CONVERSION FACTORS

Length

1 m = 39.37 in. = 3.281 ft
1 in. = 2.54 cm
1 km = 0.621 mi
1 mi = 5280 ft = 1.609 km
1 lightyear (ly) = 9.461×10^{15} m
1 angstrom (Å) = 10^{-10} m

Mass

1 kg = 10^3 g = 6.85×10^{-2} slug
1 slug = 14.59 kg
1 u = 1.66×10^{-27} kg

Time

1 min = 60 s
1 h = 3600 s
1 day = 8.64×10^4 s
1 yr = 365.242 days = 3.156×10^7 s

Volume

1 L = 1000 cm^3 = 3.531×10^{-2} ft^3
1 ft^3 = 2.832×10^{-2} m^3
1 gal = 3.786 L = 231 in.^{3}

Angle

$180° = \pi$ rad
1 rad = 57.30°
1° = 60 min = 1.745×10^{-2} rad

Speed

1 km/h = 0.278 m/s = 0.621 mi/h
1 m/s = 2.237 mi/h = 3.281 ft/s
1 mi/h = 1.61 km/h = 0.447 m/s = 1.47 ft/s

Force

1 N = 0.2248 lb = 10^5 dynes
1 lb = 4.448 N
1 dyne = 10^{-5} N = 2.248×10^{-6} lb

Work and energy

1 J = 10^7 erg = 0.738 ft·lb = 0.239 cal
1 cal = 4.186 J
1 ft·lb = 1.356 J
1 Btu = 1.054×10^3 J = 252 cal
1 J = 6.24×10^{18} eV
1 eV = 1.602×10^{-19} J
1 kWh = 3.60×10^6 J

Pressure

1 atm = 1.013×10^5 N/m^2 (or Pa) = 14.70 lb/in.^{2}
1 Pa = 1 N/m^2 = 1.45×10^{-4} lb/in.^{2}
1 lb/in.^{2} = 6.895×10^3 N/m^2

Power

1 hp = 550 ft·lb/s = 0.746 kW
1 W = 1 J/s = 0.738 ft·lb/s
1 Btu/h = 0.293 W

COLLEGE PHYSICS

SIXTH EDITION

COLLEGE PHYSICS

SIXTH EDITION

RAYMOND A. SERWAY

JAMES MADISON UNIVERSITY

JERRY S. FAUGHN

EASTERN KENTUCKY UNIVERSITY

CONTRIBUTING AUTHOR:
CLEMENT J. MOSES

THOMSON

BROOKS/COLE

Australia · Canada · Mexico · Singapore · Spain
United Kingdom · United States

Editor in Chief: Michelle Julet
Senior Development Editor: Susan Dust Pashos
Assistant Editor: Alyssa White
Editorial Assistant: Jessica Howard
Technology Project Manager: Samuel Subity
Marketing Manager: Kelley McAllister
Marketing Assistant: Sandra Perin
Advertising Project Manager: Stacey Purviance
Senior Project Manager, Editorial Production: Charlene Catlett
    Squibb, Teri Hyde
Print/Media Buyer: Karen Hunt
Production Service: Progressive Publishing Alternatives

Text Designer: Lisa Adamitis
Art Director: Carol Bleistine
Photo Researcher: Dena Betz
Copy Editor: Progressive Publishing Alternatives
Illustrator: Rolin Graphics
Cover Designer: Larry Didona
Cover Image: Aurora Borealis over Trees, © Yutaka Iijima/
    Photonica
Cover Printer: Transcontinental Printing
Compositor: Progressive Information Technologies
Printer: Transcontinental Printing

Printed in Canada
2   3   4   5   6   7   06   05   04   03

For more information about our products, contact us at:
Thomson Learning Academic Resource Center
1-800-423-0563
For permission to use material from this text, contact us by:
Phone: 1-800-730-2214
Fax: 1-800-730-2215
Web: http://www.thomsonrights.com

Brooks/Cole—Thomson Learning
511 Forest Lodge Road
Pacific Grove, CA 93950
USA

ISBN 0-030-35114-6
Text with passcard: ISBN 0-534-49258-4
International Student Edition: ISBN 0-534-40814-1
(Not for sale in the United States)

Asia
Thomson Learning
60 Albert Street, #15-01
Albert Complex
Singapore 189969

Australia
Nelson Thomson Learning
102 Dodds Street
South Melbourne, Victoria 3205
Australia

Canada
Nelson Thomson Learning
1120 Birchmount Road
Toronto, Ontario M1K 5G4
Canada

Europe/Middle East/Africa
Thomson Learning
Berkshire House
168-173 High Holborn
London WC1V 7AA
United Kingdom

Latin America
Thomson Learning
Seneca, 53
Colonia Polanco
11560 Mexico D.F.
Mexico

Spain
Paraninfo Thomson Learning
Calle/Magallanes, 25
28015 Madrid, Spain

Library of Congress Control Number: 2002113279

Contents Overview

Contents

ABOUT THE AUTHORS

Raymond A. Serway received his doctorate at the Illinois Institute of Technology and is Professor Emeritus at James Madison University. Dr. Serway began his teaching career at Clarkson University, where he conducted research and taught from 1967 to 1980. His second academic appointment was at James Madison University as Professor of Physics and Head of the Physics Department from 1980 to 1986. He remained at James Madison University until his retirement in 1997. He was the recipient of the Madison Scholar Award at James Madison University in 1990, the Distinguished Teaching Award at Clarkson University in 1977, and the Alumni Achievement Award from Utica College in 1985. As Guest Scientist at the IBM Research Laboratory in Zurich, Switzerland, he worked with K. Alex Müller, 1987 Nobel Prize recipient. Dr. Serway also held research appointments at Rome Air Development center from 1961 to 1963, at IIT Research Institute from 1963 to 1967, and as a visiting scientist at Argonne National Laboratory, where he collaborated with his mentor and friend, Sam Marshall. Dr. Serway is also the senior author of *Principles of Physics,* 3rd edition; *Physics for Scientists and Engineers,* 5th edition; *Modern Physics,* 2nd edition with Dr. Moses; and the high-school textbook *Physics* with Dr. Faughn (the latter published by Holt, Rinehart, & Winston). In addition, Dr. Serway has published more than 40 research papers in the field of condensed matter physics and has given more than 70 presentations at professional meetings. Dr. Serway and his wife Elizabeth enjoy traveling, golfing, and spending quality time with their four children and five grandchildren.

Jerry S. Faughn earned his doctorate at the University of Mississippi. He is Professor Emeritus and former Chair of the Department of Physics and Astronomy at Eastern Kentucky University. Dr. Faughn has also written a microprocessor interfacing text for upper-division physics students. He is co-author of a non-mathematical physics text and a physical science text for general education students, and (with Dr. Serway) the high-school textbook *Physics,* published by Holt, Rinehart, & Winston. He has taught courses ranging from the lower division to the graduate level, but his primary interest is in students just beginning to learn physics. He has been director of a number of NSF and state grants, many of which were devoted to the improvement of physics education. He believes that there is no greater calling than to be a teacher and an interpreter of physics for others. Dr. Faughn has a wide variety of hobbies, among which are reading, travel, genealogy, and old-time radio. His wife Mary Ann is an avid gardener, and he contributes to her efforts by staying out of the way. His daughter Laura is in family practice and his son David is an attorney.

Clement J. Moses is Emeritus Professor of Physics at Utica College of Syracuse University. He was born and brought up in Utica, New York, and holds an A. B. from Hamilton College, an M. S. from Cornell, and a Ph.D. from State University of New York at Binghamton. He has over 30 years of science writing and teaching experience at the college level, and is a co-author of the text *Modern Physics,* 2nd edition with Dr. Serway. His research work, in both industrial and university settings, has dealt with defects in solids, solar cells, and the dynamics of atoms at surfaces. In addition to science writing, Professor Moses likes cooking, fishing, singing, and going to operas.

PREFACE

College Physics is written for a one-year course in introductory physics usually taken by students majoring in biology, the health professions, and other disciplines including environmental, earth, and social sciences, and technical fields such as architecture. The mathematical techniques used in the book include algebra, geometry, and trigonometry, but not calculus.

The main objectives of this introductory textbook are twofold: to provide the student with a clear and logical presentation of the basic concepts and principles of physics, and to strengthen an understanding of the concepts and principles through a broad range of interesting applications to the real world. To meet these objectives, we have emphasized sound physical arguments and problem-solving methodology. At the same time, we have attempted to motivate the student through practical examples that demonstrate the role of physics in other disciplines.

This textbook, which covers the standard topics in classical physics and 20th century physics, is divided into six parts. Part I (Chapters 1–9) deals with Newtonian mechanics and the physics of fluids; Part II (Chapters 10–12) is concerned with heat and thermodynamics; Part III (Chapters 13–14) covers wave motion and sound; Part IV (Chapters 15–21) is concerned with electricity and magnetism; Part V (Chapters 22–25) treats the properties of light and the field of geometric and wave optics; Part VI (Chapters 26–30) represents an introduction to special relativity, quantum physics, atomic, and nuclear physics.

CHANGES TO THE SIXTH EDITION

A number of new features, changes, and improvements have been added to this edition. Based on comments from users of the fifth edition and reviewers' suggestions, a major effort was made to improve clarity of presentation, precision of language, and accuracy throughout. The new pedagogical features added to this edition are based on current trends in science education. The following represent the major changes in the sixth edition.

PEDAGOGICAL CHANGES

- **Quick Quizzes** Several questions labeled Quick Quizzes are now included in each chapter to provide students with opportunities to test their understanding of the physical concepts presented. The questions require students to make decisions on the basis of sound reasoning, and some are intended to help students overcome common misconceptions. Most questions are presented in multiple choice format, and can be adapted for assessing student performance in the classroom. Answers to all questions are found at the end of the textbook, while answers with detailed explanations are provided in the Instructor's Manual.
- **Tips** These new features, placed in the margins of the text, address common student misconceptions that often cause students to follow unproductive paths. The "Tips" should help students avoid common mistakes and misunderstandings.
- **Problems and Conceptual Questions** A substantial revision of the end-of-chapter problems and conceptual questions was made in this edition. Most of the new problems that have been added are intermediate in level, and all have been carefully edited and reworded where necessary. Many new problems require

students to make order-of-magnitude calculations. Solutions to approximately 12 problems per chapter are included in the *Student Solutions Manual and Study Guide*. Boxed numbers identify these problems. A smaller subset of solutions will be posted on the World Wide Web (**http://info.brookscole.com/serway**) and will be accessible to students and instructors using *College Physics*. The **web** icon identifies these problems.

- **Group Activities** This new feature at the end of each chapter is included to encourage students to engage in activities outside of the classroom. Some are simple experiments or demonstrations the student can perform individually or with the assistance of a classmate. Others are problems, frequently consisting of both conceptual and numerical parts, that will stimulate group discussion.

- **Webnotes** Useful World Wide Web addresses are provided as marginal notes to encourage students to explore extensions of material beyond what is covered in the text.

CONTENT CHANGES

Although the overall content and organization of the textbook is similar to that of the fifth edition, several changes were implemented.

- Chapter 2 now includes several new examples on kinematics and the revised presentation places an increased emphasis on qualitative explanations of equations and graphs.

- Chapter 5 was extensively rewritten, and now includes a discussion of the various ways that energy can be transferred into or out of a system. We also added a new discussion of energy and power considerations when someone jumps vertically into the air, and a new section on work done by a variable force.

- Chapter 6 includes a new section on rocket propulsion emphasizing that the operation of a rocket can be understood from the law of conservation of momentum as applied to the rocket plus its ejected fuel. The chapter also presents new applications on injury in automobile collisions and finding the impact force on landing in a jump.

- We have added a discussion of arches and the ultimate strength of materials, including biological substances, in Section 9.2.

- A number of changes were made to Chapter 12. The first law of thermodynamics is now expressed as $\Delta U = Q + W$, where positive W is defined as the work done *on* the system. This form of the first law is used in most modern chemistry textbooks, and has been recommended by a committee appointed by AAPT (the American Association of Physics Teachers). Chapter 12 places an increased emphasis on the application of the first law to internal combustion engines and cyclic processes. The chapter also discusses how the human body constantly transforms internal energy into other forms of energy.

- Chapter 16 now offers more explanation of how to relate circuit schematics to actual circuits. This chapter also discusses the operation of the defibrillator as a practical example of capacitance.

- In Chapter 17, we added a section entitled "Electrical Activity in the Heart," which is of special interest to pre-medical and life science students.

- Chapter 18 includes a section entitled "Conduction of Electrical Signals by Neurons." This discussion is a recast of an earlier essay by Paul Davidovits on the nervous system and will be of special interest to pre-medical and life science students.

- Chapter 19 now includes a new application on electric motors.

- A new application that discusses the refraction of a laser beam in a DVD player is presented in Chapter 22.

- Chapter 23 places an increased emphasis on ray diagrams.

- New to this edition is Section 24.5 entitled "Using Interference to Read CD's and DVD's." This discussion represents an interesting application of interference to modern technology.
- Chapter 26 was extensively rewritten and includes a new example on the conversion of mass to energy.
- Chapter 27 now includes a discussion of how X-ray diffraction is used to determine the structure of biologically important molecules.
- Chapter 28 was extensively revised and now includes two new sections entitled "Energy Bands in Solids" and "Semiconductor Devices." These topics are important for understanding the electronic properties of solids, and are basic to understanding the behavior of semiconductor devices such as diodes and transistors.
- Section 29.8, new to this edition, discusses various types of radiation detectors such as Geiger counters, bubble chambers, and scintillation counters. These practical applications will be of special interest to students preparing for health related professions.
- Some sections in the fifth edition were either deleted or moved to more appropriate locations. Section 1.7 on mathematical notation was moved to an appendix. Most of the section entitled "The Diffraction Grating" was moved from Chapter 25 to Chapter 24 because it logically follows the discussion of diffraction. The section entitled "Pair Production and Annihilation" was moved from Chapter 27 to Chapter 26 because the topic is a verification of the equivalence of mass and other forms of energy as predicted by the theory of relativity.
- The QuickLabs that were included in the fifth edition have been deleted, but the most popular QuickLabs are now incorporated into the new **Group Activities** feature described earlier. Multiple-choice questions that appeared in the fifth edition have been removed, but some have been rewritten as Quick Quiz questions.

TEXTBOOK FEATURES

Most instructors would agree that the textbook assigned in a course should be the student's primary guide for understanding and learning the subject matter. Furthermore, the textbook should be easily accessible and written in a style that facilitates instruction and learning. With this in mind, we have included many pedagogical features that are intended to enhance the textbook's usefulness to both students and instructors. These features are as follows:

STYLE   We have attempted to write the book in a style that is clear, logical, relaxed, and pleasing to the reader. At the same time, we have attempted to keep the presentation accurate and precise. New terms are carefully defined, and we have avoided the use of jargon.

PREVIEWS   All chapters begin with a preview that includes a brief discussion of the chapter's objectives and content.

ORGANIZATION   The book is divided into the following six parts: mechanics, thermodynamics, vibrations and wave motion, electricity and magnetism, light and optics, and modern physics. Each part includes an overview of the subject matter to be covered in that part and some historical perspectives.

UNITS   The international system of units (SI) is used throughout the book. The U.S. customary system of units is used only to a limited extent in the problem sets of the early chapters on mechanics.

MARGINAL NOTES   Comments and notes appearing in the margin can be used to locate important statements, equations, definitions, and concepts in the text.

PROBLEM-SOLVING STRATEGIES   General strategies and suggestions are included for solving the types of problems featured in both the worked examples and end-of-chapter problems. This feature, highlighted by a surrounding box, is intended to help students identify the essential steps in solving problems and increase their skills as problem solvers.

PHYSICS IN ACTION   This boxed material focuses on photographs of interesting demonstrations and phenomena in physics, accompanied by detailed explanations. The material can also serve as a source of information for initiating classroom discussions.

LIFE SCIENCE TOPICS   Many chapters include text, worked examples and problems dealing with applications of physics to the life sciences. These are identified by the DNA icon ( ).

WORKED EXAMPLES   A large number of worked examples, including many new ones, are presented as an aid in understanding and/or reinforcing physical concepts. In many cases, these examples serve as models for solving end-of-chapter problems. The examples are set off from the text for ease of location, and all examples are given titles to describe their content. Many examples include a **Reasoning** section to illustrate the underlying concepts and methodology used in arriving at a correct solution. This will help students understand the logic behind the solution and the advantage of using a particular approach to solve the problem. The solution answer is highlighted with a light blue screen. Many worked examples are followed immediately by exercises with answers. These exercises represent extensions of the worked examples and are intended to sharpen student's problem-solving skills and test their understanding of concepts. Students who work through these exercises on a regular basis should find the end-of-chapter problems less intimidating.

IMPORTANT STATEMENTS AND EQUATIONS   Most important statements and definitions are set in **boldface type** or are highlighted with a background screen for added emphasis and ease of review. Similarly, important equations are highlighted with a gold screen to facilitate location.

ILLUSTRATIONS AND PHOTOGRAPHS   The text material, worked examples, and end-of-chapter questions and problems are accompanied by numerous figures, photographs, and tables. Full color is used to add clarity to the figures and to make the visual presentation as realistic and pleasing as possible. Three-dimensional effects are rendered with the use of shaded and lightened areas, where appropriate. Vectors are color coded, and curves in xy plots are drawn in color. Color photographs have been carefully selected, and their accompanying captions have been written to serve as an added instructional tool. A complete description of the pedagogical use of color appears on the inside front cover.

SUMMARIES   Each chapter contains a summary which reviews the important concepts and equations discussed in that chapter.

CONCEPTUAL QUESTIONS   A set of conceptual questions is provided at the end of each chapter. The **Applying Physics** examples presented in the text should serve as models for students when conceptual questions are assigned or used in tests. The questions provide the student with a means of self-testing the concepts presented in the chapter. Some conceptual questions are appropriate for initiating classroom discussions. Answers to all odd-numbered conceptual questions are located in the answer section at the end of the book.

END-OF-CHAPTER PROBLEMS   An extensive set of problems is included at the end of each chapter. Answers to odd-numbered problems are given at the end of the book. For the convenience of both the student and instructor, about two thirds of the problems are keyed to specific sections of the chapter. The remaining problems, labeled "Additional Problems," are not keyed to specific sections. There are three levels of problems according to their level of difficulty. Straightforward problems are numbered in black, intermediate level problems are numbered in blue, and the most challenging problems are numbered in magenta. Those problems with a focus on the life sciences are identified by the DNA icon (🧬). The set of problems under the heading of **Group Activities,** new to this edition, have been included to encourage students to engage in scientific activities outside the classroom.

APPENDICES   Several appendices are provided at the end of the book. Most of the appendix material represents a review of mathematical techniques used in the book, such as scientific notation, algebra, geometry, and trigonometry. References to these appendices is made as needed throughout the book. Most of the mathematical review sections include worked examples and exercises with answers. Some appendices contain useful tables that supplement textual information. For easy reference, the front endpapers contain a chart explaining the use of color throughout the book and a list of often-used conversion factors.

TEACHING OPTIONS

This book contains more than enough material for a one-year course in introductory physics. This serves two purposes. First, it gives the instructor more flexibility in choosing topics for a specific course. Second, the book becomes more useful as a resource for students. On the average, it should be possible to cover about one chapter each week for a class that meets three hours per week. Those sections, examples, and end-of-chapter problems dealing with applications of physics to the life sciences are identified with the DNA icon (🧬). We offer the following suggestions for shorter courses for those instructors who choose to move at a slower pace through the year:

Option A: If you choose to place more emphasis on contemporary topics in physics, you should consider omitting all or parts of Chapter 8 (Rotational Equilibrium and Rotational Dynamics), Chapter 21 (Alternating Current Circuits and Electromagnetic Waves), and Chapter 25 (Optical Instruments).

Option B: If you choose to place more emphasis on classical physics, you could omit all or parts of Part VI of the textbook, which deals with special relativity and other topics in 20th century physics.

The Instructor's Manual offers additional suggestions for specific sections and topics that may be omitted without loss of continuity if time presses.

STUDENT ANCILLARIES

Thomson·Brooks/Cole offers several items to supplement and enhance the classroom experience. These ancillaries will allow instructors to customize the textbook to their students' needs and to their own style of instruction. One or more of these ancillaries may be shrink-wrapped with the text at a reduced price:

STUDENT SOLUTIONS MANUAL AND STUDY GUIDE by John R. Gordon, Charles Teague, and Raymond A. Serway. Now offered in two volumes, this manual features detailed solutions to approximately 12 problems per chapter. These problems are indicated in the text with boxed numbers. The manual also features a skills section, important notes from key sections of the text, and a list of important equations and concepts. Volume 1 contains Chapters 1–14 and Volume 2 contains Chapters 15–30.

***CORE CONCEPTS IN COLLEGE PHYSICS* CD-ROM, VERSION 2.0** The *Core Concepts in College Physics* CD-ROM applies the power of multimedia to the introductory physics course, offering full-motion animation and video, engaging interactive graphics, clear and concise text, and guiding narration. Drawing from topics in mechanics, thermodynamics, electromagnetism, and optics, *Core Concepts in College Physics* focuses on those concepts students typically find most difficult in the course. The CD-ROM also presents step-by-step explorations of essential mathematics, problem-solving strategies, and animations of problems to promote conceptual understanding and sharpen problem-solving skills. The accompanying *Workbook* contains practical physics problems coordinating with the CD, along with worked solutions.

PHYSICS LABORATORY MANUAL, 2nd edition by David Loyd. This manual supplements the learning of basic physical principles while introducing laboratory procedures and equipment. Each chapter of the manual includes a pre-laboratory assignment, objectives, an equipment list, the theory behind the experiment, experimental procedures, graphs, and questions. A laboratory report is provided for each experiment so the student can record data, calculations, and experimental results. Students are encouraged to apply statistical analysis to their data in order to develop their ability to judge the validity of their results.

ADDITIONAL ONLINE RESOURCES In addition to the companion Web site for this textbook (**http://info.brookscole.com/serway**), students using *College Physics* by Serway and Faughn are encouraged to visit the Brooks/Cole Physics Resource Center at the address below for features such as online quizzing and additional Weblinks. See **http://physics.brookscole.com**

INSTRUCTOR ANCILLARIES

Ancillaries offered in two volumes are split as follows: Volume 1 contains Chapters 1–14 and Volume 2 contains Chapters 15–30.

INSTRUCTOR'S MANUAL by Jerry Faughn and Charles Teague. Available in two volumes, this manual consists of complete solutions to all the problems in the text, answers to the even-numbered problems and conceptual questions, full answers with explanations to the Quick Quizzes, and a list of suggested readings from journals and other resources.

PRINTED TEST BANK by Ed Oberhofer. Available in two volumes, the comprehensive test bank contains approximately 1 750 problems and questions in multiple choice format. Answers are provided in a separate key. Instructors may duplicate pages for distribution to students.

COMPUTERIZED TEST BANK Available for Windows and Macintosh, the computerized test bank allows instructors to create, deliver and customize tests and quizzes using questions from the printed Test Bank. Instructors may rearrange and edit existing questions, or add their own.

OVERHEAD TRANSPARENCY ACETATES The collection of transparencies in two volumes consists of approximately 200 full-color figures and photographs from the text to enhance lectures. These transparencies feature large print for easy viewing in the classroom.

INSTRUCTOR'S MANUAL FOR PHYSICS LABORATORY MANUAL, 2nd edition by David Loyd. Each chapter contains a discussion of the experiment, teaching hints, answers to selected questions from the student laboratory manual, and a post-laboratory quiz with short answers and essay questions. The author has also included a list of the suppliers of scientific equipment and a summary of the equipment needed for all the experiments in the manual.

MULTIMEDIA PRESENTATION MANAGER FOR INTRODUCTORY PHYSICS 2003: A MICROSOFT® POWERPOINT® LINK TOOL This one-stop lecture tool and instructional resource makes it easy to assemble, edit, publish and present custom media-enhanced lectures for your course, using Microsoft® PowerPoint®. The two-volume cross-platform (Win/Mac) CD set includes electronic files of the ***Instructor's Manual*** and ***Test Bank,*** plus digital files of textbook art and additional video clips and animations.

Also found on the ***Multimedia Presentation Manager for Introductory Physics 2003*** are simulations for ***Interactive Physics™ 2000,*** the highly acclaimed software from MSC Software. Many simulations are keyed to specific worked examples and end-of-chapter problems in *College Physics,* while others stand alone as laboratory exercises. The instructor or department must own a multi-user license for ***Interactive Physics™*** in order to assign the simulations for student use.

WEBTUTOR™ ADVANTAGE ON WEBCT AND BLACKBOARD *WebTutor Advantage* enables the instructor to create and manage a course Web site. WebTutor's course management tool gives the instructor the ability to provide virtual office hours, post syllabi, set up threaded discussions, track student progress with the quizzing material, and much more. WebTutor also provides robust communication tools, such as a course calendar, asynchronous discussion, real time chat, a whiteboard, and an integrated e-mail system.

For students, WebTutor offers real-time access to a full array of study tools, including chapter outlines, summaries, learning objectives, glossary flashcards (with audio), practice quizzes, and Web links.

WebTutor is available for Semesters I (Chapters 1–14) and II (Chapters 15–30) in both WebCT and Blackboard.

MYCOURSE 2.0 *MyCourse 2.0* offers instructors a simple solution for a custom course Web site that allows for assignments, tracking and reporting student progress, syllabus loading, and more. Contact your Thomson·Brooks/Cole representative for details. To see a demo of *MyCourse 2.0,* visit **http://mycourse.thomsonlearning.com**

INSTRUCTOR OPTIONS FOR ONLINE HOMEWORK For detailed explanations and demonstrations, contact your Thomson·Brooks/Cole representative or visit the following Web sites:

- *WebAssign: A Web-Based Homework System*
  http://www.webassign.net or contact WebAssign at **webassign@ncsu.edu**
- *Homework Service*
  http://hw.ph.utexas.edu/hw.html or contact **moore@physics.utexas.edu**

INFOTRAC® COLLEGE EDITION InfoTrac® is an online university library offering the full text of articles from almost 4 000 scholarly and popular publications, updated daily and going back as much as 22 years. This is available bundled free with new copies of *College Physics,* 6th edition by Serway and Faughn, and both adopters and their students receive unlimited access for four months.

ACKNOWLEDGMENTS

In preparing the 6th edition of this textbook, we have been guided by the expertise of many people who have reviewed manuscript and/or provided pre-revision suggestions. We wish to acknowledge the following reviewers and express our sincere appreciation for their helpful suggestions, criticisms, and encouragement.

Sixth edition reviewers:

Marilyn Akins, *Broome Community College*

Lawrence Anderson-Huang, *University of Toledo*

Lattie F. Collins, *East Tennessee State University*

Michael Dennin, *University of California, Irvine*

N. John DiNardo, *Drexel University*

Leonard X. Finegold, *Drexel University*

Eric Ganz, *University of Minnesota*

Grant W. Hart, *Brigham Young University*

Christopher Herbert, *New Jersey City University*

George W. Kattawar, *Texas A & M University*

Ivan Kramer, *University of Maryland, Baltimore County*

Michael LoPresto, *Henry Ford Community College*

Ed Oberhofer, *Lake Sumter Community College*

J. Scott Payson, *Wayne State University*

Patrick Polley, *Beloit College*

W. Steve Quon, *Ventura College*

Barry Robertson, *Queen's University*

Larry Rowan, *University of North Carolina, Chapel Hill*

Perry A.Tompkins, *Samford University*

Bernard Whiting, *University of Florida*

We thank the following people for their suggestions and assistance during the preparation of earlier editions of this textbook:

Albert Altman, *University of Lowell*

John Anderson, *University of Pittsburgh*

Subhash Antani, *Edgewood College*

Neil W. Ashcroft, *Cornell University*

Charles R. Bacon, *Ferris State University*

Dilip Balamore, *Nassau Community College*

Ralph Barnett, *Florissant Valley Community College*

Lois Barrett, *Western Washington University*

Paul D. Beale, *University of Colorado at Boulder*

Paul Bender, *Washington State University*

David H. Bennum, *University of Nevada at Reno*

Jeffery Braun, *University of Evansville*

John Brennan, *University of Central Florida*

Michael Bretz, *University of Michigan, Ann Arbor*

Michael E. Browne, *University of Idaho*

Joseph Cantanzarite, *Cypress College*

Ronald W. Canterna, *University of Wyoming*

Clinton M. Case, *Western Nevada Community College*

Neal M. Cason, *University of Notre Dame*

Roger W. Clapp, *University of South Florida*

Giuseppe Colaccico, *University of South Florida*

Lawrence B. Coleman, *University of California, Davis*

Jorge Cossio, *Miami Dade Community College*

Terry T. Crow, *Mississippi State College*

Stephen D. Davis, *University of Arkansas at Little Rock*

John DeFord, *University of Utah*

Chris J. DeMarco, *Jackson Community College*

Robert J. Endorf, *University of Cincinnati*

Paul Feldker, *Florissant Valley Community College*

Tom French, *Montgomery County Community College*

Albert Thomas Fromhold, Jr., *Auburn University*

Lothar Frommhold, *University of Texas at Austin*

Teymoor Gedayloo, *California Polytechnic State University*

Simon George, *California State University, Long Beach*

John R. Gordon, *James Madison University*

George W. Greenlees, *University of Minnesota*

Wlodzimierz Guryn, *Brookhaven National Laboratory*

James Harmon, *Oklahoma State University*

Grant W. Hart, *Brigham Young University*

John Ho, *State University of New York at Buffalo*

Murshed Hossain, *Rowan University*

Robert C. Hudson, *Roanoke College*

Fred Inman, *Mankato State University*

Ronald E. Jodoin, *Rochester Institute of Technology*

Drasko Jovanovic, *Fermilab*

Frank Kolp, *Trenton State University*

Joan P. S. Kowalski, *George Mason University*

Ivan Kramer, *University of Maryland, Baltimore County*

Sol Krasner, *University of Chicago*

Karl F. Kuhn, *Eastern Kentucky University*

David Lamp, *Texas Tech University*

Harvey S. Leff, *California State Polytechnic University*

Joel Levine, *Orange Coast College*

Michael Lieber, *University of Arkansas*

James Linbald, *Saddleback Community College*

Bill Lochslet, *Pennsylvania State University*

Bo Lou, *Ferris State University*

Jeffery V. Mallow, *Loyola University of Chicago*

David Markowitz, *University of Connecticut*

Steven McCauley, *California State Polytechnic University, Pomona*

Joe McCauley, Jr., *University of Houston*

Ralph V. McGrew, *Broome Community College*

Bill F. Melton, *University of North Carolina at Charlotte*

H. Kent Moore, *James Madison University*

John Morack, *University of Alaska, Fairbanks*

Steven Morris, *Los Angeles Harbor College*

Carl R. Nave, *Georgia State University*

Martin Nikolo, *Saint Louis University*

Blaine Norum, *University of Virginia*

M. E. Oakes, *University of Texas at Austin*

Lewis J. Oakland, *University of Minnesota*

Ed Oberhofer, *University of North Carolina at Charlotte*

Lewis O'Kelly, *Memphis State University*

David G. Onn, *University of Delaware*

T. A. K. Pillai, *University of Wisconsin, La Crosse*

Lawrence S. Pinsky, *University of Houston*

William D. Ploughe, *Ohio State University*

Brooke M. Pridmore, *Clayton State University*

Joseph Priest, *Miami University*

James Purcell, *Georgia State University*

Michael Ram, *State University of New York at Buffalo*

Kurt Reibel, *Ohio State University*

Virginia Roundy, *California State University, Fullerton*

William R. Savage, *The University of Iowa*

Reinhard A. Schumacher, *Carnegie Mellon University*

John Simon, *University of Toledo*

Donald D. Snyder, *Indiana University at Southbend*

Carey E. Stronach, *Virginia State University*

Thomas W. Taylor, *Cleveland State University*

L. L. Van Zandt, *Purdue University*

Howard G. Voss, *Arizona State University*

Larry Weaver, *Kansas State University*

Donald H. White, *Western Oregon State College*

George A. Williams, *The University of Utah*

Jerry H. Wilson, *Metropolitan State College*

Robert M. Wood, *University of Georgia*

Clyde A. Zaidins, *University of Colorado at Denver*

We are delighted to acknowledge the creative and original new worked examples, applied content, questions, problems and illustrations contributed by Clement J. Moses. Edward McCliment of the University of Iowa and Doug Davis of Eastern Illinois University generously contributed many of the new end-of-chapter problems and questions, especially those of interest to the life sciences. Edward F. Redish of the University of Maryland graciously allowed us to list some of his problems from the Activity Based Physics Project as Group Activities. Some of the remaining Group Activities were written by Robert J. Beichner of North Carolina State University and appeared as QuickLabs in the fifth edition. Steve Quon of Ventura College wrote most of the Webnotes.

This book was carefully checked and re-checked for accuracy by John W. Jewett of the California State University at Pomona, Robert Ehrlich of George Mason University, and Ralph McGrew of Broome Community College. Though responsibility for any remaining errors rests with us, we thank them for their untiring efforts.

We are extremely grateful to the publishing team at the Brooks/Cole Publishing Company for their expertise and outstanding work in all aspects of this project. Susan Pashos carefully analyzed the reviews of the sixth edition, provided important guidance in all aspects of the project, and made many important improvements in the final manuscript. Donna King of Progressive Information Technologies managed the book through all production stages. We appreciate the fine work of Alyssa White in managing the development of the ancillary materials. Samuel Subity developed the Web site and other electronic media. Carol Bleistine managed the overall art and design program. Dena Betz acquired many new and excellent photos. We also recognize important contributions by Charlene Squibb (Senior Production Manager), Kelley McAllister (Marketing Manager), Lisa Adamitis (book designer), and Stacey Purviance (Manager of Creative Development). We thank Angus McDonald, Executive Editor, for his support and problem-solving spirit.

Finally, we dedicate this book to our wives and children, for their love, support, and long-term sacrifices.

Raymond A. Serway
Leesburg, Virginia

Jerry S. Faughn
Richmond, Kentucky

APPLICATIONS

Although physics is relevant to so much in our modern lives, this may not be obvious to students in an introductory course. In this sixth edition of *College Physics,* we continue a design feature begun in the previous edition. This feature makes the relevance of physics to everyday life more obvious by pointing out specific applications in the form of a marginal note. Some of these applications pertain to the life sciences and are marked with the DNA icon ( ). The list below is not intended to be a complete listing of all the applications of the principles of physics found in this textbook. Many other applications are to be found within the text and especially in the worked examples, Conceptual Questions, and end-of-chapter problems.

> **APPLICATION**
>
> DIRECTS YOU TO SECTIONS
> DISCUSSING APPLIED
> PRINCIPLES OF PHYSICS

TO THE STUDENT

It is appropriate to offer some words of advice that should be of benefit to you, the student. Before doing so, we assume that you have read the Preface, which describes the various features of the text that will help you through the course.

HOW TO STUDY

Very often instructors are asked, "How should I study physics and prepare for examinations?" There is no simple answer to this question, but we would like to offer some suggestions that are based on our own experiences in learning and teaching over the years.

First and foremost, maintain a positive attitude toward the subject matter, keeping in mind that physics is the most fundamental of all natural sciences. Other science courses that follow will use the same physical principles, so it is important that you understand and are able to apply the various concepts and theories discussed in the text.

CONCEPTS AND PRINCIPLES

It is essential that you understand the basic concepts and principles before attempting to solve assigned problems. You can best accomplish this goal by carefully reading the textbook before you attend your lecture on the covered material. When reading the text, you should jot down those points that are not clear to you. We've purposely left wide margins in the text to give you space for doing this. Also be sure to make a diligent attempt at answering the questions in the Quick Quizzes as you come to them in your reading. We have worked hard to prepare questions that help you judge for yourself how well you understand the material. Pay careful attention to the many Tips in the margins of the text. These will help you to avoid misconceptions, mistakes, and misunderstandings and will help you to maximize the efficiency of your time by minimizing adventures along fruitless paths. During class, take careful notes and ask questions about those ideas that are unclear to you. Keep in mind that few people are able to absorb the full meaning of scientific material after only one reading. Several readings of the text and your notes may be necessary. Your lectures and laboratory work supplement reading of the textbook and should clarify some of the more difficult material. You should minimize your memorization of material. Successful memorization of passages from the text, equations, and derivations does not necessarily indicate that you understand the material. Your understanding of the material will be enhanced through a combination of efficient study habits, discussions with other students and with instructors, and your ability to solve the problems presented in the textbook. Ask questions whenever you feel clarification of a concept is necessary.

STUDY SCHEDULE

It is important that you set up a regular study schedule, preferably one that calls for daily involvement with course work. Make sure that you to read the syllabus for the course and adhere to the schedule set by your instructor. The lectures will be much more illuminating if you read the corresponding textbook material *before* attending them. As a general rule, you should devote about two hours of study time for every hour you are in class. If you are having trouble with the course, seek the advice of the instructor or other students who have taken the course. You may find it necessary to seek further instruction from experienced students. Very often, instructors offer review sessions in addition to regular class periods. It is important that you avoid the practice of delaying study until a day or two before an exam. More often than not, this approach has disastrous results. Rather than undertake an all-night study session, briefly review the basic concepts and equations and get a good night's rest. If you feel you need additional help in understanding the concepts, in preparing for exams, or in problem-solving, we suggest that you acquire a copy of the *Student Solutions Manual and Study Guide* that accompanies this textbook; this manual should be available at your college bookstore.

USE THE FEATURES

You should make full use of the various features of the text discussed in the Preface. For example, marginal notes are useful for locating and describing important equations and concepts, and **boldfaced** type indicates important statements and definitions. Many useful tables are contained in the Appendices, but most are incorporated in the text where they are most often referenced. Appendix A is a convenient review of mathematical techniques.

Answers to odd numbered questions and problems and all Quick Quizzes are given at the end of the textbook. The exercises (with answers) that follow some worked examples represent extensions of those examples; in most of these exercises, you are expected to perform a simple calculation. Their purpose is to test your problem-solving skills as you read through the text. Problem-Solving Strategies and Hints are included in selected chapters throughout the text and give you additional information about how you should solve problems. The Table of Contents provides an overview of the entire text, while the Index enables you to locate specific material quickly. Footnotes sometimes are used to supplement the text or to cite other references on the subject discussed.

After reading a chapter, you should be able to define any new quantities introduced in that chapter and to discuss the principles and assumptions that were used to arrive at certain key relations. The chapter summaries and the review sections of the *Student Solutions Manual and Study Guide* should help you in this regard. In some cases, it may be necessary for you to refer to the index of the text to locate certain topics. You should be able to correctly associate with each physical quantity the symbol used to represent that quantity and the unit in which the quantity is specified. Furthermore, you should be able to express each important relation in a concise and accurate prose statement.

PROBLEM-SOLVING

R. P. Feynman, Nobel laureate in physics, once said, "You do not know anything until you have practiced." In keeping with this statement, we strongly advise that you develop the skills necessary to solve a wide range of problems. Your ability to solve problems will be one of the main tests of your knowledge of physics, and therefore you should try to solve as many problems as possible. It is essential that you understand basic concepts and principles before attempting to solve problems. It is good practice to try to find alternate solutions to the same problem. For example, you can solve problems in mechanics using Newton's laws, but very often an alternative method that draws on energy considerations is more direct. You should not deceive yourself into thinking that you understand a problem merely because you have seen it solved in class. You must be able to solve the problem and similar problems on your own.

The approach to solving problems should be carefully planned. A systematic plan is especially important when a problem involves several concepts. First, read the problem several times until you are confident you understand what is being asked. Look for any key words that will help you interpret the problem and perhaps allow you to make certain assumptions. Your ability to interpret a question properly is an integral part of problem-solving. Second, you should acquire the habit of writing down the information given in a problem and those quantities that need to be found; for example, you might construct a table listing both the quantities given and the quantities to be found. This procedure is sometimes used in the worked examples of the textbook. Finally, after you have decided on the method you feel is appropriate for a given problem, proceed with your solution. General problem-solving strategies of this type are included in the text and are highlighted with a surrounding box. If you follow the steps of this procedure, you will not only find it easier to come up with a solution, but you will also gain more from your efforts.

Often, students fail to recognize the limitations of certain equations or physical laws in a particular situation. It is very important that you understand and remember the assumptions that underlie a particular theory or formalism. For example, certain equations in kinematics apply only to a particle moving with constant acceleration. These equations are not valid for describing motion whose acceleration is not constant, such as the motion of an object connected to a spring or the motion of an object through a fluid.

EXPERIMENTS

Physics is a science based on experimental observations. In view of this fact, we recommend that you try to supplement the text by performing various types of "hands-on" experiments, either at home or in the laboratory. For example, the common Slinky™ toy is excellent for studying traveling waves; a ball swinging on the end of a long string can be used to investigate pendulum motion; various objects attached to the end of a vertical spring or rubber band can be used to investigate elasticity; an old pair of Polaroid sunglasses and some discarded lenses and a magnifying glass are the components of various experiments in optics; and an approximate measure of the free-fall acceleration can be determined simply by measuring with a stopwatch the time it takes for a ball to drop from a known height. The list of such experiments is endless. When physical models are not available, be imaginative and try to develop models of your own.

Some of the Group Activities at the end of each chapter can be performed on your own or with the help of a friend and will give you specific guidelines for acquiring "hands-on" experience.

CLOSING COMMENTS

Someone once said that there are only two professions in which people truly enjoy what they are doing, professional sports and physics. It is our sincere hope that you too will find physics an exciting and enjoyable experience and that you will profit from this experience, regardless of your chosen profession.

Welcome to the exciting world of physics!

> *To see the World in a Grain of Sand*
> *And a Heaven in a Wild Flower,*
> *Hold infinity in the palm of your hand*
> *And Eternity in an hour.*

> W. Blake, "Auguries of Innocence"

COLLEGE PHYSICS

SIXTH EDITION

PART One

Mechanics

Physics, the most fundamental science, is concerned with the basic principles of the Universe. It is one of the foundations on which the other physical sciences—astronomy, chemistry, and geology—are based. The beauty of physics lies in the simplicity of its fundamental theories and in the way just a small number of basic concepts, equations, and assumptions can describe our view of the world.

The myriad physical phenomena in our world are parts of one or more of the following five areas of physics:

- Mechanics, which is concerned with the effects of forces on material objects
- Thermodynamics, which deals with heat, temperature, and the behavior of large numbers of particles
- Electromagnetism, which deals with charges, currents, and electromagnetic fields
- Relativity, a theory that describes particles moving at any speed, and connects space and time
- Quantum mechanics, a theory dealing with the behavior of particles at the submicroscopic level as well as the macroscopic world

The first part of this textbook addresses mechanics, sometimes referred to as classical mechanics or Newtonian mechanics. This is an appropriate subject with which to begin an introductory book because many of the basic principles of mechanical systems can be used later to describe such natural phenomena as waves and the transfer of energy by heat. Furthermore, the laws of conservation of energy and momentum to be introduced in our study of mechanics retain their importance in the fundamental theories that follow, including the theories of modern physics.

The first serious attempts to develop a theory of motion were made by the Greek astronomers and philosophers. Although these early scientists devised a complex model to describe the motions of heavenly bodies, their model lacked correlation between celestial motions and the motions of objects on Earth. The study of mechanics was enhanced by a number of careful astronomical investigations by Copernicus, Brahe, and Kepler in the 16th century. In the 16th and 17th centuries, Galileo attempted to relate the motions of falling objects and projectiles to the motions of planetary bodies, and Sevin and Hooke studied forces and their relationship to motion. A major development in the theory of mechanics was provided by Newton in 1687 when he published his *Principia*. Newton's elegant theory, which remained unchallenged for more than 200 years, was based on his hypothesis of universal gravitation together with contributions made by Galileo and others.

Today, mechanics is of vital importance to students in all disciplines. It is highly successful in describing the motions of material objects, such as planets, rockets, and baseballs. In the first nine chapters of this book, we shall describe the laws of mechanics and examine a wide range of phenomena that can be understood through these laws.

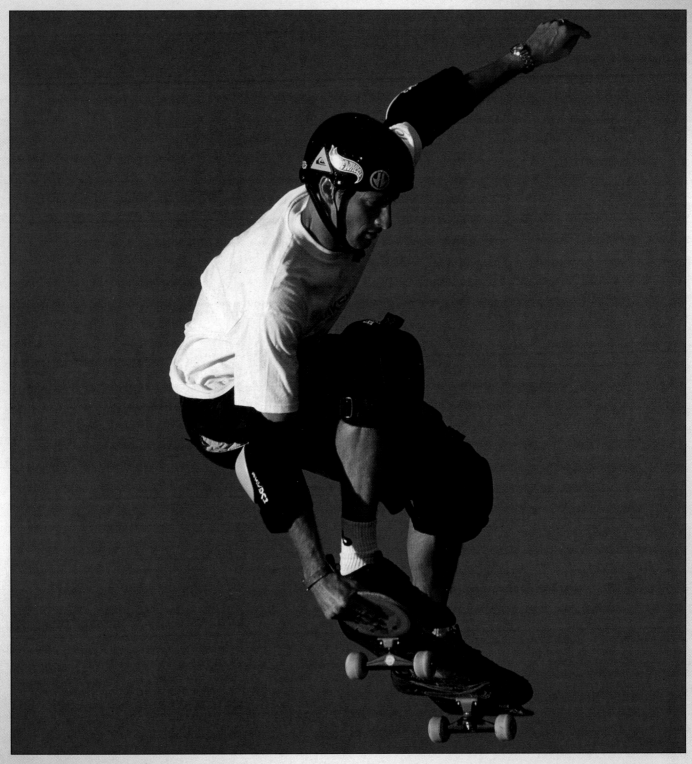

As this skateboarder moves gracefully through the air, the shape of the curved path he follows and the height he achieves depend on his velocity as he leaves the ground below. *(Allsport/Getty)*

1 Introduction

Chapter Outline

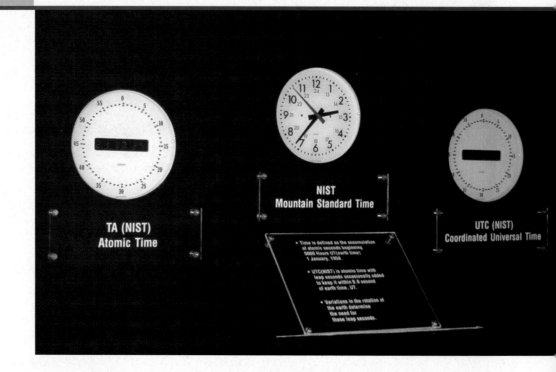

This hallway display of clocks indicates, from left to right, Atomic Time, Mountain Standard Time, and Universal Time. *(Courtesy of National Institute of Standards and Technology, Boulder Laboratories, U.S. Department of Commerce)*

The goal of physics is to provide an understanding of nature by developing theories based on experiments. The theories are usually expressed in mathematical form. Fortunately, it is possible to explain the behavior of a variety of physical systems with a limited number of fundamental laws.

Scientists continuously work at improving our comprehension of fundamental laws, and new discoveries are made every day. In many research areas, a great deal of overlap occurs among physics, chemistry, and biology. Numerous recent technological advances are the result of the cooperative efforts of many scientists, engineers, and technicians. Some of the most notable include unmanned space missions and manned Moon landings, microcircuitry and high-speed computers, and sophisticated imaging techniques used in scientific research and medicine. The impacts of such developments on our society have indeed been great, and it is likely that future discoveries will be exciting, challenging, and of great benefit to humanity.

Because the following chapters will be concerned with the laws of physics, it is necessary to provide clear definitions of the basic quantities involved in these laws. For example, such physical quantities as force, velocity, volume, and acceleration can be described in terms of more fundamental quantities. In the next several chapters we shall encounter three basic quantities: **length** (L), **mass** (M), and **time** (T). Later in the book we will add two other basic quantities, temperature and electric current. In our study of mechanics, however, we shall be concerned only with the quantities of length, mass, and time.

Magnified view of the gears in a Swiss watch. *(Dale Boyer, Tony Stone Images)*

Webnote 1.1

Find out more about standards of length, mass, and time by visiting the International Bureau of Weights and Measures and the National Institute of Standards and Technology. See
http://www.bipm.fr
http://www.nist.gov/

1.1 STANDARDS OF LENGTH, MASS, AND TIME

If we are to report the result of a measurement of a certain quantity to someone who wishes to reproduce this result, a unit for the quantity must be defined. For example, if our fundamental unit of length is defined to be 1.0 meter, and someone familiar with our system of measurement reports that a wall is 2.0 meters high, we know that the height of the wall is twice the fundamental unit of length. Likewise, if our fundamental unit of mass is defined as 1.0 kilogram, and we are told that a person has a mass of 75 kilograms, then that person has a mass 75 times as great as the fundamental unit of mass.

In 1960, an international committee agreed on a standard system of units for the fundamental quantities of science. It is called the **SI** (Système International), and its units of length, mass, and time are the meter, kilogram, and second.

LENGTH

In 1799, the legal standard of length in France became the meter, defined as one ten-millionth of the distance from the equator to the North Pole. Until 1960, the official length of the meter was the distance between two lines on a specific bar of platinum-iridium alloy stored under controlled conditions. This standard was abandoned for several reasons, the principal one being that measurements of the separation between the lines are not precise enough. In that year, the meter was defined as 1 650 763.73 wavelengths of orange-red light emitted from a krypton-86 lamp. In October 1983, this definition was abandoned also, and the **meter was redefined as the distance traveled by light in vacuum during a time interval of 1/299 792 458 second.** This latest definition establishes that the speed of light is 299 792 458 meters per second.

◁ Definition of the meter

MASS

The SI unit of mass, the kilogram, is defined as the mass of a specific platinum-iridium alloy cylinder kept at the International Bureau of Weights and Measures at Sèvres, France (similar to that shown in Figure 1.1a). As we

◁ Definition of the kilogram

(a)   (b)

FIGURE 1.1   (a) The National Standard Kilogram No. 20, an accurate copy of the International Standard Kilogram kept at Sèvres, France, is housed under a double bell jar in a vault at the National Institute of Standards and Technology. (b) The nation's primary time standard is a cesium fountain atomic clock developed at the National Institute of Standards and Technology laboratories in Boulder, Colorado. This clock will neither gain nor lose a second in 20 million years. *(Courtesy of National Institute of Standards and Technology, U.S. Dept. of Commerce)*

shall see in Chapter 4, mass is a quantity used to measure the resistance to a change in motion of an object. It is more difficult to cause a change in the motion of an object with a large mass than an object with a small mass.

TIME

Before 1960, the time standard was defined in terms of the average length of a solar day in the year 1900. (A solar day is the time interval between successive appearances of the Sun at the highest point it reaches in the sky each day.) The basic unit of time, the second, was defined to be $(1/60)(1/60)(1/24) = 1/86\ 400$ of the average solar day. In 1967, the second was redefined to take advantage of the high precision attainable with a device known as an atomic clock, which uses the characteristic frequency of the light emitted from the cesium-133 atom as its "reference clock." **The second is now defined as 9 192 631 700 times the period of oscillation of radiation from the cesium atom.** The newest type of cesium atomic clock is shown in Figure 1.1b.

▶ Definition of the second

APPROXIMATE VALUES FOR LENGTH, MASS, AND TIME INTERVALS

Approximate values of some lengths, masses, and time intervals are presented in Tables 1.1, 1.2, and 1.3, respectively. Note the wide ranges of values. Study these tables to get a feel for what is meant by a kilogram of mass (this book has a mass of about 2 kilograms) or a time interval of 10^{10} seconds (one year is about $\pi \times 10^7$ seconds). Study Appendix A if you need to learn or review the notation for powers of 10, such as the expression of the number 50 000 in the form 5×10^4.

Systems of units commonly used are the Système International, in which the units of length, mass, and time are the meter (m), kilogram (kg), and second (s), respectively; the cgs or Gaussian system, in which the units of length, mass, and time are the centimeter (cm), gram (g), and second, respectively; and the U.S. customary system, in which the units of length, mass, and time are the foot (ft),

TABLE 1.1 **Approximate Values of Some Measured Lengths**

	Length (m)
Distance from Earth to most remote known quasar	1×10^{26}
Distance from Earth to most remote known normal galaxies	4×10^{25}
Distance from Earth to nearest large galaxy (M31, the Andromeda galaxy)	2×10^{22}
Distance from Earth to nearest star (Proxima Centauri)	4×10^{16}
One lightyear	9×10^{15}
Mean orbit radius of Earth about Sun	2×10^{11}
Mean distance from Earth to Moon	4×10^{8}
Mean radius of Earth	6×10^{6}
Typical altitude of satellite orbiting Earth	2×10^{5}
Length of football field	9×10^{1}
Length of housefly	5×10^{-3}
Size of smallest dust particles	1×10^{-4}
Size of cells in most living organisms	1×10^{-5}
Diameter of hydrogen atom	1×10^{-10}
Diameter of atomic nucleus	1×10^{-14}
Diameter of proton	1×10^{-15}

TABLE 1.2

Approximate Values of Some Masses

	Mass (kg)
Observable Universe	1×10^{52}
Milky Way galaxy	7×10^{41}
Sun	2×10^{30}
Earth	6×10^{24}
Moon	7×10^{22}
Shark	1×10^{2}
Human	7×10^{1}
Frog	1×10^{-1}
Mosquito	1×10^{-5}
Bacterium	1×10^{-15}
Hydrogen atom	2×10^{-27}
Electron	9×10^{-31}

TABLE 1.3 **Approximate Values of Some Time Intervals**

	Time Interval (s)
Age of Universe	5×10^{17}
Age of Earth	1×10^{17}
Average age of college student	6×10^{8}
One year	3×10^{7}
One day (time required for one revolution of Earth about its axis)	9×10^{4}
Time between normal heartbeats	8×10^{-1}
Period[a] of audible sound waves	1×10^{-3}
Period of typical radio waves	1×10^{-6}
Period of vibration of atom in solid	1×10^{-13}
Period of visible light waves	2×10^{-15}
Duration of nuclear collision	1×10^{-22}
Time required for light to travel across a proton	3×10^{-24}

[a] A *period* is defined as the time required for one complete vibration.

TABLE 1.4

Some Prefixes for Powers of 10 Used with "Metric" (SI and cgs) Units

Power	Prefix	Abbreviation
10^{-18}	atto-	a
10^{-15}	femto-	f
10^{-12}	pico-	p
10^{-9}	nano-	n
10^{-6}	micro-	μ
10^{-3}	milli-	m
10^{-2}	centi-	c
10^{-1}	deci-	d
10^{1}	deka-	da
10^{3}	kilo-	k
10^{6}	mega-	M
10^{9}	giga-	G
10^{12}	tera-	T
10^{15}	peta-	P
10^{18}	exa-	E

slug, and second, respectively. Throughout most of this book we shall use SI units because they are almost universally accepted in science and industry. We shall make limited use of U.S. customary units in the study of mechanics.

Some of the most frequently used "metric" (SI and cgs) prefixes representing powers of 10 and their abbreviations are listed in Table 1.4. For example, 10^{-3} m is equivalent to 1 millimeter (mm), and 10^{3} m is 1 kilometer (km). Likewise, 1 kg is equal to 10^{3} g, and 1 megavolt (MV) is 10^{6} volts (V).

1.2 THE BUILDING BLOCKS OF MATTER

A 1-kg (≈ 2 pounds) cube of solid gold has a length of about 3.73 cm (≈ 1.5 in.) on a side. Is this cube nothing but wall-to-wall gold, with no empty space? If the cube is cut in half, the two resulting pieces still retain their chemical identity as solid gold. But what if the pieces of the cube are cut again and again, indefinitely? Will the smaller and smaller pieces always be the same substance, gold? Questions such as these can be traced back to early Greek philosophers. Two of them—Leucippus and Democritus—could not accept the idea that such cutting could go on forever. They speculated that the process ultimately must end when it produces a particle that can no longer be cut. In Greek, *atomos* means "not sliceable." From this comes our English word *atom,* once believed to be the smallest, ultimate particle of matter. Elementary-particle physicists still engage in speculation and experimentation concerning the ultimate building blocks of matter.

Let us review briefly what is known about the ultimate structure of the world around us. It is useful to view the atom as a miniature Solar System with a dense, positively charged nucleus occupying the position of the Sun and negatively charged electrons orbiting like planets. This model of the atom enables us to understand certain properties of the simpler atoms, such as hydrogen, but fails to explain many fine details of atomic structure.

Following the discovery of the nucleus in the early 1900s, the question arose: Does it have structure? That is, is the nucleus a single particle or a collection of particles? The exact composition of the nucleus has not been defined completely even today, but by the early 1930s a model evolved that helps us understand how the nucleus behaves. Scientists determined that occupying the nucleus are two

Webnote 1.2

For a brief overview of the atom and the history of the atom, see *http://encarta.msn.com/find/ concise.asp?mod=1&ti=761567432*

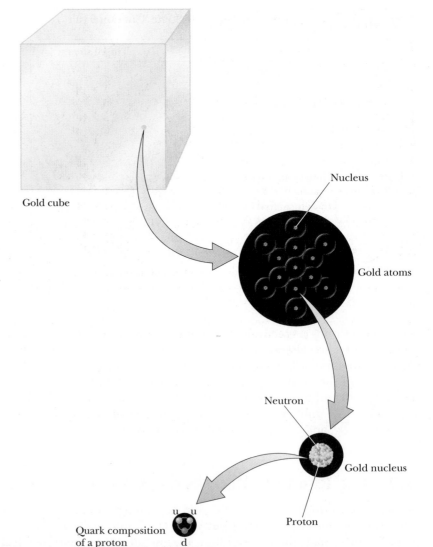

FIGURE 1.2 Levels of organization in matter. Ordinary matter consists of atoms, and at the center of each atom is a compact nucleus consisting of protons and neutrons. Protons and neutrons are composed of quarks. The quark composition of a proton is shown.

Gold cube

Nucleus

Gold atoms

Neutron

Gold nucleus

Proton

Quark composition of a proton

basic entities, protons and neutrons. The *proton* is nature's fundamental carrier of positive charge (equal in magnitude but opposite in sign to the charge on the electron), and the number of protons in a nucleus determines what element the material is. For instance, one proton in the nucleus means that the atom is an atom of hydrogen, regardless of how many neutrons may be present; two protons mean an atom of helium, and so on.

The existence of *neutrons* was verified conclusively in 1932. A neutron has no charge and a mass about equal to that of a proton. One of its primary purposes is to act as a "glue" to hold the nucleus together. If neutrons were not present, the repulsive electrical force between the positively charged particles would cause the nucleus to fly apart.

But is this where the division into smaller and smaller particles stops? As we shall explore more carefully in Chapter 30, building blocks even more elementary than protons and neutrons exist. Protons, neutrons, and a zoo of other exotic particles are now thought to be composed of six particles called **quarks** (rhymes with "forks"), which have been given the names *up, down, strange, charmed, bottom,* and *top*. The up, charmed, and top quarks have charges of $+\frac{2}{3}$ that of the proton, whereas the down, strange, and bottom quarks have charges of $-\frac{1}{3}$ that of the proton. The proton consists of two up quarks and one down quark (see Fig. 1.2), which you can easily show leads to the correct charge for the proton, $+1$. The neutron is composed of two down quarks and one up quark, giving a net charge of zero.

1.3 DIMENSIONAL ANALYSIS

The word *dimension* has a special meaning in physics. It usually denotes the physical nature of a quantity. Whether the separation between two points is measured in units of feet, meters, or furlongs, it is a distance. We say its dimension is *length*.

The symbols that we use in this section to specify length, mass, and time are L, M, and T, respectively. We often use brackets [ ] to denote the dimensions of a physical quantity. For example, in this notation the dimensions of velocity v are written $[v] = L/T$, and the dimensions of area A are $[A] = L^2$. The dimensions of area, volume, velocity, and acceleration are listed in Table 1.5, along with their units in the three common systems. The dimensions of other quantities, such as force and energy, will be described later as they are introduced.

In many situations, you may have to either derive or check the correctness of a specific formula (mathematical expression or equation). Although you may have forgotten the details of the derivation, you can use a powerful procedure called **dimensional analysis,** which makes use of the fact that **dimensions can be treated as algebraic quantities:** That is, quantities can be added or subtracted only if they have the same dimensions. Furthermore, the terms on the two sides of an equation must have the same dimensions. By following these simple rules, you can use dimensional analysis to help determine whether or not an expression has the correct form. The relationship can be correct only if the dimensions on both sides of the equation are the same.

TABLE 1.5	Dimensions and Some Units of Area, Volume, Velocity, and Acceleration			
System	**Area (L^2)**	**Volume (L^3)**	**Velocity (L/T)**	**Acceleration (L/T^2)**
SI	m^2	m^3	m/s	m/s^2
cgs	cm^2	cm^3	cm/s	cm/s^2
U.S. customary	ft^2	ft^3	ft/s	ft/s^2

To illustrate this procedure, suppose you wish to derive a formula for the distance x traveled by a car in a time t if the car starts from rest and moves with constant acceleration a. In Chapter 2 we shall find that the correct expression for this special case is $x = \frac{1}{2}at^2$. Let us check the validity of this expression from a dimensional analysis approach.

The quantity x has the dimension length. In order for the equation to be dimensionally correct, the quantity on the right side of the equals sign must also have the dimension length. We can perform a dimensional check by substituting the dimensions for acceleration, L/T^2, and time, T, into the equation. That is, the dimensional form of $\frac{1}{2}at^2$ can be written

$$\frac{L}{T^2} \cdot T^2 = L$$

The dimensions of time cancel as shown, leaving the dimension of length. Thus, the left and right sides of the equation have the same dimension, length, and the equation is correct within a numerical factor. Note that the factor $\frac{1}{2}$ is ignored because it has no dimensions. Dimensional analysis cannot give us numerical factors; this is its characteristic limitation.

Example 1.1 Analysis of an Equation

Show that the expression $v = v_0 + at$ is dimensionally correct, where v and v_0 represent velocities, a is acceleration, and t is time.

Solution Because

$$[v] = [v_0] = \frac{L}{T}$$

(see Table 1.5) and the dimensions of acceleration are L/T^2, the dimensions of at are

$$[at] = \frac{L}{T^2}\,(T) = \frac{L}{T}$$

and the expression is dimensionally correct. However, if the expression were given as $v = v_0 + at^2$, it would be dimensionally *incorrect*. Try it and see!

Webnote 1.3

For a tutorial on dimensional analysis, go to
http://www.physics.uoguelph.ca/tutorials/dimanaly/

1.4 UNCERTAINTY IN MEASUREMENT AND SIGNIFICANT FIGURES

Physics is a science in which mathematical laws are tested by experiment. For example, we may wish to test Newton's famous law relating the net force F, the mass m, and the acceleration a of an object, $F = ma$, by measuring certain quantities. Because these measured values are known only to within the limits of the experimental uncertainty, we can say only that $F = ma$ holds within certain limits. It is most important to state the value of this uncertainty, which can depend on such factors as the quality of the apparatus, the skill of the experimenter, and the number of measurements performed. There are many ways of handling uncertainties, but here we want to introduce a very basic method of keeping track of uncertainties in measured physical quantities and in calculations based on these measured quantities.

Suppose that in a laboratory experiment we are asked to measure the area of a rectangular plate with a meter stick. Let us assume that the accuracy to which we can measure a particular dimension of the plate is ± 0.1 cm. If the length of the plate is measured to be 16.3 cm, we can claim only that this length lies somewhere between 16.2 cm and 16.4 cm. In this case, we say that the measured value has three significant figures. Likewise, if the plate's width is measured to be 4.5 cm, the actual value lies between 4.4 cm and 4.6 cm. This measured value has only two significant figures. Thus, we could write the measured values as 16.3 ± 0.1 cm and 4.5 ± 0.1 cm. In general, **a significant figure is a reliably known digit** (other than a zero used to locate a decimal point).

Suppose now that we would like to find the area of the plate by multiplying the two measured values together. The value can range between $(16.3 - 0.1$ cm$)$ $(4.5 - 0.1$ cm$) = (16.2$ cm$)(4.4$ cm$) = 71$ cm^2 and $(16.3 + 0.1$ cm$)(4.5 + 0.1$ cm$) = (16.4$ cm$)(4.6$ cm$) = 75$ cm^2. If we were to claim that the area is $(16.3$ cm$)$ $(4.5$ cm$) = 73.35$ cm^2, our answer would be unjustified, because an answer of 73.35 cm^2 means we are sure of the area to within ± 0.01 cm^2, yet the results above show that the answer is about 73 cm^2 with an uncertainty of ± 2 cm^2. We must drop the 0.35 in 73.35 and keep only two significant figures, 73. Note that the answer has two significant figures, the same number of figures as the least accurately known quantity being multiplied, the 4.5-cm width. A good rule of thumb for determining the number of significant figures that can be claimed is as follows. **When multiplying two or more quantities, the number of significant figures in the final product is the same as the number of significant figures in the *least accurate* of the quantities being multiplied, where *least accurate* means *having the lowest number of significant figures*. The same rule applies to division.**

Applying this rule to the preceding multiplication example, we see that the answer for the area can have only two significant figures because the dimension 4.5 cm has only two significant figures. Thus, we must report the area to be 73 ± 2 cm^2, realizing that the value can range between $(16.2$ cm$)(4.4$ cm$) = 71$ cm^2 and $(16.4$ cm$)(4.6$ cm$) = 75$ cm^2. The 73 represents the number of significant figures and the ± 2 represents the uncertainty.

Zeros may or may not be significant figures. Zeros used to position the decimal point in such numbers as 0.03 and 0.0075 are not significant (but are useful to avoid errors). Thus, there are one and two significant figures, respectively, in these two values. When zeros are placed after other digits in a whole number, there is the possibility of misinterpretation. For example, suppose the mass of an object is given as 1 500 g. This value is ambiguous because we do not know whether the last two zeros are being used to locate the decimal point or whether they represent significant figures in the measurement. In order to remove this ambiguity, it is common to use scientific notation to indicate the number of significant figures. In this case, we would express the mass as 1.5×10^3 g if there are two significant figures in the measured value, 1.50×10^3 g if there are three significant figures, and 1.500×10^3 g if there are four. Likewise, 0.000 15 should be expressed in scientific notation as 1.5×10^{-4} if it has two significant figures or as 1.50×10^{-4} if it has three significant figures. The three zeros between the decimal point and the digit 1 in the number 0.000 15 are not counted as significant figures because they are present only to locate the decimal point. In this book, **most of the numerical examples and end-of-chapter problems will yield answers having either two or three significant figures.**

For addition and subtraction, it is best to focus on the number of decimal places in the quantities involved rather than on the number of significant figures. **When numbers are added or subtracted, the number of decimal places in the result should equal the smallest number of decimal places of any term in the sum.** For example, if we wish to compute 123 (zero decimal places) + 5.35 (two decimal places), the answer is 128 (zero decimal places) and not 128.35. If we compute the sum 1.000 1 (four decimal places) + 0.000 3 (four decimal places) = 1.000 4, the result has the correct number of decimal places, namely four; observe that the rules for multiplying significant figures do not work here because the answer has five significant figures even though one of the terms in the sum, 0.000 3, has only one significant figure. Likewise, if we perform the subtraction

This sundial is located in Majorca, Spain. The principle of the sundial is as follows. If the number 12 points to true north, the shadow gives the correct standard time. *(David R. Frazier/Photo Researchers, Inc.)*

Tip 1.1 USING CALCULATORS

Calculators were designed by engineers to yield as many digits as the memory of the calculator chip permitted. Most often, values shown on electronic calculators will give you more significant figures than you use for a given problem. Therefore, the answer must be rounded *down* to the correct number of figures. For the most accurate results, keep an extra significant figure in intermediate calculations and round to the correct number of significant figures only in the final result.

Tip 1.2 NO COMMAS IN NUMBERS WITH MANY DIGITS

In science, numbers with more than three digits are written with groups of three digits separated by spaces rather than commas. Thus, 10 000 is the same as the common American notation of 10,000. Similarly, $\pi = 3.14159265$ is written as 3.141 592 65.

$1.002 - 0.998 = 0.004$, the result has three decimal places because each term in the subtraction has three decimal places.

A short example can help you see where the decimal place rule for adding and subtracting measured quantities comes from. Look at the first example of addition above, where we added 123 and 5.35 and rewrite these numbers as 123.xxx and 5.35x. What we mean by this notation is that digits written with an x are completely unknown and can be any digit from 0 to 9. Now line up 123.xxx and 5.35x relative to the decimal point and perform the addition using the rule that an unknown digit added to a known or unknown digit yields an unknown:

$$123.\mathrm{xxx}$$
$$+\quad 5.35\mathrm{x}$$
$$128.\mathrm{xxx}$$

The answer of 128.xxx means that we are justified only in keeping the number 128 because everything after the decimal point in the sum is unknown "junk." The example shows that the controlling uncertainty is introduced into an addition or subtraction by the term with the smallest number of decimal places.

Example 1.2   Installing a Carpet

A carpet is to be installed in a room whose length is measured to be 12.71 m (four significant figures) and whose width is measured to be 3.46 m (three significant figures). Find the area of the room.

Solution   If you multiply 12.71 m by 3.46 m on your calculator, you will get the answer $43.976\,6\ \mathrm{m}^2$. How many of these numbers should you claim? Our rule of thumb for multiplication says that you can claim only the number of significant figures in the least accurate of the quantities being measured. In this example, we have three significant figures in our less accurate measurement, so we should express our final answer as $44.0\ \mathrm{m}^2$. Note that in reducing 43.976 6 to three significant figures, we used a general rule for rounding off numbers, which states that the last digit retained (the 9 in this example) is increased by 1 if the first digit dropped (here the 7) is 5 or greater. (A technique for avoiding error accumulation in a long calculation is to delay rounding off numbers until you have the final result.)

1.5   CONVERSION OF UNITS

Sometimes it is necessary to convert units from one system to another. Conversion factors between the SI and U.S. customary systems for units of length are as follows:

$$1\ \text{mile} = 1609\ \text{m} = 1.609\ \text{km} \qquad 1\ \text{ft} = 0.3048\ \text{m} = 30.48\ \text{cm}$$
$$1\ \text{m} = 39.37\ \text{in.} = 3.281\ \text{ft} \qquad 1\ \text{in.} = 0.0254\ \text{m} = 2.54\ \text{cm}$$

A more extensive list of conversion factors can be found on the inside of the front cover.

Units can be treated as algebraic quantities that can cancel each other. For example, suppose we wish to convert 15.0 in. to centimeters. Because 1 in. = 2.54 cm, we find that

$$15.0\ \text{in.} = 15.0\ \cancel{\text{in.}} \times 2.54\ \frac{\text{cm}}{\cancel{\text{in.}}} = 38.1\ \text{cm}$$

This road sign near Raleigh, North Carolina, shows distances in miles and kilometers. How accurate are the conversions? *(Billy E. Barnes/Stock Boston)*

Example 1.3 Pull Over, Buddy!

If a car is traveling at a speed of 28.0 m/s, is it exceeding the speed limit of 55.0 mi/h?

Solution We can perform the conversion from meters per second to miles per hour in two steps—meters to miles and then seconds to hours. Converting the length units, we find that

$$28.0 \text{ m/s} = \left(28.0 \frac{\text{m}}{\text{s}}\right)\left(\frac{1 \text{ mi}}{1609 \text{ m}}\right) = 1.74 \times 10^{-2} \text{ mi/s}$$

Now we finish the conversion by converting seconds to hours, as follows:

$$1.74 \times 10^{-2} \text{ mi/s} = \left(1.74 \times 10^{-2} \frac{\text{mi}}{\text{s}}\right)\left(60 \frac{\text{s}}{\text{min}}\right)\left(60 \frac{\text{min}}{\text{h}}\right) = 62.6 \text{ mi/h}$$

The car should slow down because it is exceeding the speed limit.

An alternative approach is to use the single conversion relationship 1 m/s = 2.237 mi/h:

$$28.0 \text{ m/s} = \left(28.0 \frac{\text{m}}{\text{s}}\right)\left(\frac{2.237 \text{ mi/h}}{1 \text{ m/s}}\right) = 62.6 \text{ mi/h}$$

Frequently in this book, an example will be followed by an exercise whose purpose is to test your understanding of the concept covered in the example. The answer will be provided immediately after the exercise, when appropriate. Here is your first exercise, which relates to Example 1.3.

EXERCISE Use the speedometer pictured in Figure 1.3 to see whether the foregoing result is in agreement with the speedometer dial (calibrated in miles per hour and kilometers per hour).

Webnote 1.4

For a link to an online unit converter, see the URL below. For homework and exams you still must learn how to convert units yourself, using a table of conversion factors.

http://www.digitaldutch.com/unitconverter/

FIGURE 1.3 (Example 1.3) This speedometer gives readings in both miles per hour and kilometers per hour. See if you can confirm the conversions for a few readings on the dial. *(Paul Silverman, Fundamental Photographs)*

1.6 ORDER-OF-MAGNITUDE CALCULATIONS

Often it is useful to estimate an answer to a problem in which little information is given. This answer can then be used to determine whether or not a more precise calculation is necessary. Such an approximation is usually based on certain assumptions that must be modified if greater precision is needed. Sometimes it is necessary to know a quantity only within a factor of 10. In such a case we refer to the **order of magnitude** of the quantity, by which we mean the power of 10 that is closest to the actual value of the quantity. We use the symbol $\sim$ for "is on the order of." For example, if the mass of a person is 75 kg, we would say that the person's mass is on the order of 10^2 kg, or 75 kg $\sim 10^2$ kg. Usually, when an order-of-magnitude calculation is made, the results are reliable to within a factor of 10. If a quantity increases in value by three orders of magnitude, this means that its value increases by a factor of $10^3 = 1000$.

Example 1.4   How Much Gasoline Do We Use?

Estimate the number of gallons of gasoline used by all U.S. cars each year.

Solution   Because there are about 280 million people in the United States, an estimate of the number of cars in the country is 7×10^7 (assuming one car and four people per family). We can also estimate that the average distance traveled per year is 1×10^4 miles. If we assume gasoline consumption of 0.05 gal/mi (equivalent to 20 miles per gallon), each car uses about 5×10^2 gal/year. Multiplying this by the total number of cars in the United States gives an estimated total consumption on the order of 10^{11} gal, which corresponds to a yearly consumer expenditure on the order of $100 billion at a price of $1.50 per gallon! This is a low estimate because we haven't accounted for commercial consumption.

Exercise   The 1994 Honda Civic DX gets 47 mi/gal at highway speeds of 65 mi/h. Approximately how much gasoline would be saved each year if everyone drove cars which got even 40 mi/gal?

Answer $\sim 10^{10}$ gal

Example 1.5   The High Price of Gold

All the gold that has ever been mined could be melted and molded into a cube approximately 20 m on a side. Use the fact that 1 kg of gold occupies a cube 3.73 cm on a side and the fact that gold sells for about $12 000 per kg to estimate the value of all the mined gold existing on this planet.

Solution   The volume of a block 20 m on a side is $(20 \text{ m})(20 \text{ m})(20 \text{ m}) = 8\ 000 \text{ m}^3$, and the volume of 1 kg of gold is

$$(3.73 \times 10^{-2} \text{ m})(3.73 \times 10^{-2} \text{ m})(3.73 \times 10^{-2} \text{ m}) \approx 5 \times 10^{-5} \text{ m}^3$$

Thus, our 8 000-m^3 block of gold would have a mass of

$$\frac{8\ 000 \text{ m}^3}{5 \times 10^{-5} \text{ m}^3/\text{kg}} = 1.5 \times 10^8 \text{ kg}$$

At $12 000 per kg, this mass of gold would cost $\sim \$10^{12}$.

1.7   Coordinate Systems

Many aspects of physics deal with locations in space. For example, the mathematical description of the motion of an object requires a method for describing the position of the object. Thus, it is fitting that we first discuss how to describe the position of a point in space. This is done by means of coordinates. A point on a line can be located with one coordinate; a point in a plane is located with two coordinates; and three coordinates are required to locate a point in space.

A coordinate system used to specify locations in space consists of

- A fixed reference point O, called the *origin*
- A set of specified axes, or directions, with an appropriate scale and labels on the axes
- Instructions that tell us how to label a point in space relative to the origin and axes

One convenient coordinate system that we shall use frequently is the **Cartesian coordinate system,** sometimes called the **rectangular coordinate system.** Such a system in two dimensions is illustrated in Figure 1.4. An arbitrary point in this system is labeled with the coordinates (x, y). For example, the point P in the figure has coordinates $(5, 3)$. This notation means that if we start at the origin (O), we can reach P by moving 5 meters to the right of the origin and then 3 meters above the origin. (Positive x is usually selected to the right of the origin and positive y upward from the origin. Negative x is usually to the left of the origin and negative y downward from the origin. This convention is not an absolute necessity, however. There may be instances in which you would like to take positive x to the left of the origin and negative x to the right. Feel free to do so.) In the same way, the point Q has coordinates $(-3, 4)$, which corresponds to going 3 meters to the left of the origin and 4 meters above the origin.

Sometimes it is more convenient to locate a point in space by its **plane polar coordinates** (r, θ), as in Figure 1.5. In this coordinate system, an origin O and a reference line are selected, as shown. A point is then specified by the distance r from the origin to the point and by the angle θ between the reference line and a line drawn from the origin to the point. (Frequently, the reference line is selected to be the positive x axis of a Cartesian coordinate system.) The angle θ is considered positive when measured counterclockwise from the reference line and negative when measured clockwise. For example, if a point is specified by the polar coordinates 3 m and 60°, we locate this point by moving out 3 m from the origin at an angle of 60° above (or counterclockwise from) the reference line. A point specified by polar coordinates 3 m and $-60°$ is located 3 m out from the origin and 60° below (or clockwise from) the reference line.

1.8 TRIGONOMETRY

The portion of mathematics that is based on the special properties of a right triangle is called trigonometry. Many of the concepts of this branch of mathematics are of utmost importance in the study of physics. We will now review some of the more important fundamental definitions and ideas that you will need to know.

Consider the right triangle shown in Figure 1.6, where side a is opposite the angle θ, side b is adjacent to the angle θ, and side c is the hypotenuse of the triangle. The basic trigonometric functions defined by such a triangle are the ratios of the lengths of the sides of the triangle. Specifically, the most important relationships are called the sine (sin), cosine (cos), and tangent (tan) functions. In terms of the angle θ, these functions are[1]

$$\sin \theta = \frac{\text{side opposite } \theta}{\text{hypotenuse}} = \frac{a}{c}$$

$$\cos \theta = \frac{\text{side adjacent to } \theta}{\text{hypotenuse}} = \frac{b}{c} \qquad \textbf{[1.1]}$$

$$\tan \theta = \frac{\text{side opposite } \theta}{\text{side adjacent to } \theta} = \frac{a}{b}$$

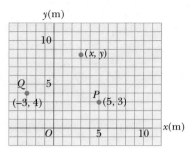

FIGURE 1.4 Designation of points in a two-dimensional Cartesian coordinate system. Every point is labeled with coordinates (x, y).

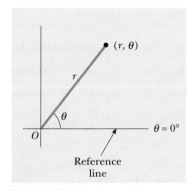

FIGURE 1.5 A polar coordinate system.

Webnote 1.5

For a link to more information on coordinate systems, see *http://www-spof.gsfc.nasa.gov/stargaze/Scelcoor.htm*

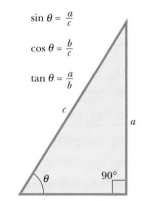

$\sin \theta = \frac{a}{c}$

$\cos \theta = \frac{b}{c}$

$\tan \theta = \frac{a}{b}$

FIGURE 1.6 Certain trigonometric functions for a right triangle.

---

[1] In order to recall these trigonometric relationships, consider using the following memory device. Remember the word *SOHCAHTOA* whose letters stand for *S*ine = *O*pposite/*H*ypotenuse, *C*osine = *A*djacent/*H*ypotenuse, and *T*angent = *O*pposite/*A*djacent. (Thanks go to Professor Don Chodrow for pointing this out.)

For example, if the angle θ is equal to 30°, it is found that the ratio of *a* to *c* is always 0.50; that is, sin 30° = 0.50. Note that the sin, cos, and tan are quantities without units because each represents the ratio of two lengths.

Another important relationship, called the **Pythagorean theorem,** exists between the lengths of the sides of a right triangle; it is

$$c^2 = a^2 + b^2 \qquad \text{[1.2]}$$

Finally, it will often be necessary to find the value of inverse relationships. For example, suppose you know the value of the sine of an angle is 0.866, but you need to know the value of the angle itself. The inverse sine function may be expressed as $\sin^{-1}(0.866)$, which is a shorthand way of asking the question "What angle has a sine of 0.866?" Punching a couple of buttons on your calculator reveals that this angle is 60.0°. Try it for yourself and show that $\tan^{-1}(0.400) = 21.8°$. (Be sure that your calculator is set for degrees and not radians.)

The following examples will give you a little practice in working with these trigonometric principles in practical situations.

Example 1.6   How High Is the Building?

A person attempts to measure the height of a building by walking out a distance of 46.0 m from its base and shining a flashlight beam toward its top. He finds that when the beam is elevated at an angle of 39.0° with respect to the horizontal, as shown in Figure 1.7, the beam just strikes the top of the building. Find the height of the building and the distance the flashlight beam has to travel before it strikes the top of the building.

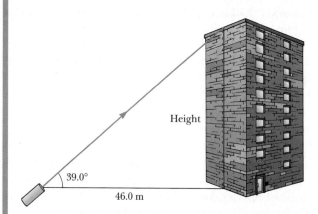

FIGURE 1.7   (Example 1.6)

Solution   Refer to the right triangle shown in the figure. We know the length of the side adjacent to the 39.0° angle, and we know the value of the angle. Thus, we can use the definition of the tangent function to find the height of the building, which is the side opposite the 39.0° angle.

$$\tan 39.0° = \frac{\text{Height}}{46.0 \text{ m}}$$

or

$$\text{Height} = (\tan 39.0°)(46.0 \text{ m}) = (0.810)(46.0 \text{ m}) = \boxed{37.3 \text{ m}}$$

Now that we know the lengths of both the side adjacent to the angle and the side opposite the angle, we can find the length of the hypotenuse, which is the distance *c* that the beam travels before it strikes the top of the building.

$$c = \sqrt{a^2 + b^2} = \sqrt{(37.3 \text{ m})^2 + (46.0 \text{ m})^2} = \boxed{59.2 \text{ m}}$$

EXERCISE   Use a different method to find the distance traveled by the light. Would another trigonometric function, such as the sin or cos, give you the answer? Try it.

Example 1.7   Things to Do When the CB Radio Goes Bad

A truck driver moves up a straight mountain highway, as shown in Figure 1.8. Elevation markers at the beginning and ending points of the trip show that he has risen vertically 0.530 km, and the mileage indicator on the truck shows that he has traveled a total distance of 3.00 km during the ascent. Find the angle of incline of the hill.

FIGURE 1.8   (Example 1.7)

Solution   This is an example in which we must use an inverse trigonometric relationship. From the sine of the angle, defined as the opposite side over the hypotenuse, we find

$$\sin \theta = \frac{0.530}{3.00} = 0.177$$

To find θ, we use the inverse sine relationship:

$$\theta = \sin^{-1}(0.177) = \boxed{10.2°}$$

1.9   PROBLEM-SOLVING STRATEGY

Most courses in general physics require the student to learn the skills of problem solving, and examinations usually include problems that test such skills. This brief section presents some useful suggestions that will help increase your success in solving problems, enhance your understanding of physical concepts, eliminate initial panic or lack of direction in approaching a problem, and organize your work. One way to accomplish these goals is to adopt a problem-solving strategy. Many chapters in this book will include a section labeled "Problem-Solving Strategy" that should help you through the rough spots.

The following steps, illustrated in Figure 1.9, are commonly used to develop a problem-solving strategy:

1. Read the problem carefully at least twice. Be sure you understand the nature of the problem before proceeding further.
2. Draw a suitable diagram with appropriate labels and coordinate axes, if needed.
3. Imagine a movie, running in your mind, of what happens in the problem.
4. As you examine what is being asked in the problem, identify the basic physical principle or principles that are involved, listing the knowns and unknowns.
5. Select a basic relationship or derive an equation that can be used to find the unknown, and algebraically solve the equation for the unknown.
6. Substitute the given values with the appropriate units into the equation.
7. Obtain a numerical value with units for the unknown. You can have confidence in your result if the following questions can be properly answered: Do the units match? Is the answer reasonable? Is the plus or minus sign proper or meaningful?

FIGURE 1.9   A guide for problem solving.

One purpose of this strategy is to increase your likelihood of getting the correct answer. Properly drawn diagrams can eliminate many sign errors. Diagrams also help to isolate the physical principles of the problem. Setting up your equations with carefully labeled knowns and unknowns will help eliminate other careless errors. Carrying through the algebra for as long as possible (substituting numbers only at the end) helps you think in terms of the physics of the problem. A check of units at the end of the problem can indicate a possible algebraic error. The physical layout and organization of your problem will make the final product more understandable and easier to follow. Once you have developed an organized system for examining problems and extracting relevant information, you will become a more confident problem solver.

The following example illustrates this procedure.

Example 1.8   A Round Trip by Air

An airplane travels 450 km due east and then travels an unknown distance due north. Finally, it returns to its starting point by traveling a distance of 525 km. How far did the airplane travel in the northerly direction?

Solution   After reading the problem carefully, draw a diagram of the situation, as in Figure 1.10. Because the path of the airplane forms a right triangle, you can use the Pythagorean theorem, which relates the three sides of the triangle:

$$c^2 = a^2 + b^2$$

FIGURE 1.10   (Example 1.8)

$a = 450$ km
$c = 525$ km
$b = ?$

In this problem, the hypotenuse c and the side a are known quantities. We must first solve symbolically for the unknown quantity b *before inserting numerical values.* Subtracting a^2 from both sides of the preceding equation and moving the unknown b^2 to the left gives

$$b^2 = c^2 - a^2$$

We now take the square root of both sides to find the solution for b:

$$b = +\sqrt{c^2 - a^2}$$

Note that the negative solution has been disregarded because it is not physically meaningful.

Finally, we substitute the given values for the quantities c and a to find a numerical value for b:

$$b = \sqrt{(525 \text{ km})^2 - (450 \text{ km})^2} = \boxed{270 \text{ km}}$$

SUMMARY

The physical quantities we shall encounter in our study of mechanics can be expressed in terms of three fundamental quantities, length, mass, and time—which have the SI units meters (m), kilograms (kg), and seconds (s), respectively. It is often helpful to use dimensional analysis to check equations and to assist in deriving equations.

When inserting numerical factors into equations, you must be sure that the dimensions of these factors are consistent throughout the equation. In many cases it will be necessary to use the table of conversion factors on the inside front cover to convert from one system of units to another.

Often it is useful to estimate an answer to a problem in which little information is given. Such estimates are called **order-of-magnitude calculations.**

The three most basic trigonometric functions for a right triangle are the sine, cosine, and tangent, defined as

$$\sin \theta = \frac{\text{side opposite } \theta}{\text{hypotenuse}}$$

$$\cos \theta = \frac{\text{side adjacent to } \theta}{\text{hypotenuse}} \qquad \text{[1.1]}$$

$$\tan \theta = \frac{\text{side opposite } \theta}{\text{side adjacent to } \theta}$$

The **Pythagorean theorem** is an important relationship between the lengths of the sides of a right triangle:

$$c^2 = a^2 + b^2 \qquad \text{[1.2]}$$

where c is the hypotenuse of the triangle and a and b are its other two sides.

CONCEPTUAL QUESTIONS

1. Estimate the order of magnitude of the length, in meters, of each of the following: (a) a mouse, (b) a pool cue, (c) a basketball court, (d) an elephant, (e) a city block.

2. Long ago, in England, the inch was defined as the length of three barley grains placed end to end. In what ways was this a good way to define a standard? In what ways was this a poor way to define a standard?

3. (a) Estimate the number of times your heart beats in a month. (b) Estimate the number of human heartbeats in an average lifetime.

4. An object with a mass of 1 kg weighs a little more than 2 lb. Use this information to estimate the mass of the following objects: (a) an apple seed; (b) the rest of the apple; (c) a football player; (d) a car.

5. Find the order of magnitude of your age in seconds.

6. Estimate the number of atoms in 1 cm^3 of a solid. (Note that the diameter of an atom is about 10^{-10} m.)

7. The height of a horse is sometimes given in units of "hands." Why is this a poor standard of length?

8. How many of the lengths or time intervals given in Tables 1.2 and 1.3 could you verify using only equipment found in a typical dormitory room?

9. An ancient unit of length called the *cubit* was equal to six palms, where a palm was the width of the four fingers of an open hand. Noah's ark was 300 cubits long, 50 cubits wide, and 30 cubits high. Estimate the volume of the ark in cubic meters. Also, estimate the volume of a typical home in cubic meters and compare it to the volume of the ark.

10. If an equation is dimensionally correct, does this mean that the equation must be true?

11. Estimate the cost per pound of a typical automobile. How does your answer compare to the cost of a good steak?

12. Figure Q1.12 is a photograph showing unit conversions on the labels of some grocery-store items. Check the accuracy of these conversions. Are the manufacturers using significant figures correctly?

FIGURE Q1.12 *(Courtesy of George Semple)*

PROBLEMS

1, 2, 3 = straightforward, intermediate, challenging   ☐ = full solution available in Student Solutions Manual/Study Guide

web = solution posted **http://info.brookscole.com/serway**   🔬 = biomedical application

Section 1.3   Dimensional Analysis

1. In a desperate attempt to come up with an equation to use during an examination, a student tries $v^2 = ax$. Use dimensional analysis to determine whether this equation might be valid.

2. The radius r of a circle inscribed in any triangle whose sides are a, b, and c is given by $r = [(s - a)(s - b)(s - c)/s]^{1/2}$, where s is an abbreviation for $(a + b + c)/2$. Check this formula for dimensional consistency.

3. The period of a simple pendulum, defined as the time for one complete oscillation, is measured in time units and is given by

$$T = 2\pi \sqrt{\frac{\ell}{g}}$$

where ℓ is the length of the pendulum and g is the acceleration due to gravity, in units of length divided by time squared. Show that this equation is dimensionally consistent. (You might want to check the formula using your keys at the end of a string and a stopwatch.)

4. Which of the equations below are dimensionally correct?
(a) $v = v_0 + ax$ (b) $y = (2 \text{ m})\cos(kx)$, where $k = 2 \text{ m}^{-1}$.

5. Newton's law of universal gravitation is represented by

$$F = G\frac{Mm}{r^2}$$

Here F is the gravitational force, M and m are masses, and r is a length. Force has the SI units kg·m/s^2. What are the SI units of the proportionality constant G?

6. (a) One of the fundamental laws of motion states that the acceleration of an object is directly proportional to the resultant force on it and inversely proportional to its mass. If the proportionality constant is defined to have no dimensions, determine the dimensions of force. (b) The newton is the SI unit of force. According to the results for (a), how can you express a force having units of newtons using the fundamental units of mass, length, and time?

Section 1.4   Uncertainty in Measurements and Significant Figures

7. How many significant figures are there in (a) 78.9 ± 0.2, (b) 3.788×10^9, (c) 2.46×10^{-6}, (d) 0.0032?

8. Two I-beams are to be joined end to end. One has been measured to be 3.456 m long; the other, 4.3 m. When they are bolted snugly together, what value can you accurately give for the total length?

9. Carry out the following arithmetic operations: (a) the sum of the measured values 756, 37.2, 0.83, and 2.5; (b) the product 0.0032×356.3; (c) the product $5.620 \times \pi$.

web

10. The value of the speed of light is now defined to be $2.997\ 924\ 58 \times 10^8$ m/s. Express the speed of light to (a) three significant figures, (b) five significant figures, and (c) seven significant figures.

11. A farmer measures the distance around a rectangular field. The length of each long side of the rectangle is found to be 38.44 m, and the length of each short side is found to be 19.5 m. What is the total distance around the field?

12. Calculate (a) the circumference of a circle of radius 3.5 cm and (b) the area of a circle of radius 4.65 cm.

13. A fisherman catches two striped bass. The smaller of the two has a measured length of 93.46 cm (two decimal places, four significant figures), and the larger fish has a measured length of 135.3 cm (one decimal place, four significant figures). What is the total length of fish caught for the day?

14. (a) Using your calculator, find, in scientific notation with appropriate rounding, the value of $(2.437 \times 10^4)(6.5211 \times 10^9)/(5.37 \times 10^4)$. (b) Find the value of $(3.14159 \times 10^2)(27.01 \times 10^4)/(1234 \times 10^6)$.

Section 1.5   Conversion of Units

15. A fathom is a method for measuring lengths, usually reserved for measuring the depth of water. A fathom is approximately 6 ft in length. Take the distance to the Moon to be 250 000 miles, and use the approximation above to find the distance in fathoms.

16. Find the height or length of these natural wonders in kilometers, meters, and centimeters. (a) The longest cave system in the world is the Mammoth Cave system in Central Kentucky. It has a mapped length of 348 miles. (b) In the United States, the waterfall with the greatest single drop is Ribbon Falls in California which drops 1 612 ft. (c) Mount McKinley in Alaska is America's highest mountain at 20 320 feet. (d) The deepest canyon in the United States is King's Canyon in California with a depth of 8 200 ft.

17. An aspirin tablet contains 325 mg of acetylsalicylic acid. Express this mass in grams.

18. Suppose your hair grows at the rate of 1/32 inch per day. Find the rate at which it grows in nanometers per second. Since the distance between atoms in a molecule is on the order of 0.1 nm, your answer suggests how rapidly atoms are assembled in this protein synthesis.

19. Find the distance to the nearest star, in feet, using the data in Table 1.1 and the appropriate conversion factors.

20. Find the age of Earth in years, using the data in Table 1.3 and the appropriate conversion factors.

21. The speed of light is about 3.00×10^8 m/s. Convert this to miles per hour.

22. A house is 50.0 ft long and 26 ft wide, and has 8.0-ft-high ceilings. What is the volume of the interior of the house in cubic meters and in cubic centimeters?

23. The amount of water in reservoirs is often measured in acre-ft. One acre-ft is a volume that covers an area of one acre to a depth of one foot. An acre is an area of 43 560 ft^2. Find the volume in SI units of a reservoir containing 25.0 acre-ft of water.

24. The base of a pyramid covers an area of 13.0 acres (1 acre = 43 560 ft^2) and has a height of 481 ft (Fig. P1.24). If the volume of a pyramid is given by the expression

$V = (1/3) bh$, where b is the area of the base and h is the height, find the volume of this pyramid in cubic meters.

FIGURE P1.24   *(© Sylvain Grandadam/Photo Researchers, Inc.)*

25. A quart container of ice cream is to be made in the form
web of a cube. What should be the length of a side, in centimeters? (Use the conversion 1 gallon = 3.786 liter.)

26. One gallon of paint (volume = 3.786×10^{-3} m^3) covers an area of 25.0 m^2. What is the thickness of the paint on the wall?

27. One cubic centimeter (1.0 cm^3) of water has a mass of 1.0×10^{-3} kg. (a) Determine the mass of 1.0 m^3 of water. (b) Assuming biological substances are 98% water, estimate the masses of a cell with a diameter of 1.0 μm, a human kidney, and a fly. Assume a kidney is roughly a sphere with a radius of 4.0 cm, and a fly is roughly a cylinder 4.0 mm long and 2.0 mm in diameter.

28. A billionaire offers to give you $1 billion if you can count it out using only $1 bills. Should you accept her offer? Assume that you can count at an average rate of one bill every second, and be sure to allow for the fact that you need about 8 hours a day for sleeping and eating.

Section 1.6   Order-of-Magnitude Calculations

Note: In developing answers to the problems in this section, you should state your important assumptions, including the numerical values assigned to parameters used in the solution.

29. Imagine that you are the equipment manager of a professional baseball team. One of your jobs is to keep baseballs on hand for games. Balls are sometimes lost when players hit them into the stands as either home runs or foul balls. Estimate how many baseballs you have to buy per season in order to make up for such losses. Assume your team plays an 81-game home schedule in a season.

30. A hamburger chain advertises that it has sold more than 50 billion hamburgers. Estimate how many pounds of hamburger meat must have been used by the restaurant chain and how many head of cattle were required to furnish the meat.

31. An automobile tire is rated to last for 50 000 miles. Estimate the number of revolutions the tire will make in its lifetime.

32. Grass grows densely everywhere on a quarter-acre plot of land. What is the order of magnitude of the number of blades of grass? Explain your reasoning. Note that 1 acre = 43 560 ft^2.

33. Estimate the number of Ping-Pong balls that would fit into a typical-size room (without being crushed).

34. Soft drinks are commonly sold in aluminum containers. Estimate the number of such containers thrown away each year by U.S. consumers. Approximately how many tons of aluminum does this represent?

Section 1.7   Coordinate Systems

35. A point is located in a polar coordinate system by the coordinates $r = 2.5$ m and $\theta = 35°$. Find the x and y coordinates of this point, assuming the two coordinate systems have the same origin.

36. A certain corner of a room is selected as the origin of a rectangular coordinate system. If a fly is crawling on an adjacent wall at a point having coordinates (2.0, 1.0), where the units are meters, what is the distance of the fly from the corner of the room?

37. Express the location of the fly in Problem 36 in polar coordinates.

38. Two points in a rectangular coordinate system have the coordinates (5.0, 3.0) and (−3.0, 4.0), where the units are centimeters. Determine the distance between these points.

Section 1.8   Trigonometry

39. For the triangle shown in Figure P1.39, what are (a) the length of the unknown side, (b) the tangent of θ, and (c) the sine of ϕ?

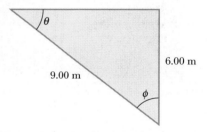

FIGURE P1.39

40. A bridge 30.0 m long crosses a chasm. If the bridge is inclined at an angle of 20.0° to the horizontal, what is the difference in height of the two ends of the bridge?

41. A pilot knows a 6.00-mi distance lies between two landmarks A and B on the level ground below. When he is over landmark A, he determines that landmark B lies 30.0° below the horizontal. What is the airplane's altitude?

42. A right triangle has a hypotenuse of length 3.00 m, and
web one of its angles is 30.0°. What are the lengths of (a) the side opposite the 30.0° angle and (b) the side adjacent to the 30.0° angle?

43. In Figure P1.43, find (a) the side opposite θ, (b) the side adjacent to ϕ, (c) cos θ, (d) sin ϕ, and (e) tan ϕ.

FIGURE P1.43

44. In a certain right triangle, the two sides that are perpendicular to each other are 5.00 m and 7.00 m long. What is the length of the third side of the triangle?

45. In Problem 44, what is the tangent of the angle for which 5.00 m is the opposite side?

46. A surveyor measures the distance across a straight river by the following method: Starting directly across from a tree on the opposite bank, she walks 100 m along the riverbank to establish a baseline. Then she sights across to the tree. The angle from her baseline to the tree is 35.0°. How wide is the river?

ADDITIONAL PROBLEMS

47. A human tissue cell has a diameter on the order of 1 μm. Estimate the number of cells in 1 cm^3 of tissue.

48. The radius of the planet Saturn is 5.85×10^7 m, and its mass is 5.68×10^{26} kg (Fig. P1.48). (a) Find the density of Saturn (its mass divided by its volume) in grams per cubic centimeter. (The volume of a sphere is given by $(4/3)\pi r^3$.) (b) Find the area of Saturn in square feet. (The surface area of a sphere is given by $4\pi r^2$.)

FIGURE P1.48   A view of Saturn. *(Courtesy of NASA)*

49. The displacement of an object moving under uniform acceleration is some function of time and the acceleration. Suppose we write this displacement as $s = ka^m t^n$ where k is a dimensionless constant. Show by dimensional analysis that this expression is satisfied if $m = 1$ and $n = 2$. Can this analysis give the value of k?

50. Compute the order of magnitude of the mass of (a) a bathtub filled with water, and (b) a bathtub filled with pennies. In your solution, list the quantities you estimate and the value you estimate for each.

51. You can obtain a rough estimate of the size of a molecule by the following simple experiment. Let a droplet of oil spread out on a smooth water surface. The resulting oil slick will be approximately one molecule thick. Given an oil droplet of mass 9.00×10^{-7} kg and density 918 kg/m^3 that spreads out into a circle of radius 41.8 cm on the water surface, what is the order of magnitude of the diameter of an oil molecule?

52. At the time of this book's printing, the U.S. national debt was about $6 trillion. (a) Assume that no interest was charged and that this debt was paid at a rate of $1000 per second. What is the order of magnitude of the time in years required to pay off the debt? (b) A dollar bill is about 15.5 cm long. If 6 trillion dollar bills were laid end to end around Earth's equator, what is the order of magnitude of the number of times they would encircle the planet? Take the radius of Earth to be about 6 378 km. (*Note:* Before doing any of the calculations, try to guess at the answers. You may be very surprised.)

53. Estimate the number of piano tuners living in New York City. This question was raised by the physicist Enrico Fermi, who was well known for making order-of-magnitude calculations.

54. Sphere 1 has surface area A_1 and volume V_1, and sphere 2 has surface area A_2 and volume V_2. If the radius of sphere 2 is double the radius of sphere 1, what is the ratio of (a) the areas A_2/A_1, and (b) the volumes V_2/V_1?

55. (a) How many seconds are there in a year? (b) If one micrometeorite (a sphere with a diameter on the order of 10^{-6} m) struck each square meter of the Moon each second, estimate the number of years would it take to cover the Moon with micrometeorites to a depth of one meter. (*Hint:* Consider a cubic box, 1 m on a side, on the Moon, and find how long it would take to fill the box.)

GROUP ACTIVITIES

Note: Following are a variety of group activities. Some are mini-experiments or demonstrations that you can carry out with the assistance of a co-worker. Others are problems, frequently consisting of both conceptual and numerical parts, that stimulate group discussion. The first step in developing both problem-solving skills and a full understanding of physics requires that you understand the basic concepts. These activities are designed to provide assistance in that endeavor.

G.1 Choose a variety of objects that range in length from a few centimeters to a few meters. Try guessing the lengths in a unit appropriate to their size, then use a meter stick supplied by your instructor to check your guesses. Keep trying this until you can estimate consistently to within 20% of the actual length.

G.2 Choose a variety of objects that range in mass from a few grams to a few kilograms. Estimate the masses by hefting the objects, then use a balance supplied by your instructor to check your guesses. Keep trying this until you can estimate consistently to within 30% of the actual mass.

G.3 A pizza company sells their product in two sizes, small with a radius of six inches and large with a radius of nine inches. A customer argues that if the small one sells for six dollars, the large should sell for nine dollars. Without doing any calculations, is the customer correct? Defend your answer. Calculate the area of each pizza to find out

how much pie you are getting in each case. If the small one costs six dollars how much should the large cost?

G.4 You know that the measurements of a typical sheet of paper are 8.50 in. by 11.0 in. Convert these measurements to mm. Use your results to calculate the length of the diagonal of the sheet of paper by using the Pythagorean theorem. Measure the diagonal with a ruler to see how well you have done. Finally, use a suitable trig function to calculate the angle that this diagonal line makes with the horizontal. Use a protractor to verify your calculation.

G.5 Each of the following equations was given by a student during an examination. Do a dimensional analysis of each equation and explain why the equation cannot be correct.

$$\tfrac{1}{2}mv^2 = \tfrac{1}{2}mv_0^2 + \sqrt{mgh} \qquad v = v_0 + at^2 \qquad ma = v^2$$

(Problem 5 courtesy of E. F. Redish. For more problems of this type, visit http://www.physics.umd.edu/perg/)

Motion in One Dimension

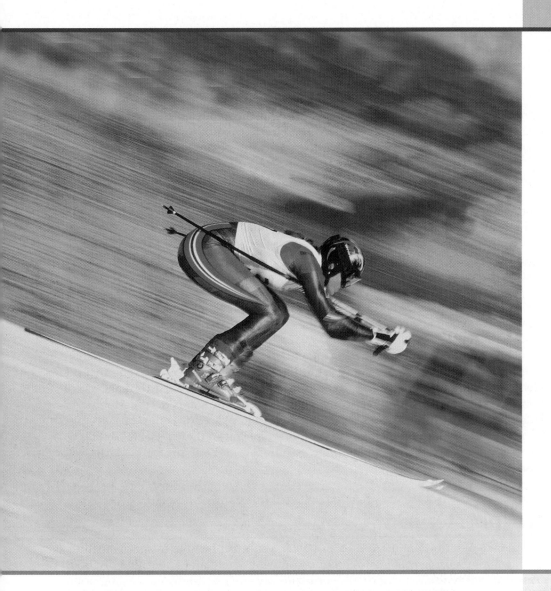

Chapter Outline

Downhill skiers can reach speeds of greater than 100 km/h. In this chapter, you will learn how to estimate the time it takes the skier to reach this speed if the skier starts from rest at the top of the slope and moves in a straight line.

(Jean Y. Ruszniewski/Tony Stone Images)

P eople have always been concerned with motion. For example, in today's world you consider motion when you describe how fast your new car will go or how much pickup it has. Likewise, prehistoric cave dwellers must have pondered motion as they devised methods for capturing a rapidly moving antelope. The branch of physics concerned with the study of the motion of an object and the relationship of this motion to such physical concepts as force and mass is called **dynamics.** The part of dynamics that describes motion without regard to its causes is called **kinematics.** In this chapter we shall focus on kinematics and on one-dimensional motion—that is, motion along a straight line. We shall start by discussing displacement, velocity, and acceleration. These concepts will enable us to study the motion of objects undergoing constant acceleration. In Chapter 3 we shall discuss the motions of objects in two dimensions.

The first recorded evidence of the study of mechanics can be traced to the people of ancient Sumeria and Egypt, whose primary concern was to understand the motions of heavenly bodies. The most systematic and detailed early studies of the heavens were conducted by the Greeks from about 300 B.C. to A.D. 300. Ancient scientists and lay people regarded Earth as the center of the Universe. This *geocentric model* was accepted by such notables as Aristotle (384–322 B.C.) and Claudius Ptolemy (about A.D. 140). Largely because of the authority of Aristotle, the geocentric model became the accepted theory of the Universe until the 17th century.

About 250 B.C., the Greek philosopher Aristarchus worked out the details of a model of the Solar System based on a moving Earth. He proposed that the sky appears to turn westward because Earth is really turning eastward. This model was not given much consideration because it was believed that if Earth turned, it would set up a great wind as it moved through the air. The critics could not see that Earth was carrying the air and everything else with it as it rotated.

The Polish astronomer Nicolaus Copernicus (1473–1543) is credited with initiating the revolution that finally replaced the geocentric model. In his system, called the *heliocentric model,* Earth and the other planets revolve in circular orbits around the Sun.

This early knowledge formed the foundation for the work of Galileo Galilei (1564–1642). Galileo stands out as the dominant facilitator of the entrance of physics into the modern era. In 1609 he became one of the first to make astronomical observations with a telescope. He observed mountains on the Moon, the larger satellites of Jupiter, spots on the Sun, and the phases of Venus. His observations convinced him of the correctness of the Copernican theory. Galileo's skills as an experimentalist and his quantitative study of motion are particularly well known.

Webnote 2.1

NASA's resilient *Galileo* project was named after the Italian scientist. For beautiful pictures generated by the mission and a summary of its many accomplishments, see *http://www.jpl.nasa.gov/galileo/*

2.1   DISPLACEMENT

In order to describe the motion of an object, you must be able to specify its position at all times using some convenient coordinate system and a specified origin. A **frame of reference** is a choice of coordinate axes that defines your starting point for measuring any quantity. For example, consider a jogger moving along the x axis as in Figure 2.1, where the initial position of the jogger is x_i and the final position is x_f. (The indices i and f refer to the initial and final locations, respectively.)

▶ Definition of displacement

The **displacement** of the jogger Δx (read as "delta" x) is defined as his *change in position.* (We shall use the Greek letter delta, Δ, to denote a change in any physical quantity.) As he moves from one position to another, his displacement is given by the difference between his final and initial positions, or $x_f - x_i$. Therefore, we write the displacement, or change in position, of the jogger as

FIGURE 2.1 A jogger moving along the x axis from x_i to x_f undergoes a displacement of $\Delta x = x_f - x_i$.

$$\Delta x \equiv x_f - x_i \qquad \text{[2.1]}$$

From this definition, we see that Δx is positive if x_f is greater than x_i and negative if x_f is less than x_i. If the jogger moves from an initial position of $x_i = 3$ m to a final position of $x_f = 15$ m, his displacement is $\Delta x = 15$ m $- 3$ m $= 12$ m.

Displacement is an example of a vector quantity. Many physical quantities in this book, including displacement, velocity, and acceleration, are vector quantities. In general, **a vector quantity is a physical quantity that must be characterized by both a magnitude and a direction.** By contrast, **a scalar quantity is one that has magnitude but not direction.** Scalar quantities, such as mass and temperature, are completely specified by a numeric value with appropriate units; no direction is involved.

A vector has both direction and magnitude

We shall usually designate vector quantities with boldface type. For example, **v** represents velocity and **a** denotes an acceleration, both vector quantities. In this chapter, which deals with one-dimensional motion, it will not be necessary to use this boldface notation. The reason is that in one-dimensional motion an object can only move in two directions, and these directions are easily specified by plus and minus signs. For example, if the truck in Figure 2.2a moves from an initial po-

$\Delta x = + 70$ m

$x_i = 10$ m

$x_f = 80$ m

(a)

FIGURE 2.2 (a) A truck moving to the right from $x_i = 10$ m to $x_f = 80$ m undergoes a displacement $\Delta x = +70$ m. (b) A truck moving to the left from $x_i = 80$ m to $x_f = 20$ m undergoes a displacement $\Delta x = -60$ m.

$\Delta x = -60$ m

$x_f = 20$ m

$x_i = 80$ m

(b)

sition of 10 m to a final position of 80 m, its displacement is

$$\Delta x = x_f - x_i = +80 \text{ m} - (+10 \text{ m}) = +70 \text{ m}$$

In this case, the displacement has a magnitude of 70 m and is directed in the positive *x* direction, as indicated by the plus sign of the result. (Sometimes the plus sign is omitted, but a result of 70 m for the displacement is understood to be the same as +70 m.) Any vector quantity is usually represented by an arrow, as is the displacement in Figure 2.2a. The length of the arrow represents the magnitude of the displacement, and the head of the arrow indicates its direction.

If the truck moves to the left from an initial position of 80 m to a final position of 20 m, as in Figure 2.2b, the displacement is

$$\Delta x = x_f - x_i = +20 \text{ m} - (+80 \text{ m}) = -60 \text{ m}$$

The minus sign in this result indicates that the displacement is a vector quantity pointing in the negative *x* direction as in Figure 2.2b.

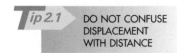

Tip 2.1   DO NOT CONFUSE DISPLACEMENT WITH DISTANCE

The displacement of an object is *not* the same as the distance it travels. For example, a model rocket fired straight upward and returning to the firing point travels some distance, but its displacement is zero.

2.2   AVERAGE VELOCITY

The motion of an object is completely known if its position is known at all times. Consider again the truck of Figure 2.2, moving along a highway (the *x* axis). Let the truck's position be x_i at some time t_i, and let its position be x_f at some time t_f. In the time interval $\Delta t = t_f - t_i$, the displacement of the truck is $\Delta x = x_f - x_i$.

 Definition of average velocity

The average velocity $\bar{v}$ during some time interval is defined as the displacement Δx divided by the time interval during which the displacement occurred:

$$\bar{v} = \frac{\Delta x}{\Delta t} = \frac{x_f - x_i}{t_f - t_i} \qquad [2.2]$$

The average velocity of an object can be either positive or negative, depending on the sign of the displacement. (The time interval Δt is always positive.) For example, in Figure 2.2b, the average velocity of the truck is negative, the minus sign indicating motion to the left along the *x* axis. Similarly, a positive average velocity for the truck in Figure 2.2a indicates that it moves to the right along the *x* axis.

In order to understand the vector nature of velocity, consider the following situation. Suppose a friend tells you that she will be taking a trip in her car and will travel at a constant rate of 55 mi/h in a straight line for 1 h. If she starts from her home, where will she be at the end of the trip? Obviously, you cannot answer this question because she did not specify the direction of her trip. All you can say is that she will be located 55 mi from her starting point. However, if she tells you that she will be driving at the rate of 55 mi/h directly northward, then her final location will be known exactly. The velocity of an object is known only if its direction of motion and its speed (a scalar) are specified.

As an example of the use of Equation 2.2, suppose the truck in Figure 2.2a moves 70 m to the right in 5.0 s. Substituting these values into Equation 2.2 gives the average velocity in this time interval as

$$\bar{v} = \frac{\Delta x}{\Delta t} = \frac{80 \text{ m} - 10 \text{ m}}{5.0 \text{ s}} = +14 \text{ m/s}$$

Note that the units of average velocity are units of length divided by units of time. These are meters per second (m/s) in SI. Other units for velocity might be feet per second (ft/s) in the U.S. customary system or centimeters per second (cm/s) in the cgs system.

FIGURE 2.3   A drag race viewed from a blimp. One car follows the red straight-line path from Ⓟ to Ⓠ, and a second car follows the blue curved path.

Let us assume we are watching a drag race from the Goodyear blimp. In one run we see a car follow the straight-line path from Ⓟ to Ⓠ shown in Figure 2.3 during the time interval Δt, and in a second run a car follows the curved path during the same time interval. If you examine the definition of average velocity (Eq. 2.2) carefully, you will see that the two cars had the same average velocity. This is because they had the same displacement $(x_f - x_i)$ during the same time interval Δt, as indicated in Figure 2.3.

We developed the idea of average velocity in this section mainly because it leads us to the extremely useful concept of instantaneous velocity (Section 2.3), that is, velocity at a specific moment.

Quick Quiz 2.1 Figure 2.4 shows the unusual path of a confused football player. He receives a kickoff at his own goal, runs downfield to within inches of a touchdown, and then reverses direction to race backward until he is tackled at the exact location where he first caught the ball. During this run, what is (a) the total distance he travels, (b) his displacement, and (c) his average velocity in the x direction?

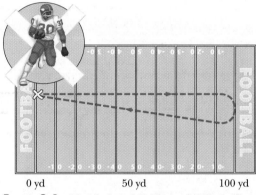

0 yd   50 yd   100 yd

FIGURE 2.4 (Quick Quiz 2.1) The path followed by a confused football player.

GRAPHICAL INTERPRETATION OF VELOCITY

If a car moves along the x axis from Ⓐ to Ⓑ to Ⓒ, and so forth, we can plot the position of these points as a function of the time elapsed since the start of motion. This gives a *position-time graph* such as the one shown in Figure 2.5. (Note that the motion is along a straight line, yet the position-time graph is not a straight line. Why?) The straight line connecting any two points on this graph, such as Ⓐ and Ⓓ, provides us with a geometric interpretation of average velocity. The slope of the line is Δx divided by the time interval for the motion, Δt. Therefore, **the average velocity of an object during the time interval Δt is equal to the**

Tip 2.2   SLOPES OF GRAPHS

The word *slope* is often used in reference to the graphs of physical data. Regardless of what data are plotted, the *slope* equals the ratio of the change in the quantity represented on the vertical axis to the change in the quantity represented on the horizontal axis. That is,

$$\text{Slope} = \frac{\text{change in } y}{\text{change in } x}$$

The slope is a quantity with units.

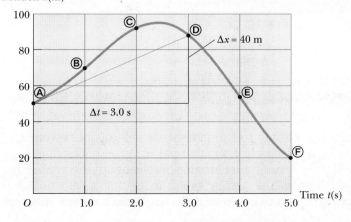

Position x(m)

$\Delta x = 40$ m

$\Delta t = 3.0$ s

Time t(s)

FIGURE 2.5 Position-time graph for the motion of a car moving along the x axis. The average velocity in the time interval Δt is the slope of the blue straight line connecting Ⓐ and Ⓓ.

Tip 2.3 AVERAGE SPEED AND AVERAGE VELOCITY

The average velocity is *not* the average speed. For example, suppose an object moves from the origin to $x = 30$ m and then back to the origin in a time interval of 5.0 s. The average velocity in this case is zero, because the object ends up at the same position at which it started—its displacement is zero. The average speed, however, is the total distance of 30 m + 30 m = 60 m divided by the time interval of 5.0 s, which is 60 m/5.0 s = 12 m/s.

slope of the straight line joining the initial and final points on a graph of the position of the object plotted versus time.

Suppose the position of the car is recorded every 1.0 s as in Figure 2.5. From the data points, we see that the car first moves in the positive x direction as it travels from Ⓐ to Ⓑ to Ⓒ, reaches a maximum position of about 95 m from the origin of motion at time $t = 2.5$ s, and then reverses its direction as it heads back toward the origin of motion. In the first 3.0 s of its motion, as it travels from Ⓐ to Ⓓ, its displacement $\Delta x = +90 - (+50) = +40$ m. Thus, its average velocity in this interval, which equals the slope of the blue line in Figure 2.5, is $\overline{v} = \Delta x/\Delta t = (+40 \text{ m})/(3.0 \text{ s}) = +13$ m/s. If we used this procedure to find the average velocity between two other points, we would most likely get a different value. For example, the average velocity between points Ⓒ and Ⓕ is negative and has a value of about -23 m/s.

AVERAGE SPEED

In day-to-day usage, the terms *speed* and *velocity* are interchangeable. In physics, however, there is a clear distinction between these two quantities. The **average speed** of an object, a scalar quantity, is defined as the total distance traveled divided by the total time it takes to travel that distance:

▶ Definition of average speed

$$\text{Average speed} = \frac{\text{total distance}}{\text{total time}}$$

The SI unit of average speed is the same as the unit of average velocity: meters per second. However, unlike average velocity, average speed has no direction and hence carries no algebraic sign. We can think of the total distance as the distance recorded by an odometer of a car.

2.3 INSTANTANEOUS VELOCITY

TABLE 2.1

Positions of a Runner at Specific Instants of Time

t (s)	x (m)
1.00	1.00
1.01	1.02
1.10	1.21
1.20	1.44
1.50	2.25
2.00	4.00
3.00	9.00

Imagine that you take a trip in your car along a perfectly straight highway. At the end of your journey, it is a relatively simple task to calculate your average velocity. The car's odometer gives you the distance traveled, a compass can give you direction, and a watch can supply the time interval. However, such a calculation would omit a great deal of information about what actually occurred on your trip. For example, if your calculation indicated that your average speed was 55 mi/h and you went east, it would be quite unlikely that your speed at every instant was 55 mi/h and you went east in a straight line. Instead, you might have traveled at 55 mi/h for a short distance, continually changed direction to follow the curves in the road, stopped for lunch and then made up some time by traveling at 70 mi/h, paused again as a police officer wrote out a ticket, and then raced forward again when the coast was clear. What was most important to the police (and to your own safety) was the speed of your car and the direction it was going at a particular instant in time—what is called the instantaneous velocity.

▶ Definition of instantaneous velocity

More precisely, **the instantaneous velocity v is defined as the limit of the average velocity as the time interval Δt becomes infinitesimally short.** In mathematical language this is written as

$$v \equiv \lim_{\Delta t \to 0} \frac{\Delta x}{\Delta t} \qquad \text{[2.3]}$$

The notation $\lim_{\Delta t \to 0}$ means that the ratio $\Delta x/\Delta t$ is to be evaluated as the time interval Δt approaches zero.

To better understand the meaning of instantaneous velocity as expressed by Equation 2.3, consider the data in Table 2.1. Assume you have been observing a runner racing along a track. One second after starting into motion, the runner has moved to a position 1.00 m from the starting point; at $t = 1.50$ s, the runner is 2.25 m from the starting point; and so on. After collecting these data, you now wish

TABLE 2.2	Calculated Values of the Time Intervals, Displacements, and Average Velocities for the Runner of Table 2.1		
Time Interval (s)	Δt **(s)**	Δx **(m)**	$\bar{v}$/ms)
1.00 to 3.00	2.00	8.00	4.00
1.00 to 2.00	1.00	3.00	3.00
1.00 to 1.50	0.50	1.25	2.50
1.00 to 1.20	0.20	0.44	2.20
1.00 to 1.10	0.10	0.21	2.10
1.00 to 1.01	0.01	0.02	2.00

to determine the velocity of the runner at the time $t = 1.00$ s. Table 2.2 presents some of the calculations you might perform to determine the velocity in question. Let us use $t = 1.00$ s as our initial time, and first consider the top row of the table. For the observed portion of the run (from 1.00 s to 3.00 s), the time interval $\Delta t = 2.00$ s and the displacement $\Delta x = +8.00$ m. Thus, the average velocity in this interval is $\Delta x/\Delta t = +4.00$ m/s. This gives only a rough approximation of the instantaneous velocity at $t = 1.00$ s. According to Equation 2.3, we can find increasingly accurate answers by letting the time interval become smaller and smaller. Therefore, consider the second entry in Table 2.2, from 1.00 s to 2.00 s, corresponding to a time interval of $\Delta t = 1.00$ s. In this interval, $\Delta x = +3.00$ m and so the average velocity is $+3.00$ m/s. This is closer to the answer we want—the runner's velocity at 1.00 s—because the time interval used in the calculation is smaller. Now let us consider a very short time interval indicated by the last entry in Table 2.2. In this case, $\Delta t = 0.01$ s and the displacement $\Delta x = +0.02$ m. Thus, the average velocity in this interval is $+2$ m/s. We could improve the reliability of our calculation by allowing the time interval to become even smaller, but we can state with some degree of confidence that the instantaneous velocity of the runner was $+2.00$ m/s at time $t = 1.00$ s.

The instantaneous speed of an object, which is a scalar quantity, is defined as the magnitude of the instantaneous velocity. As with average speed, instantaneous speed (which we shall usually call simply "speed") has no direction associated with it and hence carries no algebraic sign. For example, if one object has an instantaneous velocity of $+15$ m/s along a given line and another object has an instantaneous velocity of -15 m/s along the same line, both have an instantaneous speed of 15 m/s.

GRAPHICAL INTERPRETATION OF INSTANTANEOUS VELOCITY

Figure 2.6 shows a position-time graph for a new car trip. To find the instantaneous velocity of the car at Ⓑ, we must find the average velocity during an infini-

FIGURE 2.6 The instantaneous velocity of the car at Ⓑ equals the slope of the tangent line at that point.

tesimally short time interval. This means that point Ⓒ (or any other point) on the curve must be brought closer and closer to Ⓑ until the two points nearly overlap. From this construction we see that the line joining Ⓑ and Ⓒ is approaching the line tangent to the curve at Ⓑ. **The slope of the line tangent to the position-time curve at Ⓑ is defined to be the instantaneous velocity at that time.** For the data given in Figure 2.6, the slope at Ⓑ is about 3.4 m/s. Note that the slope at point Ⓒ is zero, which means the car stops moving at that point.

Example 2.1   A Toy Train

A toy train moves slowly along a straight portion of track according to the graph of position versus time in Figure 2.7. Find (a) the average velocity for the total trip, (b) the average velocity during the first 4.0 s of motion, (c) the average velocity during the next 4.0 s of motion, (d) the instantaneous velocity at $t = 2.0$ s, and (e) the instantaneous velocity at $t = 5.0$ s. (f) Describe in words the train's motion from the origin to Ⓐ, from Ⓐ to Ⓑ, and from Ⓑ to Ⓒ.

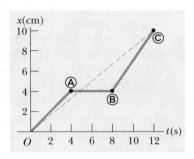

FIGURE 2.7   (Example 2.1)

Solution   (a) The slope of the line joining the origin and point *C* (the dashed blue line) provides the average velocity for the total trip. A measurement of the slope gives

$$\overline{v} = \frac{\Delta x}{\Delta t} = \frac{10 \text{ cm}}{12 \text{ s}} = \boxed{+0.83 \text{ cm/s}}$$

(b) The slope of the line joining the origin to point *A* gives us the average velocity during the first 4.0 s:

$$\overline{v} = \frac{\Delta x}{\Delta t} = \frac{4.0 \text{ cm}}{4.0 \text{ s}} = \boxed{+1.0 \text{ cm/s}}$$

(c) Following the same procedure for the next 4.0-s interval, we see that the slope of the line between points Ⓐ and Ⓑ is zero. During this time interval, the train has remained at the same location, 4.0 cm from the starting point, and its average velocity is zero.

(d) A line drawn tangent to the curve at the point corresponding to $t = 2.0$ s has the same slope as the line in part (b). Thus, the instantaneous velocity at this time is + 1.0 cm/s. This has to be true because the graph indicates that during the first 4.0 s of motion the train covers equal distances in equal intervals of time. In other words, the train moves at a constant velocity during the first 4.0 s. Under these conditions the average velocity over any interval and the instantaneous velocity at any instant are identical at all times.

(e) At $t = 5.0$ s, the slope of the position-time curve is zero. Therefore, the instantaneous velocity is zero at this instant.

(f) *Origin to* Ⓐ: The train moves at a constant speed in the positive *x* direction for the first 4.0 s because the position-time curve is rising steadily toward positive values. Ⓐ to Ⓑ: The train stops at $x = 4.0$ m for 4.0 s. Ⓑ to Ⓒ: The train travels at a higher constant speed in the positive *x* direction for the last 4.0 s. This is known from the fact that the curve from Ⓑ to Ⓒ rises steadily but more steeply than the curve from the origin to Ⓐ.

2.4 ACCELERATION

As you travel from place to place in your car, you normally do not travel long distances at a constant velocity. The velocity of the car increases when you step on the gas and decreases when you apply the brakes. Furthermore, the velocity changes when you round a curve, altering your direction of motion. When the velocity of an object changes with time, the object is said to undergo an *acceleration*. We need a more precise definition of acceleration in order to describe the motion of objects unambiguously.

AVERAGE ACCELERATION

Suppose a car moves along a straight highway as in Figure 2.8. At time t_i it has a velocity of v_i, and at time t_f its velocity is v_f. **The average acceleration during this time interval is defined as the change in velocity divided by the time interval during which that change occurred:**

◀ Definition of average acceleration

$$\bar{a} \equiv \frac{\Delta v}{\Delta t} = \frac{v_f - v_i}{t_f - t_i} \qquad [2.4]$$

For example, suppose the car shown in Figure 2.8 accelerates from an initial velocity of $v_i = +10$ m/s to a final velocity of $v_f = +30$ m/s in a time interval of 2.0 s. (Note that both velocities are toward the right, the direction selected as the positive direction.) These values can be inserted into Equation 2.4 to give the average acceleration:

$$\bar{a} = \frac{\Delta v}{\Delta t} = \frac{30 \text{ m/s} - 10 \text{ m/s}}{2.0 \text{ s}} = +10 \text{ m/s}^2$$

Acceleration is a vector quantity having dimensions of length divided by the square of time. Some common units of acceleration are meters per second per second [(m/s)/s, which is usually written m/s²] and feet per second per second (ft/s²). The acceleration calculated in the preceding case was +10 m/s². This notation means that, on average, the car accelerates in the positive x direction such that its velocity increases at a rate of 10 m/s every second.

For the case of motion in a straight line, the direction of the velocity of an object and the direction of its acceleration are related as follows. **When the object's velocity and acceleration are in the same direction, the speed of the object increases with time. When the object's velocity and acceleration are in opposite directions, the speed of the object decreases with time.**

To clarify this point, consider the following situation. Suppose the velocity of a car changes from -10 m/s to -30 m/s in a time interval of 2.0 s. The minus signs indicate that the velocity of the car is in the negative x direction. They do **not** mean that the car is slowing down! The average acceleration of the car in this time interval is

$$\bar{a} = \frac{\Delta v}{\Delta t} = \frac{-30 \text{ m/s} - (-10 \text{ m/s})}{2.0 \text{ s}} = -10 \text{ m/s}^2$$

The minus sign indicates that the acceleration vector is also in the negative x direction. Because the velocity and acceleration vectors are in the same direction, the speed of the car must increase as the car moves to the left. Note that positive and negative accelerations specify directions relative to chosen axes, not "speeding up" or "slowing down." The terms "speeding up" or "slowing down" refer to increases or decreases in speed.

FIGURE 2.8 A car moving to the right accelerates from a velocity of v_i to a velocity of v_f in the time interval $\Delta t = t_f - t_i$.

Tip 2.4 ▶ NEGATIVE ACCELERATION

Negative acceleration does not necessarily mean that an object is slowing down. If the acceleration is negative, and the velocity is negative, the object is speeding up!

Tip 2.5 ▶ DECELERATION

The word *deceleration* has a common popular connotation as *slowing down*. This is not necessarily true (see Tip 2.4), so to avoid confusion, we will not use the word deceleration in this text.

True or False? Let us define eastward as negative and westward as positive. (a) If a car is traveling eastward, its acceleration must be eastward. (b) If a car is slowing down, its acceleration may be positive. (c) A particle with constant nonzero acceleration can never stop and stay stopped.

INSTANTANEOUS ACCELERATION

In some situations, the value of the average acceleration differs in different time intervals. It is useful therefore to define instantaneous acceleration. This concept is analogous to the definition of instantaneous velocity discussed in Section 2.3. In mathematical terms, **the instantaneous acceleration is defined as the limit of the average acceleration as the time interval Δt goes to zero.** That is,

Definition of instantaneous acceleration

$$a \equiv \lim_{\Delta t \to 0} \frac{\Delta v}{\Delta t} \qquad [2.5]$$

Here again, the notation $\lim_{\Delta t \to 0}$ means that the ratio $\Delta v / \Delta t$ is to be evaluated as the time interval Δt approaches zero.

Figure 2.9 is useful for understanding the concept of instantaneous acceleration. It plots the velocity of an object versus time. This could represent, for example, the motion of a car along a busy street. The average acceleration of the car between times t_i and t_f can be found by determining the slope of the line joining points Ⓟ and Ⓠ. If we imagine that point Ⓠ is brought closer and closer to point Ⓟ, the value that we find for the average acceleration between these points approaches the value of the acceleration of the car at point Ⓟ. The instantaneous acceleration at point Ⓟ, for example, is the slope of the graph at the point. That is, **the instantaneous acceleration of an object at a certain time equals the slope of the velocity-time graph at that instant of time.** From now on we shall use the term *acceleration* to mean "instantaneous acceleration."

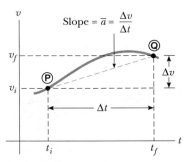

FIGURE 2.9 Velocity-time graph for an object moving in a straight line. The slope of the blue line connecting points Ⓟ and Ⓠ is defined as the average acceleration in the time interval $\Delta t = t_f - t_i$.

Parts (a), (b), and (c) of Figure 2.10 represent three graphs of the velocities of different objects moving in straight-line paths as functions of time. The possible accelerations of each object as functions of time are shown in parts (d), (e), and (f). Match each velocity-time graph with the acceleration-time graph that best describes the motion.

FIGURE 2.10 (Quick Quiz 2.3) Match each velocity-time graph to its corresponding acceleration-time graph.

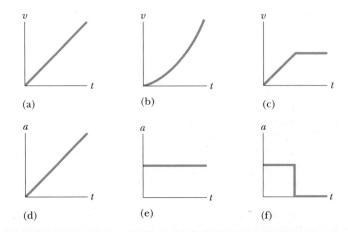

Example 2.2 A Fly Ball Is Caught

A baseball player moves in a straight-line path in order to catch a fly ball hit to the outfield. His velocity as a function of time is shown in Figure 2.11. Find his instantaneous acceleration at points Ⓐ, Ⓑ, and Ⓒ and translate the graph into a verbal description of the player's motion.

FIGURE 2.11 (Example 2.2)

Solution The instantaneous acceleration at any time is the slope of the velocity-time curve at that instant. The slope of the curve at point Ⓐ is

$$a = \frac{\Delta v}{\Delta t} = \frac{4 \text{ m/s}}{2 \text{ s}} = \boxed{+ 2 \text{ m/s}^2}$$

At point Ⓑ, the slope of the line is zero and hence the instantaneous acceleration at this time is also zero. Note that even though the instantaneous acceleration has dropped to zero at this instant, the velocity of the player is not zero. (In general, if $a = 0$, the velocity is constant in time but *not necessarily* zero.) Instead, the player is running at a constant velocity of $+ 4$ m/s. A value of zero for an instantaneous acceleration means that at a particular instant the velocity of the object is not changing.

Finally, at point Ⓒ we calculate the slope of the velocity-time curve and find that the instantaneous acceleration is

$$a = \frac{\Delta v}{\Delta t} = \frac{- 2 \text{ m/s}}{1 \text{ s}} = \boxed{- 2 \text{ m/s}^2}$$

For this situation, the minus sign and the fact that the velocities are positive indicate that the speed of the player is decreasing as he approaches the location where he will attempt to catch the baseball.

For the first 2 s the ball player is moving in the positive x direction (the velocity is positive) and steadily speeding up (the curve is rising) to a maximum speed of 4 m/s. He moves for 1 s at a steady speed of 4 m/s and then slows down in the last second (the v versus t curve is falling), still moving in the positive x direction (v is always positive).

Follow up: Do you see any way of using the v versus t curve to calculate how far he has gone in 4 s?

2.5 MOTION DIAGRAMS

The concepts of velocity and acceleration are often confused with each other, but in fact they are quite different quantities. It is instructive to use a motion diagram to describe the velocity and acceleration vectors as time progresses while an object is in motion. A **motion diagram** is a representation of a moving object at successive time intervals, with velocity and acceleration vectors sketched at each position. In order not to confuse these two vector quantities, we use red for velocity vectors and violet for acceleration vectors as in Figure 2.12. The time intervals between adjacent positions in the motion diagram are assumed to be equal.

A motion diagram is similar to the set of images resulting from a stroboscopic photograph of a moving object. Each image is made as the strobe light flashes. Figure 2.12 represents three sets of strobe photographs of cars moving along a straight roadway, from left to right. The time intervals between flashes of the stroboscope are equal in each diagram. Let us describe the motion of the car in each diagram.

In Figure 2.12a, the images of the car are equally spaced—the car moves the same distance in each time interval. Thus, the car moves with *constant positive velocity* and has *zero acceleration*. Note that the red arrows are all the same length, and there are no violet arrows.

In Figure 2.12b, the images of the car become farther apart as time progresses. In this case, the velocity vector increases in time because the car's displacement between adjacent positions increases as time progresses. Thus, the car is moving with a *positive velocity,* and a constant *positive acceleration.* The red arrows are successively longer in each image, and the violet arrows point to the right.

In Figure 2.12c, the car slows as it moves to the right because its displacement between adjacent positions decreases as time progresses. In this case, the car moves initially to the right with a constant negative acceleration. The velocity vector decreases in time (red arrows get shorter) and eventually reaches zero. (This type of motion is exhibited by a car that comes to a stop after applying its brakes.) In this diagram the acceleration and velocity vectors are *not* in the same direction. The car is moving with a *positive velocity* but with a *negative acceleration.*

You should be able to construct motion diagrams for a car that moves initially to the left with a constant positive or negative acceleration. You should also construct appropriate motion diagrams along with the mathematical solutions to kinematic problems, to see if your answers are consistent with the diagrams. In all of these exercises our convention is to choose the right-hand direction as positive.

FIGURE 2.12 (a) Motion diagram for a car moving at constant velocity (zero acceleration). (b) Motion diagram for a car whose constant acceleration is in the direction of its velocity. The velocity vector at each instant is indicated by a red arrow, and the constant acceleration vector by a violet arrow. (c) Motion diagram for a car whose constant acceleration is in the direction *opposite* the velocity at each instant.

(a)

(b)

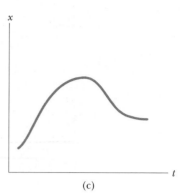
(c)

FIGURE 2.13 (Quick Quiz 2.4) Which position-time curve is impossible?

Quick Quiz 2.4 The three graphs in Figure 2.13 represent the position versus time for objects moving along the *x* axis. Which of these graphs is not physically possible?

Quick Quiz 2.5 Figure 2.14a is a diagram of a multiflash image of an airpuck moving to the right on a horizontal surface. The images sketched are separated by equal time intervals, and the first and last images show the puck at rest. (a) In Figure 2.14b, which graph best shows the puck's position as a function of time? (b) In Figure 2.14c, which graph best shows the puck's velocity as a function of time? (c) In Figure 2.14d, which graph best shows the puck's acceleration as a function of time?

(a)

(b)

(c)

FIGURE 2.14 (Quick Quiz 2.5) Choose the correct graphs.

(d)

2.6   ONE-DIMENSIONAL MOTION WITH CONSTANT ACCELERATION

Many of the mechanics applications in this book will be concerned with objects moving with *constant acceleration*. This type of motion is important because it applies to many objects in nature. For example, an object in free fall near Earth's surface moves in the vertical direction with constant acceleration, assuming that air resistance can be neglected. When an object moves with constant acceleration, *the instantaneous acceleration at any point in a time interval is equal to the value of the average acceleration over the entire time interval.* Consequently, the velocity increases or decreases at the same rate throughout the motion, and a plot of v versus t gives a straight line with either positive or negative slope.

Because the average acceleration equals the instantaneous acceleration when a is constant, we can eliminate the bar used to denote average values from our defining equation for acceleration. That is, because $\overline{a} = a$, we can write Equation 2.4 as

$$a = \frac{v_f - v_i}{t_f - t_i}$$

For convenience, let $t_i = 0$ and t_f be any arbitrary time t. Also, let $v_i = v_0$ (the initial velocity at $t = 0$) and $v_f = v$ (the velocity at any arbitrary time t). With this notation, we can express the acceleration as

$$a = \frac{v - v_0}{t}$$

or

$$v = v_0 + at \qquad \text{(for constant } a\text{)} \qquad \text{[2.6]}$$

Equation 2.6 just states that the velocity after t seconds of constant acceleration is equal to what it was at the start plus the amount added after t seconds of acceleration at a. For example, if a car starts with a velocity of $+2.0$ m/s to the right and accelerates to the right with $a = +6.0$ m/s^2, it will have a velocity of $+14$ m/s after 2.0 s have elapsed:

$$v = v_0 + at = +2.0 \text{ m/s} + (6.0 \text{ m/s}^2)(2.0 \text{ s}) = +14 \text{ m/s}$$

One of the features of one-dimensional motion with constant acceleration is the manner in which the initial, final, and average velocities are related. Because the velocity is increasing or decreasing *uniformly* with time, we can express the average velocity in any time interval as the arithmetic average of the initial velocity v_0 and the final velocity v:

$$\overline{v} = \frac{v_0 + v}{2} \qquad \text{(for constant } a\text{)} \qquad \text{[2.7]}$$

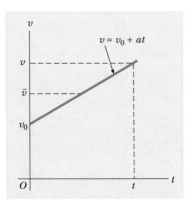

FIGURE 2.15 The velocity varies linearly with time for a particle moving with constant acceleration. The average velocity is the mean value of the initial and final velocities.

Note that this expression is valid only when the acceleration is constant—that is, when the velocity changes uniformly with time. The graphical interpretation of $\overline{v}$ is shown in Figure 2.15. As you can see, the velocity varies linearly with time according to Equation 2.6.

We can now use this result along with the defining equation for average velocity, Equation 2.2, to obtain an expression for the displacement of an object as a function of time. Again we choose $t_i = 0$ and $t_f = t$, and for convenience we write $\Delta x = x_f - x_i = x - x_0$. This gives

$$\Delta x = \overline{v}t = \left(\frac{v_0 + v}{2}\right)t$$

$$\Delta x = \tfrac{1}{2}(v_0 + v)t \qquad \text{[2.8]}$$

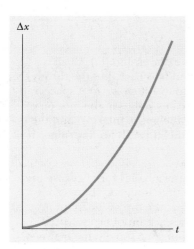

FIGURE 2.16 The displacement versus time for an object moving with constant acceleration. Note that the displacement varies as t^2. The constantly increasing slope indicates that the velocity increases with time.

We can obtain another useful expression for displacement by substituting the equation for v (Eq. 2.6) into Equation 2.8:

$$\Delta x = \tfrac{1}{2}(v_0 + v_0 + at)t$$

$$\Delta x = v_0 t + \tfrac{1}{2}at^2 \qquad \text{(for constant } a\text{)} \qquad \textbf{[2.9]}$$

Figure 2.16 shows a plot of Δx versus t for this equation. You should be able to show that the area under the curve in Figure 2.15 is equal to $v_0 t + \tfrac{1}{2}at^2$, which is equal to the displacement Δx. In fact, **the area under any curve of v versus t can be shown to be equal to the displacement of the object.**

Finally, we can obtain an expression that does not contain time by substituting the value of t from Equation 2.6 into Equation 2.8. This gives

$$\Delta x = \tfrac{1}{2}(v + v_0)\left(\frac{v - v_0}{a}\right) = \frac{v^2 - v_0^2}{2a}$$

$$v^2 = v_0^2 + 2a\,\Delta x \qquad \text{(for constant a)} \qquad \textbf{[2.10]}$$

Equations 2.6–2.10 may be used to solve any problem in one-dimensional motion with constant acceleration. For convenience, we list the four equations that are used most often in Table 2.3.

The best way to gain confidence in the use of these equations is to work a number of problems. Many times you will discover that there is more than one method for solving a given problem.

At appropriate places within each chapter we shall include interludes that we hope will motivate you to think critically about the material you have been reading. These interesting interludes, called "Applying Physics," are conceptual rather than mathematical and test your understanding of concepts. The answers serve as models for responding to the conceptual questions at the end of each chapter.

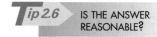

Tip 2.6   IS THE ANSWER REASONABLE?

When solving problems, you should think about your answer and determine if it seems reasonable. For example, if you are calculating the mass of a housefly and arrive at a value of 100 kg, this value is *unreasonable*—there is an error somewhere. If you are calculating the length of a spacecraft on a launch pad and end up with a value of 10 cm, this is *unreasonable*—look for a mistake.

TABLE 2.3	Equations for Motion in a Straight Line Under Constant Acceleration	
Equation	**Information Given by Equation**	
$v = v_0 + at$	Velocity as a function of time	
$\Delta x = \tfrac{1}{2}(v_0 + v)t$	Displacement as a function of velocity and time	
$\Delta x = v_0 t + \tfrac{1}{2}at^2$	Displacement as a function of time	
$v^2 = v_0^2 + 2a\,\Delta x$	Velocity as a function of displacement	

Note: Motion is along the x axis. At $t = 0$, the velocity of the particle is v_0.

APPLYING PHYSICS 2.1

Consider the following motions of an object in one dimension: (a) A ball is thrown directly upward, rises to a highest point and falls back into the thrower's hand. (b) A race car starts from rest and speeds up to 100 m/s. (c) A spacecraft drifts through space at constant velocity. Are there any points in the motion of these particles at which the average velocity (over the entire interval) and the instantaneous velocity (at an instant of time within the interval) are the same? If so, identify the point(s).

Explanation (a) The average velocity for the thrown ball is zero because the ball returns to the starting point at the end of the time interval. There is one point at which the instantaneous velocity is zero—at the top of the motion. (b) The average velocity for the motion of the race car cannot be evaluated unambiguously with the information given, but it must be some value between 0 and 100 m/s. Because the car will have every instantaneous velocity between 0 and 100 m/s at some time during the interval, there must be some instant at which the instantaneous velocity is equal to the average velocity. (c) Because the instantaneous velocity of the spacecraft is constant, its instantaneous velocity over *any* time interval and its average velocity at *any* time are the same.

PROBLEM-SOLVING STRATEGY   Accelerated Motion

The following procedure is recommended for solving problems involving accelerated motion.

1. Make sure all the units in the problem are consistent. That is, if displacements are measured in meters, be sure that velocities have units of meters per second and accelerations have units of meters per second squared.
2. Choose a coordinate system. Choose instants to call the initial and final points. Make a labeled diagram of the problem including the direction of all displacements, velocities, and accelerations.
3. Make a list of all the quantities given in the problem and a separate list of those to be determined.
4. Think about the physical concepts involved, any restrictions that apply, and select from the list of kinematic equations the one or ones that will enable you to determine the unknowns.
5. Check to see if your answers are reasonable and consistent with the diagram.

Example 2.3   The Indianapolis 500

A race car starting from rest accelerates at a rate of 5.00 m/s². What is the velocity of the car after it has traveled 100 ft?

Reasoning and Solution You should refer to the preceding Problem-Solving Strategy to see how we apply the steps indicated there to this example.

(Step 1) Be sure that the units you use are consistent. The units in this problem are *not* consistent. If we choose to leave the distance in feet, we must change the length dimension of the units of acceleration from meters to feet. Alternatively, we can leave the units of acceleration as they are and convert the distance traveled to meters. Let's do the latter. The table of conversion factors on the inside front cover gives 1 m = 3.281 ft, so we find that 1 ft = 0.305 m and 100 ft = 30.5 m.

(Step 2) You must choose a coordinate system. A convenient one for this problem is shown in Figure 2.17. The origin of the coordinate system is at the initial location of the car, and the positive direction is to the right. Using this convention, we require that velocities, accelerations, and displacements to the right are positive, and vice versa.

$v_0 = 0$

$v = ?$

$x = 0$

$x = 30.5$ m

$+ x$

FIGURE 2.17   (Example 2.3)

(Step 3) Make a list of the quantities given in the problem and a separate list of those to be determined.

Given	To Be Determined
$v_0 = 0$	v
$a = + 5.00 \text{ m/s}^2$	
$\Delta x = + 30.5 \text{ m}$	

(Step 4) Select from Table 2.3 those that will allow you to determine the unknowns. In the present case, the equation

$$v^2 = v_0{}^2 + 2a\,\Delta x$$

is our best choice because it will give us a value for v directly:

$$v^2 = (0)^2 + 2(5.00 \text{ m/s}^2)(30.5 \text{ m}) = 305 \text{ m}^2/\text{s}^2$$

$$v = \sqrt{305 \text{ m}^2/\text{s}^2} = \pm 17.5 \text{ m/s}$$

Because the car is moving to the right, we choose $+ 17.5 \text{ m/s}$ as the correct solution for v. Alternatively, the problem can be solved by using $\Delta x = v_0 t + \frac{1}{2}at^2$ to find t and then using the expression $v = v_0 + at$ to find v. Try it!

Example 2.4   Watch Out for the Speed Limit

A car traveling at a constant speed of 30.0 m/s (≈ 67 mi/h) passes a trooper hidden behind a billboard as in Figure 2.18. One second after the speeding car passes the billboard, the trooper sets off in chase with a constant acceleration of 3.00 m/s^2. (a) How long does it take the trooper to overtake the speeding car? (b) How fast is the trooper going when she overtakes the car?

$v_{car} = 30.0 \text{ m/s}$

$a_{car} = 0$

$a_{trooper} = 3.00 \text{ m/s}^2$

$t = 0$

$t = ?$

FIGURE 2.18   (Example 2.4) A speeding car passes a hidden trooper. When does the trooper catch up to the car?

Reasoning and Solution   To solve this problem algebraically, we write an expression for the position of each vehicle as a function of time. It is convenient to choose the origin at the position of the billboard and take $t = 0$ as the time the trooper begins moving. At that instant, the speeding car has already traveled a distance of 30.0 m, because it travels at a constant speed of 30.0 m/s. Thus, the initial position of the speeding car is $x_0 = +30.0$ m. (Draw a diagram for your own benefit.)

A Because the car moves with constant speed, its acceleration is zero, and applying Equation 2.9 gives

$$\Delta x_{car} = x_{car} - x_0 = v_0 t + \tfrac{1}{2} a t^2$$

In this situation, $v_0 = v$, and $a = 0$, so

$$x_{car} = x_0 + vt = 30.0 \text{ m} + (30.0 \text{ m/s})t$$

For the trooper, who starts from the origin at $t = 0$, we have $x_0 = 0$, $v_0 = 0$, and $a = 3.00$ m/s^2. Hence, the position of the trooper as a function of time is

$$x_{trooper} = \tfrac{1}{2} a t^2 = \tfrac{1}{2}(3.00 \text{ m/s}^2) t^2$$

The trooper overtakes the car at the instant that $x_{trooper} = x_{car}$, or

$$\tfrac{1}{2}(3.00 \text{ m/s}^2) t^2 = 30.0 \text{ m} + (30.0 \text{ m/s})t$$

This gives the quadratic equation

$$(1.50 \text{ m/s}^2) t^2 - (30.0 \text{ m/s})t - 30.0 \text{ m} = 0$$

Since the final point is after the original point, only the positive root is physically meaningful: $t = \boxed{21.0 \text{ s}}$.

B Note that in the time interval of 21.0 s, the trooper travels a distance of about 660 m:

$$\Delta x_{trooper} = v_0 t + \tfrac{1}{2} a t^2 = 0 + \tfrac{1}{2}(3.00 \text{ m/s}^2)(21.0 \text{ s})^2 = 6.59 \times 10^2 \text{ m}$$

Because the trooper starts from rest and accelerates at 3.00 m/s^2 for 21.0 s, when she overtakes the car she is traveling at a speed of

$$v_{trooper} = v_0 + at = 0 + (3.00 \text{ m/s}^2)(21.0 \text{ s}) = \boxed{62.9 \text{ m/s}}$$

Note that this is about twice the speed of the car.

EXERCISE   This problem also can be easily solved graphically. On the *same* graph, plot the position versus time for each vehicle, and from the intersection of the two curves determine the time at which the trooper overtakes the speeding car.

Example 2.5   Stop That 747!

A typical jetliner lands at a speed of 160 mi/h and brakes at the rate of 10 mi/h/s. If the jetliner travels at a constant speed of 160 mi/h for 1.0 s after landing before applying the brakes, what is the total displacement of the jetliner between touchdown on the runway and coming to rest?

Solution   First we convert the given quantities to SI units. Because 1.00 mi/h = 0.447 m/s, we have

$$v_0 = (160 \text{ mi/h})\left(\frac{0.447 \text{ m/s}}{1.00 \text{ mi/h}}\right) = 71.5 \text{ m/s}$$

$$a = (10.0 \text{ mi/h/s})\left(\frac{0.447 \text{ m/s}}{1.00 \text{ mi/h}}\right) = 4.47 \text{ m/s}^2$$

Next we choose our coordinate system and sketch the problem as shown in Figure 2.19.

FIGURE 2.19 (Example 2.5) Coasting and braking distances for a landing jetliner.

From the sketch we see that the problem has two parts. In part A, the plane coasts with a constant velocity and so the acceleration is zero. Part B involves a braking distance over which the acceleration is constant at -4.47 m/s^2 and the velocity drops from 71.5 m/s to 0 m/s. Listing the given quantities and those to be determined, we find:

For the Coasting Displacement—Part A

Given	To Be Determined
$v = 71.5$ m/s	Δx
$a = 0$	
$t = 1.0$ s	

Selecting $\Delta x = \overline{v}t$ from our list of equations for motion with constant acceleration ($a = 0$) gives the coasting displacement, Δx_A.

$$\Delta x_A = \overline{v}t = (71.5 \text{ m/s})(1.0 \text{ s}) = 71.5 \text{ m}$$

Note that using $\Delta x = v_0 t + \frac{1}{2}at^2$ would also give the same result, but $v^2 = v_0^2 + 2a\,\Delta x$ simply yields $v = v_0$.

For the Braking Displacement—Part B

Given	To Be Determined
$v_0 = 71.5$ m/s	Δx
$v = 0$	
$a = -4.47$ m/s^2	

Because the acceleration is again steady, we use the equations in Table 2.3. The only equation which connects v, a, and Δx is $v^2 = v_0^2 + 2a\,\Delta x$. Solving for the displacement during braking, Δx_B,

$$\Delta x_B = \frac{v^2 - v_0^2}{2a} = \frac{0^2 - (71.5 \text{ m/s})^2}{2(-4.47 \text{ m/s}^2)} = 572 \text{ m}$$

The total displacement of the plane on the runway is

$$\Delta x_A + \Delta x_B = 72 \text{ m} + 572 \text{ m} = 644 \text{ m} = \boxed{6.4 \times 10^2 \text{ m}}$$

to the correct number of significant figures. Does the approximate answer of 640 m or 2100 ft seem reasonable?

Example 2.6   The Porsche of American Trains, the Acela

The sleek high-speed electric train known as the Acela (pronounced ahh-sell-ah) is currently in service on the Washington-New York-Boston run and is shown in Figure 2.20a. It consists of two power cars and six coaches and can carry 304 passengers at speeds up to 170 mi/h. In order to negotiate curves comfortably at high speed, the train carriages tilt as much as six degrees from the vertical to prevent passengers from being pushed to the side. A velocity-time graph for the Acela is shown in Figure 2.20b. (a) Describe the motion of the train in each interval. (b) Find the peak acceleration of Acela in miles per hour per second (mi/h/s) as it speeds up from 45 to 170 mi/h. (c) Find the train's displacement in miles between $t = 0$ and $t = 200$ s. (d) Find the average acceleration of Acela and its displacement in miles in the interval from 200 s to 300 s. (The train has regenerative braking, which means that it feeds energy back into the utility lines each time it stops!) (e) Find the total displacement in the interval from 0 to 400 s.

Solution   **A** As we see in Figure 2.20b, from about -50 s to 50 s Acela is cruising at a constant positive velocity in the $+x$ direction. From 50 s to 200 s, Acela accelerates in the $+x$ direction reaching a top speed of about 170 mi/h. Around 200 s, the engineer applies the brakes, and the train, still traveling in the $+x$ direction, slows down and then stops at 350 s. Just after 350 s, Acela reverses direction (v becomes negative) and steadily gains speed in the $-x$ direction. **B** Because the acceleration is changing in this problem, we cannot use the equations for motion with constant a, but must resort to the most general graphical methods. The peak acceleration between 45 and 170 mi/h is given by the slope of the steepest tangent to the v versus t curve in this interval. From the tangent of the blue line shown in Figure 2.20c, we find

$$a = \text{slope} = \frac{\Delta v}{\Delta t} = \frac{(150 - 50)\ \text{mi/h}}{(100 - 50)\ \text{s}} = \boxed{2.0\ \text{mi/h/s}}$$

Note that if we work with *extreme caution* and treat units like algebraic quantities, we can manage mixed units of mi/h/s. There are two reasons to do this. One is that an acceleration of 2.0 mi/h/s is easy to visualize for most people, while 0.89 m/s^2 is not. The other is that it is easier to work in mixed units in this problem instead of converting all values to m/s and re-plotting the graph. At any rate, we urge extreme caution in using mixed units and point out they can have a hazardous effect on your physics grade.

C Let us use the fact that the area under the v versus t curve equals the displacement. The train's displacement between 0 and 200 s is equal to the area of the gray shaded region in Figure 2.20d, which we have approximated with a series of triangles and rectangles.

$$\Delta x_{0 \to 200\ \text{s}} = \text{area}_1 + \text{area}_2 + \text{area}_3 + \text{area}_4 + \text{area}_5$$

$$\approx (50\ \text{mi/h})(50\ \text{s}) + (50\ \text{mi/h})(50\ \text{s}) + (160\ \text{mi/h})(100\ \text{s})$$

$$+ \tfrac{1}{2}(50\ \text{s})(100\ \text{mi/h}) + \tfrac{1}{2}(100\ \text{s})(170\ \text{mi/h} - 160\ \text{mi/h}) = 24\,000\ (\text{mi/h})(\text{s})$$

Now, at the end of our calculation, we can find the displacement in miles by converting hours to seconds. As 1 h = 3 600 s,

$$\Delta x_{0 \to 200\ \text{s}} \approx \left(\frac{24\,000\ \text{mi}}{3\,600\ \text{s}}\right)(\text{s}) = \boxed{6.7\ \text{mi}}$$

D The average acceleration is just the slope of the green line from 200 s to 300 s, shown in Figure 2.20c:

$$\bar{a} = \text{slope} = \frac{\Delta v}{\Delta t} = \frac{(10 - 170)\ \text{mi/h}}{100\ \text{s}} = \boxed{-1.6\ \text{mi/h/s}}$$

The negative sign means the train's velocity in the $+x$ direction is decreasing and it is going 1.6 mi/h slower each second. Acela's displacement is given by the area of the yellow shaded region of Figure 2.20d, composed of a large triangle and a thin rectangle.

(a)

(b)

(c)

(d)

FIGURE 2.20 (Example 2.6) (a) The Acela—1 171 000 lb of cold steel thundering along at 150 mi/h. *(Courtesy Amtrak NEC Media Relations)* (b) Velocity versus time graph for the Acela. (c) The slope of the steepest tangent blue line gives the peak acceleration, while the slope of the green line is the average acceleration between 200 s and 300 s. (d) The area under the velocity-time graph in some time interval gives the displacement of Acela in that time interval.

$$\Delta x_{200 \to 300 \text{ s}} \approx \tfrac{1}{2}(100 \text{ s})(170 - 10) \text{ mi/h} + (10 \text{ mi/h})(100 \text{ s})$$

$$= 9\,000 \text{ (mi/h)(s)} = \boxed{2.5 \text{ mi}}$$

E To get the total displacement, we simply add all the individual displacements. We need the displacements for the time intervals from 300 s to 350 s and from 350 s to 400 s:

$$\Delta x_{300 \to 350 \text{ s}} \approx \tfrac{1}{2}(50 \text{ s})(10 \text{ mi/h}) = 250 \text{ (mi/h)(s)}$$

$$\Delta x_{350 \to 400 \text{ s}} \approx \tfrac{1}{2}(50 \text{ s})(-50 \text{ mi/h}) = -1\,250 \text{ (mi/h)(s)}$$

Finally we find

$$\Delta x_{0 \to 400 \text{ s}} \approx (24\,000 + 9\,000 + 250 - 1\,250)(\text{mi/h})(\text{s}) = \boxed{8.9 \text{ mi}}$$

2.7 FREELY FALLING OBJECTS

It is now well known that, in the absence of air resistance, all objects dropped near Earth's surface fall toward Earth with the same constant acceleration. It was not until about 1600 that this conclusion was accepted. Prior to that time, the teachings of the great philosopher Aristotle (384–322 B.C.) had held that heavier objects fell faster than lighter ones.

It was Galileo who originated our present-day ideas concerning falling objects. There is a legend that he discovered the law of falling objects by observing that two different weights dropped simultaneously from the Leaning Tower of Pisa hit the ground at approximately the same time. Although it is doubtful that this particular experiment was carried out, we know that Galileo performed many systematic experiments on objects moving on inclined planes. In his experiments he rolled balls down a slight incline and measured the distances they covered in successive time intervals. The purpose of the incline was to reduce the acceleration and enable Galileo to make accurate measurements of the time intervals. (Some people refer to this experiment as "diluting gravity.") By gradually increasing the slope of the incline, he was finally able to draw mathematical conclusions about freely falling objects, because a falling ball is equivalent to a ball falling down a vertical incline. Galileo's achievements in the science of mechanics paved the way for Newton in his development of the laws of motion, which we shall study in Chapter 4.

You might want to try the following experiment. Simultaneously drop a coin and a crumpled piece of paper from the same height. If the effects of air friction are negligible, both objects will have the same motion and hit the floor at the same time. In the idealized case, where air resistance is absent, such motion is referred to as free fall. If this same experiment could be conducted in a vacuum, in which air friction is zero, the paper and coin would fall with the same acceleration, regardless of the shape of the paper. (See the photographs in Physics in Action.) On August 2, 1971, such a demonstration was conducted on the Moon by astronaut David Scott. He simultaneously released a hammer and a feather, and they fell with the same acceleration to the lunar surface. This demonstration surely would have pleased Galileo!

When we use the expression *freely falling object,* we do not necessarily mean an object dropped from rest. **A freely falling object is any object moving under the influence of gravity only, regardless of its initial motion. Objects thrown upward or downward and those released from rest are all falling freely once they are released. Any freely falling object experiences an acceleration directed** *downward,* **regardless of the direction of its motion at any instant.**

We shall denote the magnitude of the **free-fall acceleration** by the symbol g. The value of g decreases with increasing altitude. Furthermore, slight variations in g occur in latitude. At Earth's surface the value of g is approximately 9.80 m/s².

GALILEO GALILEI, ITALIAN PHYSICIST AND ASTRONOMER (1564–1642)

Galileo formulated the laws that govern the motion of objects in free fall. He also investigated the motion of an object on an inclined plane, established the concept of relative motion, invented the thermometer, and discovered that the motion of a swinging pendulum could be used to measure time intervals. After designing and constructing his own telescope, he discovered four of Jupiter's moons, found that the Moon's surface is rough, discovered sunspots and the phases of Venus, and showed that the Milky Way consists of an enormous number of stars. Galileo publicly defended Nicholaus Copernicus's assertion that the Sun is at the center of the Universe (the heliocentric system). He published *Dialogue Concerning Two New World Systems* to support the Copernican model, a view which the Church declared to be heretical. After being taken to Rome in 1633 on a charge of heresy, he was sentenced to life imprisonment, and later was confined to his villa at Arcetri, near Florence, where he died in 1642. *(North Wind)*

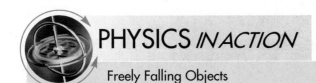

PHYSICS *IN ACTION*

Freely Falling Objects

On the left is a multiflash photograph of a falling billiard ball. As the ball falls, the spacing between successive images increases, indicating that the ball accelerates downward. The motion diagram shows that the ball's velocity (red arrows) increases with time, and its acceleration (violet arrows) remains constant.

The middle photograph shows a multiflash image of two freely falling balls released simultaneously. The ball on the left side is solid, and the ball on the right is a hollow Ping-Pong ball. The time interval between flashes is $\frac{1}{30}$ s and the scale is in centimeters. Can you determine g from these data? Note that the effect of air resistance is greater for the smaller, hollow ball on the right, as indicated by the smaller separation between successive images.

In the photograph on the right, a feather and an apple are released from rest in a 4-ft vacuum chamber. The trap door that released the two objects was opened with an electronic switch at the same instant the camera shutter was opened. The two objects fell at approximately the same rate, as indicated by the horizontal alignment of the multiple images. When air resistance is negligible, all objects fall with the same acceleration regardless of their masses.

(Richard Megna 1990, Fundamental Photographs)   *Education Development Center, Newton, MA*   *© James Sugar, Black Star*

Unless stated otherwise, we will use this value for g when doing calculations. For making quick estimates, use $g \approx 10 \text{ m/s}^2$.

If we neglect air resistance and assume that the free-fall acceleration does not vary with altitude over short vertical distances, then the motion of a freely falling object is equivalent to motion in one dimension under constant acceleration. Therefore, the equations developed in Section 2.6 for objects moving with constant acceleration can be applied. It is common to define "up" as the $+y$ direction and to use y as the position variable in the kinematic equations. The only modification that we need to make in these equations for freely falling objects is to note that the motion is in the vertical direction (the y direction) rather than the horizontal (x) direction, and that the acceleration is downward and has a magnitude of 9.80 m/s^2. Thus, we always take $a = -g = -9.80 \text{ m/s}^2$, where the minus sign means that the acceleration of a freely falling object is downward. In Chapter 7 we shall study how to deal with variations in g with altitude.

Webnote 2.2

For more on the life and work of Galileo Galilei, visit any of these sites:

http://galileo.imss.firenze.it/museo/b/ egalilg.html
http://es.rice.edu/ES/humsoc/Galileo/ index.html
http://www-groups.dcs.st-and.ac.uk/~history/Mathematicians/ Galileo.html

Quick Quiz 2.6   A ball is thrown vertically upward. While the ball is in free fall, does its acceleration (a) increase, (b) decrease, or (c) remain constant?

Quick Quiz 2.7   After a ball is thrown vertically upward and is in the air, its speed (a) increases, (b) decreases, (c) decreases and then increases, or (d) remains the same.

Quick Quiz 2.8   A skydiver jumps out of a helicopter. A few seconds later, another skydiver jumps out, so that they both fall along the same vertical line. Ignore air resistance, so that both skydivers fall with the same acceleration. Does the vertical distance between them (a) increase, (b) decrease, or (c) stay the same? Does the difference in their velocities (d) increase, (e) decrease, or (f) stay the same?

Example 2.7   Look Out Below!

A golf ball is released from rest at the top of a very tall building. Neglecting air resistance, calculate the position and velocity of the ball after 1.00, 2.00, and 3.00 s.

Solution   We choose our coordinates so that the starting point of the ball is at the origin ($y_0 = 0$ at $t = 0$) and remember that we have defined y to be positive upward. (You should make a simple sketch of the problem for your own benefit.) Because $y_0 = 0$, $v_0 = 0$, and $a = -g = -9.80 \text{ m/s}^2$, Equation 2.6 becomes

$$v = at = (-9.80 \text{ m/s}^2)t \tag{1}$$

Equation 2.9 expressed in the notation of the problem becomes

$$\Delta y = v_0 t + \tfrac{1}{2}at^2$$

where $\Delta y = y - y_0$. Using $y_0 = 0$, $v_0 = 0$, and $a = -g$ gives

$$y = \tfrac{1}{2}at^2 = \tfrac{1}{2}(-9.80 \text{ m/s}^2)t^2 \tag{2}$$

where t is in seconds, v is in meters per second, and y is in meters. Equations (1) and (2) give the velocity and position at any time t after the ball is released. Therefore, at $t = 1.00$ s,

$$v = (-9.80 \text{ m/s}^2)(1.00 \text{ s}) = \boxed{-9.80 \text{ m/s}}$$

$$y = \tfrac{1}{2}(-9.80 \text{ m/s}^2)(1.00 \text{ s})^2 = \boxed{-4.90 \text{ m}}$$

Likewise, at $t = 2.00$ s, we find that $v = -19.6$ m/s and $y = -19.6$ m. Finally, at $t = 3.00$ s, $v = -29.4$ m/s and $y = -44.1$ m. The minus sign for v indicates that the velocity vector is directed downward, and the minus sign for y indicates a position below the origin. By comparing the positions and velocities for 1, 2, and 3 s of fall, note that the velocity of a falling object is directly proportional to the time and the positions is proportional to the time squared. These two results were first proven by Galileo.

EXERCISE   Calculate the position and velocity of the ball after 4.00 s.

ANSWER -78.4 m, -39.2 m/s.

Example 2.8 Not a Bad Throw for a Rookie!

A stone is thrown from the top of a building with an initial velocity of 20.0 m/s straight upward. The building is 50.0 m high, and the stone just misses the edge of the roof on its way down, as in Figure 2.21. Determine (a) the time needed for the stone to reach its maximum height, (b) the maximum height, (c) the time needed for the stone to return to the level of the thrower, (d) the velocity of the stone at this instant, and (e) the velocity and position of the stone at $t = 5.00$ s.

Solution **A** The diagram in Figure 2.21 establishes a coordinate system with $y_0 = 0$ at the top of the building, and we take y to be positive upward.

Given	To Be Determined
$v_0 = +20.0$ m/s	t
$a = -g = -9.80$ m/s^2	

To find the time necessary for the stone to reach the maximum height, we use Equation 2.6, $v = v_0 + at$, noting that $v = 0$ at maximum height:

$$t = \frac{v - v_0}{a} = \frac{0 - 20.0 \text{ m/s}}{-9.80 \text{ m/s}^2} = \boxed{2.04 \text{ s}}$$

B This value of time can be substituted into Equation 2.9, $\Delta y = v_0 t + \frac{1}{2}at^2$, to give the maximum height as measured from the position of the thrower. In this case, $\Delta y = y_{max} - y_0$ and $y_0 = 0$, so

$$y_{max} = \Delta y = (20.0 \text{ m/s})(2.04 \text{ s}) + \frac{1}{2}(-9.80 \text{ m/s}^2)(2.04 \text{ s})^2 = \boxed{20.4 \text{ m}}$$

C When the stone is back at the height of the thrower, the y coordinate is zero. From the expression $\Delta y = v_0 t + \frac{1}{2}at^2$ (Eq. 2.9), with $\Delta y = y - y_0 = 0 - 0 = 0$, we find

$$\Delta y = 0 = 20.0t - 4.90t^2$$

This is a quadratic equation and has two solutions for t. The equation can be factored to give

$$t(20.0 - 4.90t) = 0$$

One solution is $t = 0$, corresponding to the time at which the stone starts its motion. The other solution is $t = 4.08$ s, which is the solution we are after.

D The value for t found in (c) can be inserted into $v = v_0 + at$ (Eq. 2.6) to give

$$v = 20.0 \text{ m/s} + (-9.80 \text{ m/s}^2)(4.08 \text{ s}) = \boxed{-20.0 \text{ m/s}}$$

Note that the velocity of the stone when it arrives back at its original height is equal in magnitude to the stone's initial velocity but opposite in direction. This indicates that the motion is symmetric; that is, at any given height above the origin, the magnitude of the velocity going up is the same as that going down.

E From $v = v_0 + at$ (Eq. 2.6), the velocity after 5.00 s is

$$v = 20.0 \text{ m/s} + (-9.80 \text{ m/s}^2)(5.00 \text{ s}) = \boxed{-29.0 \text{ m/s}}$$

We can use $\Delta y = v_0 t + \frac{1}{2}at^2$ (Eq. 2.9) to find the position of the particle at $t = 5.00$ s. Noting that because $y_0 = 0$, $\Delta y = y - y_0 = y$, so

$$\Delta y = (20.0 \text{ m/s})(5.00 \text{ s}) + \frac{1}{2}(-9.80 \text{ m/s}^2)(5.00 \text{ s})^2 = \boxed{-22.5 \text{ m}}$$

EXERCISE Find the velocity of the stone just before it hits the ground.

ANSWER -37.0 m/s

FIGURE 2.21 (Example 2.8) Position and velocity versus time for a freely falling object thrown initially upward with a velocity of $v_0 = +20.0$ m/s.

Labels on figure:
$t = 2.04$ s
$y_{max} = 20.4$ m
$v = 0$

$t = 0$, $y_0 = 0$
$v_0 = 20.0$ m/s

$t = 4.08$ s
$\Delta y = 0$
$v = -20.0$ m/s

$t = 5.00$ s
$\Delta y = -22.5$ m
$v = -29.0$ m/s

50.0 m

$t = 5.80$ s
$\Delta y = -50.0$ m
$v = -37.0$ m/s

Example 2.9   **A Rocket Goes Ballistic**

A rocket moves straight upward, starting from rest with an acceleration of $+29.4 \text{ m/s}^2$ for 4.00 s. It runs out of fuel at the end of this 4.00 s and continues to move upward for a while. How high does it rise above its original starting point? What is its velocity the instant before it crashes on the ground?

Reasoning   Because the units are already consistent, we begin by sketching the problem, taking $y = 0$ at the launch point and y positive upward as in Figure 2.22. The problem consists of three phases. For phase 1, the rocket has an *upward* acceleration of 29.4 m/s^2 and we can use our equations for motion with constant a to find the height and velocity of the rocket at the end of phase 1 when the fuel is burned up. In phase 2, which ends at the highest point reached by the rocket, the rocket is in free fall and has an acceleration of -9.80 m/s^2. This phase can be treated by using the final y and v values of phase 1 as input for phase 2. Phase 3 is a free-fall problem for an object dropped from rest from the height y_{max}.

FIGURE 2.22   (Example 2.9) Three linked phases of motion for a rocket that is launched, uses up its fuel, and crashes.

Solution   We can organize our thoughts by listing known quantities and quantities to be determined.

Phase 1: Going Up, Under Power	
Given	**To Be Determined**
$y_0 = 0$	$y_{\text{burn out}}$
$v_0 = 0$	$v_{\text{burn out}}$
$a = 29.4 \text{ m/s}^2$	
$t = 4.00 \text{ s}$	

$$\Delta y_{\text{burn out}} = v_0 t + \tfrac{1}{2}at^2 = 0 + \tfrac{1}{2}(29.4 \text{ m/s}^2)(4.00 \text{ s})^2 = 235.2 \text{ m}$$

$$v_{\text{burn out}} = v_0 + at = 0 + (29.4 \text{ m/s}^2)(4.00 \text{ s}) = 117.6 \text{ m/s}$$

Note that we keep four significant figures for now, but we shall round off to three significant figures at the end of the problem.

Phase 2: Coasting, Still Going Up

Given	To Be Determined
$y_0 = 235.2$ m	y_{max}
$v_0 = 117.6$ m/s	
$a = -9.80$ m/s^2	
$v_f = 0$	

Rearranging $v^2 = v_0{}^2 + 2a\,\Delta y$ yields $\Delta y = \dfrac{v^2 - v_0{}^2}{2a}$ or

$$y_{\text{max}} = y_0 + \left(\frac{v^2 - v_0{}^2}{2a}\right) = 235.2 \text{ m} + \left(\frac{0^2 - (117.6 \text{ m/s})^2}{2(-9.80 \text{ m/s}^2)}\right)$$

$$= 235.2 \text{ m} + 705.6 \,\frac{\text{m}^2/\text{s}^2}{\text{m/s}^2} = \boxed{941 \text{ m}}$$

Phase 3: Falling

Given	To Be Determined
$y_0 = 941$ m	v
$y = 0$	
$v_0 = 0$	
$a = -9.80$ m/s^2	

We can use Equation 2.10 to find v:

$$v^2 = v_0{}^2 + 2a\,\Delta y = 0^2 + 2a\,\Delta y$$

We take the negative solution to describe the downward motion:

$$v = -\sqrt{2a\,\Delta y} = -\sqrt{2(-9.80 \text{ m/s}^2)(-941 \text{ m})} = \boxed{-136 \text{ m/s}}$$

SUMMARY

The **displacement** of an object moving along the x axis is defined as the change in position of the object, and is given by

$$\Delta x \equiv x_f - x_i \tag{2.1}$$

where x_i is the initial position of the object and x_f is its final position.

A **vector** quantity is any quantity that is characterized by both a magnitude and a direction. A **scalar** quantity has a magnitude only.

The **average velocity** of an object moving along the x axis during some time interval is equal to the displacement of the object Δx divided by the time interval Δt during which the displacement occurred.

$$\bar{v} \equiv \frac{\Delta x}{\Delta t} = \frac{x_f - x_i}{t_f - t_i} \tag{2.2}$$

The average velocity is equal to the slope of the straight line joining the initial and final points on a graph of the position of the object versus time.

The slope of the line tangent to the position-time curve at some point is equal to the **instantaneous velocity** at that time. The **instantaneous speed** of an object is defined as the magnitude of the instantaneous velocity.

The **average acceleration** of an object during some time interval is defined as the change in velocity Δv divided by the time interval Δt during which the change occurred.

$$\bar{a} \equiv \frac{\Delta v}{\Delta t} = \frac{v_f - v_i}{t_f - t_i} \qquad [2.4]$$

The **instantaneous acceleration** of an object at a certain time equals the slope of a velocity-time graph at that instant.

The equations that describe the motion of an object moving with constant acceleration along the x axis are

$$v = v_0 + at \qquad [2.6]$$

$$\Delta x = \tfrac{1}{2}(v + v_0)t \qquad [2.8]$$

$$\Delta x = v_0 t + \tfrac{1}{2}at^2 \qquad [2.9]$$

$$v^2 = v_0^2 + 2a\,\Delta x \qquad [2.10]$$

An object falling in the presence of Earth's gravity experiences a free-fall acceleration directed toward Earth's center. If air friction is neglected and if the altitude of the falling object is small compared with Earth's radius, then we can assume that the free-fall acceleration g is constant over the range of motion, where the magnitude of g is equal to 9.80 m/s².

CONCEPTUAL QUESTIONS

1. If the velocity of a particle is nonzero, can its acceleration be zero? Explain.
2. If the velocity of a particle is zero, can its acceleration be zero? Explain.
3. If a car is traveling eastward, can its acceleration be westward? Explain.
4. The speed of sound in air is 331 m/s. During the next thunderstorm, try to estimate your distance from a lightning bolt by measuring the time lag between the flash and the thunderclap. You can ignore the time it takes for the light flash to reach you. Why?
5. An object is thrown upward. Discuss the sign of the free-fall acceleration relative to the object's velocity while the object is in the air.
6. If the average velocity of an object is zero in some time interval, what can you say about the displacement of the object for that interval?
7. A child throws a marble into the air with an initial speed v_0. Another child drops a ball at the same instant. Compare the accelerations of the two objects while they are in flight.
8. The strobe photographs in Figure Q2.8 were taken of a disk moving from left to right under different conditions, and the time interval between images is constant. Taking the direction to the right to be positive, describe the motion of the disk in each case. For which case is (a) the acceleration positive? (b) the acceleration negative? (c) the velocity constant?
9. Can the instantaneous velocity of an object at an instant of time ever be greater in magnitude than the average velocity over a time interval containing the instant? Can it ever be less?

(a)

(b)

(c)

FIGURE Q2.8   *(Courtesy David Rogers)*

10. Car A, traveling from New York to Miami, has a speed of 25 m/s. Car B, traveling from New York to Chicago, also has a speed of 25 m/s. Are their velocities equal? Explain.
11. A ball is thrown vertically upward. (a) What are its velocity and acceleration when it reaches its maximum altitude? (b) What is its acceleration just before it hits the ground?

B.C.

By permission of John Hart and Field Enterprises Inc.

FIGURE Q2.17

12. A rule of thumb for driving is that a separation of one car length for each 10 mi/h of speed should be maintained between moving vehicles. Assuming a constant reaction time, discuss the relevance of this rule for (a) motion with constant velocity and (b) motion with constant acceleration.

13. Two cars are moving in the same direction in parallel lanes along a highway. At some instant, the velocity of car A exceeds the velocity of car B. Does this mean that the acceleration of A is greater than that of B? Explain.

14. Consider the following combinations of signs and values for velocity and acceleration of a particle with respect to a one-dimensional *x* axis:

Velocity	Acceleration
a. Positive	Positive
b. Positive	Negative
c. Positive	Zero
d. Negative	Positive
e. Negative	Negative
f. Negative	Zero
g. Zero	Positive
h. Zero	Negative

Describe what a particle is doing in each case, and give a real-life example for an automobile on an east-west one-dimensional axis, with east considered the positive direction.

15. A student at the top of a building of height *h* throws one ball upward with a speed of v_0 and then throws a second ball downward with the same initial speed, v_0. How do the final velocities of the balls compare when they reach the ground?

16. You drop a ball from a window on an upper floor of a building. It strikes the ground with velocity *v*. You now repeat the drop, but you have a friend down on the street who throws another ball upward at velocity *v*. Your friend throws the ball upward at exactly the same time that you drop yours from the window. At some location, the balls pass each other. Is this location *at* the halfway point between window and ground, *above* this point, or *below* this point?

17. A pebble is dropped into a water well, and the splash is heard 16 s later, as illustrated in the cartoon strip shown in Figure Q2.17. Estimate the distance from the rim of the well to the water's surface.

18. A ball rolls in a straight line along the horizontal direction. Using motion diagrams (or multiflash photographs), describe the velocity and acceleration of the ball for each of the following situations: (a) The ball moves to the right at a constant speed. (b) The ball moves from right to left and continually slows down. (c) The ball moves from right to left and continually speeds up. (d) The ball moves to the right, first speeding up at a constant rate, and then slowing down at a constant rate.

19. What would happen if a moving car came to rest with constant acceleration and the acceleration were to remain constant after the car came to rest?

PROBLEMS

1, 2, 3 = straightforward, intermediate, challenging   ☐ = full solution available in Student Solutions Manual/Study Guide

web = solution posted **http://info.brookscole.com/serway**   = biomedical application

Section 2.2 Average Velocity

Section 2.3 Instantaneous Velocity

1. A person travels by car from one city to another with different constant speeds between pairs of cities. She drives for 30.0 min at 80.0 km/h, 12.0 min at 100 km/h, and 45.0 min at 40.0 km/h, and spends 15.0 min eating lunch and buying gas. (a) Determine the average speed for the trip. (b) Determine the distance between the initial and final cities along this route.

2. (a) Sand dunes in a desert move over time as sand is swept up the windward side to settle in the lee side. Such "walking" dunes have been known to walk 20 feet in a year and can travel as much as 100 feet per year in particularly windy times. Calculate the average speed in each case in m/s. (b) Fingernails grow at the rate of drifting conti-

nents, about 10 mm/yr. Approximately how long did it take for North America to separate from Europe, a distance of about 3 000 mi?

3. Two boats start together and race across a 60-km-wide lake and back. Boat A goes across at 60 km/h and returns at 60 km/h. Boat B goes across at 30 km/h and its crew, realizing how far behind it is getting, returns at 90 km/h. Turnaround times are negligible, and the boat that completes the round trip first wins. (a) Which boat wins and by how much? (Or is it a tie?) (b) What is the average velocity of the winning boat?

4. (a) A bristlecone pine tree has been known to take 4 000 years to grow to a height of 20 ft. Find the average speed of growth in m/s. (b) In contrast, the fastest growing plant is the giant kelp, which can grow at a rate of 2 feet in one day. Find the average speed of growth of this plant in m/s.

5. A motorist drives north for 35.0 minutes at 85.0 km/h and then stops for 15.0 minutes. He then continues north, traveling 130 km in 2.00 h. (a) What is his total displacement? (b) What is his average velocity?

6. The position versus time for a certain particle moving along the *x* axis is shown in Figure P2.6. Find the average velocity in the time intervals (a) 0 to 2.00 s, (b) 0 to 4.00 s, (c) 2.00 s to 4.00 s, (d) 4.00 s to 7.00 s, (e) 0 to 8.00 s.

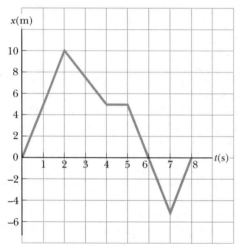

FIGURE P2.6 (Problems 6 and 15)

7. A tennis player moves in a straight-line path as shown in Figure P2.7. Find her average velocities in the time intervals (a) 0 to 1.0 s, (b) 0 to 4.0 s, (c) 1.0 s to 5.0 s, and (d) 0 to 5.0 s.

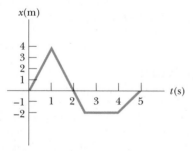

FIGURE P2.7 (Problems 7 and 17)

8. Two cars travel in the same direction along a straight highway, one at a constant speed of 55 mi/h and the other at 70 mi/h. (a) Assuming that they start at the same point, how much sooner does the faster car arrive at a destination 10 mi away? (b) How far must the faster car travel before it has a 15-min lead on the slower car?

9. Figure P2.9 shows a position-time graph for several trains traveling along a straight track. Identify the graphs that correspond to (a) forward motion only, (b) backward motion only, (c) constant velocity, (d) greatest constant velocity, (e) no movement.

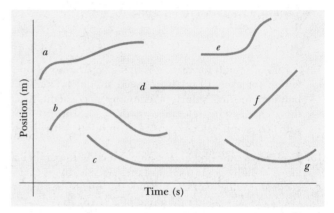

FIGURE P2.9

10. If the average speed of an orbiting space shuttle is 19 800 mi/h, determine the time required for it to circle Earth. Make sure you consider the fact that the shuttle is orbiting about 200 mi above Earth's surface, and assume that Earth's radius is 3 963 miles.

11. A person takes a trip, driving with a constant speed of 89.5 km/h except for a 22.0-min rest stop. If the person's average speed is 77.8 km/h, how much time is spent on the trip and how far does the person travel?

12. A tortoise can run with a speed of 0.10 m/s, and a hare can run 20 times as fast. In a race, they both start at the same time, but the hare stops to rest for 2.0 minutes. The tortoise wins by a shell (20 cm). (a) How long does the race take? (b) What is the length of the race?

13. In order to qualify for the finals in a racing event, a race car must achieve an average speed of 250 km/h on a track with a total length of 1 600 m. If a particular car covers the first half of the track at an average speed of 230 km/h, what minimum average speed must it have in the second half of the event in order to qualify?

14. Runner A is initially 4.0 mi west of a flagpole and is running with a constant velocity of 6.0 mi/h due east. Runner B is initially 3.0 mi east of the flagpole and is running with a constant velocity of 5.0 mi/h due west. How far are the runners from the flagpole when they meet?

15. The position versus time for a certain particle moving along the *x* axis is shown in Figure P2.6. Find the instantaneous velocity at the instants (a) *t* = 1.00 s, (b) *t* = 3.00 s, (c) *t* = 4.50 s, and (d) *t* = 7.50 s.

16. A race car moves such that its position fits the relationship

$$x = (5.0 \text{ m/s})t + (0.75 \text{ m/s}^3)t^3$$

where *x* is measured in meters and *t* in seconds. (a) Plot a graph of position versus time. (b) Determine the instantaneous velocity at *t* = 4.0 s, using time intervals of 0.40 s, 0.20 s, and 0.10 s. (c) Compare the average velocity during the first 4.0 s with the results of (b).

17. Find the instantaneous velocities of the tennis player of Figure P2.7 at (a) 0.50 s, (b) 2.0 s, (c) 3.0 s, (d) 4.5 s.

Section 2.4 Acceleration

18. Secretariat ran the Kentucky Derby with times for the quarter mile of 25.2 s, 24.0 s, 23.8 s, and 23.0 s. (a) Find his average speed during each quarter-mile segment. (b) Assuming that Secretariat's instantaneous speed at the finish line was the same as the average speed during the final quarter mile, find his average acceleration for the entire race. (*Hint:* Recall that horses in the Derby start from rest.)

19. A particle is moving with a velocity of 60.0 m/s in the positive *x* direction at *t* = 0. Between *t* = 0 and *t* = 15.0 s the velocity decreases uniformly to zero. What was the acceleration during this 15.0-s interval? What is the significance of the sign of your answer?

20. A tennis ball with a speed of 10.0 m/s is thrown perpendicularly at a wall. After striking the wall, the ball rebounds in the opposite direction with a speed of 8.0 m/s. If the ball is in contact with the wall for 0.012 s, what is the average acceleration of the ball while it is in contact with the wall?

21. A certain car is capable of accelerating at a rate of **web** + 0.60 m/s². How long does it take for this car to go from a speed of 55 mi/h to a speed of 60 mi/h?

22. The velocity-time graph for an object moving along a straight path is shown in Figure P2.22. (a) Find the average accelerations of this object during the time intervals 0 to 5.0 s, 5.0 s to 15 s, and 0 to 20 s. (b) Find the instantaneous accelerations at 2.0 s, 10 s, and 18 s.

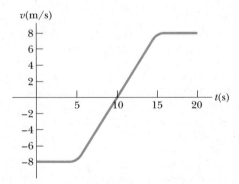

FIGURE P2.22

23. The engine of a model rocket accelerates the rocket vertically upward for 2.0 s as follows: At *t* = 0, its speed is zero; at *t* = 1.0 s, its speed is 5.0 m/s; at *t* = 2.0 s, its speed is 16 m/s. Plot a velocity-time graph for this motion, and from it determine (a) the average acceleration during the 2.0-s interval and (b) the instantaneous acceleration at *t* = 1.5 s.

Section 2.6 One-Dimensional Motion with Constant Acceleration

24. The French National Railroad holds the world's speed record for passenger trains in regular service. A TGV (*tres grand vitesse,* or very great speed) train traveling at a speed of 300 km/h requires 1.20 km to come to an emergency stop. Find the braking acceleration for this train, assuming constant acceleration.

25. Jules Verne in 1865 proposed sending men to the Moon by firing a space capsule from a 220-m-long cannon with final speed of 10.97 km/s. What would have been the unrealistically large acceleration experienced by the space travelers during launch? (A human can stand an acceleration of 15*g* for a short time.) Compare your answer with the free-fall acceleration, 9.80 m/s².

26. A truck covers 40.0 m in 8.50 s while smoothly slowing down to final speed 2.80 m/s. (a) Find its original speed. (b) Find its acceleration.

27. The minimum distance required to stop a car moving at 35.0 mi/h is 40.0 ft. What is the minimum stopping distance for the same car moving at 70.0 mi/h, assuming the same rate of acceleration?

28. A racing car reaches a speed of 40 m/s. At this instant, it begins a uniform negative acceleration, using a parachute and a braking system, and comes to rest 5.0 s later. (a) Determine the acceleration of the car. (b) How far does the car travel after the acceleration starts?

29. A Cessna aircraft has a lift-off speed of 120 km/h. (a) What minimum constant acceleration does this require if the aircraft is to be airborne after a takeoff run of 240 m? (b) How long does it take the aircraft to become airborne?

30. A truck on a straight road starts from rest and accelerates at 2.0 m/s² until it reaches a speed of 20 m/s. Then the truck travels for 20 s at constant speed until the brakes are applied, stopping the truck in a uniform manner in an additional 5.0 s. (a) How long is the truck in motion? (b) What is the average velocity of the truck for the motion described?

31. A drag racer starts her car from rest and accelerates at 10.0 m/s² for the entire distance of 400 m ($\frac{1}{4}$ mile). (a) How long did it take the race car to travel this distance? (b) What is the speed of the race car at the end of the run?

32. A jet plane lands with a speed of 100 m/s and can accelerate at a maximum rate of − 5.00 m/s² as it comes to rest. (a) From the instant the plane touches the runway, what is the minimum time needed before it can come to rest? (b) Can this plane land on a small tropical island airport where the runway is 0.800 km long?

33. A driver in a car traveling at a speed of 60 mi/h sees a deer 100 m away on the road. Calculate the minimum constant acceleration that is necessary for the car to stop without hitting the deer (assuming that the deer does not move in the meantime).

34. A record of travel along a straight path is as follows.

   1. Start from rest with constant acceleration of 2.77 m/s² for 15.0 s
   2. Constant velocity for the next 2.05 min
   3. Constant negative acceleration − 9.47 m/s² for 4.39 s

(a) What was the total displacement for the complete trip? (b) What were the average speeds for legs 1, 2, and 3 of the trip as well as for the complete trip?

35. A train is traveling down a straight track at 20 m/s when the engineer applies the brakes, resulting in an acceleration of -1.0 m/s^2 as long as the train is in motion. How far does the train move during a 40-s time interval starting at the instant the brakes are applied?

36. A car accelerates uniformly from rest to a speed of 40.0 mi/h in 12.0 s. (a) Find the distance the car travels during this time and (b) the constant acceleration of the car.

37. A car starts from rest and travels for 5.0 s with a uniform acceleration of $+1.5$ m/s^2. The driver then applies the brakes, causing a uniform acceleration of -2.0 m/s^2. If the brakes are applied for 3.0 s, (a) how fast is the car going at the end of the braking period, and (b) how far has it gone?

38. A train 400 m long is moving on a straight track with a speed of 82.4 km/h. The engineer applies the brakes at a crossing, and later the last car passes the crossing with a speed of 16.4 km/h. Assuming constant acceleration, determine how long the train blocked the crossing. Disregard the width of the crossing.

39. A hockey player is standing on his skates on a frozen pond when an opposing player, moving with a uniform speed of 12 m/s, skates by with the puck. After 3.0 s, the first player makes up his mind to chase his opponent. If he accelerates uniformly at 4.0 m/s^2, (a) how long does it take him to catch his opponent, and (b) how far has he traveled in this time? (Assume the player with the puck remains in motion at constant speed.)

40. A car, capable of a constant acceleration of 2.5 m/s^2, is stopped at a traffic light. When the light turns green, the car starts from rest with this acceleration. Also, as the light turns green, a truck traveling with a constant velocity of 40 km/h passes the car. Clearly, the car will eventually travel faster than the truck and will overtake it. Where will the car catch up with the truck?

41. A car has an initial velocity v_0 when the driver sees an obstacle in the road in front of him. His reaction time is t_r, and the braking acceleration of the car is a. Show that the total stopping distance is

$$s_{stop} = v_0 t_r - \frac{v_0{}^2}{2a}$$

Note that a is negative in this situation.

42. The yellow caution light on a traffic signal should stay on long enough to allow a driver to either pass through the intersection or safely stop before reaching the intersection. That is, if a car is more than the stopping distance of the previous problem away from the intersection, it can stop. If the car is less than this stopping distance from the intersection, the yellow light should stay on long enough to allow the car to pass entirely through the intersection. (a) Show that the yellow light should stay on for a time

$$t_{light} = t_r - \left(\frac{v_0}{2a}\right) + \left(\frac{s_i}{v_0}\right)$$

where t_r is the driver's reaction time, v_0 is the velocity of

the car approaching the light at the speed limit, a is the braking acceleration, and s_i is the length of the intersection. (b) As city traffic planner, you expect cars to approach an intersection 16 m wide with a speed of 60 km/h. Be cautious and assume a reaction time of 1.1 s to allow for a driver's indecision. Find the length of time the yellow light should remain on. Use a braking acceleration of -2.0 m/s^2.

Section 2.7   Freely Falling Objects

43. A ball is thrown vertically upward with a speed of 25.0 m/s. (a) How high does it rise? (b) How long does it take to reach its highest point? (c) How long does it take to hit the ground after it reaches its highest point? (d) What is its velocity when it returns to the level from which it started?

44. In the mid-1960s, McGill University launched high-altitude weather sensors by firing them from a cannon made from two World War II Navy cannons bolted together for a total length of 18 m. It was proposed that this arrangement be used to launch a satellite. The orbital speed of a satellite is about 29 000 km/h. What would the average acceleration throughout the 18-m length of the cannon have to be in order to have a muzzle velocity of 29 000 km/h for this satellite?

45. Killer whales at Sea World routinely jump 7.5 m above the water. What is their speed as they leave the water to achieve this?

46. A peregrine falcon dives at a pigeon. The falcon starts downward from rest and falls with free-fall acceleration. If the pigeon is 76.0 m below the initial position of the falcon, how long does it take the falcon to reach the pigeon? Assume that the pigeon remains at rest.

47. A small mailbag is released from a helicopter that is descending steadily at 1.50 m/s. After 2.00 s, (a) what is the speed of the mailbag, and (b) how far is it below the helicopter? (c) What are your answers to parts (a) and (b) if the helicopter is rising steadily at 1.50 m/s?

48. A ball thrown vertically upward is caught by the thrower after 2.00 s. Find (a) the initial velocity of the ball and (b) the maximum height it reaches.

49. A model rocket is launched straight upward with an initial speed of 50.0 m/s. It accelerates with a constant upward acceleration of 2.00 m/s^2 until its engines stop at an altitude of 150 m. (a) What is the maximum height reached by the rocket? (b) How long after lift-off does the rocket reach its maximum height? (c) How long is the rocket in the air?

50. A parachutist with a camera, both descending at a speed of 10 m/s, releases that camera at an altitude of 50 m. (a) How long does it take the camera to reach the ground? (b) What is the velocity of the camera just before it hits the ground?

51. A student throws a set of keys vertically upward to her sorority sister, who is in a window 4.00 m above. The keys are caught 1.50 s later by the sister's outstretched hand. (a) With what initial velocity were the keys thrown? (b) What was the velocity of the keys just before they were caught?

52. A ball is thrown directly downward, with an initial speed of 8.00 m/s, from a height of 30.0 m. After what interval does the ball strike the ground?

ADDITIONAL PROBLEMS

53. A truck tractor pulls two trailers, one behind the other, at a constant speed of 100 km/h. It takes 0.600 s for the big rig to completely pass onto a bridge 400 m long. For what duration of time is all or part of the truck-trailer combination on the bridge?

54. A speedboat moving at 30.0 m/s approaches a no-wake buoy marker 100 m ahead. The pilot slows the boat with a constant acceleration of -3.50 m/s^2 by reducing the throttle. (a) How long does it take the boat to reach the buoy? (b) What is the velocity of the boat when it reaches the buoy?

55. A bullet is fired through a board 10.0 cm thick in such a way that the bullet's line of motion is perpendicular to the face of the board. If the initial speed of the bullet is 400 m/s and it emerges from the other side of the board with a speed of 300 m/s, find (a) the acceleration of the bullet as it passes through the board and (b) the total time the bullet is in contact with the board.

56. An indestructible bullet 2.00 cm long is fired straight through a board that is 10.0 cm thick. The bullet strikes the board with a speed of 420 m/s and emerges with a speed of 280 m/s. (a) What is the average acceleration of the bullet through the board? (b) What is the total time that the bullet is in contact with the board? (c) What thickness of boards (calculated to 0.1 cm) would it take to stop the bullet assuming the acceleration through all boards is the same?

57. A ball is thrown upward from the ground with an initial speed of 25 m/s; at the same instant, a ball is dropped from a building 15 m high. After how long will the balls be at the same height?

58. A ranger in a national park is driving at 35.0 mi/h when a **web** deer jumps into the road 200 ft ahead of the vehicle. After a reaction time of t, the ranger applies the brakes to produce an acceleration of $a = -9.00$ ft/s^2. What is the maximum reaction time allowed if she is to avoid hitting the deer?

59. Two students are on a balcony 19.6 m above the street. One student throws a ball vertically downward at 14.7 m/s; at the same instant the other student throws a ball vertically upward at the same speed. The second ball just misses the balcony on the way down. (a) What is the difference in their time in air? (b) What is the velocity of each ball as it strikes the ground? (c) How far apart are the balls 0.800 s after they are thrown?

60. The driver of a truck slams on the brakes when he sees a tree blocking the road. The truck slows uniformly with acceleration -5.60 m/s^2 for 4.20 s, making skid marks 62.4 m long that end at the tree. With what speed does the truck then strike the tree?

61. A young woman named Kathy Kool buys a sports car that can accelerate at the rate of 4.90 m/s^2. She decides to test the car by drag racing with another speedster, Stan Speedy. Both start from rest, but experienced Stan leaves the starting line 1.00 s before Kathy. If Stan moves with a constant acceleration of 3.50 m/s^2 and Kathy maintains an acceleration of 4.90 m/s^2, find (a) the time it takes Kathy to overtake Stan, (b) the distance she travels before she catches him, and (c) the speeds of both cars at the instant she overtakes him.

62. A mountain climber stands at the top of a 50.0-m cliff that overhangs a calm pool of water. He throws two stones vertically downward 1.00 s apart and observes that they cause a single splash. The first stone has an initial velocity of -2.00 m/s. (a) How long after release of the first stone will the two stones hit the water? (b) What initial velocity must the second stone have if they are to hit simultaneously? (c) What will the velocity of each stone be at the instant they hit the water?

63. Using a rocket pack with full throttle, a lunar astronaut accelerates upward from the Moon's surface with a constant acceleration of 2.00 m/s^2. At a height of 5.00 m, a bolt comes loose. (The free-fall acceleration on the Moon's surface is about 1.67 m/s^2.) (a) How fast is the astronaut moving at that time? (b) When will the bolt hit the Moon's surface? (c) How fast will it be moving then? (d) How high will the astronaut be when the bolt hits? (e) How fast will the astronaut be traveling then?

64. In Bosnia, the ultimate test of a young man's courage once was to jump off a 400-year-old bridge (now destroyed) into the River Neretva, 23 m below the bridge. (a) How long did the jump last? (b) How fast was the diver traveling upon impact with the river? (c) If the speed of sound in air is 340 m/s, how long after the diver took off did a spectator on the bridge hear the splash?

65. A person sees a lightning bolt pass close to an airplane that is flying in the distance. The person hears thunder 5.0 s after seeing the bolt, and sees the airplane overhead 10 s after hearing the thunder. The speed of sound in air is 1 100 ft/s. (a) Find the distance of the airplane from the person at the instant of the bolt. (Neglect the time it takes the light to travel from the bolt to the eye.) (b) Assuming that the plane travels with a constant speed toward the person, find the velocity of the airplane. (c) Look up the speed of light in air, and defend the approximation used in (a).

66. Another scheme to catch the roadrunner has failed. A safe falls from rest from the top of a 25.0-m-high cliff toward Wile E. Coyote, who is standing at the base. Wile first notices the safe after it has fallen 15.0 m. How long does he have to get out of the way?

67. A stunt woman sitting on a tree limb wishes to drop verti- **web** cally onto a horse galloping under the tree. The constant speed of the horse is 10.0 m/s, and the woman is initially 3.00 m above the level of the saddle. (a) What must be the horizontal distance between the saddle and limb when the woman makes her move? (b) How long is she in the air?

68. A hard rubber ball, released at chest height, falls to the pavement and bounces back to nearly the same height. When it is in contact with the pavement, the lower side of the ball is temporarily flattened. Before this dent in the ball pops out, suppose that its maximum depth is on the order of 1 cm. Compute an order-of-magnitude estimate for the maximum acceleration of the ball. State your assumptions, the quantities you estimate, and the values you estimate for them.

GROUP ACTIVITIES

Note: Following are a variety of group activities. Some are mini-experiments or demonstrations you can carry out with the assistance of a co-worker. Others are problems, frequently consisting of both conceptual and numerical parts, that stimulate group discussion. The first step in developing problem-solving skills and in developing a full understanding of physics is understanding the basic concepts. These activities are designed to provide assistance in that endeavor.

G.1 Estimate a few speeds in metric units, using a stopwatch or wristwatch chronometer. For example, roll a ball across a table and estimate the number of centimeters it moves each second to find its speed. Other speeds you might try are for someone walking across the room, a jogger running, a car moving through some distance, and so forth. How closely do your estimates match those of a co-worker? To see how well you did, make some actual measurements for those situations in which it is feasible to do so.

G.2 Use what you know about falling objects to measure your reaction time. Hold the index finger and thumb of your dominant hand about 2.5 cm apart, and then have your co-worker hold a ruler vertically in the space between your finger and thumb, as shown in Figure GA2.2. Note the position of the ruler relative to your index finger. Your co-worker must release the ruler and you must catch it (without moving your hand downward) as quickly as you can. The ruler (a freely falling object) falls through a distance $d = \frac{1}{2}gt^2$, where t is the reaction time and $g = 9.80$ m/s². Repeat this measurement of d five times, average your results and calculate an average value of t. Now measure your co-worker's reaction time using the same procedure. Compare your results. For most people, the reaction time is at best about 0.2 s. As an extension to this experiment, replace the ruler with a crisp dollar bill. Hold the bill such that your thumb and index finger are just at the level of Washington's face. Unless you are anticipating the time of release, you will not be able to catch the bill when it is released because the time for the top to pass out of your hand is less than the typical 0.2-s reaction time.

FIGURE GA2.2

G.3 Galileo studied accelerated motion by allowing objects to roll down inclined planes so that their motion would be slow enough to make reasonable observations. Try a similar procedure. Make a mark at the top of an inclined plane as the starting point for the motion and use a metal barrier at the end as a sound cue for stopping a stopwatch. Measure the length of the plane. Record the average time for several trials of a ball rolling down the plane at a measured angle. From this information calculate the acceleration. Repeat the experiment for a larger angle of inclination. Do this for several trials until you can plot a graph of acceleration versus angle. From your graph can you guess what the acceleration would be if the inclined plane were vertical? Would the results of your experiment be different if you had used a significantly more massive ball? If you are unsure, repeat the experiment to see if there is a difference.

G.4 A ball is thrown straight upward and moves in free fall. Choose a coordinate system with its origin at the release point of the ball and the positive direction upward. What is the sign of the following quantities? (a) The velocity of the ball just before it reaches its maximum height, just after it reaches its maximum height, and at its maximum height. (b) The acceleration of the ball at just before it reaches its maximum height, just after its maximum height, and at its maximum height. (c) The displacement of the ball just before it reaches its maximum height, just after its maximum height, and at its maximum height. (d) If it takes time t_1 to reach its maximum height, how long will it take to return to ground level? (e) If it is thrown upward with a velocity of $+v_0$, what will be its velocity upon returning to ground level? (f) Assume a ball is thrown upward with a velocity of 12 m/s. Find the various answers for the above questions to test your answers.

G.5 Figure GA2.5 is a graph of position versus time for the motion of two balls, A and B, moving on parallel tracks. Sketch this figure on a sheet of paper and answer the following questions.
(a) Mark with the symbol t_A along the t axis any instant or instants at which one ball is passing the other.
(b) Which ball is moving faster at clock reading t_B?
(c) Mark with the symbol t_C along the t axis any instant or instants at which the balls have the same velocity.
(d) Over the period of time shown in the diagram, which of the following is true of ball B? Explain your answer.
  (i) It is speeding up all the time.
  (ii) It is slowing down all the time.
  (iii) It is speeding up part of the time and slowing down part of the time.

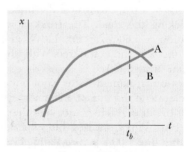

FIGURE GA2.5

G.6 Figure GA2.6 is a graph of the velocity as a function of time for a boy riding his new tricycle. Write a story describing the boy's actions that leads to this graph (keep it short!) and pose a physics problem that could be solved using the graph.

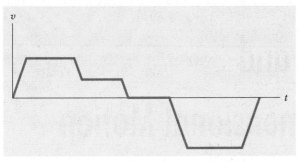

FIGURE GA2.6

G.7 Perform the following activities to verify that all objects fall with the same acceleration. First, try dropping a coffee filter oriented horizontally and a pencil. Then repeat the experiment with the filter in a loose ball, a tight ball, and finally in a compacted wad. You should note that compacting the filter tends to reduce the effects of air resistance and make the two objects fall more nearly at the same rate.

(Problems 5 and 6 are courtesy of E. F. Redish. For more problems of this type, visit http://www.physics.umd.edu/perg/)

The Wizard of Id   by Parker and Hart

By permission of John Hart and Field Enterprises, Inc.

3

Vectors and Two-Dimensional Motion

The circus stuntmen being shot out of cannons are human projectiles. In the absence of air resistance, they move in parabolic paths until they land in a strategically placed net. *(Ringling Brothers Circus)*

*I*n our discussion of one-dimensional motion in Chapter 2, we used the concept of vectors only to a limited extent. As we progress in our study of motion, the ability to manipulate vector quantities will become increasingly important. For that reason, much of this chapter is devoted to techniques for manipulating vectors. We will then apply our newfound skills to a special case of two-dimensional motion—projectiles. We shall also see that an understanding of vector manipulation is necessary in order to work with and understand relative motion.

3.1 VECTORS AND SCALARS REVISITED

Each of the physical quantities we shall encounter in this book can be categorized as either a *vector quantity* or a *scalar quantity*. As we noted in Chapter 2, a vector is a physical quantity that requires the specification of both direction and magnitude (size). A scalar is a quantity that can be completely specified by its magnitude with appropriate units; it has no direction. An example of each kind of quantity is shown in Figure 3.1.

As described in Chapter 2, displacement, velocity and acceleration are vector quantities. Temperature is an example of a scalar quantity. If someone tells you that the temperature of an object is $-5°C$, that information completely specifies the temperature of the object; no direction is required. Other examples of scalars are masses, time intervals, and the number of pages in this textbook. Scalar quantities are manipulated using the rules of ordinary arithmetic. For vectors, special procedures are needed. Most of the following section deals with various methods of adding vectors.

3.2 SOME PROPERTIES OF VECTORS

When a vector quantity is handwritten, it is often represented with an arrow over the letter ($\vec{A}$). As mentioned in Section 2.1, a vector quantity in this book will be represented by boldface type (for example, **A**). The magnitude of a vector such as **A** will be represented by italic type such as *A*. Likewise, italic type will be used to represent scalars.

Equality of Two Vectors. Two vectors **A** and **B** are equal if they have the same magnitude and the same direction. This property allows us to translate a vector parallel to itself in a diagram without affecting the vector. In fact, for most purposes, any vector can be moved parallel to itself without being affected.

Adding Vectors. When two or more vectors are added, they must all have the same units. It would be meaningless to add a velocity vector to a displacement vector because they are different physical quantities. Scalars also obey this rule. For example, it would be meaningless to add temperatures and volumes.

Here we show a geometric or graphical method for adding vectors. (Later we shall develop an algebraic technique that is much more convenient and will be used throughout the remainder of the text.) To add vector **B** to vector **A** geometrically, first draw **A** on a piece of graph paper to some scale, such as 1 cm = 1 m. Draw **A** so that its direction is specified relative to a coordinate system. Then draw vector **B** to the same scale and with the tail of **B** starting at the tip of **A**, as in Figure 3.2a. Vector **B** must be drawn along the direction that makes the proper angle relative to vector **A**. The **resultant vector** **R** = **A** + **B** is the vector drawn from the tail of **A** to the tip of **B**. This is known as the **triangle method of addition.**

When two vectors are added, the sum is independent of the order of the addition. That is, **A** + **B** = **B** + **A**. This can be seen from the geometric construction in Figure 3.2b, and is called the commutative law of addition.

(a)

(b)

FIGURE 3.1 (a) The number of grapes in this bunch ripe for picking is one example of a scalar quantity. Can you think of other examples? *(Mack Henley/Visuals Unlimited)* (b) This helpful person pointing in the right direction tells us to travel 5 blocks north to reach the courthouse. A vector is a physical quantity that must be specified by both magnitude and direction. *(George Semple)*

FIGURE 3.2 (a) When vector **B** is added to vector **A**, the vector sum **R** is the vector that runs from the tail of **A** to the tip of **B**. (b) Here the resultant runs from the tail of **B** to the tip of **A**. These constructions prove that **A** + **B** = **B** + **A**.

(a)   (b)

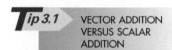

Tip 3.1 VECTOR ADDITION VERSUS SCALAR ADDITION

Keep in mind that **A** + **B** = **C** is very different from $A + B = C$. The first is a vector sum, which must be handled carefully, such as with the graphical method described here. The second is a simple algebraic addition of numbers that is handled with the normal rules of arithmetic.

Webnote 3.1

For more on the properties of vectors, see
http://www.physics.uoguelph.ca/ tutorials/vectors/vectors.html
To visualize the addition of vectors, see (best with Internet Explorer)
http://www.mcasco.com/p1agva.html

An alternative graphical procedure for adding two vectors, known as the **parallelogram method of addition,** is shown in Figure 3.3. In this construction, the tails of vectors **A** and **B** are joined together, and the resultant vector **R** is the diagonal of the parallelogram formed with **A** and **B** as its sides.

This same general approach can also be used to add more than two vectors, as in Figure 3.4 for four vectors. The resultant vector sum **R** = **A** + **B** + **C** + **D** is the vector drawn from the tail of the first vector to the tip of the last vector. Again, the order in which the vectors are added is unimportant.

Negative of a Vector. The negative of the vector **A** is defined as the vector that when added to **A** gives zero for the vector sum. This means that **A** and −**A** have the same magnitude but opposite directions.

Subtracting Vectors. Vector subtraction makes use of the definition of the negative of a vector. We define the operation **A** − **B** as vector −**B** added to vector **A**:

$$\mathbf{A} - \mathbf{B} = \mathbf{A} + (-\mathbf{B}) \qquad [3.1]$$

Thus, vector subtraction is really a special case of vector addition. The geometric construction for subtracting two vectors is shown in Figure 3.5.

Multiplying or Dividing a Vector by a Scalar. The multiplication or division of a vector by a scalar gives a vector. For example, if vector **A** is multiplied by the scalar number 3, the result, written 3**A**, is a vector with a magnitude three times that of **A**, pointing in the same direction as **A**. If we multiply vector **A** by the scalar −3, the result is −3**A**, a vector with a magnitude three times that of **A**, pointing in the direction opposite **A** (because of the negative sign).

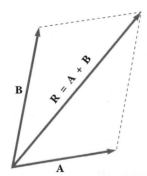

FIGURE 3.3 In this construction, the resultant **R** is the diagonal of a parallelogram with sides **A** and **B**.

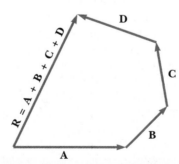

FIGURE 3.4 A geometric construction for summing four vectors. The resultant vector **R** is the vector that completes the polygon.

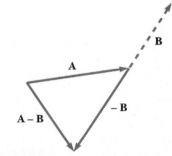

FIGURE 3.5 This construction shows how to subtract vector **B** from vector **A**. The vector −**B** has the same magnitude as the vector **B** but points in the opposite direction.

The magnitudes of two vectors **A** and **B** are $A = 12$ units and $B = 8$ units. The largest and smallest possible values for the magnitude of the resultant vector $\mathbf{R} = \mathbf{A} + \mathbf{B}$ are (a) 14.4 and 4, (b) 12 and 8, (c) 20 and 4, (d) none of these answers.

If vector **B** is added to vector **A**, the resultant vector $\mathbf{A} + \mathbf{B}$ has magnitude $A + B$ when **A** and **B** are (a) perpendicular to each other, (b) oriented in the same direction, (c) oriented in opposite directions, or (d) in any direction relative to each other.

Example 3.1   Taking a Trip

A car travels 20 km due north and then 35 km in a direction 60° west of north, as in Figure 3.6. Find the magnitude and direction of a single vector that gives the net effect of the car's trip. This vector is called the car's resultant displacement.

FIGURE 3.6   (Example 3.1) A graphical method for finding the resultant displacement vector $\mathbf{R} = \mathbf{A} + \mathbf{B}$.

Solution   The problem can be solved geometrically using graph paper and a protractor, as shown in Figure 3.6. The resultant displacement **R** is the sum of the two individual displacements **A** and **B**. We suggest that you draw the diagram much larger, with a scale of 1 cm = 2 km.

The length of **R**, drawn to the same scale as **A** and **B**, indicates that the displacement of the car is 48 km, and a measurement of the angle β shows that the displacement is approximately 39° west of north. Notice that ordinary arithmetic does not work here because the correct answer of 48 km is not equal to 20 km + 35 km = 55 km!

3.3   COMPONENTS OF A VECTOR

One method of adding vectors makes use of the projections of a vector along the axes of a rectangular coordinate system. These projections are called **components.** Any vector can be completely described by its components.

Consider a vector **A** in a rectangular coordinate system, as shown in Figure 3.7. Note that **A** can be expressed as the sum of two vectors, $\mathbf{A}_x$ parallel to the x axis, and $\mathbf{A}_y$ parallel to the y axis. That is,

$$\mathbf{A} = \mathbf{A}_x + \mathbf{A}_y$$

where $\mathbf{A}_x$ and $\mathbf{A}_y$ are the component vectors of **A**. The projection of **A** along the x axis, A_x, is called the x component of **A**, and the projection of **A** along the y axis, A_y, is called the y component of **A**. These components can be either positive or

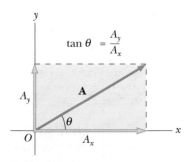

FIGURE 3.7   Any vector **A** lying in the xy plane can be represented by its rectangular components A_x and A_y.

Tip 3.2   X AND Y COMPONENTS

Equation 3.2 for the x and y components of a vector associates the cosine of the angle with the x component and the sine of the angle with the y component. This is due *solely* to the fact that we chose to measure the angle θ with respect to the x axis. You will most likely face a problem in the future in which θ is measured with respect to the y axis. In this case, the components would be given by $A_x = A \sin \theta$ and $A_y = A \cos \theta$.

Webnote 3.2

For a review of trigonometric functions, angle measurements, and circles, see
http://www.sosmath.com/trig/trig.html

Tip 3.3   TANGENTS ON CALCULATORS

Generally, the inverse tangent function on calculators provides an angle between $-90°$ and $+90°$. As a consequence, if the vector you are studying lies in the second or third quadrant, the angle measured from the positive x axis will be the angle your calculator returns plus $180°$.

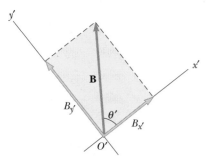

FIGURE 3.8   The components of vector **B** in a tilted coordinate system.

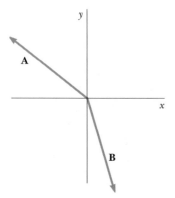

FIGURE 3.9   (Quick Quiz 3.4)

negative numbers with units. From the definitions of sine and cosine, we see that $\cos \theta = A_x/A$ and $\sin \theta = A_y/A$. Hence, the components of **A** are

$$A_x = A \cos \theta$$

$$A_y = A \sin \theta$$

[3.2]

These components form two sides of a right triangle, the hypotenuse of which has the magnitude A. Thus, it follows that **A**'s magnitude and direction are related to its components through the Pythagorean theorem and the definition of the tangent:

$$A = \sqrt{A_x^2 + A_y^2}$$

[3.3]

$$\tan \theta = \frac{A_y}{A_x}$$

[3.4]

To solve for the angle θ, we can write Equation 3.4 in the form

$$\theta = \tan^{-1}\left(\frac{A_y}{A_x}\right)$$

If a coordinate system other than the one shown in Figure 3.7 is chosen, the components of the vector must be modified accordingly. In many applications it is more convenient to express the components of a vector in a coordinate system having axes that are not horizontal and vertical but are still perpendicular to each other. Suppose a vector **B** makes an angle θ' with the x' axis defined in Figure 3.8. The rectangular components of **B** along the axes of Figure 3.8 are given by $B_{x'} = B \cos \theta'$ and $B_{y'} = B \sin \theta'$, as in Equations 3.2. The magnitude and direction of **B** are obtained from expressions equivalent to Equations 3.3 and 3.4. Thus, we can express the components of a vector in any coordinate system that is convenient for the situation.

   As you commute to work or school in the morning, which is most likely the larger? (a) the magnitude of your displacement, (b) the distance you travel, or (c) not enough information to say.

   Figure 3.9 shows two vectors lying in the xy plane. Determine the signs of the x and y components of **A**, **B**, and **A + B** and place your answers in the following table.

Vector	x component	y component
A		
B		
A + B		

APPLYING PHYSICS 3.1

You may have asked someone directions to a destination in a city and have been told something like, "Walk 3 blocks east and then 5 blocks south." If so, are you experienced with vector components?

Explanation Yes, you are! Although you may not have thought of vector component language when you heard these directions, this is exactly what the directions represent. The perpendicular streets of the city reflect an *xy* coordinate system—we can assign the *x* axis to the east-west streets and the *y* axis to the north-south streets. Thus, the comment of the person giving you directions can be translated as, "Undergo a displacement vector that has an *x* component of + 3 blocks and a *y* component of − 5 blocks." You would arrive at the same destination by undergoing the *y* component first, followed by the *x* component, demonstrating the commutative law of addition.

ADDING VECTORS ALGEBRAICALLY

In order to add vectors algebraically rather than graphically, the following procedure is used. First, find the components of all the vectors in some coordinate system that is appropriate for the problem. Next, add all the *x* components to find the resultant component in the *x* direction. Then add all the *y* components to find the resultant component in the *y* direction. Because the resultant *x* and *y* components are at right angles to each other, you can make a sketch and use the Pythagorean theorem to determine the magnitude of the resultant vector. Finally, use a suitable trigonometric function and the components of the resultant vector to find the angle that the resultant vector makes with the *x* axis. These steps are summarized as a Problem-Solving Strategy and are illustrated in the worked examples that follow.

PROBLEM-SOLVING *STRATEGY*	Adding Vectors

When two or more vectors are to be added, the following steps are used:

1. Select a coordinate system.
2. Draw a sketch of the vectors to be added (or subtracted), with a label on each vector.
3. Find the *x* and *y* components of all vectors.
4. Find the resultant components (the algebraic sum of the components) in both the *x* and *y* directions and sketch them so that you can see the actual direction of the resultant in the problem.
5. Use the Pythagorean theorem to find the magnitude of the resultant vector.
6. Use a suitable trigonometric function to find the angle the resultant vector makes with the positive *x* axis.

Example 3.2 Help Is on the Way!

Find the horizontal and vertical components of the 100-m displacement of a superhero who flies from the top of a tall building along the path shown in Figure 3.10a.

Solution The triangle formed by the displacement and its components is shown in Figure 3.10b. Because $A = 100$ m and $\theta = -30.0°$ (θ is negative because it is measured clockwise from the *x* axis), we have

$$A_y = A \sin \theta = (100 \text{ m})\sin(-30.0°) = \boxed{-50.0 \text{ m}}$$

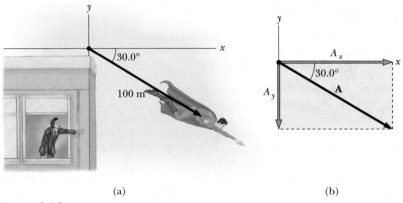

(a)   (b)

FIGURE 3.10 (Example 3.2)

Note that $\sin(-\theta) = -\sin\theta$. The negative sign for A_y reflects the fact that displacement in the y direction is *downward* from the origin.

The x component of displacement is

$$A_x = A\cos\theta = (100\text{ m})\cos(-30.0°) = \boxed{+86.6\text{ m}}$$

Note that $\cos(-\theta) = \cos\theta$. Also, from an inspection of the figure, you should be able to see that A_x is positive in this case.

QUESTION: What effect would choosing $\theta = 330°$ have on your answer?

Example 3.3 Taking a Hike

A hiker begins a trip by first walking 25.0 km southeast from her base camp. On the second day she walks 40.0 km in a direction 60.0° north of east, at which point she discovers a forest ranger's tower.

A Determine the components of the hiker's displacements on the first and second days.

Solution If we denote the displacement vectors on the first and second days by **A** and **B**, respectively, and use the camp as the origin of coordinates, we get the vectors shown in Figure 3.11a. Displacement **A** has a magnitude of 25.0 km and is 45.0° south of east. Its components are

$$A_x = A\cos(-45.0°) = (25.0\text{ km})(0.707) = \boxed{17.7\text{ km}}$$

$$A_y = A\sin(-45.0°) = -(25.0\text{ km})(0.707) = \boxed{-17.7\text{ km}}$$

(a)

(b)

FIGURE 3.11 (Example 3.3) (a) Hiker's path and the resultant vector. (b) Components of the hiker's total displacement from camp.

The negative value of A_y indicates that the y coordinate decreased in this displacement. The signs of A_x and A_y are also evident from Figure 3.11a.

Displacement **B** has a magnitude of 40.0 km and is 60.0° north of east. Its components are

$$B_x = B \cos 60.0° = (40.0 \text{ km})(0.500) = \boxed{20.0 \text{ km}}$$

$$B_y = B \sin 60.0° = (40.0 \text{ km})(0.866) = \boxed{34.6 \text{ km}}$$

B Determine the components of the hiker's total displacement for the trip.

Solution   The resultant displacement for the trip, $\mathbf{R} = \mathbf{A} + \mathbf{B}$, has components

$$R_x = A_x + B_x = 17.7 \text{ km} + 20.0 \text{ km} = \boxed{37.7 \text{ km}}$$

$$R_y = A_y + B_y = -17.7 \text{ km} + 34.6 \text{ km} = \boxed{16.9 \text{ km}}$$

Figure 3.11b shows a sketch of the components and their actual directions in space. The magnitude and direction of the resultant may be easily and correctly determined from such a sketch.

EXERCISE   Determine the magnitude and direction of the total displacement.

ANSWER   41.3 km, 24.1° north of east from the base camp.

3.4   DISPLACEMENT, VELOCITY, AND ACCELERATION IN TWO DIMENSIONS

In our discussion of one-dimensional motion in Chapter 2, the vector nature of displacement, velocity, and acceleration was taken into account through the use of positive and negative signs. To completely describe the motion of an object in two or three dimensions, we must make use of vectors.

Consider an object moving through space as shown in Figure 3.12. When the object is at some point Ⓟ at time t_i, its position is described by the position vector $\mathbf{r}_i$, drawn from the origin to Ⓟ. When the object has moved to some other point Ⓠ at time t_f, its position vector is $\mathbf{r}_f$. As you can see from the vector diagram in Figure 3.12, the final position vector is the sum of the initial position vector and $\Delta\mathbf{r}$. Because $\mathbf{r}_f = \mathbf{r}_i + \Delta\mathbf{r}$, **the displacement of the object is defined as the change in the position vector:**

$$\Delta\mathbf{r} \equiv \mathbf{r}_f - \mathbf{r}_i \qquad [3.5]$$

The average velocity of a particle during the time interval Δt is the ratio of the displacement to the time interval for this displacement.

◀ Average velocity

$$\bar{\mathbf{v}} \equiv \frac{\Delta\mathbf{r}}{\Delta t} \qquad [3.6]$$

Because the displacement $\Delta\mathbf{r}$ is a vector quantity and the time interval Δt is a scalar quantity, we conclude that the average velocity is a *vector* quantity directed along $\Delta\mathbf{r}$.

The instantaneous velocity v is defined as the limit of the average velocity $\Delta\mathbf{r}/\Delta t$ as Δt goes to zero:

◀ Instantaneous velocity

$$\mathbf{v} \equiv \lim_{\Delta t \to 0} \frac{\Delta\mathbf{r}}{\Delta t} \qquad [3.7]$$

The direction of the instantaneous velocity vector is along a line that is tangent to the path of the particle and in the direction of motion.

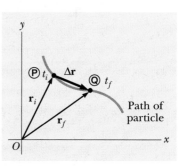

FIGURE 3.12   An object moving along some curved path between points Ⓟ and Ⓠ. The displacement vector $\Delta\mathbf{r}$ is the difference in the position vectors. That is, $\Delta\mathbf{r} = \mathbf{r}_f - \mathbf{r}_i$.

Average acceleration

The average acceleration of an object whose velocity changes by Δv in the time interval Δt is a vector defined as the ratio $\Delta v/\Delta t$.

$$\overline{\mathbf{a}} \equiv \frac{\Delta \mathbf{v}}{\Delta t} \qquad\qquad [3.8]$$

Instantaneous acceleration

The instantaneous acceleration vector a is defined as the limit of the average acceleration vector as Δt goes to zero.

It is important to recognize that a particle can accelerate in several ways. First, the magnitude of the velocity vector (the speed) may change with time. Second, the direction of the velocity vector may change with time (this happens when the motion is along a curved path) even though the speed is constant. Third, both the magnitude and the direction of the velocity vector may change.

Quick Quiz 3.5 True or false: (a) An object cannot accelerate if its speed is constant. (b) A particle can have a constant velocity and a varying speed.

APPLYING **PHYSICS** 3.2

The gas pedal in an automobile is called the *accelerator.* Are there any other controls that could also be considered an accelerator?

Explanation   The gas pedal is called the accelerator because the common usage of the word *acceleration* refers to an *increase in speed.* The scientific definition, however, is that **acceleration occurs whenever the velocity changes in any way.** Thus, the *brake pedal* can be considered an accelerator because it causes the car to slow down. The *steering wheel* is also an accelerator because it changes the direction of the velocity vector.

3.5   PROJECTILE MOTION

In the situations we considered in Chapter 2, objects moved along straight-line paths, such as the *x* axis. Now let us look at some cases in which an object moves in a plane. By this we mean that the object moves in both the *x* and *y* directions simultaneously, which is another way of saying it moves in two dimensions. The particular form of two-dimensional motion we shall concentrate on is called **projectile motion.**

Projectile motion

Anyone who has observed a baseball in motion has observed projectile motion. It is surprisingly simple to analyze if we neglect the effect of air resistance and the rotation and curvature of Earth. With these assumptions, we find that the path of a projectile, the baseball, is curved, as shown in Figure 3.13. Such a curve is called a parabola.

Let us choose our coordinate system so that the *y* direction is vertical and positive upward. In this case, **the acceleration in the y direction is $-g$, just as in free fall, and the acceleration in the x direction is 0 (because air resistance is neglected).** Furthermore, let us assume that at $t = 0$, the projectile leaves the origin with a velocity $\mathbf{v}_0$, as in Figure 3.13. If the velocity vector makes an angle θ_0 with the horizontal, where θ_0 is called the *projection angle,* then from the definitions of cosine and sine functions and Figure 3.13, we have

$$v_{x0} = v_0 \cos \theta_0 \qquad \text{and} \qquad v_{y0} = v_0 \sin \theta_0$$

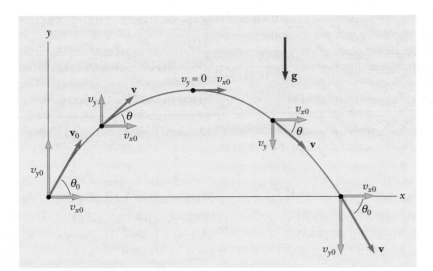

FIGURE 3.13 The parabolic trajectory of a particle that leaves the origin with a velocity of $\mathbf{v}_0$. Note that the velocity $\mathbf{v}$ changes with time. However, the x component of the velocity v_x remains constant in time. Also, $v_y = 0$ at the peak, but the acceleration is always equal to the free-fall acceleration and acts vertically downward.

In order to analyze projectile motion, we separate the motion into two parts, the x (or horizontal) motion and the y (or vertical) motion, and solve each part separately. Experiment shows that the horizontal and vertical motions are independent of each other. This means that neither the x motion nor the y motion is changed by the presence of the other. Let's look first at the x motion. As noted before, motion in the x direction occurs with $a_x = 0$. This means that *the velocity component along the x direction remains constant*. Thus, if the initial value of the velocity component in the x direction is $v_{x0} = v_0 \cos \theta_0$, this is also the value of the velocity in the x direction at any later time. That is,

$$v_x = v_{x0} = v_0 \cos \theta_0 = \text{constant} \qquad \text{[3.9]}$$

Equation 3.9 can be substituted into the defining equation for velocity (Eq. 2.2) to give us an expression for the horizontal displacement of the projectile as a function of time:

$$\Delta x = v_{x0} t = (v_0 \cos \theta_0) t \qquad \text{[3.10]}$$

These equations tell us all we need to know about the motion in the x direction. Let us now consider the y motion. Because the projectile has constant acceleration in this direction, the equations developed in Section 2.6 can be used. In these equations, v_{y0} shall denote the initial velocity in the y direction and $-g$, the free-fall acceleration. The negative sign for g indicates that positive y for the vertical motion is assumed to be upward. With this choice of signs, we have

$$v_y = v_{y0} - gt \qquad \text{[3.11]}$$

$$\Delta y = v_{y0} t - \tfrac{1}{2} g t^2 \qquad \text{[3.12]}$$

$$v_y^2 = v_{y0}^2 - 2g \, \Delta y \qquad \text{[3.13]}$$

where $v_{y0} = v_0 \sin \theta_0$. The speed v of the projectile at any instant can be calculated from the components of velocity at that instant, using the Pythagorean theorem:

$$v = \sqrt{v_x^2 + v_y^2}$$

The angle θ that the velocity vector makes with the x axis is given by

$$\theta = \tan^{-1} \left(\frac{v_y}{v_x} \right)$$

Before we examine some numerical examples dealing with projectile motion, let us pause to summarize what we have learned so far about this kind of motion:

An evening view of the illuminated water fountain in La Place de la Concorde, Paris. The individual water streams follow parabolic trajectories. The horizontal range and maximum height of a given stream of water depend on the elevation angle of its initial velocity as well as its initial speed. (© *The Telegraph Color Library/FPG*)

Tip 3.4   ACCELERATION AT THE HIGHEST POINT

Many people have trouble with the top of the trajectory of a projectile, thinking that its acceleration at the top is zero. This wrong interpretation arises from confusion between zero vertical velocity and zero vertical acceleration. At the highest point, the projectile has zero vertical component of velocity, but is still moving horizontally, and is still accelerating, because gravity is still acting. If the projectile were to experience zero acceleration at the highest point, then the velocity would not change—the projectile would move horizontally once it reached the top! We don't see that happening, because the acceleration is indeed not equal to zero.

1. Projectile motion can be described as a superposition of two independent motions in the x and y directions.
2. Provided air resistance is negligible, the horizontal component of velocity v_x remains constant because there is no horizontal component of acceleration.
3. The vertical component of acceleration is equal to the free-fall acceleration, g.
4. The vertical component of velocity v_y and the displacement in the y direction are identical to those of a freely falling body.

Quick Quiz 3.6   If you are carrying a ball and running at constant speed and wish to throw the ball so that you can catch it as it comes back down, should you (a) throw the ball at an angle of about 45° above the horizontal and maintain the same speed, (b) throw the ball straight in the air and slow down to catch it, or (c) throw the ball straight in the air and maintain the same speed?

Quick Quiz 3.7   As a projectile moves in its parabolic path, the velocity and acceleration vectors are perpendicular to each other (a) everywhere along its path, (b) at the peak of its path, (c) nowhere along its path, or (d) not enough information is given.

Quick Quiz 3.8   A home run is hit in a baseball game. The ball is hit from home plate into the center field stands, along a parabolic path. What is the acceleration of the ball (a) while it is rising, (b) at the highest point of the trajectory, and (c) while it is descending after reaching the highest point? Ignore air resistance.

PROBLEM-SOLVING STRATEGY   **Projectile Motion**

We suggest that you use the following approach to solving projectile motion problems:

1. Select a coordinate system and sketch the path of the projectile, including initial and final positions, velocities, and accelerations.
2. Resolve the initial velocity vector into x and y components.
3. Treat the horizontal motion and the vertical motion independently.
4. Follow the techniques for solving problems with constant velocity to analyze the horizontal motion of the projectile.
5. Follow the techniques for solving problems with constant acceleration to analyze the vertical motion of the projectile.

Example 3.4   The Stranded Explorers

An Alaskan rescue plane drops a package of emergency rations to stranded hikers as shown in Figure 3.14. The plane is traveling horizontally at 40.0 m/s at a height of 100 m above the ground.

A Where does the package strike the ground relative to the point at which it was released?

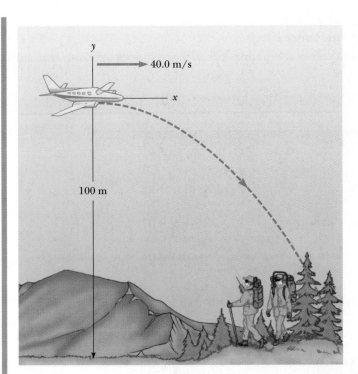

FIGURE 3.14 (Example 3.4) From the point of view of an observer on the ground, a package released from the rescue plane travels along the path shown.

Solution The coordinate system for this problem is selected as shown in Figure 3.14, with the positive x direction to the right and the positive y direction upward.

Consider first the horizontal motion of the package. The only equation available to us is $\Delta x = v_{x0}t$. The initial x component of the package velocity is the same as the velocity of the plane when the package was released, v_{x0}. Because $x_0 = 0$, we find

$$x = v_{x0}t$$

If we know t, the time at which the package strikes the ground, we can determine x, the final position of the package along the horizontal. To find t, we move to the equations for the vertical motion of the package. We know that at the instant the package hits the ground its y coordinate is -100 m. We also know that the initial velocity of the package in the vertical direction, v_{y0}, is zero because at the moment of release, the package had only a horizontal component of velocity. From Equation 3.12 we have

$$y = -\tfrac{1}{2}gt^2$$
$$-100 \text{ m} = -\tfrac{1}{2}(9.80 \text{ m/s}^2)t^2$$
$$t = 4.52 \text{ s}$$

Substitution of this value for the time of flight into the expression for the x coordinate gives

$$x = v_{x0}t = (40.0 \text{ m/s})(4.52 \text{ s}) = \boxed{181 \text{ m}}$$

B What are the horizontal and vertical components of the velocity of the package just before it hits the ground?

Solution We already know the horizontal component of the velocity of the package just before it hits because the velocity in the horizontal direction remains constant at 40.0 m/s throughout the flight.

The vertical component of the velocity just before the package hits the ground can be found by using Equation 3.11, with $v_{y0} = 0$:

$$v_y = v_{y0} - gt = 0 - (9.80 \text{ m/s}^2)(4.52 \text{ s}) = \boxed{-44.3 \text{ m/s}}$$

EXERCISE Find the speed of the package and the direction of impact just before it hits the ground.

ANSWER 59.7 m/s; 47.9° below the horizontal

APPLICATION

LONG JUMPING

Example 3.5   The Long Jump

A long jumper (Fig. 3.15) leaves the ground at an angle of 20.0° to the horizontal and at a speed of 11.0 m/s.

A How far does he jump? (Assume that his motion is equivalent to that of a particle. In other words, disregard the motion of his arms and legs.)

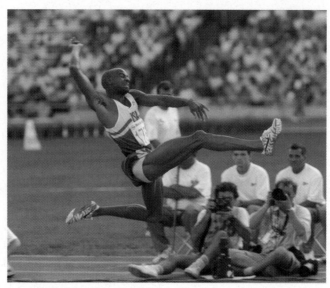

FIGURE 3.15   (Example 3.5) Mike Powell, current holder of the world long jump record of 8.95 m. *(Mike Powell/Allsport/Getty)*

Solution   The horizontal motion is described by Equation 3.10:

$$\Delta x = (v_0 \cos \theta_0) t = (11.0 \text{ m/s})(\cos 20.0°) t$$

The value of Δx can be found if t, the duration of the jump, is known. We can find t with $v_y = v_0 \sin \theta_0 - gt$ (Eq. 3.11) by noting that at the top of the jump the vertical component of velocity goes to zero:

$$t_1 = \frac{v_0 \sin \theta_0 - v_y}{g} = \frac{(11.0 \text{ m/s})(\sin 20.0°) - 0}{9.80 \text{ m/s}^2} = 0.384 \text{ s}$$

Note that t_1 is the time interval to the *top* of the jump. Because of the symmetry of the vertical motion, an identical time interval passes before the jumper returns to the ground. Therefore, the *total time* in the air is $t = 2t_1 = 0.768$ s. Taking $x_0 = 0$ at $t = 0$, the distance jumped is

$$x = (11.0 \text{ m/s})(\cos 20.0°)(0.768 \text{ s}) = \boxed{7.94 \text{ m}}$$

B What is the maximum height reached?

Solution   The maximum height reached is found using $y = (v_0 \sin \theta_0) t - \frac{1}{2} gt^2$ (Eq. 3.12) with $t = t_1 = 0.384$ s.

$$y_{\text{max}} = (11.0 \text{ m/s})(\sin 20.0°)(0.384 \text{ s}) - \frac{1}{2}(9.80 \text{ m/s}^2)(0.384 \text{ s})^2$$
$$= \boxed{0.722 \text{ m}}$$

Although the assumption that the long jumper's motion is that of a projectile is an over-simplification of the situation, the values obtained are reasonable.

Example 3.6   That's Quite an Arm

A stone is thrown upward from the top of a building at an angle of 30.0° to the horizontal and with an initial speed of 20.0 m/s, as in Figure 3.16. The height of the building is 45.0 m.

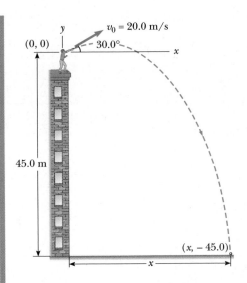

FIGURE 3.16   (Example 3.6)

A How long is the stone "in flight"?

Solution   The initial x and y components of the velocity are

$$v_{x0} = v_0 \cos \theta_0 = (20.0 \text{ m/s})(\cos 30.0°) = +17.3 \text{ m/s}$$

$$v_{y0} = v_0 \sin \theta_0 = (20.0 \text{ m/s})(\sin 30.0°) = +10.0 \text{ m/s}$$

To find t, we can use $y = v_{y0}t - \frac{1}{2}gt^2$ (Eq. 3.12) with $y_0 = 0$, $y = -45.0$ m and $v_{y0} = 10.0$ m/s (we have chosen the point of release as the origin, as shown in Fig. 3.16).

To solve any quadratic equation in the standard form $at^2 + bt + c = 0$, where a, b, and c are constants, we use the quadratic formula (see Appendix A)

$$t = \frac{-b \pm \sqrt{b^2 - 4ac}}{2a}$$

Putting our equation into standard form and using $a = 4.90 \text{ m/s}^2$, $b = -10.0$ m/s, and $c = -45.0$ m gives

$$(4.90 \text{ m/s}^2)\, t^2 - (10.0 \text{ m/s})\, t - 45.0 \text{ m} = 0$$

The positive root of this equation is $t = $ 4.22 s. Does the negative root have any physical meaning?

B What is the speed of the stone just before it strikes the ground?

Solution   The y component of the velocity just before the stone strikes the ground can be obtained using Equation 3.11 with $t = 4.22$ s:

$$v_y = v_{y0} - gt = 10.0 \text{ m/s} - (9.80 \text{ m/s}^2)(4.22 \text{ s}) = -31.4 \text{ m/s}$$

Because $v_x = v_{x0} = 17.3$ m/s, the required speed is given by

$$v = \sqrt{v_x^2 + v_y^2} = \sqrt{(17.3)^2 + (-31.4)^2} \text{ m/s} = \boxed{35.9 \text{ m/s}}$$

EXERCISE   Where does the stone strike the ground?

ANSWER   73.0 m from the base of the building.

3.6   RELATIVE VELOCITY

Recall from Chapter 2 that a frame of reference is a coordinate system that defines a starting point for measuring any quantity. Often we describe the motion of an object from a **stationary frame of reference** attached to Earth, but occasionally it is useful to use a **moving frame of reference** attached to a bus, car, or plane moving with a constant velocity. It is always important to specify the reference frame

PHYSICS *IN ACTION*

Parabolic Paths

On the left is a multiflash photograph of a popular lecture demonstration in which a projectile is fired at a target that is being held by a magnet in the device at the top right of the photograph. The conditions of the experiment are that the gun is aimed at the target and the projectile leaves the gun at the instant the target is released from rest. Under these conditions, the projectile will hit the target no matter what the initial speed of the projectile is. The reason is that the projectile and target both experience the same downward acceleration and hence their velocities change by the same amount in the same time interval. Note that the velocity of the projectile (red arrows) changes in direction and magnitude, and its downward acceleration (violet arrows) remains constant.

On the right, a welder cuts holes through a heavy metal construction beam with a hot torch. The sparks generated in the process follow parabolic paths.

(Central Scientific Company)

(© The Telegraph Colour Library/FPG)

being used because observers in different reference frames can measure different displacements or velocities for the same moving object.

For example, if two cars were moving in the same direction with speeds of 50 mi/h and 60 mi/h, a passenger in the slower car would measure the speed of the faster car relative to a frame attached to his car as 10 mi/h. Of course, a stationary observer using a reference frame rigidly attached to Earth would measure the speed of the faster car as 60 mi/h. This simple example demonstrates that velocity measurements differ in different frames of reference.

We solved the two-moving-cars problem with a minimum of thought and effort, but you will encounter many situations where velocities are not all in the same direction. If this is the case, a more systematic method for attacking such problems is beneficial. To develop this method, let us write down all the information we are given and that which we want to know in the form of velocities with

subscripts. We have

$$\mathbf{v}_{se} = +50 \text{ mi/h} \qquad \text{(The subscript "se" means the velocity of the } slower \text{ car relative to } Earth.)$$

$$\mathbf{v}_{fe} = +60 \text{ mi/h} \qquad \text{(The subscript "fe" means the velocity of the } faster \text{ car relative to } Earth.)$$

We want to know $\mathbf{v}_{fs}$, which is the velocity of the *faster* car relative to the *slower* car. To find this, we write an equation for $\mathbf{v}_{fs}$ in terms of the other two velocities:

$$\mathbf{v}_{\boxed{fs}} = \mathbf{v}_{\boxed{f}e} + \mathbf{v}_{e\boxed{s}} \qquad\qquad \textbf{[3.14]}$$

Note the pattern of the subscripts in this expression. The first subscript f on the left side of the equation is also the first subscript on the right side of the equation. The second subscript s on the left side of the equation is also the last subscript on the right side of the equation. Boxes are used in the subscripts to emphasize this pattern. The boldface notation, of course, indicates that velocity is a vector quantity. As we shall see in the following examples, this vector nature of the velocity is of paramount importance in certain instances.

We know that $\mathbf{v}_{es} = -\mathbf{v}_{se}$, so

$$\mathbf{v}_{fs} = +60 \text{ mi/h} - 50 \text{ mi/h} = +10 \text{ mi/h}$$

There is no general equation to memorize in order to solve relative velocity problems; instead, you should develop the necessary equations on your own by following the technique already demonstrated for writing subscripts. We suggest that you practice this technique as in the examples that follow.

Example 3.7   **Where Does the Ball Land?**

A passenger at the rear of a train traveling at 15 m/s relative to Earth throws a baseball with a speed of 15 m/s in the direction opposite the motion of the train. What is the velocity of the baseball relative to Earth?

Solution   We first write our knowns and unknowns with appropriate subscripts:

$$\mathbf{v}_{te} = +15 \text{ m/s} \qquad \text{(velocity of the } train \text{ relative to } Earth)$$

$$\mathbf{v}_{bt} = -15 \text{ m/s} \qquad \text{(velocity of the } baseball \text{ relative to the } train.)$$

We need $\mathbf{v}_{be}$, the velocity of the *baseball* relative to *Earth*. Following the strategy for watching the subscripts while we write the equation for our unknown, we have

$$\mathbf{v}_{\boxed{be}} = \mathbf{v}_{\boxed{b}t} + \mathbf{v}_{t\boxed{e}}$$

$$\mathbf{v}_{\boxed{be}} = +15 \text{ m/s} - 15 \text{ m/s} = \boxed{0}$$

Example 3.8   **Crossing a River**

The boat in Figure 3.17 is heading due north as it crosses a wide river with a velocity of 10.0 km/h relative to the water. The river has a uniform velocity of 5.00 km/h due east. Determine the velocity of the boat with respect to an observer on the riverbank.

Solution   We have

$$\mathbf{v}_{br} = 10.0 \text{ km/h due north (velocity of the } boat \text{ relative to the } river)$$

$$\mathbf{v}_{re} = 5.00 \text{ km/h due east (velocity of the } river \text{ relative to } Earth)$$

and we want $\mathbf{v}_{be}$, the velocity of the *boat* relative to *Earth*. Our equation becomes

$$\mathbf{v}_{\boxed{be}} = \mathbf{v}_{\boxed{b}r} + \mathbf{v}_{r\boxed{e}}$$

FIGURE 3.17   (Example 3.8)

Because the velocities are not along the same direction, the terms in the equation must be manipulated as vector quantities, as shown in Figure 3.17. The quantity $\mathbf{v}_{br}$ is due north, $\mathbf{v}_{re}$ is due east, and the vector sum of the two, $\mathbf{v}_{be}$, is at the angle θ. Thus, the speed of the boat relative to Earth can be found from the Pythagorean theorem as

$$v_{be} = \sqrt{(v_{br})^2 + (v_{re})^2}$$

$$= \sqrt{(10.0 \text{ km/h})^2 + (5.00 \text{ km/h})^2} = \boxed{11.2 \text{ km/h}}$$

and the direction of $\mathbf{v}_{be}$ is

$$\theta = \tan^{-1}\left(\frac{v_{re}}{v_{br}}\right) = \tan^{-1}\left(\frac{5.00}{10.0}\right) = \boxed{26.6°}$$

The boat travels at a speed of 11.2 km/h in the direction 63.4° north of east with respect to Earth.

Example 3.9   Which Way Should the Boat Head?

If the boat of the preceding example moves with the same speed of 10.0 km/h relative to the water but now travels due north, as in Figure 3.18, what heading should it take? That is, in what direction should it move relative to the water?

Solution   We know that

$$\mathbf{v}_{br} = \text{velocity of the } boat \text{ relative to the } river$$

$$\mathbf{v}_{re} = \text{velocity of the } river \text{ relative to } Earth$$

and we need $\mathbf{v}_{be}$, the velocity of the *boat* relative to *Earth*. We have

$$\mathbf{v}_{\boxed{be}} = \mathbf{v}_{\boxed{b}r} + \mathbf{v}_{r\boxed{e}}$$

The relationship among these three vectors is shown in Figure 3.18; it agrees with our intuitive guess that the boat must head upstream in order to be pushed directly northward across the water. The speed v_{be} can be found from the Pythagorean theorem:

$$v_{be} = \sqrt{(v_{br})^2 - (v_{re})^2} = \sqrt{(10.0 \text{ km/h})^2 - (5.00 \text{ km/h})^2} = 8.66 \text{ km/h}$$

and the direction that the boat must head in order to travel due north is

$$\theta = \tan^{-1}\left(\frac{v_{re}}{v_{be}}\right) = \tan^{-1}\left(\frac{5.00}{8.66}\right) = \boxed{30.0°}$$

where θ is west of north.

FIGURE 3.18   (Example 3.9)

SUMMARY

Two vectors **A** and **B** can be added geometrically by either the triangle method or the parallelogram method. In the **triangle method,** the two vectors are drawn to scale on graph paper with the tail of the second vector located at the tip of the first. The **resultant** vector is the vector drawn from the tail of the first to the tip of the second. In the **parallelogram method,** the vectors are drawn to scale on graph paper with the tails of the two vectors joined. The resultant vector is the diagonal of the parallelogram formed with the two vectors as its sides.

The negative of a vector **A** is a vector with the same magnitude as **A** but pointing in the opposite direction.

The *x* component of a vector is equivalent to the projection of the vector along the *x* axis of a coordinate system. The *y* component is the projection of the vector along the *y* axis of this coordinate system. The resultant of two or more vectors can be found algebraically by resolving all vectors into their x and y components, finding the resultant x and y components, and then using the Pythagorean theorem to find the resultant vector. The angle of the resultant vector with respect to the x axis can be found by use of a suitable trigonometric function.

An object moving above Earth's surface in both the x and y directions simultaneously is said to be undergoing projectile motion. Neglecting air resistance, the object moves along the horizontal (x) direction so that its velocity v_x in this direction is a constant. The object also moves in the vertical (y) direction with a constant downward free-fall acceleration of magnitude $g = 9.80 \text{ m/s}^2$. The equations describing the motion of a projectile are

$$\Delta x = v_{x0} t \qquad\qquad\qquad \textbf{[3.10]}$$

$$v_y = v_{y0} - gt \qquad\qquad\qquad \textbf{[3.11]}$$

$$\Delta y = v_{y0} t - \tfrac{1}{2} g t^2 \qquad\qquad\qquad \textbf{[3.12]}$$

$$v_y^2 = v_{y0}^2 - 2g \, \Delta y \qquad\qquad\qquad \textbf{[3.13]}$$

where $v_{y0} = v_0 \sin \theta_0$ is the initial vertical component of the velocity and $v_{x0} = v_0 \cos \theta_0$ is the initial horizontal component of the velocity.

Observations made in different frames of reference can be related to one another through the techniques of the transformation of velocities. There is no general equation for you to memorize in order to work relative velocity problems. Instead, you should follow the technique for writing subscripts on each velocity under consideration as described in the text.

CONCEPTUAL QUESTIONS

1. Vector **A** lies in the xy plane. For what orientations of vector **A** will both of its components be negative? When will the components have opposite signs?

2. If **B** is added to **A**, under what conditions does the resultant vector have a magnitude equal to $A + B$? Under what conditions is the resultant vector equal to zero?

3. A wrench is dropped from the top of a 10-m mast on a sailing ship while the ship is traveling at 20 knots. Where will the wrench hit the deck? (Galileo posed this problem.)

4. Two vectors have unequal magnitudes. Can their sum be zero? Explain.

5. A projectile is fired on Earth with some initial velocity. Another projectile is fired on the Moon with the same initial velocity. If air resistance is neglected, which projectile has the greater range? Which reaches the greater altitude? (Note that the free-fall acceleration on the Moon is about 1.6 m/s^2.)

6. Can a vector **A** have a component greater than its magnitude A?

7. Is it possible to add a vector quantity to a scalar quantity?

8. Under what circumstances would a vector have components that are equal in magnitude?

9. As a projectile moves in its path, is there any point along the path where the velocity and acceleration vectors are (a) perpendicular to each other? (b) parallel to each other?

10. A rock is dropped at the same instant that a ball, at the same initial elevation, is thrown horizontally. Which will have the greater speed when it reaches ground level?

11. Explain whether or not the following particles have an acceleration: (a) a particle moving in a straight line with constant speed and (b) a particle moving around a curve with constant speed.

12. Correct the following statement: "The racing car rounds the turn at a constant velocity of 90 miles per hour."

13. A spacecraft drifts through space at a constant velocity. Suddenly a gas leak in the side of the spacecraft causes a constant acceleration of the spacecraft in a direction perpendicular to the initial velocity. The orientation of the spacecraft does not change, so that the acceleration remains perpendicular to the original direction of the velocity. What is the shape of the path followed by the spacecraft in this situation?

14. A ball is projected horizontally from the top of a building. One second later another ball is projected horizontally from the same point with the same velocity. At what point in the motion will the balls be closest to each other? Will the first ball always be traveling faster than the second ball? What will be the time difference between when the balls hit the ground? Can the horizontal projection velocity of the second ball be changed so that the balls arrive at the ground at the same time?

15. Two projectiles are thrown with the same magnitude of initial velocity, one at an angle θ with respect to the level ground and the other at angle $90° - \theta$. Both projectiles will strike the ground at the same distance from the projection point. Will both projectiles be in the air for the same time interval?

16. A baseball is thrown such that its initial x and y components of velocity are known. Neglecting air resistance, describe how you would calculate, at the instant the ball reaches the top of its trajectory, (a) its coordinates, (b) its

velocity, and (c) its acceleration. How would these results change if air resistance were taken into account?

17. A projectile is fired at some angle to the horizontal with some initial speed v_0, and air resistance is neglected. Is the projectile a freely falling body? What is its acceleration in the vertical direction? What is its acceleration in the horizontal direction?

18. Determine which of the following moving objects obey the equations of projectile motion developed in this chapter. (a) A ball is thrown in an arbitrary direction. (b) A jet airplane crosses the sky with its engines thrusting the plane forward. (c) A rocket leaves the launch pad. (d) A rocket moving through the sky after its engines have failed. (e) A stone is thrown under water.

19. How can you throw a projectile so that it has zero speed at the top of its trajectory? So that it has nonzero speed at the top of its trajectory?

20. A ball is thrown upward in the air by a passenger on a train that is moving with constant velocity. (a) Describe the path of the ball as seen by the passenger. Describe the path as seen by a stationary observer outside the train. (b) How would these observations change if the train were accelerating along the track?

PROBLEMS

1, 2, 3 = straightforward, intermediate, challenging   ☐ = full solution available in Student Solutions Manual/Study Guide

web = solution posted at **http://info.brookscole.com/serway**   🔬 = biomedical application

Section 3.2   Some Properties of Vectors

1. A dog searching for a bone walks 3.50 m south, then 8.20 m at an angle 30.0° north of east, and finally 15.0 m west. Find the dog's resultant displacement vector, using graphical techniques.

2. An airplane flies 200 km due west from city A to city B and then 300 km in the direction of 30.0° north of west from city B to city C. (a) In straight-line distance, how far is city C from city A? (b) Relative to city A, in what direction is city C?

3. A man lost in a maze makes three consecutive displacements so that at the end of the walks he is right back where he started. The first displacement is 8.00 m westward, and the second is 13.0 m northward. Find the magnitude and direction of the third displacement, using the graphical method.

4. A jogger runs 100 m due west, then changes direction for the second leg of the run. At the end of the run, she is 175 m away from the starting point at an angle of 15.0° north of west. What were the direction and length of her second displacement? Use graphical techniques.

5. A plane flies from base camp to lake A, a distance of 280 km at a direction of 20.0° north of east. After dropping off supplies it flies to lake B, which is 190 km and 30.0° west of north from lake A. Graphically determine the distance and direction from lake B to the base camp.

6. Vector **A** has a magnitude of 8.00 units and makes an angle of 45.0° with the positive x axis. Vector **B** also has a magnitude of 8.00 units and is directed along the negative x axis. Using graphical methods, find (a) the vector sum **A** + **B** and (b) the vector difference **A** − **B**.

7. Vector **A** is 3.00 units in length and points along the positive x axis. Vector **B** is 4.00 units in length and points along the negative y axis. Use graphical methods to find the magnitude and direction of the vectors (a) **A** + **B** and (b) **A** − **B**.

8. The displacement vectors **A** and **B** shown in Figure P3.8 each have a magnitude of 3.00 m. Graphically find (a) **A** + **B**, (b) **A** − **B**, (c) **B** − **A**, (d) **A** − 2**B**

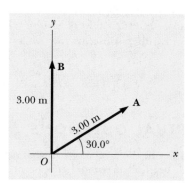

FIGURE P3.8

Section 3.3   Components of a Vector

9. A golfer takes two putts to get his ball into the hole once he is on the green. The first putt displaces the ball 6.00 m east, and the second, 5.40 m south. What displacement would have been needed to get the ball into the hole on the first putt?

10. A person walks 25.0° north of east for 3.10 km. How far would a person walk due north and due east to arrive at the same location?

11. A girl delivering newspapers covers her route by traveling 3.00 blocks west, 4.00 blocks north, then 6.00 blocks east. (a) What is her resultant displacement? (b) What is the total distance she travels?

12. While exploring a cave, a spelunker starts at the entrance and moves the following distances. She goes 75.0 m north, 250 m east, 125 m at an angle 30.0° north of east, and 150 m south. Find the resultant displacement from the cave entrance.

13. A vector has an x component of − 25.0 units and a y component of 40.0 units. Find the magnitude and direction of this vector.

14. A quarterback takes the ball from the line of scrimmage, runs backward for 10.0 yards, and then runs to the right parallel to the line of scrimmage for 15.0 yards. At this point, he throws a 50.0-yard forward pass straight down-

field, perpendicular to the line of scrimmage. What is the magnitude of the football's resultant displacement?

15. The eye of a hurricane passes over Grand Bahama Island. It is moving in a direction 60.0° north of west with a speed of 41.0 km/h. Three hours later, the course of the hurricane suddenly shifts due north, and its speed slows to 25.0 km/h. How far from Grand Bahama is the hurricane 4.50 h after it passes over the island?

16. A small map shows Atlanta to be 730 miles in a direction of 5.0° north of east from Dallas. The same map shows that Chicago is 560 miles in a direction of 21° west of north from Atlanta. Assume a flat Earth and use this information to find the displacement from Dallas to Chicago.

17. A commuter airplane starts from an airport and takes the route shown in Figure P3.17. It first flies to city A located at 175 km in a direction 30.0° north of east. Next, it flies 150 km 20.0° west of north to city B. Finally, it flies 190 km due west to city C. Find the location of city C relative to the location of the starting point.

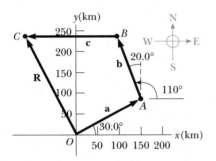

FIGURE P3.17

18. Two people pull on a stubborn mule, as seen from a helicopter in Figure P3.18. Find (a) the single force that is equivalent to the two forces shown, and (b) the force that a third person would have to exert on the mule to make the net force equal to zero.

FIGURE P3.18

19. A man pushing a mop across a floor causes the mop to undergo two displacements. The first has a magnitude of 150 cm and makes an angle of 120° with the positive x axis. The resultant displacement has a magnitude of 140 cm and is directed at an angle of 35.0° to the positive x axis. Find the magnitude and direction of the second displacement.

20. An airplane starting from airport A flies 300 km east, then 350 km at 30.0° west of north, and then 150 km north to arrive finally at airport B. (a) The next day, another plane flies directly from A to B in a straight line. In what direction should the pilot travel in this direct flight? (b) How far will the pilot travel in this direct flight? Assume there is no wind during these flights.

21. Long John Silver, a pirate, has buried his treasure on an island with five trees located at the following points: A (30.0 m, -20.0 m), B (60.0 m, 80.0 m), C (-10.0 m, -10.0 m), D (40.0 m, -30.0 m), and E (-70.0 m, 60.0 m), all measured relative to some origin, as in Figure P3.21. His map instructs you to start at A and move toward B, but cover only one-half the distance between A and B. Then move toward C, covering one-third the distance between your current location and C. Then move toward D, covering one-fourth the distance between where you are and D. Finally move toward E, covering one-fifth the distance between you and E, stop and dig. (a) What are the coordinates of the point where his treasure is buried? (b) Rearrange the order of the trees, for instance B (30 m, -20 m), A (60 m, 80 m), E (-10 m, -10 m), C (40 m, -30 m), and D (-70 m, 60 m), and repeat the calculation to show that the answer does not depend on the order of the trees.

FIGURE P3.21

Section 3.4 Displacement, Velocity and Acceleration in Two Dimensions

Section 3.5 Projectile Motion

22. One of the fastest recorded pitches in major-league baseball, thrown by Nolan Ryan in 1974, was clocked at 100.8 mi/h [Fig. P3.22]. If a pitch were thrown horizontally with this velocity, how far would the ball fall vertically by the time it reached home plate, 60.0 ft away?

23. The peregrine falcon is the fastest bird, flying at a speed of 200 mi/h (Fig. P3.23). Nature has adapted it to reach such speed by placing baffles in its nostrils to prevent air from rushing in and slowing it. Also, its eyes adjust focus faster than any other creature so it can focus quickly on its prey. Assume it is moving horizontally at this speed at a height

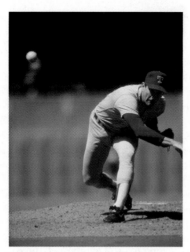

FIGURE P3.22   Nolan Ryan hurls the baseball. *(Rick Stewart/Allsport/Getty)*

FIGURE P3.24

of 100 m above the ground when it brings its wings into its sides and begins to drop in free fall. How far will the bird fall vertically while traveling horizontally a distance of 100 m?

FIGURE P3.23   Notice the structure within the peregrine falcon's nostrils. *(Joseph Kayne/ Dembinsky Photo Associates)*

24. A student stands at the edge of a cliff and throws a stone horizontally over the edge with a speed of 18.0 m/s. The cliff is 50.0 m above a flat, horizontal beach, as shown in Figure P3.24. How long after being released does the stone strike the beach below the cliff? With what speed and angle of impact does it land?

25. The best leaper in the animal kingdom is the puma, which can jump to a height of 12 ft when leaving the ground at an angle of 45°. With what speed, in SI units, must it leave the ground to reach this height?

26. Tom the cat is chasing Jerry the mouse across a table surface 1.5 m above the floor. Jerry steps out of the way at the last second, and Tom slides off the edge of the table at a speed of 5.0 m/s. Where will Tom strike the floor, and what velocity components will he have just before he hits?

27. A tennis player standing 12.6 m from the net hits the ball **web** at 3.00° above the horizontal. To clear the net, the ball must rise at least 0.330 m. If the ball just clears the net at the apex of its trajectory, how fast was the ball moving when it left the racquet?

28. An artillery shell is fired with an initial velocity of 300 m/s at 55.0° above the horizontal. To clear an avalanche, it ex-

plodes on a mountainside 42.0 s after firing. What are the *x* and *y* coordinates of the shell where it explodes, relative to its firing point?

29. A brick is thrown upward from the top of a building at an angle of 25° to the horizontal and with an initial speed of 15 m/s. If the brick is in flight for 3.0 s, how tall is the building?

30. A place kicker must kick a football from a point 36.0 m (about 39 yd) from the goal, and the ball must clear the crossbar, which is 3.05 m high. When kicked, the ball leaves the ground with a speed of 20.0 m/s at an angle of 53.0° to the horizontal. (a) By how much does the ball clear or fall short of clearing the crossbar? (b) Does the ball approach the crossbar while still rising or while falling?

31. A car is parked on a cliff overlooking the ocean on an incline that makes an angle of 24.0° below the horizontal. The negligent driver leaves the car in neutral, and the emergency brakes are defective. The car rolls from rest down the incline with a constant acceleration of 4.00 m/s² for a distance of 50.0 m to the edge of the cliff. The cliff is 30.0 m above the ocean. Find (a) the car's position relative to the base of the cliff when the car lands in the ocean, and (b) the length of time the car is in the air.

32. A fireman, 50.0 m away from a burning building, directs a stream of water from a ground level fire hose at an angle of 30.0° above the horizontal. If the speed of the stream as it leaves the hose is 40.0 m/s, at what height will the stream of water strike the building?

33. A projectile is launched with an initial speed of 60.0 m/s at an angle of 30.0° above the horizontal. The projectile lands on a hillside 4.00 s later. Neglect air friction. (a) What is the projectile's velocity at the highest point of its trajectory? (b) What is the straight-line distance from where the projectile was launched to where it hits?

34. A soccer player kicks a rock horizontally off a 40.0-m-high cliff into a pool of water. If the player hears the sound of

the splash 3.00 s later, what was the initial speed given to the rock? Assume the speed of sound in air to be 343 m/s.

Section 3.6 Relative Velocity

35. A jet airliner moving initially at 300 mi/h due east enters a region where the wind is blowing at 100 mi/h in a direction 30.0° north of east. What is the new velocity of the aircraft relative to the ground?

36. A boat moves through the water of a river at 10 m/s relative to the water, regardless of the boat's direction. If the water in the river is flowing at 1.5 m/s, how long does it take the boat to make a round trip consisting of a 300-m displacement downstream followed by a 300-m displacement upstream?

37. The pilot of an airplane notes that the compass indicates a heading due west. The airplane's speed relative to the air is 150 km/h. If there is a wind of 30.0 km/h toward the north, find the velocity of the airplane relative to the ground.

38. A river flows due east at 1.50 m/s. A boat crosses the river from the south shore to the north shore by maintaining a constant velocity of 10.0 m/s due north relative to the water. (a) What is the velocity of the boat relative to shore? (b) If the river is 300 m wide, how far downstream has the boat moved by the time it reaches the north shore?

39. A rowboat crosses a river with a velocity of 3.30 mi/h at an
web angle 62.5° north of west relative to the water. The river is 0.505 mi wide and carries an eastward current of 1.25 mi/h. How far upstream is the boat when it reaches the opposite shore?

40. The pilot of an aircraft wishes to fly due west in a 50.0-km/h wind blowing toward the south. If the speed of the aircraft relative to the air is 200 km/h, (a) in what direction should the aircraft head, and (b) what will be its speed relative to the ground?

41. How long does it take an automobile traveling in the left lane at 60.0 km/h to overtake (become even with) another car that is traveling in the right lane at 40.0 km/h, when the cars' front bumpers are initially 100 m apart?

42. A science student is riding on a flatcar of a train traveling along a straight horizontal track at a constant speed of 10.0 m/s. The student throws a ball along a path that she judges to make an initial angle of 60.0° with the horizontal and to be in line with the track. The student's professor, who is standing on the ground nearby, observes the ball to rise vertically. How high does the ball rise?

ADDITIONAL PROBLEMS

43. A particle undergoes two displacements. The first has a magnitude of 150 cm and makes an angle of 120.0° with the positive x axis. The resultant of the two displacements is 140 cm directed at an angle of 35.0° to the positive x axis. Find the magnitude and direction of the second displacement.

44. Find the sum of these four vector forces: 12.0 N to the right at 35.0° above the horizontal, 31.0 N to the left at 55.0° above the horizontal, 8.40 N to the left at 35.0° below the horizontal, and 24.0 N to the right at 55.0° below the horizontal. (*Hint:* N stands for newton, the SI unit of force. The component method allows addition of any vectors, forces as well as displacements and velocities. Make a

drawing of this situation and select the best axes for x and y so you have the least number of components.)

45. A car travels due east with a horizontal speed of 50.0 km/h. Rain is falling vertically with respect to Earth. The traces of the rain on the side windows of the car make an angle of 60.0° with the vertical. Find the velocity of the rain with respect to (a) the car and (b) Earth.

46. A stone is kicked so that it leaves the ground with an initial velocity of 10.0 m/s at an angle of 37.0°. The stone is kicked toward a loading dock whose near edge is 3.00 m away and 1.00 m high. (a) What are the components of the stone's initial velocity? (b) How high does it go? (c) How far from the edge does it strike the dock? (d) How fast is it moving just before it hits the dock?

47. Towns A and B in Figure P3.47 are 80.0 km apart. A couple arranges to drive from town A and meet a couple driving from town B at the lake, L. The two couples leave simultaneously and drive for 2.50 h in the directions shown. Car 1 has a speed of 90.0 km/h. If the cars arrive simultaneously at the lake, what is the speed of car 2?

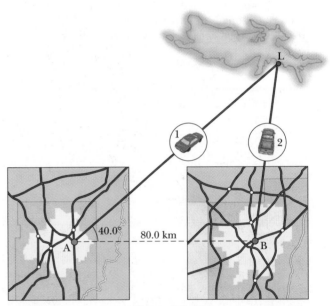

FIGURE P3.47

48. Three displacements are **A** = 200 m, due south; **B** = 250 m, due west; **C** = 150 m, 30° east of north. Construct a separate diagram for each of the following possible ways of adding these vectors:

$$\mathbf{R}_1 = \mathbf{A} + \mathbf{B} + \mathbf{C} \qquad \mathbf{R}_2 = \mathbf{B} + \mathbf{C} + \mathbf{A} \qquad \mathbf{R}_3 = \mathbf{C} + \mathbf{B} + \mathbf{A}$$

49. A rocket is launched at an angle of 53.0° above the horizontal with an initial speed of 100 m/s. It moves for 3.00 s along its initial line of motion with an acceleration of 30.0 m/s^2. At this time its engines fail and the rocket proceeds to move as a projectile. Find (a) the maximum altitude reached by the rocket, (b) its total time of flight, and (c) its horizontal range.

50. Two canoeists in identical canoes exert the same effort paddling and hence maintain the same speed relative to the water. One paddles directly upstream (and moves upstream), whereas the other paddles directly downstream.

With downstream as the positive direction, an observer on shore determines the velocities of the two canoes to be -1.2 m/s and $+2.9$ m/s, respectively. (a) What is the speed of the water relative to shore? (b) What is the speed of each canoe relative to the water?

51. If a person can jump a maximum horizontal distance (by using a 45° projection angle) of 3.0 m on Earth, what would be his maximum range on the Moon, where the free-fall acceleration is $g/6$ and $g = 9.80$ m/s^2? Repeat for Mars, where the acceleration due to gravity is $0.38g$.

52. A daredevil decides to jump a canyon. Its walls are equally **web** high and 10 m apart. He takes off by driving a motorcycle up a short ramp sloped at an angle of 15°. What minimum speed must he have in order to clear the canyon?

53. Cliff divers at Acapulco jump into the sea from a cliff 36.0 m high. At the level of the sea, a rock sticks out a horizontal distance of 6.00 m. With what minimum horizontal velocity must the cliff divers leave the top of the cliff if they are to miss the rock?

54. A mountain climber is stranded on a ledge 30 m above the ground (Fig. P3.54). Rescuers on the ground want to shoot a projectile to him with a rope attached to it. If the projectile is directed upward at an initial angle of 55° from a horizontal distance of 50 m, determine the initial speed the projectile must have in order to land on the ledge.

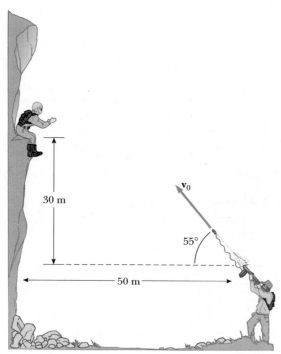

FIGURE P3.54

55. A home run is hit in such a way that the baseball just clears a wall 21 m high, located 130 m from home plate. The ball is hit at an angle of 35° to the horizontal, and air resistance is negligible. Find (a) the initial speed of the ball, (b) the time it takes the ball to reach the wall, and (c) the velocity components and the speed of the ball when it reaches the wall. (Assume the ball is hit at a height of 1.0 m above the ground.)

56. A ball is thrown straight upward and returns to the thrower's hand after 3.00 s in the air. A second ball is thrown at an angle of 30.0° with the horizontal. At what speed must the second ball be thrown so that it reaches the same height as the one thrown vertically?

57. A quarterback throws a football toward a receiver with an initial speed of 20 m/s, at an angle of 30° above the horizontal. At that instant, the receiver is 20 m from the quarterback. In what direction and with what constant speed should the receiver run in order to catch the football at the level at which it was thrown?

58. A 2.00-m-tall basketball player wants to make a basket from a distance of 10.0 m, as in Figure P3.58. If he shoots the ball at a 45.0° angle, at what initial speed must he throw the basketball so that it goes through the hoop without striking the backboard?

FIGURE P3.58

59. In a very popular lecture demonstration, a projectile is fired at a falling target as in Figure P3.59. The projectile leaves the gun at the same instant that the target is dropped from rest. Assuming that the gun is initially aimed at the target, show that the projectile will hit the target. (One restriction on this experiment is that the projectile must reach the target before the target strikes the floor.)

FIGURE P3.59

60. Figure P3.60 illustrates the difference in proportions between the male and female anatomies. The displacements $\mathbf{d}_{1m}$ and $\mathbf{d}_{1f}$ from the bottom of the feet to the navel have magnitudes of 104 cm and 84.0 cm, respectively. The dis-

placements $\mathbf{d}_{2m}$ and $\mathbf{d}_{2f}$ have magnitudes of 50.0 cm and 43.0 cm, respectively. (a) Find the vector sum of the displacements $\mathbf{d}_1$ and $\mathbf{d}_2$ in each case. (b) The male figure is 180 cm tall, the female 168 cm. Normalize the displacements of each figure to a common height of 200 cm, and re-form the vector sums as in part (a). Then find the vector difference between the two sums.

FIGURE P3.60

61. By throwing a ball at an angle of 45°, a boy can throw it a maximum horizontal distance of R on a level field. How far can he throw the same ball vertically upward? Assume that his muscles give the ball the same speed in each case. (Is this assumption valid?)

62. A projectile is fired with an initial speed of v_0 at an angle of θ_0 to the horizontal, as in Figure 3.13. When it reaches its peak, it has (x, y) coordinates given by $(R/2, h)$, and when it strikes the ground, its coordinates are $(R, 0)$, where R is called the horizontal range. (a) Show that it reaches a maximum height h given by

$$h = \frac{v_0^2 \sin^2 \theta_0}{2g}$$

(b) Show that its horizontal range is given by

$$R = \frac{v_0^2 \sin 2\theta_0}{g}$$

63. A hunter wishes to cross a river that is 1.5 km wide and flows with a speed of 5.0 km/h parallel to its banks. The hunter uses a small powerboat that moves at a maximum speed of 12 km/h with respect to the water. What is the minimum time necessary for crossing?

64. A water insect maintains a constant average position on the surface of a stream by darting upstream (against the current), then drifting downstream (with the current) to its original position. The current in the stream is 0.500 m/s relative to the shore, and the insect darts upstream 0.560 m (relative to a spot on shore) in 0.800 s during the first part of its motion. Take upstream as the positive direction. (a) Determine the velocity of the insect relative to the water (i) during its dash upstream and (ii) during its drift downstream. (b) How far upstream relative to the water does the insect move during one cycle of this motion? (c) What is the average velocity of the insect relative to the water?

65. A daredevil is shot out of a cannon at 45.0° to the horizontal with an initial speed of 25.0 m/s. A net is positioned a horizontal distance of 50.0 m from the cannon. At what height above the cannon should the net be placed in order to catch the daredevil?

66. A projectile is fired with an initial velocity of 15.0 m/s at 53.0° above the horizontal from the foot of a ramp inclined 20.0° above the horizontal, as in Figure P3.66. How far up the ramp does the projectile strike the ramp?

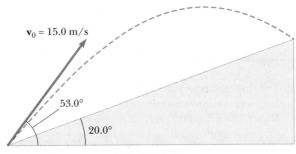

FIGURE P3.66

67. A student decides to measure the muzzle velocity of a pellet from his gun. He points the gun horizontally. He places a target on a vertical wall a distance x away from the gun. The pellet hits the target a vertical distance y below the gun. (a) Show that the position of the pellet when traveling through the air is given by $y = Ax^2$, where A is a constant. (b) Express the constant A in terms of the initial (muzzle) velocity and the free-fall acceleration. (c) If $x = 3.00$ m and $y = 0.210$ m, what is the initial speed of the pellet?

68. In ancient mythology, King Theseus of Athens is trapped in the Labyrinth (a maze) and finds his way out by following a thread given to him by Ariadne. He walks 10.0 m, makes a 90.0° right turn, walks 5.00 m, makes another 90.0° right turn, and walks 7.00 m. What is Theseus's displacement from his initial position?

69. Instructions for finding a buried treasure include the following: Go 75.0 paces at 240°, turn to 135° and walk 125 paces, then travel 100 paces at 160°. Determine the resultant displacement from the starting point.

70. When baseball outfielders throw the ball, they usually allow it to take one bounce on the theory that the ball arrives sooner this way. Suppose that after the bounce the ball rebounds at the same angle θ as it had when released (as in Fig. P3.70) but loses half its speed. (a) Assuming the ball is always thrown with the same initial speed, at what angle θ should the ball be thrown in order to go the same distance D with one bounce as one thrown upward at 45.0°

FIGURE P3.70

with no bounce? (b) Determine the ratio of the times for the one-bounce and no-bounce throws.

71. One strategy in a snowball fight is to throw a snowball at a high angle over level ground. While your opponent is watching the first one, you throw a second snowball at a low angle timed to arrive before or at the same time as the first one. Assume both snowballs are thrown with a speed of 25.0 m/s. The first one is thrown at an angle of 70.0° with respect to the horizontal. (a) At what angle should the second snowball be thrown to arrive at the same point as the first? (b) How many seconds later should the second snowball be thrown after the first to arrive at the same time?

GROUP ACTIVITIES

G.1 Take three steps, turn 90°, and then walk four steps. Now count the number of steps it takes to walk in a straight line back to your starting point. Verify your result mathematically.

G.2 For this investigation, you need to be outside with a small ball such as a tennis ball and a wristwatch with a second hand. Throw the ball vertically upward as hard as you can and determine the initial speed of your throw and the approximate maximum height of the ball using only your wristwatch. What happens when you throw the ball at some angle other than 90°? Does this change the time of flight? Can you still determine the maximum height and initial speed? Give careful explanations for your answers. Your co-worker can eyeball the maximum height by standing at a distance and noting the angle.

G.3 Using as projectiles the drops of water spraying from a garden hose at ground level, test the statement that the maximum range occurs when the angle of projection is 45 degrees. As an additional part to this experiment, hold the nozzle horizontally above the ground and have your co-worker position a marker at the location where the water strikes the ground. Increase the angle of inclination by about 10° and again record the strike position. Repeat until you have reached an angle of about 75°. Note the pattern produced. Are there two angles at which the range is the same? Explain the reason behind your observations.

G.4 Use a chessboard as a coordinate system with the intersection of the lines on the board as the positions of the coordinates. Select an origin for your coordinate system and on index cards write down several vector displacements that are at right angles to one another. For example, displacement (1) might be a movement of four units to the right, displacement (2) an upward movement of six units, and so forth. Continue this for a total of seven or eight movements until you end up at some particular location on the chessboard. Let your co-worker start at the origin and follow your vector directions to see if he/she arrives at the expected final location. Now, shuffle the cards and repeat the experiment. Does the order of the displacements make any difference as to where you eventually end up?

G.5 (a) A vector **A** is in the first quadrant of a Cartesian coordinate system. What is the sign of the *x* component of **A**? What is the sign of the *y* component? (b) A vector **B** is in the second quadrant of a Cartesian coordinate system. What is the sign of the *x* component of **B**? What is the sign of the *y* component? (c) The vector sum **A** + **B** _____. Choose the correct fill-in-the blank answer from: (i) must be in either the first or second quadrant. (ii) could be in any quadrant. (d) Let **A** = 30 m at an angle of 30° from the positive *x* axis, and **B** = 20 m at an angle of 40° from the negative *x* axis. Test your predictions of parts (a) through (c).

G.6 Roll a ball off a table. At the very instant the rolling ball leaves the table, drop a second ball from the same height above the floor. (This will require a sharp eye and good reflexes!) Do the two balls hit the floor at the same time? Try varying the speed at which you roll the ball off the table. Does this change affect the time at which the balls strike the floor? Finally, roll one of the balls down an incline and drop the other ball from the base of the incline at the instant the first ball leaves the slope. Which of these balls hits the floor first in this situation? Explain the reason behind your observations.

G.7 You can use any coordinate system you like in order to solve a projectile motion problem. To demonstrate this, let us consider a case in which we throw a ball off the top of a building with a velocity **v** at an angle θ with respect to the horizontal. We choose the origin of the coordinate system at the base of the building directly below the point where the ball is released. (a) When the ball reaches its highest point, which of the following are true? (i) The *x* component of the velocity is zero. (ii) The *y* component of the velocity is zero. (iii) Neither component of velocity is necessarily zero. (b) What is the displacement when the ball strikes the ground? (c) Let the building be 50.0 m tall, the initial horizontal velocity be 9.00 m/s, and the initial vertical velocity be 12.0 m/s. Find the maximum height and the time to reach the maximum height. Also, find the time to reach ground level. Use your choices in (a) and (b) to find your answers.

G.8 A boy and a girl are tossing an apple back and forth between them. Figure GA3.8 shows the path the apple follows when watched by an observer looking on from the side. The apple is moving from left to right. Five points are marked on the path. Ignore air resistance. (a) Make a copy of this figure. At each of the marked points, draw an arrow that indicates the magnitude and direction of the apple's velocity when it passes through that point. (b) Make a second copy of the figure. This time, at each marked point, place an arrow indicating the magnitude and direction of any acceleration the apple experiences at that point.

FIGURE GA3.8

(Problem 8 is courtesy of E. F. Redish. For more problems of this type, visit http://www.physics.umd.edu/perg/)

The Laws of Motion

4

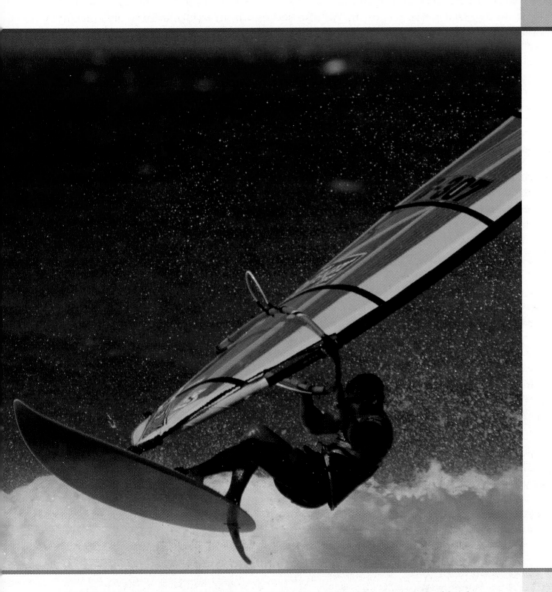

This surf boarder is lifted into the air under the action of forces exerted by the wind on the sail and by water waves on the surf board. *Peter Sterling/FPG/Getty Images*

Classical mechanics describes the relationship between the motion of objects found in our everyday world and the forces acting on them. There are conditions under which classical mechanics does not apply or applies in only a limited way. Most often these conditions are encountered when dealing either with very tiny objects (of a size comparable to that of atoms or smaller, about 10^{-10} m) or with objects moving at speeds near that of light (3×10^8 m/s). Our study of relativity and quantum mechanics in later chapters will enable us to handle such situations. However, for large, slowly moving objects, we can make accurate calculations using the laws of classical mechanics discussed in this chapter.

We learn in this chapter that an object remains in motion with constant velocity if no external force acts on it. We also see that if an external force acts on an object, the object accelerates in response to this force. The acceleration can be determined if the net force acting on the object is known and if the object's mass is known. Finally we learn that there is no such thing as a single isolated force, but that forces arise in pairs from the interaction of two entities, each pushing or pulling on the other with forces of equal magnitude and opposite direction.

4.1 THE CONCEPT OF FORCE

When we think of a **force,** we usually imagine a push or a pull exerted on some object. For instance, you exert a force on a ball when you throw it or kick it, and you exert a force on a chair when you sit down on it. What happens to an object when it is acted on by a force depends on the magnitude and the direction of the force. Force is a vector quantity; thus, we denote it with a directed arrow, just as we do velocity and acceleration.

If you pull on a spring, as in Figure 4.1a, the spring stretches. If a child pulls hard enough on a wagon (Fig. 4.1b), the wagon moves. When a football is kicked (Fig. 4.1c), it is deformed and set in motion. These are all examples of **contact forces,** so named because they result from physical contact between two objects.

Another class of forces does not involve physical contact between two objects. Early scientists, including Newton, were uneasy with the concept of forces that act between two disconnected objects. To overcome this conceptual difficulty, Michael Faraday (1791–1867) introduced the concept of a *field*. The corresponding forces are called **field forces.** According to this approach, when an object of mass m is placed at some point P near a second object of mass M, we say that the first object interacts with the second by virtue of the gravitational field that exists at P. Thus, the force of gravitational attraction between two objects, illustrated in Figure 4.1d, is an example of a field force. This force keeps objects bound to Earth and gives rise to what we commonly call the *weight* of the object.

Another common example of a field force is the electric force that one electric charge exerts on another (Fig. 4.1e). A third example is the force exerted by a bar magnet on a piece of iron (Fig. 4.1f).

The known fundamental forces in nature are all field forces. These are, in order of decreasing strength, (1) strong nuclear forces between subatomic particles; (2) electromagnetic forces between electric charges; (3) weak nuclear forces, which arise in certain radioactive decay processes; and (4) gravitational attractions between objects. In classical physics, we are concerned only with gravitational and electromagnetic forces.

Whenever a force is exerted on an object, the object's shape can change. For example, when you kick a soccer ball or strike a tennis ball with a racquet, as in Figure 4.2, the objects deform to some extent. Even objects we usually think of as rigid and inflexible are deformed under the action of external forces. Often the deformations are permanent, as in the case of a collision between automobiles.

Contact forces | Field forces

(a) | (d)

(b) | (e)

(c) | Iron | N S | (f)

FIGURE 4.1 **Examples of forces applied to various objects.** In each case a force acts on the object surrounded by the dashed lines. Something in the environment external to the boxed area exerts this force.

(a)

(b)

FIGURE 4.2 (a) A soccer ball is set in motion by the contact force **F** exerted on it by the kicker's foot. The ball is deformed during the short time in contact with the foot. *(Loren Winters/Visuals Unlimited)* (b) A tennis ball being struck by a racquet. The ball experiences a large contact force as it is struck and a correspondingly large acceleration. *(Amoz Eckerson/Visuals Unlimited)*

4.2 NEWTON'S FIRST LAW

Consider the following simple experiment. A book is lying on a table. Obviously, the book remains at rest if left alone. Now imagine that you push the book with a horizontal force great enough to overcome the force of friction between book and table, so that the book is set in motion. Because the magnitude of your applied force exceeds the magnitude of the friction force, the book accelerates. If you stop applying the force, the book stops sliding after traveling a short distance because the force of friction retards its motion. Now imagine pushing the book across a smooth, waxed floor. The book again comes to rest once the force is no longer applied, but not as soon as before. Finally, imagine that the book is moving on a horizontal, frictionless surface. In this situation, the book continues to move in a straight line with constant velocity until it hits a wall or some other object.

Before about 1600, scientists felt that the natural state of matter was the state of rest. Galileo was the first to take a different approach. He devised thought experiments—such as an object moving on a frictionless surface, as just described—and concluded that it is not the nature of an object to *stop,* once set in motion; rather, it is an object's nature to *continue in its original state of motion.* This approach to motion was later formalized by Newton in a form that has come to be known as **Newton's first law of motion:**

> **If the net force ΣF exerted on an object is zero, the object continues in its original state of motion. That is, if $\Sigma F = 0$, an object at rest remains at rest and an object moving with some velocity continues with that same velocity.**

◄ Newton's first law

By *net force* we mean *the vector sum of all external forces exerted on the object.* An *external force* is any force that results from the interaction between the object and its envi-

Unless acted on by an external force, an object at rest will remain at rest and an object in motion will continue in motion with constant velocity. In this case, the wall of the building did not exert a large enough external force on the moving train to stop it. *(Roger Viollet, Mill Valley, CA, University Science Books, 1982)*

ronment or surroundings, such as the force exerted on an object when it is lifted.

As a special case, let us see what happens to an object if there are no external forces exerted on it. For example, consider a spaceship traveling in space, far from any planets or other matter. The spaceship requires a propulsion system to change its velocity. However, if the propulsion system is turned off when a velocity **v** is reached, the spaceship coasts in space with a constant velocity and the astronauts get a free ride.

MASS AND INERTIA

A bowling ball and a golf ball lie side by side on the ground. Newton's first law tells us that both remain at rest as long as no external force acts on them. Now imagine supplying a net force by striking each ball with a golf club. Both balls resist your attempt to change their state of motion. But you know from everyday experience that if the two are struck with equal force, the golf ball will travel much farther than the bowling ball. That is, the bowling ball is more successful in maintaining its original state of motion. The tendency of an object to continue in its original motion is called **inertia.**

While inertia is the tendency for an object to continue its motion in the absence of a force, **mass** is a measure of the resistance of an object to changes in its motion due to a force. The greater the mass of a body, the less it accelerates under the action of an applied force. The SI unit of mass is the kilogram. Mass is a scalar quantity that obeys the rules of ordinary arithmetic.

Inertia can be used to explain the operation of one type of seat belt mechanism. In the event of an accident, the purpose of the seat belt is to hold the passenger firmly in place relative to the car to prevent serious injury. Figure 4.3 illustrates how one type of shoulder harness operates. Under normal conditions, the ratchet turns freely to allow the harness to wind on or unwind from the pulley as the passenger moves. When an accident occurs, the car undergoes a large acceleration and rapidly comes to rest. The large block under the seat, because of its inertia, continues to slide forward along the tracks. The pin connection between the block and the rod causes the rod to pivot about its center and engage the ratchet wheel. At this point the ratchet wheel locks in place, and the harness no longer unwinds.

APPLICATION

SEAT BELTS

Seat belt

Pulley

Rachet

Rod

Pivot

Tracks

Pin connection

Large block

FIGURE 4.3 A mechanical arrangement for an automobile seat belt.

ISAAC NEWTON, ENGLISH PHYSICIST AND MATHEMATICIAN (1642–1727)

Newton was one of the most brilliant scientists in history. Before the age of 30, he formulated the basic concepts and laws of mechanics, discovered the law of universal gravitation, and invented the mathematical methods of calculus. As a consequence of his theories, Newton was able to explain the motions of the planets, the ebb and flow of the tides, and many special features of the motions of the Moon and Earth. He also interpreted many fundamental observations concerning the nature of light. His contributions to physical theories dominated scientific thought for two centuries and remain important today. *(Giraudon/Art Resource)*

4.3 NEWTON'S SECOND LAW

Newton's first law explains what happens to an object when no force acts on it: The object either remains at rest or continues moving in a straight line with constant speed. Newton's second law answers the question of what happens to an object that has a net force acting on it.

Imagine pushing a block of ice across a frictionless horizontal surface. When you exert some horizontal force on the block, it moves with an acceleration of, say, 2 m/s^2. If you apply a force twice as large, the acceleration doubles. Pushing three times as hard triples the acceleration, and so on. From such observations, we conclude that **the acceleration of an object is directly proportional to the net force acting on it.**

Common experience with pushing objects tells you that mass also affects acceleration. Imagine that you stack identical blocks of ice on top of each other while pushing the stack with constant force. If the force when you push one block produces an acceleration of 2 m/s^2, the acceleration will drop to half that value when two blocks are pushed, one-third that initial value when three blocks are pushed, and so on. We conclude that **the acceleration of an object is inversely proportional to its mass.** These observations are summarized in **Newton's second law:**

> **The acceleration of an object is directly proportional to the net force acting on it and inversely proportional to its mass.**
>
> $$a \propto \frac{\sum \mathbf{F}}{m}$$

In vector equation form, and with the units of all quantities appropriately chosen, we can state Newton's second law as

$$\sum \mathbf{F} = m\mathbf{a} \qquad [4.1]$$

where **a** is the acceleration of the object, m is its mass, and $\sum \mathbf{F}$ represents the vec-

Webnote 4.1

Find out more about Isaac Newton at:
http://www.newton.cam.ac.uk/newtlife.html
http://userwww.sfsu.edu/~rsauzier/Newton.html

 Newton's second law

 FORCE CAUSES *CHANGES* **IN MOTION**

Many times, students make the mistake of thinking that whenever there is motion, there must be an underlying force. But motion can occur in the absence of forces, as described in Newton's first law. Force is the cause of *changes* in motion.

tor sum of all forces acting on the object. You should note that, because this is a vector equation, it is equivalent to the following three component equations:

$$\sum F_x = ma_x \qquad \sum F_y = ma_y \qquad \sum F_z = ma_z \qquad \text{[4.2]}$$

UNITS OF FORCE AND MASS

◀ Definition of newton ▶

The SI unit of force is the **newton,** defined as the force that, when acting on an object that has a mass of 1 kg, produces an acceleration of 1 m/s². From this definition and Newton's second law, we see that the newton can be expressed in terms of the fundamental units of mass, length, and time:

$$1 \text{ N} \equiv 1 \text{ kg} \cdot \text{m/s}^2 \qquad \text{[4.3]}$$

In the U.S. customary system, the unit of force is the **pound.** The following conversion from pounds to newtons will be useful to you in many problems:

$$1 \text{ N} \equiv 0.225 \text{ lb} \qquad \text{[4.4]}$$

The units of mass, acceleration, and force in the SI and U.S. customary systems are summarized in Table 4.1.

Tip 4.2 MA IS NOT A FORCE

Equation 4.1 does *not* say that the product *m***a** is a force. All forces exerted on an object are added vectorially to generate the net force on the left side of the equation. This net force is then equated to the product of the mass of the object and the acceleration that results from the net force. Do *not* include an "*m***a** force" in your analysis. Also, be sure to use consistent units when applying Newton's second law.

Quick Quiz 4.1 True or false: (a) It is possible to have motion in the absence of a force. (b) If an object is not moving, no external force acts on it.

Quick Quiz 4.2 True or false: (a) If a single force acts on an object, the object accelerates. (b) If an object experiences an acceleration, a force acts on it. (c) If an object experiences no acceleration, no external force acts on it.

Quick Quiz 4.3 True or false: If the net force acting on an object is in the *x* direction, the object moves in the *x* direction.

Tip 4.3 USING NEWTON'S SECOND LAW

The procedure for applying Newton's second law is to add *all* of the forces on the object vectorially and then find the resultant acceleration. Do *not* find individual accelerations for each force and then add the accelerations vectorially.

THE GRAVITATIONAL FORCE

The **gravitational force** is the mutual force of attraction between any two objects in the Universe. It is interesting and rather curious that, although the gravitational force can be very strong between macroscopic objects, it is the weakest of the fundamental forces. For example, the gravitational force between the electron

TABLE 4.1	**Units of Mass, Acceleration, and Force**		
System	Mass	Acceleration	Force
SI	kg	m/s²	N = kg·m/s²
U.S. customary	slug	ft/s²	lb = slug·ft/s²

Note: 1 N = 0.225 lb.

and proton in the hydrogen atom has a magnitude on the order of 10^{-47} N, whereas the electromagnetic force between these same two particles is on the order of 10^{-7} N.

In addition to his contributions to the understanding of motion, Newton studied gravity extensively. **Newton's law of universal gravitation** states that every particle in the Universe attracts every other particle with a force that is directly proportional to the product of the masses of the particles and inversely proportional to the square of the distance between them. If the particles have masses m_1 and m_2 and are separated by a distance r, as in Figure 4.4, the magnitude of the gravitational force is

◀ Law of universal gravitation

$$F_g = G \frac{m_1 m_2}{r^2} \qquad [4.5]$$

where $G = 6.67 \times 10^{-11}$ N·m²/kg² is the **universal gravitational constant.** We examine the gravitational force in more detail in Chapter 7.

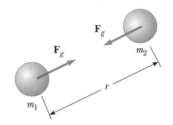

FIGURE 4.4 The gravitational force between two particles is attractive.

WEIGHT

The magnitude of the gravitational force acting on an object of mass m near Earth's surface is called the weight of the object w. Since objects of any mass fall with the same acceleration g at Earth's surface, (neglecting air resistance), we can express the second law as "weight equals mass times acceleration" or

$$w = mg \qquad [4.6]$$

Because the magnitude of the gravitational force is also given by Equation 4.5, we see that $w = F_g$, and it follows that

$$w = G \frac{M_E m}{r^2} \qquad [4.7]$$

where M_E is the mass of Earth and r is the distance from the object to Earth's center. If the object is at rest on Earth's surface, then r is equal to Earth's radius, R_E. As r increases, Equation 4.7 shows that the weight of the object decreases because r is in the denominator. Thus, the weight of an object high in the atmosphere is less than the weight of the same object at Earth's surface.

Comparing Equations 4.6 and 4.7, we see that g is given by

$$g = G \frac{M_E}{r^2} \qquad [4.8]$$

Because it depends on g, weight varies with geographic location. Objects weigh less at higher altitudes than at sea level because, as noted in Section 2.7, g decreases with increasing distance above Earth's surface. Hence, weight, unlike mass, is not an inherent property of an object. For example, if an object has a mass of 70.0 kg, then its weight at a location where $g = 9.80$ m/s² is $mg = 686$ N. At the top of a mountain, where g might be 9.76 m/s², the object's weight would be 683 N. Therefore, if you want to lose weight quickly without going on a diet, move to the top of a mountain or weigh yourself at an altitude of 30 000 ft during a flight on a jet airplane!

Another interesting feature of Equation 4.8 is that it is a general result that can be used to calculate the acceleration of an object falling at the surface of any massive object if the more massive object's radius and mass are known. You should be able to show that $g_{Sun} = 274$ m/s² and $g_{Moon} = 1.67$ m/s² using the values in Table 7.3.

Astronaut Edwin E. Aldrin, Jr., walking on the Moon after the *Apollo 11* **lunar landing.** Aldrin's weight on the Moon is less than it is on Earth, but his mass is the same in both places. *(NASA)*

Quick Quiz 4.4 You are talking by interplanetary telephone to a friend who lives on the Moon. She tells you she has just won 1 newton of gold in a contest. Excitedly, you tell her that you entered Earth's version of the same contest and also won 1 newton of gold! Who is richer? (a) Your friend, (b) you, or (c) it's a draw.

Quick Quiz 4.5 A baseball of mass m is thrown upward with some initial speed. If air resistance is neglected, the force acting on the ball when it reaches its peak is (a) mg and upward, (b) mg and downward, (c) zero, or (d) none of these.

4.4 NEWTON'S THIRD LAW

In Section 4.1 we found that a force is exerted on an object when that object comes into contact with some other object. For example, consider the task of driving a nail into a block of wood, as illustrated in Figure 4.5a. To accelerate the nail and drive it into the block, a net force must be exerted by the hammer on the nail. However, Newton recognized that a single isolated force (such as the force exerted by the hammer on the nail) cannot exist. Instead, **forces in nature always exist in pairs.** According to Newton, the hammer exerts a force on the nail, and the nail exerts a force on the hammer. You can tell that there is clearly a force exerted by the nail on the hammer, because the hammer rapidly slows down after coming into contact with the nail.

Newton described this type of situation in terms of his **third law:**

Newton's third law ▶

> If two objects interact, the force $\mathbf{F}_{12}$ exerted by object 1 on object 2 is equal in magnitude but opposite in direction to the force $\mathbf{F}_{21}$ exerted by object 2 on object 1.

This law, which is illustrated in Figure 4.5b, is equivalent to stating that **a single isolated force cannot exist.** The force $\mathbf{F}_{12}$ exerted by object 1 on object 2 is sometimes called the *action force,* and the force $\mathbf{F}_{21}$ exerted by object 2 on object 1 is called the *reaction force.* In reality, either force can be labeled the action or reac-

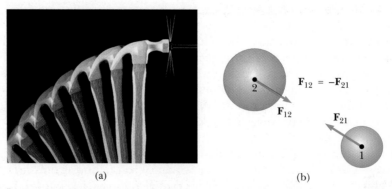

(a)   (b)

FIGURE 4.5 **Newton's third law.** (a) The force exerted by the hammer on the nail is equal in magnitude and opposite in direction to the force exerted by the nail on the hammer. *(John Gillmoure, The Stock Market)* (b) The force $\mathbf{F}_{12}$ exerted by object 1 on object 2 is equal in magnitude and opposite in direction to the force $\mathbf{F}_{21}$ exerted by object 2 on object 1.

tion force. **The action force is equal in magnitude to the reaction force and opposite in direction. In all cases, the action and reaction forces act on different objects.** For example, the force acting on a freely falling projectile is the force exerted by Earth on the projectile, $\mathbf{F}_g$, and the magnitude of this force is mg. The reaction to this force is the force exerted by the projectile on Earth, $\mathbf{F}_g' = -\mathbf{F}_g$. The reaction force $\mathbf{F}_g'$ must accelerate Earth toward the projectile just as the action force $\mathbf{F}_g$ accelerates the projectile toward Earth. However, because Earth has such a large mass, its acceleration due to this reaction force is negligibly small.

You directly experience Newton's third law whenever you slam your fist against a wall or kick a football with your bare foot. You should be able to identify the action and reaction forces in these cases.

As another example of Newton's third law, consider the helicopter. Most helicopters have a large set of blades rotating in a horizontal plane above the body and another, smaller set rotating in a vertical plane at the back. Other helicopters have two large sets of blades above the body rotating in opposite directions. Why do helicopters always have two sets of blades? In the first type of helicopter, the engine applies a force to the blades, causing them to change their rotational motion. According to Newton's third law, however, the blades must exert a force on that engine of equal magnitude and in the opposite direction. This force would cause the body of the helicopter to rotate in the direction opposite the blades. A rotating helicopter would be impossible to control, so a second set of blades is used. The small blades in the back provide a force opposite to that tending to rotate the body of the helicopter, keeping the body oriented in a stable position. In helicopters with two sets of large, counter-rotating blades, engines apply forces in opposite directions so that there is no net force rotating the helicopter.

As we mentioned earlier, Earth exerts a force $\mathbf{F}_g$ on any object. If the object is a TV at rest on a table, as in Figure 4.6a, the reaction force to $\mathbf{F}_g$ is the force the TV exerts on Earth, $\mathbf{F}_g'$. The TV does not accelerate downward, because it is held up by the table. The table, therefore, exerts on the TV an upward force $\mathbf{n}$ called the **normal force.** (It is called the normal force because it is normal, or perpendicular, to the table's surface.) This is the force that prevents the TV from falling through the table; it can have any value needed, up to the point of breaking the

Tip 4.4 ACTION-REACTION PAIRS

When applying Newton's third law, note that action and reaction forces act on *different* objects. Two external forces acting on the same object, even if they are equal in magnitude and opposite in direction, *cannot* be an action-reaction pair.

APPLICATION

HELICOPTER FLIGHT

(a)   (b)

FIGURE 4.6 When a TV set is sitting on a table, the forces acting on the set are the normal force $\mathbf{n}$ exerted by the table and the force of gravity $\mathbf{F}_g$, as illustrated in (b). The reaction to $\mathbf{n}$ is the force exerted by the TV set on the table, $\mathbf{n}'$. The reaction to $\mathbf{F}_g$ is the force exerted by the TV set on Earth, $\mathbf{F}_g'$.

table. The normal force is an elastic force arising from the sag of the table's surface, a sag caused by the heavy TV. The normal force balances the weight. The reaction to **n** is the force exerted by the TV on the table, **n**′. Therefore, we conclude that

$$\mathbf{F}_g = -\mathbf{F}_g' \quad \text{and} \quad \mathbf{n} = -\mathbf{n}'$$

The forces **n** and **n**′ have the same magnitude, which is the same as $\mathbf{F}_g$ unless the table has broken. Note that the forces acting on the TV are $\mathbf{F}_g$ and **n**, as shown in Figure 4.6b. The two reaction forces, $\mathbf{F}_g'$ and **n**′, are exerted by the TV on objects other than the TV. Remember, the two forces in an action-reaction pair always act on two different objects.

From Newton's second law, we see that, because the TV is not accelerating in any direction (**a** = 0), it follows that $F_g = n = mg$.

Webnote 4.2

Rocket propulsion is another excellent application of Newton's third law. See, for example: *http://www.howstuffworks.com/ rocket.htm*

APPLICATION

COLLIDING VEHICLES

Quick Quiz 4.6 A small sports car collides head-on with a massive truck. Which vehicle experiences the greater impact force (in magnitude)? (a) the car, (b) the truck, (c) they experience the same force. Which vehicle experiences the greater acceleration? (d) the car, (e) the truck, (f) they experience the same acceleration.

APPLYING PHYSICS 4.1

A horse pulls on a sled with a horizontal force, causing it to accelerate as in Figure 4.7a. Newton's third law says that the sled exerts a force of equal magnitude and opposite direction on the horse. In view of this, how can the sled accelerate—don't these forces cancel?

Explanation When applying Newton's third law, it is important to remember that the forces involved act on different objects. Notice that the force exerted by the horse is *on the sled,* while the force exerted by the sled is *on the horse.* Because these forces act on different objects, they cannot cancel.

The horizontal forces exerted on the *sled* are the forward force **F** exerted by the horse and the backward force of friction $\mathbf{f}_{snow, sled}$ exerted by the snow on the sled (Fig. 4.7b). When **F** exceeds $\mathbf{f}_{snow, sled}$, the sled accelerates to the right.

The horizontal forces exerted on the *horse* are the forward friction force $\mathbf{f}_{snow, horse}$ exerted by the snow, which arises as a reaction to the force exerted by the horse pushing against the ground, and the backward force **F** exerted by the sled acting through the harness. This latter force arises as a reaction to the force exerted by the horse pulling the sled forward (Fig. 4.7c). The resultant of these two forces causes the horse to accelerate. When $\mathbf{f}_{snow, horse}$ exceeds **F**, the horse accelerates to the right.

If we consider the horse and sled as a system, the two forces described (exerted by horse on sled and by sled on horse) form an action-reaction force pair that are *internal* to the system and therefore cannot affect the motion of the system.

(a)   (b)   (c)

FIGURE 4.7 (Applying Physics 4.1)

4.5 SOME APPLICATIONS OF NEWTON'S LAWS

This section applies Newton's laws to objects moving under the actions of constant external forces. We assume that objects behave as particles, and so we need not worry about rotational motion. We also neglect any friction effects. Finally, we neglect the masses of any ropes or strings involved; in these approximations, the magnitude of the force exerted along a rope (the *tension*) is the same at all points in the rope. This is illustrated by the rope in Figure 4.8, showing forces **T** and **T**′ acting on it. If the rope has mass m, then Newton's second law applied to the rope gives $T - T' = ma$. However, if $m = 0$, as it will be in the upcoming examples, then $T = T'$.

When we apply Newton's law to an object, we are interested only in those forces that act *on the object*. For example, in Figure 4.6b, the only external forces acting on the TV are **n** and **F**$_g$. The reactions to these forces, **n**′ and **F**$_g$′, act on the table and on Earth, respectively, and do not appear in Newton's second law applied to the TV.

Consider a crate being pulled to the right on a frictionless, horizontal surface, as in Figure 4.9a. Suppose you are asked to find the acceleration of the crate and the force the surface exerts on it. The horizontal force exerted on the crate acts through the rope. The force that the rope exerts on the crate is denoted by **T** (because it is a tension force). The magnitude of **T** is equal to the tension in the rope. What we mean by the words "tension in the rope" is just the force read by a spring scale when the rope in question has been cut and the scale inserted! A dashed circle is drawn around the crate in Figure 4.9a to remind you to isolate the crate from its surroundings.

Because we are interested only in the motion of the crate, we must be able to identify all forces acting on it. These are illustrated in Figure 4.9b. In addition to the force **T**, the force diagram for the crate includes the force of gravity **F**$_g$ exerted by Earth and the normal force **n** exerted by the floor. Such a force diagram is referred to as a **free-body diagram.** We call it a "free-body diagram"

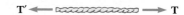

FIGURE 4.8 Newton's second law applied to the rope gives $T - T' = ma$. However, if $m = 0$, then $T = T'$. Thus, the tension in a massless rope is the same at all points in the rope.

Tip 4.5   FREE-BODY DIAGRAMS

The *most important* step in solving a problem using Newton's second law is to draw the free-body diagram. Be sure to draw only those forces that act on the object that you are isolating. An incorrect diagram will most likely lead to an incorrect solution.

FIGURE 4.9 (a) A crate being pulled to the right on a frictionless surface. (b) The free-body diagram that represents the forces exerted on the crate.

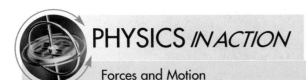

PHYSICS *IN ACTION*

Forces and Motion

On the left is an athlete running at sunset. The external forces acting on the athlete, as described by the blue vectors, are (a) the force $\mathbf{F}_g$ exerted by the Earth (b) the force $\mathbf{F}$ exerted by the ground, and (c) the force $\mathbf{R}$ due to air resistance.

On the right is a multiflash exposure of a golfer striking a golf ball. The force exerted by the golf club on the ball is equal in magnitude and opposite in direction to the force exerted by the ball on the golf club.

(Mitchell Funk/The Image Bank)

(Courtesy of Michael Hans/Photo Researchers, Inc.)

Webnote 4.3

For more on free-body diagrams including a tutorial and a self-test, see:

http://eta.physics.uoguelph.ca/ tutorials/fbd/Q.fbd.html

because the environment is replaced by a series of forces on an otherwise free body. The construction of a correct free-body diagram is an essential step in applying Newton's laws; *its importance cannot be overemphasized.*

The *reactions* to the forces we have listed—namely, the force exerted by the rope on the hand doing the pulling, the force exerted by the crate on Earth, and the force exerted by the crate on the floor—are not included in the free-body diagram because they act on other objects and not on the crate. Consequently, they do not determine the crate's motion.

Now let us apply Newton's second law to the crate. First we must choose an appropriate coordinate system. In this case it is convenient to use the one shown in Figure 4.9b, with the x axis horizontal and the y axis vertical. We can apply Newton's second law in the x direction, y direction, or both, depending on what we are asked to find in a problem. In addition, we may be able to use the equations of motion for constant acceleration that are found in Chapter 2. However, you should use these equations *only when the acceleration is constant*.

OBJECTS IN EQUILIBRIUM

Objects that are either at rest or moving with constant velocity are said to be in *equilibrium.* In equation form, because $\mathbf{a} = 0$, this condition of equilibrium can be expressed as

$$\sum \mathbf{F} = 0 \qquad \qquad [4.9]$$

This statement signifies that the *vector* sum of all the forces (the net force) acting on an object in equilibrium is zero.

Usually, equilibrium problems are solved more easily if we work with Equation 4.9 in terms of the components of the external forces acting on an object. By this we mean that, in a two-dimensional problem, the sum of all the external forces in the x and y directions must separately equal zero; that is,

$$\sum F_x = 0 \quad \text{and} \quad \sum F_y = 0 \qquad [4.10]$$

We shall not consider three-dimensional problems in this book, but the extension of Equation 4.10 to a three-dimensional situation can be made by adding a third equation, $\sum F_z = 0$.

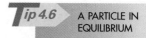

Tip 4.6   A PARTICLE IN EQUILIBRIUM

Zero net force on a particle does *not* mean that the particle is not moving. It means that the particle is not *accelerating*. If the particle has an initial velocity and experiences zero net force, it continues to move with the same velocity.

Quick Quiz 4.7   Consider the two situations shown in Figure 4.10, in which there is no acceleration. In both cases, all individuals pull with a force of magnitude F. Is the reading on the scale in part (i) of the figure (a) greater than, (b) less than, or (c) equal to the reading in part (ii)?

PROBLEM-SOLVING STRATEGY   Objects in Equilibrium

The following procedure is recommended for problems involving objects in equilibrium:

1. Make a sketch of the situation described in the problem statement.
2. Draw a free-body diagram for the *isolated* object under consideration, and label all forces acting on the object.
3. Resolve all forces into x and y components, choosing a convenient coordinate system.
4. Use the equations $\sum F_x = 0$ and $\sum F_y = 0$. Keep track of the signs of the various force components.
5. Application of Step 4 leads to a set of equations with several unknowns. Solve the simultaneous equations for the unknowns in terms of the known quantities.

(i)

(ii)

FIGURE 4.10   (Quick Quiz 4.7) (i) A person pulls with a force of magnitude F on a spring scale attached to a wall. (ii) Two people pull with forces of magnitude F in opposite directions on a spring scale attached between two ropes.

Example 4.1   A Traffic Light at Rest

A traffic light weighing 100 N hangs from a vertical cable tied to two other cables that are fastened to a support, as in Figure 4.11a. The upper cables make angles of 37.0° and 53.0° with the horizontal. Find the tension in each of the three cables.

Reasoning   We must construct two free-body diagrams. The first is for the traffic light, shown in Figure 4.11b; the second is for the knot that holds the three cables together (Fig. 4.11c). The knot is a convenient point to choose because all forces in question act on this point.

Solution   From Figure 4.11b, we see that $\sum F_y = 0$ yields $T_3 - F_g = 0$, or $T_3 = F_g = 100$ N. Next, we choose the coordinate axes shown in Figure 4.10c and resolve all forces into their x and y components:

Force	x component	y component
$\mathbf{T}_1$	$-T_1 \cos 37.0°$	$T_1 \sin 37.0°$
$\mathbf{T}_2$	$T_2 \cos 53.0°$	$T_2 \sin 53.0°$
$\mathbf{T}_3$	0	-100 N

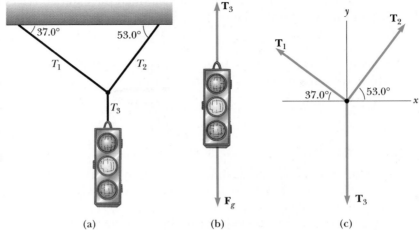

FIGURE 4.11   (Example 4.1) (a) A traffic light suspended by cables. (b) A free-body diagram for the traffic light. (c) A free-body diagram for the knot joining the cables.

The condition for equilibrium gives us the equations

$$\sum F_x = T_2 \cos 53.0° - T_1 \cos 37.0° = 0 \tag{1}$$

$$\sum F_y = T_1 \sin 37.0° + T_2 \sin 53.0° - 100 \text{ N} = 0 \tag{2}$$

(In this case, equilibrium means a motionless light.) From (1) we see that the horizontal components of $\mathbf{T}_1$ and $\mathbf{T}_2$ must be equal in magnitude, and from (2) we see that the sum of the vertical components of $\mathbf{T}_1$ and $\mathbf{T}_2$ must balance the force of gravity acting on the light. We can solve (1) for T_2 in terms of T_1 to give

$$T_2 = T_1 \left(\frac{\cos 37.0°}{\cos 53.0°} \right) = T_1 \left(\frac{0.799}{0.602} \right) = 1.33 T_1$$

This value for T_2 can be substituted into (2) to give

$$T_1 \sin 37.0° + (1.33 T_1)(\sin 53.0°) - 100 \text{ N} = 0$$

$$T_1 = 60.1 \text{ N}$$

$$T_2 = 1.33 T_1 = 1.33(60.0 \text{ N}) = 79.9 \text{ N}$$

EXERCISE   When will $T_1 = T_2$?

ANSWER   When the supporting cables make equal angles with the horizontal support.

EXERCISE   Solve this problem using the rules of vector addition and the law of cosines and sines found in Appendix A.4.

Example 4.2   Sled on a Frictionless Hill

A child holds a sled at rest on a frictionless, snow-covered hill, as shown in Figure 4.12a. If the sled weighs 77.0 N, find the force **T** exerted by the rope on the sled and the force **n** exerted by the hill on the sled.

Reasoning   Figure 4.12b shows the forces acting on the sled and a convenient coordinate system to use for this type of problem. Note that **n**, the force exerted by the hill on the sled, is perpendicular (normal) to the hill. The hill can exert a component of force along the incline only if there is friction between the sled and the hill. Because the sled is at rest, we are able to apply the condition for equilibrium as $\Sigma F_x = 0$ and $\Sigma F_y = 0$.

Solution   First, we replace the force of gravity $\mathbf{F}_g$ acting on the sled with its x and y components. The x component has magnitude $mg \sin \theta = (77.0 \text{ N})(\sin 30.0°)$

(a)

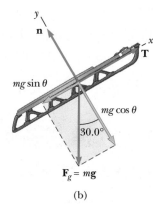

(b)

Figure 4.12   (Example 4.2) (a) A child holding a sled on a frictionless hill. (b) A free-body diagram for the sled.

and points along the negative x axis. The y component has magnitude $mg \cos \theta = (77.0 \text{ N})(\cos 30.0°)$ and points in the negative y direction. Applying the condition for equilibrium to the sled, we find that

$$\sum F_x = T - (77.0 \text{ N})(\sin 30.0°) = 0$$

$$T = 38.5 \text{ N}$$

$$\sum F_y = n - (77.0 \text{ N})(\cos 30.0°) = 0$$

$$n = 66.7 \text{ N}$$

Note that n is less than the weight of the sled in this situation. This is so because the sled is on an incline and **n** is equal in magnitude and opposite in direction to the component of the force of gravity perpendicular to the incline.

Exercise   What happens to the normal force as the angle of incline increases?

Answer   It decreases.

Exercise   When is the magnitude of the normal force equal to the weight of the sled?

Answer   When the sled is on a horizontal surface and the force exerted by the rope on the sled is either zero or along the horizontal.

Quick Quiz 4.8

Consider the child being pulled on the toboggan in Figure 4.13. Is the magnitude of the normal force exerted by the ground on the toboggan (a) equal to the total weight of the child plus the toboggan, (b) greater than the total weight, (c) less than the total weight, or (d) possibly greater than or less than the total weight depending on the size of the weight relative to the tension in the rope?

FIGURE 4.13   (Quick Quiz 4.8)

ACCELERATING OBJECTS AND NEWTON'S SECOND LAW

In a situation in which a net force acts on an object, the object accelerates, and we use Newton's second law in order to analyze the motion. The examples and suggestions that follow should help you solve problems of this kind.

PROBLEM-SOLVING STRATEGY   **Newton's Second Law**

The following procedure is recommended for working problems that involve the application of Newton's second law:

1. Draw a diagram of the system.
2. Isolate the object of interest whose motion is being analyzed. Draw a free-body diagram for this object, showing all external forces acting on it. For systems containing more than one object, draw a *separate* free-body diagram for each object.
3. Establish convenient coordinate axes for each object, and find the components of the forces along these axes. It is convenient to choose a coordinate system where one of the axes is parallel to the motion of the objects. Apply Newton's second law in the *x* and *y* directions for each object.
4. Solve the component equations for the unknowns. Remember that, in order to obtain a complete solution, you must have as many independent equations as you have unknowns.
5. If necessary, use the equations of kinematics (motion with constant acceleration) from Chapter 2 to find all the unknowns.

n

F = 20.0 N

$F_g = w = 300$ N

FIGURE 4.14   (Example 4.3)

Example 4.3   Moving a Crate

The combined weight of the crate and dolly in Figure 4.14 is 300 N. If the person pulls on the rope with a constant force of 20.0 N, what is the acceleration of the system (crate plus dolly), and how far will it move in 2.00 s? Assume that the system starts from rest and that there are no friction forces opposing its motion.

Reasoning   We can find the acceleration of the system from Newton's second law. Because the force exerted on the system is constant, its acceleration is constant. Therefore, we can apply the equations of motion with constant acceleration to find the distance traveled in 2.00 s.

Solution   In order to apply Newton's second law to the system, we must first know the system's mass:

$$m = \frac{w}{g} = \frac{300 \text{ N}}{9.80 \text{ m/s}^2} = 30.6 \text{ kg}$$

Now we can find the acceleration of the system from the second law:

$$a_x = \frac{F_x}{m} = \frac{20.0 \text{ N}}{30.6 \text{ kg}} = \boxed{0.654 \text{ m/s}^2}$$

Because the acceleration is constant, we can find the distance the system moves in 2.00 s using the relation $x = v_0 t + \frac{1}{2}at^2$ with $v_0 = 0$:

$$x = \frac{1}{2}at^2 = \frac{1}{2}(0.654 \text{ m/s}^2)(2.00 \text{ s})^2 = \boxed{1.31 \text{ m}}$$

It is important to note that the constant applied force of 20.0 N is assumed to act on the system at all times during its motion. If the force were removed at some instant, the system would continue to move with constant velocity and hence zero acceleration.

Example 4.4   The Run-Away Car

A car of mass m is on an icy driveway inclined at an angle $\theta = 20.0°$, as in Figure 4.15a. Determine the acceleration of the car, assuming the incline is frictionless.

(a)   (b)

FIGURE 4.15   (Example 4.4)

Reasoning   The free-body diagram for the car is shown in Figure 4.15b. The only forces exerted on the car are the normal force **n** acting perpendicular to the driveway surface and the force of gravity $\mathbf{F}_g$ acting vertically downward. It is convenient to choose the coordinate axes with the x axis along the incline and the y axis perpendicular to it. Then we replace the force of gravity with a component of magnitude $mg \sin \theta$ along the positive x axis and a component of magnitude $mg \cos \theta$ along the negative y axis.

Solution   Applying Newton's second law in component form, with $a_y = 0$, gives

$$\sum F_x = mg \sin \theta = ma_x \tag{1}$$

$$\sum F_y = n - mg \cos \theta = 0 \tag{2}$$

From (1) we see that the acceleration of the car along the driveway is provided by the component of the force of gravity directed down the incline:

$$a_x = g\sin\theta \qquad\qquad (3)$$

Note that the acceleration given by (3) is *constant* and *independent of the mass* of the car—it depends only on the angle of inclination and on *g*. In our example, $\theta = 20.0°$, and so we find that

$$a_x = \boxed{3.35 \text{ m/s}^2}$$

EXERCISE　A compact car and a large luxury sedan are at rest at the top of an ice-covered (frictionless) driveway. If they both slide down the driveway, which reaches the bottom first?

ANSWER　Because the acceleration is independent of the mass and thus is the same for both vehicles, they arrive at the bottom simultaneously.

EXERCISE　If the length of the driveway is 25.0 m and a car starts from rest at the top, how long does it take to travel to the bottom? What is the car's speed at the bottom?

ANSWER　3.86 s; 12.9 m/s

Example 4.5　Weighing a Fish in an Elevator

A person weighs a fish of mass *m* on a spring scale attached to the ceiling of an elevator, as shown in Figure 4.16. Show that if the elevator accelerates in either direction, the spring scale gives a reading different from the true weight of the fish.

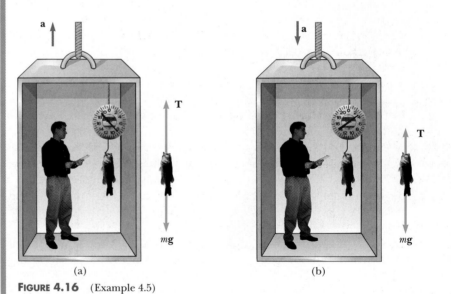

(a)　(b)

FIGURE 4.16　(Example 4.5)

Solution　The forces acting on the fish are the downward force of gravity $\mathbf{F}_g$ and the upward force **T** exerted by the spring scale. By Newton's third law, the tension *T* is also the reading of the scale. If the elevator is either at rest or moving at constant velocity, the fish is not accelerating, so $\Sigma F_y = T - F_g = 0$ or $T = F_g = mg$.

If the elevator moves with an acceleration **a**, then Newton's second law applied to the fish gives the net force exerted on the fish:

$$\sum F_y = T - mg = ma_y \qquad\qquad (1)$$

where we have chosen upward as the positive direction. Thus, we conclude from (1) that the scale reading *T* is *greater* than the weight *mg* if **a** is *upward*, so that a_y is positive as in

Figure 4.16a. The reading is *less* than mg if **a** is *downward*, so that a_y is negative as in Figure 4.16b.

For example, if the weight of the fish is 40.0 N and **a** is *upward*, so that $a_y = +2.00$ m/s², then the scale reading from (1) is

$$T = ma_y + mg = mg\left(\frac{a_y}{g} + 1\right) \qquad (2)$$

$$= (40.0 \text{ N})\left(\frac{+2.00 \text{ m/s}^2}{9.80 \text{ m/s}^2} + 1\right) = \boxed{48.2 \text{ N}}$$

If **a** is *downward*, so that $a_y = -2.00$ m/s², then (2) gives us

$$T = (40.0 \text{ N})\left(\frac{-2.00 \text{ m/s}^2}{9.80 \text{ m/s}^2} + 1\right) = \boxed{31.8 \text{ N}}$$

Hence, if you buy a fish by weight in an elevator, make sure the fish is weighed while the elevator is either at rest or else accelerating downward!

Special Cases　If the elevator cable breaks, the elevator falls freely and $a_y = -g$. We see from (2) that the scale reading T is zero in this case; that is, the fish appears to be weightless. If the elevator accelerates downward with an acceleration greater than g, the fish (along with the person in the elevator) eventually hits the ceiling because the acceleration of fish and person is still that of a freely falling object.

4.6　FORCES OF FRICTION

When an object is in motion either on a surface or through a viscous medium, such as air or water, there is resistance to the motion because the object interacts with its surroundings. We call such resistance a **force of friction.** Forces of friction are very important in our everyday lives. They allow us to walk or run and are necessary for the motion of wheeled vehicles.

Imagine you are working in your garden and have filled a plastic trash can with yard clippings. You then try to drag the trash can across the surface of your concrete patio. If you apply an external horizontal force **F** to the trash can, acting to the right, as shown in Figure 4.17a, the trash can remains stationary if **F** is small. The force that counteracts **F** and keeps the trash can from moving acts to the left and is called the **force of static friction** $\mathbf{f}_s$. As long as the trash can is not moving, $f_s = F$. Thus, if **F** is increased, $\mathbf{f}_s$ also increases. Likewise, if **F** decreases, $\mathbf{f}_s$ also decreases. Experiments show that the friction force arises from the nature of the two surfaces: Because of their roughness, contact is made only at a few points, as shown in the magnified view of the surfaces in Figure 4.17a.

If we increase the magnitude of **F**, as in Figure 4.17b, the trash can eventually slips. When the trash can is on the verge of slipping, f_s is a maximum as shown in Figure 4.17c. When F exceeds $f_{s,\,max}$, the trash can moves and accelerates to the right. When the trash can is in motion, the friction force is less than $f_{s,\,max}$ (Fig. 4.17c). We call the friction force for an object in motion the **force of kinetic friction** $\mathbf{f}_k$. The net force $F - f_k$ in the x direction produces an acceleration to the right, according to Newton's second law. If $F = f_k$, the acceleration is zero, and the trash can moves to the right with constant speed. If the applied force is removed, the friction force acting to the left provides an acceleration of the trash can in the $-x$ direction and eventually brings it to rest, again consistent with Newton's second law.

Experimentally, we find that, to a good approximation, both $f_{s,\,max}$ and f_k for an object on a surface are proportional to the normal force exerted by the surface on the object. Thus, the experimental observations can be summarized as follows:

FIGURE 4.17 (a) The force of friction $\mathbf{f}_s$ exerted by a concrete surface on a trash can is directed opposite the force $\mathbf{F}$ that you exert on the can. As long as the trash can is not moving, the magnitude of the force of static friction equals that of the applied force $\mathbf{F}$. (b) When the magnitude of $\mathbf{F}$ exceeds the magnitude of $\mathbf{f}_k$, the force of kinetic friction, the trash can accelerates to the right. (c) A graph of the magnitude of the friction force versus that of the applied force. Note that $f_{s,\,max} > f_k$.

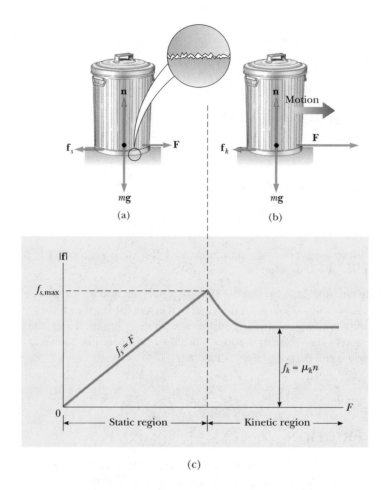

- The magnitude of the force of static friction between any two surfaces in contact can have the values

$$f_s \leq \mu_s n \qquad\qquad \textbf{[4.11]}$$

where the dimensionless constant μ_s is called the **coefficient of static friction** and n is the magnitude of the normal force exerted by one surface on the other. Equation 4.11 holds when an object is on the verge of slipping, that is, when $f_s = f_{s,\,max} \equiv \mu_s n$. This situation is called *impending motion*. The inequality holds when the component of the applied force parallel to the surfaces is less than this value.

- The magnitude of the force of kinetic friction acting between two surfaces is

$$f_k = \mu_k n \qquad\qquad \textbf{[4.12]}$$

where μ_k is the **coefficient of kinetic friction.**

- The values of μ_k and μ_s depend on the nature of the surfaces, but μ_k is generally less than μ_s. Table 4.2 lists some reported values.

- The direction of the friction force exerted by a surface on an object is opposite the actual motion (kinetic friction) or the impending motion (static friction) of the object relative to the surface.

- The coefficients of friction are nearly independent of the area of contact between the surfaces.

Although the coefficient of kinetic friction varies with speed, we shall neglect any such variations. The approximate nature of Equations 4.11 and 4.12 is easily demonstrated by trying to get an object to slide down an incline at constant speed. Especially at low speeds, the motion is likely to be characterized by alternate stick and slip episodes.

Tip 4.7   USE THE EQUAL SIGN IN LIMITED SITUATIONS

In Equation 4.11, the equal sign is used *only* when the surfaces are just about to break free and begin sliding. Do not fall into the common trap of using $f_s = \mu_s n$ in *any* static situation.

TABLE 4.2	Coefficients of Friction[a]	
	μ_s	μ_k
Steel on steel	0.74	0.57
Aluminum on steel	0.61	0.47
Copper on steel	0.53	0.36
Rubber on concrete	1.0	0.8
Wood on wood	0.25–0.5	0.2
Glass on glass	0.94	0.4
Waxed wood on wet snow	0.14	0.1
Waxed wood on dry snow	—	0.04
Metal on metal (lubricated)	0.15	0.06
Ice on ice	0.1	0.03
Teflon on Teflon	0.04	0.04
Synovial joints in humans	0.01	0.003

[a] All values are approximate.

Quick Quiz 4.9 You press your physics textbook flat against a vertical wall with your hand. What is the direction of the friction force exerted by the wall on the book? (a) downward (b) upward (c) out from the wall (d) into the wall.

Quick Quiz 4.10 A crate is sitting in the center of a flatbed truck. The truck accelerates toward the east, and the crate moves with it, not sliding on the bed of the truck. What is the direction of the friction force exerted by the bed of the truck on the crate? (a) To the west. (b) To the east. (c) There is no friction force because the crate is not sliding.

Quick Quiz 4.11 You are playing with your younger sister in the snow. She is sitting on a sled and asks you to move her across a flat, horizontal field. You have a choice of (a) pushing her from behind by applying a force downward on her shoulders at 30° below the horizontal (Fig. 4.18a), or (b) attaching a rope to the front of the sled and pulling with a force at 30° above the horizontal (Fig 4.18b). Which would be easier for you and why?

(a)

(b)

FIGURE 4.18 (Quick Quiz 4.11)

FIGURE 4.19   (Example 4.6) *After the puck is given an initial velocity to the right, the external forces acting on it are the force of gravity $\mathbf{F}_g$, the normal force $\mathbf{n}$, and the force of kinetic friction $\mathbf{f}_k$.*

Example 4.6   The Sliding Hockey Puck

The hockey puck in Figure 4.19 is given an initial speed of 20.0 m/s on a frozen pond. The puck remains on the ice and slides 120 m before coming to rest. Determine the coefficient of kinetic friction between puck and ice.

Reasoning   The puck slides to rest with a constant acceleration along the horizontal. Thus, we can use the kinematic equation $v^2 = v_0^2 + 2a\Delta x$ to find a. Newton's second law applied in the horizontal direction is $-f_k = -\mu_k n = ma$. To find μ_k, we first find the normal force $\mathbf{n}$ by applying $\Sigma F_y = 0$ in the vertical direction.

Solution   With the final speed $v = 0$; the initial speed $v_0 = 20.0$ m/s; and the displacement, $\Delta x = 120$ m:

$$v^2 = v_0^2 + 2a\Delta x$$

$$a = \frac{v^2 - v_0^2}{2\Delta x}$$

$$= \frac{0 - (20.0 \text{ m/s})^2}{2(120 \text{ m})} = -1.67 \text{ m/s}^2$$

The negative sign means that the acceleration is to the left in Figure 4.19, *opposite* the direction of the velocity.

  The magnitude of the force of kinetic friction is found from $f_k = \mu_k n$, and n is found from $\Sigma F_y = 0$ as follows:

$$\Sigma F_y = n - F_g = 0$$

$$n = F_g = mg$$

Thus,

$$f_k = \mu_k n = \mu_k mg$$

Now we apply Newton's second law along the horizontal direction, taking the positive direction toward the right:

$$\Sigma F_x = -f_k = ma$$

$$-\mu_k mg = ma$$

$$\mu_k = -\frac{a}{g} = \frac{1.67 \text{ m/s}^2}{9.80 \text{ m/s}^2} = \boxed{0.170}$$

Example 4.7   Connected Objects

Two objects are connected by a light string that passes over a frictionless pulley, as in Figure 4.20a. The coefficient of kinetic friction between the cube and the surface is 0.300. Find the acceleration of the two objects and the tension in the string.

Reasoning   Connected objects are best handled by applying Newton's second law separately to each. The free-body diagrams for the cube and the ball are shown in Figure 4.20b. Assuming the string connecting the two objects does not stretch, the magnitude of the acceleration for both objects has the same value, a. We shall obtain two equations involving the unknowns T and a that can be solved simultaneously.

Solution   With the positive x direction to the right and the positive y direction upward, Newton's second law applied to the cube of mass m_1 gives

$$\Sigma F_x = T - f_k = m_1 a$$

$$\Sigma F_y = n - m_1 g = 0$$

FIGURE 4.20   (Example 4.7) (a) Two objects connected by a light string that passes over a frictionless pulley. (b) Free-body diagrams for the objects.

Because $f_k = \mu_k n$ and $n = m_1 g$, we have $f_k = \mu_k m_1 g$ and the first equation above becomes

$$T - \mu_k m_1 g = m_1 a \qquad \textbf{(1)}$$

Now we apply Newton's second law to the ball, which is moving in the vertical direction and has mass m_2. Because the ball moves downward when the cube moves in the positive x direction (to the right), we define down as the positive direction for the ball:

$$\sum F_y = m_2 g - T = m_2 a \qquad \textbf{(2)}$$

Adding (1) and (2) eliminates T, leaving a single equation for a:

$$m_2 g - \mu_k m_1 g = (m_1 + m_2) a$$

$$a = \frac{m_2 g - \mu_k m_1 g}{m_1 + m_2}$$

Substituting the known values into this equation gives

$$a = \frac{(7.00 \text{ kg})(9.80 \text{ m/s}^2) - (0.300)(4.00 \text{ kg})(9.80 \text{ m/s}^2)}{(4.00 \text{ kg} + 7.00 \text{ kg})} = \boxed{5.17 \text{ m/s}^2}$$

When this value for the acceleration is substituted into (1), we get

$$T = \boxed{32.4 \text{ N}}$$

It is interesting that the result for a can be more directly obtained (and checked) by considering the two connected objects as a single system. In this approach, the total system mass is $m_1 + m_2$ because both objects are accelerated; the net external force exerted by the outside world on the two-object system is the difference between the gravitational force on m_2 and the friction force retarding m_1. That is, $\sum F_{\text{ext}} = m_2 g - \mu_k m_1 g = (m_1 + m_2) a$. Thus

$$a = \frac{m_2 g - \mu_k m_1 g}{m_1 + m_2}$$

Unfortunately the single system approach gives no information about **internal forces,**

Webnote 4.4

Pulleys such as the one in Example 4.7 are involved in several end-of-chapter problems. Practical pulley configurations are often more complicated than those shown in this book. If you are interested, you can study pulley support systems thoroughly at:

http://www.howstuffworks.com/pulley.htm

which are forces between adjacent parts of a system. Here tension is an internal force that does not contribute to the net external force on the system.

EXERCISE   If the coefficient of kinetic friction were very large, the ball and cube could move with constant velocity. What coefficient of friction is required for this situation to exist?

ANSWER   1.75

FIGURE 4.21   The horizontal forces acting on the car are the *forward* forces **f** exerted by the road on each tire and the force of air resistance **R**, which acts *opposite* the car's velocity. (The car's tires exert a rearward force on the road, not shown in the diagram.)

FRICTION AND THE MOTION OF A CAR

Forces of friction are important in the analysis of the motion of cars and other wheeled vehicles. There are several types of friction forces to consider, the main ones being the force of friction between tires and road surface and the retarding force produced by air resistance.

Assuming the car is a four-wheel-drive vehicle of mass *m*, as each wheel turns to propel the car forward, the tire exerts a rearward force on the road. The reaction to this rearward force is a forward force **f** exerted by the road on the tire (Fig. 4.21). If we assume that the same forward force **f** is exerted on each tire, the net forward force on the car is 4**f**, and the car's acceleration is therefore $a = 4f/m$.

When the car is in motion, we must also consider the force of air resistance **R**, which acts in the direction opposite the velocity of the car. The net force exerted on the car is therefore $4f - R$, and so the car's acceleration is $a = (4f - R)/m$. At

APPLYING PHYSICS 4.2

Consider a sky diver falling through air before reaching her terminal speed, as in Figure 4.22. As her speed increases, what happens to her acceleration? What is her acceleration once she reaches terminal speed?

Explanation   The forces exerted on the skydiver are the downward force of gravity *m***g** and an upward force of air resistance **R**. Before she reaches terminal speed, the magnitude of **R** is less than her weight. As her downward speed increases, the force of air resistance increases. The vector sum of the force of gravity and the force of air resistance gives a total force that decreases with time, so her acceleration decreases. Once the two forces balance each other so that the net force is zero, the acceleration is consequently zero, and she has reached terminal speed.

FIGURE 4.22   (Applying Physics 4.2) *(Guy Sauvage, Photo Researchers, Inc.)*

normal driving speeds, the magnitude of **R** is proportional to the first power of the speed. That is, $R = bv$, where b is a constant. Thus, the force of air resistance increases with increasing speed. When R is equal to $4f$, the acceleration is zero, and the car moves at a constant speed.

A similar situation occurs when an object falls through air. When the upward force exerted by the air on the object balances the downward force of gravity exerted on the object, the net force on the object is zero and hence its acceleration is zero. Once this condition is reached, the object continues to move downward with some constant maximum speed called the **terminal speed.**

SUMMARY

Newton's first law states that, if the net force exerted on an object is zero, an object at rest remains at rest and an object in motion continues in motion with a constant velocity.

The tendency for an object to maintain its original state of motion is called **inertia. Mass** is the physical quantity that measures the resistance of an object to changes in its velocity.

Newton's second law states that the acceleration of an object is directly proportional to the net force acting on it and inversely proportional to its mass. In equation form, this is expressed as:

$$\sum \mathbf{F} = m\mathbf{a} \qquad [4.1]$$

The magnitude of the force of gravity exerted on an object is called the **weight** of an object, denoted by w. The weight of an object of mass m is equal to the product mg:

$$w = mg \qquad [4.6]$$

Newton's third law states that, if two objects interact, the force $\mathbf{F}_{12}$ exerted by object 1 on object 2 is equal in magnitude and opposite in direction to the force $\mathbf{F}_{21}$ exerted by object 2 on object 1. Thus, an isolated force can never occur in nature.

An **object in equilibrium** has no net external force acting on it, and the second law, in component form, implies that $\Sigma F_x = 0$ and $\Sigma F_y = 0$.

The magnitude of the maximum force of static friction, $f_{s,\,max}$, between an object and a surface is proportional to the magnitude of the normal force acting on the object. This maximum force occurs when the object is on the verge of slipping. In general,

$$f_s \leq \mu_s n \qquad [4.11]$$

where μ_s is the **coefficient of static friction.** When an object slides over a surface, the direction of the force of kinetic friction on the object $\mathbf{f}_k$ is opposite the direction of the motion of the object relative to the surface, and the magnitude is proportional to that of the normal force. The magnitude of $\mathbf{f}_k$ is

$$f_k = \mu_k n \qquad [4.12]$$

where μ_k is the **coefficient of kinetic friction.** In general, $\mu_k < \mu_s$.

CONCEPTUAL QUESTIONS

1. A ball is held in a person's hand. (a) Identify all the external forces acting on the ball and the reaction to each. (b) If the ball is dropped, what force is exerted on it while it is falling? Identify the reaction force in this case. (Neglect air resistance.)

2. If a car is traveling westward with a constant speed of 20 m/s, what is the resultant force acting on it?

3. If a car moves with a constant acceleration, can you conclude there are no forces acting on it?

4. A rubber ball is dropped onto the floor. What force causes the ball to bounce?

5. If you push on a heavy box that is at rest, you must exert some force to start its motion. However, once the box is sliding, you can apply a smaller force to maintain that motion. Why?

6. If gold were sold by weight, would you rather buy it in Denver or in Death Valley? If it were sold by mass, in which of the two locations would you prefer to buy it? Why?

7. A passenger sitting in the rear of a bus claims that she was injured as the driver slammed on the brakes, causing a suitcase to come flying toward her from the front of the

bus. If you were the judge in this case, what disposition would you make? Why?

8. A space explorer is moving through space far from any planet or star. She notices a large rock, taken as a specimen from an alien planet, floating around the cabin of the ship. Should she push it gently or kick it toward the storage compartment? Why?

9. What force causes an automobile to move? A propeller-driven airplane? A rowboat?

10. Analyze the motion of a rock dropped in water in terms of its speed and acceleration as it falls. Assume that a resistive force is acting on the rock that increases as the velocity increases.

11. In the motion picture *It Happened One Night* (Columbia Pictures, 1934), Clark Gable is standing inside a stationary bus in front of Claudette Colbert, who is seated. The bus suddenly starts moving forward and Clark falls into Claudette's lap. Why did this happen?

12. A weightlifter stands on a bathroom scale. He pumps a barbell up and down. What happens to the reading on the bathroom scale as this is done? Suppose he is strong enough to actually *throw* the barbell upward. How does the reading on the scale vary now?

13. In a tug-of-war between two athletes, each pulls on the rope with a force of 200 N. What is the tension in the rope? If the rope does not move, what horizontal force does each athlete exert against the ground?

14. As a rocket is fired from a launching pad, its speed and ac-celeration increase with time as its engines continue to op-erate. Explain why this occurs even though the thrust of the engines remains constant.

15. Identify the action-reaction pairs in the following situa-tions: (a) A man takes a step, (b) a snowball hits a girl in the back, (c) a baseball player catches a ball, (d) a gust of wind strikes a window.

16. The driver of a speeding empty truck slams on the brakes and skids to a stop through a distance *d*. (a) If the truck car-ried a load that doubled its mass, what would be the truck's "skidding distance"? (b) If the initial speed of the truck were halved, what would be the truck's "skidding distance"?

17. Suppose you are driving a car at a high speed. Why should you avoid "slamming on" your brakes when you want to stop in the shortest possible distance? (Newer cars have antilock brakes that avoid this problem.)

18. A truck loaded with sand accelerates along a highway. If the driving force on the truck remains constant, what hap-pens to the truck's acceleration if its trailer leaks sand at a constant rate through a hole in its bottom?

19. A large crate is placed on the bed of a truck but not tied down. (a) As the truck accelerates forward, the crate re-mains at rest relative to the truck. What force causes the crate to accelerate forward? (b) If the driver slams on the brakes, what could happen to the crate?

20. Describe a few examples in which the force of friction ex-erted on an object is in the direction of motion of the object.

PROBLEMS

1, 2, 3 = straightforward, intermediate, challenging   ☐ = full solution available in Student Solutions Manual/Study Guide

web = solution posted at **http://info.brookscole.com/serway**   🧬 = biomedical application

Section 4.1   The Concept of Force

Section 4.2   Newton's First Law

Section 4.3   Newton's Second Law

Section 4.4   Newton's Third Law

1. A 6.0-kg object undergoes an acceleration of 2.0 m/s^2. (a) What is the magnitude of the resultant force acting on it? (b) If this same force is applied to a 4.0-kg object, what acceleration is produced?

2. A football punter accelerates a football from rest to a speed of 10 m/s during the time in which his toe is in con-tact with the ball (about 0.20 s). If the football has a mass of 0.50 kg, what average force does the punter exert on the ball?

3. The heaviest invertebrate is the giant squid, which is esti-mated to have a weight of about 2 tons spread out over its length of 70 feet. What is its weight in newtons?

4. The heaviest flying bird is the trumpeter swan, which weighs in at about 38 pounds at its heaviest. What is its weight in newtons?

5. A bag of sugar weighs 5.00 lb on Earth. What should it weigh in newtons on the Moon, where the free-fall acceler-ation is $\frac{1}{6}$ that on Earth? Repeat for Jupiter, where *g* is 2.64 times that on Earth. Find the mass of the bag of sugar in kilograms at each of the three locations.

6. A freight train has a mass of 1.5×10^7 kg. If the locomo-tive can exert a constant pull of 7.5×10^5 N, how long does it take to increase the speed of the train from rest to 80 km/h?

7. The air exerts a forward force of 10 N on the propeller of a 0.20-kg model airplane. If the plane accelerates forward at 2.0 m/s^2, what is the magnitude of the resistive force ex-erted by the air on the airplane?

8. A 5.0-g bullet leaves the muzzle of a rifle with a speed of 320 m/s. What total force (assumed constant) is exerted on the bullet while it is traveling down the 0.82-m-long barrel of the rifle?

9. A performer in a circus is fired from a cannon as a "hu-man cannonball" and leaves the cannon with a speed of 18.0 m/s. The performer's mass is 80.0 kg. The cannon barrel is 9.20 m long. Find the average net force exerted on the performer while he is being accelerated inside the cannon.

10. To lift a patient, four nurses grip the sheet on which the patient is lying and lift upward. If each nurse exerts an up-ward force of 240 N and the patient has an upward accel-eration of 0.504 m/s^2, what is the weight of the patient?

11. A boat moves through the water with two forces acting on it. One is a 2000-N forward push by the water on the pro-peller, and the other is an 1800-N resistive force due to the water around the bow. (a) What is the acceleration of the 1000-kg boat? (b) If it starts from rest, how far will it move in 10.0 s? (c) What will its velocity be at the end of this time?

12. Two forces are applied to a car in an effort to move it, as shown in Figure P4.12. (a) What is the resultant of these two forces? (b) If the car has a mass of 3000 kg, what acceleration does it have? Ignore friction.

FIGURE P4.12

13. The pelvis in Figure P4.13 has a mass of 30.0 kg. Determine its acceleration.

FIGURE P4.13

14. The force exerted by the wind on the sails of a sailboat is 390 N north. The water exerts a force of 180 N east. If the boat (including crew) has a mass of 270 kg, what are the magnitude and direction of its acceleration?

Section 4.5 Some Applications of Newton's Laws

15. Find the tension in each cable supporting the 600-N cat burglar in Figure P4.15.

FIGURE P4.15

16. Find the tension in the two wires that support the 100-N light fixture in Figure P4.16.

FIGURE P4.16

17. A 150-N bird feeder is supported by three cables as shown web in Figure P4.17. Find the tension in each cable.

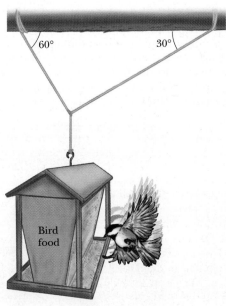

FIGURE P4.17

18. The leg and cast in Figure P4.18 weigh 220 N (w_1). Determine the weight w_2 and the angle α needed so that no force is exerted on the hip joint by the leg plus cast.

110 N 40° α w_2

$w_1 = 220$ N

FIGURE P4.18

19. A person with a broken forearm has the arm in a sling, as in Figure P4.19. The cast and the forearm together weigh 98.0 N. Assuming the upper arm exerts a horizontal force of 24.0 N to the right on the forearm, as shown, determine the force exerted by the sling on the neck.

24.0 N →

98.0 N

FIGURE P4.19

20. Two people are pulling a boat through the water as in Figure P4.20. Each exerts a force of 600 N directed at a 30.0° angle relative to the forward motion of the boat. If the boat moves with constant velocity, find the resistive force **F** exerted by the water on the boat.

600 N

F

30.0°

30.0°

600 N

FIGURE P4.20

21. The distance between two telephone poles is 50.0 m. When a 1.00-kg bird lands on the telephone wire midway between the poles, the wire sags 0.200 m. Draw a free-body diagram of the bird. How much tension does the bird produce in the wire? Ignore the weight of the wire.

22. You are a judge in a children's kite-flying contest, and two children will win prizes for the kites that pull most strongly and least strongly on their strings. To measure string tensions, you borrow a weight hanger, some slotted weights, and a protractor from your physics teacher, and use the following protocol, illustrated in Figure P4.22: Wait for a child to get her kite well controlled, hook the hanger onto the kite string about 30 cm from her hand, pile on weight until that section of string is horizontal, record the mass required, and record the angle between the horizontal and the string running up to the kite. (a) Explain how this method works. As you construct your explanation, imagine that the children's parents ask you about your method, that they might make false assumptions about your ability without concrete evidence, and that your explanation is an opportunity to give them confidence in your evaluation technique. (b) Find the string tension if the mass is 132 g and the angle is 46.3°.

FIGURE P4.22

23. A 5.0-kg bucket of water is raised from a well by a rope. If the upward acceleration of the bucket is 3.0 m/s^2, find the force exerted by the rope on the bucket.

24. A shopper in a supermarket pushes a loaded cart with a horizontal force of 10 N. The cart has a mass of 30 kg. (a) How far will it move in 3.0 s, starting from rest? (Ignore friction.) (b) How far will it move in 3.0 s if the shopper places her 30-N baby in the cart before she begins to push it?

25. A 2000-kg car is slowed down uniformly from 20.0 m/s to 5.00 m/s in 4.00 s. (a) What average total force acted on the car during this time, and (b) how far did the car travel during this time?

26. Two packing crates of masses 10.0 kg and 5.00 kg are connected by a light string that passes over a frictionless pulley as in Figure P4.26. The 5.00-kg crate lies on a smooth incline of angle 40.0°. Find the acceleration of the 5.00-kg crate and the tension in the string.

27. Assume that the three blocks in Figure P4.27 move on a frictionless surface and that a 42-N force acts as shown on the 3.0-kg block. Determine (a) the acceleration given this

FIGURE P4.26

FIGURE P4.27

FIGURE P4.31

FIGURE P4.32

system, (b) the tension in the cord connecting the 3.0-kg and the 1.0-kg blocks, and (c) the force exerted by the 1.0-kg block on the 2.0-kg block.

28. An object of mass 2.0 kg starts from rest and slides down an inclined plane 80 cm long in 0.50 s. What *net force* is acting on the object along the incline?

29. A 40.0-kg wagon is towed up a hill inclined at 18.5° with respect to the horizontal. The tow rope is parallel to the incline and has a tension of 140 N in it. Assume that the wagon starts from rest at the bottom of the hill, and neglect friction. How fast is the wagon going after moving 80.0 m up the hill?

30. An object with mass m_1 = 5.00 kg, resting on a frictionless horizontal table, is connected to a cable that passes over a pulley and then is fastened to a hanging object with mass m_2 = 10.0 kg, as in Figure P4.30. Find the acceleration of each object and the tension in the cable.

FIGURE P4.30   (Problems 30, 36, and 45)

31. The forearm (including the hand) in Figure P4.31 has a mass of 4.26 kg. The upper arm exerts the force shown on the forearm. If the forearm has an acceleration of 2.54 m/s², determine the mass of the object being pulled by the hand. (Assume the object slides over a frictionless surface.)

32. Determine the magnitude and direction of the force **F** exerted by the heel of the hand on the forehead in order that the head will exert no force on the neck in Figure P4.32. Disregard the weight of the head and assume it remains at rest.

33. A 1000-kg car is pulling a 300-kg trailer. Together the car **web** and trailer have an acceleration of 2.15 m/s² in the forward direction. Neglecting frictional forces on the trailer, determine (a) the net force on the car; (b) the net force on the trailer; (c) the force exerted by the trailer on the car; (d) the resultant force exerted by the car on the road.

34. Two objects with masses of 3.00 kg and 5.00 kg are connected by a light string that passes over a frictionless pulley as in Figure P4.34. Determine (a) the tension in the string, (b) the acceleration of each object, and (c) the distance each object will move in the first second of motion if both objects start from rest.

FIGURE P4.34

Section 4.6   Forces of Friction

35. A dockworker loading crates on a ship finds that a 20-kg crate, initially at rest on a horizontal surface, requires a 75-N horizontal force to set it in motion. However, after the crate is in motion, a horizontal force of 60 N is required to keep it moving with a constant speed. Find the coefficients of static and kinetic friction between crate and floor.

36. In Figure P4.30, m_1 = 10 kg and m_2 = 4.0 kg. The coefficient of static friction between m_1 and the horizontal surface is 0.50 while the coefficient of kinetic friction is 0.30. (a) If the system is released from rest, what will its acceleration be? (b) If the system is set in motion with m_2 moving downward, what will be the acceleration of the system?

37. A 1000-N crate is being pushed across a level floor at a constant speed by a force **F** of 300 N at an angle of 20.0° below the horizontal as shown in Figure P4.37a. (a) What is the coefficient of kinetic friction between the crate and the floor? (b) If the 300-N force is instead pulling the block at an angle of 20.0° above the horizontal as shown in Figure P4.37b, what will be the acceleration of the crate? Assume that the coefficient of friction is the same as found in (a).

(a)	(b)

FIGURE P4.37

38. A hockey puck is hit on a frozen lake and starts moving with a speed of 12.0 m/s. Five seconds later, its speed is 6.00 m/s. (a) What is its average acceleration? (b) What is the average value of the coefficient of kinetic friction between puck and ice? (c) How far does the puck travel during this 5.00-s interval?

39. The coefficient of static friction is 0.800 between the soles of a sprinter's running shoes and the level track surface on which she is running. Determine the maximum acceleration she can achieve. Do you need to know that her mass is 60.0 kg?

40. A woman at an airport is towing her 20.0-kg suitcase at constant speed by pulling on a strap at an angle of θ above the horizontal (Fig. P4.40). She pulls on the strap with a 35.0-N force, and the friction force on the suitcase is 20.0 N. Draw a free-body diagram of the suitcase. (a) What angle does the strap make with the horizontal? (b) What normal force does the ground exert on the suitcase?

41. The coefficient of static friction between the 3.00-kg crate and the 35.0° incline of Figure P4.41 is 0.300. What minimum force **F** must be applied to the crate perpendicular to the incline to prevent the crate from sliding down the incline?

42. A box of books weighing 300 N is shoved across the floor of an apartment by a force of 400 N exerted downward at an angle of 35.2° below the horizontal. If the coefficient of kinetic friction between box and floor is 0.570, how long does it take to move the box 4.00 m, starting from rest?

43. An object falling under the pull of gravity experiences a frictional force of air resistance. The magnitude of this

FIGURE P4.40

FIGURE P4.41

force is approximately proportional to the speed of the object, $f = bv$. Assume that b = 15 kg/s and m = 50 kg. (a) What is the terminal speed that the object reaches while falling? (b) Does your answer to part (a) depend on the initial speed of the object? Explain.

44. A student decides to move a box of books into her dormitory room by pulling on a rope attached to the box. She pulls with a force of 80.0 N at an angle of 25.0° above the horizontal. The box has a mass of 25.0 kg, and the coefficient of kinetic friction between box and floor is 0.300. (a) Find the acceleration of the box. (b) The student now starts moving the box up a 10.0° incline, keeping her 80.0-N force directed at 25.0° above the line of the incline. If the coefficient of friction is unchanged, what is the new acceleration of the box?

45. Objects with masses m_1 = 10.0 kg and m_2 = 5.00 kg are **web** connected by a light string that passes over a frictionless pulley as in Figure P4.30. If, when the system starts from rest, m_2 falls 1.00 m in 1.20 s, determine the coefficient of kinetic friction between m_1 and the table.

46. A car is traveling at 50.0 km/h on a flat highway. (a) If the coefficient of friction between road and tires on a rainy day is 0.100, what is the minimum distance in which the car will stop? (b) What is the stopping distance when the surface is dry and the coefficient of friction is 0.600?

47. A 2.00-kg block is given an initial speed of 2.50 m/s up a plane inclined 15.0° with the horizontal. The coefficient of kinetic friction between block and plane is 0.250. What speed does it have just as it returns to its initial position?

48. The driver of a 600-kg sports car, heading directly for a railroad crossing 250 m away, applies the brakes in a panic stop. The car is moving (illegally) at 40 m/s, and the brakes can supply a friction force of 1200 N. (a) How fast is the car moving when it reaches the crossing? (b) Will the driver escape collision with a freight train that is 80.0 m away from the intersection and traveling at 23 m/s?

49. Find the acceleration experienced by each of the two objects shown in Figure P4.49 if the coefficient of kinetic friction between the 7.00-kg object and the plane is 0.250.

FIGURE P4.49

50. A 2.00-kg block is held in equilibrium on an incline of angle $\theta = 60.0°$ by a horizontal force **F** applied in the direction shown in Figure P4.50. If the coefficient of static friction between block and incline is $\mu_s = 0.300$, determine (a) the minimum value of **F** and (b) the normal force of the incline on the block.

FIGURE P4.50

51. The person in Figure P4.51 weighs 170 lb. The crutches each make an angle of 22.0° with the vertical (as seen from the front). Half of his weight is supported by the crutches. The other half is supported by the vertical forces exerted by the ground on his feet. Assuming he is at rest and the force exerted by the ground on the crutches acts along the crutches, determine (a) the smallest possible coefficient of friction between crutches and ground and (b) the magnitude of the compression force supported by each crutch.

52. A small block of mass $m = 2.00$ kg rests on the left edge of a block of length $L = 3.00$ m and mass $M = 8.00$ kg. The coefficient of kinetic friction between the two blocks is $\mu_k = 0.300$, and the surface on which the 8.00-kg block rests is frictionless. A constant horizontal force of magnitude $F = 10.0$ N is applied to the 2.00-kg block, setting it in motion as shown in Figure P4.52a. (a) How long will it take before this block makes it to the right side of the 8.00-kg block, as shown in Figure P4.52b? (*Note:* Both blocks are set in motion when **F** is applied.) (b) How far does the 8.00-kg block move in the process?

ADDITIONAL PROBLEMS

53. Find the net force exerted by the cables on the leg in Fig. P4.53. The cables are horizontal before passing over the pulleys. Assume the leg remains at rest.

FIGURE P4.51

(a)

(b)

FIGURE P4.52

FIGURE P4.53

54. One of the great dangers to mountain climbers is an avalanche, in which a large mass of snow and ice breaks loose and goes on an essentially frictionless ride down a mountainside on a cushion of compressed air. If you were on a 30.0° mountain slope and an avalanche started 400 m

up the slope, how much time would you have to get out of the way?

55. (a) What is the resultant force exerted by the two cables supporting the traffic light in Figure P4.55? (b) What is the weight of the light?

FIGURE P4.55

56. As a protest against the umpire's calls, a baseball pitcher throws a ball straight up into the air at a speed of 20.0 m/s. In the process, he moves his hand through a distance of 1.50 m. If the ball has a mass of 0.150 kg, find the force he exerts on the ball to give it this upward speed.

57. A girl coasts down a hill on a sled, reaching a level surface at the bottom with a speed of 7.0 m/s. If the coefficient of friction between runners and snow is 0.050 and the girl and sled together weigh 600 N, how far does the sled travel on the level surface before coming to rest?

58. (a) What is the minimum force of friction required to hold the system of Figure P4.58 in equilibrium? (b) What coefficient of static friction between the 100-N block and the table ensures equilibrium? (c) If the coefficient of kinetic friction between the 100-N block and the table is 0.250, what hanging weight should replace the 50.0-N weight to allow the system to move at a constant speed once it is set in motion?

FIGURE P4.58

59. A box rests on the back of a truck. The coefficient of static
web friction between box and truck bed is 0.300. (a) When the truck accelerates forward, what force accelerates the box? (b) Find the maximum acceleration the truck can have before the box slides.

60. A 4.00-kg block is pushed along the ceiling with a constant applied force of 85.0 N that acts at an angle of 55.0° with the horizontal, as in Figure P4.60. The block accelerates to the right at 6.00 m/s². Determine the coefficient of kinetic friction between block and ceiling.

FIGURE P4.60

61. A frictionless plane is 10.0 m long and inclined at 35.0°. A sled starts at the bottom with an initial speed of 5.00 m/s up the incline. When it reaches the point at which it momentarily stops, a second sled is released from the top of this incline with an initial speed v_i. Both sleds reach the bottom of the incline at the same moment. (a) Determine the distance that the first sled traveled up the incline. (b) Determine the initial speed of the second sled.

62. Three objects are connected by light strings as shown in Figure P4.62. The string connecting the 4.00-kg object and the 5.00-kg object passes over a light frictionless pulley. Determine (a) the acceleration of each object and (b) the tension in the two strings.

FIGURE P4.62

63. A 3.00-kg block starts from rest at the top of a 30.0° incline and slides 2.00 m down the incline in 1.50 s. Find (a) the acceleration of the block, (b) the coefficient of kinetic friction between it and the incline, (c) the frictional force acting on the block, and (d) the speed of the block after it has slid 2.00 m.

64. A 5.0-kg penguin sits on a 10-kg sled, as in Figure P4.64. A horizontal force of 45 N is applied to the sled, but the penguin attempts to impede the motion by holding onto a cord attached to a wall. The coefficient of kinetic friction between the sled and snow as well as that between the sled and the penguin is 0.20. (a) Draw a free-body diagram for the penguin and one for the sled, and identify the reaction force for each force you include. Determine (b) the tension in the cord and, (c) the acceleration of the sled.

FIGURE P4.64

65. Two boxes of fruit on a frictionless horizontal surface are connected by a light string as in Figure P4.65, where $m_1 = 10$ kg and $m_2 = 20$ kg. A force of 50 N is applied to the 20-kg box. (a) Determine the acceleration of each box and the tension in the string. (b) Repeat the problem for the case where a coefficient of kinetic friction between each box and the surface is 0.10.

FIGURE P4.65

66. A high diver of mass 70.0 kg jumps off a board 10.0 m above the water. If his downward motion is stopped 2.00 s after he enters the water, what average upward force did the water exert on him?

67. Two people pull as hard as they can on ropes attached to a 200-kg boat. If they pull in the same direction, the boat has an acceleration of 1.52 m/s² to the right. If they pull in opposite directions, the boat has an acceleration of 0.518 m/s² to the left. What is the force exerted by each person on the boat? (Disregard any other forces on the boat.)

68. A 3.0-kg object hangs at one end of a rope that is attached to a support on a railroad car. When the car accelerates to the right, the rope makes an angle of 4.0° with the vertical, as shown in Figure P4.68. Find the acceleration of the car.

FIGURE P4.68

69. The three blocks of masses 10.0 kg, 5.00 kg, and 3.00 kg are connected by light strings that pass over frictionless pulleys as shown in Figure P4.69. The acceleration of the 5.00-kg block is 2.00 m/s² to the left, and the surfaces are rough. Find (a) the tension in each string and (b) the coefficient of kinetic friction between blocks and surfaces.

(Assume the same μ_k for both blocks in contact with surfaces.)

FIGURE P4.69

70. An inquisitive physics student, wishing to combine pleasure with scientific inquiry, rides on a roller coaster sitting on a bathroom scale. (Do not try this yourself on a roller coaster that forbids loose heavy packages.) The bottom of the seat in the roller coaster car is in a plane parallel to the track. The seat has a perpendicular back and a seat belt that fits around the student's chest in a plane parallel to the bottom of the seat. The student lifts his feet from the floor, so that the scale reads his weight, 200 lb, when the car is horizontal. At one point during the ride, the car zooms with negligible friction down a straight slope inclined at 30.0° below the horizontal. What does the scale read there?

71. A van accelerates down a hill (Fig. P4.71), going from rest to 30.0 m/s in 6.00 s. During the acceleration, a toy ($m = 0.100$ kg) hangs by a string from the van's ceiling. The acceleration is such that the string remains perpendicular to the ceiling. Determine (a) the angle θ and (b) the tension in the string.

FIGURE P4.71

72. An 80-kg stuntman jumps from the top of a building 30 m above a catching net. Assuming that air resistance exerts a 100-N force on the stuntman as he falls, determine his velocity just before he hits the net.

73. The parachute on a race car of weight 8800 N opens at the end of a quarter-mile run when the car is traveling at 35 m/s. What total retarding force must be supplied by the parachute to stop the car in a distance of 1000 m?

74. On takeoff, the combined action of the air around the engines and wings of an airplane exerts an 8000-N force on the plane, directed upward at an angle of 65.0° above the horizontal. The plane rises with constant velocity in the vertical direction while continuing to accelerate in the horizontal direction. (a) What is the weight of the plane? (b) What is its horizontal acceleration?

75. A 72-kg man stands on a spring scale in an elevator. Starting from rest, the elevator ascends, attaining its maximum speed of 1.2 m/s in 0.80 s. It travels with this constant

speed for 5.0 s, undergoes a uniform *negative* acceleration for 1.5 s, and comes to rest. What does the spring scale register (a) before the elevator starts to move? (b) during the first 0.80 s? (c) while the elevator is traveling at constant speed? (d) during the negative acceleration?

76. A sled weighing 60.0 N is pulled horizontally across snow so that the coefficient of kinetic friction between sled and snow is 0.100. A penguin weighing 70.0 N rides on the sled, as in Fig. P4.76. If the coefficient of static friction between penguin and sled is 0.700, find the maximum horizontal force that can be exerted on the sled before the penguin begins to slide off.

FIGURE P4.76

77. The board sandwiched between two other boards in Figure P4.77 weighs 95.5 N. If the coefficient of friction between the boards is 0.663, what must be the magnitude of the compression forces (assume horizontal) acting on both sides of the center board to keep it from slipping?

FIGURE P4.77

78. Determine the magnitude of the net force exerted by the cable on the leg in Figure P4.78.

79. An inventive child named Chris wants to reach an apple in a tree without climbing the tree. Sitting in a chair connected to a rope that passes over a frictionless pulley (Fig. P4.79), Chris pulls on the loose end of the rope with such a force that the spring scale reads 250 N. Chris's true weight is 320 N, and the chair weighs 160 N. (a) Show that the acceleration of the system is *upward* and find its magnitude. (b) Find the force Chris exerts on the chair.

FIGURE P4.78

FIGURE P4.79

80. A fire helicopter carries a 620-kg bucket of water at the end of a 20.0-m long cable. When the helicopter is flying back from a fire at a constant speed of 40.0 m/s, the cable makes an angle of 40.0° with respect to the vertical. Determine the force exerted by air resistance on the bucket.

GROUP ACTIVITIES

G.1 There is a simple method for measuring the coefficients of static and kinetic friction between an object and some surface. For this investigation, you will need a few coins, your textbook or some other flat surface that can be inclined, a protractor, and some double-stick tape. Place a coin at one edge of the book as it lies on a table, and lift

that edge of the book until the coin just slips down the incline at constant speed after you give it a push to get it started as shown in Figure GA4.1. When this occurs, measure the incline angle using your protractor. Repeat the measurement five times, and find the average value of this critical angle θ_c. The coefficient of static friction between the coin and book's surface is $\mu_s = \tan \theta_c$. (You should prove this as an exercise.) Calculate the average value of μ_s. To measure the coefficient of kinetic friction, find the angle θ_c' at which the coin moves down the incline with constant speed after you give it a push to get it started. This angle should be less than θ_c. Measure this new angle five times, and get its average value. Calculate the average value of μ_k using the fact that $\mu_k = \tan \theta_c'$, where $\theta_c' < \theta_c$. Repeat these measurements using two or three stacked coins, with double-stick tape between them. You should get the same results as with one coin. Why?

Coin

θ

FIGURE GA4.1

G.2 Borrow a spring scale from your instructor and use it to study some of the properties of the force of friction. (1) Attach the scale to a block of wood resting on the surface of a table and note the force required to start the block into motion. You and your co-worker should each take at least five trials and average your results. This measured force is the maximum value of the force of static friction between the block and surface. (2) Now use the spring scale to measure the force required to keep the block moving at constant velocity. Again, perform several trials to find the average value for this force. The force you find is the force of kinetic friction. (3) Turn the block so that a side with a different surface area is in contact with the table. Repeat the experiments above to see if the area of contact between the surfaces produces different values for the forces of friction.

G.3 Get a bathroom scale and stand on it while riding on an elevator. Watch carefully what happens to your apparent weight (the scale reading) as the elevator moves upward or downward as a function of time. What does this tell you about the acceleration of the elevator during the ride?

G.4 (a) An elevator moving upward has two forces acting on it, the tension in the cable and its weight. When the elevator is accelerating upward, which is greater, T or w? (b) When the elevator is moving at a constant velocity upward, which is greater, T or w? (c) When the elevator is moving upward, but the acceleration is downward, which is greater, T or w? (d) Let the elevator have a mass of 1500 kg and an upward acceleration of 2.5 m/s². Find T. Is your answer consistent with the answer to part (a)? (e) The elevator of part (d) now moves with a constant velocity upward of 10 m/s. Find T. Is your answer consistent with your answer to part (b)? (e) The elevator now initially moving upward begins to accelerate downward at 1.50 m/s². Find T. Is your answer consistent with your answer to part (c)?

G.5 You are pushing on a block that is resting on a table. (a) You press on the block, but not quite hard enough to start it moving. What are the forces acting on the block? (Be sure to specify the type of force and the object causing each force.) Wherever you can, compare the magnitudes of forces. Give a brief explanation for how you know each of the results you state. (b) You press a bit harder and the block begins to move. After a moment of starting up, you press so that the block is moving with a constant velocity. For the time while the block is moving with a constant velocity have any of the comparisons you stated in (a) changed? Which ones? Have any of the forces changed? Which ones? (c) Suppose the block has a mass of 0.4 kg and the coefficient of kinetic friction between the block and the table is 0.3. What force will you have to use to keep it going at a constant speed of 0.2 m/s?

G.6 Consider a metal sphere (S) a few centimeters in diameter and a feather (F). For each quantity in the list below indicate whether the quantity is larger for S than F, the same size for both, or larger for F. Explain each of your answers.

(a) The gravitational force.

(b) The time it will take to fall a given distance in air.

(c) The time it will take to fall a given distance in vacuum.

(d) The total force on the object when falling in vacuum.

(e) The total force on the object when falling in air.

(Problems 5 and 6 are courtesy of E. F. Redish. For more problems of this type, visit http://www.physics.umd. edu/perg/)

5

Energy

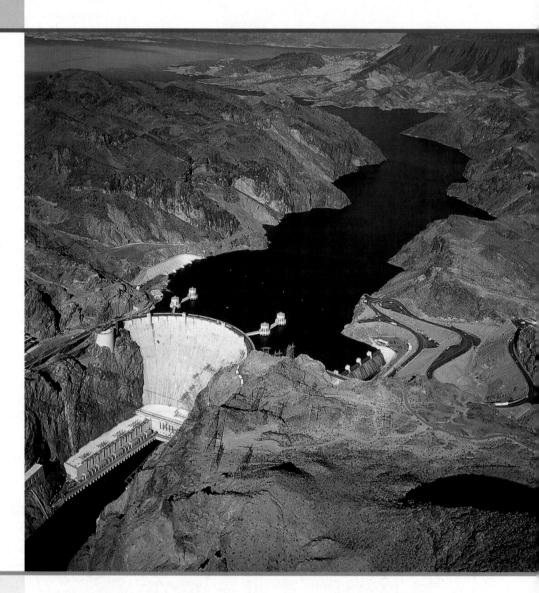

Aerial view of the Hoover Dam, a hydroelectric power facility on the Colorado River. This power plant harnesses the energy in flowing water stored in the reservoir. Falling water rotates turbines that drive electrical generators, producing about 4 billion kilowatt-hours of energy every year. *(Bernhard Edmaier/SPL/Photo Researchers, Inc.)*

T he concept of energy is one of the most important in the world of science. In everyday use, the term *energy* has to do with the cost of fuel for transportation and heating, electricity for lights and appliances, and the foods we consume. However, these ideas do not really define energy. They tell us only that fuels are needed to do a job and that those fuels provide us with something we call energy.

Energy is present in the Universe in a variety of forms, including mechanical energy, chemical energy, electromagnetic energy, and nuclear energy. Although energy can be transformed from one form to another, the principle of conservation of energy says that the total amount of energy in the Universe never changes. This is also true for an isolated system, which is a collection of objects that can exchange energy with each other but not with the rest of the Universe. If one form of energy in an isolated system decreases, then another form of energy in the system must increase. For example, if the system consists of a motor connected to a battery, the battery converts chemical energy to electrical energy, and the motor converts electrical energy to mechanical energy. Understanding how energy is transformed from one form into another is essential in the study of physics, chemistry, biology, geology, and astronomy.

In this chapter we are concerned primarily with mechanical energy. We introduce the concept of *kinetic energy*, which is defined as the energy associated with motion, and the concept of *potential energy*, the energy associated with position. Mechanical energy is the sum of these two quantities. We shall see that the ideas of work and energy can be used in place of Newton's laws to solve certain problems more simply.

We begin by defining *work,* a concept that provides a link between force and energy. With this as a foundation, we can find expressions for different forms of potential energy and then discuss the principle of conservation of energy and apply it to problems.

5.1 WORK

Almost all the terms we have used thus far in this book have the same meaning in physics as in everyday life. Now, however, we encounter a term whose meaning in physics is distinctly different from its meaning in our day-to-day affairs. This new term is **work.** As a start, imagine pushing a heavy sofa across your living room floor. If you push on the sofa *and* it undergoes a displacement, then you have done work on the sofa.

Work can be more precisely defined with the help of Figure 5.1. Here we see an object that undergoes a displacement $\Delta \mathbf{x}$ along a straight line while acted on by a constant force $\mathbf{F}$ that makes an angle of θ with $\Delta \mathbf{x}$. (We have assumed a horizontal displacement; we could have chosen any arbitrary direction instead.) The force has accomplished something—it has moved the object. We say that work was done by the force $\mathbf{F}$ on the object.

FIGURE 5.1 If an object undergoes a displacement $\Delta \mathbf{x}$, the work done by the force $\mathbf{F}$ on the object is $(F \cos \theta) \Delta x$.

The work W done by a constant force on an object is defined as the product of the component of the force along the direction of displacement and the magnitude of the displacement:

$$W \equiv (F \cos \theta) \Delta x \qquad [5.1]$$

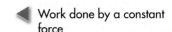 Work done by a constant force

Notice that this expression tells us only about the applied force and the displacement of the object. We have no information about how long it took for this displacement to occur, nor about the object's velocity or acceleration. We do not

TABLE 5.1	Units of Work in the Two Common Systems of Measurement	
System	**Unit of Work**	**Name of Combined Unit**
SI	newton-meter (N·m)	joule (J)
U.S. customary	foot-pound (ft·lb)	foot-pound

need these values to find the work done. That is, for given values of F, θ, and Δx, the amount of work done by a force on an object is the same regardless of the amount of time required to do the work.

Work is a scalar quantity—no direction is associated with it. Its units are those of force times length. Therefore, the SI unit of work is the **newton-meter** (N·m). Another name for the newton-meter is the **joule** (J) (rhymes with "pool"). The U.S. customary unit of work is given in Table 5.1.

As an example of the distinction between *work* in physics and our everyday understanding of the word, consider holding a heavy chair at arm's length for 10 min. At the end of this time interval, your tired arms may lead you to think that you have done a considerable amount of work. According to the definition above, however, you have done no work on the chair.[1] You have exerted a force on the chair in order to support it, but you have not moved it through a displacement. A force does no work on an object if the object does not move. This can be seen by noting that if $\Delta x = 0$, Equation 5.1 gives $W = 0$. The work is also zero if the applied force is zero—if you don't push, you don't do any work.

Also note from Equation 5.1 that the work done by a force is zero when the direction of the force is perpendicular to that of the displacement. That is, if $\theta = 90°$, then $W = 0$ because $\cos 90° = 0$. For example, no work is done on a bucket of water when it is carried horizontally at constant velocity because the upward force exerted to support the bucket is perpendicular to the displacement of the bucket, as shown in Figure 5.2. Likewise, the work done by the force of gravity during the horizontal displacement is also zero, for the same reason.

In general, an object may be moving under the influence of several external forces. In that case, the total work done as the object undergoes some displacement is the algebraic sum of the amount of work done by each force.

Work can be either positive or negative. The sign depends on the direction of **F** relative to the direction of $\Delta\mathbf{x}$. The work done by the applied force is positive when the vector associated with the component $F\cos\theta$ is in the *same direction* as the displacement. For example, when you lift a box as in Figure 5.3, the work done by the force you exert on the box is positive because that force is upward, in the same direction as the displacement. When the vector associated with the component $F\cos\theta$ is in the direction *opposite* the displacement, W is negative. Thus, when you lower the box to the floor, you do negative work on the box because you still exert an upward force on it, but the displacement is now downward. In general, the work done by a force on an object is negative when the direction in which the force acts is opposite the direction of motion of the object.

If a constant applied force acts along the direction of the displacement, then $\theta = 0°$ and $\cos 0° = 1$. In this case, Equation 5.1 gives

$$W = F\Delta x \qquad \text{[5.2]}$$

Equations 5.1 and 5.2 are special cases of a more general definition of work, because both equations assume a force constant in size and direction. The more general case of the work done by a varying force is treated briefly in Section 5.8.

Tip 5.1 WORK IS A SCALAR QUANTITY

Note that *work is a scalar quantity*—there is no direction associated with it. In fact, all the types of energy and energy transfer discussed in this textbook are scalar quantities. This is one of the nice features of the energy approach—it is easier to deal with scalar quantities than with vector quantities!

Tip 5.2 WORK IS DONE BY . . . ON . . .

When discussing work, try to use the phrase "the work done by the . . . on the. . . ." For example, "the work done by the *hammer* on the nail" identifies the nail as the system, and the force exerted by the hammer represents the nail's interaction with its environment.

Webnote 5.1

Further discussion as well as animations on the subject of work can be found at:
http://id.mind.net/~zona/mstm/physics/mechanics/energy/work/work.html
Additional explanation of positive and negative work can be found at:
http://theory.uwinnipeg.ca/mod_tech/node28.html

---

[1] Actually, you do expend energy while holding the chair at arm's length because your muscles are continuously contracting and relaxing while the chair is being supported. Thus, work is being done within your body (something moves within your muscles with each contraction and relaxation) but not on the chair (because the chair does not undergo a displacement).

FIGURE 5.2 No work is done on a bucket when it is moved horizontally because the applied force **F** is perpendicular to the displacement.

FIGURE 5.3 Positive work is done by this person when the box is lifted because the applied force **F** is in the same direction as the displacement. When the box is lowered to the floor, the work done by the person is negative because his applied force is still directed upward but the displacement is downward.

Quick Quiz 5.1

Figure 5.4 shows four situations in which a force is exerted on an object. In all four cases, the force has the same magnitude, and the displacement of the object is to the right and of the same magnitude. Rank the situations in order of the work done by the force on the object, from most positive to most negative.

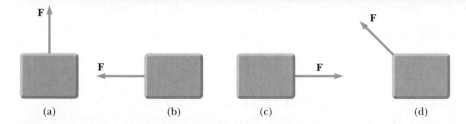

(a) (b) (c) (d)

FIGURE 5.4 (Quick Quiz 5.1) A force **F** is exerted on an object that undergoes a displacement to the right. The magnitudes of the force and displacement are the same in all four cases.

APPLYING PHYSICS 5.1

A student lifts a heavy box of mass m a vertical height h and then walks horizontally a distance d while holding the box, as in Figure 5.5. Determine the work done by the student and by the force of gravity on the box.

Explanation Assuming that he lifts the box with a force of magnitude equal to the weight of the box mg, the work done by the student on the box during the vertical displacement is $W = F\Delta y = (mg)(h) = mgh$ because the force is in the direction of the displacement. For the horizontal displacement, we assume that the acceleration of the box is approximately zero. Hence, the horizontal force he exerts on the box is approximately zero. As a result, the work he does on the box as it moves horizontally is zero because the force he applies (which is holding up the box) is perpendicular to the displacement. Thus, the net work he does during the complete process is mgh.

The work done by the force of gravity on the box during the box's vertical displacement is $-mgh$, negative because the direction of this force is opposite that of the displacement. The work done by the force of gravity is zero during the horizontal displacement because this force is perpendicular to the displacement. Hence, the net work done by the force of gravity for the complete process is $-mgh$. The net work done by *all* forces on the box is zero because $+mgh + (-mgh) = 0$.

FIGURE 5.5 (Applying Physics 5.1) A student lifts a heavy box of mass m a distance h and then walks a horizontal distance d.

Example 5.1   Mr. Clean

A man cleaning his apartment pulls the canister of a vacuum cleaner with a force of magnitude $F = 55.0$ N. The force makes an angle of $30.0°$ with the horizontal, as shown in Figure 5.6. The canister is displaced 3.00 m to the right. Calculate the work done by the 55.0-N force on the vacuum cleaner.

FIGURE 5.6   (Example 5.1) A vacuum cleaner being pulled at an angle of $30.0°$ above the horizontal.

Solution   Using the definition of work (Eq. 5.1), we have

$$W = (F\cos\theta)\,\Delta x = (55.0 \text{ N})(\cos 30.0°)(3.00 \text{ m}) = \boxed{143 \text{ J}}$$

Note that the normal force **n**, the gravitational force $m\mathbf{g}$, and the upward component of the applied force do *no* work because they are perpendicular to the displacement.

EXERCISE   Find the work done by the man on the canister if he pulls it 3.00 m with a horizontal force of magnitude $F = 32.0$ N.

ANSWER   96.0 J

EXERCISE   A 65.0-kg woman climbs a flight of 20 stairs, each 23.0 cm high. How much work is done by the force of gravity on the woman in this process?

ANSWER -2.93 kJ

5.2   KINETIC ENERGY AND THE WORK-KINETIC ENERGY THEOREM

Now that we have seen how to evaluate the work done by a force on an object, let us explore the power of this approach. Solving problems using Newton's second law can be difficult if the forces involved are complicated. An alternative approach to such problems is to relate the speed of an object to the work done on the object by some net external force. If the work done by the net force on the object can be calculated for a given displacement, the change in the object's speed is easy to evaluate. Let's see how this is done.

Figure 5.7 shows an object of mass m moving to the right under the action of a constant net force $\mathbf{F}_{\text{net}}$, also directed to the right. Because the force is constant, we know from Newton's second law that the object moves with constant acceleration $\mathbf{a}$. If the object is displaced a distance of Δx, the work done by $\mathbf{F}_{\text{net}}$ on the object is

$$W_{\text{net}} = F_{\text{net}}\,\Delta x = (ma)\,\Delta x \qquad [5.3]$$

In Chapter 2 we found that the following relationship holds when an object undergoes constant acceleration:

$$v^2 = v_0^2 + 2a\,\Delta x \qquad \text{or} \qquad a\,\Delta x = \frac{v^2 - v_0^2}{2}$$

FIGURE 5.7   An object undergoing a displacement and a change in velocity under the action of a constant net force $\mathbf{F}_{\text{net}}$.

We can substitute this expression into Equation 5.3 to get

$$W_{net} = m\left(\frac{v^2 - v_0^2}{2}\right)$$

$$W_{net} = \tfrac{1}{2}mv^2 - \tfrac{1}{2}mv_0^2 \qquad [5.4]$$

The quantity $\tfrac{1}{2}mv^2$ has a special name in physics: **kinetic energy.** The kinetic energy KE of an object of mass m moving with a speed v is defined as

$$KE \equiv \tfrac{1}{2}mv^2 \qquad [5.5]$$

◀ Kinetic energy

We can think of kinetic energy as the energy associated with the motion of an object. Kinetic energy is a scalar quantity and has the same units as work. This means that the unit $kg \cdot m^2/s^2$ (from $\tfrac{1}{2}mv^2$) can be replaced with the more streamlined unit, the joule: $1\ kg \cdot m^2/s^2 = 1\ J$. This conversion factor can be used to simplify units in energy expressions.

It is often convenient to write Equation 5.4 in the form

$$W_{net} = KE_f - KE_i = \Delta KE \qquad [5.6]$$

◀ Work-kinetic energy theorem

Equation 5.6 is an important result known as the **work-kinetic energy theorem.** According to this result, the work done by a *net* force acting on an object is equal to the change in the kinetic energy of the object. Thus, we conclude that

> **When work is done by a net force on an object and the only change in the object is its speed, the work done is equal to the change in the object's kinetic energy.**

The work-kinetic energy theorem indicates that the speed of the object increases if the net work done on it is positive because the final kinetic energy is greater than the initial kinetic energy. The speed decreases if the net work is negative because now the final kinetic energy is less than the initial kinetic energy.

From the relationship between the work done on an object and the change in its kinetic energy as expressed by Equations 5.4 and 5.6, we can understand why a golf ball initially at rest is given a large kinetic energy after being struck by a golf club. We can also think of kinetic energy as the work a moving object can do in coming to rest. For example, suppose a hammer is on the verge of striking a nail, as in Figure 5.8. The moving hammer has kinetic energy and can therefore do work on the nail. The work done on the nail is $F\Delta x$, where F is the average force exerted by the hammer on the nail and Δx is the distance the nail is driven into the wall.

For convenience, Equation 5.4 was derived under the assumption that the net force acting on the object is constant. A more general derivation would show that this equation is valid under all circumstances, including that of a variable force.

FIGURE 5.8 The moving hammer has kinetic energy and thus can do work on the nail, driving it into the wall.

APPLYING **PHYSICS** 5.2

A car traveling at a speed v skids a distance d after its brakes lock. Estimate how far it will skid if it is traveling at a speed $2v$ when its brakes lock.

Explanation Let us assume that the force of kinetic friction between car and road surface is constant and the same at both speeds. From the work-kinetic energy theorem, the net force exerted on the car times the displacement of the car is equal to its initial kinetic energy. (Watch your negative signs and remember the definition of negative work.) When the speed is doubled, the kinetic energy of the car is quadrupled. So for a given applied friction force, the distance traveled must increase fourfold when the initial speed is doubled. The estimated distance the car skids is $4d$.

Example 5.2   Towing a Car

A car of mass 1.40×10^3 kg being towed has a net forward force of 4.50×10^3 N applied to it as in Figure 5.9. The car starts from rest and travels down a horizontal highway. Find the speed of the car after it has traveled 100.0 m. (Ignore kinetic energy losses caused by friction and air resistance.)

FIGURE 5.9   (Example 5.2)

Reasoning   Three external forces act on the car. Neither the gravitational force nor the normal force does work on the car because these forces are vertical and the displacement is horizontal. Because we are ignoring friction and air resistance, the only external force we include in the calculation is the forward force of 4.50×10^3 N.

Solution   The work done by the horizontal forward force on the car is

$$W = F\Delta x = (4.50 \times 10^3 \text{ N})(100.0 \text{ m}) = 4.50 \times 10^5 \text{ J}$$

Using the work-kinetic energy theorem and noting that the initial kinetic energy of the car is zero, we find the final speed of the car as follows:

$$W = KE_f - KE_i = \tfrac{1}{2}mv_f^2 - 0$$

$$v_f = \sqrt{\frac{2W}{m}} = \sqrt{\frac{2(4.50 \times 10^3 \text{ J})}{1.40 \times 10^3 \text{ kg}}} = \boxed{25.4 \text{ m/s}}$$

Notice that an energy calculation such as this gives only the speed of the object, not the velocity. In many cases, this is all you need. If you need the direction of the velocity vector, you must either analyze the diagram or perform other calculations.

5.3   POTENTIAL ENERGY

In the preceding section we saw that an object that has kinetic energy (energy of motion) can do work on another object, as illustrated by the moving hammer driving the nail into the wall. Now we introduce **potential energy,** the form of mechanical energy associated with the position of an object within some system. We consider potential energy to be a property of the system rather than a property of the object. In this chapter, we define a **system** as a collection of objects or particles interacting via forces or processes that are internal to the system. We shall see that the potential energy associated with a system can also result in work being done on objects.

GRAVITATIONAL POTENTIAL ENERGY

The first systems we consider consist of Earth and small objects positioned near its surface. The objects interact with Earth through the gravitational force. (The objects' gravitational interaction with each other is negligibly small.) As an object falls freely in the gravitational field, the field exerts a force on it, doing positive work on it and thereby increasing its kinetic energy. Consider Figure 5.10, which shows a brick held at some height above a nail in a board lying on the ground. When the brick is released, it falls toward the ground, gaining speed and thereby gaining kinetic energy. As a result of the brick's position relative to the ground,

Brick

FIGURE 5.10   Because of its position in space, the brick can do work on the nail.

the potential energy associated with this system is converted to kinetic energy as the brick falls. When the brick reaches the nail, it does work on the nail, driving it into the board. The energy associated with the relative position of the brick in space near Earth's surface is called **gravitational potential energy.**

Let us now derive an expression for the gravitational potential energy at a given location above Earth's surface. Consider a new system in which a book of mass m falls from at an initial height y_i to a final height y_f above the ground, as in Figure 5.11, and let's ignore air resistance. The downward displacement has a magnitude of $\Delta y = y_i - y_f$. As the book falls, the only force that does work on it is the gravitational force, $m\mathbf{g}$. To find the work done by the gravitational force, we can use $W = (F \cos \theta)\, \Delta y$ with $F = mg$ and $\Delta y = y_i - y_f$. Because the force and displacement are in the same direction, $\theta = 0°$. Therefore,

$$W_g = (mg \cos 0°)(y_i - y_f) = mgy_i - mgy_f \qquad [5.7]$$

We define the quantity mgy to be the gravitational potential energy PE of the book-Earth system near Earth's surface:

$$PE \equiv mgy \qquad [5.8]$$

Thus, the gravitational potential energy associated with an object located above Earth's surface is the product of the object's weight and its vertical coordinate. This relationship is valid only for objects near Earth's surface, where g is approximately constant. Furthermore, the gravitational potential energy is actually the potential energy of the object-Earth system.

If we substitute this expression for PE into the expression for W_g, we have

$$W_g = PE_i - PE_f \qquad [5.9]$$

The work done on any object by the force of gravity is equal to the difference between the initial and final values of the gravitational potential energy.

The units of gravitational potential energy are the same as those of work and kinetic energy—joules in SI. Potential energy, like work and kinetic energy, is a scalar quantity.

REFERENCE LEVELS FOR GRAVITATIONAL POTENTIAL ENERGY

In solving problems involving gravitational potential energy, you must choose a location at which to set the gravitational potential energy equal to zero. The choice of zero level is completely arbitrary because the important quantity is the *difference* in potential energy, and this difference is independent of the choice of zero level.

It is often convenient to choose Earth's surface as the reference position for zero potential energy, but again this is not essential. Often the statement of a problem will suggest a convenient level to use. As an example, consider a book at several possible locations as in Figure 5.12. When the book is at Ⓐ, above the surface

FIGURE 5.11 The work done by the gravitational force as the book falls from y_i to y_f is equal to $mgy_i - mgy_f$.

 Gravitational potential energy

Tip 5.3 POTENTIAL ENERGY BELONGS TO A SYSTEM

Potential energy is always associated with a *system* of two or more interacting objects. One example is the gravitational potential energy of the Earth-Moon system.

FIGURE 5.12 Any reference level can be used for representing the zero gravitational potential energy for the book-Earth system.

of the desk, a logical zero level for potential energy would be the surface of the desk. When the book is at Ⓑ, however, the floor might be a more appropriate zero reference level. Finally, a location such as Ⓒ, where the book is held out a window, would suggest choosing Earth's surface as the zero level of potential energy. The choice, however, makes no difference—any of the reference levels mentioned could be used as the zero level, regardless of whether the book is at Ⓐ, Ⓑ, or Ⓒ. Example 5.3 illustrates this important point.

APPLYING *PHYSICS* 5.3

A common carnival scene is the Ring-the-Bell attraction, in which the player swings a hammer downward in an attempt to project an object upward to ring a bell. What technique should the player use to raise the object to its maximum height?

Explanation In physics terms, the goal is to provide the object with enough initial kinetic energy so that it will rise at least as high as the bell. This kinetic energy is converted to gravitational potential energy as the object moves upward. The initial kinetic energy of the object is related to how much work was done by the player in striking the device with the hammer. To maximize this work and, ultimately, the height of the object, the player must strike with as much force as possible and move the hammer through as much distance as possible while it is in contact with the device.

Example 5.3 Wax Your Skis

A 60.0-kg skier is at the top of a slope, as shown in Figure 5.13a. At the initial point Ⓐ, she is 10.0 m vertically above point Ⓑ.

10.0 m

(a)　　　　　　(b)

FIGURE 5.13 (Example 5.3) *(b, Gregg Adams/Stone/Getty Images)*

Ⓐ Setting the zero level for gravitational potential energy at Ⓑ, find the gravitational potential energy of this system when the skier is at Ⓐ and then at Ⓑ, and then find the difference in potential energy between these two configurations.

Solution   Call the gravitational potential energy PE_i when the skier is at Ⓐ and PE_f when she is at Ⓑ. PE_f is zero by choice. Hence, the potential energy when the skier is at Ⓐ is

$$PE_i = mgy_i = (60.0 \text{ kg})(9.80 \text{ m/s}^2)(10.0 \text{ m}) = 5.88 \times 10^3 \text{ J}$$

Because $PE_f = 0$, the difference in potential energy is

$$PE_i - PE_f = 5.88 \times 10^3 \text{ J} - 0 \text{ J} = \boxed{5.88 \times 10^3 \text{ J}}$$

B   Repeat this problem with the zero level at point Ⓐ.

Solution   In this case, PE_i is zero because of the chosen reference level. The final potential energy is

$$PE_f = mgy_f = (60.0 \text{ kg})(9.80 \text{ m/s}^2)(-10.0 \text{ m}) = -5.88 \times 10^3 \text{ J}$$

The position y_f is negative because the final point is 10.0 m below the zero reference level. The difference in potential energy is now

$$PE_i - PE_f = 0 \text{ J} - (-5.88 \times 10^3 \text{ J}) = \boxed{5.88 \times 10^3 \text{ J}}$$

These calculations show that the gravitational potential energy when the skier is at the top of the slope is 5 880 J greater than it is when she is at the bottom, *regardless of the zero level selected.*

EXERCISE   If the zero level for gravitational potential energy is selected to be midway down the slope, at a height of 5.00 m, find the initial potential energy, the final potential energy, and the difference in potential energy between points Ⓐ and Ⓑ, in Figure 5.13a.

ANSWER   2.94 kJ, −2.94 kJ, 5.88 kJ

5.4   CONSERVATIVE AND NONCONSERVATIVE FORCES

If you sled down a very slick hill, your speed, and hence your kinetic energy, increase dramatically. However, if the hill is not very icy, your speed and kinetic energy do not increase as rapidly. What has happened to this "lost" kinetic energy? To answer this question, we shall examine the properties of two categories of forces that exist in nature—conservative and nonconservative forces.

CONSERVATIVE FORCES

A force is conservative if the work it does on an object moving between two points is independent of the path the object takes between the points. In other words, the work done on an object by a conservative force depends only on the initial and final positions of the object.

◀ Definition of a conservative force

The force of gravity is conservative. As we learned in the preceding section, the work done by the gravitational force on an object moving between any two points near Earth's surface is $W_g = mgy_i - mgy_f$. From this we see that W_g depends only on the initial and final vertical coordinates of the object and hence is independent of path.

We can associate a potential energy function with any conservative force. In the preceding section, the potential energy function associated with the gravitational force was found to be $PE = mgy$. Potential energy functions can be defined only for conservative forces. Some other examples of conservative forces are the spring force and the electric force between charged objects. These conservative forces have different potential energy functions.

In general, the work W_c done on a moving object by a conservative force is equal to the initial value of the potential energy minus the final value. We can generalize Equation 5.9 as follows:

$$W_c = PE_i - PE_f \qquad \text{[5.9a]}$$

For example, when an object falls through some vertical displacement, the work done by the gravitational force on the object appears as a decrease in the gravitational potential energy. From everyday experience, we recognize that the decrease in potential energy is accompanied by an increase in the kinetic energy of the object. We shall study these situations in Section 5.5.

NONCONSERVATIVE FORCES

▶ Definition of a nonconservative force

A force is nonconservative if the work it does on an object depends on the path taken by the object between its final and starting points. Some common examples of nonconservative forces are kinetic friction, air drag, and propulsive forces, such as the force exerted by a jet engine on a plane and the force exerted by a propeller on a submarine.

To understand this more clearly, suppose you displace a book between two points on a table as in Figure 5.14. If the book is displaced in a straight line along the blue path between points Ⓐ and Ⓑ, you do a certain amount of work against the kinetic friction force to keep the book moving at a constant speed. Now, imagine that you push the book along the red semicircular path in Figure 5.14. You perform more work against friction along this longer path than along the straight path. The work done depends on the path, so the friction force is nonconservative.

What has happened to the kinetic energy of an initially sliding book that comes to a stop because of friction? The friction force has transformed the kinetic energy of the book into a type of energy associated with temperature—the book and table are warmer than they were before the book moved. We use the phrase **internal energy** for the energy associated with an object's temperature. We shall return to the concept of internal energy in Section 5.6.

In any realistic situation, both conservative and nonconservative forces act on an object at the same time. As we shall see in the next sections, the way to do energy calculations on such problems is to separate the total work done into the two categories: work done by conservative forces, W_c, and work done by nonconservative forces, W_{nc}.

FIGURE 5.14 The work done against the force of friction depends on the path taken as the book is moved from Ⓐ to Ⓑ. The work required is greater along the red path than the blue path. Hence, friction is a nonconservative force.

As this skillful rock climber moves cautiously up the mountain, her internal (biochemical) energy is transformed into gravitational potential energy. Can you identify other forms of energy transformations if she descends the mountain with the help of supporting ropes? *(P. H. Royer/ Explorer/Photo Researchers, Inc.)*

5.5   CONSERVATION OF MECHANICAL ENERGY

Conservation principles play a very important role in physics, and you will encounter a number of them as you continue in this course. Before we describe its mathematical details, let us examine what is meant by conservation. When we say that a physical quantity is *conserved,* we simply mean that the numeric value of the quantity remains constant. Although the form of the quantity may change in some manner, its final value is the same as its initial value. For example, the energy in an isolated system may change from gravitational potential energy to kinetic energy or to one of a variety of other forms, but the system never loses energy. The total quantity of energy in the system remains constant.

When an object is dropped, it falls to the floor; as it falls, its speed and thus its kinetic energy increase, while the potential energy of the object-Earth system decreases. If nonconservative forces such as air resistance are ignored, whatever potential energy is lost as the object moves downward appears as kinetic energy. In other words, the sum of the kinetic and potential energies—called the **total mechanical energy *E***—remains constant. This is an example of the principle of **conservation of mechanical energy.** For the case of an object in free fall, this principle tells us that any increase (or decrease) in potential energy is accompanied by an equal decrease (or increase) in kinetic energy. **In any isolated system of objects that interact only through conservative forces, the total mechanical energy of the system remains constant.**

Because the total mechanical energy $E \equiv KE + PE$, we can state the principle of conservation of mechanical energy as $E_i = E_f$, and so we have

$$KE_i + PE_i = KE_f + PE_f \qquad \text{[5.10]}$$

If the force of gravity is the *only* force doing work within a system, then the total mechanical energy of the system remains constant, and the principle of conservation of mechanical energy takes the form

$$\tfrac{1}{2}mv_i^2 + mgy_i = \tfrac{1}{2}mv_f^2 + mgy_f \qquad \text{[5.11]}$$

Because there are other forms of energy besides kinetic and gravitational potential energy, we will modify Equation 5.11 to include these new forms as we learn about them.

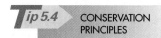 Conservation of mechanical energy

Webnote 5.2

A short but informative discussion of potential and kinetic energy in the context of a freely falling object may be found at:
http://theory.uwinnipeg.ca/mod_tech/node32.html

Tip 5.4 CONSERVATION PRINCIPLES

Equation 5.10 represents conservation of the mechanical energy of an isolated system. In later chapters, we will generate conservation laws for such quantities as momentum, angular momentum, and electric charge. Conservation principles essentially say that a conserved quantity that either changes in form or moves out of one part of an isolated system must appear in exactly the same amount in either another form or in another part of the system.

Quick Quiz 5.2

Three identical balls are thrown from the top of a building, all with the same initial speed. The first ball is thrown horizontally, the second at some angle above the horizontal, and the third at some angle below the horizontal as in Figure 5.15. Neglecting air resistance, rank the speeds of the balls as they reach the ground.

FIGURE 5.15 (Quick Quiz 5.2) Three identical balls are thrown with the same initial speed from the top of a building.

Quick Quiz 5.3

Bobby, of mass m, drops from a tree limb at the same time that Sally, also of mass m, begins her descent down a frictionless slide. The slide is in the shape of a quadrant of a circle. If they both start at the same height above the ground, which of the following is true about their kinetic energies as they reach the ground?
 (a) The kinetic energy of Bobby is greater than that of Sally.
 (b) The kinetic energy of Sally is greater than that of Bobby.
 (c) They have the same kinetic energy.
 (d) It is impossible to say from the information given.

Webnote 5.3

You can practice designing a roller coaster at:
http://www.learner.org/exhibits/ parkphysics/coaster.html

APPLYING PHYSICS 5.4

You have graduated from college and now design roller coasters for a living. Your boss asks you to design a roller coaster that is supposed to release a car from rest at the top of a hill of height h and then roll freely down the hill and reach the peak of the next hill of height $1.1h$. What would you tell your boss in this situation?

Explanation   You should politely tell your boss to study some physics. It is impossible to design such a roller coaster because it would violate the principle of conservation of mechanical energy. At the starting point, the roller coaster has no kinetic energy and the Earth-coaster system has gravitational potential energy mgh relative to ground level. If the roller coaster were able to reach the top of the second hill, the system would have gravitational potential energy $1.1\, mgh$, which is greater than the initial potential energy. If this roller coaster were actually built, the car would move up the second hill to a height h (neglecting friction), stop short of the peak, and then start rolling backward, becoming trapped between the two peaks.

PROBLEM-SOLVING STRATEGY	Applying Conservation of Mechanical Energy

Take the following steps in applying the principle of conservation of mechanical energy:

1. Define your system.
2. Select a location for the zero of gravitational potential energy and *do not change* this location while solving the problem.
3. Determine whether or not nonconservative forces are present.
4. If mechanical energy is conserved (that is, if only conservative forces are present), you can write the total initial mechanical energy of the system $KE_i + PE_i$ at some point. Then, write an expression for the total final energy of the system $KE_f + PE_f$ at the final point of interest. Because mechanical energy is conserved, you can equate the two total energies and solve for the unknown.

Example 5.4   The Daring Diver

A diver of mass m drops from a board 10.0 m above the water surface, as in Figure 5.16.

A   Use conservation of mechanical energy to find his speed 5.00 m above the water surface. Neglect air resistance.

Reasoning   The system consists of the diver and Earth. As the diver falls, only one force acts on him: the force of gravity. Therefore, only the force of gravity does any work on him, and the mechanical energy of the system is conserved. In order to find the diver's speed at the 5.00-m mark, we will choose the water surface as the zero level for potential energy. Also, note that the diver drops from the board; that is, he leaves with zero velocity and therefore zero kinetic energy.

Solution   Because mechanical energy is conserved and $v_i = 0$, Equation 5.11 gives

$$\tfrac{1}{2}mv_i^2 + mgy_i = \tfrac{1}{2}mv_f^2 + mgy_f$$

$$0 + mgy_i = \tfrac{1}{2}mv_f^2 + mgy_f$$

$$v_f = \sqrt{2g(y_i - y_f)}$$

$$= \sqrt{2(9.80 \text{ m/s}^2)(10.0 \text{ m} - 5.00 \text{ m})} = \boxed{9.90 \text{ m/s}}$$

Note that the result does not depend on the mass of the diver.

10.0 m

$KE_i = 0$
$PE_i = mgy_i$

5.00 m

$KE_f = \tfrac{1}{2}mv_f^2$
$PE_f = 0$

0

FIGURE 5.16   (Example 5.4) The zero of gravitational potential energy is taken to be when the diver is at the water surface.

B Find the speed of the diver just before he strikes the water.

Solution   With the final position of the diver at the surface of the pool, $y_f = 0$ and Equation 5.11 gives

$$0 + mgy_i = \tfrac{1}{2}mv_f^2 + 0$$
$$v_f = \sqrt{2gy_i} = \sqrt{2(9.80 \text{ m/s}^2)(10.0 \text{ m})} = \boxed{14.0 \text{ m/s}}$$

EXERCISE   If the diver pushes off so that he leaves the board with an initial speed of 2.00 m/s, find his speed when he strikes the water. Use conservation of mechanical energy.

ANSWER   14.1 m/s

Example 5.5   Sliding Down a Frictionless Hill

A sled and its rider together weigh 800 N. They move down a frictionless hill through a vertical distance of 10.0 m, as shown in Figure 5.17a. Use conservation of mechanical energy to find the speed of the sled at the bottom of the hill, assuming the rider pushes off with an initial speed of 5.00 m/s. Neglect air resistance.

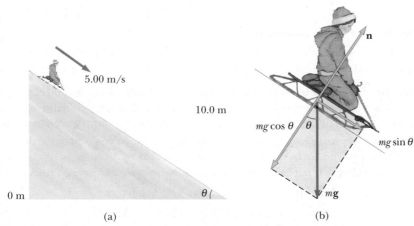

(a)   (b)

FIGURE 5.17   (Example 5.5) (a) A sled and rider start from the top of a frictionless hill with a speed of 5.00 m/s. (b) A free-body diagram for the sled plus rider.

Reasoning   The forces acting on the sled and rider as they move down the hill are shown in Figure 5.17b. In the absence of a frictional force, the only forces acting are the normal force **n** and the gravitational force. At all points along the path, **n** is perpendicular to the direction of travel and hence does no work. Likewise, the component of weight perpendicular to the incline ($mg \cos \theta$) does no work. The only force that does any work is the component of the gravitational force, $mg \sin \theta$, along the slope of the hill. As a result, we are justified in using Equation 5.11.

Solution   The initial energy of the sled-rider-Earth system includes kinetic energy because of the initial speed:

$$\tfrac{1}{2}mv_i^2 + mgy_i = \tfrac{1}{2}mv_f^2 + mgy_f$$

or, after canceling m throughout the equation,

$$\tfrac{1}{2}v_i^2 + gy_i = \tfrac{1}{2}v_f^2 + gy_f$$

If we set the origin of our coordinates at the bottom of the incline, the initial and final y

coordinates of the sled are $y_i = 10.0$ m and $y_f = 0$. Thus, we get

$$\tfrac{1}{2}v_i^2 + gy_i = \tfrac{1}{2}v_f^2 + 0$$

$$v_f^2 = v_i^2 + 2gy_i$$

$$= (5.00 \text{ m/s})^2 + 2(9.80 \text{ m/s}^2)(10.0 \text{ m})$$

$$v_f = \boxed{14.9 \text{ m/s}}$$

Notice that this solution does not depend on the shape of the hill. The only things we need to know are the rider's initial speed and the vertical distance traveled. If the hill had a varying curvature with bumps and troughs, the conservation of energy approach would yield the same result. If you try to solve the problem using Newton's laws, the varying slope of the hill would cause the acceleration to vary along the hill, making this approach very complicated. As you progress further with your study of physics, you will find many problems that are difficult or impossible to solve using Newton's laws, whereas conservation of energy provides a simple, yet powerful, alternative.

EXERCISE If the sled and rider start at the bottom of the incline and are given an initial speed of 5.00 m/s up the incline, how high will they rise vertically, and what will their speed be when they return to the bottom of the hill?

ANSWER 1.28 m; 5.00 m/s

Example 5.6 The Jumping Bug

A powerful grasshopper launches itself at an angle of 45° above the horizontal and rises vertically about 1.0 m. With what speed v does it leave the ground? Neglect air resistance.

Solution Because we are ignoring the nonconservative force of air resistance, we can use the principle of conservation of mechanical energy. We sketch the initial state of the grasshopper as it takes off and its final state when it is at a height of 1.0 m and its vertical speed is zero, as shown in Figure 5.18. The initial energy of the grasshopper-Earth system is all kinetic energy. We take the ground as the zero level for the gravitational potential energy. The final energy is partly kinetic energy and partly potential energy. Because mechanical energy is conserved,

$$KE_i + PE_i = KE_f + PE_f$$

$$\tfrac{1}{2}mv^2 + 0 = \tfrac{1}{2}mv_x^2 + mgh$$

$$v^2 = v_x^2 + 2gh \tag{1}$$

We can eliminate v_x from (1) by noting that initially $v^2 = v_x^2 + v_y^2$ and that $v_x = v_y$ for 45°; these conditions yield $v^2 = 2v_x^2$ or $v_x^2 = v^2/2$. Substituting this result into (1) gives

$$v = 2\sqrt{gh}$$

FIGURE 5.18
(Example 5.6)

Inserting numbers we find a surprisingly high value!

$$v = 2\sqrt{(9.8 \text{ m/s}^2)(1.0 \text{ m})} = 6.3 \text{ m/s}$$

Where does the grasshopper get the energy to acquire this large initial speed?

POTENTIAL ENERGY STORED IN A SPRING

The concept of potential energy is of tremendous value in describing certain types of mechanical motion. One of these, the motion of an object attached to a stretched or compressed spring, is discussed in more detail in Chapter 13, but it is instructive to examine it here.

Consider Figure 5.19a, which shows a spring in its equilibrium position–that is, the spring is neither compressed nor stretched. If we push a block against the spring as in Figure 5.19b, the spring is compressed a distance x. In order to compress the spring, we must exert on the block a force of magnitude $F = kx$, where k (the **spring constant** or **force constant**) is a constant for a particular spring. Because the block is in equilibrium in Figure 5.19b, the spring must be exerting a force on it given by

$$F_s = -kx \qquad\qquad [5.12]$$

This equation for F_s, describing the force exerted by a stretched or compressed spring, is also called **Hooke's law** after Sir Robert Hooke, who discovered the relationship between F_s and x. The force F_s is often called a *restoring force* because the spring exerts a force whose direction is opposite the displacement of its end. For a flexible spring, k is a small number (about 100 N/m), whereas for a stiff spring, k is large (about 10 000 N/m). The value of the spring constant k is determined by how the spring was formed, its material composition, and the thickness of the

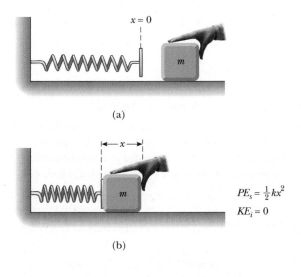

(a)

$PE_s = \frac{1}{2}kx^2$
$KE_i = 0$

(b)

$PE_s = 0$
$KE_f = \frac{1}{2}mv^2$

(c)

FIGURE 5.19 (a) A spring at equilibrium–neither compressed nor stretched. (b) A block of mass m on a frictionless surface is pushed against the spring. If x is the compression in the spring, the potential energy stored in the spring is $\frac{1}{2}kx^2$. (c) When the block is released, this energy is transferred to the block in the form of kinetic energy.

spring wire. The equation $F = kx$ says that the more we compress the spring, the larger the force we must exert.

To find an expression for the potential energy associated with the spring force, called the **elastic potential energy**, let us determine the work required to compress a spring from its equilibrium position to some final arbitrary position x. In this situation, we cannot use the expression $W = Fx$ to calculate the work done on the spring because the force is not constant as the spring is being compressed. The force that must be applied to compress the spring varies from $F = 0$ to $F = kx$ at maximum compression. Therefore, we use the average force $\overline{F}$ during the displacement. Because the force increases linearly with position (that is, $F \propto x$), the average force that must be applied is

$$\overline{F} = \frac{F_0 + F_x}{2} = \frac{0 + kx}{2} = \tfrac{1}{2}kx$$

Therefore, the work *done by the applied force* is

$$W = \overline{F}x = \tfrac{1}{2}kx^2$$

This work is stored in the compressed spring as elastic potential energy. Thus, we define the elastic potential energy PE_s associated with the spring force as

$$PE_s = \tfrac{1}{2}kx^2 \qquad\qquad\qquad \textbf{[5.13]}$$

Figure 5.19c shows how this stored elastic potential energy can be recovered. When the block is released, the spring snaps back to its original length and the stored elastic potential energy is converted to kinetic energy of the block. The elastic potential energy stored in the spring is zero when the spring is in equilibrium ($x = 0$). Note that potential energy as given by Equation 5.13 is also stored in the spring when it is stretched. Furthermore, the elastic potential energy is a maximum when the spring has reached its maximum compression or extension. Finally, the potential energy is always positive when the spring is not at the equilibrium position, because PE_s is proportional to x^2.

This new form of energy is included as another term in our equation for conservation of mechanical energy:

$$(KE + PE_g + PE_s)_i = (KE + PE_g + PE_s)_f$$

where PE_g is the gravitational potential energy.

Webnote 5.4

For an animation of an object on a hanging spring and accompanying discussion of conservation of energy, see:
http://id.mind.net/~zona/mstm/physics/ mechanics/energy/massOnASpring/ massOnASpring.html

Example 5.7   A Block Projected Up a Frictionless Incline

A 0.50-kg block rests on a horizontal, frictionless surface as in Figure 5.20. The block is pressed against a light spring having a spring constant of $k = 80.0$ N/m. The spring is compressed a distance of 2.0 cm to point Ⓐ and released. Consider the system to be the spring, block, and Earth and find (a) the speed of the block when it is at the bottom of the incline, position Ⓑ.

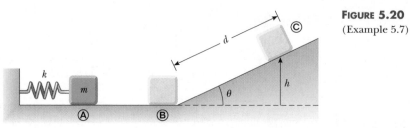

FIGURE 5.20
(Example 5.7)

Ⓐ **Reasoning**   Applied to positions Ⓐ and Ⓑ, our expression for the conservation of mechanical energy of the system is

$$(KE + PE_g + PE_s)_A = (KE + PE_g + PE_s)_B$$

The initial kinetic energy of the block at Ⓐ is zero, and because the block and spring both remain at the same level throughout the motion, the gravitational potential energy terms on the two sides of the equation are equal. PE_{sB} is zero because after the block leaves the spring, the spring is no longer compressed. Thus,

$$PE_{sA} = KE_B$$

Solution From the last expression we have

$$\tfrac{1}{2}kx^2 = \tfrac{1}{2}mv_B^{\,2}$$

$$v_B = \sqrt{\frac{k}{m}}\,x = \sqrt{\frac{80.0\text{ N/m}}{0.50\text{ kg}}}\,(2.0 \times 10^{-2}\text{ m}) = \boxed{0.25\text{ m/s}}$$

B Find the maximum distance d the block travels up the frictionless incline if the incline angle θ is 25°.

Solution Conservation of mechanical energy of our system applied to positions Ⓑ and Ⓒ gives

$$(KE + PE_g + PE_s)_B = (KE + PE_g + PE_s)_C$$

The kinetic energy is zero at Ⓒ when the block stops at its highest position. The potential energy stored in the spring is zero at both Ⓑ and Ⓒ. The gravitational potential energy is zero at Ⓑ and mgh at Ⓒ if we choose the reference level at the level of the horizontal surface. Thus, we have

$$\tfrac{1}{2}mv_B^{\,2} = mgh = mgd \sin \theta$$

or

$$d = \frac{v_B^{\,2}}{2g \sin \theta} = \frac{(0.25\text{ m/s})^2}{2(9.8\text{ m/s}^2)\sin(25°)} = \boxed{7.5 \times 10^{-3}\text{ m}}$$

Example 5.8 Dropping a Block onto a Spring

A spring that has a force constant of 1.00×10^3 N/m is placed on a table in a vertical position as in Figure 5.21. A block of mass 1.60 kg is held 1.00 m above the free end of the spring. The block is dropped from rest so that it falls vertically onto the spring. By what distance does the spring compress?

FIGURE 5.21 (Example 5.8) An object is dropped onto a vertical spring, causing it to compress.

Solution Once again, consider the system to be the spring, block, and Earth. Conservation of mechanical energy applied to the system at positions Ⓐ and Ⓑ gives

$$(KE + PE_g + PE_s)_A = (KE + PE_g + PE_s)_B \qquad (1)$$

The initial kinetic energy is zero because the block is released from rest. The final kinetic energy is also zero because the block comes to a stop when the spring is fully compressed. The gravitational potential energy when the block is at position Ⓐ is $mg(h + d)$, where d is the distance by which the spring is compressed. Due to our chosen reference level of gravitational potential energy, this quantity is zero when the block is at position

Ⓑ. The potential energy stored in the spring at position Ⓐ is zero, while the potential energy stored in the spring at position Ⓑ is $\frac{1}{2}kd^2$. Therefore (1) gives

$$0 + mg(h + d) + 0 = 0 + 0 + \tfrac{1}{2}kd^2$$

$$mg(h + d) = \tfrac{1}{2}kd^2$$

$$(1.60 \text{ kg})(9.80 \text{ m/s}^2)(1.00 \text{ m} + d) = \tfrac{1}{2}(1.00 \times 10^3 \text{ N/m})d^2$$

$$(-5.00 \times 10^2)d^2 + 15.7d + 15.7 = 0$$

This quadratic equation can be solved (use the quadratic formula, Equation A.8); the solutions are $d = 0.19$ m and $d = -0.16$ m. Because we have chosen the value of d as a positive number, by claiming that $y = -d$ is *below* the initial position of the end of the spring, we need to choose the positive root, $d = 0.19$ m. (The negative root gives the position for the final condition as $y = -d = -(-0.16 \text{ m}) = +0.16$ m; this is the position above the *initial* $y = 0$ at which the spring again comes to rest in its oscillation, assuming that the block remains attached to the spring.)

5.6   NONCONSERVATIVE FORCES, NONISOLATED SYSTEMS, AND CONSERVATION OF ENERGY

In realistic situations, nonconservative forces such as friction act on parts of a system. In such situations, the total mechanical energy of the system is not constant, and we cannot apply Equation 5.10. In order to account for friction forces, let us return to the work-kinetic energy theorem:

$$W_{\text{net}} = \tfrac{1}{2}mv_f^2 - \tfrac{1}{2}mv_i^2$$

We will separate the net work into two parts, that done by the nonconservative forces W_{nc} and that done by the conservative forces W_c. The work-kinetic energy relationship then becomes

$$W_{nc} + W_c = \tfrac{1}{2}mv_f^2 - \tfrac{1}{2}mv_i^2 \qquad \text{[5.14]}$$

Solving for W_{nc} and substituting the expression for W_c provided by Equation 5.9a, we have

$$W_{nc} = \tfrac{1}{2}mv_f^2 - \tfrac{1}{2}mv_i^2 - (PE_i - PE_f)$$

$$W_{nc} = (KE_f + PE_f) - (KE_i + PE_i) \qquad \text{[5.15]}$$

The work done by all nonconservative forces acting on parts of a system equals the change in mechanical energy of the system. This means that energy either crosses the boundary of the system and enters the remainder of the Universe (so that the system is not isolated) or that energy is transformed within the system into a form we have not yet accounted for.

A simple nonisolated system might be an object acted on by a net external force. We covered this in Section 5.2 (see Equation 5.6). If the work-kinetic energy theorem is applied to a nonisolated system, it tells us that the work done by external forces changes the system's kinetic energy. In this case, energy is transferred across the system-environment boundary. If positive work is done on the system, energy is transferred from the environment to the system. If negative work is done on the system, energy is transferred from the system to the environment.

Friction is an example of a nonconservative force that can transform kinetic energy into internal energy, as mentioned in Section 5.4. If a book slides across a horizontal table (so that the gravitational potential energy does not change) and comes to a stop, kinetic energy has decreased but the book and table surfaces are warmer.

We have now seen three methods of storing energy in a system—kinetic energy, potential energy, and internal energy. We have seen only one way to transfer energy into the system so far—work. We shall introduce other ways to transfer energy into or out of a system in later chapters. We illustrate these in Figure 5.22 and summarize them as follows:

- **Work,** in the mechanical sense as we have learned in this chapter, is the process of transferring energy to a system through the displacement of the system by an applied force.
- **Heat** is the process of transferring energy by means of microscopic collisions between atoms or molecules. For example, a metal spoon resting in a cup of coffee becomes hot because some of the kinetic energy of molecules in the liquid coffee is transferred to the spoon as internal energy.
- **Mechanical waves** are a means of transferring energy by allowing a disturbance to propagate through air or another medium. This is the method by which energy (which you detect as sound) leaves your stereo system through the loudspeakers and enters your ears to stimulate the hearing process. Other examples of mechanical waves are seismic waves and ocean waves.

FIGURE 5.22 Energy-transfer mechanisms. (a) Energy is transferred from the environment to the block by *work*. (b) Energy transfers up the handle of the spoon by *heat*. (c) Energy leaves the radio by *mechanical waves*. (d) Energy enters the hair dryer by *electrical transmission*. (e) Energy leaves the light bulb by *electromagnetic radiation*. (*George Semple*)

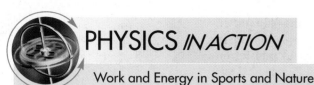

PHYSICS *IN ACTION*

Work and Energy in Sports and Nature

On the left, these cyclists are working hard and expending energy as they pedal uphill in Marin County, California.

On the right, Twin Falls on the Island of Kauai, Hawaii. The potential energy of the water-Earth system at the top of the falls is converted to kinetic energy at the bottom. In many locations, this mechanical energy is used to produce electrical energy.

(David Madison/Stone/Getty Images)

(Bruce Byers/FPG/Getty Images)

- **Electrical transmission** involves energy transfer by means of electric currents. This is how energy enters your stereo system or any other electrical device.
- **Electromagnetic radiation** refers to electromagnetic waves such as light, microwaves, and radio waves. Examples of this method of transfer include cooking your baked potato in your microwave oven and light energy traveling from the Sun to Earth through space.

CONSERVATION OF ENERGY IN GENERAL

▶ Total energy is always conserved

One of the central features of the energy approach is the notion that **we can neither create nor destroy energy—energy is always *conserved*. Thus, if the total amount of energy in a system changes, it can *only* be due to the fact that energy has crossed the boundary of the system by a transfer mechanism such as one of the methods listed above.** This is a general statement of the principle of **conservation of energy.**

The principle of conservation of energy is not solely confined to physics. In biology, energy transformations take place in myriad ways inside all living organisms. One example is the transformation of chemical energy to mechanical energy that causes flagella to move and propel an organism. Some bacteria use chemical energy to produce light (see Fig. 5.23). Although the mechanisms that produce these light emissions are not well understood, living creatures often rely on this light for their existence. For example, certain fish have sacs beneath their eyes filled with

APPLICATION

FLAGELLAR MOVEMENT;
BIOLUMINESCENCE

light-emitting bacteria. The emitted light attracts other smaller creatures that become food for the fish.

A book of mass m is projected with a speed v across a horizontal surface. The book slides until it stops due to the friction force between the book and the surface. The surface is now tilted 30°, and the book is projected up the surface with the same initial speed v. When the book has come to rest, how does the decrease in mechanical energy of the book-Earth system compare to that when the book slid over the horizontal surface? (a) It is the same. (b) It is larger. (c) It is smaller. (d) The relationship cannot be determined.

FIGURE 5.23 This small plant, found in warm southern waters, exhibits bioluminescence, a process in which chemical energy is converted to light. The red areas are chlorophyll, which glows when excited by blue light. (*Jan Hinsch/Science Photo Library/Photo Researchers, Inc.*)

APPLYING PHYSICS 5.5

An automobile with kinetic energy plus potential energy carried within the gasoline strikes a tree and comes to rest. The total energy of the system (the automobile) is now less than before. How did the energy leave the system, and what is the form of the energy remaining in the system?

Explanation The energy left the system in several ways. The automobile did some *work* on the tree during the collision, causing the tree to deform. There was a large crash during the collision, representing transfer of energy by *sound*. After the automobile came to rest, there was no more kinetic energy. *Potential energy* is left in any gas remaining in the tank. There is also more *internal energy* in the automobile, as its temperature is likely to be higher after the collision than before.

PROBLEM-SOLVING STRATEGY | Conservation of Energy Revisited

If nonconservative forces such as friction act within a system, mechanical energy is not conserved. Hence, the problem-solving strategy given immediately before Example 5.4 must be modified as follows. First, write expressions for the total initial and total final mechanical energies. In this case, the difference between the two total energies is equal to the work done by the nonconservative force(s). These steps are summarized in Equation 5.15.

Example 5.9 Fun on the Water Slide

A child of mass $m = 20.0$ kg takes a ride on an irregularly curved water slide of height 2.00 m, as in Figure 5.24. He starts from rest at the top.

A Determine the speed of the child at the bottom, assuming no friction is present.

Reasoning The normal force does no work on the child because this force is always perpendicular to the displacement. Furthermore, because there is no friction, $W_{nc} = 0$ and we can apply the principle of conservation of mechanical energy for the child-Earth system.

FIGURE 5.24 (Example 5.9) If the slide is frictionless, the speed of the child at the bottom depends only on the height of the slide.

Solution If we measure the y coordinate for the child from the bottom of the slide, then $y_i = h = 2.00$ m, $y_f = 0$, and we have for the system,

$$KE_i + PE_i = KE_f + PE_f$$

$$0 + mgh = \tfrac{1}{2}mv_f^2 + 0$$

$$v_f = \sqrt{2gh}$$

$$= \sqrt{2(9.80 \text{ m/s}^2)(2.00 \text{ m})} = \boxed{6.26 \text{ m/s}}$$

Note that this speed is the same as if the child fell vertically through a distance $h = 2.00$ m!

B If a friction force acts on the child, and he arrives at the bottom of the slide with a speed $v_f = 3.00$ m/s, by how much does the mechanical energy of the system decrease due to this force?

Solution We define the system as the child, Earth, and the slide. In this case, a nonconservative force acts within the system, and therefore mechanical energy is *not* conserved. We can find the change in mechanical energy ΔE due to friction, given that the final speed at the bottom is known:

$$\Delta E = KE_f + PE_f - KE_i - PE_i = \tfrac{1}{2}mv_f^2 + 0 - 0 - mgh$$

$$= \tfrac{1}{2}(20.0 \text{ kg})(3.00 \text{ m/s})^2 - (20.0 \text{ kg})(9.80 \text{ m/s}^2)(2.00 \text{ m})$$

$$= \boxed{-302 \text{ J}}$$

The change in mechanical energy is negative because friction reduces the mechanical energy of the system. The "lost" mechanical energy appears as an increase in internal energy in the system, $+302$ J.

Example 5.10 Let's Go Skiing

A skier starts from rest at the top of a frictionless incline of height 20.0 m as in Figure 5.25. At the bottom of the incline, she encounters a horizontal surface where the coefficient of kinetic friction between skis and snow is 0.210. How far does she travel on the horizontal surface before coming to rest?

FIGURE 5.25 (Example 5.10) The skier slides down the slope and onto a level surface, stopping after a distance d from the bottom of the hill.

Solution First, let's calculate her speed at the bottom of the incline, which we choose as our zero point of gravitational potential energy. Because the incline is frictionless, the mechanical energy of the skier-Earth system remains constant, and we find, as we did in the previous example,

$$v_\text{B} = \sqrt{2gh} = \sqrt{2(9.80 \text{ m/s}^2)(20.0 \text{ m})} = 19.8 \text{ m/s}$$

Now we apply Equation 5.15 as the skier moves along the rough horizontal surface from Ⓑ to Ⓒ. Note that the work done by the nonconservative force of friction is $W_{nc} = -f_k d$, so Equation 5.15 gives

$$-f_k d = \tfrac{1}{2}mv_\text{C}^2 - \tfrac{1}{2}mv_\text{B}^2$$

where d is the horizontal distance the skier travels before coming to rest. Because $v_\text{B} = 19.8$ m/s, $v_\text{C} = 0$, and the friction force is given by $f_k = \mu_k n = \mu_k mg$, we get

$$-\mu_k mgd = -\tfrac{1}{2}mv_\text{B}^2$$

$$d = \frac{v_\text{B}^2}{2\mu_k g} = \frac{(19.8 \text{ m/s})^2}{2(0.210)(9.80 \text{ m/s}^2)} = \boxed{95.2 \text{ m}}$$

EXERCISE Find the horizontal distance the skier travels before coming to rest if the incline also has a coefficient of kinetic friction equal to 0.210.

ANSWER 40.3 m

5.7 POWER

From a practical viewpoint, it is interesting to know not only the amount of energy transferred to or from a system, but also the rate at which the transfer occurred. This is a particularly interesting issue for living creatures because the maximum work per second, or power output, of an animal varies greatly with output duration. To handle such problems in a systematic way, **power is defined as the time rate of energy transfer.**

If an external force is applied to an object and if the work done by this force is W in the time interval Δt, then the **average power** during this interval is defined as the ratio of the work done to the time interval:

$$\overline{\mathcal{P}} \equiv \frac{W}{\Delta t} \qquad\qquad [5.16] \qquad \blacktriangleleft \text{ Average power}$$

It is sometimes useful to rewrite Equation 5.16 by substituting $W = F\,\Delta x$ and noting that $\Delta x/\Delta t$ is the average speed of the object during the time Δt:

$$\overline{\mathcal{P}} = \frac{W}{\Delta t} = \frac{F\,\Delta x}{\Delta t} = F\overline{v} \qquad\qquad [5.17]$$

That is, the average power delivered to an object is equal to the product of the force acting on the object during some time interval and the object's average speed during this time interval. In Equation 5.17, F is the component of force in the direction of the average velocity.

A result of the same form as Equation 5.17 is obtained for instantaneous values. That is, the instantaneous power delivered to an object is equal to the product of the force on the object at that instant and the instantaneous speed, or $\mathcal{P} = Fv$.

The SI units of power are joules per second, which are also called **watts** (W) after James Watt:

$$1 \text{ W} = 1 \text{ J/s} = 1 \text{ kg} \cdot \text{m}^2/\text{s}^3 \qquad\qquad [5.18]$$

The unit of power in the U.S. customary system is the horsepower (hp), where

$$1 \text{ hp} \equiv 550 \, \frac{\text{ft} \cdot \text{lb}}{\text{s}} = 746 \text{ W} \qquad \text{[5.19]}$$

The horsepower was first defined by Watt, who needed a large power unit to rate the power output of his new invention, the steam engine.

The watt is commonly used in electrical applications, but it can be used in other scientific areas as well. For example, European sports-car engines are now rated in kilowatts.

We now define units of energy (or work) in terms of the unit of power, the watt. One kilowatt-hour (kWh) is the energy transferred in 1 h at the constant rate of 1 kW = 1 000 J/s. Therefore,

$$1 \text{ kWh} = (10^3 \text{ W})(3 \, 600 \text{ s}) = (10^3 \text{ J/s})(3 \, 600 \text{ s}) = 3.60 \times 10^6 \text{ J}$$

It is important to realize that a kilowatt-hour is a unit of energy, *not* power. When you pay your electric bill, you are buying energy, and that's why your bill lists a charge for electricity (no pun intended) of about 10 cents/kWh. The amount of electricity used by an appliance can be calculated by multiplying its power rating (usually expressed in watts and valid only for normal household electrical circuits) by the length of time the appliance is operated. For example, an electric bulb rated at 100 W (= 0.100 kW) "consumes" 3.6×10^5 J of energy in 1 h.

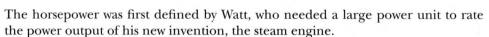

Tip 5.5   WATTS THE DIFFERENCE?

Do not confuse the nonitalic symbol W for the unit watt with the italic symbol *W* for work. Also, remember that the watt represents a rate of energy transfer—the watt is *the same as* a joule per second.

APPLYING **PHYSICS** 5.6

A lightbulb is described by some individuals as "having 60 watts." What is wrong with this phrase?

Reasoning   The number represents the power of the lightbulb, which is the rate at which energy passes through it, entering as electrical transmission from the power source. It is not a value that the lightbulb "possesses," however. It is the approximate power rating of the bulb when it is connected to a normal 120-volt household supply of electricity. If the bulb were connected to a 12-volt supply, energy would enter the bulb at a rate other than 60 watts. Thus, the 60 watts is not an intrinsic property of the bulb, such as density is for a given solid material.

Example 5.11   **Power Delivered by an Elevator Motor**

A 1 000-kg elevator carries a maximum load of 800 kg. A constant frictional force of 4 000 N retards its motion upward, as in Figure 5.26. What minimum power, in kilowatts

FIGURE 5.26   (Example 5.11) The motor exerts an upward force **T** on the elevator. A frictional force **f** and the force of gravity *M***g** act downward.

and in horsepower, must the motor deliver to lift the fully loaded elevator at a constant speed of 3.00 m/s?

Solution   The motor must supply the force **T** that pulls the elevator upward. From Newton's second law and from the fact that **a** = 0 because **v** is constant, we get

$$T - f - Mg = 0$$

where M is the total mass (elevator plus load), equal to 1800 kg. Therefore,

$$
\begin{aligned}
T &= f + Mg \\
&= 4.00 \times 10^3 \, \text{N} + (1.80 \times 10^3 \, \text{kg})(9.80 \, \text{m/s}^2) \\
&= 2.16 \times 10^4 \, \text{N}
\end{aligned}
$$

From $\mathcal{P} = Fv$ and the fact that **T** is in the same direction as **v**, we have

$$
\begin{aligned}
\mathcal{P} &= Tv \\
&= (2.16 \times 10^4 \, \text{N})(3.00 \, \text{m/s}) = 6.48 \times 10^4 \, \text{W} \\
&= \boxed{64.8 \, \text{kW} = 86.9 \, \text{hp}}
\end{aligned}
$$

ENERGY AND POWER IN A VERTICAL JUMP

The stationary jump consists of two parts, extension and free flight.[2] In the extension phase the person jumps up from a crouch, straightening the legs and throwing up the arms; the free-flight phase occurs when the jumper leaves the ground. Because the body is an extended object and different parts move with different speeds, we describe the motion of the jumper in terms of the position and velocity of the **center of mass (CM),** which is the point in the body at which all the mass may be considered to be concentrated. Figure 5.27 shows the position and velocity of the CM at different stages of the jump.

Using the principle of the conservation of mechanical energy, we can find H, the maximum increase in height of the CM, in terms of the velocity v_{CM} of the CM at liftoff. Taking PE_i, the gravitational potential energy of the jumper-Earth system just as the jumper lifts off from the ground to be zero, and noting that the kinetic

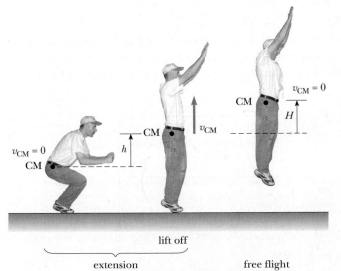

FIGURE 5.27   Extension and free flight in the vertical jump.

[2] For more information on this topic, see E. J. Offenbacher, *American Journal of Physics,* **38**, 829 (1969).

energy KE_f of the jumper at the peak is zero, we have

$$PE_i + KE_i = PE_f + KE_f$$

$$\tfrac{1}{2}mv_{CM}^2 = mgH \qquad \text{or} \qquad H = \frac{v_{CM}^2}{2g}$$

We can estimate v_{CM} by assuming that the acceleration of the CM is constant during the extension phase. If the depth of the crouch is h and the time for extension is Δt, we find $v_{CM} = 2\bar{v} = 2h/\Delta t$. Measurements on a group of male college students show typical values of $h = 0.40$ m and $\Delta t = 0.25$ s, the latter value being set by the fixed speed with which muscle can contract. Substituting, we find $v_{CM} = 2(0.40 \text{ m})/(0.25 \text{ s}) = 3.2$ m/s and

$$H = \frac{v_{CM}^2}{2g} = \frac{(3.2 \text{ m/s})^2}{2(9.80 \text{ m/s}^2)} = 0.52 \text{ m}$$

Measurements on this same group of students found that H was between 0.45 m and 0.61 m in all cases, confirming the basic validity of our simple calculation.

In order to relate the physical concepts of energy, power, and efficiency to humans, it is interesting to calculate these values for the vertical jump. The kinetic energy given to the body in a jump is $KE = \tfrac{1}{2}mv_{CM}^2$ and for a person of mass 68 kg, the kinetic energy is

$$KE = \tfrac{1}{2}(68 \text{ kg})(3.2 \text{ m/s})^2 = 3.5 \times 10^2 \text{ J}$$

Although this may seem like a large expenditure of energy, we can make a simple calculation to show that jumping and exercise in general, are not good ways to lose weight, in spite of their many health benefits. Since the muscles are at most 25% efficient at producing work from chemical energy (muscles always warm up a person as well as do work—that's why you sweat when you work out), they use up four times the 350 J (or 1 400 J) of chemical energy in one jump. This chemical energy ultimately comes from the food we eat, whose energy content is given in units of food calories, with 1 food calorie equal to 4 200 J. So the total energy supplied by the body in a vertical jump is only about $\tfrac{1}{3}$ of a food calorie! You are a lot better off not eating that piece of cheesecake than trying to work it off by jumping.

Finally, it is interesting to calculate the mechanical power that can be generated by the body in strenuous activity for brief periods. Here we find

$$\mathcal{P} = \frac{KE}{\Delta t} = \frac{3.5 \times 10^2 \text{ J}}{0.25 \text{ s}} = 1.4 \times 10^3 \text{ W}$$

or $(1\ 400 \text{ W})(1 \text{ hp}/746 \text{ W}) = 1.9$ hp. Thus, humans can produce about 2 hp of mechanical power for periods on the order of seconds. Table 5.2 shows the maximum power outputs from humans for various time intervals based on bicycling and rowing, activities in which it is possible to measure power output accurately.

TABLE 5.2	Maximum Power Output from Humans for Various Time Periods
Power	**Time**
2 hp or 1 500 W	6 s
1 hp or 750 W	60 s
0.35 hp or 260 W	35 min
0.2 hp or 150 W	5 h
0.1 hp or 75 W (Safe daily level)	8 h

5.8 WORK DONE BY A VARYING FORCE

Suppose an object is being displaced along the x axis under the action of a force F_x that acts in the x direction and varies with position, as shown in Figure 5.28. The object is displaced in the direction of increasing x from $x = x_i$ to $x = x_f$. In such a situation, we cannot use Equation 5.1 to calculate the work done by the force because this relationship applies only when **F** is constant in magnitude and direction. As shown in Figure 5.28, we do not have a *single* value of the force to substitute into Equation 5.1. However, if we imagine that the object undergoes the *small* displacement Δx shown in Figure 5.28a, then the x component F_x of the force is approximately constant over this interval. Thus, we can approximate the work done by the force for this small displacement as

$$W_1 \cong F_x \Delta x \qquad \text{[5.20]}$$

This quantity is just the area of the shaded rectangle in Figure 5.28a. If we imagine that the F_x versus x curve is divided into a large number of such intervals, then the total work done for the displacement from x_i to x_f is approximately equal to the sum of a large number of small rectangles:

$$W \cong F_1 \Delta x_1 + F_2 \Delta x_2 + F_3 \Delta x_3 + \cdots \qquad \text{[5.21]}$$

If the displacements Δx are allowed to approach zero, then the number of terms in the sum increases without limit, but the value of the sum approaches a definite value equal to the area under the curve bounded by F_x and the x axis in Figure 5.28b. In other words, **the work done by a variable force acting on an object that undergoes a displacement is equal to the area under the graph of F_x versus x.**

A common physical system discussed in Section 5.5 for which the force varies with position consists of a block on a horizontal, frictionless surface connected to a spring. When the spring is stretched or compressed a small distance x from its equilibrium position $x = 0$, the spring exerts a force on the block given by $F_s = -kx$, where k is the force constant of the spring.

Now let us determine the work done by an *external agent* on the block as the spring is stretched *very slowly* from $x_i = 0$ to $x_f = x_{max}$, as in Figure 5.29a. This work can be easily calculated by noting that at any value of the displacement, Newton's third law tells us that the applied force $\mathbf{F}_{app}$ is equal in magnitude to the spring force $\mathbf{F}_s$ and acts in the opposite direction, so that $F_{app} = -(-kx) = kx$. A plot of F_{app} versus x is a straight line, as shown in Figure 5.29b. Therefore, the work done by this applied force in stretching the spring from $x = 0$ to $x = x_{max}$ is the area under the straight line in Figure 5.29b, which in this case is the area of the shaded triangle:

$$W_{F_{app}} = \tfrac{1}{2}kx_{max}^2$$

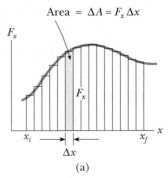

Area = $\Delta A = F_x \Delta x$

(a)

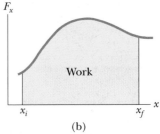

(b)

FIGURE 5.28 (a) The work done by the force component F_x for the small displacement Δx is $F_x \Delta x$, which equals the area of the shaded rectangle. The total work done for the displacement from x_i to x_f is approximately equal to the sum of the areas of all the rectangles. (b) The work done by the component F_x of the varying force as the particle moves from x_i to x_f is exactly equal to the area under this curve.

(a)

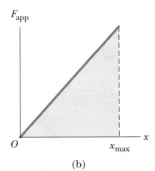

(b)

FIGURE 5.29 (a) A block being pulled from $x_i = 0$ to $x_f = x_{max}$ on a frictionless surface by a force $\mathbf{F}_{app}$. If the process is carried out very slowly, the applied force is equal in magnitude and opposite in direction to the spring force at all times. (b) A graph of F_{app} versus x.

Example 5.12 Work Required to Stretch a Spring

One end of a horizontal spring ($k = 80$ N/m) is held fixed while an external force is applied to the free end, stretching it slowly from $x_A = 0$ to $x_B = 4.0$ cm.

A Find the work done by the applied force on the spring.

Solution We place the zero reference of the coordinate axis at the free end of the unstretched spring. The applied force is $F_{app} = kx = (80 \text{ N/m})(x)$. The work done by F_{app} is the area of the triangle from 0 to 4.0 cm in Figure 5.30:

$$W = \tfrac{1}{2}kx_B^2 = \tfrac{1}{2}(80 \text{ N/m})(0.040 \text{ m})^2 = \boxed{0.064 \text{ J}}$$

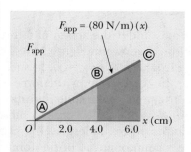

FIGURE 5.30 (Example 5.12) A graph of the external force required to stretch a spring that obeys Hooke's law versus the elongation of the spring.

B Find the additional work done in stretching the spring from $x_B = 4.0$ cm to $x_C = 7.0$ cm.

Solution The work done in stretching the spring the additional amount is the darker-shaded area between these limits in Figure 5.30. Geometrically, it is the difference in area between the large and small triangles.

$$W = \tfrac{1}{2}kx_C^2 - \tfrac{1}{2}kx_B^2 = \tfrac{1}{2}(80 \text{ N/m})[(0.070 \text{ m})^2 - (0.040 \text{ m})^2] = \boxed{0.13 \text{ J}}$$

SUMMARY

The **work** done by a *constant* force **F** acting on an object is defined as the product of the component of the force in the direction of the object's displacement and the magnitude of the displacement. If the force makes an angle θ with the displacement $\Delta \mathbf{x}$, the work done by **F** for this displacement is

$$W \equiv (F \cos \theta)\, \Delta x \qquad\qquad \textbf{[5.1]}$$

The **kinetic energy** of an object of mass m moving with speed v is defined as

$$KE \equiv \tfrac{1}{2}mv^2 \qquad\qquad \textbf{[5.5]}$$

The **work–kinetic energy theorem** states that the net work done on an object equals the change in kinetic energy of the object:

$$W_{\text{net}} = KE_f - KE_i = \Delta KE \qquad\qquad \textbf{[5.6]}$$

The **gravitational potential energy** associated with an object of mass m that is elevated a distance of y above Earth's surface is given by

$$PE \equiv mgy \qquad\qquad \textbf{[5.8]}$$

A force is **conservative** if the work it does on an object depends only on the initial and final positions of the object and not on the path taken between those positions. A force for which this is not true is said to be **nonconservative.**

The **principle of conservation of mechanical energy** states that if the only forces acting within a system are conservative, the total mechanical energy of the system remains constant.

$$KE_i + PE_i = KE_f + PE_f \qquad\qquad \textbf{[5.10]}$$

The **elastic potential energy** stored in a spring is given by

$$PE_s \equiv \tfrac{1}{2}kx^2 \qquad\qquad \textbf{[5.13]}$$

where k is the spring constant and x is the distance the spring is compressed or extended from its unstretched (equilibrium) position.

Energy can never be created or destroyed. Energy may be transformed from one form to another, but the total energy of an isolated system is always constant. The work done by all nonconservative forces acting within a system equals the change in the total mechanical energy of the system:

$$W_{nc} = (KE_f + PE_f) - (KE_i + PE_i) \qquad [5.15]$$

Average power is the ratio of energy transfer to the time interval during which the transfer occurs. For work, this expression is

$$\overline{\mathcal{P}} \equiv \frac{W}{\Delta t} \qquad [5.16]$$

If a force **F** acts on an object moving with an average speed of $\overline{v}$, the average power delivered to the object is

$$\overline{\mathcal{P}} = F\overline{v} \qquad [5.17]$$

CONCEPTUAL QUESTIONS

1. Consider a tug-of-war as in Fig. Q5.1 in which two teams pulling on a rope are evenly matched, so that no motion takes place. Is work done on the rope? On the pullers? On the ground? Is work done on anything?

FIGURE Q5.1 *(Arthur Tilley/FPG/Getty Images)*

2. Discuss whether any work is being done by each of the following agents and, if so, whether the work is positive or negative: (a) a chicken scratching the ground, (b) a person studying, (c) a crane lifting a bucket of concrete, (d) the force of gravity on the bucket in part (c), (e) the leg muscles of a person in the act of sitting down.

3. If the height of a playground slide is kept constant, will it make any difference in the final speed of children playing on it as to how long the slide is or whether it has bumps? Assume the slide is slick enough to be considered frictionless. Repeat this question without the assumption of no friction.

4. (a) Can the kinetic energy of a system be negative? (b) Can the gravitational potential energy of a system be negative? Explain.

5. Roads going up mountains are constructed with switchbacks with the road weaving back and forth along the face of the slope, such that there is only a gentle rise on any portion of the roadway. Does this require any less work to be done by an automobile climbing the mountain, compared to driving on a roadway that is straight up the slope? Why are switchbacks used?

6. (a) If the speed of a particle is doubled, what happens to its kinetic energy? (b) If the net work done on a particle is zero, what can be said about its speed?

7. As a simple pendulum swings back and forth, the forces acting on the suspended object are the force of gravity, the tension in the supporting cord, and air resistance. (a) Which of these forces, if any, does no work on the pendulum? (b) Which of these forces does negative work at all times during its motion? (c) Describe the work done by the force of gravity while the pendulum is swinging.

8. A bowling ball is suspended from the ceiling of a lecture hall by a strong cord. The ball is drawn away from its equilibrium position and released from rest at the tip of the demonstrator's nose as in Figure Q5.8. If the demonstrator remains stationary, explain why the ball does not strike her on its return swing. Would she be safe if the ball were given a push from its starting position at her nose?

FIGURE Q5.8

9. An older model car accelerates from 0 to speed v in 10 s. A newer, more powerful sports car accelerates from 0 to $2v$ in the same time interval. What is the ratio of powers expended by the two cars? Consider the energy coming from the engine to appear only as kinetic energy of the cars.

10. During a stress test of the cardiovascular system, a patient walks and runs on a treadmill. (a) Is the energy expended by the patient equivalent to the energy of walking and running on the ground? Explain. (b) If the treadmill is tilted upward, what effect does this have? Discuss.

11. When a punter kicks a football, is he doing any work on the ball while the toe of his foot is in contact with it? Is he doing any work on the ball after it loses contact with his toe? Are any forces doing work on the ball while it is in flight?

12. As a sled moves across a flat snow-covered field at constant velocity, is any work done? How does air resistance enter into the picture?

13. A weight is connected to a spring that is suspended vertically from the ceiling. If the weight is displaced downward from its equilibrium position and released, it will oscillate up and down. If air resistance is neglected, will the total mechanical energy of the system (weight plus Earth plus spring) be conserved? How many forms of potential energy are there for this situation?

14. The driver of a car slams on her brakes to avoid colliding with a deer crossing the highway. What happens to the car's kinetic energy as it comes to rest?

15. You are reshelving books in a library. You lift a book from the floor to the top shelf. The kinetic energy of the book on the floor was zero, and the kinetic energy of the book on the top shelf is zero, so there is no change in kinetic energy. Yet you did some work in lifting the book. Is the work-kinetic energy theorem violated?

16. A ball is thrown straight up into the air. At what position is its kinetic energy a maximum? At what position is the gravitational potential energy a maximum?

17. An Earth satellite is in a circular orbit at an altitude of 500 km. Explain why the work done by the gravitational force acting on the satellite is zero. Using the work-kinetic energy theorem, what can you say about the speed of the satellite?

18. In most circumstances, the normal force acting on an object and the force of static friction do no work on the object. However, the reason that the work is zero is different for the two cases. Explain why each does zero work.

PROBLEMS

1, 2, 3 = straightforward, intermediate, challenging ☐ = full solution available in Student Solutions Manual/Study Guide

web = solution posted at **http://info.brookscole.com/serway** 🎵 = biomedical application

Section 5.1 Work

1. A weight lifter lifts a 350-N set of weights from ground level to a position over his head, a vertical distance of 2.00 m. How much work does the weight lifter do, assuming he moves the weights at constant speed?

2. If a man lifts a 20.0-kg bucket from a well and does 6.00 kJ of work, how deep is the well? Assume that the speed of the bucket remains constant as it is lifted.

3. A tugboat exerts a constant force of 5.00×10^3 N on a ship moving at constant speed through a harbor. How much work does the tugboat do on the ship if each moves a distance of 3.00 km?

4. A shopper in a supermarket pushes a cart with a force of 35 N directed at an angle of 25° downward from the horizontal. Find the work done by the shopper as she moves down a 50-m length of aisle.

5. Starting from rest, a 5.00-kg block slides 2.50 m down a **web** rough 30.0° incline. The coefficient of kinetic friction between the block and the incline is $\mu_k = 0.436$. Determine (a) the work done by the force of gravity, (b) the work done by the friction force between block and incline, and (c) the work done by the normal force.

6. A scraper is drawn over a tooth 20 times, each time moving a distance of 0.75 cm. The scraper is held against the tooth with a normal force of 5.0 N. Assuming a coefficient of kinetic friction of 0.90 between the scraper and the tooth, determine the work done to clean the tooth.

7. A sledge loaded with bricks has a total mass of 18.0 kg and is pulled at constant speed by a rope. The rope is inclined at 20.0° above the horizontal, and the sledge moves a distance of 20.0 m on a horizontal surface. The coefficient of kinetic friction between the sledge and surface is 0.500. (a) What is the tension of the rope? (b) How much work is done by the rope on the sledge? (c) What is the mechanical energy lost due to friction?

8. A block of mass 2.50 kg is pushed 2.20 m along a frictionless horizontal table by a constant 16.0 N force directed 25.0° below the horizontal. Determine the work done by (a) the applied force, (b) the normal force exerted by the table, (c) the force of gravity, and (d) the net force on the block.

Section 5.2 Kinetic Energy and the Work-Kinetic Energy Theorem

9. A mechanic pushes a 2.50×10^3-kg car from rest to a speed of v, doing 5 000 J of work in the process. During this time, the car moves 25.0 m. Neglecting friction between car and road, find (a) v and (b) the horizontal force exerted on the car.

10. A 7.00-kg bowling ball moves at 3.00 m/s. How fast must a 2.45-g Ping-Pong ball move so that the two balls have the same kinetic energy?

11. A person doing a chin-up weighs 700 N exclusive of the arms. During the first 25.0 cm of the lift, each arm exerts an upward force of 355 N on the torso. If the upward movement starts from rest, what is the person's velocity at this point?

12. A crate of mass 10.0 kg is pulled up a rough incline with an initial speed of 1.50 m/s. The pulling force is 100 N parallel to the incline, which makes an angle of 20.0° with the horizontal. The coefficient of kinetic friction is 0.400, and the crate is pulled 5.00 m. (a) How much work is done by gravity? (b) How much mechanical energy is lost due to friction? (c) How much work is done by the 100-N force? (d) What is the change in kinetic energy of the crate? (e) What is the speed of the crate after being pulled 5.00 m?

13. A 70-kg base runner begins his slide into second base when moving at a speed of 4.0 m/s. The coefficient of friction between his clothes and Earth is 0.70. He slides so

that his speed is zero just as he reaches the base. (a) How much mechanical energy is lost due to friction acting on the runner? (b) How far does he slide?

14. An outfielder throws a 0.150-kg baseball at a speed of 40.0 m/s and an initial angle of 30.0°. What is the kinetic energy of the ball at the highest point of its motion?

15. A 2.0-g bullet leaves the barrel of a gun at a speed of 300 m/s. (a) Find its kinetic energy. (b) Find the average force exerted by the expanding gases on the bullet as the bullet moves the length of the 50-cm-long barrel.

16. A 0.60-kg particle has a speed of 2.0 m/s at point *A* and kinetic energy of 7.5 J at point *B*. What is (a) its kinetic energy at *A*? (b) its speed at point *B*? (c) the total work done on the particle as it moves from *A* to *B*?

17. A 2 000-kg car moves down a level highway under the actions of two forces. One is a 1 000-N forward force exerted on the drive wheels by the road; the other is a 950-N resistive force. Use the work–kinetic energy theorem to find the speed of the car after it has moved a distance of 20 m, assuming it starts from rest.

18. On a frozen pond, a 10-kg sled is given a kick that imparts to it an initial speed of $v_0 = 2.0$ m/s. The coefficient of kinetic friction between sled and ice is $\mu_k = 0.10$. Use the work–kinetic energy theorem to find the distance the sled moves before coming to rest.

Section 5.3 Potential Energy

Section 5.4 Conservative and Nonconservative Forces

19. A person's heart and head are 1.3 m and 1.8 m above the feet, respectively. Determine the potential energy associated with 0.50 kg of blood in the heart relative to (a) the feet, (b) the head.

20. A 2.00-kg ball is attached to a ceiling by a 1.00-m-long string. The height of the room is 3.00 m. What is the gravitational potential energy associated with the ball relative to (a) the ceiling? (b) the floor? (c) a point at the same elevation as the ball?

21. The arm, including the hand, in Figure P5.21 has a mass of 7.0 kg. Treating the arm as if it were a single point mass *m* attached to a rigid massless rod at point *A* as shown, determine the work that must be done by the deltoid muscle to raise the arm from position 1 to position 2.

FIGURE P5.21

22. The spring attached to the cart in Figure P5.22 has a spring constant $k = 250$ N/m. How much work is done to straighten the legs? Disregard gravity and assume the spring is initially unstretched.

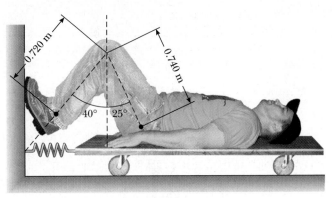

FIGURE P5.22

23. The triceps muscle can be modeled as a spring of such a strength that a force of 105 N is required to stretch it 2.00 cm. Determine the work done to stretch this muscle 2.00 cm.

24. A softball pitcher rotates a 0.250-kg ball around a vertical circular path of radius 0.600 m before releasing it. The pitcher exerts a 30.0-N force directed parallel to the motion of the ball around the complete circular path. The speed of the ball at the top of the circle is 15.0 m/s. If the ball is released at the bottom of the circle, what is its speed upon release?

25. A 40-N toy is placed in a light swing that is attached to ropes 2.0 m long. Find the gravitational potential energy associated with the toy relative to its lowest position (a) when the ropes are horizontal, (b) when the ropes make a 30° angle with the vertical, and (c) at the bottom of the circular arc.

Section 5.5 Conservation of Mechanical Energy

26. A 50-kg pole-vaulter running at 10 m/s vaults over the bar. Her speed when she is above the bar is 1.0 m/s. Neglect air resistance, as well as any energy absorbed by the pole, and determine her altitude as she crosses the bar.

27. A child and sled with a combined mass of 50.0 kg slide down a frictionless slope. If the sled starts from rest and has a speed of 3.00 m/s at the bottom, what is the height of the hill?

28. A 0.400-kg bead slides on a curved wire, starting from rest at point Ⓐ in Figure P5.28. If the wire is frictionless, find the speed of the bead (a) at Ⓑ and (b) at Ⓒ.

FIGURE P5.28 (Problems 28 and 36)

29. A gymnast swings on the high bar as shown in Figure P5.29. Starting from rest directly over the bar, he swings

around the bar while keeping his arms and legs out-stretched. Treating the gymnast as though his entire mass were concentrated at a point 1.20 m from the bar, determine his speed as he passes under the bar at position Ⓐ.

FIGURE P5.29

30. A bead of mass $m = 5.00$ kg is released from point Ⓐ and slides on the frictionless track shown in Figure P5.30. Determine (a) the bead's speed at points Ⓑ and Ⓒ and (b) the net work done by the force of gravity in moving the bead from Ⓐ to Ⓒ.

FIGURE P5.30

31. Tarzan swings on a 30.0-m-long vine initially inclined at an angle of 37.0° with the vertical. What is his speed at the bottom of the swing (a) if he starts from rest? (b) if he pushes off with a speed of 4.00 m/s?

32. Three objects with masses, $m_1 = 5.0$ kg, $m_2 = 10$ kg, and $m_3 = 15$ kg, are attached by strings over frictionless pulleys as indicated in Figure P5.32. The horizontal surface is frictionless, and the system is released from rest. Using energy concepts, find the speed of m_3 after it moves down 4.0 m.

33. The launching mechanism of a toy gun consists of a spring of unknown spring constant, as shown in Figure P5.33a. If the spring is compressed a distance of 0.120 m and the gun fired vertically as shown, the gun can launch a 20.0-g projectile from rest to a maximum height of 20.0 m above the starting point of the projectile. Neglecting all resistive

FIGURE P5.32

forces, determine (a) the spring constant and (b) the speed of the projectile as it moves through the equilibrium position of the spring (where $x = 0$), as shown in Figure P5.33b.

FIGURE P5.33

34. A projectile is launched with a speed of 40 m/s at an angle of 60° above the horizontal. Find the maximum height reached by the projectile during its flight by using conservation of energy.

35. A 0.250-kg block is placed on a light vertical spring ($k = 5.00 \times 10^3$ N/m) and pushed downward, compressing the spring 0.100 m. After the block is released, it leaves the spring and continues to travel upward. What height above the point of release will the block reach if air resistance is negligible?

Section 5.6 Nonconservative Forces, Nonisolated Systems, and Conservation of Energy

36. If the wire in Problem 28 (Fig. P5.28) is frictionless between points A and B and rough between B and C, and if the 0.400-kg bead starts from rest at A, (a) find its speed at

B. (b) If the bead comes to rest at *C*, find the loss in mechanical energy as it goes from *B* to *C*.

37. A morsel of food with a mass of 4.2 g is injected into the esophagus with an initial speed of 2.5 cm/s. On the way down to the stomach, the walls of the esophagus exert an upward resistive force of 2.7×10^{-3} N on the morsel. If the esophagus is 20.0 cm long, with what speed does the morsel of food enter the stomach?

38. An attendant pushes a patient on a gurney 20.7 m down a hall with a constant speed of 0.88 m/s, doing 2 000 J of work on the gurney. If the gurney is released, it travels 0.48 m before coming to rest. Assuming the attendant exerts a horizontal force on the gurney, determine the total weight of the gurney plus patient.

39. A 70-kg diver steps off a 10-m tower and drops, from rest, **web** straight down into the water. If he comes to rest 5.0 m beneath the surface, determine the average resistive force exerted on him by the water.

40. An airplane of mass 1.5×10^4 kg is moving at 60 m/s. The pilot then revs up the engine so that the forward thrust exerted by the air around the propeller becomes 7.5×10^4 N. If the force exerted by air resistance on the body of the airplane has a magnitude of 4.0×10^4 N, find the speed of the airplane after it has traveled 500 m. Assume that the airplane is in level flight throughout this motion.

41. A 2.1×10^3-kg car starts from rest at the top of a 5.0-m-long driveway that is sloped at 20° with the horizontal. If an average friction force of 4.0×10^3 N impedes the motion, find the speed of the car at the bottom of the driveway.

42. A 25.0-kg child on a 2.00-m-long swing is released from rest when the ropes of the swing make an angle of 30.0° with the vertical. (a) Neglecting friction, find the child's speed at the lowest position. (b) If the actual speed of the child at the lowest position is 2.00 m/s, what is the mechanical energy lost due to friction?

43. Starting from rest, a 10.0-kg block slides 3.00 m down a frictionless ramp (inclined at 30.0° from the floor) to the bottom. The block then slides an additional 5.00 m along the floor before coming to a stop. Determine (a) the speed of the block at the bottom of the ramp, (b) the coefficient of kinetic friction between block and floor, and (c) the mechanical energy lost due to friction.

44. A child slides without friction from a height *h* along a curved water slide (Fig. P5.44). She is launched from a height *h*/5 into the pool. Determine her maximum airborne height *y* in terms of *h* and *θ*.

FIGURE P5.44

45. A skier starts from rest at the top of a hill that is inclined at 10.5° with the horizontal. The hillside is 200 m long, and the coefficient of friction between snow and skis is 0.075 0. At the bottom of the hill, the snow is level and the coefficient of friction is unchanged. How far does the skier glide along the horizontal portion of the snow before coming to rest?

46. In a circus performance, a monkey is strapped to a sled and both are given an initial speed of 4.0 m/s up a 20° inclined track. The combined mass of monkey and sled is 20 kg, and the coefficient of kinetic friction between sled and incline is 0.20. How far up the incline do the monkey and sled move?

47. An 80.0-kg sky diver jumps out of a balloon at an altitude of 1 000 m and opens the parachute at an altitude of 200.0 m. (a) Assuming that the total retarding force on the diver is constant at 50.0 N with the parachute closed and constant at 3 600 N with the parachute open, what is the speed of the diver when he lands on the ground? (b) Do you think the sky diver will get hurt? Explain. (c) At what height should the parachute be opened so that the final speed of the sky diver when he hits the ground is 5.00 m/s? (d) How realistic is the assumption that the total retarding force is constant? Explain.

Section 5.7 Power

48. A skier of mass 70 kg is pulled up a slope by a motor-driven cable. (a) How much work is required to pull him 60 m up a 30° slope (assumed frictionless) at a constant speed of 2.0 m/s? (b) What power must a motor have to perform this task?

49. A 50.0-kg student climbs a 5.00-m-long rope and stops at the top. (a) What must her average speed be in order to match the power output of a 200-W light bulb? (b) How much work does she do?

50. While running, a person dissipates about 0.60 J of mechanical energy per step per kilogram of body mass. If a 60-kg person develops a power of 70 W during a race, how fast is the person running? Assume a running step is 1.5 m long.

51. Water flows over a section of Niagara Falls at the rate of 1.2×10^6 kg/s and falls 50 m (Fig. P5.51). How much power is generated by the falling water?

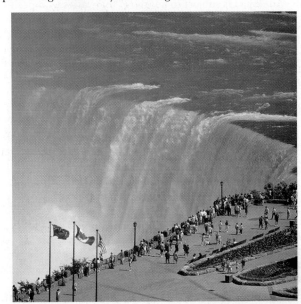

FIGURE P5.51 Some of the energy released by this large natural waterfall (and popular tourist attraction) is converted to electrical energy at a nearby hydroelectric plant. *(CORBIS Stock Market)*

52. A 1.50×10^3-kg car accelerates uniformly from rest to 10.0 m/s in 3.00 s. Find (a) the work done on the car in this time interval, (b) the average power delivered by the engine in this time interval, and (c) the instantaneous power delivered by the engine at $t = 2.00$ s.

53. A 1.50×10^3 kg car starts from rest and accelerates uniformly to 18.0 m/s in 12.0 s. Assume that air resistance remains constant at 400 N during this time. Find (a) the average power developed by the engine and (b) the instantaneous power output of the engine at $t = 12.0$ s just before the car stops accelerating.

54. A 650-kg elevator starts from rest. It moves upward for 3.00 s with constant acceleration until it reaches its cruising speed, 1.75 m/s. (a) What is the average power of the elevator motor during this interval? (b) How does this compare with its power during an upward cruise with constant speed?

Section 5.8 Work Done by a Varying Force

55. The force acting on a particle varies as in Figure P5.55. Find the work done by the force as the particle moves (a) from $x = 0$ to $x = 8.00$ m, (b) from $x = 8.00$ m to $x = 10.0$ m, and (c) from $x = 0$ to $x = 10.0$ m.

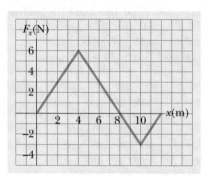

FIGURE P5.55

56. An object is subject to a force F_x that varies with position as in Figure P5.56. Find the work done by the force on the object as it moves (a) from $x = 0$ to $x = 5.00$ m, (b) from $x = 5.00$ m to $x = 10.0$ m, and (c) from $x = 10.0$ m to $x = 15.0$ m. (d) What is the total work done by the force over the distance $x = 0$ to $x = 15.0$ m?

FIGURE P5.56

57. The force acting on an object is given by $F_x = (8x - 16)$ N, where x is in meters. (a) Make a plot of this force versus x from $x = 0$ to $x = 3.00$ m. (b) From your graph, find the net work done by this force as the object moves from $x = 0$ to $x = 3.00$ m.

ADDITIONAL PROBLEMS

58. A 98.0-N grocery cart is pushed 12.0 m by a shopper who exerts a constant horizontal force of 40.0 N. If all frictional forces are neglected and the cart starts from rest, what is its final speed?

59. An archer pulls her bowstring back 0.400 m by exerting a force that increases uniformly from zero to 230 N. (a) What is the equivalent spring constant of the bow? (b) How much work does the archer do in pulling the bow?

60. A block of mass 12.0 kg slides from rest down a frictionless 35.0° incline and is stopped by a strong spring with $k = 3.00 \times 10^4$ N/m. The block slides 3.00 m from the point of release to the point where it comes to rest against the spring. When the block comes to rest, how far has the spring been compressed?

61. (a) A 75-kg man steps out a window and falls (from rest) 1.0 m to a sidewalk. What is his speed just before his feet strike the pavement? (b) If the man falls with his knees and ankles locked, the only cushion for his fall is an approximately 0.50-cm give in the pads of his feet. Calculate the average force exerted on him by the ground in this situation. This average force is sufficient to cause cartilage damage in the joints or to break bones.

62. A toy gun uses a spring to project a 5.3-g soft rubber sphere horizontally. The spring constant is 8.0 N/m, the barrel of the gun is 15 cm long, and a constant frictional force of 0.032 N exists between barrel and projectile. With what speed does the projectile leave the barrel if the spring was compressed 5.0 cm for this launch?

63. Two objects are connected by a light string passing over a light, frictionless pulley as in Figure P5.63. The 5.00-kg object is released from rest at a point 4.00 m above the floor. (a) Determine the speed of each object when the two pass each other. (b) Determine the speed of each object at the moment the 5.00-kg object hits the floor. (c) How much higher does the 3.00-kg object travel after the 5.00-kg object hits the floor?

FIGURE P5.63

64. Two blocks, *A* and *B* (with mass 50 kg and 100 kg, respectively), are connected by a string, as shown in Figure P5.64. The pulley is frictionless and of negligible mass. The coefficient of kinetic friction between block *A* and the incline is $\mu_k = 0.25$. Determine the change in the kinetic energy of block *A* as it moves from © to ©, a distance of 20 m up the incline if the system starts from rest.

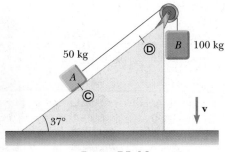

FIGURE P5.64

65. A 700-N Marine in basic training climbs a 10.0-m vertical rope at a constant speed in 8.00 s. What is his power output?

66. Energy is conventionally measured in Calories as well as in joules. One Calorie in nutrition is one kilocalorie, which we define in Chapter 11 as 1 kcal = 4 186 J. Metabolizing one gram of fat can release 9.00 kcal. A student decides to try to lose weight by exercising. She plans to run up and down the stairs in a football stadium as fast as she can and as many times as necessary. Is this in itself a practical way to lose weight? To evaluate the program, suppose she runs up a flight of 80 steps, each 0.150 m high, in 65.0 s. For simplicity ignore the energy she uses in coming down (which is small). Assume that a typical efficiency for human muscles is 20.0%. This means that when your body converts 100 J from metabolizing fat, 20 J goes into doing mechanical work (here, climbing stairs). The remainder goes into internal energy. Assume the student's mass is 50.0 kg. (a) How many times must she run the flight of stairs to lose one pound of fat? (b) What is her average power output, in watts and in horsepower, as she is running up the stairs?

67. For saving energy, bicycling and walking are far more efficient means of transportation than is travel by automobile. For example, when riding at 10.0 mi/h, a cyclist uses food energy at a rate of about 400 kcal/h above what he would use if merely sitting still. (In exercise physiology, power is often measured in kcal/h rather than in watts. Here 1 kcal = 1 nutritionist's Calorie = 4 186 J.) Walking at 3.00 mi/h requires about 220 kcal/h. It is interesting to compare these values with the energy consumption required for travel by car. Gasoline yields about 1.30×10^8 J/gal. Find the fuel economy in equivalent miles per gallon for a person (a) walking, and (b) bicycling.

68. A catcher "gives" with the ball when he catches a 0.15-kg baseball moving at 25 m/s. (a) If he moves his glove a distance of 2.0 cm, what is the average force acting on his hand? (b) Repeat for the case in which his glove and hand move 10 cm.

69. A ski jumper starts from rest 50.0 m above the ground on a frictionless track, and flies off the track at an angle of 45.0° above the horizontal and at a height of 10.0 m above the level ground. Neglect air resistance. (a) What is his speed when he leaves the track? (b) What is the maximum altitude he attains after leaving the track? (c) Where does he land relative to the end of the track?

70. A 5.0-kg block is pushed 3.0 m up a vertical wall with constant speed by a constant force of magnitude F applied at an angle of $\theta = 30°$ with the horizontal, as shown in Figure P5.70. If the coefficient of kinetic friction between block and wall is 0.30, determine the work done by (a) **F**, (b) the force of gravity, and (c) the normal force between block and wall. (d) By how much does the gravitational potential energy increase during this motion?

FIGURE P5.70

71. The ball launcher in a pinball machine has a spring that has a force constant of 1.20 N/cm (Fig. P5.71). The surface on which the ball moves is inclined 10.0° with respect to the horizontal. If the spring is initially compressed 5.00 cm, find the launching speed of a 0.100-kg ball when the plunger is released. Friction and the mass of the plunger are negligible.

FIGURE P5.71

72. The masses of the javelin, discus, and shot are 0.80 kg, 2.0 kg, and 7.2 kg, respectively, and record throws in the corresponding track events are about 89 m, 69 m, and 21 m, respectively. Neglecting air resistance, (a) calculate the minimum initial kinetic energies that would produce these throws, and (b) estimate the average force exerted on each object during the throw, assuming the force acts over a distance of 2.0 m. (c) Do your results suggest that air resistance is an important factor?

73. Jane, whose mass is 50.0 kg, needs to swing across a river filled with man-eating crocodiles in order to rescue Tarzan. However, she must swing into a *constant* horizontal wind force **F** on a vine that is initially at an angle of θ with the vertical (see Fig. P5.73). $D = 50.0$ m, $F = 110$ N, $L = 40.0$ m, and $\theta = 50.0°$. (a) With what minimum speed must Jane begin her swing in order to just make it to the other side? (*Hint:* First determine the potential energy that can be associated with the wind force. Because the wind force is constant, use an analogy with the constant gravitational force.) (b) Once the rescue is complete, Tarzan and Jane must swing back across the river. With what minimum speed must they begin their swing? Assume that Tarzan has a mass of 80.0 kg.

74. As it plows a parking lot, a snowplow pushes an ever-growing pile of snow in front of it. Suppose a car moving through air is similarly modeled as a cylinder pushing a growing plug of air in front of it. The originally stationary air is set into motion at the constant speed v of the

FIGURE P5.73

FIGURE P5.75

FIGURE P5.76

cylinder, as in Figure P5.74. In a time Δt, a new disk of air of area A and mass Δm must be moved a distance $v\Delta t$ and hence must be given kinetic energy $\frac{1}{2}(\Delta m)v^2$. Using this model, show that the power loss due to air resistance is $\rho A v^3/2$ and the resistive force is $\rho A v^2/2$, where ρ is the density of air.

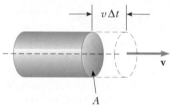

FIGURE P5.74

75. A child's pogo stick (Fig. P5.75) stores energy in a spring ($k = 2.50 \times 10^4$ N/m). At position Ⓐ ($x_1 = -0.100$ m), the spring compression is a maximum and the child is momentarily at rest. At position Ⓑ ($x = 0$), the spring is relaxed and the child is moving upward. At position Ⓒ, the child is again momentarily at rest at the top of the jump. Assuming that the combined mass of child and pogo stick is 25.0 kg, (a) calculate the total energy of the system if both potential energies are zero at $x = 0$, (b) determine x_2, (c) calculate the speed of the child at $x = 0$, (d) determine the value of x for which the kinetic energy of the system is a maximum, and (e) obtain the child's maximum upward speed.

76. A 2.00-kg block situated on a rough incline is connected to a spring of negligible mass having a spring constant of 100 N/m (Fig. P5.76). The block is released from rest when the spring is unstretched, and the pulley is frictionless. The block moves 20.0 cm down the incline before coming to rest. Find the coefficient of kinetic friction between the block and incline.

77. In the dangerous "sport" of bungee jumping, a daring student jumps from a balloon with a specially designed elastic cord attached to his waist, as shown in Figure P5.77. The

unstretched length of the cord is 25.0 m, the student weighs 700 N, and the balloon is 36.0 m above the surface of a river below. Calculate the required force constant of the cord if the student is to stop safely 4.00 m above the river.

FIGURE P5.77 Bungee jumping. (*Gamma*)

78. An object of mass m is suspended from the top of a cart by a string of length L as in Figure P5.78a. The cart and object are initially moving to the right at constant speed v_0. The cart comes to rest after colliding and sticking to a bumper, as in Figure P5.78b, and the suspended object swings through an angle θ. (a) Show that the initial speed is $v_0 = \sqrt{2gL(1 - \cos \theta)}$. (b) If $L = 1.20$ m and $\theta = 35.0°$, find the initial speed of the cart. (*Hint:* The force exerted by the string on the object does no work on the object.)

(a)   (b)

FIGURE P5.78

GROUP ACTIVITIES

G.1 Suspend a rubber band from a support and borrow some weights from your instructor to measure the rubber band's spring constant for small extensions. Calculate how much elastic potential energy is stored in the rubber band for a given extension. Use conservation of energy to predict how high a paper wad will go into the air when released from a given extension of the band. Try it to test your prediction.

G.2 Tightly wrap a rubber band around a tennis ball. Now fasten one end of a string through the rubber band and the other end to the top of a door frame to construct a pendulum. Pull the pendulum to the side at a variety of angles to observe that the energy of the pendulum-Earth system is always conserved as the pendulum swings (almost) to the same height of its arc as the height from which it was released. The word "almost" in the last sentence occurs because some energy is lost to friction at the point of support and to air resistance. You can observe this slight loss of energy by pulling the ball to the side and letting it go from a point about half an inch from your chin. Let it go (but don't push it) and test your belief in conservation of energy by seeing if you can avoid flinching when it swings back toward your chin.

G.3 While you have your pendulum of the preceding activity set up, predict what will happen in the following situation and then test your guess. When a pendulum is released from a given height, it swings to the same height at the other end of its arc as you noted above. However, suppose you place a meter stick across the door opening such that the pendulum string strikes the stick about halfway up the string when it moves through the door opening. How high will the ball swing in this case? Will it return to the same height as that at which it started, swing to a lower height, or swing to a greater height? Explain your answer.

G.4 Measure your pulse rate while at rest. Now slowly walk up a flight of stairs and measure your pulse rate at the top of the stairs. Repeat this activity starting with about the same rest pulse rate at the bottom of the stairs. This time run up the stairs. Based on your pulse rate readings, what can you conclude about the amount of work and power expended in each case? Repeat this experiment with a series of 10 pushups.

G.5 (a) A block with a mass m is pulled along a horizontal surface for a distance x by a constant force $\mathbf{F}$ at an angle θ with respect to the horizontal. The coefficient of kinetic friction between block and table is μ_k. Is the force exerted by friction equal to $\mu_k mg$? If not, what is the force exerted by friction? (b) How much work is done by the friction force and by $\mathbf{F}$? Don't forget the signs. (c) Identify all the forces that do no work on the block. (d) Let $m = 2.00$ kg, $x = 4.00$ m, $\theta = 37.0°$, $F = 15.0$ N, and $\mu_k = 0.400$ to find the answers to the questions posed above. Do the answers agree with your initial judgments?

G.6 (a) A child slides down a waterslide at an amusement park from an initial height h. The slide can be considered frictionless because of the water flowing down it. Can the equation for conservation of mechanical energy be used for the child? Defend your answer. (b) Is the mass of the child a factor in determining the speed at the bottom of the slide? Defend your answer. (c) The child drops straight down rather than following the curved ramp of the slide. In which case will he be traveling faster at ground level? Defend your answer. (d) If friction is present, how would the conservation of energy equation be modified? Defend your choice for the equation to be used. (e) Find the maximum speed of the child if the initial height of the slide is 25.0 m.

G.7 A child's toy consists of a base that is clamped firmly to a table, a spring and launching ramp attached to the base, and a ball. By compressing the spring, the child can launch the ball up the ramp to fire it off the table. The spring has a spring constant k, the ball has a mass m, and the ramp rises a height h. The spring is compressed a distance Δr in order to launch the ball. When the ball leaves the launching ramp, its velocity makes an angle θ with respect to the horizontal. (a) Assuming that friction and air resistance can be ignored for the purposes of this problem, describe the changes in the forms of energy in the system from the time the spring is compressed until the ball hits the ground. (b) Calculate the vector velocity of the ball when it leaves the launching ramp. Be sure to specify your coordinate system. (c) The spring constant is 32 N/m, the spring's compression is 5.0 cm, the ball's mass is 20 g, the height of the ramp is 10 cm, and the top of the table is 1.0 m above the floor. With what speed will the ball hit the floor?

(Problem 7 is courtesy of E. F. Redish. For more problems of this type, visit http://www.physics.umd.edu/ripe/perg/)

6
Momentum and Collisions

A bowling ball rolling down an alley has momentum, a quantity that is proportional to the ball's velocity. As the ball collides with the pins, part of the ball's momentum before the collision is transferred to the pins after the collision.

(Mark Cooper/Corbis Stock Market)

*C*onsider what happens when you strike a golf ball with a club. As you well know, the ball undergoes a large change in velocity and travels a long distance because of the force exerted by the club on that ball. But why doesn't the club also undergo a large change in velocity? It has the same force acting on it. Why does a good follow-through on the swing help the ball to go farther? Does the deformation of the ball upon contact have any effect on its subsequent motion?

As a first step in answering such questions, we shall introduce the term *momentum,* a term that has a slightly different meaning in physics than it does in everyday life. For example, a candidate moving rapidly in the opinion polls is said to be gaining momentum, but from the standpoint of physics, is he? As another example, a very massive football player is said to have a great deal of momentum as he runs down the field. Can a much less massive player have the same momentum? What must he do to achieve this momentum? Can a Ping-Pong ball have the same momentum as the expertly released bowling ball in the chapter opening photo?

The concept of momentum will lead us to a second conservation law: conservation of momentum. This law is especially useful for treating problems that involve collisions between objects because it gives us important information about the objects' motion without having to know the complicated variation of collision forces with time.

6.1 MOMENTUM AND IMPULSE

The linear momentum of an object of mass *m* moving with a velocity v is defined as the product of the mass and the velocity:

$$\mathbf{p} \equiv m\mathbf{v} \qquad \text{[6.1]}$$

 Linear momentum

As its definition shows, momentum is a vector quantity, with its direction matching that of the velocity. Momentum has dimensions ML/T, and its SI units are kilogram-meters per second (kg·m/s).

Often we find it advantageous to work with the components of momentum. For two-dimensional motion, these are

$$p_x = mv_x \qquad p_y = mv_y$$

where p_x represents the momentum of an object in the *x* direction and p_y its momentum in the *y* direction.

Quick Quiz 6.1 Two objects have equal kinetic energy. How do the magnitudes of their momenta compare? (a) $p_1 < p_2$, (b) $p_1 = p_2$, (c) $p_1 > p_2$, (d) not enough information to determine the answer.

The definition of momentum in Equation 6.1 coincides with our everyday use of the word. When we think of a massive object moving with a high velocity, we often say that the object has a large momentum, in accordance with Equation 6.1. Likewise, a small object moving slowly is said to have a small momentum. On the other hand, a small object moving with a high velocity can have a large momentum.

In order to change the momentum of an object, a force must be applied to it. In modern terms, Newton's second law of motion can be stated as follows:

The time rate of change of momentum of an object is equal to the net force acting on the object.

Newton's second law ▶

$$\mathbf{F}_{\text{net}} = \frac{\text{change in momentum}}{\text{time interval}} = \frac{\Delta \mathbf{p}}{\Delta t} \qquad [6.2]$$

where Δt is the time interval during which the momentum changes by $\Delta \mathbf{p}$ and $\mathbf{F}_{\text{net}}$ is the net force acting on the object. To see that this is equivalent to $\mathbf{F}_{\text{net}} = m\mathbf{a}$ for an object of constant mass, consider a single constant force $\mathbf{F}$ acting on a particle and producing a constant acceleration. In this case, we can write Equation 6.2 as

$$\mathbf{F} = \frac{\Delta \mathbf{p}}{\Delta t} = \frac{m\mathbf{v}_f - m\mathbf{v}_i}{\Delta t} = \frac{m(\mathbf{v}_f - \mathbf{v}_i)}{\Delta t} \qquad [6.3]$$

Now recall that the average acceleration of an object as given by Equation 3.8 is $\mathbf{a} = \Delta \mathbf{v}/\Delta t$. Thus, we see that Equation 6.3 reduces to the familiar equation $\mathbf{F} = m\mathbf{a}$.

Note from Equation 6.2 that if the net force is zero, the momentum of the object does not change. In other words, the linear momentum and velocity of a particle are conserved when $\mathbf{F}_{\text{net}} = 0$.

When a single constant force $\mathbf{F}$ acts on a particle, Equation 6.3 can be written as $\mathbf{F}\,\Delta t = \Delta \mathbf{p}$ or

Impulse-momentum theorem ▶

$$\mathbf{F}\,\Delta t = \Delta \mathbf{p} = m\mathbf{v}_f - m\mathbf{v}_i \qquad [6.4]$$

The product $\mathbf{F}\,\Delta t$ is defined to be the **impulse** of a constant force for the time interval Δt:

$$\text{Impulse} \equiv \mathbf{F}\,\Delta t \qquad [6.5]$$

Note that impulse is a vector quantity whose direction is the same as the constant force on the object. Equation 6.4 is called the **impulse-momentum theorem** and shows that **the impulse of the force acting on an object equals the change in momentum of that object.** This is true even if the force is not constant, although we need to modify the definition of impulse in this case, as discussed below.

The concept of impulse is most useful in describing collisions that last for a short time, for example, a bat hitting a ball, but a word of caution is in order here. If you tried to solve a problem such as this using Newton's second law, you would encounter some difficulty in measuring the value to be used for $\mathbf{F}$, because the force exerted on the ball is large, of short duration, and not constant. Instead, it might be represented by a curve like that in Figure 6.1a. The force starts out small as the bat comes in contact with the ball, rises to a maximum value when they are firmly in contact, and then drops off as the ball leaves the bat. In such instances, it is necessary to define an **average force** $\overline{\mathbf{F}}$ shown as a dashed line in Figure 6.1b. This average force can be thought of as the constant force that would give the same impulse to the object in the time interval Δt as the actual time-varying force

FIGURE 6.1   (a) A force acting on an object may vary in time. The impulse is the area under the force-time curve. (b) The average force (horizontal dashed line) gives the same impulse to the object in the time interval Δt as the real time-varying force described in (a).

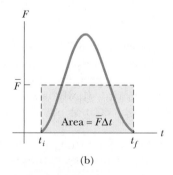

gives in this interval. In this situation, we can write the impulse-momentum theorem as

$$\overline{\mathbf{F}}\,\Delta t = \Delta \mathbf{p} \qquad [6.6]$$

The impulse imparted by a force during the time interval Δt is equal to the area under the force-time graph from the beginning to the end of the time interval or, equivalently, to $\overline{F}\,\Delta t$ as shown in Figure 6.1b. The brief collision between a bullet and an apple is illustrated in Figure 6.2.

APPLYING **PHYSICS** 6.1

In boxing matches of the 19th century, bare fists were used. In modern boxing, fighters wear padded gloves. How does this protect the brain of the boxer from injury? Boxers often "roll with the punch." How does this protect their health?

Explanation   The brain is immersed in a cushioning fluid inside the skull. If the head is struck suddenly by a bare fist, there is a rapid acceleration of the skull. The brain matches this acceleration only because of the large impulsive force exerted by the skull on the brain. This large and sudden force (large $\overline{F}$ and small Δt) can cause severe brain injury. Padded gloves extend the time Δt over which the force is applied to the head. Thus, for a given impulse $\overline{F}\,\Delta t$, the gloves provide a longer time interval than the bare fist, decreasing the average force. Because the average force is decreased, the acceleration of the skull is decreased, reducing (but not eliminating) the chance for brain injury. The same argument can be made for "rolling with the punch." If the head is held steady while being struck, the time interval over which the force is applied is relatively short and the average force is large. If the head is allowed to move in the same direction of the punch, the time interval is lengthened, and the average force reduced.

FIGURE 6.2   An apple being pierced by a 30-caliber bullet traveling at a supersonic speed of 900 m/s. This collision was photographed with a microflash stroboscope using an exposure time of 0.33 μs. Shortly after the photograph was taken, the apple disintegrated completely. Note that the points of entry and exit of the bullet are visually explosive. (© *Harold and Esther Edgerton Foundation, 2002, courtesy of Palm Press, Inc.*)

APPLICATION

BOXING AND BRAIN INJURY

Example 6.1   Teeing Off

A 50-g golf ball is struck with a club as in Figure 6.3. The force on the ball varies from zero when contact is made up to some maximum value (when the ball is deformed) and then back to zero when the ball leaves the club. Thus, the force-time graph is somewhat like that in Figure 6.1. Assume that the ball leaves the club face with a velocity of $+44$ m/s.

A Estimate the size and direction of the impulse due to the collision.

Solution   Before the club hits the ball, the ball is at rest on the tee, and as a result its initial momentum is zero. Taking the positive direction in Figure 6.3 to be that of the motion of the ball (to the right), the momentum of the ball immediately after the collision is

$$\mathbf{p}_f = m\mathbf{v}_f = (50 \times 10^{-3}\,\text{kg})(44\,\text{m/s}) = +2.2\,\text{kg}\cdot\text{m/s}$$

The positive sign means the final momentum is directed to the right in Figure 6.3. Thus, the impulse imparted to the ball, which equals its change in momentum, is

$$\Delta \mathbf{p} = m\mathbf{v}_f - m\mathbf{v}_i = \boxed{+2.2\,\text{kg}\cdot\text{m/s}}$$

where the positive sign means that the change in momentum and the impulse are both directed to the right.

B Estimate the duration of the collision and the average force on the ball.

Solution   From Figure 6.3, it appears that a reasonable estimate of the distance the ball travels while in contact with the club is the radius of the ball, about 2.0 cm. The time

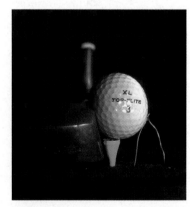

FIGURE 6.3   (Example 6.1) A golf ball being struck by a club. (© *Harold and Esther Edgerton Foundation, 2002, courtesy of Palm Press, Inc.*)

it takes the club to move this distance (the contact time) is then

$$\Delta t = \frac{\Delta x}{\overline{v}} = \frac{2.0 \times 10^{-2} \text{ m}}{22 \text{ m/s}} = \boxed{9.1 \times 10^{-4} \text{ s}}$$

Finally, the average force is estimated to be

$$\overline{\mathbf{F}} = \frac{\Delta \mathbf{p}}{\Delta t} = \frac{2.2 \text{ kg} \cdot \text{m/s}}{9.1 \times 10^{-4} \text{ s}} = \boxed{+2.4 \times 10^{3} \text{ N}}$$

and the direction of $\overline{\mathbf{F}}$ is in the same direction as $\Delta \mathbf{p}$, the positive x direction.

Example 6.2   How Good Are the Bumpers?

In a particular crash test, a car of mass 1.50×10^{3} kg collides with a wall as in Figure 6.4a. The initial and final velocities of the car are $\mathbf{v}_i = -15.0$ m/s and $\mathbf{v}_f = +2.60$ m/s, respectively. If the collision lasts for 0.150 s, find the impulse on the car due to the collision and the size and direction of the average force exerted on the car.

Before

−15.0 m/s

After

+2.60 m/s

(a)

(b)

FIGURE 6.4   (Example 6.2) (a) This car's momentum changes as a result of its collision with the wall. (b) In a crash test (an inelastic collision), much of the car's initial kinetic energy is transformed into the energy it took to damage the vehicle. *(b, courtesy of the Insurance Institute for Highway Safety)*

Solution   The initial and final momenta of the car are

$$\mathbf{p}_i = m\mathbf{v}_i = (1.50 \times 10^{3} \text{ kg})(-15.0 \text{ m/s}) = -2.25 \times 10^{4} \text{ kg} \cdot \text{m/s}$$

$$\mathbf{p}_f = m\mathbf{v}_f = (1.50 \times 10^{3} \text{ kg})(+2.60 \text{ m/s}) = +0.390 \times 10^{4} \text{ kg} \cdot \text{m/s}$$

where the minus sign means the initial momentum is to the left in Figure 6.4a and the plus sign means the final momentum is to the right. Hence, the impulse, which equals the change in momentum, is

$$\overline{\mathbf{F}} \Delta t = \Delta \mathbf{p} = \mathbf{p}_f - \mathbf{p}_i$$
$$= +0.390 \times 10^{4} \text{ kg} \cdot \text{m/s} - (-2.25 \times 10^{4} \text{ kg} \cdot \text{m/s})$$
$$= \boxed{+2.64 \times 10^{4} \text{ kg} \cdot \text{m/s}}$$

The average force exerted on the car is

$$\overline{\mathbf{F}} = \frac{\Delta \mathbf{p}}{\Delta t} = \frac{2.64 \times 10^{4} \text{ kg} \cdot \text{m/s}}{0.150 \text{ s}} = \boxed{+1.76 \times 10^{5} \text{ N}}$$

where the plus sign indicates the average force is in the positive x direction.

EXERCISE A 0.150-kg baseball is thrown with a speed of 40.0 m/s. It is hit straight back at the pitcher with a speed of 50.0 m/s. (a) What is the impulse delivered to the baseball? (b) Find the magnitude of the average force exerted by the bat on the ball if the two are in contact for 2.00×10^{-3} s.

ANSWER (a) 13.5 kg·m/s (b) 6.75 kN

INJURY IN CAR COLLISIONS

The main injuries that occur to a person hitting the interior of a car in a crash are brain damage, bone fracture, and trauma to the skin, blood vessels, and internal organs. Here we shall compare the rather imprecisely known thresholds for human injury to typical forces and accelerations experienced in a car crash.

APPLICATION

INJURY TO PASSENGERS IN
CAR COLLISIONS

A force of about 90 kN (20 000 lb) compressing the tibia can cause fracture. Although the breaking force varies with the bone considered, we may take this value as the threshold force for fracture. It is well known that rapid acceleration of the head, even without skull fracture, can be fatal. Estimates show that head accelerations of $150g$ experienced for about 4 ms or $50g$ for 60 ms are fatal 50% of the time. Such injuries from rapid acceleration often result in spinal cord nerve damage where the nerves enter the base of the brain. The threshold for damage to skin, blood vessels, and internal organs can be estimated in an approximate way from whole-body impact data, where the force is uniformly distributed over the entire front surface area of 0.7–0.9 m². These data show that if the collision lasts for less than about 70 ms, a person will survive if the whole-body impact pressure (force per unit area) is less than 1.9×10^5 N/m² (28 lb/in.²). Death results in 50% of cases where the whole-body impact pressure reaches 3.4×10^5 N/m² (50 lb/in.²).

Armed with the data above, we can estimate the forces and accelerations in a typical car crash and see how seat belts, air bags, and padded interiors can reduce the chance of death or serious injury in a collision. Consider a typical collision involving a 75-kg passenger not wearing a seat belt, traveling at 27 m/s (60 mi/h) who comes to rest in about 0.010 s after striking an unpadded dashboard. Using $\overline{F}\Delta t = mv_f - mv_i$, we find

$$\overline{F} = \frac{mv_f - mv_i}{\Delta t} = \frac{0 - (75 \text{ kg})(27 \text{ m/s})}{0.010 \text{ s}} = -2.0 \times 10^5 \text{ N}$$

and

$$a = \left|\frac{\Delta v}{\Delta t}\right| = \frac{27 \text{ m/s}}{0.010 \text{ s}} = 2\,700 \text{ m/s}^2 = \frac{2\,700 \text{ m/s}^2}{9.8 \text{ m/s}^2}g = 280g$$

If we assume the passenger crashes into the dashboard and windshield so that the head and chest with a combined surface area of 0.5 m² experience the force, we find a whole-body pressure of

$$\frac{\overline{F}}{A} = \frac{2.0 \times 10^5 \text{ N}}{0.5 \text{ m}^2} \cong 4 \times 10^5 \text{ N/m}^2$$

Thus, we see that the force, the acceleration, and the whole-body pressure all *exceed* the threshold for fatality or broken bones and that an unprotected collision at 60 mi/h is almost certainly fatal.

What can be done to reduce or eliminate the chance of dying in a car crash? The most important factor is the collision time or the time it takes the person to come to rest. If this time can be increased by 10–100 times the value of 0.01 s for a hard collision, the chances of survival in a car crash are much higher because the increase in Δt makes the contact force 10–100 times smaller. Seat belts restrain people so that they come to rest in about the same amount of time it takes to stop the car, typically about 0.15 s. This increases the effective collision time to about

FIGURE 6.5 **Force on the car versus time for the collision shown in Figure 6.4b.** *(Based on data from the Insurance Institute for Highway Safety)*

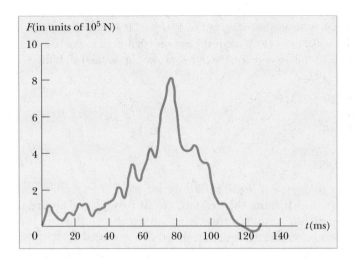

0.15 s. Figure 6.5 shows the measured force on a car versus time for a car crash, and that the car comes to rest in about 0.15 s.

Air bags also increase the collision time, absorb energy from the body as they rapidly deflate, and spread the contact force over an area of the body of about 0.5 m². preventing penetration wounds and fractures. Air bags must deploy very rapidly (in less than 10 ms) in order to stop a human traveling at 27 m/s before he or she comes to rest against the steering column about 0.3 m away. In order to achieve this rapid deployment, accelerometers send a signal to discharge a bank of capacitors (devices that store electric charge) that ignites an explosive, thereby filling the air bag with gas very quickly. The electrical charge for ignition is stored in capacitors to ensure air bag operation in the event of damage to the battery or the car's electrical system in a severe collision.

Webnote 6.1

For information on the use of air bags and seatbelts in automobiles, visit the National Safety Council at:

http://www.nsc.org/airbag.htm

The important reduction of potentially fatal forces, accelerations, and pressures to tolerable levels by the simultaneous use of seat belts and airbags is summarized as follows. If a 75-kg person traveling at 27 m/s is stopped by a seat belt in 0.15 s, the person experiences an average force of 9.8 kN, an average acceleration of 18g, and a whole-body pressure of 2.8×10^4 N/m² for a contact area of 0.5 m². These values are about one order of magnitude less than the values estimated earlier for an unprotected person and well below the thresholds for life-threatening injuries.

6.2   CONSERVATION OF MOMENTUM

An important feature of collisions is that the momentum of an isolated system in which a collision occurs is conserved. In this section we shall see how the laws of motion lead us to this important conservation law.

A collision may be the result of physical contact between two objects, as illustrated in Figure 6.6a. This is a common observation when two macroscopic objects, such as two billiard balls or a baseball and a bat, strike each other. But "contact" on a submicroscopic scale is ill defined and hence meaningless, and so the notion of *collision* must be generalized. More accurately, forces between two bodies arise from the electrostatic interaction of the electrons in the surface atoms of the bodies. To understand the distinction between macroscopic and microscopic collisions, consider the collision between two positive charges as in Figure 6.6b. Because the two particles are positively charged, they repel each other and do not touch.

FIGURE 6.6 (a) A collision between two objects resulting from direct contact. (b) A collision between two charged objects (in this case, a proton and a helium nucleus).

Figure 6.7 shows an isolated system of two particles before and after they collide. By isolated we mean that no external forces, such as the gravitational force or friction, act on the system. Before the collision, the velocities of the two particles are $\mathbf{v}_{1i}$ and $\mathbf{v}_{2i}$; after the collision, the velocities are $\mathbf{v}_{1f}$ and $\mathbf{v}_{2f}$. The impulse-momentum theorem applied to m_1 becomes

$$\overline{\mathbf{F}}_{21} \, \Delta t = m_1 \mathbf{v}_{1f} - m_1 \mathbf{v}_{1i}$$

Before collision

After collision

(b)

FIGURE 6.7 Before and after a head-on collision between two objects. The momentum of each object changes as a result of the collision, but the total momentum of the system remains constant.

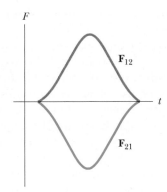

FIGURE 6.8 Force as a function of time for the two colliding particles in Figures 6.6(a) and 6.7. Note that $\mathbf{F}_{21} = -\mathbf{F}_{12}$.

Likewise, for m_2 we have

$$\overline{\mathbf{F}}_{12} \, \Delta t = m_2\mathbf{v}_{2f} - m_2\mathbf{v}_{2i}$$

where $\overline{\mathbf{F}}_{21}$ is the average force exerted by m_2 on m_1 during the collision and $\overline{\mathbf{F}}_{12}$ is the average force exerted by m_1 on m_2 during the collision (review Fig. 6.6a).

We are using average values for $\overline{\mathbf{F}}_{21}$ and $\overline{\mathbf{F}}_{12}$ even though the actual forces may vary in time in a complicated way, as is the case in Figure 6.8. Newton's third law states that at all times these two forces are equal in magnitude and opposite in direction ($\mathbf{F}_{21} = -\mathbf{F}_{12}$). In addition, the two forces act for the same time interval. Thus,

$$\overline{\mathbf{F}}_{21} \, \Delta t = -\overline{\mathbf{F}}_{12} \, \Delta t$$

or

$$m_1\mathbf{v}_{1f} - m_1\mathbf{v}_{1i} = -(m_2\mathbf{v}_{2f} - m_2\mathbf{v}_{2i})$$

from which we find

$$m_1\mathbf{v}_{1i} + m_2\mathbf{v}_{2i} = m_1\mathbf{v}_{1f} + m_2\mathbf{v}_{2f} \qquad [6.7]$$

◀ Conservation of momentum

This result is known as **conservation of momentum.**

> **The principle of conservation of momentum states that, when no external forces act on a system consisting of two objects that collide with each other, the total momentum of the system before the collision is equal to the total momentum of the system after the collision.**

Note that momentum is conserved for a *system* of objects. In the example used to derive Equation 6.7, the system was taken to be two colliding objects. In the general case of collisions, the system includes all the objects that are interacting with one another. In addition, we have assumed that the only forces acting during the collision are internal forces, meaning the forces that arise between the interacting objects of the system. For example, in the collision of two objects shown in Figure 6.6a, the internal forces are $\mathbf{F}_{21}$ and $\mathbf{F}_{12}$. If a third object outside the system (consisting of m_1 and m_2) were to exert a force on either m_1 or m_2 (or both objects) during the collision, momentum would not be conserved for the system. In all the example problems that we shall consider, the system will be assumed to be isolated.

Our derivation has assumed that only two objects interact, but the result remains valid regardless of the number involved. In its most general form, conservation of momentum can be stated as follows: **The total momentum of an isolated**

MOMENTUM CONSERVATION APPLIES TO A *SYSTEM*!

Remember that the momentum of an *isolated system* is conserved. The momentum of one particle within an isolated system is not necessarily conserved, because other particles in the system may be interacting with it. Apply conservation of momentum to an isolated system *only*.

(a)

(b)

Conservation of momentum is the principle behind these two propulsion systems. (a) The force from a nitrogen-propelled, hand-controlled device allows an astronaut to move about freely in space without restrictive tethers. *(Courtesy of NASA)*   (b) A squid propels itself by expelling water at a high velocity. *(Mike Severns/Stone/ Getty Images)*

system of objects is conserved regardless of the nature of the forces between the objects.

To understand what this statement means, imagine that you initially stand at rest and then jump upward, leaving the ground with velocity **v**. Obviously, your momentum is not conserved, because before the jump it was zero and it became *m***v** as you began to rise. However, the total momentum of the system is conserved if the system selected includes all objects that exert forces on one another. You must include Earth as part of this system because you exert a downward force on Earth when you jump. Earth in turn exerts on you an upward force of the same magnitude, as required by Newton's third law. Momentum is conserved for the system consisting of you and Earth. Thus, as you move upward with some momentum *m***v**, Earth moves in the opposite direction with momentum of the same magnitude. The recoil velocity of Earth due to this event is imperceptibly small, of course, because Earth is so massive, but its recoil momentum is not zero.

APPLYING PHYSICS 6.2

A baseball is projected into the air at an angle to the ground. As it moves through its trajectory, its velocity and therefore its momentum constantly changes. Is this a violation of conservation of momentum?

Explanation   The principle of conservation of momentum states that the momentum of an isolated system of particles is conserved. A ball projected through the air is subject to the external force of gravity, so we would not expect its momentum to be conserved. The gravitational force is in the vertical direction, so it is actually only the vertical component of the momentum that changes due to this force. In the horizontal direction, no force is exerted (ignoring air friction), so the horizontal component of the ball's momentum is conserved. In Chapter 3, we used the same idea in stating that the horizontal component of the velocity of a projectile remains constant.

If we consider the baseball and Earth as a system of two particles, the gravitational force is internal to this system. The momentum of the ball-Earth system is conserved. The downward force from the thrower's feet as the ball is thrown gives Earth an initial motion. As the ball rises and falls, Earth initially sinks and then rises (although imperceptibly!) due to the upward gravitational force from the ball, so that the total momentum of the system remains unchanged.

Example 6.3   The Recoiling Pitching Machine

A baseball player wishes to maintain his physical condition during the winter. He uses a 50.0-kg pitching machine to help him, placing the machine on the pitcher's mound as in Figure 6.9. The ground is covered with a thin layer of ice, so that there is negligible friction between the ground and the machine. The machine fires a 0.15-kg baseball horizontally with a speed of 36 m/s. What is the recoil speed of the machine?

Reasoning   We take the system to consist of the baseball and the pitching machine. Because of the gravitational force and the normal force, the system is not truly isolated. However, both of these forces are in the *y* direction, and therefore are directed perpendicularly to the motion of the system. In addition, there is no friction between the machine and the ground. Therefore, momentum is conserved in the *x* direction because there are no external forces in this direction. We will restrict our attention to the *x* direction, and use the symbol *v* for components of the velocity in the *x* direction.

Solution   The total momentum of the system before firing is zero ($m_1 v_{1i} + m_2 v_{2i} = 0$). Therefore, the total momentum after firing must be zero; that is,

$$m_1 v_{1f} + m_2 v_{2f} = 0$$

With $m_1 = 0.15$ kg, $v_{1i} = 36$ m/s, and $m_2 = 50$ kg, we solve for v_{2f} and find the recoil

FIGURE 6.9   (Example 6.3) When the baseball is fired to the right, the pitching machine recoils to the left. The momentum of the baseball-machine system is conserved.

speed of the pitching machine to be

$$v_{2f} = -\frac{m_1}{m_2} v_{1f} = -\left(\frac{0.15 \text{ kg}}{50 \text{ kg}}\right)(36 \text{ m/s}) = -0.11 \text{ m/s}$$

The negative sign for v_{2f} indicates that the pitching machine is moving to the left after firing, in the direction opposite the direction of motion of the baseball. Because the pitching machine is much more massive than the ball, the acceleration and consequent speed of the pitching machine are much smaller than the acceleration and speed of the ball.

Quick Quiz 6.2   A boy stands at one end of a floating raft that is stationary relative to the shore. He then walks to the opposite end of the raft, away from the shore. As a consequence, the raft (a) remains stationary, (b) moves away from the shore, or (c) moves toward the shore.

Quick Quiz 6.3   Two objects are at rest on a frictionless surface. Object 1 has a larger mass than object 2. When a force is applied to object 1, it accelerates through a displacement Δx. The force is removed from object 1 and is applied to object 2. At the moment when object 2 has accelerated through the same displacement Δx, which statements are true? (a) $p_1 < p_2$ (b) $p_1 = p_2$ (c) $p_1 > p_2$ (d) $KE_1 < KE_2$ (e) $KE_1 = KE_2$ (f) $KE_1 > KE_2$

Quick Quiz 6.4   An astronaut of mass M is floating in the environment of her spacecraft and is initially at rest. She decides to get to the other side of the spacecraft by throwing a notebook of mass m in the direction opposite to that in which she wishes to move. If the speed of the notebook is v, what is her recoil speed? (a) 0   (b) $mv/(M - m)$ (c) $mv/(M + m)$ (d) mv/M (e) $2mv/(M - m)$

6.3 COLLISIONS

We have seen that, for any type of collision, the total momentum of the system just before collision equals the total momentum just after collision. We can say that the total momentum is always conserved for any type of collision. However, the total **kinetic energy** is generally not conserved in a collision, because some of the kinetic energy is converted to internal energy, sound energy, and into the work needed to permanently deform an object, as in a car crash. We define an **inelastic collision** as a collision in which momentum is conserved but kinetic energy is not. The collision of a rubber ball with a hard surface is inelastic, because some of the kinetic energy is lost when the ball is deformed during contact with the surface. When two objects collide and stick together, the collision is called **perfectly inelastic.** For example, if two pieces of putty collide, they stick together and move with some common velocity after the collision. If a meteorite collides head-on with Earth, it becomes buried in Earth and the collision is considered perfectly inelastic. Not all of the initial kinetic energy is necessarily lost in a perfectly inelastic collision.

An **elastic collision** is defined as one in which both momentum and kinetic energy are conserved. Billiard ball collisions and the collisions of air molecules with the walls of a container at ordinary temperatures are highly elastic. Macroscopic collisions such as those between billiard balls are only approximately elastic, because some permanent deformation and hence some loss of kinetic energy takes place. Perfectly elastic collisions do occur, however, between atomic and subatomic particles. Elastic and perfectly inelastic collisions are *limiting* cases; most actual collisions fall into a range in between them.

A practical example of an inelastic collision is used to detect glaucoma, a disease in which the pressure inside the eye builds up and leads to blindness by damaging the cells of the retina. In this application, medical professionals use a device called a *tonometer* to measure the pressure inside the eye. This device releases a puff of air against the outer surface of the eye and measures the speed of the air after reflection from the eye. At normal pressure, the eye is slightly spongy, and the pulse is reflected at low speed. As the pressure inside the eye increases, the outer surface becomes more rigid, and the speed of the reflected pulse increases. Thus, the speed of the reflected puff of air is used to measure the internal pressure of the eye.

We can summarize the types of collisions as follows:

- **An elastic collision is one in which both momentum and kinetic energy are conserved.**
- **An inelastic collision is one in which momentum is conserved but kinetic energy is not.**
- **A perfectly inelastic collision is an inelastic collision in which the two objects stick together after the collision, so that their final velocities are the same and the momentum of the system is conserved.**

In the remainder of this section, we shall treat perfectly inelastic and elastic collisions in one dimension.

Tip 6.2 MOMENTUM AND KINETIC ENERGY IN COLLISIONS

The momentum of an isolated system is conserved in all collisions. However, the kinetic energy of an isolated system is conserved only when the collisions within the system are elastic. This is due to the fact that kinetic energy can be transferred out of the system or transformed into other types of energy, but there is only one type of linear momentum.

APPLICATION

GLAUCOMA TESTING

Elastic collision

Inelastic collision

Quick Quiz 6.5 Suppose you are blindfolded in a game room where your friends are playing billiards. If you only hear a clicking sound as one ball collides with another, you would conclude that the collisions are (a) elastic, (b) inelastic, or (c) not enough information is given.

PERFECTLY INELASTIC COLLISIONS

Consider two objects having masses m_1 and m_2 moving with known initial velocity components v_{1i} and v_{2i} along a straight line, as in Figure 6.10. If the two objects collide head-on, stick together, and move with some common velocity component v_f after the collision, the collision is perfectly inelastic. Because the total momentum of the two-object isolated system before the collision equals the total momentum of the combined-object system after the collision, we can solve for the final velocity from

Before collision

$$m_1 v_{1i} + m_2 v_{2i} = (m_1 + m_2) v_f \qquad \text{[6.8]}$$

$$v_f = \frac{m_1 v_{1i} + m_2 v_{2i}}{m_1 + m_2} \qquad \text{[6.9]}$$

After collision

It is important to note that v_{1i}, v_{2i}, and v_f represent the x components of the vectors $\mathbf{v}_{1i}$, $\mathbf{v}_{2i}$, and $\mathbf{v}_f$, so we must be careful to enter the known values of v_{1i} and v_{2i} with the correct signs. For example, in Figure 6.10, v_{1i} would have a positive value (m_1 moving to the right), whereas v_{2i} would have a negative value (m_2 moving to the left). Once these values are entered into Equation 6.9, this equation can be solved for both the correct size and direction of the final velocity as shown in Example 6.5.

FIGURE 6.10 (a) Before and (b) after a perfectly inelastic head-on collision between two objects.

Tip 6.3 INELASTIC VS. PERFECTLY INELASTIC COLLISIONS

If the colliding particles stick together, the collision is perfectly inelastic. If they bounce off each other (and kinetic energy is not conserved), the collision is inelastic.

Example 6.4 The Cadillac Versus the Compact Car

An 1 800-kg car stopped at a traffic light is struck from the rear by a 900-kg car and the two become entangled. If the smaller car was moving at 20 m/s before the collision, what is the speed of the entangled cars after the collision? Take the positive x direction to be in the direction of motion of the smaller car.

Reasoning The total momentum of the two cars before the collision equals the total momentum of the system after the collision if we ignore friction and assume the system is isolated. This is called the impulse approximation, where we assume that the collision force is much larger than any other forces present, such as friction, and acts for such a short time that these other forces may be neglected during the collision. Thus, the result we obtain for the final speed will only be approximately true just after the collision. For longer time intervals after the collision, we would use Newton's second law to describe the slowing down of the system due to friction. Because the cars "become entangled," this is a perfectly inelastic collision.

Solution The magnitude of the total momentum of the system before the collision is equal to that of only the smaller car because the larger car is initially at rest:

$$p_i = m_1 v_i = (900 \text{ kg})(20 \text{ m/s}) = 1.80 \times 10^4 \text{ kg·m/s}$$

After the collision, the mass that moves is the sum of the masses of the cars. The magnitude of the momentum of the combination is

$$p_f = (m_1 + m_2) v_f = (2\,700 \text{ kg}) v_f$$

Equating the initial momentum to the final momentum and solving for v_f, the speed of the entangled cars just after the collision, we have

$$v_f = \frac{p_i}{m_1 + m_2} = \frac{1.80 \times 10^4 \text{ kg·m/s}}{2\,700 \text{ kg}} = 6.67 \text{ m/s}$$

Example 6.5 Here's Mud in Your Eye

Two balls of mud collide head-on in a perfectly inelastic collision, as in Figure 6.10. Suppose $m_1 = 0.500$ kg, $m_2 = 0.250$ kg, $v_{1i} = +4.00$ m/s, and $v_{2i} = -3.00$ m/s.

A Find the velocity of the composite ball of mud after the collision.

Solution Writing Equation 6.9 for conservation of momentum with the positive direction for velocity to the right, we can find the velocity of the combined mass after the collision:

$$m_1 v_{1i} + m_2 v_{2i} = (m_1 + m_2) v_f$$

$$(0.500 \text{ kg})(4.00 \text{ m/s}) + (0.250 \text{ kg})(-3.00 \text{ m/s}) = (0.750 \text{ kg})(v_f)$$

$$v_f = \boxed{+1.67 \text{ m/s}}$$

Note that the positive sign of v_f tells us that the final velocity is in the positive x direction.

B What is the change in kinetic energy in this collision?

Solution The kinetic energy of the system *before* the collision is

$$KE_i = KE_1 + KE_2 = \tfrac{1}{2} m_1 v_{1i}{}^2 + \tfrac{1}{2} m_2 v_{2i}{}^2$$

$$= \tfrac{1}{2}(0.500 \text{ kg})(4.00 \text{ m/s})^2 + \tfrac{1}{2}(0.250 \text{ kg})(-3.00 \text{ m/s})^2$$

$$= 5.13 \text{ J}$$

The kinetic energy of the system *after* the collision is

$$KE_f = \tfrac{1}{2}(m_1 + m_2) v_f{}^2 = \tfrac{1}{2}(0.750 \text{ kg})(1.67 \text{ m/s})^2 = 1.05 \text{ J}$$

Hence, the *change* in kinetic energy is

$$KE_f - KE_i = \boxed{-4.08 \text{ J}}$$

The minus sign tells us that kinetic energy is lost as a result of the collision. This energy is converted to internal energy, sound, and mechanical vibrations as the two objects collide and deform.

Example 6.6 The Ballistic Pendulum

The ballistic pendulum (Fig. 6.11) is a device used to measure the speed of a fast-moving projectile such as a bullet. The bullet is fired into a large block of wood suspended from some light wires. The bullet is stopped by the block, and the entire system swings through the vertical distance h. It is possible to obtain the initial speed of the bullet by measuring h and the two masses. As an example of the technique, assume that the mass of the bullet m_1 is 5.00 g, the mass of the pendulum m_2 is 1.00 kg, and h is 5.00 cm. Find the initial speed of the bullet, v_{1i}.

Solution The collision between the bullet and the block is perfectly inelastic. Writing the conservation of momentum for the collision in the form of Equation 6.8, we have

$$m_1 v_{1i} = (m_1 + m_2) v_f$$

$$(5.00 \times 10^{-3} \text{ kg})(v_{1i}) = (1.005\ 0 \text{ kg})(v_f) \tag{1}$$

There are two unknowns in this equation, v_{1i} and v_f; the latter is the speed of the block plus embedded bullet *immediately after the collision*. We must look for additional information if we are to complete the problem. Kinetic energy of the bullet-block system is not conserved during an inelastic collision. However, the mechanical energy of the bullet-block-Earth system is conserved after the collision, and so the kinetic energy of this system when the block is at the bottom is transformed into the potential energy of the sys-

(a)   (b)

FIGURE 6.11 (Example 6.6) (a) Diagram of a ballistic pendulum. Note that $\mathbf{v}_f$ is the velocity of the system just *after* the perfectly inelastic collision. (b) Multiflash photograph of a laboratory ballistic pendulum. *(Courtesy of Central Scientific Company)*

tem when the block is at height h:

$$\tfrac{1}{2}(m_1 + m_2)\,v_f^2 = (m_1 + m_2)gh$$

$$v_f = \sqrt{2gh} = \sqrt{2(9.80 \text{ m/s}^2)(5.00 \times 10^{-2} \text{ m})}$$

$$= 0.990 \text{ m/s}$$

With v_f now known, (1) yields v_{1i}:

$$v_{1i} = \frac{(1.005\,0 \text{ kg})(0.990 \text{ m/s})}{5.00 \times 10^{-3} \text{ kg}} = \boxed{199 \text{ m/s}}$$

EXERCISE Explain why it would be incorrect to equate the initial kinetic energy of the incoming bullet to the final gravitational potential energy associated with the bullet-block combination.

Quick Quiz 6.6 An object of mass m moves to the right with a speed v. It collides head-on with an object of mass $3\,m$ moving with speed $v/3$ in the opposite direction. If the two objects stick together, what is the speed of the combined object of mass $4\,m$ after the collision? (a) 0 (b) $v/2$ (c) v (d) $2v$

Quick Quiz 6.7 A skater is using very low friction rollerblades. A friend throws a Frisbee at her, along the straight line along which she is coasting. Describe each of the following events as an elastic, an inelastic, or a perfectly inelastic collision between the skater and the Frisbee: (a) She catches the Frisbee and holds it. (b) She tries to catch the Frisbee but it bounces off her hands and falls to the ground in front of her. (c) She catches the Frisbee and immediately throws it back with the same speed (relative to the ground) to her friend.

In a perfectly inelastic head-on collision between two identical objects, all of the initial kinetic energy of the system of two objects is transformed to forms other than kinetic energy (a) when the final momentum of the system is not zero; (b) when both objects have the same speed; (c) when they are initially moving toward each other along the same line with different speeds.

ELASTIC COLLISIONS

Before collision

(a)

After collision

(b)

FIGURE 6.12 (a) Before and (b) after an elastic head-on collision between two hard spheres.

Now consider two objects that undergo an elastic head-on collision (Fig. 6.12). In this situation, *both momentum and kinetic energy of the system of two objects are conserved;* we can write these conditions as

$$m_1 v_{1i} + m_2 v_{2i} = m_1 v_{1f} + m_2 v_{2f}$$ [6.10]

$$\tfrac{1}{2} m_1 v_{1i}^2 + \tfrac{1}{2} m_2 v_{2i}^2 = \tfrac{1}{2} m_1 v_{1f}^2 + \tfrac{1}{2} m_2 v_{2f}^2$$ [6.11]

where v is positive if an object moves to the right and negative if it moves to the left.

In a typical problem involving elastic collisions, there are two unknown quantities, and Equations 6.10 and 6.11 can be solved simultaneously to find them. An alternative approach, employing a little mathematical manipulation of Equation 6.11, often simplifies this process. To see this, let's cancel the factor $\tfrac{1}{2}$ in Equation 6.11 and rewrite the equation as

$$m_1(v_{1i}^2 - v_{1f}^2) = m_2(v_{2f}^2 - v_{2i}^2)$$

Here we have moved the terms containing m_1 to one side of the equation and those containing m_2 to the other. Next, let us factor both sides of the equation:

$$m_1(v_{1i} - v_{1f})(v_{1i} + v_{1f}) = m_2(v_{2f} - v_{2i})(v_{2f} + v_{2i})$$ [6.12]

We now separate the terms containing m_1 and m_2 in the equation for the conservation of momentum (Eq. 6.10) to get

$$m_1(v_{1i} - v_{1f}) = m_2(v_{2f} - v_{2i})$$ [6.13]

To obtain our final result, we divide Equation 6.12 by Equation 6.13 and find

$$v_{1i} + v_{1f} = v_{2f} + v_{2i}$$

Gathering initial and final values on opposite sides of the equation gives

$$v_{1i} - v_{2i} = -(v_{1f} - v_{2f})$$ [6.14]

This equation, in combination with the condition for conservation of momentum, Equation 6.10, will be used to solve problems dealing with perfectly elastic head-on collisions. According to Equation 6.14, the relative velocity of the two objects before the collision, $v_{1i} - v_{2i}$, equals the negative of the relative velocity of the two objects after the collision, $-(v_{1f} - v_{2f})$. To better understand Equation 6.14, imagine that you are riding along on one of the objects. As you measure the velocity of the other object from your vantage point, you will be measuring the relative velocity of the two objects. In your view of the collision, the other object comes toward you and bounces off, leaving the collision with the same speed, but in the opposite direction. This is just what Equation 6.14 states.

APPLYING **PHYSICS** 6.3

A great deal of expense is involved in repairing automobiles after collisions. It seems that much of this money could be saved if cars were built from hard rubber, so that they made approximately elastic collisions and simply bounced off each other without denting. Why isn't this done?

Explanation Making cars out of hard rubber would save on automobile repair bills, but the results for the drivers and passengers would be disastrous. Imagine that two identical cars traveling toward each other at identical speeds collide head-on. In the approximation that the collision is elastic, the cars bounce off each other and leave the collision with their velocities reversed in direction. Thus, the inhabitants of the cars would suffer a large change in momentum in a short period of time. As a result, the force on the inhabitants would become very large, causing significant bodily injury. For real cars in this situation, the cars would come to rest during the collision, as the front sections of the cars crumpled. Thus, the change in momentum is only half what it was in the elastic case, and the change occurs over the longer time interval during which the cars crumpled. The resulting smaller force on the inhabitants will result in fewer injuries.

◄ **APPLICATION**

WHY NOT USE RUBBER FOR CAR BUMPERS?

PROBLEM-SOLVING *STRATEGY* | One-Dimensional Collisions

The following procedure is recommended for one-dimensional problems involving collisions between two objects:

1. Set up a coordinate axis and define your velocities with respect to this axis. That is, objects moving in the direction selected as positive are considered to have a positive velocity, and those moving in the negative direction have a negative velocity. It is convenient to make your axis coincide with one of the initial velocities.
2. In your sketch, draw all velocity vectors with labels and include all the given information.
3. Write expressions for the momentum of each object before and after the collision. Remember to include the appropriate signs for the velocity vectors.
4. Now write expressions for the *total* momentum of the system of two objects *before* and *after* the collision and equate the two. Remember, it is the momentum of the *system* (the two colliding objects) that is conserved, *not* the momenta of the individual objects.
5. If the collision is *inelastic*, the kinetic energy of the system is not conserved. Proceed to solve the momentum equations for the unknown quantities.
6. If the collision is *elastic*, the kinetic energy of the system is conserved, so you can equate the total kinetic energies before and after the collision. This gives an additional relationship between the velocities.

See p. 171 for the case of two-dimensional collisions.

Webnote 6.2

You can review definitions relating to linear momentum and practice applying conservation of momentum at:

http://id.mind.net/~zona/mstm/ physics/mechanics/momentum/ momentum.html

Example 6.7 **Let's Play Pool**

Two billiard balls move toward one another as in Figure 6.12. The balls have identical masses, and assume that the collision between them is perfectly elastic. If the initial velocities of the balls are $+30$ cm/s and -20 cm/s, what is the velocity of each ball after the collision?

Solution We turn first to Equation 6.10. The equal masses cancel on each side, and after substituting the appropriate values for the initial velocities, we have

$$30 \text{ cm/s} + (-20 \text{ cm/s}) = v_{1f} + v_{2f}$$

$$10 \text{ cm/s} = v_{1f} + v_{2f} \qquad (1)$$

Because kinetic energy is also conserved, we can apply Equation 6.14, which gives

$$30 \text{ cm/s} - (-20 \text{ cm/s}) = v_{2f} - v_{1f}$$

$$50 \text{ cm/s} = v_{2f} - v_{1f} \qquad (2)$$

Solving (1) and (2) simultaneously, we find

$$v_{1f} = \boxed{-20 \text{ cm/s}} \qquad \text{and} \qquad v_{2f} = \boxed{+30 \text{ cm/s}}$$

That is, the balls *exchange velocities!* This is always the case when two objects of equal mass collide elastically head-on.

EXERCISE Find the final velocity of the two balls if the ball with initial velocity −20 cm/s has a mass equal to half that of the ball with initial velocity +30 cm/s.

ANSWER $v_{1f} = -3.0$ cm/s; $v_{2f} = +47$ cm/s

Example 6.8 Two Blocks and a Spring

A block of mass $m_1 = 1.60$ kg, initially moving to the right with a speed of 4.00 m/s on a frictionless horizontal track, collides with a massless spring attached to a second block of mass $m_2 = 2.10$ kg moving to the left with a speed of −2.50 m/s, as in Figure 6.13a. The spring has a spring constant of 600 N/m.

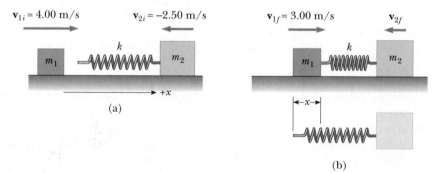

FIGURE 6.13 (Example 6.8)

A At the instant when m_1 is moving to the right with a speed of 3.00 m/s, as in Figure 6.13b, determine the speed of m_2.

Solution We identify the system as the two blocks and the spring. Because the total momentum of the system is conserved, we have

$$m_1 v_{1i} + m_2 v_{2i} = m_1 v_{1f} + m_2 v_{2f}$$

$$v_{2f} = \frac{m_1 v_{1i} + m_2 v_{2i} - m_1 v_{1f}}{m_2}$$

$$= \frac{(1.60 \text{ kg})(4.00 \text{ m/s}) + (2.10 \text{ kg})(-2.50 \text{ m/s}) - (1.60 \text{ kg})(3.00 \text{ m/s})}{2.10 \text{ kg}}$$

$$= \boxed{-1.74 \text{ m/s}}$$

Note that the initial velocity component of m_2 is −2.50 m/s because its direction is to the left. The negative value for v_{2f} means that m_2 is still moving to the left at the instant we are considering.

B Determine the distance the spring is compressed at that instant.

Solution   Because this is an isolated system, and there are no friction or other noncon-servative forces acting within the system, the mechanical energy of the system is conserved:

$$E_i = E_f$$

$$\tfrac{1}{2}m_1 v_{1i}^2 + \tfrac{1}{2}m_2 v_{2i}^2 = \tfrac{1}{2}m_1 v_{1f}^2 + \tfrac{1}{2}m_2 v_{2f}^2 + \tfrac{1}{2}kx^2$$

Substituting the given values and the result of part (a) into this expression gives

$$x = \boxed{0.173 \text{ m}}$$

EXERCISE   Find the velocity of m_1 and the compression in the spring at the instant that m_2 is at rest.

ANSWER   0.719 m/s to the right; 0.251 m

6.4   GLANCING COLLISIONS

In Section 6.2 we showed that the total linear momentum of a system is conserved when the system is isolated (that is, when no external forces act on the system). For a general collision of two objects in three-dimensional space, the conservation of momentum principle implies that the total momentum of the system in each direction is conserved. However, an important subset of collisions takes place in a plane. The game of billiards is a familiar example involving multiple collisions of objects moving on a two-dimensional surface. Let us restrict our attention to a single two-dimensional collision between two objects that takes place in a plane. For such collisions, we obtain two component equations for the conservation of momentum:

$$m_1 v_{1ix} + m_2 v_{2ix} = m_1 v_{1fx} + m_2 v_{2fx}$$

$$m_1 v_{1iy} + m_2 v_{2iy} = m_1 v_{1fy} + m_2 v_{2fy}$$

where we must use three subscripts in this general equation to represent, respectively, (1) the identification of the object, (2) initial and final values, and (3) the velocity component.

Consider a two-dimensional problem in which an object of mass m_1 collides with an object of mass m_2 that is initially at rest, as in Figure 6.14. After the collision, m_1 moves at an angle θ with respect to the horizontal, and m_2 moves at an angle ϕ with respect to the horizontal. This is called a *glancing* collision. Applying the law of conservation of momentum in component form, and noting that the initial y component of momentum is zero, we have

x component:	$m_1 v_{1i} + 0 = m_1 v_{1f}\cos\theta + m_2 v_{2f}\cos\phi$	**[6.15]**
y component:	$0 + 0 = m_1 v_{1f}\sin\theta - m_2 v_{2f}\sin\phi$	**[6.16]**

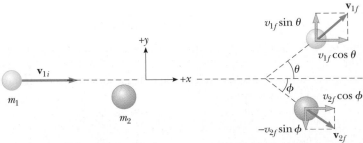

(a) Before the collision   (b) After the collision

FIGURE 6.14   (a) Before and (b) after a glancing collision between two balls.

If the collision is elastic, we can write a third equation for kinetic energy, in the form

$$\tfrac{1}{2}m_1 v_{1i}{}^2 = \tfrac{1}{2}m_1 v_{1f}{}^2 + \tfrac{1}{2}m_2 v_{2f}{}^2 \qquad [6.17]$$

If we know the initial velocity v_{1i} and the masses, we are left with four unknowns (v_{1f}, v_{2f}, θ, and ϕ). Because we have only three equations, one of the four remaining quantities must be given to determine the motion after the collision from conservation principles alone.

If the collision is inelastic, the kinetic energy of the system is *not* conserved, and Equation 6.17 does *not* apply.

PROBLEM-SOLVING STRATEGY   **Two-Dimensional Collisions**

The following procedure is recommended when dealing with two-dimensional problems involving collisions between two objects.

1. Set up coordinate axes and define your velocities with respect to these axes. It is convenient to have the *x* axis coincide with one of the initial velocities.

2. In your sketch, draw and label all velocity vectors and include all the given information. Write expressions for the *x* and *y* components of the momentum of each object before and after the collision. Remember to include the appropriate signs for the components of the velocity vectors. For example, if an object is moving in the negative *x* direction, the *x* component of its velocity must be negative. It is essential that you pay careful attention to signs.

3. Write expressions for the *total* momentum of the system of two objects in the *x* direction *before* and *after* the collision, and equate the two. Repeat this procedure for the total momentum in the *y* direction.

4. Proceed to solve the momentum equations for the unknown quantities. If the collision is inelastic, kinetic energy of the system is *not* conserved, and additional information is probably required. If the collision is perfectly inelastic, the final velocities of the two objects are equal.

5. If the collision is elastic, kinetic energy of the system is conserved, and you can equate the total kinetic energy before the collision to the total kinetic energy after the collision. This provides an additional relationship between the velocity magnitudes.

FIGURE 6.15 (Example 6.9) A top view of a perfectly inelastic collision between a car and a van.

Example 6.9   Collision at an Intersection

A 1 500-kg car traveling east with a speed of 25.0 m/s collides at an intersection with a 2 500-kg van traveling north at a speed of 20.0 m/s as shown in Figure 6.15. Find the direction and magnitude of the velocity of the wreckage after the collision, assuming that the vehicles undergo a perfectly inelastic collision (that is, they stick together) and assuming that friction between the vehicles and the road can be neglected.

Solution   Let us choose east to be along the positive *x* direction and north to be along the positive *y* direction as in Figure 6.15. Before the collision, the only object having momentum in the *x* direction is the car. Thus, the magnitude of the total initial momentum of the system (car plus van) in the *x* direction is

$$\sum p_{xi} = m_{car} v_{i\,car} = (1\,500\text{ kg})(25.0\text{ m/s}) = 3.75 \times 10^4\text{ kg·m/s}$$

The wreckage moves at an angle θ and speed v after the collision. The magnitude of the total momentum in the x direction after the collision is

$$\sum p_{xf} = (m_{car} + m_{van})v \cos \theta = (4\ 000\ \text{kg})v \cos \theta$$

Because the total momentum in the x direction is conserved, we can equate these two expressions to obtain

$$(3.75 \times 10^4)\ \text{kg·m/s} = (4\ 000\ \text{kg})v \cos \theta \qquad \textbf{(1)}$$

Similarly, the total initial momentum of the system in the y direction is that of the van, whose magnitude is equal to $(2\ 500\ \text{kg})(20.0\ \text{m/s})$. Applying conservation of momentum to the y direction, we have

$$\sum p_{yi} = \sum p_{yf}$$

$$(2\ 500\ \text{kg})(20.0\ \text{m/s}) = (4\ 000\ \text{kg})v \sin \theta$$

$$5.00 \times 10^4\ \text{kg·m/s} = (4\ 000\ \text{kg})v \sin \theta \qquad \textbf{(2)}$$

If we divide (2) by (1), we find

$$\tan \theta = \frac{5.00 \times 10^4}{3.75 \times 10^4} = 1.33$$

$$\theta = \boxed{53.1°}$$

Substituting this angle into (2), the value of v is found to be

$$v = \frac{5.00 \times 10^4\ \text{kg·m/s}}{(4\ 000\ \text{kg})\sin 53.1°} = \boxed{15.6\ \text{m/s}}$$

EXERCISE A 3.00-kg object with an initial velocity of $+5.00$ m/s in the positive x direction collides with and sticks to a 2.00-kg object with an initial velocity of -3.00 m/s in the negative y direction. Find the final components of velocity of the composite object.

ANSWER $v_{xf} = 3.00$ m/s; $v_{yf} = -1.20$ m/s

6.5 ROCKET PROPULSION

When ordinary vehicles such as cars and locomotives are propelled, the driving force for the motion is friction. In the case of the car, this driving force is exerted by the road on the car. A locomotive "pushes" against the tracks; hence, the driving force is the reaction force exerted by the tracks on the locomotive. However, a rocket moving in space has no road or tracks to "push" against. Therefore, the source of rocket propulsion must be different from that of earthbound vehicles. **The operation of a rocket depends on the law of conservation of momentum as applied to a system, where the system is the rocket plus its ejected fuel.**

The propulsion of a rocket can be understood by first considering the pitching machine on ice in Example 6.3. As a baseball is fired from the machine, the baseball receives momentum $m\mathbf{v}$ in one direction, and the machine receives a momentum of equal magnitude in the opposite direction. As each baseball is fired, the machine moves faster, so that a large velocity can be established by firing many baseballs.

In a similar manner, as a rocket moves in free space (a vacuum), its momentum changes when some of its mass is released in the form of ejected gases. Because the ejected gases acquire some momentum, the rocket receives a compensating momentum in the opposite direction. Therefore, the rocket is accelerated as a result of the "push," or thrust, from the exhaust gases. Note that the rocket represents the *inverse* of an inelastic collision; that is, momentum is conserved, but the kinetic energy of the system is *increased* (at the expense of energy stored in the fuel of the rocket).

$$M + \Delta m$$

$$\mathbf{p}_i = (M + \Delta m)\mathbf{v}$$

(a)

$$M$$

$$\Delta m$$

$$\mathbf{v} + \Delta \mathbf{v}$$

(b)

FIGURE 6.16 Rocket propulsion. (a) The initial mass of the rocket and fuel is $M + \Delta m$ at a time t, and its speed is v. (b) At a time $t + \Delta t$, the rocket's mass has been reduced to M, and an amount of fuel Δm has been ejected. The rocket's speed increases by an amount Δv.

Suppose that at some time t, the momentum of the rocket plus the fuel is $(M + \Delta m)v$ (Fig. 6.16a). During a short time interval Δt, the rocket ejects fuel of mass Δm and the rocket's speed therefore increases to $v + \Delta v$ (Fig. 6.16b). If the fuel is ejected with speed v_e *relative to the rocket*, the speed of the fuel relative to a stationary frame of reference is $v - v_e$. Thus, if we equate the total initial momentum of the system with the total final momentum, we have

$$(M + \Delta m)v = M(v + \Delta v) + \Delta m(v - v_e)$$

Simplifying this expression gives

$$M \Delta v = \Delta m(v_e)$$

Furthermore, the increase Δm in the exhaust mass corresponds to an equal decrease in the rocket mass, so that $\Delta m = -\Delta M$. Using this fact, we have

$$M \Delta v = -v_e \Delta M \qquad [6.18]$$

This result, together with the methods of calculus, can be used to obtain the following important equation:

$$v_f - v_i = v_e \ln\left(\frac{M_i}{M_f}\right) \qquad [6.19]$$

where M_i is the initial mass of the rocket plus fuel and M_f is the final mass of the rocket plus its remaining fuel. This is the basic expression for rocket propulsion. It tells us that the increase in speed is proportional to the exhaust speed v_e and to the natural log of M_i/M_f. Because the maximum ratio of M_i to M_f for a single-stage rocket is about 10:1, the increase in speed can reach $v_e \ln 10 = 2.3 v_e$ or about twice the exhaust speed! Therefore, the exhaust speed should be very high. Currently, typical rocket exhaust speeds are several km/s.

The **thrust** on the rocket is defined as the force exerted on the rocket by the ejected exhaust gases. We can obtain an expression for the instantaneous thrust from Equation 6.18:

Rocket thrust

$$\text{Instantaneous thrust} = Ma = M\frac{\Delta v}{\Delta t} = \left|v_e \frac{\Delta M}{\Delta t}\right| \qquad [6.20]$$

Here we see that the thrust increases as the exhaust speed increases and as the rate of change of mass $\Delta M/\Delta t$ (the burn rate) increases.

APPLYING **PHYSICS** 6.4

When Robert Goddard proposed the possibility of rocket-propelled vehicles, the *New York Times* agreed that such vehicles would be useful and successful within Earth's atmosphere ("Topics of the Times," the *New York Times*, January 13, 1920, p. 12). But the *Times* balked at the idea of using such a rocket in the vacuum of space, noting that ". . . its flight would be neither accelerated nor maintained by the explosion of the charges it then might have left. To claim that it would be is to deny a fundamental law of dynamics, and only Dr. Einstein and his chosen dozen, so few and fit, are licensed to do that . . . That Professor Goddard, with his 'chair' in Clark College and the countenancing of the Smithsonian Institution, does not know the relation of action to reaction, and of the need to have something better than a vacuum against which to react—to say that would be absurd. Of course, he only seems to lack the knowledge ladled out daily in high schools." What did the writer of these passages overlook?

Explanation The writer of these passages was making a common mistake in believing that a rocket works by expelling gases which push on something, propelling the rocket forward. With this belief, it is impossible to see how a rocket fired in empty space would work. Gases do not need to push on anything—it is the act of expelling the gases itself that pushes the rocket forward. This can be argued from Newton's third law—the rocket pushes the gases backward, resulting in the gases pushing the rocket forward. It can also be argued from conservation of momentum—as the gases gain momentum in one direction, the rocket must gain momentum in the opposite direction to conserve the original momentum of the rocket-gas system.

The *New York Times* did publish a retraction 49 years later ("A Correction," *New York Times,* July 17, 1969, p. 43), while the *Apollo 11* astronauts were on their way to the Moon. It appeared on a page with two other articles entitled, "Fundamentals of Space Travel" and "Spacecraft, Like Squid, Maneuver by 'Squirts,'" and contained the following passages: ". . . an editorial feature of the *New York Times* dismissed the notion that a rocket could function in a vacuum and commented on the ideas of Robert H. Goddard . . . Further investigation and experimentation have confirmed the findings of Isaac Newton in the 17th century, and it is now definitely established that a rocket can function in a vacuum as well as in an atmosphere. The *Times* regrets the error."

Example 6.10 A Rocket in Space

A rocket moving in free space has a speed of 3.0×10^3 m/s relative to Earth. Its engines are turned on, and fuel is ejected in a direction opposite the rocket's motion at a speed of 5.0×10^3 m/s relative to the rocket.

A What is the speed of the rocket relative to Earth once its mass is reduced to one-half its mass before ignition?

Solution Applying Equation 6.19, we have

$$v_f = v_i + v_e \ln\left(\frac{M_i}{M_f}\right)$$

$$= 3.0 \times 10^3 \text{ m/s} + (5.0 \times 10^3 \text{ m/s}) \ln\left(\frac{M_i}{0.5 \, M_i}\right)$$

$$= \boxed{6.5 \times 10^3 \text{ m/s}}$$

B What is the thrust on the rocket if it burns fuel at the rate of 50 kg/s?

Solution Using Equation 6.20, we have

$$\text{Thrust} = \left| v_e \frac{\Delta M}{\Delta t} \right| = (5.0 \times 10^3 \text{ m/s})(50 \text{ kg/s}) = \boxed{2.5 \times 10^5 \text{ N}}$$

SUMMARY

The **linear momentum** of an object of mass m moving with a velocity $\mathbf{v}$ is defined to be

$$\mathbf{p} \equiv m\mathbf{v} \qquad \text{[6.1]}$$

The **impulse** of a constant force $\mathbf{F}$ acting on an object is equal to the product of the force and the time interval during which the force acts:

$$\text{Impulse} \equiv \mathbf{F} \, \Delta t \qquad \text{[6.5]}$$

The **impulse-momentum theorem** states that the impulse of a constant force on an object is equal to the change in momentum of the object:

$$\mathbf{F} \, \Delta t = \Delta \mathbf{p} = m\mathbf{v}_f - m\mathbf{v}_i \qquad \text{[6.4]}$$

Conservation of momentum applied to a system of two interacting objects states that, when no external forces act on the system, the total momentum of the system before the collision is equal to the total momentum of the system after the collision.

$$m_1\mathbf{v}_{1i} + m_2\mathbf{v}_{2i} = m_1\mathbf{v}_{1f} + m_2\mathbf{v}_{2f}$$ [6.7]

An **inelastic collision** is one in which momentum of the system is conserved but kinetic energy is not. A **perfectly inelastic collision** is one in which the colliding objects stick together after the collision. An **elastic collision** is one in which both momentum and kinetic energy of the system are conserved.

In glancing collisions, conservation of momentum can be applied along two perpendicular directions, that is, along an x axis and a y axis.

CONCEPTUAL QUESTIONS

1. Use conservation of momentum to discuss what happens to a passenger in a car crash. Discuss the use of a seat belt in bringing the passenger to rest. Discuss accidents that cause whiplash neck injuries.

2. Two objects initially traveling along the same straight line meet in a perfectly elastic collision. After they collide, will their final motion necessarily be along the same straight line as their initial motion?

3. In perfectly inelastic collisions between two objects, there are some instances in which all of the original kinetic energy is transformed to forms other than kinetic. Give an example of such a situation.

4. If two objects collide and one is initially at rest, is it possible for both to be at rest after the collision? Is it possible for one to be at rest after the collision? Explain.

5. A ball of clay of mass m is thrown with a speed v against a brick wall. The clay sticks to the wall and stops. Is the principle of conservation of momentum violated in this example?

6. A skater is standing still on a frictionless ice rink. Her friend throws a Frisbee straight at her. In which of the following cases is the largest momentum transferred to the skater? (a) The skater catches the Frisbee and holds onto it. (b) The skater catches the Frisbee momentarily, but then drops it vertically downward. (c) The skater catches the Frisbee, holds it momentarily, and throws it back to her friend.

7. You are standing perfectly still and then you take a step forward. Before the step your momentum was zero, but afterward you have some momentum. Is the conservation of momentum violated in this case?

8. If two particles have equal kinetic energies, are their momenta necessarily equal? Explain.

9. A more ordinary example of conservation of momentum than a rocket ship occurs in a kitchen dishwashing machine. In this device, water at high pressure is forced out of small holes on the spray arms. Show how conservation of momentum explains why the arms rotate, directing water to all the dishes.

10. If two automobiles collide, they usually do not stick together. Does this mean the collision is elastic? Explain why a head-on collision is likely to be more dangerous than other types of collisions.

11. Why are rockets often built with several stages so that the fuel tanks and other assemblies can be discarded once they are no longer needed?

12. Consider a perfectly inelastic collision between a car and a large truck. Which vehicle loses more kinetic energy as a result of the collision?

13. Your physical education teacher, who knows something about physics, throws you a tennis ball at a certain velocity, and you catch it. You are now given the following choice: The teacher can throw you a medicine ball (which is much more massive than the tennis ball) with the same velocity as the tennis ball, the same momentum, or the same kinetic energy. Which choice would you make, in order to make the easiest catch, and why?

14. You are watching a movie about a superhero and notice that the superhero hovers in the air and throws a piano at some bad guys while remaining stationary in the air. What's wrong with this scenario?

15. In golf, novice players are often advised to be sure to "follow through" with their swing. Why does this make the ball travel a longer distance? If a shot is taken near the green, very little follow-through is required. Why?

16. An airbag inflates when a collision occurs, which protects the passenger (the dummy, in this case) from serious injury (Fig. Q6.16). Why does the airbag soften the blow? Discuss the physics involved in this dramatic photograph.

FIGURE Q6.16   *(Courtesy of Saab)*

17. A sharpshooter fires a rifle while standing with the butt of the gun against his shoulder. If the forward momentum of a bullet is the same as the backward momentum of the gun, why isn't it as dangerous to be hit by the gun as by the bullet?

18. A large bedsheet is held vertically by two students. A third student, who happens to be the star pitcher on the baseball team, throws a raw egg at the sheet. Explain why the egg does not break when it hits the sheet, regardless of its

initial speed. (If you try this, make sure the pitcher hits the sheet near its center, and do not allow the egg to fall on the floor after being caught.)

PROBLEMS

1, 2, 3 = straightforward, intermediate, challenging   ☐ = full solution available in Student Solutions Manual/Study Guide

web = solution posted on **http://info.brookscole.com/serway**   🔥 = biomedical application

Section 6.1   Momentum and Impulse

1. Show that the kinetic energy of a particle of mass m is related to the magnitude of the momentum p of that particle by $KE = p^2/2m$. [*Note:* This expression is invalid for relativistic particles (those traveling at speeds near that of light).]

2. A tennis player receives a shot with the ball (0.060 0 kg) traveling horizontally at 50.0 m/s and returns the shot with the ball traveling horizontally at 40.0 m/s in the opposite direction. (a) What is the impulse delivered to the ball by the racquet? (b) What work does the racquet do on the ball?

3. Calculate the magnitude of the linear momentum for the following cases: (a) a proton with mass 1.67×10^{-27} kg, moving with a speed of 5.00×10^6 m/s; (b) a 15.0-g bullet moving with a speed of 300 m/s; (c) a 75.0-kg sprinter running with a speed of 10.0 m/s; (d) Earth (mass = 5.98×10^{24} kg) moving with an orbital speed equal to 2.98×10^4 m/s.

4. A 0.10-kg ball is thrown straight up into the air with an initial speed of 15 m/s. Find the momentum of the ball (a) at its maximum height and (b) halfway to its maximum height.

5. A pitcher claims he can throw a 0.145-kg baseball with as much momentum as a 3.00-g bullet moving with a speed of 1.50×10^3 m/s. (a) What must the baseball's speed be if the pitcher's claim is valid? (b) Which has greater kinetic energy, the ball or the bullet?

6. A stroboscopic photo of a club hitting a golf ball like that shown in Figure 6.3 was made by Harold Edgerton in 1933. The ball was initially at rest and the club was shown to be in contact with the club for about 0.0020 s. Also, the ball was found to end up with a speed of 2.0×10^2 ft/s. Assuming that the golf ball had a mass of 55 g, find the average force exerted by the club on the ball.

7. A professional diver performs a dive from a platform 10 m above the water surface. Estimate the order of magnitude of the average impact force she experiences in her collision with the water. State the quantities you take as data and their values.

8. A 60.0-kg woman jumps from a burning building and falls 10.0 m before making contact with a safety net, which stops her in 0.120 s. What is the average force exerted by the net on her?

9. A car is stopped for a traffic signal. When the light turns green, the car accelerates, increasing its speed from 0 to 5.20 m/s in 0.832 s. What are the magnitudes of the linear impulse and the average total force experienced by a 70.0-kg passenger in the car during this time?

10. A friend claims that he can hold on to a 12-kg child in a 60-mi/h collision lasting for 0.05 s as long as he has his seat belt on. Prove that the violent forces during a collision will tear the child from his arms. (A child should always be in a toddler seat secured with a seat belt in the back seat of a car.)

11. The force shown in the force-time diagram in Figure P6.11 acts on a 1.5-kg object. Find (a) the impulse of the force, (b) the final velocity of the object if it is initially at rest, and (c) the final velocity of the object if it is initially moving along the x axis with a velocity of -2.0 m/s.

FIGURE P6.11

12. The force of magnitude F_x acting in the x direction on a 2.00-kg particle varies in time as shown in Figure P6.12. Find (a) the impulse of the force, (b) the final velocity of the particle if it is initially at rest, and (c) the final velocity of the particle if it is initially moving along the x axis with a velocity of -2.00 m/s.

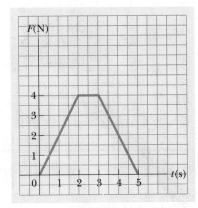

FIGURE P6.12

13. The forces shown in the force-time diagram in Figure P6.13 act on a 1.5-kg particle. Find (a) the impulse for the interval $t = 0$ to $t = 3.0$ s and (b) the impulse for the interval $t = 0$ to $t = 5.0$ s. (c) If the forces act on a 1.5-kg particle that is initially at rest, find the particle's speed at $t = 3.0$ s and at $t = 5.0$ s.

CONCEPTUAL QUESTIONS

19. If you jump from a table and land with your knees locked, why are you more likely to be hurt than if you land with your legs relaxed?

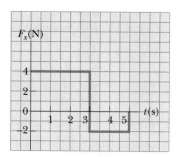

FIGURE P6.13

14. A fire hose sends 1 000 kg of water per minute against a burning building. The water strikes the building at 20.0 m/s and does not bounce back. (a) What is the rate of change of momentum of the water? (b) What force does the building exert on the water? (c) What force does the water exert on the building?

15. The front of a 1 400 kg car is designed to absorb the shock of a collision by having a "crumple zone" in which the front 1.20 m of the car collapses in absorbing the shock of a collision. If a car traveling 25.0 m/s stops uniformly in 1.20 m, (a) how long does the collision last, (b) what is the magnitude of the average force on the car, and (c) what is the acceleration of the car? Express the acceleration as a multiple of the acceleration of gravity.

16. A pitcher throws a 0.15-kg baseball so that it crosses home plate horizontally with a speed of 20 m/s. It is hit straight back at the pitcher with a final speed of 22 m/s. (a) What is the impulse delivered to the ball? (b) Find the average force exerted by the bat on the ball if the two are in contact for 2.0×10^{-3} s.

17. A fire hose sends 20.0 kg of water per second onto a burning building. The water strikes the roof horizontally at 40.0 m/s and is deflected as shown in Figure P6.17. What are the magnitude and direction of the force exerted by the water on the roof? [*Hint:* Treat the horizontal and vertical components separately.]

FIGURE P6.17

Section 6.2 Conservation of Momentum

18. A 730-N man stands in the middle of a frozen pond of radius 5.0 m. He is unable to get to the other side because of a lack of friction between his shoes and the ice. To overcome this difficulty, he throws his 1.2-kg physics textbook horizontally toward the north shore, at a speed of 5.0 m/s. How long does it take him to reach the south shore?

19. High-speed stroboscopic photographs show that the head of a 200-g golf club is traveling at 55 m/s just before it strikes a 46-g golf ball at rest on a tee. After the collision, the club head travels (in the same direction) at 40 m/s. Find the speed of the golf ball just after impact.

20. A rifle with a weight of 30 N fires a 5.0-g bullet with a speed of 300 m/s. (a) Find the recoil speed of the rifle. (b) If a 700-N man holds the rifle firmly against his shoulder, find the recoil speed of man and rifle.

21. A 45.0-kg girl is standing on a 150-kg plank. The plank, originally at rest, is free to slide on a frozen lake, which is a flat, frictionless surface. The girl begins to walk along the plank at a constant velocity of 1.50 m/s to the right relative to the plank. (a) What is her velocity relative to the ice surface? (b) What is the velocity of the plank relative to the ice surface?

22. A 65.0-kg person throws a 0.045 0-kg snowball forward with a ground speed of 30.0 m/s. A second person, with a mass of 60.0 kg, catches the snowball. Both people are on skates. The first person is initially moving forward with a speed of 2.50 m/s, and the second person is initially at rest. What are the velocities of the two people after the snowball is exchanged? Disregard the friction between the skates and the ice.

23. In Section 6.2 we implied that the kinetic energy of Earth can be ignored when considering the energy of a system consisting of Earth and a dropped ball of mass m_b. Verify this by first setting up a ratio of Earth's kinetic energy to that of the ball as they collide. Then use conservation of momentum to show that

$$\frac{v_E}{v_b} = -\frac{m_b}{m_E}$$

and

$$\frac{KE_E}{KE_b} = \frac{m_b}{m_E}$$

Find the order of magnitude of the ratio of kinetic energies, based on data that you specify.

24. An amoeba of mass 1.0×10^{-12} kg propels itself through the water by blowing a jet of water through a tiny orifice. Suppose the amoeba ejects water with a speed of 1.0×10^{-4} m/s and at a rate of 1.0×10^{-13} kg/s. Assume the water is continuously replenished so the mass of the amoeba stays the same. (a) If there were no force on the amoeba other than the reaction force caused by the emerging jet, what would be the acceleration of the amoeba? (b) If the amoeba moves with constant velocity through the water, what is the force exerted by the surrounding water (exclusive of the jet) on the amoeba?

Section 6.3 Collisions

Section 6.4 Glancing Collisions

25. A 7.00-kg bowling ball collides head-on with a 2.00-kg bowling pin, which was originally at rest. The pin flies forward with a speed of 3.00 m/s. If the ball continues forward with a speed of 1.80 m/s, what was the initial speed of the ball? Ignore rotation of the ball.

26. A 75.0-kg ice skater, moving at 10.0 m/s, crashes into a stationary skater of equal mass. After the collision, the two skaters move as a unit at 5.00 m/s. Suppose the average force a skater can experience without breaking a bone is 4 500 N. If the impact time is 0.100 s, does a bone break?

27. A railroad car of mass 2.00×10^4 kg moving at 3.00 m/s collides and couples with two coupled railroad cars, each of the same mass as the single car and moving in the same direction at 1.20 m/s. (a) What is the speed of the three coupled cars after the collision? (b) How much kinetic energy is lost in the collision?

28. A 7.0-g bullet is fired into a 1.5-kg ballistic pendulum. The bullet emerges from the block with a speed of 200 m/s, and the block rises to a maximum height of 12 cm. Find the initial speed of the bullet.

29. A 0.030-kg bullet is fired vertically at 200 m/s into a **web** 0.15-kg baseball that is initially at rest. How high does the combination rise after the collision, assuming the bullet embeds itself in the ball?

30. An 8.00-g bullet is fired into a 250-g block that is initially at rest at the edge of a table of height 1.00 m (Fig. P6.30). The bullet remains in the block, and after the impact the block lands 2.00 m from the bottom of the table. Determine the initial speed of the bullet.

8.00 g

250 g

1.00 m

2.00 m

FIGURE P6.30

31. Gayle runs at a speed of 4.00 m/s and dives on a sled, initially at rest on the top of a frictionless snow-covered hill. After she has descended a vertical distance of 5.00 m, her brother, who is initially at rest, hops on her back and together they continue down the hill. What is their speed at the bottom of the hill if the total vertical drop is 15.0 m? Gayle's mass is 50.0 kg, the sled has a mass of 5.00 kg, and her brother has a mass of 30.0 kg.

32. A 1 200-kg car traveling initially with a speed of 25.0 m/s in an easterly direction crashes into the rear end of a 9 000-kg truck moving in the same direction at 20.0 m/s (Fig. P6.32). The velocity of the car right after the collision is 18.0 m/s to the east. (a) What is the velocity of the truck right after the collision? (b) How much mechanical energy is lost in the collision? Account for this loss in energy.

33. A 12.0-g bullet is fired horizontally into a 100-g wooden block that is initially at rest on a frictionless horizontal surface and connected to a spring having spring constant 150 N/m. The bullet becomes embedded in the block. If the bullet-block system compresses the spring by a maximum of 80.0 cm, what was the speed of the bullet at impact with the block?

34. (a) Three carts of masses 4.0 kg, 10 kg, and 3.0 kg move on a frictionless horizontal track with speeds of 5.0 m/s, 3.0 m/s, and 4.0 m/s, as shown in Figure P6.34. The carts stick together after colliding. Find the final velocity of the three carts. (b) Does your answer require that all carts collide and stick together at the same time?

5.0 m/s   3.0 m/s   −4.0 m/s

4.0 kg   10 kg   3.0 kg

FIGURE P6.34

35. A 5.00-g object moving to the right at 20.0 cm/s makes an elastic head-on collision with a 10.0-g object that is initially at rest. Find (a) the velocity of each object after the collision and (b) the fraction of the initial kinetic energy transferred to the 10.0-g object.

36. A 10.0-g object moving to the right at 20.0 cm/s makes an elastic head-on collision with a 15.0-g object moving in the opposite direction at 30.0 cm/s. Find the velocity of each object after the collision.

37. A 25.0-g object moving to the right at 20.0 cm/s overtakes and collides elastically with a 10.0-g object moving in the same direction at 15.0 cm/s. Find the velocity of each object after the collision.

38. Four railroad cars, each of mass 2.50×10^4 kg, are coupled together and coasting along horizontal tracks at speed v_i toward the south. A very strong but foolish movie actor, riding on the second car, uncouples the front car and gives it a big push, increasing its speed to 4.00 m/s south. The remaining three cars continue moving south, now at 2.00 m/s. (a) Find the initial speed of the cars. (b) How much work did the actor do?

39. A 7.00-g bullet, when fired from a gun into a 1.00-kg block of wood held in a vise, penetrates the block to a depth of 8.00 cm. This block of wood is placed on a frictionless horizontal surface, and a second 7.00-g bullet is fired from the gun into the block. To what depth does the bullet penetrate the block in this case?

40. A billiard ball rolling across a table at 1.50 m/s makes a head-on elastic collision with an identical ball. Find the speed of each ball after the collision (a) when the second ball is initially at rest, (b) when the second ball is moving toward the first at a speed of 1.00 m/s, and (c) when the second ball is moving away from the first at a speed of 1.00 m/s.

25.0 m/s   20.0 m/s

Before

18.0 m/s   **v**

After

FIGURE P6.32

41. A 90-kg fullback moving east with a speed of 5.0 m/s is tackled by a 95-kg opponent running north at 3.0 m/s. If the collision is perfectly inelastic, calculate (a) the velocity of the players just after the tackle and (b) the kinetic energy lost as a result of the collision. Can you account for the missing energy?

42. An 8.00-kg object moving east at 15.0 m/s on a frictionless horizontal surface collides with a 10.0-kg object that is initially at rest. After the collision, the 8.00-kg object moves south at 4.00 m/s. (a) What is the velocity of the 10.0-kg object after the collision? (b) What percentage of the initial kinetic energy is lost in the collision?

43. A 2 000-kg car moving east at 10.0 m/s collides with a 3 000-kg car moving north. The cars stick together and move as a unit after the collision, at an angle of 40.0° north of east and at a speed of 5.22 m/s. Find the speed of the 3 000-kg car before the collision.

44. Two automobiles of equal mass approach an intersection. One vehicle is traveling with velocity 13.0 m/s toward the east and the other is traveling north with speed v_{2i}. Neither driver sees the other. The vehicles collide in the intersection and stick together, leaving parallel skid marks at an angle of 55.0° north of east. The speed limit for both roads is 35 mi/h, and the driver of the northward-moving vehicle claims he was within the speed limit when the collision occurred. Is he telling the truth?

45. A billiard ball moving at 5.00 m/s strikes a stationary ball of the same mass. After the collision, the first ball moves at 4.33 m/s at an angle of 30° with respect to the original line of motion. (a) Find the velocity (magnitude and direction) of the second ball after collision. (b) Was this an inelastic collision or an elastic collision?

ADDITIONAL PROBLEMS

46. In research in cardiology and exercise physiology it is often important to know the mass of blood pumped by a person's heart in one stroke. This information can be obtained by means of a *ballistocardiograph*. The instrument works as follows. The subject lies on a horizontal pallet floating on a film of air. Friction on the pallet is quite negligible. Initially the momentum of the system is zero. When the heart beats, it expels a mass m of blood into the aorta with speed v, and the body and platform move in the opposite direction with speed V. The blood speed can be determined independently (for example, by observing the Doppler shift of ultrasound). Assume it is 50.0 cm/s in one typical trial. The mass of the subject plus the pallet is 54.0 kg. The pallet moves 6.00×10^{-5} m in 0.160 s after one heartbeat. Calculate the mass of blood that leaves the heart. Assume the mass of blood is negligible compared to the total mass of the person. This simplified example illustrates the principle of ballistocardiography, but in practice a more sophisticated model of heart function is used.

47. A 0.400-kg soccer ball approaches a player horizontally with a speed of 15.0 m/s. The player illegally strikes the ball with her hand and causes it to move in the opposite direction with a speed of 22.0 m/s. What impulse was delivered to the ball by the player?

48. Consider a frictionless track as shown in Figure P6.48. A block of mass $m_1 = 5.00$ kg is released from Ⓐ. It makes a head-on elastic collision at Ⓑ with a block of mass

$m_2 = 10.0$ kg that is initially at rest. Calculate the maximum height to which m_1 rises after the collision.

FIGURE P6.48

49. Most of us know intuitively that in a head-on collision between a large dump truck and a subcompact car, you are better off being in the truck than in the car. Why is this? Many people imagine that the collision force exerted on the car is much greater than that experienced by the truck. To substantiate this view, they point out that the car is crushed, whereas the truck is only dented. This idea of unequal forces, of course, is false. Newton's third law tells us that both objects experience forces of the same magnitude. The truck suffers less damage because it is made of stronger metal. But what about the two drivers? Do they experience the same forces? To answer this question, suppose that each vehicle is initially moving at 8.00 m/s and that they undergo a perfectly inelastic head-on collision. Each driver has mass 80.0 kg. Including the drivers, the total vehicle masses are 800 kg for the car and 4 000 kg for the truck. If the collision time is 0.120 s, what force does the seat belt exert on each driver?

50. As shown in Figure P6.50, a bullet of mass m and speed v passes completely through a pendulum bob of mass M. The bullet emerges with a speed of $v/2$. The pendulum bob is suspended by a stiff rod of length ℓ and negligible mass. What is the minimum value of v such that the pendulum bob will barely swing through a complete vertical circle?

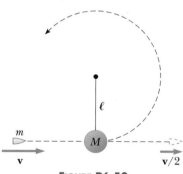

FIGURE P6.50

51. A 2.0-g particle moving at 8.0 m/s makes a perfectly elastic head-on collision with a resting 1.0-g object. (a) Find the speed of each after the collision. (b) If the stationary particle has a mass of 10 g, find the speed of each particle after the collision. (c) Find the final kinetic energy of the incident 2.0-g particle in the situations described in (a) and (b). In which case does the incident particle lose more kinetic energy?

52. A 0.400-kg bead slides on a curved frictionless wire, starting from rest at point Ⓐ in Figure P6.52. At point Ⓑ the bead collides elastically with a 0.600-kg ball at rest. Find the distance the ball rises as it moves up the wire.

FIGURE P6.56

high up the track does the 0.500-kg block travel back after the collision? (c) How far away from the bottom of the table does the 1.00-kg block land, given that the table is 2.00 m high? (d) How far away from the bottom of the table does the 0.500-kg block eventually land?

$m_1 = 0.500$ kg
$m_2 = 1.00$ kg
$h_1 = 2.50$ m
$h_2 = 2.00$ m

FIGURE P6.57

58. Tarzan, whose mass is 80.0 kg, swings from a 3.00-m vine that is horizontal when he starts. At the bottom of his arc, he picks up 60.0-kg Jane in a perfectly inelastic collision. What is the height of the highest tree limb they can reach on their upward swing?

59. A small block of mass $m_1 = 0.500$ kg is released from rest at the top of a curved wedge of mass $m_2 = 3.00$ kg, which sits on a frictionless horizontal surface as in Figure P6.59a. When the block leaves the wedge, its velocity is measured to be 4.00 m/s to the right, as in Figure P6.59b. (a) What is the velocity of the wedge after the block reaches the horizontal surface? (b) What is the height h of the wedge?

(a)   (b)

FIGURE P6.59

60. Two carts of equal mass $m = 0.250$ kg are placed on a frictionless track that has a light spring of force constant $k = 50.0$ N/m attached to one end of it, as in Figure P6.60. The red cart is given an initial velocity of

---

Ⓐ

1.50 m

Ⓑ

FIGURE P6.52

53. An 80-kg man, standing erect, steps off a 3.0-m high diving platform and begins to fall from rest. The man comes to rest 2.0 s after reaching the water. What average force did the water exert on him?

54. A 12.0-g bullet is fired horizontally into a 100-g wooden
web block initially at rest on a horizontal surface. After impact, the block slides 7.5 m before coming to rest. If the coefficient of kinetic friction between block and surface is 0.650, what was the speed of the bullet immediately before impact?

55. A 60.0-kg man running at an initial speed of 4.00 m/s jumps onto a 120-kg cart initially at rest (Figure P6.55). He slides on the cart's top surface and finally comes to rest relative to the cart. The coefficient of kinetic friction between the man and the cart is 0.400. Friction between the cart and ground can be neglected. (a) Find the final speed of the man and cart relative to the ground. (b) Find the frictional force acting on the men while he is sliding across the top surface of the cart. (c) How long does the frictional force act on him? (d) Find the change in momentum of the man and the change in momentum of the cart. (e) Determine the displacement of the man relative to the ground while he is sliding on the cart. (f) Determine the displacement of the cart relative to the ground while he is sliding. (g) Find the change in the man's kinetic energy. (h) Find the change in kinetic energy of the cart. (i) Explain why the answers to (g) and (h) differ. (What kind of collision is this, and what accounts for the loss of mechanical energy?)

60.0 kg   4.00 m/s

120 kg

FIGURE P6.55

56. Two blocks of masses $m_1 = 2.00$ kg and $m_2 = 4.00$ kg are each released from rest at a height of 5.00 m on a frictionless track, as shown in Figure P6.56, and undergo an elastic head-on collision. (a) Determine the velocity of each block just before the collision. (b) Determine the velocity of each block immediately after the collision. (c) Determine the maximum heights to which m_1 and m_2 rise after the collision.

57. A 0.500-kg block is released from rest at the top of a frictionless track 2.50 m above the top of a table. It then collides elastically with a 1.00-kg block that is initially at rest on the table, as shown in Figure P6.57. (a) Determine the velocities of the two blocks just after the collision. (b) How

$v_0 = 3.00$ m/s to the right, and the blue cart is initially at rest. If the carts collide elastically, find (a) the velocity of the carts just after the first collision and (b) the maximum compression of the spring.

FIGURE P6.60

61. A cannon is rigidly attached to a carriage, which can move along horizontal rails but is connected to a post by a large spring, initially unstretched and with force constant $k = 2.00 \times 10^4$ N/m, as in Figure P6.61. The cannon fires a 200-kg projectile at a velocity of 125 m/s directed 45.0° above the horizontal. (a) If the mass of the cannon and its

FIGURE P6.61

carriage is 5 000 kg, find the recoil speed of the cannon. (b) Determine the maximum extension of the spring. (c) Find the maximum force the spring exerts on the carriage. (d) Consider the system consisting of the cannon, carriage, and shell. Is the momentum of this system conserved during the firing? Why or why not?

62. Two objects of masses m and $3m$ are moving toward each other along the x axis with the same initial speeds v_0. The object with mass m is traveling to the left, and the object with mass $3m$ is traveling to the right. They undergo an elastic glancing collision such that m is moving downward after the collision at right angles from its initial direction. (a) Find the final speeds of the two objects. (b) What is the angle θ at which the object with mass $3m$ is scattered?

63. A neutron in a reactor makes an elastic head-on collision with a carbon atom that is initially at rest. (The mass of the carbon nucleus is about 12 times that of the neutron.) (a) What fraction of the neutron's kinetic energy is transferred to the carbon nucleus? (b) If the neutron's initial kinetic energy is 1.6×10^{-13} J, find its final kinetic energy and the kinetic energy of the carbon nucleus after the collision.

64. A cue ball traveling at 4.00 m/s makes a glancing, elastic collision with a target ball of equal mass that is initially at rest. The cue ball is deflected so that it makes an angle of 30.0° with its original direction of travel. Find (a) the angle between the velocity vectors of the two balls after the collision and (b) the speed of each ball after the collision.

65. A block of mass m lying on a rough horizontal surface is given an initial velocity of v_0. After traveling a distance d, it makes a head-on elastic collision with a block of mass $2m$. How far does the second block move before coming to rest? (Assume the coefficient of friction μ_k is the same for both blocks.)

GROUP ACTIVITIES

G.1 Hold a Ping-Pong ball or tennis ball on top of a basketball as in Figure GA6.1. Drop them both at the same time so that the basketball hits the floor first, bounces up, and hits the smaller falling ball. What happens and why?

FIGURE GA6.1

G.2 Tie a string to two tennis balls to make identical pendulums. Support the two such that they are of the same length and just touching. Now, model an inelastic collision by attaching a piece of putty or double-stick tape to one of the balls so that when the balls collide they will

stick together and move as a unit. Move one pendulum upward to some starting amplitude, and observe the amplitude to which the combination rises after the collision. Discuss and explain the result. Don't forget that *after* the collision, the energy of the system (the two tennis balls and Earth) is conserved, and you may need to use this fact to determine the height to which the combination rises.

G.3 A fun device that illustrates conservation of momentum and conservation of kinetic energy is the so-called executive stress-reliever shown in Figure GA6.3. It consists of five identical hard balls supported by strings of equal lengths. When one ball is pulled out and released, an almost elastic collision causes one ball to move out on the opposite side with the same speed as the incoming ball. If two balls are pulled out and released, two balls swing out on the opposite side, and so forth—but how do the balls know? For example, is it possible that on occasion when one ball is released, two will swing out on the opposite side traveling with half the speed of the incoming ball? If you have access to this toy, try the above experiment to see if you can get the latter collision to occur. When you have convinced yourself that it is not going to happen, find the kinetic energy of the system with one incoming

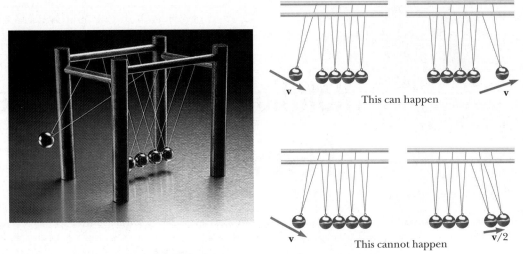

FIGURE GA6.3   **An executive stress-reliever.** *(Photo courtesy of CENCO)*

ball of mass m and speed v and the kinetic energy of two outgoing balls with mass m and speed $v/2$ to show that kinetic energy is not conserved in such a collision.

G.4 Two ice skaters are holding hands at the center of a frozen pond when an argument ensues. Skater A shoves skater B along a horizontal direction. Identify (a) the horizontal forces acting on A and (b) those acting on B. (c) Which force is greater, the force on A or the force on B? (d) Can conservation of momentum be used for the system of A and B? Defend your answer. (e) If A has a mass of 0.900 that of B, and B begins to move away with a speed of 2.00 m/s, find the speed of A.

G.5 (a) A car traveling due east strikes a car traveling due north at an intersection, and the two move together as a unit. A property owner on the southeast corner of the intersection claims his fence was torn down in the collision. Should he be awarded damages by the insurance company? Defend your answer. (b) Let the eastward-moving car have a mass of 1 300 kg, a speed of 30 km/h, and the northward-moving car a mass of 1 100 kg and a speed of 20 km/h. Find the velocity after the collision. Are the results consistent with your answer to part (a)?

G.6 A professor of physics is going ice skating for the first time. He has gotten himself into the middle of an ice rink and cannot figure out how to make the skates work. Every motion he makes simply slips on the ice and leaves him in

the place he started. He decides that he can get off the ice by throwing his gloves in the opposite direction. (a) Suppose he has a mass M and his gloves have a mass m. He throws the gloves as hard as he can away from him, and they leave his hand with a velocity **v** relative to Earth. Explain whether or not he moves. If he does move, calculate his velocity **V** relative to Earth. (b) Discuss his motion from the point of view of the forces acting on him. (c) If the ice rink is 10 m in diameter and the professor starts in the center, estimate how long it will take him to reach the edge, assuming there is no friction at all.

G.7 Two blocks collide on a frictionless surface. After the collision, the two blocks stick together. Block A has a mass M and is initially moving at speed v. Block B has a mass $2M$ and is initially at rest. System C is composed of both blocks. (a) Draw a free-body diagram for each block at an instant *during* the collision. (b) Rank the magnitudes of the horizontal forces in your diagram. Explain your reasoning. (c) Calculate the change in momentum of block A, block B, and system C. Show all of your work. (d) Is kinetic energy conserved in this collision? Explain your answer.

(Problems 6 and 7 are courtesy of Edward F. Redish. For more problems of this type, visit http://www.physics.umd.edu/perg)

7

Rotational Motion and the Law of Gravity

Chapter Outline

Astronauts F. Story Musgrave and Jeffrey A. Hoffman, along with the Hubble Space Telescope and the space shuttle *Endeavor,* are all "falling" around Earth.
(Courtesy NASA)

*I*n this chapter we investigate circular motion, a specific type of two-dimensional motion. We shall encounter such terms as *angular speed, angular acceleration,* and *centripetal acceleration.* The results derived will help you understand the motions of a diverse range of objects in our environment, from a car moving around a circular race track to clusters of galaxies orbiting a common center.

We shall also introduce Newton's law of universal gravitation, one of the fundamental laws in nature, and show how this law, together with Newton's laws of motion, enables us to understand a variety of familiar phenomena.

Circular motion and the law of universal gravitation are related historically in that Newton discovered the law of gravity when he attempted to explain the circular motion of the Moon about Earth and the motions of the planets about the Sun. Thus, it is appropriate to consider these two important physical topics together. We shall also extend the concept of gravitational potential energy first introduced in Chapter 5 and develop the concept of escape velocity.

7.1 ANGULAR SPEED AND ANGULAR ACCELERATION

We began our study of linear motion by defining the terms *displacement, velocity,* and *acceleration.* We follow the same basic approach here as we turn to a study of rotational motion. Consider Figure 7.1a, a top view of a rotating compact disc. The axis of rotation is at the center of the disc, at O. A point P on the disc is at the distance r from the origin and moves about O in a circle of radius r. In fact, every point on the disc undergoes circular motion about O. To analyze such motion, it is convenient to set up a *fixed* reference line, as shown in Figure 7.1a. Let us assume that at time $t = 0$, the point P is on the reference line as in Figure 7.1a and a line is drawn on the disc from the origin out to P. After a time interval Δt has elapsed, P has advanced to a new position (Fig. 7.1b). In this time interval, the line OP has moved through the angle θ with respect to the reference line. Likewise, P has moved a distance s measured along the circumference of the circle; s is called an *arc length.*

In the situations we have encountered thus far, angles have been measured in degrees. However, in scientific work angles are often measured in *radians* (rad) rather than degrees, in order to simplify certain equations. In fact, almost all of the equations derived in this chapter and the next require that angles be measured in radians. With reference to Figure 7.1b, when the arc length s is equal to the radius r, the angle θ swept out by r is equal to one radian. In general, any angle θ measured in radians is defined by the relation

$$\theta \equiv \frac{s}{r} \qquad \text{[7.1]}$$

The radian is a pure number with no dimensions. This can be seen from Equation 7.1, because θ is the ratio of an arc length to the radius of the circle (also a length).

To convert degrees to radians, note that when point P in Figure 7.1 moves through an angle of 360° (one revolution), the arc length s is equal to the circumference of the circle, $2\pi r$. From Equation 7.1 we see that the corresponding angle in radians is $2\pi r/r = 2\pi$ rad. Hence, 2π rad = 360° or

$$1 \text{ rad} \equiv \frac{360°}{2\pi} = 57.3°$$

From this definition, it follows that any angle in degrees can be converted to its equivalent in radians with the expression

$$\theta \text{ (rad)} = \frac{\pi}{180°}\theta \text{ (deg)}$$

For example, 60° equals $\pi/3$ rad and 45° equals $\pi/4$ rad.

(a)

(b)

FIGURE 7.1 (a) The point P on a rotating compact disc at $t = 0$. (b) As the disc rotates, the point P moves through an arc length s.

Tip 7.1 REMEMBER THE RADIAN

Equation 7.1 defines an angle expressed in *radians.* Do not use this equation for angles measured in degrees. Also, be sure to set your calculator in radian mode when doing problems in rotation.

We see from Figure 7.2 that, as a point on the rotating disc moves from Ⓐ to Ⓑ in a time Δt, the angle through which the disc rotates is $\Delta\theta = \theta_f - \theta_i$. We define $\Delta\theta$ as the **angular displacement.** Note that we use angular variables to describe the rotating disc because each point on the disc undergoes the same angular displacement in any given time interval. As we shall see, other variables are common to each point on the rotating object.

Average angular speed ▶

The average angular speed $\overline{\omega}$ (ω is the Greek letter omega) of a rotating rigid object is the ratio of the angular displacement $\Delta\theta$ to the time interval Δt:

$$\overline{\omega} = \frac{\theta_f - \theta_i}{t_f - t_i} = \frac{\Delta\theta}{\Delta t} \qquad [7.2]$$

Instantaneous angular speed ▶

By analogy with linear speed, **the instantaneous angular speed ω is defined as the limit of the average speed $\Delta\theta/\Delta t$ as the time interval Δt approaches zero:**

$$\omega \equiv \lim_{\Delta t \to 0} \frac{\Delta\theta}{\Delta t} \qquad [7.3]$$

Angular speed has the units radians per second (rad/s). We shall take ω to be positive when θ is increasing (counterclockwise motion) and negative when θ is decreasing (clockwise motion). When the angular speed is constant, the instantaneous angular speed at any instant in a time interval is equal to the average angular speed over the time interval.

FIGURE 7.2 As a point on the compact disc moves from Ⓐ to Ⓑ, the disc rotates through the angle $\Delta\theta = \theta_f - \theta_i$.

Example 7.1   Whirlybirds

The rotor on a helicopter turns at an angular speed of 320 revolutions per minute. (In this book, we sometimes use the abbreviation rpm, but in most cases we will use rev/min.) Express this in radians per second.

Solution   We use the conversion factors 1 rev $= 2\pi$ rad and 60 s $= 1$ min, to give

$$320 \frac{\text{rev}}{\text{min}} = 320 \frac{\text{rev}}{\text{min}} \left(\frac{2\pi \, \text{rad}}{\text{rev}}\right) \left(\frac{1 \, \text{min}}{60 \, \text{s}}\right) = \boxed{33.5 \, \text{rad/s}}$$

Quick Quiz 7.1   A rigid body is rotating counterclockwise about a fixed axis. The following pairs of quantities represent an initial angular position and a final angular position of the rigid body. Which of the sets can *only* occur if the rigid body rotates through more than 180°? (a) 3 rad, 6 rad;   (b) -1 rad, 1 rad;   (c) 1 rad, 5 rad.

Quick Quiz 7.2   Suppose that the change in angular position for each pair of values in Quick Quiz 7.1 occurred in 1 s. Which choice represents the lowest average angular speed?

Figure 7.3 shows a bicycle turned upside down so that a repair technician can work on the rear wheel. The bicycle pedals are turned so that at time t_i the wheel has angular speed ω_i (Fig. 7.3a) and at a later time t_f it has angular speed ω_f (Fig. 7.3b). **The average angular acceleration $\overline{\alpha}$ (α is the Greek letter alpha) of an object is defined as the ratio of the change in the angular speed to the time**

Average angular acceleration ▶

FIGURE 7.3 An accelerating bicycle wheel rotates with (a) angular speed ω_i at time t_i and (b) angular speed ω_f at time t_f.

Δt it takes the object to undergo the change:

$$\overline{\alpha} \equiv \frac{\omega_f - \omega_i}{t_f - t_i} = \frac{\Delta\omega}{\Delta t} \qquad [7.4]$$

The instantaneous angular acceleration is defined as the limit of the ratio $\Delta\omega/\Delta t$ as Δt approaches zero. Angular acceleration has the units radians per second per second (rad/s^2). **When a rigid object rotates about a fixed axis, as does the bicycle wheel, every portion of the object has the same angular speed and the same angular acceleration.** This, in fact, is precisely what makes these variables so useful for describing rotational motion. Note that the linear speed and acceleration have a different value at every point along a radius.

◀ Instantaneous angular acceleration

The following argument should convince you that ω and α are the same for every point on the wheel. If a point on the rim of the wheel had a greater angular speed than a point nearer the center, the shape of the wheel would be changing. The wheel remains circular (symmetrically distributed about the axle) only if all points have the same angular speed and the same angular acceleration.

7.2 ROTATIONAL MOTION UNDER CONSTANT ANGULAR ACCELERATION

Let us pause for a moment to consider some similarities between the equations we have found thus far for rotational motion and those we found for linear motion in earlier chapters. For example, compare the defining equation for average angular speed,

$$\overline{\omega} \equiv \frac{\theta_f - \theta_i}{t_f - t_i} = \frac{\Delta\theta}{\Delta t}$$

with the defining equation for average linear speed,

$$\overline{v} \equiv \frac{x_f - x_i}{t_f - t_i} = \frac{\Delta x}{\Delta t}$$

The equations are similar in the sense that θ replaces x and ω replaces v. Take careful note of such similarities as you study rotational motion, because virtually every linear quantity we have encountered thus far has a corresponding "twin" in rotational motion. Once you are adept at recognizing such analogies, you will find it unnecessary to memorize many of the equations in this chapter. In addition, the techniques for solving rotational motion problems are quite similar to those you have already learned for linear motion. For example, problems concerned with objects that rotate with constant angular acceleration can be solved in much the same manner as those dealing with linear motion under constant acceleration. If you understand problems involving objects that move with constant linear acceleration, these rotational motion problems should be little more than a review for you.

An additional analogy between linear motion and rotational motion is revealed when we compare the defining equation for average angular acceleration,

$$\overline{\alpha} \equiv \frac{\omega_f - \omega_i}{t_f - t_i} = \frac{\Delta \omega}{\Delta t}$$

with the defining equation for average linear acceleration,

$$\overline{a} \equiv \frac{v_f - v_i}{t_f - t_i} = \frac{\Delta v}{\Delta t}$$

In light of the analogies between variables in linear motion and those in rotational motion, it should not surprise you that the equations of rotational motion involve the variables θ, ω, and α. In Section 2.6, we developed a set of kinematic equations for linear motion under constant acceleration. The same procedure can be used to derive a similar set of equations for rotational motion under constant angular acceleration. The resulting equations of rotational kinematics, along with the corresponding equations for linear motion under constant acceleration, are as follows:

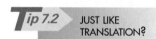

Tip 7.2 JUST LIKE TRANSLATION?

Equations 7.5–7.7 suggest that rotational kinematics is just like translational kinematics. That is almost true, but remember the following two differences. (1) In rotational kinematics, you must specify a rotation axis. (2) In rotational motion, the object keeps returning to its original orientation—thus, you may be asked for the number of revolutions made by a rigid body, which is something that has no meaning in translational motion.

Rotational Motion About a Fixed Axis with α Constant (Variables: θ and ω)		Linear Motion with a Constant (Variables: x and v)
$\omega = \omega_i + \alpha t$	**[7.5]**	$v = v_i + at$
$\Delta \theta = \omega_i t + \frac{1}{2}\alpha t^2$	**[7.6]**	$\Delta x = v_i t + \frac{1}{2}at^2$
$\omega^2 = \omega_i^2 + 2\alpha \Delta \theta$	**[7.7]**	$v^2 = v_i^2 + 2a \Delta x$

Again, note the one-to-one correspondence between the rotational equations involving the angular variables θ, ω, and α and the equations of linear motion involving the variables x, v, and a.

Example 7.2 A Rotating Wheel

A wheel rotates with a constant angular acceleration of 3.50 rad/s^2.

A If the angular speed of the wheel is 2.00 rad/s at $t_i = 0$, through what angle does the wheel rotate between $t = 0$ and $t = 2.00$ s?

Solution Because the angular acceleration in the problem is given as constant, we can use the rotational kinematic equations. We use Equation 7.6, setting $\theta_i = 0$,

$$\Delta \theta = \omega_i t + \frac{1}{2}\alpha t^2$$
$$= (2.00 \text{ rad/s})(2.00 \text{ s}) + \frac{1}{2}(3.50 \text{ rad/s}^2)(2.00 \text{ s})^2$$
$$= \boxed{11.0 \text{ rad}}$$

This is equivalent to 11.0 rad/(2π rad/rev) = 1.75 rev.

B What is the angular speed of the wheel at $t = 2.00$ s?

Solution   We use Equation 7.5:

$$\omega = \omega_i + \alpha t = 2.00 \text{ rad/s} + (3.50 \text{ rad/s}^2)(2.00 \text{ s}) = \boxed{9.00 \text{ rad/s}}$$

We could also obtain this result using Equation 7.7 and the results of part A. Try it!

EXERCISE   Find the angle through which the wheel rotates between $t = 2.00$ s and $t = 3.00$ s.

ANSWER   10.8 rad

7.3 RELATIONS BETWEEN ANGULAR AND LINEAR QUANTITIES

In this section we derive some useful relations between the angular speed and acceleration of a rotating object and the linear speed and acceleration of an arbitrary point in the object.

Consider the arbitrarily shaped object in Figure 7.4, rotating about the z axis through the point O. Assume that the object rotates through the angle $\Delta\theta$, and hence point P moves through the arc length Δs in the interval Δt. We know from the defining equation for radian measure that

$$\Delta\theta = \frac{\Delta s}{r}$$

Let us now divide both sides of this equation by Δt, the interval during which the rotation occurred:

$$\frac{\Delta\theta}{\Delta t} = \frac{1}{r}\frac{\Delta s}{\Delta t}$$

If Δt is very small, then the angle $\Delta\theta$ through which the object rotates is small and the ratio $\Delta\theta/\Delta t$ is the instantaneous angular speed ω. Also, Δs is very small when Δt is very small, and the ratio $\Delta s/\Delta t$ equals the instantaneous linear speed v for small values of Δt. Hence, the preceding equation is equivalent to

$$\omega = \frac{v}{r}$$

Figure 7.4 allows us to interpret this equation. The distance Δs is traversed along an arc of the circular path followed by the point P as it moves during the time Δt. Thus, v must be the linear speed of a point lying along this arc, where the direction of the point's vector velocity **v** is *tangent to the circular path*. As a result, we often refer to this linear speed as the *tangential speed* of a particle moving in a circular path, and write

$$v_t = r\omega \qquad\qquad [7.8]$$

◀ Tangential speed

That is, **the tangential speed of a point on a rotating object equals the distance of that point from the axis of rotation multiplied by the angular speed.** Equation 7.8 shows that the linear speed of a point on the rotating object increases with movement outward from the center of rotation toward the rim, as we would intuitively expect. **Although every point on the rotating object has the same angular speed, not every point has the same linear (tangential) speed.**

Exercise caution when you use Equation 7.8. It has been derived using the defining equation for radian measure and hence is valid only when ω is measured

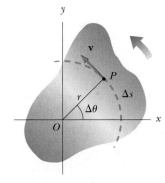

FIGURE 7.4   Rotation of an object about an axis through O that is perpendicular to the plane of the figure (the z axis). Note that a point P on the object rotates in a circle of radius r centered at O.

in radians per unit time. Other measures of angular speed, such as degrees per second and revolutions per second, will not give correct answers in Equation 7.8.

To find a second equation relating linear and angular quantities, imagine that an object rotating about a fixed axis (Fig. 7.4) changes its angular speed by $\Delta \omega$ in the interval Δt. At the end of this interval, the speed of a point on the object, such as P, has changed by the amount Δv_t. From Equation 7.8 we have

$$\Delta v_t = r \Delta \omega$$

Dividing by Δt gives

$$\frac{\Delta v_t}{\Delta t} = r \frac{\Delta \omega}{\Delta t}$$

If the time interval Δt is very small, then the ratio $\Delta v_t / \Delta t$ is the tangential acceleration of that point and $\Delta \omega / \Delta t$ is the angular acceleration. Therefore, we see that

Tangential acceleration

$$a_t = r\alpha \qquad\qquad [7.9]$$

That is, **the tangential acceleration of a point on a rotating object equals the distance of that point from the axis of rotation multiplied by the angular acceleration.** Again, radian measure must be used for the angular acceleration in this equation.

There is one more equation that relates linear quantities to angular quantities, but we shall defer its derivation to the next section.

Quick Quiz 7.3	When a wheel of radius R rotates about a fixed axis as in Figure 7.3, which of the following statements are false? (a) All points on the wheel have the same angular speed. (b) All points on the wheel have the same tangential speed. (c) All points on the wheel have the same angular acceleration. (d) All points on the wheel have the same tangential acceleration.

APPLYING **PHYSICS** 7.1

The launch area for the European Space Agency is not in Europe—it is in South America. Why?

Explanation   Placing a satellite in Earth orbit requires providing a large tangential speed to the satellite. This is the task of the rocket propulsion system. Anything that reduces the requirements on the propulsion system is a welcome contribution. The surface of Earth is already traveling toward the east at a high speed, due to Earth's rotation. Thus, if rockets are launched toward the east, the rotation of Earth provides some initial tangential speed, reducing somewhat the requirements on the propulsion system. If rockets were launched from Europe, which is at a relatively large latitude, the contribution of Earth's rotation is relatively small, because the distance between Europe and the rotation axis of Earth is relatively small. The ideal place for launching is at the Equator, which is as far as one can be from the rotation axis of Earth and still be on Earth's surface. This results in the largest possible tangential speed due to Earth's rotation. The European Space Agency exploits this advantage by launching from French Guiana, which is only a few degrees north of the Equator.

A second advantage of this launch location is that launching toward the east takes the spacecraft over water. In the event of an accident or a failure, the wreckage will fall into the ocean rather than falling into populated areas as it would if launched to the east from Europe. This is the primary reason why the United States launches spacecraft from Florida rather than California, despite the more favorable weather conditions in California.

APPLYING PHYSICS 7.2

When a tall smokestack falls over, it often breaks somewhere along its length before it hits the ground as in Figure 7.5. Why does this happen?

Explanation As the smokestack rotates around its base, each higher portion of the smokestack falls with an increasing tangential acceleration. The tangential acceleration of a given point on the smokestack is proportional to the distance of that portion from the base. As the acceleration increases, eventually higher portions of the smokestack will experience an acceleration greater than that resulting from gravity alone. This can only happen if these portions are being pulled downward by a force in addition to the gravitational force. The force that causes this to occur is the cohesive force from lower portions of the smokestack. Eventually, the cohesive force that provides this acceleration is larger than the smokestack can withstand, and it breaks.

FIGURE 7.5 (Applying Physics 7.2) A falling smokestack.

Example 7.3 Compact Discs

A compact disc in a computer rotates from rest up to an angular speed of 31.4 rad/s in a time of 0.892 s.

A What is the angular acceleration of the disc, assuming the angular acceleration is uniform?

Solution If we use $\omega = \omega_i + \alpha t$ and the fact that $\omega_i = 0$ at $t = 0$, we get

$$\alpha = \frac{\omega}{t} = \frac{31.4 \text{ rad/s}}{0.892 \text{ s}} = \boxed{35.2 \text{ rad/s}^2}$$

B How many rotations does the disc make while coming up to speed?

Solution Equation 7.6 enables us to find the angular displacement during the 0.892-s time interval. Taking $\omega_i = 0$ and $\theta_i = 0$ gives

$$\Delta\theta = \omega_i t + \tfrac{1}{2}\alpha t^2 = \tfrac{1}{2}(35.2 \text{ rad/s}^2)(0.892 \text{ s})^2 = \boxed{14.0 \text{ rad}}$$

Because 2π rad = 1 rev, this angular displacement corresponds to 2.23 rev.

C If the radius of the disc is 4.45 cm, find the final linear speed of a microbe riding on the rim of the disc.

Solution The relation $v_t = r\omega$ and the given values lead to

$$v_t = r\omega = (0.044\ 5 \text{ m})(31.4 \text{ rad/s}) = \boxed{1.40 \text{ m/s}}$$

D What is the magnitude of the tangential acceleration of the microbe at this time?

Solution We use $a_t = r\alpha$, which gives

$$a_t = r\alpha = (0.044\ 5 \text{ m})(35.2 \text{ rad/s}^2) = \boxed{1.57 \text{ m/s}^2}$$

EXERCISE What is the angular speed and angular displacement of the disc 0.300 s after it begins to rotate?

ANSWER 10.6 rad/s; 1.58 rad

Long before compact discs became the medium of choice for recorded music, phonographs were popular. There are similarities between the rotational motion of phonograph records and compact discs, and there are some interesting differences. A phonograph record, for instance, rotates at a <u>constant angular speed</u>. Popular angular speeds were $33\frac{1}{3}$ rev/min for long-playing albums (hence the nickname "LP"), 45 rev/min for "singles," and 78 rev/min used in very early

> **APPLICATION**
>
> PHONOGRAPH RECORDS AND COMPACT DISCS

recordings. At the outer edge of the record, the pickup needle (stylus) was moving over the vinyl material at a faster tangential speed than when the needle was close to the center of the record. As a result, the sound information was compressed into a smaller length of track near the center of the record than near the outer edge.

Compact discs, on the other hand, are designed such that the disc moves past the laser pickup at a constant tangential speed. Because the pickup moves radially as it follows the tracks of information, this requires that the angular speed of the compact disc must vary according to the radial position of the laser. Because the tangential speed is fixed, the information density (per length of track) anywhere on the disc is the same. Example 7.4 demonstrates numerical calculations for both compact discs and phonograph records.

Example 7.4   Length of a CD Track

A compact disc player is designed such that as the read head moves out from the center of the disc, the angular speed of the disc changes so that the linear speed at the position of the head will always be at a constant value of about 1.3 m/s.

A Find the angular speed of the disc when the read head is at a distance of 2.0 cm from the center of the disc and at 5.6 cm.

Solution   From $v_t = r\omega$, we have at $r = 2.0$ cm,

$$\omega = \frac{v_t}{r} = \frac{1.3 \text{ m/s}}{2.0 \times 10^{-2} \text{ m}} = \boxed{65 \text{ rad/s}}$$

and at 5.6 cm,

$$\omega = \frac{v_t}{r} = \frac{1.3 \text{ m/s}}{5.6 \times 10^{-2} \text{ m}} = \boxed{23 \text{ rad/s}}$$

B On the other hand, an old-fashioned record player rotated at a constant angular speed so that the linear speed of the record groove moving under the detector (the stylus) changed. Find the linear speed of a 45.0-rpm record at points 2.0 and 5.6 cm from the center.

Solution   Because 45.0 rev/min = 4.7 rad/s, the linear speed at $r = 2.0$ cm is

$$v_t = r\omega = (2.0 \times 10^{-2} \text{ m})(4.7 \text{ rad/s}) = \boxed{0.094 \text{ m/s}}$$

At $r = 5.6$ cm, we have

$$v_t = r\omega = (5.6 \times 10^{-2} \text{ m})(4.7 \text{ rad/s}) = \boxed{0.26 \text{ m/s}}$$

C In both the CD and phonograph record, information is recorded in a continuous spiral track as one moves over the surface of the disc from the edge. Use the information above to calculate the total length of the track for a CD designed to play for 1 hour.

Solution   The CD moves under the read head with a constant linear speed of 1.3 m/s. Thus, the total length of the track that moves under it in 3 600 seconds is

$$\Delta x = v_t t = (1.3 \text{ m/s})(3 \, 600 \text{ s}) = \boxed{4 \, 680 \text{ m}}$$

7.4   CENTRIPETAL ACCELERATION

Figure 7.6a shows a car moving in a circular path with *constant linear speed v*. Students are often surprised to find that **even though the car moves at a constant speed, it still has an acceleration.** To see why this occurs, consider the defining equation for average acceleration:

$$\bar{\mathbf{a}} = \frac{\mathbf{v}_f - \mathbf{v}_i}{t_f - t_i} \qquad \text{[7.10]}$$

Note that the average acceleration depends on the *change in the velocity vector*. Because velocity is a vector, there are two ways in which an acceleration can be produced: by a change in the *magnitude* of the velocity and by a change in the *direction* of the velocity. It is the latter situation that occurs for the car moving in a circular path with constant speed (see Fig. 7.6b). We shall show that the acceleration vector in this case is perpendicular to the path and always points toward the center of the circle. An acceleration of this nature is called a **centripetal** (center-seeking) **acceleration.** Its magnitude is given by

$$a_c = \frac{v^2}{r} \qquad \text{[7.11]}$$

To derive Equation 7.11, consider Figure 7.7a. Here an object is seen first at point Ⓐ with velocity $\mathbf{v}_i$ at time t_i and then at point Ⓑ with velocity $\mathbf{v}_f$ at a later time t_f. Let us assume that here $\mathbf{v}_i$ and $\mathbf{v}_f$ differ only in direction; their magnitudes are the same (that is, $v_i = v_f = v$). To calculate the acceleration, we begin with Equation 7.10, which indicates that we must vectorially subtract $\mathbf{v}_i$ from $\mathbf{v}_f$:

$$\bar{\mathbf{a}} = \frac{\mathbf{v}_f - \mathbf{v}_i}{t_f - t_i} = \frac{\Delta\mathbf{v}}{\Delta t} \qquad \text{[7.12]}$$

where $\Delta\mathbf{v} = \mathbf{v}_f - \mathbf{v}_i$ is the change in velocity. This can be accomplished graphically, as shown by the vector triangle in Figure 7.7b. Note that when Δt is very small, Δs and $\Delta\theta$ will also be very small. In this case, $\mathbf{v}_f$ will be almost parallel to $\mathbf{v}_i$, and the vector $\Delta\mathbf{v}$ will be approximately perpendicular to them, pointing toward the center of the circle. In the limiting case where Δt becomes vanishingly small, $\Delta\mathbf{v}$ will point exactly toward the center of the circle. Furthermore, in this limiting case the acceleration will also be directed toward the center of the circle because it is in the direction of $\Delta\mathbf{v}$.

Now consider the triangle in Figure 7.7a, which has sides Δs and r. This triangle and the one formed by the vectors in Figure 7.7b are similar. This enables us to write a relationship between the lengths of the sides:

$$\frac{\Delta v}{v} = \frac{\Delta s}{r}$$

or

$$\Delta v = \frac{v}{r}\Delta s$$

This can be substituted into Equation 7.12 for Δv to give

$$a = \frac{v}{r}\frac{\Delta s}{\Delta t} \qquad \text{[7.13]}$$

In this situation, Δs is a small distance measured along the arc of the circle (a tangential distance), so $v = \Delta s/\Delta t$, where v is the tangential speed. Therefore, Equation 7.13 reduces to Equation 7.11:

$$a_c = \frac{v^2}{r}$$

Because the tangential speed is related to the angular speed through the relation $v_t = r\omega$ (Eq. 7.8), an alternative form of Equation 7.11 is

$$a_c = \frac{r^2\omega^2}{r} = r\omega^2 \qquad \text{[7.14]}$$

(a)

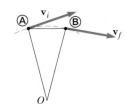

(b)

FIGURE 7.6 (a) Circular motion of a car moving with constant speed. (b) As the car moves along the circular path from Ⓐ to Ⓑ, the direction of its velocity vector changes, so the car undergoes a centripetal acceleration.

(a)

(b)

FIGURE 7.7 (a) As the particle moves from Ⓐ to Ⓑ, the direction of its velocity vector changes from $\mathbf{v}_i$ to $\mathbf{v}_f$. (b) The construction for determining the direction of the change in velocity $\Delta\mathbf{v}$, which is toward the center of the circle.

Thus, we conclude that

Centripetal acceleration ▶

In circular motion the centripetal acceleration is directed inward toward the center of the circle and has a magnitude given by either v^2/r or $r\omega^2$.

You should show that the dimensions of a_c are L/T^2, as required.

Let us clear up any misconceptions that might exist concerning centripetal and tangential acceleration by considering a car moving around a circular race track. In order to move in a circular path, the car always has a centripetal component of acceleration because its direction of travel, and hence the direction of its velocity, is continuously changing. If the car's speed is increasing or decreasing, it also has a tangential component of acceleration. To summarize, the tangential component of acceleration is due to changing speed; the centripetal component of acceleration is due to changing direction.

When both components of acceleration exist simultaneously, the component vector associated with the tangential acceleration is tangent to the circular path, and the component vector associated with the centripetal acceleration points toward the center of the circular path. Because these component vectors are perpendicular to each other, we can find the magnitude of the **total acceleration** using the Pythagorean theorem:

Total acceleration ▶

$$a = \sqrt{a_t^2 + a_c^2} \qquad [7.15]$$

Quick Quiz 7.4

A race track is constructed such that two arcs of radius 80 m at Ⓐ and 40 m at Ⓑ are joined by two stretches of straight track as in Figure 7.8. In a particular trial run, a driver traveled at a constant speed of 50 m/s for one complete lap.

1. The ratio of the tangential acceleration at Ⓐ to that at Ⓑ is
   (a) $\frac{1}{2}$ (b) $\frac{1}{4}$ (c) 2   (d) 4
   (e) The tangential acceleration is zero at both points.
2. The ratio of the centripetal acceleration at Ⓐ to that at Ⓑ is
   (a) $\frac{1}{2}$ (b) $\frac{1}{4}$ (c) 2   (d) 4
   (e) The centripetal acceleration is zero at both points.
3. The angular speed is greatest at
   (a) Ⓐ   (b) Ⓑ   (c) It is equal at both Ⓐ and Ⓑ.

FIGURE 7.8   (Quick Quiz 7.4)

Quick Quiz 7.5

An object moves in a circular path with constant speed v. Which of the following statements is true concerning the object? (a) Its velocity is constant but its acceleration is changing. (b) Its acceleration is constant but its velocity is changing. (c) Both its velocity and acceleration are changing. (d) Its velocity and acceleration remain constant.

Example 7.5   Let's Go for a Spin

A test car moves at a constant speed of 10 m/s around a circular road of radius 50 m. Find the car's (a) centripetal acceleration and (b) angular speed.

Solution

A From Equation 7.11, the magnitude of the centripetal acceleration of the car is found to be

$$a_c = \frac{v^2}{r} = \frac{(10 \text{ m/s})^2}{50 \text{ m}} = \boxed{2.0 \text{ m/s}^2}$$

By definition, the direction of $\mathbf{a}_c$ is toward the center of curvature of the road.

B The angular speed of the car can be found using the expression $v_t = r\omega$ which gives

$$\omega = \frac{v_t}{r} = \frac{10 \text{ m/s}}{50 \text{ m}} = \boxed{0.20 \text{ rad/s}}$$

Note that we can also find the centripetal acceleration by using this value of ω, in an alternative method to part A. From Equation 7.14 we have

$$a_c = r\omega^2 = (50 \text{ m})(0.20 \text{ rad/s})^2 = 2.0 \text{ m/s}^2$$

As expected, the two methods for finding the centripetal acceleration give the same answer.

EXERCISE   Find the car's tangential acceleration and total acceleration.

ANSWER   The tangential acceleration is zero because the speed of the car remains constant. Because $a_t = 0$, the total acceleration in this example equals the centripetal acceleration, 2.0 m/s², found in part A.

7.5   THE VECTOR NATURE OF ANGULAR QUANTITIES

Before continuing our discussion of centripetal acceleration, a brief digression is in order. In Sections 7.1–7.4 we treated angular displacement, speed, and acceleration as though they were scalar quantities with linear analogs. You should not be surprised to learn that, like their linear analogs, these rotational concepts generalize to vector quantities that have both magnitude and direction. The first hint of this came just before Example 7.1, when we assigned positive and negative values to angular speed ω. That assignment was analogous to our practice in Chapter 2 of choosing a positive or negative direction for one-dimensional linear motion.

Let us describe how to find the direction of the **angular velocity** vector $\boldsymbol{\omega}$. When a rigid object such as a disk rotates about a fixed axis, the angular velocity $\boldsymbol{\omega}$ points along this axis. A convenient right-hand rule, illustrated in Figure 7.9a, can

FIGURE 7.9 (a) The right-hand rule for determining the direction of the angular velocity vector **ω**. (b) The direction of **ω** is in the direction of advance of a right-handed screw.

FIGURE 7.10 An end view of a disk rotating about an axis through its center perpendicular to the page. (a) When the disk rotates clockwise, **ω** is into the page. (b) When the disk rotates counterclockwise, **ω** is out of the page.

be used to determine the direction of **ω**. Grasp the axis of rotation with your right hand so that your fingers wrap in the direction of rotation. Your extended thumb points in the direction of **ω**. Figure 7.9b illustrates that **ω** is also in the direction of advance of a rotating right-handed screw.

We can apply this rule to a rotating disk when viewed along the axis of rotation as in Figure 7.10. When the disk rotates clockwise as in Figure 7.10a, applying the right-hand rule shows that the direction of **ω** is into the page. When the disk rotates counterclockwise as in Figure 7.10b, the direction of **ω** is out of the page.

Finally, the direction of the angular acceleration vector **α** is in the direction of **ω** if the angular speed (the magnitude of **ω**) is increasing in time and antiparallel to **ω** if the angular speed is decreasing in time.

Although we will not work problems involving vector angular velocity and acceleration, the vector nature of rotational quantities plays an important role in a number of phenomena we shall consider later in this book.

7.6 FORCES CAUSING CENTRIPETAL ACCELERATION

Consider a ball of mass m tied to a string of length r and being whirled in a horizontal circular path, as in Figure 7.11. Let us assume that the ball moves with constant speed. Because the velocity vector **v** changes its direction continuously during the motion, the ball experiences a centripetal acceleration directed toward the center of motion, as described in Section 7.4, with magnitude

$$a_c = \frac{v_t^2}{r}$$

The inertia of the ball tends to maintain motion in a straight-line path; however, the string exerts a force on the ball that makes it follow a circular path instead. This force (equal to the force of tension) is directed along the length of the string toward the center of the circle in Figure 7.11. Thus, the equation for Newton's second law along the radial direction is

$$F = ma_c = m\frac{v_t^2}{r} \qquad [7.16]$$

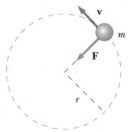

FIGURE 7.11 A ball attached to a string of length r, rotating in a circular path at constant speed.

Here F stands for any force—tension in a string, gravity, the force of friction—acting in a given situation to keep an object following a circular path.

Because it acts at right angles to the motion, a force that produces a centripetal acceleration causes a change in the direction of the velocity. Beyond this, such a force is no different from any of the other forces we have studied. For example, friction between tires and road provides the force that enables a race car to travel in a circular path on a flat road, and the gravitational force exerted by Earth on the Moon provides the force necessary to keep the Moon in its orbit. (We shall discuss the gravitational force in more detail in Section 7.8.)

FIGURE 7.12   Front view of a car rounding a banked roadway.

Regardless of the example used, if the force that produces a centripetal acceleration vanishes, the object does not continue to move in its circular path; instead, it moves along a straight-line path tangent to the circle. As the race car in the preceding paragraph travels faster and faster on the curved, flat road, the centripetal acceleration increases, as does the friction force needed to cause this acceleration. At some speed, the friction force needed to keep the car on a circular path will be greater than the maximum possible static friction force allowed by the static coefficient of friction. At this point, the car will skid off the road.

In order to prevent this from happening, roadways are often banked instead of being built with flat curves. Banking the roadway toward the inside of the circular path as in Figure 7.12 creates a component of the normal force exerted by the road on the car that is directed toward the center of the path. This component of the normal force adds to the centrally directed friction force. Thus, the car can negotiate a banked curve at a higher speed than it can an unbanked curve. The banking of roadways is especially evident at high-speed race tracks, in bicycle tracks in velodromes, and in bobsled tracks.

Tip 7.3   CENTRIPETAL FORCE

The force causing centripetal acceleration is called a *centripetal force* in some textbooks. Giving the force causing circular motion a name—centripetal force—leads many students to consider this as a new *kind* of force. Furthermore, a common mistake in drawing force diagrams is to include all of the usual forces and then add another vector for the centripetal force. But it is not a separate force—*it is simply one of our familiar forces that causes circular motion.* For Earth's motion around the Sun, the "centripetal force" is the force of *gravity.* For a rock whirled on the end of a string, the "centripetal force" is the *tension* in the string. For an amusement park patron pressed against the inner wall of a rapidly rotating circular room, the "centripetal force" is the *normal force* exerted by the wall.

◄ APPLICATION

BANKED ROADWAYS

PROBLEM-SOLVING STRATEGY   **Forces That Cause Centripetal Acceleration**

Use the following steps when dealing with centripetal accelerations and the forces that produce them:

1. Draw a free-body diagram of the object(s) under consideration, showing and labeling all forces that act on it (them).
2. Choose a coordinate system that has one axis perpendicular to the circular path followed by the object and one axis tangent to the circular path.
3. Find the net force toward the center of the circular path. This is the force that causes the centripetal acceleration.
4. From here on the steps are virtually identical to those encountered when solving Newton's second-law problems with $\Sigma F_x = ma_x$ and $\Sigma F_y = ma_y$. However, instead of the x and y directions, we apply Newton's second law in the radial and tangential directions by writing $\Sigma F_{\text{along radius}} = ma_c$ and $\Sigma F_t = ma_t$. Also, note that the magnitude of the centripetal acceleration can always be written $a_c = v_t^2/r$.

(a)

(b)

FIGURE 7.13   (Example 7.6) (a) Top view of a car on a curved path. (b) A free-body diagram for the car, showing an end view.

Example 7.6   Buckle Up for Safety

A car travels at a constant speed of 30.0 mi/h (13.4 m/s) on a level circular turn of radius 50.0 m, as shown in the bird's-eye view in Figure 7.13a. What is the minimum coefficient of static friction between the tires and roadway so that the car makes the circular turn without sliding?

Reasoning   The force in the radial direction acting on the car is the force of static friction directed toward the center of the circular path as in Figure 7.13a. Thus, our objective in this example is to find an expression for the maximum frictional force $f = \mu_s n$ and use it in Equation 7.16.

Solution   Equation 7.16 applied to this situation becomes

$$f = m\frac{v_t^2}{r} \tag{1}$$

All the forces acting on the car are shown in Figure 7.13b. Because the car is in equilibrium in the vertical direction, the normal force upward is balanced by the force of gravity downward, so that

$$n = mg$$

From this expression we can find the maximum force of static friction, as follows:

$$f = \mu_s n = \mu_s mg \tag{2}$$

By setting the right-hand sides of (1) and (2) equal to each other, we find that

$$\mu_s m\!\!\!/g = m\!\!\!/\frac{v_t^2}{r}$$

or

$$\mu_s = \frac{v_t^2}{rg} \tag{3}$$

Therefore, the minimum coefficient of static friction required for the car to make the turn without sliding outward is

$$\mu_s = \frac{v_t^2}{rg} = \frac{(13.4 \text{ m/s})^2}{(50.0 \text{ m})(9.80 \text{ m/s}^2)} = \boxed{0.366}$$

The value of μ_s for rubber on dry concrete is very close to 1; thus, the car can negotiate the curve with ease. If the road were wet or icy, however, the value for μ_s could be 0.2 or lower. Under such conditions, the radial force provided by static friction would not be great enough to enable the car to follow the circular path, and it would slide off the roadway.

Example 7.7   Fun with a Yo-Yo

A child swings a yo-yo of weight mg in a horizontal circle so that the cord makes an angle of θ with the vertical, as in Figure 7.14a. Find the centripetal acceleration of the yo-yo.

Reasoning   Two forces act on the yo-yo: the force of gravity and the force **T** associated with the tension in the cord, as in Figure 7.14b. The condition for equilibrium from Chapter 4 (Equation 4.9) applied in the y direction, $\Sigma F_y = 0$, enables us to find T. The component of **T** toward the center of the circular path is the force that produces the centripetal acceleration, and we can therefore find a_c from $F = ma_c$.

Solution   Applying ΣF_y gives

$$T\cos\theta - mg = 0$$

$$T = \frac{mg}{\cos\theta}$$

The *horizontal component* of **T** causes the centripetal acceleration.

$$F = T\sin\theta = \frac{mg\sin\theta}{\cos\theta} = mg\tan\theta$$

(a)

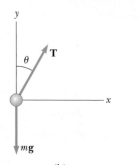

(b)

FIGURE 7.14   (Example 7.7) (a) A yo-yo swinging in a horizontal circle so that the cord makes a constant angle θ with the vertical. (b) A free-body diagram for the yo-yo.

The centripetal acceleration can now be found by applying Newton's second law along the horizontal direction:

$$F = ma_c$$

$$a_c = \frac{F}{m} = \frac{mg \tan \theta}{m} = g \tan \theta = (9.80 \text{ m/s}^2)(\tan 30.0) = \boxed{5.66 \text{ m/s}^2}$$

Example 7.8   Riding the Tracks

Figure 7.15a shows a roller coaster moving around an almost circular loop of radius R. What speed must the car have at the bottom of the track so that it just makes it over the top of the loop?

FIGURE 7.15   (Example 7.8) (a) A roller coaster traveling around a nearly circular track. (b) Passengers on this roller coaster experience the thrill of a loop-the-loop. *(Ernst Haas/Stone/Getty Image)*

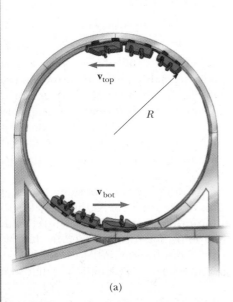

(a)

(b)

Solution   If the car is to just make it over the top and complete the loop, the force exerted by the track on the car must become zero at the top of the loop. Therefore, the only force exerted on the car at the top of the loop is the force of gravity mg, and this force provides the centripetal acceleration of the car. Thus, we have

$$mg = m\frac{v_{\text{top}}^2}{R}$$

where v_{top} is the minimum speed the car must have at the top to stay in contact with the tracks. Solving this equation for v_{top} gives

$$v_{\text{top}} = \sqrt{gR}$$

At the top of the loop, the car has both kinetic energy and potential energy. If we select the zero level for potential energy to be at the bottom of the loop, the mechanical energy of the car-Earth system at the *top* is

$$E_{\text{top}} = \tfrac{1}{2}mv_{\text{top}}^2 + mg2R = \tfrac{1}{2}mgR + 2mgR = 2.5mgR$$

The mechanical energy of the system at the *bottom* of the loop is solely kinetic energy:

$$E_{\text{bot}} = \tfrac{1}{2}mv_{\text{bot}}^2$$

Because the mechanical energy of the system is conserved, the energy at the bottom must equal the energy at the top:

$$\tfrac{1}{2}mv_{\text{bot}}^2 = 2.5mgR$$

Solving for v_{bot} gives

$$v_{\text{bot}} = \sqrt{5gR}$$

For example, if the radius of the loop is 10.0 m, the speed at the bottom must be at least

$$v_{\text{bot}} = \sqrt{5(9.80 \text{ m/s}^2)(10.0 \text{ m})} = 22.1 \text{ m/s}$$

Such a high speed can make the trip very uncomfortable for the passengers. The required speed can be reduced by changing the shape of the loop from a circular shape to one that closes down slightly at the top as shown in Figure 7.15b.

7.7 DESCRIBING FORCES IN ACCELERATED REFERENCE FRAMES

We have seen that an object moving in a circle of radius r with constant speed v has a centripetal acceleration whose magnitude is v_t^2/r and whose direction is toward the center of rotation. The force necessary to maintain this centripetal acceleration must also act toward the center of rotation. In the case of a ball rotating at the end of a string, the force exerted by the string on the ball (equal in magnitude to the tension in the string) is the force causing the centripetal acceleration. For a satellite in a circular orbit around Earth, the force of gravity plays that role; for a car rounding a curve on a level road, it is the force of friction between the tires and the pavement; and so forth.

To better understand the motion of a rotating system, consider a car approaching a curved exit ramp at high speed as in Figure 7.16. As the car takes the sharp left turn onto the ramp, a person sitting in the passenger seat slides to the right across the seat and hits the door. At that point, the force exerted by the door on the passenger keeps her from being ejected from the car. What causes the passenger to move toward the door? A popular but *erroneous* explanation is that some mysterious force pushes her outward.

The phenomenon is correctly explained as follows. Before the car enters the ramp, the passenger is moving in a straight-line path. As the car enters the ramp and travels a curved path, the passenger, because of inertia, tends to move along the original straight-line path. This is in accordance with Newton's first law: The natural tendency of an object in motion is to continue moving in a straight line. However, if a sufficiently large force toward the center of curvature acts on the passenger, she moves in a curved path along with the car. The origin of this force is the force of friction between the passenger and the car seat. If this frictional force is not great enough, the passenger slides across the seat as the car turns under her. Eventually the passenger encounters the door, which provides a large enough force to enable her to follow the same curved path as the car. The passenger slides toward the door, not because of some mysterious outward force but because there is no force great enough to enable her to travel along the circular path followed by the car.

FIGURE 7.16 A car approaching a curved exit ramp.

Tip 7.4 CENTRIFUGAL FORCE

We often hear the phrase "centrifugal force," described as a force pulling *outward* on an object moving in a circular path. It is likely that you have felt this effect, perhaps on an amusement park ride, or during a sharp turn in your car. This is not a real force, however. Forces always represent interactions between objects. If you feel a "centrifugal force," you cannot identify another object with which you are interacting because this is not a real force. This is a fictitious "force" that occurs because you are moving in a noninertial (accelerating) reference frame.

As a second example, consider what happens when you run clothes through the rinse cycle of a washing machine. In the last phase of this cycle, the drum spins rapidly to remove water from the clothes. Why is the water thrown off? An *erroneous* explanation is that the rotating system creates some mysterious outward force on each drop of water, and this force causes the water to be hurled to the outer drum of the machine. The correct explanation is as follows. When the clothes are at rest in the machine, water is held to them by molecular forces between the water and the fabric. During the spin cycle, the clothes rotate and the molecular forces are not great enough to provide the necessary force to keep the water molecules moving in a circular path along with the clothes. Hence, the drops of water, because of their inertia, move in straight-line paths until they encounter the sides of the spinning drum.

In summary, in describing motion in an accelerating reference frame, we must be very careful to distinguish real forces from fictitious ones. An observer in a car rounding a curve is in an accelerating frame and invents a fictitious outward force to explain why he or she is thrown outward. A stationary observer outside the car, however, considers only real forces on the passenger. To the outside observer, the mysterious outward force *does not exist!* The only real external force on the passenger is the force exerted by friction or the contact force exerted by the door.

> ◄ **APPLICATION**
>
> THE SPIN-DRY CYCLE OF A WASHING MACHINE

7.8   NEWTON'S LAW OF UNIVERSAL GRAVITATION

Prior to 1686 a great amount of data had been collected on the motions of the Moon and the planets, but a clear understanding of the forces that cause these celestial bodies to move the way they do was not available. In that year, Isaac Newton provided the key that unlocked the secrets of the heavens. He knew from the first law that a net force had to be acting on the Moon. If it were not, the Moon would move in a straight-line path rather than in its almost circular orbit around Earth. Newton reasoned that this force arises as a result of an attractive force between Moon and Earth, which we call the force of gravity. He also concluded that there could be nothing special about the Earth-Moon system or the Sun and its planets that would cause gravitational forces to act on them alone. In other words, he saw that the same force of attraction that causes the Moon to follow its path also causes an apple to fall to Earth from a tree. He wrote, "I deduced that the forces which keep the planets in their orbs must be reciprocally as the squares of their distances from the centers about which they revolve; and thereby compared the force requisite to keep the Moon in her orb with force of gravity at the surface of Earth; and found them answer pretty nearly."

In 1687 Newton published his work on the law of universal gravitation, which states that

> **Every particle in the Universe attracts every other particle with a force that is directly proportional to the product of their masses and inversely proportional to the square of the distance between them.**

If the particles have masses m_1 and m_2 and are separated by the distance r, the magnitude of the gravitational force between them is

$$F = G \frac{m_1 m_2}{r^2}$$ [7.17]   ◄ Law of universal gravitation

where G is a universal constant called the **constant of universal gravitation,** which has been measured experimentally. Its value in SI units is

$$G = 6.673 \times 10^{-11} \, \text{N} \cdot \text{m}^2/\text{kg}^2$$ [7.18]

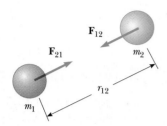

FIGURE 7.17 The gravitational force between two particles is attractive and acts along the line joining the two particles. Note that, according to Newton's third law, $\mathbf{F}_{12} = -\mathbf{F}_{21}$.

Webnote 7.1

For a perspective on science during the time of Newton, a brief overview of Newton's *Principia*, and more on the law of universal gravitation, explore:
http://chemweb.richmond.edu/~rubin/pedagogy/131/131notes/Newton.html

This force law is an example of an **inverse-square law** in that it varies as one over the square of the separation. The force acts so that the objects are always attracted to one another. From Newton's third law, we also know that the force exerted by m_1 on m_2, designated $\mathbf{F}_{12}$ in Figure 7.17, is equal in magnitude to the force $\mathbf{F}_{21}$ exerted by m_2 on m_1, but in the opposite direction. That is, these forces form an action-reaction pair.

Several features of the law of universal gravitation deserve some attention:

1. The gravitational force is a field force that always exists between two particles regardless of the medium that separates them.
2. The force varies as one over the square of the distance between the particles and therefore decreases rapidly with increasing separation.
3. The force is proportional to the product of the particles' masses.
4. The gravitational force is actually very weak as shown by the size of G. It is easily observed for a falling ball of mass m because the mass of Earth M_E is so large in the product mM_E.

Another important fact is that **the gravitational force exerted by a spherical object on a particle outside the sphere is the same as if the entire mass of the sphere were concentrated at its center.** For example, the force on a particle of mass m at Earth's surface has the magnitude

$$F = G\frac{M_E m}{R_E^2}$$ [7.19]

where M_E is Earth's mass and R_E is its radius. This force is directed toward the center of Earth.

Quick Quiz 7.6 As a ball drops and falls to the ground, which statements are false? (a) The force that the ball exerts on Earth is equal in magnitude to the force that Earth exerts on the ball. (b) The ball undergoes the same acceleration as Earth. (c) Earth is much more massive than the ball, so Earth pulls much harder on the ball than the ball pulls on Earth. Thus, the ball falls while Earth remains stationary.

MEASUREMENT OF THE GRAVITATIONAL CONSTANT

The gravitational constant was first measured in an important experiment by Henry Cavendish in 1798. His apparatus consisted of two small spheres, each of mass m, fixed to the ends of a light horizontal rod suspended by a thin metal wire, as in Figure 7.18a. Two large spheres, each of mass M, were placed near the smaller spheres. The attractive force between the smaller and larger spheres caused the rod to rotate in a horizontal plane and the wire to twist. The angle through which the suspended rod rotated was measured with a light beam reflected from a mirror attached to the vertical suspension. (Such a moving spot of light is an effective technique for amplifying the motion.) The experiment was carefully repeated with different masses at various separations. In addition to providing a value for G, the results showed that the force is attractive, proportional to the product mM, and inversely proportional to the square of the distance r.

Mirror

Light source

M r

m

(a)

(b)

FIGURE 7.18 (a) The Cavendish apparatus for measuring *G*. The smaller spheres of mass *m* are attracted to the large spheres of mass *M*, and the rod rotates through a small angle. A light beam reflected from a mirror on the rotating apparatus measures the angle of rotation. (b) A student Cavendish apparatus. (*Courtesy of PASCO Scientific*)

> **Quick Quiz 7.7**
>
> Superman circles Earth at a radius of $2R$ where R is Earth's radius. He then moves to a radius of $4R$. The gravitational force on him at this second orbit as compared to the first orbit is (a) the same, (b) twice as great, (c) four times as great, (d) half as great, (e) one-fourth as great.

Example 7.9 Billiards, Anyone?

Three 0.300-kg billiard balls are placed on a table at the corners of a right triangle, as shown from overhead in Figure 7.19. Find the net gravitational force on the cue ball (designated as m_1) resulting from the other two balls.

Solution To find the net gravitational force on the cue ball m_1, we first calculate the force exerted by m_2 on m_1. Then we find the force exerted by m_3 on m_1. Finally, we add these two forces *vectorially* to obtain the net force **F** on the cue ball.

The force exerted by m_2 on m_1, denoted by $\mathbf{F}_{21}$ in Figure 7.19, is upward. Its magnitude is calculated using Equation 7.17:

$$F_{21} = G\frac{m_2 m_1}{r_{21}^2} = (6.67 \times 10^{-11}\ \text{N} \cdot \text{m}^2/\text{kg}^2)\ \frac{(0.300\ \text{kg})(0.300\ \text{kg})}{(0.400\ \text{m})^2}$$

$$= 3.75 \times 10^{-11}\ \text{N}$$

This result shows that gravitational forces between the common objects that surround us have extremely small magnitudes.

Now let us calculate the gravitational force exerted by m_3 on m_1, denoted by $\mathbf{F}_{31}$ in Figure 7.19. It is directed to the right, and its magnitude is

$$F_{31} = G\frac{m_3 m_1}{r_{31}^2} = (6.67 \times 10^{-11}\ \text{N} \cdot \text{m}^2/\text{kg}^2)\ \frac{(0.300\ \text{kg})(0.300\ \text{kg})}{(0.300\ \text{m})^2}$$

$$= 6.67 \times 10^{-11}\ \text{N}$$

The net gravitational force **F** exerted on the cue ball is found by adding $\mathbf{F}_{21}$ and $\mathbf{F}_{31}$ as *vectors*. The magnitude of this net force is given by

$$F = \sqrt{F_{21}^2 + F_{31}^2} = \sqrt{(3.75)^2 + (6.67)^2} \times 10^{-11}\ \text{N} = \boxed{7.65 \times 10^{-11}\ \text{N}}$$

EXERCISE Find the direction of **F**.

ANSWER The vector **F** makes an angle of 29.3° with the line joining m_1 and m_3.

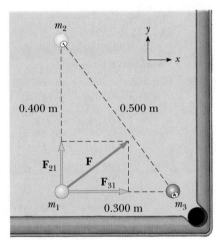

m_2

y

x

0.400 m

0.500 m

$\mathbf{F}_{21}$ **F**

$\mathbf{F}_{31}$

m_1 0.300 m m_3

FIGURE 7.19 (Example 7.9)

FIGURE 7.20   (Example 7.10)

Example 7.10   The Mass of Earth

Use the law of universal gravitation to find an approximate value for the mass of Earth.

Solution   Consider a baseball of mass m falling toward Earth at a location where the free-fall acceleration is g as in Figure 7.20. We know that the magnitude of the gravitational force exerted by Earth on the baseball is the same as the weight of the ball and that this is given by $F_g = mg$. Because the magnitude of the gravitational force is given by Equation 7.19, we find that

$$mg = G\frac{M_E m}{R_E^{\,2}}$$

We can divide each side of this equation by m_b and solve for the mass of Earth M_E:

$$M_E = \frac{gR_E^{\,2}}{G}$$

The falling baseball is close enough to Earth so that the distance of separation between the center of the ball and the center of Earth can be taken as Earth's radius, 6.38×10^6 m. Thus, the Earth's mass is

$$M_E = \frac{(9.80 \text{ m/s}^2)(6.38 \times 10^6 \text{ m})^2}{6.67 \times 10^{-11} \text{ N} \cdot \text{m}^2/\text{kg}^2} = \boxed{5.98 \times 10^{24} \text{ kg}}$$

TABLE 7.1

Free-Fall Acceleration g at Various Altitudes

Altitude (km)[a]	g (m/s^2)
1 000	7.33
2 000	5.68
3 000	4.53
4 000	3.70
5 000	3.08
6 000	2.60
7 000	2.23
8 000	1.93
9 000	1.69
10 000	1.49
50 000	0.13

[a] All values are distances above Earth's surface.

Example 7.11   Gravity and Altitude

Derive an expression that shows how the free-fall acceleration varies with distance from the center of Earth at an *exterior* point.

Solution   The falling baseball of Example 7.10 can be used here also. Now, however, assume that the ball is located at some arbitrary distance r from Earth's center. The first equation in Example 7.10, with r replacing R_E and m removed from both sides, becomes

$$g = G\frac{M_E}{r^2}$$

This indicates that the free-fall acceleration at an exterior point decreases as the inverse square of the distance from the center of Earth. Our assumption in Chapter 2 that objects fall with constant acceleration is obviously incorrect in light of the present example. For short falls, however, this change in g is so small that neglecting the variation does not introduce a significant error in the result.

Because the weight of an object is mg, we see that a change in the value of g produces a change in the weight of the object. For example, if you weigh 800 N at Earth's surface, you will weigh only 200 N at a height above Earth equal to the radius of Earth. Also, we see that if the distance of an object from Earth becomes infinitely large, the weight approaches zero. Values of g at various altitudes are listed in Table 7.1.

EXERCISE   If an object weighs 270 N at Earth's surface, what will it weigh at an altitude above the surface of Earth equal to twice the radius of Earth?

ANSWER   30.0 N

7.9   GRAVITATIONAL POTENTIAL ENERGY REVISITED

In Chapter 5 we introduced the concept of gravitational potential energy and found that the potential energy associated with an object can be calculated using the equation $PE = mgh$, where h is the height of the object above or below some reference level. This equation is actually valid only when the object is near Earth's surface. For objects high above Earth's surface, such as a satellite, an alternative expression must be used to compute the gravitational potential energy. The general expression for the gravitational potential energy associated with an object of mass m at a distance r from the center of Earth can be shown (using integral calculus) to be

$$PE = -G\frac{M_E m}{r}$$ [7.20]

◀ General form of gravitational potential energy

where M_E is Earth's mass.

This equation assumes that the zero level for potential energy is at an infinite distance from Earth's center. This point is necessary because the gravitational force goes to zero when r is set equal to infinity. The negative sign in Equation 7.20 means that the work done by an external force in moving an object of mass m in from infinity to a distance r away from Earth's center is negative, or the gravitational potential energy **decreases** in such a process. Alternatively, we can say that the work is negative because the external force has to hold the object back against the attractive force of gravity.

APPLYING PHYSICS 7.3

Why is the Sun hot?

Explanation   The Sun was formed when a cloud of gas and dust coalesced, due to gravitational attraction, into a massive astronomical object. Before this occurred, the particles were widely scattered, representing a large amount of gravitational potential energy. As the particles came together to form the Sun, the gravitational potential energy decreased. According to the principle of conservation of energy, this potential energy had to be transformed to another form. The form to which it transformed was internal energy, representing an increase in temperature. If enough particles come together, the temperature rises to a point at which nuclear fusion occurs, and the object becomes a star. If there are not enough particles, the temperature rises, but not to a point at which fusion occurs—the object is a planet, if it is in orbit around a star. Jupiter is an example of a large planet that might have been a star had more particles been available to be collected.

Example 7.12   Does the Potential Energy Reduce to *mgh*?

A Find expressions for the gravitational potential energy associated with an object at Earth's surface and for the same object when at a height h above Earth's surface (Fig. 7.21).

Solution   Equation 7.20 gives the potential energy at the surface of Earth as

$$PE_1 = -G\frac{M_E m}{R_E}$$

We can also use Equation 7.20 to find the potential energy when the object is at the height h above the surface of Earth. Taking $r = R_E + h$, we find

$$PE_2 = -G\frac{M_E m}{(R_E + h)}$$

B From the answers to part A, show that the difference in potential energy between these two points reduces to the familiar expression $PE = mgh$ when h is small compared to Earth's radius.

FIGURE 7.21   (Example 7.12)

Solution   The difference in potential energy between the two configurations is found as follows:

$$PE_2 - PE_1 = -G\frac{M_E m}{(R_E + h)} - \left(-G\frac{M_E m}{R_E}\right)$$

$$= -GM_E m\left[\frac{1}{(R_E + h)} - \frac{1}{R_E}\right]$$

After finding a common denominator and applying some algebra to the equation above, we find

$$PE_2 - PE_1 = \frac{GM_E m h}{R_E(R_E + h)}$$

When the height h is very small compared to R_E, the denominator in the preceding expression is approximately equal to R_E^2. Thus, we have

$$PE_2 - PE_1 \cong \frac{GM_E}{R_E^2} mh$$

Now note that the free-fall acceleration at Earth's surface is given by $g = GM_E/R_E^2$ (see Example 7.11). Thus,

$$PE_2 - PE_1 \cong mgh$$

ESCAPE SPEED

If an object is projected upward from Earth's surface with a large enough speed, it can go soaring off into space and never return. This particular speed is called the **escape speed** of an object from Earth. We shall now proceed to derive an expression that enables us to calculate the escape speed.

Suppose an object of mass m is projected vertically upward from Earth's surface with an initial speed v_i, as in Figure 7.22. The initial mechanical energy (kinetic plus potential energy) of the object–Earth system is given by

$$KE_i + PE_i = \tfrac{1}{2}mv_i^2 - G\frac{M_E m}{R_E}$$

We neglect air resistance and assume that the initial speed is just large enough to allow the object to reach infinity with a speed of zero. This value of v_i is the escape speed v_{esc}. When the object is at an infinite distance from Earth, its kinetic energy is zero because $v_f = 0$, and the gravitational potential energy is also zero because our zero level of potential energy was selected at $r = \infty$. Hence, the total mechanical energy is zero, and the law of conservation of energy gives

and

$$\tfrac{1}{2}mv_{\text{esc}}^2 - G\frac{M_E m}{R_E} = 0$$

$v_f = 0$

h

v_i

m

R_E

M_E

FIGURE 7.22   An object of mass m projected upward from Earth's surface with an initial speed v_i reaches a maximum altitude h (where $M_E \gg m$).

Escape speed ▷

$$v_{\text{esc}} = \sqrt{\frac{2GM_E}{R_E}}   \qquad [7.21]$$

From this equation, we will find that the escape speed for an object projected from Earth is about 11.2 km/s, which corresponds to about 25 000 mi/h (see Example 7.13). Note that this expression for v_{esc} is independent of the mass of the object projected from Earth. Thus, a spacecraft has the same escape speed as a molecule. A list of escape speeds for the planets, the Moon, and the Sun is given in Table 7.2. The data presented in this table help us understand why some planets have atmospheres and others do not. For example, on the very hot planet Mercury, gas molecules often reach speeds that are greater than the escape speed of 4.3 km/s. Consequently, any gases that might have been present on the surface of the planet at its formation have long since wandered off into space. Likewise, in the atmosphere of our own Earth, hydrogen and helium molecules through collisions often gain speeds greater than 11.2 km/s, the escape speed on Earth. Therefore, these gases were not retained in Earth's atmosphere. On the other hand, heavier gases such as oxygen and nitrogen in Earth's atmosphere have average speeds of less than 11.2 km/s and have not escaped.

TABLE 7.2	
Escape Speeds for the Planets and the Moon	
Planet	v_{esc} **(km/s)**
Mercury	4.3
Venus	10.3
Earth	11.2
Moon	2.3
Mars	5.0
Jupiter	60.0
Saturn	36.0
Uranus	22.0
Neptune	24.0
Pluto	1.1

Example 7.13   Escape Speed of a Rocket

Calculate the escape speed from Earth for a 5 000-kg spacecraft.

Solution   Using Equation 7.21 with $M_E = 5.98 \times 10^{24}$ kg and $R_E = 6.38 \times 10^6$ m gives

$$v_{esc} = \sqrt{\frac{2GM_E}{R_E}} = \sqrt{\frac{2(6.67 \times 10^{-11})(5.98 \times 10^{24})}{6.38 \times 10^6}} = 1.12 \times 10^4 \text{ m/s}$$

This corresponds to about 25 000 mi/h or about 7 mi/s. Note that the mass of the spacecraft was not required for this calculation.

7.10 KEPLER'S LAWS

The movements of the planets, stars, and other celestial bodies have been observed for thousands of years. In early history, scientists regarded Earth as the center of the Universe. This *geocentric model* was developed extensively by the Greek astronomer Claudius Ptolemy in the second century A.D. and was accepted for the next 1 400 years. In 1543 the Polish astronomer Nicolaus Copernicus (1473–1543) showed that Earth and the other planets revolve in circular orbits about the Sun (the *heliocentric model*).

The Danish astronomer Tycho Brahe (pronounced Brah or BRAH-huh; 1546–1601) made accurate astronomical measurements over a period of 20 years and provided the data for the currently accepted model of the Solar System. Brahe's precise observations of the planets and 777 stars were carried out with nothing more elaborate than a large sextant and compass; the telescope had not yet been invented.

The German astronomer Johannes Kepler, who was Brahe's assistant, acquired Brahe's astronomical data and spent about 16 years trying to deduce a mathematical model for the motions of the planets. After many laborious calculations, he found that Brahe's precise data on the motion of Mars about the Sun provided the answer. Kepler's analysis first showed that the concept of circular orbits about the Sun had to be abandoned. He eventually discovered that the orbit of Mars could be accurately described by an ellipse with the Sun at one focus. He then generalized this analysis to include the motions of all planets. The complete analysis is summarized in three statements known as **Kepler's laws:**

1. **All planets move in elliptical orbits with the Sun at one of the focal points.**
2. **A line drawn from the Sun to any planet sweeps out equal areas in equal time intervals.**
3. **The square of the orbital period of any planet is proportional to the cube of the average distance from the planet to the Sun.**

Newton later demonstrated that these laws are consequences of the gravitational force that exists between any two objects. Newton's law of universal

JOHANNES KEPLER, GERMAN ASTRONOMER (1571–1630)

Kepler is best known for developing the laws of planetary motion based on the careful observations of Tycho Brahe. Throughout his life, Kepler was sidetracked by mystic ideas dating back to the ancient Greeks. For example, he believed in the notion of the "music of the spheres" proposed by Pythagoras, in which each planet in its motion sounds out an exact musical note. After spending several years trying to work out a "regular-solid theory" of the planets, he concluded that the Copernican view of circular planetary orbits had to be abandoned for the view that the planetary orbits are ellipses with the Sun always at one of the foci. *(Art Resource)*

◄ Kepler's laws

Webnote 7.2

For more on the lives of Johannes Kepler, Tycho Brahe, and Nicolaus Copernicus, visit:
http://www.phys.virginia.edu/classes/109N/1995/lectures/kepler.html
http://www.phys.virginia.edu/classes/109N/1995/lectures/tychob.html
http://www.phy.bg.ac.yu/web_projects/giants/copernicus.html

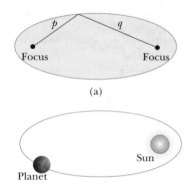

FIGURE 7.23   (a) The sum $p + q$ is the same for every point on the ellipse. (b) In the Solar System, the Sun is at one focus of the elliptical orbit of each planet and the other focus is empty.

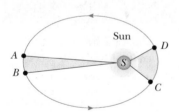

FIGURE 7.24   The two areas swept out by the planet in its elliptical orbit about the Sun are equal if the time interval between points A and B is equal to the time interval between points C and D.

Kepler's third law ▶

Webnote 7.3

For a complete look at Kepler's laws of planetary motion, explore:
http://csep10.phys.utk.edu/astr161/lect/history/kepler.html

gravitation, together with his laws of motion, provides the basis for a full mathematical description of the motions of planets and satellites.

KEPLER'S FIRST LAW

We shall not attempt to derive Kepler's first law. It can be shown that the first law arises as a natural consequence of the inverse-square nature of Newton's law of gravitation. That is, any object bound to another by a force that varies as $1/r^2$ will move in an elliptical orbit. As shown in Figure 7.23a, an ellipse is a curve drawn so that the sum of the distances from any point on the curve to two internal points called focal points or foci (singular, focus) is always the same. For the Sun-planet configuration (Fig. 7.23b), the Sun is at one focus and the other focus is empty. Because the orbit is an ellipse, the distance from the Sun to the planet continuously changes.

KEPLER'S SECOND LAW

Kepler's second law states that a line drawn from the Sun to any planet sweeps out equal areas in equal time intervals. Consider a planet in an elliptical orbit about the Sun, as in Figure 7.24. Imagine that at some instant of time we draw a line from the Sun to a planet, tracing out the line AS where S represents the position of the Sun. Exactly 30 days later, we repeat the process and draw the line BS. The area swept out is shown in color in the diagram. When the planet is at point C, we repeat the process, drawing the line CS from the planet to the Sun. We now wait 30 days, exactly the same amount of time we waited before, and draw the line DS. The area swept out in this interval is also shown in color in the figure. The two colored areas are equal. As we shall see in Chapter 8, Kepler's second law is related to a physical principle known as conservation of angular momentum.

KEPLER'S THIRD LAW

The derivation of Kepler's third law is simple enough to carry out here. Consider a planet of mass M_p moving about the Sun, which has a mass of M_S, in a circular orbit (Fig. 7.25). (The assumption of a circular rather than an elliptical orbit will not introduce serious error into our approach, because the orbits of all planets except Mercury and Pluto are very nearly circular.) Because the gravitational force on the planet is the force that causes the planet to move in a circular path, Newton's second law gives

$$G\frac{M_S M_p}{r^2} = \frac{M_p v^2}{r}$$

The speed v of the planet in its orbit is equal to the circumference of the orbit divided by the time required for one revolution T, called the **period** of the planet. That is, $v = 2\pi r/T$, and the preceding expression becomes

$$G\frac{M_S}{r^2} = \frac{(2\pi r/T)^2}{r}$$

$$T^2 = \left(\frac{4\pi^2}{GM_S}\right) r^3 = K_S r^3 \qquad \text{[7.22]}$$

where K_S is a constant given by

$$K_S = \frac{4\pi^2}{GM_S} = 2.97 \times 10^{-19} \text{ s}^2/\text{m}^3$$

Equation 7.22 is Kepler's third law. Note that K_S is independent of the mass of

PHYSICS *IN ACTION*

Views of Planets and Moons

(Right) Olympus Mons, a giant volcano on Mars, is taller than three Mount Everests and wider than the entire Hawaiian Island chain. This spectacular wide-angle photograph was obtained by the *Mars Orbiter* camera on April 25, 1998, during the 263rd orbit of the Mars Global Surveyor's mission. Images of polar caps on Mars taken by this camera over the last few years suggests that Mars may have major global climate changes that are occurring on the same time scales as Earth's most recent climate shifts. For more recent photographs and information, visit http://mars.jpl.nasa.gov/gallery.

(Below) Images taken by the spacecraft *Galileo* of two tall volcanic plumes and rings of red material deposited on the surface of Io, one of the moons of Jupiter and one of the most active objects in the Solar System. (A second spacecraft, *Cassini*, acquired images of Io at ultraviolet wavelengths to reveal actively erupting plumes of gas and dust.) One distinctive feature in the high-resolution *Galileo* image at the center is a giant red ring (the Pele ring) with a diameter of about 1 400 km (870 mi.). A more recent *Galileo* image on the far right shows a second giant red ring on Io centered around Tvashtar. This ring shows that Tvashtar must be the vent for north polar plumes imaged by *Cassini*, producing plumes that range in height between 355 km and 415 km. For other more recent photographs and information, visit http://photojournal.jpl.nasa.gov/cgi-bin/PIAGenCatalog.pl?PIA02588.

(NASA/JPL)

(NASA/JPL/Malin Space Science Systems)

---

the planet. Therefore, Equation 7.22 is valid for any planet. If we consider the orbit of a satellite such as the Moon about Earth, then the constant has a different value, with the mass of the Sun replaced by the mass of Earth. In this case, K_E equals $4\pi^2/GM_E$.

Kepler's third law gives us a method for measuring the mass of the Sun or of any celestial object around which at least one other object moves in orbit. Notice that the constant K_S in Equation 7.22 includes the mass of the Sun. The value of this constant can be found by substituting the values of the period and orbital radius and solving for K_S. The Sun's mass is then

$$M_S = \frac{4\pi^2}{GK_S}$$

This same process can be used to calculate Earth's mass (by considering the period and orbital radius of the Moon) and also the mass of other planets in the Solar System that possess moons.

A collection of useful planetary data is presented in Table 7.3, where the last column verifies that T^2/r^3 is a constant.

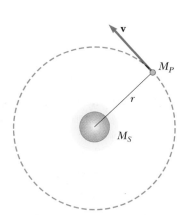

FIGURE 7.25 A planet of mass M_p moving in a circular orbit about the Sun. The orbits of all planets except Mercury and Pluto are nearly circular.

TABLE 7.3   **Useful Planetary Data**

Body	Mass (kg)	Mean Radius (m)	Period (s)	Mean Distance from Sun (m)	$\dfrac{T^2}{r^3}$ $(10^{-19}$ s^2/m$^3)$
Mercury	3.18×10^{23}	2.43×10^6	7.60×10^6	5.79×10^{10}	2.97
Venus	4.88×10^{24}	6.06×10^6	1.94×10^7	1.08×10^{11}	2.99
Earth	5.98×10^{24}	6.38×10^6	3.156×10^7	1.496×10^{11}	2.97
Mars	6.42×10^{23}	3.37×10^6	5.94×10^7	2.28×10^{11}	2.98
Jupiter	1.90×10^{27}	6.99×10^7	3.74×10^8	7.78×10^{11}	2.97
Saturn	5.68×10^{26}	5.85×10^7	9.35×10^8	1.43×10^{12}	2.99
Uranus	8.68×10^{25}	2.33×10^7	2.64×10^9	2.87×10^{12}	2.95
Neptune	1.03×10^{26}	2.21×10^7	5.22×10^9	4.50×10^{12}	2.99
Pluto	$\approx 1 \times 10^{23}$	$\approx 3 \times 10^6$	7.82×10^9	5.91×10^{12}	2.96
Moon	7.36×10^{22}	1.74×10^6	—	—	—
Sun	1.991×10^{30}	6.96×10^8	—	—	—

Quick Quiz 7.8   A comet is in a highly elliptical orbit around the Sun. The period of the comet's orbit is 90 days. Which of the following statements is true about the possibility of a collision between this comet and Earth? (a) There is no possible danger of a collision. (b) There is a possibility of a collision. (c) There is not enough information to determine whether there is danger of a collision.

APPLYING **PHYSICS** 7.4

Some science fiction stories fantasize about a sister planet to Earth in the same orbit as Earth, but 180° ahead of us. Thus, we could never see this planet from Earth because it would always be behind the Sun, hidden from our point of view. How would you plan a visit to this proposed planet if you start from an Earth orbit?

Explanation   The quickest way to arrive at the planet is to fire your rocket so that you slow down in your orbit around the Sun. This would transfer you to an elliptical orbit around the Sun with a smaller major axis than that of Earth's orbit. Thus, you will complete this orbit in a time less than one year. If you design your orbit so that the period is 0.5 year, then when you return to your original point, the sister planet will also be there.

A calculation using Kepler's third law shows that your spacecraft will dip inside the orbit of Venus, so you must be sure not to collide with Venus and that you have the appropriate thermal protection to deal with increased radiation from the Sun.

Another possibility is to fire your rockets so that you speed up in your orbit around the Sun, transferring to an elliptical orbit with a period of 1.5 years. This avoids the thermal problems of going near the Sun, but takes three times as long.

Example 7.14   A Free Ride

NASA employs a method called either "slingshotting" or "gravitational boosting" to add velocity to a spacecraft as it passes a planet. To understand how this works, consider a spacecraft moving with a speed of 10.0 km/s approaching Jupiter at a point where the

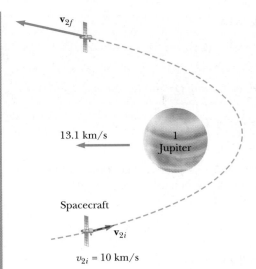

FIGURE 7.26 (Example 7.14)

gravitational pull of Jupiter is negligible (Fig. 7.26). Jupiter moves in its orbit at an average speed of about 13.1 km/s and all speeds are measured relative to a reference frame fixed in the Sun. Find the speed of the spacecraft after the "collision" with Jupiter, when the spacecraft is almost free of Jupiter's pull. Assume that the spacecraft's velocity after the collision is directly opposite its initial velocity.

Solution We can treat this event as an elastic collision. Because Jupiter is so massive, its velocity will be relatively unaffected, so before and after the collision it will have a velocity of -13.1 km/s. To find the velocity of the spacecraft after the collision, we call Jupiter object 1 and the spacecraft object 2 and apply Equation 6.14:

$$v_{1i} - v_{2i} = -(v_{1f} - v_{2f})$$

$$-13.1 \text{ km/s} - 10.0 \text{ km/s} = -(-13.1 \text{ km/s} - v_{2f})$$

$$v_{2f} = \boxed{-36.2 \text{ km/s}}$$

Thus, the spacecraft leaves the vicinity of Jupiter traveling more than three times faster than it did upon its encounter with the planet.

EXERCISE The extra kinetic energy of the spacecraft comes from Jupiter's kinetic energy, which is transferred to the spacecraft by the force of gravity; that is, Jupiter "drags" the spacecraft along with it in its orbital motion and slows down a bit. Using Table 7.3, calculate the approximate decrease in orbital speed of Jupiter for a 500-kg spacecraft.

ANSWER $\approx 1 \times 10^{-20}$ m/s—which is immeasurably small!

Example 7.15 Geosynchronous Orbit

Satellite dishes do not have to change directions in order to stay focused on a signal from a satellite. This means that the satellite always has to be found at the same location with respect to the surface of Earth. For this to occur, the satellite must be at a height such that its revolution period is the same as that of Earth, 24 h. At what height must a satellite be to achieve this?

APPLICATION

SATELLITE DISHES

Reasoning and Solution The force that produces the centripetal acceleration of the satellite is the gravitational force, so

$$G\frac{M_E m}{r^2} = \frac{mv^2}{r} \tag{1}$$

where M_E is Earth's mass and r is the satellite's distance from the center of Earth.

Also, we find the speed of the satellite to be

$$v = \frac{d}{T} = \frac{2\pi r}{T} \tag{2}$$

where T is the orbital period of the satellite. Solving (1) and (2) simultaneously for r yields

$$r = \left(\frac{T^2}{4\pi^2} GM_E \right)^{1/3}$$

EXERCISE   Taking $T = 24\,\text{h} = 86\,400\,\text{s}$ and $M_E = 5.98 \times 10^{24}\,\text{kg}$, show that the height above Earth is approximately 22 000 miles. (Be careful; will $r = 22\,000$ miles?)

Example 7.16   An Earth Satellite

A satellite of mass m moves in a circular orbit about Earth with a constant speed v and a height $h = 1000$ km above Earth's surface, as in Figure 7.27. (For clarity, this figure is not drawn to scale.) Find the orbital speed of the satellite. Earth's radius is 6.38×10^6 m, and its mass is 5.98×10^{24} kg.

Reasoning   The only force acting on the satellite is the gravitational force directed toward the center of the circular orbit. This can be equated to mv^2/r to find v. (The distance r is the distance from the center of Earth to the satellite.)

Solution   The gravitational force exerted by Earth produces the centripetal acceleration of the satellite. Because the magnitude of the gravitational force is GM_Em/r^2, we find that

$$F = G\frac{M_Em}{r^2} = m\frac{v^2}{r}$$

$$v^2 = \frac{GM_E}{r}$$

In this expression, the distance r is Earth's radius plus the height of the satellite. That is, $r = R_E + h = 7.38 \times 10^6$ m, and so

$$v^2 = \frac{(6.67 \times 10^{-11}\,\text{N} \cdot \text{m}^2/\text{kg}^2)(5.98 \times 10^{24}\,\text{kg})}{7.38 \times 10^6\,\text{m}} = 5.40 \times 10^7\,\text{m}^2/\text{s}^2$$

Therefore,

$$v = 7.35 \times 10^3\,\text{m/s} \approx \boxed{16\,400\,\text{mi/h}}$$

Note that v *is independent of the mass of the satellite!*

EXERCISE   Calculate the period T of the satellite (the time required for one revolution about Earth).

ANSWER   105 min

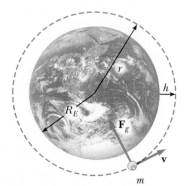

FIGURE 7.27   (Example 7.16) A satellite of mass m moving in a circular orbit of radius r and with constant speed v around Earth (not drawn to scale). The force causing the centripetal acceleration is the gravitational force acting on the satellite.

SUMMARY

The **average angular speed** $\overline{\omega}$ of a rigid object is defined as the ratio of the angular displacement $\Delta\theta$ to the time interval Δt:

$$\overline{\omega} = \frac{\theta_f - \theta_i}{t_f - t_i} = \frac{\Delta\theta}{\Delta t} \tag{7.2}$$

where $\overline{\omega}$ is in radians per second (rad/s).

The **average angular acceleration** $\bar{\alpha}$ of a rotating object is defined as the ratio of the change in angular speed $\Delta\omega$ to the time interval Δt.

$$\bar{\alpha} \equiv \frac{\omega_f - \omega_i}{t_f - t_i} = \frac{\Delta\omega}{\Delta t} \qquad [7.4]$$

where $\bar{\alpha}$ is in radians per second per second (rad/s^2).

If an object undergoes rotational motion about a fixed axis under constant angular acceleration α, we can describe its motion with the following set of equations:

$$\omega = \omega_i + \alpha t \qquad [7.5]$$

$$\Delta\theta = \omega_i t + \tfrac{1}{2}\alpha t^2 \qquad [7.6]$$

$$\omega^2 = \omega_i^2 + 2\alpha\,\Delta\theta \qquad [7.7]$$

When an object rotates about a fixed axis, the angular speed and angular acceleration are related to the tangential speed and tangential acceleration through the relationships

$$v_t = r\omega \qquad [7.8]$$

$$a_t = r\alpha \qquad [7.9]$$

Any object moving in a circular path has an acceleration directed toward the center of the circular path, called a **centripetal acceleration.** Its magnitude is given by

$$a_c = \frac{v^2}{r} = r\omega^2 \qquad [7.11, 7.14]$$

Any object moving in a circular path must have a net force exerted on it that is directed toward the center of the circular path. From Newton's second law, this force and the centripetal acceleration are related by

$$F = m\frac{v_t^2}{r} \qquad [7.16]$$

Some examples of forces that cause centripetal acceleration are the force of gravity (as in the motion of a satellite) and the force of tension in a string.

Newton's law of universal gravitation states that every particle in the Universe attracts every other particle with a force that is directly proportional to the product of their masses and inversely proportional to the square of the distance r between them:

$$F = G\frac{m_1 m_2}{r^2} \qquad [7.17]$$

where $G = 6.673 \times 10^{-11}$ N·m^2/kg^2 is the **constant of universal gravitation.**

Kepler's laws of planetary motion state that

1. All planets move in elliptical orbits with the Sun at one of the focal points.
2. A line drawn from the Sun to any planet sweeps out equal areas in equal time intervals.
3. The square of the orbital period of a planet is proportional to the cube of the average distance from the planet to the Sun:

$$T^2 = \left(\frac{4\pi^2}{GM_S}\right) r^3 \qquad [7.22]$$

CONCEPTUAL QUESTIONS

1. In a race like the Indianapolis 500, a driver circles the track counterclockwise and feels his head pulled toward one shoulder. To relieve his neck muscles from having to hold his head erect, the driver fastens a strap to one wall of the car and the other to his helmet. The length of the strap is adjusted to keep his head vertical. (a) Which shoulder does his head tend to lean toward? What force(s) produce the centripetal acceleration (b) when there is no strap, and (c) when there is a strap?

2. Consider a compact disc rotating at constant angular speed. Does a scratch on the rim have a radial component of velocity? A tangential component? A radial component

of acceleration? A tangential component? Repeat these questions for the brief interval of time during which the CD moves from rest to its final angular speed.

3. An object executes circular motion with a constant speed whenever a net force of constant magnitude acts perpendicular to the velocity. What happens to the speed if the force is not perpendicular to the velocity?

4. Explain why Earth is not spherical in shape, but bulges at the Equator.

5. If a car's wheels are replaced with wheels of greater diameter, will the reading of the speedometer change? Explain.

6. At night, you are farther from the Sun than during the day. What's more, the force exerted by the Sun on you is downward into Earth at night, and upward into the sky during the day. If you had a sensitive enough bathroom scale, would you appear to weigh more at night than during the day?

7. Correct the following statement: "The race car rounds the turn at a constant velocity of 90 miles per hour."

8. Why does an astronaut in a spacecraft orbiting Earth experience a feeling of weightlessness?

9. Explain why it is easier to determine the mass of a planet when it has a moon than to find its mass if it has no moon.

10. Because of Earth's rotation about its axis, you will weigh slightly less at the Equator than at the poles. Why?

11. It has been suggested that rotating cylinders about 10 miles long and 5 miles in diameter be placed in space for colonies. The purpose of their rotations is to simulate gravity for the inhabitants. Explain the concept behind this proposal.

12. Describe the path of a moving object in the event that the object's acceleration is constant in magnitude at all times and (a) perpendicular to its velocity; (b) parallel to its velocity.

13. A pail of water can be whirled in a vertical circular path such that no water is spilled. Why does the water remain in the pail, even when the pail is upside down above your head?

14. The orbital planes of all the planets must pass through the center of the Sun. Why?

15. Is it possible for a car to move in a circular path in such a way that it has a tangential acceleration but no centripetal acceleration?

16. Use Kepler's second law to convince yourself that Earth moves faster in its orbit during the northern-hemisphere winter, when it is closest to the Sun, than during the summer, when it is farthest from the Sun.

17. Why does a pilot sometimes black out when pulling out of a steep dive?

PROBLEMS

1, 2, 3 = straightforward, intermediate, challenging ☐ = full solution available in Student Solutions Manual/Study Guide

web = solution posted at **http://info.brookscole.com/serway** [icon] = biomedical application

Section 7.1 Angular Speed and Angular Acceleration

1. The tires on a new compact car have a diameter of 2.0 ft and are warranted for 60 000 miles. (a) Determine the angle (in radians) through which one of these tires will rotate during the warranty period. (b) How many revolutions of the tire are equivalent to your answer in (a) above?

2. A wheel has a radius of 4.1 m. How far (path length) does a point on the circumference travel if the wheel is rotated through angles of 30°, 30 rad, and 30 rev, respectively?

3. Find the angular speed of Earth about the Sun in radians per second and degrees per day.

4. A potter's wheel moves from rest to an angular speed of 0.20 rev/s in 30 s. Find its angular acceleration in radians per second per second.

Section 7.2 Rotational Motion Under Constant Angular Acceleration

Section 7.3 Relations Between Angular and Linear Quantities

5. A dentist's drill starts from rest. After 3.20 s of constant angular acceleration, it turns at a rate of 2.51×10^4 rev/min. (a) Find the drill's angular acceleration. (b) Determine the angle (in radians) through which the drill rotates during this period.

6. An electric motor rotating a workshop grinding wheel at a rate of 100 rev/min is switched off. Assume constant negative angular acceleration of magnitude 2.00 rad/s². (a) How long does it take for the grinding wheel to stop? (b) Through how many radians has the wheel turned during the interval found in (a)?

7. A car is traveling with a velocity of 17.0 m/s on a straight horizontal highway. The wheels of the car have a radius of 48.0 cm. If the car then speeds up with an acceleration of 2.00 m/s² for 5.00 s, find the number of revolutions of the wheels during this period.

8. The car in the preceding problem runs out of gas and the rotation of the wheels slows with an angular acceleration of magnitude 1.35 rad/s². If the car has a speed of 24.0 m/s when this happens, through how many revolutions do the wheels turn as the car comes to rest?

9. The diameters of the main rotor and tail rotor of a single-engine helicopter are 7.60 m and 1.02 m, respectively. The respective rotational speeds are 450 rev/min and 4 138 rev/min. Calculate the speeds of the tips of both rotors. Compare these speeds with the speed of sound, 343 m/s.

10. The tub of a washer goes into its spin-dry cycle, starting from rest and reaching an angular speed of 5.0 rev/s in 8.0 s. At this point the person doing the laundry opens the lid, and a safety switch turns off the washer. The tub slows to rest in 12.0 s. Through how many revolutions does the tub turn during this 20-s interval? Assume constant angular acceleration while it is starting and stopping.

11. A standard cassette tape is placed in a standard cassette player. Each side lasts for 30 minutes. The two tape wheels of the cassette fit onto two spindles in the player. Suppose

that a motor drives one spindle at constant angular velocity ~ 1 rad/s and the other spindle is free to rotate at any angular speed. Find the order of magnitude of the tape's thickness. Specify any other quantities you estimate and the values you take for them.

12. A coin with a diameter of 2.40 cm is dropped on edge onto a horizontal surface. The coin starts out with an initial angular speed of 18.0 rad/s and rolls in a straight line without slipping. If the rotation slows with an angular acceleration of magnitude 1.90 rad/s^2, how far does the coin roll before coming to rest?

13. A pulsar is a celestial object that emits light in short bursts. A pulsar in the Crab Nebula flashes at a rate of 30 times/s. Suppose the light pulses are caused by the rotation of a spherical object that emits light from a pair of diametrically opposed "flashlights" on its equator. What is the maximum radius of the pulsar if no part of its surface can move faster than the speed of light (3.00×10^8 m/s)?

Section 7.4 Centripetal Acceleration

14. As noted in Conceptual Question 11, it has been suggested that rotating cylinders about 10 mi long and 5.0 mi in diameter be placed in space and used as colonies. What angular speed must such a cylinder have so that the centripetal acceleration at its surface equals the free-fall acceleration?

15. Find the centripetal accelerations of (a) a point on the Equator and (b) the North Pole, due to the rotation of Earth about its axis.

16. In civil aviation, a "standard turn" for level flight of a propeller-driven airplane is one in which the airplane makes a complete circular turn in 2.0 min. If the plane's speed is 180 m/s, what is the radius of the circle? What is the centripetal acceleration of the plane?

17. (a) What is the tangential acceleration of a bug on the rim of a 10-in.-diameter disk if the disk moves from rest to an angular speed of 78 rev/min in 3.0 s? (b) When the disk is at its final speed, what is the tangential velocity of the bug? (c) One second after the bug starts from rest, what are its tangential acceleration, centripetal acceleration, and total acceleration?

18. A race car starts from rest on a circular track of radius 400 m. The car's speed increases at the constant rate of 0.500 m/s^2. At the point where the magnitudes of the centripetal and tangential accelerations are equal, determine (a) the speed of the race car, (b) the distance traveled, and (c) the elapsed time.

Section 7.6 Forces Causing Centripetal Acceleration

Section 7.7 Describing Forces in Accelerated Reference Frames

19. A 55.0-kg ice skater is moving at 4.00 m/s when she grabs the loose end of a rope, the opposite end of which is tied to a pole. She then moves in a circle of radius 0.800 m around the pole. (a) Determine the force exerted by the horizontal rope on her arms. (b) Compare this force with her weight.

20. A sample of blood is placed in a centrifuge of radius 15.0 cm. The mass of a red blood cell is 3.0×10^{-16} kg, and the magnitude of the force acting on it as it settles out

of the plasma is 4.0×10^{-11} N. At how many revolutions per second should the centrifuge be operated?

21. A 2 000-kg car rounds a circular turn of radius 20 m. If the road is flat and the coefficient of friction between tires and road is 0.70, how fast can the car go without skidding?

22. The cornering performance of an automobile is evaluated on a skidpad, where the maximum speed that a car can maintain around a circular path on a dry, flat surface is measured. Then the centripetal acceleration, also called the lateral acceleration, is calculated as a multiple of the free-fall acceleration *g*. The main factors affecting the performance are the tire characteristics and the suspension system of the car. A Dodge Viper GTS can negotiate a skidpad of radius 61.0 m at 86.5 km/h. Calculate its maximum lateral acceleration.

23. A 50.0-kg child stands at the rim of a merry-go-round of radius 2.00 m, rotating with an angular speed of 3.00 rad/s. (a) What is the child's centripetal acceleration? (b) What is the minimum force between his feet and the floor of the carousel that is required to keep him in the circular path? (c) What minimum coefficient of static friction is required? Is the answer you found reasonable? In other words, is he likely to stay on the merry-go-round?

24. An engineer wishes to design a curved exit ramp for a toll road in such a way that a car will not have to rely on friction to round the curve without skidding. She does so by banking the road in such a way that the force causing the centripetal acceleration will be supplied by the component of the normal force toward the center of the circular path. (a) Show that for a given speed *v* and a radius *r*, the curve must be banked at the angle θ such that $\tan \theta = v^2/rg$. (b) Find the angle at which the curve should be banked if a typical car rounds it at a 50.0-m radius and a speed of 13.4 m/s.

25. An air puck of mass 0.25 kg is tied to a string and allowed to revolve in a circle of radius 1.0 m on a frictionless horizontal table. The other end of the string passes through a hole in the center of the table, and a mass of 1.0 kg is tied to it (Fig. P7.25). The suspended mass remains in equilibrium while the puck on the tabletop revolves. (a) What is the tension in the string? (b) What is the horizontal force acting on the puck? (c) What is the speed of the puck?

FIGURE P7.25

26. Tarzan (*m* = 85 kg) tries to cross a river by swinging from a 10-m-long vine. His speed at the bottom of the swing (as he just clears the water) is 8.0 m/s. Tarzan doesn't know

that the vine has a breaking strength of 1 000 N. Does he make it safely across the river? Justify your answer.

27. A 40.0-kg child takes a ride on a Ferris wheel that rotates four times each minute and has a diameter of 18.0 m. (a) What is the centripetal acceleration of the child? (b) What force (magnitude and direction) does the seat exert on the child at the lowest point of the ride? (c) What force does the seat exert on the child at the highest point of the ride? (d) What force does the seat exert on the child when she is halfway between the top and bottom?

28. A roller-coaster vehicle has a mass of 500 kg when fully loaded with passengers (Fig. P7.28). (a) If the vehicle has a speed of 20.0 m/s at Ⓐ, what is the magnitude of the force that the track exerts on the vehicle at this point? (b) What is the maximum speed the vehicle can have at Ⓑ in order for gravity to hold it on the track?

FIGURE P7.28

Section 7.8   Newton's Law of Universal Gravitation

Section 7.9   Gravitational Potential Energy Revisited

Section 7.10   Kepler's Laws

29. The average distance separating Earth and the Moon is 384 000 km. Use the data in Table 7.3 to find the net gravitational force exerted by Earth and the Moon on a spaceship with mass 3.00×10^4 kg located halfway between them.

30. During a solar eclipse, the Moon, Earth, and Sun all lie on the same line, with the Moon between Earth and the Sun. (a) What force is exerted by the Sun on the Moon? (b) What force is exerted by Earth on the Moon? (c) What force is exerted by the Sun on Earth? (See Table 7.3 and Problem 29 above.)

31. A coordinate system (in meters) is constructed on the surface of a pool table, and three objects are placed on the coordinate system as follows: a 2.0-kg object at the origin, a 3.0-kg object at (0, 2.0), and a 4.0-kg object at (4.0, 0). Find the resultant gravitational force exerted by the other two objects on the object at the origin.

32. Use the data of Table 7.3 to find the point between Earth and the Sun at which an object can be placed so that the net gravitational force exerted by Earth and Sun on this object is zero.

33. Objects with masses of 200 kg and 500 kg are separated by 0.400 m. (a) Find the net gravitational force exerted by these objects on a 50.0-kg object placed midway between them. (b) At what position (other than infinitely remote ones) can the 50.0-kg object be placed so as to experience a net force of zero?

34. Two objects attract each other with a gravitational force of magnitude 1.00×10^{-8} N when separated by 20.0 cm. If the total mass of the two objects is 5.00 kg, what is the mass of each?

35. A satellite moves in a circular orbit around Earth at a speed of 5 000 m/s. Determine (a) the satellite's altitude above Earth's surface and (b) the period of the satellite's orbit.

36. A 600-kg satellite is in a circular orbit about Earth at a height above Earth equal to Earth's mean radius. Find (a) the satellite's orbital speed, (b) the period of its revolution, and (c) the gravitational force acting on it.

37. Io, a satellite of Jupiter, has an orbital period of 1.77 days and an orbital radius of 4.22×10^5 km. From these data, determine the mass of Jupiter.

38. A satellite has a mass of 100 kg and is located at 2.00×10^6 m above Earth's surface. (a) What is the potential energy associated with the satellite at this location? (b) What is the magnitude of the gravitational force on the satellite?

39. A satellite of mass 200 kg is launched from a site on the Equator into an orbit at 200 km above Earth's surface. (a) If the orbit is circular, what is the orbital period of this satellite? (b) What is the satellite's speed in orbit? (c) What is the minimum energy necessary to place this satellite in orbit, assuming no air friction?

40. Neutron stars are extremely dense objects that are formed from the remnants of supernova explosions. Many rotate very rapidly. Suppose that the mass of a certain spherical neutron star is twice the mass of the Sun and its radius is 10.0 km. Determine the greatest possible angular speed it can have so that the matter at the surface of the star on its equator is just held in orbit by the gravitational force.

ADDITIONAL PROBLEMS

41. (a) Find the angular speed of Earth's rotation on its axis. As Earth turns toward the east, we see the sky turning toward the west at this same rate.

(b) *The rainy Pleiads wester*
   And seek beyond the sea
   The head that I shall dream of
   That shall not dream of me.
   —A. E. Housman (© Robert E. Symons)

Cambridge, England, is at longitude 0° and Saskatoon, Saskatchewan, is at longitude 107° west. How much time elapses after the Pleiades set in Cambridge until these stars fall below the western horizon in Saskatoon?

42. The Mars probe *Pathfinder* is designed to drop the instrument package from a height of 20 m above the surface, after the speed of the probe has been brought to zero by a combination parachute-rocket system at that height. To cushion the landing, giant airbags surround the package. The mass of Mars is 0.107 4 times that of Earth and the radius of Mars is 0.528 2 that of Earth. Find (a) the acceleration due to gravity at the surface of Mars and (b) how long it takes for the instrument package to fall the last 20 meters.

43. An athlete swings a 5.00-kg ball horizontally on the end of a rope. The ball moves in a circle of radius 0.800 m at an angular speed of 0.500 rev/s. What are (a) the tangential speed of the ball and (b) its centripetal acceleration?

(c) If the maximum tension the rope can withstand before breaking is 100 N, what is the maximum tangential speed the ball can have?

44. A digital audio compact disc carries data, with each bit occupying 0.6 μm, along a continuous spiral track from the inner circumference of the disc to the outside edge. A CD player turns the disc to carry the track counterclockwise above a lens at a constant speed of 1.30 m/s. Find the required angular speed (a) at the beginning of the recording, where the spiral has a radius of 2.30 cm and (b) at the end of the recording, where the spiral has a radius of 5.80 cm. (c) A full-length recording lasts for 74 min 33 s. Find the average angular acceleration of the disc. (d) Assuming that the acceleration is constant, find the total angular displacement of the disc as it plays. (e) Find the total length of the track.

45. The Solar Maximum Mission Satellite was placed in a circular orbit about 150 mi above Earth. Determine (a) the orbital speed of the satellite and (b) the time required for one complete revolution.

46. A car rounds a banked curve where the radius of curvature of the road is R, the banking angle is θ, and the coefficient of static friction is μ. (a) Determine the range of speeds the car can have without slipping up or down the road. (b) What is the range of speeds possible if $R = 100$ m, $\theta = 10°$, and $\mu = 0.10$ (slippery conditions)?

47. A car moves at speed v across a bridge made in the shape **web** of a circular arc of radius r. (a) Find an expression for the normal force acting on the car when it is at the top of the arc. (b) At what minimum speed will the normal force become zero (causing occupants of the car to seem weightless) if $r = 30.0$ m?

48. A 0.400-kg pendulum bob passes through the lowest part of its path at a speed of 3.00 m/s. (a) What is the tension in the pendulum cable at this point if the pendulum is 80.0 cm long? (b) When the pendulum reaches its highest point, what angle does the cable make with the vertical? (c) What is the tension in the pendulum cable when the pendulum reaches its highest point?

49. Because of Earth's rotation about its axis, a point on the Equator experiences a centripetal acceleration of 0.034 0 m/s^2 while a point at the poles experiences no centripetal acceleration. (a) Show that at the Equator the gravitational force on an object (the true weight) must exceed the object's apparent weight. (b) What are the apparent weights at the Equator and at the poles of a 75.0-kg person? (Assume Earth is a uniform sphere, and take $g = 9.800$ m/s^2.)

50. A stunt man whose mass is 70 kg swings from the end of a 4.0-m-long rope along the arc of a vertical circle. Assuming he starts from rest when the rope is horizontal, find the tensions in the rope that are required to make him follow his circular path, (a) at the beginning of his motion, (b) at a height of 1.5 m above the the bottom of the circular arc, and (c) at the bottom of his arc.

51. In a popular amusement park ride, a rotating cylinder of radius 3.00 m is set in rotation at an angular speed of 5.00 rad/s, as in Figure P7.51. The floor then drops away, leaving the riders suspended against the wall in a vertical position. What minimum coefficient of friction between a rider's clothing and the wall is needed to keep the rider from slipping? (*Hint:* Recall that the magnitude of the maximum force of static friction is equal to μn, where n is the normal force—in this case, the force causing the centripetal acceleration.

FIGURE P7.51

52. A 0.50-kg ball that is tied to the end of a 1.5-m light cord is revolved in a horizontal plane with the cord making a 30° angle with the vertical (see Fig. 7.14). (a) Determine the ball's speed. (b) If instead the ball is revolved so that its speed is 4.0 m/s, what angle does the cord make with the vertical? (c) If the cord can withstand a maximum tension of 9.8 N, what is the highest speed at which the ball can move?

53. A skier starts at rest at the top of a large hemispherical hill (Fig. P7.53). Neglecting friction, show that the skier will leave the hill and become airborne at a distance of $h = R/3$ below the top of the hill. (*Hint:* At this point, the normal force goes to zero.)

FIGURE P7.53

54. After consuming all of its nuclear fuel, a massive star can collapse to form a black hole, which is an immensely dense object for which the escape speed is greater than the speed of light. Newton's law of universal gravitation still describes the force that a black hole exerts on objects

outside it. A spacecraft in the shape of a long cylinder has a length of 100 m and its mass with occupants is 1 000 kg. It has strayed too close to a 1.0-m-radius black hole having a mass 100 times that of the Sun (Fig. P7.54). (a) If the nose of the spacecraft points toward the center of the black hole, and if distance between the nose of the spacecraft and the black hole's center is 10 km, determine the total force on the spacecraft. (b) What is the difference in the force per kilogram of mass felt by the occupants in the nose of the ship and those in the rear of the ship farthest from the black hole?

FIGURE P7.54

55. In Robert Heinlein's *The Moon Is a Harsh Mistress,* the colonial inhabitants of the Moon threaten to launch rocks down onto Earth if they are not given independence (or at least representation). Assuming that a gun could launch a rock of mass *m* at twice the lunar escape speed, calculate the speed of the rock as it enters Earth's atmosphere.

56. Show that the escape speed from the surface of a planet of **web** uniform density is directly proportional to the radius of the planet.

57. A massless spring of constant $k = 78.4$ N/m is fixed on the left side of a level track. A block of mass $m = 0.50$ kg is pressed against the spring and compresses it a distance *d*, as in Figure P7.57. The block (initially at rest) is then released and travels toward a circular loop-the-loop of radius $R = 1.5$ m. The entire track and the loop-the-loop are frictionless except for the section of track between points *A* and *B*. Given that the coefficient of kinetic friction between the block and the track along *AB* is $\mu_k = 0.30$, and that the length of *AB* is 2.5 m, determine the minimum compression *d* of the spring that enables the block to just make it through the loop-the-loop at point *C*. (*Hint:* The force exerted by the track on the block will be zero if the block barely makes it through the loop-the-loop.)

FIGURE P7.57

58. A small block of mass $m = 0.50$ kg is fired with an initial speed of $v_0 = 4.0$ m/s along a horizontal section of frictionless track, as shown in the top portion of Figure P7.58. The block then moves along the frictionless semicircular, *vertical* tracks of radius $R = 1.5$ m. (a) Determine the force exerted by the track on the block at points Ⓐ and Ⓑ. (b) The bottom of the track consists of a section ($L = 0.40$ m) with friction. Determine the coefficient of kinetic friction between the block and that portion of the bottom track if the block just makes it to point Ⓒ on the first trip. (*Hint:* If the block just makes it to point Ⓒ, the force of contact exerted by the track on the block at that point should be zero.)

FIGURE P7.58

59. A frictionless roller coaster is given an initial velocity of v_0 at height *h*, as in Figure P7.59. The radius of curvature of the track at point Ⓐ is *R*. (a) Find the maximum value of v_0 so that the roller coaster stays on the track at Ⓐ solely because of gravity. (b) Using the value of v_0 calculated in (a), determine the value of *h'* that is necessary if the roller coaster just makes it to point Ⓑ.

FIGURE P7.59

GROUP ACTIVITIES

G.1 Experiment with a bicycle wheel, like the one in Figure 7.3. Wrap tape around the middle of one spoke so that you can easily see it and draw a chalk mark on a point on the rim at the end of the spoke. Make the tire rotate by always gently and steadily turning the crank by hand. Compare the linear speed of the tape with that of the chalk mark. Why do these two points move at different speeds?

What happens to the speed of the tape as you slide it closer to the center of the wheel?

With the tape at a given distance from the center, measure the time it takes for the tape to make five rotations. From this, find the angular speed and the linear speed. Repeat at least five times and average your results. Move the tape to a different point and repeat the observations.

G.2 Tie a tennis ball to string of about 1.0 m length. At a safe distance from other students, whirl the ball in a horizontal circle. Note the increasing tension in the string as you whirl it faster. Now, whirl the ball in a vertical circle and observe the difference in tension at the top of the path and at the bottom. Why does this occur? While whirling the ball, release it and observe that it flies off in a direction tangent to its circular path.

G.3 On a clear night spend a few hours outside observing the stars. Choose a pattern of stars toward the north, and others toward the east, south, and west. As time passes, notice how these patterns move. Obtain a star chart of the constellations from your instructor and also a list of the planets visible for your date of observation. Locate as many of these as possible.

G.4 A roller coaster travels in a circular path in a vertical plane. (a) Identify the forces on a passenger at the top of the circular loop that cause centripetal acceleration. Show the direction of all forces in a sketch. (b) Identify the forces on the passenger at the bottom of the loop that produce centripetal acceleration. Show these in a sketch. (c) Based on your answers to (a) and (b) at what point, top or bottom, should the seat exert the greatest force on the passenger? (d) Assume the speed of the roller coaster is 12.0 m/s at the top of the loop, which has radius 8.00 m. Find the force exerted by the seat on a 70.0-kg passenger. Then, assume the speed remains the same at the bottom of the path and find the force exerted by the seat on the passenger at this point. Are your answers consistent with your choice of answers for (a) and (b)?

G.5 (a) If Earth's mass were doubled at the same time its radius is doubled, what would happen to free-fall acceleration? Would it increase, decrease, or stay the same? (b) Use the mass of Earth and its radius as given in Table 7.3 to find precise values for g at these values of M and R. Are your answers consistent with your predictions?

G.6 Figure GA7.6 shows the elliptical orbit of a spacecraft around Earth. Take the origin of your coordinate system to be at the center of Earth.
(a) On a copy of the figure (enlarged if necessary), draw vectors representing
   (i) the position of the spacecraft when it is at Ⓐ and Ⓑ;
   (ii) the velocity of the spacecraft when it is at Ⓐ and Ⓑ;

   (iii) the acceleration of the spacecraft when it is at Ⓐ and Ⓑ.
Make sure that each type of vector can be distinguished. (You might use the color scheme found on the inside front cover of this book.) Provide a legend that shows how each type is represented.
(b) Have you drawn the velocity vector at Ⓐ longer, shorter, or the same length as the one at Ⓑ? Explain why. Have you drawn the acceleration vector at Ⓐ longer, shorter, or the same length as the one at Ⓑ? Explain why you have done so.

FIGURE GA7.6

G.7 Two students are driving a Porsche of mass M at a constant speed v around the traffic circle in front of the physics building. Assume that the road is level and has a radius R. The coefficient of friction between the car's wheels and the road is μ.
(a) In terms of the quantities given, what is the magnitude and direction of the car's acceleration?
(b) At the instant when it passes directly in front of the physics building, is there a net force on the car? If there is, what is responsible for it and how big must it be? Express your answer in terms of the quantities given and explain your reasoning.
(c) The mass of the car and passengers is 1 700 kg, the circle has a radius $R = 25$ m, and the coefficient of friction between rubber and a dry road is $\mu = 0.60$. What is the fastest speed the car can go without slipping?

(Problems 6 and 7 are courtesy of E. F. Redish. For more problems of this type, visit http://www.physics.umd.edu/perg/)

8

Rotational Equilibrium and Rotational Dynamics

Chapter Outline

This adventurous child shifts his body weight around the center of the steering wheel to change the direction of the ship. As the center of his body shifts to the left of the wheel's rotation axis, the wheel rotates counterclockwise, causing the ship to move toward the left as indicated by the tilt from the horizontal. If he were to shift his body to the right of the wheel's axis, the ship would move to the **right.** (*Christoph Wilhelm/FPG/Getty Images*)

T his chapter completes our study of mechanical equilibrium begun in Chapter 4. An understanding of equilibrium problems is important in a variety of fields. For example, students of architecture or industrial technology benefit from understanding the forces that act on buildings or large machines, and biology students should understand the forces at work in muscles and on bones and joints. In order to fully understand objects in equilibrium, we must consider torque. This concept will also play a key role in completing our discussion of rotational motion.

In this chapter we build on the definitions of angular speed and angular acceleration encountered in Chapter 7 by examining the relationship between these concepts and the forces that produce rotational motion. Specifically, we shall find the rotational analog of Newton's second law and define a term that needs to be added to our equation for conservation of mechanical energy: rotational kinetic energy. One of the central purposes of this chapter is to develop the concept of angular momentum, a quantity that plays a key role in rotational motion. Finally, just as we found that linear momentum is conserved for an isolated system, we shall also find that the angular momentum of any isolated system is always conserved.

8.1 TORQUE

Consider Figure 8.1, an overhead view of a door hinged at point O. From this viewpoint, the door is free to rotate about a line perpendicular to the page and passing through O. When the force **F** is applied at the outer edge, as shown, the door easily rotates counterclockwise; that is, the rotational effect of the force is quite large. On the other hand, the same force applied at a point nearer the hinges has a smaller rotational effect on the door. (Observe for yourself that a half-open door rotates with difficulty if you push it near the inner edge, close to the hinges, but turns easily if you push near the doorknob.)

The tendency of a force to rotate an object about some axis is measured by a quantity called the torque τ. The magnitude of the torque due to a force **F** is given by

$$\tau = Fd \qquad \text{[8.1]}$$

In this equation, τ (the Greek letter tau) is the torque, and the distance d is the **lever arm** (or moment arm) of the force **F**. **The lever arm is the perpendicular distance from the axis of rotation to a line drawn along the direction of the force.** Note that **the value of τ depends on the axis of rotation.** Torque is a vector perpendicular to the plane determined by the lever arm and the force. For the two-dimensional problems used in this book, the torque is either into or out of the plane of the paper. We shall use the clockwise or counterclockwise turning tendency of an applied force to determine whether the torque is positive or negative.

As an example, consider the wrench pivoted about the axis O in Figure 8.2a.

FIGURE 8.1 A bird's-eye view of a door hinged at O, with a force applied perpendicularly to the door.

(a) (b)

FIGURE 8.2 (a) A force **F** acting at an angle ϕ with the horizontal produces a torque of magnitude Fd about the pivot O. (b) The component $F \sin \phi$ tends to rotate the system about O. The component $F \cos \phi$ produces no torque about O.

In this case, the applied force **F** acts at an angle ϕ with the horizontal. If you examine the definition of lever arm just given, you will see that the lever arm is the distance d shown in the figure and not L, the length of the wrench. That is, d is the perpendicular distance from the axis of rotation to the line along which the applied force acts. Here, d is related to L by the expression $d = L \sin \phi$. Thus, the net torque produced by the force **F** is given by $\tau = FL \sin \phi$. You can also find the torque in situations such as this by resolving the force into components, as shown in Figure 8.2b. The torque about the axis of rotation produced by the component $F \cos \phi$ is zero because the lever arm of this component is zero. That is, the distance from the pivot to the line along which the force acts is zero, because the line along which the force acts passes through the axis of rotation. The component $F \sin \phi$ has a lever arm of L, and the torque produced by this component is, as before, $\tau = FL \sin \phi$.

If two or more forces act on an object, as in Figure 8.3, then each has a tendency to produce a rotation about the pivot O. In this example, **F**$_2$ tends to rotate the object clockwise and **F**$_1$ tends to rotate the object counterclockwise. We shall use the convention that **the sign of the torque is positive if its turning tendency is counterclockwise and negative if its turning tendency is clockwise.** In Figure 8.3, the torque associated with **F**$_1$, which has a moment arm of d_1, is positive and equal to $F_1 d_1$; the torque associated with **F**$_2$ is negative and equal to $-F_2 d_2$. The *net torque* acting on the object about O is found by summing the torques:

$$\sum \tau = \tau_1 + \tau_2 = F_1 d_1 - F_2 d_2$$

Notice that the units of torque are units of force times length, such as the newton-meter (N·m) or the foot-pound (ft·lb).

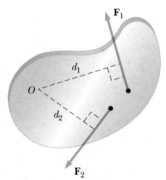

FIGURE 8.3   The force **F**$_1$ tends to rotate the object counterclockwise about O and **F**$_2$ tends to rotate the object clockwise.

Example 8.1   The Spinning Crate

Figure 8.4 is a top view of a packing crate being pushed by two forces of equal magnitude acting in opposite directions as shown. Find the net torque exerted on the crate if its width is 1.0 m. Assume an axis of rotation through the center of the crate.

FIGURE 8.4   (Example 8.1) A top view of a packing crate being pushed by forces of equal magnitude and opposite direction.

Solution   Because each force produces clockwise rotation, the torques are both negative. The torque produced by **F**$_1$ is

$$\tau_1 = F_1 d_1 = -(500 \text{ N})(0.50 \text{ m}) = -250 \text{ N} \cdot \text{m}$$

and the torque produced by **F**$_2$ is

$$\tau_2 = F_2 d_2 = -(500 \text{ N})(0.50 \text{ m}) = -250 \text{ N} \cdot \text{m}$$

Thus, the net torque is exerted on the crate $-500 \text{ N} \cdot \text{m}$.

Example 8.2   The Swinging Door

Find the torque produced by the 300-N force applied at an angle of 60° to the door of Figure 8.5a.

Reasoning   In Figure 8.5b, the 300-N force has been replaced by its horizontal and vertical components:

$$F_x = (300 \text{ N})\cos 60° = 150 \text{ N}$$

$$F_y = (300 \text{ N})\sin 60° = 260 \text{ N}$$

A convenient and obvious location for the axis of rotation is the hinge of the door.

Solution   In this case, the 150-N force produces zero torque about the axis of rotation, because the line along which the force acts passes through the axis of rotation and hence the lever arm is zero. The 260-N force has a lever arm of 2.0 m and thus produces a torque of

$$\tau = (260 \text{ N})(2.0 \text{ m}) = \boxed{+520 \text{ N} \cdot \text{m}}$$

where the plus sign indicates the door rotates in the counterclockwise direction.

FIGURE 8.5   (Example 8.2) (a) A top view of a door being pulled by a 300-N force. (b) The components of the 300-N force.

8.2   TORQUE AND THE TWO CONDITIONS FOR EQUILIBRIUM

In Chapter 4 we examined some situations in which an object has no net external force acting on it; that is, $\Sigma F_x = 0$ and $\Sigma F_y = 0$. We found there that if such an object is initially at rest with respect to an observer, it remains at rest. If the object is initially moving with respect to the observer, the object's velocity remains constant (does not change). Such an object was said to be in equilibrium. However, the condition that $\Sigma \mathbf{F} = 0$ (Equation 4.9), although necessary, is not sufficient to ensure that an object is in complete mechanical equilibrium. This can be understood by considering Figure 8.4. In this situation, the two applied forces acting on the crate are equal in magnitude and opposite in direction. Nevertheless, the crate will still accelerate—it will begin to rotate clockwise because the forces do not act through a common point.

This case illustrates that, to fully understand the effect of a force or group of forces on an object, we must know not only the magnitude and direction of the force(s) but also their point(s) of application. That is, the net torque acting on an object must be considered.

We shall call the requirement that $\Sigma \mathbf{F} = 0$ the first condition for equilibrium. **The second condition for equilibrium asserts that if an object is in rotational equilibrium, the net torque acting on it about any axis must be zero.** That is,

$$\sum \tau = 0$$

In summary, an object in mechanical equilibrium must satisfy both conditions:

1. **The net external force must be zero.** $\sum \mathbf{F} = 0$

    $\quad$ [8.2]

2. **The net external torque must be zero.** $\sum \tau = 0$

◀ The two conditions for mechanical equilibrium

The first condition is a statement of translational equilibrium; the second is a statement of rotational equilibrium. We shall discuss torque and its relation to rotational motion in more detail later in the chapter, but for now we simply point out

This large balanced rock at the Garden of the Gods in Colorado Springs, Colorado is in mechanical equilibrium. *(David Serway)*

Tip 8.1   SPECIFY YOUR AXIS

In solving rotation problems, you must specify an axis of rotation. The choice is arbitrary, but once you make it, you must maintain that choice consistently throughout the problem. For example, there is no unique value of the torque. Its value depends on your choice of rotation axis.

Webnote 8.1

To see the classic case of equilibrium for a seesaw, go to:
http://theory.uwinnipeg.ca/mod_tech/node48.html

FIGURE 8.7   The center of gravity of an object is the point where all the weight of the object can be considered to be concentrated.

that just as an object in translational equilibrium has $\mathbf{a} = 0$, an object in rotational equilibrium has $\boldsymbol{\alpha} = 0$, meaning no rotational acceleration.

POSITION OF THE AXIS OF ROTATION

In the cases we have described so far, the axes of rotation for calculating torques were selected without explanation. Often the nature of a problem suggests a convenient location for the axis, but just as often no single location stands out as being preferable. You might ask, "What axis should I choose in calculating the net torque?" The answer is as follows: **If the object is in equilibrium, it does not matter where you put the axis of rotation for calculating the net torque; the location of the axis is completely arbitrary.**

APPLYING **PHYSICS** *8.1*

When an automobile driver steps on the accelerator, the nose of the car moves upward. When the driver brakes, the nose moves downward. Why do these effects occur?

Explanation   When the driver steps on the accelerator, there is an increased force exerted by the roadway on the tires. This force is parallel to the roadway and directed toward the front of the automobile, as suggested in Figure 8.6a. This force provides a torque that causes the car to rotate in the clockwise direction. The result of this rotation is a "nosing up" of the car. When the driver steps on the brake, there is also an increased force exerted by the roadway on the tires, but now directed toward the rear of the automobile, as suggested in Figure 8.6b. This force results in a torque that causes a counterclockwise rotation and the subsequent "nosing down" of the automobile.

(a)   (b)

FIGURE 8.6   (Applying Physics 8.1)

8.3   THE CENTER OF GRAVITY

One of the forces that must be considered in dealing with a rigid object is that of gravity acting on the object. To compute the torque due to the force of gravity, all of the weight can be thought of as concentrated at a single point.

Consider an object of arbitrary shape lying in the xy plane, as in Figure 8.7. The object is divided into a large number of very small particles of weight m_1g, m_2g, m_3g, . . . having coordinates (x_1, y_1), (x_2, y_2), (x_3, y_3), ... If the object is free to rotate about the origin, each particle contributes a torque about the origin that is equal to its weight multiplied by its lever arm. For example, the torque due to the weight m_1g is m_1gx_1, and so forth.

We wish to locate the point of application of the single force of magnitude $w = F_g$ (the total weight of the object) whose effect on the rotation of the object is the same as that of the individual particles. This point is called the **center of gravity** of the object. Equating the torque exerted by w at the center of gravity to the sum of the torques acting on the individual particles gives

$$(m_1g + m_2g + m_3g + \cdots)x_{cg} = m_1gx_1 + m_2gx_2 + m_3gx_3 + \cdots$$

If we assume that g is uniform over the object (which will be the case in all the situations we examine), then the g terms in the preceding equation cancel and we get

$$x_{cg} = \frac{m_1 x_1 + m_2 x_2 + m_3 x_3 + \cdots}{m_1 + m_2 + m_3 + \cdots} = \frac{\sum m_i x_i}{\sum m_i} \qquad [8.3]$$

Similarly, the y coordinate of the center of gravity of the system can be found from

$$y_{cg} = \frac{\sum m_i y_i}{\sum m_i} \qquad [8.4]$$

The center of gravity of a homogeneous, symmetric body must lie on the axis of symmetry. For example, the center of gravity of a homogeneous rod must lie midway between the ends of the rod. The center of gravity of a homogeneous sphere or a homogeneous cube must lie at the geometric center of the object. We can determine the center of gravity of an irregularly shaped object, such as a wrench, experimentally by suspending the wrench from two different arbitrary points (Fig. 8.8). The wrench is first hung from point A, and a vertical line AB (which can be established with a plumb bob) is drawn when the wrench is in equilibrium. The wrench is then hung from point C, and a second vertical line CD is drawn. The center of gravity coincides with the intersection of these two lines. In fact, if the wrench is hung freely from any point, the center of gravity always lies straight below the point of support. Furthermore, the vertical line through that point must pass through the center of gravity.

In several examples in Section 8.4, we shall be concerned with homogeneous, symmetric objects whose centers of gravity coincide with their geometric centers. A rigid object in a uniform gravitational field can be balanced by a single force equal in magnitude to the weight of the object, as long as the force is directed upward through the object's center of gravity.

FIGURE 8.8 An experimental technique for determining the center of gravity of a wrench. The wrench is hung freely from two different pivots, A and C. The intersection of the two lines, AB and CD, locates the center of gravity.

Example 8.3   Where Is the Center of Gravity?

Three particles are located in a coordinate system as shown in Figure 8.9. Find the center of gravity.

Reasoning and Solution   The y coordinate of the center of gravity is zero because all the particles are on the x axis. To find the x coordinate of the center of gravity, we use Equation 8.3:

$$x_{cg} = \frac{\sum m_i x_i}{\sum m_i}$$

For the numerator, we find

$$\sum m_i x_i = m_1 x_1 + m_2 x_2 + m_3 x_3$$
$$= (5.00 \text{ kg})(-0.500 \text{ m}) + (2.00 \text{ kg})(0 \text{ m}) + (4.00 \text{ kg})(1.00 \text{ m})$$
$$= 1.50 \text{ kg} \cdot \text{m}$$

The denominator is $\sum m_i = 11.0$ kg; therefore,

$$x_{cg} = \frac{1.50 \text{ kg} \cdot \text{m}}{11.0 \text{ kg}} = \boxed{0.136 \text{ m}}$$

EXERCISE   If a fourth particle of mass 2.00 kg is placed at $x = 0$, $y = 0.250$ m, find the x and y coordinates of the center of gravity for this system of four particles.

ANSWER $x_{cg} = 0.115$ m; $y_{cg} = 0.038\ 0$ m

FIGURE 8.9   (Example 8.3) Locating the center of gravity for a system of three particles.

Example 8.4   Locating Your Lab Partner's Center of Gravity

In this example we show how to find the location of a person's center of gravity. Suppose your lab partner has a height L of 173 cm (5 ft 8 in) and a weight w of 715 N (160 lb). You can easily determine the position of his center of gravity by having him stretch out on a uniform board supported at one end by a scale as shown in Figure 8.10. If the board's weight w_b is 49 N and the scale reading F is 350 N, find the distance of your lab partner's center of gravity from the left end of the board.

FIGURE 8.10   (Example 8.4) Determining your lab partner's center of gravity.

Reasoning   The center of gravity must lie somewhere on the long symmetry axis of the body. To find the position x_{cg} of the center of gravity, we set the sum of the torques about point O equal to zero.

Solution   Referring to Figure 8.10,

$$\sum \tau_O = 0$$

$$-wx_{cg} - w_b(L/2) + FL = 0$$

$$x_{cg} = \frac{FL - w_b(L/2)}{w} = \frac{(350 \text{ N})(173 \text{ cm}) - (49 \text{ N})(86.5 \text{ cm})}{715 \text{ N}} = \boxed{79 \text{ cm}}$$

Thus x_{cg}, measured from the head, is about 46% of the person's height.

8.4   EXAMPLES OF OBJECTS IN EQUILIBRIUM

In Chapter 4 we discussed some techniques for solving problems concerned with objects in equilibrium. Recall that when the objects are treated as geometric points, it is sufficient to simply apply the condition that the net force on the object must be zero. In this chapter we have shown that for objects of finite dimensions, a second condition for equilibrium must be satisfied—namely that the net torque on the object must also be zero. The following general procedure is recommended for solving problems that involve objects in equilibrium.

Tip 8.2   ZERO TORQUE

Zero net torque does not mean an absence of rotational motion. An object that rotates at a constant angular speed can be under the influence of zero net torque. This is analogous to the translational situation—zero net force on an object does not necessarily mean that it is not in motion.

PROBLEM-SOLVING STRATEGY   **Objects in Equililbrium**

1. Draw a simple, neat diagram of the system.
2. Isolate the object that is being analyzed. Draw a free-body diagram showing all external forces that are acting on this object. For systems that contain more than one object, draw a *separate* diagram for each object. Do not include forces that the object exerts on its surroundings.
3. Establish convenient coordinate axes for each object and find the components of the forces along these axes. Now apply the first condition of equilibrium (the net force on the object in the x and y directions must be zero) for each object under consideration.

4. Choose a convenient rotation axis for calculating the net torque on the object. Now apply the second condition of equilibrium (the net torque on the object about any axis must be zero). Remember that the choice of the axis for the torque equation is arbitrary; therefore, choose an axis that will simplify your calculation as much as possible. Note that a force that acts along a line passing through the axis of rotation gives zero contribution to the torque.

5. The first and second conditions for equilibrium will give a set of simultaneous equations with several unknowns. All that is left to complete your solution is to solve for the unknowns in terms of the known quantities.

Example 8.5 Walking a Horizontal Beam

A uniform horizontal 300-N beam, 5.00 m long, is attached to a wall by a pin connection that allows the beam to rotate. Its far end is supported by a cable that makes an angle of 53.0° with the horizontal (Fig. 8.11a). If a 600-N person stands 1.50 m from the wall, find the tension in the cable and the force exerted by the wall on the beam.

Reasoning   First we must identify all the external forces acting on the beam and sketch them on a free-body diagram. This is shown in Figure 8.11b. The forces on the beam consist of the downward force of gravity, which has a magnitude of 300 N; the downward force exerted by the man, which is equal in magnitude to his weight, 600 N; the force **T** exerted by the cable; and the force **R** exerted by the wall. We then resolve the forces **T** and **R** into their horizontal and vertical components, as shown in Figure 8.11c. Note that the x component of the force exerted by the cable ($T\cos 53.0°$) is to the left, whereas the y component ($T\sin 53.0°$) is upward. The horizontal and vertical components of **R** are denoted by R_x and R_y, respectively. The first condition for equilibrium can now be applied in the x and y directions to give us two equations in terms of our unknowns R_x, R_y, and T. The necessary third equation can be found from the second condition of equilibrium.

Solution   From the first condition for equilibrium, we find

$$\sum F_x = R_x - T\cos 53.0° = 0 \tag{1}$$

$$\sum F_y = R_y + T\sin 53.0° - 600\ \text{N} - 300\ \text{N} = 0 \tag{2}$$

The unknowns are R_x, R_y, and T. Because there are three unknowns and only two equations, we cannot find a solution from just the first condition of equilibrium.

Now let us use the second condition of equilibrium. The axis that passes through the pivot at the wall is a convenient one to choose for the torque equation because the components R_x, R_y, and $T\cos 53.0°$ all have lever arms of zero and hence have zero torque about this axis. Recalling our sign convention for the torque about an axis and noting that the lever arms of the 600-N, 300-N, and $T\sin 53.0°$ components are 1.50 m, 2.50 m, and 5.00 m, respectively, we get

$$\sum \tau_O = (T\sin 53.0°)(5.00\ \text{m}) - (300\ \text{N})(2.50\ \text{m}) - (600\ \text{N})(1.50\ \text{m}) = 0 \tag{3}$$

$$T = \boxed{413\ \text{N}}$$

Thus, the torque equation using this axis gives us one of the unknowns immediately! This value for T is then substituted into (1) and (2) to give

$$R_x = \boxed{249\ \text{N}} \qquad R_y = \boxed{570\ \text{N}}$$

If we selected some other axis for the torque equation, the solution would be the same. For example, if the axis were to pass through the center of gravity of the beam, the torque equation would involve both T and R_y; together with (1) and (2), however, it could still be solved for the unknowns. Try it!

(a)

(b)

(c)

FIGURE 8.11   (Example 8.5) (a) A uniform beam attached to a wall and supported by a cable. (b) A free-body diagram for the beam. (c) The component form of the free-body diagram.

FIGURE 8.12   (Example 8.6) (a) A weight held with the forearm horizontal. (b) The mechanical model for the system.

Example 8.6   A Weighted Forearm

A 50.0-N (11-lb) weight is held in a person's hand with the forearm horizontal, as in Figure 8.12a. The biceps muscle is attached 0.030 0 m from the joint, and the weight is 0.350 m from the joint. Find the upward force exerted by the biceps on the forearm (the ulna) and the downward force exerted by the humerus on the forearm, acting at the joint. Neglect the weight of the forearm.

Solution   The forces acting on the forearm are equivalent to those acting on a bar of length 0.350 m, as shown in Figure 8.12b, where **F** is the upward force exerted by the biceps and **R** is the downward force at the joint. From the first condition for equilibrium, we have

$$\sum F_y = F - R - 50.0 \text{ N} = 0 \tag{1}$$

From the second condition for equilibrium, we know that the sum of the torques about any axis must be zero. With the joint O as the axis, we have

$$F(0.030\ 0 \text{ m}) - (50.0 \text{ N})(0.350 \text{ m}) = 0$$

$$F = \boxed{583 \text{ N (131 lb)}}$$

This value for F can be substituted into (1) to give $R = \boxed{533 \text{ N (120 lb)}}$. Clearly, the forces at joints and in muscles can be extremely large compared to the weight of the lifted object.

Example 8.7   Don't Climb the Ladder

A uniform 10-m-long, 50-N ladder rests against a smooth vertical wall as in Figure 8.13a. If the ladder is just on the verge of slipping when it makes a 50° angle with the ground, find the coefficient of static friction between the ladder and ground.

Reasoning   Figure 8.13b is the free-body diagram for the ladder, showing all external forces acting on it. At the base of the ladder, Earth exerts an upward normal force **n**, and a force of static friction **f** acts to the right. The wall exerts the force **P** to the left. Note that **P** is horizontal because the wall is smooth. (If the wall were rough, an upward frictional force would be exerted on the ladder.) The first condition for equilibrium can now be applied in the x and y directions to give us two equations in terms of our unknowns f, P, and n. The necessary third equation can be found from the second condition of equilibrium.

Solution   From the first condition for equilibrium applied to the ladder, we have

$$\sum F_x = f - P = 0 \tag{1}$$

$$\sum F_y = n - 50 \text{ N} = 0 \tag{2}$$

From (2) we see that $n = 50$ N. Furthermore, when the ladder is on the verge of slipping, the magnitude of the force of static friction must be maximum and given by the relation $f_{s,\text{max}} = \mu_s n = \mu_s(50 \text{ N})$. Thus, (1) reduces to

$$\mu_s(50 \text{ N}) = P \tag{3}$$

Let us now apply the second condition of equilibrium and take the torques about the axis through O at the bottom of the ladder, as in Figure 8.13c. The force **P** and the force of gravity acting on the ladder are the only forces that contribute to the torque about this axis, and their lever arms are shown in Figure 8.13c. Note that, because the length of the ladder is 10 m, the lever arm for **P** is $d_1 = (10 \text{ m})\sin 50°$. Likewise, the lever arm for

FIGURE 8.13   (Example 8.7) (a) A ladder leaning against a frictionless wall. (b) A free-body diagram for the ladder. (c) Lever arms for the force of gravity and **P**.

the 50-N force of gravity is $d_2 = (5.0 \text{ m}) \cos 50°$, where the force of gravity acts through the center because the ladder is uniform. Thus, we find that

$$\sum \tau_O = P(10 \text{ m})\sin 50° - (50 \text{ N})(5.0 \text{ m})\cos 50° = 0$$

$$P = 21 \text{ N}$$

Now that P is known, we can substitute its value into (3) to find μ_s:

$$\mu_s = \frac{P}{50 \text{ N}} = \frac{21 \text{ N}}{50 \text{ N}} = \boxed{0.42}$$

EXERCISE   Determine the direction of the total force acting on the ladder at its base. Does this force act along the ladder?

ANSWER $\tan^{-1}(n/P) = 67°$;   no. The force at the base of the ladder is generally *not* in the direction parallel to the ladder.

8.5   RELATIONSHIP BETWEEN TORQUE AND ANGULAR ACCELERATION

Earlier in this chapter we considered the situation in which both the net force and the net torque acting on an object are zero. Objects in this situation are said to be in equilibrium. We now examine the behavior of an object when the net torque acting on it is not zero. As you shall see, when a rigid object is subject to a net torque, it undergoes an angular acceleration. Furthermore, the angular acceleration is directly proportional to the net torque. The end result of our investigation will be an expression that is analogous to $\Sigma \mathbf{F} = m\mathbf{a}$ in translational motion.

Let us begin by considering the system shown in Figure 8.14, which consists of an object of mass m connected to a very light rod of length r. The rod is pivoted at the point O, and its movement is confined to rotation on a frictionless *horizontal* table. Now let us assume that a force $\mathbf{F}_t$ perpendicular to the rod and hence tangent to the circular orbit is acting on m. Because there is no force to oppose this tangential force, the object undergoes a tangential acceleration according to Newton's second law:

$$F_t = ma_t$$

Multiplying the left and right sides of this equation by r gives

$$F_t r = mra_t$$

In Chapter 7 (Equation 7.9) we found that the tangential acceleration and angular acceleration for a particle rotating in a circular path are related by the expression

$$a_t = r\alpha$$

FIGURE 8.14   An object of mass m attached to a light rod of length r moves in a circular path on a frictionless horizontal surface while the tangential force $\mathbf{F}_t$ acts on it.

so now we find that

$$F_t r = m r^2 \alpha \qquad \text{[8.5]}$$

The left side of Equation 8.5, which should be familiar to you, is the torque acting on the object about its axis of rotation. That is, the torque is equal in magnitude to the force on the object multiplied by the perpendicular distance from the pivot to the line along which the force acts, or $\tau = F_t r$. Hence, we can write Equation 8.5 as

$$\tau = m r^2 \alpha \qquad \text{[8.6]}$$

Equation 8.6 shows that the torque on the system is proportional to angular acceleration, where the constant of proportionality $m r^2$ is called the **moment of inertia** of the object of mass m. (Because the rod is very light, its moment of inertia can be neglected.)

> **Quick Quiz 8.1**
>
> Using a screwdriver, you try to loosen a stubborn screw from a piece of wood and you fail. In order to succeed, you should find a screwdriver that (a) is longer, (b) is shorter, (c) has a narrower handle, or (d) has a fatter handle.

TORQUE ON A ROTATING OBJECT

Now consider a solid disk rotating about its axis as in Figure 8.15a. The disk consists of many particles at various distances from the axis of rotation, as in Figure 8.15b. The torque on each one of these particles is given by Equation 8.6. The *total* torque on the disk is given by the sum of the individual torques on all the particles:

$$\sum \tau = \left(\sum m r^2 \right) \alpha \qquad \text{[8.7]}$$

Note that, because the disk is rigid, all particles have the *same* angular acceleration, so α is not involved in the sum. If the masses and distances of the particles are labeled with subscripts as in Figure 8.15b, then

$$\sum m r^2 = m_1 r_1^2 + m_2 r_2^2 + m_3 r_3^2 + \cdots$$

This quantity is the moment of inertia of the whole body and is given the symbol I:

Moment of inertia ▷

$$I \equiv \sum m r^2 \qquad \text{[8.8]}$$

The moment of inertia has the SI units $\text{kg} \cdot \text{m}^2$. Using this result in Equation 8.7, we see that the total torque on a rigid body rotating about a fixed axis is given by

Relationship between net torque and angular acceleration ▷

$$\sum \tau = I \alpha \qquad \text{[8.9]}$$

FIGURE 8.15   (a) A solid disk rotating about its axis. (b) The disk consists of many particles, all with the same angular acceleration.

(a)   (b)

The angular acceleration of an extended rigid object is proportional to the net torque acting on it. The proportionality constant I between the net torque and angular acceleration is the object's moment of inertia.

Note that the equation $\Sigma\tau = I\alpha$ (Eq. 8.9) is the rotational counterpart to Newton's second law, $\Sigma F = ma$ as stated in Equation 4.1. Thus, the correspondence between rotational motion and linear motion continues. Recall from Chapter 7 that the linear variables x, v, and a are replaced in rotational motion by the variables θ, ω, and α. Likewise, we now see that **the force and mass in linear motion correspond to torque and moment of inertia in rotational motion.** The moment of inertia may be viewed as a kind of "rotational mass"—for a fixed applied torque acting on an object, the larger its moment of inertia, the smaller its angular acceleration. This is analogous to the case of linear motion of an object when a fixed force is applied to it—the larger its mass, the smaller its linear acceleration. Note, however, this major difference between m and I: **m depends only on the quantity of matter in an object whereas I depends on both the quantity of matter and its distribution (through the r^2 term in $I = \Sigma\ mr^2$) in the rigid object.** In this chapter we shall develop other equations for the rotational kinetic energy and the angular momentum of an object rotating about a fixed axis. Based on the analogies already presented, you should be able to predict the form of these equations.

Quick Quiz 8.2

A constant net torque is applied to an object. One of the following will definitely not be a constant. It is the object's (a) angular acceleration, (b) angular velocity, (c) moment of inertia, or (d) center of gravity.

Quick Quiz 8.3

The two rigid objects shown in Figure 8.16 have the same mass, radius, and angular speed. If the same braking torque is applied to each, which takes longer to stop? (a) A, (b) B, (c) not enough information to say. (*Hint:* See Table 8.1 on page 234 for the moment of inertia of these objects.)

FIGURE 8.16 (Quick Quiz 8.3)

The gear system on a bicycle provides an easily visible example of the relationship between torque and angular acceleration. Consider first a five-speed gear system in which the drive chain can be adjusted to wrap around any of five gears attached to the back wheel (Fig. 8.17). The gears are concentric with the wheel hub and are of different radii. When the cyclist begins from rest, the chain is attached to the largest gear. Because it has the largest radius, this gear provides the largest torque to the drive wheel. Large torque is required for starting, because the bicycle must change its motion from rest to some speed. As the bicycle rolls faster, the tangential speed of the chain increases and becomes too fast for the cyclist to maintain by pushing the pedals. The chain is then moved to the gear with the next smallest radius, and the chain now has a smaller tangential speed, which the cyclist

FIGURE 8.17 The drive wheel and gears of a bicycle. (*George Semple*)

APPLICATION

BICYCLE GEARS

can maintain by pushing the pedals. This gear does not provide as much torque as the first, but the cyclist does not need as much—he or she only needs to accelerate from some speed to a somewhat higher speed. This process continues as the bicycle moves faster and faster, and the cyclist shifts through all five gears. The fifth gear supplies the lowest torque, but now the main function of the supplied torque is to counter the frictional torque from the rolling tires, which will tend to slow the bicycle. The small radius of this gear allows the cyclist to keep up with the chain's movement by pushing the pedals, despite the rapid movement of the bicycle.

A 15-speed bicycle has the same gear structure on the drive wheel, but has three gears on the sprocket connected to the pedals. By combining different positions of the chain on the rear gears and the sprocket gears, 15 different torques are available.

MORE ON THE MOMENT OF INERTIA

Because the moment of inertia of an object—as defined by $I = \Sigma \, mr^2$ (Eq. 8.8)—is used throughout the remainder of this chapter, it will be useful to examine it in more detail before discussing other aspects of rotational motion. As seen earlier, a small object (or particle) has a moment of inertia equal to mr^2 about some axis. When you are dealing with a composite object, the total moment of inertia about some axis is just the sum of the moments of inertia of the object's component pieces. For example, suppose a majorette twirls a baton as in Figure 8.18. Let us assume that the baton can be modeled as a very light rod of length 2ℓ with a heavy object at each end. (The rod of a real baton has significant mass relative to its ends.) Because we are neglecting the mass of the rod, the moment of inertia of the baton about an axis through its center and perpendicular to its length is given by Equation 8.8:

$$I = \sum mr^2$$

Because in this system there are two objects with equal masses equidistant from the axis of rotation, we see that $r = \ell$ for each object, and the sum is

$$I = \sum mr^2 = m\ell^2 + m\ell^2 = 2m\ell^2$$

If the mass of the rod is not neglected, we would have to include its moment of inertia to find the total moment of inertia of the baton.

We pointed out earlier that I is the rotational counterpart of m. However, there are some important distinctions between the two. For example, mass is an intrinsic property of an object that does not change, whereas **the moment of inertia of a system depends on how the mass is distributed and on the location of the axis of rotation.** Examples 8.8 and 8.9 illustrate this point.

FIGURE 8.18   A baton of length 2ℓ and mass $2m$ (the mass of the connecting rod is neglected). The moment of inertia about the axis through the baton's center and perpendicular to its length is $2m\ell^2$.

0.20 kg   0.30 kg

0.30 kg   0.20 kg

0.50 m

FIGURE 8.19   (Example 8.8) Four objects connected to light rods rotating in the plane of the page.

Example 8.8   The Baton Twirler—Act 1

In an effort to be the star of the half-time show, a majorette twirls a highly unusual baton made up of four spheres fastened to the ends of light rods (Fig. 8.19). Each rod is 1.0 m long. Find the moment of inertia of the system about an axis perpendicular to the page and passing through the point where the rods cross.

Solution   Applying Equation 8.8, we get

$$I = \sum mr^2 = m_1 r_1^2 + m_2 r_2^2 + m_3 r_3^2 + m_4 r_4^2$$
$$= (0.20 \text{ kg})(0.50 \text{ m})^2 + (0.30 \text{ kg})(0.50 \text{ m})^2 + (0.20 \text{ kg})(0.50 \text{ m})^2$$
$$+ (0.30 \text{ kg})(0.50 \text{ m})^2$$
$$= \boxed{0.25 \text{ kg} \cdot \text{m}^2}$$

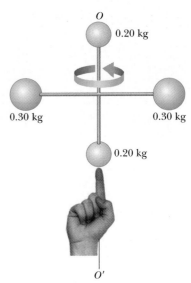

Example 8.9   The Baton Twirler—Act 2

Not satisfied with the crowd reaction from the baton twirling of Example 8.8, the majorette tries spinning her strange baton about the axis OO', as shown in Figure 8.20. Calculate the moment of inertia about this axis.

Solution   Again applying Equation 8.8, we have

$$I = \sum mr^2$$
$$= (0.20 \text{ kg})(0)^2 + (0.30 \text{ kg})(0.50 \text{ m})^2 + (0.20 \text{ kg})(0)^2$$
$$+ (0.30 \text{ kg})(0.50 \text{ m})^2$$
$$= 0.15 \text{ kg} \cdot \text{m}^2$$

Note that I is less in this case than in Example 8.8 because some of the mass is closer to the axis of rotation.

FIGURE 8.20   (Example 8.9) A double baton rotating about the axis OO'.

CALCULATION OF MOMENTS OF INERTIA FOR EXTENDED OBJECTS

The method used for calculating moments of inertia in Examples 8.8 and 8.9 is simple enough when you have only a few small objects rotating about an axis. The situation becomes much more complex when the object is an extended mass, such as a sphere, a cylinder, or a cone. One type of extended object that is amenable to a simple solution is a hoop rotating about an axis perpendicular to its plane and passing through its center, as shown in Figure 8.21. A bicycle tire, for example, fits in this category.

To evaluate the moment of inertia of the hoop, we can still use the equation $I = \sum mr^2$ (Eq. 8.8) and imagine that the hoop is divided into a number of small segments having masses m_1, m_2, m_3, . . . , as in Figure 8.21. This approach is simply an extension of the baton problem described in the preceding examples, except that now we have a large number of small masses in rotation instead of only four.

We can express the sum for I as

$$I = \sum mr^2 = m_1 r_1^2 + m_2 r_2^2 + m_3 r_3^2 + \cdots$$

All of the segments around the hoop are at the *same distance* R from the axis of rotation; thus, we can drop the subscripts on the distances and factor out the common factor R^2:

$$I = (m_1 + m_2 + m_3 + \cdots)R^2$$

We know, however, that the sum of the masses of all the segments must equal the total mass of the hoop, M:

$$M = m_1 + m_2 + m_3 + \cdots$$

and so we can express I as

$$I = MR^2 \qquad\qquad \text{[8.10]}$$

This expression can be used for the moment of inertia of any ring-shaped object rotating about an axis through its center and perpendicular to its plane. Note that the result is strictly valid only if the thickness of the ring is small relative to its inner radius.

The hoop we selected as an example is unique in that we were able to find an expression for its moment of inertia using only simple algebra. Unfortunately, most extended objects are more difficult to work with because all of the mass elements are not located at the same distance from the axis, and the methods of integral calculus are required. The moments of inertia for some other common shapes

FIGURE 8.21   A uniform hoop can be divided into a large number of small segments that are equidistant from the center of the hoop.

Tip 8.3   NO SINGLE MOMENT OF INERTIA

Moment of inertia is analogous to mass, but there are major differences. Mass is an inherent property of an object. The moment of inertia of an object depends on the shape of the object and the choice of rotation axis.

TABLE 8.1	**Moments of Inertia for Various Rigid Objects of Uniform Composition**

Hoop or thin cylindrical shell
$I = MR^2$

Solid sphere
$I = \frac{2}{5} MR^2$

Solid cylinder or disk
$I = \frac{1}{2} MR^2$

Thin spherical shell
$I = \frac{2}{3} MR^2$

Long thin rod with rotation axis through center
$I = \frac{1}{12} ML^2$

Long thin rod with rotation axis through end
$I = \frac{1}{3} ML^2$

Webnote 8.2

Torque, center of gravity, and moment of inertia are all involved in the design of golf clubs and tennis racquets. See:
http://clubmaker-online.com/ physics1.html
http://tennisone.com/Product/ racquet.kp/racquet.custom2.kp.htm

are given without proof in Table 8.1. When the need arises, you can use this table to determine the moment of inertia of a body having any one of the listed shapes.

If mass elements in an object are redistributed parallel to the axis of rotation, the moment of inertia of the object does not change. Therefore the expression $I = MR^2$ can be used equally well to find the axial moment of inertia of an embroidery hoop or of a long sewer pipe. Likewise, a door turning on its hinges is described by the same moment-of-inertia expression as that tabulated for a long thin rod rotating about its end.

0.350 m

FIGURE 8.22   (Example 8.10) A ball being tossed by a pitcher. The forearm is being used to accelerate the ball.

Example 8.10   Warming Up

A baseball player loosening up his arm before a game tosses a 0.150-kg baseball using only the rotation of his forearm to accelerate the ball (Fig. 8.22). The forearm has a mass of 1.50 kg and a length of 0.350 m. The ball starts at rest and is released with a speed of 30.0 m/s in 0.300 s.

A Find the constant angular acceleration of the arm and ball.

Solution   During its acceleration, the ball moves through an arc of a circle with a radius of 0.350 m. We can determine the angular acceleration using Equation 7.5, $\omega = \omega_i + \alpha t$. Because $\omega_i = 0$, however, $\omega = \alpha t$, or

$$\alpha = \frac{\omega}{t}$$

We also know that $v = r\omega$, and so we get

$$\alpha = \frac{\omega}{t} = \frac{v}{rt} = \frac{30.0 \text{ m/s}}{(0.350 \text{ m})(0.300 \text{ s})} = \boxed{286 \text{ rad/s}^2}$$

B Find the torque exerted on the ball to give it this angular acceleration.

Solution The moment of inertia of the ball about an axis that passes through the elbow, perpendicularly to the arm, is

$$I_{ball} = mr^2 = (0.150 \text{ kg})(0.350 \text{ m})^2 = 1.84 \times 10^{-2} \text{ kg} \cdot \text{m}^2$$

If we model the forearm as a uniform bar of length L, its moment of inertia about an axis that passes through the elbow, perpendicularly to the arm, is

$$I_{forearm} = \tfrac{1}{3}ML^2 = \tfrac{1}{3}(1.50 \text{ kg})(0.350 \text{ m})^2 = 6.13 \times 10^{-2} \text{ kg} \cdot \text{m}^2$$

The total moment of inertia of the composite system (ball plus forearm) about the elbow is

$$I_{system} = I_{ball} + I_{forearm} = 7.97 \times 10^{-2} \text{ kg} \cdot \text{m}^2$$

Thus, the required torque is

$$\tau = I_{system}\alpha = (7.97 \times 10^{-2} \text{ kg} \cdot \text{m}^2)(286 \text{ rad/s}^2) = \boxed{22.8 \text{ N} \cdot \text{m}}$$

Example 8.11 The Falling Bucket

A solid, frictionless cylindrical reel of mass $M = 3.00$ kg and radius $R = 0.400$ m is used to draw water from a well (Fig. 8.23a). A bucket of mass $m = 2.00$ kg is attached to a cord that is wrapped around the cylinder. If the bucket starts from rest at the top of the well and falls for 3.00 s before hitting the water, how far does it fall?

Reasoning Figure 8.23b shows the two forces on the bucket as it falls: **T** is the force exerted by the cord, and $m\mathbf{g}$ is the force of gravity. We shall choose downward as the positive direction (an atypical but convenient choice) and write Newton's second law for the bucket as

$$mg - T = ma \qquad \text{(bucket)} \qquad \text{(1)}$$

With one equation and two unknowns, T and a, we must develop an additional equation to complete the problem. To obtain this second equation, let us consider the cylinder's rotational motion. Equation 8.9 applied to the cylinder gives the necessary expression:

$$\sum \tau = I\alpha = \tfrac{1}{2}MR^2\alpha \qquad \text{(cylinder)} \qquad \text{(2)}$$

Figure 8.23c shows that the only force producing a torque on the cylinder as it rotates about an axis through its center is **T**, the force exerted by the cord. The magnitude of the torque due to this force about the axis of rotation is $\tau = TR$. Thus, we can express (2) as

$$TR = \tfrac{1}{2}MR^2\alpha$$

or

$$T = \tfrac{1}{2}MR\alpha \qquad \text{(3)}$$

Actually, two other forces act on the cylinder—the force of gravity $M\mathbf{g}$ and the upward force **n** exerted by the axle, but we do not have to consider them here because the lever arm of each about the axis of rotation is zero.

At this point, it is important to recognize that the downward acceleration of the bucket is equal to the tangential acceleration of a point on the rim of the cylinder. Therefore, the angular acceleration of the cylinder and the linear acceleration of the bucket are related by $a_t = R\alpha$. When this relation is used in (3), we get

$$T = \tfrac{1}{2}Ma_t \qquad \text{(4)}$$

(a)

(b)　　　(c)

FIGURE 8.23 (Example 8.11) (a) A water bucket attached to a rope passing over a frictionless reel. (b) A free-body diagram for the bucket. (c) The tension produces a torque on the cylinder about its axis of rotation.

Solution Equations (1) and (4) can now be solved simultaneously to find a_t and T. This procedure gives

$$a_t = 5.60 \text{ m/s}^2 \qquad T = 8.40 \text{ N}$$

Finally, we turn to the equations for motion with constant linear acceleration to find the distance d that the bucket falls in 3.00 s. Because $v_0 = 0$, we get

$$d = v_0 t + \tfrac{1}{2} a t^2 = \tfrac{1}{2}(5.60 \text{ m/s}^2)(3.00 \text{ s})^2 = \boxed{25.2 \text{ m}}$$

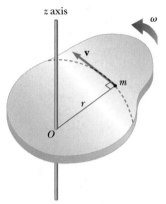

FIGURE 8.24 A rigid plate rotating about the z axis with angular speed ω. The kinetic energy of a particle of mass m is $\tfrac{1}{2} m v^2$. The total kinetic energy of the plate is $\tfrac{1}{2} I \omega^2$.

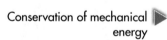

Rotational kinetic energy ▶

8.6 ROTATIONAL KINETIC ENERGY

In Chapter 5 we defined the kinetic energy of a particle moving through space with a speed v as the quantity $\tfrac{1}{2} m v^2$. Analogously, **an object rotating about some axis with an angular speed ω has rotational kinetic energy given by $\tfrac{1}{2} I \omega^2$.** To prove this, consider a thin rigid plate rotating about some axis perpendicular to its plane, as in Figure 8.24. The plate consists of many small particles, each of mass m. All these particles rotate in circular paths about the axis. If r is the distance of one of the particles from the axis of rotation, the speed of this particle is $v = r\omega$. Because the *total* kinetic energy of the plate's rotation is the sum of all the kinetic energies associated with all of its particles, we have

$$KE_r = \sum \left(\tfrac{1}{2} m v^2\right) = \sum \left(\tfrac{1}{2} m r^2 \omega^2\right) = \tfrac{1}{2}\left(\sum m r^2\right)\omega^2$$

In the last step, the ω^2 term is factored out because it is the same for every particle. This leads to the expression for rotational kinetic energy,

$$KE_r = \tfrac{1}{2} I \omega^2 \qquad \text{[8.11]}$$

where $I = \sum m r^2$ is the moment of inertia of the object.

We found the energy concept to be extremely useful for describing the linear motion of a system. It can be equally useful for simplifying the analysis of rotational motion. Consider a system such as a bowling ball rolling down a ramp that is described by three types of energy: **gravitational potential energy PE_g, translational kinetic energy KE_t** and **rotational kinetic energy KE_r.** We must include all these forms of energy in our equation for conservation of mechanical energy for an isolated system.

Conservation of mechanical ▶
energy

$$(KE_t + KE_r + PE_g)_i = (KE_t + KE_r + PE_g)_f \qquad \text{[8.12]}$$

where i and f refer to initial and final values, respectively. This relation is *only* true if we ignore dissipative forces such as friction.

FIGURE 8.25 (Example 8.12) A ball starts from rest at the top of an incline and rolls to the bottom without slipping.

Example 8.12 A Ball Rolling Down an Incline

A ball of mass M and radius R starts from rest at a height of 2.00 m and rolls down a 30.0° slope, as shown in Figure 8.25. What is the linear speed of the ball when it leaves the incline? Assume that the ball rolls without slipping.

Reasoning and Solution The initial energy associated with the ball is gravitational potential energy, and when the ball reaches the bottom of the ramp, this potential energy has been converted to translational and rotational kinetic energy. The conservation of mechanical energy equation becomes

$$(PE_g)_i = (KE_t + KE_r)_f$$

$$Mgh = \tfrac{1}{2} M v^2 + \tfrac{1}{2}\left(\tfrac{2}{5} M R^2\right)\omega^2$$

where h is the distance through which the ball's center of gravity falls and where we have used $I = \frac{2}{5}MR^2$ from Table 8.1 as the moment of inertia of the ball. If the ball rolls without slipping, the ball's center of gravity must advance a linear distance $2\pi R$ in one revolution or in a time of $2\pi/\omega$. This means that the linear speed of the ball is related to its angular speed by

$$v = \frac{2\pi R}{2\pi/\omega} = R\omega$$

This expression can be used to eliminate ω from the equation for conservation of mechanical energy:

$$Mgh = \tfrac{1}{2}Mv^2 + \tfrac{1}{5}Mv^2$$

Solving for v gives

$$v = \sqrt{\frac{10gh}{7}} = \sqrt{\frac{10(9.80 \text{ m/s}^2)(2.00 \text{ m})}{7}} = \boxed{5.29 \text{ m/s}}$$

EXERCISE Repeat this example for a solid cylinder of the same mass and radius as the ball and released from the same height. In a race between the two objects on the incline, which one would win?

ANSWER $v = \sqrt{4gh/3} = 5.11$ m/s; the ball would win.

Quick Quiz 8.4 Two spheres, one hollow and one solid, are rotating with the same angular speed about an axis through their centers. Both spheres have the same mass and radius. Which sphere, if either, has the higher rotational kinetic energy? (a) The hollow sphere. (b) The solid sphere. (c) They have the same kinetic energy.

Quick Quiz 8.5 Two children, Nathan and Kaitlyn, are playing on a hill, rolling automobile tires down the slope. Nathan claims that the tire will roll faster if one of them curls up inside the tire and rides down inside it. Kaitlyn claims that this will cause the tire to roll more slowly. Which child is correct? (a) Nathan, (b) Kaitlyn, (c) not enough information is given. Assume that the children release the tires from rest.

Quick Quiz 8.6 Which arrives at the bottom first? (a) A ball rolling without sliding down incline A, (b) a solid cylinder rolling without sliding down incline A, or (c) a box of the same mass as the ball sliding down a frictionless incline B having the same dimensions as A. Assume that each object is released from rest at the top of its incline.

8.7 ANGULAR MOMENTUM

In Figure 8.26, an object of mass m is positioned a distance r away from a center of rotation. Under the action of a constant torque on the object, its angular speed will increase from the value ω_0 to the value ω in a time interval Δt. Thus, we can write

$$\tau = I\alpha = I\left(\frac{\omega - \omega_0}{\Delta t}\right) = \frac{I\omega - I\omega_0}{\Delta t}$$

FIGURE 8.26 An object of mass m rotating in a circular path under the action of a constant torque.

If we define the product

Definition of angular ▶
momentum

$$L \equiv I\omega \qquad \text{[8.13]}$$

as the **angular momentum** of the object, then we can write

$$\tau = \frac{\text{change in angular momentum}}{\text{time interval}} = \frac{\Delta L}{\Delta t} \qquad \text{[8.14]}$$

Equation 8.14 is the rotational analog of Newton's second law, $F = \Delta p / \Delta t$, and states that **the torque acting on an object is equal to the time rate of change of the object's angular momentum.**

When the net external torque $(\Sigma \tau)$ acting on the system is zero, we see from Equation 8.14 that $\Delta L / \Delta t = 0$. In this case, the rate of change of the system's angular momentum is zero. Therefore, **the angular momentum remains constant in time.** That is,

$$L_i = L_f \qquad \text{if} \qquad \sum \tau = 0 \qquad \text{[8.15]}$$

Conservation of angular ▶
momentum

The angular momentum of a system is conserved when the net external torque acting on the system is zero. That is, when $\Sigma \ \tau = 0$, the initial angular momentum equals the final angular momentum.

Equation 8.15 gives us a third conservation law to add to our list: **conservation of angular momentum.** We can now state that **the energy, linear momentum, and angular momentum of an isolated system all remain constant.**

If an isolated rotating system experiences a change in the moment of inertia, its angular speed will change. In this situation, conservation of angular momentum requires that

$$I_i \omega_i = I_f \omega_f \qquad \text{if} \qquad \sum \tau = 0 \qquad \text{[8.16]}$$

Note that conservation of angular momentum applies to macroscopic objects such as planets and people, as well as to atoms and molecules. There are many examples of conservation of angular momentum, some of which should be familiar to you. You may have observed a figure skater spinning in the finale of his act. The skater's angular speed increases when he pulls his hands and feet close to the trunk of his body, as in Figure 8.27. That is, $\omega_f > \omega_i$. Neglecting friction between skater and ice, we see that no external torques act on the skater. The moment of inertia of his body decreases as his hands and feet are brought in. The resulting increase in

APPLICATION

FIGURE SKATING

FIGURE 8.27 Todd Eldridge's angular speed increases when he pulls his arms in close to his body, demonstrating that angular momentum is conserved. (© *1998 David Madison*)

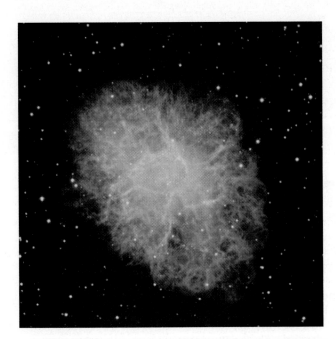

FIGURE 8.28 The Crab Nebula in the constellation Taurus. This nebula is the remnant of a supernova explosion, which was seen on Earth in A.D. 1054. It is located some 6 300 lightyears away and is approximately 6 lightyears in diameter, still expanding outward. *(Max Planck Institute for Astronomy and Calar Alto Observatory, K. Meisenheimer and A. Quetz)*

Webnote 8.3

To see a single, double, and triple axel in skating, visit:
http://btc.montana.edu/olympics/ physbio/biomechanics/cam03.html

APPLICATION

AERIAL SOMERSAULTS

APPLICATION

ROTATING NEUTRON STARS

angular speed is accounted for as follows. Because angular momentum must be conserved, the product $I\omega$ is constant, and a decrease of the skater's moment of inertia must be compensated by a corresponding increase in his angular speed.

Similarly, when a diver or acrobat wishes to make several somersaults, she pulls her hands and feet close to the trunk of her body in order to rotate at a greater angular speed. In this case, the external force due to gravity acts through her center of gravity and hence exerts no torque about the axis of rotation, and the angular momentum about the center of gravity is conserved. For example, when a diver wishes to double her angular speed, she must reduce her moment of inertia to half its initial value.

An interesting astrophysical example of conservation of angular momentum occurs when, at the end of its lifetime, a massive star uses up all its fuel and collapses under the influence of gravitational forces, causing a gigantic outburst of energy called a supernova explosion. The best-studied example of a remnant of a supernova explosion is the Crab Nebula, a chaotic, expanding mass of gas (Fig. 8.28). In a supernova, part of the star's mass is released into space, where it eventually condenses into new stars and planets. Most of what is left behind typically collapses into a **neutron star,** an extremely dense sphere of matter with a diameter of about 10 km (a great reduction from the 10^6-km diameter of the original star), and containing a large fraction of the star's original mass. As the moment of inertia of the system decreases during the collapse, the star's rotational speed increases. More than 700 rapidly rotating neutron stars have been identified since their first discovery in 1967, with periods of rotation ranging from a millisecond to several seconds. The neutron star is a most dramatic system—an object with a mass greater than that of the Sun, rotating about its axis many times each second!

Quick Quiz 8.7

A horizontal disk with moment of inertia I_1 rotates with angular speed ω_0 about a vertical frictionless axle. A second horizontal disk, with moment of inertia I_2 and initially not rotating, drops onto the first. Because the surfaces are rough, the two eventually reach the same angular speed ω. Is the ratio ω/ω_0 equal to (a) I_1/I_2, (b) I_2/I_1, (c) $I_1/(I_1 + I_2)$, or (d) $I_2/(I_1 + I_2)$?

(a)

(b)

FIGURE 8.29 (Applying Physics 8.2) (a) This person is given an initial angular speed while holding two objects as shown. (b) When the objects are pulled in close to the body, the angular speed of the system increases. Why?

> **Quick Quiz 8.8**
>
> If global warming continues to occur, it is likely that some ice from Earth's polar ice caps will melt and the water will be distributed closer to the Equator. If this occurs, would the length of the day (one revolution) (a) increase, (b) decrease, or (c) remain the same?

APPLYING PHYSICS 8.2

A common classroom demonstration is to spin on a rotating stool with heavy weights in your outstretched hands, and bring the weights close to your body (Fig. 8.29). The result is an increase in the angular speed, in agreement with the conservation of angular momentum. As discussed previously, ice skaters use this effect to produce high-speed spins when they bring their arms in close to the rotation axis. Let us imagine we perform such a demonstration, and moving the weights inward results in halving the moment of inertia, and, therefore, doubling the angular speed. If we consider the rotational kinetic energy, we see that the energy is doubled in this situation. Thus, angular momentum is conserved, but kinetic energy is not. Where does this extra energy come from?

Explanation As you spin with the weights in your hands, your arms provide the tension force that causes the weights to move in a circular path. As you bring the weights in, this force from your arms does work on the weights, because it is moving them through a radial displacement. What's more, as your angular speed increases in response to the weights moving inward, you must apply more and more force to keep the weights moving in the circular path of smaller radius. Thus, the work that you do in bringing the weights in keeps increasing as they are brought in. This work appears as the increase in rotational kinetic energy of the system. You can perform this work due to the chemical energy stored in your body from previous meals, so you are transforming potential energy stored in your body into rotational kinetic energy.

PROBLEM-SOLVING STRATEGY · Rotational Motion

Keep the following facts and procedures in mind when solving rotational motion problems.

1. Very few new techniques must be learned in order to solve rotational motion problems. For example, problems involving the equation $\Sigma \tau = I\alpha$ are very similar to those encountered in Newton's second law problems, $\Sigma F = ma$. Note the correspondences between linear and rotational quantities, in that F is replaced by τ, m by I, and a by α.

2. Other analogs between rotational quantities and linear quantities include the replacement of x by θ and v by ω. These are helpful as memory devices for such rotational motion quantities as rotational kinetic energy $KE_r = \frac{1}{2}I\omega^2$ and angular momentum $L = I\omega$.

3. With the analogs mentioned in the previous step, techniques for conservation of energy are the same as those examined in Chapter 5, except for the fact that any rotational kinetic energy in the system must be included in the expression for the conservation of energy.

4. Likewise, the techniques for solving problems in conservation of angular momentum are essentially the same as those for solving problems in conservation of linear momentum, except that total initial angular momentum is equated to total final angular momentum as $L_i = L_f$.

Example 8.13 The Spinning Stool

A student sits on a pivoted stool while holding a pair of objects (see Fig. 8.29). The stool is free to rotate about a vertical axis with negligible friction. The moment of inertia of student, objects, and stool is $2.25 \text{ kg} \cdot \text{m}^2$. The student is set in rotation with an initial angular speed of 5.00 rad/s, with arms outstretched. As he rotates, he pulls the objects inward so that the new moment of inertia of the system (student, objects, and stool) becomes $1.80 \text{ kg} \cdot \text{m}^2$. What is the new angular speed of the system?

Reasoning We will apply the principle of conservation of angular momentum to find the new angular speed. The initial angular momentum $I_i\omega_i$ will be equated to the final angular momentum $I_f\omega_f$.

Solution The initial angular momentum of the system is

$$L_i = I_i\omega_i = (2.25 \text{ kg} \cdot \text{m}^2)(5.00 \text{ rad/s}) = 11.3 \text{ kg} \cdot \text{m}^2/\text{s}$$

When the objects are pulled in, they are closer to the axis of rotation, and as a result the moment of inertia of the system is reduced. The new angular momentum is

$$L_f = I_f\omega_f = (1.80 \text{ kg} \cdot \text{m}^2)\omega_f$$

Because the net external torque on the system is zero, angular momentum is conserved. Thus, we find that

$$(11.3 \text{ kg} \cdot \text{m}^2/\text{s}) = (1.80 \text{ kg/m}^2)\omega_f$$

$$\omega_f = \boxed{6.28 \text{ rad/s}}$$

Example 8.14 The Merry-Go-Round

You have finally coaxed your young nephew to leave the playground and come home for lunch. You notice that he has left his sandals on the still-spinning merry-go-round, and you must retrieve them. Assume that the merry-go-round is a simple circular platform of mass $M = 100$ kg and radius $R = 2.00$ m, rotating in a horizontal plane about a frictionless vertical axle (Fig. 8.30). Just after you jump onto the merry-go-round, the angular speed of the system is 2.00 rad/s. If your mass is $m = 60.0$ kg and you walk slowly from the edge toward the center to get the sandals, find the angular speed when you reach a point 0.500 m from the center. Neglect the mass of the sandals.

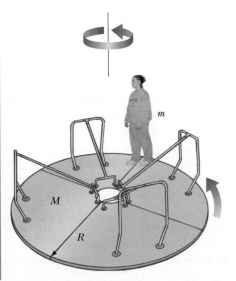

FIGURE 8.30 (Example 8.14) If you walk toward the center of the rotating platform, the angular speed of the system increases because the angular momentum of the system (student + platform) must remain constant.

Reasoning   We will use the principle of conservation of angular momentum. The initial angular momentum of the system is the sum of the angular momentum of the platform plus your angular momentum when you are at the edge of the merry-go-round. The final angular momentum is the sum of the angular momentum of the platform plus your angular momentum when you are 0.500 m from the center.

Solution   The moment of inertia of the platform I_p is

$$I_p = \tfrac{1}{2}MR^2 = \tfrac{1}{2}(100 \text{ kg})(2.00 \text{ m})^2 = 200 \text{ kg} \cdot \text{m}^2$$

Assuming that you (the student) are a particle, your initial moment of inertia is

$$I_s = mR^2 = (60.0 \text{ kg})(2.00 \text{ m})^2 = 240 \text{ kg} \cdot \text{m}^2$$

Thus, the initial angular momentum of the platform plus student is

$$L_i = (I_p + I_s)(\omega_i) = (200 \text{ kg} \cdot \text{m}^2 + 240 \text{ kg} \cdot \text{m}^2)(2.00 \text{ rad/s}) = 880 \text{ kg} \cdot \text{m}^2/\text{s}$$

When you have walked to the position 0.500 m from the center, your moment of inertia is

$$I_s' = mr_f^2 = (60.0 \text{ kg})(0.500 \text{ m})^2 = 15.0 \text{ kg} \cdot \text{m}^2$$

No change occurs in the moment of inertia of the platform. Because there are no external torques on the *system* (you plus the platform) about the axis of rotation, we can apply the law of conservation of angular momentum:

$$L_i = L_f$$

$$880 \text{ kg} \cdot \text{m}^2/\text{s} = 200\omega_f + 15.0\omega_f = 215\omega_f$$

$$\omega_f = \boxed{4.09 \text{ rad/s}}$$

Your angular speed has doubled. You would have done well to stop the merry-go-round before retrieving the sandals.

EXERCISE   Calculate the change in kinetic energy of the system (platform plus student) for this situation. What accounts for this change in energy?

ANSWER $KE_f - KE_i = 918 \text{ J}$. You must perform positive work in order to walk toward the center of the platform.

SUMMARY

The tendency of a force to rotate an object about some axis is measured by a quantity called **torque,** τ. The magnitude of the torque is given by

$$\tau = Fd \tag{8.1}$$

In this equation, d is the **lever arm** — the perpendicular distance from the axis of rotation to a line drawn along the direction of the force. The sign of the torque is negative if the turning tendency of the corresponding force is clockwise and positive if the turning tendency is counterclockwise.

An object is in **equilibrium** when the following conditions are satisfied: (1) the net external force must be zero, and (2) the net external torque must be zero about any axis. That is,

$$\sum \mathbf{F} = 0 \qquad \sum \tau = 0 \tag{8.2}$$

The first equation is called the **first condition for equilibrium.** When this equation holds, an object is either at rest or moving with a constant velocity. The second equation is called the **second condition for equilibrium.** When it holds, an object is said to be in rotational equilibrium.

The **moment of inertia** of a group of particles is

$$I \equiv \sum mr^2 \qquad \text{[8.8]}$$

If a rigid object free to rotate about a fixed axis has a net external torque acting on it, the object undergoes an angular acceleration α, where

$$\sum \tau = I\alpha \qquad \text{[8.9]}$$

If a rigid object rotates about a fixed axis with angular speed ω, its **rotational kinetic energy** is

$$KE_r = \tfrac{1}{2}I\omega^2 \qquad \text{[8.11]}$$

where I is the moment of inertia about the axis of rotation.

The **angular momentum** of a rotating object is

$$L \equiv I\omega \qquad \text{[8.13]}$$

If the net external torque acting on a system is zero, the total angular momentum of the system is constant.

$$L_i = L_f \qquad \text{[8.15]}$$

Applying the law of conservation of angular momentum to an object whose moment of inertia changes with time gives

$$I_i\omega_i = I_f\omega_f \qquad \text{[8.16]}$$

CONCEPTUAL QUESTIONS

1. Why can't you put your heels firmly against a wall and then bend over without falling?

2. Why does a tall athlete have an advantage over a smaller one when competing in the high jump?

3. Both torque and work are products of force and distance. How are they different? Do they have the same units?

4. Is it possible to calculate the torque acting on a rigid object without specifying the origin? Is the torque independent of the location of the origin?

5. Can an object be in equilibrium when only one force acts on it? If you believe the answer is yes, give an example to support your conclusion.

6. Often when a high diver wants to turn a flip in midair, she draws her legs up against her chest. Why does this make her rotate faster? What should she do when she wants to come out of her flip?

7. In some motorcycle races, the riders drive over small hills, and the motorcycle becomes airborne for a short time. If the motorcycle racer keeps the throttle open while leaving the hill and going into the air, the motorcycle tends to nose upward. Why does this happen?

8. Why does a long pole help a tightrope walker stay balanced?

9. Suppose you are designing a car for a coasting race—the cars in this race have no engines, they simply coast downhill. Do you want large wheels or small wheels? Do you want solid, disk-like wheels or hoop-like wheels? Should the wheels be heavy or light?

10. If you toss a textbook into the air, rotating it each time about one of the three axes perpendicular to the textbook, you will find that it will not rotate smoothly about one of these axes. (Try placing a strong rubber band around the book before the toss so it will stay closed.) Its rotation is stable about those axes having the largest and smallest moments of inertia but unstable about the axis of intermediate moment. Try this on your own to find the axis which has this intermediate moment of inertia.

11. Stars originate as large bodies of slowly rotating gas. Because of gravity, these clumps of gas slowly decrease in size. What happens to the angular speed of a star as it shrinks? Explain.

12. If a high jumper positions his body correctly when going over the bar, the center of gravity of the athlete may actually pass under the bar. Explain how this is possible.

FIGURE Q8.12 *(© Eye Ubiquitous/Corbis)*

13. In a tape recorder, the tape is pulled past the read/write heads at a constant speed by the drive mechanism. Consider the reel from which the tape is pulled—as the tape is pulled off, the radius of the roll of remaining tape

decreases. How does the torque on the reel change with time? How does the angular speed of the reel change with time? If the tape mechanism is suddenly turned on so that the tape is quickly pulled with a large force, is the tape more likely to break when pulled from a nearly full reel or a nearly empty reel?

14. (a) Give an example in which the net force acting on an object is zero and yet the net torque is nonzero. (b) Give an example in which the net torque acting on an object is zero and yet the net force is nonzero.

15. A mouse is initially at rest on a horizontal turntable mounted on a frictionless vertical axle. If the mouse begins to walk clockwise around the perimeter, what happens to the turntable? Explain.

16. A cat usually lands on its feet regardless of the position from which it is dropped. A slow-motion film of a cat falling shows that the upper half of its body twists in one direction while the lower half twists in the opposite direction. (See Figure Q8.16.) Why does this type of rotation occur?

17. A ladder rests inclined against a wall. Would you feel safer climbing up the ladder if you were told that the floor is frictionless but the wall is rough, or that the wall is frictionless but the floor is rough? Justify your answer.

18. Neutral equilibrium is a special condition of equilibrium in which an object's center of gravity moves neither up nor down if disturbed slightly. How do you think an object in neutral equilibrium will react to a slight disturbance?

FIGURE Q8.16   A falling, twisting cat. *(Gerard Lacz/NHPA)*

PROBLEMS

1, 2, 3 = straightforward, intermediate, challenging   ☐ = full solution available in Student Solutions Manual/Study Guide

web = solution posted at **http://info.brookscole.com/serway**   🧬 = biomedical application

Section 8.1   Torque

1. If the torque required to loosen a nut that is holding a flat tire in place on a car has a magnitude of 40.0 N·m, what *minimum* force must be exerted by the mechanic at the end of a 30.0-cm lug wrench to accomplish the task?

2. A steel band exerts a horizontal force of 80.0 N on a tooth at point *B* in Figure P8.2. What is the torque on the root of the tooth about point *A*?

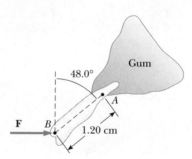

FIGURE P8.2

3. The person in Figure P8.3 weighs 800 N. He is exercising by bending back and forth as he pushes against a wall. At one moment, the forces **F**₁ and **F**₂ have magnitudes of 100 N and 900 N, respectively. Assume the force of gravity acts downward through point *A* as shown. Determine the

FIGURE P8.3

net torque on the person about axes through points *A*, *B*, and *C* perpendicular to the plane of the paper.

4. As part of a physical therapy program following a knee operation, a 10-kg object is attached to an ankle and leg lifts are done as sketched in Figure P8.4. Calculate the torque about the knee due to this weight for the four positions shown.

FIGURE P8.4

5. A simple pendulum consists of a small object of mass 3.0 kg hanging at the end of a 2.0-m-long light string that is connected to a pivot point. Calculate the magnitude of the torque (due to the force of gravity) about this pivot point when the string makes a 5.0° angle with the vertical.

6. A fishing pole is 2.00 m long and inclined to the horizontal at an angle of 20.0° (Fig. P8.6). What is the torque exerted by the fish about an axis perpendicular to the page and passing through the hand of the person holding the pole?

FIGURE P8.6

Section 8.2 Torque and the Two Conditions for Equilibrium

Section 8.3 The Center of Gravity

Section 8.4 Examples of Objects in Equilibrium

7. The arm in Figure P8.7 weighs 41.5 N. The force of gravity acting on the arm acts through point *A*. Determine the magnitudes of the tension force $\mathbf{F}_t$ in the deltoid muscle and the force $\mathbf{F}_s$ exerted by the shoulder on the humerus (upper-arm bone) to hold the arm in the position shown.

FIGURE P8.7

8. A water molecule consists of an oxygen atom with two hydrogen atoms bound to it as shown in Figure P8.8. The bonds are 0.100 nm in length and the angle between the two bonds is 106°. Use the *xy* axes shown and determine the location of the center of gravity of the molecule. Take the mass of an oxygen atom to be 16 times the mass of a hydrogen atom.

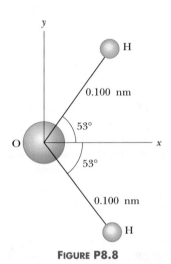

FIGURE P8.8

9. A cook holds a 2.00-kg carton of milk at arm's length (Fig. P8.9). What force $\mathbf{F}_B$ must be exerted by the biceps muscle? (Ignore the weight of the forearm.)

FIGURE P8.9

10. The sailor in Figure P8.10 weighs 750 N. The force $\mathbf{F}_1$ exerted by the wind on the sail is horizontal and acts through point *B*. The weight of the boat is 1 250 N and acts through point *O*, which is 0.800 m from the point *A* along the line *OA*. The force $\mathbf{F}_2$ exerted by the water acts through point *A*. Determine the net force exerted by the wind on the sail.

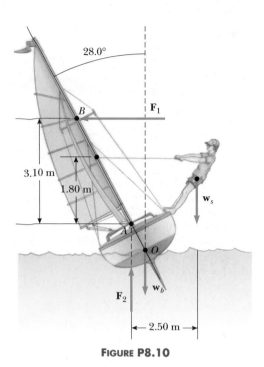

FIGURE P8.10

11. Figure P8.11 is a crude model of the back (spine) as a rigid rod supported by a guy wire W (back muscles). The rod supporting the weight is free to pivot about the point P. Compare the tension in the wire W necessary to pick up the weight in the two positions shown. Assume the spine is rigidly attached to the pelvis and that the compression force in the rod representing the spine acts along the rod.

(a)   (b)

FIGURE P8.11

12. Consider the following mass distribution where the xy coordinates are given in meters: 5.0 kg at (0.0, 0.0) m, 3.0 kg at (0.0, 4.0) m, and 4.0 kg at (3.0, 0.0) m. Where should a fourth object of 8.0 kg be placed so the center of gravity of the four-object arrangement will be at (0.0, 0.0) m?

13. The approximate mass distributions of a female at ages 10 and 20 are shown in Figure P8.13. The masses quoted for portions of the arms and legs are for each arm and each leg separately. Using the midpoint X as your origin, determine the vertical position of the center of gravity at age 10

and at age 20. Express each result as a fraction of the person's height.

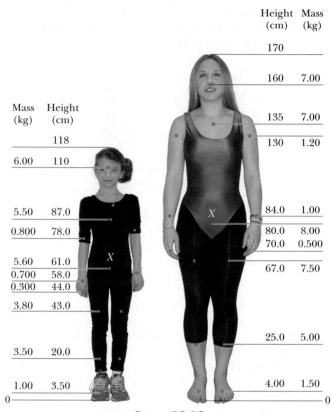

FIGURE P8.13

14. The normal center of gravity of a woman is shifted forward during pregnancy, particularly in the third trimester. To compensate, the upper part of the body bends backward, as shown in Figure P8.14. The weight distribution, including the extra weight of the fetus at position C, is as follows: 65% at A, 25% at B, 10% at C. The support of the feet is centered on the dashed line through A and D. (a) Determine the angle β. (b) Find the vertical position of the

FIGURE P8.14

woman's center of gravity. (*Hint:* Imagine rotating the figure through 90°.)

15. What is the tension *T* exerted by the hamstring muscles in the back of the thigh and the compressive force F_c in the knee joint due to the application of a horizontal force of 100 N to the ankle as shown in Figure P8.15?

38.0 cm

100 N

4.00 cm   T

F_c

FIGURE P8.15

16. The principal forces acting on a foot when a person is squatting are shown in Figure P8.16. Determine the magnitude of the force $\mathbf{F}_H$ exerted by the Achilles tendon on the heel at point *H* and the magnitude of the force $\mathbf{F}_J$ exerted on the ankle joint at point *J*.

$\mathbf{F}_J$

$\mathbf{F}_H$

d_1

52°

H   *J*

d_H d_2

3.20 cm   4.00 cm

$\mathbf{F}_n$

350 N

FIGURE P8.16

17. Figure P8.17 shows a person using both hands to lift a 30.0-kg barbell. When the lifter's back is horizontal, what is the magnitude of the tension **T** in the back muscles and the hinge force $\mathbf{F}_c$ which the hips exert on the base of the spine? The weight $\mathbf{w}_1 = 380$ N is the weight of the upper torso, $\mathbf{w}_2 = 60$ N is the weight of the head, and $\mathbf{w}_3 = 394$ N is the weight of the arms (100 N) plus the weight being lifted (294 N). The tension **T** in the back muscles is directed 12° above the horizontal.

18. A window washer is standing on a scaffold supported by a vertical rope at each end. The scaffold weighs 200 N and is

36 cm 12 cm 12 cm 18 cm

T

12°

$\mathbf{F}_c$

$\mathbf{w}_1$ $\mathbf{w}_2$

$\mathbf{w}_3$

FIGURE P8.17

3.00 m long. What is the tension in each rope when the 700-N worker stands 1.00 m from one end?

19. The chewing muscle, the masseter, is one of the strongest in the human body. It is attached to the mandible (lower jawbone) as shown in Figure P8.19a. The jawbone is pivoted about a socket just in front of the auditory canal. The forces acting on the jawbone are equivalent to those acting on the curved bar in Figure P8.19b: $\mathbf{F}_c$ is the force exerted by the food being chewed against the jawbone, **T** is the tension in the masseter, and **R** is the force exerted by the socket on the mandible. Find **T** and **R** if you bite down on a piece of steak with a force of 50.0 N.

Masseter

Mandible

(a)

F_C T   R

3.50 cm

7.50 cm

(b)

FIGURE P8.19

20. A hungry 700-N bear walks out on a beam in an attempt to retrieve some "goodies" hanging at the end (Fig. P8.20). The beam is uniform, weighs 200 N, and is 6.00 m long; the goodies weigh 80.0 N. (a) Draw a free-body diagram for the beam. (b) When the bear is at $x = 1.00$ m, find the tension in the wire and the components of the reaction force at the hinge. (c) If the wire can withstand a maximum tension of 900 N, what is the maximum distance the bear can walk before the wire breaks?

60.0°

x   Goodies

FIGURE P8.20

21. A uniform semicircular sign 1.00 m in diameter and of weight w is supported by two wires as shown in Figure P8.21. What is the tension in each of the wires supporting the sign?

0.25 m

0.75 m

BuBu's Boutique

FIGURE P8.21

22. A 20.0-kg floodlight in a park is supported at the end of a horizontal beam of negligible mass that is hinged to a pole, as shown in Figure P8.22. A cable at an angle of 30.0° with the beam helps to support the light. Find (a) the tension in the cable and (b) the horizontal and vertical forces exerted on the beam by the pole.

23. A uniform plank of length 2.00 m and mass 30.0 kg is supported by three ropes, as indicated by the blue vectors in Figure P8.23. Find the tension in each rope when a 700-N person is 0.500 m from the left end.

24. A 15.0-m, 500-N uniform ladder rests against a frictionless wall, making an angle of 60.0° with the horizontal. (a) Find the horizontal and vertical forces exerted on the base of the ladder by Earth when an 800-N fire fighter is

30.0°

FIGURE P8.22

$\mathbf{T}_2$

$\mathbf{T}_3$

$\mathbf{T}_1$

40.0°

0.500 m

2.00 m

FIGURE P8.23

4.00 m from the bottom. (b) If the ladder is just on the verge of slipping when the fire fighter is 9.00 m up, what is the coefficient of static friction between ladder and ground?

25. An 8.00-m, 200-N uniform ladder rests against a smooth **web** wall. The coefficient of static friction between the ladder and the ground is 0.600, and the ladder makes a 50.0° angle with the ground. How far up the ladder can an 800-N person climb before the ladder begins to slip?

26. A 1 200-N uniform boom is supported by a cable perpendicular to the boom as in Figure P8.26. The boom is hinged at the bottom, and a 2 000-N weight hangs from its top. Find the tension in the supporting cable and the components of the reaction force exerted on the boom by the hinge.

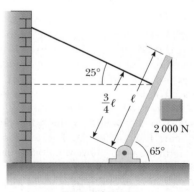

25°

$\frac{3}{4}\ell$

ℓ

2 000 N

65°

FIGURE P8.26

27. The large quadriceps muscle in the upper leg terminates at its lower end in a tendon attached to the upper end of the tibia (Fig. P8.27a). The forces on the lower leg when the leg is extended are modeled as in Figure P8.27b, where **T** is the tension in the tendon, **w** is the force of gravity acting on the lower leg, and **F** is the weight of the foot. Find **T** when the tendon is at an angle of 25.0° with the tibia, assuming that $w = 30.0$ N, $F = 12.5$ N, and the leg is extended at an angle of 40.0° with the vertical ($\theta = 40.0°$). Assume that the center of gravity of the lower leg is at its center, and that the tendon attaches to the lower leg at a point one fifth of the way down the leg.

(a)

(b)

FIGURE P8.27

28. One end of a uniform 4.0-m-long rod of weight w is supported by a cable. The other end rests against the wall, where it is held by friction (see Fig. P8.28). The coefficient of static friction between the wall and the rod is $\mu_s = 0.50$. Determine the minimum distance x from point A at which an additional weight w (same as the weight of the rod) can be hung without causing the rod to slip at point A.

FIGURE P8.28

Section 8.5 Relationship Between Torque and Angular Acceleration

29. Four objects are held in position at the corners of a rectangle by light rods as shown in Figure P8.29. Find the moment of inertia of the system about (a) the x axis, (b) the y axis, and (c) an axis through O and perpendicular to the page.

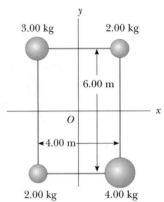

FIGURE P8.29   (Problems 29 and 30)

30. If the system shown in Figure P8.29 is set in rotation about each of the axes mentioned in Problem 29, find the torque that will produce an angular acceleration of 1.50 rad/s² in each case.

31. A model airplane with mass 0.750 kg is tethered by a wire so that it flies in a circle 30.0 m in radius. The airplane engine provides a net thrust of 0.800 N perpendicular to the tethering wire. (a) Find the torque the net thrust produces about the center of the circle. (b) Find the angular acceleration of the airplane when it is in level flight. (c) Find the linear acceleration of the airplane tangent to its flight path.

32. A potter's wheel having a radius of 0.50 m and a moment of inertia of 12 kg·m² is rotating freely at 50 rev/min. The potter can stop the wheel in 6.0 s by pressing a wet rag against the rim and exerting a radially inward force of 70 N. Find the effective coefficient of kinetic friction between the wheel and the wet rag.

33. A cylindrical fishing reel has a moment of inertia of $I = 6.8 \times 10^{-4}$ kg·m² and a radius of 4.0 cm. A friction clutch in the reel exerts a restraining torque of 1.3 N·m if a fish pulls on the line. The fisherman gets a bite, and the reel begins to spin with an angular acceleration of 66 rad/s². (a) What is the force exerted by the fish on the line? (b) How much line unwinds in 0.50 s?

34. A bicycle wheel has a diameter of 64.0 cm and a mass of 1.80 kg. Assume that the wheel is a hoop with all the mass concentrated on the outside radius. The bicycle is placed on a stationary stand and a resistive force of 120 N is applied tangent to the rim of the tire. (a) What force must be applied by a chain passing over a 9.00-cm-diameter sprocket in order to give the wheel an acceleration of 4.50 rad/s²? (b) What force is required if you shift to a 5.60-cm-diameter sprocket?

35. A 150-kg merry-go-round in the shape of a uniform, solid, horizontal disk of radius 1.50 m is set in motion by wrapping a rope about the rim of the disk and pulling on the rope. What constant force must be exerted on the rope to bring the merry-go-round from rest to an angular speed of 0.500 rev/s in 2.00 s?

36. A cylindrical 5.00-kg reel, with a radius of 0.600 m and a frictionless axle, starts from rest and speeds up uniformly as a 3.00-kg bucket falls into a well, making a light rope unwind from the reel (Fig. P8.36). The bucket starts from rest and falls for 4.00 s. (a) What is the linear acceleration of the falling bucket? (b) How far does it drop? (c) What is the angular acceleration of the reel?

5.00 kg

0.600 m

3.00 kg

FIGURE P8.36

37. An airliner lands with a speed of 50.0 m/s. Each wheel of the plane has a radius of 1.25 m and a moment of inertia of 110 kg·m². At touchdown the wheels begin to spin under the action of friction. Each wheel supports a weight of 1.40×10^4 N, and the wheels attain their angular speed while rolling without slipping in 0.480 s. What is the coefficient of kinetic friction between the wheels and the runway? Assume that the speed of the plane is constant.

Section 8.6   Rotational Kinetic Energy

38. A constant torque of 25.0 N·m is applied to a grindstone whose moment of inertia is 0.130 kg·m². Using energy principles, and neglecting friction, find the angular speed after the grindstone has made 15.0 revolutions. (*Hint:* the angular equivalent of $W_{net} = F\Delta x = \frac{1}{2}mv_f^2 - \frac{1}{2}mv_i^2$ is $W_{net} = \tau\Delta\theta = \frac{1}{2}I\omega_f^2 - \frac{1}{2}I\omega_i^2$. You should convince yourself that this is correct.)

39. A 10.0-kg cylinder rolls without slipping on a rough surface. At an instant when its center of gravity has a speed of 10.0 m/s, determine (a) the translational kinetic energy of its center of gravity, (b) the rotational kinetic energy about its center of gravity, and (c) its total kinetic energy.

40. The net work done in accelerating a propeller from rest to an angular speed of 200 rad/s is 3 000 J. What is the moment of inertia of the propeller?

41. A horizontal 800-N merry-go-round of radius 1.50 m is started from rest by a constant horizontal force of 50.0 N applied tangentially to the merry-go-round. Find the kinetic energy of the merry-go-round after 3.00 s. (Assume it is a solid cylinder.)

42. A car is designed to get its energy from a rotating flywheel with a radius of 2.00 m and a mass of 500 kg. Before a trip, the flywheel is attached to an electric motor, which brings the flywheel's rotational speed up to 5 000 rev/min. (a) Find the kinetic energy stored in the flywheel. (b) If the flywheel is to supply energy to the car as would a 10.0-hp motor, find the length of time the car could run before the flywheel would have to be brought back up to speed.

43. The top in Figure P8.43 has a moment of inertia of 4.00×10^{-4} kg·m² and is initially at rest. It is free to rotate about a stationary axis AA'. A string, wrapped around a peg along the axis of the top, is pulled in such a manner as to maintain a constant tension of 5.57 N in the string. If the string does not slip while wound around the peg, what is the angular speed of the top after 80.0 cm of string has been pulled off the peg? (*Hint:* Consider the work done.)

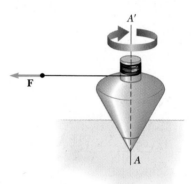

A'

F

A

FIGURE P8.43

44. A 240-N sphere 0.20 m in radius rolls, without slipping, 6.0 m down a ramp that is inclined at 37° with the horizontal. What is the angular speed of the sphere at the bottom of the slope if it starts from rest?

Section 8.7   Angular Momentum

45. (a) Calculate the angular momentum of Earth that arises from its spinning motion on its axis and (b) the angular momentum of Earth that arises from its orbital motion about the Sun.

46. Halley's comet moves about the Sun in an elliptical orbit, with its closest approach to the Sun being 0.59 A.U. and its greatest distance being 35 A.U. (1 A.U. = Earth-Sun distance). If the comet's speed at closest approach is 54 km/s, what is its speed when it is farthest from the Sun? You may neglect any change in the comet's mass and assume that its angular momentum about the Sun is conserved.

47. The system of small objects shown in Figure P8.47 is rotating at an angular speed of 2.0 rev/s. The objects are connected by light, flexible spokes that can be lengthened or shortened. What is the new angular speed if the spokes are shortened to 0.50 m? (An effect similar to that illustrated in this problem occurred in the early stages of the formation of our Galaxy. As the massive cloud of dust and gas that was the source of the stars and planets contracted, an initially small angular speed increased with time.)

48. A playground merry-go-round of radius $R = 2.00$ m has a moment of inertia $I = 250$ kg·m² and is rotating at

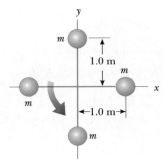

FIGURE P8.47

10.0 rev/min about a frictionless vertical axle. Facing the axle, a 25.0-kg child hops onto the merry-go-round, and manages to sit down on the edge. What is the new angular speed of the merry-go-round?

49. A solid, horizontal cylinder of mass 10.0 kg and radius 1.00 m rotates with an angular speed of 7.00 rad/s about a fixed vertical axis through its center. A 0.250-kg piece of putty is dropped vertically onto the cylinder at a point 0.900 m from the center of rotation, and sticks to the cylinder. Determine the final angular speed of the system.

50. A student sits on a rotating stool holding two 3.0-kg objects. When his arms are extended horizontally, the objects are 1.0 m from the axis of rotation, and he rotates with an angular speed of 0.75 rad/s. The moment of inertia of the student plus stool is 3.0 kg·m^2 and is assumed to be constant. The student then pulls the objects horizontally to 0.30 m from the rotation axis. (a) Find the new angular speed of the student. (b) Find the kinetic energy of the student before and after the objects are pulled in.

51. The puck in Figure P8.51 has a mass of 0.120 kg. Its original distance from the center of rotation is 40.0 cm, and the puck is moving with a speed of 80.0 cm/s. The string is pulled downward 15.0 cm through the hole in the frictionless table. Determine the work done on the puck. (*Hint:* Consider the change of kinetic energy of the puck.)

web

FIGURE P8.51

52. A merry-go-round rotates at the rate of 0.20 rev/s with an 80-kg man standing at a point 2.0 m from the axis of rotation. (a) What is the new angular speed when the man walks to a point 1.0 m from the center? Assume that the merry-go-round is a solid 25-kg cylinder of radius 2.0 m.

(b) Calculate the change in kinetic energy due to this movement. How do you account for this change in kinetic energy?

53. A 60.0-kg woman stands at the rim of a horizontal turntable having a moment of inertia of 500 kg·m^2 and a radius of 2.00 m. The turntable is initially at rest and is free to rotate about a frictionless, vertical axle through its center. The woman then starts walking around the rim clockwise (as viewed from above the system) at a constant speed of 1.50 m/s relative to Earth. (a) In what direction and with what angular speed does the turntable rotate? (b) How much work does the woman do to set herself and the turntable into motion?

54. A space station shaped like a giant wheel has a radius 100 m and a moment of inertia of 5.00×10^8 kg·m^2. A crew of 150 are living on the rim, and the station is rotating so that the crew experience an apparent acceleration of 1g (Fig. P8.54). When 100 people move to the center of the station for a union meeting, the angular speed changes. What apparent acceleration is experienced by the managers remaining at the rim? Assume an average mass of 65.0 kg for all the inhabitants.

FIGURE P8.54

ADDITIONAL PROBLEMS

55. A cylinder with moment of inertia I_1 rotates with angular velocity ω_0 about a frictionless vertical axle. A second cylinder, with moment of inertia I_2, initially not rotating, drops onto the first cylinder (Fig. P8.55). Since the surfaces are rough, the two eventually reach the same angular speed ω. (a) Calculate ω. (b) Show that kinetic energy is lost in this situation, and calculate the ratio of the final to the initial kinetic energy.

56. A 0.100-kg meter stick is supported at its 40.0-cm mark by a string attached to the ceiling. A 0.700-kg object hangs vertically from the 5.00-cm mark. An object of mass m is attached somewhere on the meter stick to keep it horizontal and in rotational and translational equilibrium. If the tension in the string attached to the ceiling is 19.6 N, determine (a) the value of m and (b) its point of attachment on the stick.

Before   After

FIGURE P8.55

FIGURE P8.59

57. Show that the kinetic energy of an object rotating about a fixed axis with angular momentum $L = I\omega$ can be written as $KE = L^2/2I$.

58. Figure P8.58 shows a claw hammer as it is being used to pull a nail out of a horizontal board. If a force of magnitude 150 N is exerted horizontally as shown, find (a) the force exerted by the hammer claws on the nail and (b) the force exerted by the surface on the point of contact with the hammer head. Assume that the force the hammer exerts on the nail is parallel to the nail.

F

30.0 cm

30.0° | Single point
        | of contact

5.00 cm

FIGURE P8.58

59. An electric motor turns a flywheel through a drive belt that joins a pulley on the motor and a pulley that is rigidly attached to the flywheel, as shown in Figure P8.59. The flywheel is a solid disk with a mass of 80.0 kg and a diameter of 1.25 m. It turns on a frictionless axle. Its pulley has much smaller mass and a radius of 0.230 m. If the tension in the upper (taut) segment of the belt is 135 N and the flywheel has a clockwise angular acceleration of 1.67 rad/s², find the tension in the lower (slack) segment of the belt.

60. A 12.0-kg object is attached to a cord that is wrapped around a wheel of radius $r = 10.0$ cm (Fig. P8.60). The acceleration of the object down the frictionless incline is measured to be 2.00 m/s². Assuming the axle of the wheel to be frictionless, determine (a) the tension in the rope,

(b) the moment of inertia of the wheel, and (c) the angular speed of the wheel 2.00 s after it begins rotating, starting from rest.

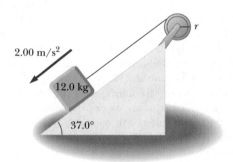

2.00 m/s²

12.0 kg

37.0°

FIGURE P8.60

61. A uniform ladder of length L and weight w is leaning against a vertical wall. The coefficient of static friction between the ladder and the floor is the same as that between the ladder and the wall. If this coefficient of static friction is $\mu_s = 0.500$, determine the smallest angle the ladder can make with the floor without slipping.

62. A uniform 10.0-N picture frame is supported as shown in Figure P8.62. Find the tension in the cords and the magnitude of the horizontal force at P that are required to hold the frame in the position shown.

T$_2$ **T**$_1$

30.0 cm   50.0°

15.0 cm

P **F**

10.0 N

FIGURE P8.62

63. A solid 2.0-kg ball of radius 0.50 m starts at a height of 3.0 m above the surface of Earth and *rolls* down a 20° slope. A solid disk and a ring start at the same time and the same height. The ring and disk each have the same mass and radius as the ball. Which of the three wins the race to the bottom if all roll without slipping?

64. A common physics demonstration (Figure P8.64) consists of a ball resting at the end of a board of length ℓ that is elevated at an angle θ with the horizontal. A light cup is attached to the board at r_c so that it will catch the ball when the support stick is suddenly removed. (a) Show that the ball will lag behind the falling board when $\theta < 35.3°$, and (b) the ball will fall into the cup when the board is supported at this limiting angle and the cup is placed at

$$r_c = \frac{2\ell}{3\cos\theta}$$

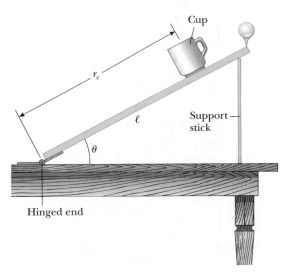

FIGURE P8.64

65. In Figure P8.65 the sliding block has a mass of 0.850 kg, the counterweight has a mass of 0.420 kg, and the pulley is a uniform solid cylinder with a mass of 0.350 kg and an outer radius of 0.0300 m. The coefficient of kinetic friction between the block and the horizontal surface is 0.250. The pulley turns without friction on its axle. The light cord does not stretch and does not slip on the pulley. The block has a velocity of 0.820 m/s toward the pulley when it passes through a photogate. (a) Use energy methods to predict its speed after it has moved to a second photogate, 0.700 m away. (b) Find the angular speed of the pulley at the same moment.

FIGURE P8.65

66. (a) Without the wheels, a bicycle frame has a mass of 8.44 kg. Each of the wheels can be roughly modeled as a

uniform solid disk with a mass of 0.820 kg and a radius of 0.343 m. Find the kinetic energy of the whole bicycle when it is moving forward at 3.35 m/s. (b) Before the invention of a wheel turning on an axle, ancient people moved heavy loads by placing rollers under them. (Modern people use rollers too. Any hardware store will sell you a roller bearing for a lazy susan.) A stone block of mass 844 kg moves forward at 0.335 m/s, supported by two uniform cylindrical tree trunks, each of mass 82.0 kg and radius 0.343 m. There is no slipping between the block and the rollers or between the rollers and the ground. Find the total kinetic energy of the moving objects.

67. In exercise physiology studies it is sometimes important to determine the location of a person's center of gravity. This can be done with the arrangement shown in Figure P8.67. A light plank rests on two scales, which read $F_{g1} = 380$ N and $F_{g2} = 320$ N. The scales are separated by a distance of 2.00 m. How far from the woman's feet is her center of gravity?

FIGURE P8.67

68. Two astronauts (Fig. P8.68), each having a mass of 75.0 kg, are connected by a 10.0-m rope of negligible mass. They are isolated in space, moving in circles around the point halfway between them at speeds of 5.00 m/s. Treating the astronauts as particles, calculate (a) the magnitude of the angular momentum and (b) the rotational energy of the system. By pulling on the rope, the astronauts shorten the distance between them to 5.00 m. (c) What is the new angular momentum of the system? (d) What are their new speeds? (e) What is the new rotational energy of the system? (f) How much work is done by the astronauts in shortening the rope?

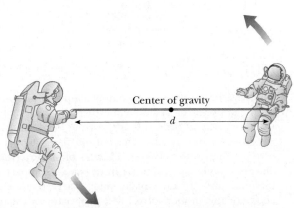

FIGURE P8.68 (Problems 68 and 69)

69. Two astronauts (Fig. P8.68), each having a mass *M*, are connected by a rope of length *d* having negligible mass. They are isolated in space, moving in circles around the point halfway between them at speeds *v*. Calculate (a) the magnitude of the angular momentum of the system by treating the astronauts as particles and (b) the rotational energy of the system. By pulling on the rope, the astronauts shorten the distance between them to *d*/2. (c) What is the new angular momentum of the system? (d) What are their new speeds? (e) What is the new rotational energy of the system? (f) How much work is done by the astronauts in shortening the rope?

70. Two window washers, Bob and Joe, are on a 3.00-m-long, 345-N scaffold supported by two cables attached to its ends. Bob weighs 750 N and stands 1.00 m from the left end, as shown in Figure P8.70. Two meters from the left end is the 500-N washing equipment. Joe is 0.500 m from the right end and weighs 1000 N. Given that the scaffold is in rotational and translational equilibrium, what are the forces on each cable?

—1.00 m— |
——— 2.00 m ——— | 0.500 m |
———————— 3.00 m ————————

FIGURE P8.70

71. We have all complained that there aren't enough hours in a day. In an attempt to change that, suppose that all the people in the world lined up at the Equator, and all started running east at 2.5 m/s relative to the surface of Earth. By how much would the length of a day increase? (Assume that there are 5.5×10^9 people in the world with an average mass of 70 kg each, and that Earth is a solid homogeneous sphere. In addition, you may use the result $1/(1 - x) \approx 1 + x$ for small *x*.)

72. In a circus performance, a large 5.0-kg hoop of radius 3.0 m rolls without slipping. If the hoop is given an angular speed of 3.0 rad/s while rolling on the horizontal and allowed to roll up a ramp inclined at 20° with the horizontal, how far (measured along the incline) does the hoop roll?

73. A uniform, solid cylinder of mass *M* and radius *R* rotates on a frictionless horizontal axle (Fig. P8.73). Two objects with equal masses hang from light cords wrapped around the cylinder. If the system is released from rest, find (a) the tension in each cord and (b) the acceleration of each object after the objects have descended a distance *h*.

FIGURE P8.73

74. Figure P8.74 shows a vertical force applied tangentially to a uniform cylinder of weight *w*. The coefficient of static friction between the cylinder and all surfaces is 0.500. Find, in terms of *w*, the maximum force **F** that can be applied without causing the cylinder to rotate. (*Hint:* When the cylinder is on the verge of slipping, both friction forces are at their maximum values. Why?)

FIGURE P8.74

GROUP ACTIVITIES

G.1 Compare the motion of an empty soup can and a filled soup can down the same incline such as a tilted table. If they are released from rest at the same height on the incline, which one reaches the bottom first? Repeat your observations with different kinds of soup (tomato, chicken noodle, etc.) and a can of beans. Compare their motions, and try to explain your observations. Finally, compare the motion of a filled soup can with a tennis ball, and explain your results to a friend.

G.2 Before attempting this exercise, review Example 8.13 dealing with the spinning stool. The techniques used here are similar to those used there. (a) First, make an estimate of the moment of inertia of your body. One way to do this would be to model your body as a solid cylinder and find *I* from $I = \frac{1}{2}MR^2$. You would have to determine your mass and to estimate your average "radius" for this approach. Can you think of an alternative way to estimate *I*? (b) Now, use the approach of Example 8.13 to measure

I. Sit on a rotating stool holding two weights (say two books), and determine the angular speed of rotation with the books extended and after they are pulled in. The angular speed can be found by counting the time for a given number of rotations. Use conservation of angular momentum to determine *I.* Do this five times to determine an average value for *I.* How well do your results for (a) and (b) compare, and if they differ greatly, what might cause the discrepancy?

G.3 This experiment demonstrates a simple way to find the center of gravity of an irregularly shaped object. Cut out an irregular shape from a piece of cardboard, and punch three to five holes around the edge of the shape. Put a pushpin through one of the holes, and tack the shape to a corkboard so that the shape can rotate freely. Now tie a weight to one end of a string and hang the other end of the string from the pushpin. When the string stops moving, trace a line on the cardboard that follows the string. Repeat this for each hole in the cardboard. You will find that there is a point where all the lines intersect. This is the center of gravity.

G.4 A painter climbs a ladder leaning against a smooth wall. At a certain height, the ladder is on the verge of slipping. (a) Explain why the force exerted by the vertical wall on the ladder is horizontal. (b) If the ladder of length *L* leans at an angle θ with the horizontal, what is the lever arm for this horizontal force with the axis of rotation taken at the base of the ladder? (c) If the ladder is uniform, what is the lever arm for the weight of the ladder? (d) Let the mass of the painter be 80 kg, $L = 4.0$ m, the ladder's mass be 30 kg, $\theta = 53°$, and the coefficient of friction between ground and ladder be 0.45. Find the maximum distance the painter can climb up the ladder.

G.5 A horizontal cylinder is fixed in place and allowed to rotate about an axis through its center. The cylinder has a radius of 1.00 m and moment of inertia 350 kg·m². It is given an angular acceleration by wrapping several turns of rope around it and pulling on the rope with a constant force of 120 N. Then, the rope is again wrapped around the cylinder, but this time the rope is passed over a pulley and attached to a hanging weight of 120 N. The weight is released and the cylinder again receives an angular acceleration. (a) Without doing any calculations, in which case is the angular acceleration the greatest, or will it be the same in both cases? Defend your answer. (b) Do the calculations to find the acceleration in each case.

G.6 The polar ice caps contain about 2.3×10^{19} kg of ice. This mass contributes essentially nothing to the moment of inertia of Earth because it is located at the poles, close to the axis of rotation. Estimate the change in the length of the day that would be expected if the polar ice caps were to melt and the water were distributed uniformly over the surface of Earth. (Note that the moment of inertia of a thin spherical shell of radius *r* and mass *m* is $\frac{2}{3}mr^2$.)

G.7 (a) Estimate the angular momentum your body has as a result of Earth's daily rotation about its north-south axis. (b) Estimate the angular momentum an automobile tire has about its axis of rotation while the car is driving along a highway at 55 miles per hour.

G.8 In the movie *Jurassic Park,* there is a scene in which some members of the visiting group are trapped in the kitchen with dinosaurs outside. The paleontologist is pressing against the center of the door, trying to keep out the dinosaurs who are on the other side. The botanist throws herself against the door at the edge near the hinge. A pivotal point in the film is that she cannot reach a gun on the floor because she is trying to hold the door closed. If the paleontologist is pressing at the center of the door, and the botanist at the edge about 8 cm from the hinge, estimate how far the paleontologist would have to relocate in order to have a greater effect on keeping the door closed than both of them pushing together have in their original positions.

(Problems 6, 7, and 8 are courtesy of Edward F. Redish. For more problems of this type, see http://www.physics.umd.edu/perg/)

9

Solids and Fluids

Chapter Outline

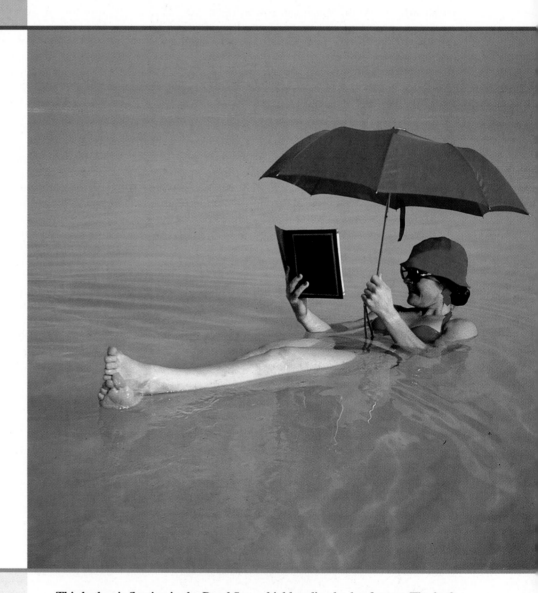

This bather is floating in the Dead Sea, a highly saline body of water. The bather floats in the Dead Sea with ease, because the upward buoyant force exerted by the salt water on the bather is much greater than it would be if the bather were floating in fresh water. The Dead Sea is actually a landlocked lake located between Israel and Jordan. At 395 m below sea level, it is the lowest body of water on Earth. Its salinity near the surface is about 30% and increases to about 33% at depths of 150 ft (46 m). The salinity of the Dead Sea is so high that except for bacteria, it cannot sustain animal or vegetable life. (© *Carl Purcell/Photo Researchers, Inc.*)

*I*n this chapter we consider some properties of solids and fluids (both liquids and gases). We spend some time looking at properties that are peculiar to solids, but much of our emphasis is on the properties of fluids. We take this approach because an understanding of the behavior of fluids is fundamentally important to students in the life sciences. We open the fluids part of the chapter with a study of fluids at rest and finish with a discussion of the properties of fluids in motion. Additional topics discussed in this chapter include surface tension, viscosity, and transport phenomena (diffusion and osmosis).

Crystals of natural quartz (SiO_2), one of the most common minerals on Earth. Quartz crystals are used to make special lenses and prisms and in certain electronic applications. *(Charles Winters)*

9.1 STATES OF MATTER

Matter is normally classified as being in one of three states—solid, liquid, or gas. Often this classification system is extended to include a fourth state, referred to as a plasma. When matter is heated to high temperatures, many of the electrons surrounding each atom are freed from the nucleus. The resulting substance is a collection of free, electrically charged particles—negatively charged electrons and positively charged ions. Such a highly ionized substance containing equal amounts of positive and negative charges is a **plasma.** Plasmas exist inside stars, for example. If we were to take a grand tour of our Universe, we would find that there is far more matter in the plasma state than in the more familiar solid, liquid, and gaseous states because there are far more stars around than any other form of celestial matter. In this chapter, however, we ignore plasmas and concentrate on the more familiar solid, liquid, and gaseous forms that make up the environment on our planet.

Everyday experience tells us that a solid has definite volume and shape. A brick, for example, maintains its familiar shape and size day in and day out. We also know that a liquid has a definite volume but no definite shape. For instance, when you fill the tank on a lawn mower, the gasoline changes its shape from that of the original container to that of the tank on the mower. If there is a gallon of gasoline before you pour, however, there still is a gallon after. Finally, a gas has neither definite volume nor definite shape.

All matter consists of some distribution of atoms or molecules. The atoms in a solid are held, by forces that are mainly electrical, at specific positions with respect to one another and vibrate about these equilibrium positions. At low temperatures, however, the vibrating motion is slight and the atoms can be considered essentially fixed. As energy is added to the material, the amplitude of the vibrations increases. A vibrating atom can be viewed as being bound in its equilibrium position by springs attached to neighboring atoms. A collection of such atoms and imaginary springs is shown in Figure 9.1. If a solid is compressed by external forces, we can picture the forces as compressing these tiny internal springs. When the external forces are removed, the solid tends to return to its original shape and size. Consequently, a solid is said to have *elasticity*.

Solids can be classified as either crystalline or amorphous. A **crystalline solid** is one in which the atoms have an ordered structure. For example, in the sodium chloride crystal (common table salt), sodium and chlorine atoms occupy alternate corners of a cube, as in Figure 9.2a. In an **amorphous solid,** such as glass, the atoms are arranged randomly, as in Figure 9.2b.

For any given substance, the liquid state exists at a higher temperature than the solid state. The intermolecular forces in a liquid are not strong enough to keep the molecules in fixed positions, and they wander through the liquid in a random fashion (Fig. 9.2c). Solids and liquids have the following property in common: When an attempt is made to compress them, strong repulsive atomic forces act internally to resist compression.

In the gaseous state, molecules are in constant random motion and exert only weak forces on each other. The average distance between the molecules of a gas is

FIGURE 9.1 A model of a portion of a solid. The atoms (spheres) are imagined as being attached to each other by springs, which represent the elastic nature of the interatomic forces. A solid consists of trillions of segments like this with springs connecting all of them.

FIGURE 9.2   (a) The NaCl structure, with the Na^+ (red) and Cl^- (blue) ions at alternate corners of a cube. (b) In an amorphous solid, the atoms are arranged randomly. (c) Erratic motion of a molecule in a liquid.

Webnote 9.1

For a good visualization of solids, liquids, and gases from a macroscopic and microscopic viewpoint, take a look at
http://intro.chem.okstate.edu/1225/ Lecture/Chapter13/State.html
http://intro.chem.okstate.edu/1225/ Lecture/Chapter13/Microstates.html

quite large compared with the size of the molecules. Occasionally the molecules collide with each other, but most of the time they move as nearly free, noninteracting particles. We shall say more about gases in subsequent chapters.

9.2   THE DEFORMATION OF SOLIDS

As mentioned in Section 9.1, we usually think of a solid as an object having definite shape and volume. In our study of mechanics, to keep things simple, we assumed that objects remain undeformed when external forces act on them. In reality, all objects are deformable. That is, it is possible to change the shape or size of an object (or both) through the application of external forces. When the forces are removed, the object tends to return to its original shape and size, which means that the deformation exhibits an elastic behavior.

The elastic properties of solids are discussed in terms of stress and strain. **Stress** is related to the force causing a deformation; **strain** is a measure of the degree of deformation. For sufficiently small stresses, stress is proportional to strain, and the constant of proportionality depends on the material being deformed and on the nature of the deformation. We call this proportionality constant the **elastic modulus:**

$$\text{Elastic modulus} = \frac{\text{stress}}{\text{strain}} \qquad [9.1]$$

The elastic modulus can be thought of as the stiffness of a material: A material having a large elastic modulus is very stiff and is difficult to deform.

YOUNG'S MODULUS: ELASTICITY IN LENGTH

FIGURE 9.3   A long bar clamped at one end is stretched by the amount ΔL under the action of a force **F**.

Consider a long bar of cross-sectional area A and length L_0, clamped at one end (Fig. 9.3). When an external force **F** is applied along the bar, perpendicularly to the cross section, internal forces in the bar resist the distortion ("stretching") that **F** tends to produce, but the bar nevertheless attains an equilibrium in which (1) its length is greater than L_0 and (2) the external force is balanced by internal forces. In such a situation the bar is said to be stressed. We define the **tensile stress** as the ratio of the magnitude of the external force F to the cross-sectional area A. The word "tensile" has the same root as the word "tension" and is used because the bar is under tension. The SI units of stress are newtons per square meter (N/m^2), which are also given a special name, the **pascal** (Pa):

The pascal

$$1\ Pa \equiv 1\ N/m^2 \qquad [9.2]$$

TABLE 9.1	Typical Values for Elastic Moduli		
Substance	Young's Modulus (Pa)	Shear Modulus (Pa)	Bulk Modulus (Pa)
Aluminum	7.0×10^{10}	2.5×10^{10}	7.0×10^{10}
Bone	1.8×10^{10}	8.0×10^{10}	—
Brass	9.1×10^{10}	3.5×10^{10}	6.1×10^{10}
Copper	11×10^{10}	4.2×10^{10}	14×10^{10}
Steel	20×10^{10}	8.4×10^{10}	16×10^{10}
Tungsten	35×10^{10}	14×10^{10}	20×10^{10}
Glass	$6.5–7.8 \times 10^{10}$	$2.6–3.2 \times 10^{10}$	$5.0–5.5 \times 10^{10}$
Quartz	5.6×10^{10}	2.6×10^{10}	2.7×10^{10}
Rib Cartilage	1.2×10^{7}	—	—
Rubber	0.1×10^{7}	—	—
Tendon	2×10^{7}	—	—
Water	—	—	0.21×10^{10}
Mercury	—	—	2.8×10^{10}

The **tensile strain** in this case is defined as the ratio of the change in length ΔL to the original length L_0 and is therefore a dimensionless quantity. Thus, we can use Equation 9.1 to define **Young's modulus** Y:

$$Y \equiv \frac{\text{tensile stress}}{\text{tensile strain}} = \frac{F/A}{\Delta L / L_0} = \frac{FL_0}{A\,\Delta L} \qquad \text{[9.3]}$$

A material having a large Young's modulus is difficult to stretch. This quantity is typically used to characterize a rod or wire stressed under *either tension or compression*. Note that because the strain is a dimensionless quantity, Y is in pascals. Typical values are given in Table 9.1. Experiments show that (1) the change in length for a fixed external force is proportional to the original length and (2) the force necessary to produce a given strain is proportional to the cross-sectional area.

It is possible to exceed the *elastic limit* of a substance by applying a sufficiently great stress (Fig. 9.4). At the *elastic limit*, the stress-strain curve departs from a straight line. A material subjected to a stress beyond this level ordinarily does not return to its original length when the external force is removed. As the stress is increased further, the material ultimately breaks.

SHEAR MODULUS: ELASTICITY OF SHAPE

Another type of deformation occurs when an object is subjected to a force **F** parallel to one of its faces while the opposite face is held fixed (Fig. 9.5a). If the object is originally a rectangular block, such a parallel force results in a shape whose cross section is a parallelogram. In this situation, the stress is called a shear stress. A book pushed sideways as in Figure 9.5b is under a shear stress. There is no change in volume with this deformation. We define the **shear stress** as F/A, the ratio of the magnitude of the parallel force to the area A of the face being sheared. The **shear strain** is the ratio $\Delta x/h$, where Δx is the horizontal distance the sheared face moves and h is the height of the object. In terms of these quantities, the **shear modulus** S is defined as

$$S \equiv \frac{\text{shear stress}}{\text{shear strain}} = \frac{F/A}{\Delta x/h} \qquad \text{[9.4]}$$

A material having a large shear modulus is difficult to bend.

Shear moduli for some representative materials are listed in Table 9.1. Note that the units of shear modulus are force per unit area (Pa).

FIGURE 9.4 Stress-versus-strain curve for an elastic solid.

FIGURE 9.5 (a) A shear deformation in which a rectangular block is distorted by a force applied tangent to one of its faces. (b) A book under shear stress.

BULK MODULUS: VOLUME ELASTICITY

Bulk modulus characterizes the response of a substance to uniform squeezing. Suppose that the external forces acting on an object are all perpendicular to the surface on which the force acts and distributed uniformly over the surface of the object (Fig. 9.6). This occurs when an object is immersed in a fluid. An object subject to this type of deformation undergoes a change in volume but no change in shape. The **volume stress** ΔP is defined as the ratio of the magnitude of the change in the applied force ΔF to the surface area A. (When dealing with fluids, we shall refer to the quantity F/A as the **pressure.**) The volume strain is equal to the change in volume ΔV divided by the original volume V. Thus, from Equation 9.1 we can characterize a volume compression in terms of the **bulk modulus** B, defined as

Bulk modulus ▶

FIGURE 9.6 When a solid is under uniform pressure, it undergoes a change in volume but no change in shape. This cube is compressed on all sides by forces normal to its six faces.

$$B \equiv \frac{\text{volume stress}}{\text{volume strain}} = -\frac{\Delta F/A}{\Delta V/V} = -\frac{\Delta P}{\Delta V/V} \qquad [9.5]$$

A material having a large bulk modulus does not compress easily. Note that a negative sign is included in this defining equation so that B is always positive. An increase in pressure (positive ΔP) causes a decrease in volume (negative ΔV), and vice versa.

Table 9.1 lists bulk modulus values for some materials. If you look up such values in a different source, you will often find that the reciprocal of the bulk modulus, called the **compressibility** of the material, is listed. Note from Table 9.1 that both solids and liquids have bulk moduli. However, there is no shear modulus and no Young's modulus for liquids because a liquid will not sustain a shearing stress or a tensile stress. (It flows instead.)

Example 9.1   Built to Last

A vertical steel beam in a building supports a load of 6.0×10^4 N. If the length of the beam is 4.0 m and its cross-sectional area is 8.0×10^{-3} m^2, find the distance it is compressed along its length.

Solution   Because the beam is under compression, we can use Equation 9.3. Taking Young's modulus for steel, $Y = 20 \times 10^{10}$ Pa, from Table 9.1, we have

$$Y = \frac{FL_0}{A \, \Delta L}$$

or

$$\Delta L = \frac{FL_0}{YA} = \frac{(6.0 \times 10^4 \text{ N})(4.0 \text{ m})}{(20 \times 10^{10} \text{ Pa})(8.0 \times 10^{-3} \text{ m}^2)} = \boxed{1.5 \times 10^{-4} \text{ m}}$$

Example 9.2   Squeezing a Lead Sphere

A solid lead sphere of volume 0.50 m^3 is dropped in the ocean to a depth of about 2 000 m (about 1 mile) where the pressure increases by 2.0×10^7 Pa. Lead has a bulk modulus of 7.7×10^9 Pa. What is the change in volume of the sphere?

Solution   From the definition of bulk modulus (Eq. 9.5), we have

$$B = -\frac{\Delta P}{\Delta V/V}$$

$$\Delta V = -\frac{V \Delta P}{B}$$

When the sphere is at the surface, where its volume is 0.50 m³, the pressure exerted on it is atmospheric pressure. The increase in pressure ΔP when the sphere is submerged is 2.0×10^7 Pa. Therefore, the change in its volume when it is submerged is

$$\Delta V = -\frac{(0.50 \text{ m}^3)(2.0 \times 10^7 \text{ Pa})}{7.7 \times 10^9 \text{ Pa}} = \boxed{-1.3 \times 10^{-3} \text{ m}^3}$$

The negative sign indicates a *decrease* in volume.

Arches and the Ultimate Strength of Materials

◀ **APPLICATION**

ARCH STRUCTURES IN BUILDINGS

The **ultimate strength** of a material is the maximum force per unit area the material can withstand before it breaks or fractures. Such values are of great importance, particularly in the construction of buildings, bridges, and roads. Table 9.2 gives the ultimate strength of a variety of materials under both tension and compression. Note that both bone and a variety of building materials (concrete, brick and marble) are stronger in compression than in tension. The greater ability of brick and stone to resist compression is the basis of the semicircular arch, developed and used extensively by the Romans in everything from memorial arches, to expansive temples, to aqueduct supports.

Before the development of the arch, the principal method of spanning a space was the simple post-and-beam construction (Fig. 9.7a), in which a horizontal beam is supported by two columns. This type of construction was used to build the great Greek temples. The columns of these temples are closely spaced because of the limited length of available stones and also because of the low ultimate tensile strength of a sagging stone beam.

TABLE 9.2	Ultimate Strength of Materials	
Material	**Tensile Strength (N/m²)**	**Compressive Strength (N/m²)**
Iron	1.7×10^8	5.5×10^8
Steel	5.0×10^8	5.0×10^8
Aluminum	2.0×10^8	2.0×10^8
Bone	1.2×10^8	1.5×10^8
Marble	—	8.0×10^7
Brick	1×10^6	3.5×10^7
Concrete	2×10^6	2×10^7

Post-and-beam
(a)

Semicircular arch (Roman)
(b)

Gothic arch

Flying buttress

Flying buttress

Pointed arch (Gothic)
(c)

FIGURE 9.7 (a) A simple post-and-beam structure. (b) The semicircular arch developed by the Romans. (c) Gothic arch with flying buttresses to provide lateral support.

The semicircular arch (Fig. 9.7b) developed by the Romans was a great technological achievement in architectural design. It effectively allowed the heavy load of a wide roof span to be channeled into horizontal and vertical forces on narrow supporting columns. The stability of this arch depends on the compression between its wedge-shaped stones. The stones are forced to squeeze against each other by the uniform loading as shown in Figure 9.7b. This results in horizontal outward forces at the base of the arch where it starts curving away from the vertical, which must be balanced by the stone walls shown on the sides of the arch. It is common to use very heavy walls (buttresses) on either side of the arch to provide horizontal stability. If the foundation of the arch should move, the compressive forces between the wedge-shaped stones may decrease to the extent that the arch collapses. The stone surfaces used in the arches constructed by the Romans were cut to make a very tight joint; it is interesting to note that mortar was usually not used in these joints. The resistance to slipping between stones was provided by the compression force and the friction between the stone faces.

Another important architectural innovation was the pointed Gothic arch shown in Figure 9.7c. This type of structure was first used in Europe beginning in the 12th century, followed by the construction of several magnificent Gothic cathedrals in France in the 13th century. One of the most striking features of these cathedrals is their extreme height. For example, the cathedral at Chartres rises to 118 ft and the one at Reims has a height of 137 ft. Such magnificent buildings evolved over a very short period of time, without the benefit of any mathematical theory of structures. However, Gothic arches required flying buttresses to prevent the spreading of the arch supported by the tall, narrow columns.

9.3   DENSITY AND PRESSURE

The density of a substance of uniform composition is defined as its mass per unit volume.

In symbolic form, a substance of mass M and volume V has a density ρ (Greek rho), given by

Density ▷

$$\rho \equiv \frac{M}{V} \qquad\qquad [9.6]$$

The units of density are kilograms per cubic meter in SI units and grams per cubic centimeter in the cgs system. Table 9.3 lists the densities of some substances.

TABLE 9.3	Density of Some Common Substances		
Substance	$\rho(\mathrm{kg/m^3})^a$	**Substance**	$\rho(\mathrm{kg/m^3})^a$
Ice	0.917×10^3	Water	1.00×10^3
Aluminum	2.70×10^3	Glycerin	1.26×10^3
Iron	7.86×10^3	Ethyl alcohol	0.806×10^3
Copper	8.92×10^3	Benzene	0.879×10^3
Silver	10.5×10^3	Mercury	13.6×10^3
Lead	11.3×10^3	Air	1.29
Gold	19.3×10^3	Oxygen	1.43
Platinum	21.4×10^3	Hydrogen	8.99×10^{-2}
Uranium	18.7×10^3	Helium	1.79×10^{-1}

^{a} All values are at standard atmospheric temperature and pressure
   (STP), defined as 0°C (273 K) and 1 atm (1.013×10^5 Pa). To convert
   to grams per cubic centimeter, multiply by 10^{-3}.

The densities of most liquids and solids vary slightly with changes in temperature and pressure; the densities of gases vary greatly with such changes. Note that under normal conditions the densities of solids and liquids are about 1 000 times greater than the densities of gases. This difference implies that the average spacing between molecules in a gas under these conditions is about ten times greater than that in a solid or liquid.

The **specific gravity** of a substance is the ratio of its density to the density of water at 4°C, which is 1.0×10^3 kg/m^3. (Note that the size of the kilogram was originally defined to make the density of water 1.0×10^3 kg/m^3 at 4°C.) By definition, specific gravity is a dimensionless quantity. For example, if the specific gravity of a substance is 3.0, its density is $3.0(1.0 \times 10^3$ kg/m$^3) = 3.0 \times 10^3$ kg/m^3.

We have seen that fluids do not sustain shearing stresses, and thus the only stress that can exist on an object submerged in a fluid is one that tends to compress the object. The force exerted by the fluid on the object is always perpendicular to the surfaces of the object, as shown in Figure 9.8a.

The pressure at a specific point in a fluid can be measured with the device pictured in Figure 9.8b—an evacuated cylinder enclosing a light piston connected to a spring. As the device is submerged in a fluid, the fluid presses down on the top of the piston and compresses the spring until the inward force exerted by the fluid is balanced by the outward force exerted by the spring. The fluid pressure can be measured directly if the spring is calibrated in advance. This is accomplished by applying a known force to the spring to compress it a given distance. If F is the magnitude of the force exerted by the fluid on the piston and A is the area of the piston, then **the average pressure P in the fluid at the level to which the device has been submerged is defined as the ratio of force to area:**

$$P \equiv \frac{F}{A} \qquad [9.7]$$

Because pressure is defined as force per unit area, it has units of pascals (newtons per square meter).

As you can see from this definition, we need to know the magnitude of the force exerted on a surface *and* the area over which that force is applied in order to determine the effect of pressure. For example, a 700-N man can stand on a vinyl-covered floor in regular street shoes without damaging the surface. If he wears golf shoes with numerous metal cleats protruding from each sole, he does considerable damage to the floor. In both cases the net force applied to the floor is 700 N. However, when the man wears ordinary shoes, the area of his contact with the floor is considerably larger than when he wears golf shoes. (In the latter case, the only area in contact with the floor is the sum of the small cross-sectional areas of the ends of the metal cleats.) Hence, the *pressure* on the floor is much smaller when he wears ordinary shoes.

Snowshoes use this principle (Fig. 9.9). The snow exerts an upward normal force on the shoes to support the weight of the person. According to Newton's third law, this upward force is accompanied by a downward force exerted by the

(b)

FIGURE 9.8 (a) The force exerted by a fluid on a submerged object at any point is perpendicular to the surface of the object. The force exerted by the fluid on the walls of the container is perpendicular to the walls at all points and increases with depth. (b) A simple device for measuring pressure in a fluid.

◀ Average pressure

Tip 9.1 FORCE AND PRESSURE

Equation 9.7 makes a clear distinction between force and pressure. Another important distinction is that *force is a vector* and *pressure is a scalar.* There is no direction associated with pressure, but the direction of the force associated with the pressure is perpendicular to the surface of interest.

FIGURE 9.9 Snowshoes prevent the person from sinking into the soft snow because the force on the snow is spread over a larger area, which reduces the pressure on the snow's surface. (*Earl Young/FPG/Getty Images*)

shoes on the snow. If the person is wearing snowshoes, this force is distributed over the very large areas of the snowshoes so that the pressure at any given point is relatively low and the person does not penetrate very deeply into the snow.

Quick Quiz 9.1

Suppose someone in a crowded room accidentally stomps on your foot with the heel of one shoe. Would you be better off if that person were (a) a basketball player wearing sneakers or (b) a petite woman wearing spike-heeled shoes?

APPLYING PHYSICS 9.1

The daring physics professor, after a long lecture, stretches out for a nap on a bed of nails, as in Figure 9.10. How is this possible?

FIGURE 9.10   (Applying Physics 9.1) Does anyone have a pillow?

Explanation   If you try to support your entire weight on a single nail, the pressure on your body is your weight divided by the very small area of the end of the nail. This pressure is sufficiently large to penetrate the skin. However, if you distribute your weight over several hundred nails, as the professor is doing, the pressure is considerably reduced because the area that supports your weight is the total area of all nails in contact with your body. (Note that lying on a bed of nails is much more comfortable than sitting on the bed. Why? Extend the logic to show that it would be more uncomfortable yet to stand on a bed of nails without shoes.)

Example 9.3   The Water Bed

A water bed is 2.00 m on a side and 30.0 cm deep.

A Find its weight.

Solution   Because the density of water is 1 000 kg/m^3 (Table 9.3) and the bed's volume is $(2.00 \times 2.00 \times 0.300)$ m^3 = 1.20 m^3, the mass of the bed is

$$M = \rho V = (1\ 000 \text{ kg/m}^3)(1.20 \text{ m}^3) = 1.20 \times 10^3 \text{ kg}$$

and its weight is

$$w = Mg = (1.20 \times 10^3 \text{ kg})(9.80 \text{ m/s}^2) = \boxed{1.18 \times 10^4 \text{ N}}$$

This is equivalent to approximately 2 640 lb. In order to support such a heavy load, you would be well advised to keep your water bed in the basement or on a sturdy, well-supported floor.

B Find the pressure that the water bed exerts on the floor. Assume that the entire lower surface of the bed makes contact with the floor.

Solution The weight of the water bed is 1.18×10^4 N. Its cross-sectional area is 4.00 m². The pressure exerted on the floor is therefore

$$P = \frac{1.18 \times 10^4 \text{ N}}{4.00 \text{ m}^2} = \boxed{2.95 \times 10^3 \text{ Pa}}$$

EXERCISE Calculate the pressure exerted by the bed on the floor if the bed rests on its side.

ANSWER Because the area of the bed's side is 0.600 m², the pressure is 1.96×10^4 Pa.

9.4 VARIATION OF PRESSURE WITH DEPTH

If a fluid is at rest in a container, *all portions* of the fluid must be in static equilibrium (at rest with respect to the observer). Furthermore, **all points at the same depth must be at the same pressure.** If this were not the case, a given portion of the fluid would not be in equilibrium. For example, consider the small block of fluid shown in Figure 9.11a. If the pressure were greater on the left side of the block than on the right, $\mathbf{F}_1$ would be greater than $\mathbf{F}_2$, and the block would accelerate and thus would not be in equilibrium.

Now let us examine the portion of the fluid (assumed to be liquid) contained within the volume indicated by the darker region in Figure 9.11b. This region has a cross-sectional area A and extends to a depth h below the surface of the liquid. Three external forces act on this volume of liquid: the force of gravity Mg; the upward force PA exerted by the liquid below it; and a downward force P_0A exerted by the atmosphere, where P_0 is atmospheric pressure. Because this volume of liquid is in equilibrium, these forces must add to zero, and so we get

$$PA - Mg - P_0A = 0 \qquad \text{[9.8]}$$

From the relation $M = \rho V = \rho Ah$, the weight of the fluid in the volume is

$$w = Mg = \rho gAh \qquad \text{[9.9]}$$

When Equation 9.9 is substituted into Equation 9.8, we get

$$\boxed{P = P_0 + \rho gh} \qquad \text{[9.10]}$$

where normal atmospheric pressure at sea level is $P_0 = 1.013 \times 10^5$ Pa (equivalent to 14.7 lb/in.²). According to Equation 9.10, **the pressure P at a depth h below the surface of a liquid open to the atmosphere is greater than atmospheric pressure by the amount ρgh.** Moreover, the pressure is not affected by the shape of the vessel, as shown in Figure 9.12.

For example, we can use Equation 9.10 to calculate the pressure at an ocean depth of 1 000 m as follows:

$$P = P_0 + \rho gh$$
$$= 1.013 \times 10^5 \text{ Pa} + (1.0 \times 10^3 \text{ kg/m}^3)(9.80 \text{ m/s}^2)(1.00 \times 10^3 \text{ m})$$
$$\approx 9.9 \times 10^6 \text{ Pa}$$

This is approximately 100 times greater than atmospheric pressure! Obviously, the design and construction of vessels that can withstand such enormous pressures are not trivial matters.

In view of the fact that the pressure in a fluid depends on depth and on the value of P_0, any increase in pressure at the surface must be transmitted to every

FIGURE 9.11 (a) If the block of fluid is to be in equilibrium, the force $\mathbf{F}_1$ must balance the force $\mathbf{F}_2$. (b) The net force on the volume of liquid within the darker region must be zero.

FIGURE 9.12 This photograph illustrates that the pressure in a liquid is the same at all points having the same elevation. For example, the pressure is the same at points *A, B, C,* and *D*. Note that the shape of the vessel does not affect the pressure. *(Courtesy of Central Scientific Company)*

(a)

(b)

FIGURE 9.13 (a) Diagram of a hydraulic press. Because the pressure is the same at the left and right sides, a small force $\mathbf{F}_1$ at the left produces a much larger force $\mathbf{F}_2$ at the right. (b) A vehicle under repair is supported by a hydraulic lift in a garage. *(David Frazier)*

point in the fluid. This was first recognized by the French scientist Blaise Pascal (1623–1662) and is called **Pascal's principle:**

> **A change in pressure applied to an enclosed fluid is transmitted undiminished to every point of the fluid and to the walls of the container.**

APPLICATION

HYDRAULIC LIFTS

An important application of Pascal's principle is the hydraulic press (Fig. 9.13a). A downward force $\mathbf{F}_1$ is applied to a small piston of area A_1. The pressure is transmitted through a fluid to a larger piston of area A_2. Because the pressure is the same on both sides, we see that $P = F_1/A_1 = F_2/A_2$. Therefore, the magnitude of the force $\mathbf{F}_2$ is larger than the magnitude of $\mathbf{F}_1$ by the factor A_2/A_1. That is why a large load, such as a car, can be supported on the large piston by a much smaller force on the smaller piston. Hydraulic brakes, car lifts, hydraulic jacks, forklifts, and other machines make use of this principle.

Webnote 9.2

For more on the life of Blaise Pascal, go to
http://www.eingang.org/Lecture/pascal.html
http://norfacad.pvt.k12.va.us/project/pascal/blaise3.htm

APPLYING **PHYSICS** 9.2

A corollary to the statement that pressure in a fluid increases with depth is that water always seeks its own level. This means that if a vessel is filled with water, then regardless of the vessel's shape the surface of the water is perfectly flat and at the same height at all points. The ancient Egyptians used this fact to level the pyramids. Devise a scheme showing how this could be done.

(a)   (b)   (c)   (d)

FIGURE 9.14 (Applying Physics 9.2)

Explanation There are many ways this could have been done, but Figure 9.14 shows the scheme used by the Egyptians. The builders cut grooves in the base of the pyramid as in (a) and partially filled the grooves with water. The height of the water was marked as in (b) and the rock was chiseled down to this line (c). Finally, the groove was filled with crushed rock and gravel, as in (d).

9.5 PRESSURE MEASUREMENTS

A simple device for measuring pressure is the open-tube manometer (Fig. 9.15a). One end of a U-shaped tube containing a liquid is open to the atmosphere, and the other end is connected to a system of unknown pressure P. The pressure at point B equals $P_0 + \rho g h$, where ρ is the density of the fluid. The pressure at B, however, equals the pressure at A, which is also the unknown pressure P. Therefore, we conclude that

$$P = P_0 + \rho g h$$

The pressure P is called the *absolute pressure*, and $P - P_0$ is called the *gauge pressure*. Thus, if P in the system is greater than atmospheric pressure, h is positive. If P is less than atmospheric pressure (a partial vacuum), h is negative, meaning that the right-hand column is lower than the left-hand column.

Another instrument that is used to measure pressure is the *barometer* (Fig. 9.15b), invented by Evangelista Torricelli (1608–1647). A long tube that is closed at one end is filled with mercury and then inverted into a dish of mercury. The closed end of the tube is nearly a vacuum, and so its pressure can be taken as zero. It follows that $P_0 = \rho g h$, where ρ is the density of the mercury and h is the height of the mercury column. Note that the barometer measures the pressure of the atmosphere, whereas the manometer measures pressure in an enclosed fluid.

One atmosphere of pressure is defined to be the pressure equivalent of a column of mercury that is exactly 0.76 m in height at 0°C with $g = 9.806\ 65$ m/s². At this temperature mercury has a density of 13.595×10^3 kg/m³; therefore,

$$P_0 = \rho g h = (13.595 \times 10^3 \text{ kg/m}^3)(9.806\ 65 \text{ m/s}^2)(0.760\ 0 \text{ m})$$
$$= 1.013 \times 10^5 \text{ Pa} = 1 \text{ atm}$$

It is interesting to note that the force of the atmosphere on our bodies (assuming a body area of 2 000 in.²) is extremely large, on the order of 30 000 lb! A natural question to raise is, How can we exist under such great forces attempting to collapse our bodies? The answer is that our body cavities and tissues are permeated with fluids that are pushing outward with this same atmospheric pressure. Consequently, our bodies are in equilibrium under the force of the atmosphere pushing in and an equal internal force pushing out.

BLOOD PRESSURE MEASUREMENTS

A specialized manometer (called a sphygmomanometer) is often used to measure blood pressure. In this application, a rubber bulb forces air into a cuff wrapped tightly around the upper arm and simultaneously into a manometer as in Figure 9.16. The pressure in the cuff is increased until the flow of blood through the brachial artery in the arm is stopped. A valve on the bulb is then opened, and the measurer listens with a stethoscope to the artery at a point just below the cuff. When the pressure in the cuff and brachial artery is just below the maximum value produced by the heart (the systolic pressure), the artery opens momentarily on each beat of the heart. At this point, the velocity of the blood is high and turbulent, and the flow is noisy and can be heard with the stethoscope. The manometer is calibrated to read the pressure in millimeters of mercury, and the value obtained is about 120 mm for a normal heart. Values of 140 mm or above are considered to be high, and medication to lower the blood pressure is often prescribed.

(a)

(b)

FIGURE 9.15 Two devices for measuring pressure: (a) an open-tube manometer and (b) a mercury barometer.

Manometer

Rubber bulb

Stethoscope

Cuff

FIGURE 9.16 A sphygmomanometer can be used to measure blood pressure.

APPLICATION

MEASURING BLOOD PRESSURE

Webnote 9.3
You can make a water barometer using some simple items found at home. Visit
http://kids.earth.nasa.gov/archive/ air_pressure/barometer.html

As the pressure in the cuff is lowered further, intermittent sounds are still heard until the pressure falls just below the minimum heart pressure (the diastolic pressure). At this point, continuous sounds are heard. In the normal heart, this transition occurs at about 80 mm of mercury, and values above 90 are considered to need medical intervention. Blood pressure readings are usually expressed as the ratio systolic/diastolic or 120/80 for a healthy heart.

Quick Quiz 9.2

Blood pressure is normally measured with the cuff of the sphygmomanometer around the arm. Suppose that the blood pressure were measured with the cuff around the calf of the leg of a standing person. Would the reading of the blood pressure be (a) the same here as it is for the arm? (b) less than it is for the arm? or (c) greater than it is for the arm?

APPLYING PHYSICS 9.3

In a ball-point pen, ink moves down a tube to the tip where it is spread on a sheet of paper by a rolling stainless steel ball. Near the top of the ink cartridge, there is a small hole open to the atmosphere. If you seal this hole, you will find that the pen no longer works. Use your knowledge of how a barometer works to explain this behavior.

APPLICATION

BALL-POINT PENS

Explanation If the hole is sealed, or if it were not present, the pressure of the air above the ink would decrease as some ink is used. Consequently, atmospheric pressure exerted against the ink at the bottom of the cartridge would prevent some of the ink from flowing out. The hole allows the pressure above the ink to remain at atmospheric pressure. Why does a ball-point pen seem to run out of ink when you write on a vertical surface?

Example 9.4   The Car Lift

In a car lift used in a service station, compressed air exerts a force on a small piston of circular cross section having a radius of 5.00 cm. This pressure is transmitted by an incompressible liquid to a second piston of radius 15.0 cm. (a) What force must the compressed air exert in order to lift a car weighing 13 300 N? (b) What air pressure will produce this force? (c) Show that the input energy transfer is equal in magnitude to the output energy transfer.

FIGURE 9.17   (Example 9.4)

Solution

A Because the pressure exerted by the compressed air is transmitted undiminished throughout the liquid, we have

$$F_1 = \left(\frac{A_1}{A_2}\right) F_2 = \frac{\pi(5.00 \times 10^{-2}\,\text{m})^2}{\pi(15.0 \times 10^{-2}\,\text{m})^2}\,(1.33 \times 10^4\,\text{N}) = \boxed{1.48 \times 10^3\,\text{N}}$$

B The air pressure that will produce this force is

$$P = \frac{F_1}{A_1} = \frac{1.48 \times 10^3\,\text{N}}{\pi(5.00 \times 10^{-2}\,\text{m})^2} = \boxed{1.88 \times 10^5\,\text{Pa}}$$

This pressure is approximately twice atmospheric pressure.

C The energy transfer input and output are by means of work done by the forces as the piston moves. To determine the work done, we must find the magnitude of the displacement through which each force acts. Because we assume the liquid to be incompressible, the volume of the cylinder through which the input piston moves must equal that through which the output piston moves (see Fig. 9.17). The lengths of these cylinders are the magnitudes Δx_1 and Δx_2 of the displacements of the forces. Setting the volumes equal, we have $V_1 = V_2$, or $A_1\,\Delta x_1 = A_2\,\Delta x_2$. Therefore,

$$\frac{A_1}{A_2} = \frac{\Delta x_2}{\Delta x_1}$$

Evaluating the ratio of the input work to the output work, we find

$$\frac{W_1}{W_2} = \frac{F_1\,\Delta x_1}{F_2\,\Delta x_2} = \left(\frac{F_1}{F_2}\right)\left(\frac{\Delta x_1}{\Delta x_2}\right) = \left(\frac{A_1}{A_2}\right)\left(\frac{A_2}{A_1}\right) = 1$$

which verifies that the work input and output are the same, as they must be to conserve energy.

9.6 BUOYANT FORCES AND ARCHIMEDES'S PRINCIPLE

A fundamental principle affecting objects submerged in fluids was discovered by the Greek mathematician Archimedes. **Archimedes's principle** can be stated as follows:

> **Any object completely or partially submerged in a fluid is buoyed up by a force whose magnitude is equal to the weight of the fluid displaced by the object.**

Everyone has experienced Archimedes's principle. For example, recall that it is relatively easy to lift someone if you are both standing in a swimming pool, whereas lifting that same individual on dry land would be a difficult task. Water provides partial support to any object placed in it. We say that an object placed in a fluid is buoyed up by the fluid, and we call this upward force the **buoyant force.** The buoyant force is *not* a mysterious new force that arises in fluids. As we shall see, the physical cause of the buoyant force is the pressure difference between the top and bottom of the object. According to Archimedes's principle, **the magnitude of the buoyant force always equals the weight of the fluid displaced by the object.** The buoyant force acts vertically upward through what was the center of gravity of the fluid before the fluid was displaced.

Archimedes's principle can be verified in the following manner. Suppose we focus our attention on the cube of water that is colored red in the container of Figure 9.18. This cube of water is in equilibrium under the action of the forces on it.

ARCHIMEDES , GREEK MATHEMATICIAN, PHYSICIST, AND ENGINEER (287 – 212 B.C.)

Archimedes was perhaps the greatest scientist of antiquity. He is well known for discovering the nature of the buoyant force and was a gifted inventor. According to legend, Archimedes was asked by King Hieron to determine whether the king's crown was made of pure gold or merely a gold alloy. The task was to be performed without damaging the crown. Archimedes allegedly arrived at a solution while taking a bath, noting a partial loss of weight after submerging his arms and legs in the water. As the story goes, he was so excited about his great discovery that he ran naked through the streets of Syracuse shouting, "Eureka!" which is Greek for "I have found it."

◀ Archimedes's principle

FIGURE 9.18 The external forces on a cube of water are the force of gravity $\mathbf{w} = \mathbf{F}_g$ and the buoyant force $\mathbf{B}$. Under equilibrium conditions, $B = w$.

Tip 9.2 | BUOYANT FORCE IS EXERTED BY THE FLUID

Note that the buoyant force is exerted by the fluid. If several objects of different densities but the same volume are immersed in a fluid, they will all experience the *same* buoyant force. Whether they sink or float is determined by the relationship between the buoyant force and the weight.

Hot-air balloons. Because hot air is less dense than cold air, there is a net upward force on the balloons.
(© Sylvester Allred/Visuals Unlimited)

One of those forces is the force of gravity. What cancels that downward force? Apparently, the water beneath the cube is buoying it up and holding it in equilibrium. Thus, the buoyant force $\mathbf{B}$ on the cube of water must be exactly equal in magnitude to the weight of the water inside the cube: $B = w$.

Now imagine that the cube of water is replaced by a cube of steel of the same dimensions. What is the buoyant force on the steel? The water surrounding a cube behaves in the same way whether the cube is made of water or steel; therefore, **the buoyant force acting on the steel is the same as the buoyant force acting on a cube of water of the same dimensions.** This result applies for a totally submerged object of any shape, size, or density.

Let us show explicitly that the buoyant force is equal in magnitude to the weight of the displaced fluid. The pressure at the bottom of the cube in Figure 9.18 is greater than the pressure at the top by the amount $\rho_{\text{fluid}}gh$, where ρ_{fluid} is the density of the fluid and h is the height of the cube. Because the pressure difference ΔP is equal to the buoyant force per unit area, or, $\Delta P = B/A$, we see that $B = (\Delta P)(A) = (\rho_{\text{fluid}}gh)(A) = \rho_{\text{fluid}}gV$, where V is the volume of the cube. Because the mass of the fluid in the cube is $M = \rho_{\text{fluid}}V$, we see that

$$B = \rho_{\text{fluid}}Vg = Mg = w_{\text{fluid}} \qquad [9.11]$$

where w_{fluid} is the weight of the displaced fluid.

The weight of the submerged object is $w_{\text{obj}} = mg = \rho_{\text{obj}}Vg$, where ρ_{obj} is the density of the object. Because $w_{\text{fluid}} = \rho_{\text{fluid}}Vg$ and $w_{\text{obj}} = \rho_{\text{obj}}Vg$, we see that, if the density of the object is greater than the density of the fluid, the unsupported object sinks. If the density of the object is less than that of the fluid, the unsupported submerged object accelerates upward and ultimately floats. When a *floating* object is in equilibrium, only part of it is submerged. In this case, the magnitude of the buoyant force equals the weight of the object.

It is instructive to compare the forces on a totally submerged object with those on a floating object.

Case I: A Totally Submerged Object When an object is *totally* submerged in a fluid of density ρ_{fluid}, the upward buoyant force has a magnitude of $B = \rho_{\text{fluid}}V_{\text{obj}}g$, where V_{obj} is the volume of the object. If the object has density ρ_{obj}, the downward gravitational force has a magnitude equal to $w = mg = \rho_{\text{obj}}V_{\text{obj}}g$, and the net force on it is $B - w = (\rho_{\text{fluid}} - \rho_{\text{obj}})V_{\text{obj}}g$. Hence, if the density of the object is less than the density of the fluid, as in Figure 9.19a, the net force is positive (upward) and the unsupported object accelerates upward. If the density of the object is greater than the density of the fluid, as in Figure 9.19b, the net force is negative and the unsupported object accelerates downward.

Case II: A Floating Object Now consider an object in static equilibrium floating on a fluid—that is, a partially submerged object. In this case, the upward buoyant force is balanced by the downward force of gravity acting on the object. If V_{fluid} is the volume of the fluid displaced by the object (which corresponds to the volume

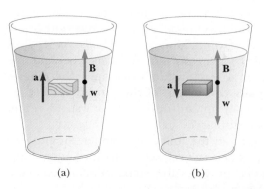

(a)   (b)

FIGURE 9.19 (a) A totally submerged object that is less dense than the fluid in which it is submerged experiences a net upward force. (b) A totally submerged object that is denser than the fluid sinks.

of the part of the object beneath the fluid level), then the magnitude of the buoyant force is given by $B = \rho_{\text{fluid}} V_{\text{fluid}} g$. Because the weight of the object is $w = mg = \rho_{\text{obj}} V_{\text{obj}} g$, and $w = B$, we see that $\rho_{\text{fluid}} V_{\text{fluid}} g = \rho_{\text{obj}} V_{\text{obj}} g$, or

$$\frac{\rho_{\text{obj}}}{\rho_{\text{fluid}}} = \frac{V_{\text{fluid}}}{V_{\text{obj}}} \qquad [9.12]$$

Under normal conditions, the average density of a fish is slightly greater than the density of water. This being the case, a fish would sink if it did not have a mechanism for adjusting its density: the internal regulation of the size of the swim bladder. In this manner, fish maintain neutral buoyancy as they swim to various depths.

The human brain is immersed in a fluid (the cerebrospinal fluid) of density 1 007 kg/m³, which is slightly less than the average density of the brain, 1 040 kg/m³. Consequently, most of the weight of the brain is supported by the buoyant force of the surrounding fluid. In some clinical procedures, it is necessary to remove a portion of this fluid for diagnostic purposes. During such procedures, the nerves and blood vessels in the brain are placed under great strain, which in turn can cause extreme discomfort and pain. Great care must be exercised with such patients until the initial brain fluid volume has been restored by the body.

When service station attendants check the antifreeze in your car or the condition of your battery, they often use devices that apply Archimedes's principle. Figure 9.20 shows a common device that is used to check the antifreeze in a car radiator. The small balls in the enclosed tube vary in density so that all of them float when the tube is filled with pure water, none float in pure antifreeze, one floats in a 5% mixture, two in a 10% mixture, and so forth. The number of balls that float thus serves as a measure of the percentage of antifreeze in the mixture, which in turn is used to determine the lowest temperature the mixture can withstand without freezing.

Similarly, the charge of some car batteries can be determined with a so-called "magic-dot" process that is built into the battery (Fig. 9.21). When we look down into a viewing port in the top of the battery, a red dot indicates that the battery is sufficiently charged; a black dot indicates that the battery has lost its charge. This is because, if the battery has sufficient charge, the density of the battery fluid is high enough to cause the red ball to float. As the battery loses its charge, the density of the battery fluid decreases and the ball sinks beneath the surface of the fluid, where the dot appears black.

APPLICATION

SWIM BLADDERS IN FISH

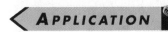
APPLICATION

CEREBROSPINAL FLUID

APPLICATION

TESTING YOUR CAR'S ANTIFREEZE

APPLICATION

CHECKING THE BATTERY CHARGE

Tip 9.3   HEAVIER THAN AIR?

Whether an object sinks or floats is determined by a comparison of the fluid and object *densities*. In the case of aircraft, we often hear descriptive phrases such as "heavier-than-air" (for an airplane) or "lighter than air" (for a blimp). Despite the popularity of these phrases, try to train yourself to say "denser than air" and "less dense than air."

FIGURE 9.20 The number of balls that float in this device is a measure of the density of the antifreeze solution in a vehicle's radiator, and consequently a measure of the temperature at which freezing will occur.

Charged battery   Discharged battery

FIGURE 9.21 The red ball in the plastic tube inside the battery serves as an indicator of whether the battery is (a) charged or (b) discharged. As the battery loses its charge, the density of the battery fluid decreases, and the ball sinks out of sight.

Most of the volume of this iceberg is beneath the water. Can you determine what fraction of the total volume is underwater? *(Geraldine Prentice/Tony Stone Images)*

Quick Quiz 9.3

Suppose an office party is taking place on the top floor of a tall building. Carrying an iced soft drink, you step on the elevator, which begins to accelerate downward. What happens to the level of the ice in the drink? Does it (a) rise farther out of the liquid? (b) sink deeper into the liquid? or (c) remain unaffected by the motion?

Quick Quiz 9.4

Atmospheric pressure varies from day to day. What happens to the level of a floating ship on a high-pressure day compared to a low-pressure day? (a) It is higher. (b) It is lower. (c) Its level doesn't change.

Quick Quiz 9.5

A person in a boat floating in a small pond throws an anchor overboard. Does the level of the pond (a) rise, (b) fall, or (c) remain the same?

Quick Quiz 9.6

Lead has a greater density than iron, and both are denser than water. Is the buoyant force on a solid lead object (a) greater than, (b) less than, or (c) equal to the buoyant force on a solid iron object of the same dimensions?

Example 9.5   A Red Tag Special on Crowns

A bargain hunter purchases a "gold" crown at a flea market. After she gets home, she hangs it from a scale and finds its weight to be 7.84 N (Fig. 9.22a). She then weighs the crown while it is immersed in water of density 1 000 kg/m^3, as in Figure 9.22b, and now the scale reads 6.86 N. Is the crown made of pure gold?

(a)   (b)

FIGURE 9.22   (Example 9.5) (a) When the crown is suspended in air, the scale reads the true weight w. (b) When the crown is immersed in water, the buoyant force **B** reduces the scale reading to the apparent weight, $T_2 = w - B$.

Reasoning and Solution   When the crown is suspended in air, the scale reads the weight w (neglecting the buoyant force exerted by the air). When it is immersed in water, the buoyant force **B** from the water reduces the tension in the string (and the scale reading) to $T_2 = w - B$. Hence, the magnitude of the buoyant force on the crown is the difference between the first scale reading in air (the weight w) and the second scale reading when the crown is immersed in water:

$$B = w - T_2 = 7.84 \text{ N} - 6.86 \text{ N} = 0.98 \text{ N}$$

Because the magnitude of the buoyant force is equal to the weight of the displaced fluid w_f, we have

$$w_f = \rho_f g V_f = 0.98 \text{ N}$$

where V_f is the volume of the displaced fluid, and ρ_f is its density (1 000 kg/m^3). Also, the volume of the crown V_c is equal to the volume of the displaced fluid (because the crown is completely submerged). We find that

$$V_c = V_f = \frac{0.98 \text{ N}}{g\rho_f} = \frac{0.98 \text{ N}}{(9.8 \text{ m/s}^2)(1\,000 \text{ kg/m}^3)} = 1.0 \times 10^{-4} \text{ m}^3$$

Finally, the density of the crown is

$$\rho_c = \frac{m_c}{V_c} = \frac{w_c}{gV_c} = \frac{7.84 \text{ N}}{(9.8 \text{ m/s}^2)(1.0 \times 10^{-4} \text{ m}^3)} = 8.0 \times 10^3 \text{ kg/m}^3$$

From Table 9.3 we see that the density of gold is 19.3×10^3 kg/m^3. Thus, either the crown is hollow or it is made of an alloy. This was not a good day for the bargain hunter!

Example 9.6   Floating Down the River

A raft is constructed of wood having a density of 600 kg/m^3. Its surface area is 5.7 m^2, and its volume is 0.60 m^3. When the raft is placed in fresh water of density 1 000 kg/m^3, as in Figure 9.23, to what depth does the raft sink in the water?

FIGURE 9.23   (Example 9.6) A raft partially submerged in water.

Reasoning   The magnitude of the upward buoyant force acting on the raft must equal the weight of the raft if the raft is to float. In addition, from Archimedes's principle the magnitude of the buoyant force is equal to the weight of the displaced water.

Solution   The magnitude of the upward buoyant force acting on the raft equals the weight of the displaced water, which in turn must equal the weight of the raft:

$$B = \rho_{\text{water}} g V_{\text{water}} = \rho_{\text{water}} g A h$$

Because the area A and density ρ_{water} are known, we can find the depth h to which the raft sinks in the water:

$$h = \frac{w_{\text{raft}}}{\rho_{\text{water}} g A} \qquad (1)$$

The weight of the raft is

$$w_{\text{raft}} = \rho_{\text{raft}} g V_{\text{raft}} = (600 \text{ kg/m}^3)(9.8 \text{ m/s}^2)(0.60 \text{ m}^3) = 3.5 \times 10^3 \text{ N}$$

Therefore, substitution into (1) gives

$$h = \frac{3.5 \times 10^5 \text{ N}}{(1\,000 \text{ kg/m}^3)(9.8 \text{ m/s}^2)(5.7 \text{ m}^2)} = \boxed{0.060 \text{ m}}$$

9.7 FLUIDS IN MOTION

When a fluid is in motion, its flow can be characterized in one of two ways. The flow is said to be **streamline,** or **laminar,** if every particle that passes a particular point moves exactly along the smooth path followed by particles that passed that point earlier. The path is called a *streamline* (Fig. 9.24). Different streamlines cannot cross each other under this steady-flow condition, and the streamline at any point coincides with the direction of fluid velocity at that point.

In contrast, the flow of a fluid becomes irregular, or **turbulent,** above a certain velocity or under any conditions that can cause abrupt changes in velocity. Irregular motions of the fluid, called *eddy currents,* are characteristic in turbulent flow as shown in Figure 9.25.

In discussions of fluid flow, the term *viscosity* is used for the degree of internal friction in the fluid. This internal friction is associated with the resistance between two adjacent layers of the fluid moving relative to each other. A fluid such as kerosene has a lower viscosity than does crude oil or molasses.

Many features of fluid motion can be understood by considering the behavior of an **ideal fluid,** which satisfies the following conditions:

1. *The fluid is nonviscous;* that is, there is no internal friction force between adjacent layers.
2. *The fluid is incompressible,* which means that its density is constant.
3. *The fluid motion is steady,* meaning that the velocity, density, and pressure at each point in the fluid do not change in time.
4. *The fluid moves without turbulence.* This implies that each element of the fluid has zero angular velocity about its center; that is, there can be no eddy currents present in the moving fluid. A small wheel placed in the fluid would translate but not rotate.

EQUATION OF CONTINUITY

Figure 9.26a represents a fluid flowing through a pipe of nonuniform size. The particles in the fluid move along the streamlines in steady-state flow. In a small

FIGURE 9.24 An illustration of streamline flow around an automobile in a test wind tunnel. The streamlines in the airflow are made visible by smoke particles. *(Andy Sacks/Stone)*

FIGURE 9.25 Turbulent flow: The tip of a rotating blade (the dark region at the top) forms a vortex in air that is being heated by an alcohol lamp (the wick is at the bottom). Note the air turbulence on both sides of the rotating blade. *(Kim Vandiver and Harold Edgerton. © Harold and Esther Edgerton Foundation, 2002, courtesy of Palm Press, Inc.)*

FIGURE 9.26 (a) A fluid moving with streamline flow through a pipe of varying cross-sectional area. The volume of fluid flowing through A_1 in a time interval Δt must equal the volume flowing through A_2 in the same time interval. Therefore, $A_1 v_1 = A_2 v_2$. (b) Water flowing slowly out of a faucet. The width of the stream narrows as the water falls and speeds up in accord with the continuity equation. (*George Semple*)

time interval Δt, the fluid entering the bottom end of the pipe moves a distance of $\Delta x_1 = v_1 \Delta t$, where v_1 is the speed of the fluid at this location. If A_1 is the cross-sectional area in this region, then the mass contained in the bottom blue region is $\Delta M_1 = \rho_1 A_1 \Delta x_1 = \rho_1 A_1 v_1 \Delta t$, where ρ_1 is the density of the fluid at A_1. Similarly, the fluid that moves out of the upper end of the pipe in the same time interval Δt has a mass of $\Delta M_2 = \rho_2 A_2 v_2 \Delta t$. However, **because mass is conserved** and because the flow is steady, the mass that flows into the bottom of the pipe through A_1 in the time Δt must equal the mass that flows out through A_2 in the same interval. Therefore, $\Delta M_1 = \Delta M_2$, or

$$\rho_1 A_1 v_1 = \rho_2 A_2 v_2 \qquad \text{[9.13]}$$

For the case of an incompressible fluid, $\rho_1 = \rho_2$ and Equation 9.13 reduces to

$$A_1 v_1 = A_2 v_2 \qquad \text{[9.14]}$$

 Equation of continuity

This expression is called the **equation of continuity.** From this result, we see that **the product of the cross-sectional area of the pipe and the fluid speed at that cross section is a constant.** Therefore, the speed is high where the tube is constricted and low where the tube has a larger diameter. The product Av, which has dimensions of volume per unit time, is called the **flow rate. The condition Av = constant is equivalent to the fact that the amount of fluid that enters one end of the tube in a given time interval equals the amount of fluid leaving the tube in the same interval, assuming that the fluid is incompressible and that there are no leaks.** Figure 9.26b is a one example of an application of the equation of continuity. As the stream of water flows continuously from a faucet, the width of the stream narrows as it falls and speeds up.

There are many instances in your everyday experiences when you have seen the equation of continuity in action. For example, you place your thumb over the open end of a garden hose to make the water spray out farther and with greater speed. As you reduce the cross-sectional area of the nozzle, the water drops leave the nozzle at a higher speed, causing them to move a longer distance. Similar reasoning explains why rising smoke from a smoldering piece of wood behaves as it does. The smoke first rises in a streamline pattern, getting thinner as it rises, and eventually breaks up into a swirling, turbulent pattern. The smoke rises because it is less dense than air, and the buoyant force of the air accelerates it upward. As the

 CONTINUITY EQUATIONS

Equation 9.14 states that the rate of flow of fluid into the system equals the rate of flow out of the system, because fluid cannot be stored in the system once it is full of fluid. Example 9.4, which analyzed a car lift, represents an analogous result for energy conservation—the energy transferred into the lift is equal to that transferred out.

speed of the smoke stream increases, the cross-sectional area of the stream decreases, as seen from the equation of continuity. However, the stream soon reaches a speed so great that streamline flow is not possible. We shall study the relationship between speed of fluid flow and turbulence in a later discussion of the Reynolds number.

Example 9.7 Watering a Garden

A water hose 2.50 cm in diameter is used by a gardener to fill a 30.0-liter bucket (1 liter = 1 000 cm^3, exactly). The gardener notes that it takes 1.00 min to fill the bucket. A nozzle with an opening of cross-sectional area 0.500 cm^2 is then attached to the hose. The nozzle is held so that water is projected horizontally from a point 1.00 m above the ground. Over what horizontal distance can the water be projected?

Solution We identify point 1 within the hose and point 2 at the exit of the nozzle. We first use the continuity equation for fluids to find the speed $v_2 = v_{x0}$ with which the water exits the nozzle:

$$A_1 v_1 = A_2 v_2 = A_2 v_{x0}$$

$$v_{x0} = \frac{A_1 v_1}{A_2}$$

The numerator of this expression, $A_1 v_1$, is the volume flow rate in the hose, which can be evaluated from the bucket-filling information. Thus,

$$v_{x0} = \frac{A_1 v_1}{A_2} = \left(\frac{30.0 \text{ L/min}}{0.500 \text{ cm}^2}\right)\left(\frac{10^3 \text{ cm}^3}{1 \text{ L}}\right)\left(\frac{1 \text{ min}}{60 \text{ s}}\right)\left(\frac{1 \text{ m}}{100 \text{ cm}}\right) = 10.0 \text{ m/s}$$

Because the water is in free fall once it exits the nozzle, it exhibits projectile motion after leaving the hose. A particle of the water falls through a vertical distance of 1.00 m starting from rest. Setting $y_0 = 0$ and $v_{y0} = 0$ at $t = 0$, the time required to fall this distance can be calculated from Equation 3.12:

$$\Delta y = v_{y0} t - \tfrac{1}{2} g t^2$$

$$t = \sqrt{\frac{2y}{g}}$$

$$= \sqrt{\frac{2(1.00 \text{ m})}{9.80 \text{ m/s}^2}} = 0.452 \text{ s}$$

In the horizontal direction, we apply Equation 3.10 to a particle of water to find the horizontal distance, setting $x_0 = 0$ at $t = 0$:

$$x = v_{x0} t = (10.0 \text{ m/s})(0.452 \text{ s}) = \boxed{4.52 \text{ m}}$$

EXERCISE The Garfield Thomas water tunnel at Pennsylvania State University has a circular cross-section that constricts from a diameter of 3.6 m to the test section, which is 1.2 m in diameter. If the speed of flow is 3.0 m/s in the larger-diameter pipe, determine the speed of flow in the test section.

ANSWER 27 m/s

DANIEL BERNOULLI, SWISS PHYSICIST AND MATHEMATICIAN (1700–1782)

In his most famous work, *Hydrodynamica*, Bernoulli showed that, as the velocity of fluid flow increases, its pressure decreases. In this same publication, Bernoulli also attempted the first explanation of the behavior of gases with changing pressure and temperature; this was the beginning of kinetic theory of gases. *(Corbis-Bettmann)*

BERNOULLI'S EQUATION

As a fluid moves through a pipe of varying cross section and elevation, the pressure changes along the pipe. In 1738 the Swiss physicist Daniel Bernoulli (1700–1782) derived an expression that relates pressure to fluid speed and elevation. Bernoulli's equation is not a freestanding law of physics. As we shall see in this section, **Bernoulli's equation is a consequence of energy conservation as applied to an ideal fluid.**

In deriving Bernoulli's equation, we again assume that the fluid is incompressible and nonviscous and flows in a nonturbulent, steady-state manner. Consider the flow through a nonuniform pipe in the time Δt, as illustrated in Figure 9.27. The force on the lower end of the fluid is $P_1 A_1$, where P_1 is the pressure at the lower end. The work done on the lower end of the fluid by the fluid behind it is

$$W_1 = F_1 \Delta x_1 = P_1 A_1 \Delta x_1 = P_1 V$$

where V is the volume of the lower blue region in Figure 9.27. In a similar manner, the work done on the fluid on the upper portion in the time Δt is

$$W_2 = -P_2 A_2 \Delta x_2 = -P_2 V$$

(Remember that the volume of fluid that passes through A_1 in the time Δt equals the volume that passes through A_2 in the same interval.) The work W_2 is negative because the force on the fluid at the top is opposite its displacement. The net work done by these forces in the time Δt equals the net work done by the fluid:

$$W_{\text{fluid}} = P_1 V - P_2 V$$

Part of this work goes into changing the fluid's kinetic energy, and part goes into changing the gravitational potential energy of the fluid–earth system. If m is the mass of the fluid passing through the pipe in the time interval Δt, then the change in kinetic energy of the volume of fluid is

$$\Delta KE = \tfrac{1}{2} m v_2{}^2 - \tfrac{1}{2} m v_1{}^2$$

The change in the gravitational potential energy is

$$\Delta PE = mgy_2 - mgy_1$$

Because the net work done by the fluid on the segment of fluid shown in Figure 9.27 changes the kinetic energy and the potential energy of the nonisolated system, we have

$$W_{\text{fluid}} = \Delta KE + \Delta PE$$

The three terms in this equation are those we have just evaluated. Substituting expressions for each of the terms gives

$$P_1 V - P_2 V = \tfrac{1}{2} m v_2{}^2 - \tfrac{1}{2} m v_1{}^2 + mgy_2 - mgy_1$$

If we divide each term by V and recall that $\rho = m/V$, this expression reduces to

$$P_1 - P_2 = \tfrac{1}{2}\rho v_2{}^2 - \tfrac{1}{2}\rho v_1{}^2 + \rho gy_2 - \rho gy_1$$

Let us move those terms that refer to point ① to one side of the equation and those that refer to point ② to the other side:

$$P_1 + \tfrac{1}{2}\rho v_1{}^2 + \rho gy_1 = P_2 + \tfrac{1}{2}\rho v_2{}^2 + \rho gy_2 \qquad \text{[9.15]}$$

This is **Bernoulli's equation.** It is often expressed as

$$P + \tfrac{1}{2}\rho v^2 + \rho gy = \text{constant} \qquad \text{[9.16]}$$

> Bernoulli's equation states that the sum of the pressure (P), the kinetic energy per unit volume ($\tfrac{1}{2}\rho v^2$), and the potential energy per unit volume (ρgy) has the same value at all points along a streamline.

An important consequence of Bernoulli's equation can be demonstrated by considering Figure 9.28, which shows water flowing through a horizontal constricted pipe from a region of large cross-sectional area into a region of smaller cross-sectional area. This device, called a *Venturi tube,* can be used to measure the

FIGURE 9.27 A fluid flowing through a constricted pipe with streamline flow. The fluid in the section with a length of Δx_1 moves to the section with a length of Δx_2. The volumes of fluid in the two sections are equal.

◀ Bernoulli's equation

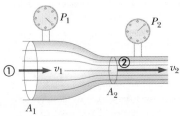

FIGURE 9.28 The pressure P_1 is greater than the pressure P_2, because $v_1 < v_2$. This device can be used to measure the speed of fluid flow.

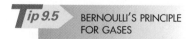

speed of fluid flow. Let us compare the pressure at point ① to the pressure at point ②. Because the pipe is horizontal, $y_1 = y_2$ and Equation 9.15 applied to points ① and ② gives

$$P_1 + \tfrac{1}{2}\rho v_1{}^2 = P_2 + \tfrac{1}{2}\rho v_2{}^2 \qquad \text{[9.17]}$$

Because the water is not backing up in the pipe, its speed v_2 in the constriction must be greater than its speed v_1 in the region of greater diameter. From Equation 9.17, $v_2 > v_1$ means that P_2 must be less than P_1. This result is often expressed by the statement that **swiftly moving fluids exert less pressure than do slowly moving fluids.** As we shall see in the next section, this important result enables us to understand a wide range of everyday phenomena.

Quick Quiz 9.7

You observe two helium balloons floating next to each other at the ends of strings secured to a table. The facing surfaces of the balloons are separated by 1–2 cm. You blow through the opening between the balloons. What happens to the balloons? (a) They move toward each other. (b) They move away from each other. (c) They are unaffected.

Example 9.8 Shoot-Out at the Old Water Tank

A nearsighted sheriff fires at a cattle rustler with his trusty six-shooter. Fortunately for the cattle rustler, the bullet misses him and penetrates the town water tank and causes a leak (Fig. 9.29). If the top of the tank is open to the atmosphere, determine the speed at which the water leaves the hole when the water level is 0.500 m above the hole.

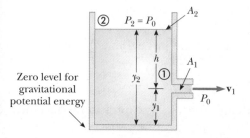

FIGURE 9.29 (Example 9.8) The water speed v_1 from the hole in the side of the container is given by $v_1 = \sqrt{2gh}$.

Reasoning If we assume that the cross-sectional area of the tank is large relative to that of the hole ($A_2 \gg A_1$), then the water level drops very slowly and we can assume $v_2 \approx 0$. Let us apply Bernoulli's equation to points ① and ②. If we note that $P_1 = P_0$ at the hole, we get

$$P_0 + \tfrac{1}{2}\rho v_1{}^2 + \rho g y_1 = P_0 + \rho g y_2$$

$$v_1 = \sqrt{2g(y_2 - y_1)} = \sqrt{2gh}$$

Solution This says that the speed of the water emerging from the hole is equal to the speed acquired by a body falling freely through the vertical distance h. This is known as **Torricelli's law.** If the height h is 0.500 m, for example, the speed of the stream is

$$v = \sqrt{2(9.80 \text{ m/s}^2)(0.500 \text{ m})} = \boxed{3.13 \text{ m/s}}$$

EXERCISE If the head of the cattle rustler is 3.00 m below the level of the hole in the tank, where must he stand to get doused with water?

ANSWER 2.45 m from the base of the tank

9.8 OTHER APPLICATIONS OF FLUID DYNAMICS

In this section we describe some common phenomena that can be explained, at least in part, by Bernoulli's equation.

In general, an object moving through a fluid experiences a net upward force as the result of any effect that causes the fluid to change its direction as it flows past the object. For example, a golf ball struck with a club is given a rapid backspin, as shown in Figure 9.30. The dimples on the ball help "entrain" the air to follow the curvature of the ball's surface. Figure 9.30 shows a thin layer of air wrapping part way around the ball and being deflected downward as a result. Because the ball pushes the air down, by Newton's third law the air must push up on the ball and cause it to rise. Without the dimples, the air is not as well entrained and the golf ball does not travel as far. For the same reason, a tennis ball's fuzz helps the spinning ball "grab" the air rushing by and helps deflect it.

Many devices operate in the manner illustrated in Figure 9.31. A stream of air passing over an open tube reduces the pressure above the tube. This causes the liquid to rise into the airstream. The liquid is then dispersed into a fine spray of droplets. You might recognize that this so-called atomizer is used in perfume bottles and paint sprayers. The same principle is used in the carburetor of a gasoline engine. In that case, the low-pressure region in the carburetor is produced by air drawn in by the piston through the air filter. The gasoline vaporizes, mixes with the air, and enters the cylinder of the engine for combustion.

◀ APPLICATION

LIFT EXPERIENCED BY A SPINNING BALL

◀ APPLICATION

"ATOMIZERS" IN PERFUME BOTTLES AND PAINT SPRAYERS

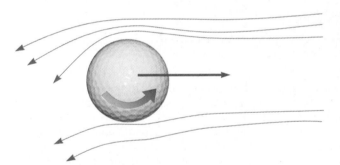

FIGURE 9.30 A spinning golf ball experiences a lifting force that allows it to travel much farther than it would if it were not spinning.

FIGURE 9.31 A stream of air passing over a tube dipped in a liquid causes the liquid to rise in the tube as shown. This effect is used in perfume bottles and paint sprayers.

In a person with advanced arteriosclerosis, the Bernoulli effect produces a symptom called *vascular flutter*. In this situation, the artery is constricted as a result of accumulated plaque on its inner walls, as in Figure 9.32. To maintain a constant flow rate, the blood must travel faster than normal through the constriction. If the blood speed is sufficiently high in the constricted region, its pressure is low, and the artery may collapse under external pressure, causing a momentary interruption in blood flow. At this moment there is no Bernoulli effect, and the vessel reopens under arterial pressure. As the blood rushes through the constricted artery, the internal pressure drops and again the artery closes. Such variations in blood flow can be heard with a stethoscope. If the plaque becomes dislodged and ends up in a smaller vessel that delivers blood to the heart, the person can suffer a heart attack.

An aneurysm is a weakened spot on an artery where the artery walls have ballooned outward. Blood flows more slowly through this region, as can be seen from the equation of continuity, resulting in an increase in pressure in the vicinity of the aneurysm relative to the pressure in other parts of the artery. This condition is dangerous because the excess pressure can cause the artery to rupture.

The lift on an aircraft wing can also be explained, in part, by the Bernoulli effect. Airplane wings are designed so that the air speed above the wing is greater than that below. As a result, the air pressure above the wing is less than the pressure below, and there is a net upward force on the wing, called the "lift." Another

◀ APPLICATION

VASCULAR FLUTTER AND ANEURYSMS

FIGURE 9.32 Blood must travel faster than normal through a constricted region of an artery.

◀ APPLICATION

LIFT ON AIRCRAFT WINGS

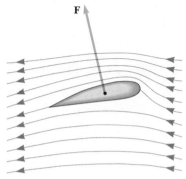

FIGURE 9.33 Streamline flow around an airplane wing. The pressure above is less than the pressure below, and there is a dynamic lift force upward.

factor influencing the lift on a wing is shown in Figure 9.33. The wing has a slight upward tilt that causes air molecules striking the bottom to be deflected downward. The air molecules bouncing off the wing at the bottom produce an upward force on the wing and a significant lift on the aircraft. Finally, turbulence also has an effect. If the wing is tilted too much, the flow of air across the upper surface becomes turbulent, and the pressure difference across the wing is not as great as that predicted by the Bernoulli effect. In an extreme case, this turbulence may cause the aircraft to stall.

APPLYING **PHYSICS** 9.4

The Bernoulli effect can be used to partially explain how a sailboat can accomplish the seemingly impossible task of sailing into the wind. How can this be done?

FIGURE 9.34 (Applying Physics 9.4)

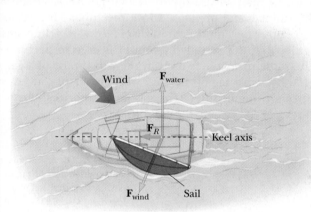

Explanation As shown in Figure 9.34, the wind blowing in the direction of the arrow causes the sail to billow out, taking on a shape similar to that of an airplane wing. For the same reason as for an airplane wing, there is a force acting on the sail in the direction shown. The component of force perpendicular to the boat tends to make the boat move sideways in the water, but the keel prevents this sideways motion. The component of the force in the forward direction drives the boat almost against the wind. The word *almost* is used because a sailboat can move forward only when the wind direction is about 10–15° with respect to the forward direction. This means that in order to sail directly against the wind, a boat must follow a zigzag path, a procedure called *tacking*, so that the wind is always at some angle with respect to the direction of travel.

APPLYING **PHYSICS** 9.5

Consider the portion of a home plumbing system shown in Figure 9.35. The water trap in the pipe below the sink captures a plug of water that prevents sewer gas from finding its way from the sewer pipe, up the sink drain, and into the home. Suppose the dishwasher is draining, so that water is moving to the left in the sewer pipe. What is the purpose of the vent, which is open to the air above the roof of the house? In which direction is air moving at the opening of the vent, upward or downward?

Explanation Let us imagine that the vent is not present, so that the drain pipe for the sink is simply connected through the trap to the sewer pipe. As water from the dishwasher moves to the left in the sewer pipe, the pressure in the sewer pipe is reduced below atmospheric pressure, according to Bernoulli's principle. The pressure at the drain in the sink is still at atmospheric pressure. Thus, this pressure differential can push the plug of water in the water trap of the sink down the drain pipe into the sewer pipe, removing it as a barrier to sewer gas. With the addition of the vent to the roof, the reduced pressure of the dishwasher

FIGURE 9.35 (Applying Physics 9.5)

water results in air entering the vent pipe at the roof. This keeps the pressure in the vent pipe and the right-hand side of the sink drain pipe at a pressure close to atmospheric, so that the plug of water in the water trap remains in place.

9.9 SURFACE TENSION, CAPILLARY ACTION, AND VISCOUS FLUID FLOW

If you look closely at a dewdrop sparkling in the morning sunlight, you will find that the drop is spherical. The drop takes this shape because of a property of liquid surfaces called **surface tension.** In order to understand the origin of surface tension, consider a molecule at point A in a container of water, as in Figure 9.36. Although nearby molecules exert forces on this molecule, the net force on it is zero because it is completely surrounded by other molecules and hence is attracted equally in all directions. The molecule at B, however, is not attracted equally in all directions. Because there are no molecules above it to exert upward forces, the molecule at B is pulled toward the interior of the liquid. The contraction at the surface of the liquid ceases when the inward pull exerted on the surface molecules is balanced by the outward repulsive forces that arise from collisions with molecules in the interior of the liquid. **The net effect of this pull on all the surface molecules is to make the surface of the liquid contract and consequently to make the surface area of the liquid as small as possible.** Drops of water take on a spherical shape because a sphere has the smallest surface area for a given volume.

If you place a sewing needle very carefully on the surface of a bowl of water, you will find that the needle floats even though the density of steel is about eight times that of water. This can also be explained by surface tension. A close examination of the needle shows that it actually rests in a depression in the liquid surface, as shown in Figure 9.37. The water surface acts like an elastic membrane under tension. The weight of the needle produces a depression, thus increasing the surface area of the film. Molecular forces now act at all points along the depression in an attempt to restore the surface to its original horizontal position. The vertical components of these forces act to balance the force of gravity acting on the needle.

The **surface tension** γ in a film of liquid is defined as the ratio of the magnitude of the surface tension force **F** to the length L along which the force acts:

$$\gamma \equiv \frac{F}{L} \qquad [9.18]$$

The SI units of surface tension are newtons per meter, and values for a few representative materials are given in Table 9.4.

The concept of surface tension can be thought of as the energy content of the fluid at its surface per unit surface area. To see that this is reasonable, we can

FIGURE 9.36 The net force on a molecule at A is zero because such a molecule is completely surrounded by other molecules. The net force on a surface molecule at B is downward because it is not completely surrounded by other molecules.

FIGURE 9.37 End view of a needle resting on the surface of water. The components of surface tension balance the force of gravity.

TABLE 9.4	Surface Tensions for Various Liquids	
Liquid	$T(°C)$	Surface Tension (N/m)
Ethyl alcohol	20	0.022
Mercury	20	0.465
Soapy water	20	0.025
Water	20	0.073
Water	100	0.059

FIGURE 9.38 An apparatus for measuring the surface tension of liquids. The force on the wire ring is measured just before it breaks free of the liquid.

Webnote 9.4

Surface tension plays an important part in the formation of bubbles. Read more at *http://www.exploratorium.edu/ronh/bubbles/bubbles.html*

manipulate the units of surface tension γ as

$$\frac{N}{m} = \frac{N \cdot m}{m^2} = \frac{J}{m^2}$$

In general, **any equilibrium configuration of an object is one in which the energy is a minimum.** Consequently, a fluid will take on a shape such that its surface area is as small as possible. For a given volume, a spherical shape is the one that has the smallest surface area; therefore, a drop of water takes on a spherical shape.

An apparatus used to measure the surface tension of liquids is shown in Figure 9.38. A circular wire with a circumference L is lifted from a body of liquid. The surface film clings to the inside and outside edges of the wire, holding back the wire and causing the spring to stretch. If the spring is calibrated, we can measure the force required to overcome the surface tension of the liquid. In this case, the surface tension is given by

$$\gamma = \frac{F}{2L}$$

We must use $2L$ for the length because the surface film exerts forces on the inside and outside of the ring.

The surface tension of liquids decreases with increasing temperature. This occurs because the faster moving molecules of a hot liquid are not bound together as strongly as are those in a cooler liquid. Furthermore, certain ingredients called surfactants decrease surface tension when added to liquids. For example, soap or detergent decreases the surface tension of water. This reduction in surface tension makes it easier for soapy water to penetrate the cracks and crevices of your clothes to clean them better than plain water. An effect similar to this occurs in our lungs. The surface tissue of the air sacs in the lungs contains a fluid that has a surface tension of about 0.050 N/m. A liquid with a surface tension this high would make it very difficult for the lungs to expand as we inhale. However, as the area of the lungs increases with inhalation, the body secretes into the tissue a substance that gradually reduces the surface tension of the liquid. At full expansion, the surface tension of the lung fluid can drop to as low as 0.005 N/m.

Example 9.9   Walking on Water

In this example, we illustrate how an insect is supported on the surface of water by surface tension. Let us assume that the insect's "foot" is spherical. When the insect steps onto the water with all six legs, a depression is formed in the water around each foot, as shown in Figure 9.39a. The surface tension of the water produces upward forces on the water that tend to restore the water surface to its normally flat shape. If the insect has a mass of 2.0×10^{-5} kg and if the radius of each foot is 1.5×10^{-4} m, find the angle θ.

Solution   From the definition of surface tension, we can find the magnitude of the net force **F** directed tangential to the depressed part of the water surface:

$$F = \gamma L$$

The length L along which this force acts is equal to the distance around the insect's foot, $2\pi r$. (It is assumed that the insect depresses the water surface such that the radius of the depression is equal to the radius of the foot.) Thus, $F = \gamma(2\pi r)$, and the net vertical force is

$$F_v = \gamma(2\pi r)(\cos \theta)$$

Because the insect has six legs, this upward force must equal one sixth the weight of the insect, assuming its weight is equally distributed on all 6 feet. Thus,

$$\gamma(2\pi r)(\cos\theta) = \tfrac{1}{6}w = \tfrac{1}{6}mg$$

$$\cos\theta = \frac{mg}{12\pi r\gamma} = \frac{(2.0 \times 10^{-5}\,\text{kg})(9.80\,\text{m/s}^2)}{12\pi(1.5 \times 10^{-4}\,\text{m})(0.073\,\text{N/m})} \qquad \textbf{(1)}$$

$$\theta = \boxed{62°}$$

Note that if the weight of the insect were great enough to make the right side of (1) greater than unity, a solution for θ would be impossible because the cosine can never be greater than unity. Under these conditions, the insect would sink.

(a)   **(b)**

FIGURE 9.39 (Example 9.9) (a) One foot of an insect resting on the surface of water. (b) This water strider resting on the surface of a lake remains on the surface, rather than sinking, because an upward surface tension force acts on each leg, which balances the force of gravity acting on the insect. *(Herman Eisenbeiss/Photo Researchers, Inc.)*

THE SURFACE OF LIQUIDS

If you have ever closely examined the surface of water in a glass container, you may have noticed that the surface of the liquid near the walls of the glass curves upward as you move from the center to the edge, as shown in Figure 9.40a. However, if mercury is placed in a glass container, the mercury surface curves downward, as in Figure 9.40b. These surface effects can be explained by considering the forces between molecules. In particular, we must consider the forces that the molecules

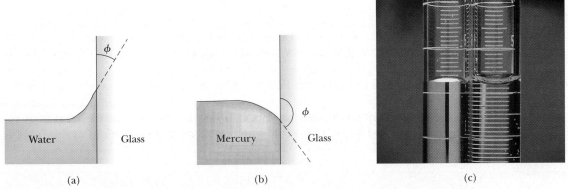

(a)   **(b)**   **(c)**

FIGURE 9.40 A liquid in contact with a solid surface. (a) For water, the adhesive force is greater than the cohesive force. (b) For mercury, the adhesive force is less than the cohesive force. (c) The surface of mercury *(left)* curves downward in a glass container, whereas the surface of water *(right)* curves upward as you move from the center to the edge. *(Charles D. Winters)*

FIGURE 9.41 (a) The contact angle between water and paraffin is about 107°. In this case, the cohesive force is greater than the adhesive force. (b) When a chemical called a wetting agent is added to the water, it wets the paraffin surface, and $\phi < 90°$. In this case, the adhesive force is greater than the cohesive force.

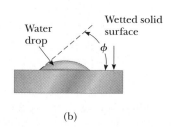

(a)   (b)

of the liquid exert on one another and the forces that the molecules of the glass surface exert on those of the liquid. In general terms, forces between like molecules, such as the forces between water molecules, are called *cohesive forces* and forces between unlike molecules, such as those exerted by glass on water, are called *adhesive forces*.

Water tends to cling to the walls of the glass because the adhesive forces between the liquid molecules and the glass molecules are *greater* than the cohesive forces between the liquid molecules. In effect, the liquid molecules cling to the surface of the glass rather than fall back into the bulk of the liquid. When this condition prevails, the liquid is said to "wet" the glass surface. The surface of the mercury curves downward near the walls of the container because the cohesive forces between the mercury atoms are greater than the adhesive forces between mercury and glass. That is, a mercury atom near the surface is pulled more strongly toward other mercury atoms than toward the glass surface; hence mercury does not wet the glass surface.

The angle ϕ between the solid surface and a line drawn tangent to the liquid at the surface is called *the angle of contact* (Fig. 9.40). Note that ϕ is less than 90° for any substance in which adhesive forces are stronger than cohesive forces and greater than 90° if cohesive forces predominate. For example, if a drop of water is placed on paraffin, the contact angle is approximately 107° (Fig. 9.41a). If certain chemicals, called wetting agents or detergents, are added to the water, the contact angle becomes less than 90°, as shown in Figure 9.41b. The addition of such substances to water is of value when one wants to ensure that water makes intimate contact with a surface and penetrates it. For this reason, detergents are added to water to wash clothes or dishes. On the other hand, it is often necessary to keep water from making intimate contact with a surface, as in waterproofing clothing, where a situation somewhat the reverse of that shown in Figure 9.41 is called for. The clothing is sprayed with a waterproofing agent, which changes ϕ from less than 90° to greater than 90°. Thus, the water beads up on the surface and does not easily penetrate the clothing.

APPLICATION >

DETERGENTS AND WATERPROOFING AGENTS

CAPILLARY ACTION

Capillary tubes are tubes in which the diameter of the opening is very small, on the order of a hundredth of a centimeter. In fact, the word *capillary* means "hair-like." If such a tube is inserted into a fluid for which adhesive forces dominate over cohesive forces, the liquid will rise into the tube, as shown in Figure 9.42. The rising of the liquid in the tube can be explained in terms of the shape of the surface of the liquid and in terms of the surface tension effects in the liquid. At the point of contact between liquid and solid, the upward force of surface tension is directed as shown in Figure 9.42. From Equation 9.18, the magnitude of this force is

$$F = \gamma L = \gamma(2\pi r)$$

We use $L = 2\pi r$ here because the liquid is in contact with the surface of the tube at all points around its circumference. The vertical component of this force due to surface tension is

$$F_v = \gamma(2\pi r)(\cos \phi) \qquad [9.19]$$

FIGURE 9.42 A liquid rises in a narrow tube because of capillary action, a result of surface tension and adhesive forces.

For the liquid in the capillary tube to be in equilibrium, this upward force must be equal to the weight of the cylinder of water of height h inside the capillary tube. The weight of this water is

$$w = Mg = \rho Vg = \rho g \pi r^2 h \qquad [9.20]$$

Equating F_v in Equation 9.19 to w in Equation 9.20, we have

$$\gamma(2\pi r)(\cos \phi) = \rho g \pi r^2 h$$

Thus, the height to which water is drawn into the tube is

$$h = \frac{2\gamma}{\rho g r} \cos \phi \qquad [9.21]$$

FIGURE 9.43 When cohesive forces between molecules of the liquid exceed adhesive forces, the liquid level in the capillary tube is below the surface of the surrounding fluid.

If a capillary tube is inserted into a liquid in which cohesive forces dominate over adhesive forces, the level of the liquid in the capillary tube will be below the surface of the surrounding fluid, as shown in Figure 9.43. An analysis similar to that done above would show that the distance h the surface is depressed is given by Equation 9.21.

Capillary tubes are often used to draw small samples of blood from a needle prick in the skin. Capillary action must also be considered in the construction of concrete-block buildings because water seepage through capillary pores in the blocks or the mortar may cause damage to the inside of the building. To prevent this, the blocks are usually coated with a waterproofing agent either outside or inside the building. Water seepage through a wall is an undesirable effect of capillary action, but paper towels use capillary action in a useful manner to absorb spilled fluids.

(a)

(b)

FIGURE 9.44 (a) The particles in an ideal (nonviscous) fluid all move through the pipe with the same velocity. (b) In a viscous fluid, the velocity of the fluid particles is zero at the surface of the pipe and increases to its maximum value at the center of the pipe.

Example 9.10 How High Does the Water Rise?

Find the height to which water would rise in a capillary tube with a radius equal to 5.0×10^{-5} m. Assume that the angle of contact between the water and the material of the tube is small enough to be considered zero.

Solution From Table 9.4, the surface tension of water is 0.073 N/m. For a contact angle of $0°$, we have $\cos \phi = \cos 0° = 1$, so that Equation 9.21 gives

$$h = \frac{2\gamma}{\rho g r} = \frac{2(0.073 \text{ N/m})}{(1.00 \times 10^3 \text{ kg/m}^3)(9.80 \text{ m/s}^2)(5.0 \times 10^{-5} \text{ m})} = \boxed{0.29 \text{ m}}$$

VISCOUS FLUID FLOW

It is considerably easier to pour water out of a container than to pour honey. This is because honey has a higher viscosity than water. In a general sense, *viscosity refers to the internal friction of a fluid.* It is very difficult for layers of a viscous fluid to slide past one another. Likewise, it is difficult for one solid surface to slide past another if there is a highly viscous fluid, such as soft tar, between them.

When an ideal (nonviscous) fluid flows through a pipe, the fluid layers slide past one another with no resistance. If the pipe has a uniform cross section, each layer has the same velocity as in Figure 9.44a. In contrast, the layers of a viscous fluid have different velocities as in Figure 9.44b. The fluid has the greatest velocity at the center of the pipe; the layer next to the wall does not move because of cohesive forces between molecules and the wall surface.

To better understand the concept of viscosity, consider a liquid layer placed between two solid surfaces, as in Figure 9.45. The lower surface is fixed in position, and the top surface moves to the right with a velocity $\mathbf{v}$ under the action of an external force $\mathbf{F}$. Because of this motion, a portion of the liquid is distorted

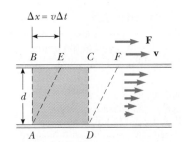

FIGURE 9.45 A layer of liquid between two solid surfaces in which the lower surface is fixed and the upper surface moves to the right with a velocity $\mathbf{v}$.

from its original shape, *ABCD*, at one instant to the shape *AEFD* a moment later. The force required to move the upper plate and distort the liquid depends on several factors. The force is proportional to both the area *A* in contact with the fluid and the speed *v* of the fluid. Furthermore, the force is inversely proportional to the distance *d* between the two plates. We can express these proportionalities as $F \propto Av/d$. Thus, the force required to move the upper plate at a fixed speed *v* is

$$F = \eta \frac{Av}{d} \qquad \text{[9.22]}$$

Coefficient of viscosity

where η (lowercase Greek letter *eta*) is the **coefficient of viscosity** of the fluid.

The SI units of viscosity are $N \cdot s/m^2$. The units of viscosity in many reference sources are often expressed in $dyne \cdot s/cm^2$, called 1 **poise** in honor of the French scientist J. L. Poiseuille (1799–1869). The relationship between the SI unit of viscosity and the poise is

$$1 \text{ poise} = 10^{-1} \text{ N} \cdot \text{s/m}^2 \qquad \text{[9.23]}$$

Small viscosities are often expressed in centipoise (cp), where 1 cp = 10^{-2} poise. The coefficients of viscosity for some common substances are listed in Table 9.5.

POISEUILLE'S LAW

Figure 9.46 shows a section of a tube of length *L* and radius *R* containing a fluid under a pressure P_1 at the left end and a pressure P_2 at the right. Because of this pressure difference, the fluid flows through the tube. The rate of flow (volume per unit time) depends on the pressure difference $(P_1 - P_2)$, the dimensions of the tube, and the viscosity of the fluid. The result, known as **Poiseuille's law,** is

FIGURE 9.46 Velocity profile of a fluid flowing through a uniform pipe of circular cross section. The rate of flow is given by Poiseuille's law. Note that the fluid velocity is greatest at the middle of the pipe.

Poiseuille's law

$$\text{Rate of flow} = \frac{\Delta V}{\Delta t} = \frac{\pi R^4 (P_1 - P_2)}{8\eta L} \qquad \text{[9.24]}$$

where η is the coefficient of viscosity of the fluid. We shall not attempt to derive this equation here because the methods of integral calculus are required. However, you should note that the equation does agree with common sense. That is, it is reasonable that the rate of flow should increase if the pressure difference across the tube or the tube radius increases. Likewise, the flow rate should decrease if the viscosity of the fluid or the length of the tube increases. Thus, the presence of *R* and the pressure difference in the numerator of Equation 9.24 and of *L* and η in the denominator makes sense.

From Poiseuille's law, we see that in order to maintain a constant flow rate, the pressure difference across the tube has to increase if the viscosity of the fluid increases. This is important when we consider the flow of blood through the circulatory system. The viscosity of blood increases as the number of red blood cells rises. Thus, blood with a high concentration of red blood cells requires greater pumping pressure from the heart to keep it circulating than does blood of lower red blood cell concentration.

Webnote 9.5

A calculator version of Poiseuille's law can be found at the following URL. Make sure you understand how to use Equation 9.24 on your own at exam time.
http://hyperphysics.phy-astr.gsu.edu/hbase/ppois.html
If you want to know how Poiseuille came up with his somewhat complicated relationship, work through the site below. Keep going at least until you get to the animation.
http://grad.math.arizona.edu/~walton/biomath/history.htm

TABLE 9.5	The Viscosities of Various Fluids	
Fluid	$T(°C)$	**Viscosity** $\eta(N \cdot s/m^2)$
Water	20	1.0×10^{-3}
Water	100	0.3×10^{-3}
Whole blood	37	2.7×10^{-3}
Glycerin	20	$1\,500 \times 10^{-3}$
10-wt motor oil	30	250×10^{-3}

Note that the flow rate varies as the radius of the tube raised to the fourth power. Consequently, if a constriction occurs in a vein or artery, the heart will have to work considerably harder in order to produce a higher pressure drop and hence to maintain the required flow rate.

Example 9.11 A Blood Transfusion

A patient receives a blood transfusion through a needle of radius 0.20 mm and length 2.0 cm. The density of blood is 1 050 kg/m³. The bottle supplying the blood is 0.50 m above the patient's arm. What is the rate of flow through the needle?

Solution The pressure differential between the level of the blood and the patient's arm is

$$P_1 - P_2 = \rho g h = (1\ 050\ \text{kg/m}^3)(9.80\ \text{m/s}^2)(0.50\ \text{m}) = 5.15 \times 10^3\ \text{Pa}$$

The viscosity of whole blood is given in Table 9.5. Thus, the rate of flow, from Poiseuille's law, is

$$\frac{\Delta V}{\Delta t} = \frac{\pi R^4 (P_1 - P_2)}{8 \eta L}$$

$$= \frac{\pi (2.0 \times 10^{-4}\ \text{m})^4 (5.15 \times 10^3\ \text{Pa})}{8(2.7 \times 10^{-3}\ \text{N} \cdot \text{s/m}^2)(2.0 \times 10^{-2}\ \text{m})} = \boxed{6.0 \times 10^{-8}\ \text{m}^3/\text{s}}$$

Exercise How long will it take to inject 1 pint (500 cm³) of blood into the patient?

Answer 140 min

REYNOLDS NUMBER

As we mentioned earlier, at sufficiently high velocities, fluid flow changes from simple streamline flow to turbulent flow, that is, flow characterized by a highly irregular motion of the fluid. It is found experimentally that the onset of turbulence in a tube is determined by a dimensionless factor called the **Reynolds number** (*RN*), given by

$$RN = \frac{\rho v d}{\eta} \qquad\qquad [9.25]$$

◀ Reynolds number

where ρ is the density of the fluid, v is the average speed of the fluid along the direction of flow, d is the diameter of the tube, and η is the viscosity of the fluid. If *RN* is below about 2 000, the flow of fluid through a tube is streamline; turbulence occurs if *RN* is above 3 000. In the region between 2 000 and 3 000, the flow is unstable, meaning that the fluid can move in streamline flow but any small disturbance will cause its motion to change to turbulent flow.

Example 9.12 Turbulent Flow of Blood

Determine the speed at which blood flowing through an artery of diameter 0.20 cm would become turbulent. Assume that the density of blood is 1.05×10^3 kg/m³ and that its viscosity is 2.7×10^{-3} N·s/m².

Solution At the onset of turbulence, the Reynolds number is 3 000. Thus, the speed of the blood would have to be

$$v = \frac{\eta(RN)}{\rho d} = \frac{(2.7 \times 10^{-3}\ \text{N} \cdot \text{s/m}^2)(3\ 000)}{(1.05 \times 10^3\ \text{kg/m}^3)(0.20 \times 10^{-2}\ \text{m})} = \boxed{3.9\ \text{m/s}}$$

9.10 TRANSPORT PHENOMENA

When a fluid flows through a tube, the basic mechanism that produces the flow is a difference in pressure across the ends of the tube. This pressure difference is responsible for the transport of a mass of fluid from one location to another. The fluid may also move from place to place because of a second mechanism, one that depends on a concentration difference between two points in the fluid, as opposed to a pressure difference. When the concentration (the number of molecules per unit volume) is higher at one location than at another, molecules will flow from the point where the concentration is high to the point where it is lower. The two fundamental processes involved in fluid transport resulting from concentration differences are called *diffusion* and *osmosis*. The following sections examine the nature and importance of these processes.

DIFFUSION

In a diffusion process, molecules move from a region where their concentration is high to a region where their concentration is lower. To understand why diffusion occurs, consider Figure 9.47, which represents a container in which a high concentration of molecules has been introduced into the left side. The dashed line in Figure 9.47 represents an imaginary barrier separating the region of high concentration from the region of lower concentration. Because the molecules are moving with high speeds in random directions, many of them cross the imaginary barrier moving from left to right. Very few molecules pass through this area moving from right to left simply because very few of them are on the right side of the container at any instant. Thus, there will always be a *net* movement from the region with many molecules to the region with fewer molecules. For this reason, the concentration on the left side of the container decreases in time and that on the right side increases. Once a concentration equilibrium has been reached, there will be no *net* movement across the cross-sectional area. That is, when the concentration is the same on both sides, the number of molecules diffusing from right to left in a given time interval equals the number moving from left to right in the same time interval.

The basic equation for diffusion is **Fick's law,** which in equation form is

Fick's law ▶

$$\text{Diffusion rate} = \frac{\text{mass}}{\text{time}} = \frac{\Delta M}{\Delta t} = DA\left(\frac{C_2 - C_1}{L}\right) \qquad [9.26]$$

where D is a constant of proportionality. The left side of this equation is called the diffusion rate and is a measure of the mass being transported per unit time. This equation says that the rate of diffusion is proportional to the cross-sectional area A and to the change in concentration per unit distance, $(C_2 - C_1)/L$, which is called the concentration gradient. The concentrations C_1 and C_2 are measured in kilograms per cubic meter. The proportionality constant D is called the **diffusion coefficient** and has units of square meters per second. Table 9.6 lists diffusion coefficients for a few substances.

FIGURE 9.47   When the concentration of gas molecules on the left side of the container exceeds the concentration on the right side, there will be a net motion (diffusion) of molecules from left to right.

TABLE 9.6	Diffusion Coefficients for Various Substances at 20°C

Substance	D (m²/s)
Oxygen through air	6.4×10^{-5}
Oxygen through tissue	1×10^{-11}
Oxygen through water	1×10^{-9}
Sucrose through water	5×10^{-10}
Hemoglobin through water	76×10^{-11}

THE SIZE OF CELLS AND OSMOSIS

Diffusion through cell membranes is vital for carrying oxygen to the cells of the body and for removing carbon dioxide and other waste products from them. Cells require oxygen for those metabolic processes in which substances are either synthesized or broken down. In such metabolic processes, the cell uses up oxygen and produces carbon dioxide as a by-product. A fresh supply of oxygen diffuses from the blood, where its concentration is high, into the cell, where its concentration is low. Likewise, carbon dioxide diffuses from the cell into the blood, where it is in lower concentration. Water, ions, and other nutrients also pass into and out of cells by diffusion.

A common characteristic of cells in all plants and animals is their extremely small size. The adult human body contains literally trillions of cells. In order to understand why cells are so small, we must consider the relationship between the surface area of an object and its volume.

Let us consider a cube 2 cm on a side. The area of one of its faces is 2 cm $\times$ 2 cm = 4 cm², and because a cube has six sides, the total surface area is 24 cm². Its volume is 2 cm $\times$ 2 cm $\times$ 2 cm = 8 cm³. Hence, the ratio of surface area to volume is 24/8 = 3. Now consider a larger cube, one measuring 3 cm on a side. Repeating the calculations gives us a surface area of 54 cm² and a volume of 27 cm³. In this case, the ratio of surface area to volume is 54/27 = 2. Thus, we see that as the size of an object decreases, the ratio of its surface area to its volume increases. This, of course, says that a small cell has a larger surface-area-to-volume ratio than a large cell. But how does this pertain to the operation of a cell?

A cell can function properly only if it can (1) rapidly receive vital substances such as oxygen and (2) rapidly eliminate waste products. If such substances are to move readily into and out of cells, the cells should have a large surface area. However, if the volume of the cell is too large, it could take a considerable period of time for the nutrients to diffuse into the interior of the cell where they are needed. Under optimum conditions, the surface area of the cell should be large enough so that the exposed membrane area can exchange materials effectively, while at the same time the volume should be small enough so that materials can reach or leave particular locations rapidly. To reach these optimum conditions, a small cell with its high surface-area-to-volume ratio is necessary.

As we have seen, the movement of material through cell membranes is necessary for the efficient functioning of cells. The diffusion of material through a membrane is partially determined by the size of the pores (holes) in the membrane wall. That is, small molecules, such as water, pass through the pores easily and larger molecules, such as sugar, pass through only with difficulty or not at all. A membrane that allows passage of some molecules but not others is called a **selectively permeable** membrane.

Osmosis is defined as the movement of water from a region where its concentration is high, across a selectively permeable membrane, into a region where its concentration is lower. As in the case of diffusion, osmosis continues until the concentrations on the two sides of the membrane are equal. Osmosis is often described simply as the diffusion of water across a membrane.

To understand the effect of osmosis on living cells, let us consider a particular cell in the body that contains a sugar concentration of 1%. (That is, 1 g of sugar is dissolved in enough water to make 100 mL of solution; "mL" is the abbreviation for milliliters, 10^{-3} L = 1 cm^3.) Now assume that this cell is immersed in a 5% sugar solution (5 g of sugar dissolved in enough water to make 100 mL). In such a situation, water diffuses from inside the cell, where its concentration is higher, across the cell wall membrane, to the outside solution, where the concentration of water is lower. This loss of water from the cell causes it to shrink and perhaps become damaged through dehydration. If the concentrations are reversed, water diffuses into the cell, causing it to swell and perhaps burst. It should be obvious from this description that normal osmotic relationships must be maintained in the body. If solutions are introduced into the body intravenously, care must be taken to ensure that these solutions do not disturb the osmotic balance of the body because such a disturbance can lead to cell damage. For example, if 9% saline solution surrounds a red blood cell, the cell will shrink. On the other hand, if the saline solution is about 1%, the cell will eventually burst.

In the body, blood is cleansed of impurities by osmosis as it flows through the kidneys (see Fig. 9.48a). Arterial blood first passes through a bundle of capillaries known as a glomerulus where most of the waste products along with some essential salts and minerals are removed. From the glomerulus, a narrow tube emerges that is in intimate contact with other capillaries throughout its length. As blood passes through the tubules, most of the essential elements are returned to it; waste products are not allowed to reenter and are eventually removed as urine.

If the kidneys fail, an artificial kidney or a dialysis machine can filter the blood. Figure 9.48b shows how this is done. Blood from an artery in the arm is mixed with heparin, a blood thinner, and allowed to pass through a tube covered with a semipermeable membrane. This tubing is immersed in a bath of a dialysate fluid with the same chemical composition as purified blood. Waste products from the blood enter the dialysate by diffusion through the membrane walls. The filtered blood is then returned to a vein.

APPLICATION

KIDNEY FUNCTION AND DIALYSIS

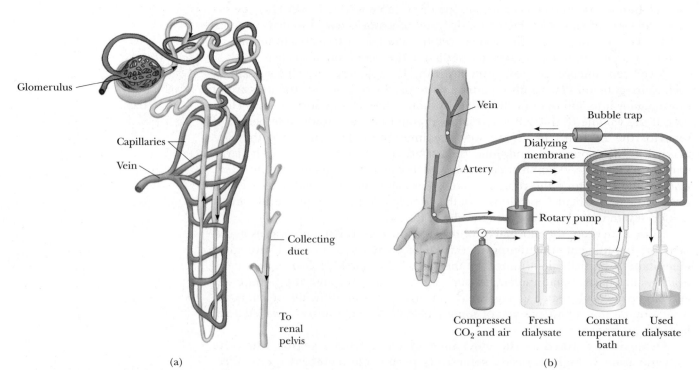

Glomerulus

Capillaries

Vein

Collecting duct

To renal pelvis

(a)

Vein

Bubble trap

Dialyzing membrane

Artery

Rotary pump

Compressed CO_2 and air

Fresh dialysate

Constant temperature bath

Used dialysate

(b)

FIGURE 9.48 (a) Diagram of a single nephron in the human excretory system. (b) An artificial kidney.

MOTION THROUGH A VISCOUS MEDIUM

When an object falls through air, its motion is impeded by the force of air resistance. In general, this force is dependent on the shape of the falling object and on its velocity. This viscous drag acts on all falling objects, but the exact details of the motion can be calculated only for a few cases in which the object has a simple shape, such as a sphere. In this section, we shall examine the motion of a tiny spherical object falling slowly through a viscous medium.

In 1845 a scientist named George Stokes found that the magnitude of the resistive force on a very small spherical object of radius r falling slowly through a fluid of viscosity η with speed v is given by

$$F_r = 6\pi\eta r v \qquad \textbf{[9.27]}$$

This equation, called **Stokes's law,** has many important applications. For example, it describes the sedimentation of particulate matter in blood samples. It was used by Robert Millikan (1886–1953) to calculate the radius of charged oil droplets falling through air. From this, Millikan was ultimately able to determine the charge on the electron. Millikan was awarded the Nobel prize in 1923 for this pioneering work on elemental charge.

As a sphere falls through a viscous medium, three forces act on it, as shown in Figure 9.49: $\mathbf{F}_r$ is the force of frictional resistance, $\mathbf{B}$ is the buoyant force of the fluid, and $\mathbf{w}$ is the force of gravity acting on the sphere, whose magnitude is given by

$$w = \rho g V = \rho g \left(\tfrac{4}{3}\pi r^3\right)$$

where ρ is the density of the sphere and $\tfrac{4}{3}\pi r^3$ is its volume. According to Archimedes's principle, the magnitude of the buoyant force is equal to the weight of the fluid displaced by the sphere:

$$B = \rho_f g V = \rho_f g\left(\tfrac{4}{3}\pi r^3\right)$$

where ρ_f is the density of the fluid.

At the instant the sphere begins to fall, the force of frictional resistance is zero because the speed of the sphere is zero. As the sphere accelerates, its speed increases and so does $\mathbf{F}_r$. Finally, at a speed called the **terminal speed** v_t, *the net force goes to zero.* This occurs when the net upward force balances the downward force of gravity. Hence, the sphere reaches terminal speed when

$$F_r + B = w$$

or

$$6\pi\eta r v_t + \rho_f g\left(\tfrac{4}{3}\pi r^3\right) = \rho g\left(\tfrac{4}{3}\pi r^3\right)$$

When this is solved for v_t, we get

$$v_t = \frac{2r^2 g}{9\eta}\,(\rho - \rho_f) \qquad \textbf{[9.28]}$$

SEDIMENTATION AND CENTRIFUGATION

If an object is not spherical, we can still use the basic approach just described to determine its terminal speed. The only difference is that we cannot use Stokes's law for the resistive force. Instead, let us assume that the resistive force has a magnitude given by $F_r = kv$, where k is a coefficient of frictional resistance that must be determined experimentally. As we discussed previously, the object reaches its terminal speed when the downward force of gravity is balanced by the net upward force, or

$$w = B + F_r \qquad \textbf{[9.29]}$$

where B is the buoyant force, given by $B = \rho_f g V$. We can use the fact that volume V

FIGURE 9.49 A sphere falling through a viscous medium. The forces acting on the sphere are the resistive frictional force $\mathbf{F}_r$, the buoyant force $\mathbf{B}$, and the force of gravity $\mathbf{w}$.

 Terminal speed

By spreading their arms and legs out from their bodies while keeping the planes of their bodies parallel to the ground, skydivers experience maximum air drag. The forces acting on this skydiver are the force of gravity and the force of air resistance. When these two forces balance each other, the skydiver has zero acceleration and reaches a terminal speed of about 60 m/s. (© *Steve Fitchett/FPG/Getty Images*)

of the displaced fluid is related to the density ρ of the falling object by $V = m/\rho$. Hence, we can express the buoyant force as

$$B = \frac{\rho_f}{\rho} mg$$

Let us substitute this expression for B and $F_r = kv_t$ into Equation 9.29 (terminal speed condition):

$$mg = \frac{\rho_f}{\rho} mg + kv_t$$

or

$$v_t = \frac{mg}{k}\left(1 - \frac{\rho_f}{\rho}\right) \qquad [9.30]$$

The terminal speed for particles in biological samples is usually quite small. For example, the terminal speed for blood cells falling through plasma is about 5 cm/h in Earth's gravitational field. The terminal speeds for the molecules that make up a cell are many orders of magnitude smaller than this because of their much smaller mass. The speed at which materials fall through a fluid is called the *sedimentation rate*. This number is often important in clinical analysis.

It is often desired to increase the sedimentation rate in a fluid. A common method used to accomplish this is to increase the effective acceleration g that appears in Equation 9.30. A fluid containing various biological molecules is placed in a centrifuge and whirled at very high angular speeds (Fig. 9.50). Under these conditions, the particles experience a large radial acceleration, $a_c = v^2/r = \omega^2 r$, which is much greater than the free-fall acceleration, and so we can replace g in Equation 9.30 by $\omega^2 r$:

$$v_t = \frac{m\omega^2 r}{k}\left(1 - \frac{\rho_f}{\rho}\right) \qquad [9.31]$$

ω

FIGURE 9.50 Simplified diagram of a centrifuge (top view).

This equation indicates that the sedimentation rate is enormously speeded up in a centrifuge ($\omega^2 r \gg g$) and that those particles having the greatest mass also have the largest terminal speed. Therefore, the most massive particles settle out on the bottom of a test tube first.

Example 9.13   The Spinning Test Tube

A centrifuge rotates at 50 000 rev/min, which corresponds to an angular speed of 5 240 rad/s (a typical speed). A test tube placed in this device has its top 5.0 cm from the axis of rotation and its bottom 13 cm from this axis. Find the effective value of g at the midpoint of the test tube, which corresponds to a distance 9.0 cm from the axis of rotation.

Solution   The acceleration experienced by the particles of the tube at a distance $r = 9.0$ cm from the axis of rotation is given by

$$a_c = \omega^2 r = (5\ 240\ \text{rad/s})^2(9.0 \times 10^{-2}\ \text{m}) = \boxed{2.5 \times 10^6\ \text{m/s}^2}$$

EXERCISE   If the mass of the contents of the test tube is 15 g, find the force that the bottom of the tube must exert on the contents of the tube to provide this centripetal acceleration. Assume a centripetal acceleration equal to that found at the midpoint of the tube.

ANSWER 3.7×10^4 N, or about 8 000 lb! (Because of such large forces, the base of the tube in a centrifuge must be rigidly supported to keep the glass from shattering.)

SUMMARY

Matter is normally classified as being in one of three states: solid, liquid, or gaseous.

The elastic properties of a solid can be described using the concepts of stress and strain. **Stress** is related to the force producing a deformation; **strain** is a measure of the degree of deformation. Stress is proportional to strain, and the constant of proportionality is the **elastic modulus:**

$$\text{Elastic modulus} \equiv \frac{\text{stress}}{\text{strain}} \qquad [9.1]$$

Three common types of deformation are: (1) the resistance of a solid to elongation or compression, characterized by **Young's modulus** Y; (2) the resistance to displacement of the faces of a solid sliding past each other, characterized by the **shear modulus** S; (3) the resistance of a solid or liquid to a volume change, characterized by the **bulk modulus** B.

In the SI, pressure is expressed in pascals (Pa), where $1\ \text{Pa} \equiv 1\ \text{N/m}^2$.

The **density** ρ of a substance of uniform composition is its mass per unit volume—kilograms per cubic meter (kg/m^3) in the SI:

$$\rho \equiv \frac{M}{V} \qquad [9.6]$$

The **pressure** P in a fluid is the force per unit area that the fluid exerts on an object immersed in it:

$$P \equiv \frac{F}{A} \qquad [9.7]$$

The pressure in an incompressible fluid varies with depth h according to the expression

$$P = P_0 + \rho g h \qquad [9.10]$$

where P_0 is atmospheric pressure $(1.013 \times 10^5\ \text{Pa})$ and ρ is the density of the fluid.

Pascal's principle states that, when pressure is applied to an enclosed fluid, the pressure is transmitted undiminished to every point of the fluid and to the walls of the containing vessel.

When an object is partially or fully submerged in a fluid, the fluid exerts an upward force, called the **buoyant force,** on the object. According to **Archimedes's principle,** the magnitude of the buoyant force is equal to the weight of the fluid displaced by the object.

Certain aspects of a fluid in motion can be understood by assuming that the fluid is nonviscous and incompressible and that its motion is in a steady state with no turbulence.

1. The flow rate through the pipe is a constant, which is equivalent to stating that the product of the cross-sectional area A and the speed v at any point is constant.

$$A_1 v_1 = A_2 v_2 \qquad [9.14]$$

This relation is referred to as the **equation of continuity.**

2. The sum of the pressure, the kinetic energy per unit volume, and the potential energy per unit volume has the same value at all points along a streamline:

$$P + \tfrac{1}{2}\rho v^2 + \rho g y = \text{constant} \qquad [9.16]$$

This is known as **Bernoulli's equation.**

CONCEPTUAL QUESTIONS

1. Explain the effect of an elastic stocking on a leg.
2. Baseball homerun hitters like to play in Denver, but curveball pitchers do not. Why?
3. A woman wearing high-heeled shoes is invited into a home in which the kitchen has vinyl floor covering. Why should the homeowner be concerned?
4. Figure Q9.4 shows aerial views from directly above two dams. Both dams are equally long (the vertical dimension in the diagram) and equally deep (into the page in the diagram). The dam on the left holds back a very large lake, whereas the dam on the right holds back a narrow river. Which dam has to be built more strongly?

FIGURE Q9.4

5. A typical silo on a farm has many bands wrapped around its perimeter, as shown in Figure Q9.5. Why is the spacing between successive bands smaller at the lower portions of the silo?

FIGURE Q9.5

6. During inhalation, the pressure in the lungs is slightly less than external pressure, and the muscles controlling exhalation are relaxed. Underwater, the body equalizes internal and external pressures. Discuss the condition of the muscles if a person underwater is breathing through a snorkel. Would a snorkel work in deep water?

7. Colloquially, we say a nurse uses a syringe to "draw" blood. Why is it more accurate to say that a syringe is used to "accept" blood?

8. Many people believe that a vacuum created inside a vacuum cleaner causes particles of dirt to be drawn in. Actually, however, the dirt is pushed in. Explain.

9. Suppose a damaged ship just barely floats in the ocean after a hole in its hull has been sealed. It is pulled by a tug-

boat toward shore and into a river, heading toward a dry-dock for repair. As it is pulled up the river, it sinks. Why?

10. Will an ice cube float higher in water or in an alcoholic beverage?

11. A pound of Styrofoam and a pound of lead have the same weight. If they are placed on a sensitive equal arm balance, will it balance?

12. An ice cube is placed in a glass of water. What happens to the level of the water as the ice melts?

13. Place two cans of soft drinks, one regular and one diet, in a container of water. You will find that the diet drink floats while the regular one sinks. Use Archimedes's principle to devise an explanation. (*Broad Hint:* The artificial sweetener used in diet drinks is less dense than sugar.)

14. Prairie dogs live in underground burrows with at least two entrances. They ventilate their burrows by building a mound over one entrance, as in the photograph. It is open to a stream of air when a breeze blows from any direction. A second entrance at ground level is open to almost stagnant air. How does this construction create an air flow through the burrow?

FIGURE Q9.14 *(Pamela Zilly/The Image Bank)*

15. When you are driving a small car on the freeway and a truck passes you at high speed, you feel pulled toward the truck. Why?

16. A barge is carrying a load of gravel along a river. It approaches a low bridge, and the captain realizes that the top of the pile of gravel is not going to make it under the bridge. The captain orders the crew to quickly shovel gravel from the pile into the water. Is this a good decision?

17. Tornadoes and hurricanes often lift the roofs of houses. Use the Bernoulli effect to explain why. Why should you keep your windows open under these conditions?

PROBLEMS

1, 2, 3 = straightforward, intermediate, challenging   ☐ = full solution available in Student Solutions Manual/Study Guide

web = solution posted at **http://info.brookscole.com/serway**   🔬 = biomedical application

Section 9.2   The Deformation of Solids

1. A steel wire of diameter 1 mm can support tension 0.2 kN. A cable to support tension 20 kN should have diameter of what order of magnitude?

2. If the shear stress in steel exceeds about 4.00×10^8 N/m^2, the steel ruptures. Determine the shearing force necessary to (a) shear a steel bolt 1.00 cm in diameter and (b) punch a 1.00-cm-diameter hole in a steel plate 0.500 cm thick.

3. The heels on a pair of women's shoes have radii of 0.50 cm at the bottom. If 30% of the weight of a woman weighing 480 N is supported by each heel, find the stress on each heel.

4. When water freezes, it expands about 9.00%. What would be the pressure increase inside your automobile engine block if the water in it froze? (The bulk modulus of ice is $2.00 \times 10^9 \, \text{N/m}^2$.)

5. For safety in climbing, a mountaineer uses a nylon rope that is 50 m long and 1.0 cm in diameter. When supporting a 90-kg climber, the rope elongates 1.6 m. Find its Young's modulus.

6. A stainless steel orthodontic wire is applied to a tooth, as in Figure P9.6. The wire has an unstretched length of 3.1 cm and a diameter of 0.22 mm. If the wire is stretched 0.10 mm, find the magnitude and direction of the force on the tooth. Disregard the width of the tooth and assume the Young's modulus for stainless steel is $18 \times 10^{10} \, \text{Pa}$.

30°   30°

FIGURE P9.6

7. Bone has a Young's modulus of about $18 \times 10^9 \, \text{Pa}$. Under compression, it can withstand a stress of about $160 \times 10^6 \, \text{Pa}$ before breaking. Assume that a femur (thigh bone) is 0.50 m long and calculate the amount of compression this bone can withstand before breaking.

8. The distortion of Earth's crustal plates is an example of shear on a large scale. A particular crustal rock has a shear modulus of $1.5 \times 10^{10} \, \text{Pa}$. What shear stress is involved when a 10-km layer of this rock is sheared through a distance of 5.0 m?

9. A child slides across a floor in a pair of rubber-soled shoes. The friction force acting on each foot is 20 N, the footprint area of each foot is 14 cm², and the thickness of the soles is 5.0 mm. Find the horizontal distance traveled by the sheared face of the sole. The shear modulus of the rubber is $3.0 \times 10^6 \, \text{Pa}$.

10. A high-speed lifting mechanism supports an 800-kg object with a steel cable 25.0 m long and 4.00 cm² in cross-sectional area. (a) Determine the elongation of the cable. (b) By what additional amount does the cable increase in length if the object is accelerated upward at a rate of $3.0 \, \text{m/s}^2$? (c) What is the greatest mass that can be accelerated upward at $3.0 \, \text{m/s}^2$ if the stress in the cable is not to exceed the elastic limit of the cable, which is $2.2 \times 10^8 \, \text{Pa}$?

11. The stainless-steel hip pin in Figure P9.11 has a radius of 0.25 cm. A total upward force of 300 N is exerted by the leg on the pin. Determine the deformation of the pin in the gap between the bones. Assume that the shear modulus is equal to that given for steel in Table 9.1.

12. A person weighing 800 N stands on the ball of one foot. The tibia is 36 cm long; other dimensions are given in Figure P9.12. Find (a) the stress in the tibia, (b) the strain in the tibia, and (c) the change in length of the tibia. Young's modulus for the tibia is $1.80 \times 10^{10} \, \text{Pa}$.

Section 9.3 Density and Pressure

13. A 50.0-kg ballet dancer stands on her toes during a performance with 4 in.² (26.0 cm²) in contact with the floor.

0.20 mm

FIGURE P9.11

Tibia

Outside radius 12 mm

Inside radius 6.0 mm

Achilles tendon

T   F

A

5.0 cm   14 cm

800 N

FIGURE P9.12

What is the pressure exerted by the floor over the area of contact (a) if the dancer is stationary, and (b) if the dancer is leaping upwards with an acceleration of $4.00 \, \text{m/s}^2$?

14. The four tires of an automobile are inflated to a gauge pressure of $2.0 \times 10^5 \, \text{Pa}$. Each tire has an area of 0.024 m² in contact with the ground. Determine the weight of the automobile.

15. A specimen of urea dissolved in water has a volume of 100 cm³ and a mass of 133 g. Determine the density of the solution.

16. A 70-kg man in a 5.0-kg chair tilts back so that all the weight is balanced on two legs of the chair. Assume that each leg makes contact with the floor over a circular area with a radius of 1.0 cm, and find the pressure exerted by each leg on the floor.

17. If 1.0 m³ of concrete weighs $5.0 \times 10^4 \, \text{N}$, what is the height of the tallest cylindrical concrete pillar that will not collapse under its own weight? The compression strength of concrete (the maximum pressure that can be exerted on the base of the structure) is $1.7 \times 10^7 \, \text{Pa}$.

Section 9.4 Variation of Pressure with Depth

Section 9.5 Pressure Measurements

18. Water is to be pumped to the top of the Empire State Building, which is 1 200 ft high. What gauge pressure is

needed in the water line at the base of the building to raise the water to this height?

19. A collapsible plastic bag (Fig. P9.19) contains a glucose solution. If the average gauge pressure in the vein is 1.33×10^4 Pa, what must be the minimum height h of the bag in order to infuse glucose into the vein? Assume that the specific gravity of the solution is 1.02.

Glucose solution

h

FIGURE P9.19

20. (a) A very powerful vacuum cleaner has a hose 2.86 cm in diameter. With no nozzle on the hose, what is the weight of the heaviest brick it can lift? (b) A very powerful octopus uses one sucker, of diameter 2.86 cm, on each of the two shells of a clam, in an attempt to pull the shells apart. Find the greatest force the octopus can exert in salt water 32.3 m deep.

21. For the cellar of a new house, a hole is dug in the ground, with vertical sides going down 2.40 m. A concrete foundation wall is built all the way across the 9.60-m width of the excavation. This foundation wall is 0.183 m away from the front of the cellar hole. During a rainstorm, drainage from the street fills up the space in front of the concrete wall, but not the cellar behind the wall. The water does not soak into the clay soil. Find the force the water causes on the foundation wall. For comparison, the weight of the water is given by

$$2.40 \text{ m} \times 9.60 \text{ m} \times 0.183 \text{ m} \times 1\,000 \text{ kg/m}^3$$
$$\times 9.80 \text{ m/s}^2 = 41.3 \text{ kN}$$

22. Blaise Pascal duplicated Torricelli's barometer using a red Bordeaux wine, of density 984 kg/m³, as the working liquid (Fig. P9.22). What was the height h of the wine column for normal atmospheric pressure? Would you expect the vacuum above the column to be as good as for mercury?

23. A container is filled to a depth of 20.0 cm with water. On **web** top of the water floats a 30.0-cm-thick layer of oil with specific gravity 0.700. What is the absolute pressure at the bottom of the container?

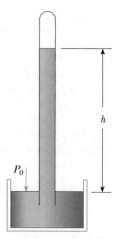

P_0

h

FIGURE P9.22

24. Piston ① in Figure P9.24 has a diameter of 0.25 in.; piston ② has a diameter of 1.5 in. In the absence of friction, determine the force **F** necessary to support the 500-lb weight.

500 lb

2.0 in.

10 in.

②

①

F

FIGURE P9.24

25. Figure P9.25 shows the essential parts of a hydraulic brake system. The area of the piston in the master cylinder is 6.4 cm², and that of the piston in the brake cylinder is 1.8 cm². The coefficient of friction between shoe and wheel drum is 0.50. If the wheel has a radius of 34 cm, determine the frictional torque about the axle when a force of 44 N is exerted on the brake pedal.

Wheel drum

Pedal

Shoe

Master cylinder

Brake cylinder

FIGURE P9.25

Section 9.6 Buoyant Forces and Archimedes's Principle

26. A frog in a hemispherical pod finds that he just floats without sinking in a fluid of density 1.35 g/cm^3. If the pod has a radius of 6.00 cm and negligible mass, what is the mass of the frog? (See Fig. P9.26.)

FIGURE P9.26

27. A small ferry boat is 4.00 m wide and 6.00 m long. When a loaded truck pulls onto it, the boat sinks an additional 4.00 cm into the river. What is the weight of the truck?

28. The density of ice is 920 kg/m^3, and that of sea water is 1 030 kg/m^3. What fraction of the total volume of an iceberg is exposed?

29. A kidney weighs 5.7 N in air. If the kidney is completely submerged in water, its apparent weight is 1.6 N. Determine the specific gravity of the kidney.

30. A 10.0-kg block of metal is suspended from a scale and immersed in water as in Figure P9.30. The dimensions of the block are 12.0 cm × 10.0 cm × 10.0 cm. The 12.0-cm dimension is vertical, and the top of the block is 5.00 cm below the surface of the water. (a) What are the forces exerted by the water on the top and bottom of the block? (Take $P_0 = 1.013\,0 \times 10^5$ N/m^2.) (b) What is the reading of the spring scale? (c) Show that the buoyant force equals the difference between the forces at the top and bottom of the block.

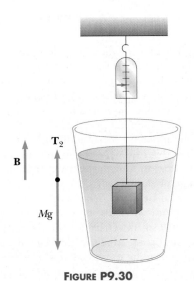

FIGURE P9.30

31. A bathysphere used for deep sea exploration has a radius of 1.50 m and a mass of 1.20×10^4 kg. In order to dive, this submarine takes on mass in the form of sea water. De-

termine the mass the submarine must take on so that it can descend at a constant speed of 1.20 m/s, when the resistive force on it is 1 100 N upward. The density of sea water is 1.03×10^3 kg/m^3.

32. The United States possesses the eight largest warships in the world, aircraft carriers of the *Nimitz* class, and is building one more. Suppose one of the ships bobs up to float 11.0 cm higher in the water when 50 fighters take off from it in 25 minutes, at a location where $g = 9.78$ m/s^2. Bristling with bombs and missiles, the planes have average mass 29 000 kg. Find the horizontal area enclosed by the waterline of the $4-billion ship. By comparison, its flight deck has area 18 000 m^2. Below decks are passageways hundreds of meters long, so narrow that two large men cannot pass each other.

33. An empty rubber balloon has a mass of 0.012 0 kg. The balloon is filled with helium at a density of 0.181 kg/m^3. At this density the balloon has a radius of 0.500 m. If the filled balloon is fastened to a vertical line, what is the tension in the line?

34. A light spring of constant $k = 160$ N/m rests vertically on the bottom of a large beaker of water (Fig. P9.34a). A 5.00-kg block of wood (density = 650 kg/m^3) is connected to the spring, and the block-spring system is allowed to come to static equilibrium (Fig. P9.34b). What is the elongation ΔL of the spring?

(a)   (b)

FIGURE P9.34

35. A sample of an unknown material appears to weigh 300 N in air and 200 N when immersed in alcohol of specific gravity 0.700. What are (a) the volume and (b) the density of the material?

36. An object weighing 300 N in air is immersed in water after being tied to a string connected to a balance. The scale now reads 265 N. Immersed in oil, the object appears to weigh 275 N. Find (a) the density of the object and (b) the density of the oil.

37. A thin spherical shell of mass 0.400 kg and diameter 0.200 m is filled with alcohol ($\rho = 806$ kg/m^3). It is then released from rest at the bottom of a pool of water. Find the acceleration of the alcohol-filled shell as it starts to rise toward the surface of the water.

38. A rectangular air mattress is 2.0 m long, 0.50 m wide, and 0.08 m thick. If it has a mass of 2.0 kg, what additional mass can it support in water?

39. A 1.00-kg beaker containing 2.00 kg of oil (density = 916 kg/m^3) rests on a scale. A 2.00-kg block of iron is suspended from a spring scale and completely submerged in the oil (Fig. P9.39). Find the equilibrium readings of both scales.

FIGURE P9.39

Section 9.7 Fluids in Motion

Section 9.8 Other Applications of Fluid Dynamics

40. A cowboy at a dude ranch fills a horse trough that is 1.5 m long, 60 cm wide, and 40 cm deep. He uses a 2.0-cm-diameter hose from which water emerges at 1.5 m/s. How long does it take him to fill the trough?

41. (a) Calculate the mass flow rate (in grams per second) of blood ($\rho = 1.0$ g/cm^3) in an aorta with a cross-sectional area of 2.0 cm^2 if the flow speed is 40 cm/s. (b) Assume that the aorta branches to form a large number of capillaries with a combined cross-sectional area of 3.0×10^3 cm^2. What is the flow speed in the capillaries?

42. A liquid ($\rho = 1.65$ g/cm^3) flows through two horizontal sections of tubing joined end to end. In the first section the cross-sectional area is 10.0 cm^2, the flow speed is 275 cm/s, and the pressure is 1.20×10^5 Pa. In the second section the cross-sectional area is 2.50 cm^2. Calculate the smaller section's (a) flow speed and (b) pressure.

43. A hypodermic syringe contains a medicine with the density of water (Fig. P9.43). The barrel of the syringe has a cross-sectional area of 2.50×10^{-5} m^2. In the absence of a force on the plunger, the pressure everywhere is 1.00 atm. A force **F** of magnitude 2.00 N is exerted on the plunger, making medicine squirt from the needle. Determine the medicine's flow speed through the needle. Assume that the pressure in the needle remains equal to 1.00 atm and that the syringe is horizontal.

FIGURE P9.43

44. When a person inhales, air moves down the bronchus (windpipe) at 15 cm/s. The average flow speed of the air doubles through a constriction in the bronchus. Assuming incompressible flow, determine the pressure drop in the constriction.

45. A jet of water squirts out horizontally from a hole near the bottom of the tank in Figure P9.45. If the hole has a diam-

eter of 3.50 mm, what is the height h of the water level in the tank?

FIGURE P9.45

46. A large storage tank, open to the atmosphere at the top and filled with water, develops a small hole in its side at a point 16.0 m below the water level. If the rate of flow from the leak is 2.50×10^{-3} m^3/min, determine (a) the speed at which the water leaves the hole and (b) the diameter of the hole.

47. The inside diameters of the larger portions of the horizontal pipe in Figure P9.47 are 2.50 cm. Water flows to the right at a rate of 1.80×10^{-4} m^3/s. Determine the inside diameter of the constriction.

FIGURE P9.47

48. Through a pipe of diameter 15.0 cm, water is pumped from the Colorado River up to Grand Canyon Village, on the rim of the canyon. The river is at 564 m elevation and the village is at 2 096 m. (a) At what minimum pressure must the water be pumped to arrive at the village? (b) If 4 500 m^3 are pumped per day, what is the speed of the water in the pipe? (c) What additional pressure is necessary to deliver this flow? (*Note:* You may assume that the free-fall acceleration and the density of air are constant over this range of elevations.)

49. Old Faithful Geyser in Yellowstone Park erupts at approximately 1-hour intervals, and the height of the fountain reaches 40.0 m. (a) Consider the rising stream as a series of separate drops. Analyze the free-fall motion of one of them to determine the speed at which the water leaves the ground. (b) Treat the rising stream as an ideal fluid in streamline flow. Use Bernoulli's equation to determine the speed of the water as it leaves ground level. (c) What is the pressure (above atmospheric) in the heated underground chamber 175 m below the vent? You may assume the chamber is large compared to the geyser vent.

50. An airplane is cruising at altitude 10 km. The pressure outside the craft is 0.287 atm; within the passenger compartment the pressure is 1.00 atm and the temperature is 20°C. The density of air is 1.20 kg/m^3 at 20°C and 1 atm of pressure. A small leak occurs in one of the window seals in the passenger compartment. Model the air as an ideal fluid to find the speed of the stream of air flowing through the leak.

51. A siphon is a device that allows a fluid to seemingly defy gravity (Fig. P9.51). The flow must be initiated by a partial vacuum in the tube, as in a drinking straw. (a) Show that the speed at which the water emerges from the siphon is given by $v = \sqrt{2gh}$. (b) For what values of y will the siphon work?

FIGURE P9.51

Section 9.9 Surface Tension, Capillary Action, and Viscous Fluid Flow

52. In order to lift a wire ring of radius 1.75 cm from the surface of a container of blood plasma, a vertical force of 1.61×10^{-2} N greater than the weight of the ring is required. Calculate the surface tension of blood plasma from this information.

53. A square metal sheet 3.0 cm on a side and of negligible thickness is attached to a balance and inserted into a container of fluid. The contact angle is found to be zero, as shown in Figure P9.53a and the balance to which the metal sheet is attached reads 0.40 N. A thin veneer of oil is then spread over the metal sheet and the contact angle becomes 180°, as shown in Figure P9.53b. The balance now reads 0.39 N. What is the surface tension of the fluid?

(a)   (b)

FIGURE P9.53

54. Whole blood has a surface tension of 0.058 N/m and a density of 1 050 kg/m^3. To what height can whole blood rise in a capillary blood vessel that has a radius of 2.0×10^{-6} m if the contact angle is zero?

55. A certain fluid has a density of 1 080 kg/m^3 and is observed to rise to a height of 2.1 cm in a 1.0-mm-diameter tube. The contact angle between the wall and the fluid is zero. Calculate the surface tension of the fluid.

56. A staining solution used in a microbiology laboratory has a surface tension of 0.088 N/m and a density 1.035 times the density of water. What must be the diameter of a capillary tube so that this solution will rise to a height of 5 cm? (Assume a contact angle of zero.)

57. The block of ice (temperature 0°C) shown in Figure P9.57 is drawn over a level surface lubricated by a layer of water 0.10 mm thick. Determine the magnitude of the force **F** needed to pull the block with a constant speed of 0.50 m/s. At 0°C, the viscosity of water has the value $\eta = 1.79 \times 10^{-3}$ N·s/m^2.

FIGURE P9.57

58. A thin 1.5-mm coating of glycerine has been placed between two microscope slides of width 1.0 cm and length 4.0 cm. Find the force required to pull one of the microscope slides at a constant speed of 0.30 m/s relative to the other.

59. A straight horizontal pipe with a diameter of 1.0 cm and a length of 50 m carries oil with a coefficient of viscosity of 0.12 N·s/m^2. At the output of the pipe, the flow rate is 8.6×10^{-5} m^3/s and the pressure is 1.0 atm. Find the gauge pressure at the pipe input.

60. The pulmonary artery, which connects the heart to the lungs, has an inner radius of 2.6 mm and is 8.4 cm long. If the pressure drop between the heart and lungs is 400 Pa, what is the average speed of blood in the pulmonary artery?

61. Spherical particles of a protein of density 1.8 g/cm^3 are shaken up in a solution of 20°C water. The solution is allowed to stand for 1.0 h. If the depth of water in the tube is 5.0 cm, find the radius of the largest particles that remain in solution at the end of the hour.

62. Blood, during transfusion, flows from a bottle through a needle that is 3.5 cm long and has an inside diameter of 0.60 mm. How high above the needle must the bottle be held in order to transfuse 0.50 L of blood in 40 min? The density of whole blood is 1 050 kg/m^3. Assume the patient's blood is nearly at atmospheric pressure.

63. Glucose, during transfusion, flows from a bag through a needle that is 3.0 cm long and has an inside diameter of 0.50 mm. If the transfusion rate is to be 0.50 L in 40 min, how high above the needle must the bag be held? Assume a density of 1 040 kg/m^3 and a coefficient of viscosity of 2.25×10^{-3} N·s/m^2 for the glucose. Assume the patient's blood is nearly at atmospheric pressure.

64. (a) If cholesterol builds up in major arteries so that their diameter is reduced by 10%, what effect will the shrinkage have on blood flow? That is, by what percentage will blood

flow be reduced? (b) If cholesterol builds up enough to reduce the diameter by 20%, what effect does this have?

65. The aorta in humans has a diameter of about 2.0 cm and, at certain times, the blood speed through it is about 55 cm/s. Is the blood flow turbulent? The density of whole blood is 1 050 kg/m^3, and its coefficient of viscosity is 2.7×10^{-3} N·s/m^2.

66. A pipe carrying 20°C water has a diameter of 2.5 cm. Estimate the maximum flow speed if the flow must be streamline.

Section 9.10 Transport Phenomena

67. Sucrose is allowed to diffuse along a 10-cm length of tubing filled with water. The tube is 6.0 cm^2 in cross-sectional area. The diffusion coefficient is equal to 5.0×10^{-10} m^2/s, and 8.0×10^{-14} kg is transported along the tube in 15 s. What is the difference in the concentration levels of sucrose at the two ends of the tube?

68. Glycerin in water diffuses along a horizontal column that has a cross-sectional area of 2.0 cm^2. The concentration gradient is 3.0×10^{-2} kg/m^4, and the diffusion rate is found to be 5.7×10^{-15} kg/s. Determine the diffusion coefficient.

69. The viscous force on an oil drop is measured to be 3.0×10^{-13} N when the drop is falling through air with a speed of 4.5×10^{-4} m/s. If the radius of the drop is 2.5×10^{-6} m, what is the viscosity of air?

70. Small spheres of diameter 1.00 mm fall through 20°C water with a terminal speed of 1.10 cm/s. Calculate the density of the spheres.

ADDITIONAL PROBLEMS

71. Leatherback turtles can dive to a depth of about 3 000 feet. Calculate the absolute pressure at this depth. Assume the density of sea water is 1 024 kg/m^3 and that the air above exerts a pressure of 101.3 kPa.

72. Take the density of blood to be ρ and the distance between the feet and the heart to be h_H. Ignore the flow of blood. (a) Show that the difference in blood pressure between the feet and heart is given by $P_F - P_H = \rho g h_H$. (b) Take the density of blood to be 1.05×10^3 kg/m^3 and the distance between the heart and the feet to be 1.20 m. Find the difference in blood pressure between these two points. This problem indicates that pumping blood from the extremities is very difficult for the heart. The veins in the legs have valves in them that open when blood is pumped toward the heart and close when blood flows away from the heart. Also, pumping action produced by actions such as walking and breathing assists the heart.

73. The approximate inside diameter of the aorta is 0.50 cm; that of a capillary is 10 μm. The approximate average blood flow speed is 1.0 m/s in the aorta and 1.0 cm/s in the capillaries. If all the blood in the aorta eventually flows through the capillaries, estimate the number of capillaries in the circulatory system.

74. The small piston of a hydraulic lift has a cross-sectional area of 3.00 cm^2, and its large piston has a cross-sectional area of 200 cm^2. What force must be applied to the small piston for the lift to raise a load of 15.0 kN? (In service stations, this force is usually exerted by compressed air.)

75. The human brain and spinal cord are immersed in the cerebrospinal fluid. The fluid is normally continuous between the cranial and spinal cavities. It normally exerts a pressure of 100 to 200 mm of H$_2$O above the prevailing atmospheric pressure. In medical work, pressures are often measured in units of mm of H$_2$O because body fluids, including the cerebrospinal fluid, typically have nearly the same density as water. The pressure of the cerebrospinal fluid can be measured by means of a *spinal tap*. A hollow tube is inserted into the spinal column, and the height to which the fluid rises is observed, as shown in Figure P9.75. If the fluid rises to a height of 160 mm, we write its gauge pressure as 160 mm H$_2$O. (a) Express this pressure in pascals, in atmospheres and in millimeters of mercury. (b) Sometimes it is necessary to determine if an accident victim has suffered a crushed vertebra that is blocking flow of the cerebrospinal fluid in the spinal column. In other cases, a physician may suspect a tumor or other growth is blocking the spinal column and inhibiting flow of cerebrospinal fluid. Such conditions can be investigated by means of the *Queckensted test*. In this procedure, the veins in the patient's neck are compressed, to make the blood pressure rise in the brain. The increase in pressure in the blood vessels is transmitted to the cerebrospinal fluid. What should be the normal effect on the height of the fluid in the spinal tap? (c) Suppose that compressing the veins has no effect on the fluid level. What might account for this?

FIGURE P9.75

76. Determining the density of a fluid has many important applications. A car battery contains sulfuric acid, and the battery will not function properly if the acid density is too low. Similarly, the effectiveness of antifreeze in your car's engine coolant depends on the density of the mixture (usually ethylene glycol and water). When you donate blood to a blood bank, its screening includes determination of the density of the blood, since higher density correlates with higher hemoglobin content. A *hydrometer* is an instrument used to determine liquid density. A simple one is sketched in Figure P9.76. The bulb of a syringe is squeezed and released to lift a sample of the liquid of interest into a tube containing a calibrated rod of known density. (Assume the rod is cylindrical.) The rod, of length L and average density ρ_0, floats partially immersed in the liquid of density ρ. A length h of the rod protrudes above the surface of the liquid. Show that the density of the liquid is given by

$$\rho = \frac{\rho_0 L}{L - h}$$

FIGURE P9.76

77. A 600-kg weather balloon is designed to lift a 4 000-kg package. What volume should the balloon have after being inflated with helium at standard temperature and pressure (see Table 9.3) in order that the total load can be lifted?

78. A helium-filled balloon is tied to a 2.0-m-long, 0.050-kg string. The balloon is spherical with a radius of 0.40 m. When released, it lifts a length (h) of the string and then remains in equilibrium as in Figure P9.78. Determine the value of h. When deflated, the balloon has a mass of 0.25 kg. (*Hint:* Only that part of the string above the floor contributes to the load being held up by the balloon.)

FIGURE P9.78

79. A small sphere 0.60 times as dense as water is dropped from a height of 10 m above the surface of a smooth lake. Determine the maximum depth to which the sphere will sink. Neglect any kinetic energy transferred to the water during impact and sinking.

80. A U-tube open at both ends is partially filled with water (Fig. P9.80a). Oil ($\rho = 750$ kg/m³) is then poured into the right arm and forms a column $L = 5.00$ cm high (Fig. P9.80b). (a) Determine the difference h in the

heights of the two liquid surfaces. (b) The right arm is then shielded from any air motion while air is blown across the top of the left arm until the surfaces of the two liquids are at the same height (Fig. 9.80c). Determine the speed of the air being blown across the left arm. Assume that the density of air is 1.29 kg/m³.

FIGURE P9.80

81. Figure P9.81 shows a water tank with a valve at the bottom. If this valve is opened, what is the maximum height attained by the water stream coming out of the right side of the tank? Assume that $h = 10.0$ m, $L = 2.00$ m, and $\theta = 30.0°$ and that the cross-sectional area at A is very large compared with that at B.

FIGURE P9.81

82. A solid copper ball with a diameter of 3.00 m at sea level is placed at the bottom of the ocean, at a depth of 10.0 km. If the density of the sea water is 1 030 kg/m³, how much does the diameter of the ball decrease when it reaches bottom?

83. A 1.0-kg hollow ball with a radius of 0.10 m, filled with air, is released from rest at the bottom of a 2.0-m-deep pool of water. How high above the water does the ball shoot upward? Neglect all frictional effects, and neglect changes in the ball's motion when it is only partially submerged.

84. In about 1657 Otto von Guericke, inventor of the air pump, evacuated a sphere made of two brass hemispheres (Fig. P9.84). Two teams of eight horses each could pull the hemispheres apart only on some trials, and then with the greatest difficulty. (a) Show that the force required to pull the evacuated hemispheres apart is $\pi R^2(P_0 - P)$, where R is the radius of the hemispheres and P is the pressure inside the sphere, which is much less than

atmospheric pressure P_0. (b) Determine the required force if $P = 0.10 \, P_0$ and $R = 0.30$ m.

FIGURE P9.84 The colored engraving, dated 1672, illustrates Otto von Guericke's demonstration of the force due to air pressure as performed before Emperor Ferdinand III in 1657. (*The Granger Collection*)

85. A 2.0-cm-thick bar of soap is floating on a water surface so that 1.5 cm of the bar is underwater. Bath oil of specific gravity 0.60 is poured into the water and floats on top of the water. What is the depth of the oil layer when the top of the soap is just level with the upper surface of the oil?

86. A cube of ice whose edge is 20.0 mm is floating in a glass of ice-cold water with one of its faces parallel to the water surface. (a) How far below the water surface is the bottom face of the block? (b) Ice-cold ethyl alcohol is gently poured onto the water surface to form a layer 5.00 mm thick above the water. When the ice cube attains hydrostatic equilibrium again, what will be the distance from the top of the water to the bottom face of the block? (c) Additional cold ethyl alcohol is poured onto the water surface until the top surface of the alcohol coincides with the top surface of the ice cube (in hydrostatic equilibrium). How thick is the required layer of ethyl alcohol?

87. A water tank open to the atmosphere at the top has two small holes punched in its side, one above the other. The holes are 5.00 cm and 12.0 cm above the floor. How high does water stand in the tank if the two streams of water hit the floor at the same place?

88. Oil having a density of 930 kg/m³ floats on water. A rec-
web tangular block of wood 4.00 cm high and with a density of 960 kg/m³ floats partly in the oil and partly in the water. The oil completely covers the block. How far below the interface between the two liquids is the bottom of the block?

89. A hollow object with an average density of 900 kg/m³ floats in a pan containing 500 cm³ of water. Ethanol is added to the water and mixed into it until the object is just on the verge of sinking. What volume of ethanol has been added? Disregard the loss of volume caused by mixing.

GROUP ACTIVITIES

G.1 You will need a large spring scale, a clear container partially filled with water and a few cylinders of the same size but made of different materials. Your instructor may be able to supply these items. Measure the volume of each cylinder. Hang a cylinder on the scale, record the reading of the scale, and start lowering the cylinder into the water. What happens to the scale reading as more of the cylinder is submerged? Why does the reading behave as it does? Is the effect independent of the type of material used for the cylinder? Record the reading for a cylinder when it is not submerged and when it is completely immersed. How can you use these two readings to verify Archimedes's principle?

G.2 Place an egg at the bottom of a container of fresh water. Now use a funnel to slowly add a salt solution to the water. You will observe that the egg begins to rise to the surface. Use Archimedes's principle to explain your observation.

G.3 An iron block of volume 0.20 m³ is suspended from a spring scale and immersed in a flask of water. Then the iron block is removed and an aluminum block of the same volume replaces it. (a) In which case is the buoyant force the greatest, for the iron block or the aluminum block? (b) In which case does the spring scale read the largest value? (c) Use the known densities of these materials to find the answers to the above. Are your calculations consistent with your answers to part (a) and (b)?

G.4 (a) Assume Superman attempts to drink through a very long vertical straw. Is there a length of the straw such that he is unable to lift the fluid to his mouth? Defend your answer. (b) Assume the fluid is water, and find the length of the straw. (c) Based on your answer to (b), consider what would happen if Superman were on the Moon where there is no atmosphere. Could he drink through a straw of any length there?

G.5 The density of air is 1.3 kg/m³ at sea level. From your knowledge of air pressure at ground level, estimate the height of the atmosphere. As a simplifying assumption, take the atmosphere to be of uniform density up to some height after which the density rapidly falls to zero. (In reality, the density of the atmosphere decreases as we go up.)

G.6 Water is poured to the same level in each of the three vessels shown in Figure GA9.6. Each vessel has the same base area. Because the water fills each vessel to the same depth, each vessel has the same pressure at the bottom. Because the area and pressure at each base is the same, each liquid should exert the same force on the base of the vessel. Yet, if the vessels are weighed, three different values are obtained. (The one in the center clearly holds less liquid than the one at the left, so it will weigh less.) How can you resolve this apparent contradiction?

FIGURE GA9.6

G.7 Suppose you have the following collection of objects: a pencil, a coin, an empty plastic box for a tape cassette with its edges taped shut, a needle, an unopened can of soda pop, and an empty can of soda pop. Which of these objects do you expect will float and which will sink in water? Will it make a difference if you carefully place the object with its largest surface on the surface of the water? In which cases? Explain your reasoning. After you have written your answer, perform the experiments and compare the results with your predictions.

(Problems 5, 6, and 7 are courtesy of Edward F. Redish. For more problems of this type, see http://www.physics.umd.edu/perg/)

Thermodynamics

As we saw in the first part of this book, Newtonian mechanics explains a wide range of phenomena, such as the motions of baseballs, rockets, and planets. We now turn to the study of thermodynamics, which is concerned with the concepts of energy transfers between a system and its environment and the resulting temperature variations. As we shall see, thermodynamics is very successful in explaining the bulk properties of matter and the correlation between those properties and the mechanics of atoms and molecules.

Historically, the development of thermodynamics paralleled the development of the atomic theory of matter. By the middle of the 19th century, chemical experiments provided solid evidence for the existence of atoms. At that time, scientists recognized that there must be a connection between thermodynamics and the structure of matter. In 1827 the botanist Robert Brown reported that grains of pollen suspended in a liquid moved erratically from one place to another, as if under constant agitation. In 1905 Albert Einstein developed a theory about the cause of this erratic motion, which today is called *Brownian motion*. Einstein explained the phenomenon by assuming that the grains of pollen are under constant bombardment by "invisible" molecules in the liquid, which themselves are moving erratically. Einstein's insight gave scientists a means of discovering vital information concerning molecular motion. It also helped to confirm the concept of the atomic constituents of matter.

Have you ever wondered how refrigerators cool or what types of transformations occur in automobile engines or why a bicycle pump becomes warm as you inflate a tire? The laws of thermodynamics and the concepts of heat and temperature enable us to answer such questions.

Many things can happen to an object when its temperature increases. Its size generally changes slightly, but it can also melt, boil, ignite, or even explode. The outcome depends on the composition of the object and the amount by which its temperature changes. Thermodynamics must therefore concern itself with the physical and chemical transformations of matter in all of its forms: solid, liquid, and gas.

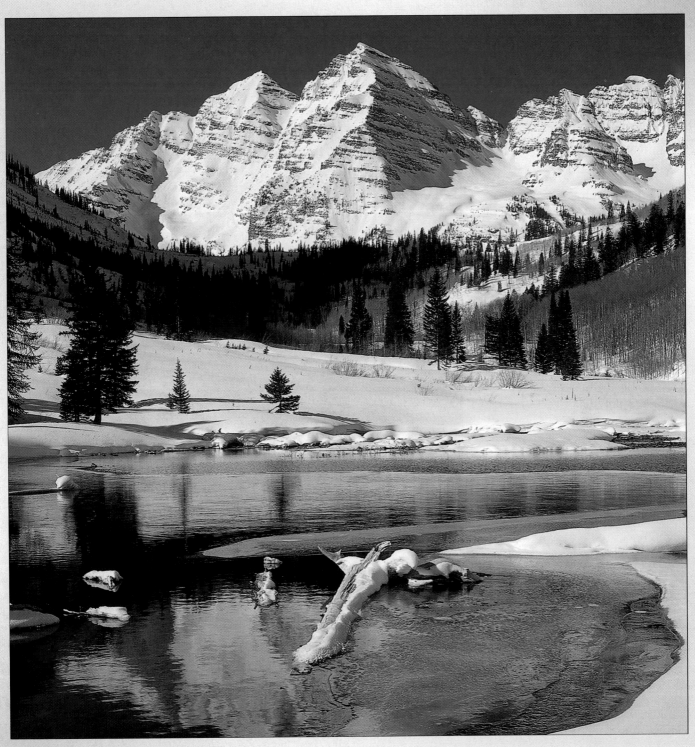

This winter photograph of Maroon Lake in the foreground near Aspen, Colorado, shows evidence of water in two of its three phases. In the lake is liquid water, and solid water in the form of snow appears in the mountains. On a cloudy day, the clouds would consist of water droplets that have condensed from gaseous water vapor in the air. (© Jeff Lepore/Photo Researchers, Inc.)

10 Thermal Physics

This dramatic photograph shows molten lava flowing from the Kilauea volcano in Hawaii and entering the Pacific Ocean at two locations. Note that as the lava is cooled by the ocean, tall steam plumes are created and large quantities of ocean water are converted to water vapor. Violent eruptions can cause both local and global (atmospheric) damage, as in the case of the Mt. St. Helens eruption in 1980.

(© Brad Lewis/Visuals Unlimited)

*O*ur study thus far has focused exclusively on mechanics. Such concepts as mass, force, and kinetic energy have been carefully defined in order to make the subject quantitative. We now move to a new branch of physics, **thermal physics.** Here we shall find that quantitative descriptions of thermal phenomena require careful definitions of the concepts of temperature, heat, and internal energy. We will seek answers to such questions as: What happens to an object when energy is added to it or removed from it by heat? What physical changes occur in an object when its temperature increases or decreases? One familiar outcome of a temperature increase or decrease for an object is a change in size. We examine the details of this process in our discussion of linear expansion.

This chapter concludes with a study of ideal gases. We approach this study on two levels. The first examines ideal gases on the macroscopic scale. Here we are concerned with the relationships among such quantities as pressure, volume, and temperature. On the second level we examine gases on a microscopic scale, using a model that pictures the components of a gas as small particles. This latter approach, called the kinetic theory of gases, helps us understand what happens on the atomic level to affect such macroscopic properties as pressure and temperature.

10.1 TEMPERATURE AND THE ZEROTH LAW OF THERMODYNAMICS

We often associate the concept of temperature with how hot or cold an object feels when we touch it. Thus, our senses provide us with qualitative indications of temperature. However, our senses are unreliable and often misleading. For example, if we remove a metal ice tray and a package of frozen vegetables from the freezer, the ice tray feels colder to the hand than the vegetables even though the two are at the same temperature. This is because the properties of metal are such that the transfer of energy (by heat) to the ice tray from your hand is more rapid than the transfer of energy to the cardboard package. The larger the difference in temperature between the object and your hand, the faster the energy is transferred. Therefore, temperature and your sense of touch are related in some way. What we need is a reliable and reproducible method of making quantitative measurements to establish the relative "hotness" or "coldness" of objects that is solely related to the temperature of the object. Scientists have developed a variety of thermometers for making such quantitative measurements.

We are all familiar with the fact that two objects at different initial temperatures may eventually reach some intermediate temperature when placed in contact with each other. For example, if a cup of hot water is combined with a cup of cold water, the combined water reaches an equilibrium temperature somewhere between the initial temperatures of the hot and cold water. Likewise, if a cup of hot coffee is cooled with an ice cube, the ice rises in temperature and eventually melts, and the temperature of the coffee decreases.

In order to understand the concept of temperature, it is useful to first define two often-used phrases, *thermal contact* and *thermal equilibrium*. To grasp the meaning of thermal contact, imagine two objects placed in an insulated container so that they interact with each other but not with the rest of the world. If the objects are at different temperatures, energy can be exchanged between them by, for example, heat or electromagnetic radiation. The process by which energy is exchanged between objects because of differences in their temperature is called **heat.** We examine the concept of heat in more detail in Chapter 11. For purposes of our current discussion, we shall assume that two objects are in **thermal contact** with each other if energy can be exchanged between them. **Thermal equilibrium**

FIGURE 10.1   The zeroth law of thermodynamics. (a) and (b) If the temperatures of A and B are found to be the same by placing them in contact with a thermometer (object C), no energy will be exchanged between them when they are placed in thermal contact with each other (c).

(a)   (b)   (c)

is the situation in which two objects in thermal contact with each other cease to exchange energy.

Using these ideas, we can develop a formal definition of temperature. Consider two objects, A and B, that are not in thermal contact with each other, and a third object, C, that acts as a **thermometer**—a device calibrated to measure the temperature of an object. We wish to determine whether A and B will be in thermal equilibrium if they are placed in thermal contact. The thermometer (object C) is first placed in thermal contact with A until thermal equilibrium is reached as in Figure 10.1a. The thermometer is then placed in thermal contact with B, and its reading is recorded (Fig. 10.1b). If the two readings are the same, then A and B are in thermal equilibrium with each other. If A and B are placed in thermal contact with each other as in Figure 10.1c, there is no transfer of energy between them.

We can summarize these results in a statement known as the **zeroth law of thermodynamics (the law of equilibrium):**

> **If objects A and B are separately in thermal equilibrium with a third object C, then A and B are in thermal equilibrium with each other.**

Zeroth law of thermodynamics ▶

This statement, insignificant and obvious as it may seem, is easily proved experimentally and is important because it makes it possible to define temperature. We can think of **temperature** as the property that determines whether or not an object is in thermal equilibrium with other objects. **Two objects in thermal equilibrium with each other are at the same temperature.**

10.2   THERMOMETERS AND TEMPERATURE SCALES

Thermometers are devices used to measure the temperature of an object or a system with which the thermometer is in thermal equilibrium. All thermometers make use of some physical property exhibiting a change with temperature that can be calibrated in order to make the temperature measurable. Some of the physical properties used are (1) the volume of a liquid, (2) the length of a solid, (3) the pressure of a gas held at constant volume, (4) the volume of a gas held at constant pressure, (5) the electric resistance of a conductor, and (6) the color of a very hot object.

One common thermometer in everyday use consists of a mass of liquid—usually mercury or alcohol—that expands into a glass capillary tube when its temperature rises (Fig. 10.2). In this case, the physical property that changes is the volume of a liquid. Because the cross-sectional area of the capillary tube is uniform, the change in volume of the liquid varies linearly with its length along the tube. We can then define a temperature to be related to the length of the liquid col-

FIGURE 10.2   Diagram of a mercury thermometer. Because of thermal expansion, the level of the mercury rises as the temperature of the mercury changes from 0°C (the ice point) to 100°C (the steam point).

umn. The thermometer can be calibrated by placing it in thermal contact with some environments that remain at constant temperature. One such environment is a mixture of water and ice in thermal equilibrium at atmospheric pressure. Another commonly used system is a mixture of water and steam in thermal equilibrium at atmospheric pressure. Once we mark the ends of the liquid column for our chosen environment on our thermometer, we need to define a scale of numbers associated with various temperatures. An example of such a scale is the **Celsius temperature scale.** On the Celsius scale, the temperature of the ice-water mixture is defined to be zero degrees Celsius, written 0°C; this temperature is called the **ice point** or **freezing point** of water. Likewise, on the Celsius scale, the temperature of the water-steam mixture is defined as 100°C, called the **steam point** or **boiling point** of water. Once the ends of the liquid column in the thermometer are marked at these two points, the distance between marks is divided into 100 equal segments, each corresponding to a change in temperature of one degree Celsius.

Thermometers calibrated in this way present problems when extremely accurate readings are needed. For instance, an alcohol thermometer calibrated at the ice and steam points of water might agree with a mercury thermometer only at the calibration points. Because mercury and alcohol have different thermal expansion properties, when one indicates a temperature of 50°C, say, the other may indicate a slightly different value. The discrepancies between different types of thermometers are especially large when the temperatures to be measured are far from the calibration points.

THE CONSTANT-VOLUME GAS THERMOMETER AND THE KELVIN SCALE

We can construct practical thermometers such as the mercury thermometer, but these types of thermometers do not define temperature in a fundamental way. One thermometer, however, is more fundamental, and it offers a way to define temperature and relate it to internal energy directly—the **gas thermometer.** In a gas thermometer, the temperature readings are nearly independent of the substance used in the thermometer. One type of gas thermometer is the constant-volume unit shown in Figure 10.3. The behavior observed in this device is the pressure variation with temperature of a fixed volume of gas. When the constant-volume gas thermometer was developed, it was calibrated using the ice and steam points of water, as follows. (A different calibration procedure, to be discussed shortly, is now used.) The gas flask was inserted into an ice-water bath, and mercury reservoir B was raised or lowered until the volume of the confined gas was at some value, indicated by the zero point on the scale. The height h, the difference between the levels in the reservoir and column A, indicated the pressure in the flask at 0°C. The flask was inserted into water at the steam point, and reservoir B was readjusted until the height in column A was again brought to zero on the scale, ensuring that the gas volume was the same as it had been in the ice bath (hence the designation "constant volume"). A measure of the new value for h gave a value for the pressure at 100°C. These pressure and temperature values were then plotted on a graph, as in Figure 10.4. The line connecting the two points served as a calibration curve for measuring unknown temperatures. If we want to measure the temperature of a substance, we place the gas flask in thermal contact with the substance and adjust the column of mercury until the level in column A returns to zero. The height of the mercury column tells us the pressure of the gas, and we then find the temperature of the substance from the calibration curve.

Now suppose that temperatures are measured with various gas thermometers containing different gases. Experiments show that the thermometer readings are nearly independent of the type of gas used, as long as the gas pressure is low and the temperature is well above the point at which the gas liquifies.

We can also perform the temperature measurements with the gas in the flask at different starting pressures at 0°C. As long as the pressure is low, we generate

FIGURE 10.3 A constant-volume gas thermometer measures the pressure of the gas contained in the flask immersed in the bath. The volume of gas in the flask is kept constant by raising or lowering reservoir B to keep the mercury level constant.

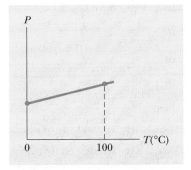

FIGURE 10.4 A typical graph of pressure versus temperature taken with a constant-volume gas thermometer. The dots represent known reference temperatures (the ice point and the steam point).

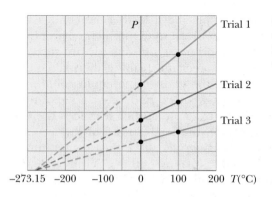

FIGURE 10.5 **Pressure versus temperature for dilute gases.** Note that, for all three experimental trials, the pressure extrapolates to zero at the unique temperature of $-273.15°$C.

straight-line calibration curves for each starting pressure, as shown for three experimental trials (solid lines) in Figure 10.5.

If the curves in Figure 10.5 are extended back toward negative temperatures, we find a startling result. In every case, regardless of the type of gas or the value of the low starting pressure, **the pressure extrapolates to zero when the temperature is $-273.15°$C.** This suggests that this particular temperature is universal in its importance, because it does not depend on the substance used in the thermometer. In addition, because the lowest possible pressure is $P = 0$, which would be a perfect vacuum, this temperature must represent a lower bound for physical processes. Thus, we define this temperature as **absolute zero.**

This significant temperature is used as the basis for the **Kelvin temperature scale,** which sets $-273.15°$C as its zero point (0 K). The size of a "degree" on the Kelvin scale (called a kelvin) is chosen to be identical to the size of a degree on the Celsius scale. Thus, the relationship that enables us to convert between these temperatures is

$$T_C = T - 273.15 \qquad \text{[10.1]}$$

where T_C is the Celsius temperature and T is the Kelvin temperature (sometimes called the **absolute temperature**).

Early gas thermometers made use of ice and steam points according to the procedure just described. However, these points are experimentally difficult to duplicate because they are pressure-sensitive. Consequently, a procedure based on two new points was adopted in 1954 by the International Committee on Weights and Measures. The first point is absolute zero. The second point is **the triple point of water, which is the single temperature and pressure at which water, water vapor, and ice can coexist in equilibrium.** This point is a convenient and reproducible reference temperature for the Kelvin scale. It occurs at a temperature of $0.01°$C and a pressure of 4.58 mm of mercury. The temperature at the triple point of water on the Kelvin scale occurs at 273.16 K. Thus, **the SI unit of temperature, the kelvin, is defined as 1/273.16 of the temperature of the triple point of water.**

Figure 10.6 shows the Kelvin temperatures for various physical processes and structures. As the figure reveals, absolute zero has never been achieved, although laboratory experiments have come very close to absolute zero.

What would happen to a substance if its temperature could reach 0 K? As Figure 10.5 indicates, the substance would exert zero pressure on the walls of its container (assuming that the gas does not liquefy or solidify all the way to absolute zero). In Section 10.6 we will show that the pressure of a gas is proportional to the kinetic energy of the molecules of that gas. Thus, according to classical physics, the kinetic energy of the gas would go to zero, and there would be no motion at all of the individual components of the gas; hence, the molecules would settle out on the bottom of the container. Quantum theory, to be discussed in Chapter 27, modifies this statement to indicate that there would be some residual energy, called the *zero-point energy*, at this low temperature.

Webnote 10.1

The SI unit of temperature is described at *http://physics.nist.gov/cuu/Units/kelvin.html*

The kelvin ▶

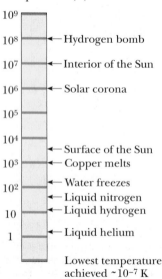

Temperature (K)

10^9

10^8 ← Hydrogen bomb

10^7 ← Interior of the Sun

10^6 ← Solar corona

10^5

10^4 ← Surface of the Sun

10^3 ← Copper melts

10^2 ← Water freezes
  ← Liquid nitrogen

10 ← Liquid hydrogen

1 ← Liquid helium

Lowest temperature achieved ~10^{-7} K

FIGURE 10.6 Absolute temperatures at which various selected physical processes take place. Note that the scale is logarithmic.

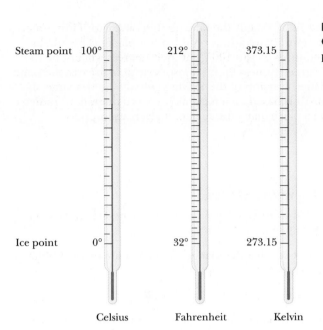

FIGURE 10.7 A comparison of the Celsius, Fahrenheit, and Kelvin temperature scales.

THE CELSIUS, KELVIN, AND FAHRENHEIT TEMPERATURE SCALES

Equation 10.1 shows that the Celsius temperature T_C is shifted from the absolute (Kelvin) temperature T by 273.15. Because the size of a Celsius degree is the same as that of a kelvin, a temperature difference of 5°C is equal to a temperature difference of 5 K. The two scales differ only in the choice of zero point. Thus, the ice point (273.15 K) corresponds to 0.00°C, and the steam point (373.15 K) is equivalent to 100.00°C.

The most common temperature scale in use in the United States is the Fahrenheit scale. It sets the temperature of the ice point at 32°F and the temperature of the steam point at 212°F. The relationship between the Celsius and Fahrenheit temperature scales is

$$T_F = \tfrac{9}{5}T_C + 32 \qquad\qquad [10.2]$$

For example, you should show that a temperature of 50.0°F corresponds to a Celsius temperature of 10.0°C and an absolute temperature of 283 K.

Equation 10.2 can easily be used to find a relationship between changes in temperature on the Celsius and Fahrenheit scales. In an end-of-chapter problem you will be asked to show that if the Celsius temperature changes by ΔT_C, the Fahrenheit temperature changes by the amount ΔT_F, given by

$$\Delta T_F = \tfrac{9}{5}\Delta T_C \qquad\qquad [10.3]$$

Figure 10.7 compares the three temperature scales that we have discussed.

APPLYING PHYSICS 10.1

A group of future astronauts lands on an inhabited planet. They strike up a conversation with the life forms about temperature scales. It turns out that the inhabitants of this planet have a temperature scale based on the freezing and boiling points of water. Would these two temperatures on this planet be the same as those on Earth? Would the planet's inhabitants use degrees that are the same size as ours? Suppose that the inhabitants have also devised a scale similar to the Kelvin scale. Would their absolute zero be the same as ours?

Explanation The values of 0°C and 100°C for the freezing and boiling points of water are defined at atmospheric pressure. On another planet, it is unlikely

Webnote 10.2

For a convenient calculator that converts temperature readings between all three scales, go to the URL below. (You should be able to convert temperatures on your own during exams!)
http://hyperphysics.phy-str.gsu.edu/ hbase/thermo/temper.html#c3

that atmospheric pressure will be exactly the same as that on Earth. Thus, water will freeze and boil at different temperatures on the other planet. The inhabitants may call these temperatures 0° and 100°, but their degrees will not be the same size as our Celsius degrees (unless their atmospheric pressure was the same as ours). However, this planet's version of the Kelvin scale will use the same absolute zero as ours, because it is based on a natural, universal definition, rather than being associated with a particular substance or a given atmospheric pressure.

Example 10.1 Heating a Pan of Water

A pan of water is heated from 25°C to 80°C. What is the change in its temperature on the Kelvin scale and on the Fahrenheit scale?

Solution From Equation 10.1 we see that the change in temperature on the Celsius scale equals the change on the Kelvin scale. Therefore,

$$\Delta T = \Delta T_C = 80 - 25 = 55°C = \boxed{55 \text{ K}}$$

From Equation 10.3 we find that the change in temperature on the Fahrenheit scale is $\frac{9}{5}$ as great as the change on the Celsius scale. That is,

$$\Delta T_F = \frac{9}{5}\Delta T_C = \frac{9}{5}(80 - 25) = \boxed{99°F}$$

EXERCISE On a hot summer day, the temperature is reported as 30.0°C. What is this temperature in Fahrenheit degrees and in kelvins?

ANSWER 86.0°F; 303 K

10.3 THERMAL EXPANSION OF SOLIDS AND LIQUIDS

Our discussion of the liquid thermometer made use of one of the best-known changes that occur in most substances: as temperature increases, volume increases. This phenomenon, known as **thermal expansion,** plays an important role in numerous applications. For example, thermal expansion joints must be included in buildings, concrete highways, and bridges to compensate for changes in dimensions with temperature variations (Figure 10.8).

The overall thermal expansion of an object is a consequence of the change in the average separation between its constituent atoms or molecules. To understand this, consider how the atoms in a solid substance behave. These atoms are located at fixed equilibrium positions; if an atom is pulled away from its position, a restoring force pulls it back. We can imagine that the atoms are particles connected by springs to their neighboring atoms. (See Figure 9.1.) If an atom is pulled away from its equilibrium position, the distortion of the springs provides a restoring force.

At ordinary temperatures, the atoms vibrate about their equilibrium positions with an amplitude (maximum distance from the center of vibration) of about 10^{-11} m, and the average spacing between the atoms is about 10^{-10} m. As the temperature of the solid increases, the atoms vibrate with greater amplitudes and the average separation between them increases. Consequently, the solid as a whole expands.

If the thermal expansion of an object is sufficiently small compared with the object's initial dimensions, then the change in any dimension is, to a good approximation, proportional to the first power of the temperature change. Suppose an object has an initial length L_0 along some direction at some temperature. The length increases by ΔL for the change in temperature ΔT. Thus, for small changes

FIGURE 10.8 Thermal expansion joints are used to separate sections of roadways on bridges. Without these joints, the surfaces would buckle due to thermal expansion on very hot days or crack due to contraction on very cold days. (© *Frank Siteman, Stock/Boston*)

TABLE 10.1	Average Coefficients of Expansion for Some Materials Near Room Temperature		
Material	Average Coefficient of Linear Expansion $[(°C)^{-1}]$	Material	Average Coefficient of Volume Expansion $[(°C)^{-1}]$
Aluminum	24×10^{-6}	Ethyl alcohol	1.12×10^{-4}
Brass and bronze	19×10^{-6}	Benzene	1.24×10^{-4}
Copper	17×10^{-6}	Acetone	1.5×10^{-4}
Glass (ordinary)	9×10^{-6}	Glycerin	4.85×10^{-4}
Glass (Pyrex®)	3.2×10^{-6}	Mercury	1.82×10^{-4}
Lead	29×10^{-6}	Turpentine	9.0×10^{-4}
Steel	11×10^{-6}	Gasoline	9.6×10^{-4}
Invar (Ni-Fe alloy)	0.9×10^{-6}	Air	3.67×10^{-3}
Concrete	12×10^{-6}	Helium	3.665×10^{-3}

in temperature,

$$\Delta L = \alpha L_0 \, \Delta T \qquad \qquad \textbf{[10.4]}$$

or

$$L - L_0 = \alpha L_0 (T - T_0)$$

where L is the final length, T is the final temperature, and the proportionality constant α is called the **coefficient of linear expansion** for a given material and has units of $(°C)^{-1}$.

Table 10.1 lists the coefficients of linear expansion for various materials. Note that for these materials α is positive, indicating an increase in length with increasing temperature.

It is helpful to picture a thermal expansion as a magnification or a photographic enlargement. For example, as the temperature of a metal washer increases (Fig. 10.9), all dimensions, including the radius of the hole, increase according to Equation 10.4.

One practical application of thermal expansion is the common technique of using hot water to loosen a metal lid that is stuck on a glass jar. This succeeds because the circumference of the lid expands more than that of the glass jar.

Because the linear dimensions of an object change with temperature, it follows that surface area and volume also change with temperature. Consider a square having an initial length L_0 on a side and therefore an initial area $A_0 = L_0^2$. As the temperature is increased, the length of each side increases to

$$L = L_0 + \alpha L_0 \, \Delta T$$

We are now able to calculate the change in the area of the square object as follows. The new area $A = L^2$ is

$$L^2 = (L_0 + \alpha L_0 \, \Delta T)(L_0 + \alpha L_0 \, \Delta T) = L_0^2 + 2\alpha L_0^2 \, \Delta T + \alpha^2 L_0^2 (\Delta T)^2$$

The last term in this expression contains the quantity $\alpha \, \Delta T$ raised to the second power. Because $\alpha \, \Delta T$ is much less than unity, squaring it makes it even smaller. Therefore, we can neglect this term to get a simpler expression:

$$A = L^2 = L_0^2 + 2\alpha L_0^2 \, \Delta T$$
$$= A_0 + 2\alpha A_0 \, \Delta T$$

or

$$\Delta A = A - A_0 = \gamma A_0 \, \Delta T \qquad \qquad \textbf{[10.5]}$$

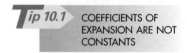

Tip 10.1 COEFFICIENTS OF EXPANSION ARE NOT CONSTANTS

Note that the coefficients of expansion can vary somewhat with temperature, so the coefficients are actually averages.

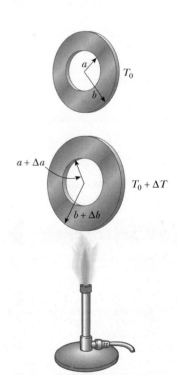

FIGURE 10.9 **Thermal expansion of a homogeneous metal washer.** As the washer is heated, all dimensions increase. (Note that the expansion is exaggerated in this figure.)

FIGURE 10.10 (a) A bimetallic strip bends as the temperature changes because the two metals have different expansion coefficients. (b) A bimetallic strip used in a thermostat to break or make electrical contact. (c) The interior of a thermostat, showing the coiled bimetallic strip. Why do you suppose the strip is coiled? *(c, Charles D. Winters)*

where $\gamma = 2\alpha$. The quantity γ (Greek letter gamma) is called the **coefficient of area expansion.**

By a similar procedure we can show that the *increase in volume* of an object accompanying a change in temperature is

$$\Delta V = \beta V_0 \, \Delta T \qquad \text{[10.6]}$$

where β, the **coefficient of volume expansion,** is equal to 3α. (Note that $\gamma = 2\alpha$ and $\beta = 3\alpha$ only if the coefficient of linear expansion of the object is the same in all directions.)

As Table 10.1 indicates, each substance has its own characteristic coefficients of expansion. For example, when the temperatures of a brass rod and a steel rod of equal length are raised by the same amount from some common initial value, the brass rod expands more than the steel rod because brass has a larger coefficient of expansion than steel. A simple device that uses this principle is a *bimettalic strip*. Certain thermostats in home furnace systems, for example, make use of a bimetallic strip. The strip is made by securely bonding two different metals together. As the temperature of the strip increases, the two metals expand by different amounts and the strip bends, as in Figure 10.10.

APPLICATION

THERMOSTATS

Thermal expansion affects the choice of glassware used in kitchens and laboratories. If hot liquid is poured into a cold container made of ordinary glass, the container may well break due to thermal stress. The inside surface of the glass becomes hot and expands while the outside surface remains at room temperature, and ordinary glass may not withstand the differential expansion without breaking.

APPLICATION

PYREX GLASS

Pyrex glass has a coefficient of linear expansion of about one third that of ordinary glass, so that the thermal stresses are smaller. Kitchen measuring cups and laboratory beakers are often made of Pyrex so that they can be used with hot liquids.

APPLICATION

RISING SEA LEVELS

The thermal expansion of water also has profound influence on the rising ocean levels. At current rates of global warming, scientists predict that about one half of the expected rise in sea level will be caused by thermal expansion; the remainder will be due to the melting of polar ice.

Quick Quiz 10.1

If you quickly plunge a room-temperature mercury thermometer into very hot water, the mercury level will (a) go up briefly before reaching a final reading, (b) go down briefly before reaching a final reading, or (c) not change.

Quick Quiz 10.2

If you were to make a very sensitive glass thermometer, which of the following working liquids would you choose? (a) mercury (b) alcohol (c) gasoline (d) glycerin

Quick Quiz 10.3

Two spheres are made of the same metal and have the same radius, but one is hollow and the other is solid. The spheres are taken through the same temperature increase. Which sphere expands more? (a) solid sphere, (b) hollow sphere, (c) they expand by the same amount, or (d) not enough information to say.

APPLYING **PHYSICS** *10.2*

A homeowner is painting the ceiling, and a drop of paint from the brush falls onto an operating incandescent light bulb. The bulb breaks. Why?

Explanation The glass envelope of an incandescent light bulb receives energy on the inside surface by radiation from the very hot filament and convection of the gas filling the bulb. Thus, the glass becomes very hot. When the drop of paint falls onto the glass, that portion of the glass envelope suddenly becomes cold, and the contraction of this region creates thermal stresses that cause the glass to shatter.

Example 10.2 **Expansion of a Railroad Track**

A steel railroad track has a length of 30.000 m when the temperature is 0°C. What is its length on a hot day when the temperature is 40°C?

Solution If we use Table 10.1 and Equation 10.4, and note that the change in temperature is 40°C, we find that the *increase* in length is

$$\Delta L = \alpha L_0 \, \Delta T = [11 \times 10^{-6} \, (°C)^{-1}](30.000 \text{ m})(40°C) = 0.013 \text{ m}$$

Therefore, the track's length at 40°C is

$$L = L_0 + \Delta L = \boxed{30.013 \text{ m}}$$

EXERCISE What is the length of the same railroad track on a cold winter day when the temperature is 0°F?

ANSWER 29.994 m

(Example 10.2) Thermal expansion: The extreme heat of a July day in Asbury Park, New Jersey, caused these railroad tracks to buckle. (*Wide World Photos*)

Example 10.3 **Does the Hole Get Bigger or Smaller?**

A hole of cross-sectional area 100.00 cm² is cut in a piece of steel at 20°C. What is the area of the hole if the steel is heated from 20°C to 100°C?

Solution A hole in a substance expands in exactly the same way as would a piece of the substance having the same shape as the hole (see Fig. 10.9). The change in area of the hole can be found by using Equation 10.5.

$$\Delta A = \gamma A_0\, \Delta T = [22 \times 10^{-6}\ (°C)^{-1}](100.00\ cm^2)(80°C) = 0.18\ cm^2$$

Therefore, the area of the hole at 100°C is

$$A = A_0 + \Delta A = \boxed{100.18\ cm^2}$$

EXERCISE In the derivation of Equation 10.5, the term $\alpha^2 L_0{}^2(\Delta T)^2 = \alpha^2 A_0(\Delta T)^2$ was neglected because it was said to be negligibly small. Use the values found in Example 10.3 to verify that this term is, indeed, extremely small.

THE UNUSUAL BEHAVIOR OF WATER

Liquids generally increase in volume with increasing temperature and have volume expansion coefficients about ten times greater than those of solids. Water is an exception to this rule over a small temperature range, as we can see from its density-versus-temperature curve in Figure 10.11. As the temperature increases from 0°C to 4°C, water contracts and thus its density increases. Above 4°C, water exhibits the expected expansion with increasing temperature. Thus, the density of water reaches its maximum value of 1 000 kg/m³ at 4°C.

We can use this unusual thermal expansion behavior of water to explain why a pond freezes slowly from the top down. When the atmospheric temperature drops from 7°C to 6°C, for example, the water at the surface of the pond also cools and consequently decreases in volume. This means that the surface water is denser then the water below it, which has not cooled and decreased in volume. As a result, the surface water sinks and warmer water from below is forced to the surface to be cooled, a process called upwelling. When the atmospheric temperature is between 4°C and 0°C, however, the surface water expands as it cools, becoming less dense than the water below it. The sinking process stops, and eventually the surface water freezes. As the water freezes, the ice remains on the surface because ice is less dense than water. The ice continues to build up on the surface, and water near the bottom of the pool remains at 4°C. If this did not happen, fish and other forms of marine life would not survive through the winter.

The same peculiar thermal expansion properties sometimes cause water pipes to burst in winter. As energy leaves the water through the pipe by heat and is transferred to the outside cold air, the outer layers of water in the pipe freeze first. Con-

APPLICATION

BURSTING PIPES IN WINTER

FIGURE 10.11 The density of water as a function of temperature. The inset at the right shows that the maximum density of water occurs at 4°C.

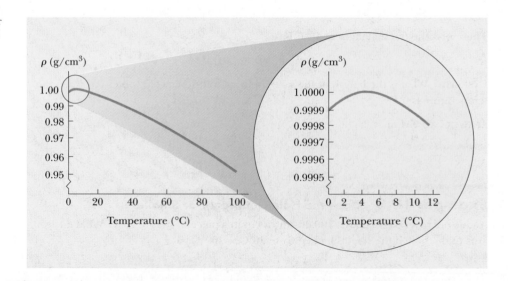

tinuing energy transfer causes ice formation to move ever closer to the center of the pipe. As long as there is still an opening through the ice, the water can expand as its temperature approaches 0°C or as it freezes into more ice, pushing itself into another part of the pipe. Eventually, however, the ice freezes to the center somewhere along the pipe's length, forming a plug of ice there. If there is still liquid water between this plug and some other obstruction, such as another ice plug or a spigot, then there is no additional volume available for further expansion and freezing. It is possible for the pressure buildup due to this frustrated expansion to burst the pipe.

APPLYING PHYSICS 10.3

Suppose a solid object is at the bottom of a container of water. Can the temperature of the system be changed in any way to make the object float to the surface?

Explanation   Let us imagine lowering the temperature. Liquids tend to have larger coefficients of volume expansion than solids, so the density of water increases more rapidly with the dropping temperature than that of the solid. If the density of the solid is initially only slightly larger than that of water, it would in principle be possible to make it float by lowering the temperature, although we then face the possible problem of reaching 4°C, the maximum density point, before this happens.

10.4   MACROSCOPIC DESCRIPTION OF AN IDEAL GAS

The properties of gases are important in a number of thermodynamic processes. Our weather is a good example of the types of processes that depend on the behavior of gases.

   If we introduce a gas into a container, it expands to fill the container uniformly. The gas does not have a fixed volume or pressure. Its volume is that of the container, and its pressure depends on the size of the container. In this section we will be concerned with the properties of a gas of pressure P, temperature T, confined to a container of volume V. It is useful to know how these quantities are related. In general, the equation that interrelates them, called the *equation of state,* can be complicated. If the gas is maintained at a low pressure (or low density), however, the equation of state is found experimentally to be relatively simple. Such a low-density gas approximates what is called an **ideal gas.** Most gases at room temperature and atmospheric pressure behave approximately as ideal gases. An ideal gas is a **collection of atoms or molecules that move randomly, exert no long-range forces on one another, and occupy a negligible fraction of the volume of their container.**

   It is convenient to express the amount of gas in a given volume in terms of the number of moles, n. The number of moles of a substance is related to its mass m by the expression

$$n = \frac{m}{\text{molar mass}} \qquad [10.7]$$

where the molar mass of the substance is usually expressed in grams per mole.

   In general, **one mole (mol) of any substance is that amount of the substance that contains as many particles (atoms, molecules, or other particles) as there are atoms in 12 g of the carbon-12 isotope.** One mole of substance A contains the same number of particles as there are in one mole of any other substance B. For example, 1 mol of aluminum contains the same number of atoms as 1 mol of lead. Experiments have shown that this number, known as Avogadro's number N_A, is

$$N_A = 6.02 \times 10^{23} \text{ particles/mol}$$

FIGURE 10.12 A gas confined to a cylinder whose volume can be varied with a movable piston.

Avogadro's number is defined so that 1 mol of carbon-12 atoms has a mass of exactly 12 g. In general, the mass contained in 1 mol of any element is the element's atomic mass expressed in grams. (For molecular substances, the mass of 1 mol of the substance is its molecular mass expressed in grams.) For example, 1 mol of aluminum (atomic mass = 27 u) has a mass of 27 g (we say its *molar mass* is 27 g/mol) and 1 mol of hydrogen gas (H_2, molecular mass = 2.0 u) has a mass of 2.0 g (its molar mass is 2.0 g/mol). Because there are 6.02×10^{23} particles in 1 mol of *any* element, the mass per atom for a given element is

$$m_{atom} = \frac{\text{molar mass}}{N_A}$$

For example, the mass of a helium atom is

$$m_{He} = \frac{4.00 \text{ g/mol}}{6.02 \times 10^{23} \text{ atoms/mol}} = 6.64 \times 10^{-24} \text{ g/atom}$$

Now suppose an ideal gas is confined to a cylindrical container whose volume can be varied by means of a movable piston, as in Figure 10.12. Let us assume that the cylinder does not leak, and so the number of moles remains constant. For such a system, experiments provide the following information. First, when the gas is kept at a constant temperature, its pressure is inversely proportional to the volume. (This is known as Boyle's law.) Second, when the pressure of the gas is kept constant, the volume is directly proportional to the temperature. (This is known as Charles' law.) Third, when the volume is held constant, the pressure is directly proportional to the temperature. (This is known as Gay-Lussac's law.) These observations can be summarized by the following equation of state, known as the **ideal gas law:**

▶ Equation of state for an ideal gas

$$PV = nRT \qquad \text{[10.8]}$$

In this expression, R is a constant for a specific gas that can be determined from experiments, and T is the temperature in kelvins. Each point on a P-versus-V diagram represents a different state of the system. Experiments on several gases show that, as the pressure approaches zero, the quantity PV/nT approaches the same value of R for all gases. For this reason R is called the **universal gas constant.** In SI units, where pressure is expressed in pascals and volume in cubic meters, R has the value

▶ The universal gas constant

$$R = 8.31 \text{ J/mol} \cdot \text{K} \qquad \text{[10.9]}$$

If the pressure is expressed in atmospheres and the volume is given in liters (recall that $1 \text{ L} = 10^3 \text{ cm}^3 = 10^{-3} \text{ m}^3$), then R has the value

$$R = 0.0821 \text{ L} \cdot \text{atm/mol} \cdot \text{K}$$

Using this value of R and Equation 10.8, we find that the volume occupied by 1 mol of any gas at atmospheric pressure and at 0°C (273 K) is 22.4 L.

 Tip 10.2 "STANDARD" TEMPERATURE AND PRESSURE

Chemists often define standard temperature and pressure (STP) to be 20°C and 1.0 atm. We choose STP to be 0°C and 1.0 atm. See Table 9.3.

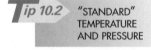 **Quick Quiz 10.4** A common material for cushioning objects in packages is made by trapping bubbles of air between sheets of plastic. This material is more effective at cushioning the package contents on (a) a hot day, (b) a cold day, or (c) it is effective on both hot and cold days.

Quick Quiz 10.5   A helium-filled balloon is released into the atmosphere. As the balloon rises, it (a) expands, (b) contracts, or (c) remains unchanged in size.

APPLYING PHYSICS 10.4

On some canned foods, you can test the seal by observing the lid of the container when it is opened. When sealed, the lid is indented, but when opened, the lid springs back out. How does this work?

Explanation   The food is sealed inside the container while it is still hot. As the food and the air above it cools, the pressure of the trapped air is lowered. Thus, the atmospheric pressure pressing against the outside of the lid causes it to indent inward. When you open the container, breaking the seal allows outside air to rush in, and the lid snaps back to its original shape.

Example 10.4   How Much Gas Is in a Container?

An ideal gas occupies a volume of 100 cm^3 at 20°C and a pressure of 100 Pa. Determine the number of moles of gas in the container.

Solution   The quantities given are volume, pressure, and temperature:

$$V = 100 \text{ cm}^3 = 1.00 \times 10^{-4} \text{ m}^3; \qquad P = 100 \text{ Pa}; \qquad T = 20°\text{C} = 293 \text{ K}$$

From the ideal gas law $PV = nRT$, we solve for n and find

$$n = \frac{PV}{RT} = \frac{(100 \text{ Pa})(1.00 \times 10^{-4} \text{ m}^3)}{(8.31 \text{ J/mol} \cdot \text{K})(293 \text{ K})} = \boxed{4.11 \times 10^{-6} \text{ mol}}$$

EXERCISE   Show that the universal gas constant $R = 0.082\ 1 \text{ L} \cdot \text{atm/mol} \cdot \text{K}$ is also equal to the value $R = 8.31 \text{ J/mol K}$.

Example 10.5   Squeezing a Tank of Gas

Pure helium gas is admitted into a leakproof cylinder containing a movable piston. The initial volume, pressure, and temperature of the gas are 15 L, 2.0 atm, and 300 K. If the volume is decreased to 12 L and the pressure increased to 3.5 atm, find the final temperature of the gas. (Assume that helium behaves as an ideal gas.)

Solution   From the ideal gas law, $PV = nRT$, we divide both sides by T to find

$$\frac{PV}{T} = nR \tag{1}$$

Because no gas escapes from the cylinder, the number of moles remains constant, so the right side of (1) is a constant. Applying (1) at the initial and final points gives

$$\frac{P_i V_i}{T_i} = \frac{P_f V_f}{T_f} \tag{2}$$

where i and f refer to the initial and final values. Solving for T_f, we get

$$T_f = \left(\frac{P_f V_f}{P_i V_i}\right)(T_i) = \frac{(3.5 \text{ atm})(12 \text{ L})}{(2.0 \text{ atm})(15 \text{ L})}(300 \text{ K}) = \boxed{420 \text{ K}}$$

Example 10.6 Heating a Bottle of Air

A sealed glass bottle at 27°C contains air at atmospheric pressure and has a volume of 30 cm³. The bottle is then tossed into an open fire. When the temperature of the air in the bottle reaches 200°C, what is the pressure inside the bottle? Assume that any volume changes of the bottle are small enough to be negligible.

Solution As in the previous example, the number of moles of gas remains constant, so we can start with Equation (2) of Example 10.5:

$$\frac{P_i V_i}{T_i} = \frac{P_f V_f}{T_f}$$

Because the initial and final volumes of the gas are assumed equal, this expression reduces to

$$\frac{P_i}{T_i} = \frac{P_f}{T_f}$$

Before evaluating the final pressure, we must convert the given temperatures to kelvins: $T_i = 27°C = 300$ K and $T_f = 200°C = 473$ K. Thus,

$$P_f = \left(\frac{T_f}{T_i}\right)(P_i) = \left(\frac{473 \text{ K}}{300 \text{ K}}\right)(1.0 \text{ atm}) = \boxed{1.6 \text{ atm}}$$

Obviously, the higher the temperature, the higher the pressure exerted by the trapped air. Of course, if the pressure rises high enough, the bottle will shatter.

EXERCISE In this example we neglected the change in volume of the bottle. If the coefficient of volume expansion for glass is 27×10^{-6} (°C)$^{-1}$, find the magnitude of this volume change.

ANSWER 0.14 cm³

10.5 Avogadro's Number and the Ideal Gas Law

In the early 1800s an important field of experimental investigation was created to determine the relative masses of molecules. The equipment used in one such investigation is shown in Figure 10.13. The lower section of the glass vessel contains two electrodes that are connected to a battery so that an electric current can be passed through lightly salted water. (The salt improves the electrical conductivity of the water.) Bubbles of gas are produced at each electrode and become trapped in the column above it. An analytical examination of the bubbles reveals that the column above the positive electrode contains oxygen gas and the column above the negative electrode contains hydrogen gas. Obviously, the current decomposes the water into its constituent parts.

Experimentally it is found that the volume of hydrogen gas collected is always exactly twice the volume of oxygen gas collected, as we now know in light of the chemical composition of water, H_2O. In addition, it is found that if 9 g of water is decomposed, 8 g of oxygen and 1 g of hydrogen are collected. From this information it is possible to determine the relative masses of oxygen and hydrogen molecules.

Following such an experimental investigation, Amedeo Avogadro in 1811 stated the following hypothesis:

Equal volumes of gas at the same temperature and pressure contain the same numbers of molecules.

FIGURE 10.13 When water is decomposed by an electric current, the volume of hydrogen gas collected is twice that of oxygen.

Oxygen gas

Hydrogen gas

Water

Battery

Based on this hypothesis and the fact that the ratio of hydrogen to oxygen molecules collected in the experiment is 2:1, the mass of an oxygen molecule must be 16 times that of a hydrogen molecule.

It can be shown that single moles of all gases occupy the same volume at 1 atm and 0°C. Thus, a corollary to Avogadro's hypothesis is as follows:

> **At standard temperature and pressure, one-mole quantities of all gases contain the same numbers of molecules.**

Specifically, the number of molecules contained in 1 mol of any gas is Avogadro's number, $N_A = 6.02 \times 10^{23}$ particles/mol. It is of historical interest that Avogadro never knew, even approximately, the value for this number.

◀ Avogadro's number

Because the number of moles of a gas in a container is equal to the total number N of molecules present divided by Avogadro's number N_A, we have

$$n = \frac{N}{N_A}$$

[10.10]

With this expression we can rewrite the ideal gas law in terms of the total number of molecules as

$$PV = nRT = \frac{N}{N_A} RT$$

or

$$PV = Nk_B T$$

[10.11]   ◀ Ideal gas law

where k_B is **Boltzmann's constant** and has the value

$$k_B = \frac{R}{N_A} = 1.38 \times 10^{-23} \, \text{J/K}$$

[10.12]   ◀ Boltzmann's constant

10.6 THE KINETIC THEORY OF GASES

In Section 10.4 we discussed the macroscopic properties of an ideal gas, using such quantities as pressure, volume, number of moles, and temperature. In this section we consider the ideal gas model from the microscopic point of view. We will show that the macroscopic properties can be understood on the basis of what is happening on the atomic scale. In addition, we reexamine the ideal gas law in terms of the behavior of the individual molecules that make up the gas.

Using the model of an ideal gas, we describe what is known as the **kinetic theory of gases.** With this theory, we can interpret the pressure and temperature of an ideal gas in terms of microscopic variables. The kinetic theory of gases model makes the following assumptions:

1. **The number of molecules in the gas is large, and the average separation between them is large compared with their dimensions.** The fact that the number of molecules is large allows us to use statistical analysis of their behavior. The large separation between molecules means that the molecules occupy a negligible volume in the container. This is consistent with the ideal gas model, in which we imagine the molecules to be point-like.

2. **The molecules obey Newton's laws of motion, but as a whole they move randomly.** By "randomly" we mean that any molecule can move in any direction with equal probability, with a wide distribution of speeds.

3. **The molecules interact only by short-range forces during elastic collisions.** This is consistent with the ideal gas model, in which the molecules exert no long-range forces on each other.

◀ Assumptions of kinetic theory for an ideal gas

4. **The molecules make elastic collisions with the walls.**
5. **The gas under consideration is a pure substance; that is, all molecules are identical.**

Although we often picture an ideal gas as consisting of single atoms, molecular gases exhibit equally good approximations to ideal behavior at low pressures. On average, effects associated with molecular structure have no effect on the motions considered. Thus, we can apply the results of the following development to molecular gases as well as to monatomic gases.

MOLECULAR MODEL FOR THE PRESSURE OF AN IDEAL GAS

For our first application of kinetic theory, let us derive an expression for the pressure of an ideal gas in a container in terms of microscopic quantities. The pressure of the gas is the result of collisions between the gas molecules and the walls of the container. During their random motion, the gas molecules collide with the walls of the container and undergo a change of momentum as a result of the force exerted by the walls on the molecules.

Now we derive an expression for the pressure of an ideal gas consisting of N molecules in a container of volume V. In this section, we use m to represent the mass of one molecule. The container is a cube with edges of length d (Fig. 10.14). Consider the collision of one molecule moving with a velocity $-v_x$ toward the left-hand face of the box (Fig. 10.15). After colliding elastically with the wall, it moves in the positive x direction with a velocity $+v_x$. Because the momentum of the molecule is $-mv_x$ before the collision and $+mv_x$ afterward, the change in momentum of the molecule is

$$\Delta p_x = mv_x - (-mv_x) = 2mv_x$$

If F_1 is the magnitude of the average force exerted by a molecule on the *wall* in the time Δt, applying Newton's second law to the wall gives

$$F_1 = \frac{\Delta p_x}{\Delta t} = \frac{2mv_x}{\Delta t}$$

For the molecule to make two collisions with the same wall, it must travel a distance $2d$ along the x direction in a time Δt. Therefore, the time interval between two collisions with the same wall is $\Delta t = 2d/v_x$ and the force imparted to the wall by a single collision is

$$F_1 = \frac{2mv_x}{\Delta t} = \frac{2mv_x}{2d/v_x} = \frac{mv_x^2}{d}$$

The total force F exerted by all the molecules on the wall is found by adding the forces exerted by the individual molecules:

$$F = \frac{m}{d}(v_{x1}^2 + v_{x2}^2 + \cdots)$$

In this equation, v_{x1} is the x component of velocity of molecule 1, v_{x2} is the x component of velocity of molecule 2, and so on. The summation terminates when we reach N molecules because there are N molecules in the container.

To proceed further, note that the average value of the square of the velocity in the x direction for N molecules is

$$\overline{v_x^2} = \frac{v_{x1}^2 + v_{x2}^2 + \cdots + v_{xN}^2}{N}$$

Thus, the total force on the wall can be written

$$F = \frac{Nm}{d}\overline{v_x^2}$$

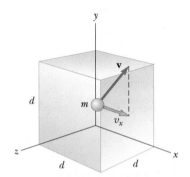

FIGURE 10.14   **A cubical box with sides of length *d* containing an ideal gas.** The molecule shown moves with velocity **v**.

Before collision

After collision

FIGURE 10.15   A molecule moving along the *x* axis makes elastic collisions with the walls of the container. After colliding with a wall the molecule's momentum is reversed, and the molecule exerts a force on the wall.

Now let us focus on one molecule in the container traveling in some arbitrary direction with velocity **v**, whose components are v_x, v_y, and v_z. In this case, we must express the total force on the wall in terms of the speed of the molecules rather than just a single component. The Pythagorean theorem relates the square of the speed to the square of these components according to the expression $v^2 = v_x^2 + v_y^2 + v_z^2$. Hence, the average value of v^2 for all the molecules in the container is related to the average values $\overline{v_x^2}$, $\overline{v_y^2}$, and $\overline{v_z^2}$ according to the expression $\overline{v^2} = \overline{v_x^2} + \overline{v_y^2} + \overline{v_z^2}$. Because the motion is completely random, the average values $\overline{v_x^2}$, $\overline{v_y^2}$, and $\overline{v_z^2}$ are equal to each other. Using this fact and the above result, we find that

$$\overline{v_x^2} = \tfrac{1}{3}\overline{v^2}$$

Thus, the total force on the wall is

$$F = \frac{N}{3}\left(\frac{m\overline{v^2}}{d}\right)$$

This expression allows us to find the total pressure exerted on the wall:

$$P = \frac{F}{A} = \frac{F}{d^2} = \frac{1}{3}\left(\frac{N}{d^3}\,m\overline{v^2}\right) = \frac{1}{3}\left(\frac{N}{V}\right)m\overline{v^2}$$

$$P = \frac{2}{3}\left(\frac{N}{V}\right)\left(\frac{1}{2}\,m\overline{v^2}\right) \qquad \textbf{[10.13]}$$

The glass vessel contains dry ice (solid carbon dioxide). The white cloud is carbon dioxide vapor, which is denser than air and hence falls from the vessel as shown. *(© R. Folwell/Science Photo Library)*

This result shows that **the pressure is proportional to the number of molecules per unit volume and to the average translational kinetic energy of a molecule,** $\frac{1}{2}m\overline{v^2}$. With this simplified model of an ideal gas, we have arrived at an important result that relates the large-scale quantity of pressure to an atomic quantity, the average value of the square of the molecular speed. Thus, we have a key link between the atomic world and the large-scale world.

◀ Pressure of an ideal gas

You should note that Equation 10.13 verifies some features of pressure that are probably familiar to you. One way to increase the pressure inside a container is to increase the number of molecules per unit volume in the container. You do this when you add air to a tire. The pressure in the tire can also be increased by increasing the average translational kinetic energy of the molecules in the tire. As we shall see shortly, this can be accomplished by increasing the temperature of the gas inside the tire. That is why the pressure inside a tire increases as the tire warms up during long trips. The continuous flexing of the tires as they move along the road surface transfers energy to the air inside the tires, increasing the air's temperature, which in turn produces an increase in pressure.

MOLECULAR INTERPRETATION OF TEMPERATURE

We have related the pressure of a gas to the average kinetic energy of the gas molecules; let us now relate temperature to a microscopic description of the gas. We obtain some insight into the meaning of temperature by first writing Equation 10.13 in the form

$$PV = \tfrac{2}{3}N(\tfrac{1}{2}\,m\overline{v^2})$$

If we compare this with the equation of state for an ideal gas in the form of Equation 10.11,

$$PV = Nk_B T$$

we can see that the left-hand sides of these two equations are identical. Equating the right sides of these expressions, we find that

$$T = \frac{2}{3k_B}\,(\tfrac{1}{2}\,m\overline{v^2}) \qquad \textbf{[10.14]}$$

◀ Temperature is proportional to average kinetic energy

That is, **the temperature of a gas is a direct measure of average molecular kinetic energy.** As the temperature of a gas increases, the molecules move with higher average kinetic energy.

By rearranging Equation 10.14, we can relate the translational molecular kinetic energy to the temperature:

▷ Average kinetic energy per molecule

$$\frac{1}{2}\,m\overline{v^2} = \frac{3}{2}\,k_B T \qquad \text{[10.15]}$$

That is, the average translational kinetic energy per molecule is $\frac{3}{2}k_B T$.

The total translational kinetic energy of N molecules of gas is simply N times the average energy per molecule, which is given by Equation 10.15:

▷ Total kinetic energy of N molecules

$$KE_{\text{total}} = N\left(\frac{1}{2}m\overline{v^2}\right) = \frac{3}{2}Nk_B T = \frac{3}{2}nRT \qquad \text{[10.16]}$$

where we have used $k_B = R/N_A$ for Boltzmann's constant and $n = N/N_A$ for the number of moles of gas. From this result, we see that **the total translational kinetic energy of a system of molecules is proportional to the absolute temperature of the system.**

For a monatomic gas, translational kinetic energy is the only type of energy the molecules can have. Thus, Equation 10.16 gives the **internal energy U for a monatomic gas:**

$$U = \frac{3}{2}nRT \qquad \text{(monatomic gas)} \qquad \text{[10.17]}$$

For diatomic and polyatomic molecules, additional possibilities for energy storage are available in the vibration and rotation of the molecule.

The square root of $\overline{v^2}$ is called the **root-mean-square (rms) speed** of the molecules. From Equation 10.15 we get, for the rms speed,

▷ Root-mean-square speed

$$v_{\text{rms}} = \sqrt{\overline{v^2}} = \sqrt{\frac{3k_B T}{m}} = \sqrt{\frac{3RT}{M}} \qquad \text{[10.18]}$$

where M is the molar mass in kilograms per mole. This expression shows that, at a given temperature, lighter molecules move faster, on average, than heavier molecules. For example, hydrogen gas (H_2), with a molar mass of 2.0×10^{-3} kg/mol, has molecules moving four times as fast as oxygen gas (O_2), whose molar mass is 32×10^{-3} kg/mol. If we calculate the rms speed for hydrogen at room temperature (~ 300 K), we find

$$v_{\text{rms}} = \sqrt{\frac{3RT}{M}} = \sqrt{\frac{3(8.31 \text{ J/mol} \cdot \text{K})(300 \text{ K})}{2.0 \times 10^{-3} \text{ kg/mol}}} = 1.9 \times 10^3 \text{ m/s}$$

This value is about 17% of the escape speed for Earth, which we calculated in Chapter 7. Because this is an average speed, a large number of molecules have speeds much higher than this, and therefore can escape from Earth's atmosphere.

TABLE 10.2	Some rms Speeds	
Gas	**Molar Mass (kg/mol)**	**v_{rms} at 20°C (m/s)**
H_2	2.02×10^{-3}	1902
He	4.0×10^{-3}	1352
H_2O	18×10^{-3}	637
Ne	20.1×10^{-3}	603
N_2 and CO	28×10^{-3}	511
NO	30×10^{-3}	494
CO_2	44×10^{-3}	408
SO_2	48×10^{-3}	390

This is why Earth's atmosphere does not presently contain hydrogen—it has all bled off into space. Table 10.2 lists the rms speeds for various molecules at 20°C.

Quick Quiz 10.6 One container is filled with argon gas and another with helium gas. Both containers are at the same temperature. Which molecules have the higher rms speed? (a) argon, (b) helium, (c) they have the same speed, or (d) not enough information to say.

APPLYING PHYSICS 10.5

Imagine a gas in an insulated cylinder with a movable piston. The piston has been pushed inward, compressing the gas, and is now released. As the molecules of the gas strike the piston, they move it outward. From the point of view of the kinetic theory, explain how the expansion of this gas causes its temperature to drop.

Explanation From the point of view of kinetic theory, a molecule colliding with the piston causes the piston to rebound with some velocity. According to the conservation of momentum then, the molecule must rebound with less speed than it had before the collision. Thus, as these collisions occur, the average speed of the collection of molecules is reduced. Because temperature is related to the average speed of the molecules, the temperature of the gas drops.

Example 10.7 A Tank of Helium

A tank contains 2.0 mol of helium gas at 20°C. Assume that the helium behaves as an ideal gas.

A Find the total internal energy of the system.

Solution Using Equation 10.17 with $n = 2.0$ and $T = 293$ K, we get

$$U = \tfrac{3}{2}nRT$$

$$= \tfrac{3}{2}(2.0 \text{ mol})(8.31 \text{ J/mol} \cdot \text{K})(293 \text{ K}) = \boxed{7.3 \times 10^3 \text{ J}}$$

B What is the average kinetic energy per molecule?

Solution From Equation 10.15 we see that the average kinetic energy per molecule is

$$\tfrac{1}{2}m\overline{v^2} = \tfrac{3}{2}k_B T = \tfrac{3}{2}(1.38 \times 10^{-23} \text{ J/K})(293 \text{ K}) = \boxed{6.1 \times 10^{-21} \text{ J}}$$

EXERCISE Using the fact that the molar mass of helium is 4.0×10^{-3} kg/mol, determine the rms speed of the atoms at 20°C.

ANSWER 1.4×10^3 m/s

Webnote 10.3

To see a simulation of a gaseous mixture reaching thermodynamic equilibrium, run the virtual laboratory found in
http://jersey.uoregon.edu/vlab/Thermodynamics/

Webnote 10.4

An excellent general resource on temperature can be found at
http://www.temperatureworld.com/
This site contains information on current events, applications and technology, including global warming and climate perspectives.

SUMMARY

The **zeroth law of thermodynamics** states that if two objects, A and B, are separately in thermal equilibrium with a third object, then A and B are in thermal equilibrium with each other.

The relationship between the **Celsius** temperature T_C and the **Kelvin (absolute)** temperature T is

$$T_C = T - 273.15 \qquad\qquad \textbf{[10.1]}$$

The relationship between the **Fahrenheit** and **Celsius** temperatures is

$$T_F = \tfrac{9}{5}T_C + 32 \qquad \text{[10.2]}$$

Generally, when a substance is heated, it expands. If an object has an initial length L_0 at some temperature and undergoes a change in temperature ΔT, this linear dimension changes by the amount ΔL, which is proportional to the object's initial length and the temperature change:

$$\Delta L = \alpha L_0 \, \Delta T \qquad \text{[10.4]}$$

The parameter α is called the **coefficient of linear expansion.**
The change in area of a substance with temperature is given by

$$\Delta A = \gamma A_0 \, \Delta T \qquad \text{[10.5]}$$

where γ is the **coefficient of area expansion** and is equal to 2α.
The change in volume with temperature of most substances is proportional to the initial volume V_0 and the temperature change ΔT:

$$\Delta V = \beta V_0 \, \Delta T \qquad \text{[10.6]}$$

where β is the **coefficient of volume expansion** and is equal to 3α.
An **ideal gas** is one that obeys the equation

$$PV = nRT \qquad \text{[10.8]}$$

where P is the pressure of the gas, V is its volume, n is the number of moles of gas, R is the universal gas constant ($8.31 \, \text{J/mol} \cdot \text{K}$), and T is the absolute temperature in kelvins. A real gas at very low pressures behaves approximately as an ideal gas.
The **pressure** of N molecules of an ideal gas contained in a volume V is given by

$$P = \frac{2}{3}\left(\frac{N}{V}\right)(\tfrac{1}{2}\,\overline{mv^2}) \qquad \text{[10.13]}$$

where $\tfrac{1}{2}\overline{mv^2}$ is the **average kinetic energy per molecule.**
The average kinetic energy of the molecules of a gas is directly proportional to the absolute temperature of the gas:

$$\tfrac{1}{2}\overline{mv^2} = \tfrac{3}{2}k_{\mathrm{B}}T \qquad \text{[10.15]}$$

where k_{B} is **Boltzmann's constant** ($1.38 \times 10^{-23}\,\text{J/K}$).
The internal energy of n moles of a monatomic ideal gas is

$$U = \tfrac{3}{2}nRT \qquad \text{[10.17]}$$

The **root-mean-square (rms) speed** of the molecules of a gas is

$$v_{\mathrm{rms}} = \sqrt{\frac{3k_{\mathrm{B}}T}{m}} = \sqrt{\frac{3RT}{M}} \qquad \text{[10.18]}$$

CONCEPTUAL QUESTIONS

1. Why does an ordinary glass dish usually break when placed on a hot stove? Dishes made of Pyrex glass do not break so easily. What characteristic of Pyrex prevents this from occurring?

2. The balance wheel of a mechanical watch governs the frequency with which the watch ticks. A wheel cut from a single piece of metal will expand upon heating, thus increasing its moment of inertia. Will this cause the watch to speed up or slow down?

3. In an astronomy class, the temperature at the core of a star is given by the teacher as 1.5×10^7 degrees. A student asks if this is Kelvin or Celsius. How would you respond?

4. If a car engine overheats, you are warned not to remove the radiator cap to add cold water until there is time for the engine to cool down. Is this good advice? Why or why not?

5. Common thermometers are made of a mercury column in a glass tube. Based on the operation of these common thermometers, which has the larger coefficient of linear expansion—glass or mercury? (Don't answer this by looking in a table.)

6. A steel wheel bearing is 1 mm smaller in diameter than an axle. How can the bearing be fit onto the axle without removing any material?

7. Objects deep below the surface of the ocean are subjected to extremely high pressures, as we saw in Chapter 9. Some bacteria in these environments have adapted to pressures as much as a thousand times atmospheric. How might these bacteria be affected if they were rapidly moved to the surface of the ocean?

8. Why is a power line more likely to break in winter than in summer even if it is loaded with the same weight?

9. Although the average speed of gas molecules in thermal equilibrium at some temperature is greater than zero, the average velocity is zero. Explain.

10. After food is cooked in a pressure cooker, why is it very important to cool the container with cold water before attempting to remove the lid?

11. Some picnickers stop at a convenience store to buy food, including bags of potato chips. They then drive up into the mountains to their picnic site. When they unload the food, they notice that the chip bags are puffed up like balloons. Why did this happen?

12. Markings to indicate length are placed on a steel tape in a room that is at a temperature of 22°C. Measurements are then made with the same tape on a day when the temperature is 27°C. Are the measurements too long, too short, or accurate?

13. Why do vapor bubbles in a pot of boiling water get larger as they approach the surface?

14. Small planets tend to have little or no atmosphere. Why is this?

PROBLEMS

1, 2, 3 = straightforward, intermediate, challenging   ☐ = full solution available in Student Solutions Manual/Study Guide

web = solution posted at **http://info.brookscole.com/serway**   🔬 = biomedical application

Section 10.1   Temperature and the Zeroth Law of Thermodynamics

Section 10.2   Thermometers and Temperature Scales

1. For each of the following temperatures, find the equivalent temperature on the indicated temperature scale: (a) −273.15°C on the Fahrenheit scale, (b) 98.6°F on the Celsius scale, and (c) 100 K on the Fahrenheit scale.

2. The highest recorded temperature on Earth was 136°F, at Azizia, Libya, in 1922. The lowest recorded temperature was −127°F, at Vostok Station, Antarctica, in 1960. Express these temperature extremes in degrees Celsius.

3. Convert the following temperatures to their values on the Fahrenheit and Kelvin scales. (a) the boiling point of liquid hydrogen −252.87°C; (b) the temperature of a room at 20°C.

4. Death Valley holds the record for the highest recorded temperature in the United States. On July 10, 1913, at a place called Furnace Creek Ranch, the temperature rose to 134°F. The lowest occurred at Prospect Creek Camp in Alaska on January 23, 1971, when the temperature plummeted to −79.8° F. Convert these temperatures to the Celsius scale.

5. Show that the temperature −40° is unique in that it has the same numerical value on the Celsius and Fahrenheit scales.

6. A constant-volume gas thermometer is calibrated in dry ice (−80.0°C) and in boiling ethyl alcohol (78.0°C). The two pressures are 0.900 atm and 1.635 atm. (a) What value of absolute zero does the calibration yield? (b) What pressures would be found at the freezing and boiling points of water? (Note that we have a linear relationship between P and T as $P = A + BT$, where A and B are constants.)

7. Show that if the temperature on the Celsius scale changes by ΔT_C, the Fahrenheit temperature changes by the amount $\Delta T_F = \frac{9}{5}\Delta T_C$.

8. The temperature difference between the inside and the outside of an automobile engine is 450°C. Express this temperature difference on the (a) Fahrenheit scale (b) Kelvin scale.

9. The melting point of gold is 1 064°C, and the boiling point is 2 660°C. (a) Express these temperatures in kelvins. (b) Compute the difference between these temperatures in Celsius degrees and kelvins.

Section 10.3   Thermal Expansion of Solids and Liquids

10. A copper telephone wire has essentially no sag between poles 35.0 m apart on a winter day when the temperature is −20.0°C. How much longer is the wire on a summer day when $T_C = 35.0°C$? Assume that the thermal coefficient of copper is constant throughout this range at its room temperature value.

11. The New River Gorge bridge in West Virginia is a 518-m-long steel arch. How much will its length change between temperature extremes of −20°C and 35°C?

12. A grandfather clock is controlled by a swinging brass pendulum that is 1.3 m long at a temperature of 20°C. (a) What is the length of the pendulum rod when the temperature drops to 0.0°C? (b) If a pendulum's period is given by $T = 2\pi\sqrt{L/g}$, where L is its length, does the change in length of the rod cause the clock to run fast or slow?

13. A pair of eyeglass frames are made of epoxy plastic [coefficient of linear expansion = 130×10^{-6} (°C)$^{-1}$]. At room temperature (assume 20.0°C) the frames have circular lens holes 2.20 cm in radius. To what temperature must the frames be heated in order to insert lenses 2.21 cm in radius?

14. A cube of solid aluminum has a volume of 1.00 m³ at 20°C. What temperature change is required to produce a 100-cm³ increase in the volume of the cube?

15. A brass ring of diameter 10.00 cm at 20.0°C is heated and slipped over an aluminum rod of diameter 10.01 cm at 20.0°C. Assuming the average coefficients of linear expansion are constant, (a) to what temperature must this combination be cooled to separate them? Is this attainable? (b) What if the aluminum rod is 10.02 cm in diameter?

16. Show that the coefficient of volume expansion β is related to the coefficient of linear expansion α through the expression $\beta = 3\alpha$.

17. The inside diameter of a steel lid and the outside diameter of a glass peanut butter jar are both precisely 11.500 cm at room temperature, 21.0°C. If the lid is stuck and you run 80.0°C hot water over it until the lid and the jar both come to 80.0°C, what will the new diameters be?

18. A construction worker uses a steel tape to measure the length of an aluminum support column. If the measured length is 18.700 m when the temperature is 21.2°C, what is the measured length when the temperature rises to 29.4°C? (*Note:* Don't neglect the expansion of the steel tape.)

19. The band in Figure P10.19 is stainless steel (coefficient of linear expansion = 17.3×10^{-6} $(°C)^{-1}$; Young's modulus = 18×10^{10} N/m^2). It is essentially circular with an initial mean radius of 5.0 mm, a height of 4.0 mm, and a thickness of 0.50 mm. If the band just fits snugly over the tooth when heated to 80°C, what is the tension in the band when it cools to 37°C?

FIGURE P10.19

20. The Trans-Alaskan pipeline is 1 300 km long, reaching from Prudhoe Bay to the port of Valdez, and experiences temperatures from $-73°C$ to $+35°C$. How much does the steel pipeline expand due to the difference in temperature? How can this expansion be compensated for?

21. An automobile fuel tank is filled to the brim with 45 L (12 gal) of gasoline at 10°C. Immediately afterward, the vehicle is parked in the Sun, where the temperature is 35°C. How much gasoline overflows from the tank as a result of the expansion? (Neglect the expansion of the tank.)

22. An underground gasoline tank at 54°F can hold 1 000 gallons of gasoline. If the driver of a tanker truck fills the underground tank on a day when the temperature is 90°F, how many gallons, according to his measure on the truck, can he pour in? Assume that the temperature of the gasoline cools to 54°F upon entering the tank.

23. A disk has moment of inertia I_0. Its temperature increases by ΔT. Show that its moment of inertia increases by $\Delta I = 2\alpha I_0 \Delta T$.

24. A concrete walk is poured on a day when the temperature is 20.0°C in such a way that the ends are unable to move. (a) What is the stress in the cement when its temperature is 50.0°C on a hot sunny day? (b) Does the concrete fracture? Take Young's modulus for concrete to be 7.00×10^9 N/m^2 and the compressive strength to be 2.00×10^7 N/m^2.

25. Figure P10.25 shows a circular steel casting with a gap. If the casting is heated, (a) does the width of the gap increase or decrease? (b) The gap width is 1.600 cm when the temperature is 30.0°C. Determine the gap width when the temperature is 190°C.

FIGURE P10.25

26. A hollow aluminum cylinder 20.0 cm deep has an internal capacity of 2.000 L at 20.0°C. It is completely filled with turpentine and then warmed to 80.0°C. (a) How much turpentine overflows? (b) If it is then cooled back to 20.0°C, how far below the surface of the cylinder's rim is the turpentine surface?

Section 10.4   Macroscopic Description of an Ideal Gas

27. The ideal gas constant in the SI is 8.315 J/mol·K. In chemistry classes, the mixed units L·atm/mol·K are used for this constant. Find the ideal gas constant in the mixed units to four significant figures.

28. Gas is contained in an 8.0-L vessel at a temperature of 20°C and a pressure of 9.0 atm. (a) Determine the number of moles of gas in the vessel. (b) How many molecules are in the vessel?

29. (a) An ideal gas occupies a volume of 1.0 cm^3 at 20°C and atmospheric pressure. Determine the number of molecules of gas in the container. (b) If the pressure of the 1.0-cm^3 volume is reduced to 1.0×10^{-11} Pa (an extremely good vacuum) while the temperature remains constant, how many moles of gas remain in the container?

30. A tank having a volume of 0.100 m^3 contains helium gas at 150 atm. How many balloons can the tank blow up if each filled balloon is a sphere 0.300 m in diameter at an absolute pressure of 1.20 atm?

31. Gas is confined in a tank at a pressure of 10.0 atm and a temperature of 15.0°C. If half of the gas is withdrawn and the temperature is raised to 65.0°C, what is the new pressure in the tank?

32. A rigid tank contains 0.40 mol of oxygen (O_2). Determine the mass (in kg) of oxygen that must be withdrawn from the tank to lower the pressure of the gas from 40 atm to 25 atm. Assume that the volume of the tank and the temperature of the oxygen are constant during this operation.

33. A weather balloon is designed to expand to a maximum radius of 20 m when in flight at its working altitude where the air pressure is 0.030 atm and the temperature is 200 K. If the balloon is filled at atmospheric pressure and 300 K, what is its radius at lift-off?

34. A cylindrical diving bell, 3.00 m in diameter and 4.00 m tall with an open bottom, is submerged to a depth of 220 m in the ocean. The surface temperature is 25.0°C, and the temperature 220 m down is 5.00°C. The density of sea water is 1 025 kg/m^3. How high does the sea water rise in the bell when it is submerged?

35. An air bubble has a volume of 1.50 cm^3 when it is released by a submarine 100 m below the surface of a lake. What is the volume of the bubble when it reaches the surface? Assume that the temperature and the number of air molecules in the bubble remains constant during ascent.

Section 10.5 Avogadro's Number and the Ideal Gas Law

Section 10.6 The Kinetic Theory of Gases

36. A sealed cubical container 20.0 cm on a side contains three times Avogadro's number of molecules at a temperature of 20.0°C. Find the force exerted by the gas on one of the walls of the container.

37. What is the average kinetic energy of a molecule of oxygen at a temperature of 300 K?

38. (a) What is the total random kinetic energy of all the molecules in 1 mol of hydrogen at a temperature of 300 K? (b) With what speed would a mole of hydrogen have to move so that the kinetic energy of the mass as a whole would be equal to the total random kinetic energy of its molecules?

39. Use Avogadro's number to find the mass of a helium atom.

40. The temperature near the top of the atmosphere on Venus is 240 K. (a) Find the rms speed of hydrogen (H$_2$) at this point in the atmosphere. (b) Repeat for carbon dioxide (CO$_2$). (c) It has been found that if the rms speed exceeds one sixth of the planet's escape velocity, the gas eventually leaks out of the atmosphere and into outer space. If the escape velocity on Venus is 10.3 km/s, does hydrogen escape? Does carbon dioxide?

41. A cylinder contains a mixture of helium and argon gas in equilibrium at a temperature of 150°C. (a) What is the average kinetic energy of each type of molecule? (b) What is the rms speed of each type of molecule?

42. If 2.0 mol of a gas is confined to a 5.0-L vessel at a pressure of 8.0 atm, what is the average kinetic energy of a gas molecule?

43. **web** Superman leaps in front of Lois Lane to save her from a volley of bullets. In a 1-min interval, an automatic weapon fires 150 bullets, each of mass 8.0 g, at 400 m/s. The bullets strike his mighty chest, which has an area of 0.75 m^2. Find the average force exerted on Superman's chest if the bullets bounce back after an elastic, head-on collision.

44. In a period of 1.0 s, 5.0 × 10^{23} nitrogen molecules strike a wall of area 8.0 cm^2. If the molecules move at 300 m/s and strike the wall head-on in a perfectly elastic collision, find the pressure exerted on the wall. (The mass of one N$_2$ molecule is 4.68 × 10^{-26} kg.)

ADDITIONAL PROBLEMS

45. Inside the wall of a house, an L-shaped section of hot water pipe consists of a straight horizontal piece 28.0 cm long, an elbow, and a straight vertical piece 134 cm long (Figure P10.45). A stud and a second-story floorboard hold stationary the ends of this section of copper pipe. Find the magnitude and direction of the displacement of the pipe elbow when the water flow is turned on, raising the temperature of the pipe from 18.0°C to 46.5°C.

FIGURE P10.45

46. The active element of a certain laser is an ordinary glass rod 20 cm long and 1.0 cm in diameter. If the temperature of the rod increases by 75°C, find its increases in (a) length, (b) diameter, and (c) volume.

47. A popular brand of cola contains 6.50 g of carbon dioxide dissolved in 1.00 L of soft drink. If the evaporating carbon dioxide is trapped in a cylinder at 1.00 atm and 20.0°C, what volume does the gas occupy?

48. A 1.5-m-long glass tube, closed at one end, is weighted and lowered to the bottom of a freshwater lake. When the tube is recovered, an indicator mark shows that water rose to within 0.40 m of the closed end. Determine the depth of the lake. Assume constant temperature.

49. Long-term space missions require reclamation of the oxygen in the carbon dioxide exhaled by the crew. In one method of reclamation, 1.00 mol of carbon dioxide produces 1.00 mol of oxygen and 1.00 mol of methane as a byproduct. The methane is stored in a tank under pressure and is available to control the attitude of the spacecraft by controlled venting. A single astronaut exhales 1.09 kg of carbon dioxide each day. If the methane generated in the respiration recycling of three astronauts during one week of flight is stored in an originally empty 150-L tank at − 45.0°C, what is the final pressure in the tank?

50. A vertical cylinder of cross-sectional area 0.050 m^2 is fitted with a tight-fitting, frictionless piston of mass 5.0 kg (Fig. P10.50). If there is 3.0 mol of an ideal gas in the cylinder at 500 K, determine the height h at which the piston will be in equilibrium under its own weight.

51. A liquid with coefficient of volume expansion β just fills a spherical flask of volume V_0 at temperature T (Fig. P10.51). The flask is made of a material that has a coefficient of linear expansion α. The liquid is free to expand into a capillary of cross-sectional area A at the top. (a) If the temperature increases by ΔT, show that the liquid rises in the capillary by the amount $\Delta h = (V_0/A)(\beta - 3\alpha)\Delta T$. (b) For a typical system, such as a mercury thermometer, why is it a good approximation to neglect the expansion of the flask?

FIGURE P10.50

FIGURE P10.51

52. A hollow aluminum cylinder is to be fitted over a steel piston. At 20°C the inside diameter of the cylinder is 99% of the outside diameter of the piston. To what common temperature should the two pieces be heated in order that the cylinder just fit over the piston?

53. A steel measuring tape was designed to read correctly at 20°C. A parent uses the tape to measure the height of a 1.1-m-tall child. If the measurement is made on a day when the temperature is 25°C, is the tape reading longer or shorter than the actual height, and by how much?

54. Before beginning a long trip on a hot day, a driver inflates an automobile tire to a gauge pressure of 1.80 atm at 300 K. At the end of the trip the gauge pressure has increased to 2.20 atm. (a) Assuming the volume has remained constant, what is the temperature of the air inside the tire? (b) What percentage of the original mass of air in the tire should be released so the pressure returns to the original value? Assume the temperature remains at the value found in (a), and the volume of the tire remains constant as air is released.

55. Two concrete spans of a 250-m-long bridge are placed end to end so that no room is allowed for expansion (Fig. P10.55a). If the temperature increases by 20.0°C, what is

the height *y* to which the spans rise when they buckle (Fig. P10.55b)?

(a)   (b)

FIGURE P10.55

56. A copper rod and steel rod are heated. At 0°C the copper rod has a length L_C and the steel one has a length L_S. When the rods are being heated or cooled, a difference of 5.00 cm is maintained between their lengths. Determine the values of L_C and L_S.

57. If 9.00 g of water is placed in a 2.00-L pressure cooker and heated to 500°C, what is the pressure inside the container?

58. An expandable cylinder has its top connected to a spring of constant 2.00×10^3 N/m (see Fig. P10.58). The cylinder is filled with 5.00 L of gas with the spring relaxed at a pressure of 1.00 atm and a temperature of 20.0°C. (a) If the lid has a cross-sectional area of 0.010 0 m^2 and negligible mass, how high will the lid rise when the temperature is raised to 250°C? (b) What is the pressure of the gas at 250°C?

FIGURE P10.58

59. A swimmer has 0.820 L of dry air in his lungs when he dives into a lake. Assuming the pressure of the dry air is 95% of the external pressure at all times, what is the volume of the dry air at a depth of 10.0 m? Assume that atmospheric pressure at the surface is 1.013×10^5 Pa.

60. Two small containers of equal volume, 100 cm^3, each contain helium gas at 0°C and 1.00 atm pressure. The two containers are joined by a small open tube of negligible volume, allowing gas to flow from one container to the other. What common pressure will exist in the two containers if the temperature of one container is raised to 100°C while the other container is kept at 0°C?

61. A bimetallic bar is made of two thin strips of dissimilar metals bonded together. As they are heated, the one with the larger average coefficient of expansion expands more than the other, forcing the bar into an arc, with the outer

radius having a larger circumference (see Fig. P10.61). (a) Derive an expression for the angle of bending θ as a function of the initial length of the strips, their average coefficients of linear expansion, the change in temperature, and the separation of the centers of the strips ($\Delta r = r_2 - r_1$). (b) Show that the angle of bending goes to zero when ΔT goes to zero or when the two coefficients of expansion become equal. (c) What happens if the bar is cooled?

FIGURE P10.61

GROUP ACTIVITIES

G.1 Fill one basin with hot tap water (not to exceed about 100°F). Fill another basin with cold tap water, and add ice until about one third of the mixture is ice. Fill the third basin with an equal mixture of hot and cold tap water. Place your left hand in the hot water and your right hand in the cold water for about 15 s. Then place both hands in the basin of lukewarm water for 15 s. Describe whether the water feels hot or cold to either of your hands and why this effect occurs.

G.2 Tape two plastic straws tightly together along their entire length but with a 2-cm offset. Hold them in a stream of very hot water from a faucet so that the water pours through one straw but not the other. Quickly hold the straws up and sight along their length. You should be able to see a very slight curvature in the tape. The effect is small, so look closely. Running cold water through the same straw and again sighting along the length will help you see the small change in shape more clearly. Explain these observations.

G.3 You can study the thermal expansion of air with a Florence flask and a balloon. To do so, boil a small amount of water in the flask. Then, quickly remove the flask from the heat source and place the mouth of the balloon over the mouth of the flask. Observe what happens to the balloon when you place the flask in a basin of cold water. Explain your observation. Reheat the water in the flask with the balloon still in place and explain your observations.

G.4 An ideal gas is contained in an airtight box. Complete the following five statements to show the quantitative change that will occur. For example, if you want to say that the initial volume V quadruples, complete the statement with "$4V$."

  (a) If the absolute temperature of the gas is halved, the average speed $\bar{v}$ of a gas molecule becomes ____.

  (b) If the average speed of a gas molecule doubles, the pressure P on a wall of the box becomes ____.

  (c) If the absolute temperature of the gas is halved, the pressure P on a wall of the box becomes ____.

  (d) If the number of gas molecules inside the box is doubled, but the temperature is kept the same, the pressure P on the wall of the box becomes ____.

G.5 The diameter of an atom is on the order of 1 angstrom (1 Å $= 10^{-10}$ m). In order to develop a feel for a gas on the molecular scale, consider one liter of air (mostly N_2 and O_2) at room temperature and a pressure of 10^5 Pa.

  (a) Calculate the number of molecules in the sample of gas.

  (b) Estimate the average spacing between the molecules.

  (c) Suppose that the gas is rescaled upwards so that each atom is the size of a tennis ball. Now what is the average spacing between molecules?

Problems 4 and 5 are courtesy of Edward F. Redish. For more problems of this type, see http://www.physics.umd.edu/perg/)

11

Energy in Thermal Processes

Chapter Outline

Water can undergo phase changes as suggested by this beautiful winter photograph of icicles on the branches of a tree. The icicles form on rainy days as the temperature drops below freezing (a liquid to solid phase change). When the temperature rises above freezing, the icicles melt into water (a solid to liquid phase change). Both processes require that energy be transferred between the icicles and the environment. *(Gary Holscher/Stone/Getty Images)*

When you place two objects with different temperatures in thermal contact, the temperature of the warmer object decreases while the temperature of the cooler object increases. With time, they reach a common equilibrium temperature that is intermediate between their initial temperatures. During this process, we say that energy is transferred from the warmer object to the cooler one.

Until about 1850, the subjects of thermodynamics and mechanics were considered to be two distinct branches of science, and the principle of conservation of energy seemed to describe only certain kinds of mechanical systems. Experiments performed by the English physicist James Joule (1818–1889) and others showed that the ratio of the decrease in mechanical energy (kinetic plus potential) of an isolated system to the increase in internal energy of the system was always the same. Today, internal energy is treated as a form of energy that can be transformed into mechanical energy, and vice versa. Once the concept of energy was broadened to include internal energy, the law of conservation of energy emerged as a universal law of nature.

The focus of this chapter is to describe some of the processes that enable energy to be transferred between a system and its surroundings.

11.1 HEAT AND INTERNAL ENERGY

A major distinction must be made between internal energy and heat, particularly because "heat" is used very loosely in everyday speech. You should read the following descriptions carefully and make efforts to use these terms correctly, because they are not interchangeable—they mean very different things when used in physics.

The following is a formal definition of internal energy:

> **Internal energy U is the energy associated with the microscopic components of a system—atoms and molecules. It includes kinetic and potential energy associated with the random translational, rotational, and vibrational motion of the atoms or molecules that make up the system as well as intermolecular potential energy.**

In the preceding chapter we showed that the internal energy of a monatomic ideal gas is associated with the translational motion of its atoms. In this special case, the internal energy is the total translational kinetic energy of the atoms; the higher the temperature of the gas, the greater the kinetic energy of the atoms and the greater the internal energy of the gas. For more complicated diatomic and polyatomic gases, internal energy includes other forms of molecular energy, such as rotational kinetic energy and the kinetic and potential energy associated with molecular vibrations. Internal energy is also associated with the intermolecular potential energy ("bond energy") between molecules in a liquid or solid.

Heat was introduced in Chapter 5 as one possible method of transferring energy between a system and its environment, and we provide a formal definition here:

> **Heat is a mechanism by which energy is transferred between a system and its environment because of a temperature difference between them.**

The symbol Q is used to represent the amount of energy transferred by heat between a system and its environment. For brevity, we will often use the phrase "the energy Q transferred to a system . . ." rather than "the energy Q transferred by heat to a system . . .."

JAMES PRESCOTT JOULE, BRITISH PHYSICIST (1818–1889)

Joule received some formal education in mathematics, philosophy, and chemistry from John Dalton but was in large part self-educated. Joule's most active research period, from 1837 through 1847, led to the establishment of the principle of conservation of energy and the relationship between heat and other forms of energy transfer. His study of the quantitative relationship among electrical, mechanical, and chemical effects of heat culminated in his announcement in 1843 of the amount of work required to produce a unit of internal energy, called the mechanical equivalent of heat. *(By kind permission of the President and Council of the Royal Society)*

◀ Internal energy

◀ Heat

Suppose a pan of water is placed on the burner of a stove. Energy enters the water from the hot burner and the internal energy of the water increases as a result. It is *incorrect* to say that there is more heat in the water as time goes by.

As further clarification of our use of the word *heat*, consider the distinction between work and energy discussed in Chapter 5. The work done on (or by) a system is a measure of the amount of energy transferred between the system and its surroundings, whereas the mechanical energy of the system (kinetic and/or potential) is a consequence of its motion and coordinates. Thus, when a person does work on a system, energy is transferred from the person to the system. It makes no sense to talk about the work *in* a system—we refer only to the work done *on* or *by* a system when some process has occurred in which energy has been transferred to or from the system. Likewise, it makes no sense to use the term *heat* unless energy has been transferred as a result of a temperature difference.

UNITS OF HEAT

Early in the development of thermodynamics, before scientists realized the connection between thermodynamics and mechanics, heat was defined in terms of the temperature changes it produced in an object, and a separate unit of energy, the **calorie,** was used for heat. The calorie (cal) was defined as **the energy necessary to raise the temperature of 1 g of water from 14.5°C to 15.5°C.** (The "Calorie," with a capital C, used in describing the energy content of foods, is actually a kilocalorie.) Likewise, the unit of heat in the U.S. customary system, the **British thermal unit** (Btu), was defined as **the energy required to raise the temperature of 1 lb of water from 63°F to 64°F.**

> ◀ Definition of the calorie

In 1948, scientists agreed that because heat (like work) is a measure of the transfer of energy, its SI unit should be the joule. The calorie is now defined to be exactly 4.186 J:

> ◀ The mechanical equivalent of heat

$$1 \text{ cal} \equiv 4.186 \text{ J} \qquad \text{[11.1]}$$

Note that this definition makes no reference to raising the temperature of water. The calorie is a general energy unit. It is introduced here for historical reasons, but we shall make little use of it. The definition in Equation 11.1 is known, from the historical background we have discussed, as the **mechanical equivalent of heat.**

Example 11.1   Losing Weight the Hard Way

A student eats a dinner containing 2 000 Calories of energy. He wishes to do an equivalent amount of work in the gymnasium by lifting a 50.0-kg object. How many times must he raise the object to expend this much energy? Assume that he raises it a distance of 2.00 m each time.

Solution   The student desires to transfer 2.00×10^3 Calories of energy from his body by doing work on the object-Earth system. Because 1 Calorie = 1.00×10^3 cal, the total work required is 2.00×10^6 cal. Converting this to joules, we have for the total work required

$$W = (2.00 \times 10^6 \text{ cal})(4.186 \text{ J/cal}) = 8.37 \times 10^6 \text{ J}$$

The work done in lifting an object of mass m a distance h is equal to mgh, and the work done in lifting it n times is $nmgh$. We equate this to the total work required:

$$W = nmgh$$

Solving for n,

$$n = \frac{W}{mgh} = \frac{8.37 \times 10^6 \text{ J}}{(50.0 \text{ kg})(9.80 \text{ m/s}^2)(2.00 \text{ m})} = 8.54 \times 10^3 \text{ times}$$

If the student is in good shape and lifts the weight once every 5 s, it will take him about 12 h to perform this feat. Clearly, it is much easier to lose weight by dieting than by lifting weights!

Comments: This problem is somewhat misleading in that it assumes that the student's body is perfectly efficient at converting food energy into work. In reality, our physiology is such that we convert about one sixth of the food energy we consume into useful work. When we tense our muscles without moving, we do no work even though our muscles consume food energy. The food energy that is not used to do work ends up as internal energy in our muscles, which results in an increase in our body temperature and a release of energy to our surroundings. This explains why we become warmer while exercising—an important factor in the process of maintaining body temperature.

11.2 SPECIFIC HEAT

The historical definition of the calorie indicates the amount of energy necessary to raise the temperature of one gram of a specific substance—water—by 1°, which is 4.186 J. To raise the temperature of 1 kg of water by 1°, we must transfer 4 186 J of energy. The amount of energy required to raise the temperature of 1 kg of an arbitrary substance by 1° varies with the substance. For example, the energy required to raise the temperature of 1 kg of copper by 1.0°C is 387 J, which is significantly less than that required for water. Every substance requires a unique amount of energy per unit mass to change the temperature of that substance by 1.0°C.

Suppose that a quantity of energy Q is transferred to a mass m of a substance, thereby changing its temperature by $\Delta T = T_f - T_i$. The **specific heat** c of the substance is defined as

$$c \equiv \frac{Q}{m\,\Delta T}$$

[11.2] ◀ Specific heat

The units of specific heat are J/kg·°C. Table 11.1 lists specific heats for several substances. From the definition of the calorie, the specific heat of water is 4 186 J/kg·°C.

TABLE 11.1	Specific Heats of Some Materials at Atmospheric Pressure	
Substance	**J/kg·°C**	**cal/g·°C**
Aluminum	900	0.215
Beryllium	1 820	0.436
Cadmium	230	0.055
Copper	387	0.092 4
Germanium	322	0.077
Glass	837	0.200
Gold	129	0.030 8
Ice	2 090	0.500
Iron	448	0.107
Lead	128	0.030 5
Mercury	138	0.033
Silicon	703	0.168
Silver	234	0.056
Steam	2 010	0.480
Water	4 186	1.00

Tip 11.1 FINDING ΔT

In Equation 11.3, be sure to remember that ΔT is *always* the final temperature minus the initial temperature, $\Delta T = T_f - T_i$.

APPLICATION

SEA BREEZES AND THERMALS

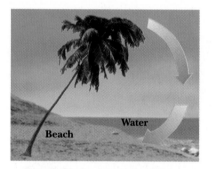

FIGURE 11.1 Circulation of air at the beach. On a hot day, the air above the sand warms faster than the air above the cooler water. The warmer air floats upward due to Archimedes's principle, resulting in the movement of cooler air toward the beach.

From this definition, we can express the energy Q needed to raise the temperature of a system of mass m by ΔT as

$$Q = mc\,\Delta T \qquad [11.3]$$

For example, the energy required to raise the temperature of 0.500 kg of water by 3.00°C is $Q = (0.500 \text{ kg})(4\ 186 \text{ J/kg}\cdot°\text{C})(3.00°\text{C}) = 6.28 \times 10^3$ J. Note that when the temperature increases, ΔT and Q are taken to be *positive*, corresponding to energy flowing *into* the system. When the temperature decreases, ΔT and Q are *negative* and energy flows *out* of the system.

Table 11.1 shows that water has the highest specific heat relative to most other common substances. This high specific heat is responsible for the moderate temperatures found in regions near large bodies of water. As the temperature of a body of water decreases during winter, the water transfers energy to the air, which carries the energy landward when prevailing winds are toward the land. For example, the prevailing winds off the western coast of the United States are toward the land, and the energy liberated by the Pacific Ocean as it cools keeps coastal areas much warmer than they would be otherwise. This explains why the western coastal states generally have more favorable winter weather than the eastern coastal states, where the winds do not carry the energy toward land.

The fact that the specific heat of water is higher than that of sand is responsible for the pattern of air flow at a beach. During the day, the Sun adds roughly equal amounts of energy to beach and water, but the lower specific heat of sand causes the beach to reach a higher temperature than the water. As a result, the air above the land reaches a higher temperature than the air above the water. The denser cold air pushes the less dense hot air upward (due to Archimedes's principle). This results in a breeze from ocean to land during the day. Because the hot air gradually cools as it rises, it subsequently sinks, setting up the circulation pattern shown in Figure 11.1.

A similar effect produces rising layers of air, called *thermals,* that help eagles to soar higher and hang gliders to stay in flight longer. A thermal is created when a portion of Earth reaches a higher temperature than neighboring regions. This often happens to plowed fields, which are warmed by the Sun to higher temperatures than nearby fields shaded by vegetation. The cooler, denser air over the vegetation-covered fields pushes the expanding air over the plowed field upward, and a thermal is formed.

Quick Quiz 11.1 Imagine you have 1 kg each of iron, glass, and water, and that all three samples are at 10°C. (a) Rank the samples from lowest to highest temperature after 100 J of energy is added to each by heat. (b) Rank them from least to greatest amount of energy transferred by heat if enough energy is transferred so that each increases in temperature by 20°C.

Quick Quiz 11.2 The specific heat of substance A is greater than that for substance B. If equal amounts of energy are added by heat to both these substances, the one reaching the higher temperature, assuming no melting, freezing, or evaporation occurs, will be (a) substance A or (b) substance B, or (c) there will be no difference in the final temperatures, or (d) could be either A or B.

11.3 CALORIMETRY

One technique for measuring the specific heat of a solid or liquid is to raise the temperature of the substance to some value, place it in a vessel containing cold water of known mass and temperature, and measure the temperature of the combination after equilibrium is reached. Let us define the system as the substance and the water. If the vessel is assumed to be a good insulator, so that energy does not leave the system, then we can assume the system is isolated. Vessels having this property are called **calorimeters,** and analysis performed using such vessels is called **calorimetry.**

The principle of conservation of energy for this isolated system requires that the energy that leaves the warmer substance (of unknown specific heat) equals the energy that enters the water.[1] Thus, we can write

$$Q_{cold} = -Q_{hot} \qquad [11.4]$$

To see how to set up a calorimetry problem, suppose that m_x is the mass of a substance whose specific heat we wish to determine, c_x its specific heat, and T_x its initial temperature. Let m_w, c_w, and T_w represent corresponding values for the water. If T is the final equilibrium temperature after the substance and the water are combined, from Equation 11.3 we find that the energy gained by the water is $m_w c_w(T - T_w)$ and the energy lost by the substance of unknown specific heat is $m_x c_x(T - T_x)$. Substituting these values into Equation 11.4, we have

$$m_w c_w(T - T_w) = -m_x c_x(T - T_x)$$

Solving for c_x gives

$$c_x = \frac{m_w c_w(T - T_w)}{m_x(T_x - T)}$$

By substituting the known values in the right-hand side, we can calculate the specific heat of the substance.

Tip 11.2 SIGN CONVENTION FOR ENERGY TRANSFER

The negative sign in Equation 11.4 is necessary for consistency with our sign convention for energy transfer. The energy transfer Q_{hot} is negative because energy is leaving the hot substance. The negative sign in the equation assures that the right-hand side is a positive number, consistent with the left-hand side, which is positive because energy is entering the cold water.

Tip 11.3 CELSIUS VERSUS KELVIN

In equations in which T appears, such as the ideal gas law, the Kelvin temperature *must* be used. In equations involving ΔT, such as calorimetry equations, it is *possible* to use Celsius temperatures, because a change in temperature is the same on both scales. It is *safest*, however, to use Kelvin temperatures *consistently* in all equations involving T or ΔT.

Example 11.2 Cooling a Hot Ingot

The temperature of a 0.050 0-kg ingot of metal is raised to 200.0°C, and the ingot is then dropped into a light, insulated beaker containing 0.400 kg of water initially at 20.0°C. If the final equilibrium temperature of the mixed system is 22.4°C, find the specific heat of the metal.

Solution Using Equation 11.4, we can write $Q_{cold} = -Q_{hot}$ for the system, or

$$m_w c_w(T - T_w) = -m_x c_x(T - T_x)$$

Solving for c_x and substituting numerical values, we have

$$c_x = \frac{m_w c_w(T - T_w)}{m_x(T_x - T)} = \frac{(0.400 \text{ kg})(4\,186 \text{ J/kg} \cdot °C)(22.4°C - 20.0°C)}{(0.0\,500 \text{ kg})(200.0°C - 22.4°C)}$$

$$= \boxed{453 \text{ J/kg} \cdot °C}$$

The ingot is most likely iron, as can be seen by comparing this result with the data in Table 11.1.

EXERCISE What is the total energy transferred to the water in this example as the ingot is cooled?

ANSWER 4.02 kJ

---

[1] For precise measurements, the container holding the water should be included in our calculations because it also changes temperature. However, if the mass of the water is large compared with that of the container, we can ignore the energy gained by the container.

EXERCISE A cowboy fires a silver bullet of mass 2.00 g with a muzzle speed of 2.00×10^2 m/s into the pine wall of a saloon, where the bullet stops. Assume that all the internal energy generated by the impact remains within the bullet. What is the temperature change of the bullet?

ANSWER 85.5°C

11.4 LATENT HEAT AND PHASE CHANGE

A substance often undergoes a change in temperature when energy is transferred between the substance and its environment. In some situations, however, the transfer of energy does not result in a change in temperature. This can occur when the physical characteristics of the substance change from one form to another, commonly referred to as a **phase change.** Some common phase changes are solid to liquid (melting), liquid to gas (boiling), and a change in crystalline structure of a solid. All such phase changes involve a change in internal energy, but no change in temperature.

The energy required to change the phase of a given mass *m* of a pure substance is

$$Q = \pm mL \qquad [11.5]$$

where *L* is called the **latent heat** of the substance and depends on the nature of the phase change as well as on the substance. The word *latent* is from the Latin *latere*, meaning *hidden* or *concealed*. The proper sign in Equation 11.5 is chosen according to the direction of the energy flow. For example, when an ice cube is melting, we are adding energy to the ice, and we express Equation 11.5 as $Q = +mL$. However, when water is freezing into ice, we are removing energy from the water, so we express Equation 11.5 with the negative sign: $Q = -mL$.

Latent heat of fusion L_f is the term used when the phase change occurs during melting or freezing, and **latent heat of vaporization** L_v is the term used when the phase change occurs during boiling or condensing.[2] For example, the latent heat of fusion for water at atmospheric pressure is 3.33×10^5 J/kg, and the latent heat of vaporization for water is 2.26×10^6 J/kg. The latent heats of different substances vary considerably, as is seen in Table 11.2.

With our knowledge of latent heat, we can understand the full behavior of a substance as energy is added to it. Consider, for example, the addition of energy to a 1-g cube of ice at $-30.0°C$ in a container held at constant pressure. Suppose that this energy results in the ice turning to steam (water vapor) at 120.0°C. Figure 11.2 indicates the experimental measurement of temperature as energy is added to the system. Let us examine each portion of the curve separately.

Part A During this portion of the curve, the temperature of the ice changes from $-30.0°C$ to 0.0°C. Because the specific heat of ice is 2 090 J/kg·°C, we can calculate the amount of energy added from Equation 11.3:

$$Q = mc_{\text{ice}} \Delta T = (1.00 \times 10^{-3} \text{ kg})(2\,090 \text{ J/kg·°C})(30.0°C) = 62.7 \text{ J}$$

Part B When the ice reaches 0.0°C, the ice-water mixture remains at this temperature—even though energy is being added—until all the ice melts to become water at 0.0°C. According to Equation 11.5, the energy required to melt 1.00 g of ice at 0.0°C is

$$Q = mL_f = (1.00 \times 10^{-3} \text{ kg})(3.33 \times 10^5 \text{ J/kg}) = 333 \text{ J}$$

Tip 11.4 SIGNS ARE CRITICAL

For phase changes, use the correct explicit sign in Equation 11.5, depending on whether you are adding or removing energy from the substance.

---

[2] When a gas cools, it eventually returns to the liquid phase, or *condenses.* The energy per unit mass given up during the process is called the *heat of condensation,* and it equals the heat of vaporization. When a liquid cools, it eventually solidifies, and the *heat of solidification* equals the heat of fusion.

TABLE 11.2	**Latent Heats of Fusion and Vaporization**					
		Latent Heat of Fusion			**Latent Heat of Vaporization**	
Substance	**Melting Point (°C)**	**(J/kg)**	**(cal/g)**	**Boiling Point (°C)**	**(J/kg)**	**(cal/g)**
Helium	−269.65	5.23×10^3	(1.25)	−268.93	2.09×10^4	(4.99)
Nitrogen	−209.97	2.55×10^4	(6.09)	−195.81	2.01×10^5	(48.0)
Oxygen	−218.79	1.38×10^4	(3.30)	−182.97	2.13×10^5	(50.9)
Ethyl alcohol	−114	1.04×10^5	(24.9)	78	8.54×10^5	(204)
Water	0.00	3.33×10^5	(79.7)	100.00	2.26×10^6	(540)
Sulfur	119	3.81×10^4	(9.10)	444.60	3.26×10^5	(77.9)
Lead	327.3	2.45×10^4	(5.85)	1 750	8.70×10^5	(208)
Aluminum	660	3.97×10^5	(94.8)	2 450	1.14×10^7	(2 720)
Silver	960.80	8.82×10^4	(21.1)	2 193	2.33×10^6	(558)
Gold	1 063.00	6.44×10^4	(15.4)	2 660	1.58×10^6	(377)
Copper	1 083	1.34×10^5	(32.0)	1 187	5.06×10^6	(1 210)

Part C   Between 0.0°C and 100°C, no phase change occurs. The energy added to the water is used to increase its temperature, as it was in part A. The amount of energy necessary to increase the temperature from 0.0°C to 100°C is

$$Q = mc_{water} \, \Delta T = (1.00 \times 10^{-3} \, \text{kg})(4.19 \times 10^3 \, \text{J/kg} \cdot °\text{C})(100°\text{C}) = 4.19 \times 10^2 \, \text{J}$$

Part D   At 100°C, another phase change occurs as the water changes to steam at 100°C. Similar to the behavior in part B, the water-steam mixture remains at constant temperature, this time at 100°C—even though energy is being added—until all the liquid has been converted to steam. The energy required to convert 1.00 g of water to steam at 100°C is

$$Q = mL_v = (1.00 \times 10^{-3} \, \text{kg})(2.26 \times 10^6 \, \text{J/kg}) = 2.26 \times 10^3 \, \text{J}$$

Part E   During this portion of the curve, as in parts A and C, no phase change occurs, so all the energy added is used to increase the temperature of the steam. The energy that must be added to raise the temperature of the steam to 120.0°C is

$$Q = mc_{steam} \, \Delta T = (1.00 \times 10^{-3} \, \text{kg})(2.01 \times 10^3 \, \text{J/kg} \cdot °\text{C})(20°\text{C}) = 40.2 \, \text{J}$$

The total amount of energy that must be added to change 1 g of ice at −30.0°C to steam at 120°C is the sum of the results from all five parts of the curve, 3.11×10^3 J. Conversely, to cool 1 g of steam at 120.0°C down to the point at which we have ice at −30.0°C, we must remove 3.11×10^3 J of energy.

FIGURE 11.2   A plot of temperature versus energy added when 1.00 g of ice, initially at −30.0°C, is converted to steam at 120°C.

Phase changes can be described in terms of rearrangements of molecules when energy is added to or removed from a substance. Consider first the liquid-to-gas phase change. The molecules in a liquid are close together, and the forces between them are stronger than those between the more widely separated molecules of a gas. Therefore, work must be done on the liquid against these attractive molecular forces in order to separate the molecules. The latent heat of vaporization is the amount of energy that must be added to the one kilogram of liquid to accomplish this.

Similarly, at the melting point of a solid, we imagine that the amplitude of vibration of the atoms about their equilibrium positions becomes great enough to allow the atoms to pass the barriers of adjacent atoms and move to their new positions. The new locations are, on average, less symmetrical and therefore have higher energy. The latent heat of fusion is equal to the work required at the molecular level to transform the mass from the ordered solid phase to the disordered liquid phase.

The average distance between atoms is much greater in the gas phase than in either the liquid or the solid phase. Each atom or molecule is removed from its neighbors, without the compensation of attractive forces to new neighbors. Therefore, more work is required at the molecular level to vaporize a given mass of substance than to melt it; thus, the latent heat of vaporization is much greater than the latent heat of fusion (Table 11.2).

> **Quick Quiz 11.3**
>
> An amount of energy is added to ice, raising its temperature from $-10°C$ to $-5°C$. A larger amount of energy is added to the same mass of liquid water, raising its temperature from $15°C$ to $20°C$. From these results, we can conclude that (a) overcoming the latent heat of fusion of ice requires an input of energy, (b) the latent heat of fusion of ice delivers some energy to the system, (c) the specific heat of ice is less than that of water, (d) the specific heat of ice is greater than that of water.

PROBLEM-SOLVING STRATEGY   Calorimetry Problems

If you are having difficulty with calorimetry problems, be sure to consider the following points:

1. Use consistent units. For instance, if you are using specific heats in $cal/g \cdot °C$, be sure masses are in grams and temperatures in Celsius degrees.
2. Transfers of energy are given by $Q = mc \, \Delta T$ only for those processes in which no phase changes occur. The equations $Q = \pm mL_f$ and $Q = \pm mL_v$ are to be used only when phase changes *are* taking place.
3. Often sign errors occur in the equation $Q_{cold} = -Q_{hot}$. Remember to use the negative sign in the equation, and that ΔT is always the final temperature minus the initial temperature.

Example 11.3   Cooling the Steam

What mass of steam initially at $130°C$ is needed to warm 200 g of water in a 100-g glass container from $20.0°C$ to $50.0°C$?

Solution The steam loses energy in three stages. In the first stage, the steam is cooled to 100°C. The energy transfer in the process is

$$Q_1 = m_s c_s \Delta T = m_s (2.01 \times 10^3 \, \text{J/kg} \cdot °\text{C})(-30.0°\text{C}) = -m_s(6.03 \times 10^4 \, \text{J/kg})$$

where m_s is the unknown mass of the steam.

In the second stage, the steam is converted to water. To find the energy transfer during this phase change, we use $Q = -mL_v$, where the negative sign indicates that energy is leaving the steam.

$$Q_2 = -m_s(2.26 \times 10^6 \, \text{J/kg})$$

In the third stage, the temperature of the water created from the steam is reduced to 50.0°C. This requires an energy transfer of

$$Q_3 = m_s c_w \Delta T = m_s(4.19 \times 10^3 \, \text{J/kg} \cdot °\text{C})(-50.0°\text{C}) = -m_s(2.09 \times 10^5 \, \text{J/kg})$$

Adding the energy transfers in these three stages gives us

$$Q_{\text{hot}} = Q_1 + Q_2 + Q_3 = -m_s(6.03 \times 10^4 \, \text{J/kg} + 2.26 \times 10^6 \, \text{J/kg} + 2.09 \times 10^5 \, \text{J/kg})$$
$$= -m_s(2.53 \times 10^6 \, \text{J/kg})$$

Now, we turn our attention to the temperature increase of the water and the glass. Using Equation 11.3,

$$Q_{\text{cold}} = (0.200 \, \text{kg})(4.19 \times 10^3 \, \text{J/kg} \cdot °\text{C})(30.0°\text{C})$$
$$+ (0.100 \, \text{kg})(837 \, \text{J/kg} \cdot °\text{C})(30.0°\text{C})$$
$$= 2.77 \times 10^4 \, \text{J}$$

Using Equation 11.4, we can solve for the unknown mass:

$$Q_{\text{cold}} = -Q_{\text{hot}}$$
$$2.77 \times 10^4 \, \text{J} = -[-m_s(2.53 \times 10^6 \, \text{J})]$$
$$m_s = 1.09 \times 10^{-2} \, \text{kg} = \boxed{10.9 \, \text{g}}$$

Example 11.4 Boiling Liquid Helium

Liquid helium has a very low boiling point, 4.2 K, and a very low latent heat of vaporization, $2.09 \times 10^4 \, \text{J/kg}$. If energy is transferred to a container of liquid helium from an immersed electric heater at a rate of 10.0 W, how long does it take to boil away 1.00 kg of the liquid?

Solution Because $L_v = 2.09 \times 10^4 \, \text{J/kg}$, we must supply $2.09 \times 10^4 \, \text{J}$ of energy to boil away 1.00 kg. Because 10.0 W = 10.0 J/s, 10.0 J of energy is transferred to the helium each second. Therefore, the time interval required to transfer $2.09 \times 10^4 \, \text{J}$ of energy is

$$\Delta t = \frac{2.09 \times 10^4 \, \text{J}}{10.0 \, \text{J/s}} = 2.09 \times 10^3 \, \text{s} \approx \boxed{35 \, \text{min}}$$

EXERCISE If 10.0 W of power is supplied to 1.00 kg of water at 100°C, how long will it take for the water to completely boil away?

ANSWER 62.8 h

The ceramic material being heated by a flame is a poor conductor of energy but is able to withstand a large temperature gradient; its left side is buried in ice (0°C) and its right side is extremely hot. *(Courtesy of Corning Glass Works)*

11.5 ENERGY TRANSFER BY THERMAL CONDUCTION

For some applications we must know the rate at which energy is transferred between a system and its surroundings and the mechanisms responsible for the transfer. There are several ways in which energy (not classified as work) can be

FIGURE 11.3 Energy reaches the hand by conduction through the copper rod.

transferred from one location to another. These include conduction, convection, and radiation. These processes can result in a change in internal energy of a system. Regardless of the process, however, no net energy transfer takes place between a system and its surroundings when the two are at the same temperature. In this section we discuss energy transfer by conduction. Convection and radiation are discussed in Sections 11.6 and 11.7.

The energy transfer process that is most clearly associated with a temperature difference is called **thermal conduction** or simply **conduction.** In this process, the transfer can be viewed on an atomic scale as an exchange of kinetic energy between microscopic particles—molecules, atoms, and electrons, where less energetic particles gain energy in collisions with more energetic particles. For example, if you hold one end of a long metal bar and insert the other end into a flame as in Figure 11.3, the temperature of the metal in your hand soon increases. The energy reaches your hand through conduction. How this happens can be understood by examining what is happening to the microscopic particles in the metal. Before the rod is inserted into the flame, the microscopic particles are vibrating about their equilibrium positions. As the rod is brought in the vicinity of the flame, those particles near the flame begin to vibrate with larger and larger amplitudes. These, in turn, collide with their neighbors and transfer some of their energy in the collisions. Slowly, metal atoms and electrons farther and farther from the flame increase their amplitude of vibration, until eventually those in the metal near your hand are affected. This increased vibration represents an increase in temperature of the metal (and possibly a burned hand).

Although the transfer of energy through a substance can be partially explained by atomic vibrations, the rate of conduction depends on the properties of the substance. For example, it is possible to hold a piece of asbestos in a flame indefinitely. This implies that very little energy is being conducted through the asbestos. In general, metals are good thermal conductors because they contain large numbers of electrons that are relatively free to move through the metal and can transport energy from one region to another. Thus, in a good conductor such as copper, conduction takes place via the vibration of atoms and via the motion of free electrons. Materials such as asbestos, cork, paper, and fiberglass are poor conductors. Gases are also poor conductors because of the large distance between the molecules.

Conduction occurs only if there is a difference in temperature between two parts of the conducting medium. The temperature difference drives the flow of energy. Consider a slab of material of thickness Δx and cross-sectional area A with its opposite faces at different temperatures T_c and T_h, where $T_h > T_c$ (Fig. 11.4). The slab allows energy to transfer from the region of higher temperature to that of lower temperature by thermal conduction. The rate of energy transfer, $\mathcal{P} = Q/\Delta t$, is proportional to the cross-sectional area of the slab and the temperature differ-

Melted snow pattern on a parking lot surface indicates the presence of underground hot-water pipes used to aid snow removal. Energy from the water is conducted from the pipes to the pavement, where it causes the snow to melt. *(Courtesy of Dr. Albert A. Bartlett, University of Colorado, Boulder, CO)*

FIGURE 11.4 Energy transfer through a conducting slab of cross-sectional area A and thickness Δx. The opposite faces are at different temperatures, T_c and T_h.

FIGURE 11.5 Conduction of energy through a uniform, insulated rod of length L. The opposite ends are in thermal contact with energy reservoirs at different temperatures.

ence, and inversely proportional to the thickness of the slab:

$$\mathscr{P} = \frac{Q}{\Delta t} \propto A \frac{\Delta T}{\Delta x}$$

Note that $\mathscr{P}$ has units of watts when Q is in joules and Δt is in seconds.

Suppose a substance is in the shape of a long, uniform rod of length L as in Figure 11.5. We shall assume that the rod is insulated so that energy cannot escape by conduction from its surface except at the ends. One end is in thermal contact with an energy reservoir at temperature T_c and the other end is in thermal contact with a reservoir at temperature $T_h > T_c$. When a steady state is reached, the temperature at each point along the rod is constant in time. In this case, $\Delta T = T_h - T_c$ and $\Delta x = L$, so that

$$\frac{\Delta T}{\Delta x} = \frac{T_h - T_c}{L}$$

Thus, the actual rate of energy transfer by conduction through the rod is given by the equation

$$\mathscr{P} = kA\left(\frac{T_h - T_c}{L}\right) \qquad [11.6]$$

where k is a proportionality constant, depending on material, called the **thermal conductivity.** Substances that are good conductors have large thermal conductivity values, whereas good insulators have low thermal conductivity values. Table 11.3 lists thermal conductivities for various substances.

TABLE 11.3

Thermal Conductivities

Substance	Thermal Conductivity ($J/s \cdot m \cdot °C$)
Metals (at 25°C)	
Aluminum	238
Copper	397
Gold	314
Iron	79.5
Lead	34.7
Silver	427
Gases (at 20°C)	
Air	0.023 4
Helium	0.138
Hydrogen	0.172
Nitrogen	0.023 4
Oxygen	0.023 8
Nonmetals	
Asbestos	0.25
Concrete	1.3
Glass	0.84
Ice	1.6
Rubber	0.2
Water	0.60
Wood	0.10

Quick Quiz 11.4 Will an ice cube wrapped in a wool blanket remain frozen for (a) less time, (b) the same length of time, or (c) a longer time than an identical ice cube exposed to air at room temperature?

Tip 11.5 BLANKETS AND COATS IN COLD WEATHER

When you sleep under a blanket in the winter or wear a warm coat outside, the blanket or coat is not a source of energy. It serves as a layer of material with low thermal conductivity in order to reduce the transfer of energy away from your body by heat. The primary insulating medium is the air trapped in small pockets within the material.

Example 11.5 Energy Transfer Through a Concrete Wall

Find the energy transferred in 1.00 h by conduction through a concrete wall 2.0 m high, 3.65 m long, and 0.20 m thick if one side of the wall is held at 20°C and the other side is at 5°C.

Solution Equation 11.6 gives the rate of energy transfer by conduction in joules per second. Because $\mathscr{P} = Q/\Delta t$, the amount of energy Q transferred in a time interval Δt is

$$Q = \mathscr{P}\,\Delta t = kA\,\Delta t\left(\frac{T_h - T_c}{L}\right)$$

If we substitute the values given and consult Table 11.3, we find that

$$Q = (1.3\,J/s \cdot m \cdot °C)(7.3\ m^2)(3\ 600\ s)\left(\frac{15°C}{0.20\ m}\right) = 2.6 \times 10^6\,J$$

This is an enormous amount of energy—enough to raise the temperature of 5.4×10^3 kg of water by 1°C!

Early houses were insulated by the material of which their walls were constructed—thick masonry blocks. Masonry restricts energy loss by conduction because its k is relatively low. The large thickness L also decreases energy loss by conduction, as shown by Equation 11.6.

HOME INSULATION

If you would like to determine whether or not to add insulation to a ceiling or to some other portion of a building, you need to slightly extend what you have just learned about conduction, for two reasons.

1. The insulating properties of materials used in buildings are usually expressed in engineering (U.S. customary) rather than SI units. For example, measurements stamped on a package of fiberglass insulating board will be in units such as British thermal units, feet, and degrees Fahrenheit.
2. In dealing with the insulation of a building, we must consider conduction through a compound slab, with each portion of the slab having a different thickness and a different thermal conductivity. For example, a typical wall in a house consists of an array of materials, such as wood paneling, dry wall, insulation, sheathing, and wood siding.

It is found that the rate of energy transfer by conduction through a compound slab is

$$\frac{Q}{\Delta t} = \frac{A(T_h - T_c)}{\sum_i L_i/k_i}$$

[11.7]

where T_h and T_c are the temperatures of the *outer extremities* of the slab and the summation is over all portions of the slab. For example, if the slab consists of three different materials, the denominator is the sum of three terms. In engineering practice, the term L/k for a particular substance is referred to as the **R value** of the material. Thus, Equation 11.7 reduces to

$$\frac{Q}{\Delta t} = \frac{A(T_h - T_c)}{\sum_i R_i}$$

[11.8]

The R values for a few common building materials are listed in Table 11.4. Note the units of R and that the R values are defined for specific thicknesses.

TABLE 11.4	*R* Values for Common Building Materials

Material	R value (ft^2·°F·h/Btu)
Hardwood siding (1.0 in. thick)	0.91
Wood shingles (lapped)	0.87
Brick (4.0 in. thick)	4.00
Concrete block (filled cores)	1.93
Styrofoam (1.0 in. thick)	5.0
Fiber glass batting (3.5 in. thick)	10.90
Fiber glass batting (6.0 in. thick)	18.80
Fiber glass board (1.0 in. thick)	4.35
Cellulose fiber (1.0 in. thick)	3.70
Flat glass (0.125 in. thick)	0.89
Insulating glass (0.25-in. space)	1.54
Vertical air space (3.5 in. thick)	1.01
Air film	0.17
Dry wall (0.50 in. thick)	0.45
Sheathing (0.50 in. thick)	1.32

Next to any vertical outside surface is a very thin, stagnant layer of air that must be considered when the total R value for a wall is figured. The thickness of this stagnant layer depends on the speed of the wind. As a result, energy loss by conduction from a house on a day when the wind is blowing hard is greater than energy loss on a day when the wind speed is zero. A representative R value for a stagnant air layer is given in Table 11.4.

Example 11.6   The *R* Value of a Typical Wall

Calculate the total R value for a wall constructed as shown in Figure 11.6a. Starting outside the house (to the left in the figure) and moving inward, the wall consists of 4.0-in. brick, 0.50 in. sheathing, an air space 3.5 in. thick, and 0.50-in. drywall. Do not forget the stagnant air layers inside and outside the house.

Solution   Referring to Table 11.4, we find

$$\begin{aligned}
\text{Outside stagnant air layer:} \quad & R_1 = 0.17 \text{ ft}^2 \cdot {}^\circ\text{F} \cdot \text{h/Btu} \\
\text{Brick:} \quad & R_2 = 4.00 \\
\text{Sheathing:} \quad & R_3 = 1.32 \\
\text{Air space:} \quad & R_4 = 1.01 \\
\text{Drywall:} \quad & R_5 = 0.45 \\
\text{Inside stagnant air layer:} \quad & R_6 = 0.17 \\
\hline
& R_{\text{total}} = \boxed{7.12 \text{ ft}^2 \cdot {}^\circ\text{F} \cdot \text{h/Btu}}
\end{aligned}$$

EXERCISE   If a layer of fiberglass insulation 3.5 in. thick is placed inside the wall to replace the air space, as in Figure 11.6b, what is the new total R value? By what factor is the energy loss reduced?

ANSWER $R = 17 \text{ ft}^2 \cdot {}^\circ\text{F} \cdot \text{h/Btu}$; 2.4

FIGURE 11.6   (Example 11.6) A cross-sectional view of an exterior wall containing (a) an air space and (b) insulation.

11.6 ENERGY TRANSFER BY CONVECTION

At one time or another you have probably warmed your hands by holding them over an open flame as shown in Figure 11.7. In this situation, the air directly above the flame is being warmed and expands. As a result, the density of this air decreases and the air rises. This mass of air warms your hands as it flows by. **Energy transferred by the movement of a substance is said to have been transferred by convection.** When the movement results from differences in density, as with air around a fire, it is referred to as *natural convection*. Air flow at a beach is an example of natural convection, as is the mixing that occurs as surface water in a lake cools and sinks. When the substance is forced to move by a fan or pump, as in some hot air and hot water heating systems, the process is called *forced convection*.

Were it not for convection currents, boiling water would be very difficult. As water is being warmed in a teakettle, the lower layers are warmed first. The warmed water at the bottom expands and rises to the top because its density is lowered. At the same time, the denser, cool water at the surface sinks to the bottom of the kettle and is warmed.

The same process occurs when a radiator raises the temperature of a room. The hot radiator warms the air in the lower regions of the room. The warm air expands and rises to the ceiling because of its lower density. The denser, cooler air

FIGURE 11.7   Warming a hand by convection.

Photograph of a teakettle showing steam and turbulent convection air currents. *(Gary Settles/Science Source/Photo Researchers)*

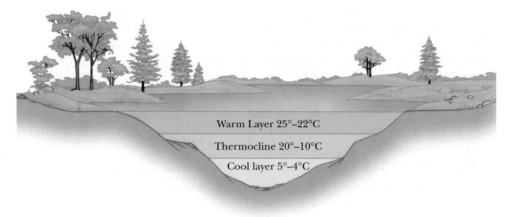

Warm Layer 25°–22°C

Thermocline 20°–10°C

Cool layer 5°–4°C

(a) Summer layering of water

FIGURE 11.8 Convection currents are set up in a room warmed by a radiator.

(b) Fall and spring upwelling

FIGURE 11.9 (a) During the summer, the warm upper layer of water is separated from a cooler lower layer by a thermocline. (b) Convection currents during the spring or fall mix the water and can cause algal blooms.

APPLICATION

COOLING AUTOMOBILE ENGINES

APPLICATION

ALGAL BLOOMS IN PONDS AND LAKES

from above sinks, setting up the continuous air current pattern shown in Figure 11.8.

Your automobile engine is maintained at a safe operating temperature by a combination of conduction and forced convection. Water (actually, a mixture of water and antifreeze) circulates in the interior of the engine. As the metal of the engine block increases in temperature, energy passes from the hot metal to the cooler water by thermal conduction. The water pump forces water out of the engine and into the radiator, carrying energy along with it (by forced convection). In the radiator, the hot water passes through metal pipes that are in contact with the cooler outside air, and energy passes into the air by conduction. The cooled water is then returned to the engine by the water pump to absorb more energy. The process of air being pulled past the radiator by the fan is also forced convection.

The algal blooms often seen in temperate lakes and ponds during spring or fall are caused by convection currents in the water. To understand this process, examine Figure 11.9. During the summer, bodies of water develop temperature gradients such that a warm upper layer of water is separated from a cold lower layer by a buffer zone called a thermocline. In the spring or fall, the temperature changes in the water break down this thermocline, setting up convection currents that mix the water. This mixing process transports nutrients from the bottom to the surface. The nutrient-rich water forming at the surface can cause a rapid, temporary increase in the population of algae.

The body temperature of mammals ranges from about 35°C to 38°C, whereas that of birds ranges from about 40°C to 43°C. To maintain this temperature in cold weather, it is necessary to provide insulation to prevent energy loss from the surface of the skin. How is this accomplished?

Explanation A natural method of accomplishing this is via layers of fat beneath the skin. Fat protects against both conduction and convection in that it has a low thermal conductivity, and there are few blood vessels in fat to carry blood to the surface where energy losses by convection can occur. Birds ruffle their feathers in cold weather such that a layer of air with a low thermal conductivity is trapped between the feathers and the skin. Bristling the fur produces the same effect in fur-bearing animals.

11.7 ENERGY TRANSFER BY RADIATION

Another way to transfer energy is through **radiation.** You have most likely experienced energy transfer by radiation when sitting in front of a fireplace. Figure 11.10 shows how we warm our hands at an open flame using energy transfer by radiation. Our hands are placed to one side of the flame; they are not in physical contact with the flame either directly or indirectly, and therefore conduction cannot account for the energy transfer. Furthermore, convection is not important in this situation because our hands are not above the flame in the path of convection currents. The important process in this case is the transfer of energy by radiation.

FIGURE 11.10 Warming hands by radiation.

All objects radiate energy continuously in the form of electromagnetic waves due to thermal vibrations of the molecules. You also encounter electromagnetic radiation in the form of the orange glow from an electric stove burner, an electric space heater, or the coils of a toaster.

The rate at which an object radiates energy is proportional to the fourth power of its absolute temperature. This is known as **Stefan's law** and is expressed in equation form as

$$\mathscr{P} = \sigma A e T^4 \qquad \text{[11.9]}$$

◄ Stefan's law

where $\mathscr{P}$ is the power in watts (or joules per second) radiated by the object, σ is a parameter equal to $5.669\,6 \times 10^{-8}$ W/m$^2 \cdot$K^4, A is the surface area of the object in square meters, e is a constant called the **emissivity,** and T is the object's Kelvin temperature. The value of e can vary between zero and unity, depending on the properties of the surface of the object.

Approximately 1 340 J of electromagnetic radiation from the Sun passes perpendicularly through each 1 m^2 at the top of Earth's atmosphere every second. This radiation is primarily visible light, accompanied by significant amounts of infrared and ultraviolet radiation. We shall study these types of radiation in detail in Chapter 21. Some of this energy is reflected back into space, and some is absorbed by the atmosphere, but enough arrives at Earth's surface each day to supply all our energy needs hundreds of times over—if it could be captured and used efficiently. The growth in the number of solar houses in this country is one example of our attempts to make use of this abundant energy. Radiant energy from the Sun affects our day-to-day existence in a number of ways. It influences Earth's average temperature, ocean currents, agriculture, rain patterns, and so on.

As another example of the effects of energy transfer by radiation, consider what happens to the atmospheric temperature at night. If there is cloud cover above Earth, the water vapor in the clouds absorbs part of the infrared radiation emitted by Earth and reemits it back to the surface. Consequently, the temperature at the surface remains at moderate levels. In the absence of cloud cover, there

Webnote 11.1

Radiation makes solar heating possible. To learn more about solar heating from an enthusiastic proponent, visit
http://www.flasolar.com/overview. htm#The%20Big%20Picture

FIGURE 11.11 This thermogram of a home, made during cold weather, shows colors ranging from white and yellow (areas of greatest energy loss) to blue and purple (areas of least energy loss). *(Daedalus Enterprises, Inc./Peter Arnold, Inc.)*

is nothing to prevent this radiation from escaping into space, and thus the temperature drops more on a clear night than when it is cloudy.

As an object radiates energy at a rate given by Equation 11.9, it also absorbs electromagnetic radiation. If the latter process did not occur, an object would eventually radiate all its energy and its temperature would reach absolute zero. The energy an object absorbs comes from its surroundings, which consists of other objects that radiate energy. If an object is at a temperature T and its surroundings are at a temperature T_0, the net energy gained or lost each second by the object as a result of radiation is

$$\mathcal{P}_{\text{net}} = \sigma Ae(T^4 - T_0{}^4) \tag{11.10}$$

When an object is in equilibrium with its surroundings, it radiates and absorbs energy at the same rate, and so its temperature remains constant. When an object is hotter than its surroundings, it radiates more energy than it absorbs and so cools.

An **ideal absorber** is defined as an object that absorbs all the energy incident on it, and for such a body, $e = 1$. Such an object is referred to as a **black body** because a room-temperature black body would look black—it reflects no light to the eye of the viewer. An ideal absorber is also an ideal radiator of energy. In contrast, an object for which $e = 0$ absorbs none of the energy incident on it. Such an object reflects all the incident energy and so is an **ideal reflector.**

APPLICATION

LIGHT-COLORED SUMMER CLOTHING

White clothing is more comfortable to wear in the summer than black clothing. Black fabric acts as a good absorber of incoming sunlight and as a good emitter of this absorbed energy. However, about half of the emitted energy travels toward the body, causing the person wearing the garment to feel uncomfortably warm. In contrast, white or light-colored clothing reflects away much of the incoming energy.

APPLICATION

THERMOGRAPHY

The amount of energy radiated by an object can be measured with temperature-sensitive recording equipment, using a technique called **thermography.** The amount of radiation emitted by an object is greatest in the object's warmest regions. An image of the pattern formed by varying radiation levels, called a **thermogram,** is brightest in the warmest areas. Figure 11.11 reproduces a thermogram of a house. The center portions of the door and windows are yellow, signifying temperatures higher than those of surrounding areas. A higher temperature usually means that energy is escaping. Thermograms can be useful for purposes of energy conservation. For example, the owners of this house could conserve energy and reduce their heating costs by adding insulation to the attic area and by installing thermal draperies over the windows.

APPLICATION

RADIATION THERMOMETERS FOR MEASURING BODY TEMPERATURE

Figure 11.12 shows a recently developed radiation thermometer that has removed most of the risk of taking the temperature of young children or the aged with a traditional thermometer—risks such as bowel perforation or bacterial contamination. The instrument measures the intensity of the infrared radiation leaving the eardrum and surrounding tissues and converts this information to a standard numerical reading. The eardrum is a particularly good location to measure body temperature because it is near the hypothalamus—the body's temperature control center.

FIGURE 11.12 A radiation thermometer measures a patient's temperature by monitoring the intensity of infrared radiation leaving the ear. *(Photodisc/Getty Images)*

Quick Quiz 11.5 Star A has twice the radius and twice the absolute temperature of star B. What is the ratio of the power output of star A to that of star B due to electromagnetic radiation? The emissivity of both stars can assumed to be 1. (a) 4 (b) 8 (c) 16 (d) 32 (e) 64

APPLYING **PHYSICS** 11.2

If you sit in front of a fire in a fireplace with your eyes closed, you feel significant warmth in your eyelids. If you now put on a pair of eyeglasses and repeat this activity, your eyelids will not feel as warm. Why?

Explanation Much of the warm feeling that we have while sitting in front of a fireplace is due to radiation from the fire. A large fraction of this radiation is in the infrared part of the spectrum. Our eyelids are particularly sensitive to infrared radiation. However, glass is partially opaque to infrared radiation. Thus, when you put on the glasses, you block much of the radiation from reaching your eyelids, and they feel cooler.

Example 11.7   **Who Turned Down the Thermostat?**

A student is trying to decide what to wear. The surroundings (his bedroom) are at 20.0°C. If the skin temperature of the unclothed student is 35°C, what is the net energy loss from his body in 10.0 min by radiation? Assume the emissivity of skin is 0.900 and the surface area of the student is 1.50 m^2.

Solution Using Equation 11.10, the net rate of energy loss from the skin by radiation is

$$\mathscr{P}_{net} = \sigma A e (T^4 - T_0{}^4)$$
$$= (5.67 \times 10^{-8}\ \text{W/m}^2 \cdot \text{K}^4)(1.50\ \text{m}^2)$$
$$\times (0.90)[(308\ \text{K})^4 - (293\ \text{K})^4] = 125\ \text{W}$$

(Why is the temperature given in kelvins?) At this loss rate, the total energy lost by the skin by radiation in 10 min is

$$Q = \mathscr{P}_{net} \times \Delta t = (125\ \text{W})(600\ \text{s}) = \boxed{7.5 \times 10^4\ \text{J}}$$

This is enough energy to light a 100-watt light bulb for about 13 minutes!

11.8   **RESISTING ENERGY TRANSFER**

The Thermos bottle, called a *Dewar flask* (after its inventor) in scientific applications, is designed to minimize energy transfer by conduction, convection, and radiation. It is used to store either cold or hot liquids for long periods of time. The standard vessel (Fig. 11.13) is a double-walled Pyrex glass with silvered walls. The space between the walls is evacuated to minimize energy transfer by conduction and convection. The silvered surface minimizes energy transfer by radiation because silver is a very good reflector and has very low emissivity. A further reduction

Vacuum

Silvered surfaces

Hot or cold liquid

FIGURE 11.13 A cross-sectional view of a Thermos bottle designed to store hot or cold liquids.

APPLICATION

THERMOS BOTTLES

in energy loss is achieved by reducing the size of the neck. Dewar flasks are commonly used to store liquid nitrogen (boiling point 77 K) and liquid oxygen (boiling point 90 K).

To confine liquid helium (boiling point 4.2 K), which has a very low heat of vaporization, it is often necessary to use a double Dewar system in which the Dewar flask containing the liquid is surrounded by a second Dewar flask. The space between the two flasks is filled with liquid nitrogen.

Some of the principles of the Thermos bottle are used to protect sensitive electronic instruments in orbiting space satellites. In half of its orbit about Earth, a satellite is exposed to intense radiation from the Sun, and in the other half it is in Earth's cold shadow. Without protection, its interior would thus be subjected to tremendous extremes of temperature. The interior of the satellite is wrapped with blankets of highly reflective aluminum foil. The foil's shiny surface reflects away much of the Sun's radiation while the satellite is in the unshaded part of the orbit, and helps retain interior energy while the satellite is in Earth's shadow.

APPLICATION

DRESSING WARMLY IN WINTER

Wool sweaters and down jackets keep us warm by trapping the warmer air in regions close to our bodies and hence reducing energy loss by convection and conduction. In other words, what keeps us warm is not the clothing itself but the air trapped in the clothing (see Tip 11.5).

11.9 GLOBAL WARMING AND GREENHOUSE GASES

Many of the principles of energy transfer, and opposition to it, can be understood by studying the operation of a glass greenhouse. During the day, sunlight passes into the greenhouse and is absorbed by the walls, earth, plants, and so on. This absorbed visible light is subsequently re-radiated as infrared radiation, which causes the temperature of the interior to rise.

In addition, convection currents are inhibited in a greenhouse. As a result, warmed air cannot rapidly pass over the surfaces of the greenhouse that are exposed to the outside air and thereby cause energy loss by conduction through those surfaces. Most experts now consider this to be a more important warming effect than that of any trapped infrared radiation. In fact, experiments have shown that when the glass over a greenhouse is replaced by a special glass known to transmit infrared light, the temperature inside is lowered only slightly. Based on this evidence, the primary mechanism that raises the temperature of a greenhouse is not the trapping of infrared radiation but the inhibition of air flow that occurs under any roof (in an attic, for example).

A phenomenon formerly known as the *greenhouse effect* appears to also play a major role in determining Earth's temperature. First note that Earth's atmosphere is a good transmitter (and hence a poor absorber) of visible radiation and a good absorber of infrared radiation. The visible light that reaches Earth's surface is absorbed and re-radiated as infrared light, which in turn is absorbed (trapped) by Earth's atmosphere. An extreme case is the warmest planet, Venus, which has a carbon dioxide-rich atmosphere and temperatures approaching 850°F.

As fossil fuels (coal, oil, and natural gas) are burned, large amounts of carbon dioxide are released into the atmosphere, causing it to retain more energy. This is of great concern to scientists and governments throughout the world. Many scientists are convinced that the 10% increase in the amount of atmospheric carbon dioxide in the past 30 years could lead to drastic changes in world climate. The increase in concentration of atmospheric carbon dioxide in the latter part of the 20th century is shown in Figure 11.14. According to one estimate, doubling the carbon dioxide content in the atmosphere will cause temperatures to increase by 2°C! In temperate regions, such as Europe and the United States, the temperature rise would save billions of dollars per year in fuel costs. Unfortunately, the temperature increase would also melt a large amount of ice from the polar ice caps, which could cause flooding and destroy many coastal areas. It would also increase the frequency

FIGURE 11.14 The concentration of atmospheric carbon dioxide in parts per million (ppm) of dry air as a function of time during the latter part of the 20th century. These data were recorded at Mauna Loa Observatory in Hawaii. The yearly variations (red curve) coincide with growing seasons because vegetation absorbs carbon dioxide from the air. The steady increase (black curve) is of concern to scientists.

of droughts, and consequently decrease already low crop yields in tropical and subtropical countries. Even slightly higher average temperatures might make it impossible for certain plants and animals to survive in their customary ranges.

At present, about 3.5×10^{11} tons of CO_2 are released into the atmosphere each year. Most of this gas results from human activities such as the burning of fossil fuels, the cutting of forests, and manufacturing processes. Other greenhouse gases are also increasing in concentration in the atmosphere. One of these is methane, CH_4, which is released in the digestive process of cows and other ruminants. This gas originates from that part of the animal's stomach called the *rumen* where cellulose is digested. Termites are also major producers of this gas. Finally, greenhouse gases such as nitrous oxide (N_2O) and sulfur dioxide (SO_2) are increasing due to automobile and industrial pollution.

Whether the increasing greenhouse gases are responsible or not, there is convincing evidence that global warming is certainly underway. The evidence comes from melting of ice in Antarctica and the retreat of glaciers at widely scattered sites throughout the world. For example, satellite images of Antarctica show James Ross Island completely surrounded by water for the first time since maps were made, about 100 years ago. Previously the island was connected to the mainland by an ice bridge. In addition, at various places across the continent, ice shelves are retreating, some at a rapid rate.

Perhaps at no place in the world are glaciers monitored with greater interest than in Switzerland. There, it has been found that the Alps have lost about 50% of their glacial ice compared to 130 years ago. In 1818 the citizens of the town of Brig became concerned about the advance of the Aletsch glacier, the largest glacier in Switzerland, because it was threatening their summer pastures and forests. A religious procession was held, and a cross was placed against the glacier as a symbolic blockade. Today, the glacier is about a mile farther back in the mountains away from the cross, and it continues to retreat. This retreat is evidence of the loss of glacial ice in the Swiss Alps.

The retreat of glaciers on high-altitude peaks in the tropics is even more severe than in Switzerland. The Lewis glacier on Mount Kenya and the snows of Kilimanjaro are two examples. However, in certain regions of the planet where glaciers are near large bodies of water and are fed by large and frequent snows, glaciers continue to advance, so the overall picture of a catastrophic global warming scenario may be premature. However, in about 50 years, the amount of carbon dioxide in the atmosphere is expected to be about twice what it was in the pre-Industrial Era. Because of this, most scientists voice the concern that reductions in greenhouse gas emissions need to be made now.

Webnote 11.2

Find out more about global warming at this excellent EPA site: *http://www.epa.gov/globalwarming/*

SUMMARY

Internal energy is all of a system's energy that is associated with the system's microscopic components. Internal energy includes kinetic energy of translation, rotation, and vibration of molecules.

Heat is a transfer of energy across the boundary of a system resulting from a temperature difference between the system and its surroundings. We use the symbol Q for the amount of energy transferred by this process.

The **calorie** is the amount of energy necessary to raise the temperature of 1 g of water from 14.5°C to 15.5°C. The **mechanical equivalent of heat** is 4.186 J/cal.

The energy required to change the temperature of a mass m of a substance by an amount ΔT is

$$Q = mc \, \Delta T \qquad \text{[11.3]}$$

where c is the **specific heat** of the substance.

The energy required to change the phase of a pure substance of mass m is

$$Q = \pm \, mL \qquad \text{[11.5]}$$

where L is the **latent heat** of the substance and depends on the nature of the phase change and the properties of the substance.

Energy may be transferred by several processes including work, which we discussed in Chapter 5, and by conduction, convection, and radiation. **Conduction** can be viewed as an exchange of kinetic energy between colliding molecules or electrons. The rate at which energy transfers by conduction through a slab of area A and thickness L is

$$\mathscr{P} = kA \left(\frac{T_h - T_c}{L} \right) \qquad \text{[11.6]}$$

where k is the **thermal conductivity** of the material making up the slab.

Energy is transferred by **convection** as a substance moves from one place to another.

All objects emit **radiation** from the surface in the form of electromagnetic waves at the rate

$$\mathscr{P} = \sigma A e T^4 \qquad \text{[11.9]}$$

An object that is hotter than its surroundings radiates more energy than it absorbs, whereas a body that is cooler than its surroundings absorbs more energy than it radiates.

CONCEPTUAL QUESTIONS

1. Dogs pant when they are hot. How does this help them to cool off?

2. A person cannot remain long in a swimming pool if its temperature is above 37.0°C. Explain. Apply your explanation to the effect of thermal pollution on fish in a stream.

3. Rub the palm of your hand on a metal surface for 30–45 seconds. Place the palm of your other hand on an unrubbed portion of the surface and then the rubbed portion. The rubbed portion will feel warmer. Now repeat this process on a wooden surface. You should notice that the difference in temperature between the rubbed and unrubbed portions of the wood surface seems larger than that of the metal surface. Why is this?

4. Pioneers stored fruits and vegetables in underground cellars. Discuss fully this choice for a storage site.

5. In usually warm climates that experience an occasional hard freeze, fruit growers spray the fruit trees with water, hoping that a layer of ice will form on the fruit. Why is this advantageous?

6. In winter, the pioneers mentioned in Question 4 stored an open barrel of water alongside their produce. Why?

7. Cups of water for coffee or tea can be warmed with an immersion coil, which is immersed in the water and raised to a high temperature by means of electricity. The instructions warn the user not to operate the coils when they are out of water. Why? Can the immersion coil be used to warm up a cup of stew?

8. The U.S. penny is now made of copper-coated zinc. Can a calorimetric experiment be devised to test for the metal content in a collection of pennies? If so, describe the procedure you would use.

9. On a clear, cold night, why does frost tend to form on the tops of mailboxes and cars rather than the sides?

10. A warning sign often seen on highways just before a bridge is "Caution—Bridge Surface Freezes Before Road Surface." Of the three energy transfer processes discussed in Sections 11.5–11.7, which is most important in causing a bridge surface to freeze before the road surface on very cold days?

11. A tile floor feels uncomfortably cold to your bare feet, but a carpeted floor in an adjoining room at the same temperature feels warm. Why?

12. On a very hot day it is possible to cook an egg on the hood of a car. Would you select a black car or a white car on which to cook your egg? Why?

13. Concrete has a higher specific heat than does soil. Use this fact to explain (partially) why cities have higher average temperatures than the surrounding countryside. Would you expect evening breezes to blow from city to country or from country to city? Explain.

14. You need to pick up a very hot cooking pot in your kitchen. You have a pair of hot pads. Should you soak them in cold water or keep them dry in order to pick up the pot most comfortably?

15. In a daring lecture demonstration, a professor dips her wetted fingers into molten lead (327°C) and withdraws them quickly without getting burned. How is this possible?

16. The air temperature above coastal areas is profoundly influenced by the large specific heat of water. One reason is that the energy released when 1 m³ of water cools by 1.0°C raises the temperature of an enormously larger volume of air by 1.0°C. Estimate the volume of air. The specific heat of air is approximately 1.0 kJ/kg·°C. Take the density of air to be 1.3 kg/m³.

17. Ethyl alcohol has about one half the specific heat of water. If equal masses of alcohol and water in separate beakers are supplied with the same amount of energy, compare the temperature increases of the two liquids.

18. Why are the handles of some fireplace pokers made in the shape of a spring?

19. If you fill a brown paper bag with water and carefully set it on an outdoor barbecue grill, you can often boil the water without burning the bag. Explain how this can be. (Do NOT try this on an indoor stove.)

PROBLEMS

1, 2, 3 = straightforward, intermediate, challenging ☐ = full solution available in Student Solutions Manual/Study Guide

web = solution posted at **http://info.brookscole.com/serway** 🐍 = biomedical application

Section 11.1 Heat and Internal Energy

Section 11.2 Specific Heat

1. How many joules of energy are required to raise the temperature of 100 g of gold from 20.0°C to 100°C?

2. A 50-g sample of copper is at 25°C. If 1 200 J of energy is added to the copper by heat, what is its final temperature?

3. Lake Erie contains roughly 4.00×10^{11} m³ of water. (a) How much energy is required to raise the temperature of that volume of water from 11.0°C to 12.0°C? (b) How many years would it take to supply this amount of energy by using the 1 000-MW exhaust energy of an electric power plant?

4. An aluminum rod is 20.0 cm long at 20°C and has a mass of 350 g. If 10 000 J of energy is added to the rod by heat, what is the change in length of the rod?

5. While on their honeymoon in the French Alps, James Prescott Joule and his bride carefully measured the temperature of water at the top and bottom of a waterfall. If the water fell a distance of 110 m and if all of its initial potential energy was converted to internal energy as a result of the friction forces involved in hitting the bottom, what temperature rise should the new Mr. and Mrs. Joule have found? (Actually, they found no rise because the water loses energy through evaporation.)

6. As part of an exercise routine a 50.0-kg person climbs 10.0 m up a vertical rope. How many (food) Calories are expended in a single climb up the rope? (1 food Calorie = 10³ calories)

7. A 75.0-kg weight-watcher wishes to climb a mountain to work off the equivalent of a large piece of chocolate cake rated at 500 (food) Calories. How high must the person climb? (1 food Calorie = 10³ calories)

8. A person's body converts chemical energy into internal energy at a metabolic rate of 100 kcal/h. The person stands neck-deep in an insulated tub containing 0.800 m³ of water at a temperature of 20.0°C. Disregarding energy loss from the tub, determine the change of water temperature in 45.0 min.

9. A 5.00-g lead bullet traveling at 300 m/s is stopped by a large tree. If half the kinetic energy of the bullet is transformed into internal energy and remains with the bullet while the other half is transmitted to the tree, what is the increase in temperature of the bullet?

10. A 1.5-kg copper block is given an initial speed of 3.0 m/s on a rough horizontal surface. Because of friction, the block finally comes to rest. (a) If the block absorbs 85% of its initial kinetic energy as internal energy, calculate its increase in temperature. (b) What happens to the remaining energy?

11. A 200-g aluminum cup contains 800 g of water in thermal equilibrium with the cup at 80°C. The combination of cup and water is cooled uniformly so that the temperature decreases by 1.5°C per minute. At what rate is energy being removed? Express your answer in watts.

Section 11.3 Calorimetry

12. Lead pellets, each of mass 1.00 g, are heated to 200°C. How many pellets must be added to 500 g of water that is initially at 20.0°C to make the equilibrium temperature 25.0°C? Neglect any energy transfer to or from the container.

13. A 0.40-kg iron horseshoe that is initially at 500°C is dropped into a bucket containing 20 kg of water at 22°C. What is the final equilibrium temperature? Neglect any energy transfer to or from the surroundings.

14. In a showdown on the streets of Laredo, the good guy drops a 5.0-g silver bullet, at a temperature of 20°C, into a 100-cm³ cup of water at 90°C. Simultaneously, the bad guy drops a 5.0-g copper bullet, at the same initial temperature, into an identical cup of water. Which one ends the showdown with the coolest cup of water in the West? Neglect any energy transfer into or away from the container.

15. An aluminum cup contains 225 g of water and a 40-g copper stirrer, all at 27°C. A 400-g sample of silver at an initial temperature of 87°C is placed in the water. The stirrer is used to stir the mixture gently until it reaches its final equilibrium temperature of 32°C. Calculate the mass of the aluminum cup.

16. If 200 g of water is contained in a 300-g aluminum vessel at 10°C and an additional 100 g of water at 100°C is poured

into the container, what is the final equilibrium temperature of the mixture?

17. A 100-g aluminum calorimeter contains 250 g of water. The two substances are in thermal equilibrium at 10°C. Two metallic blocks are placed in the water. One is a 50-g piece of copper at 80°C. The other sample has a mass of 70 g and is originally at a temperature of 100°C. The entire system stabilizes at a final temperature of 20°C. Determine the specific heat of the unknown second sample.

18. A combination of 0.250 kg of water at 20.0°C, 0.400 kg of aluminum at 26.0°C, and 0.100 kg of copper at 100°C is mixed in an insulated container and allowed to come to thermal equilibrium. Neglect any energy transfer to or from the container and determine the final temperature of the mixture.

19. A student drops two metallic objects into a 120-g steel container holding 150 g of water at 25°C. One object is a 200-g cube of copper that is initially at 85°C, and the other is a chunk of aluminum that is initially at 5.0°C. To the student's surprise, the water reaches a final temperature of 25°C, precisely where it started. What is the mass of the aluminum chunk?

Section 11.4   Latent Heat and Phase Change

20. A 50-g ice cube at 0°C is heated until 45 g has become water at 100°C and 5.0 g has become steam at 100°C. How much energy was added to accomplish this?

21. A 100-g cube of ice at 0°C is dropped into 1.0 kg of water that was originally at 80°C. What is the final temperature of the water after the ice melts?

22. How much energy is required to change a 40-g ice cube from ice at −10°C to steam at 110°C?

23. What mass of steam that is initially at 120°C is needed to warm 350 g of water and its 300-g aluminum container from 20°C to 50°C?

24. A 3.00-g lead bullet at 30.0°C is fired at a speed of 240 m/s into a large block of ice at 0°C, in which it embeds itself. What quantity of ice melts?

25. A 75-kg cross-country skier glides over snow as in Figure P11.25. The coefficient of friction between skis and snow is 0.20. Assume all the snow beneath his skis is at 0°C and that all the internal energy generated by friction is added to snow that sticks to his skis until melted. How far would he have to ski to melt 1.0 kg of snow?

FIGURE P11.25   A cross-country skier. (*Nathan Bilow, Leo de Wys, Inc.*)

26. A 100-g ice cube at 0°C is placed in 650 g of water at 25°C. What is the final temperature of the mixture?

27. A 40-g block of ice is cooled to − 78°C. It is added to 560 g of water in an 80-g copper calorimeter at a temperature of 25°C. Determine the final temperature. (If not all the ice melts, determine how much ice is left.) Remember that the ice must first warm to 0°C, melt, and then continue warming as water. The specific heat of ice is 0.500 cal/g · °C = 2 090 J/kg · °C.

28. A 60.0-kg runner expends 300 W of power while running a marathon (Fig. P11.28). Assuming that 10.0% of the energy is delivered to the muscle tissue and that the excess energy is primarily removed from the body by sweating, determine the volume of bodily fluid (assume it is water) lost per hour. (At 37.0°C the latent heat of vaporization of water is 2.41×10^6 J/kg.)

FIGURE P11.28   Abdelkhader El Mouaziz of Morocco, winner of the New York Marathon in 2000. (*© Reuters New Media Inc./CORBIS*)

29. Two speeding lead bullets, each of mass 5.00 g and at temperature 20.0°C, collide head-on at speeds of 500 m/s each. Assuming a perfectly inelastic collision and no loss of energy to the atmosphere, describe the final state of the two-bullet system. *Hint:* Assume a constant value for the specific heat of lead in the liquid and solid state.

30. A beaker of water sits in the sun until it reaches an equilibrium temperature of 30°C. The beaker is made of 100 g of aluminum and contains 180 g of water. In an attempt to cool this system, 100 g of ice at 0°C is added to the water. (a) Determine the final temperature. If $T_f = 0$°C, determine how much ice remains. (b) Repeat this for 50 g of ice.

31. Steam at 100°C is added to ice at 0°C. (a) Find the amount of ice melted and the final temperature when the mass of steam is 10 g and the mass of ice is 50 g. (b) Repeat with steam of mass 1.0 g and ice of mass 50 g.

32. An iron nail is driven into a block of ice by a single blow of a hammer. The hammerhead has a mass of 0.50 kg and an initial speed of 2.0 m/s. Nail and hammer are at rest after the blow. How much ice melts? Assume the temperature of the nail is 0.0°C before and after.

Section 11.5   Energy Transfer by Thermal Conduction

33. (a) Find the rate of energy flow through a copper block of cross-sectional area 15 cm^2 and length 8.0 cm when a tem-

perature difference of 30°C is established across the block. Repeat the calculation assuming the material is (b) a block of stagnant air with these dimensions; (c) a block of wood with these dimensions.

34. A window has a glass surface of 1.6×10^3 cm^2 and a thickness of 3.0 mm. (a) Find the rate of energy transfer by conduction through this pane when the temperature of the inside surface of the glass is 70°F and the outside temperature is 90°F. (b) Repeat for the same inside temperature and an outside temperature of 0°F.

35. A steam pipe is covered with 1.50-cm-thick insulating material of thermal conductivity 0.200 cal/cm·°C·s. How much energy is lost every second when the steam is at 200°C and the surrounding air is at 20.0°C? The pipe has a circumference of 800 cm and a length of 50.0 m. Neglect losses through the ends of the pipe.

36. A box with a total surface area of 1.20 m^2 and a wall thickness of 4.00 cm is made of an insulating material. A 10.0-W electric heater inside the box maintains the inside temperature at 15.0°C above the outside temperature. Find the thermal conductivity k of the insulating material.

37. Determine the R value for a wall constructed as follows: The outside of the house consists of lapped wood shingles placed over 0.50-in.-thick sheathing, over 3.0 in. of cellulose fiber, over 0.50 in. of dry wall.

38. A Thermopane window consists of two glass panes, each 0.50 cm thick, with a 1.0-cm-thick sealed layer of air between. If the inside surface temperature is 23°C and the outside surface temperature is 0.0°C, determine the rate of energy transfer through 1.0 m^2 of the window. Compare this with the rate of energy transfer through 1.0 m^2 of a single 1.0-cm-thick pane of glass.

39. A copper rod and an aluminum rod of equal diameter are joined end to end in good thermal contact. The temperature of the free end of the copper rod is held constant at 100°C, and that of the far end of the aluminum rod is held at 0°C. If the copper rod is 0.15 m long, what must be the length of the aluminum rod so that the temperature at the junction is 50°C?

40. A Styrofoam box has a surface area of 0.80 m^2 and a wall thickness of 2.0 cm. The temperature of the inner surface is 5.0°C, and that outside is 25°C. If it takes 8.0 h for 5.0 kg of ice to melt in the container, determine the thermal conductivity of the Styrofoam.

Section 11.7 Energy Transfer by Radiation

41. A sphere that is a perfect blackbody radiator has a radius **web** of 0.060 m and is at 200°C in a room where the temperature is 22°C. Calculate the net rate at which the sphere radiates energy.

42. The surface temperature of the Sun is about 5 800 K. Taking the Sun's radius to be 6.96×10^8 m, calculate the total energy radiated by the Sun each second. (Assume $e = 0.965$.)

43. A large hot pizza floats in outer space. It is 70 cm in diameter and 2.0 cm thick, with a temperature of 100°C. Assume its emissivity is 0.8. What is the order of magnitude of its rate of energy loss?

44. Calculate the temperature at which a tungsten filament that has an emissivity of 0.25 and a surface area of 2.5×10^{-5} m^2 will radiate energy at the rate of 25 W in a room where the temperature is 22°C.

45. Measurements on two stars indicate that Star X has a surface temperature of 5 727°C and Star Y has a surface temperature of 11 727°C. If both stars have the same radius, what is the ratio of the luminosity (total power output) of Star Y to the luminosity of Star X? Both stars can be considered to have an emissivity of 1.0.

46. At high noon, the Sun delivers 1.00 kW to each square meter of a blacktop road. If the hot asphalt loses energy only by radiation, what is its equilibrium temperature?

ADDITIONAL PROBLEMS

47. The bottom of a copper kettle has a 10-cm radius and is 2.0 mm thick. The temperature of the outside surface is 102°C, and the water inside the kettle is boiling at 1 atm of pressure. Find the rate at which energy is being transferred through the bottom of the kettle.

48. A 1.00-m^2 solar collector collects radiation from the Sun and focuses it on 250 g of water that is initially at 23.0°C. The average intensity of radiation arriving from the Sun at Earth's surface at this location is 550 W/m^2, and we assume that this is collected with 100% efficiency. Find the time required for the collector to raise the temperature of the water to 100°C.

49. Solar energy can be the primary source of winter space heating for a typical house (with floor area 130 m^2 = 1 400 ft^2) in the north central United States. If the house has very good insulation, you may model it as losing energy by heat steadily at the rate of 1 000 W during the winter, when the average exterior temperature is −5°C. The passive solar-energy collector can consist simply of large windows facing south. Sunlight shining in during the daytime is absorbed by the floor, interior walls, and objects in the house, raising their temperature to 30°C. As the Sun goes down, insulating draperies or shutters are closed over the windows. During the period between 4 P.M. and 8 A.M., the temperature of the house will drop, and a sufficiently large "thermal mass" is required to keep it from dropping too far. The thermal mass can be a large quantity of stone (with specific heat 800 J/kg·°C) in the floor and the interior walls exposed to sunlight. What mass of stone is required if the temperature is not to drop below 18°C overnight?

50. A water heater is operated by solar power. If the solar collector has an area of 6.00 m^2, and the intensity delivered by sunlight is 550 W/m^2, how long does it take to increase the temperature of 1.00 m^3 of water from 20.0°C to 60.0°C?

51. A 40-g ice cube floats in 200 g of water in a 100-g copper cup; all are at a temperature of 0°C. A piece of lead at 98°C is dropped into the cup, and the final equilibrium temperature is 12°C. What is the mass of the lead?

52. Assume that a hailstone at 0°C falls through air at a uniform temperature of 0°C and lands on a sidewalk also at this temperature. From what initial height must the hailstone fall in order to entirely melt on impact?

53. A 200-g block of copper at a temperature of 90°C is dropped into 400 g of water at 27°C. The water is contained in a 300-g glass container. What is the final temperature of the mixture?

54. A class of 10 students taking an exam has a power output per student of about 200 W. Assume that the initial

temperature of the room is 20°C and that its dimensions are 6.0 m by 15.0 m by 3.0 m. What is the temperature of the room at the end of 1.0 h if all the energy remains in the air in the room and none is added by an outside source? The specific heat of air is 837 J/kg·°C, and its density is about 1.3×10^{-3} g/cm^3.

55. The human body must maintain its core temperature inside a rather narrow range around 37°C. Metabolic processes, notably muscular exertion, convert chemical energy into internal energy deep in the interior. From the interior, energy must flow out to the skin or lungs to be lost by heat to the environment. During moderate exercise, an 80-kg man can metabolize food energy at the rate 300 kcal/h, do 60 kcal/h of mechanical work, and put out the remaining 240 kcal/h of energy by heat. Most of the energy is carried from the body interior out to the skin by forced convection (as a plumber would say): Blood is warmed in the interior and then cooled at the skin, which is a few degrees cooler than the body core. Without blood flow, living tissue is a good thermal insulator, with thermal conductivity about 0.210 W/m·°C. Show that blood flow is essential to cool the body by calculating the rate of energy conduction in kcal/h through the tissue layer under the skin. Assume that its area is 1.40 m^2, its thickness is 2.50 cm, and it is maintained at 37.0°C on one side and at 34.0°C on the other side.

56. An aluminum rod and an iron rod are joined end to end **web** in good thermal contact. The two rods have equal lengths and radii. The free end of the aluminum rod is maintained at a temperature of 100°C, and the free end of the iron rod is maintained at 0°C. (a) Determine the temperature of the interface between the two rods. (b) If each rod is 15 cm long and each has a cross-sectional area of 5.0 cm^2, what quantity of energy is conducted across the combination in 30 min?

57. Water is being boiled in an open kettle that has a 0.500-cm-thick circular aluminum bottom with a radius of 12.0 cm. If the water boils away at rate of 0.500 kg/min, what is the temperature of the lower surface of the bottom of the kettle? Assume that the top surface of the bottom of the kettle is at 100°C.

58. A 3.00-g copper penny at 25.0°C drops 50.0 m to the ground. (a) If 60.0% of the initial potential energy associated with the penny goes into increasing the internal energy, determine the final temperature. (b) Does the result depend on the mass of the penny? Explain.

59. A bar of gold is in thermal contact with a bar of silver of the same length and area (Fig. P11.59). One end of the compound bar is maintained at 80.0°C, and the opposite end is at 30.0°C. When the energy flow reaches steady state, find the temperature at the junction.

60. An iron plate is held against an iron wheel so that a sliding frictional force of 50 N acts between the two pieces of metal. The relative speed at which the two surfaces slide over each other is 40 m/s. (a) Calculate the rate at which

FIGURE P11.59

mechanical energy is converted to internal energy. (b) The plate and the wheel have masses of 5.0 kg each, and each receives 50% of the internal energy. If the system is run as described for 10 s and each object is then allowed to reach a uniform internal temperature, what is the resultant temperature increase?

61. An automobile has a mass of 1 500 kg, and its aluminum brakes have an overall mass of 60 kg. (a) Assuming that all of the internal energy transformed by friction when the car stops is deposited in the brakes, and neglecting energy transfer, how many times could the car be braked to rest starting from 25 m/s (56 mi/h) before the brakes would begin to melt? (Assume an initial temperature of 20°C.) (b) Identify some effects that are neglected in part (a) but are likely to be important in a more realistic assessment of the temperature increase of the brakes.

62. A 1.0-m-long aluminum rod of cross-sectional area 2.0 cm^2 is inserted vertically into a thermally insulated vessel containing liquid helium at 4.2 K. The rod is initially at 300 K. If half of the rod is inserted into the helium, how many liters of helium boil off in the very short time during which the inserted half cools to 4.2 K? The density of liquid helium at 4.2 K is 122 kg/m^3.

63. A *flow calorimeter* is an apparatus used to measure the specific heat of a liquid. The technique is to measure the temperature difference between the input and output points of a flowing stream of the liquid while adding energy at a known rate. (a) Start with the equations $Q = mc\,\Delta T$ and $m = \rho V$ and show that the rate at which energy is added to the liquid is given by the expression

$$\frac{\Delta Q}{\Delta t} = \rho c\,\Delta T\left(\frac{\Delta V}{\Delta t}\right)$$

(b) In a particular experiment, a liquid of density 0.72 g/cm^3 flows through the calorimeter at the rate of 3.5 cm^3/s. At steady state, a temperature difference of 5.8°C is established between the input and output points when energy is supplied at the rate of 40 J/s. What is the specific heat of the liquid?

64. Three liquids are at temperatures of 10°C, 20°C, and 30°C, respectively. Equal masses of the first two liquids are mixed, and the equilibrium temperature is 17°C. Equal masses of the second and third are then mixed, and the equilibrium temperature is 28°C. Find the equilibrium temperature when equal masses of the first and third are mixed.

GROUP ACTIVITIES

G.1 A plot of the temperature of a substance versus time as its temperature decreases is called a cooling curve and has the same shape and basic explanation as the curve shown in Figure 11.2. You can plot such a curve using observations on some water in a container in the freezer compartment of a refrigerator. Place a thermometer in the liquid and record the reading of the thermometer every minute until about 5 minutes after the liquid has frozen completely. Explain your observations. Another material that is a little easier to work with is naphthalene (mothballs). You can plot the cooling curve in this case without a freezer. Melt a small amount of this in a container and plot a graph of temperature versus time as you did above. Again, explain your observations.

G.2 You have probably heard someone say that hot water freezes faster than cold water. Is this an urban legend or is it true? To test this, fill one container with hot water, about 200°F, and another with cooler water at about 70°F. Place the two in the freezer compartment of a refrigerator and find out for yourself. There are a number of variables that you will need to control in such an experiment. Here are a few; can you list more? (1) The two containers need to be located at similar locations in the freezer compartment. That is, one should not be near the door while another is in the back of the compartment. (2) The two containers should not be placed close together so that an exchange of energy takes place between the two. (3) If the freezer is one of the old types that forms frost on its walls, the hot container should not be allowed to melt through the frost and make intimate contact with the cold walls of the freezer.

G.3 You perhaps have heard that you can greatly reduce the baking time for potatoes in a *conventional* oven by inserting a nail through each potato. Are there any scientific reasons to believe this is true? Test the theory with a couple of similar sized potatoes, but do not bake them in a microwave oven.

G.4 Suppose you have two beakers of water. Each holds 1 kg of water. One is insulated from its surroundings and at 25°C. The other is sitting on a hot plate and boiling. You place a block of iron with a mass of 0.20 kg into the boiling water and leave it until the water starts boiling again. You then remove it from the boiling water and place it into the insulated beaker. If the specific heat of iron is 0.107 kcal/kg·°C, what will be the final temperature of the water and iron once they come to thermal equilibrium?

12

The Laws of Thermodynamics

Chapter Outline

As this youngster pumps air into his bicycle tire, positive work is done on the air in the pump. If the pumping rate is rapid enough, no energy enters or leaves the air by heat and the internal energy of the air increases. Therefore, the air in the pump gets warmer. *(George Semple)*

*I*n Chapter 5, we hinted at the relationship between energy in mechanics and energy in thermodynamics. We discussed the transformation of mechanical energy to internal energy when a nonconservative force such as friction is acting. In Chapter 10, we defined internal energy and found the relationship between internal energy and temperature for an ideal gas. In this chapter, we develop a complete treatment of energy in thermal processes.

According to the first law of thermodynamics, the internal energy of an object can be increased either by energy added to the object or by work done on it. It tells us that the internal energy of a system, which is just the sum of the molecular kinetic and potential energies, can change as a result of two separate types of energy transfer across the boundary of the system. The law essentially imposes conservation of energy for both energy added by heat and work done on a system but does not predict which of several possible energy-conserving processes actually occur in nature.

The second law of thermodynamics, which can be stated in many equivalent ways, establishes which processes allowed by the first law do occur in nature and which do not. For example, the second law tells us that energy never flows by heat spontaneously from a cold object to a hot object. One important application of this law is in the study of heat engines, such as the internal combustion engine, and the principles that limit their efficiency.

12.1 WORK IN THERMODYNAMIC PROCESSES

In the macroscopic approach to thermodynamics, we describe the *state* of a system with the use of such variables as pressure, volume, temperature, and internal energy, which are called *state variables*. The number of macroscopic state variables needed to characterize a system depends on the system's nature. For a homogeneous system, such as a gas containing only one type of molecule, usually only two variables are needed. It is important to note that a *macroscopic state* of an isolated system can be specified only if the system is in internal thermal equilibrium. In the case of a gas in a container, internal thermal equilibrium requires that every part of the gas be at the same pressure and temperature and that these variables be constant in time. If the temperature varies from one part of the gas to another, for example, we cannot specify a single temperature for the entire gas to be used in the ideal gas law.

In this section, we study the other important energy transfer mechanism besides heat for thermodynamic systems—work. We confine our discussion to the work done on a gas, which is important in understanding internal combustion engines. Consider a gas contained by a cylinder fitted with a movable piston (Fig. 12.1a) and in equilibrium. The gas occupies a volume V and exerts a uniform pressure P on the cylinder walls and the piston. Now let us assume that the gas is compressed slowly enough to allow the system to remain essentially in thermodynamic equilibrium at all times. As the piston is pushed downward by an external force F through a distance Δy, the work done on the gas is

$$W = -F\Delta y = -PA\,\Delta y$$

where we have set the magnitude F of the external force equal to PA because the piston is always in equilibrium. Note that if the piston is pushed downward, Δy is negative, so we need an explicit negative sign in the expression for W to make the work positive. Because $A\,\Delta y$ is the change in volume of the gas (that is, $\Delta V = A\,\Delta y$), we can express the **work done on the gas** as

$$W = -P\Delta V \qquad\qquad \text{[12.1]}$$

If the gas is compressed as in Figure 12.1b, ΔV is negative and the work done on the gas is positive. If the gas expands, ΔV is positive and the work done on the gas is negative. If the volume remains constant, the work done on the gas is zero.

FIGURE 12.1 (a) A gas in a cylinder occupying a volume V at a pressure P. (b) The piston is pushed downward so that the gas is compressed.

FIGURE 12.2 A gas is compressed from state i to state f. The work done on the gas equals the negative of the area under the PV curve.

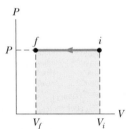

FIGURE 12.3 The PV diagram for a gas being compressed at constant pressure. The shaded area represents the work done on the gas.

Equation 12.1 can be used to calculate the work done on the system only when the pressure of the gas remains constant during the expansion or compression. A process in which the pressure remains constant is called an **isobaric process.** If the pressure changes, an average value of the pressure may be used to estimate the work done. Calculus is required to determine the exact work done, and we will not attempt such calculations here.

In general, the pressure is not constant during a process followed by a gas, but depends on the volume and temperature. If the pressure and volume are known at each step of the process, the state of the gas at each step can be plotted on a graph that is very important in thermodynamics—a ***PV* diagram,** as in Figure 12.2. This type of diagram allows us to visualize the process through which a gas is progressing. The curve on such a graph is called the *path* taken between the initial and final states.

Consider the process represented by the pressure-volume diagram, or PV diagram, in Figure 12.3. The gas is compressed from an initial volume V_i to a final volume V_f at constant pressure P. From Equation 12.1 we see that the work done on the gas in this case is $-P(V_f - V_i)$. Note that this is just the negative of the area under the pressure-volume curve. In general, **the work done on a gas that takes it from some initial state to some final state is the negative of the area under the curve on a *PV* diagram.** This statement is true whether or not the pressure remains constant during the process.

As Figure 12.3 suggests, for our process of compressing a gas in the cylinder, the work done depends on the particular path taken between the initial and final states. To illustrate this important point, consider several different paths connecting i and f (Fig. 12.4). In the process depicted in Figure 12.4a, the volume of the gas is first reduced from V_i to V_f at constant pressure P_i and the pressure of the gas then increases from P_i to P_f by heating at constant volume V_f. (A constant-volume process is called an **isovolumetric process** and consists of a vertical line on a PV diagram.) The work done on the gas along this path is $-P_i(V_f - V_i)$. In Figure 12.4b, the pressure of the gas is increased from P_i to P_f at constant volume V_i and then the volume of the gas is reduced from V_i to V_f at constant pressure P_f. The work done on the gas along this path (an isobaric process) is $-P_f(V_f - V_i)$, which is greater than that for the process described in Figure 12.4a. It is greater because the piston was displaced through the same distance by a larger force than for the situation in Figure 12.4a. Finally, for the process described in Figure 12.4c, where both P and V change continuously, the work done on the gas has some value intermediate between the values obtained in the first two processes. To evaluate the work in this case, the shape of the PV curve must be known. Thus, the work done on a system depends on the process by which the system goes from the initial to the final state. In other words, the work done depends on the initial, final, and intermediate states of the system.

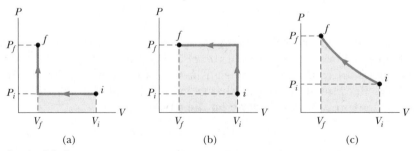

FIGURE 12.4 The work done on a gas as it is taken from an initial state to a final state depends on the path taken between these states.

APPLYING *PHYSICS* **12.1**

Imagine a gas in an insulated cylinder with a movable piston. The piston has been pushed inward, compressing the gas, and is now released. As the molecules of the gas strike the piston, they move it outward. From the point of view of energy principles, explain how this expansion causes the temperature of the gas to drop.

Explanation   From the point of view of energy principles, the molecules strike the piston and move it through a distance. Thus, the molecules do work on the piston, which represents a transfer of energy out of the gas. As a result, the internal energy of the gas drops. Because the temperature is related to internal energy, the temperature of the gas drops.

Example 12.1   Work Done by an Expanding Gas

In the system similar to that shown in Figure 12.1, the gas in the cylinder is at a pressure of 8.0×10^3 Pa and the piston has an area of 0.10 m^2. As energy is added slowly by heat to the gas, the piston is pushed up a distance of 4.0 cm. Calculate the work done by the expanding gas on the surroundings. Assume that the pressure remains constant.

Solution   The change in volume of the gas is

$$\Delta V = A \, \Delta y = (0.10 \text{ m}^2)(4.0 \times 10^{-2} \text{ m}) = 4.0 \times 10^{-3} \text{ m}^3$$

and from Equation 12.1, the work done *on* the gas is $-P\,\Delta V$, so the work done *by* the expanding gas is $+P\,\Delta V$, or

$$W_{\text{by gas}} = P\,\Delta V = (8.0 \times 10^3 \text{ Pa})(4.0 \times 10^{-3} \text{ m}^3) = \boxed{32 \text{ J}}$$

12.2   THE FIRST LAW OF THERMODYNAMICS

When the principle of conservation of energy was first introduced in Chapter 5, we stated that the mechanical energy of an isolated system is constant in the absence of nonconservative forces such as friction. That mechanical model did not encompass changes in the internal energy of the system—energy underneath the skin residing in the jiggling motion and positions of all the molecules. We now introduce a special case of the general principle of conservation of energy called the **first law of thermodynamics** that encompasses changes in internal energy. It is a universally valid law that can be applied to all kinds of processes and provides a connection between the microscopic and macroscopic worlds.

In this chapter, we have discussed two ways in which energy can be transferred between a system and its surroundings. One is by doing work, which requires a macroscopic displacement of the point of application of a force (or pressure). The other is by heat, which occurs through random molecular collisions. Both mechanisms result in a change in internal energy of the system and therefore usually result in measurable changes in the macroscopic variables of the system, such as the pressure, temperature, and volume of a gas.

◀ Change in internal energy

To express these ideas more quantitatively, suppose a system undergoes a change from an initial state to a final state. During this change, positive Q is the energy transferred *to* the system by heat, and positive W is the work done *on* the system. If we represent the internal energy of the system (the sum of the molecular energies) with the letter U, then the *change* in internal energy $\Delta U = U_f - U_i$ can

be expressed as

First law of thermodynamics

$$\Delta U = U_f - U_i = Q + W \qquad\qquad [12.2]$$

where all quantities must have the same energy units. Equation 12.2 is known as **the first law of thermodynamics.** The first law indicates that the change in the internal energy of a system is equal to the sum of the energy transferred across the system boundary by heat and the energy transferred by work.

SOME APPLICATIONS OF THE FIRST LAW OF THERMODYNAMICS

First consider an *isolated system*—that is, one that does not interact with its surroundings. In this case, no energy transfer takes place by heat and the work done is zero; hence, the internal energy remains constant. That is, because $Q = W = 0$, $\Delta U = 0$, and so $U_i = U_f$. We conclude **that the internal energy of an isolated system remains constant.**

For isolated systems, U remains constant

Next consider the case in which a system (not isolated from its surroundings) is taken through a **cyclic process**—that is, a process that originates and ends at the same state so that $U_f = U_i$. In this case, the change in the internal energy must again be *zero*, and therefore the energy added to the system by heat must equal the negative of the work done on the system in the course of one cycle. That is, in a cyclic process,

Cyclic process

$$\Delta U = 0 \qquad \text{and} \qquad Q = -W$$

Note that **the net work done per cycle by the gas on its surroundings equals the area enclosed by the path representing the process on a *PV* diagram.** As we shall see in the next section, cyclic processes are very important for describing the thermodynamics of *heat engines*—devices in which some part of the energy input exits the engine by mechanical work.

Another important type of idealized process in which $\Delta U = 0$ is an **isothermal process** (a constant-temperature process). One way to picture such a process physically is to consider an ideal monatomic gas contained in a cylinder with a movable piston as shown in Figure 12.5. Imagine the cylinder and gas are in thermal contact with a large source of energy (an energy reservoir) that can exchange energy with the gas without changing its own temperature. Now, allow energy to transfer slowly by heat into the gas, always maintaining a steady equilibrium temperature of the system and allowing the gas to expand and the pressure to fall. This system is interesting because it is also one way to model a heat engine in which gases push against a piston and provide useful work. We can find the work done by our primitive engine by recalling that $U = \frac{3}{2}nRT$ for an ideal monatomic gas (Chapter 10). Because $T_i = T_f$ for an isothermal process, $\Delta U = 0$. Thus, we have $\Delta U = Q + W = 0$ or $W = -Q$. This means that the work done on an ideal

---

Tip 12.1   DUAL SIGN CONVENTIONS

Many physics and engineering textbooks present the first law as $\Delta U = Q - W$, with a minus sign between the heat and work. The reason for this is that work is defined in these treatments as the work done *by* the gas rather than *on* the gas, as in our treatment. This form of the first law represents the original interest in applying it to steam engines, where the primary concern was the work extracted from the engine.

Isothermal expansion

Q_h

Energy reservoir at T_h

FIGURE 12.5 The gas in the cylinder expands isothermally while in contact with a reservoir at T_h.

gas in a constant-temperature process is the negative of the energy added by heat. Because Q is a positive number for an isothermal expansion (energy Q is added to the gas here), W is a negative number. Because W is negative, the work done *by* this system on its surroundings is positive, and the engine supplies work to the outside world.

If a process occurs in which the work done is zero, then the change in internal energy equals the energy entering or leaving the system by heat. If energy enters the system, Q is positive and the internal energy increases. For a gas, we can associate this increase in internal energy with an increase in the kinetic energy of the molecules. If a process occurs in which the energy transferred by heat is zero, a so-called **adiabatic process,** and work is done on the system, then the magnitude of the change in internal energy equals the work done on the system. For example, if a hot gas is allowed to expand so quickly that there is no time for energy to leave the system by heat, the work done on the gas is negative and the internal energy decreases. This is because kinetic energy is transferred from the gas molecules to the moving piston. Such an adiabatic expansion is of practical importance and is nearly realized in an internal combustion engine when a gasoline-air mixture is ignited and expands rapidly against a piston. One of the examples in this section shows how to calculate the work output from an adiabatic expansion.

On a microscopic scale, no practical distinction exists between the results of energy transfer by heat and work. Each can produce a change in the internal energy of a system. Although the macroscopic processes Q and W are *not* properties defined for a specific state of a system like P, V, T, and U, they are related to changes in internal energy of a system through the first law of thermodynamics. Once a process is defined, Q and W can be either calculated or measured, and the change in internal energy can be found from the first law of thermodynamics.

 Adiabatic process

Tip 12.2 THE FIRST LAW AND CONSERVATION OF ENERGY

The first law can be considered as the general equation for energy conservation. In this approach, the first law is applied to a closed system, energy transfer Q includes electromagnetic radiation, and work W includes electrical transmission ("electrical work") and mechanical waves ("molecular work").

Quick Quiz 12.1 Fill in the boxes with $-$, $+$, or 0 for each of the three terms in the first law of thermodynamics. For each situation, the system to be considered is identified.

Situation	System	Q	W	ΔU
(a) Rapidly pumping up a bicycle tire	Air in the pump			
(b) Pan of room-temperature water sitting on a hot stove	Water in the pan			
(c) Air quickly leaking out of a balloon	Air originally in balloon			

Example 12.2 An Isobaric Process

A gas is enclosed in a container fitted with a piston of cross-sectional area 0.10 m^2. The pressure of the gas is maintained at 8 000 Pa while energy is slowly added by heat; as a result, the piston is pushed up a distance of 4.0 cm. (Recall that any process in which the pressure remains constant is called an isobaric process.) If 42 J of energy is added to the system by heat during the expansion, what is the change in internal energy of the system?

Solution The work done on the gas is

$$W = -P\Delta V = -(8\,000 \text{ Pa})(0.10 \text{ m}^2)(4.0 \times 10^{-2} \text{ m}) = -32 \text{ N·m} = -32 \text{ J}$$

The change in internal energy is found from the first law of thermodynamics:

$$\Delta U = Q + W = 42\,\text{J} + (-32\,\text{J}) = \boxed{10\,\text{J}}$$

We see that **in an isobaric process the work done and the energy transferred by heat are both nonzero.**

EXERCISE   If 42 J of energy is added by heat to the system with the piston clamped in a *fixed* position, what is the work done by the gas? What is the change in its internal energy?

ANSWER   No work is done; $\Delta U = 42\,\text{J}$

An isobaric process is one that occurs at constant pressure

Example 12.3   Work Done in a Car Engine (Adiabatic Expansion)

In a car engine operating at 1 800 rpm, the expansion of hot, high-pressure gas against a piston occurs in about 10 ms. Because energy transfer by heat typically takes a time on the order of minutes or hours, it is safe to assume that little energy leaves the hot gas during the expansion. Estimate the work done in this adiabatic expansion by assuming the engine cylinder contains 0.10 mole of an ideal monatomic gas that goes from 1 200 K to 400 K, typical engine temperatures, during the expansion.

Solution   Because there is no time for energy to leave the system by heat, we know that $Q = 0$, and the first law becomes

$$W = \Delta U - Q = \Delta U - 0 = \Delta U$$

We find ΔU from the expression for the internal energy of an ideal monatomic gas, $U = \frac{3}{2}nRT$:

$$\Delta U = U_f - U_i = \tfrac{3}{2}nR(T_f - T_i)$$
$$= \tfrac{3}{2}(0.10\ \text{mol})(8.315\ \text{J/mol}\cdot\text{K})(400\ \text{K} - 1\,200\ \text{K}) = -1.0\ \text{kJ}$$

Finally we have

$$W = \Delta U = \boxed{-1.0\ \text{kJ}}$$

The negative sign means the gas does work on the surroundings in the expansion.

In case you are wondering, the use of $W = -P\,\Delta V$ would not apply here because P is changing, as you can see from $P = nRT/V$.

Example 12.4   An Isovolumetric Process

Water with a mass of 2.0 kg is held at constant volume in a container while 10.0 kJ of energy is slowly added by a flame. The container is not well insulated, and as a result 2.0 kJ of energy leaks out to the surroundings. What is the temperature increase of the water?

An isovolumetric process is one that takes place at constant volume

Solution   Recall that an isovolumetric process is one that takes place at constant volume. In such a process the work done is equal to zero because there is no change in volume. Thus, the first law of thermodynamics gives

$$\Delta U = Q$$

This indicates that the net energy Q added to the water goes into increasing the internal energy of the water. The net energy added to the water is

$$Q = 10.0\ \text{kJ} - 2.0\ \text{kJ} = 8.0\ \text{kJ}$$

Because $Q = mc\,\Delta T$, the temperature increase of the water is

$$\Delta T = \frac{Q}{mc} = \frac{8.0 \times 10^3\ \text{J}}{(2.0\ \text{kg})(4.186 \times 10^3\ \text{J/kg}\cdot{}^\circ\text{C})} = \boxed{0.96\,^\circ\text{C}}$$

Example 12.5   Boiling Water

At atmospheric pressure $(1.013 \times 10^5$ Pa), 1.0 g of water occupies a volume of 1.0 cm^3. When this water is boiled, it becomes 1 671 cm^3 of steam. Calculate the change in internal energy for this process.

Solution   Because the latent heat of vaporization of water is 2.26×10^6 J/kg at atmospheric pressure, the energy required to boil 1.0 g is

$$Q = mL_v = (1.0 \times 10^{-3} \text{ kg})(2.26 \times 10^6 \text{ J/kg}) = 2\ 260 \text{ J}$$

The work done on the system when 1 cm^3 of liquid expands to 1 671 cm^3 of gas is negative and equal to

$$W = -P(V_{\text{steam}} - V_{\text{water}})$$
$$= -(1.013 \times 10^5 \text{ Pa})[(1\ 671 - 1.0) \times 10^{-6} \text{ m}^3] = -169 \text{ J}$$

Hence, the change in internal energy of the system is

$$\Delta U = Q + W = 2\ 260 \text{ J} + (-169 \text{ J}) = \boxed{2.1 \times 10^3 \text{ J}}$$

The positive ΔU tells us that the internal energy of the system has increased. We see that most of the energy (93%) transferred to the liquid by heat goes into an increase in the internal energy, which corresponds to work done to overcome the attractive forces holding the water molecules close together in the liquid phase. Only a small fraction (7%) goes into external work.

Example 12.6   First Law of Thermodynamics Applied to a Cyclic Process

An ideal monatomic gas is confined in a cylinder by a movable piston. Means are provided to add or withdraw energy slowly by heat from the gas and to carry out each of the processes $A \rightarrow B$, $B \rightarrow C$, and $C \rightarrow A$ shown in Figure 12.6. The gas starts at A with $P = 1.00$ atm, $V = 5.00$ L, and $T = 300$ K. $A \rightarrow B$ is an isovolumetric process that raises P to 3.00 atm at B. $B \rightarrow C$ is an isothermal expansion to 1.00 atm. $C \rightarrow A$ is an isobaric compression at 1.00 atm that completes the cycle and returns the gas to its original state. This cycle may be thought of as a model of a peculiar kind of heat engine, which basically consists of a gas confined in a cylinder by a piston, the gas continually exchanging energy by heat and work with its surroundings.

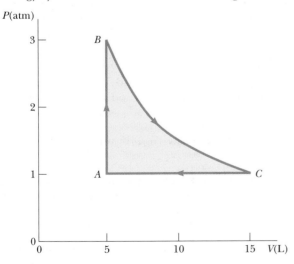

FIGURE 12.6   (Example 12.6)

A Find the number of moles of gas, the temperature at B, and the volume of the gas at C.

Solution n, T, and V can be found from the ideal gas law, which connects the equilibrium values of P, V, and T.

$$n = \frac{PV_A}{RT_A} = \frac{(1.00 \text{ atm})(5.00 \text{ L})}{(0.082\ 1 \text{ L} \cdot \text{atm/mol} \cdot \text{K})(300 \text{ K})} = \boxed{0.203 \text{ mol}}$$

$$T_B = \frac{P_BV_B}{nR} = \frac{(3.00 \text{ atm})(5.00 \text{ L})}{(0.203 \text{ mol})(0.082\ 1 \text{ L} \cdot \text{atm/mol} \cdot \text{K})} = \boxed{900 \text{ K}}$$

$$V_C = \frac{nRT_C}{P_C} = \frac{(0.203 \text{ mol})(0.082\ 1 \text{ L} \cdot \text{atm/mol} \cdot \text{K})(900 \text{ K})}{(1.00 \text{ atm})} = \boxed{15.0 \text{ L}}$$

B Find U, the internal energy of the gas, at A, B, and C. List P, V, T, and U, the state variables, at points A, B, and C.

Solution   For an ideal monatomic gas, $U = \frac{3}{2}nRT$ (Chapter 10).

$$U_A = \frac{3}{2}(0.203 \text{ mol})(8.315 \text{ J/mol} \cdot \text{K})(300 \text{ K}) = \boxed{0.760 \text{ kJ}}$$

Similarly,

$$U_B = \boxed{2.28 \text{ kJ}} \qquad \text{and} \qquad U_C = \boxed{2.28 \text{ kJ}}$$

From our calculations and from Figure 12.6, the state variables are

Point	P (atm)	V (L)	T (K)	U (kJ)
A	1.00	5.00	300	0.760
B	3.00	5.00	900	2.28
C	1.00	15.0	900	2.28

C Note that Q and W are not defined for specific states like A, B, and C, but only make sense for some process or path like $A \rightarrow B$, and so forth. State whether Q and W are +, −, or 0 for the processes $A \rightarrow B$, $B \rightarrow C$, and $C \rightarrow A$. Give your reasoning and explain how, experimentally, each process could be carried out.

Solution   For the process $A \rightarrow B$, $W_{A \rightarrow B} = -P\Delta V = 0$ because $\Delta V = 0$. As the temperature rises for this process, energy must be added by heat to the gas, so Q is +. This process can be carried out by placing the room-temperature system in an oven with the piston locked and gradually heating to 900 K.

For the isothermal process $B \rightarrow C$, T is constant and there is no change in internal energy for an ideal gas. The first law becomes $\Delta U = Q + W = 0$ or $Q = -W$. Because the gas expands and does work on the surroundings, W is a negative number. This means that Q is positive and energy is added by heat to the system. This process may be carried out by unlocking the piston and allowing the gas to expand to 15.0 L while keeping the system in the oven at 900 K.

For the process $C \rightarrow A$, T drops and the system cools as energy leaves the gas by heat, so Q is negative. W is positive because the gas contracts and work is done on the system. To carry out this process, remove the cylinder from the oven and allow the gas to cool at room temperature and contract.

D Calculate Q, W, and ΔU for each process. Estimate $W_{B \rightarrow C}$ by using an average value of $P = 1.70$ atm.

Solution   In all cases, we can calculate ΔU for the ideal gas from $\Delta U = \frac{3}{2}nR(T_f - T_i)$ and W from $W = -P\Delta V$ where W is the work done *on* the gas. We then use the first law to find Q:

$$Q = \Delta U - W$$

$$Q_{A \rightarrow B} = \Delta U_{A \rightarrow B} - W_{A \rightarrow B} = (2.28 \text{ kJ} - 0.760 \text{ kJ}) - 0 = \boxed{1.52 \text{ kJ}}$$

$$Q_{B \rightarrow C} = \Delta U_{B \rightarrow C} - W_{B \rightarrow C} = 0 - (-P\Delta V) = +P\Delta V = P(V_f - V_i)$$
$$= (1.70 \text{ atm})(15.0 \text{ L} - 5.00 \text{ L}) = 17.0 \text{ atm} \cdot \text{L}$$

An atm·L is 1.01×10^2 J, so $Q_{B \to C} =$ 1.72 kJ

$$Q_{C \to A} = \Delta U_{C \to A} - W_{C \to A}$$
$$= (0.760 \text{ kJ} - 2.28 \text{ kJ}) - (-P\Delta V) = -1.52 \text{ kJ} + P(V_f - V_i)$$
$$= -1.52 \text{ kJ} + (1.00 \text{ atm})(5.00 \text{ L} - 15.0 \text{ L}) = -1.52 \text{ kJ} - 1.01 \text{ kJ}$$
$$= -2.53 \text{ kJ}$$

EXERCISE Tabulate Q, W, and ΔU for each process and calculate Q_{net}, W_{net}, and ΔU_{net} for the entire cycle.

ANSWER $\Delta U = 0$; $Q_{net} = +0.71$ kJ; $W_{net} = -0.71$ kJ

Quick Quiz 12.2 Identify the nature of paths A, B, C, and D in Figure 12.7 as being isobaric, isothermal, isovolumetric, or adiabatic. Note that $Q = 0$ for path B.

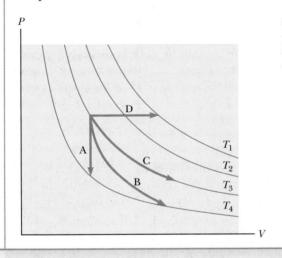

FIGURE 12.7 (Quick Quiz 12.2) Identify the nature of paths A, B, C, and D.

12.3 THE FIRST LAW AND HUMAN METABOLISM

Animals do work and give off energy by heat, and this leads us to believe the first law of thermodynamics can be applied to living organisms to describe them in a general way. The internal energy stored in humans goes into other forms needed for maintaining and repairing the major body organs and is transferred out of the body by work as a person walks or lifts a heavy object, and by heat when the body is warmer than its surroundings. Because the rates of change of internal energy, energy loss by heat, and energy loss by work vary widely with the intensity and duration of human activity, it is best to measure the time rates of change of ΔU, Q, and W. Rewriting the first law, these time rates of change are related by

$$\frac{\Delta U}{\Delta t} = \frac{Q}{\Delta t} + \frac{W}{\Delta t} \qquad \text{[12.3]}$$

On average, energy Q flows *out* of the body, and work is done *by* the body on its surroundings, so both $Q/\Delta t$ and $W/\Delta t$ are negative. This means that $\Delta U/\Delta t$ would be negative and the internal energy and body temperature would decrease with time if a human were a closed system with no way of ingesting matter or replenishing its internal energy store. Because all animals are actually open systems, animals add internal energy (chemical potential energy) to themselves in the delightful processes of eating and breathing, and thus the internal energy and the organism's temperature are kept constant. Overall, the energy from the oxidation of food ultimately supplies the work done by the body and energy lost from the body by heat, and this is the interpretation we give Equation 12.3. That is, $\Delta U/\Delta t$ is the rate at which internal energy is added to our bodies by food and this term just balances the rate of energy loss by heat, $Q/\Delta t$, and by work, $W/\Delta t$. Finally, note that if we have a way of measuring $\Delta U/\Delta t$ and $W/\Delta t$ for a human, we can calculate $Q/\Delta t$ from Equation 12.3 and gain useful information on the efficiency of the body as a machine.

MEASURING THE METABOLIC RATE $\Delta U/\Delta t$

The value of $W/\Delta t$, the work done by a person per unit time, is easily determined by measuring the power output supplied by a human (in pedaling a bike, for example). The **metabolic rate** $\Delta U/\Delta t$ is the rate at which chemical potential energy in food and oxygen are transformed into internal energy to just balance the body losses of internal energy by work and heat. Although the mechanisms of food oxidation and energy release in the body are complicated, involving many intermediate reactions and enzymes (organic compounds that speed up the chemical reactions taking place at "low" body temperatures), an amazingly simple rule summarizes these processes: **The metabolic rate is directly proportional to the rate of oxygen consumption by volume.** It is found that for an average diet, the consumption of 1 liter of oxygen releases 4.8 kcal or 20 kJ of energy. We can write this important summary rule as

$$\frac{\Delta U}{\Delta t} = 4.8 \frac{\Delta V_{O_2}}{\Delta t} \qquad [12.4]$$

where the metabolic rate $\Delta U/\Delta t$ is measured in kcal/s and $\Delta V_{O_2}/\Delta t$, the volume rate of oxygen consumption, is in L/s. Thus, by measuring the oxygen-use rate during various activities ranging from sleep to intense bicycle racing, we can effectively measure the variation of metabolic rate or the variation in the total power the body generates. A simultaneous measurement of the work per unit time done by a person along with the metabolic rate allows the efficiency of the body as a machine to be determined. Figure 12.8 shows a person monitored for oxygen consumption while riding a bike attached to a dynamometer, a device for measuring power output.

FIGURE 12.8 This bike rider is being monitored for oxygen consumption. *(©BSIP/Laurent/Science Source/Photo Researchers, Inc.)*

METABOLIC RATE, ACTIVITY, AND WEIGHT GAIN

Table 12.1 shows the measured rate of oxygen consumption in milliliters per minute per kilogram of body mass and the calculated metabolic rate for a 65-kg male engaged in various activities. A sleeping person uses about 80 W of power, the **basal metabolic rate,** just to maintain and run different body organs—heart, lungs, liver, kidneys, brain, and skeletal muscles. More intense activity increases the metabolic rate to a maximum of about 1 600 W for a superb racing cyclist, although such a high rate can only be maintained for periods of a few seconds. When we sit watching a riveting film, we give off about as much energy by heat as a bright (250-W) light bulb.

Regardless of level of activity, the daily food intake should just balance the loss in internal energy if a person is not to gain weight. Furthermore, exercise is a poor substitute for dieting as a method of losing weight. For example, the loss of 1 pound of body fat requires the muscles to expend 4 100 kcal of energy. If the goal is to lose a pound of fat in 35 days, a jogger could run an extra mile a day because a 65-kg jogger uses about 120 kcal to jog one mile (35 days × 120 kcal/day = 4 200 kcal). An easier way to lose the pound of fat would be to diet and eat two

TABLE 12.1	**Oxygen Consumption and Metabolic Rates for Various Activities for a 65-kg Male**[a]		
Activity	**O$_2$ use rate (mL/min · kg)**	**Metabolic rate (kcal/h)**	**Metabolic rate (W)**
Sleeping	3.5	70	80
Light activity(dressing, slow walking, desk work)	10	200	230
Moderate activity (walking briskly)	20	400	465
Heavy activity (basketball, (fast breast stroke)	30	600	700
Extreme activity (bicycle racing)	70	1 400	1 600

[a] Source: *A Companion to Medical Studies*, 2/e, R. Passmore, Philadelphia, F. A. Davis, 1968.

fewer slices of bread per day for 35 days, because bread has a calorie content of 60 kcal/slice (35 days × 2 slices/day × 60 kcal/slice = 4 200 kcal).

Example 12.7   Fighting Fat

In the course of 24 hours a 65-kg person spends 8 h at a desk, 2 h puttering around the house, 1 h jogging 5 miles, 5 h in moderate activity, and 8 h sleeping. What is the change in her internal energy in this period?

Solution   Using Table 12.1 and information from the previous paragraph, we find

$$\Delta U = -(10 \text{ h})(200 \text{ kcal/h}) - (1 \text{ h})(5 \text{ mi/h})(120 \text{ kcal/mi})$$
$$- (5 \text{ h})(400 \text{ kcal/h}) - (8 \text{ h})(70 \text{ kcal/h})$$
$$= \boxed{-5\ 160 \text{ kcal}}$$

Note that the calories from food that she should consume if she is to just replenish this internal energy is 5 160 kcal. Furthermore, she should consume less than 5 160 kcal if she wishes to lose weight.

PHYSICAL FITNESS AND EFFICIENCY OF THE HUMAN BODY AS A MACHINE

One measure of a person's physical fitness is their maximum capacity to use or consume oxygen. This "aerobic" fitness can be increased and maintained with regular exercise, but falls when training stops. Typical maximum rates of oxygen consumption and corresponding fitness levels are shown in Table 12.2; we see that the maximum oxygen consumption rate varies from 28 mL/min · kg of body mass for poorly conditioned subjects to 70 mL/min · kg for superb athletes.

We have already pointed out that the first law can be rewritten to relate the metabolic rate $\Delta U/\Delta t$ to the rate at which energy leaves the body by work and by heat:

$$\frac{\Delta U}{\Delta t} = \frac{Q}{\Delta t} + \frac{W}{\Delta t}$$

Now consider the body as a machine capable of supplying mechanical power to the outside world and ask for its efficiency. The body's efficiency *e* is defined as the

TABLE 12.2	Physical Fitness and Maximum Oxygen Consumption Rate[a]
Fitness Level	**Maximum Oxygen Consumption Rate (mL/min · kg)**
Very poor	28
Poor	34
Fair	42
Good	52
Excellent	70

[a] Source: *Aerobics*, K. H. Cooper, New York, Bantam Books, 1968.

ratio of the mechanical power supplied by a human to the metabolic rate or the total power input to the body:

$$e = \text{ body's efficiency} = \frac{\left|\dfrac{W}{\Delta t}\right|}{\left|\dfrac{\Delta U}{\Delta t}\right|} \qquad \text{[12.5]}$$

In this definition, absolute value signs are used to show that e is a positive number and to avoid explicitly using minus signs required by our definition of W and Q in the first law. Table 12.3 shows the efficiency of workers engaged in different activities for several hours. These values were obtained by measuring the power output and simultaneous oxygen consumption of mine workers and calculating the metabolic rate from the oxygen consumption. The table shows that a person can steadily supply mechanical power for several hours at about 100 W with an efficiency of about 17%. It also shows the dependence of efficiency on activity, and that e can drop to values as low as 3% for highly inefficient activities like shoveling, which involves many starts and stops. Finally, it is interesting in comparison to the average results of Table 12.3 that a superbly-conditioned athlete efficiently coupled to a mechanical device for extracting power (a bike!) can supply a power of around 300 W for about 30 minutes at a peak efficiency of 22%.

TABLE 12.3	Metabolic Rate, Power Output, and Efficiency for Different Activities[a]		
Activity	**Metabolic rate $\dfrac{\Delta U}{\Delta t}$ (W)**	**Power Output $\dfrac{W}{\Delta t}$ (W)**	**Efficiency e**
Cycling	505	96	0.19
Pushing loaded coal cars in a mine	525	90	0.17
Shoveling	570	17.5	0.03

[a] Source: "Inter- and Intra-Individual Differences in Energy Expenditure and Mechanical Efficiency," C. H. Wyndham et al., *Ergonomics* **9**, 17 (1966).

12.4 HEAT ENGINES AND THE SECOND LAW OF THERMODYNAMICS

The invention and improvement of the steam engine and internal combustion engine played an important role in the development of the second law of thermodynamics. These devices also made industrial development possible, greatly changing the nature of everyday life.

A **heat engine** is a device that converts internal energy to other useful forms, such as electrical and mechanical energy. In a typical process for producing electricity in a power plant, for instance, coal or some other fuel is burned, and the resulting internal energy is used to convert water to steam. This steam is directed at the blades of a turbine, setting it in rotation. Finally, the mechanical energy associated with this rotation is used to drive an electric generator. In another heat engine, the internal combustion engine in your automobile, energy enters the engine as the fuel is injected into the cylinder, and a fraction of this energy is converted to mechanical energy.

In general, a heat engine carries some working substance through cyclical processes[1] during which (1) energy is transferred by heat from a source at a high temperature, (2) work is done by the engine, and (3) energy is expelled by the engine by heat to a source at a lower temperature. As an example, consider the operation of a steam engine in which the working substance is water. The water in the engine is carried through a cycle in which it first evaporates into steam in a boiler and then expands against a piston. After the steam is condensed with cooling water, it is returned to the boiler, and the process is repeated.

It is useful to draw a heat engine schematically as in Figure 12.9. The engine absorbs energy Q_h from the hot reservoir, does work W_{eng}, and then gives up energy Q_c to the cold reservoir. (Note that *negative* work is done *on* the engine, so that $W = -W_{eng}$.) Because the working substance goes through a cycle, its initial and final internal energies are equal, so $\Delta U = 0$. Hence, from the first law of thermodynamics

$$\Delta U = Q + W \quad \text{or} \quad Q_{net} = -W = W_{eng}$$

This shows that **the work W_{eng} done by a heat engine equals the net energy absorbed by the engine.** As we can see from Figure 12.9, $Q_{net} = |Q_h| - |Q_c|$. Therefore,

$$W_{eng} = |Q_h| - |Q_c| \qquad \textbf{[12.6]}$$

If the working substance is a gas, **the work done by the engine for a cyclic process is the area enclosed by the curve representing the process on a *PV* diagram.** This is shown for an arbitrary cyclic process in Figure 12.10.

The **thermal efficiency** e of a heat engine is defined as the ratio of the work done by the engine to the energy absorbed at the higher temperature during one cycle:

$$e = \frac{W_{eng}}{|Q_h|} = \frac{|Q_h| - |Q_c|}{|Q_h|} = 1 - \frac{|Q_c|}{|Q_h|} \qquad \textbf{[12.7]}$$

We can think of the efficiency as the ratio of what you gain (work) to what you give (energy transfer at the higher temperature). Equation 12.7 shows that a heat engine has 100% efficiency ($e = 1$) only if $Q_c = 0$—that is, if no energy is expelled to the cold reservoir. In other words, a heat engine with perfect efficiency would have to expel all the input energy by mechanical work.

The Kelvin-Planck statement of the **second law of thermodynamics** can be stated as follows: **It is impossible to construct a heat engine that, operating in**

FIGURE 12.9 **A schematic representation of a heat engine.** The engine receives energy Q_h from the hot reservoir, expels energy Q_c to the cold reservoir, and does work W.

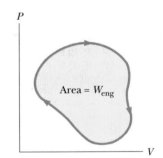

FIGURE 12.10 **The *PV* diagram for an arbitrary cyclic process.** The area enclosed by the curve equals the net work done.

◀ Thermal efficiency

◀ Second law of thermodynamics

---

[1] Strictly speaking, the automobile is not a heat engine according to the cyclic process description, because the air-fuel mixture undergoes only one cycle and is then expelled through the exhaust system.

a cycle, produces no other effect than the absorption of energy from a reservoir and the performance of an equal amount of work.

This form of the second law is useful for understanding the operation of heat engines. With reference to Equation 12.7, the second law says that, during the operation of a heat engine, W_{eng} can never be equal to Q_h or, alternatively, that some energy Q_c must be rejected to the environment. The essence of this form of the second law is that it is theoretically impossible to construct a heat engine that works with 100% efficiency.

Heat engines can, however, operate in reverse—we send in energy, modeled as work W_{eng} in Figure 12.9, resulting in energy being extracted from the cold reservoir and transferred to the hot reservoir. The system is now operating as a heat pump. Your refrigerator is a common heat pump. Energy is extracted from the cold food in the interior and delivered to the warmer air in the kitchen. The work is done in the compressor unit of the refrigerator, in which a piston compresses the refrigerant (freon, or other similar material) to a small volume, increasing its temperature. A household air conditioner is another example of a heat pump. Some homes are both heated and cooled by heat pumps. In the winter, the heat pump extracts energy from the outside air and delivers it to the inside, and the process is reversed for summer cooling. In a groundwater heat pump, energy is extracted in the winter from water deep in the ground rather than from the outside air, and energy is delivered to that water in the summer. This increases the yearround efficiency of the heating and cooling unit, because the ground water is at a higher temperature than the air in the winter and is at a cooler temperature than the air in summer.

Our assessment of the first two laws of thermodynamics can be summed up as follows: The first law says **we cannot get a greater amount of energy out of a cyclic process than we put in,** and the second law says that if we start with heat **we cannot break even.**

APPLICATION

HEAT PUMPS AND REFRIGERATORS

LORD KELVIN, BRITISH PHYSICIST AND MATHEMATICIAN (1824–1907)

Born William Thomson in Belfast, Kelvin was the first to propose the use of an absolute scale of temperature. His studies also led to the idea that energy cannot pass spontaneously from a colder object to a hotter object; this is known as the second law of thermodynamics. (*J-L. Charmet/ SPL/Photo Researchers*)

Example 12.8 The Efficiency of an Engine

An engine transfers 2.00×10^3 J of energy from a hot reservoir during a cycle and transfers 1.50×10^3 J to a cold reservoir.

A Find the efficiency of the engine.

Solution The efficiency of the engine is given by Equation 12.7 as

$$e = 1 - \frac{|Q_c|}{|Q_h|} = 1 - \frac{1.50 \times 10^3 \text{ J}}{2.00 \times 10^3 \text{ J}} = \boxed{0.250, \text{ or } 25.0\%}$$

B How much work does this engine do in one cycle?

Solution The work done by the engine is the difference between the input and output energies:

$$W_{eng} = |Q_h| - |Q_c| = 2.00 \times 10^3 \text{ J} - 1.50 \times 10^3 \text{ J} = \boxed{500 \text{ J}}$$

EXERCISE If an engine has an efficiency of 20.0% and exhausts 3.00×10^3 J to water used to cool the engine, how much work is done by the engine?

ANSWER 750 J

EXERCISE The energy absorbed by an engine is three times greater than the work it performs. (a) What is its thermal efficiency? (b) What fraction of the energy absorbed is expelled to the cold reservoir?

ANSWER (a) 33% (b) $\frac{2}{3}$

12.5 REVERSIBLE AND IRREVERSIBLE PROCESSES

In Section 12.6 we shall discuss a theoretical heat engine that is the most efficient engine possible. To understand its nature, we must first examine the meaning of reversible and irreversible processes. A **reversible** process is one in which every state along some path is an equilibrium state, and one for which the system can be returned to its initial conditions along the same path. A process that does not satisfy these requirements is **irreversible.**

Most natural processes are known to be irreversible—the reversible process is an idealization. Although real processes are always irreversible, some are *almost* reversible. If a real process occurs very slowly so that the system is virtually always in equilibrium, the process can be considered reversible. For example, imagine compressing a gas very slowly by dropping some grains of sand onto a frictionless piston, as in Figure 12.11. The temperature can be kept constant by placing the gas in thermal contact with an energy reservoir. The pressure, volume, and temperature of the gas are well defined during this isothermal compression. Each added grain of sand represents a change to a new equilibrium state. The process can be reversed by slowly removing grains of sand from the piston.

Energy reservoir

FIGURE 12.11 A gas in thermal contact with an energy reservoir is compressed slowly by grains of sand dropped onto the piston. The compression is isothermal and reversible.

12.6 THE CARNOT ENGINE

In 1824, a French engineer named Sadi Carnot (1796–1832), in an effort to understand the efficiency of real engines, described a theoretical engine, now called a *Carnot engine,* that is of great importance from both practical and theoretical viewpoints. He showed that a heat engine operating in an ideal, reversible cycle—called a **Carnot cycle**—between two energy reservoirs is the most efficient engine possible. Such an engine establishes an upper limit on the efficiencies of all real engines. **Carnot's theorem** can be stated as follows:

> **No real engine operating between two energy reservoirs can be more efficient than a Carnot engine operating between the same two reservoirs.**

To describe the Carnot cycle, we assume that the substance whose temperature varies between T_c and T_h is an ideal gas contained in a cylinder with a movable piston at one end. The cylinder walls and the piston are thermally nonconducting. Figure 12.12 shows four stages of the Carnot cycle, and Figure 12.13 is the *PV* diagram for the cycle. The cycle consists of two adiabatic and two isothermal processes, all reversible.

1. The process $A \rightarrow B$ is an isothermal expansion at temperature T_h, in which the gas is placed in thermal contact with a hot reservoir (that is, placed in a large oven) at temperature T_h (Fig. 12.12a). During the process, the gas absorbs energy Q_h from the reservoir and does work W_{AB} in raising the piston.
2. In the process $B \rightarrow C$, the base of the cylinder is replaced by a thermally nonconducting wall and the gas expands adiabatically; that is, no energy enters or leaves the system by heat (Fig. 12.12b). During the process, the temperature falls from T_h to T_c and the gas does work W_{BC} in raising the piston.
3. In the process $C \rightarrow D$, the gas is placed in thermal contact with a cold reservoir at temperature T_c (Fig. 12.12c) and is compressed isothermally at temperature T_c. During this time, the gas expels energy Q_c to the reservoir, and the work done on the gas is W_{CD}.
4. In the final process $D \rightarrow A$, the base of the cylinder is again replaced by a thermally nonconducting wall (Fig. 12.12d) and the gas is compressed adiabatically. The temperature of the gas increases to T_h, and the work done on the gas is W_{DA}.

SADI CARNOT, FRENCH ENGINEER (1796–1832)

Carnot is considered to be the founder of the science of thermodynamics. Some of his notes found after his death indicate that he was the first to recognize the relationship between work and heat. (*J-L. Charmet/Science Photo Library/Photo Researchers, Inc.*)

FIGURE 12.12 **The Carnot cycle.**
(a) In process $A \rightarrow B$, the gas expands isothermally while in contact with a reservoir at T_h. (b) In process $B \rightarrow C$, the gas expands adiabatically ($Q = 0$). (c) In process $C \rightarrow D$, the gas is compressed isothermally while in contact with a reservoir at $T_c < T_h$. (d) In process $D \rightarrow A$, the gas is compressed adiabatically. The upward arrows on the piston indicate removal of sand during the expansions, and the downward arrows indicate addition of sand during the compressions.

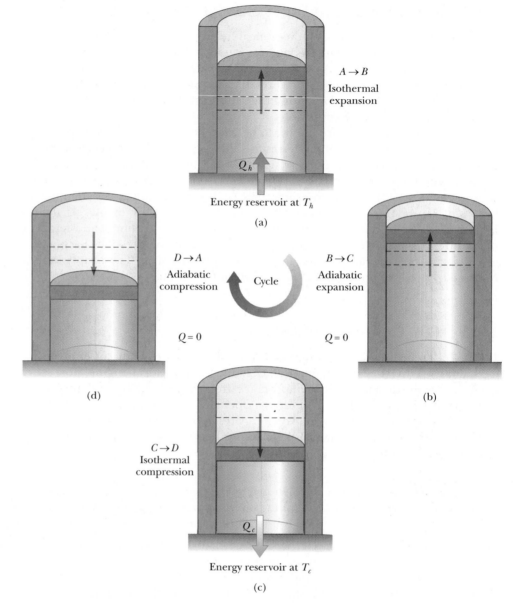

FIGURE 12.12 **The Carnot cycle.**
(a) In process $A \rightarrow B$, the gas expands isothermally while in contact with a reservoir at T_h. (b) In process $B \rightarrow C$, the gas expands adiabatically ($Q = 0$). (c) In process $C \rightarrow D$, the gas is compressed isothermally while in contact with a reservoir at $T_c < T_h$. (d) In process $D \rightarrow A$, the gas is compressed adiabatically. The upward arrows on the piston indicate removal of sand during the expansions, and the downward arrows indicate addition of sand during the compressions.

Tip 12.3 DON'T SHOP FOR A CARNOT ENGINE

The Carnot engine is an idealization—do not expect a Carnot engine to be developed for commercial use. If a Carnot engine were developed in an effort to maximize the efficiency, it would have zero power output, because, in order for all of the processes to be reversible, the engine would have to run infinitely slowly.

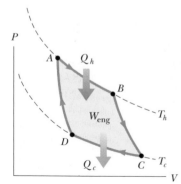

FIGURE 12.13 **The *PV* diagram for the Carnot cycle.** The net work done *W* equals the net energy received by heat in one cycle, $Q_h - Q_c$.

Carnot showed that the thermal efficiency of a Carnot engine is

$$e_c = 1 - \frac{T_c}{T_h} \qquad [12.8]$$

where *T* must be in kelvins. From this result, we see that all Carnot engines operating reversibly between the same two temperatures have the same efficiency.

Equation 12.8 can be applied to any working substance operating in a Carnot cycle between two energy reservoirs. According to this result, the efficiency is zero if $T_c = T_h$. The efficiency increases as T_c is lowered and as T_h is increased. The efficiency can be unity (100%), however, only if $T_c = 0$ K. It is impossible to reach absolute zero, so such reservoirs are not available. Thus, the maximum efficiency is always less than unity. In most practical cases, the cold reservoir is near room temperature, about 300 K. Therefore, we usually strive to increase the efficiency by raising the temperature of the hot reservoir. **All real engines are less efficient than the Carnot engine because they operate irreversibly because of friction and because they complete a cycle in a brief time period.**

PHYSICS *IN ACTION*

Devices That Convert Internal Energy into Other Forms of Energy

The device in the photograph on the left, called Hero's engine, was invented around 150 B.C. by Hero in Alexandria. When water is boiled in the flask, which is suspended by a cord, steam exits through two tubes at the sides of the flask (in opposite directions), creating a torque that rotates the flask.

The device shown in the center and right-hand photographs, called a thermoelectric converter, uses a series of semiconductor cells to convert internal energy to electrical energy. In the center photograph, the two "legs" of the device are at the same temperature, and no electrical energy is produced. However, when one leg is at a higher temperature than the other, as in the photograph on the right, electrical energy is produced as the device extracts energy from the hot reservoir and drives a small electric motor. How does this intriguing experiment demonstrate the second law of thermodynamics?

(Courtesy of Central Scientific Company)   *(Courtesy of PASCO Scientific Company)*

Quick Quiz 12.3

Three engines operate between reservoirs separated in temperature by 300 K. The reservoir temperatures are as follows:

Engine A: $T_h = 1\,000$ K, $T_c = 700$ K

Engine B: $T_h = 800$ K, $T_c = 500$ K

Engine C: $T_h = 600$ K, $T_c = 300$ K

Rank the engines in order of theoretically possible efficiency, from highest to lowest.

Example 12.9   The Steam Engine

A steam engine has a boiler that operates at 500 K. The energy from the boiler changes water to steam, which drives the piston. The temperature of the exhaust is that of the outside air, about 300 K. What is the maximum thermal efficiency of this steam engine?

Solution   From the expression for the efficiency of a Carnot engine, we find the maximum thermal efficiency for any engine operating between these temperatures:

$$e_c = 1 - \frac{T_c}{T_h} = 1 - \frac{300 \text{ K}}{500 \text{ K}} = \boxed{0.4, \text{ or } 40\%}$$

This is the highest theoretical efficiency of the engine. In practice, the efficiency is considerably lower.

EXERCISE   Determine the maximum work the engine can perform in each cycle of operation if it absorbs 200 J of energy from the hot reservoir during each cycle.

ANSWER   80 J

Example 12.10   The Carnot Efficiency

The highest theoretical efficiency of a gasoline engine, based on the Carnot cycle, is 30%. If this engine expels its gases into the atmosphere, which has a temperature of 300 K, what is the temperature in the cylinder immediately after combustion?

Solution   The Carnot efficiency is used to find T_h:

$$e_c = 1 - \frac{T_c}{T_h}$$

$$T_h = \frac{T_c}{1 - e_c} = \frac{300 \text{ K}}{1 - 0.30} = \boxed{430 \text{ K}}$$

Actual gasoline engines operate on a cycle significantly different from the Carnot cycle and therefore have lower maximum possible efficiencies.

EXERCISE   If the heat engine absorbs 837 J of energy from the hot reservoir during each cycle, how much work can it perform in each cycle?

ANSWER   251 J

RUDOLF CLAUSIUS, GERMAN PHYSICIST (1822–1888)

Born with the name Rudolf Gottlieb, he adopted the classical name of Clausius, which was a popular thing to do in his time. "I propose . . . to call S the entropy of a body, after the Greek word 'transformation.' I have designedly coined the word 'entropy' to be similar to energy, for these two quantities are so analogous in their physical significance, that an analogy of denominations seems to be helpful." *(AIP Niels Bohr Library, Lande Collection)*

12.7   ENTROPY

The zeroth law of thermodynamics involves the concept of temperature (Chapter 10), and the first law involves the concept of internal energy. Temperature and internal energy are both state variables; that is, they can be used to describe the thermodynamic state of a system. Another state variable, this one related to the second law of thermodynamics, is **entropy** S. In this section we define entropy on a macroscopic scale as the German physicist Rudolf Clausius (1822–1888) first expressed it in 1865.

Consider a reversible process between two equilibrium states. If Q_r is the energy absorbed or expelled by the system during some small interval of the path, **the change in entropy ΔS between two equilibrium states is given by the energy Q_r transferred along the reversible path divided by the absolute temperature T of the system in this interval.** That is,

$$\Delta S \equiv \frac{Q_r}{T} \qquad \text{[12.9]}$$

The subscript r on the term Q_r emphasizes that the definition applies only to a reversible path, even though the system may actually follow some irreversible path. When energy is absorbed by the system, Q_r is positive and hence the entropy increases. To calculate the change in entropy for a nonreversible process, we must model the process by a reversible process between the *same* initial and final states of the system.

When energy is expelled by the system, Q_r is negative and the entropy decreases. Note that Equation 12.9 defines not entropy but the *change* in entropy. Hence, the meaningful quantity in a description of a process is the change in entropy.

One reason the concept of entropy became useful and gained wide acceptance is because it provides another variable to describe the state of a system, to go along with pressure, volume, and temperature. The concept of entropy reached a position of even more significance when it was found that **the entropy of the Universe increases in all natural processes.** This is yet another way of stating the second law of thermodynamics.

The statement of the second law just given must be interpreted with care. Although it says that the entropy of the Universe always increases in all natural processes, there are processes in which the entropy of a system decreases. That is, there are situations in which the entropy of one system (system A) decreases while accompanied by a net increase in entropy of some other system (system B). In all cases, the change in entropy of system B is greater than the change in entropy of system A.

For centuries, individuals have attempted to build perpetual motion machines that would operate continuously without input of energy and without any net increase in entropy. Using the laws of thermodynamics, we can understand why a perpetual motion machine will never be invented. Perpetual motion machines of the first kind are those that violate the first law, putting out more energy than is put into the machine. Perpetual motion machines of the second kind are those that violate the second law. Imagine, for example, a waterfall in which water passes over a waterwheel, which turns an electric generator providing electricity to a nearby motor. The temperature of the falling water has risen after it hits the paddlewheel and the pond below. The pond is used as the high-temperature reservoir of a heat engine. The electric motor and the heat engine together are used to pump water from the pond back up to the top of the waterfall. This is a perpetual motion machine of both the first and second kinds. It violates the first law because it assumes all energy remains in the system so that it operates forever at the same rate. This would require no friction in any of the mechanisms and it requires that the mechanisms make no sound, because sound would carry energy away from the system. The machine we have described also violates the second law because it assumes that there is no exhaust energy from the heat engine, which would result in energy loss from the system.

The concept of entropy is satisfying because it enables us to present the second law of thermodynamics in the form of a mathematical statement. In the next section we will find that entropy can also be interpreted in terms of probabilities, a relationship that has profound implications for our world.

Tip 12.4   ENTROPY ≠ ENERGY

Be sure that you do not confuse energy and entropy—even though the names sound similar, they are very different.

◀ **APPLICATION**

"PERPETUAL MOTION" MACHINES

Webnote 12.1

Entropy is one of the most abstract and fascinating topics in physics. For links to the use of entropy in information theory, biology and the humanities, visit *http://www.math.psu.edu/gunesch/entropy.html*

Quick Quiz 12.4   Which of the following is true for the entropy change of a system that undergoes a reversible, adiabatic process? (a) $\Delta S < 0$ (b) $\Delta S = 0$ (c) $\Delta S > 0$

Example 12.11   Melting a Piece of Lead

Calculate the change in entropy when 300 g of lead melts at 327°C (600 K). Lead has a latent heat of fusion of 2.45×10^4 J/kg.

Solution The amount of energy added to the lead to melt it is

$$Q = mL_f = (0.300 \text{ kg})(2.45 \times 10^4 \text{ J/kg}) = 7.35 \times 10^3 \text{ J}$$

From Equation 12.9, the entropy change of the lead is

$$\Delta S = \frac{Q}{T} = \frac{7.35 \times 10^3 \text{ J}}{600 \text{ K}} = \boxed{12.3 \text{ J/K}}$$

Example 12.12 Which Way Does the Energy Flow?

A large, cold object is at 273 K, and a large, hot object is at 373 K. Show that it is impossible for a small amount of energy, say 8.00 J, to be transferred from the cold object to the hot object without decreasing the entropy of the Universe and hence violating the second law.

Solution We assume here that during the energy transfer, the two systems undergo no significant temperature change. This is not a necessary assumption; it is used to avoid a need for the techniques of integral calculus. The hot object gains energy, and its entropy change is

$$\Delta S_h = \frac{Q}{T_h} = \frac{8.00 \text{ J}}{373 \text{ K}} = 0.021 \, 4 \text{ J/K}$$

The cold reservoir loses energy, and its entropy change is

$$\Delta S_c = \frac{Q}{T_c} = \frac{-8.00 \text{ J}}{273 \text{ K}} = -0.029 \, 3 \text{ J/K}$$

The net entropy change of the Universe is

$$\Delta S_U = \Delta S_c + \Delta S_h = -0.007 \, 9 \text{ J/K}$$

This is in violation of the law that the entropy of the Universe always increases in natural processes. That is, **the spontaneous transfer of energy by heat from a cold object to a hot object cannot occur.**

EXERCISE In the preceding example, suppose that 8.00 J of energy were transferred by heat from the hot to the cold object. What would be the net change in entropy of the Universe?

ANSWER $+0.007 \, 9 \text{ J/K}$

12.8 ENTROPY AND DISORDER

As you look around at nature in all its beauty, it is easy to recognize that the events of natural processes have in them a large element of chance. For example, the spacing between trees in a natural forest is quite random; if you were to discover a forest where all the trees were equally spaced, you would conclude that it was a planted forest. Likewise, leaves fall to the ground with random arrangements. It would be highly unlikely to find the leaves laid out in perfectly straight rows. We can express the results of such observations by saying that **a disorderly arrangement is much more probable than an orderly one if the laws of nature are allowed to act without interference.**

Entropy originally found its place in thermodynamics, but its importance grew tremendously as the field of statistical mechanics developed. This analytical approach employs an alternative interpretation of entropy. In statistical mechanics, the behavior of a substance is described in terms of the statistical behavior of the atoms and molecules contained in the substance. One of the main products of this approach is the conclusion that **isolated systems tend toward greater disorder, and entropy is a measure of that disorder.**

In light of this new view of entropy, Boltzmann found an alternative method for calculating entropy through use of the relation

$$S = k_B \ln W \qquad \text{[12.10]}$$

where k_B is Boltzmann's constant ($k_B = 1.38 \times 10^{-23}$ J/K) and W is a number proportional to the probability that the system has a particular configuration. The symbol "ln" in Equation 12.10 is an abbreviation for the natural logarithm discussed in Appendix A.

Let us explore the meaning of this equation through a specific example. Imagine that you have a bag of 100 marbles, 50 red and 50 green. You are allowed to draw four marbles from the bag according to the following rules. Draw one marble, record its color, return it to the bag, and draw again. Continue this process until four marbles have been drawn. Note that because each marble is returned to the bag before the next one is drawn, the probability of drawing a red marble is always the same as the probability of drawing a green one.

The results of all possible drawing sequences are shown in Table 12.4. For example, the result RRGR means that a red marble was drawn first, a red one second, a green one third, and a red one fourth. This table indicates that there is only one possible way to draw four red marbles. There are four possible sequences that produce one green and three red marbles, six sequences that produce two green and two red, four sequences that produce three green and one red, and one sequence that produces all green. From Equation 12.10, we see that the state with the greatest disorder (two red and two green marbles) has the highest entropy because it is most probable. In contrast, the most ordered states (all red marbles and all green marbles) are least likely to occur and are states of lowest entropy.

The outcome of the draw can range between these highly ordered (lowest-entropy) and highly disordered (highest-entropy) states. Thus, entropy can be regarded as an index of how far a system has progressed from an ordered to a disordered state.

The second law of thermodynamics is really a statement of what is most probable rather than of what must be. Imagine placing an ice cube in contact with a hot piece of pizza. There is nothing in nature that absolutely forbids the transfer of energy by heat from the ice to the much warmer pizza. Statistically, it is possible for a slow-moving molecule in the ice to collide with a faster-moving molecule in the pizza so that the slow one transfers some of its energy to the faster one. However, when the great number of molecules present in the ice and pizza are considered, the odds are overwhelmingly in favor of the transfer of energy from the faster-moving molecules to the slower-moving molecules. Furthermore, this example demonstrates that a system naturally tends to move from a state of order to a state of disorder. The initial state, in which all the pizza molecules have high kinetic energy and all the ice molecules have lower kinetic energy, is much more ordered than the final state after energy transfer has taken place and the ice has melted.

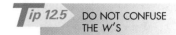

Tip 12.5 DO NOT CONFUSE THE *W*'S

The symbol *W* used here is a probability not to be confused with the same symbol used for work.

TABLE 12.4	Possible Results of Drawing Four Marbles from a Bag	
End Result	**Possible Draws**	**Total Number of Same Results**
All R	RRRR	1
1G, 3R	RRRG, RRGR, RGRR, GRRR	4
2G, 2R	RRGG, RGRG, GRRG, RGGR, GRGR, GGRR	6
3G, 1R	GGGR, GGRG, GRGG, RGGG	4
All G	GGGG	1

APPLICATION

THE DIRECTION OF TIME

Even more generally, the second law of thermodynamics defines the direction of time for all events as the direction in which the entropy of the universe increases. Although conservation of energy is not violated if energy spontaneously flows from a cold object (the ice cube) to a hot object (the pizza slice), the event violates the second law because it represents a spontaneous increase in order, and it also violates everyday experience. If the melting ice cube is filmed and the film speeded up, the difference between running the film in forward and reverse directions would be obvious to an audience. The same would be true of filming any event involving a large number of particles—such as a dish dropping to the floor and shattering.

APPLYING PHYSICS 12.2

Suppose you film the collision of two billiard balls on a pool table. Now suppose you run the film in reverse. Will the difference between forward and reverse directions be obvious? Why or why not?

Explanation The film shows nothing about the molecular arrangement of the billiard balls, so there are effectively only two particles in the event—the two billiard balls. During the collision, there is no change from order to disorder or vice versa, so the film appears to be authentic (describing a valid situation) when run in either direction.

As another example, suppose you were able to measure the velocities of all the air molecules in a room at some instant. It is very unlikely that you would find all molecules moving in the same direction with the same speed—this would be a highly ordered state, indeed. The most probable situation is a system of molecules moving haphazardly in all directions with a distribution of speeds—a highly disordered state. Let us compare this case to that of drawing marbles from a bag. If a container held 10^{23} molecules of a gas, the probability of finding all of the molecules moving in the same direction with the same speed at some instant would be similar to that of drawing a marble from the bag 10^{23} times and getting a red marble on every draw. This is clearly an unlikely set of events.

The tendency of nature to move toward a state of disorder affects a system's ability to do work. Consider a ball thrown toward a wall. The ball has kinetic energy, and its state is an ordered one; that is, all of the atoms and molecules of the ball move in unison at the same speed and in the same direction (apart from their random internal motions). When the ball hits the wall, however, part of the ball's kinetic energy is transformed into the random, disordered, internal motion of the molecules in the ball and the wall, and the temperatures of the ball and the wall both increase slightly. Before the collision, the ball was capable of doing work. It could drive a nail into the wall, for example. With the transformation of part of the ordered energy into disordered internal energy, its capability of doing work is reduced. That is, the ball rebounds with less kinetic energy than it originally had, because the collision is inelastic.

Various forms of energy can be converted to internal energy, as in the collision between the ball and the wall, but the reverse transformation is never complete. In general, if two kinds of energy, A and B, can be completely interconverted, we say that they are the *same grade*. However, if form A can be completely converted to form B and the reverse is never complete, then form A is a *higher grade* of energy than form B. In the case of a ball hitting a wall, the kinetic energy of the ball is of a higher grade than the internal energy contained in the ball and the wall after the collision. Therefore, when high-grade energy is converted to internal energy, it can never be fully recovered as high-grade energy.

This conversion of high-grade energy to internal energy is referred to as **degradation of energy.** The energy is said to be degraded because it takes on a

A full house is a very good hand in the game of poker. Can you calculate the probability of being dealt a full house (a pair and three of a kind) from a standard deck of 52 cards? (*Tom Mareschel, The IMAGE Bank*)

form that is less useful for doing work. In other words, **in all real processes, the energy available for doing work decreases.**

Finally, note once again that the statement that entropy must increase in all natural processes is true only for isolated systems. There are instances in which the entropy of some system decreases, but with a corresponding net increase in entropy for some other system. When all systems are taken together to form the Universe, *the entropy of the Universe always increases.*

Ultimately, the entropy of the Universe should reach a maximum. At this time the Universe will be in a state of uniform temperature and density. All physical, chemical, and biological processes will have ceased, because a state of perfect disorder implies no available energy for doing work. This gloomy state of affairs is sometimes referred to as an ultimate "heat death" of the Universe.

An illustration from Flammarion's novel *La Fin du Monde*, depicting the heat death of the Universe.

Quick Quiz 12.5 Suppose you are throwing two dice in a friendly game of craps. For any given throw, the two numbers that are face up can have a sum of 2, 3, 4, 5, 6, 7, 8, 9, 10, 11, or 12. Which outcome is most probable? Which is least probable?

SUMMARY

The work done on a gas at a constant pressure is

$$W = -P\Delta V \qquad [12.1]$$

The work done on the gas is positive if the gas is compressed (when ΔV is negative) and negative if the gas expands (when ΔV is positive). In general, the work done on a gas that takes it from some initial state to some final state is the negative of the area under the curve on a *PV* diagram.

From the first law of thermodynamics, when a system undergoes a change from one state to another, the **change in its internal energy** ΔU is

$$\Delta U = U_f - U_i = Q + W \qquad [12.2]$$

where Q is the energy transferred into the system by heat and W is the work done on the system. Q is positive when energy enters the system by heat and negative when the system loses energy.

The **first law of thermodynamics** is the generalization of the law of conservation of energy that includes transfer by heat.

In a **cyclic process** (one in which the system returns to its initial state), $\Delta U = 0$ and therefore $Q = W$. That is, the energy transferred into the system by heat equals the work done during the cycle.

An **adiabatic process** is one in which no energy is transferred by heat between the system and its surroundings ($Q = 0$). In this case, the first law gives $\Delta U = W$. That is, the internal energy changes as a consequence of work being done on the system.

An **isobaric process** is one that occurs at constant pressure. The work done on the system in such a process is $-P\Delta V$.

A **heat engine** is a device that takes in energy by heat and partially converts it to other forms of energy, such as mechanical and electrical energy. The work W_{eng} done by a heat engine in carrying a working substance through a cyclic process ($\Delta U = 0$) is

$$W_{\text{eng}} = |Q_h| - |Q_c| \qquad [12.6]$$

where Q_h is the energy absorbed from a hot reservoir and Q_c is the energy expelled to a cold reservoir.

The **thermal efficiency** of a heat engine is defined as the ratio of the work done by the engine to the energy absorbed per cycle:

$$e = \frac{W_{\text{eng}}}{|Q_h|} = 1 - \frac{|Q_c|}{|Q_h|} \qquad [12.7]$$

No real heat engine operating between the Kelvin temperatures T_h and T_c can exceed the efficiency of an engine operating between the same two temperatures in a **Carnot cycle,** given by

$$e_c = 1 - \frac{T_c}{T_h} \qquad [12.8]$$

Real processes proceed in an order governed by the **second law of thermodynamics,** which can be stated in two ways:

1. Energy will not flow spontaneously by heat from a cold object to a hot object.
2. No heat engine operating in a cycle can absorb energy from a reservoir and perform an equal amount of work.

The second law can also be stated in terms of a quantity called **entropy.** The **change in entropy** of a system is equal to the energy flowing by heat into (or out of) the system as the system changes from one state (A) to another (B) by a reversible process divided by the absolute temperature:

$$\Delta S \equiv \frac{Q_r}{T} \qquad [12.9]$$

One of the primary findings of statistical mechanics is that systems tend toward disorder and that entropy is a measure of this disorder. The entropy of the Universe increases in all natural processes; this is an alternative statement of the second law.

CONCEPTUAL QUESTIONS

1. The originally compressed gas from a carbon dioxide fire extinguisher emerges so rapidly that no immediate energy transfer occurs by heat between the gas and the atmosphere. Use this information to explain why the escaping gas is very cold.
2. "An adiabatic process can also be called an isentropic process ($\Delta S = 0$)." Is this statement true?
3. For an ideal gas in an isothermal process, there is no change in internal energy. Suppose the gas does work W during such a process. How much energy was transferred by heat?
4. If you shake a jar full of jelly beans of different sizes, the larger beans tend to appear near the top, and the smaller ones tend to fall to the bottom. Why does this occur? Does this process violate the second law of thermodynamics?
5. Consider the human body performing a strenuous exercise, such as lifting weights or riding a bicycle. Work is being done by the body, and energy is leaving by conduction from the skin into the surrounding air. According to the first law of thermodynamics, the temperature of the body should be steadily decreasing during the exercise. This is not what happens, however. Is the first law invalid for this situation? Explain.
6. Clearly distinguish among temperature, heat, and internal energy.
7. What is wrong with this statement? "Given any two objects, the one with the higher temperature contains more heat."
8. A steam-driven turbine is one major component of an electric power plant. Why is it advantageous to increase the temperature of the steam as much as possible?
9. When a sealed Thermos bottle full of hot coffee is shaken, what changes, if any, take place in (a) the temperature of the coffee and (b) its internal energy?

10. In solar ponds constructed in Israel, the Sun's energy is concentrated near the bottom of a salty pond. With the proper layering of salt in the water, convection is prevented, and temperatures of 100°C may be reached. Can you guess the maximum efficiency with which mechanical work can be extracted from the pond?
11. Can energy transfer into a system by heat even though the temperature of the system decreases?
12. Can the temperature of a system rise even though energy transfers out of the system by heat?
13. Is it possible to construct a heat engine that creates no thermal pollution?
14. Suppose your roommate is "Mr. Clean" and tidies up your messy room after a big party. Because more order is being created by your roommate, does this represent a violation of the second law of thermodynamics?
15. A thermodynamic process occurs in which the entropy of a system changes by -8.0 J/K. According to the second law of thermodynamics, what can you conclude about the entropy change of the environment?
16. If a supersaturated sugar solution is allowed to evaporate slowly, sugar crystals form in the container. Hence, sugar molecules go from a disordered form (in solution) to a highly ordered, crystalline form. Does this process violate the second law of thermodynamics? Explain.
17. The first law of thermodynamics says we cannot get more out of a process than we put in, but the second law says that we cannot break even. Explain this statement.
18. Give some examples of irreversible processes that occur in nature. Give an example of a process in nature that is nearly reversible.

PROBLEMS

1, **2**, **3** = straightforward, intermediate, challenging ☐ = full solution available in Student Solutions Manual/Study Guide

web = solution posted at **http://info.brookscole.com/serway** 🦠 = biomedical application

Section 12.1 Work in Thermodynamic Processes

1. The only form of energy possessed by molecules of a
web monatomic ideal gas is translational kinetic energy. Using the results from the discussion of kinetic theory in Section 10.6, show that the internal energy of a monatomic ideal gas at pressure P and occupying volume V may be written as $U = \frac{3}{2}PV$.

2. A sample of helium behaves as an ideal gas as it is heated at constant pressure from 273 K to 373 K. If 20.0 J of work is done *by* the gas during this process, what is the mass of helium present?

3. A container of volume 0.40 m³ contains 3.0 mol of argon gas at 30°C. Assuming argon behaves as an ideal gas, find the total internal energy of this gas. (*Hint:* See Problem 1 above.)

4. A 40.0-g projectile is launched by the expansion of hot gas in an arrangement shown in Figure P12.4a. The cross-sectional area of the launch tube is 1.0 cm², and the length that the projectile travels down the launch tube after starting from rest is 32 cm. As the gas expands, the pressure varies as shown in Figure P12.4b. The values for the initial pressure and volume are $P_i = 11 \times 10^5$ Pa and $V_i = 8.0$ cm³, and the final values are $P_f = 1.0 \times 10^5$ Pa and $V_f = 40.0$ cm³. Friction between the projectile and the launch tube is negligible. (a) If the projectile is launched into a vacuum, what is the speed of the projectile as it leaves the launch tube? (b) If instead the projectile is launched into air at a pressure of 1.0×10^5 Pa, what fraction of the work done by the expanding gas in the tube is spent by the projectile pushing air out of the way as it proceeds down the tube?

(a)   (b)

FIGURE P12.4

5. A gas expands from I to F along the three paths indicated in Figure P12.5. Calculate the work done *on* the gas along paths (a) *IAF*, (b) *IF*, and (c) *IBF*.

6. Sketch a *PV* diagram of the following processes. (a) A gas expands at constant pressure P_1 from volume V_1 to volume V_2. It is then kept at constant volume while the pressure is reduced to P_2. (b) A gas is reduced in pressure from P_1 and P_2 while its volume is held constant at V_1. It is then expanded at constant pressure P_2 to a final volume V_2. (c) In which of the processes is more work done *by* the gas? Why?

FIGURE P12.5 (Problems 5 and 15)

7. Gas in a container is at a pressure of 1.5 atm and a volume of 4.0 m³. What is the work done *on* the gas (a) if it expands at constant pressure to twice its initial volume? (b) if it is compressed at constant pressure to one-quarter its initial volume?

8. A movable piston having a mass of 8.00 kg and a cross-sectional area of 5.00 cm² traps 0.200 mol of an ideal gas in a vertical cylinder. If the piston slides without friction in the cylinder, how much work is done *on* the gas when its temperature is increased from 20°C to 300°C?

9. One mole of an ideal gas initially at a temperature of $T_i = 0$°C undergoes an expansion at a constant pressure of one atmosphere to four times its original volume. (a) Calculate the new temperature T_f of the gas. (b) Calculate the work done *on* the gas during the expansion.

10. (a) Determine the work done *on* a fluid that expands from i to f as indicated in Figure P12.10. (b) How much work is done *on* the fluid if it is compressed from f to i along the same path?

FIGURE P12.10

Section 12.2 The First Law of Thermodynamics

11. A container is placed in a water bath and held at constant volume as a mixture of fuel and oxygen is burned inside it. The temperature of the water is observed to rise during the burning. (The water is also held at constant volume.) (a) Consider the burning mixture to be the system. What are the signs of Q, ΔU, and W? (b) What are the signs of these quantities if the water bath is considered to be the system?

12. A quantity of a monatomic ideal gas undergoes a process in which both its pressure and volume are doubled as shown in Figure P12.12. What is the energy absorbed by heat into the gas during this process? (*Hint:* See Problem 1 above.)

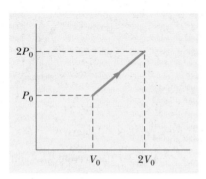

FIGURE P12.12

13. A gas is compressed at a constant pressure of 0.800 atm from 9.00 L to 2.00 L. In the process, 400 J of energy leaves the gas by heat. (a) What is the work done *on* the gas? (b) What is the change in its internal energy?

14. A monatomic ideal gas undergoes the thermodynamic process shown in the *PV* diagram of Figure P12.14. Determine whether each of the values ΔU, Q, and W for the gas is positive, negative, or zero. (*Hint:* See Problem 1 above.)

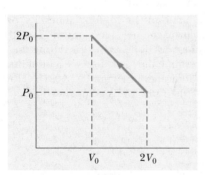

FIGURE P12.14

15. A gas expands from I to F in Figure P12.5. The energy
web added to the gas by heat is 418 J when the gas goes from I to F along the diagonal path. (a) What is the change in internal energy of the gas? (b) How much energy must be added to the gas by heat for the indirect path IAF to give the same change in internal energy?

16. A gas is taken through the cyclic process described by Figure P12.16. (a) Find the net energy transferred to the system by heat during one complete cycle. (b) If the cycle is reversed — that is, the process follows the path $ACBA$ — what is the net energy transferred by heat per cycle?

17. A thermodynamic system undergoes a process in which its internal energy decreases by 500 J. If at the same time

FIGURE P12.16 (Problems 16 and 18)

220 J of work is done on the system, find the energy transferred to or from it by heat.

18. Consider the cyclic process described by Figure P12.16. If Q is negative for the process BC and ΔU is negative for the process CA, determine the signs of Q, W, and ΔU associated with each process.

19. One gram of water changes from liquid to solid at a constant pressure of 1.00 atm and a constant temperature of $0°C$. In the process, the volume changes from 1.00 cm^3 to 1.09 cm^3. (a) Find the work done *on* the water and (b) the change in the internal energy of the water.

20. A person takes in a breath of $0°C$ air and holds it until it warms to $37.0°C$. The air has an initial volume of 0.600 L and a mass of 7.70×10^{-4} kg. Determine (a) the work done *by* the air *on* the lungs if the pressure remains constant at 1.00 atm, (b) the change in internal energy of the air, and (c) the energy added to the air by heat. Model the air as if it were a monatomic gas and use the result of Problem 1.

21. A 5.0-kg block of aluminum is heated from $20°C$ to $90°C$ at atmospheric pressure. Find (a) the work done by the aluminum, (b) the amount of energy transferred to it by heat, and (c) the increase in its internal energy.

22. One mole of gas is initially at a pressure of 2.00 atm, volume of 0.300 L and has an internal energy equal to 91.0 J. In its final state the gas is at a pressure of 1.50 atm and a volume of 0.800 L, and its internal energy equals 180 J. For the paths IAF, IBF, and IF in Figure P12.22, calculate (a) the work done *on* the gas and (b) the net energy transferred to the gas by heat in the process.

FIGURE P12.22

Section 12.4 Heat Engines and the Second Law of Thermodynamics

Section 12.5 Reversible and Irreversible Processes

Section 12.6 The Carnot Engine

23. A heat engine operates between two reservoirs at temperatures of 20°C and 300°C. What is the maximum efficiency possible for this engine?

24. The efficiency of a Carnot engine is 30.0%. The engine absorbs 800 J of energy per cycle by heat from a hot reservoir at 500 K. Determine (a) the energy expelled per cycle and (b) the temperature of the cold reservoir.

25. The energy absorbed by an engine is three times greater than the work it performs. (a) What is its thermal efficiency? (b) What fraction of the energy absorbed is expelled to the cold reservoir?

26. A particular engine has a power output of 5.00 kW and an efficiency of 25.0%. If the engine expels 8 000 J of energy as heat in each cycle, find (a) the energy absorbed in each cycle and (b) the time for each cycle.

27. One of the most efficient engines ever built is a coal-fired steam turbine in the Ohio valley, driving an electric generator as it operates between 1 870°C and 430°C. (a) What is its maximum theoretical efficiency? (b) Its actual efficiency is 42.0%. How much mechanical power does the engine deliver if it absorbs 1.40×10^5 J of energy each second from the hot reservoir?

28. A heat engine performs 200 J of work in each cycle and has an efficiency of 30%. For each cycle of operation, (a) how much energy is absorbed by heat and (b) how much energy is expelled by heat?

29. An engine absorbs 1 700 J from a hot reservoir and expels 1 200 J to a cold reservoir in each cycle. (a) What is the engine's efficiency? (b) How much work is done in each cycle? (c) What is the mechanical power output of the engine if each cycle lasts for 0.300 s?

30. A power plant has been proposed that would make use of the temperature gradient in the ocean. The system is to operate between 20.0°C (surface-water temperature) and 5.00°C (water temperature at a depth of about 1 km). (a) What is the maximum efficiency of such a system? (b) If the useful power output of the plant is 75.0 MW, how much energy is absorbed per hour? (c) In view of your answer to (a), do you think such a system is worthwhile (considering that there is no charge for fuel)?

31. In one cycle, a heat engine absorbs 500 J from the high-temperature reservoir and expels 300 J to a low-temperature reservoir. If the efficiency of this engine is 60% of the efficiency of a Carnot engine, what is the ratio of the low temperature to the high temperature in the Carnot engine?

32. A heat engine operates in a Carnot cycle between 80.0°C and 350°C. It absorbs 21 000 J of energy per cycle from the hot reservoir. The duration of each cycle is 1.00 s. (a) What is the mechanical power output of this engine? (b) How much energy does it expel in each cycle by heat?

33. A nuclear power plant has an electrical power output of 1 000 MW and operates with an efficiency of 33%. If excess energy is carried away from the plant by a river with a flow rate of 1.0×10^6 kg/s, what is the rise in temperature of the flowing water?

34. A 20.0%-efficient real engine is used to speed up a train from rest to 5.00 m/s. It is known that an ideal (Carnot) engine using the same cold and hot reservoirs would accelerate the same train from rest to a speed of 6.50 m/s using the same amount of fuel. If the engines use air at 300 K as a cold reservoir, find the temperature of the steam serving as the hot reservoir.

Section 12.7 Entropy

Section 12.8 Entropy and Disorder

35. A freezer is used to freeze 1.0 L of water completely into ice. The water and the freezer remain at a constant temperature of $T = 0°C$. Determine (a) the change in the entropy of the water and (b) the change in the entropy of the freezer.

36. What is the change in entropy of 1.00 kg of liquid water at 100°C as it changes to steam at 100°C?

37. A 70-kg log falls from a height of 25 m into a lake. If the log, the lake, and the air are all at 300 K, find the change in entropy of the Universe for this process.

38. Two 2 000-kg cars, both traveling at 20 m/s, undergo a head-on collision and stick together. Find the entropy change of the Universe resulting from the collision if the temperature is 23°C.

39. The surface of the Sun is approximately at 5 700 K, and the temperature of Earth's surface is approximately 290 K. What entropy change occurs when 1 000 J of energy is transferred by heat from the Sun to Earth?

40. Repeat the procedure used to construct Table 12.4 (a) for the case in which you draw three marbles rather than four from your bag and (b) for the case in which you draw five rather than four.

41. Prepare a table like Table 12.4 for the following occurrence. You toss four coins into the air simultaneously. Record all the possible results of the toss in terms of the numbers of heads and tails that can result. (For example, HHTH and HTHH are two possible ways in which three heads and one tail can be achieved.) (a) On the basis of your table, what is the most probable result of a toss? (b) In terms of entropy, what is the most ordered state and (c) what is the most disordered?

42. Consider a standard deck of 52 playing cards that has been thoroughly shuffled. (a) What is the probability of drawing the ace of spades in one draw? (b) What is the probability of drawing any ace? (c) What is the probability of drawing any spade?

ADDITIONAL PROBLEMS

43. A student claims that she has constructed a heat engine that operates between the temperatures of 200 K and 100 K with 60% efficiency. The professor does not give her credit for the project. Why not?

44. A Carnot heat engine uses a steam boiler at 100°C as the high-temperature reservoir. The low-temperature reservoir is the outside environment at 20.0°C. Energy is exhausted to the low-temperature reservoir at the rate of 15.4 W. (a) Determine the useful power output of the heat engine. (b) How much steam will it cause to condense in the high-temperature reservoir in one hour?

45. A Carnot heat engine extracts energy Q_h from a hot reservoir at constant temperature T_h and rejects energy Q_c to a cold reservoir at constant temperature T_c. Find the entropy changes of the (a) hot reservoir, (b) the cold reservoir, (c) the engine, and (d) the complete system.

46. One end of a copper rod is in thermal contact with a hot reservoir at $T = 500$ K, and the other end is in thermal contact with a cooler reservoir at $T = 300$ K. If 8 000 J of energy is transferred from one end to the other, with no change in the temperature distribution, find the entropy change of each reservoir and the total entropy change of the Universe.

47. Find the Celsius temperature change of a river due to the exhausted energy from a nuclear power plant. Assume that the input power to the boiler in the plant is 25×10^8 W, the efficiency of the use of this power is 30%, and the river flow rate is 9.0×10^6 kg/min.

48. Every second at Niagara Falls, some 5 000 m³ of water falls a distance of 50 m. What is the increase in entropy per second due to the falling water? (Assume a 20°C environment.)

49. One object is at a temperature of T_h and another is at a lower temperature T_c. Use the second law of thermodynamics to show that energy transfer by heat can only occur from the hotter to the colder object. Assume that the heat capacity of each object is large enough that its temperature is not measurably changed.

50. When a gas follows path 123 on the PV diagram in Figure P12.50, 418 J of energy flows into the system by heat, and − 167 J of work is done *on the gas.* (a) What is the change in the internal energy of the system? (b) How much energy Q flows into the system if the gas follows path 143? The work done on the gas along this path is − 63.0 J. What net work would be done on or by the system if the system followed (c) path 12341? (d) path 14321? (e) What is the change in internal energy of the system in the processes described in parts (c) and (d)?

FIGURE P12.50

51. A substance undergoes the cyclic process shown in Figure P12.51. Work output occurs along path AB, whereas work input is required along path BC, and no work is involved in the constant-volume process CA. Energy transfers by heat occur during each process involved in the cycle. (a) What is the work output during process AB? (b) How much work input is required during process BC? (c) What is the net energy input Q during this cycle?

52. A power plant, having a Carnot efficiency, produces 1 000 MW of electrical power from turbines that take in steam at 500 K and reject water at 300 K into a flowing river. The water downstream is 6.00 K warmer due to the output of the power plant. Determine the flow rate of the river.

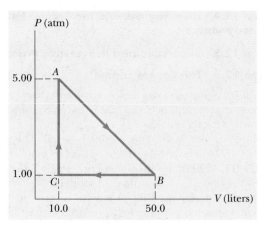

FIGURE P12.51

53. A 100-kg steel support rod in a building has a length of 2.0 m at a temperature of 20°C. The rod supports a hanging load of 6 000 kg. Find (a) the work done *on* the rod as the temperature increases to 40°C, (b) the energy Q added to the rod (assume the specific heat of steel is the same as that for iron), and (c) the change in internal energy of the rod.

54. An ideal gas initially at pressure P_0, volume V_0, and temperature T_0 is taken through the cycle described in Figure P12.54. (a) Find the net work done *by* the gas per cycle in terms of P_0 and V_0. (b) What is the net energy Q added to the system per cycle? (c) Obtain a numerical value for the net work done per cycle for 1.00 mol of gas initially at 0°C. (*Hint:* Recall that work done by the system equals the area under a PV curve.)

FIGURE P12.54

55. One mole of neon gas is heated from 300 K to 420 K at constant pressure. Calculate (a) the energy Q transferred to the gas, (b) the change in the internal energy of the gas, and (c) the work done on the gas. Note that neon has a molar specific heat of $c = 20.79$ J/mol · K for a constant-pressure process.

56. A 1.0-kg block of aluminum is heated at atmospheric pressure so that its temperature increases from 22°C to 40°C. Find (a) the work done *on* the aluminum, (b) the energy Q added to the aluminum, and (c) the change in internal energy of the aluminum.

57. Suppose a heat engine is connected to two energy reservoirs, one a pool of molten aluminum (660°C) and the other a block of solid mercury (− 38.9°C). The engine runs by freezing 1.00 g of aluminum and melting 15.0 g of mercury during each cycle. The latent heat of fusion of aluminum is 3.97×10^5 J/kg, and that of mercury is 1.18×10^4 J/kg. (a) What is the efficiency of this engine? (b) How does the efficiency compare with that of a Carnot engine?

58. One mole of an ideal gas is taken through the cycle shown in Figure P12.58. At point A, the pressure, volume, and temperature are P_0, V_0, and T_0. In terms of R and T_0, find (a) the total energy entering the system by heat per cycle, (b) the total energy leaving the system by heat per cycle, (c) the efficiency of an engine operating in this cycle, and (d) the efficiency of an engine operating in a Carnot cycle between the temperature extremes for this process. (*Hint:* Recall that work done on the gas is the negative of the area under a PV curve.)

59. An electrical power plant has an overall efficiency of 15%. The plant is to deliver 150 MW of electrical power to a city, and its turbines use coal as fuel. The burning coal produces steam at 190°C, which drives the turbines. This steam is then condensed into water at 25°C by passing through coils in contact with river water. (a) How many

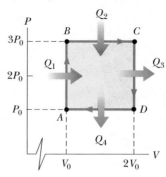

FIGURE P12.58

metric tons of coal does the plant consume each day (1 metric ton = 1×10^3 kg)? (b) What is the total cost of the fuel per year if the delivered price is $8 per metric ton? (c) If the river water is delivered at 20°C, at what minimum rate must it flow over the cooling coils in order that its temperature not exceed 25°C? (*Note:* The heat of combustion of coal is 7.8×10^3 cal/g.)

GROUP ACTIVITIES

G.1 Bend a paper clip back and forth several times and touch it to your lip. Why does it warm up? Repeat this with a rubber band after it is stretched several times. You should notice that it also warms up after stretching. Now take the stretched rubber band from your lip and wait for it to come to equilibrium with the surrounding air. Touch the stretched band to your lip, let it relax to its unstretched length, and note that it becomes cooler. Explain how the second law of thermodynamics helps you understand these results.

G.2 You should be able to locate one of the following in either the chemistry or physics department of your university: a helium tank, a nitrogen tank, or a CO_2 tank. Ask to use one of these for a short period for the following observation. Learn to open and close the valve or valves correctly. Touch the tanks and nozzle to verify that they are at room temperature. Open a tank for 2–3 seconds and then close it. Touch the nozzle again, and you will find that it is very cold. Why? As a hint, the expansion of the gas is nearly adiabatic.

G.3 You can study order and disorder and verify the answer to Quick Quiz 12.5 by rolling a pair of dice 100 times and recording the number of spots appearing on the dice for each throw. Which total comes up most frequently?

G.4 Another way to study order and disorder is to toss pennies. Predict in advance which is the more likely occurrence when tossing three pennies: ordered with all the same orientation (all heads or all tails), or disordered (not all the same). Show that the ordered result includes two out of eight possibilities, whereas the disordered re-

sult includes six out of eight. Toss the three pennies 100 times to see how close you come to this prediction. Repeat this process for four pennies. (The all-the-same result should be expected two times in sixteen throws).

G.5 For each of the situations described below, the system considered is undergoing some changes. Among the possibilities, you should consider these changes: (*Q*) The object is absorbing or giving off energy by heat. (*T*) The object's temperature is changing. (*U*) The object's internal energy is changing. (*W*) The object is doing mechanical work or having work done on it. For each of the situations described below, identify which of the four changes are taking place and write as many of the letters *Q, T, U,* or *W* (or none) as are appropriate.

(a) A cylinder with a piston on top contains a compressed gas and is sitting on an energy reservoir (a large iron block). After everything has come to thermal equilibrium, the piston is moved upward somewhat (very slowly). The system to be considered is the gas in the cylinder.

(b) A cylinder with a piston on top contains a compressed gas and is wrapped in styrofoam, a very good thermal insulator. The piston is pressed downward (again, very slowly), compressing the gas. The system to be considered is the gas in the cylinder.

(c) An ice cube sitting in the open air is melting.

(Problem 5 is courtesy of Edward F. Redish. For more problems of this type, see http://www.physics.umd.edu/perg/)

Three

Vibrations and Waves

As we look around us, we find many examples of objects that vibrate or oscillate: a pendulum, the strings of a guitar, an object suspended on a spring, the piston of an engine, the head of a drum, the reed of a saxophone. Most elastic objects vibrate when an impulse is applied to them; that is, once they are distorted, their shape tends to be restored to some equilibrium configuration. Even at the atomic level, the atoms in a solid vibrate about some position as if they were connected to their neighbors by imaginary springs.

Wave motion is closely related to the phenomenon of vibration. Sound waves, earthquake waves, waves on stretched strings, and water waves are all produced by vibrations. As a sound wave travels through a medium (such as air), the elements of the medium vibrate back and forth; as a water wave travels across a pond, the water elements vibrate up and down. When any wave travels through a medium, the elements of the medium move in repetitive cycles. Therefore, the motion of the elements bears a strong resemblance to the periodic motion of a vibrating pendulum or an object attached to a spring.

Many other natural phenomena occur whose explanations require an understanding of vibrations and waves. Although many large structures, such as skyscrapers and bridges, appear to be rigid, they actually vibrate—a fact that must be taken into account by the architects and engineers who design and build them. To understand how radio and television work, we must comprehend the origin and nature of electromagnetic waves and how they propagate through space. Finally, much of what scientists have learned about atomic structure has come from information carried by waves; therefore, to understand the concepts and theories of atomic physics, we must first study waves and vibrations.

As these water droplets fall into a pond, a circular pattern of waves spread outward from the entry point of the droplets. *(© Peter K. Ziminski/Visuals Unlimited)*

13

Vibrations and Waves

Chapter Outline

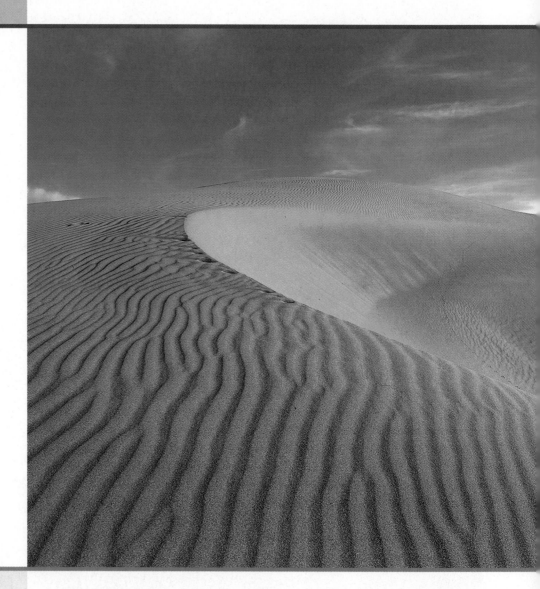

When sand is abundant and evenly distributed, transverse dunes are formed perpendicular to the direction of prevailing winds. These transverse dunes were formed in Great Sand Dunes National Monument, Colorado. *(Arthur C. Smith III/Grant Heilman Photography, Inc.)*

This chapter constitutes a brief return to the subject of mechanics as we examine various forms of periodic motion. We concentrate especially on motion that occurs when the force on an object is proportional to the displacement of the object from its equilibrium position. When such a force acts only toward the equilibrium position, the result is a back-and-forth periodic motion—an oscillation, or vibration, between two extreme positions. You may be familiar with several types of periodic motion, such as the oscillations of an object attached to a spring, the motion of a pendulum, and the vibrations of a stringed musical instrument.

Because vibrations can cause disturbances that move through a medium, we also study wave motion in this chapter. Many kinds of waves occur in nature, including sound waves, water waves, seismic waves, and electromagnetic waves. We conclude with a brief discussion of some terms and concepts that are common to all types of waves, and in later chapters we shall focus our attention on specific categories of waves.

13.1 HOOKE'S LAW

One of the simplest types of vibrational motion is that of an object attached to a spring, which we discussed briefly in Chapter 5. Let us assume that the object moves on a frictionless horizontal surface. If the spring is stretched or compressed a small distance x from its unstretched or equilibrium position and then released, it exerts a force on the object as shown in Figure 13.1. From experiment, this spring force is found to obey the equation

$$F_s = -kx \qquad \text{[13.1]}$$

◀ Hooke's law

where x is the displacement of the object from its equilibrium ($x = 0$) position and k is a positive constant called the **spring constant.** This force law for springs was discovered by Robert Hooke in 1678 and is known as **Hooke's law.** The value of k is a measure of the stiffness of the spring. Stiff springs have large k values, and soft springs have small k values.

The negative sign in Equation 13.1 signifies that the force exerted by the spring is always directed *opposite* the displacement of the object. When the object is to the right of the equilibrium position, as in Figure 13.1a, x is positive and F_s is negative. This means that the force is in the negative direction, to the left. When the object is to the left of the equilibrium position, as in Figure 13.1c, x is negative and F_s is positive, indicating that the direction of the force is to the right. Of course, when $x = 0$, as in Figure 13.1b, the spring is unstretched and $F_s = 0$. Because the spring force always acts toward the equilibrium position, it is sometimes called a restoring force. **The direction of the restoring force is such that the object is being either pulled or pushed toward the equilibrium position.**

Let us examine the motion of the object if it is initially pulled a distance A to the right and released from rest. The force exerted by the spring on the object pulls the object back toward the equilibrium position. As the object moves toward $x = 0$, the magnitude of the force decreases (because x decreases) and reaches zero at $x = 0$. However, the object gains speed as it moves toward the equilibrium position. In fact, the speed reaches its maximum value when $x = 0$. The momentum achieved by the object causes it to overshoot the equilibrium position and to compress the spring. As the object moves to the left of the equilibrium position (negative x values), the force acting on it begins to increase to the right and the speed of the object begins to decrease. The object finally comes to a stop at $x = -A$. The process is then repeated, and the object continues to oscillate back and forth over the same path. This type of motion is called **simple harmonic motion. Simple harmonic motion occurs when the net force along the direction**

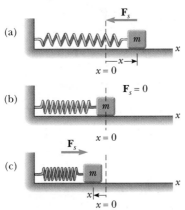

FIGURE 13.1 The force exerted by a spring on an object varies with the displacement of the object from the equilibrium position, $x = 0$.
(a) When x is positive (stretched spring), the spring force is to the left.
(b) When x is zero (unstretched spring), the spring force is zero.
(c) When x is negative (compressed spring), the spring force is to the right.

of motion is a Hooke's law type of force—that is, when the net force is proportional to the displacement and in the opposite direction.

Not all periodic motion over the same path can be classified as simple harmonic motion. For example, consider a ball being tossed back and forth between a parent and a child. The ball moves repetitively, but the motion is not simple harmonic. Unless the force acting on an object along the direction of motion has the form of Equation 13.1, the object does not exhibit simple harmonic motion.

The motion of an object suspended from a vertical spring is also simple harmonic. In this case, the force of gravity acting on the attached object stretches the spring until equilibrium is reached (the object is suspended and at rest.) This position of equilibrium establishes the position of the object for which $x = 0$. When the object is moved away from equilibrium by a distance x and then released, a net force acts (in the form of Hooke's law) toward the equilibrium position. Because the net force is proportional to x, the motion is simple harmonic.

In order to discuss the motion of an object oscillating in simple harmonic motion in some detail, it is necessary to define the following terms.

- The **amplitude** A is the magnitude of the maximum position of the object relative to its equilibrium position. In the absence of friction, an object continues in simple harmonic motion and reaches a maximum position equal to $\pm A$ on each side of the equilibrium position during each cycle.
- The **period** T is the time it takes the object to execute one complete cycle of motion—that is, from $x = A$ to $x = -A$ and back to $x = A$.
- The **frequency** f is the number of complete cycles or vibrations per unit of time.

Example 13.1   Measuring the Spring Constant

A common technique used to evaluate the spring constant is illustrated in Figure 13.2. The spring is hung vertically (Fig. 13.2a), and an object of mass m is attached to the lower end of the spring (Fig. 13.2b). The spring is stretched by a distance d from its initial position under the action of the "load" mg. Because the spring force is upward, it must balance the weight mg downward *when the system is at rest* (see Fig. 13.2c). In this case, we can apply Hooke's law to give

$$F_s = -kd = -mg$$

$$k = \frac{mg}{d}$$

For example, if a spring is displaced 2.0 cm by a mass of 0.55 kg, the spring constant is

$$k = \frac{mg}{d} = \frac{(0.55 \text{ kg})(9.80 \text{ m/s}^2)}{2.0 \times 10^{-2} \text{ m}} = \boxed{2.7 \times 10^2 \text{ N/m}}$$

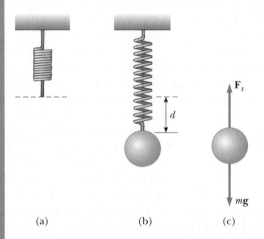

FIGURE 13.2   (Example 13.1) Determining the spring constant. The elongation d of the spring is due to the suspended weight, mg. Because the upward spring force balances the weight when the system is in equilibrium, it follows that $k = mg/d$.

(a)   (b)   (c)

Example 13.2   Simple Harmonic Motion on a Frictionless Surface

A 0.35-kg object attached to a spring of spring constant 130 N/m is free to move on a frictionless horizontal surface, as in Figure 13.1. If the object is released from rest at $x = 0.10$ m, find the force on it and its acceleration at (a) $x = 0.10$ m, (b) $x = 0.050$ m, (c) $x = 0$ m, and (d) $x = -0.050$ m.

Solution

A The point at which the object is released at rest defines the amplitude of the motion. In this case, $A = 0.10$ m. The object moves continuously between the limits 0.10 m and -0.10 m. When x is a maximum ($x = A$), the force on the object is a maximum and is calculated as follows:

$$F_s = -kx$$

$$F_{max} = -kA = -(130 \text{ N/m})(0.10 \text{ m}) = \boxed{-13 \text{ N}}$$

The negative sign indicates that the force acts to the left, in the negative x direction. We can use Newton's second law to calculate the acceleration at this position:

$$F_{max} = ma_{max}$$

$$a_{max} = \frac{F_{max}}{m} = -\frac{13 \text{ N}}{0.35 \text{ kg}} = \boxed{-37 \text{ m/s}^2}$$

Again, the negative sign indicates that the acceleration is to the left.

B We can use the same approach to find the force and acceleration at other positions. At $x = 0.05$ m, we have

$$F_s = -kx = -(130 \text{ N/m})(0.050 \text{ m}) = \boxed{-6.5 \text{ N}}$$

$$a = \frac{F}{m} = -\frac{6.5 \text{ N}}{0.35 \text{ kg}} = \boxed{-19 \text{ m/s}^2}$$

Note that the acceleration of an object moving with simple harmonic motion *is not constant* because F_s is not constant.

C At $x = 0$, the spring force is zero (because $F_s = -kx = 0$) and the acceleration is zero. In other words, when the spring is unstretched, it exerts no force on the object attached to it.

D At $x = -0.050$ m, we have

$$F_s = -kx = -(130 \text{ N/m})(-0.050 \text{ m}) = \boxed{6.5 \text{ N}}$$

$$a = \frac{F_s}{m} = \frac{6.5 \text{ N}}{0.35 \text{ kg}} = \boxed{19 \text{ m/s}^2}$$

This result shows that the force and acceleration are positive when the object is on the negative side of the equilibrium position. This indicates that the force exerted by the spring on the object is acting to the right as the spring is being compressed. At the same time, the object is slowing down as it moves from $x = 0$ to $x = -0.050$ m.

EXERCISE   Find the force and acceleration when $x = -0.10$ m.

ANSWER   13 N; 37 m/s^2

As indicated in Example 13.2, the acceleration of an object moving with simple harmonic motion can be found by using Hooke's law to specify the force in the equation for Newton's second law, $F = ma$. This gives

$$-kx = ma$$

$$a = -\frac{k}{m} x \qquad \text{[13.2]}$$

◀ Acceleration in simple harmonic motion

A NONCONSTANT ACCELERATION

The acceleration of a particle in simple harmonic motion is *not* constant. Equation 13.2 shows that *a* is proportional to the position *x*. Thus, *a* varies with *x* and we *cannot* apply the constant acceleration kinematic equations of Chapter 2 in this situation.

Because the maximum value of *x* is defined to be the amplitude, *A*, we see that the acceleration ranges over the values $-kA/m$ to $+kA/m$. Equation 13.2 enables us to find the acceleration of the object as a function of its position. In subsequent sections, we will find equations for velocity as a function of position and position as a function of time.

Earlier we stated that an object moves with simple harmonic motion when the net force acting on it is proportional to its displacement from equilibrium and is directed toward the equilibrium position. Equation 13.2 provides an alternative definition of simple harmonic motion. An object moves with simple harmonic motion if its acceleration is proportional to its displacement and is in the opposite direction to it.

Quick Quiz 13.1

A block on the end of a spring is pulled to position $x = A$ and released. Through what total distance does it travel in one full cycle of its motion? (a) $A/2$ (b) A (c) $2A$ (d) $4A$

Quick Quiz 13.2

Which of the following combination of quantities cannot be in the same direction for a simple harmonic oscillator? (a) position and velocity, (b) velocity and acceleration, (c) position and acceleration.

13.2 ELASTIC POTENTIAL ENERGY

For the most part, we have worked with three types of mechanical energy: gravitational potential energy, translational kinetic energy, and rotational kinetic energy. In Chapter 5 we briefly discussed a fourth type of mechanical energy, elastic potential energy. We now consider this form of energy in more detail.

A system of interacting objects has potential energy associated with the configuration of the system. As we learned in Chapter 5, a system of an object and Earth has potential energy of *mgh* if the object is at a height *h* above Earth's surface and the zero of potential energy is chosen as that configuration in which the object is located on the surface. This potential energy can be transformed to kinetic energy of the object as the object falls after being released. Likewise, a compressed spring has potential energy. The compressed spring, when allowed to expand, can apply a force on an object such that the potential energy in the spring is transformed to kinetic energy of the object. As an example, Figure 13.3 shows a ball being projected from a spring-loaded toy gun, where the spring is compressed a distance *x*. As the gun is fired, the compressed spring does work on the ball and imparts kinetic energy to it. **The energy stored in a stretched or compressed spring or other elastic material is called elastic potential energy PE_s given by**

FIGURE 13.3 A ball projected from a spring-loaded gun. The elastic potential energy stored in the spring is transformed into the kinetic energy of the ball.

Energy = elastic PE_s

Energy = KE

$$PE_s \equiv \tfrac{1}{2}kx^2 \qquad\qquad \textbf{[13.3]} \qquad\qquad \blacktriangleleft \text{ Elastic potential energy}$$

Note that the elastic potential energy stored in a spring is zero when the spring is unstretched or uncompressed ($x = 0$). **Energy is stored in a spring only when it is either stretched or compressed. Furthermore, the elastic potential energy is a maximum when a spring has reached its maximum compression or extension.** Finally, the potential energy in a spring is always positive because it is proportional to x^2. We now include this new form of energy in our equation for conservation of mechanical energy:

$$(KE + PE_g + PE_s)_i = (KE + PE_g + PE_s)_f \qquad\qquad \textbf{[13.4]}$$

If nonconservative forces such as friction are present, then the final mechanical energy does not equal the initial mechanical energy. In this case, the difference in the two energies must equal the work W_{nc} done by the nonconservative force. Using Equation 5.15, we find that

$$W_{nc} = (KE + PE_g + PE_s)_f - (KE + PE_g + PE_s)_i \qquad\qquad \textbf{[13.5]}$$

As an example of the energy conversions that take place when a spring is included in the system, consider Figure 13.4. A block of mass m slides on a frictionless horizontal surface with constant velocity $\mathbf{v}_i$ and collides with a coiled spring. The description that follows is greatly simplified by assuming that the spring is very light and therefore has negligible kinetic energy. As the spring is compressed, it exerts a leftward force on the block. At maximum compression, the block momentarily stops (Fig. 13.4c). The initial total energy in the system (block plus spring) before the collision is the kinetic energy of the block. After the block collides with the spring and the spring is partially compressed, as in Figure 13.4b, the block has kinetic energy $\tfrac{1}{2}mv^2$ (where $v < v_i$) and the spring has potential energy $\tfrac{1}{2}kx^2$. When the block stops momentarily after colliding with the spring, the kinetic energy is zero. Because the spring force is conservative and because there are no external

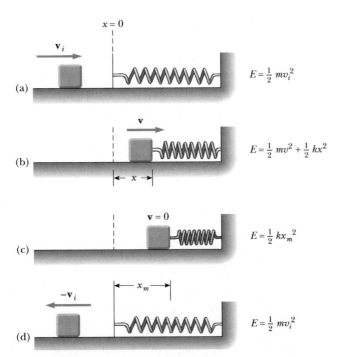

FIGURE 13.4 A block sliding on a frictionless horizontal surface collides with a light spring. (a) Initially, the mechanical energy is entirely the kinetic energy of the block. (b) The mechanical energy at some arbitrary position is the sum of the kinetic energy of the block and the elastic potential energy stored in the spring. (c) When the block comes to rest, the mechanical energy is entirely elastic potential energy stored in the compressed spring. (d) When the block leaves the spring, the mechanical energy is equal to the block's kinetic energy. The total energy remains constant.

FIGURE 13.5 Elastic potential energy is stored in this drawn bow. *(Eric Lars Baleke/Black Star)*

forces that can do work on the system, **the total mechanical energy of the system consisting of the block and spring remains constant.** Thus, energy is transformed from kinetic energy of the block into potential energy stored in the spring. As the spring expands, the block moves in the opposite direction and regains all of its initial kinetic energy, as in Figure 13.4d.

Elastic potential energy is stored in the bow of an archer when she pulls back the bowstring (Figure 13.5). This is a more complicated system than a single object on a spring, but if we consider the essential features, there is a restoring force on the arrow as a result of the stretching of the string and the bending of the bow. The potential energy stored in the system is transformed into kinetic energy of the arrow as it is released. We can make similar statements about devices such as crossbows and slingshots.

APPLICATION

ARCHERY

Quick Quiz 13.3	When an object moving in simple harmonic motion is at its maximum displacement, which of the following is at a maximum? (a) velocity, (b) acceleration, or (c) kinetic energy.

Example 13.3   Stop That Car!

A 13 000-N car starts at rest and rolls down a hill from a height of 10 m (Fig. 13.6). It then moves across a level surface and collides with a light spring-loaded guardrail. Neglecting any losses due to friction, find the maximum distance the spring is compressed. Assume a spring constant of 1.0×10^6 N/m.

FIGURE 13.6 (Example 13.3) A car starts from rest on a hill at the position shown. When the car reaches the bottom of the hill, it collides with a spring-loaded guardrail.

Solution   The initial and final values of the car's kinetic energy are zero. Thus, the initial potential energy of the car-spring-Earth system is completely converted to elastic potential energy in the spring at the end of the trip (assuming we neglect any energy losses due to friction during the collision). Conservation of mechanical energy for the system gives

$$mgh = \tfrac{1}{2}kx^2$$

Solving for x gives

$$x = \sqrt{\frac{2mgh}{k}} = \sqrt{\frac{2(13\ 000\ \text{N})(10\ \text{m})}{1.0 \times 10^6\ \text{N/m}}} = \boxed{0.50\ \text{m}}$$

Note that it was not necessary at any point to calculate the velocity of the car to obtain a solution. This demonstrates the power of the principle of conservation of energy. We work with the initial and final energy values only, without having to consider all the details in between.

EXERCISE What is the speed of the car just before it collides with the guardrail?

ANSWER 14 m/s

Example 13.4 Motion With and Without Friction

A 1.6-kg block on a horizontal surface is attached to a spring with a spring constant of 1.0×10^3 N/m (Fig. 13.1). The spring is compressed a distance of 2.0 cm, and the block is released from rest.

A Calculate the speed of the block as it passes through the equilibrium position, $x = 0$, if the surface is frictionless.

Solution From Equation 13.3, see that the initial elastic potential energy of the compressed spring is

$$PE_s = \tfrac{1}{2}kx_i^2$$

Because the block is always at the same height above Earth's surface, the gravitational potential energy of the system remains constant. Hence, the initial potential energy stored in the spring is converted to kinetic energy at $x = 0$. That is,

$$\tfrac{1}{2}kx_i^2 = \tfrac{1}{2}mv_f^2$$

Solving for v_f gives

$$v_f = \sqrt{\frac{k}{m}}\, x_i = \sqrt{\frac{1.0 \times 10^3\ \text{N/m}}{1.6\ \text{kg}}}\,(2.0 \times 10^{-2}\ \text{m}) = \boxed{0.50\ \text{m/s}}$$

B Calculate the speed of the block as it passes through the equilibrium position if a constant frictional force of 4.0 N retards its motion.

Solution Because sliding friction is present in this situation, we know that the final mechanical energy is less than the initial mechanical energy. The work done by the frictional force as the block undergoes a displacement x_i is

$$W_{nc} = W_f = -fx_i$$

Applying Equation 13.5 to this situation gives

$$-fx_i = \tfrac{1}{2}mv_f^2 - \tfrac{1}{2}kx_i^2$$

$$-(4.0\ \text{N})(2.0 \times 10^{-2}\ \text{m}) = \tfrac{1}{2}(1.6\ \text{kg})(v_f^2) - \tfrac{1}{2}(1.0 \times 10^3\ \text{N/m})(2.0 \times 10^{-2}\ \text{m})^2$$

$$v_f = \boxed{0.39\ \text{m/s}}$$

Note that this value for v_f is less than that obtained in the frictionless case. Does the result make sense?

EXERCISE How far does the block travel before coming to rest? Assume a constant friction force of 4.0 N and the same initial conditions as before.

ANSWER 5.0 cm

13.3 VELOCITY AS A FUNCTION OF POSITION

Conservation of energy provides a simple method of deriving an expression for the velocity of an object undergoing periodic motion as a function of position.

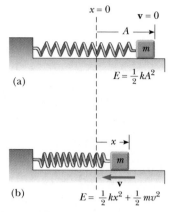

$x = 0$
$v = 0$
A
m

(a) $E = \frac{1}{2}kA^2$

x
m
v

(b) $E = \frac{1}{2}kx^2 + \frac{1}{2}mv^2$

FIGURE 13.7 (a) An object attached to a spring on a frictionless surface is released from rest with the spring extended a distance A. Just before release, the total energy is elastic potential energy $kA^2/2$. (b) When the object reaches position x, it has kinetic energy $mv^2/2$ and the elastic potential energy has decreased to $kx^2/2$.

The object in question is initially at its maximum extension A (Fig. 13.7a) and is then released from rest. In this situation, the initial energy of the system is entirely elastic potential energy stored in the spring, $\frac{1}{2}kA^2$. As the object moves toward the origin to some new position x (Fig. 13.7b), part of this energy is transformed into kinetic energy, and the potential energy stored in the spring is reduced to $\frac{1}{2}kx^2$. Because the total energy of the system is equal to $\frac{1}{2}kA^2$ (the initial energy stored in the spring), we can equate this to the sum of the kinetic and potential energies at the position x:

$$\tfrac{1}{2}kA^2 = \tfrac{1}{2}mv^2 + \tfrac{1}{2}kx^2$$

Solving for v, we get

$$v = \pm\sqrt{\frac{k}{m}\,(A^2 - x^2)} \qquad \text{[13.6]}$$

This expression shows us that the speed is a maximum at $x = 0$ and zero at the extreme positions $x = \pm A$.

The right side of Equation 13.6 is preceded by the $\pm$ sign because the square root of a number can be either positive or negative. The sign of v that is selected depends on the circumstances of the motion. If the object in Figure 13.7 is moving to the right, v must be positive; if the object is moving to the left, v must be negative.

Example 13.5 The Object-Spring System Revisited

A 0.50-kg object connected to a light spring with a spring constant of 20 N/m oscillates on a frictionless horizontal surface.

A Calculate the total energy of the system and the maximum speed of the object if the amplitude of the motion is 3.0 cm.

Solution From Equation 13.3, we have

$$E = PE_s = \tfrac{1}{2}kA^2 = \tfrac{1}{2}(20 \text{ N/m})(3.0 \times 10^{-2} \text{ m})^2 = \boxed{9.0 \times 10^{-3} \text{ J}}$$

When the object is at $x = 0$, $PE_s = 0$ and $E = \frac{1}{2}mv_{\text{max}}^2$; therefore,

$$\tfrac{1}{2}mv_{\text{max}}^2 = 9.0 \times 10^{-3}\text{ J}$$

$$v_{\text{max}} = \sqrt{\frac{18 \times 10^{-3}\text{ J}}{0.50\text{ kg}}} = \boxed{0.19 \text{ m/s}}$$

B What is the velocity of the object when the displacement is 2.0 cm?

Solution We can apply Equation 13.6 directly:

$$v = \pm\sqrt{\frac{k}{m}\,(A^2 - x^2)} = \pm\sqrt{\frac{20}{0.50}\,(3.0^2 - 2.0^2) \times 10^{-4}} = \boxed{\pm 0.14 \text{ m/s}}$$

The $\pm$ sign indicates that the object could be moving to the right or to the left at this instant.

C Compute the kinetic and potential energies of the system when the displacement is 2.0 cm.

Solution The results of part B can be used to give

$$KE = \tfrac{1}{2}mv^2 = \tfrac{1}{2}(0.50\text{ kg})(0.14\text{ m/s})^2 = \boxed{5.0 \times 10^{-3}\text{ J}}$$

$$PE_s = \tfrac{1}{2}kx^2 = \tfrac{1}{2}(20\text{ N/m})(2.0 \times 10^{-2}\text{ m})^2 = \boxed{4.0 \times 10^{-3}\text{ J}}$$

Note that the sum $KE + PE_s$ equals the total energy E found in part A.

EXERCISE For what values of x is the speed of the object 0.10 m/s?

ANSWER ± 2.5 cm

13.4 COMPARING SIMPLE HARMONIC MOTION WITH UNIFORM CIRCULAR MOTION

We can better understand and visualize many aspects of simple harmonic motion along a straight line by looking at its relationship to uniform circular motion. Figure 13.8 is a top view of an experimental arrangement that is useful for this purpose. A ball is attached to the rim of a phonograph turntable of radius A, illuminated from the side by a lamp. Rather than concentrating on the ball, let us focus our attention on the shadow that the ball casts on the screen. We find that **as the turntable rotates with constant angular speed, the shadow of the ball moves back and forth with simple harmonic motion.**

In order to understand why, let us examine Equation 13.6 more closely. This equation says that the velocity of an object moving with simple harmonic motion is related to the displacement by

$$v = C\sqrt{A^2 - x^2}$$

where C is a constant. To see that the shadow also obeys this relation, consider Figure 13.9, which shows the ball moving with a constant speed v_0 in a direction tangent to the circular path. At this instant, the velocity of the ball in the x direction is given by $v = v_0 \sin \theta$, or

$$\sin \theta = \frac{v}{v_0}$$

Likewise, the larger triangle containing the angle θ in Figure 13.9 enables us to write

$$\sin \theta = \frac{\sqrt{A^2 - x^2}}{A}$$

Equating the right-hand sides of the last two expressions, we see that v is related to the displacement x as

$$\frac{v}{v_0} = \frac{\sqrt{A^2 - x^2}}{A}$$

or

$$v = \frac{v_0}{A}\sqrt{A^2 - x^2} = C\sqrt{A^2 - x^2}$$

The velocity of the ball in the x direction is related to the displacement x in exactly the same manner as the velocity of an object undergoing simple harmonic motion. Hence, the shadow moves with simple harmonic motion.

A valuable example of the relationship between simple harmonic motion and circular motion is seen in vehicles and machines that use the back-and-forth motion of a piston to create rotational motion of a wheel. For example, consider the drive wheel of a locomotive. In Figure 13.10, the curved housing at the left of the photo contains a piston that moves back and forth in simple harmonic motion. The piston is connected to an arrangement of rods that transforms its back-and-forth motion into rotational motion of the wheels. A similar mechanism in an automobile engine transforms the back-and-forth motion of the pistons to rotational motion of the crankshaft.

PERIOD AND FREQUENCY

Note that the period T of the shadow in Figure 13.8, which represents the time required for one complete trip back and forth, is also the time it takes the ball to make one complete circular trip on the turntable. Because the ball moves through

FIGURE 13.8 An experimental setup for demonstrating the connection between simple harmonic motion and uniform circular motion. As the ball rotates on the turntable with constant angular speed, its shadow on the screen moves back and forth with simple harmonic motion.

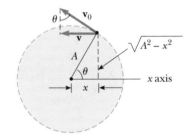

FIGURE 13.9 The ball rotates with constant speed v_0. The x component of the ball's velocity equals the projection of $\mathbf{v}_0$ on the x axis.

> **APPLICATION**
>
> PISTONS AND DRIVE WHEELS

FIGURE 13.10 The drive wheel mechanism of an old locomotive. (© *Link/Visuals Unlimited*)

Webnote 13.1

To see an excellent simulation relating uniform circular motion and simple harmonic motion, explore *http://www.phy.ntnu.edu.tw/~hwang/ shm/shm.html*

the distance $2\pi A$ (the circumference of the circle) in the time T, the speed v_0 of the ball around the circular path is

$$v_0 = \frac{2\pi A}{T}$$

$$T = \frac{2\pi A}{v_0}$$

However, for our purposes, let us consider only a fraction of the complete trip. Imagine that the ball moves from P to Q, a quarter of a revolution, in Figure 13.8. This requires a time interval equal to one fourth of the period, and the distance traveled by the ball is $2\pi A/4$. Therefore,

$$\frac{T}{4} = \frac{2\pi A}{4v_0} \qquad \text{[13.7]}$$

Note that the motion of the shadow in Figure 13.8 is equivalent to the horizontal motion of an object on the end of a spring. For this reason, the radius A of the circular motion is the same as the amplitude A of the simple harmonic motion of the shadow. During this quarter-cycle, the shadow moves from a point where its energy is solely elastic potential energy to a point where its energy is solely kinetic energy. That is,

$$\tfrac{1}{2}kA^2 = \tfrac{1}{2}mv_0^2$$

$$\frac{A}{v_0} = \sqrt{\frac{m}{k}}$$

Substituting for A/v_0 in Equation 13.7, we find that the period is

The period of an object-spring ▶ system moving with simple harmonic motion

$$T = 2\pi\sqrt{\frac{m}{k}} \qquad \text{[13.8]}$$

This expression gives the time required for an object of mass m attached to a spring constant k to complete one cycle of its motion. To see that this result agrees with common sense, picture an object hanging on a spring; if the object has a large mass, the motion is slow and a complete oscillation takes a long time. Furthermore, for a stiff spring with a large value of k, the motion is rapid and the period of oscillation is short. It is also interesting that the period does not depend on the size of the amplitude A.

The inverse of the period is the frequency of the motion. **The frequency f represents the number of oscillations that the object makes per unit time:**

$$f = \frac{1}{T} \qquad \text{[13.9]}$$

Therefore, the frequency of the periodic motion is

Frequency ▶

$$f = \frac{1}{2\pi}\sqrt{\frac{k}{m}} \qquad \text{[13.10]}$$

The units of frequency are cycles per second = s^{-1}, or **hertz** (Hz). The **angular frequency** ω is

Angular frequency ▶

$$\omega = 2\pi f = \sqrt{\frac{k}{m}} \qquad \text{[13.11]}$$

APPLYING PHYSICS 13.1

We know that the period of oscillation of an object hung on a spring is proportional to the square root of the object's mass. Thus, if we perform an experiment in which we hang objects with varying masses on the end of a spring and measure the period of oscillation, a graph of the square of the period on the vertical axis versus the mass of the suspended object on the horizontal axis will result in a straight line. But we would find that the line does not go through the origin. That is, for $m = 0$, T^2 has a finite value. How do you explain this result?

Explanation The reason that the line does not go through the origin is that the spring itself has mass, which was not considered in the derivation of Equation 13.8. Thus, the system's resistance to change in motion is a combination of the mass hung on the end of the spring and the mass of the oscillating coils of the spring. The entire mass of the spring is not oscillating, however. The bottommost coil is oscillating with the same amplitude as the object, and the attached point of the topmost coil is not oscillating at all. For a cylindrical spring, energy arguments can be used to show that the additional mass affecting the oscillations of a light spring is one third of the mass of the spring. The square of the period is proportional to the total oscillating mass. Therefore, a graph of period squared against total mass (mass hung on the spring plus the effective oscillating mass of the spring) would pass through the origin.

Tip 13.2 TWO KINDS OF FREQUENCY

We identify two kinds of frequency for a simple harmonic oscillator — f, called simply the *frequency*, is measured in hertz, and ω, the *angular frequency*, is measured in radians per second. Be sure that you are clear about which frequency is being discussed or requested in a given problem. Equation 13.11 shows the relationship between the two frequencies.

APPLYING PHYSICS 13.2

An object oscillating on the end of a horizontal spring slides back and forth over a frictionless surface. During one oscillation, you set an identical object on the first object when the system is at the maximum displacement point where it is at rest. The two objects are connected by Velcro and continue their oscillation together. Does the period of the oscillation change? Does the amplitude of the oscillation change? Does the energy of the oscillation change?

Explanation The period of the oscillation does change because the period depends on the mass of the object that is oscillating. The amplitude does not change. Because the new object was added while the original object was at rest, the combined objects are at rest at this point also, and the amplitude is the same as in the original oscillation. The energy does not change, either. At the maximum displacement point, the energy is all potential energy stored in the spring, which depends only on the spring constant and the amplitude, not the mass. The combined objects will pass through the equilibrium point with less velocity than in the original oscillation, but with the same kinetic energy. Another approach is to think about how energy could be transferred into the oscillating system — no work was done (nor was there any other form of energy transfer), so the energy in the system cannot change.

APPLYING PHYSICS 13.3

A bungee cord can be modeled as a spring. If you go bungee jumping, you will bounce up and down at the end of the cord after your daring dive off a bridge (Fig. 13.11). Suppose you perform this dive and measure the frequency of your bouncing. You then move to another bridge. You discover that the bungee cord is too long for dives off this bridge; you will hit the ground. Thus, you fold the bungee cord in half and make the dive from the doubled bungee cord. How does the frequency of your bouncing at the end of this dive compare to the frequency after the dive from the first bridge?

Explanation The force exerted by the bungee cord, modeled as a spring, is proportional to the separation of the coils as the spring is extended. Imagine that

APPLICATION

BUNGEE JUMPING

FIGURE 13.11   (Applying Physics 13.3) Bungee jumping from a bridge. *(Terje Rakke/Image Bank/Getty Images)*

we extend a spring by a given distance and measure the distance between coils. We then cut the spring in half. If one of the half-springs is now extended by the same distance, the coils will be twice as far apart as they were for the complete spring. Thus, it takes twice as much force to stretch the half-spring, from which we conclude that the half-spring has a spring constant that is twice that of the complete spring. Now consider the folded bungee cord that we model as two half-springs in parallel. Each half has a spring constant that is twice the original spring constant of the bungee cord. In addition, an object hanging on the folded bungee cord will experience two forces—one from each half-spring. As a result, the required force for a given extension will be four times as much as for the original bungee cord. The effective spring constant of the folded bungee cord is, therefore, four times as large as the original spring constant. Because the frequency of oscillation is proportional to the square root of the spring constant, your bouncing frequency on the folded cord will be twice what it is on the original cord.

Example 13.6   That Car Needs Shock Absorbers!

A 1 300-kg car is constructed on a frame supported by four springs. Each spring has a spring constant of 20 000 N/m. If two people riding in the car have a combined mass of 160 kg, find the frequency of free vibration of the car when it is driven over a pothole in the road.

Solution   We assume that the weight is evenly distributed; thus each spring supports one fourth of the load. The total mass supported by the springs is 1 460 kg, and therefore each spring supports 365 kg. Hence, the frequency of vibration is

$$f = \frac{1}{2\pi}\sqrt{\frac{k}{m}} = \frac{1}{2\pi}\sqrt{\frac{20\,000\text{ N/m}}{365\text{ kg}}} = \boxed{1.18\text{ Hz}}$$

EXERCISE   How long does it take the car to execute three complete periods?

ANSWER   2.54 s

13.5   POSITION, VELOCITY, AND ACCELERATION AS A FUNCTION OF TIME

We can obtain an expression for the position of an object moving with simple harmonic motion as a function of time by returning to the relationship between simple harmonic motion and uniform circular motion. Again, consider a ball on the

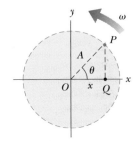

rim of a rotating turntable of radius A, as in Figure 13.12. We shall refer to the circle made by the ball as the *reference circle* for the motion. Let us assume that the turntable revolves at a *constant* angular speed ω. As the ball rotates on the reference circle, the angle θ made by the line OP with the x axis changes with time. Furthermore, as the ball rotates, the projection of P on the x axis, labeled point Q, moves back and forth along the axis with simple harmonic motion.

From the right triangle OPQ we see that $\cos \theta = x/A$. Therefore, the x coordinate of the ball is

$$x = A \cos \theta$$

Because the ball rotates with constant angular speed, it follows that $\theta = \omega t$ (see Chapter 7). Therefore,

$$x = A \cos \omega t \qquad \text{[13.12]}$$

In one complete revolution, the ball rotates through an angle of 2π rad in the period T. In other words, the motion repeats itself every T seconds. Therefore,

$$\omega = \frac{\Delta \theta}{\Delta t} = \frac{2\pi}{T} = 2\pi f \qquad \text{[13.13]}$$

where f is the frequency of the motion. The angular speed of the ball as it moves around the reference circle is the same as the angular frequency of the projected simple harmonic motion. Consequently, Equation 13.12 can be written

$$x = A \cos(2\pi ft) \qquad \text{[13.14]}$$

This equation represents the position of an object moving with simple harmonic motion as a function of time; it is graphed in Figure 13.13a. The curve should be familiar to you from trigonometry. Its shape is called sinusoidal. Note that x varies between A and $-A$ because the cosine function varies between 1 and -1.

Figures 13.13b and 13.13c represent curves for velocity and acceleration as a function of time. An end-of-chapter problem will ask you to show that the velocity and acceleration are sinusoidal functions of time. Note that when x is a maximum or minimum, the velocity is zero, and when x is zero, the magnitude of the velocity is a maximum. Furthermore, when x has its maximum positive value, the acceleration is a maximum but in the negative x direction, and when x is at its maximum negative position, the acceleration has its maximum value in the positive direction. These curves are consistent with our earlier discussion of the points at which v and a reach their maximum, minimum, and zero values.

FIGURE 13.12 A reference circle. As the ball at P rotates in a circle with uniform angular speed, its projection Q along the x axis moves with simple harmonic motion.

Tip 13.3 WHERE'S THE TRIANGLE?

Equation 13.14 includes a trigonometric function. Trigonometric functions are normally used to analyze right triangles. In this situation there are no triangles. This is because trigonometric functions are simply mathematical functions that can be used whether they refer to triangles or not. In this case, the cosine function has the correct behavior to represent position in simple harmonic motion.

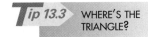

FIGURE 13.13 (a) Displacement, (b) velocity, and (c) acceleration versus time for an object moving with simple harmonic motion under the initial conditions that $x_0 = A$ and $v_0 = 0$ at $t = 0$.

Motion
of paper

FIGURE 13.14 An experimental apparatus for demonstrating simple harmonic motion. A pen attached to the oscillating object traces out a sinusoidal wave on the moving chart paper.

Figure 13.14 illustrates one experimental arrangement that demonstrates the sinusoidal nature of simple harmonic motion. An object connected to a spring has a marking pen attached to it. While the object vibrates vertically, a sheet of paper is moved horizontally with constant speed. The pen traces out a sinusoidal pattern.

Quick Quiz 13.4   If the amplitude of a system moving in simple harmonic motion is doubled, which of the following quantities does not change? (a) total energy, (b) maximum speed, (c) maximum acceleration, (d) period.

Example 13.7   The Vibrating Object-Spring System

A Find the amplitude, frequency, and period of motion for an object vibrating at the end of a spring if the equation for its position as a function of time is

$$x = (0.25 \text{ m})\cos\left(\frac{\pi}{8.0}\, t\right)$$

Solution   We can find two of our unknowns by comparing this equation with the general equation for such motion:

$$x = A\cos(2\pi ft)$$

By comparison, we see that

$$A = \boxed{0.25 \text{ m}}$$

$$2\pi f = \frac{\pi}{8.0} \text{ rad/s}$$

$$f = \boxed{\tfrac{1}{16}} \text{ Hz}$$

Because the period $T = 1/f$, it follows that $T = 1/f = \boxed{16 \text{ s}}$.

B What is the position of the object after 1.0 s has elapsed?

Solution   Substituting $t = 1.0$ s in the given expression for x gives

$$x = (0.25 \text{ m})\cos\left(\frac{\pi}{8.0} \text{ rad}\right) = \boxed{0.23 \text{ m}}$$

Note that in evaluating the cosine function, the angle is in radians. Thus, you should either set your calculator to evaluate trigonometric functions based on radian measure or convert from radians to degrees. The angle is called the *phase* of the oscillation at that instant.

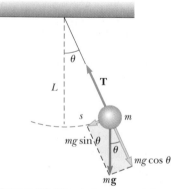

FIGURE 13.15 A simple pendulum consists of a bob of mass m suspended by a light string of length L. (L is the distance from the pivot to the center of mass of the bob.) The restoring force that causes the pendulum to undergo simple harmonic motion is the component of weight tangent to the path of motion, $mg\sin\theta$.

13.6 MOTION OF A PENDULUM

A simple pendulum is another mechanical system that exhibits periodic motion. It consists of a small bob of mass m suspended by a light string of length L fixed at its upper end, as in Figure 13.15. (By a light string, we mean that the string's mass is assumed to be very small compared to the mass of the bob and hence can be ignored.) When released, the bob swings to and fro over the same path, but is its motion simple harmonic? In order to answer this question, we must examine the force that acts as the restoring force on the pendulum. If this force is proportional to the displacement s, then the force is of the Hooke's law form, $F = -ks$, and hence the motion is simple harmonic. Furthermore, because $s = L\theta$ in this case, we see that the motion is simple harmonic if F is proportional to the angle θ.

The component of the force of gravity tangential to the circular path is the force that acts to restore the pendulum to its equilibrium position. Thus, the restoring force is

$$F_t = -mg \sin \theta$$

From this equation, we see that the restoring force is proportional to $\sin \theta$ rather than to θ. Thus, in general, the motion of a pendulum is *not* simple harmonic. However, for small angles, less than about 15 degrees, the angle θ measured in radians and the sine of the angle are approximately equal. As an example, for $\theta = 12.0° = 0.209$ rad, we find $\sin(12.0°) = 0.208$. Therefore, if we restrict the motion to small angles, the restoring force can be written as

$$F_t = -mg\theta$$

Because $s = L\theta$, we have

$$F_t = -\left(\frac{mg}{L}\right)s$$

This equation follows the general form of the Hooke's law force, given by $F = -ks$, with $k = mg/L$. Thus, we are justified in saying that a pendulum undergoes simple harmonic motion only when it swings back and forth at very small amplitudes (or, in this case, small values of θ, so that $\sin \theta \cong \theta$).

We can find the period of a pendulum by first recalling that the period of an object-spring system (Eq. 13.8) is

$$T = 2\pi \sqrt{\frac{m}{k}}$$

If we replace k with its equivalent, mg/L, we see that the period of a simple pendulum is

$$T = 2\pi \sqrt{\frac{m}{mg/L}}$$

$$T = 2\pi \sqrt{\frac{L}{g}} \qquad \text{[13.15]}$$

This equation reveals the somewhat surprising result that the period of a simple pendulum depends not on mass but only on the pendulum's length and on the free-fall acceleration. Furthermore, the amplitude of the motion is not a factor as long as it is relatively small. The analogy between the motion of a simple pendulum and the object-spring system is illustrated in Figure 13.16.

It is of historical interest to point out that it was Galileo who first noted that the period of a pendulum is independent of its amplitude. He supposedly observed this while attending church services at the cathedral in Pisa. The pendulum he studied was a swinging chandelier that was set in motion when someone bumped it while lighting candles. Galileo was able to measure its frequency, and hence its period, by timing the swings with his pulse.

The dependence of the period of a pendulum on its length and on the free-fall acceleration allows us to use a pendulum as a timekeeper for a clock. A number of clock designs employ a pendulum, with the length adjusted so that its period serves as the basis for the rate at which the clock's hands turn. Of course, these clocks are used at different locations on Earth, so there will be some variation of the free-fall acceleration. In order to compensate for this, the pendulum of a clock should have some movable mass so that the effective length can be adjusted for variations in g.

A multiflash photograph of a swinging pendulum. Is the oscillating motion simple harmonic in this case? *(Robert Mathena/Fundamental Photographs)*

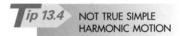

Tip 13.4 NOT TRUE SIMPLE HARMONIC MOTION

Remember that the pendulum *does not* exhibit true simple harmonic motion for *any* angle. If the angle is less than about 15°, the motion can be *modeled* as simple harmonic.

◄ The period of a simple pendulum depends only on L and g

APPLICATION

PENDULUM CLOCKS

FIGURE 13.16 Simple harmonic motion for an object-spring system and its analogy, the motion of a simple pendulum.

APPLICATION

USE OF PENDULUM IN PROSPECTING

Geologists often make use of the simple pendulum and Equation 13.15 when prospecting for oil or minerals. Deposits beneath Earth's surface can produce irregularities in the free-fall acceleration over the region being studied. A specially designed pendulum of known length is used to measure the period, which in turn is used to calculate g. Although such a measurement in itself is inconclusive, it is an important tool for geological surveys.

Quick Quiz 13.5 A simple pendulum is suspended from the ceiling of a stationary elevator, and the period is measured. When the elevator accelerates upward, what happens to the period? (a) increases, (b) decreases, or (c) remains the same. If the elevator moves with constant velocity, what happens to the period? (a) increases, (b) decreases, or (c) remains the same.

Quick Quiz 13.6 A pendulum clock depends on the period of a pendulum to keep correct time. Suppose a pendulum clock is keeping correct time and then Dennis the Menace slides the bob of the pendulum downward on the oscillating rod. Does the clock run (a) slow, (b) fast, or (c) correctly?

Quick Quiz 13.7 The period of a simple pendulum is measured on Earth to be T. If the same pendulum is taken to the Moon, its period on the Moon's surface will be (a) less than T, (b) greater than T, or (c) equal to T.

Example 13.8   What Is the Height of That Tower?

A man needs to know the height of a tower, but darkness obscures the ceiling. He knows, however, that a long pendulum extends from the ceiling almost to the floor and that its period is 12.0 s. How tall is the tower?

Solution   If we use $T = 2\pi \sqrt{L/g}$ and solve for L, we get

$$L = \frac{gT^2}{4\pi^2} = \frac{(9.80 \text{ m/s}^2)(12.0 \text{ s})^2}{4\pi^2} = \boxed{35.8 \text{ m}}$$

EXERCISE   If the length of the pendulum were halved, what would its period of vibration be?

ANSWER   8.49 s

13.7   DAMPED OSCILLATIONS

The vibrating motions we have discussed so far have taken place in ideal systems, that is, systems that *oscillate indefinitely* under the action of a linear restoring force. In all real mechanical systems, forces of friction retard the motion, and consequently the systems do not oscillate indefinitely. The friction reduces the mechanical energy of the system as time passes, and the motion is said to be **damped.**

Shock absorbers in automobiles (Fig. 13.17) make practical application of damped motion. A shock absorber consists of a piston moving through a liquid such as oil. The upper part of the shock absorber is firmly attached to the body of the car. When the car travels over a bump in the road, holes in the piston allow it to move up and down in the fluid in a damped fashion.

Damped motion varies depending on the fluid used. For example, if the fluid has a relatively low viscosity, the vibrating motion is preserved, but the amplitude of vibration decreases in time and the motion ultimately ceases. This is known as *underdamped* oscillation, and the position-time curve for an object undergoing underdamped oscillation appears in Figure 13.18. If the fluid viscosity is increased, the object returns rapidly to equilibrium after it is released and does not oscillate.

APPLICATION

SHOCK ABSORBERS

(a)   (b)

FIGURE 13.17   (a) A shock absorber consists of a piston oscillating in a chamber filled with oil. As the piston oscillates, the oil is squeezed through holes between the piston and the chamber, causing a damping of the piston's oscillations. (b) One type of automotive suspension system, in which a shock absorber is placed inside a coil spring at each wheel.

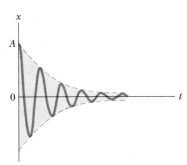

FIGURE 13.18   A graph of displacement versus time for an underdamped oscillator. Note the decrease in amplitude with time.

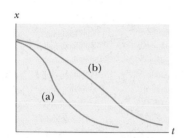

FIGURE 13.19   Plots of displacement versus time for (a) a critically damped oscillator and (b) an overdamped oscillator.

In this case, the system is said to be *critically damped* (Fig. 13.19a), and the piston returns to the equilibrium position in the shortest time possible without once overshooting the equilibrium position. If the viscosity is made greater still, the system is said to be *overdamped*. In this case, the piston returns to equilibrium without ever passing through the equilibrium point, but the time required to reach equilibrium is greater than at critical damping, as shown by Figure 13.19b.

To make automobiles more comfortable to ride in, shock absorbers are designed to be slightly underdamped. This can be demonstrated by a sharp downward push on the hood of a car. After the applied force is removed, the body of the car oscillates a few times about the equilibrium position before returning to its fixed position.

13.8   WAVE MOTION

Most of us first saw waves when, as children, we dropped a pebble into a pool of water. The disturbance created by the pebble generates water waves, which move outward until they finally reach the edge of the pool. A wide variety of physical phenomena have wave-like characteristics. The world is full of waves: sound waves, waves on a string, earthquake waves, and electromagnetic waves, such as visible light, radio waves, television signals, and x-rays. All of these waves have as their source a vibrating object. Thus, we shall use the terminology and concepts of simple harmonic motion as we move into the study of wave motion.

In the case of sound waves, the vibrations that produce waves arise from such sources as a person's vocal cords or a plucked guitar string. The vibrations of electrons in an antenna produce radio or television waves, and the simple up-and-down motion of a hand can produce a wave on a string. Regardless of the type of wave under consideration, certain concepts are common to all varieties. In the remainder of this chapter, we focus our attention on the general properties of wave motion. In later chapters we will study specific types of waves, such as sound and electromagnetic waves.

WHAT IS A WAVE?

As we noted before, when you drop a pebble into a pool of water, the disturbance produced by the pebble excites water waves, which move away from the point at which the pebble entered the water. If you carefully examined the motion of a leaf floating near the disturbance, you would see that it moves up and down and back and forth about its original position but does not undergo any net displacement attributable to the disturbance. That is, the water wave (or disturbance) moves from one place to another, *but the water is not carried with it.*

Einstein and Infeld made these remarks about wave phenomena:

> A bit of gossip starting in Washington reaches New York very quickly [by word of mouth], even though not a single individual who takes part in spreading it travels between these two cities. There are two quite different motions involved, that of the rumor, Washington to New York, and that of the persons who spread the rumor. The wind, passing over a field of grain, sets up a wave which spreads out across the whole field. Here again we must distinguish between the motion of the wave and the motion of the separate plants, which undergo only small oscillations. The particles constituting the medium perform only small vibrations, but the whole motion is that of a progressive wave. The essentially new thing here is that for the first time we consider the motion of something which is not matter, but energy propagated through matter.[1]

---

[1] A. Einstein and L. Infeld, *The Evolution of Physics*, New York, Simon and Schuster, 1961.

When we observe what is called a water wave, what we see is a rearrangement of the water's surface. Without the water there would be no wave. A wave traveling on a string would not exist without the string. Sound waves travel through air as a result of pressure variations from point to point. (Sound waves are discussed in Chapter 14.) The wave motion we consider in this chapter corresponds to the disturbance of a body or medium. Therefore, we can consider a wave to be *the motion of a disturbance*. We discuss electromagnetic waves in a later chapter; they do not require a medium.

The mechanical waves we discuss here require (1) some source of disturbance, (2) a medium that can be disturbed, and (3) some physical connection or mechanism through which adjacent portions of the medium can influence each other. All waves carry energy and momentum. The amount of energy transmitted through a medium and the mechanism responsible for the transport of energy differ from case to case. For instance, the energy carried by ocean waves during a storm is much greater than that carried by a sound wave generated by a single human voice.

Webnote 13.2

Seismic waves are a powerful reminder that waves transmit energy. For more on seismic waves, go to
http://www.geo.mtu.edu/UPSeis/waves.html
http://www.seismo.unr.edu/ftp/pub/louie/class/100/seismic-waves.html

APPLYING PHYSICS 13.4

Why is it important to be quiet in avalanche country? In *On Her Majesty's Secret Service*, a James Bond film from the 1960s, the bad guys try to stop the hero, who is escaping on skis, by firing a gun and causing an avalanche. Why did the avalanche happen?

Explanation The essence of a wave is the propagation of a disturbance through a medium. An impulsive sound, like the gunshot, can cause an acoustical disturbance that propagates through the air and can impact a ledge of snow that is just ready to break free to begin an avalanche. Such a disastrous event occurred in 1916 during World War I, when Austrian soldiers in the Alps were smothered by an avalanche caused by cannon fire.

FIGURE 13.20 A wave pulse traveling along a stretched rope. The shape of the pulse is approximately unchanged as it travels.

13.9 TYPES OF WAVES

One of the simplest ways to demonstrate wave motion is to flip one end of a long rope that is under tension and has its opposite end fixed, as in Figure 13.20. The bump (called a *pulse*) travels to the right with a definite speed. A disturbance of this type is called a **traveling wave**. Figure 13.20 shows the shape of the rope at three closely spaced times.

As a traveling wave pulse travels along the rope, **each segment of the rope that is disturbed moves perpendicularly to the wave motion.** Figure 13.21 illustrates this point for a particular tiny segment *P*. Never does the rope move in the direction of the wave. A traveling wave such as this, in which the particles of the disturbed medium move perpendicularly to the wave velocity, is called a **transverse wave.** Figure 13.22a illustrates the formation of transverse waves on a long spring.

In another class of waves, called **longitudinal waves, the elements of the medium undergo displacements parallel to the direction of wave motion.** Sound waves in air, for instance, are longitudinal. Their disturbance corresponds to a series of high- and low-pressure regions that travel through air or any material medium with a certain speed. A longitudinal pulse can be easily produced in a stretched spring, as in Figure 13.22b. The free end is pumped back and forth along the length of the spring. This action produces compressed and stretched regions of the coil that travel along the spring, parallel to the wave motion.

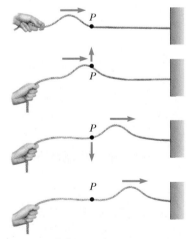

FIGURE 13.21 A pulse traveling on a stretched rope is a transverse wave. That is, any element *P* on the rope moves (blue arrows) perpendicularly to the wave motion (red arrows).

(a) Transverse wave

(b) Longitudinal wave

FIGURE 13.22   (a) A transverse wave is set up in a spring by moving one end of the spring perpendicularly to its length. (b) A longitudinal pulse along a stretched spring. The displacement of the coils is in the direction of the wave motion. For the starting motion described in the text, the compressed region is followed by a stretched region.

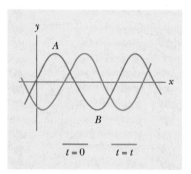

FIGURE 13.23   A one-dimensional sinusoidal wave traveling to the right with a speed v. The red curve is a snapshot of the wave at $t = 0$, and the blue curve is another snapshot at some later time t.

PICTURE OF A WAVE

Figure 13.23 shows the curved shape of a vibrating string. This pattern, sometimes called a waveform, should be familiar to you from our study of simple harmonic motion; it is a sinusoidal curve. The red curve can be thought of as a snapshot of a traveling wave taken at some instant of time, say $t = 0$; the blue curve, a snapshot of the same traveling wave at a later time. It is not difficult to imagine that this picture can as easily be used to represent a wave on water. In such a case, point A would correspond to the *crest* of the wave and point B to the low point, or *trough*, of the wave.

This same waveform can be used to describe a longitudinal wave even though no up-and-down motion is taking place. To see how this is done, consider a longitudinal wave traveling on a spring. Figure 13.24a is a snapshot of this wave at some instant, and Figure 13.24b shows the sinusoidal curve that represents the wave. Points where the coils of the spring are compressed correspond to the crests of the waveform, and stretched regions correspond to troughs.

The type of wave represented by the curve in Figure 13.24b is often referred to as a density or pressure wave. This is because the crests, where the spring coils are compressed, are regions of high density, and the troughs, where the coils are stretched, are regions of low density.

13.10   FREQUENCY, AMPLITUDE, AND WAVELENGTH

Figure 13.25 illustrates a method of producing a continuous wave or a steady stream of pulses on a very long string. One end of the string is connected to a blade that is set in vibration. As the blade oscillates vertically with simple harmonic motion, a traveling wave moving to the right is set up in the string. Figure 13.25 consists of views of the wave at intervals of one quarter of a period. Note that **each small segment of the string, such as P, oscillates vertically in the y direction with simple harmonic motion.** This must be the case, because each segment follows the simple harmonic motion of the blade. Therefore, every segment of the string can be treated as a simple harmonic oscillator vibrating with the same frequency as the blade that drives the string.

The frequencies of the waves we shall study will range from rather low values for waves on strings and waves on water, to values for sound waves between 20 Hz and 20 000 Hz (recall that 1 Hz = 1 s^{-1}), and to much higher frequencies for electromagnetic waves.

FIGURE 13.24   (a) A longitudinal wave on a spring. (b) The crests of the waveform correspond to compressed regions of the spring, and the troughs correspond to stretched regions of the spring.

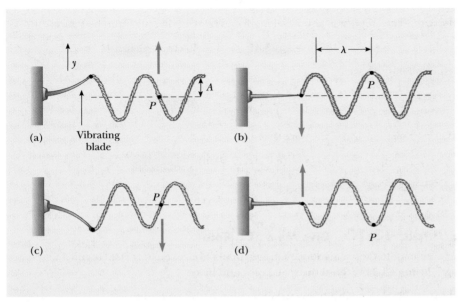

FIGURE 13.25 One method for producing traveling waves on a continuous string. The left end of the string is connected to a blade that is set in vibration. Note that every segment, such as *P*, oscillates vertically with simple harmonic motion.

The horizontal dashed line in Figure 13.25 represents the position of the string when no wave is present. The maximum distance the string moves above or below this equilibrium value is called the **amplitude *A*** of the wave. For the waves we work with, the amplitudes at the crest and the trough will be identical.

Figure 13.25b illustrates another characteristic of a wave. The horizontal arrows show the distance between two successive points that behave identically. This distance is called the **wavelength λ** (Greek letter lambda).

We can use these definitions to derive an expression for the speed of a wave. We start with the defining equation for the **wave speed *v*:**

$$v = \frac{\Delta x}{\Delta t}$$

where the wave speed is the speed at which a particular part of the wave, say a crest, moves through the medium.

A little reflection should convince you that a wave advances a distance of one wavelength in a time interval equal to one period of the vibration. Thus, taking $\Delta x = \lambda$ and $\Delta t = T$, we see that

$$v = \frac{\lambda}{T}$$

Because the frequency is equal to the reciprocal of the period, we have

$$v = f\lambda \qquad \text{[13.16]}$$

◀ Wave speed

We can apply this important general equation to many types of waves. For example, we shall use it often in our study of sound and electromagnetic waves.

Webnote 13.3

Many workers use vibrating tools or are subject to whole-body vibration, leading to occupational health issues. Read more at
http://www.18004liberty.com/research/about/1996_report/human-vibration.html
http://www.ccohs.org/oshanswers/phys_agents/vibration/vibration_measure.html

Webnote 13.4

To see interactive illustrations on the parts of a wave, visit
http://id.mind.net/~zona/mstm/physics/waves/introduction/introductionWaves.html
http://id.mind.net/~zona/mstm/physics/waves/partsOfAWave/waveParts.htm#pictureOfAWave

Example 13.9   A Traveling Wave

A wave traveling in the positive *x* direction is pictured in Figure 13.26. Find the amplitude, wavelength, period, and speed of the wave if it has a frequency of 8.0 Hz.

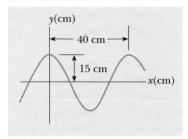

FIGURE 13.26 (Example 13.9) A harmonic wave of wavelength $\lambda = 40$ cm and amplitude $A = 15$ cm.

Solution   The amplitude and wavelength can be read directly from the figure:

$$A = 15 \text{ cm} = \boxed{0.15 \text{ m}} \qquad \lambda = 40 \text{ cm} = \boxed{0.40 \text{ m}}$$

The period of the wave is

$$T = \frac{1}{f} = \frac{1}{8.0 \text{ Hz}} = \boxed{0.13 \text{ s}}$$

and using Equation 13.16, the wave speed is

$$v = f\lambda = (8.0 \text{ Hz})(0.40 \text{ m}) = \boxed{3.2 \text{ m/s}}$$

Example 13.10   Give Me a "C" Note

The note middle C on a piano has a frequency of approximately 262 Hz and a wavelength in air of 1.31 m. Find the speed of sound in air.

Solution   By direct substitution into Equation 13.16, we find that

$$v = f\lambda = (262 \text{ s}^{-1})(1.31 \text{ m}) = \boxed{343 \text{ m/s}}$$

Example 13.11   The Speed of Radio Waves

An FM station broadcasts at a frequency of 99.9 MHz (M = mega = 10^6), with a radio wave having a wavelength of 3.00 m. Find the speed of the radio wave.

Solution   As in the last example, we use Equation 13.16:

$$v = f\lambda = (99.9 \times 10^6 \text{ s}^{-1})(3.00 \text{ m}) = \boxed{3.00 \times 10^8 \text{ m/s}}$$

This, in fact, is the speed of *any* electromagnetic wave, including light, traveling through empty space.

EXERCISE   Find the wavelength of an electromagnetic wave whose frequency is 9.0 GHz = 9.0×10^9 Hz (G = giga = 10^9), which is in the microwave range.

ANSWER   0.033 m

13.11   THE SPEED OF WAVES ON STRINGS

In this section we focus our attention on the speed of a wave on a stretched string. Rather than deriving the equation, we use dimensional analysis to verify that the expression can be valid.

It is easy to understand why wave speed depends on the tension in a string. If a string under tension is pulled sideways and released, the tension is responsible for accelerating a particular segment back toward its equilibrium position. The acceleration and wave speed increase with increasing tension in the string. Likewise, the wave speed is inversely dependent on the mass per unit length of the string. This is because it is more difficult to accelerate (and impart a large wave speed to) segments of a dense string than the lighter segments of a less dense string. Thus, wave speed is directly dependent on the tension and inversely dependent on the mass per unit length of the string. The exact relationship of the wave speed v, the tension F, and the mass per length μ is

$$v = \sqrt{\frac{F}{\mu}} \qquad\qquad \text{[13.17]}$$

From this expression we see that the speed of a mechanical wave, such as a wave on a string, depends only on the properties of the medium through which the disturbance travels.

Now let us verify that this expression is dimensionally correct. The dimensions of F are ML/T^2, and the dimensions of μ are M/L. Therefore, the dimensions of F/μ are L^2/T^2, and those of $\sqrt{F/\mu}$ are L/T, which are indeed the dimensions of speed. No other combination of F and μ is dimensionally correct. If we assume that these are the only variables relevant to the situation, Equation 13.17 seems reasonable.

Equation 13.17 indicates that we can increase the speed of a wave on a stretched string by increasing the tension in the string. It also shows that if we wrap a string with a metallic winding, as is done to the bass strings of pianos and guitars, we decrease the speed of a transmitted wave because the mass per unit length is increased.

> **APPLICATION**
>
> BASS GUITAR STRINGS

Example 13.12   A Pulse Traveling on a String

A uniform string has a mass M of 0.300 kg and a length L of 6.00 m. Tension is maintained in the string by suspending a 2.00-kg block from one end (Fig. 13.27). Find the speed of a pulse on this string.

Solution   The tension F in the string is equal to the mass m of the block multiplied by the free-fall acceleration. That is, $F = mg$. The mass per unit length of the string is $\mu = M/L$. Therefore, the speed of the pulse is

$$v = \sqrt{\frac{F}{\mu}} = \sqrt{\frac{mg}{M/L}}$$

$$= \sqrt{\frac{(2.00 \text{ kg})(9.80 \text{ m/s}^2)}{(0.300 \text{ kg})/(6.00 \text{ m})}} = \sqrt{\frac{19.6 \text{ N}}{0.050\ 0 \text{ kg/m}}} = \boxed{19.8 \text{ m/s}}$$

EXERCISE   Find the time it takes the pulse to travel from the wall to the pulley.

ANSWER   0.253 s

FIGURE 13.27   (Example 13.12) The tension F in the string is maintained by the suspended block. The wave speed v is given by the expression $v = \sqrt{F/\mu}$.

13.12   INTERFERENCE OF WAVES

Many interesting wave phenomena in nature are impossible to describe with a single moving wave. Instead, we must analyze what happens when two or more waves pass through the same region of space at the same time. We find that **two traveling waves can meet and pass through each other without being destroyed or even altered.** For instance, when two pebbles are thrown into a pond, the expanding circular waves do not destroy each other. In fact, the ripples pass through each other. Likewise, when sound waves from two sources move through air, they pass through each other. However, in the region of overlap, the resultant wave is found by adding the displacements of individual waves. For such analyses we can use the **superposition principle:**

> If two or more traveling waves are moving through a medium, the resultant wave is found by adding together the displacements of the individual waves point by point.

Experiments show that the superposition principle is valid only when the individual waves have small amplitudes of displacement—an assumption we make in all our examples.

Webnote 13.5

To see an excellent simulation showing the superposition principle, visit
http://www.phy.ntnu.edu.tw/ ~hwang/waveSuperposition/ waveSuperposition.html

(a)

(b)

(c)

FIGURE 13.28 Constructive inter-ference. If two waves having the same frequency and amplitude are in phase, as in (a) and (b), the resultant wave when they combine (c) has the same frequency as the individual waves but twice their amplitude.

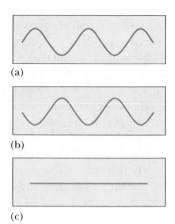

(a)

(b)

(c)

FIGURE 13.29 Destructive interfer-ence. When two waves with the same frequency and amplitude are 180° out of phase, as in (a) and (b), the result when they combine (c) is com-plete cancellation.

FIGURE 13.30 Interference pat-terns produced by outward-spreading waves from many drops of liquid falling into a body of water. *(Martin Dohrn/SPL/Photo Researchers)*

Figures 13.28a and 13.28b show two waves of the same amplitude and fre-quency. If at some instant of time these two waves attempted to travel through the same region of space, the resultant wave at that instant would have a shape like Figure 13.28c. For example, suppose these are water waves of amplitude 1 m. At the instant they overlap so that crest meets crest and trough meets trough, the re-

(a)

(b)

(c)

(d)

(e)

FIGURE 13.31 Two wave pulses traveling on a stretched string in op-posite directions pass through each other. When the pulses overlap, as in (b), (c), and (d), the net displace-ment of the string equals the sum of the displacements produced by each pulse.

(a)

(b)

(c)

(d)

(e)

FIGURE 13.32 Two wave pulses traveling in opposite directions with displacements that are inverted rela-tive to each other. When the two over-lap as in (c), their displacements sub-tract from each other.

sultant wave has an amplitude of 2 m. Waves coming together like this are said to be *in phase* and to undergo **constructive interference.**

Figures 13.29a and 13.29b show two similar waves. In this case, however, the crest of one coincides with the trough of the other; that is, one wave is *inverted* relative to the other. The resultant wave, shown in Figure 13.29c, is seen to be a state of complete cancellation. If these were water waves coming together, one of the waves would be trying to pull an individual drop of water upward at the same instant the other wave was trying to pull it downward. The result would be no motion of the water at all. In such a situation, the two waves are said to be 180° out of phase and to undergo **destructive interference.** Figure 13.30 illustrates the interference of water waves produced by drops of water falling into a pond.

Figure 13.31 shows constructive interference in two pulses moving toward each other along a stretched string; Figure 13.32 shows destructive interference in two pulses. Notice in each case that, when the two pulses separate, their shapes are unchanged, as if they had never met!

13.13 REFLECTION OF WAVES

In our discussion so far, we have assumed that the waves being analyzed could travel indefinitely without striking anything. Often, such conditions are not realized in practice. Whenever a traveling wave reaches a boundary, part or all of the wave is reflected. For example, consider a pulse traveling on a string that is fixed at one end (Fig. 13.33). When the pulse reaches the wall, it is reflected.

Note that the reflected pulse is inverted. This can be explained as follows. When the pulse meets the wall, the string exerts an upward force on the wall. According to Newton's third law, the wall must exert an equal and opposite (downward) reaction force on the string. This downward force causes the pulse to invert on reflection.

Now consider a case in which the pulse arrives at the string's end, which is attached to a ring of negligible mass that is free to slide along the post without friction (Fig. 13.34). Again the pulse is reflected, but this time it is not inverted. On

FIGURE 13.33 The reflection of a traveling wave at the fixed end of a stretched string. Note that the reflected pulse is inverted, but its shape remains the same.

FIGURE 13.34 The reflection of a traveling wave at the free end of a stretched string. In this case, the reflected pulse is not inverted.

reaching the post, the pulse exerts a force on the ring, causing it to accelerate upward. The ring is then returned to its original position by the downward component of the tension.

An alternative method of showing that a pulse is reflected without inversion when it strikes a free end of a string is to send a pulse down a string hanging vertically. When the pulse hits the free end, it is reflected without inversion, similarly to the pulse in Figure 13.34.

Finally, when a pulse reaches a boundary, it is partly reflected and partly transmitted past the boundary into the new medium. This effect is easy to observe in the case of two ropes of different density joined at some boundary.

SUMMARY

Simple harmonic motion occurs when the net force along the direction of motion is a **Hooke's law** type of force—that is, when the net force is proportional to the displacement and in the opposite direction:

$$F_s = -kx \qquad [13.1]$$

The time required for one complete vibration is called the **period** of the motion. The inverse of the period is the **frequency** of the motion, which is the number of oscillations per second.

When an object is moving with simple harmonic motion, its **acceleration** as a function of location is

$$a = -\frac{k}{m}x \qquad [13.2]$$

The energy stored in a stretched or compressed spring or other elastic material is called **elastic potential energy:**

$$PE_s \equiv \tfrac{1}{2}kx^2 \qquad [13.3]$$

The **velocity** of an object as a function of position, when the object is moving with simple harmonic motion, is

$$v = \pm\sqrt{\frac{k}{m}(A^2 - x^2)} \qquad [13.6]$$

The **period** of an object of mass m moving with simple harmonic motion while attached to a spring of spring constant k is

$$T = 2\pi\sqrt{\frac{m}{k}} \qquad [13.8]$$

where T is independent of the amplitude A.

The **frequency** of an object-spring system is $f = 1/T$. The **angular frequency** ω of the system in rad/s is

$$\omega = 2\pi f = \sqrt{\frac{k}{m}} \qquad [13.11]$$

The **position** of an object as a function of time, when the object is moving with simple harmonic motion, is

$$x = A\cos(2\pi ft) \qquad [13.14]$$

A **simple pendulum** of length L moves with simple harmonic motion for small angular displacements from the vertical, with a period of

$$T = 2\pi\sqrt{\frac{L}{g}} \qquad [13.15]$$

The period is independent of the mass of the suspended object.

A **transverse wave** is one in which the elements of the medium move perpendicularly to the direction of the wave velocity. An example is a wave on a stretched string.

A **longitudinal wave** is one in which the elements of the medium move parallel to the direction of the wave velocity. An example is a sound wave.

The relationship of the speed, wavelength, and frequency of a wave is

$$v = f\lambda \qquad [13.16]$$

The speed of a wave traveling on a stretched string of mass per unit length μ and under tension F is

$$v = \sqrt{\frac{F}{\mu}} \qquad [13.17]$$

The **superposition principle** states that if two or more traveling waves are moving through a medium, the resultant wave is found by adding the individual waves together point by point. When waves meet crest to crest and trough to trough, they undergo **constructive interference.** When crest meets trough, the waves undergo **destructive interference.**

When a wave pulse reflects from a rigid boundary, the pulse is inverted. When the boundary is free, the reflected pulse is not inverted.

CONCEPTUAL QUESTIONS

1. If one end of a heavy rope is attached to one end of a light rope, the speed of a wave will change as the wave goes from the heavy rope to the light one. Will the speed increase or decrease? What happens to the frequency? To the wavelength?

2. If a spring is cut in half, what happens to its spring constant?

3. An object-spring system undergoes simple harmonic motion with an amplitude A. Does the total energy change if the mass is doubled but the amplitude is not changed? Are the kinetic and potential energies at a given point in its motion affected by the change in mass? Explain.

4. The speed of sound in air as given in this chapter (343 m/s) is an enormous speed compared to the speed of common objects. Yet the speed of sound is of the same order of magnitude as the rms speed of air molecules at one atmosphere and 20°C as given by the kinetic theory of gases. Is this just a remarkable coincidence? Explain.

5. An object is hung on a spring and the frequency of the oscillation of the system f is measured. The object, a second identical object, and the spring are carried in the Space Shuttle to space. The two objects are attached to the ends of the spring, and the system is taken out into space on a space walk. The spring is extended, and the system is released to oscillate while floating in space. The coils of the spring do not bump into one another. What is the frequency of oscillation for this system, in terms of f?

6. If an object-spring system is hung vertically and set into oscillation, why does the motion eventually stop?

7. Is a bouncing ball an example of simple harmonic motion? Is the daily movement of a student from home to school and back simple harmonic motion?

8. If a pendulum clock keeps perfect time at the base of a mountain, will it also keep perfect time when moved to the top of the mountain? Explain.

9. A pendulum bob is made from a sphere filled with water. What would happen to the frequency of vibration of this pendulum if the sphere had a hole in it that allowed the water to leak out slowly?

10. If a grandfather clock were running slow, how could we adjust the length of the pendulum to correct the time?

11. A grandfather clock depends on the period of a pendulum to keep correct time. Suppose a grandfather clock is calibrated correctly and then the temperature of the room in which it resides increases. Does the grandfather clock run slow, fast, or correctly? (*Hint:* A material expands when its temperature increases.)

12. If you stretch a rubber hose and pluck it, you can observe a pulse traveling up and down the hose. What happens to the speed of the pulse if you stretch the hose more tightly? What happens to the speed if you fill the hose with water?

13. In a long line of people waiting to buy tickets at a movie theater, when the first person leaves, a pulse of motion occurs as people step forward to fill in the gap. The gap moves through the line of people. What determines the speed of this pulse? Is it transverse or longitudinal? How

FIGURE Q13.13 *(Gregg Adams/Stone/Getty Images)*

about the "wave" at a baseball game, where people in the stands stand up and shout as the wave arrives at their location (Fig. Q13.13), and this pulse moves around the stadium—what determines the speed of this pulse? Is it transverse or longitudinal?

14. How would you create a longitudinal wave in a stretched spring? Would it be possible to create a transverse wave in a spring?

15. In mechanics, massless strings are often assumed. Why is this not a good assumption when discussing waves on strings?

16. What happens to the wavelength of a wave on a string when the frequency is doubled? Assume that the tension in the string remains the same.

17. Explain why the kinetic and potential energies of an object-spring system can never be negative.

18. What happens to the speed of a wave on a string when the frequency is doubled? Assume that the tension in the string remains the same.

19. By what factor would you have to multiply the tension in a stretched spring in order to double the wave speed?

20. The motion of Earth going around the Sun is periodic with a period of 1 year. Is this motion simple harmonic? Explain.

PROBLEMS

1, **2**, **3** = straightforward, intermediate, challenging   ☐ = full solution available in Student Solutions Manual/Study Guide

web = solution posted at **http://info.brookscole.com/serway**   🧬 = biomedical application

Section 13.1   Hooke's Law

1. A 0.40-kg object is attached to a spring with a spring constant 160 N/m so that the object is allowed to move on a horizontal frictionless surface. The object is released from rest when the spring is compressed 0.15 m. Find (a) the force on the object and (b) its acceleration at this instant.

2. A load of 50 N attached to a spring hanging vertically stretches the spring 5.0 cm. The spring is now placed horizontally on a table and stretched 11 cm. (a) What force is required to stretch the spring by this amount? (b) Plot a graph of force (on the *y* axis) versus spring displacement from the equilibrium position along the *x* axis.

3. A ball dropped from a height of 4.00 m makes a perfectly elastic collision with the ground. Assuming no mechanical energy is lost due to air resistance, (a) show that the motion is periodic and (b) determine the period of the motion. (c) Is the motion simple harmonic? Explain.

4. The four springs of the device shown in Figure P13.4 each have a spring constant of 65.0 N/m. The hands holding the unstretched device are 43.5 cm apart. If the person cannot move his hands more than 80.0 cm apart, what is the maximum lateral force each hand can exert?

FIGURE P13.4

5. The springs 1 and 2 in Figure P13.5 have spring constants of 40.0 N/cm and 25.0 N/cm, respectively. The object *A* remains at rest, and both springs are stretched equally. Determine the stretch.

FIGURE P13.5

6. (a) The spring in Figure P13.6 has a force constant of 5.60×10^4 N/m. If the tension in either side of the rope is 210 N, how far is the spring stretched? (b) Suppose the rope is replaced by two springs which make the same V shape as the rope. If the stretch of each of these springs is twice that of the lower spring, what must be the spring constant of these springs?

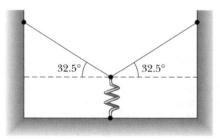

FIGURE P13.6

Section 13.2   Elastic Potential Energy

7. A slingshot consists of a light leather cup, containing a stone, that is pulled back against two parallel rubber bands. It takes a force of 15 N to stretch either one of these bands 1.0 cm. (a) What is the potential energy stored in the two bands together when a 50-g stone is placed in the cup and pulled back 0.20 m from the equilibrium position? (b) With what speed does the stone leave the slingshot?

8. An archer pulls her bow string back 0.400 m by exerting a force that increases uniformly from zero to 230 N. (a) What is the equivalent spring constant of the bow? (b) How much work is done in pulling the bow?

9. A child's toy consists of a piece of plastic attached to a
web spring (Fig. P13.9). The spring is compressed against the
floor a distance of 2.00 cm, and the toy is released. If the
toy has a mass of 100 g and rises to a maximum height of
60.0 cm, estimate the force constant of the spring.

FIGURE P13.9

10. An automobile having a mass of 1 000 kg is driven into a
brick wall in a safety test. The bumper behaves like a
spring of constant 5.00×10^6 N/m and compresses
3.16 cm as the car is brought to rest. What was the speed
of the car before impact, assuming no energy is lost during impact with the wall?

11. A simple harmonic oscillator has a total energy of *E*.
(a) Determine the kinetic and potential energies when the
displacement is one half the amplitude. (b) For what value
of the displacement does the kinetic energy equal the potential energy?

12. A 1.50-kg block at rest on a tabletop is attached to a horizontal spring having constant 19.6 N/m as in Figure
P13.12. The spring is initially unstretched. A constant
20.0-N horizontal force is applied to the object causing the
spring to stretch. (a) Determine the speed of the block after it has moved 0.300 m from equilibrium if the surface
between the block and tabletop is frictionless. (b) Answer
part (a) if the coefficient of kinetic friction between block
and tabletop is 0.200.

20.0 N

FIGURE P13.12

13. A 10.0-g bullet is fired into and embeds itself in a 2.00-kg
block attached to a spring with a spring constant of
19.6 N/m and whose mass is negligible. How far is the
spring compressed if the bullet has a speed of 300 m/s just
before it strikes the block, and the block slides on a frictionless surface? (*Note:* You must use conservation of momentum in this problem. Why?)

14. A 1.5-kg block is attached to a spring with a spring constant of 2 000 N/m. The spring is then stretched a distance of 0.30 cm and the block is released from rest.
(a) Calculate the speed of the block as it passes through
the equilibrium position if no friction is present. (b) Calculate the speed of the block as it passes through the equilibrium position if a constant frictional force of 2.0 N retards its motion. (c) What would be the strength of the
frictional force if the block reached the equilibrium position the first time with zero velocity?

Section 13.3 Velocity as a Function of Position

15. A 0.40-kg object connected to a light spring with a spring
constant of 19.6 N/m oscillates on a frictionless horizontal
surface. If the spring is compressed 4.0 cm and released
from rest, determine (a) the maximum speed of the object, (b) the speed of the object when the spring is compressed 1.5 cm, and (c) the speed of the object when the
spring is stretched 1.5 cm. (d) For what value of *x* does the
speed equal one half the maximum speed?

16. An object-spring system oscillates with an amplitude of
3.5 cm. If the spring constant is 250 N/m and the object
has a mass of 0.50 kg, determine (a) the mechanical energy of the system, (b) the maximum speed of the object,
and (c) the maximum acceleration.

17. At an outdoor market, a bunch of bananas is set into oscil-
latory motion with an amplitude of 20.0 cm on a spring
with a spring constant of 16.0 N/m. It is observed that the
maximum speed of the bunch of bananas is 40.0 cm/s.
What is the weight of the bananas in newtons?

18. A 50.0-g object is attached to a horizontal spring with a
spring constant of 10.0 N/m and released from rest with
an amplitude of 25.0 cm. What is the velocity of the object
when it is halfway to the equilibrium position if the surface
is frictionless?

Section 13.4 Comparing Simple Harmonic Motion with Uniform Circular Motion

19. While riding behind a car traveling at 3.00 m/s, you notice
that one of the car's tires has a small hemispherical bump
on its rim, as in Figure P13.19. (a) Explain why the bump,
from your viewpoint behind the car, executes simple harmonic motion. (b) If the radius of the car's tires is 0.30 m,
what is the bump's period of oscillation?

Bump

FIGURE P13.19

20. An object moves uniformly around a circular path of radius 20.0 cm, making one complete revolution every
2.00 s. What are the (a) translational speed of the object,
(b) the frequency of motion in hertz, and (c) the angular
speed of the object?

21. Consider the simplified single-piston engine in Figure
13.21. If the wheel rotates at a constant angular speed ω,
explain why the piston rod oscillates in simple harmonic
motion.

22. A 200-g object is attached to a spring and executes simple
harmonic motion with a period of 0.250 s. If the total energy of the system is 2.00 J, find (a) the force constant of
the spring and (b) the amplitude of the motion.

FIGURE P13.21

23. A spring stretches 3.9 cm when a 10-g object is hung from it. The object is replaced with a block of mass of 25 g which oscillates in simple harmonic motion. Calculate the period of motion.

24. When four people with a combined mass of 320 kg sit down in a car, they find that the car drops 0.80 cm lower on its springs. Then they get out of the car and bounce it up and down. What is the frequency of the car's vibration if its mass (empty) is 2.0×10^3 kg?

25. A 5.50-g object is suspended from a cylindrical sample of collagen 3.50 cm long and 2.00 mm in diameter. If the object vibrates up and down with a frequency of 36.0 Hz, what is the Young's modulus of the collagen?

Section 13.5   Position, Velocity, and Acceleration as a Function of Time

26. The motion of an object is described by the equation

$$x = (0.30 \text{ m}) \cos(\pi t/3)$$

Find (a) the position of the object at $t = 0$ and $t = 0.60$ s, (b) the amplitude of the motion, (c) the frequency of the motion, and (d) the period of the motion.

27. A 2.00-kg object on a frictionless horizontal track is attached to the end of a horizontal spring whose force constant is 5.00 N/m. The object is displaced 3.00 m to the right from its equilibrium position and then released, which initiates simple harmonic motion. (a) What is the force (magnitude and direction) acting on the object 3.50 s after it is released? (b) How many times does the object oscillate in 3.50 s?

28. A spring of negligible mass stretches 3.00 cm from its relaxed length when a force of 7.50 N is applied. A 0.500-kg particle rests on a frictionless horizontal surface and is attached to the free end of the spring. The particle is pulled horizontally so that it stretches the spring 5.00 cm and is then released from rest at $t = 0$. (a) What is the force constant of the spring? (b) What are the angular frequency ω, the frequency, and the period of the motion? (c) What is the total energy of the system? (d) What is the amplitude of the motion? (e) What are the maximum velocity and the maximum acceleration of the particle? (f) Determine the displacement x of the particle from the equilibrium position at $t = 0.500$ s.

29. Given that $x = A \cos \omega t$ is a sinusoidal function of time, **web** show that v (velocity) and a (acceleration) are also sinusoidal functions of time. (*Hint:* Use Equations 13.6 and 13.2.)

Section 13.6   Motion of a Pendulum

30. A man enters a tall tower, needing to know its height. He notes that a long pendulum extends from the ceiling almost to the floor and that its period is 15.5 s. (a) How tall is the tower? (b) If this pendulum is taken to the Moon, where the free-fall acceleration is 1.67 m/s², what is the period there?

31. A "seconds" pendulum is one that moves through its equilibrium position once each second. (The period of the pendulum is 2.000 s.) The length of a seconds pendulum is 0.992 7 m at Tokyo and 0.994 2 m at Cambridge, England. What is the ratio of the free-fall accelerations at these two locations?

32. An aluminum clock pendulum having a period of 1.00 s keeps perfect time at 20.0°C. (a) When placed in a room at a temperature of − 5.0°C, will it gain time or lose time? (b) How much time will it gain or lose every hour? (*Hint:* See Chapter 10.)

33. A pendulum clock that works perfectly on Earth is taken to the Moon. (a) Does it run fast or slow there? (b) If the clock is started at 12:00 midnight, what will it read after one Earth-day (24.0 h)? Assume that the free-fall acceleration on the Moon is 1.63 m/s².

34. A simple pendulum is 5.00 m long. (a) What is the period of simple harmonic motion for this pendulum if it is located in an elevator accelerating upward at 5.00 m/s²? (b) What is its period if the elevator is accelerating downward at 5.00 m/s²? (c) What is the period of simple harmonic motion for this pendulum if it is placed in a truck that is accelerating horizontally at 5.00 m/s²?

35. The free-fall acceleration on Mars is 3.7 m/s². (a) What length pendulum has a period of 1 s on Earth? What length pendulum would have a 1-s period on Mars? (b) An object is suspended from a spring with spring constant 10 N/m. Find the mass suspended from this spring that would result in a period of 1 s on Earth and on Mars.

36. Two playground swings start out together. After ten complete oscillations the swings are out of step by a half-cycle. Find the percentage difference in the length of the swings.

Section 13.10   Frequency, Amplitude, and Wavelength

37. A wave traveling in the positive x direction has a frequency of 25.0 Hz as in Figure P13.37. Find the (a) amplitude, (b) wavelength, (c) period, and (d) speed of the wave.

FIGURE P13.37

38. A bat can detect small objects such as an insect whose size is approximately equal to one wavelength of the sound the bat makes. If bats emit a chirp at a frequency of 60.0 kHz, and if the speed of sound in air is 340 m/s, what is the smallest insect a bat can detect?

39. If the frequency of oscillation of the wave emitted by an FM radio station is 88.0 MHz, determine the wave's (a) period of vibration and (b) wavelength. (Radio waves travel at the speed of light, 3.00×10^8 m/s.)

40. A piano emits sound waves with frequencies that range from a low of about 28 Hz to a high of about 4 200 Hz. Find the range of wavelengths spanned by this instrument. The speed of sound in air is approximately 343 m/s.

41. A harmonic wave is traveling along a rope. It is observed that the oscillator that generates the wave completes 40.0 vibrations in 30.0 s. Also, a given maximum travels 425 cm along the rope in 10.0 s. What is the wavelength?

42. Ocean waves are traveling to the east at 4.0 m/s with a distance of 20 m between crests. With what frequency do the waves hit the front of a boat (a) when the boat is at anchor and (b) when the boat is moving westward at 1.0 m/s?

Section 13.11 The Speed of Waves on Strings

43. A phone cord is 4.00 m long. The cord has a mass of 0.200 kg. A transverse wave pulse is produced by plucking one end of the taut cord. The pulse makes four trips down and back along the cord in 0.800 s. What is the tension in the cord?

44. A circus performer stretches a tightrope between two towers. He strikes one end of the rope and sends a wave along it toward the other tower. He notes that it takes the wave 0.800 s to reach the opposite tower, 20.0 m away. If a 1-m length of the rope has a mass of 0.350 kg, find the tension in the tightrope.

45. Transverse waves with a speed of 50.0 m/s are to be produced on a stretched string. A 5.00-m length of string with a total mass of 0.060 0 kg is used. (a) What is the required tension in the string? (b) Calculate the wave speed in the string if the tension is 8.00 N.

46. An astronaut on the Moon wishes to measure the local value of g by timing pulses traveling down a wire that has a large object suspended from it. Assume a wire of mass 4.00 g is 1.60 m long and has a 3.00-kg object suspended from it. A pulse requires 36.1 ms to traverse the length of the wire. Calculate g_{Moon} from these data. (You may neglect the mass of the wire when calculating the tension in it.)

47. A simple pendulum consists of a ball of mass 5.00 kg hanging from a uniform string of mass 0.060 0 kg and length L. If the period of oscillation for the pendulum is 2.00 s, determine the speed of a transverse wave in the string when the pendulum hangs vertically.

48. A string is 50.0 cm long and has a mass of 3.00 g. A wave travels at 5.00 m/s along this string. A second string has the same length but half the mass of the first. If the two strings are under the same tension, what is the speed of a wave along the second string?

49. Tension is maintained in a string as in Figure P13.49. The observed wave speed is 24 m/s when the suspended mass is 3.0 kg. (a) What is the mass per unit length of the string? (b) What is the wave speed when the suspended mass is 2.0 kg?

3.0 kg

FIGURE P13.49

50. The elastic limit of a piece of steel wire is 2.70×10^9 Pa. What is the maximum speed at which transverse wave pulses can propagate along this wire without exceeding this stress? (The density of steel is 7.86×10^3 kg/m³.)

51. Transverse waves travel at 20.0 m/s on a string that is under a tension of 6.00 N. What tension is required for a wave speed of 30.0 m/s in the same string?

Section 13.12 Interference of Waves

Section 13.13 Reflection of Waves

52. A series of pulses of amplitude 0.15 m are sent down a string that is attached to a post at one end. The pulses are reflected at the post and travel back along the string without loss of amplitude. What is the amplitude at a point on the string where two pulses are crossing, (a) if the string is rigidly attached to the post? (b) if the end at which reflection occurs is free to slide up and down?

53. A wave of amplitude 0.30 m interferes with a second wave of amplitude 0.20 m traveling in the same direction. What are the (a) largest and (b) smallest resultant amplitudes that can occur, and under what conditions will these maxima and minima occur?

ADDITIONAL PROBLEMS

54. The position of a 0.30-kg object attached to a spring is described by

$$x = (0.25 \text{ m})\cos(0.4\pi t)$$

Find (a) the amplitude of the motion, (b) the spring constant, (c) the position at $t = 0.30$ s, and (d) the object's speed at $t = 0.30$ s.

55. A large block P executes horizontal simple harmonic motion as it slides across a frictionless surface with a frequency $f = 1.50$ Hz. Block B rests on it, as shown in Figure P13.55, and the coefficient of static friction between the two is $\mu_s = 0.600$. What maximum amplitude of oscillation can the system have if block B is not to slip?

μ_s

B

P

FIGURE P13.55

56. A 500-g block is released from rest and slides down a frictionless track that begins 2.00 m above the horizontal, as shown in Figure P13.56. At the bottom of the track, where the surface is horizontal, the block strikes and sticks to a light spring with a spring constant of 20.0 N/m. Find the maximum distance the spring is compressed.

500 g

2.00 m

k

FIGURE P13.56

57. A 3.00-kg object is fastened to a light spring with the intervening cord passing over a pulley (Fig. P13.57). The pulley is frictionless, and its inertia may be neglected. The object is released from rest when the spring is unstretched. If the object drops 10.0 cm before stopping, find (a) the spring constant of the spring and (b) the speed of the object when it is 5.00 cm below its starting point.

web

3.00 kg

k

FIGURE P13.57

58. A 5.00-g bullet moving with an initial speed of 400 m/s is fired into and passes through a 1.00-kg block, as in Figure P13.58. The block, initially at rest on a frictionless horizontal surface, is connected to a spring with a spring constant of 900 N/m. If the block moves 5.00 cm to the right after impact, find (a) the speed at which the bullet emerges from the block and (b) the mechanical energy lost in the collision.

59. A 25-kg block is connected to a 30-kg block by a light string that passes over a frictionless pulley. The 30-kg block is connected to a light spring of force constant 200 N/m, as in Figure P13.59. The spring is unstretched when the system is as shown in the figure, and the incline is smooth. The 25-kg block is pulled 20 cm down the incline (so that the 30-kg block is 40 cm above the floor) and is released from rest. Find the speed of each block when the 30-kg block is 20 cm above the floor (that is, when the spring is unstretched).

400 m/s

5.00 cm

v

FIGURE P13.58

25 kg

30 kg

20 cm

40°

FIGURE P13.59

60. A 30.0-m steel wire and a 20.0-m copper wire, both with 1.00-mm diameters, are connected end to end and stretched to a tension of 150 N. How long does it take a transverse wave to travel the entire length of the two wires?

61. A 2.00-kg block hangs without vibrating at the end of a spring ($k = 500$ N/m) that is attached to the ceiling of an elevator car. The car is rising with an upward acceleration of $g/3$ when the acceleration suddenly ceases (at $t = 0$). (a) What is the angular frequency of oscillation of the block after the acceleration ceases? (b) By what amount is the spring stretched during the time that the elevator car is accelerating? This distance will be the amplitude of the ensuing oscillation of the block.

62. An object of mass m is connected to two rubber bands of length L, each under tension F, as in Figure P13.62. The object is displaced vertically by a small distance y. Assuming the tension does not change, show that (a) the restoring force is $-(2F/L)y$ and (b) the system exhibits simple harmonic motion with an angular frequency $\omega = \sqrt{2F/mL}$.

y

L   L

FIGURE P13.62

63. A light balloon filled with helium of density 0.180 kg/m^3 is tied to a light string of length $L = 3.00$ m. The string is tied to the ground, forming an "inverted" simple pendulum (Fig. P13.63a). If the balloon is displaced slightly from equilibrium, as in Figure P13.63b, show that the motion is simple harmonic, and determine the period of the

motion. Take the density of air to be 1.29 kg/m³. (*Hint:* Use an analogy with the simple pendulum discussed in the text, and see Chapter 9.)

(a)

(b)

FIGURE P13.63

64. A light string of mass 10.0 g and length $L = 3.00$ m has its ends tied to two walls that are separated by the distance $D = 2.00$ m. Two objects, each of mass $m = 2.00$ kg, are suspended from the string as in Figure P13.64. If a wave pulse is sent from point A, how long does it take to travel to point B?

FIGURE P13.64

65. Assume that a hole is drilled through the center of Earth. It can be shown that an object of mass m at a distance r from Earth's center is pulled toward the center of the Earth only by the material in the shaded portion of Figure P13.65. Assume Earth has a uniform density ρ. Write down Newton's law of gravitation for an object at distance r from the center of Earth, and show that the force on it is of Hooke's law form, $F = -kr$, with an effective force constant of $k = \frac{4}{3}\pi\rho Gm$, where G is the gravitational constant.

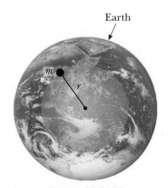

FIGURE P13.65

66. A 60.0-kg firefighter slides down a pole while a constant frictional force of 300 N retards his motion. A horizontal

20.0-kg platform is supported by a spring at the bottom of the pole to cushion the fall. The firefighter starts from rest 5.00 m above the platform, and the spring constant is 2 500 N/m. Find (a) the firefighter's speed just before he collides with the platform and (b) the maximum distance the spring is compressed. Assume that the frictional force acts during the entire motion. (*Hint:* The collision between the firefighter and the platform is perfectly inelastic.)

67. An object of mass $m_1 = 9.0$ kg is in equilibrium while connected to a light spring of constant $k = 100$ N/m that is fastened to a wall as in Figure P13.67a. A second object of mass $m_2 = 7.0$ kg is slowly pushed up against m_1, compressing the spring by the amount $A = 0.20$ m, as shown in Figure P13.67b. The system is then released, causing both objects to start moving to the right on the frictionless surface. (a) When m_1 reaches the equilibrium point, m_2 loses contact with m_1 (Fig. P13.67c) and moves to the right with velocity v. Determine the magnitude of v. (b) How far apart are the objects when the spring is fully stretched for the first time (Fig. P13.67d)? (*Hint:* First determine the period of oscillation and the amplitude of the m_1-spring system after m_2 loses contact with m_1.)

FIGURE P13.67

68. An 8.00-kg block travels on a rough horizontal surface and collides with a spring. The speed of the block just before the collision is 4.00 m/s. As it rebounds to the left with the spring uncompressed, the block travels at 3.00 m/s. If the coefficient of kinetic friction between the block and the surface is 0.400, determine (a) the loss in mechanical energy due to friction while the block is in contact with the spring and (b) the maximum distance the spring is compressed.

GROUP ACTIVITIES

G.1 Construct a simple pendulum by tying a metal bolt to one end of a string and taping the other end to the top of a doorframe. Adjust the length of the pendulum to be about 1.0 m and measure it precisely. Obtain the period by timing 25 complete oscillations, making sure the string always makes small angles with the vertical. Repeat the measurements for precisely measured pendulum lengths ranging from 0.4 m to 1.6 m in increments of 0.2 m. Plot the square of the period versus the length of the pendulum and measure the slope of the line best fitting your data points. Does your slope agree with that predicted by $T^2 = (4\pi^2/g)L$? What value of g do you obtain from your data?

While you have your pendulum in position, use a procedure similar to the above to verify that the period is independent of the amplitude for small angles and that the period is independent of the mass.

G.2 Attach one end of a rope (or a spring such as a "Slinky") to a wall, stretch it taut and use this system to study several aspects of wave motion discussed below.

(a) Send a pulse down the rope by striking it sharply from the side. An observer watching from the side can measure the time elapsed during 3 to 5 trips of the pulse from one end to the other. Dividing the total distance traveled by the elapsed time yields the wave speed. To get reliable results, take a number of readings and average the results.

(b) Use the setup above to test whether the initial amplitude of the pulse changes the wave speed.

(c) Have two people hold opposite ends of the rope (or spring). Then at the same instant, have both people hit the rope sharply from the side. Observe what happens when the two pulses meet. The superposition lasts only for the short time that the pulses overlap, and you must look carefully to see the effect.

(d) While using the setup above, devise a way to check whether pulses traveling toward one another pass through when they collide or reflect off one another.

(e) Tie one end of a rope to a doorknob and send a pulse down it. Observe what happens when the pulse reflects from the door. Does it return on the same side of the rope or does it invert?

G.3 A 50-g object is hanging from a spring whose unstretched length is 10 cm and whose spring constant is 2.5 N/m. In the list below are described five situations. In some of the

situations, the object is at rest and remains at rest. In other situations, at the instant described, the object is in the middle of an oscillation initiated by a person pulling the object downward 5 cm from its equilibrium position and releasing it. Ignore both air resistance and internal damping of the spring.

At the time the situation occurs, indicate whether the force vector requested is up (U), down (D), or has magnitude zero (0).

(a) The force on the object exerted by the spring when the object is at its equilibrium position and is at rest.

(b) The force on the object exerted by the spring when the object is at its equilbrium position and is moving downward.

(c) The net force on the object when it is at its equilibrium position and is moving upward.

(d) The force on the object exerted by the spring when it is at the top of its oscillation.

(e) The net force on the object when it is at the top of its oscillation.

G.4 One end of a spring is attached to a wall, and the other to a cart. The cart has a mass of 250 g, the rest length of the spring is 6 cm and it has a spring constant $k = 0.5$ N/cm. An experimenter pulls the cart to the right a distance of 2 cm and holds it there.

(a) What forces are felt by the spring? Be sure to identify the type of force and the agent causing it in each case. What is the total force on the spring?

(b) What forces are felt by the cart? Be sure to identify the type of force and the agent causing it in each case. What is the total force on the cart?

(c) If the cart is released, with what period will it oscillate?

G.5 As part of a Physics Open House, a department sets up a bungee jump from the top of their physics building. Assume that one end of the elastic band will be firmly attached to the top of the building and the other to the waist of a courageous participant. The participant will step off the edge of the building to be slowed down and brought back up by the elastic band before hitting the ground. Estimate the length and spring constant of the elastic you would recommend.

(Problem 5 is courtesy of Edward F. Redish. For more problems of this type, see http://www.physics.umd.edu/perg/)

Sound

14

The characteristic sound of any instrument is referred to as the quality of that sound. What is it about the sound from the tuba that allows us to distinguish between it and the sound from a flute? *(Patrick Ward/CORBIS)*

*S*ound waves are the most important example of longitudinal waves. In this chapter we discuss the characteristics of sound waves—how they are produced, what they are, and how they travel through matter. We then investigate what happens when sound waves interfere with each other. The insights gained in this chapter will help you understand how we hear.

14.1 PRODUCING A SOUND WAVE

Whether it conveys the shrill whine of a jet engine or the soft melodies of a crooner, any sound wave has its source in a vibrating object. Musical instruments produce sounds in a variety of ways. For example, the sound from a clarinet is produced by a vibrating reed, the sound from a drum by the vibration of the taut drum head, the sound from a piano by vibrating strings, and the sound from a singer by vibrating vocal cords.

Sound waves are longitudinal waves traveling through a medium, such as air. In order to investigate how sound waves are produced, we focus our attention on the tuning fork, a common device for producing pure musical notes. A tuning fork consists of two metal prongs, or tines, that vibrate when struck. Their vibration disturbs the air near them, as shown in Figure 14.1. (The amplitude of vibration of the tine in Figure 14.1 has been greatly exaggerated for clarity.) When a tine swings to the right, as in Figure 14.1a, the molecules in an element of air in front of its movement are forced closer together than normal. Such a region of high molecular density and high air pressure is called a **compression.** This compression moves away from the fork like a ripple on a pond. When the tine swings to the left, as in Figure 14.1b, the molecules in an element of air to the right of the tine spread apart, and the density and air pressure in this region are then lower than normal. Such a region of lower-than-normal density is called a **rarefaction** (pronounced "rare a fak′ shun".) Molecules to the right of the rarefaction in the figure move to the left. Hence, the rarefaction itself moves to the right, following the previously produced compression.

As the tuning fork continues to vibrate, a succession of compressions and rarefactions forms and spreads out from it. The resultant pattern in the air is somewhat like that pictured in Figure 14.2a. We can use a sinusoidal curve to represent a sound wave, as in Figure 14.2b. Notice that there are crests in the sinusoidal wave at the points where the sound wave has compressions and troughs where the sound wave has rarefactions. The compressions and rarefactions of the sound waves are superposed on the random thermal motion of the atoms and molecules, (discussed in Chapter 10), and so sound waves in gases travel at about the molecular rms speed.

FIGURE 14.1 A vibrating tuning fork. (a) As the right tine of the fork moves to the right, a high-density region (compression) of air is formed in front of its movement. (b) As the right tine moves to the left, a low-density region (rarefaction) of air is formed behind it. (c) A tuning fork. (© *Larry Stepanowicz/Visuals Unlimited*)

FIGURE 14.2 (a) As the tuning fork vibrates, a series of compressions and rarefactions moves outward, away from the fork. (b) The crests of the wave correspond to compressions, and the troughs correspond to rarefactions.

14.2 CHARACTERISTICS OF SOUND WAVES

As already noted, the general motion of elements of air near a vibrating object is back and forth between regions of compression and rarefaction. Back-and-forth motion of elements of the medium in the direction of the disturbance is characteristic of a longitudinal wave. **The motion of the elements of the medium in a longitudinal sound wave is back and forth along the direction in which the wave travels.** In contrast, **in a transverse wave the vibrations of the elements of the medium are at right angles to the direction of travel of the wave.**

CATEGORIES OF SOUND WAVES

Sound waves fall into three categories that cover different ranges of frequencies. **Audible waves** are longitudinal waves that lie within the range of sensitivity of the human ear, approximately 20 to 20 000 Hz. **Infrasonic waves** are longitudinal waves with frequencies below the audible range. Earthquake waves are an example. **Ultrasonic waves** are longitudinal waves with frequencies above the audible range for humans. For example, certain types of whistles produce ultrasonic waves. Some animals, such as dogs, can hear the waves emitted by these whistles, even though humans cannot.

APPLICATIONS OF ULTRASOUND

Ultrasonic waves are sound waves with frequencies greater than 20 kHz. Because of their high frequency and corresponding short wavelengths, ultrasonic waves can be used to produce images of small objects and are currently in wide use in medical applications, both as a diagnostic tool and in certain treatments. Internal organs are examined via the images produced by the reflection and absorption of ultrasonic waves. Although ultrasonic waves are far safer than x-rays, their images do not always have as much detail. Certain organs, however, such as the liver and the spleen, are invisible to x-rays but can be imaged with ultrasonic waves.

Medical workers can measure the speed of the blood flow in the body using a device called an ultrasonic flow meter, which makes use of the Doppler effect. The Doppler effect is discussed in Section 14.6. By comparing the frequency of the waves scattered by flowing blood with the incident frequency, we can obtain the flow speed.

Figure 14.3 illustrates the technique used to produce ultrasonic waves for clinical use. Electrical contacts are made to the opposite faces of a crystal, such as quartz or strontium titanate. If an alternating voltage of high frequency is applied to these contacts, the crystal vibrates at the same frequency as the applied voltage, emitting a beam of ultrasonic waves. At one time, this was how almost all headphones produced sound. This method of transforming electrical energy into mechanical energy, called the **piezoelectric effect,** is also reversible. If some external source causes the crystal to vibrate, an alternating voltage is produced across the crystal. Hence, a single crystal both generates and receives ultrasonic waves.

The primary physical principle that makes ultrasound imaging possible is the fact that a sound wave is partially reflected whenever it is incident on a boundary between two materials having different densities. If a sound wave is traveling in a material of density ρ_i and strikes a material of density ρ_t, the percentage of the incident sound wave intensity reflected, *PR*, is given by

$$PR = \left(\frac{\rho_i - \rho_t}{\rho_i + \rho_t} \right)^2 \times 100$$

This equation assumes that the incident sound wave travels perpendicularly to the boundary and that the speed of sound is approximately the same in the two

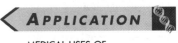

APPLICATION

MEDICAL USES OF ULTRASOUND

Direction of vibration

Electrical connections

Crystal

FIGURE 14.3 An alternating voltage applied to the faces of a piezoelectric crystal causes the crystal to vibrate.

Webnote 14.1

The March of Dimes maintains a Web site related to the obstetrical use of ultrasound:
http://www.modimes.org/ HealthLibrary2/factsheets/Ultrasound. htm

An ultrasound image of a human fetus in the womb, showing the head, right arm, and hand. *(Howard Sochurek/Woodfin Camp & Associates, Inc.)*

APPLICATION

ULTRASONIC RANGE UNIT FOR CAMERAS

Webnote 14.2

For some interesting information on autofocus cameras, go to *http://www.howstuffworks.com/ autofocus.htm*

materials. The latter assumption holds very well for the human body because the speed of sound does not vary much in the organs of the body.

Physicians commonly use ultrasonic waves to observe fetuses. This technique presents far less risk than x-rays, which deposit more energy in cells and can produce birth defects. First the abdomen of the mother is coated with a liquid, such as mineral oil. If this were not done, most of the incident ultrasonic waves from the piezoelectric source would be reflected at the boundary between the air and the mother's skin. Mineral oil has a density similar to that of skin, and a very small fraction of the incident ultrasonic wave is reflected when $\rho_i \approx \rho_t$. The ultrasound energy is emitted in pulses rather than as a continuous wave, so the same crystal can be used as a detector as well as a transmitter. An image of the fetus is obtained using an array of transducers placed on the abdomen. The reflected sound waves picked up by the transducers are converted to an electric signal, which is used to form an image on a fluorescent screen. Difficulties such as the likelihood of spontaneous abortion or of breech birth are easily detected with this technique. Also, such fetal abnormalities as spina bifida and water on the brain are readily observable.

A relatively new medical application of ultrasonics is the cavitron ultrasonic surgical aspirator (CUSA). This device has made it possible to surgically remove brain tumors that were previously inoperable. The probe of this device emits ultrasonic waves (about 23 kHz) at its tip. When the tip of the probe touches a tumor, the part of the tumor near the probe is shattered, and the residue can be sucked up (aspirated) through the hollow probe. Using this technique, neurosurgeons are able to remove brain tumors without causing serious damage to healthy surrounding tissue.

Another interesting application of ultrasound is the ultrasonic ranging unit used in some cameras to provide an almost instantaneous measurement of the distance between the camera and object to be photographed. The principal component of this device is a crystal that acts as both a loudspeaker and a microphone. A pulse of ultrasonic waves is transmitted from the transducer to the object to be photographed. The object reflects part of the signal, producing an echo that is detected by the device. The time interval between the outgoing pulse and the detected echo is then electronically converted to a distance value, because the speed of sound is a known quantity.

14.3 THE SPEED OF SOUND

The speed of a sound wave in a liquid depends on the liquid's compressibility and inertia. If the liquid has a bulk modulus B and an equilibrium density ρ, the speed of sound is

Speed of sound in a liquid ▶

$$v = \sqrt{\frac{B}{\rho}}$$
[14.1]

Recall from Chapter 9 that bulk modulus is defined as the ratio of the change in pressure ΔP to the resulting fractional change in volume $\Delta V/V$:

$$B \equiv -\frac{\Delta P}{\Delta V/V}$$
[14.2]

Note that B is always positive because an increase in pressure (positive ΔP) results in a decrease in volume. Hence the ratio $\Delta P/\Delta V$ is always negative.

It is interesting to compare Equation 14.1 with Equation 13.17 for the speed of transverse waves on a string $v = \sqrt{F/\mu}$ discussed in Chapter 13. In both cases, the wave speed depends on an elastic property of the medium (B or F) and on an inertial property of the medium (ρ or μ). In fact, the speed of all mechanical waves follows an expression of the general form

$$v = \sqrt{\frac{\text{elastic property}}{\text{inertial property}}}$$

Another example of this general form is the **speed of a longitudinal wave in a solid rod,** which is

$$v = \sqrt{\frac{Y}{\rho}}$$

[14.3]

where Y is the Young's modulus of the solid, defined as the longitudinal stress divided by the longitudinal strain (Equation 9.3), and ρ is the density of the solid.

Table 14.1 lists the speeds of sound in various media. As you can see, the speed of sound is much higher in solids than in gases. This makes sense because the molecules in a solid interact more strongly with each other than those in a gas. In general, sound travels more slowly in liquids than in solids because liquids are more compressible and hence have smaller bulk moduli. To see how sound travels faster in a solid than in air, suppose that a student places her ear against a long metal object, such as a long steel rail, and listens as her distant friend strikes the rail with a hammer. The student will hear two sounds. The first is produced by the sound traveling rapidly through the rail, and the second is the sound taking the slower path through the air.

The speed of sound also depends on the temperature of the medium. For sound traveling through air, the relationship between the speed of sound and temperature is

$$v = (331 \text{ m/s}) \sqrt{\frac{T}{273 \text{ K}}}$$

[14.4]

where 331 m/s is the speed of sound in air at 0°C and T is the absolute (Kelvin) temperature. Using this equation, we find that at 293 K, the speed of sound in air is approximately 343 m/s.

Quick Quiz 14.1 The speed of sound in air is affected by changes in (a) wavelength, (b) frequency, (c) temperature, (d) amplitude, or (e) none of these.

TABLE 14.1 Speeds of Sound in Various Media

Medium	v (m/s)
Gases	
Air (0°C)	331
Air (100°C)	386
Hydrogen (0°C)	1 290
Oxygen (0°C)	317
Helium (0°C)	972
Liquids at 25°C	
Water	1 490
Methyl alcohol	1 140
Sea water	1 530
Solids	
Aluminum	5 100
Copper	3 560
Iron	5 130
Lead	1 320
Vulcanized rubber	54

APPLYING **PHYSICS** 14.1

APPLICATION

THE SOUNDS HEARD DURING
A STORM

Why does thunder produce an extended "rolling" sound? And, how does lightning produce thunder in the first place?

Explanation   Let us assume that we are at ground level and neglect ground reflections. When lightning strikes, a channel of ionized air carries a large electric current from a cloud to the ground. This results in a rapid temperature increase of this channel of air as the current moves through it. The temperature increase causes a sudden expansion of the air. This expansion is so sudden and so intense that a tremendous disturbance — thunder — is produced in the air. The thunder rolls due to the fact that the lightning channel is a long extended source — the entire length of the channel produces the sound at essentially the same instant of time. Sound produced at the bottom of the channel reaches you first, if you are on the ground, because that is the point closest to you, and then sounds from progressively higher portions of the channel reach you. If the lightning channel were a perfectly straight line, the resulting sound might be a steady roar, but the zigzagged shape of the path results in the rolling variation in loudness, with sound from some portions of the channel arriving at your location simultaneously. The result is a particularly loud sound, interspersed with other instants when the net sound reaching you is low in intensity.

Example 14.1   Sound Waves in a Solid Bar

If a solid bar is struck at one end with a hammer, a longitudinal pulse propagates down the bar. Find the speed of sound in a bar of aluminum, which has a Young's modulus of 7.0×10^{10} Pa and a density of 2.7×10^3 kg/m³.

Solution   From Equation 14.3 we find that

$$v_{Al} = \sqrt{\frac{Y}{\rho}} = \sqrt{\frac{7.0 \times 10^{10} \text{ Pa}}{2.7 \times 10^3 \text{ kg/m}^3}} \approx \boxed{5\ 100 \text{ m/s or about } 11\ 000 \text{ mi/h!}}$$

This is a typical value for the speed of sound in solids (see Table 14.1).

14.4   ENERGY AND INTENSITY OF SOUND WAVES

As the tines of a tuning fork move back and forth through the air, they exert a force on a layer of air and cause it to move. In other words, the tines do work on the layer of air. The fact that the fork pours energy into the air as sound is one of the reasons that the vibration of the fork slowly dies out. (Other factors, such as the energy lost to friction as the tines bend, also are responsible for the diminution of movement.)

> We define the **intensity I** of a wave to be the rate at which energy flows through a unit area, A, oriented perpendicular to the direction of travel of the wave.

In equation form this is

$$I \equiv \frac{1}{A} \frac{\Delta E}{\Delta t} \qquad \text{[14.5]}$$

Equation 14.5 can be written in an alternative form if you recall that the rate of transfer of energy is defined as power. Thus,

Intensity of a wave ▶

$$I \equiv \frac{\text{power}}{\text{area}} = \frac{\mathcal{P}}{A} \qquad \text{[14.6]}$$

where $\mathscr{P}$ is the sound power passing through A, measured in watts, and the intensity has units of watts per square meter.

The faintest sounds the human ear can detect at a frequency of 1 000 Hz have an intensity of about 1×10^{-12} W/m^2. This intensity is called the **threshold of hearing.** The loudest sounds the ear can tolerate have an intensity of about 1 W/m^2 (the **threshold of pain**). At the threshold of hearing, the increase in pressure in the ear is approximately 3×10^{-5} Pa over normal atmospheric pressure. Because atmospheric pressure is about 1×10^5 Pa, this means the ear can detect pressure fluctuations as small as about 3 parts in 10^{10}! Also, at the threshold of hearing, the maximum displacement of an air molecule is about 1×10^{-11} m. This is a remarkably small number! If we compare this result with the diameter of a molecule (about 10^{-10} m), we see that the ear is an extremely sensitive detector of sound waves.

In a similar manner, we find that the loudest sounds the human ear can tolerate at 1 kHz correspond to a pressure variation of about 29 Pa away from normal atmospheric pressure, with a maximum displacement of air molecules of 1×10^{-5} m.

INTENSITY LEVEL IN DECIBELS

As was just mentioned, the human ear can detect a wide range of intensities, with the loudest tolerable sounds having intensities about 1.0×10^{12} times greater than those of the faintest detectable sounds. However, the most intense sound is not perceived as being 1.0×10^{12} times louder than the faintest sound. This is because the sensation of loudness is approximately logarithmic in the human ear. (If the word *logarithmic* strikes fear into your heart, look at Section A.3, heading G in Appendix A.) The relative intensity of a sound is called the **intensity level** or **decibel level β,** and is defined as

$$\beta \equiv 10 \log \left(\frac{I}{I_0} \right) \qquad\qquad [14.7]$$

◀ Intensity level

The constant I_0 is the reference intensity, taken to be the sound intensity at the threshold of hearing ($I_0 = 1.0 \times 10^{-12}$ W/m^2), I is any intensity, and β is the corresponding intensity level measured in decibels (dB). (The word *decibel*, which is one tenth of a *bel*, comes from the name of the inventor of the telephone, Alexander Graham Bell, 1847–1922.) On this scale, the threshold of pain ($I = 1.0$ W/m^2) corresponds to an intensity level of $\beta = 10 \log(1/1 \times 10^{-12}) = 10 \log(10^{12}) = 120$ dB. Nearby jet airplanes can create intensity levels of 150 dB, and subways and riveting machines have levels of 90–100 dB. The electronically amplified sound heard at rock concerts can be at levels of up to 120 dB, the threshold of pain. Exposure to such high intensity levels can seriously damage the ear. Earplugs are recommended whenever prolonged intensity levels exceed 90 dB. Recent evidence suggests that noise pollution, which is common in most large cities and in some industrial environments, may be a contributing factor to high blood pressure, anxiety, and nervousness. Table 14.2 gives some idea of the intensity levels of various sounds.

Webnote 14.3

To see a decibel calculator, try the following Web site. (You must be able to do calculations on your own for exams!)
http://hyperphysics.phy-astr.gsu.edu/ hbase/sound/db.html#c3

Quick Quiz 14.2 A violin plays a melody line and is then joined by nine other violins, all playing at the same intensity as the first violin, in a repeat of the same melody. (a) When all of the violins are playing together, by how many decibels does the sound level increase? (b) If ten more violins join in, by how many decibels does the sound increase over that for the single violin?

TABLE 14.2	Intensity Levels in Decibels for Different Sources
Source of Sound	**β (dB)**
Nearby jet airplane	150
Jackhammer, machine gun	130
Siren, rock concert	120
Subway, power mower	100
Busy traffic	80
Vacuum cleaner	70
Normal conversation	50
Mosquito buzzing	40
Whisper	30
Rustling leaves	10
Threshold of hearing	0

Example 14.2   Intensity Levels of Sound

Calculate the intensity level of a sound wave having an intensity of
(a) 1.0×10^{-12} W/m^2; (b) 1.0×10^{-11} W/m^2; (c) 1.0×10^{-10} W/m^2.

Solution

A For an intensity of 1.0×10^{-12} W/m^2, the intensity level, in decibels, is

$$\beta = 10 \log \left(\frac{1.0 \times 10^{-12} \text{ W/m}^2}{1.0 \times 10^{-12} \text{ W/m}^2} \right) = 10 \log(1) = \boxed{0 \text{ dB}}$$

This answer should have been obvious without calculation, because an intensity of 1.0×10^{-12} W/m^2 corresponds to the threshold of hearing.

B In this case, the intensity is exactly ten times that in part (a). The intensity level is

$$\beta = 10 \log \left(\frac{1.0 \times 10^{-11} \text{ W/m}^2}{1.0 \times 10^{-12} \text{ W/m}^2} \right) = 10 \log(10) = \boxed{10 \text{ dB}}$$

C Here the intensity is 100 times greater than at the threshold of hearing, and the intensity level is

$$\beta = 10 \log \left(\frac{1.0 \times 10^{-10} \text{ W/m}^2}{1.0 \times 10^{-12} \text{ W/m}^2} \right) = 10 \log(100) = \boxed{20 \text{ dB}}$$

Note the pattern in these answers. A sound with an intensity level of 10 dB is ten times more intense than the 0-dB sound, and a sound with an intensity level of 20 dB is 100 times more intense than a 0-dB sound. This pattern is continued throughout the decibel scale. In short, on the decibel scale **an increase of 10 dB means that the intensity of the sound is multiplied by a factor of 10.** For example, a 50-dB sound is 10 times as intense as a 40-dB sound and a 60-dB sound is 100 times as intense as a 40-dB sound.

EXERCISE Determine the intensity level of a sound wave with an intensity of 5.0×10^{-7} W/m^2.

ANSWER 57 dB

Example 14.3   A Noisy Grinding Machine

A noisy grinding machine in a factory produces a sound intensity of 1.0×10^{-5} W/m^2. Find the decibel level of this machine, and calculate the new intensity level when a second, identical machine is added to the factory.

Solution   The intensity level of the single grinder is

$$\beta = 10 \log \left(\frac{1.0 \times 10^{-5} \text{ W/m}^2}{1.0 \times 10^{-12} \text{ W/m}^2} \right) = 10 \log(10^7) = \boxed{70 \text{ dB}}$$

Adding the second grinder doubles the energy input into sound and hence doubles the intensity. The new intensity level is

$$\beta = 10 \log \left(\frac{2.0 \times 10^{-5} \text{ W/m}^2}{1.0 \times 10^{-12} \text{ W/m}^2} \right) = \boxed{73 \text{ dB}}$$

Federal regulations now demand that no office or factory worker be exposed to noise levels that average more than 90 dB over an 8-h day. The results in this example read like one of the old jokes that start "There is some good news and some bad news." First the good news. Imagine that you are a manager analyzing the noise conditions in your factory. One machine in the factory produces a noise level of 70 dB. When you add a second machine, the noise level increases by only 3 dB. Because of the logarithmic nature of intensity levels, doubling the intensity does not double the intensity level; in fact, it alters it by a surprisingly small amount. This means that additional equipment can be added to the factory without appreciably altering the intensity level of the environment.

Now the bad news. The results also work in reverse. As you remove noisy machinery, the intensity level is not lowered appreciably. For example, consider a factory with 60 machines producing a noise level of 93 dB, which is 3 dB above the maximum allowed. In order to reduce the noise level by 3 dB, half the machines would have to be removed! That is, you would have to remove 30 machines to reduce the noise level to 90 dB. To reduce the level another 3 dB, you would have to remove half of the remaining machines, and so on.

APPLICATION

OSHA NOISE LEVEL
REGULATIONS

14.5 SPHERICAL AND PLANE WAVES

If a small spherical object oscillates so that its radius changes periodically with time, a spherical sound wave is produced (Fig. 14.4). The wave moves outward from the source at a constant speed.

Because all points on the vibrating sphere behave in the same way, we conclude that the energy in a spherical wave propagates equally in all directions. That is, no one direction is preferred over any other. If $\mathcal{P}_{av}$ is the average power emitted

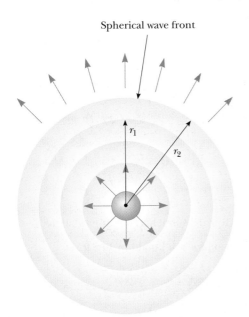

Spherical wave front

FIGURE 14.4 A spherical wave propagating radially outward from an oscillating sphere. The intensity of the spherical wave varies as $1/r^2$.

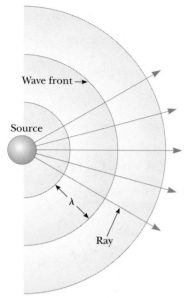

FIGURE 14.5   Spherical waves emitted by a point source. The circular arcs represent the spherical wavefronts concentric with the source. The rays are radial lines pointing outward from the source, perpendicular to the wavefronts.

Webnote 14.4

For a diagram clarifying the inverse-square behavior of sound intensity, see
http://hyperphysics.phy-astr.gsu.edu/hbase/acoustic/invsqs.html#c1

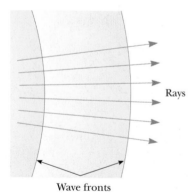

Rays

Wave fronts

FIGURE 14.6   Far away from a point source, the wavefronts are nearly parallel planes and the rays are nearly parallel lines perpendicular to the planes. Hence, a small segment of a spherical wavefront is approximately a plane wave.

by the source, then at any distance r from the source, this power must be distributed over a spherical surface of area $4\pi r^2$, assuming no absorption in the medium. (Recall that $4\pi r^2$ is the surface area of a sphere.) Hence, the **intensity** of the sound at a distance of r from the source is

$$I = \frac{\text{average power}}{\text{area}} = \frac{\mathscr{P}_{av}}{A} = \frac{\mathscr{P}_{av}}{4\pi r^2}. \qquad [14.8]$$

This shows that the intensity of a wave decreases with increasing distance from its source, as you might expect. The fact that I varies as $1/r^2$ is a result of the assumption that the small source (sometimes called a **point source**) emits a spherical wave. (In fact, light waves also obey this so-called inverse-square relationship.) Because the average power is the same through any spherical surface centered at the source, we see that the intensities at distances r_1 and r_2 (Fig. 14.4) from the center of the source are

$$I_1 = \frac{\mathscr{P}_{av}}{4\pi r_1^2} \qquad I_2 = \frac{\mathscr{P}_{av}}{4\pi r_2^2}$$

Therefore, the ratio of intensities at these two spherical surfaces is

$$\frac{I_1}{I_2} = \frac{r_2^2}{r_1^2}$$

It is useful to represent spherical waves graphically with a series of circular arcs (lines of maximum intensity) concentric with the source representing part of a spherical surface, as in Figure 14.5. We call such an arc a **wave front.** The distance between adjacent wave fronts equals the wavelength, λ. The radial lines pointing outward from the source and cutting the arcs perpendicularly are called **rays.**

Now consider a small portion of a wave front that is at a *great* distance (great relative to λ) from the source, as in Figure 14.6. In this case, the rays are nearly parallel to each other and the wave fronts are very close to being planes. Therefore, at distances from the source that are great relative to the wavelength, we can approximate the wave front with parallel planes. We call such waves plane waves. Any small portion of a spherical wave that is far from the source can be considered a **plane wave.** Figure 14.7 illustrates a plane wave propagating along the x axis. If x is taken to be the direction of the wave motion (or ray) in this figure, then the wave fronts are parallel to the plane containing the y and z axes.

Example 14.4   **Intensity Variations of a Point Source**

A small source emits sound waves with a power output of 80 W.

A Find the intensity 3.0 m from the source.

Solution   A small source emits energy in the form of spherical waves (see Fig. 14.4). Let $\mathscr{P}_{av}$ be the average power output of the source. At a distance of r from the source, the power is distributed over the surface area of a sphere, $4\pi r^2$. Therefore, the intensity at distance r from the source is given by Equation 14.8. Because $\mathscr{P}_{av} = 80$ W and $r = 3.0$ m, we find that

$$I = \frac{\mathscr{P}_{av}}{4\pi r^2} = \frac{80 \text{ W}}{4\pi(3.0 \text{ m})^2} = \boxed{0.71 \text{ W/m}^2}$$

which is close to the threshold of pain.

B At what distance would the intensity be one fourth as much as it is at $r = 3.0$ m?

Solution   Taking $I = (0.71 \text{ W/m}^2)/4$, and solving for r in Equation 14.8 gives

$$r = \left(\frac{\mathscr{P}_{av}}{4\pi I}\right)^{1/2} = \left(\frac{80 \text{ W}}{4\pi(0.71 \text{ W/m}^2)/4}\right)^{1/2} = \boxed{6.0 \text{ m}}$$

C Find the distance at which the sound level is 40 dB.

Solution We can find the intensity at the 40-dB intensity level by using Equation 14.7 with $I_0 = 1.0 \times 10^{-12}$ W/m^2:

$$40 = 10 \log(I/I_0)$$

$$4 = \log(I/I_0)$$

$$10^4 = I/I_0$$

$$I = 10^4 I_0 = 1.0 \times 10^{-8} \text{ W/m}^2$$

When this value for I is used in Equation 14.8, solving for r gives

$$r = \left(\frac{\mathscr{P}_{av}}{4\pi I}\right)^{1/2} = \left(\frac{80 \text{ W}}{4\pi \times 10^{-8} \text{ W/m}^2}\right)^{1/2} = \boxed{2.5 \times 10^4 \text{ m}}$$

which is approximately 15 mi!

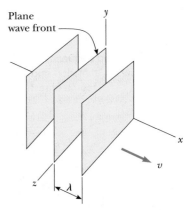

FIGURE 14.7 A representation of a plane wave moving in the positive x direction with a speed of v. The wave fronts are planes parallel to the yz plane.

14.6 THE DOPPLER EFFECT

If a car or truck is moving while its horn is blowing, the frequency of the sound you hear is higher as the vehicle approaches you and lower as it moves away from you. This is one example of the Doppler effect, named for the Austrian physicist Christian Doppler (1803–1853), who discovered it.

> In general, a **Doppler effect** is experienced whenever there is relative motion between a source of waves and an observer. When the source and observer are moving toward each other, the observer hears a frequency higher than the frequency of the source in the absence of relative motion. When the source and observer are moving away from each other, the observer hears a frequency lower than the source frequency.

Although the Doppler effect is most commonly experienced with sound waves, it is a phenomenon common to all waves. For example, the relative motion of source and observer also shifts the frequencies of light waves.

First let us consider the case in which the observer is moving and the sound source is stationary. For simplicity, we assume that the air is also stationary and that all velocity measurements are made relative to this stationary medium. Figure 14.8 describes the situation in which the observer is moving with a speed of v_O toward the source (considered a point source), which is at rest ($v_S = 0$).

We shall take the frequency of the source to be f, the wavelength to be λ, and the speed of sound in air to be v. Clearly, if both observer and source were stationary, the observer would detect f wave fronts per second. (That is, when $v_O = 0$ and $v_S = 0$, the observed frequency equals the source frequency.) When the observer is moving toward the source, he or she moves a distance of $v_O t$ in t seconds. During this interval, **the observer detects an additional number of wave fronts.** The number of extra wave fronts detected is equal to the distance traveled, $v_O t$, divided by the wavelength, λ. Thus,

$$\text{Additional wave fronts detected} = \frac{v_O t}{\lambda}$$

FIGURE 14.8 An observer moving with a speed of v_O *toward* a stationary point source (S) hears a frequency, f', that is *greater* than the source frequency, f.

The number of additional wave fronts detected *per second* is v_O/λ. Hence, the frequency f' heard by the observer is *increased* to

$$f' = f + \frac{v_O}{\lambda}$$

Substituting $\lambda = v/f$, into this expression for f' we find

Observed frequency—
observer in motion and
source at rest

$$f' = f\left(\frac{v + v_O}{v}\right) \qquad [14.9]$$

An observer traveling *away* from the source, as in Figure 14.9, **detects fewer wave fronts per second.** In this situation, we simply take v_O to be *negative* in Equation 14.9. Thus, it follows that the observed frequency in this case is *lower* than f.

In general, when an observer is moving with velocity v_O relative to a stationary source, we take v_O to be positive when the observer is moving toward the source, and negative when the observer is moving away from the source.

Now consider the situation in which the source is in motion and the observer is at rest. If the source is moving directly toward observer A in Figure 14.10a, the wave fronts heard by A are closer together because the source is moving in the direction of the outgoing wave. As a result, the wavelength λ' measured by observer A is shorter than the wavelength λ of the source at rest. During each vibration, which lasts for an interval of T (the period), the source moves a distance of $v_S T = v_S/f$ and **the wavelength is shortened by this amount.** Therefore, the observed wavelength λ' is given by

$$\lambda' = \lambda - \frac{v_S}{f}$$

Because $\lambda = v/f$, the frequency heard by observer A is

$$f' = \frac{v}{\lambda'} = \frac{v}{\lambda - \dfrac{v_S}{f}} = \frac{v}{\dfrac{v}{f} - \dfrac{v_S}{f}}$$

or

Observed frequency—
source in motion and
observer at rest

$$f' = f\left(\frac{v}{v - v_S}\right) \qquad [14.10]$$

FIGURE 14.9 An observer moving with a speed of v_O *away* from a stationary source hears a frequency, f', that is *lower* than the source frequency.

(a)

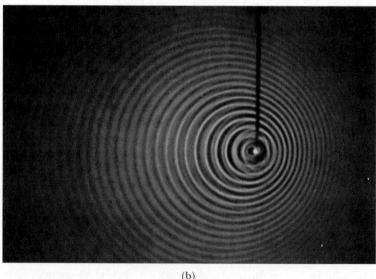

(b)

FIGURE 14.10　(a) A source S moving with speed v_S toward stationary observer A and away from stationary observer B. Observer A hears an *increased* frequency, and observer B hears a *decreased* frequency. (b) The Doppler effect in water, observed in a ripple tank. The source producing the water waves is moving to the right. (*Courtesy Educational Development Center, Newton, Mass.*)

That is, **the observed frequency increases when the source is moving toward the observer.**

In Figure 14.10a, the source is moving away from observer B, who is at rest to the left of the source. In this situation, we take v_O to be negative in Equation 14.10. Thus, observer B hears a frequency that is lower than f.

Finally, if both the source and the observer are in motion, we find the following general relationship for the observed frequency:

$$f' = f\left(\frac{v + v_O}{v - v_S}\right)$$ [14.11]

◀ Observed frequency—observer and source in motion

In this expression, the signs for the values substituted for v_O and v_S depend on the direction of the velocity. A positive value is used for motion of the observer or the source *toward* the other, and a negative sign for motion of one *away* from the other.

When working with any Doppler-effect problem, remember the following rules concerning signs. The word *toward* is associated with an *increase* in the observed frequency. The words *away from* are associated with a *decrease* in the observed frequency.

Tip 14.1 DOPPLER EFFECT DOES NOT DEPEND ON DISTANCE

The Doppler effect does not depend on the distance between the source and the observer. Although the intensity of a sound varies as the distance changes, the observed frequency does not—the frequency depends only on the speed. As you listen to an approaching source, you will hear a gradual increase in intensity but a constant frequency. As the source passes, you will hear a sudden decrease in frequency to a new constant value and a gradual decrease in intensity.

Quick Quiz 14.3

You are a passenger on a hot-air balloon that is rising with constant velocity, and you are carrying a buzzer that emits a sound of frequency f. If you accidentally drop the buzzer from the balloon, what can you conclude about the sound you hear as it falls towards the ground? (a) The frequency and intensity increase, (b) the frequency decreases and the intensity increases, (c) the frequency decreases and the intensity decreases, or (d) the frequency remains the same but the intensity decreases.

APPLYING PHYSICS 14.2

Suppose you place your stereo speakers far apart and run past them from right to left or left to right. If you run rapidly enough and have excellent pitch discrimination, you may notice that the music that is playing seems to be out of tune when you are between the speakers. Why?

Explanation When you are between the speakers, you are running away from one of them and toward the other. Thus, there is a Doppler shift downward for the sound from the speaker behind you and a Doppler shift upward for the sound from the speaker ahead of you. As a result, the sound from the two speakers will not be in tune. A calculation shows that a world-class sprint runner could run fast enough to generate about a semitone difference in the sound from the two speakers.

Example 14.5   Listen, but Don't Stand on the Track

A train moving at a speed of 40.0 m/s sounds its whistle, which has a frequency of 500 Hz. Determine the frequency heard by a stationary observer as the train approaches the observer.

Solution We can use Equation 14.10 to get the observed frequency as the train approaches the observer. Taking $v = 345$ m/s as the speed of sound in air, we have

$$f' = f\left(\frac{v}{v - v_S}\right) = (500 \text{ Hz})\left(\frac{345 \text{ m/s}}{345 \text{ m/s} - 40.0 \text{ m/s}}\right) = \boxed{566 \text{ Hz}}$$

EXERCISE   Determine the frequency heard by the stationary observer as the train recedes from the observer.

ANSWER   448 Hz

Example 14.6   The Noisy Siren

An ambulance travels down a highway at a speed of 75.0 mi/h, its siren emitting sound at a frequency of 400 Hz. What frequency is heard by a passenger in a car traveling at 55.0 mi/h in the opposite direction as the car (a) approaches? (b) moves away from the ambulance?

Solution   Let us take the velocity of sound in air to be $v = 345$ m/s and use the conversion 1.00 mi/h = 0.447 m/s. Therefore, $v_S = 75.0$ mi/h = 33.5 m/s and $v_O = 55.0$ mi/h = 24.6 m/s. We can use Equation 14.11 in both cases.

A   As the ambulance and car approach each other, the observed frequency is

$$f' = f\left(\frac{v + v_O}{v - v_S}\right) = (400 \text{ Hz})\left(\frac{345 \text{ m/s} + 24.6 \text{ m/s}}{345 \text{ m/s} - 33.5 \text{ m/s}}\right) = \boxed{475 \text{ Hz}}$$

B   As the two vehicles recede from each other, we must take $v_O = -24.6$ m/s and $v_S = -33.5$ m/s in Equation 14.11. Thus, the passenger in the car hears a frequency of

$$f' = f\left(\frac{v + v_O}{v - v_S}\right) = (400 \text{ Hz})\left[\frac{345 \text{ m/s} + (-24.6 \text{ m/s})}{345 \text{ m/s} - (-3.5 \text{ m/s})}\right] = \boxed{339 \text{ Hz}}$$

SHOCK WAVES

Now let us consider what happens when the source speed v_S *exceeds* the wave velocity v. Figure 14.11a describes this situation graphically. The circles represent spherical wave fronts emitted by the source at various times during its motion. At $t = 0$, the source is at point S_0, and at some later time t, the source is at point S_n. In the interval t, the wave front centered at S_0 reaches a radius of vt. In this same interval, the source travels to S_n, a distance of $v_S t$. At the instant the source is at S_n, the waves just beginning to be generated at this point have wave fronts of zero radius. The line drawn from S_n to the wave front centered on S_0 is tangent to all other wave fronts generated at intermediate times. All such tangent lines lie on the surface of a cone. The angle θ between one of these tangent lines and the direction of travel is given by

$$\sin \theta = \frac{v}{v_S}$$

The ratio v_S/v is referred to as the **Mach number.** The conical wave front produced when $v_S > v$ (supersonic speeds) is known as a **shock wave.** Figure 14.11b is a photograph of a bullet traveling at supersonic speed through the hot air rising above a candle. Note the shock waves in the vicinity of the bullet. Another interesting example of shock waves is the V-shaped wave front produced by a canoe (the bow wave) when the canoe's speed exceeds the speed of the water waves (Fig. 14.12).

Jet aircraft and space shuttles traveling at supersonic speeds produce shock waves, which are responsible for the loud explosion, or sonic boom, heard on the

Webnote 14.5

Take a look at an amazing and varied gallery of shock wave images. Explore *http://www.eng.vt.edu/fluids/msc/gallery/gall.htm* Start by clicking on "Shock Waves."

Webnote 14.6

To interact with an excellent animation illustrating both the Doppler effect and shock waves, visit *http://www.phy.ntnu.edu.tw/~hwang/Doppler/Doppler.html*

Conical
wave front

(a)

(b)

FIGURE 14.11 (a) A representation of a shock wave, produced when a source moves from S_0 to S_n with a speed, v_S, that is *greater* than the wave speed, v, in that medium. The envelope of the wave fronts forms a cone whose half-angle is $\sin \theta = v/v_S$. (b) A stroboscopic photograph of a bullet moving at supersonic speed through the hot air above a candle. *(Harold Edgerton, Courtesy of Palm Press, Inc.)*

ground. A shock wave carries a great deal of energy concentrated on the surface of the cone, with correspondingly great pressure variations. Shock waves are unpleasant to hear and can damage buildings when aircraft fly supersonically at low altitudes. In fact, an airplane flying at supersonic speeds produces a double boom because two shock waves are formed, one from the nose of the plane and one from the tail (Fig. 14.13).

◀ **APPLICATION**

SONIC BOOMS

Quick Quiz 14.4	As an airplane flying with constant velocity moves from a cold air mass into a warm air mass, does the Mach number (a) increase, (b) decrease, or (c) remain the same?

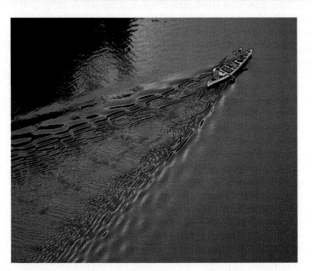

FIGURE 14.12 The V-shaped bow wave of a canoe is formed because the canoe travels at a speed greater than the speed of water waves. A bow wave is analogous to a shock wave formed by an airplane traveling faster than sound. *(© 1994, Comstock)*

FIGURE 14.13 (a) The two shock waves produced by the nose and tail of a jet airplane traveling at supersonic speed. (b) A shock wave due to a jet traveling at the speed of sound is made visible as a fog of water vapor. The large pressure variation in the shock wave causes the water in the air to condense into water droplets. *(U.S. Navy. Photo by Ensign John Gay)*

14.7 INTERFERENCE OF SOUND WAVES

Sound waves can be made to interfere with each other. This can be demonstrated with the device shown in Figure 14.14. Sound from a loudspeaker at S is sent into a tube at *P*, where there is a T-shaped junction. The sound splits and follows two separate pathways indicated by the red arrows. Half of the sound travels upward, and half downward. Finally, the two sounds merge at an opening where a listener places her ear. If the two paths r_1 and r_2 have the same length, the waves that enter the junction will separate into two halves, travel the two paths, and then combine again at the ear. This reuniting of the two waves produces constructive interference, and thus the listener hears a loud sound. If the upper path is adjusted to be one full wavelength longer than the lower path, constructive interference of the two waves occurs again, and a loud sound is detected at the receiver. In general, **if the path difference $r_2 - r_1$ is zero or some integer multiple of wavelengths, constructive interference occurs:**

◀ Condition for constructive interference

$$r_2 - r_1 = n\lambda \qquad (n = 0, 1, 2, \ldots) \qquad \text{[14.12]}$$

Suppose, however, that path length r_2 is adjusted so that the upper path is half a wavelength *longer* than the lower path r_1. In this case, an entering sound wave splits and travels the two paths as before, but now the wave along the upper path must travel a distance equivalent to half a wavelength farther than the wave traveling along the lower path. As a result, the crest of one wave meets the trough of the other when they merge at the receiver. Because this is the condition for destructive

FIGURE 14.14 **An acoustical system for demonstrating interference of sound waves.** Sound from the speaker enters the tube and splits into two parts at *P*. The two waves combine at the opposite side and are detected at *R*. The upper path length is varied by the sliding section.

interference, no sound is detected at the receiver. In general, **if the path difference $r_2 - r_1$ is $\frac{1}{2}$, $1\frac{1}{2}$, $2\frac{1}{2}$, . . . wavelengths, destructive interference occurs:**

$$r_2 - r_1 = (n + \tfrac{1}{2})\lambda \qquad (n = 0, 1, 2, \ldots) \qquad \textbf{[14.13]}$$

◀ Condition for destructive interference

Nature provides many other examples of interference phenomena. In Chapter 24 we shall describe several interesting interference effects involving light waves.

When connecting the wires between your stereo system and your loudspeakers, you may notice that the wires are color-coded and that the speakers have positive and negative signs on the connections. The reason for this is that the speakers need to be connected with the same "polarity." If they are not, then the same electrical signal fed to both speakers will result in one speaker cone moving outward at the same time that the other speaker cone is moving inward. In this case, the sound leaving the two speakers will be 180° out of phase with each other. If you are sitting midway between the speakers, the sounds from both speakers travel the same distance and preserve the phase difference they had when they left. In an ideal situation, for a 180° phase difference you would get complete destructive interference and no sound! In reality, the cancellation is not complete and is much more significant for bass notes (which have a long wavelength) than for the shorter wavelength treble notes. Nevertheless, to avoid significant reduction in the intensity of bass notes, the color-coded wires and the signs on the speaker connections should be noted carefully.

APPLICATION

CONNECTING YOUR STEREO SPEAKERS

Tip 14.2 DO WAVES REALLY INTERFERE?

In popular usage, the term *interfere* implies that an agent affects the outcome of some event. For example, in American football, *pass interference* means that a defending player has made contact with the receiver so that he is unable to catch the ball. This is very different than its use in physics, where waves pass through each other and interfere, but they do not affect each other in any way.

Example 14.7 Two Speakers Driven by the Same Source

Two speakers placed 3.00 m apart are driven by the same oscillator (Fig. 14.15). A listener is originally at point *O*, which is located 8.00 m from the center of the line connecting the two speakers. The listener then walks to point *P*, which is a perpendicular distance 0.350 m from *O*, before reaching the *first minimum* in sound intensity. What is the frequency of the oscillator?

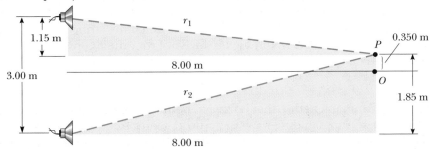

FIGURE 14.15 (Example 14.7) Two loudspeakers driven by the same source can produce interference.

Solution In order to find the frequency of the oscillator, we need to know the wavelength of the sound coming from the speakers. This information, combined with our knowledge of the speed of sound, allows us find the frequency. We can determine the wavelength from the information given in the problem about the interference of the sound waves from the speakers. The first minimum occurs when the two waves reaching the listener at *P* are 180° out of phase — in other words, when their path difference Δr equals $\lambda/2$. In order to calculate the path difference, we must first find the path lengths r_1 and r_2. Figure 14.15 shows the physical arrangement of the speakers, along with two shaded right triangles, drawn using the lengths described in the problem. Making use of these triangles in Figure 14.15, we find the path lengths to be

$$r_1 = \sqrt{(8.00 \text{ m})^2 + (1.15 \text{ m})^2} = 8.08 \text{ m}$$

$$r_2 = \sqrt{(8.00 \text{ m})^2 + (1.85 \text{ m})^2} = 8.21 \text{ m}$$

Hence, the path difference is $r_2 - r_1 = 0.13$ m. Because we require that this path difference be equal to $\lambda/2$ for the first minimum, we find that $\lambda = 0.26$ m.

To obtain the oscillator frequency, we use $v = \lambda f$, where v is the speed of sound in air, 343 m/s:

$$f = \frac{v}{\lambda} = \frac{343 \text{ m/s}}{0.26 \text{ m}} = \boxed{1.3 \text{ kHz}}$$

EXERCISE If the oscillator frequency is adjusted such that the first location at which a listener hears no sound is at a distance of 0.75 m from O, what is the new frequency?

ANSWER 0.63 kHz

14.8 STANDING WAVES

Standing waves can be set up in a stretched string by connecting one end of the string to a stationary clamp and connecting the other end to a vibrating object, such as the end of a tuning fork, or by shaking your hand up and down at a steady rate (Fig. 14.16). In this situation, traveling waves reflect from the ends, creating waves traveling in both directions on the string. The incident and reflected waves combine according to the **superposition principle.** (See section 13.12). If the string is vibrated at exactly the right frequency, the wave appears to stand still—hence its name, **standing wave.** A **node** occurs where the two traveling waves always have the same magnitude of displacement but have opposite sign, so that the net displacement is zero at that point. There is no motion in the string at the nodes but midway between two adjacent nodes, at an **antinode,** the string vibrates with the largest amplitude.

Figure 14.17 shows snapshots of the oscillation of a standing wave during half of a cycle. Notice that **all points on the string oscillate together vertically with the same frequency but that different points have different amplitudes of motion.** The nodes are stationary and the points at which the string is attached to the wall are also nodes, labeled N in Figure 14.17a. From the figure, observe that the distance between adjacent nodes is one half of the wavelength of the wave:

$$d_{NN} = \tfrac{1}{2}\lambda$$

Consider a string of length L that is fixed at both ends, as in Figure 14.18a. For a string, we can set up standing-wave patterns at many frequencies—the more loops, the higher the frequency. Three of these are pictured in Figures 14.18b, 14.18c, and 14.18d. Each has a characteristic frequency, which we will now calculate.

First, note that **the ends of the string must be nodes because these points are fixed.** If the string is displaced at its midpoint and released, the vibration shown in Figure 14.18b can be produced, in which case the center of the string is an antinode. For this standing-wave pattern, the length of the string equals $\lambda/2$ (the distance between nodes). Thus,

$$L = \frac{\lambda_1}{2} \quad \text{or} \quad \lambda_1 = 2L$$

and the frequency of this vibration is

$$f_1 = \frac{v}{\lambda_1} = \frac{v}{2L} \tag{14.14}$$

In Chapter 13 (Equation 13.17), the speed of a wave on a string was given as $v = \sqrt{F/\mu}$, where F is the tension in the string and μ is its mass per unit length.

Vibrating blade

FIGURE 14.16 Standing waves can be set up in a stretched string by connecting one end of the string to a vibrating blade. When the blade vibrates at one of the natural frequencies of the string, large-amplitude standing waves are created.

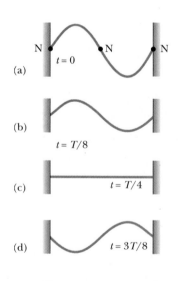

(a) $t = 0$ N N N
(b) $t = T/8$
(c) $t = T/4$
(d) $t = 3T/8$
(e) $t = T/2$

FIGURE 14.17 A standing-wave pattern in a stretched string, shown by snapshots of the string during one half of a cycle.

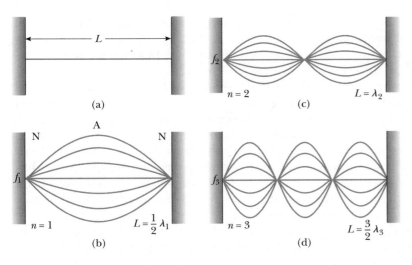

FIGURE 14.18 (a) Standing waves in a stretched string of length L, fixed at both ends. The characteristic frequencies of vibration form a harmonic series: (b) the fundamental frequency, or first harmonic; (c) the second harmonic; and (d) the third harmonic.

Thus, we can express Equation 14.14 as

$$f_1 = \frac{1}{2L}\sqrt{\frac{F}{\mu}}$$

[14.15]

This lowest frequency of vibration is called the **fundamental frequency** of the vibrating string.

The next natural frequency of vibration occurs when the length of the string equals one wavelength λ_2—that is, when $L = \lambda_2$ (Fig. 14.18c). Hence,

$$f_2 = \frac{v}{L} = \frac{2v}{2L} = 2f_1$$

[14.16]

Note that this frequency is equal to *twice* the fundamental frequency. You should convince yourself that the next highest frequency of vibration, shown in Figure 14.18d, is

$$f_3 = \frac{3v}{2L} = 3f_1$$

[14.17]

In general, the characteristic frequencies are given by

$$f_n = nf_1 = \frac{n}{2L}\sqrt{\frac{F}{\mu}}$$

[14.18]

Tip 14.3 JUMPING ROPE

If you played jump rope as a child, you set up a pattern on the rope something like that in Figure 14.18b, although the elements of the rope also exhibited a circular motion, rather than a simple up-and-down motion. If you had moved your hand faster, you could have set up a pattern something like Figure 14.18c, and you could have had two friends jumping rope—one in each of the two loops!

Multiflash photographs of standing-wave patterns in a cord driven by a vibrator at the left end. The single-loop pattern at the top left represents the fundamental ($n = 1$), the two-loop pattern at the right represents the second harmonic ($n = 2$), and the three-loop pattern at the lower left represents the third harmonic ($n = 3$).
(© *Richard Megna, Fundamental Photographs, NYC*)

where $n = 1, 2, 3, \ldots$. In other words, the frequencies are integral multiples of the fundamental frequency. The frequencies f_1, $2f_1$, $3f_1$, and so on form a **harmonic series.** The fundamental f_1 corresponds to the **first harmonic;** the frequency $f_2 = 2f_1$, to the **second harmonic;** and so on.

When a stretched string is distorted to a shape that corresponds to any one of its harmonics, after being released it vibrates only at the frequency of that harmonic. If the string is struck or bowed, however, the resulting vibration includes different amounts of various harmonics, including the fundamental. Waves not in the harmonic series are quickly damped out on a string fixed at both ends. In effect, when disturbed, the string "selects" the standing-wave frequencies. As we shall see later, the presence of several harmonics on a string gives stringed instruments their characteristic sound, which enables us to distinguish one from another even when they produce identical fundamental frequencies.

The frequency of a string on a musical instrument can be changed either by varying the tension or by changing the length. For example, the tension in guitar and violin strings is varied by turning pegs on the neck of the instrument. As the tension is increased, the frequency of the harmonic series increases according to Equation 14.18. Once the instrument is tuned, the musician varies the frequency by pressing the strings against the neck at a variety of positions, thereby changing the effective lengths of the vibrating portions of the strings. As the length is reduced, the frequency increases, as Equation 14.18 indicates.

Finally, Equation 14.18 shows that a string of fixed length can be made to vibrate at a lower fundamental frequency by increasing its mass per unit length. This is achieved in the bass strings of guitars and pianos by wrapping them with metal windings.

APPLICATION

TUNING A MUSICAL
INSTRUMENT

A concert-style harp. *(Lyon & Healy Harps, Chicago)*

Example 14.8   Harmonics of a Stretched String

Find the first four harmonics of a 1.0-m-long string if the string has a mass per unit length of 2.0×10^{-3} kg/m and is under a tension of 80 N.

Solution   The speed of the wave on the string is

$$v = \sqrt{\frac{F}{\mu}} = \sqrt{\frac{80 \text{ N}}{2.0 \times 10^{-3} \text{ kg/m}}} = 200 \text{ m/s}$$

The fundamental frequency can be found using Equation 14.14:

$$f_1 = \frac{v}{2L} = \frac{200 \text{ m/s}}{2(1.0 \text{ m})} = \boxed{100 \text{ Hz}}$$

The frequencies of the next three modes are $f_2 = 2f_1$, $f_3 = 3f_1$, and $f_4 = 4f_1$. Thus, $f_2 = \boxed{200 \text{ Hz}}$, $f_3 = \boxed{300 \text{ Hz}}$, and $f_4 = \boxed{400 \text{ Hz}}$.

EXERCISE   Find the tension in the string if the fundamental frequency is increased to 120 Hz.

ANSWER   120 N

14.9   FORCED VIBRATIONS AND RESONANCE

In Chapter 13 we learned that the energy of a damped oscillator decreases in time because of friction. It is possible to compensate for this energy loss by applying an external force that does positive work on the system.

For example, suppose an object-spring system having some natural frequency of vibration f_0 is pushed back and forth with a periodic force whose frequency is f. The system vibrates at the frequency f of the driving force. This type of motion is referred to as a **forced vibration.** Its amplitude reaches a maximum when the frequency of the driving force equals the natural frequency of the system f_0 called the **resonant frequency** of the system. Under this condition, the system is said to be in **resonance.**

In Section 14.8 we learned that a stretched string can vibrate in one or more of its natural modes. Here again, if a periodic force is applied to the string, the amplitude of vibration increases as the frequency of the applied force approaches one of the natural frequencies of vibration.

Resonance vibrations occur in a wide variety of circumstances. Figure 14.19 illustrates one experiment that demonstrates a resonance condition. Several pendulums of different lengths are suspended from a flexible beam. If one of them, such as *A*, is set in motion, the others begin to oscillate because of vibrations in the flexible beam. Pendulum *C*, the same length as *A*, oscillates with the greatest amplitude because its natural frequency matches that of pendulum *A* (the driving force).

Another simple example of resonance is a child being pushed on a swing, which is essentially a pendulum with a natural frequency that depends on the length. The swing is kept in motion by a series of appropriately timed pushes. For its amplitude to be increased, the swing must be pushed each time it returns to the person's hands. This corresponds to a frequency equal to the natural frequency of the swing. If the energy put into the system per cycle of motion exactly equals the energy lost due to friction, the amplitude remains constant.

Opera singers have been known to set crystal goblets in audible vibration with their powerful voices, as shown in Figure 14.20. This is yet another example of resonance: The sound waves emitted by the singer can set up large-amplitude vibrations in the glass. If a highly amplified sound wave has the right frequency, the amplitude of forced vibrations in the glass increases to the point where the glass becomes heavily strained and shatters.

A striking example of structural resonance occurred in 1940, when the Tacoma Narrows Bridge in the state of Washington was set in oscillation by the wind (Fig. 14.21). The amplitude of the oscillations increased rapidly and reached a high value until the bridge ultimately collapsed (probably because of metal fatigue).

A more recent example of destruction by structural resonance occurred during the Loma Prieta earthquake near Oakland, California, in 1989. In one section—almost a mile long—of the double-decker Nimitz Freeway, the upper deck collapsed onto the lower deck, killing several people. The collapse occurred because this particular section was built on mudfill, whereas other parts were built on bedrock. As seismic waves pass through mudfill or other loose soil, their wave speed decreases and the amplitude of the wave increases. Thus, the section of the freeway that collapsed oscillated at the same frequency as other sections but at a much larger amplitude.

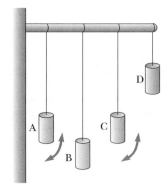

FIGURE 14.19 **Resonance.** If pendulum A is set in oscillation, only pendulum C, whose length matches that of A, will eventually oscillate with a large amplitude, or resonate. The arrows indicate motion perpendicular to the page.

◄ APPLICATION

SHATTERING GOBLETS WITH THE VOICE

◄ APPLICATION

STRUCTURAL RESONANCE IN BRIDGES AND BUILDINGS

FIGURE 14.20 *(Left)* Standing-wave pattern in a vibrating wine glass. The wine glass will shatter if the amplitude of vibration becomes too large. *(Courtesy of Professor Thomas D. Rossing, Northern Illinois University)* *(Right)* A wine glass shattered by the amplified sound of a human voice. *(Ben Rose/The IMAGE Bank)*

FIGURE 14.21 The collapse of the Tacoma Narrows suspension bridge in 1940 was a vivid demonstration of mechanical resonance. High winds set up standing waves in the bridge, causing it to oscillate at one of its natural frequencies. Once established, this resonance condition led to the bridge's collapse. *(United Press International Photo)*

14.10 STANDING WAVES IN AIR COLUMNS

Standing longitudinal waves can be set up in a tube of air, such as an organ pipe, as the result of interference between sound waves traveling in opposite directions. The relationship between the incident wave and the reflected wave depends on whether the reflecting end of the tube is open or closed. A portion of the sound wave is reflected back into the tube even at an open end. **If one end is closed, a node must exist at this end because the movement of air is restricted. If the end is open, the elements of air have complete freedom of motion, and an antinode exists.**

Figure 14.22a shows the first three modes of vibration of a pipe open at both ends. When air is directed against an edge at the left, longitudinal standing waves are formed and the pipe vibrates at its natural frequencies. Note that for the fundamental frequency the wavelength is twice the length of the pipe and hence

Tip 14.4 SOUND WAVES ARE NOT TRANSVERSE

The standing longitudinal waves in Figure 14.22 are drawn as transverse waves. This is because it is difficult to draw longitudinal displacements—they are in the same direction as the propagation. Thus, in Figure 14.22, the vertical axis represents either pressure or horizontal displacement of the elements of the medium.

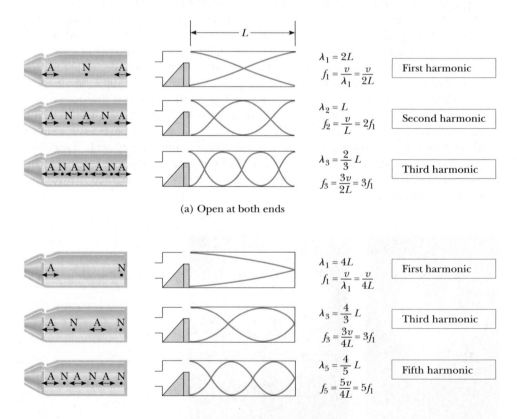

(a) Open at both ends

$\lambda_1 = 2L$
$f_1 = \dfrac{v}{\lambda_1} = \dfrac{v}{2L}$ — First harmonic

$\lambda_2 = L$
$f_2 = \dfrac{v}{L} = 2f_1$ — Second harmonic

$\lambda_3 = \dfrac{2}{3}L$
$f_3 = \dfrac{3v}{2L} = 3f_1$ — Third harmonic

(b) Closed at one end, open at the other

$\lambda_1 = 4L$
$f_1 = \dfrac{v}{\lambda_1} = \dfrac{v}{4L}$ — First harmonic

$\lambda_3 = \dfrac{4}{3}L$
$f_3 = \dfrac{3v}{4L} = 3f_1$ — Third harmonic

$\lambda_5 = \dfrac{4}{5}L$
$f_5 = \dfrac{5v}{4L} = 5f_1$ — Fifth harmonic

FIGURE 14.22 (a) Standing longitudinal waves in an organ pipe open at both ends. The natural frequencies f_1, $2f_1$, $3f_1$, . . . form a harmonic series. (b) Standing longitudinal waves in an organ pipe closed at one end. Only *odd* harmonics are present, and the natural frequencies are f_1, $3f_1$, $5f_1$, and so on.

$f_1 = v/2L$. Similarly, we find that the frequencies of the second and third harmonics are $2f_1$, $3f_1$, ... Thus, **in a pipe open at both ends, the natural frequencies of vibration form a series in which all harmonics are present. These harmonics are equal to integral multiples of the fundamental frequency.** We can express this harmonic series as

$$f_n = n\frac{v}{2L} \qquad n = 1, 2, 3, \ldots \qquad [14.19]$$

◀ Pipe open at both ends; all harmonics present

where v is the speed of sound in air.

If a pipe is closed at one end and open at the other, the closed end is a node (Fig. 14.22b). In this case, the wavelength of the fundamental mode is four times the length of the tube. Hence, $f_1 = v/4L$ and the frequencies of the third and fifth harmonics are $3f_1$, $5f_1$, ... That is, **in a pipe closed at one end and open at the other, only odd harmonics are present.** These are given by

$$f_n = n\frac{v}{4L} \qquad n = 1, 3, 5, \ldots \qquad [14.20]$$

◀ Pipe closed at one end; odd harmonics present

APPLYING PHYSICS 14.3

Passing ocean waves sometimes cause the water in a harbor to undergo very large oscillations, called a *seiche* (pronounced *saysh*). Why does this happen?

Explanation Water in a harbor is enclosed and possesses a natural frequency based on the size of the harbor. This is similar to the natural frequency of the enclosed air in a bottle, which can be excited by blowing across the edge of the opening. Ocean waves pass by the opening of the harbor at a certain frequency. If this frequency matches that of the enclosed harbor, then a large standing wave can be set up in the water by resonance. This can be simulated by carrying a fish tank with water. If your walking frequency matches the natural frequency of the water as it sloshes back and forth, a large standing wave in the fish tank can be established.

◀ **APPLICATION**

OSCILLATIONS IN A HARBOR

APPLYING PHYSICS 14.4

If an orchestra doesn't warm up before a performance, the strings go flat and the wind instruments go sharp during the performance. Why?

Explanation Without warming up, all the instruments will be at room temperature at the beginning of the concert. As the wind instruments are played, they fill with warm air from the player's exhalation. The increase in temperature of the air in the instrument causes an increase in the speed of sound, which raises the resonance frequencies of the air columns. As a result, the instruments go sharp. The strings on the stringed instruments also increase in temperature due to the friction of rubbing with the bow. This results in thermal expansion, which causes a decrease in tension in the strings. With a decrease in tension, the wave speed on the strings drops, and the fundamental frequencies decrease. Thus, the stringed instruments go flat.

◀ **APPLICATION**

WHY ARE INSTRUMENTS WARMED UP?

APPLYING PHYSICS 14.5

A bugle has no valves, keys, slides, or finger holes—how can it play a song?

Explanation Songs for the bugle are limited to harmonics of the fundamental frequency, because there is no control over frequencies without valves, keys, slides, or finger holes. The player obtains different notes by changing the tension

◀ **APPLICATION**

HOW DO BUGLES WORK?

in the lips as the bugle is played, in order to excite different harmonics. The normal playing range of a bugle is among the third, fourth, fifth, and sixth harmonics of the fundamental. For example, "Reveille" is played with just the three notes G, C, and F. And "Taps" is played with these three notes and the G one octave above the lower G.

Example 14.9   Harmonics of a Pipe

A pipe is 2.46 m long.

A Determine the frequencies of the first three harmonics if the pipe is open at both ends. Take 345 m/s as the speed of sound in air.

Solution   The fundamental frequency of a pipe open at both ends can be found from Equation 14.19, with $n = 1$.

$$f_1 = \frac{v}{2L} = \frac{345 \text{ m/s}}{2(2.46 \text{ m})} = \boxed{70.0 \text{ Hz}}$$

Because all harmonics are present in a pipe open at both ends, the second and third harmonics have frequencies of $f_2 = 2f_1 = \boxed{140 \text{ Hz}}$ and $f_3 = 3f_1 = \boxed{210 \text{ Hz}}$.

B What are the three lowest possible frequencies if the pipe is closed at one end and open at the other?

Solution   The fundamental frequency of a pipe closed at one end can be found from Equation 14.20, with $n = 1$.

$$f_1 = \frac{v}{4L} = \frac{345 \text{ m/s}}{4(2.46 \text{ m})} = \boxed{35.0 \text{ Hz}}$$

In this case, only odd harmonics are present, and so the third and fifth harmonics have frequencies of $f_3 = 3f_1 = \boxed{105 \text{ Hz}}$ and $f_5 = 5f_1 = \boxed{175 \text{ Hz}}$.

EXERCISE   If the pipe is open at one end, how many harmonics are possible in the normal hearing range, 20 to 20 000 Hz?

ANSWER   286

Example 14.10   Resonance in a Tube of Variable Length

Figure 14.23 shows a simple apparatus for demonstrating resonance in a tube. A long tube open at both ends is partially submerged in a beaker of water, and a vibrating tuning fork of unknown frequency is placed near the top of the tube. The length of the air column L is adjusted by moving the tube vertically. The sound waves generated by the

FIGURE 14.23   (Example 14.10) (a) Apparatus for demonstrating the resonance of sound waves in a tube closed at one end. The length, L, of the air column is varied by moving the tube vertically while it is partially submerged in water. (b) The first three resonances of the system.

fork are reinforced when the length of the air column corresponds to one of the resonant frequencies of the tube. The smallest value of L for which a peak occurs in the sound intensity is 9.00 cm. For this measurement, determine the frequency of the tuning fork and the value of L for the next two resonant vibrations.

Reasoning and Solution   Once the tube is in the water, this setup represents a pipe closed at one end, and the fundamental has a frequency of $v/4L$ (Fig. 14.23b). If we take $v = 345$ m/s for the speed of sound in air, and $L = 0.0900$ m, we get

$$f_1 = \frac{v}{4L} = \frac{345 \text{ m/s}}{4(0.090\ 0 \text{ m})} = \boxed{958 \text{ Hz}}$$

The fundamental wavelength of the pipe is given by $\lambda = 4L = 0.360$ m. Because the frequency of the source is constant, we see that the next resonance positions (Fig. 14.23b) correspond to lengths of $3\lambda/4 = \boxed{0.270 \text{ m}}$ and $5\lambda/4 = \boxed{0.450 \text{ m}}$.

  This arrangement is often used to measure the speed of sound, in which case the frequency of the tuning fork and the lengths at which resonance occurs must be known.

14.11 BEATS

The interference phenomena we have discussed so far involved the superposition of two or more waves with the same frequency, traveling in opposite directions. Let us now consider another type of interference effect that results from the superposition of two waves with slightly different frequencies. In this situation, the waves at some fixed point are periodically in and out of phase, corresponding to an alternation in time between constructive and destructive interference. In order to understand this phenomenon, consider Figure 14.24. The two waves in Figure 14.24a were emitted by two tuning forks having slightly different frequencies; Figure 14.24b shows the superposition of these two waves. At some time t_a, the two waves are in phase, and constructive interference occurs, as demonstrated by the resultant curve in Figure 14.24b. At some later time, however, the vibrations of the two forks move out of step with each other. At time t_b, one fork emits a compression while the other emits a rarefaction, and destructive interference occurs, as demonstrated by the curve in Figure 14.24b. As time passes, the vibrations of the two forks move out of phase, then into phase, and so on. As a consequence, a listener at some fixed point hears an alternation in loudness, known as **beats.** The number of beats per second, or beat frequency, equals the difference in frequency between the two

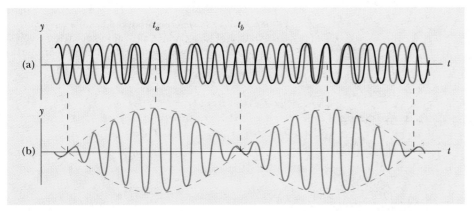

FIGURE 14.24 Beats are formed by the combination of two waves of slightly different frequencies traveling in the same direction. (a) The individual waves heard by an observer at a fixed point in space. (b) The combined wave has an amplitude (dashed line) that oscillates in time.

sources. We can therefore tune stringed instruments, such as a piano, by beating a note on the instrument against a note of known frequency. The string can then be tuned to the desired frequency by adjusting the tension until no beats are heard.

Example 14.11 Sour Notes

A particular piano string is supposed to vibrate at a frequency of 440 Hz. In order to check its frequency, a tuning fork known to vibrate at a frequency of 440 Hz is sounded at the same time the piano key is struck, and a beat frequency of 4 beats per second is heard. Find the possible frequencies at which the string could be vibrating.

Solution The number of beats per second is equal to the difference in frequency between the two sound sources. In this case, because one of the source frequencies is 440 Hz, 4 beats per second would be heard if the frequency of the string (the second source) were either $444\ \text{Hz}$ or $436\ \text{Hz}$.

14.12 QUALITY OF SOUND

The sound-wave patterns produced by most musical instruments are very complex. Figure 14.25 shows characteristic waveforms, (pressure is plotted on the vertical axis and time on the horizontal axis), produced by a tuning fork, a flute, and a clarinet, each playing the same steady note. Although each instrument has its own characteristic pattern, the figure reveals that each waveform is periodic. Note that the tuning fork produces only one harmonic (the fundamental frequency), but the two instruments emit mixtures of harmonics. Figure 14.26 graphs the harmonics of the waveforms of Figure 14.25. When this note is played on the flute (Fig. 14.26b), part of the sound consists of a vibration at the fundamental frequency, an even higher intensity is contributed by the second harmonic, the fourth harmonic produces about the same intensity as the fundamental, and so on. These sounds add together according to the principle of superposition to give the complex waveform shown. The clarinet emits a certain intensity at a frequency of the first harmonic, about half as much intensity at the frequency of the second harmonic, and so forth. The resultant superposition of these frequencies produces the pattern shown in Figure 14.25c. The tuning fork (Figs. 14.25a and 14.26a) emits sound only at the frequency of the first harmonic.

In music, the characteristic sound of any instrument is referred to as the *quality,* or *timbre,* of the sound. The quality depends on the mixture of harmonics in the sound. We say that the note C on a flute differs in quality from the same C on a

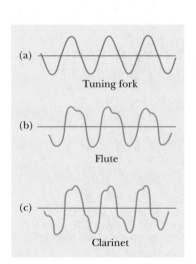

FIGURE 14.25 Waveforms produced by (a) a tuning fork, (b) a flute, and (c) a clarinet, all at approximately the same frequency. Pressure is plotted vertically and time horizontally. *(Adapted from C. A. Culver, Musical Acoustics, 4th ed., New York, McGraw-Hill, 1956)*

(a) Tuning fork

(b) Flute

(c) Clarinet

FIGURE 14.26 Harmonics of the waveforms in Figure 14.25. Note their variation in intensity. *(Adapted from C. A. Culver, Musical Acoustics, 4th ed., New York, McGraw-Hill, 1956)*

clarinet. Instruments such as the bugle, trumpet, violin, and tuba are rich in harmonics. Musicians who play wind instruments can emphasize one or another of these harmonics by changing the configuration of their lips, and can thus play different musical notes with the same valve openings.

APPLYING **PHYSICS** *14.6*

A professor performs a demonstration in which he breathes helium and then speaks with a comical voice. One student explains, "The velocity of sound in helium is higher than in air, so the fundamental frequency of the standing waves in the mouth is increased." Another student says, "No, the fundamental frequency is determined by the vocal folds and cannot be changed. Only the quality of the voice has changed." Which student is correct?

Explanation The second student is correct. The fundamental frequency of the complex tone from the voice is determined by the vibration of the vocal folds and is not changed by substituting a different gas in the mouth. The introduction of the helium into the mouth results in harmonics of higher frequencies being excited more than in the normal voice, but the fundamental frequency of the voice is the same—only the quality has changed. The unusual inclusion of the higher-frequency harmonics results in a common description of this effect as a "high-pitched" voice, but this is incorrect (it is really a "quacky" timbre).

◀ **APPLICATION**

WHY DOES THE PROFESSOR SOUND LIKE DONALD DUCK?

Each musical instrument has its own characteristic sound and mixture of harmonics. Instruments shown are (a) the violin, (b) the saxophone, and (c) the trumpet. *(Photographs courtesy of (a) © 1989 Gary Buss/FPG; (b) and (c) © 1989 Richard Laird/FPG)*

(a)   (b)   (c)

14.13 THE EAR

The human ear is divided into three regions: the outer ear, the middle ear, and the inner ear (Fig. 14.27). The *outer ear* consists of the ear canal (open to the atmosphere), that terminates at the eardrum (tympanum). Sound waves travel down the ear canal to the eardrum, which vibrates in and out in phase with the pushes and pulls caused by the alternating high and low pressures of the sound wave. Behind the eardrum are three small bones of the *middle ear,* called the hammer, the anvil, and the stirrup because of their shapes. These bones transmit the vibration to the *inner ear,* which contains the cochlea, a snail-shaped tube about 2 cm long. The cochlea makes contact with the stirrup at the oval window and is divided along its length by the basilar membrane, which consists of small hairs (cilia) and nerve fibers. This membrane varies in mass per unit length and in tension along its length, and different portions of it resonate at different frequencies. (Recall that the natural frequency of a string depends on its mass per unit length and on the tension on it.) Along the basilar membrane are numerous nerve endings, which sense the vibration of the membrane and in turn transmit impulses to the brain. The brain interprets the impulses as sounds of varying frequency, depending on the locations along the basilar membrane of the impulse-transmitting nerves and on the rates at which the impulses are transmitted. This is a simplified description of the complex functioning of the ear.

Figure 14.28 shows the frequency response curves of an average human ear for sounds of equal loudness, ranging from 0 to 120 dB. To interpret this series of graphs, take the bottom curve as the threshold of hearing. Compare the intensity level on the vertical axis for the two frequencies 100 Hz and 1 000 Hz. The vertical axis shows that the 100-Hz sound must be about 38 dB greater than the 1 000-Hz sound to be at the threshold of hearing. Thus, we see that the threshold of hearing is very strongly dependent on frequency. The easiest frequencies to hear are around 3 300 Hz; those above 12 000 Hz or below about 50 Hz must be relatively intense to be heard.

Now consider the curve labeled 80. This curve uses as its reference a 1 000-Hz tone at an intensity level of 80 dB. The curve shows that a tone of frequency 100 Hz would have to be about 4 dB louder than the 80-dB, 1 000-Hz tone in order to sound as loud. Notice that the curves flatten out as the intensity levels of the

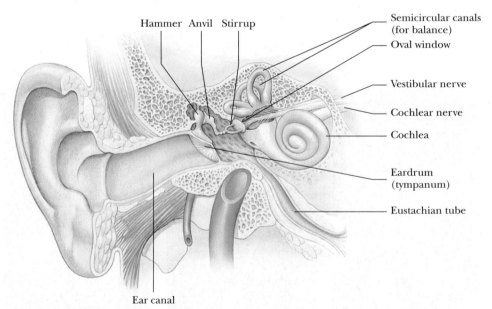

Hammer Anvil Stirrup

Semicircular canals (for balance)

Oval window

Vestibular nerve

Cochlear nerve

Cochlea

Eardrum (tympanum)

Eustachian tube

Ear canal

FIGURE 14.27 The structure of the human ear. The three tiny bones (ossicles) that connect the eardrum to the window of the cochlea act as a double-lever system to decrease the amplitude of vibration and hence increase the pressure on the fluid in the cochlea.

FIGURE 14.28 Curves of intensity level versus frequency for sounds that are perceived to be of equal loudness. Note that the ear is most sensitive at a frequency of about 3 300 Hz. The lowest curve corresponds to the threshold of hearing for only about 1% of the population.

sounds increase. That is, when sounds are loud, all frequencies can be heard equally well.

The small bones in the middle ear represent an intricate lever system that increases the force on the oval window. The pressure is greatly magnified because the surface area of the eardrum is about 20 times that of the oval window (in analogy with a hydraulic press). The middle ear, together with the eardrum and oval window, in effect acts as a matching network between the air in the outer ear and the liquid in the inner ear. The overall energy transfer between the outer ear and inner ear is highly efficient, with pressure amplification factors of several thousand. In other words, pressure variations in the inner ear are much greater than those in the outer ear.

The ear has its own built-in protection against loud sounds. The muscles connecting the three middle-ear bones to the walls control the volume of the sound by changing the tension on the bones as sound builds up, thus hindering their ability to transmit vibrations. In addition, the eardrum becomes stiffer as the sound intensity increases. These two occurrences make the ear less sensitive to loud incoming sounds. There is a time delay between the onset of loud sound and the ear's protective reaction, however, so a very sudden loud sound can still damage the ear.

The complex structure of the human ear is believed to be related to the fact that mammals have evolved from sea-going creatures. In comparison, insect ears are considerably simpler in design because they have always been land residents. A typical insect ear consists of an eardrum exposed directly to the air on one side and to an air filled cavity on the other side. Nerve cells communicate directly with the cavity and the brain without the need for the complex intermediary of an inner and middle ear. This simple design allows the ear to be placed virtually anywhere on the body. For example, a grasshopper has its ears on its legs. One advantage of the simple insect ear is that the distance and orientation of the ears can be varied so that it is easier to locate sources of sound such as other insects.

One of the most amazing medical advances in recent decades is the cochlear implant, allowing the deaf to hear. Deafness can occur when the hair-like sensors (cilia) in the cochlea break off over a lifetime or sometimes because of prolonged exposure to loud sounds. Because the cilia do not grow back, the ear loses sensitivity to certain frequencies of sound. The nerves can sometimes be stimulated electronically by a small electronic device implanted in the ear to restore this hearing loss.

Webnote 14.7

Read more about the physiology of the human ear at the following URL. Scroll to the bottom and click on "Java Animation" to see the vibration of the middle ear bones as a function of frequency.
http://www.infj.ulst.ac.uk/~pnic/ HumanEar/Andy%27s%20Stuff/ MScProject/workingcode_Local/ EarChapter.html

SUMMARY

Sound waves are longitudinal waves. **Audible waves** are sound waves with frequencies between 20 and 20 000 Hz. **Infrasonic waves** have frequencies below the audible range, and **ultrasonic waves** have frequencies above the audible range.

The speed of sound in a medium of bulk modulus B and density ρ is

$$v = \sqrt{\frac{B}{\rho}} \qquad \text{[14.1]}$$

The speed of sound also depends on the temperature of the medium. The relationship between temperature and speed for sound in air is

$$v = (331 \text{ m/s}) \sqrt{\frac{T}{273 \text{ K}}} \qquad \text{[14.4]}$$

where T is the absolute (Kelvin) temperature and 331 m/s is the speed of sound in air at 0°C.

The **intensity level** of a sound wave, in decibels, is given by

$$\beta \equiv 10 \log \left(\frac{I}{I_0} \right) \qquad \text{[14.7]}$$

The constant I_0 is a reference intensity, usually taken to be at the threshold of hearing ($I_0 = 1.0 \times 10^{-12}$ W/m^2), and I is the intensity with level β, where β is measured in **decibels** (dB).

The **intensity** of a *spherical wave* produced by a point source is proportional to the average power emitted and inversely proportional to the square of the distance from the source:

$$I = \frac{\mathcal{P}_{av}}{4\pi r^2} \qquad \text{[14.8]}$$

The change in frequency heard by an observer whenever there is relative motion between the source and observer is called the **Doppler effect.** If the observer is moving with velocity v_O and the source is at rest, the observed frequency f' is

$$f' = f \left(\frac{v + v_O}{v} \right) \qquad \text{[14.9]}$$

The sign of v_O is positive when the observer is moving toward the source, and negative when the observer is moving away from the source.

If the source is moving with velocity v_S and the observer is at rest, the observed frequency is

$$f' = f \left(\frac{v}{v - v_S} \right) \qquad \text{[14.10]}$$

The sign of v_S is positive when the source is moving toward the observer, and negative when the source is moving away from the observer.

When the observer and source are both moving, the observed frequency is

$$f' = f \left(\frac{v + v_O}{v - v_S} \right) \qquad \text{[14.11]}$$

When waves interfere, the resultant wave is found by adding the individual waves together point by point. When crest meets crest and trough meets trough, the waves undergo **constructive interference.** When crest meets trough, **destructive interference** occurs.

Standing waves are formed when two waves having the same frequency, amplitude, and wavelength travel in opposite directions through a medium. We can set up standing waves of specific frequencies in a stretched string. The natural frequencies of vibration of a stretched string of length L, fixed at both ends, are

$$f_n = \frac{n}{2L} \sqrt{\frac{F}{\mu}} \qquad n = 1, 2, 3, \ldots \qquad \text{[14.18]}$$

where F is the tension in the string and μ is its mass per unit length. The natural frequencies of vibration form a **harmonic series;** that is, the frequencies are integral multiples of the fundamental (lowest) frequency.

A system capable of oscillating is said to be in **resonance** with some driving force whenever the frequency of the driving force matches one of the natural frequencies of the system. When the system is resonating, it oscillates with maximum amplitude.

Standing waves can be produced in a tube of air. If the reflecting end of the tube is *open*, all harmonics are present and the natural frequencies of vibration are

$$f_n = n \frac{v}{2L} \qquad n = 1, 2, 3, \ldots \qquad \text{[14.19]}$$

If the tube is *closed* at the reflecting end, only the *odd* harmonics are present, and the natural frequencies of vibration are

$$f_n = n \frac{v}{4L} \qquad n = 1, 3, 5, \ldots \qquad \text{[14.20]}$$

The phenomenon of **beats** is an interference effect that occurs when two waves with slightly different frequencies combine at a fixed point in space. For sound waves, the intensity of the resultant sound changes periodically with time.

CONCEPTUAL QUESTIONS

1. To avoid overcrowding of the AM radio band, the range of audio frequencies transmitted by the radio signal of any AM radio station is limited by law to 5 000 Hz. Discuss the effect of this limitation on the quality of the sound.

2. A crude model of the human throat is that of a pipe open at both ends with a vibrating source to introduce the sound into the pipe at one end. Assuming the vibrating source produces a range of frequencies, discuss the effect of changing the pipe's length.

3. An autofocus camera sends out a pulse of sound and measures the time for the pulse to reach an object, reflect off of it and return to be detected. Can the temperature affect your focus for such a camera?

4. To keep animals away from their cars, some people mount short, thin pipes on the fenders. The pipes give out a high-pitched wail when the cars are moving. How do they create the sound?

5. Secret agents in the movies always want to get to a secure phone where a voice scrambler is in use. How do these devices work?

6. When a bell is rung, standing waves are set up around the bell's circumference. What boundary conditions must be satisfied by the resonant wavelengths? How does a crack in the bell, such as in the Liberty Bell, affect the satisfying of the boundary conditions and the sound emanating from the bell?

7. How does air temperature affect the tuning of a wind instrument?

8. Explain how the distance to a lightning bolt can be determined by counting the seconds between the flash and the sound of thunder.

9. You are driving toward a cliff and you honk your horn. Is there a Doppler shift of the sound when you hear the echo? Is it like a moving source or moving observer? What if the reflection occurs not from a cliff but from the forward edge of a huge alien spacecraft that is moving toward you as you drive?

10. Of the following sounds, state which is most likely to have an intensity level of 60 dB: a rock concert, the turning of a page in this text, a normal conversation, a cheering crowd at a football game, or background noise at a church?

11. Guitarists sometimes play a "harmonic" by lightly touching a string at the exact center and plucking the string. The result is a clear note one octave higher than the fundamental of the string, even though the string is not pressed to the fingerboard. Why does this happen?

12. By listening to a band or orchestra, how can you determine that the speed of sound is the same for all frequencies?

13. An archer shoots an arrow from a bow. Does the string of the bow exhibit standing waves after the arrow leaves? If so, and if the bow is perfectly symmetric so that the arrow leaves from the center of the string, what harmonics are excited?

14. The radar systems used by police to detect speeders are sensitive to the Doppler shift of a pulse of radio waves. Discuss how this sensitivity can be used to measure the speed of a car.

15. As oppositely moving pulses of the same shape (one upward, one downward) on a string pass through each other, there is one instant at which the string shows no displacement from the equilibrium position at any point. Has the energy carried by the pulses disappeared at this instant of time? If not, where is it?

16. A soft-drink bottle resonates as air is blown across its top. What happens to the resonant frequency as the level of fluid in the bottle decreases?

17. If a child continuously blows a whistle while on a merry-go-round, explain what you hear as the child comes by each time.

18. Despite a reasonably steady hand, we often spill our coffee when carrying it to our seat. Discuss resonance as a possible cause of this difficulty, and devise a means for solving the problem.

19. An airplane mechanic notices that the sound from a twin-engine aircraft rapidly varies in loudness when both engines are running. What could be causing this variation from loud to soft?

20. Why does a vibrating guitar string sound louder when placed on the instrument than it would if allowed to vibrate in the air while off the instrument?

PROBLEMS

1, **2**, **3** = straightforward, intermediate, challenging   ☐ = full solution available in Student Solutions Manual/Study Guide

web = solution posted at **http://info.brookscole.com/serway**   🐍 = biomedical application

Section 14.2 Characteristics of Sound Waves

Section 14.3 The Speed of Sound

Unless otherwise stated, use 345 m/s as the speed of sound in air.

1. Suppose that you hear a clap of thunder 16.2 s after seeing the associated lightning stroke. The speed of sound waves in air is 343 m/s and the speed of light in air is 3.00×10^8 m/s. How far are you from the lightning stroke?

2. A dolphin located in sea water at a temperature of 25°C emits a sound directed toward the bottom of the ocean 150 m below. How much time passes before it hears an echo?

3. A sound wave has a frequency of 700 Hz in air and a wavelength of 0.50 m. What is the temperature of the air?

4. The range of human hearing extends from approximately 20 Hz to 20 000 Hz. Find the wavelengths of these extremes at a temperature of 27°C.

5. Find the speed of sound in mercury, which has a bulk modulus of approximately 2.80×10^{10} N/m² and a density of 13 600 kg/m³.

6. A jet passes overhead at an altitude of 500 m. When the plane is directly overhead, as in Fig. P14.6, the sound of its engines appears to come from point *A*. If the average temperature of the air is 10.0°C, how fast is the plane moving?

7. You are watching a pier being constructed on the far shore of a saltwater inlet when some blasting occurs. You hear the sound in the water 4.50 s before it reaches you through the air. How wide is the inlet? (*Hint:* See Table 14.1. Assume the air temperature is 20°C.)

8. A rescue plane flies horizontally at a constant speed searching for a disabled boat. When the plane is directly above the boat, the boat's crew blows a loud horn. By the time the plane's sound detector receives the horn's sound, the plane has traveled a distance equal to one half of its altitude above the ocean. If it takes the sound 2.00 s to reach the plane, determine (a) the speed of the plane, and (b) its altitude. (Take the speed of sound to be 343 m/s.)

Section 14.4 Energy and Intensity of Sound Waves

9. The toadfish makes use of resonance in a closed tube to produce very loud sounds. The tube is its swim bladder used as an amplifier. The sound level of this creature has been measured as high as 100 dB. (a) Calculate the intensity of the sound wave emitted. (b) What is the intensity level if three of these fish try to imitate three frogs by saying "Budweiser" at the same time?

10. The area of a typical eardrum is about 5.0×10^{-5} m². Calculate the sound power (the energy per second) incident on an eardrum at (a) the threshold of hearing and (b) the threshold of pain.

11. A microphone in the ocean is sensitive to sounds emitted by porpoises. To produce a usable signal, sound waves striking the microphone must have an intensity of 10 dB. If porpoises emit sound waves with a power of 0.050 W, how far can a porpoise be from the microphone and still be heard? Disregard absorption of sound waves by the water.

12. The intensity level of an orchestra is 85 dB. A single violin achieves a level of 70 dB. How does the intensity of the sound of the full orchestra compare with that of the violin's sound?

13. A noisy machine in a factory produces sound with a level of 80 dB. How many identical machines could you add to the factory without exceeding the 90-dB limit?

14. A family ice show is held at an enclosed arena. The skaters perform to music with level 80.0 dB. This is too loud for your baby, who yells at 75.0 dB. (a) What total sound in-

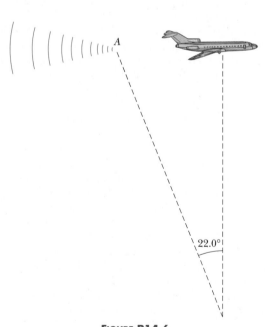

FIGURE P14.6

tensity engulfs you? (b) What is the combined sound level?

15. A person wears a hearing aid that uniformly increases the intensity level of all audible frequencies of sound by 30.0 dB. The hearing aid picks up sound having a frequency of 100 Hz and an intensity of 1.0×10^{-10} W/m². What is the intensity delivered to the eardrum?

Section 14.5 Spherical and Plane Waves

16. An outside loudspeaker (considered a small source) emits sound waves with a power output of 100 W. (a) Find the intensity 10.0 m from the source. (b) Find the intensity level, in decibels, at this distance. (c) At what distance would you experience the sound at the threshold of pain, 120 dB?

17. A train sounds its horn as it approaches an intersection. The horn can just be heard at a level of 50 dB by an observer 10 km away. (a) What is the average power generated by the horn? (b) What intensity level of the horn's sound is observed by someone waiting at an intersection 50 m from the train? Treat the horn as a point source and neglect any absorption of sound by the air.

18. A skyrocket explodes 100 m above the ground (Fig. P14.18). Three observers are spaced 100 m apart, with observer A directly under the point of the explosion. (a) What is the ratio of sound intensities heard by observers A and B? (b) What is the ratio of intensities heard by observers A and C?

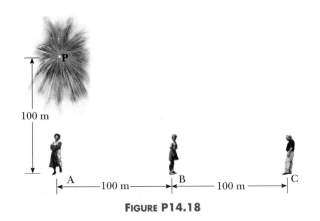

FIGURE P14.18

19. Show that the difference in decibel levels β_1 and β_2 of a sound source is related to the ratio of its distances r_1 and r_2 from the receivers by

$$\beta_2 - \beta_1 = 20 \log\left(\frac{r_1}{r_2}\right)$$

Section 14.6 The Doppler Effect

20. A train at rest emits a sound at a frequency of 1 000 Hz. An observer in a car travels away from the sound source at a speed of 30.0 m/s. What is the frequency heard by the observer?

21. A commuter train passes a passenger platform at a constant speed of 40.0 m/s. The train horn is sounded at its characteristic frequency of 320 Hz. (a) What overall change in frequency is detected by a person on the platform as the train moves from approaching to receding? (b) What wavelength is detected by a person on the platform as the train approaches?

22. At rest, a car's horn sounds the note A (440 Hz). The horn is sounded while the car moves down the street. A bicyclist moving in the same direction with one third of the car's speed hears a frequency of 415 Hz. What is the speed of the car? Is the cyclist ahead of or behind the car?

23. Two trains on separate tracks move toward one another. Train 1 has a speed of 130 km/h and train 2 a speed of 90.0 km/h. Train 2 blows its horn, emitting a frequency of 500 Hz. What is the frequency heard by the engineer on train 1?

24. A bat flying at 5.0 m/s emits a chirp at 40 kHz. If this sound pulse is reflected by a wall, what is the frequency of the echo received by the bat?

25. An alert physics student stands beside the tracks as a train rolls slowly past. He notes that the frequency of the train whistle is 442 Hz when the train is approaching him and 441 Hz when the train is receding from him. From this he can find the speed of the train. What value does he find?

26. Expectant parents are thrilled to hear their unborn baby's heartbeat, revealed by an ultrasonic motion detector. Suppose the fetus's ventricular wall moves in simple harmonic motion with amplitude 1.80 mm and frequency 115 per minute. (a) Find the maximum linear speed of the heart wall. Suppose the motion detector in contact with the mother's abdomen produces sound at precisely 2 MHz, which travels through tissue at 1.50 km/s. (b) Find the maximum frequency at which sound arrives at the wall of the baby's heart. (c) Find the maximum frequency at which reflected sound is received by the motion detector. (By electronically "listening" for echoes at a frequency different from the broadcast frequency, the motion detector produces beeps of audible sound in synchronization with the fetal heartbeat.)

27. A tuning fork vibrating at 512 Hz falls from rest and accelerates at 9.80 m/s². How far below the point of release is the tuning fork when waves of frequency 485 Hz reach the release point? Take the speed of sound in air to be 340 m/s.

28. A supersonic jet traveling at Mach 3 at an altitude of 20 000 m is directly overhead at time $t = 0$, as in Figure P14.28. (a) How long will it be before the ground observer encounters the shock wave? (b) Where will the plane be when it is finally heard? (Assume an average value of 330 m/s for the speed of sound in air.)

29. The Concorde flies at Mach 1.5, which means the speed of the plane is 1.5 times the speed of sound in air. What is the angle between the direction of propagation of the shock wave and the direction of the plane's velocity?

Section 14.7 Interference of Sound Waves

30. The acoustical system shown in Figure 14.14 is driven by a speaker emitting a 400-Hz note. If *destructive* interference occurs at a particular instant, how much must the path length in the U-shaped tube be increased in order to hear (a) constructive interference and (b) destructive interference once again?

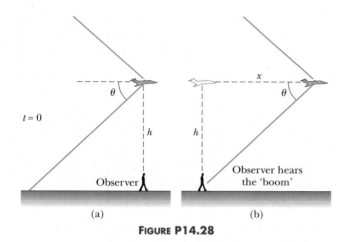

FIGURE P14.28

31. The ship in Figure P14.31 travels along a straight line parallel to the shore and 600 m from the shore. The ship's radio receives simultaneous signals of the same frequency from antennas at points *A* and *B*. The signals interfere constructively at point *C*, which is equidistant from *A* and *B*. The signal goes through the first minimum at point *D*. Determine the wavelength of the radio waves.

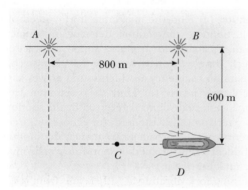

FIGURE P14.31

32. Two loudspeakers are placed above and below one another, as in Figure 14.15, and driven by the same source at a frequency of 500 Hz. (a) What minimum distance should the top speaker be moved back in order to create destructive interference between the two speakers? (b) If the top speaker is moved back twice the distance calculated in part (a), will constructive or destructive interference occur?

33. A pair of speakers separated by 0.700 m are driven by the same oscillator at a frequency of 690 Hz. An observer, originally positioned at one of the speakers, begins to walk along a line perpendicular to the line joining the two speakers. (a) How far must the observer walk before reaching a relative maximum in intensity? (b) How far will the observer be from the speaker when the first relative minimum is detected in the intensity?

Section 14.8   Standing Waves

34. A steel wire in a piano has a length of 0.700 0 m and a mass of 4.300×10^{-3} kg. To what tension must this wire

be stretched in order that the fundamental vibration correspond to middle C ($f_C = 261.6$ Hz on the chromatic musical scale)?

35. A stretched string fixed at each end has a mass of 40.0 g and a length of 8.00 m. The tension in the string is 49.0 N. (a) Determine the positions of the nodes and antinodes for the third harmonic. (b) What is the vibration frequency for this harmonic?

36. A 0.300-g wire is stretched between two points 70.0 cm apart. If the tension in the wire is 600 N, find the frequencies of the wire's first, second, and third harmonics.

37. Two speakers are driven by a common oscillator at 800 Hz and face each other at a distance of 1.25 m. Locate the points along a line joining the two speakers where relative minima of pressure amplitude be expected. (Use $v = 343$ m/s.)

38. A cello A string vibrates in its fundamental mode with a frequency of 220 vibrations/s. The vibrating segment is 70.0 cm long and has a mass of 1.20 g. (a) Find the tension in the string. (b) Determine the frequency of the string when it vibrates in three segments.

39. A 12-kg object hangs in equilibrium from a string of total length $L = 5.0$ m and linear mass density $\mu = 0.001\ 0$ kg/m. The string is wrapped around two light, frictionless pulleys that are separated by the distance $d = 2.0$ m (Fig. P14.39a). (a) Determine the tension in the string. (b) At what frequency must the string between the pulleys vibrate in order to form the standing-wave pattern shown in Figure P14.39b?

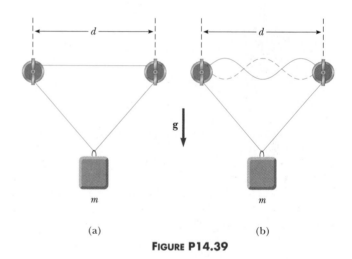

FIGURE P14.39

40. In the arrangement shown in Figure P14.40, an object of mass $m = 5.0$ kg hangs from a cord around a light pulley. The length of the cord between point *P* and the pulley is $L = 2.0$ m. (a) When the vibrator is set to a frequency of 150 Hz, a standing wave with six loops is formed. What must be the linear mass density of the cord? (b) How many loops (if any) will result if *m* is changed to 45 kg? (c) How many loops (if any) will result if *m* is changed to 10 kg?

41. A 60.000-cm guitar string under a tension of 50.000 N has a mass per unit length of 0.100 00 g/cm. What is the high-

FIGURE P14.40

est resonant frequency that can be heard by a person capable of hearing frequencies up to 20 000 Hz?

Section 14.9 Forced Vibrations and Resonance

42. Standing-wave vibrations are set up in a crystal goblet with four nodes and four antinodes equally spaced around the 20.0-cm circumference of its rim. If transverse waves move around the glass at 900 m/s, an opera singer would have to produce a high harmonic with what frequency in order to shatter the glass with a resonant vibration?

Section 14.10 Standing Waves in Air Columns

43. The windpipe of a typical whooping crane is about 5.0 feet long. What is the lowest resonant frequency of this pipe assuming it is a pipe closed at one end? Assume a temperature of 37°C.

44. The overall length of a piccolo is 32.0 cm. The resonating air column vibrates as in a pipe open at both ends. (a) Find the frequency of the lowest note a piccolo can play, assuming the speed of sound in air is 340 m/s. (b) Opening holes in the side effectively shortens the length of the resonant column. If the highest note a piccolo can sound is 4 000 Hz, find the distance between adjacent antinodes for this mode of vibration.

45. The human ear canal is about 2.8 cm long. If it is regarded as a tube open at one end and closed at the eardrum, what is the fundamental frequency around which we would expect hearing to be most sensitive? (Take the speed of sound to be 340 m/s.)

46. A shower stall measures 86.0 cm × 86.0 cm × 210 cm. When you sing in the shower, which frequencies will sound the richest (because of resonance)? Assume the stall acts as a pipe closed at both ends, with nodes at opposite sides. Assume that the voices of various singers range from 130 Hz to 2 000 Hz. (Let the speed of sound in the hot shower stall be 355 m/s.)

47. A pipe open at both ends has a fundamental frequency of **web** 300 Hz when the temperature is 0°C. (a) What is the length of the pipe? (b) What is the fundamental frequency at a temperature of 30°C?

48. A 2.00-m-long air column is open at both ends. The frequency of a certain harmonic is 410 Hz, and the frequency of the next higher harmonic is 492 Hz. Determine the speed of sound in the air column.

Section 14.11 Beats

49. Two identical mandolin strings under 200 N of tension are sounding tones with frequencies of 523 Hz. The peg of one string slips slightly, and the tension in it drops to 196 N. How many beats per second are heard?

50. The G string on a violin has a fundamental frequency of 196 Hz. It is 30.0 cm long and has a mass of 0.500 g. While this string is sounding, a nearby violinist effectively shortens (by sliding her finger down the string) the G string on her identical violin until a beat frequency of 2.00 Hz is heard between the two strings. When this occurs, what is the effective length of her string?

51. Two train whistles have identical frequencies of 180 Hz. When one train is at rest in the station, sounding its whistle, a beat frequency of 2 Hz is heard from a moving train. What two possible speeds and directions can the moving train have?

52. Two pipes, equal in length, are each open at one end. Each has a fundamental frequency of 480 Hz at 300 K. In one pipe the air temperature is increased to 305 K. If the two pipes are sounded together, what beat frequency results?

53. A student holds a tuning fork oscillating at 256 Hz. He walks toward a wall at a constant speed of 1.33 m/s. (a) What beat frequency does he observe between the tuning fork and its echo? (b) How fast must he walk away from the wall to observe a beat frequency of 5.00 Hz?

Section 14.13 The Ear

54. If a human ear canal can be thought of as resembling an organ pipe, closed at one end, that resonates at a fundamental frequency of 3 000 Hz, what is the length of the canal? (Use normal body temperature 37°C for your determination of the speed of sound in the canal.)

55. Some studies suggest that the upper frequency limit of hearing is determined by the diameter of the eardrum. The wavelength of the sound wave and the diameter of the eardrum are approximately equal at this upper limit. If this is precisely true, what is the diameter of the eardrum of a person capable of hearing 20 000 Hz? (Assume a body temperature of 37°C.)

ADDITIONAL PROBLEMS

56. Two cars are traveling in the same direction, both at a speed of 55.0 mi/h (80.7 ft/s). The driver of the trailing car sounds his horn, which has a frequency of 300 Hz. If the speed of sound is 1 100 ft/s, what frequency sound is heard by the driver of the leading car? (Hint: Consider the relative motion between the source and observer in this case.)

57. A quartz watch contains a crystal oscillator in the form of a block of quartz which vibrates by contracting and expanding. Two opposite faces of the block, 7.05 mm apart, are antinodes, moving alternately toward each other and away from each other. The plane halfway between these two faces is a node of the vibration. The speed of sound in quartz is 3.70 km/s. Find the frequency of the vibration. An oscillating electric voltage accompanies the mechanical

oscillation—the quartz is described as *piezoelectric*. An electric circuit feeds in energy to maintain the oscillation and also counts the voltage pulses to keep time.

58. A flowerpot is knocked off a balcony 20.0 m above the sidewalk and falls toward an unsuspecting 1.75-m-tall man who is standing below. How close to the sidewalk can the flower pot fall before it is too late for a warning shouted from the balcony to reach the man in time? Assume that the man below requires 0.300 s to respond to the warning.

59. On a workday the average decibel level of a busy street is
web 70 dB, with 100 cars passing a given point every minute. If the number of cars is reduced to 25 every minute on a weekend, what is the decibel level of the street?

60. A variable-length air column is placed just below a vibrating wire that is fixed at both ends. The length of the air column, open at one end, is gradually increased from zero until the first position of resonance is observed at $L = 34.0$ cm. The wire is 120 cm long and is vibrating in its third harmonic. If the speed of sound in air is 340 m/s, what is the speed of transverse waves in the wire?

61. A block with a speaker bolted to it is connected to a spring having spring constant $k = 20.0$ N/m, as shown in Figure P14.61. The total mass of the block and speaker is 5.00 kg, and the amplitude of this unit's motion is 0.500 m. If the speaker emits sound waves of frequency 440 Hz, determine the lowest and highest frequencies heard by the person to the right of the speaker.

FIGURE P14.61

62. A flute is designed so that it plays a frequency of 261.6 Hz, middle C, when all the holes are covered and the temperature is 20.0°C. (a) Consider the flute to be a pipe open at both ends, and find its length, assuming that the middle-C frequency is the fundamental. (b) A second player, nearby in a colder room, also attempts to play middle C on an identical flute. A beat frequency of 3.00 beats/s is heard. What is the temperature of the room?

63. When at rest, two trains have sirens that emit a frequency of 300 Hz. The two trains travel toward one another and toward an observer stationed between them. One of the trains moves at 30.0 m/s, and the observer hears a beat frequency of 3.0 beats per second. What is the velocity of the second train, which travels faster than the first?

64. Many artists sing very high notes in *ad lib* ornaments and cadenzas. The highest note written for a singer in a pub-

lished score was F-sharp above high C, 1.480 kHz, sung by Zerbinetta in the original version of Richard Strauss's opera *Ariadne auf Naxos*. (a) Find the wavelength of this sound in air. (b) In response to complaints, Strauss later transposed the note down to F above high C, 1.397 kHz. By what increment did the wavelength change?

65. A speaker at the front of a room and an identical speaker at the rear of the room are being driven at 456 Hz by the same sound source. A student walks at a uniform rate of 1.50 m/s away from one speaker and toward the other. How many beats does the student hear per second?

66. Two identical speakers separated by 10.0 m are driven by the same oscillator with a frequency of $f = 21.5$ Hz (Fig. P14.66). Explain why a receiver at A records a minimum in sound intensity from the two speakers. (b) If the receiver is moved in the plane of the speakers, what path should it take so that the intensity remains at a minimum? That is, determine the relationship between x and y (the coordinates of the receiver) that causes the receiver to record a minimum in sound intensity. Take the speed of sound to be 344 m/s.

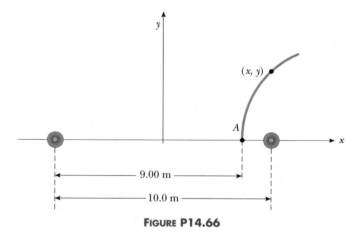

FIGURE P14.66

67. By proper excitation, it is possible to produce both longitudinal and transverse waves in a long metal rod. In a particular case, the rod is 150 cm long and 0.200 cm in radius and has a mass of 50.9 g. Young's modulus for the material is 6.80×10^{10} Pa. Determine the required tension in the rod so that the ratio of the speed of longitudinal waves to the speed of transverse waves is 8.

68. A student stands several meters in front of a smooth reflecting wall, holding a board on which a wire is fixed at each end. The wire, vibrating in its third harmonic, is 75.0 cm long, has a mass of 2.25 g, and is under a tension of 400 N. A second student, moving toward the wall, hears 8.30 beats per second. What is the speed of the student approaching the wall? (Use 340 m/s as the speed of sound in air.)

69. Two ships are moving along a line due east. The trailing vessel has a speed of 64.0 km/h relative to a land-based observation point, and the leading ship has a speed of 45.0 km/h relative to the same station. The trailing ship transmits a sonar signal at a frequency of 1 200 Hz.

What frequency is monitored by the leading ship? (Use 1 520 m/s as the speed of sound in ocean water.)

70. The Doppler equation presented in the text is valid when the motion between the observer and the source occurs on a straight line, so that the source and observer are moving either directly toward or directly away from each other. If this restriction is relaxed, we must use the more general Doppler equation

$$f' = \left[\frac{v + v_O \cos(\theta_O)}{v - v_S \cos(\theta_S)}\right] f$$

where θ_O and θ_S are defined in Figure P14.70a. (a) If both observer and source are moving away from each other along a straight line, show that the preceding equation yields the same result as Equation 14.11 in the text. (b) Use the preceding equation to solve the following problem. A train moves at a constant speed of 25.0 m/s toward the intersection shown in Figure P14.70b. A car is stopped near the intersection, 30.0 m from the tracks. If the train's horn emits a frequency of 500 Hz, what is the frequency heard by the passengers in the car when the train is 40.0 m to the left of the intersection? (Take the speed of sound to be 343 m/s.)

(a)   (b)

FIGURE P14.70

GROUP ACTIVITIES

G.1 Use an empty 1-liter soft-drink container, blow over the open end, and listen to the sound that is produced. Add some water to the container to change the height of the air column and repeat. How does the frequency that you hear change with the height of the air column?

If you want to investigate this phenomenon in more detail, construct a musical instrument made up of several soft-drink bottles with different amounts of water in each. You can play your instrument as a wind instrument by blowing over the mouths of the bottles.

G.2 Beats can easily be heard on a guitar. When a finger is placed at the fifth fret of the second string on a guitar, the note produced when the string is plucked should be identical to the note from the first string when it is played without fingering. With your finger in position on the second string, pluck the two simultaneously. If one string is slightly out of tune, a very pronounced beat frequency will be heard. What happens to the beat frequency as the string tension is changed in small increments from too low for the intended tuning to too high?

G.3 Attach a rope to a door and shake the other end to see how many of the standing-wave patterns in Figure 14.18 you can produce. When a pattern is formed, note that the amplitude of the rope's vibration is very much larger than the movement of your hand.

G.4 Snip off the corners of one end of a paper straw so that the end tapers to a point, as shown in Figure GA14.4. Chew on this end to flatten it, and you have created a double-reed instrument. Put your lips around the tapered end of the straw, press them together slightly, and blow through the straw. When you hear a steady tone, snip off a piece of the straw at the other end. Be careful to keep about the same amount of pressure with your lips. How does the frequency of the sound change as the straw becomes shorter? Why does this change occur? You may be able to produce more than one tone for any given length of the straw. Why?

FIGURE GA14.4

G.5 Inflate a balloon just enough to form a small sphere. Measure its diameter. Use a marker to color in a 1-cm square on its surface. Now continue inflating the balloon until it reaches twice the original diameter. Measure the size of the square now. Also, note how the color of the marked area has changed. Use the information in Section 14.5 to explain these results.

G.6 (a) Use the definition of the decibel level of a sound to predict the intensity of a 50-dB sound. (b) Repeat for an 80-dB sound. (c) Will two seperate 50-dB sounds constitute together a 100-dB sound? (d) Verify your answers to part (c) by a calculation using the definition of decibel level.

G.7 (a) You are driving down the highway in your car when a police car sounding its siren overtakes you and passes you. If its frequency at rest is f_0, is the frequency you hear while the car is catching up to you higher or lower than f_0? (b) What about the frequency you hear after the car has passed you? (c) Find the specific frequencies you hear in the situation where your speed = 70.0 km/h, f_0 = 600 Hz, and speed of police car = 100 km/h.

G.8 A string on a guitar has a length L and a mass per unit length of μ. It vibrates with a fundamental frequency f_1. (a) Is the wavelength of the fundamental standing wave greater than L, equal to L, or less than L? (b) Is the wavelength of the next normal mode greater than L, equal to

L, or less than *L*? (c) Is the speed of a wave on the string about the same as the speed of sound in air? Answer with true or false and defend your answer. (d) Let the length of the guitar string be 63.5 cm, and $f_1 = 110$ Hz to find precise values for all the quantities in parts (a), (b) and (c). Evaluate your conceptual predictions.

"I love hearing that lonesome wail of the train whistle as the magnitude of the frequency of the wave changes due to the Doppler effect."

Welcome to Volume 2 of Serway and Faughn's

COLLEGE PHYSICS
SIXTH EDITION

As explained in the Preface, we have materials available to assist you in this second term.

Volume 2 of the *Student Solutions Manual and Study Guide* (ISBN 0-03-034818-8) by Gordon, Teague and Serway contains detailed solutions to approximately twelve problems for each of Chapters 15–30 to help you learn by example. These problems are indicated in the textbook with boxed problem numbers. The manual also features a skills section, important notes from key sections of the text, and a list of important equations and concepts.

In addition, you might be interested in the *Core Concepts in College Physics* CD-ROM Version 2.0, ISBN 0-03-033701-1. This three-disc CD-ROM is a multimedia presentation of the most important principles of a college physics course. Content screens provide in-depth coverage of abstract and often difficult principles, building connections between physical concepts and mathematics. The presentation contains more than 350 movies—both animated and live video—including laboratory demonstrations, "real world" examples, graphic models, and step-by-step explanations of essential mathematics. The free accompanying workbook contains practical physics problems directly related to the presentation, along with worked solutions.

If you would like one or both of these publications and they are not in stock in your college or university bookstore, please see the manager about placing a special order or call (800) 354-9706 toll-free.

We wish you success in the remainder of your physics course.

THOMSON

BROOKS/COLE

PART Four

Electricity and Magnetism

We now begin a study of the branch of physics concerned with electric and magnetic phenomena. The laws of electricity and magnetism play central roles in the operation of many devices such as DVDs, VCRs, electric motors, computers, high-energy accelerators, and a host of electronic devices used in medicine. More fundamentally, we know that the interatomic and intermolecular forces that are responsible for the formation of solids and liquids and most basic life processes, such as the replication of chromosomes, are electric in origin. Furthermore, such forces as the pushes and pulls between objects and the elastic force in a spring arise from electric forces at the atomic level.

The ancient Greeks observed electric and magnetic phenomena as early as 700 B.C. They found that a piece of amber, when rubbed, became electrified and attracted pieces of straw or feathers. The existence of magnetic forces was known as a result of observations that a naturally occurring stone called *magnetite* (Fe_2O_3) was attracted to iron. (The word *electric* comes from the Greek word for amber, *elecktron*. The word *magnetic* comes from the name of the country where magnetite was found, *Magnesia*, now Turkey.)

In 1600 William Gilbert discovered that electrification was not limited to amber but is a general phenomenon. In 1785 Charles Coulomb described experiments that confirmed the inverse-square force law for electricity.

It was not until the early part of the 19th century that scientists established that electricity and magnetism are, in fact, related phenomena. In 1820 Hans Oersted discovered that a compass needle is deflected when placed near a wire carrying an electric current. A few years later, Michael Faraday showed that when a wire is moved near a magnet (or, equivalently, when a magnet is moved near a wire), an electric current is observed in the wire. From 1865 to 1873 James Clerk Maxwell used these observations and other experimental facts as bases for formulating the laws of electromagnetism as we now know them. (*Electromagnetism* is a name given to the combined subjects of electricity and magnetism.) Shortly thereafter, Heinrich Hertz verified Maxwell's predictions by producing electromagnetic waves in the laboratory. Guglielmo Marconi, inspired by the clever experiments of Hertz and others, invented the radio and filed the first patent for wireless telegraphy in 1896. The rest is history as today we sit bathed in electromagnetic waves from hundreds of practical devices ranging from hand-held computers to pagers.

Maxwell's contributions to the science of electromagnetism were especially significant because the laws he formulated are basic to *all* forms of electromagnetic phenomena including light. His work is comparable in importance to Newton's discovery of the laws of motion and the theory of gravitation.

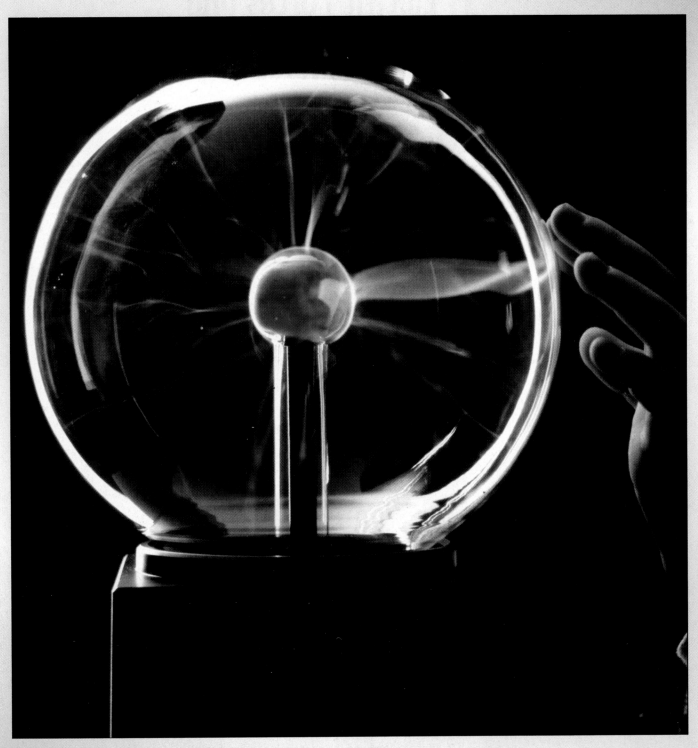

In plasma globes such as this, a high voltage (typically 5 kV) is applied between the center of the globe and the outer grounded glass surface. The interior of the globe contains a mixture of inert gases, and the applied voltage partially ionizes the gases. The ionized gases create filaments that conduct current as shown by the colorful paths in the photograph. When the outer surface is touched, the electrical discharge is drawn to your hand because your body is the path of least re-sistance to Earth compared to that of the grounded glass globe. *(Richard Megna/Fundamental Photographs)*

15

Electric Forces and Electric Fields

This nighttime view of multiple bolts of lightning was photographed in Tucson, Arizona. During a thunderstorm, a high concentration of electrical charge in a thundercloud creates a higher than normal electric field between the thundercloud and the negatively charged Earth's surface. This strong electric field creates an electric discharge between the charged cloud and ground—an enormous spark. Other discharges that are observed in the sky include cloud-to-cloud discharges and the more frequent intracloud discharges. *(© Keith Kent/Science Photo Library/Photo Researchers, Inc.)*

T he earliest known study of electricity was conducted by the Greeks about 700 B.C. By modern standards, their contributions to the field were modest. However, from those roots have sprung the enormous electrical distribution systems and sophisticated electronic instruments that are so much a part of our world today. It all began, apparently, when someone noticed that a fossil material called amber attracted small objects after being rubbed with wool. Since then we have learned that this phenomenon is not restricted to amber and wool but occurs (to some degree) when almost any two nonconducting substances are rubbed together.

In this chapter we use this effect—charging by friction—to begin an investigation of electric forces. We then discuss Coulomb's law, which is the fundamental law of force between any two stationary charged particles. The concept of an electric field associated with charges is then introduced and its effects on other charged particles described. We include a brief discussion of the Van de Graaff generator, and end by describing a way to calculate electric fields in certain situations.

15.1 PROPERTIES OF ELECTRIC CHARGES

A number of simple experiments demonstrate the existence of electrostatic forces. For example, after running a plastic comb through your hair, you will find that the comb attracts bits of paper. The attractive force is often strong enough to suspend the paper from the comb. The same effect occurs with other rubbed materials, such as glass and hard rubber.

Another simple experiment is to rub an inflated balloon with wool (or across your hair). On a dry day, the rubbed balloon will then stick to the wall of a room, often for hours. When materials behave in this way, they are said to have become **electrically charged.** You can give your body an electric charge by vigorously rubbing your shoes on a wool rug or by sliding across a car seat. You can then feel— and remove—the charge on your body by lightly touching another person. Under the right conditions, a visible spark can be seen when you touch, and a slight tingle is felt by both parties. (Experiments such as these work best on a dry day because excessive moisture can provide a pathway for charge to leak off a charged object.)

Experiments also demonstrate that there are two kinds of electric charge, which Benjamin Franklin (1706–1790) named **positive** and **negative.** Figure 15.1

BENJAMIN FRANKLIN (1706–1790)

Franklin was a printer, author, physical scientist, inventor, diplomat, and a founding father of the United States. This original and innovative thinker invented the Franklin stove, bifocals, the lightning rod, and the glass harmonica. He never patented any of his inventions, preferring that they be used for the benefit of all people. Franklin's clear thoughts, economy of expression, and quintessential optimism are evident in both his diplomatic and scientific works.

Franklin's work on electricity in the late 1740s changed a jumbled, unrelated set of observations into a connected science of electricity. Briefly, Franklin believed that electricity consisted of a single fluid and that an excess of fluid charged a body "positively" and a deficit "negatively." He conducted careful experiments to show that the natural state of an object was one of electrical neutrality. However, if an object was charged positively by rubbing it with a donor material, the donor was found to have an equal negative charge. He interpreted this to mean that the electric fluid added to the first object was just equal to that lost by the donor. Thus, electric fluid could be collected, transferred, and circulated, but it was *always* conserved. Franklin proposed several experiments involving lightning in order to "ascertain its sameness with the electric fluid." During a lightning storm, Franklin performed the dangerous kite experiment, and later used its positive outcome to practical use with the invention of the lightning rod. *(Science VU/Visuals Unlimited)*

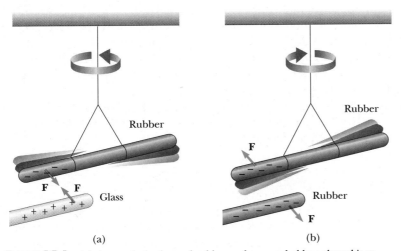

(a)　　　　　　　　(b)

FIGURE 15.1　(a) A negatively charged rubber rod suspended by a thread is attracted to a positively charged glass rod. (b) A negatively charged rubber rod is repelled by another negatively charged rubber rod.

illustrates the interaction of the two charges. A hard rubber (or plastic) rod that has been rubbed with fur is suspended by a piece of string. When a glass rod that has been rubbed with silk is brought near the rubber rod, the rubber rod is attracted toward the glass rod (Fig. 15.1a). If two charged rubber rods (or two charged glass rods) are brought near each other, as in Figure 15.1b, the force between them is repulsive. This observation demonstrates that the rubber and glass have different kinds of charge. We use the convention suggested by Franklin, where the electric charge on the glass rod is called positive and that on the rubber rod is called negative. On the basis of observations such as these, we conclude that **like charges repel one another and unlike charges attract one another.** Note that every object usually contains equal amounts of positive and negative charge—electrical forces occur when excess charge is transferred from one object to another.

Nature's basic carrier of positive charge is the proton, a particle which, along with neutrons, is located in the nucleus of an atom. Because the nucleus of an atom is held firmly in place inside a solid, protons are never moved from one material to another. Thus, when an object becomes charged, it does so because it has either gained or lost nature's basic carrier of negative charge, the electron. It is convenient to picture the electron circling about the nucleus of an atom like a planet revolving about the Sun. The usual state of an atom is neutral; that is, for every positive charge in the nucleus there is an electron orbiting the nucleus to cancel the charge. Any uncharged object in our large-scale world contains an enormous number of electrons and protons—about 10^{23} of each per cubic centimeter. However, for every negative electron a positively charged proton is present.

Charge has a natural tendency to be transferred between unlike materials. Rubbing the two materials together serves to increase the area of contact and thus to enhance the charge transfer process.

An important characteristic of charge is that **electric charge is always conserved.** That is, when two initially neutral objects are charged by being rubbed together, charge is not created in the process. The objects become charged because **negative charge is transferred from one object to the other.** One object gains some amount of negative charge while the other loses an equal amount of negative charge and hence is left with a positive charge. For example, when a glass rod is rubbed with silk, as in Figure 15.2, the silk obtains a negative charge that is equal in magnitude to the positive charge on the glass rod as negatively charged electrons are transferred from the glass to the silk in the rubbing process. Likewise, when rubber is rubbed with fur, electrons are transferred from the fur to the rubber.

In 1909 Robert Millikan (1886–1953) discovered that if an object is charged, its charge is always a multiple of a fundamental unit of charge, which we designate with the symbol e. In modern terms, the charge is said to be **quantized.** This means that charge occurs as discrete bundles in nature. Thus, an object may have a charge of $\pm e$, or $\pm 2e$, or $\pm 3e$, and so on, but never,[1] say, a fractional charge of $\pm 1.5e$. Other experiments in Millikan's time showed that the electron has a charge of $-e$ and the proton has an equal and opposite charge, $+e$. Some particles, such as a neutron, have no net charge. A neutral atom (one with no net charge) contains as many protons as electrons. The value of e is now known to be $1.602\ 19 \times 10^{-19}$ C. (The SI unit of electric charge, the **coulomb** [C], will be defined more precisely in a later section.)

15.2   INSULATORS AND CONDUCTORS

It is convenient to classify substances in terms of their ability to conduct electric charge.

Like charges repel; unlike charges attract ▶

Webnote 15.1

For a more all-inclusive view of the electron and its function in the atom, read the article at *http://encarta.msn.com/find/concise. asp?z=1&pg=2&ti=01A8A000#s3*

Charge is conserved; charge is quantized ▶

FIGURE 15.2 When a glass rod is rubbed with silk, electrons are transferred from the glass to the silk. Because of conservation of charge, each electron adds negative charge to the silk, and an equal positive charge is left behind on the rod. Also, because the charges are transferred in discrete bundles, the charges on the two objects are $\pm e$ or $\pm 2e$ or $\pm 3e$, and so on.

---

[1] Recent developments have shown the existence of fundamental particles called **quarks** that have charges of $\pm e/3$ or $\pm 2e/3$. Note that charge is *still* quantized, but in units of $\pm e/3$. A more complete discussion of quarks and their properties is presented in Chapter 30.

(a) Before

(b) After

FIGURE 15.3 Charging a metal object by conduction. (a) The charged rubber rod is placed in contact with the insulated metal sphere. Some electrons move from the rod onto the sphere. (b) When the rod is removed, the sphere is left with a negative charge.

Conductors are materials in which electric charges move freely, and insulators are materials in which electric charges do not move freely.

Glass and rubber are insulators. When such materials are charged by rubbing, only the rubbed area becomes charged, and there is no tendency for the charge to move into other regions of the material. In contrast, materials such as copper, aluminum, and silver are good conductors. When such materials are charged in some small region, the charge readily distributes itself over the entire surface of the material. If you hold a copper rod in your hand and rub the rod with wool or fur, it will not attract a piece of paper. This might suggest that a metal cannot be charged. However, if you hold the copper rod with an insulator and then rub it with wool or fur, the rod remains charged and attracts the paper. In the first case, the electric charges produced by rubbing readily move from the copper through your body and finally to ground. In the second case, the insulating handle prevents the flow of charge to ground.

Semiconductors are a third class of materials, and their electrical properties are somewhere between those of insulators and those of conductors. Silicon and germanium are well-known semiconductors that are widely used in the fabrication of a variety of electronic devices.

CHARGING BY CONDUCTION

Consider a negatively charged rubber rod brought into contact with a neutral conducting sphere that is insulated so that there is no conducting path for charges to leave the sphere. Some electrons on the rubber rod are now able to move onto the sphere, as in Figure 15.3. When the rubber rod is removed, the sphere is left with a negative charge. This process is referred to as charging by **conduction.** The object being charged in such a process (the sphere) is always left with a charge having the same sign as the object doing the charging (the rubber rod).

CHARGING BY INDUCTION

When an object is connected to a conducting wire or copper pipe buried in Earth, it is said to be **grounded.** Earth can be considered an infinite reservoir for electrons; this means that it can accept or supply an unlimited number of electrons. With this in mind, we can understand the charging of a conductor by a process known as **induction.**

Consider a negatively charged rubber rod brought near a neutral (uncharged) conducting sphere that is insulated so that there is no conducting path to ground (Fig. 15.4a). The repulsive force between the electrons in the rod and those in the sphere causes a redistribution of charge on the sphere so that some

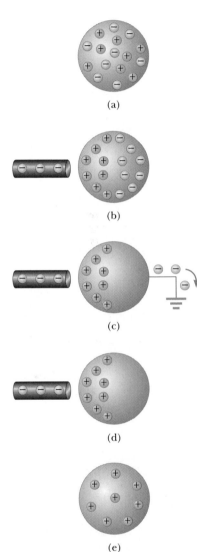

(a)

(b)

(c)

(d)

(e)

FIGURE 15.4 Charging a metal object by induction. (a) A neutral metallic sphere, with equal numbers of positive and negative charge. (b) The charge on the neutral sphere is redistributed when a charged rubber rod is placed near the sphere. (c) When the sphere is grounded, some of its electrons leave through the ground wire. (d) When the ground connection is removed, the sphere has excess positive charge that is nonuniformly distributed. (e) When the rubber rod is removed, the excess positive charge becomes uniformly distributed over the surface of the sphere.

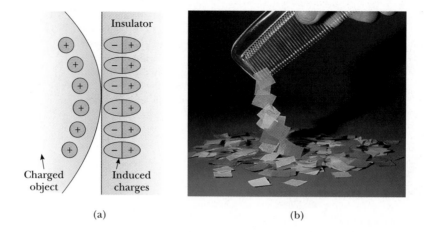

(a)   (b)

electrons move to the side of the sphere farthest away from the rod (Fig. 15.4b). The region of the sphere nearest the negatively charged rod has an excess of positive charge because of the migration of electrons away from this location. If a grounded conducting wire is then connected to the sphere, as in Figure 15.4c, some of the electrons leave the sphere and travel to ground. If the wire to ground is then removed (Fig. 15.4d), the conducting sphere is left with an excess of induced positive charge. Finally, when the rubber rod is removed from the vicinity of the sphere (Fig. 15.4e), the induced positive charge remains on the ungrounded sphere. Even though the positively charged atomic nuclei remain fixed, this excess positive charge becomes uniformly distributed over the surface of the ungrounded sphere because of the repulsive forces among the like charges and the high mobility of electrons in a metal.

In the process of inducing a charge on the sphere, the charged rubber rod loses none of its negative charge because it never came in contact with the sphere. Furthermore, the sphere is left with a charge opposite to that of the rubber rod. Note that **charging an object by induction requires no contact with the object inducing the charge.**

A process very similar to charging by induction in conductors also takes place in insulators. In most neutral atoms or molecules, the center of positive charge coincides with the center of negative charge. However, in the presence of a charged object, these centers may separate slightly, resulting in more positive charge on one side of the molecule than on the other. This effect is known as **polarization.** The realignment of charge within individual molecules produces an induced charge on the surface of the insulator, as shown in Figure 15.5a. Knowing about induction in insulators, we can explain why a comb that has been rubbed through hair attracts bits of electrically neutral paper (Fig. 15.5b) and why a balloon that has been rubbed against our clothing is able to stick to an electrically neutral wall.

APPLYING **PHYSICS** *15.1*

A positively charged ball hanging from a nonconducting string is brought near a nonconducting object. Based on the behavior of the ball-string combination, the ball is seen to be attracted to the object. From this experiment, it is not possible to determine whether the object is negatively charged or neutral. Why not? What additional experiment would help you decide between these two possibilities?

Explanation   The attraction between the ball and the object could be an attraction of unlike charges, or it could be an attraction between a charged object and a neutral object as a result of polarization of the molecules of the neutral object. Two additional experiments help us determine whether the object is charged. First, a known neutral ball is brought near the object—if there is an attraction, the object is negatively charged. Another possibility is to bring a known negatively charged ball near the object—if there is a repulsion, then the object is negatively charged. If there is no attraction, then the object is neutral.

If a suspended object A is attracted to object B, which is charged, we can conclude that (a) object A is uncharged, (b) object A is charged, (c) object B is positively charged, or (d) object A may be either charged or uncharged.

CHARLES COULOMB (1736–1806)

Coulomb's major contribution to science was in the field of electrostatics and magnetism. During his lifetime, he also investigated the strengths of materials and determined the forces that affect objects on beams, thereby contributing to the field of structural mechanics. *(Photo courtesy of AIP Niels Bohr Library, E. Scott Barr Collection)*

15.3 COULOMB'S LAW

In 1785 Charles Coulomb (1736–1806) established the fundamental law of electric force between two stationary charged particles. Experiments show that

An electric force has the following properties:

1. **It is inversely proportional to the square of the separation r between the two particles and is along the line joining them.**
2. **It is proportional to the product of the magnitudes of the charges $|q_1|$ and $|q_2|$ on the two particles.**
3. **It is attractive if the charges are of opposite sign and repulsive if the charges have the same sign.**

From these observations, we can express the magnitude of the electric force between two charges separated by a distance r as

$$F = k_e \frac{|q_1||q_2|}{r^2} \qquad [15.1]$$

◄ Coulomb's law

where k_e is a constant called the *Coulomb constant*. Note that Equation 15.1, known as **Coulomb's law,** applies only to point charges and to spherical distributions of charges.

The value of the Coulomb constant in Equation 15.1 depends on the choice of units. The SI unit of charge is the **coulomb** (C). From experiment, we know that the **Coulomb constant** in SI units has the value

$$k_e = 8.987\,5 \times 10^9 \text{ N} \cdot \text{m}^2/\text{C}^2 \qquad [15.2]$$

To simplify our calculations, we shall use the approximate value

$$k_e \approx 8.99 \times 10^9 \text{ N} \cdot \text{m}^2/\text{C}^2 \qquad [15.3]$$

The charge on the proton has a magnitude of $e = 1.6 \times 10^{-19}$ C. Therefore, it would take $1/e = 6.3 \times 10^{18}$ protons to create a total charge of $+1$ C. Likewise, 6.3×10^{18} electrons would have a total charge of -1 C. Compare this with the number of free electrons in 1 cm^3 of copper, which is on the order of 10^{23}. Even so, 1 C is a large amount of charge. In typical electrostatic experiments, where a rubber or glass rod is charged by friction, a net charge on the order of 10^{-6} C ($= 1\ \mu$C) is obtained. In other words, only a very small fraction of the total available charge is transferred between the rod and rubbing material. Table 15.1 lists the charges and masses of the electron, proton, and neutron.

When dealing with Coulomb's force law, remember that force is a vector quantity and must be treated accordingly. Figure 15.6a shows the electric force of repulsion between two positively charged particles. Electric forces obey Newton's third law, and hence the forces $\mathbf{F}_{12}$ and $\mathbf{F}_{21}$ are equal in magnitude but opposite in direction. (The notation $\mathbf{F}_{12}$ denotes the force exerted by particle 1 on particle 2. Likewise, $\mathbf{F}_{21}$ is the force exerted by particle 2 on particle 1.) It bears repeating

FIGURE 15.6 Two point charges separated by a distance r exert a force on each other given by Coulomb's law. The force on q_1 is equal in magnitude and opposite in direction to the force on q_2. (a) When the charges are of the same sign, the force is repulsive. (b) When the charges are of opposite sign, the force is attractive.

TABLE 15.1	Charge and Mass of the Electron, Proton, and Neutron	
Particle	Charge (C)	Mass (kg)
Electron	-1.60×10^{-19}	9.11×10^{-31}
Proton	$+1.60 \times 10^{-19}$	1.67×10^{-27}
Neutron	0	1.67×10^{-27}

that F_{12} and F_{21} are always equal regardless of whether q_1 and q_2 have the same magnitude.

The Coulomb force is the second example we have seen of a *field force*—a force exerted by one object on another even though there is no physical contact between them. Recall that another example of a field force is gravitational attraction. The mathematical form of the Coulomb force is the same as that of the gravitational force. That is, they are both inversely proportional to the square of the distance of separation. However, there are some important differences between electric and gravitational forces. Electric forces can be either attractive or repulsive, but gravitational forces are always attractive.

Quick Quiz 15.2

Object A has a charge of $+2\ \mu\text{C}$, and object B has a charge of $+6\ \mu\text{C}$. Which statement is true:
(a) $\mathbf{F}_{AB} = -3\mathbf{F}_{BA}$, (b) $\mathbf{F}_{AB} = -\mathbf{F}_{BA}$, or (c) $3\mathbf{F}_{AB} = -\mathbf{F}_{BA}$

Example 15.1   The Electric Force and the Gravitational Force

The electron and proton of a hydrogen atom are separated (on the average) by a distance of about 5.3×10^{-11} m. Find the magnitudes of the electric force and the gravitational force that each particle exerts on the other.

Solution   From Coulomb's law, we find that the attractive electric force has the magnitude

$$F_e = k_e \frac{|e|^2}{r^2} = \left(8.99 \times 10^9\ \frac{\text{N} \cdot \text{m}^2}{\text{C}^2}\right) \frac{(1.6 \times 10^{-19}\ \text{C})^2}{(5.3 \times 10^{-11}\ \text{m})^2} = 8.2 \times 10^{-8}\ \text{N}$$

From Newton's law of universal gravitation and Table 15.1, we find that the gravitational force has the magnitude

$$F_g = G \frac{m_e m_p}{r^2} = \left(6.67 \times 10^{-11}\ \frac{\text{N} \cdot \text{m}^2}{\text{kg}^2}\right) \frac{(9.11 \times 10^{-31}\ \text{kg})(1.67 \times 10^{-27}\ \text{kg})}{(5.3 \times 10^{-11}\ \text{m})^2}$$

$$= 3.6 \times 10^{-47}\ \text{N}$$

Because $F_e/F_g \approx 2 \times 10^{39}$, the gravitational force between the charged atomic particles is negligible compared with the electric force.

THE SUPERPOSITION PRINCIPLE

Frequently, more than two charges are present and it is necessary to find the resultant electric force on one of them. This can be accomplished by noting that the electric force between any pair of charges is given by Equation 15.1. Therefore, the resultant force on any one charge equals the vector sum of the forces exerted by the other individual charges that are present. This is another example of the **superposition principle.** For example, if you have three charges and you want to find the force exerted by charges 2 and 3 on charge 1, you first find the force exerted by charge 2 on charge 1 and the force exerted by charge 3 on charge 1. You then add these two forces together *vectorially* to get the resultant force on charge 1. The following example illustrates this procedure.

Webnote 15.2

You can check your answer to a Coulomb's law calculation at *http://hyperphysics.phy-astr.gsu.edu/hbase/electric.elefor.html#c1*

Example 15.2   Using the Superposition Principle

Consider three point charges at the corners of a triangle, as in Figure 15.7, where $q_1 = 6.00 \times 10^{-9}$ C, $q_3 = 5.00 \times 10^{-9}$ C, $q_2 = -2.00 \times 10^{-9}$ C, and the distances of separation are as shown in the figure. Find the resultant force on q_3.

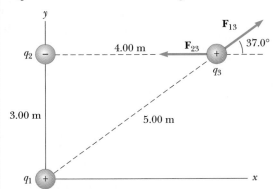

FIGURE 15.7 (Example 15.2) The force exerted by q_1 on q_3 is $\mathbf{F}_{13}$. The force exerted by q_2 on q_3 is $\mathbf{F}_{23}$. The *resultant force* exerted on q_3 is the *vector* sum $\mathbf{F}_{13} + \mathbf{F}_{23}$.

Reasoning   It is first necessary to find the direction of the forces exerted by q_1 and q_2 on q_3. The force $\mathbf{F}_{23}$ exerted by q_2 on q_3 is attractive because q_2 and q_3 have opposite signs. The force $\mathbf{F}_{13}$ exerted by q_1 on q_3 is repulsive because both q_1 and q_3 are positive. To find the resultant force on q_3, it is necessary to find the magnitudes of $\mathbf{F}_{23}$ and $\mathbf{F}_{13}$ by use of Coulomb's law and then add the two forces vectorially.

Solution   The magnitude of the force exerted by q_2 on q_3 is

$$F_{23} = k_e \frac{|q_2||q_3|}{r^2} = (8.99 \times 10^9 \text{ N} \cdot \text{m}^2/\text{C}^2) \frac{(2.00 \times 10^{-9} \text{ C})(5.00 \times 10^{-9} \text{ C})}{(4.00 \text{ m})^2}$$

$$= 5.62 \times 10^{-9} \text{ N}$$

The magnitude of the force exerted by q_1 on q_3 is

$$F_{13} = k_e \frac{|q_1||q_3|}{r^3} = (8.99 \times 10^9 \text{ N} \cdot \text{m}^2/\text{C}^2) \frac{(6.00 \times 10^{-9} \text{ C})(5.00 \times 10^{-9} \text{ C})}{(5.00 \text{ m})^2}$$

$$= 1.08 \times 10^{-8} \text{ N}$$

The force $\mathbf{F}_{13}$ makes an angle of 37.0° with the x axis and is directed along the line connecting q_1 and q_3. Therefore, the x component of this force has the magnitude $F_{13} \cos 37.0° = 8.63 \times 10^{-9}$ N, and the y component of $\mathbf{F}_{13}$ has the magnitude $F_{13} \sin 37.0° = 6.50 \times 10^{-9}$ N. The force $\mathbf{F}_{23}$ is in the negative x direction. Hence, the x and y components of the resultant force on q_3 are

$$F_x = 8.63 \times 10^{-9} \text{ N} - 5.62 \times 10^{-9} \text{ N} = 3.01 \times 10^{-9} \text{ N}$$

$$F_y = 6.50 \times 10^{-9} \text{ N}$$

The magnitude of the resultant force on the charge q_3 is therefore

$$\sqrt{(3.01 \times 10^{-9} \text{ N})^2 + (6.50 \times 10^{-9} \text{ N})^2} = \boxed{7.16 \times 10^{-9} \text{ N}}$$

and the force vector makes an angle with the x axis of

$$\theta = \tan^{-1}\left(\frac{F_y}{F_x}\right) = \tan^{-1}\left(\frac{6.50 \times 10^{-9} \text{ N}}{3.01 \times 10^{-9} \text{ N}}\right) = \boxed{65.2°}$$

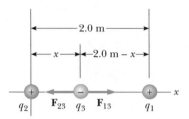

FIGURE 15.8   (Example 15.3) Three point charges are placed along the *x* axis. The charge q_3 is negative, whereas q_1 and q_2 are positive. If the resultant force on q_3 is zero, then the force $\mathbf{F}_{13}$ exerted by q_1 on q_3 must be equal in magnitude and opposite the force $\mathbf{F}_{23}$ exerted by q_2 on q_3.

Example 15.3   Where Is the Resultant Force Zero?

Three charges lie along the *x* axis as in Figure 15.8. The positive charge $q_1 = 15 \ \mu C$ is at $x = 2.0$ m, and the positive charge $q_2 = 6.0 \ \mu C$ is at the origin. Where must a *negative* charge q_3 be placed on the *x* axis so that the resultant force on it is zero?

Reasoning   Because q_3 is negative and q_1 and q_2 are positive, the forces $\mathbf{F}_{13}$ and $\mathbf{F}_{23}$ are both attractive, as indicated in Figure 15.8. The only location where the force exerted by q_2 on q_3 is opposite the force exerted by q_1 on q_3 lies on the *x* axis between q_1 and q_2. Because we require that the resultant force on q_3 be zero, then F_{13} must equal F_{23}.

Solution   If we let *x* be the coordinate of q_3, then the forces $\mathbf{F}_{13}$ and $\mathbf{F}_{23}$ have the magnitudes

$$F_{13} = k_e \frac{(15 \times 10^{-6} \ C)|q_3|}{(2.0 - x)^2} \quad \text{and} \quad F_{23} = k_e \frac{(6.0 \times 10^{-6} \ C)|q_3|}{x^2}$$

If the resultant force on q_3 is zero, then $F_{13} = F_{23}$, or

$$k_e \frac{(15 \times 10^{-6} \ C)|q_3|}{(2.0 - x)^2} = k_e \frac{(6.0 \times 10^{-6} \ C)|q_3|}{x^2}$$

Because k_e, 10^{-6}, and q_3 are common to both sides, they can be cancelled from the equation, and we have (after some reduction)

$$(2.0 - x)^2(6.0) = x^2(15)$$

This can be expanded to a quadratic equation, which can then be solved for *x*. An easier approach is first to take the positive square root of both sides:

$$(2.0 - x)\sqrt{6.0} = x\sqrt{15}$$

$$2.0 - x = x(1.58)$$

$$x = \boxed{0.78 \text{ m}}$$

15.4   THE ELECTRIC FIELD

Two different field forces have been introduced into our discussions so far: the gravitational force and the electrostatic force. As pointed out earlier, these forces are capable of acting through space, producing an effect even when there is no physical contact between the objects involved. Field forces can be discussed in a variety of ways, but an approach developed by Michael Faraday (1791–1867) is of such practical value that we shall devote much attention to it in the next few chapters. In this approach, an **electric field** is said to exist in the region of space around a charged object. When another charged object enters this electric field, the field is what exerts a force on the second charged object. As an example, consider Figure 15.9, which shows an object with a small positive charge q_0 placed near a second object with a larger positive charge Q.

We define the electric field at the location of the small "test" charge to be the electric force acting on it divided by the charge q_0 of the test charge:

$$\mathbf{E} \equiv \frac{\mathbf{F}}{q_0} \qquad [15.4]$$

Note that this is the electric field at the location of q_0 produced by the charge Q, not the field produced by q_0. The electric field is a vector quantity having the SI units newtons per coulomb (N/C). **The direction of E at a point is defined to be the direction of the electric force that would be exerted on a small posi-**

FIGURE 15.9   A small object with a positive charge q_0 placed near an object with a larger positive charge Q experiences an electric field $\mathbf{E}$ directed as shown. The magnitude of the electric field at the location of q_0 is defined as the electric force on q_0 divided by the charge q_0.

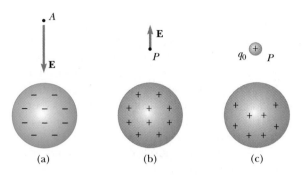

FIGURE 15.10   (a) The electric field at A due to the negatively charged sphere is downward, toward the negative charge. (b) The electric field at P due to the positively charged conducting sphere is upward, away from the positive charge. (c) A test charge q_0 placed at P will cause a rearrangement of charge on the sphere unless q_0 is very small compared with the charge on the sphere.

tive test charge placed at that point. Thus, in Figure 15.9, the direction of the electric field is horizontal and to the right. The electric field at point A in Figure 15.10a is vertical and downward because at this point a positive charge would experience a force of attraction toward the negatively charged sphere. Note that the test charge q_0 is required to be a small charge, in order to cause no significant rearrangement of the charge on the large sphere as shown in Figure 15.10c.

Once the electric field is known at some point, the force on *any* particle with charge q placed at that point can be calculated from a rearrangement of Equation 15.4:

$$\mathbf{F} = q\mathbf{E} \qquad [15.5]$$

Consider a point charge q located a distance r from a test charge q_0. According to Coulomb's law, the *magnitude* of the force on the test charge is

$$F = k_e \frac{|q||q_0|}{r^2}$$

Because the magnitude of the electric field at the position of the test charge is defined as $E = F/q_0$, we see that the *magnitude* of the electric field due to the charge q at the position of q_0 is

$$E = k_e \frac{|q|}{r^2} \qquad [15.6]$$

◀ Electric field due to a charge q

If q is *positive*, as in Figure 15.11a, the field at P due to this charge is *radially outward* from q. If q is *negative*, as in Figure 15.11b, the field at P is directed *toward* q. Equation 15.6 points out an important property of electric fields that makes them useful quantities for describing electrical phenomena. As the equation indicates, an electric field at a given point depends only on the charge q on the object setting up the field and the distance r from that object to a specific point in space. As a result, we can say that an electric field exists at point P in Figure 15.11 whether or not there is a test charge at P.

The principle of superposition holds when the electric field due to a group of point charges is calculated. We first use Equation 15.6 to calculate the electric field produced by each charge individually at a point, and then add these electric fields together as vectors.

PROBLEM-SOLVING STRATEGY   **Electric Forces and Fields**

1. **Units.** When performing calculations that use the Coulomb constant k_e, charges must be in coulombs and distances in meters. If they are given in other units, you must convert them.
2. **Applying Coulomb's law to point charges.** It is important to use the superposition principle properly when dealing with a collection of

FIGURE 15.11   A test charge q_0 at P is a distance r from a point charge q. (a) If q is positive, the electric field at P points radially *outward* from q. (b) If q is negative, the electric field at P points radially *inward* toward q.

interacting point charges. If several charges are present, the resultant force on any one of them is found by determining the individual force exerted on it by every other charge and then determining the vector sum of all these forces. The force that any charged object exerts on another is given by Coulomb's law, and the direction of the force is found by noting that the forces are repulsive between like charges and attractive between unlike charges.

3. **Calculating the electric field of point charges.** Because the electric field is a vector quantity, the superposition principle can be applied. To find the resultant electric field at a given point due to several charges, first calculate the electric field at the point due to each individual charge. The vector sum of the fields due to all the individual charges is the resultant field at the point.

APPLYING **PHYSICS** 15.2

An electron moving horizontally passes between two horizontal plates, the upper charged negatively, the lower positively. A uniform, upward-directed electric field exists in this region, and this field exerts an electric force downward on the electron. Describe the movement of the electron in this region.

Explanation The magnitude of the electric force on the electron of charge e due to a uniform electric field $\mathbf{E}$ is $F = eE$. Thus, the force is constant. Compare this to the force on a projectile of mass m moving in the gravitational field of Earth. The magnitude of the gravitational force is mg. In both cases, the particle is subject to a constant force in the vertical direction and has an initial velocity in the horizontal direction. Thus, the path will be the same in each case—the electron will move as a projectile with an acceleration in the vertical direction and constant velocity in the horizontal direction. Once the electron leaves the region between the plates, the electric field disappears, and the electron continues moving in a straight line according to Newton's first law.

Example 15.4 Electric Force on a Proton

Find the electric force on a proton placed in an upward electric field of magnitude 2.0×10^4 N/C.

Solution Because the charge on a proton is $+q = +1.6 \times 10^{-19}$ C, the electric force acting on the proton is

$$F = qE = (1.6 \times 10^{-19}\,\text{C})(2.0 \times 10^4\,\text{N/C}) = \boxed{3.2 \times 10^{-15}\,\text{N}}$$

where the force is upward, in the positive y direction. The magnitude of the gravitational force acting downward on the proton has the value $m_p g = (1.67 \times 10^{-27}\,\text{kg})(9.80\,\text{m/s}^2) = 1.64 \times 10^{-26}$ N. Hence, the magnitude of the gravitational force is negligible compared with that of the electric force.

Example 15.5 Electric Field Due to Two Point Charges

Charge $q_1 = 7.00\ \mu$C is at the origin, and charge $q_2 = -5.00\ \mu$C is on the x axis, 0.300 m from the origin (Fig. 15.12). Find the electric field at point P, which has coordinates (0, 0.400) m.

Reasoning It is first necessary to find the direction of the field at P set up by each charge. The field $\mathbf{E}_1$ at P due to q_1 is vertically upward as in Figure 15.12 because $\mathbf{E}_1$ is in the direction that a positive test charge would move if it were placed at P. Likewise, the field $\mathbf{E}_2$ at P due to q_2 is directed toward q_2 as in Figure 15.12. The magnitudes of the fields can be found from $E = k_e |q|/r^2$ and then added together vectorially.

Solution The magnitudes of $\mathbf{E}_1$ and $\mathbf{E}_2$ are

$$E_1 = k_e \frac{|q_1|}{r_1^2} = (8.99 \times 10^9 \text{ N} \cdot \text{m}^2/\text{C}^2) \frac{(7.00 \times 10^{-6} \text{ C})}{(0.400 \text{ m})^2} = 3.93 \times 10^5 \text{ N/C}$$

$$E_2 = k_e \frac{|q_2|}{r_2^2} = (8.99 \times 10^9 \text{ N} \cdot \text{m}^2/\text{C}^2) \frac{(5.00 \times 10^{-6} \text{ C})}{(0.500 \text{ m})^2} = 1.80 \times 10^5 \text{ N/C}$$

The vector $\mathbf{E}_1$ has an x component of zero. The vector $\mathbf{E}_2$ has an x component given by $E_2 \cos \theta = \frac{3}{5} E_2 = 1.08 \times 10^5$ N/C and a negative y component given by $-E_2 \sin \theta = -\frac{4}{5} E_2 = -1.44 \times 10^5$ N/C. Hence, the resultant component in the x direction is

$$E_x = 1.08 \times 10^5 \text{ N/C}$$

and the resultant component in the y direction is

$$E_y = E_{y1} + E_{y2} = 3.93 \times 10^5 \text{ N/C} - 1.44 \times 10^5 \text{ N/C} = 2.49 \times 10^5 \text{ N/C}$$

From the Pythagorean theorem ($E = \sqrt{E_x^2 + E_y^2}$), we find that $\mathbf{E}$ has a magnitude 2.72×10^5 N/C and makes an angle ϕ with the positive x axis given by

$$\phi = \tan^{-1}\left(\frac{E_y}{E_x}\right) = \tan^{-1}\left(\frac{2.49 \times 10^5 \text{ N/C}}{1.08 \times 10^5 \text{ N/C}}\right) = \boxed{66.5°}$$

EXERCISE Find the force on a positive test charge of 2.00×10^{-8} C placed at P.

ANSWER 5.44×10^{-3} N in the same direction as $\mathbf{E}$.

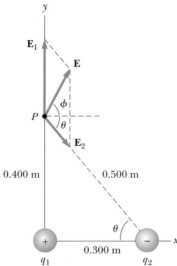

FIGURE 15.12 (Example 15.5) The resultant electric field $\mathbf{E}$ at P equals the vector sum $\mathbf{E}_1 + \mathbf{E}_2$, where $\mathbf{E}_1$ is the field due to the positive charge q_1 and $\mathbf{E}_2$ is the field due to the negative charge q_2.

Quick Quiz 15.3 A test charge of $+3\ \mu\text{C}$ is at a point P where the electric field due to other charges is directed to the right and has a magnitude of 4×10^6 N/C. If the test charge is replaced with a charge of $-3\ \mu\text{C}$, the electric field at P (a) has the same magnitude but changes direction, (b) increases in magnitude and changes direction, (c) remains the same, or (d) decreases in magnitude and changes direction.

Quick Quiz 15.4 A Styrofoam ball covered with a conducting paint has a mass of 5.0×10^{-3} kg and a charge of $4.0\ \mu\text{C}$. What electric field directed upward will produce an electric force on the ball that will balance the weight of the ball? (a) 8.2×10^2 N/C (b) 1.2×10^4 N/C (c) 2.0×10^{-2} N/C (d) 5.1×10^6 N/C

Quick Quiz 15.5 A circular ring of radius b has a total charge q uniformly distributed around it. The magnitude of the electric field at the center of the ring is (a) 0, (b) $k_e q/b^2$, (c) $k_e q^2/b^2$, (d) $k_e q^2/b$, or (e) none of these.

A "free" electron and "free" proton are placed in an identical electric field. Which of the following statements are true? (a) Each particle experiences the same electric force and the same acceleration. (b) The electric force on the proton is greater in magnitude than the force on the electron but in the opposite direction. (c) The electric force on the proton is equal in magnitude to the force on the electron, but in the opposite direction. (d) The magnitude of the acceleration of the electron is greater than that of the proton. (e) Both particles experience the same acceleration.

15.5 ELECTRIC FIELD LINES

A convenient aid for visualizing electric field patterns is to draw lines pointing in the direction of the electric field vector at any point. These lines introduced by Michael Faraday—called **electric field lines**—are related to the electric field in any region of space in the following manner:

1. The electric field vector **E** is tangent to the electric field lines at each point.
2. The number of lines per unit area through a surface perpendicular to the lines is proportional to the strength of the electric field in a given region.

Thus, **E** is large when the field lines are close together and small when they are far apart.

Figure 15.13a shows some representative electric field lines for a single positive point charge. Note that this two-dimensional drawing contains only the field lines that lie in the plane containing the point charge. The lines are actually directed radially outward from the charge in *all* directions, somewhat like the quills of an angry porcupine. Because a positive test charge placed in this field would be repelled by the charge *q*, the lines are directed radially away from the positive charge. In a similar way, the electric field lines for a single negative point charge are directed toward the charge (Fig. 15.13b). In either case, the lines are radial and extend all the way to infinity. Note that the lines are closer together as they get near the charge, indicating that the strength of the field is increasing. Equation 15.6 verifies that this should indeed be the case.

The rules for drawing electric field lines for any charge distribution are as follows:

1. The lines for a group of point charges must begin on positive charges and end on negative charges. In the case of an excess of charge, some lines will begin or end infinitely far away.

FIGURE 15.13 The electric field lines for a point charge. (a) For a positive point charge, the lines radiate outward. (b) For a negative point charge, the lines converge inward. Note that the figures show only those field lines that lie in the plane containing the charge. (c) The dark lines are small pieces of thread suspended in oil, which align with the electric field produced by a small charged conductor at the center. *(Photo courtesy of Harold M. Waage, Princeton University)*

(a)   (b)   (c)

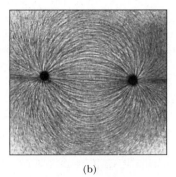

(a)   (b)

FIGURE 15.14 (a) The electric field lines for two equal and opposite point charges (an electric dipole). Note that the number of lines leaving the positive charge equals the number terminating at the negative charge. (b) The dark lines are small pieces of thread suspended in oil, which align with the electric field produced by two charged conductors. *(Photo courtesy of Harold M. Waage, Princeton University)*

2. The number of lines drawn leaving a positive charge or ending on a negative charge is proportional to the magnitude of the charge.

3. No two field lines can cross each other.

Figure 15.14 shows the beautifully symmetric electric field lines for two point charges of equal magnitude but opposite sign. This charge configuration is called an **electric dipole.** In this case, the number of lines that begin at the positive charge must equal the number that terminate at the negative charge. At points very near either charge, the lines are nearly radial. The high density of lines between the charges indicates a strong electric field in this region.

Figure 15.15 shows the electric field lines in the vicinity of two equal positive point charges. Again, close to either charge the lines are nearly radial. The same number of lines emerges from each charge because the charges are equal in magnitude. At great distances from the charges, the field is approximately equal to that of a single point charge of magnitude $2q$. The bulging out of the electric field lines between the charges indicates the repulsive nature of the electric force between like charges. Also, the low density of field lines between the charges indicates a weak field in this region, unlike the dipole.

Finally, Figure 15.16 is a sketch of the electric field lines associated with the positive charge $+2q$ and the negative charge $-q$. In this case, the number of lines leaving charge $+2q$ is twice the number terminating on charge $-q$. Hence, only

Tip 15.1 ELECTRIC FIELD LINES ARE NOT PATHS OF PARTICLES

Electric field lines are *not* material objects. They are used only as a pictorial representation of the electric field at various locations. Except in special cases, they *do not* represent the path of a charged particle released in an electric field.

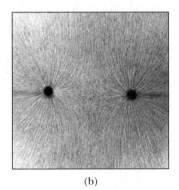

(a)   (b)

FIGURE 15.15 (a) The electric field lines for two positive point charges. The points *A*, *B*, and *C* will be discussed in Quick Quiz 15.7. (b) The dark lines are small pieces of thread suspended in oil, which align with the electric field produced by two charged conductors. *(Photo courtesy of Harold M. Waage, Princeton University)*

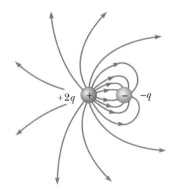

FIGURE 15.16 The electric field lines for a point charge $+2q$ and a second point charge $-q$. Note that two lines leave the charge $+2q$ for every line that terminates on $-q$.

half of the lines that leave the positive charge end at the negative charge. The remaining half terminate on negative charges that we assume to be located at infinity. At great distances from the charges (great compared with the charge separation), the electric field lines are equivalent to those of a single charge + *q*.

> **Quick Quiz 15.7** Rank the magnitudes of the electric field at points *A*, *B*, and *C* in Figure 15.15, largest magnitude first.

APPLYING **PHYSICS** 15.3

The electric field near Earth's surface in fair weather is about 100 N/C downward. Under a thundercloud, the electric field can be very large, on the order of 20 000 N/C. How are these electric fields measured?

Explanation A device for measuring these fields is called the *field mill*. Figure 15.17 shows the fundamental components of a field mill—two metal plates parallel to the ground. Each plate is connected to ground with a wire, with an ammeter (a low-resistance device for measuring flow of charge, to be discussed in Section 19.6) in one path. Consider first just the lower plate. Because it is connected to ground, and the ground happens to carry a negative charge, the plate is negatively charged. Thus, electric field lines, directed downward, end on the plate, as in Figure 15.17a. Now, imagine that the upper plate is suddenly moved over the lower plate, as in Figure 15.17b. This plate is also connected to ground, and is also negatively charged, so the field lines now end on the upper plate. The negative charges in the lower plate are repelled by those on the upper plate and must pass through the ammeter, registering a flow of charge. The amount of charge that was on the lower plate is related to the strength of the electric field. Thus, the flow of charge through the ammeter can be calibrated to measure the electric field. The plates are normally designed like the blades of a fan, with the upper plate rotating so that the lower plate is alternately covered and uncovered. As a result, charges flow back and forth continually through the ammeter, and the reading can be related to the electric field strength.

(a)   (b)

FIGURE 15.17 (Applying Physics 15.3) In (a), electric field lines end on negative charges on the lower plate. In (b), the second plate is moved above the lower plate. Electric field lines now end on the upper plate, and the negative charges in the lower plate are repelled through the ammeter.

15.6 CONDUCTORS IN ELECTROSTATIC EQUILIBRIUM

A good electric conductor, such as copper, contains charges (electrons) that are not bound to any atom and are free to move about within the material. When no net motion of charge occurs within a conductor, the conductor is said to be in

electrostatic equilibrium. An isolated conductor (one that is insulated from ground) has the following properties:

1. **The electric field is zero everywhere inside the conducting material.**
2. **Any excess charge on an isolated conductor resides entirely on its surface.**
3. **The electric field just outside a charged conductor is perpendicular to the conductor's surface.**
4. **On an irregularly shaped conductor, the charge accumulates at locations where the radius of curvature of the surface is smallest—that is, at sharp points.**

Properties of an isolated conductor

The first property can be understood by examining what would happen if it were *not* true. If there were an electric field inside a conductor, the free charge there would move and a flow of charge, or current, would be created. However, if there were a net movement of charge, the conductor would no longer be in electrostatic equilibrium.

Property 2 is a direct result of the $1/r^2$ repulsion between like charges described by Coulomb's law. If by some means an excess of charge is placed inside a conductor, the repulsive forces between the like charges push them as far apart as possible, causing them to quickly migrate to the surface. (We do not prove it here, but it is of interest to note that the excess charge resides on the surface due to the fact that Coulomb's law is an inverse-square law. With any other power law, an excess of charge would exist on the surface, but there would be a distribution of charge, of either the same or opposite sign, inside the conductor.)

Property 3 can be understood by again considering what would happen if it were not true. If the electric field in Figure 15.18a were not perpendicular to the surface, the electric field would have a component along the surface, which would cause the free charges of the conductor to move (to the left in the figure). If the charges moved, however, a current would be created and the conductor would no longer be in electrostatic equilibrium. Hence, **E** must be perpendicular to the surface.

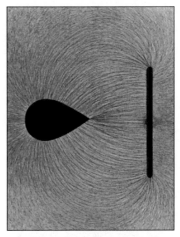

Electric field pattern of a charged conducting plate near an oppositely charged pointed conductor. Small pieces of thread suspended in oil align with the electric field lines. Note that the electric field is most intense near the pointed part of the conductor where the radius of curvature is the smallest. Also, the lines are perpendicular to the conductors. *(Courtesy of Harold M. Waage, Princeton University)*

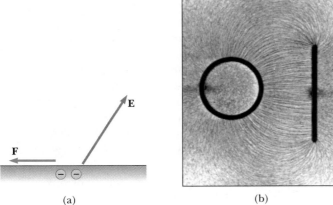

(a)   (b)

FIGURE 15.18 (a) Negative charges at the surface of a conductor. If the electric field were at an angle to the surface as shown, an electric force would be exerted on the charges along the surface and they would move to the left. Because the conductor is assumed to be in electrostatic equilibrium, **E** cannot have a component along the surface and hence must be perpendicular to it. (b) The electric field pattern of a charged conducting plate near an oppositely charged conducting cylinder. Small pieces of thread suspended in oil align with the electric field lines. Note that (1) the electric field lines are perpendicular to the conductors and (2) there are no lines inside the cylinder (**E** = 0). *(Courtesy of Harold M. Waage, Princeton University)*

FIGURE 15.19 (a) A conductor with a flatter end A and a relatively sharp end B. Excess charge placed on a conductor resides entirely at its surface and is distributed so that (b) there is less charge per unit area on the flatter end and (c) there is a large charge per unit area on the sharper end.

FIGURE 15.20 An experiment showing that any charge transferred to a conductor resides on its surface in electrostatic equilibrium. The hollow conductor is insulated from ground, and the small metal ball is supported by an insulating thread.

To see why property 4 must be true, consider Figure 15.19a, which shows a conductor that is fairly flat at one end and relatively pointed at the other. Any excess charge placed on the object moves to its surface. Figure 15.19b shows the forces between two such charges at the flatter end of the object. These forces are predominantly directed parallel to the surface. Thus, the charges move apart until repulsive forces from other nearby charges create an equilibrium situation. At the sharp end, however, the forces of repulsion between two charges are directed predominantly away from the surface, as in Figure 15.19c. As a result, there is less tendency for the charges to move apart along the surface here, and the amount of charge per unit area is greater than at the flat end. The cumulative effect of many such outward forces from nearby charges at the sharp end produces a large resultant force directed away from the surface that can be great enough to cause charges to leap from the surface into the surrounding air.

Many experiments have shown that the net charge on a conductor resides on its surface. The experiment described here was first performed by Michael Faraday. A metal ball having a negative charge was lowered at the end of a silk thread (an insulator) into an uncharged hollow conductor insulated from ground, as in Figure 15.20a. (This experiment is referred to as **Faraday's ice-pail experiment** because he used a metal ice pail as the hollow conductor.) As the ball was lowered into the pail, the needle on an electrometer attached to the outer surface of the pail was observed to deflect. (An electrometer is a device used to measure charge.) The needle deflected because the charged ball induced a positive charge on the inner wall of the pail, which left an equal negative charge on the outer wall (Fig. 15.20b).

Faraday next touched the inner surface of the pail with the ball and noted that the needle deflection did not change, either when the ball touched the inner surface of the pail (Fig. 15.20c) or when it was removed (Fig. 15.20d). Furthermore, he found that the ball was now completely uncharged. Apparently, when the ball touched the inside of the pail, the excess negative charge on the ball was neutralized by the induced positive charge on the inner surface of the pail. Thus, Faraday discovered the useful result that *all* the excess charge on an object can be transferred to an already charged metal shell if the object is touched to the *inside* of the shell. As we shall shortly see, this is the principle of operation of the Van de Graff generator.

Faraday concluded that because the electrometer deflection did not change when the charged ball touched the inside of the pail, the positive charge induced on the inside surface of the pail was just enough to neutralize the negative charge on the ball. As a result of his investigations, he concluded that a charged object suspended inside a metal container causes a rearrangement of charge on the container in such a manner that the sign of the charge on the inside surface of the container is *opposite* the sign of the charge on the suspended object. This produces a charge on the outside surface of the container of the same sign as that on the suspended object.

Faraday also found that if the electrometer was connected to the inside surface of the pail after the experiment was run, the needle showed no deflection. Thus, the *excess* charge acquired by the pail when contact was made between ball and pail appeared on the outer surface of the pail.

If a metal rod having sharp points is attached to a house, most of any charge on the house passes through these points, thus eliminating the induced charge on the house produced by storm clouds. In addition, a lightning discharge striking the house passes through the metal rod and is safely carried to the ground through wires leading from the rod to Earth. Lightning rods using this principle were first developed by Benjamin Franklin. It is an interesting sidelight to American history to note that some European countries could not accept the fact that such a worthwhile idea could have originated in the New World. As a result, they "improved" the design by eliminating the sharp points.

> **APPLICATION**
>
> LIGHTNING RODS

APPLYING PHYSICS 15.4

Suppose a point charge $+ Q$ is in empty space. Wearing rubber gloves, you proceed to surround the charge with a concentric spherical conducting shell. What effect does this have on the field lines from the charge?

Explanation When the spherical shell is placed around the charge, the charges in the shell rearrange to satisfy the rules for a conductor in equilibrium. A net charge of $- Q$ moves to the interior surface of the conductor, so that the electric field inside the conductor becomes zero. That is, the field lines originating on the $+ Q$ charge terminate on the $- Q$ charges. The movement of the $- Q$ charges to the inner surface of the sphere leaves a net charge of $+ Q$ on the outer surface of the sphere. Thus, the only change in the field lines from the initial situation will be the absence of field lines within the conductor.

APPLYING PHYSICS 15.5

Why is it safe to stay inside an automobile during a lightning storm?

Explanation Although many people believe that this is safe because of the insulating characteristics of the rubber tires, this is not true. Lightning is able to travel through several kilometers of air, so it can certainly penetrate a centimeter of rubber. The interior of the car is safe because charges on the car's metal shell reside on its outer surface, as noted in property 2 discussed earlier. Thus, an occupant in the automobile touching the inner surfaces is not in danger.

> **APPLICATION**
>
> DRIVER SAFETY DURING
> ELECTRICAL STORMS

15.7 THE MILLIKAN OIL-DROP EXPERIMENT

From 1909 to 1913, Robert Andrews Millikan (1868–1953) performed a brilliant set of experiments at the University of Chicago in which he measured the elementary charge e of the electron and demonstrated the quantized nature of the electronic charge. The apparatus he used, diagrammed in Figure 15.21, contains two parallel metal plates. Oil droplets that have been charged by friction in an atomizer are allowed to pass through a small hole in the upper plate. A horizontal light beam illuminates the oil droplets, which are viewed by a telescope whose axis is at right angles to the beam. When the droplets are viewed in this manner, they appear as shining stars against a dark background, and the rate of fall of individual drops can be determined.

Let us assume that a single drop having a mass m and carrying a charge q is being viewed, and that its charge is negative. If no electric field is present between the plates, the two forces acting on the charge are the force of gravity $m\mathbf{g}$ acting

> **Webnote 15.3**
>
> You can read the 1923 Nobel Prize presentation speech and download Millikan's Nobel lecture at *http://nobel.sdsc.edu/physics/ laureates/1923/millikan-bio.html*

FIGURE 15.21 A schematic view of the Millikan oil-drop apparatus.

downward and an upward viscous drag force **D** (Fig. 15.22a). The drag force is proportional to the speed of the drop. When the drop reaches its terminal speed v, the two forces balance each other ($mg = D$).

Now suppose that an electric field is set up between the plates by a battery connected so that the upper plate is positively charged. In this case, a third force $q\mathbf{E}$ acts on the charged drop. Because q is negative and **E** is downward, the electric force is *upward* as in Figure 15.22b. If this force is great enough, the drop moves upward and the drag force **D**′ acts downward. When the upward electric force $q\mathbf{E}$ balances the sum of the force of gravity and the drag force, both acting downward, the drop reaches a new terminal speed v'.

With the field turned on, a drop moves slowly upward, typically at rates of *hundredths* of a centimeter per second. The rate of fall in the absence of a field is comparable. Hence, a single droplet with constant mass and radius can be followed for hours as it alternately rises and falls, by simply turning the electric field on and off.

After making measurements on thousands of droplets, Millikan and his coworkers found that every drop, to within about 1% precision, had a charge equal to some positive or negative integer multiple of the elementary charge e. That is,

$$q = ne \qquad n = 0, \pm 1, \pm 2, \pm 3, \ldots \qquad \text{[15.7]}$$

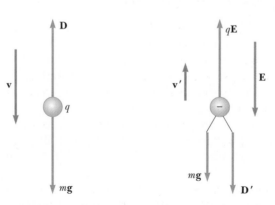

(a) Field off

(b) Field on

FIGURE 15.22 The forces on a charged oil droplet in the Millikan experiment.

where $e = 1.60 \times 10^{-19}$ C. It was later established that positive integer multiples of e arose when an oil droplet lost one or more electrons. Likewise, negative integer multiples of e occurred when a drop gained one or more electrons. Gains or losses in integral numbers provide conclusive evidence that charge is quantized. In 1923, Millikan was awarded the Nobel Prize in physics for this work.

Webnote 15.4

Interesting historical photos of Van de Graaff generators built for research (including a giant one on railroad tracks) can be found at *http://www.mos.org/sln/toe/history.html*

15.8 THE VAN DE GRAAFF GENERATOR

In 1929 Robert J. Van de Graaff designed and built an electrostatic generator that is used extensively in nuclear physics research. The principles of its operation can be understood with the help of the properties of electric fields and charges already presented in this chapter. Figure 15.23 shows the basic construction details of this device. A motor-driven pulley P moves a belt past positively charged comb-like metallic needles positioned at A. Negative charges are attracted to these needles from the belt, leaving the left side of the belt with a net positive charge. The positive charges attract electrons onto the belt as it moves past a second comb of needles at B, increasing the excess positive charge on the dome. Because the electric field inside the metal dome is negligible, the positive charge on it can easily be increased regardless of how much charge is already present. The result is that the dome is left with a large amount of positive charge.

This accumulation of charge on the dome cannot continue indefinitely, because eventually an electric discharge through the air takes place. To understand why, consider that, as more and more charge appears on the surface of the dome, the magnitude of the electric field at the surface of the dome is also increasing. Finally, the strength of the field becomes great enough to partially ionize the air near the surface, thus making the air partially conducting. Charges on the dome now have a pathway to leak off into the air, which can produce some spectacular "lightning bolts" as the discharge occurs. As noted earlier, charges find it easier to leap off a surface at points where the curvature is great. As a result, one way to inhibit the electric discharge and to increase the amount of charge that can be stored on the dome is to increase its radius. Another method for inhibiting discharge is to place the entire system in a container filled with a high-pressure gas, which is significantly more difficult to ionize than air at atmospheric pressure.

If protons (or other charged particles) are introduced into a tube attached to the dome, the large electric field outside the dome exerts a repulsive force on the protons, causing them to accelerate to energies high enough to initiate nuclear reactions between the protons and various target nuclei.

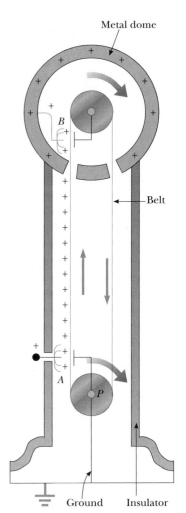

FIGURE 15.23 A diagram of a Van de Graaff generator. Charge is transferred to the dome by means of a rotating belt. The charge is deposited on the belt at point A and transferred to the dome at point B.

15.9 ELECTRIC FLUX AND GAUSS'S LAW

In this section, we introduce the concept of electric flux and then show that it can be used to calculate the electric field of a charged object. This technique for calculating electric fields was developed by Karl Friedrich Gauss (1777–1855). Gauss's law provides a direct connection between the electric flux through a closed surface and the total charge inside that surface producing the electric field. It is basically a statement that the number of electric field lines coming out of a closed surface is proportional to the positive charge inside the surface, a consequence of field lines starting on positive charges.

ELECTRIC FLUX

Consider an electric field that is uniform in both magnitude and direction as in Figure 15.24. The electric field lines penetrate a surface of area A, which is perpendicular to the field. The technique used for drawing a figure such as Figure 15.24 is that the number of lines per unit area N/A is proportional to the

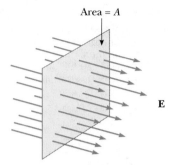

FIGURE 15.24 Field lines of a uniform electric field penetrating a plane of area *A* perpendicular to the field. The electric flux Φ_E through this area is equal to *EA*.

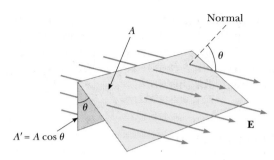

FIGURE 15.25 Field lines for a uniform electric field through an area *A* that is at an angle to the field. Because the number of lines that go through the shaded area *A'* is the same as the number that go through *A*, we conclude that the flux through *A'* is equal to the flux through *A* and is given by $\Phi_E = EA \cos \theta$.

magnitude of the electric field, or $E \propto N/A$. We can rewrite this as $N \propto EA$, which means that the number of field lines is proportional to the *product* of *E* and *A*. This product of *E* and *A* is called the **electric flux** and is represented by the symbol Φ_E, where

$$\Phi_E = EA \tag{15.8}$$

Note that Φ_E has SI units of $N \cdot m^2/C$ and is essentially the number of field lines that pass through some area *A* oriented perpendicular to the field. (It is called flux by analogy with the term flux in fluid flow, which is just the volume of liquid flowing through a perpendicular area per second.). If the surface under consideration is not perpendicular to the field as in Figure 15.25, the expression for the electric flux is

▶ Electric flux

$$\Phi_E = EA \cos \theta \tag{15.9}$$

where the perpendicular ("normal") to the area *A* is at an angle θ with respect to the field. The number of lines that cross this area is equal to the number that cross the projected area *A'*, which is perpendicular to the field. We see that the two areas are related by $A' = A \cos \theta$. From Equation 15.9 we see that the flux through a surface of fixed area has the maximum value *EA* when the surface is perpendicular to the field (when $\theta = 0°$) and that the flux is zero when the surface is parallel to the field (when $\theta = 90°$). **When the area is constructed such that a closed surface is formed, we shall adopt the convention that flux lines passing into the interior of the volume are negative and those passing out of the interior of the volume are positive.**

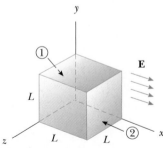

FIGURE 15.26 (Example 15.6) A hypothetical surface in the shape of a cube in a uniform electric field parallel to the *x* axis. The net flux through the surface is zero when the net charge inside the cube is zero.

Example 15.6 Flux Through a Cube

Consider a uniform electric field oriented in the *x* direction. Find the net electric flux through the surface of a cube of edges *L* oriented as shown in Figure 15.26.

Solution The net flux can be evaluated by summing up the fluxes through each face of the cube. First, note that the flux through four of the faces is zero, because **E** is parallel to the surface in these areas. These surfaces are those that are parallel to the *xy* and *xz* planes. For these surfaces, $\theta = 90°$, so $\Phi_E = EA \cos 90° = 0$. For surface 1, which lies in the *yz* plane in Figure 15.26, the flux lines pass into the interior of the cube, and the flux is taken to be negative. We have

$$\Phi_{E1} = -EA = -EL^2$$

For surface 2 at the right face of the cube, the flux is positive and given by

$$\Phi_{E2} = EA = EL^2$$

The net flux through the surface of the cube is

$$\Phi_{E,\text{net}} = \sum EA = \Phi_{E1} + \Phi_{E2} = -EL^2 + EL^2 = \boxed{0}$$

GAUSS'S LAW

Consider a point charge q surrounded by a spherical surface of radius r centered on the charge, as in Figure 15.27. The magnitude of the electric field everywhere on the surface of the sphere is

$$E = k_e \frac{q}{r^2}$$

Note that the electric field is perpendicular to the spherical surface at all points on the surface. Therefore, the electric flux through the surface is EA, where $A = 4\pi r^2$ is the surface area of the sphere:

$$\Phi_E = EA = k_e \frac{q}{r^2} (4\pi r^2) = 4\pi k_e q$$

It is convenient to express k_e in terms of another constant ϵ_0, where $k_e = 1/(4\pi\epsilon_0)$. The constant ϵ_0 is called the **permittivity of free space** and has the value

$$\epsilon_0 = \frac{1}{4\pi k_e} = 8.85 \times 10^{-12} \text{ C}^2/\text{N} \cdot \text{m}^2 \qquad \textbf{[15.10]}$$

The electric flux through the closed spherical surface that surrounds the charge q can now be expressed as

$$\Phi_E = 4\pi k_e q = \frac{q}{\epsilon_0}$$

This result says that the electric flux through a sphere that surrounds a charge q is equal to the charge divided by the constant ϵ_0. Using calculus, we can show that this simple result is true for *any* closed surface that surrounds the charge q. For example, if the surface surrounding q is irregular, the flux through that surface is also q/ϵ_0. This result is a special case of Gauss's law and the closed surface is referred to as a *gaussian surface*.

In general, **Gauss's law** states that **the electric flux through any closed surface is equal to the net charge Q inside the surface divided by ϵ_0:**

$$\Phi_E = \frac{Q}{\epsilon_0} \qquad \textbf{[15.11]}$$

 Gauss's law

It is the fundamental law describing how charges create electric fields. In principle, Gauss's law can always be used to calculate the electric field of a system of charges or a continuous distribution of charge. In practice, the technique is useful only in a limited number of situations in which there is a high degree of symmetry.

FIGURE 15.27 The flux through the spherical surface of radius r surrounding a point charge q is $\Phi_E = q/\epsilon_0$. The flux through any arbitrary surface surrounding the charge is also equal to q/ϵ_0.

Tip 15.2 GAUSSIAN SURFACES ARE NOT REAL

Note that a gaussian surface is an imaginary surface. It does not have to coincide with the surface of a physical object.

APPLYING PHYSICS 15.6

A spherical surface surrounds a point charge q. Describe what happens to the total flux through the surface if (a) the charge is tripled, (b) the volume of the sphere is doubled, (c) the surface is changed to a cube, (d) the charge is moved to another location inside the surface, and (e) the charge is moved outside the surface.

Explanation (a) If the charge is tripled, the flux through the surface is also tripled, because the net flux is proportional to the charge inside the surface. (b) The flux remains constant when the volume changes because the surface surrounds the same amount of charge, regardless of its volume. (c) The flux does not change when the shape of the closed surface changes. (d) The flux through the closed surface remains unchanged as the charge inside the surface is moved to another location inside that surface. (e) The flux is zero because the charge inside the surface is zero. All of these conclusions are arrived at through an understanding of Gauss's law.

For a surface through which the net flux is zero, the following four statements *could* be true. Which of the statements *must* be true? (a) There are no charges inside the surface. (b) The net charge inside the surface is zero. (c) The electric field is zero everywhere on the surface. (d) The number of electric field lines entering the surface equals the number leaving the surface.

Example 15.7   The Electric Field of a Charged Thin Spherical Shell

A thin spherical shell of radius R has a total charge Q distributed uniformly over its surface (Fig. 15.28). Find the electric field at points (a) outside and (b) inside the shell.

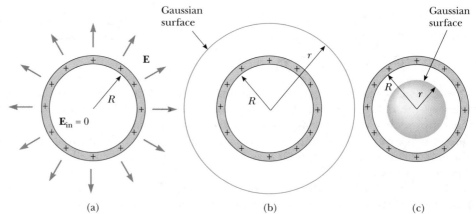

(a)   (b)   (c)

FIGURE 15.28   (Example 15.7) (a) The electric field inside a uniformly charged spherical shell is *zero*. The field outside is the same as that of a point charge having a total charge Q located at the center of the shell. (b) The construction of a gaussian surface for calculating the electric field *outside* a spherical shell. (c) The construction of a gaussian surface for calculating the electric field *inside* a spherical shell.

Solution

A The calculation of the field outside the shell is identical to that carried out in the body of the text for a point charge. Because the charge distribution is spherically symmetric, we imagine a spherical gaussian surface of radius $r > R$, concentric with the shell, as in Figure 15.28b. Following the same line of reasoning as that for a point charge, the flux through this surface is $\Phi_E = E(4\pi r^2) = Q/\epsilon_0$. Thus, the magnitude of the electric field everywhere on this surface is

$$E = \frac{Q}{4\pi r^2 \epsilon_0} = k_e \frac{Q}{r^2}$$

Note that the field at a point outside the shell is equivalent to that of a point charge located at the center of the shell.

B The electric field inside the spherical shell is *zero*. This follows from Gauss's law applied to a spherical gaussian surface of radius $r < R$, placed inside the shell, as shown in Figure 15.28c. Because the net charge inside this surface is zero, $Q = 0$, and we see that $\mathbf{E}_{in} = 0$ in the region inside the shell.

Example 15.8 A Nonconducting Plane Sheet of Charge

Find the electric field due to a nonconducting infinite plane sheet of charge with uniform charge per unit area σ.

Solution The symmetry of the situation shows that the electric field must be perpendicular to the plane and that the direction of the field on one side of the plane must be opposite its direction on the other side, as shown in Figure 15.29. It is convenient to choose for our gaussian surface a small cylinder whose axis is perpendicular to the plane and whose ends each have an area A. Note that no electric field lines pass through the curved surface of the cylinder. Field lines pass only through the two ends of the cylinder. Hence, the total flux through the cylinder is $\Phi_E = E(2A)$. Because the total charge *inside* the gaussian surface is $Q = \sigma A$, Gauss's law gives

$$\Phi_E = 2EA = \frac{\sigma A}{\epsilon_0}$$

or

$$E = \frac{\sigma}{2\epsilon_0}$$

Because the distance of the surfaces from the plane does not appear in our result, we conclude that $E = \sigma/2\epsilon_0$ at *any* distance from the infinite plane. That is, the electric field is uniform everywhere, depending only on the charge per unit area σ.

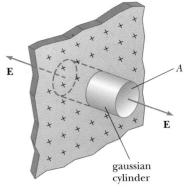

FIGURE 15.29 (Example 15.8) A cylindrical gaussian surface penetrating an infinite sheet of charge. The flux through each end of the gaussian surface is *EA*. There is no flux through the cylinder's curved surface.

SUMMARY

Electric charges have the following important properties:

1. Unlike charges attract one another and like charges repel one another.
2. Electric charge is always conserved.
3. Charge is quantized — that is, it exists in discrete packets that are integral multiples of the electronic charge.
4. The force between charged particles varies as the inverse square of their separation.

    Conductors are materials in which charges move freely. **Insulators** are materials that do not readily transport charge.

    Coulomb's law states that the electric force between two stationary charged particles separated by a distance r has the magnitude

$$F = k_e \frac{|q_1||q_2|}{r^2} \qquad \text{[15.1]}$$

where $|q_1|$ and $|q_2|$ are the magnitudes of the charges on the particles in coulombs and k_e is the **Coulomb constant,** which has the approximate value

$$k_e \approx 8.99 \times 10^9 \ \text{N} \cdot \text{m}^2/\text{C}^2 \qquad \text{[15.3]}$$

    An electric field **E** exists at some point in space if a small positive test charge q_0 placed at that point experiences an electric force **F**. The electric field is defined as

$$\mathbf{E} \equiv \frac{\mathbf{F}}{q_0} \qquad \text{[15.4]}$$

The **direction** of the electric field at a point in space is defined to be the direction of the electric force that would be exerted on a small positive charge placed at that point.

The magnitude of the electric field due to a point charge q at distance r from the point charge is

$$E = k_e \frac{|q|}{r^2} \qquad\qquad \text{[15.6]}$$

Electric field lines are useful for describing the electric field in any region of space. The electric field vector **E** is tangent to the electric field lines at every point. Furthermore, the number of electric field lines per unit area through a surface perpendicular to the lines is proportional to the strength of the electric field in that region.

A **conductor in electrostatic equilibrium** has the following properties:

1. The electric field is zero everywhere inside the conducting material.
2. Any excess charge on an isolated conductor must reside entirely on its surface.
3. The electric field just outside a charged conductor is perpendicular to the conductor's surface.
4. On an irregularly shaped conductor, charge accumulates where the radius of curvature of the surface is smallest—that is, at sharp points.

Gauss's law states that the electric flux through any closed surface is equal to the net charge Q inside the surface divided by the permittivity of free space, ϵ_0:

$$\Phi_E = \frac{Q}{\epsilon_0} \qquad\qquad \text{[15.11]}$$

CONCEPTUAL QUESTIONS

1. Because of a higher moisture content, air is a better conductor of electricity in the summer than in the winter. Would you expect the shocks from static electricity to be more severe in summer or winter? Explain.

2. Hospital personnel must wear special conducting shoes when they work around oxygen in an operating room. Why? Contrast with what might happen when personnel wear rubber-soled shoes.

3. Two insulated rods are oppositely charged on the ends. They are mounted at the centers so that they are free to rotate, and then held in the position shown in a view from above (Fig. Q15.3). The rods rotate in the plane of the paper. Will the rods stay in the positions when released? If not, into what position(s) will they move? Will the final configuration(s) be stable?

FIGURE Q15.3

4. Explain from an atomic viewpoint why charge is usually transferred by electrons.

5. Explain how a positively charged object can leave another metallic object with a net negative charge. Discuss the motion of charges during this process.

6. Why are electrostatic effects often seen on clothes just out of a dryer?

7. If a metal object receives a positive charge, does its mass appreciably increase, decrease, or stay the same? What happens to its mass if the object receives a negative charge?

8. When defining the electric field, why is it necessary to specify that the magnitude of the test charge be very small?

9. In fair weather, there is an electric field at Earth's surface, pointing down into the ground. What is the electric charge on the ground in this situation?

10. A student stands on a thick piece of insulating material, places her hand on top of a Van de Graaff generator, and then turns on the generator. Does she receive a shock?

11. An uncharged, metal-coated Styrofoam ball is suspended in the region between two vertical metal plates. If the two plates are charged, one positive and one negative, describe the motion of the ball after it is brought into contact with one of the plates.

12. Is it possible for an electric field to exist in empty space? Explain.

13. There are great similarities between electric and gravitational fields. A room can be electrically shielded so that there are no electric fields in the room by surrounding it with a conductor. Can a room be gravitationally shielded? Why or why not? (*Hint:* There are two kinds of charge in nature, but only one kind of mass.)

14. Would life be different if the electron were positively charged and the proton were negatively charged? Does the choice of signs have any bearing on physical and chemical interactions? Explain.

15. Explain why Gauss's law cannot be used to calculate the electric field of (a) a polar molecule consisting of a positive and a negative charge separated by a very small distance, (b) a charged disk, and (c) three point charges at the corner of a triangle.

16. Why should a ground wire be connected to the metal support rod for a television antenna?

17. A balloon is negatively charged by rubbing and then clings to a wall. Does this mean that the wall is positively charged? Why does the balloon eventually fall?

18. Why is it not a good idea to seek shelter under a tree during a lightning storm?

19. A charged comb often attracts small bits of dry paper that then fly away when they touch the comb. Explain.

PROBLEMS

1, **2**, **3** = straightforward, intermediate, challenging   ☐ = full solution available in Student Solutions Manual/Study Guide

web = solution posted at **http://info.brookscole.com/serway**   🔬 = biomedical application

Section 15.3 Coulomb's Law

1. A 4.5×10^{-9} C charge is located 3.2 m from a -2.8×10^{-9} C charge. Find the electrostatic force exerted by one charge on the other.

2. Two neighboring cells are ionized (charged) by x-radiation. (a) If the charge on each cell is equal to 1.60×10^{-14} C and the cells are 2.5 μm apart, what is the force exerted by each cell? (b) If the distance between the cells is doubled, what happens to the force between them?

3. An alpha particle (charge = $+2.0e$) is sent at high speed toward a gold nucleus (charge = $+79e$). What is the electrical force acting on the alpha particle when it is 2.0×10^{-14} m from the gold nucleus?

4. Determine what the mass of a proton would be if the gravitational force between two of them were equal to the electrical force between them.

5. The nucleus of ^{8}Be, which consists of 4 protons and 4 neutrons, is very unstable and spontaneously breaks into two alpha particles (helium nuclei, each consisting of 2 protons and 2 neutrons). (a) What is the force between the two alpha particles when they are 5.00×10^{-15} m apart, and (b) what will be the magnitude of the acceleration of the alpha particles due to this force? Note that the mass of an alpha particle is 4.0026 u.

6. A molecule of DNA (deoxyribonucleic acid) is 2.17 μm long. The ends of the molecule become singly ionized—negative on one end, positive on the other. The helical molecule acts like a spring and compresses 1.00% upon becoming charged. Determine the effective spring constant of the molecule.

7. Suppose that 1.00 g of hydrogen is separated into electrons and protons. Suppose also that the protons are placed at Earth's North Pole and the electrons are placed at the South Pole. What is the resulting compressional force on Earth?

8. An electron is released a short distance above Earth's surface. A second electron directly below it exerts an electrostatic force on the first electron just great enough to cancel the gravitational force on it. How far below the first electron is the second?

9. Two identical conducting spheres are placed with their centers 0.30 m apart. One is given a charge of 12×10^{-9} C and the other a charge of -18×10^{-9} C. (a) Find the electrostatic force exerted by one sphere on the other. (b) The spheres are connected by a conducting wire. After equilibrium has occurred, find the electrostatic force between the two.

10. Calculate the magnitude and direction of the Coulomb force on each of the three charges in Figure P15.10.

FIGURE P15.10   (Problems 10 and 18)

11. Three charges are arranged as shown in Figure P15.11.
web Find the magnitude and direction of the electrostatic force on the charge at the origin.

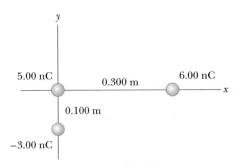

FIGURE P15.11

12. Three charges are arranged as shown in Figure P15.12. Find the magnitude and direction of the electrostatic force on the 6.00-nC charge.

FIGURE P15.12

13. Three point charges are located at the corners of an equilateral triangle as in Figure P15.13. Calculate the net electric force on the 7.00-μC charge.

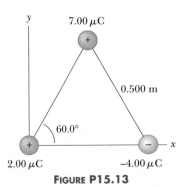

FIGURE P15.13

14. Two small beads having positive charges $3q$ and q are fixed at the opposite ends of a horizontal, insulating rod, extending from the origin to the point $x = d$. As in Figure

P15.14, a third small charged bead is free to slide on the rod. At what position is the third bead in equilibrium? Can it be in stable equilibrium?

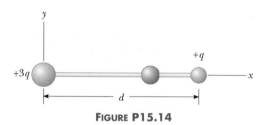

FIGURE P15.14

15. Two small metallic spheres, each of mass 0.20 g, are suspended as pendulums by light strings from a common point as shown in Figure P15.15. The spheres are given the same electric charge, and it is found that the two come to equilibrium when each string is at an angle of 5.0° with the vertical. If each string is 30.0 cm long, what is the magnitude of the charge on each sphere?

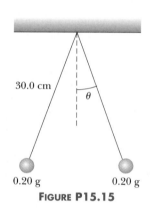

FIGURE P15.15

16. A charge of 6.00×10^{-9} C and a charge of -3.00×10^{-9} C are separated by a distance of 60.0 cm. Find the position at which a third charge of 12.0×10^{-9} C can be placed so that the net electrostatic force on it is zero.

Section 15.4   The Electric Field

17. In a hydrogen atom, what are the magnitude and direction of the electric field set up by the proton at the location of the electron $(0.51 \times 10^{-10}$ m away from the proton)?

18. (a) Determine the electric field strength at a point 1.00 cm to the left of the middle charge shown in Figure P15.10. (b) If a charge of -2.00 μC is placed at this point, what are the magnitude and direction of the force on it?

19. An airplane is flying through a thundercloud at a height of 2 000 m. (This is a very dangerous thing to do because of updrafts, turbulence, and the possibility of electric discharge.) If there are charge concentrations of $+40.0$ C at height 3 000 m within the cloud and -40.0 C at height 1 000 m, what is the electric field **E** at the aircraft?

20. An electron is accelerated by a constant electric field of magnitude 300 N/C. (a) Find the acceleration of the elec-

tron. (b) Use the equations of motion with constant acceleration to find the electron's speed after 1.00×10^{-8} s, assuming it starts from rest.

21. A piece of aluminum foil of mass 5.00×10^{-2} kg is suspended by a string in an electric field directed vertically upward. If the charge on the foil is 3.00 μC, find the strength of the field that will reduce the tension in the string to zero.

22. An electron with a speed of 3.00×10^6 m/s moves into a uniform electric field of 1 000 N/C. The field is parallel to the electron's motion. How far does the electron travel before it is brought to rest?

23. A proton accelerates from rest in a uniform electric field of 640 N/C. At some later time, its speed is 1.20×10^6 m/s. (a) Find the magnitude of the acceleration of the proton. (b) How long does it take the proton to reach this speed? (c) How far has it moved in this interval? (d) What is its kinetic energy at the later time?

24. Positive charges are situated at three corners of a rectangle, as shown in Figure P15.24. Find the electric field at the fourth corner.

FIGURE P15.24

25. Three identical charges $(q = -5.0$ μC$)$ are along a circle of 2.0-m radius at angles of 30°, 150°, and 270°, as shown in Figure P15.25. What is the resultant electric field at the center of the circle?

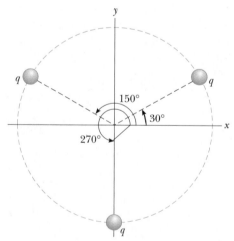

FIGURE P15.25

26. Two point charges lie along the y axis. A charge of $q_1 = -9.0$ μC is at $y = 6.0$ m, and a charge of $q_2 = -8.0$ μC is at $y = -4.0$ m. Locate the point (other than infinity) at which the total electric field is zero.

27. In Figure P15.27, determine the point (other than infinity) at which the total electric field is zero.

FIGURE P15.27

Section 15.5 Electric Field Lines

Section 15.6 Conductors in Electrostatic Equilibrium

28. Figure P15.28 shows the electric field lines for two point charges separated by a small distance. (a) Determine the ratio q_1/q_2. (b) What are the signs of q_1 and q_2?

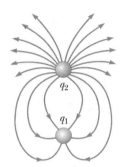

FIGURE P15.28

29. (a) Sketch the electric field lines around an isolated point charge, $q > 0$. (b) Sketch the electric field pattern around an isolated negative point charge of magnitude $-2q$.

30. (a) Sketch the electric field pattern around two positive point charges of magnitude 1 μC placed close together. (b) Sketch the electric field pattern around two negative point charges of -2 μC placed close together. (c) Sketch the pattern around two point charges of $+1$ μC and -2 μC placed close together.

31. Two point charges are a small distance apart. (a) Sketch the electric field lines for the two if one has a charge four times that of the other and both charges are positive. (b) Repeat for the case in which both charges are negative.

32. (a) Sketch the electric field pattern set up by a positively charged hollow sphere. Include the regions both inside and outside the sphere. (b) A conducting cube is given a positive charge. Sketch the electric field pattern both inside and outside the cube.

33. Refer to Figure 15.20. The charge lowered into the center of the hollow conductor has a magnitude of 5 μC. Find the magnitude and sign of the charge on the inside and outside of the hollow conductor when the charge is as shown in (a) Figure 15.20a, (b) Figure 15.20b, (c) Figure 15.20c, and (d) Figure 15.20d.

Section 15.8 The Van de Graaff Generator

34. The dome of a Van de Graaff generator receives a charge of 2.0×10^{-4} C. Find the strength of the electric field (a) inside the dome; (b) at the surface of the dome, assuming it has a radius of 1.0 m; and (c) 4.0 m from the center of the dome. (*Hint:* See Section 15.6 to review properties of conductors in electrostatic equilibrium. Also

use the fact that the points on the surface are outside a spherically symmetric charge distribution; the total charge may be considered as located at the center of the sphere.)

35. If the electric field strength in air exceeds 3.0×10^6 N/C, the air becomes a conductor. Using this fact, determine the maximum amount of charge that can be carried by a metal sphere 2.0 m in radius. (See the hint in Problem 34.)

36. Air breaks down (loses its insulating quality) and sparking results if the field strength is increased to about 3.0×10^6 N/C. (a) What acceleration does an electron experience in such a field? (b) If the electron starts from rest, in what distance does it acquire a speed equal to 10% of the speed of light?

37. A Van de Graaff generator is charged so that the electric field at its surface is 3.0×10^4 N/C. Find (a) the electric force exerted on a proton released at its surface and (b) the acceleration of the proton at this instant of time.

Section 15.9 Electric Flux and Gauss's Law

38. A flat surface having an area of 3.2 m² is rotated in a uniform electric field of magnitude $E = 6.2 \times 10^5$ N/C. Determine the electric flux through this area (a) when the electric field is perpendicular to the surface and (b) when the electric field is parallel to the surface.

39. An electric field of intensity 3.50 kN/C is applied along the x axis. Calculate the electric flux through a rectangular plane 0.350 m wide and 0.700 m long if (a) the plane is parallel to the yz plane; (b) the plane is parallel to the xy plane; and (c) the plane contains the y axis and its normal makes an angle of 40.0° with the x axis.

40. The electric field everywhere on the surface of a thin spherical shell of radius 0.750 m is measured to be equal to 890 N/C and points radially toward the center of the sphere. (a) What is the net charge within the sphere's surface? (b) What can you conclude about the nature and distribution of the charge inside the spherical shell?

41. A 40-cm-diameter loop is rotated in a uniform electric field until the position of maximum electric flux is found. The flux in this position is measured to be 5.2×10^5 N·m²/C. Calculate the electric field strength in this region.

42. A point charge of $+5.00$ μC is located at the center of a sphere with a radius of 12.0 cm. Determine the electric flux through the surface of the sphere.

43. A point charge q is located at the center of a spherical shell of radius a, which has a charge $-q$ uniformly distributed on its surface. Find the electric field (a) for all points outside the spherical shell and (b) for a point inside the shell a distance r from the center.

44. Use Gauss's law and the fact that the electric field inside any closed conductor in electrostatic equilibrium is zero to show that any excess charge placed on the conductor must reside on its surface.

45. An infinite plane conductor has charge spread out on its surface as shown in Figure P15.45. Use Gauss's law to show that the electric field at any point outside the conductor is given by $E = \sigma/\epsilon_0$, where σ is the charge per unit area on the conductor. (*Hint:* Choose a gaussian surface in the

shape of a cylinder with one end inside the conductor and one end outside the conductor.)

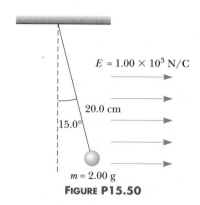

$E = 1.00 \times 10^3$ N/C

20.0 cm

15.0°

$m = 2.00$ g

FIGURE P15.50

FIGURE P15.45

46. Show that the electric field just outside the surface of a good conductor of any shape is given by $E = \sigma/\epsilon_0$, where σ is the charge per unit area on the conductor. (*Hint:* The electric field just outside the surface of a charged conductor is perpendicular to its surface.)

ADDITIONAL PROBLEMS

47. Two protons in an atomic nucleus are typically separated by a distance of 2×10^{-15} m. The electric repulsion force between the protons is huge, but the attractive nuclear force is even stronger, and keeps the nucleus from bursting apart. What is the magnitude of the electrical force between two protons separated by 2.00×10^{-15} m?

48. In the Bohr theory of the hydrogen atom, an electron moves in a circular orbit about a proton, where the radius of the orbit is 0.53×10^{-10} m. (a) Find the electrostatic force acting on each particle. (b) If this force causes the centripetal acceleration of the electron, what is the speed of the electron?

49. Three point charges are aligned along the x axis as shown in Figure P15.49. Find the electric field at the position $x = +2.0$ m, $y = 0$.

y

|← 0.50 m →|← 0.80 m →|

− 4.0 nC   5.0 nC   3.0 nC

x

FIGURE P15.49

50. A small 2.00-g plastic ball is suspended by a 20.0-cm-long string in a uniform electric field, as shown in Figure P15.50. If the ball is in equilibrium when the string makes a 15.0° angle with the vertical as indicated, what is the net charge on the ball?

51. (a) Two identical point charges $+ q$ are located on the y axis at $y = + a$ and $y = - a$. What is the electric field along the x axis at $x = b$? (b) A circular ring of charge of radius a has a total positive charge Q distributed uniformly around it. The ring is in the $x = 0$ plane with its center

at the origin. What is the electric field along the x axis at $x = b$ due to the ring of charge? (*Hint:* Consider the charge Q to consist of very many pairs of identical point charges positioned at ends of diameters of the ring.)

52. A positively charged bead having a mass of 1.00 g falls from rest in a vacuum from a height of 5.00 m in a uniform vertical electric field with a magnitude of 1.00×10^4 N/C. The bead hits the ground at a speed of 21.0 m/s. Determine (a) the direction of the electric field (up or down), and (b) the charge on the bead.

53. A solid conducting sphere of radius 2.00 cm has a charge of 8.00 μC. A conducting spherical shell of inner radius 4.00 cm and outer radius 5.00 cm is concentric with the solid sphere and has a charge of $- 4.00$ μC. Find the electric field at (a) $r = 1.00$ cm, (b) $r = 3.00$ cm, (c) $r = 4.50$ cm, and (d) $r = 7.00$ cm from the center of this charge configuration.

54. Two small silver spheres, each with a mass of 100 g, are separated by 1.00 m. Calculate the fraction of the electrons in one sphere that must be transferred to the other in order to produce an attractive force of 1.00×10^4 N (about a ton) between the spheres. (The number of electrons per atom of silver is 47, and the number of atoms per gram is Avogadro's number divided by the molar mass of silver, 107.87.)

55. A vertical electric field of magnitude 2.00×10^4 N/C exists above Earth's surface on a day when a thunderstorm is brewing. A car with a rectangular size of 6.00 m by 3.00 m is traveling along a roadway sloping downward at 10.0°. Determine the electric flux through the bottom of the car.

56. A 2.00-μC charged 1.00-g cork ball is suspended vertically on a 0.500-m-long light string in the presence of a uniform downward-directed electric field of magnitude $E = 1.00 \times 10^5$ N/C. If the ball is displaced slightly from the vertical, it oscillates like a simple pendulum. (a) Determine the period of this oscillation. (b) Should gravity be included in the calculation for part (a)? Explain.

57. Two 2.0-g spheres are suspended by 10.0-cm-long light strings (Fig. P15.57). A uniform electric field is applied in the x direction. If the spheres have charges of $- 5.0 \times 10^{-8}$ C and $+ 5.0 \times 10^{-8}$ C, determine the electric field intensity that enables the spheres to be in equilibrium at $\theta = 10°$.

58. Two point charges like those in Figure P15.58 are called an electric dipole. Show that the electric field at a distant point along the x axis is given by the expression $E_x = 4 k_e q a / x^3$.

FIGURE P15.57

FIGURE P15.58

59. A charged cork ball of mass 1.00 g is suspended on a light string in the presence of a uniform electric field as in Figure P15.59. When the electric field has an *x* component of 3.00×10^5 N/C and a *y* component of 5.00×10^5 N/C, the ball is in equilibrium at $\theta = 37.0°$. Find (a) the charge on the ball and (b) the tension in the string.

FIGURE P15.59

GROUP ACTIVITIES

G.1 Listed below are a number of experiments that you can perform to investigate static electricity.

(a) Attach two inflated balloons to the ends of a light string having a length of about 2 m. Tape the center of the string to the top of an open doorway as in Fig. GA15.1. Note that the balloons touch each other as they hang freely. Now rub each balloon several times with a wool cloth and let them hang freely once again. Why are the balloons no longer touching each other but are now separated?

(b) Rub an inflated balloon with a piece of wool and press it against a wall. Note that the balloon adheres to the wall. Why?

(c) Rub nylon hose with a plastic dry cleaner bag and observe how the hose expands like a balloon. Why?

(d) Tear some paper into very small pieces. Comb your hair and then bring the comb close to the paper pieces. Notice that they are accelerated toward the comb. How does the magnitude of the electric force compare with the magnitude of the gravitational force exerted on the paper? Keep watching and you might see a few pieces jump away from the comb. They do not fall away; they are repelled. What causes this?

G.2 For this experiment, you will need two 20-cm strips of transparent tape (mass of each is about 65 mg). Fold about 1 cm of tape over at one end of each strip to create a handle. Press both pieces of tape side by side onto a table top, rubbing your finger back and forth across the strips. Quickly pull the strips off the surface so that they become charged. Hold the tape handles together and the strips will repel each other, forming an inverted V shape. Measure the angle between the pieces and estimate the excess charge on each strip. Assume that the charges act as if they were located at the center of mass of each strip.

G.3 Rub an inflated balloon with a piece of wool cloth and place the balloon near a fine stream of water falling from a faucet. The stream of water deflects toward the balloon. Why? Vary the distance between the balloon and the stream and observe the displacement of the water stream for different distances. What is the relationship between the displacement of the stream and the distance of separation?

G.4 If you have access to a Van de Graaff generator, here are a few interesting things to try. (a) Stack several aluminum pie plates on top of the generator, then turn the generator on. What do you think is going to happen? Why? Try it. (b) Tape one pie plate on top of the generator and pour in some puffed rice or paper pieces. What do you think will happen when you turn the generator on? Why? Try it and see. (c) Tape a glass beaker to the top of the generator and pour in some puffed rice. What do you

FIGURE GA15.1

think will happen when you turn the generator on? Try it. (d) Bring a fluorescent bulb close to a charged generator and observe what happens. Why?

G.5 A point charge of magnitude 5.00 μC is at the origin of a coordinate system while a charge of -4.00 μC is at the point $x = 1.00$ m. There is a point on the x axis at x less than infinity where the electric field goes to zero. (a) Show by conceptual arguments that this point cannot be located between the charges. (b) Show by conceptual arguments that the point cannot be at any location between $x = 0$ and negative infinity. (c) Show by conceptual arguments that the point must be between $x = 1.00$ m and $x =$ positive infinity. (d) Use the values given to find the point and show that it is consistent with your conceptual argument.

G.6 Fill in the blanks using $+Q$, 0, or $-Q$ and defend your answer. A spherical conducting object A with a charge of $+Q$ is lowered through a hole into a metal container B, which is initially uncharged.

(a) When A is at the center of B but not touching it, the charge on the inner surface of B is _____.

(b) The charge on the outer surface of B is _____.

(c) Object A now is allowed to touch the inner surface of B. The charge on A is now _____.

(d) The charge on the inner surface of B is now _____.

(e) The charge on the outer surface of B is now _____.

G.7 In the Millikan oil-drop experiment, an atomizer (a sprayer with a fine nozzle) is used to introduce many tiny droplets of oil between two oppositely charged parallel metal plates. Some of the droplets pick up one or more excess electrons. The charge on the plates is adjusted so that the electric force on the excess electrons exactly balances the weight of the droplet. The idea is to look for a droplet that has the smallest electric force and assume that it has only one excess electron. This lets the observer measure the charge on the electron. Suppose we are using an electric field of 3×10^4 N/C. The charge on one electron is about 1.6×10^{-19} C. Estimate the radius of an oil drop whose weight could be balanced by the electric force of this field on one electron.

G.8 Two hard rubber spheres of mass 10 g are rubbed vigorously with fur on a dry day. They are then suspended from a rod with two insulating strings of length 5.0 cm. They are observed to hang at equilibrium as shown in Fig. GA15.8, each at an angle of 10° with the vertical. Estimate the amount of charge that is found on each sphere.

FIGURE GA15.8

(Problems 7 and 8 are courtesy of E. F. Redish. For more problems of this type, visit **http://www.physics.umd.edu/perg/**)

16

Electrical Energy and Capacitance

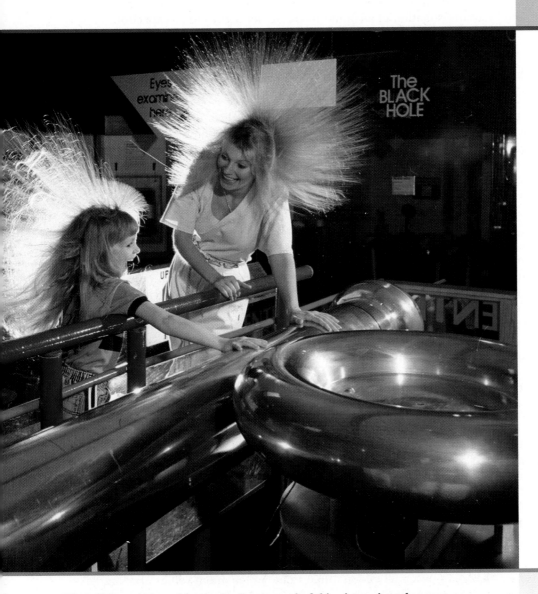

High Voltage! Everything in the foreground of this picture is at the same electrical potential of many kilovolts. With no differences in potential, no charge is moving and no one gets a shock. *(Courtesy of Resonance Research Corporation)*

*T*he concept of potential energy was first introduced in Chapter 5. A potential energy function can be defined for any conservative force, such as the force of gravity. By using the principle of conservation of energy, we were often able to avoid working directly with forces when solving problems. In this chapter we discover that the potential energy concept is also useful in the study of electricity. Because the Coulomb force is conservative, we can define an electrical potential energy corresponding to the Coulomb force. This concept of potential energy is of value, but perhaps even more valuable is a quantity called electric potential, defined as potential energy per unit charge.

Electric potential is of great practical value for dealing with electric circuits. For example, when we speak of a voltage applied between two points, we are actually referring to an electric potential difference between those points. We take our first steps toward understanding circuits with a discussion of electric potential, followed by an investigation of a common circuit element called a capacitor.

16.1   POTENTIAL DIFFERENCE AND ELECTRIC POTENTIAL

FIGURE 16.1 When a charge q moves in a uniform electric field **E** from point A to point B, the work done by the electric force on the charge is qEd.

In Chapter 5 we showed that the gravitational force is conservative. This means that the work done by this force on an object depends only on the initial and final positions of the object and not on the path connecting the two positions. Furthermore, because the gravitational force is conservative, it is possible to define a potential energy function, which we call gravitational potential energy. Because the Coulomb force law is of the same form as the universal law of gravity, it follows that **the electrostatic force is also conservative.** Therefore, it is possible to define an electrical potential energy function associated with this force.

Let us consider electrical potential energy from the point of view of the particular situation shown in Figure 16.1. Imagine a small positive charge placed at point A in a *uniform* electric field of magnitude E. (The electric field between equally and oppositely charged parallel metal plates is uniform.) As the charge moves from point A to point B, under the influence of the electric force of magnitude qE exerted on it, the work done by the electric field on the charge is positive, and given by $W_{AB} = Fd = qEd$. Therefore, the charge accelerates to the right, gaining kinetic energy. **As the charged particle gains kinetic energy, it loses an equal amount of potential energy.**

By definition, **the work done by a conservative force equals the negative of the change in potential energy, ΔPE.** The change in electrical potential energy is therefore

Change in electrical potential ▶
energy

$$\Delta PE = -W_{AB} = -qEd \qquad \text{[16.1]}$$

Note that although potential energy can be defined for any electric field, **Equation 16.1 is valid only for the case of a uniform electric field.** In subsequent sections we shall examine situations in which the electric field is not uniform.

In the coming pages we often have occasion to use electrical potential energy, but of even more practical importance in the study of electricity is the concept of potential difference.

> The potential difference between points A and B, $V_B - V_A$, is defined as the change in potential energy (final value minus initial value) of a charge q moved from A to B divided by the size of the charge.

Potential difference between ▶
two points

$$\Delta V \equiv V_B - V_A = \frac{\Delta PE}{q} \qquad \text{[16.2]}$$

Potential difference should not be confused with potential energy. The change in electric potential between two points is proportional to the change in electrical potential energy of a charge as it moves between the points, and we see from Equation 16.2 that the two are related as $\Delta PE = q\,\Delta V$. Because potential energy is a scalar quantity, **electric potential is also a scalar quantity.** From Equation 16.2 we see that electric potential difference is a measure of energy per unit charge. Alternatively, electrical potential difference is the work done to move a charge from a point A to a point B divided by the magnitude of the charge. Thus, the SI units of electric potential are[1] joules per coulomb, called volts (V):

$$1\,\text{V} \equiv 1\,\text{J/C} \qquad \text{[16.3]}$$

This says that 1 J of work must be done to move a 1-C charge between two points that are at a potential difference of 1 V. In the process of moving through a potential difference of 1 V, the 1-C charge gains (or loses) 1 J of energy.

For the special case of a uniform electric field such as that between charged parallel plates, dividing Equation 16.1 by q gives

$$\frac{\Delta PE}{q} = V_B - V_A = -Ed \qquad \text{[16.4]}$$

This equation shows that potential difference also has units of electric field times distance. From this, it follows that the SI units of electric field, newtons per coulomb, can also be expressed as volts per meter:

$$1\,\text{N/C} = 1\,\text{V/m}$$

Because Equation 16.4 is directly related to Equation 16.1, remember that it is valid only for the case of a uniform electric field.

Let us examine the changes in energy associated with movements of charge in the electric field pictured in Figure 16.2a. Because the positive charge q tends to move in the direction of the electric field, we must apply an upward external force on the charge to move it from B to A. Work is done by the external force on the charge, and this means that **a positive charge gains electrical potential energy when it is moved in a direction opposite the electric field.** This is analogous to an object gaining gravitational potential energy when it rises to higher elevations in the presence of gravity, as in Figure 16.2b. If a positive charge is released from rest at point A, it experiences a force $q\mathbf{E}$ in the direction of the field (downward in Figure 16.2a). Therefore, it accelerates downward, gaining kinetic energy. **As it gains kinetic energy, it loses an equal amount of electrical potential energy.** Also, as Equation 16.4 shows, if a positive charge moves from A to B, its electric potential decreases.

By contrast, if the test charge q is negative, the situation is reversed. **A negative charge loses electrical potential energy when it moves in the direction**

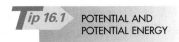

Tip 16.1 POTENTIAL AND POTENTIAL ENERGY

Note that electric potential is characteristic of the field only, independent of a test charge that may be placed in that field. On the other hand, potential energy is a characteristic of the charge-field system due to an interaction between the field and a charge placed in the field.

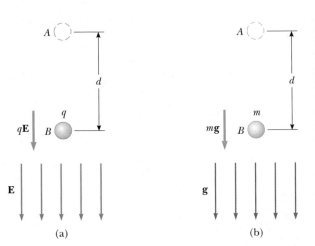

FIGURE 16.2 (a) When the electric field $\mathbf{E}$ is directed downward, point B is at a lower electric potential than point A. A positive test charge that moves from A to B loses electric potential energy. (b) An object of mass m moving downward in the direction of the gravitational field $\mathbf{g}$ loses gravitational potential energy.

---

[1] Note that the symbol V (italic) represents potential, whereas V (roman) is the symbol for the unit of this quantity—volts. Do not confuse these two symbols.

opposite the electric field. That is, a negative charge released from rest in the field **E** accelerates in a direction *opposite* the field.

Thus, when a positive charge is placed in an electric field, it moves in the direction of the field, from a point of high potential to a point of lower potential. In the process, its electrical potential energy decreases and its kinetic energy increases.

When a negative charge is placed in an electric field, it moves opposite to the direction of the field, from a point of low potential to a point of higher potential. In the process, it also undergoes a decrease in electrical potential energy and an increase in kinetic energy.

APPLICATION

AUTOMOBILE BATTERIES

Let us pause briefly to discuss a situation that illustrates the concept of electric potential difference. Consider the common 12-V automobile battery. Such a battery maintains a potential difference across its terminals, where the positive terminal is 12 V higher in potential than the negative terminal. In practice, the negative terminal is usually connected to the metal body of the car, which can be considered at a potential of zero volts. The battery becomes a useful device when it is connected by conducting wires to such things as headlights, a radio, power windows, motors, and so forth. Now consider a charge of + 1 C, to be moved around a circuit that contains the battery connected to some of these external devices. As the charge is moved inside the battery from the negative terminal (at 0 V) to the positive terminal (at 12 V), the work done by the battery on the charge is 12 J. Thus, every coulomb of positive charge that leaves the positive terminal of the battery carries an energy of 12 J. As this charge moves through the external circuit toward the negative terminal, it gives up its 12 J of electrical energy to the external devices. When the charge reaches the negative terminal, its electrical energy is zero. At this point, the battery takes over and restores 12 J of energy to the charge as it is moved from the negative to the positive terminal, enabling it to make another transit of the circuit. The actual amount of charge that leaves the battery each second and traverses the circuit depends on the properties of the external devices, as we shall see in the next chapter.

> **Quick Quiz 16.1**
>
> If an electron is released from rest in a uniform electric field, the electric potential energy of the charge-field system (a) increases, (b) decreases, or (c) remains the same.

Example 16.1   The Field Between Two Parallel Plates of Equal and Opposite Charge

Figure 16.3 illustrates a situation in which a constant electric field can be set up. A 12-V battery is connected between two parallel metal plates separated by 0.30 cm. Find the magnitude of the electric field.

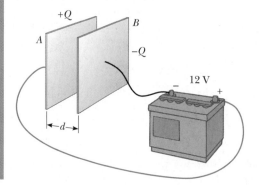

FIGURE 16.3   (Example 16.1) A 12-V battery connected to two parallel plates. The electric field between the plates has a magnitude given by the potential difference divided by the plate separation d.

Reasoning   The electric field is uniform (except near the edges of the metal plates), and thus the relationship between potential difference and the magnitude of the field is given by Equation 16.4:

$$V_B - V_A = -Ed$$

Chemical forces inside a battery maintain one electrode, called the positive terminal, at a higher potential than a second electrode, the negative terminal. Thus, in Figure 16.3, plate *B*, which is connected to the negative terminal, must be at a lower potential than plate *A*, which is connected to the positive terminal.

Solution   We have

$$V_B - V_A = -12 \text{ V}$$

This gives a value for *E* of

$$E = -\frac{(V_B - V_A)}{d} = -\frac{(-12 \text{ V})}{0.30 \times 10^{-2} \text{ m}} = \boxed{4.0 \times 10^3 \text{ V/m}}$$

The direction of this field is from the positive plate to the negative plate. A device consisting of two plates separated by a small distance is called a parallel-plate capacitor (to be discussed later in this chapter).

Example 16.2   Motion of a Proton in a Uniform Electric Field

A proton is released from rest in a uniform electric field of magnitude 8.0×10^4 V/m, directed along the positive *x* axis (Fig. 16.4). The proton undergoes a displacement of 0.50 m in the direction of the field.

A Find the change in electric potential of the proton as a result of this displacement.

Solution   From Equation 16.4, we have

$$\Delta V = V_B - V_A = -Ed = -(8.0 \times 10^4 \text{ V/m})(0.50 \text{ m}) = \boxed{-4.0 \times 10^4 \text{ V}}$$

Thus, the electric potential of the proton decreases as it moves from *A* to *B*.

B Find the change in electrical potential energy of the proton for this displacement and explain the physical meaning of the sign of the change.

Solution

$$\Delta PE = q \, \Delta V = e \, \Delta V = (1.6 \times 10^{-19} \text{ C})(-4.0 \times 10^4 \text{ V}) = \boxed{-6.4 \times 10^{-15} \text{ J}}$$

The negative sign here means that the electrical potential energy of the proton decreases as it moves in the direction of the electric field. This makes sense because, as the proton *accelerates* in the direction of the field, it gains kinetic energy and at the same time loses electrical potential energy (mechanical energy is conserved).

C Find the speed of the proton after it has moved 0.50 m, starting from rest.

Solution   If no forces other than the conservative electrical force are acting on the proton, we can apply the principle of conservation of mechanical energy in the form

$$KE_i + PE_i = KE_f + PE_f$$

In our case, $KE_i = 0$; hence, the preceding expression gives

$$KE_f = PE_i - PE_f = -\Delta PE$$

With this equation and the results of part B, we find that

$$\tfrac{1}{2}m_p v_f^2 = 6.4 \times 10^{-15} \text{ J}$$

FIGURE 16.4   (Example 16.2) A proton accelerates from *A* to *B* in the direction of the uniform electric field.

and

$$v_f^2 = \frac{2(6.4 \times 10^{-15}\,\text{J})}{1.67 \times 10^{-27}\,\text{kg}} = 7.66 \times 10^{12}\,\text{m}^2/\text{s}^2$$

$$v_f = \boxed{2.8 \times 10^6\,\text{m/s}}$$

16.2 ELECTRIC POTENTIAL AND POTENTIAL ENERGY DUE TO POINT CHARGES

In electric circuits, a point of zero electric potential is often defined by grounding (connecting to Earth) some point in the circuit. For example, if the negative plate in Example 16.1 were grounded, it would be considered to have a potential of zero and the positive plate to have a potential of 12 V. In this section, we wish to define the *electric potential due to a point charge* at a point in space. In this case, the point of zero electric potential is taken to be at an infinite distance from the charge, far from its influence and the influence of any other charges. With this choice, the methods of calculus can be used to show that the electric potential created by a point charge q at any distance r from the charge is given by

Electric potential created by a point charge ▶

$$V = k_e \frac{q}{r} \qquad\qquad [16.5]$$

Equation 16.5 shows that the potential, or work per unit charge to move a test charge in from infinity to a distance r from a positive point charge q increases the closer the positive test charge is moved to q. The plot of Equation 16.5 shown in Figure 16.5 shows that the potential associated with a point charge decreases as $1/r$ as we move away from the charge in contrast to the magnitude of the electric field of a point charge, which decreases as $1/r^2$.

Equation 16.5 points out a significant property of electric potential that makes it an important quantity in the study of electricity: The potential at a given point depends only on the charge q of the object setting up the potential and the distance r from that object to a specific point in space. As a result, we can say that a potential exists at some point in space whether or not there is a test charge at that point.

Superposition principle ▶

The electric potential of two or more charges is obtained by applying the **superposition principle.** That is, **the total electric potential at some point P due to several point charges is the algebraic sum of the electric potentials**

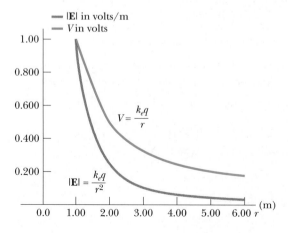

FIGURE 16.5 Electric field and potential versus distance from a point charge of 1.11×10^{-10} C. V decreases as $1/r$, whereas E decreases as $1/r^2$.

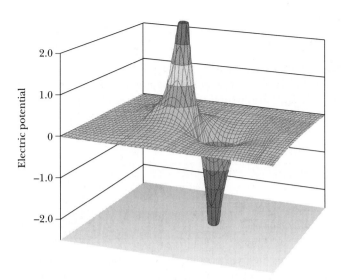

FIGURE 16.6 The electric potential (in arbitrary units) in the plane containing an electric dipole. Potential is plotted in the vertical dimension.

due to the individual charges. This is similar to the method used in Chapter 15 to find the resultant electric field at a point in space. However, note that in the case of potentials, we must evaluate an *algebraic sum* of individual potentials to obtain the total, because **potentials are scalar quantities.** Thus, it is much easier to evaluate the electric potential at some point due to several charges than to evaluate the electric field, which is a vector quantity.

Figure 16.6 is a computer-generated plot of the electric potential associated with an electric dipole. The charges (equal magnitude, opposite sign) lie in a horizontal plane at the center of the potential spikes. The value of the potential is plotted in the vertical dimension. The computer program has added the potential due to each charge to arrive at total values of the potential.

We now consider the electrical potential energy of the interaction of a system of two charged particles. If V_1 is the electric potential due to charge q_1 at a point P, then the work required to bring charge q_2 from infinity to P without acceleration is $q_2 V_1$. By definition, this work equals the potential energy PE of the two-particle system when the particles are separated by a distance r (Fig. 16.7).

Therefore, we can express the electrical potential energy of the *pair* of charges as

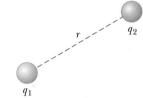

FIGURE 16.7 If two point charges are separated by the distance r, the potential energy of the pair is $k_e q_1 q_2 / r$.

$$PE = q_2 V_1 = k_e \frac{q_1 q_2}{r} \qquad [16.6]$$

◀ Potential energy of a pair of charges

Note that if the charges are of the *same* sign, *PE* is positive. This is consistent with the fact that like charges repel, and so positive work must be done on the system by an external agent to force the two charges near one another. Conversely, if the charges are of *opposite* sign, the force is attractive and *PE* is negative. This means that negative work must be done to hold back unlike charges from accelerating as they are brought close together.

Quick Quiz 16.2 If the electric potential at some point is zero, you can conclude that (a) no charges exist in the vicinity of that point, (b) some charges are positive and some are negative, or (c) all charges in the vicinity have the same sign. Choose each correct answer.

Quick Quiz 16.3

A spherical balloon contains a positively charged particle at its center. As the balloon is inflated to a larger volume while the charged particle remains at the center, which of the following changes? (a) the electric potential at the surface of the balloon, (b) the magnitude of the electric field at the surface of the balloon, (c) the electric flux through the balloon.

PROBLEM-SOLVING STRATEGY **Electric Potential**

When you solve problems involving electric potential, remember that potential is a *scalar quantity* (rather than a vector quantity, like the electric field), so there are no components to worry about. Therefore, when using the superposition principle to evaluate the electric potential at a point due to a system of point charges, simply take the algebraic sum of the potentials due to all charges. You must keep track of signs, however. The potential due to each positive charge is positive, and the potential due to each negative charge is negative. Use the basic equation $V = k_e q/r$.

Example 16.3 Finding the Electric Potential

A 5.0-μC point charge is at the origin, and a point charge $q_2 = -2.0\ \mu$C is on the x axis at (3.0, 0) m, as in Figure 16.8.

A If the electric potential is taken to be zero at infinity, find the total electric potential due to these charges at point P, with coordinates (0, 4.0) m.

Reasoning The electric potential at P due to each charge can be calculated from $V = k_e q/r$. The total electric potential is the scalar sum of these two potentials.

Solution The electric potential at P due to the 5.0-μC charge at the origin is

$$V_1 = k_e \frac{q_1}{r_1} = \left(8.99 \times 10^9\ \frac{\text{N} \cdot \text{m}^2}{\text{C}^2}\right)\left(\frac{5.0 \times 10^{-6}\ \text{C}}{4.0\ \text{m}}\right) = 1.12 \times 10^4\ \text{V}$$

and the electric potential due at P to the charge of $-2.0\ \mu$C is

$$V_2 = k_e \frac{q_2}{r_2} = \left(8.99 \times 10^9\ \frac{\text{N} \cdot \text{m}^2}{\text{C}^2}\right)\left(\frac{-2.0 \times 10^{-6}\ \text{C}}{5.0\ \text{m}}\right) = -0.360 \times 10^4\ \text{V}$$

and

$$V_P = V_1 + V_2 = \boxed{7.6 \times 10^3\ \text{V}}$$

B How much work is required to bring a third point charge of 4.0 μC from infinity to P?

Solution

$$W = q_3 V_P = (4.0 \times 10^{-6}\ \text{C})(7.6 \times 10^3\ \text{V})$$

Because 1 V = 1 J/C, W reduces to

$$W = \boxed{3.1 \times 10^{-2}\ \text{J}}$$

EXERCISE Find the magnitude and direction of the electric field at point P.

ANSWER 2.3×10^3 N/C at an angle of 79° with the x axis.

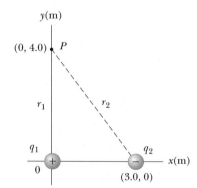

FIGURE 16.8 (Example 16.3) The electric potential at point P due to the point charges q_1 and q_2 is the algebraic sum of the potentials due to the individual charges.

16.3 POTENTIALS AND CHARGED CONDUCTORS

In order to determine the electric potential at all points on a charged conductor, let us combine Equations 16.1 and 16.2. From Equation 16.1 we see that the work done by electric forces on a charge is related to the change in electrical potential energy of the charge by

$$W = -\Delta PE$$

Furthermore, from Equation 16.2 we see that the change in electrical potential energy between two points A and B is related to the potential difference between these points by

$$\Delta PE = q(V_B - V_A)$$

Combining these two equations, we find that

$$W = -q(V_B - V_A) \qquad [16.7]$$

As we see from this result, in general, **no work is required to move a charge between two points that are at the same electric potential. That is, $W = 0$ when $V_B = V_A$.**

In Chapter 15 we found that when a conductor is in electrostatic equilibrium, a net charge placed on it resides entirely on its surface. Furthermore, we showed that the electric field just outside the surface of a charged conductor in electrostatic equilibrium is perpendicular to the surface and that the field inside the conductor is zero. We shall now show that **all points on the surface of a charged conductor in electrostatic equilibrium are at the same potential.**

Consider a surface path connecting any points A and B on a charged conductor, as in Figure 16.9. The electric field **E** is always perpendicular to the displacement along this path; therefore, no work is done by the electric field if a charge is moved between these points. From Equation 16.7 we see that if the work done is zero, the difference in electric potential $V_B - V_A$ is also zero. Therefore, **the electric potential is a constant everywhere on the surface of a charged conductor in equilibrium.** Furthermore, because the electric field inside a conductor is zero, no work is required to move a charge between two points inside the conductor. Again, Equation 16.7 shows that if the work done is zero, the difference in electric potential between any two points inside a conductor must also be zero. Thus, we conclude that the electric potential is constant everywhere inside a conductor.

Finally, because one of the points inside the conductor could be arbitrarily close to the surface of the conductor, we conclude that **the electric potential is constant everywhere inside a conductor and equal to its value at the surface.** As a consequence, no work is required to move a charge from the interior of a charged conductor to its surface. (Note that the potential inside a conductor is not necessarily zero even though the interior electric field is zero.)

◀ Properties of a charged conductor in equilibrium

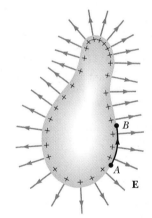

FIGURE 16.9 An arbitrarily shaped conductor with an excess positive charge. When the conductor is in electrostatic equilibrium, all of the charge resides at the surface, **E** = 0 inside the conductor, and the electric field just outside the conductor is perpendicular to the surface. The potential is constant inside the conductor and is equal to the potential at the surface.

THE ELECTRON VOLT

A unit of energy commonly used in atomic and nuclear physics because of its convenient size is the electron volt (eV). Electrons in normal atoms typically have energies of tens of electron volts, excited electrons in atoms emitting x-rays have energies of thousands of electron volts, and high-energy gamma rays (electromagnetic waves) emitted by the nucleus have energies of millions of electron volts.

The electron volt is defined as the energy that an electron (or proton) gains when accelerated through a potential difference of 1 V.

◀ Definition of the electron volt

Because 1 V = 1 J/C and because the magnitude of charge on the electron or proton is 1.60×10^{-19} C, we see that the electron volt is related to the joule by

$$1 \text{ eV} = 1.60 \times 10^{-19} \text{ C} \cdot \text{V} = 1.60 \times 10^{-19} \text{ J} \qquad \text{[16.8]}$$

APPLYING PHYSICS 16.1

Suppose scientists had chosen to measure small energies in proton volts rather than electron volts. What difference would this make?

Explanation There would be no change at all. An electron volt is the kinetic energy gained by an electron in being accelerated through a potential difference of 1 V. A proton accelerated through 1 V would have the same kinetic energy, because it carries the same charge as the electron (except for sign). The proton would be moving in the opposite direction and more slowly after accelerating through 1 V, due to its opposite charge and its larger mass, but it would still gain 1 electron volt, or 1 proton volt, of kinetic energy.

16.4 EQUIPOTENTIAL SURFACES

A surface on which all points are at the same potential is called an **equipotential surface.** The potential difference between any two points on an equipotential surface is zero. Hence, **no work is required to move a charge at constant speed on an equipotential surface.** Equipotential surfaces have a simple relationship to the electric field. **The electric field at every point on an equipotential surface is perpendicular to the surface.** If the electric field **E** had a component parallel to the surface, this component would produce an electric force on a charge placed on the surface. This force would do work on the charge as it moved from one point to another, in contradiction to the definition of an equipotential surface.

It is convenient to represent equipotential surfaces on a diagram by drawing equipotential contours, which are two-dimensional views of the intersections of the equipotential surfaces with the plane of the drawing. We shall refer to these equipotential contours as simply **equipotentials.** Figure 16.10a shows the equipotentials (in blue) associated with a positive point charge. Note that the equipotentials are perpendicular to the electric field lines (in red) at all points. Recall that the electric potential created by a point charge q is given by $V = k_e q / r$. This rela-

Webnote 16.1

For an excellent applet demonstrating electric field lines and equipotentials of point charges, visit
http://www.cco.caltech.edu/~phys1/java/phys1/EField/EField.html

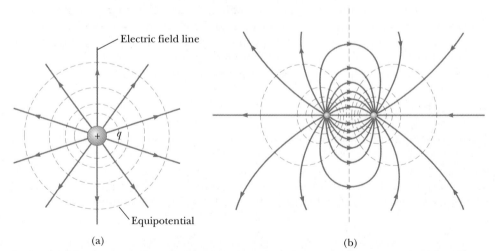

FIGURE 16.10 Equipotentials (dashed blue lines) and electric field lines (red lines) for (a) a positive point charge and (b) two point charges of equal magnitude and opposite sign. In all cases, the equipotentials are *perpendicular* to the electric field lines at every point.

tion shows that, for a single point charge, the potential is constant on any surface for which *r* is constant. Therefore, the equipotentials of a point charge are a family of spheres centered on the point charge. Figure 16.10b shows the equipotentials associated with two charges of equal magnitude but opposite sign (an electric dipole).

16.5 APPLICATIONS

THE ELECTROSTATIC PRECIPITATOR

One important application of electric discharge in gases is a device called an *electrostatic precipitator*. It is used to remove particulate matter from combustion gases, thereby reducing air pollution. It is especially useful in coal-burning power plants and in industrial operations that generate large quantities of smoke. Systems currently in use can eliminate approximately 90% by mass of the ash and dust from the smoke. Unfortunately, a very high percentage of the lighter particles still escape, and these contribute significantly to smog and haze.

Figure 16.11 illustrates the basic idea of the electrostatic precipitator. A high voltage (typically 4–100 kV) is maintained between a wire running down the center of a duct and the outer wall, which is grounded. The wire is maintained at a negative electric potential with respect to the wall, and so the electric field is directed toward the wire. The electric field near the wire reaches a high enough value to cause a discharge around the wire and the formation of positive ions, electrons, and negative ions, such as O_2^-. As the electrons and negative ions are accelerated toward the outer wall by the nonuniform electric field, the dirt particles in the streaming gas become charged by collisions and ion capture. Because most of the charged dirt particles are negative, they are also drawn to the outer wall by the electric field. When the duct is shaken, the particles fall loose and are collected at the bottom.

In addition to reducing the amounts of harmful gases and particulate matter in the atmosphere, the electrostatic precipitator recovers valuable metal oxides from the stack.

An *electrostatic air cleaner*, used in homes to relieve the discomfort of allergy sufferers, uses many of the same principles as the precipitator. Air laden with dust and pollen is drawn into the device across a positively charged mesh screen. The

> **◄ APPLICATION**
>
> THE ELECTROSTATIC PRECIPITATOR

> **◄ APPLICATION**
>
> THE ELECTROSTATIC AIR CLEANER

(a) (b) (c)

FIGURE 16.11 (a) Diagram of an electrostatic precipitator. The high voltage maintained on the central wires creates an electric discharge in the vicinity of the wire. Compare air pollution when the electrostatic precipitator is (b) operating and (c) turned off. *(b, Rei O'Hara/Black Star/PNI; c, Greig Cranna/Stock, Boston/PNI)* airborne particles become positively charged when they make intimate contact with the screen. The particles then pass through a second, negatively charged mesh screen. The electrostatic force of attraction between the positively charged particles in the air and the negatively charged screen causes the particles to precipitate out onto the surface of the screen. In this fashion, a very high percentage of contaminants are removed from the air stream.

XEROGRAPHY AND LASER PRINTERS

APPLICATION

XEROGRAPHIC COPIERS

The process of xerography is widely used for making photocopies of printed materials. The basic idea behind the process was developed by Chester Carlson, who was granted a patent for his invention in 1940. In 1947 the Xerox Corporation launched a full-scale program to develop automated duplicating machines using Carlson's process. The huge success of that development is quite evident; today, practically all offices and libraries have one or more duplicating machines, and the capabilities of these machines continue to evolve.

Some features of the xerographic process involve simple concepts from electrostatics and optics. However, the one idea that makes the process unique is the use of photoconductive material to form an image. (A photoconductor is a material that is a poor conductor of electricity in the dark but becomes a reasonably good electric conductor when exposed to light.)

Figure 16.12 illustrates the steps in the xerographic process. First, the surface of a plate or drum is coated with a thin film of the photoconductive material (usually selenium or some compound of selenium), and the photoconductive surface is given a positive electrostatic charge in the dark (Fig. 16.12a). The page to be copied is then projected onto the charged surface (Fig. 16.12b). The photocon-

(a) Charging the drum   (b) Imaging the document   (c) Applying the toner

(d) Transferring the toner to the paper   (e) Laser printer drum

FIGURE 16.12 The xerographic process. (a) The photoconductive surface is positively charged. (b) Through the use of a light source and lens, a hidden image is formed on the charged surface in the form of positive charges. (c) The surface containing the image is covered with a negatively charged powder, which adheres only to the image area. (d) A piece of paper is placed over the surface and given a charge. This transfers the image to the paper, which is then heated to "fix" the powder to the paper. (e) The image on the drum of a laser printer is produced by turning a laser beam on and off as it sweeps across the selenium-coated drum.

ducting surface becomes conducting only in areas where light strikes; there the light produces charge carriers in the photoconductor, which neutralize the positively charged surface. The charges remain on those areas of the photoconductor not exposed to light, however, leaving a hidden image of the object in the form of a positive surface charge distribution.

Next, a powder called a *toner* is negatively charged and dusted onto the photoconducting surface (Fig. 16.12c). The charged powder adheres only to the areas that contain the positively charged image. At this point, the image becomes visible. It is then transferred to the surface of a sheet of positively charged paper. Finally, the toner is "fixed" to the surface of the paper by heat (Fig. 16.12d). This results in a permanent copy of the original.

The steps for producing a document on a laser printer are similar to those used in a photocopy machine, in that parts (a), (c), and (d) of Figure 16.12 remain essentially the same. The difference between the two techniques lies in the way the image is formed on the selenium-coated drum. In a laser printer, the command to print the letter O, for instance, is sent to a laser from the memory of a computer. A rotating mirror inside the printer causes the beam of the laser to sweep across the selenium-coated drum in an interlaced pattern (Fig. 16.12e). Electrical signals generated by the printer turn the laser beam on and off in a pattern that traces out the letter O in the form of positive charges on the selenium. Toner is then applied to the drum, and the transfer to paper is accomplished as in a photocopy machine.

◀ **APPLICATION**

LASER PRINTERS

16.6 CAPACITANCE

A **capacitor** is a device used in a variety of electric circuits—for example, to tune the frequency of radio receivers, eliminate sparking in automobile ignition systems, or to store short-term energy for rapid release in electronic flash units. Figure 16.13 shows a typical design for a capacitor. It consists of two parallel metal plates separated by a distance d. When used in an electric circuit, the plates are connected to the positive and negative terminals of a battery or some other voltage source. When this connection is made, electrons are pulled off one of the plates, leaving it with a charge of $+Q$, and transferred through the battery to the other plate, leaving it with a charge of $-Q$, as shown in the figure. This charge transfer stops when the potential difference across the plates equals the potential difference of the battery. Thus, a charged capacitor is a device that stores energy that can be reclaimed when needed for a specific application.

FIGURE 16.13 A parallel-plate capacitor consists of two parallel plates, each of area A, separated by a distance d. The plates carry charges of equal magnitude and opposite sign.

> The capacitance *C* of a capacitor is defined as the ratio of the magnitude of the charge on either conductor (plate) to the magnitude of the potential difference between the conductors (plates):

$$C \equiv \frac{Q}{\Delta V}$$ [16.9]

◀ Capacitance of a pair of conductors

Capacitance has the SI units coulombs per volt, called **farads** (F) in honor of Michael Faraday. That is,

$$1\ \text{F} \equiv 1\ \text{C/V}$$

The farad is a very large unit of capacitance. In practice, most typical capacitors have capacitances ranging from microfarads (1 μF = 1 × 10^{-6} F) to picofarads (1 pF = 1 × 10^{-12} F).

For example, if a 3.0-μF capacitor is connected to a 12-V battery, the magnitude of the charge on each plate of the capacitor is

$$Q = C\,\Delta V = (3.0 \times 10^{-6}\ \text{F})(12\ \text{V}) = 36\ \mu\text{C}$$

Tip 16.2 POTENTIAL DIFFERENCE IS ΔV, NOT V

In this book, we use the symbol ΔV for the potential difference across a circuit element or a device. Note that almost every other book uses the symbol V for potential difference, without the delta sign, which can lead to confusion.

From Equation 16.9, we see that a large capacitance is needed to store a large amount of charge for a given applied voltage.

16.7 THE PARALLEL-PLATE CAPACITOR

The capacitance of a device depends on the geometric arrangement of the conductors. For example, the capacitance of a parallel-plate capacitor whose plates are separated by air (see Fig. 16.13) is

▶ Capacitance of a parallel-plate capacitor

$$C = \epsilon_0 \frac{A}{d} \qquad [16.10]$$

where A is the area of one of the plates, d is the distance between the plates, and ϵ_0 is the permittivity of free space.

Although we do not derive Equation 16.10, let us attempt to make it seem plausible. As you can see from the definition of capacitance, $C \equiv Q/\Delta V$, the amount of charge a given capacitor can store for a given potential difference across its plates increases as the capacitance increases. Therefore, it seems reasonable that a capacitor constructed from plates with large areas should be able to store a large charge. Furthermore, if the oppositely charged plates are close together, the attractive force between them will be large. In fact, for a given potential difference, the charge on the plates increases with decreasing plate separation.

APPLICATION

CAMERA FLASH ATTACHMENTS

One practical device that uses a capacitor is the flash attachment on a camera. A battery is used to charge the capacitor, and this stored charge is then released when the shutter-release button is pressed to take a picture. The stored charge is delivered to a flash tube very quickly, illuminating the subject at the instant more light is needed.

APPLICATION

COMPUTER KEYBOARDS

Computers make use of capacitors in many ways. For example, one type of computer keyboard has capacitors at the bases of its keys, as in Figure 16.14. Each key is connected to a movable plate, which represents one side of the capacitor; the fixed plate on the bottom of the keyboard represents the other side of the capacitor. When a key is pressed, the capacitor spacing decreases, causing an increase in capacitance. External electronic circuits recognize each key by the *change* in its capacitance when it is pressed.

Key ——▶

Movable plate ——▶

Dielectric ——▶

Fixed plate ——▶

FIGURE 16.14 When the key of one type of keyboard is pressed, the capacitance of a parallel-plate capacitor increases as the plate spacing decreases. The substance labeled "dielectric" is an insulating material as described in Section 16.10.

Example 16.4   **Calculating C for a Parallel-Plate Capacitor**

A parallel-plate capacitor has an area of $A = 2.00 \text{ cm}^2 = 2.00 \times 10^{-4} \text{ m}^2$ and a plate separation of $d = 1.00 \text{ mm} = 1.00 \times 10^{-3} \text{ m}$. Find its capacitance.

Solution   From $C = \epsilon_0 A/d$ we find that

$$C = \epsilon_0 \frac{A}{d} = (8.85 \times 10^{-12} \text{ C}^2/\text{N} \cdot \text{m}^2)\left(\frac{2.00 \times 10^{-4} \text{ m}^2}{1.00 \times 10^{-3} \text{ m}} \right)$$

$$= 1.77 \times 10^{-12} \text{ F} = \boxed{1.77 \text{ pF}}$$

EXERCISE   Show that $1 \text{ C}^2/\text{N} \cdot \text{m}$ equals 1 F.

SYMBOLS FOR CIRCUIT ELEMENTS AND CIRCUITS

The symbol that is commonly used to represent a capacitor in a circuit is ——│├—— , or sometimes ——│(—— . Do not confuse this with the circuit symbol ——┼│├—— used to designate a battery (or any other direct current source). The positive terminal of the battery is at the higher potential and is represented by the longer vertical line in the battery symbol. In the next chapter we

FIGURE 16.15 (a) A real circuit and (b) its equivalent circuit diagram.

The electric field pattern of two oppositely charged conducting parallel plates. Note the nonuniform nature of the electric field at the ends of the plates. *(Courtesy of Harold M. Waage, Princeton University)*

shall discuss another circuit element called a resistor, represented by the symbol ——⋀⋀⋀——. The wires in a circuit that do not have appreciable resistance compared to other elements in the circuit will be represented by straight lines.

Although we will discuss circuits in more detail in the next chapter, it is important to realize that a circuit is a collection of real objects usually containing a source of electrical energy (such as a battery) connected to elements that convert electrical energy to other forms (light, heat, sound), or store the energy in electric or magnetic fields for later retrieval. A real circuit and its diagram are sketched in Figure 16.15. The circuit symbol for a lightbulb shown in Figure 16.15b is ——(⊗)——. If you are not familiar with circuit diagrams, trace the path of the real circuit with your finger to see that it is equivalent to the geometrically regular schematic diagram.

16.8 COMBINATIONS OF CAPACITORS

Two or more capacitors can be combined in circuits in several ways. The equivalent capacitances of certain combinations can be calculated with methods described in this section. The results are of practical use, for example, if you wish to build a circuit having a specific capacitance and you have only specific values of capacitance at your disposal.

PARALLEL COMBINATION

Two capacitors connected as shown in Figure 16.16a are known as a *parallel combination* of capacitors. The left plate of each capacitor is connected by a conducting wire to the positive terminal of the battery, and the left plates are therefore at the same potential. Likewise, the right plates are connected to the negative terminal of the battery. When the capacitors are first connected in the circuit, electrons are transferred from the left plates through the battery to the right plates, leaving the left plates positively charged and the right plates negatively charged. The energy source for this charge transfer is the internal chemical energy stored in the battery, which is converted to electrical energy. The flow of charge ceases when the voltage across the capacitors equals that of the battery. The capacitors reach their maximum charge when the flow of charge ceases. Let us call the maximum charges on the two capacitors Q_1 and Q_2. Then the *total charge* Q stored by the two capacitors is

$$Q = Q_1 + Q_2 \qquad \textbf{[16.11]}$$

We can replace these two capacitors with one equivalent capacitor having a capacitance of C_{eq}. This equivalent capacitor must have exactly the same external effect on the circuit as the original two. That is, it must store Q units of charge. We also see from Figure 16.16b that **the potential differences across the capacitors**

> **Tip 16.3** VOLTAGE IS THE SAME AS POTENTIAL DIFFERENCE
>
> A voltage *across* a device, such as a capacitor, has the same meaning as the potential difference across the device. For example, if we say that the voltage across a capacitor is 12 V, we mean that the potential difference between its plates is 12 V.

FIGURE 16.16 (a) A parallel connection of two capacitors. (b) The circuit diagram for the parallel combination. (c) The potential differences across the capacitors are the same, and the equivalent capacitance is $C_{eq} = C_1 + C_2$.

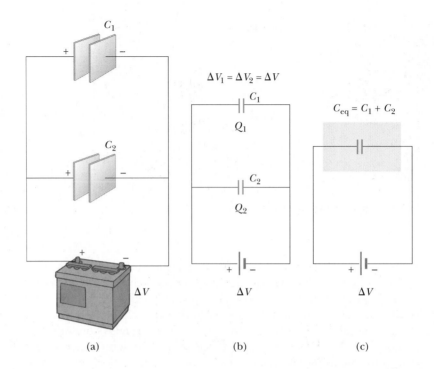

(a)   (b)   (c)

in a parallel circuit are the same; each is equal to the voltage ΔV of the battery. From Figure 16.16c, we see that the voltage across the equivalent capacitor is also ΔV. Thus, the charge on each capacitor is

$$Q_1 = C_1 \, \Delta V \qquad \text{and} \qquad Q_2 = C_2 \, \Delta V$$

The charge on the equivalent capacitor is

$$Q = C_{eq} \, \Delta V$$

Substituting these relations into Equation 16.11 gives

$$C_{eq} \, \Delta V = C_1 \, \Delta V + C_2 \, \Delta V$$

or

$$C_{eq} = C_1 + C_2 \qquad \left(\begin{array}{c} \text{parallel} \\ \text{combination} \end{array} \right) \qquad \textbf{[16.12]}$$

If we extend this treatment to three or more capacitors connected in parallel, the equivalent capacitance is found to be

$$C_{eq} = C_1 + C_2 + C_3 + \cdots \qquad \left(\begin{array}{c} \text{parallel} \\ \text{combination} \end{array} \right) \qquad \textbf{[16.13]}$$

Thus, we see that **the equivalent capacitance of a parallel combination of capacitors is greater than any of the individual capacitances.**

FIGURE 16.17 (Example 16.5) Four capacitors connected in parallel.

Example 16.5   Four Capacitors Connected in Parallel

Determine the capacitance of the single capacitor that is equivalent to the parallel combination of capacitors shown in Figure 16.17, and find the charge on the 12.0-μF capacitor.

Solution   The equivalent capacitance is found by use of Equation 16.13:

$$C_{eq} = C_1 + C_2 + C_3 + C_4$$
$$= 3.00 \; \mu\text{F} + 6.00 \; \mu\text{F} + 12.0 \; \mu\text{F} + 24.0 \; \mu\text{F} = 45.0 \; \mu\text{F}$$

The potential difference across the 12.0-μF capacitor (and all other capacitors in this case) is equal to the voltage of the battery, and so

$$Q = C\,\Delta V = (12.0 \times 10^{-6}\,\text{F})(18.0\,\text{V}) = 216 \times 10^{-6}\,\text{C} = \boxed{216\,\mu\text{C}}$$

SERIES COMBINATION

Now consider two capacitors connected in *series,* as illustrated in Figure 16.18a. **For a series combination of capacitors, the magnitude of the charge must be the same on all the plates.** To see why this must be true, let us consider the charge transfer process in some detail. We start with uncharged capacitors. When a battery is connected to the circuit, electrons are transferred from the left plate of C_1 to the right plate of C_2 through the battery. As this negative charge accumulates on the right plate of C_2, an equivalent amount of negative charge is removed from the left plate of C_2, leaving it with an excess positive charge. The negative charge leaving the left plate of C_2 accumulates on the right plate of C_1, where again an equivalent amount of negative charge is removed from the left plate. The result of this is that **all of the right plates gain charges of $-Q$ and all the left plates have charges of $+Q$.** (This is a consequence of the conservation of charge.)

    We can find an equivalent capacitor that performs the same function as the series combination. After it is fully charged, **the equivalent capacitor must end up with a charge of $-Q$ on its right plate and a charge of $+Q$ on its left plate.** By applying the definition of capacitance to the circuit in Figure 16.18b, we have

◀ Q is the same for all capacitors connected in series

$$\Delta V = \frac{Q}{C_{\text{eq}}}$$

where ΔV is the potential difference between the terminals of the battery and C_{eq} is the equivalent capacitance. From Figure 16.18a we see that

$$\Delta V = \Delta V_1 + \Delta V_2 \qquad\qquad \textbf{[16.14]}$$

where ΔV_1 and ΔV_2 are the potential differences across capacitors C_1 and C_2. (This is a consequence of the conservation of energy.) The potential difference across any number of capacitors (or other circuit elements) in series equals the sum of the potential differences across the individual capacitors. Because $Q = C\,\Delta V$ can be applied to each capacitor, the potential differences across them

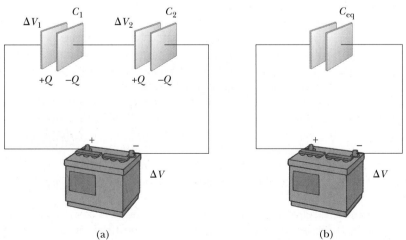

(a)   (b)

FIGURE 16.18 A series combination of two capacitors. The charges on the capacitors are the same, and the equivalent capacitance can be calculated from the reciprocal relationship $1/C_{\text{eq}} = (1/C_1) + (1/C_2)$.

are given by

$$\Delta V_1 = \frac{Q}{C_1} \qquad \Delta V_2 = \frac{Q}{C_2}$$

Substituting these expressions into Equation 16.14, and noting that $\Delta V = Q/C_{eq}$, we have

$$\frac{Q}{C_{eq}} = \frac{Q}{C_1} + \frac{Q}{C_2}$$

Canceling Q, we arrive at the relationship

$$\frac{1}{C_{eq}} = \frac{1}{C_1} + \frac{1}{C_2} \qquad \left(\begin{array}{c} \text{series} \\ \text{combination} \end{array}\right) \qquad \text{[16.15]}$$

If this analysis is applied to three or more capacitors connected in series, the equivalent capacitance is found to be

$$\frac{1}{C_{eq}} = \frac{1}{C_1} + \frac{1}{C_2} + \frac{1}{C_3} + \cdots \qquad \left(\begin{array}{c} \text{series} \\ \text{combination} \end{array}\right) \qquad \text{[16.16]}$$

As we shall demonstrate in Example 16.6, this implies that **the equivalent capacitance of a series combination is always less than any individual capacitance in the combination.**

Quick Quiz 16.4

A capacitor is designed so that one plate is large and the other is small. If the plates are connected to a battery, (a) the large plate has a greater charge than the small plate, (b) the large plate has less charge than the small plate, or (c) the plates have charges equal in magnitude but opposite in sign.

PROBLEM-SOLVING STRATEGY **Capacitors**

1. Be careful with your choice of units. To calculate the capacitance of a device in farads, make sure that distances are in meters and use the SI value of ϵ_0.

2. When two or more unequal capacitors are connected in *series*, they carry the same charge, but the potential differences across them are not the same. Their capacitances add as reciprocals, and the equivalent capacitance of the combination is always *less* than the smallest individual capacitor.

3. When two or more capacitors are connected in *parallel*, the potential differences across them are the same. The charge on each capacitor is proportional to its capacitance; hence, the capacitances add directly to give the equivalent capacitance of the parallel combination.

4. A complicated circuit consisting of capacitors can often be reduced to a simple circuit containing only one capacitor. To do this, examine your initial circuit and replace any capacitors in series or any in parallel with equivalent capacitors, using the rules in 2 and 3. After making these changes, sketch your new circuit. Examine it and replace any series or parallel combinations again. Continue this process until a single equivalent capacitor is found.

5. To find the charge on, or the potential difference across, one of the capacitors in the complicated circuit, start with the final circuit found in 4, and gradually work your way back through the circuits using $C = Q/\Delta V$ and the information in 2 and 3 above.

Example 16.6 Four Capacitors Connected in Series

Four capacitors are connected in series with a battery, as in Figure 16.19.

A Find the capacitance of the equivalent capacitor.

Solution The equivalent capacitance is found from Equation 16.16:

$$\frac{1}{C_{eq}} = \frac{1}{3.0 \ \mu F} + \frac{1}{6.0 \ \mu F} + \frac{1}{12 \ \mu F} + \frac{1}{24 \ \mu F}$$

$$C_{eq} = \boxed{1.6 \ \mu F}$$

Note that the equivalent capacitance is less than the capacitance of any of the individual capacitors in the combination.

B Find the charge on the 12-μF capacitor.

Solution We find the charge on the equivalent capacitor:

$$Q = C_{eq} \, \Delta V = (1.6 \times 10^{-6} \ F)(18 \ V) = \boxed{29 \ \mu C}$$

This is also the charge on each of the capacitors it replaced. Thus, the charge on the 12-μF capacitor in the original circuit is 29 μC.

FIGURE 16.19 (Example 16.6) Four capacitors connected in series.

Example 16.7 Equivalent Capacitance

Find the equivalent capacitance between a and b for the combination of capacitors shown in Figure 16.20a. All capacitances are in microfarads.

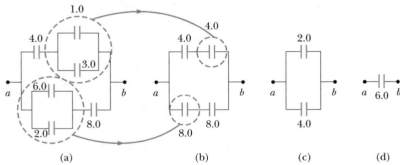

(a)   (b)   (c)   (d)

FIGURE 16.20 (Example 16.7) To find the equivalent capacitance of the circuit in (a), the circuit is reduced in steps—as indicated in (b), (c), and (d)—using the series and parallel rules described in the text.

Solution Using Equations 16.13 and 16.16, we reduce the combination step by step as indicated in the figure. The 1.0-μF and 3.0-μF capacitors are in *parallel* and combine according to $C_{eq} = C_1 + C_2$. Their equivalent capacitance is 4.0 μF. Likewise, the 2.0-μF and 6.0-μF capacitors are also in *parallel* and have an equivalent capacitance of 8.0 μF. The upper branch in Figure 16.20b now consists of two 4.0-μF capacitors in *series*, which combine according to

$$\frac{1}{C_{eq}} = \frac{1}{C_1} + \frac{1}{C_2} = \frac{1}{4.0 \ \mu F} + \frac{1}{4.0 \ \mu F} = \frac{1}{2.0 \ \mu F}$$

$$C_{eq} = 2.0 \ \mu F$$

Likewise, the lower branch in Figure 16.20b consists of two 8.0-μF capacitors in *series* and has an equivalent capacitance of 4.0 μF. Finally, the 2.0-μF and 4.0-μF capacitors in Figure 16.20c are in *parallel* and have an equivalent capacitance of 6.0 μF. Hence, the equivalent capacitance of the circuit is $\boxed{6.0 \ \mu F}$.

16.9 ENERGY STORED IN A CHARGED CAPACITOR

Almost everyone who works with electronic equipment has at some time verified that a capacitor can store energy. If the plates of a charged capacitor are connected by a conductor such as a wire, charge transfers from one plate to the other until the two are uncharged. The discharge can often be observed as a visible spark. If you accidentally touched the opposite plates of a charged capacitor, your fingers would act as a pathway by which the capacitor could discharge, inflicting an electric shock. The degree of shock would depend on the capacitance and voltage applied to the capacitor. Where high voltages and large quantities of charge are present, as in the power supply of a television set, such a shock can be fatal.

If a capacitor is initially uncharged (both plates neutral), so that the plates are at the same potential, almost no work is required to transfer a small amount of charge ΔQ from one plate to the other. However, once this charge has been transferred, a small potential difference $\Delta V = \Delta Q/C$ appears between the plates. Therefore, work must be done to transfer additional charge against this potential difference. We must look at the way the voltage changes at each moment. As more charge is transferred from one plate to the other, the potential difference increases in proportion to the charge. If the potential difference at any instant during the charging process is ΔV, the work required to move more charge ΔQ through this potential difference is $\Delta V \Delta Q$; that is,

$$\Delta W = \Delta V \Delta Q$$

We know that $\Delta V = Q/C$ for a capacitor that has a total charge of Q. Therefore, a plot of voltage versus total charge gives a straight line with a slope of $1/C$, as shown in Figure 16.21. Because the work ΔW is the area of the shaded rectangle, the total work done in charging the capacitor to a final voltage ΔV is the area under the voltage-charge curve, which in this case equals the area under the straight line. Because the area under this line is the area of a triangle (which is one half of the product of the base and height), the total work done is

$$W = \tfrac{1}{2} Q \Delta V \qquad\qquad \textbf{[16.17]}$$

Note that this is also the energy stored in the capacitor, because the work required to charge the capacitor equals the energy stored in the capacitor after it is charged. From the definition of capacitance, we find $Q = C\,\Delta V$; hence, we can express the energy stored as

$$\text{Energy stored} = \tfrac{1}{2} Q\,\Delta V = \tfrac{1}{2} C(\Delta V)^2 = \frac{Q^2}{2C} \qquad\qquad \textbf{[16.18]}$$

For example, the amount of energy stored in a 5.0-μF capacitor when it is connected across a 120-V battery is

$$\text{Energy stored} = \tfrac{1}{2} C(\Delta V)^2 = \tfrac{1}{2}(5.0 \times 10^{-6}\,\text{F})(120\,\text{V})^2 = 3.6 \times 10^{-2}\,\text{J}$$

In practice, there is a limit to the maximum energy (or charge) that can be stored in a capacitor, because electrical breakdown ultimately occurs between the plates of the capacitor at a sufficiently large value of ΔV. For this reason, capacitors are usually labeled with a maximum operating voltage.

Large capacitors can store enough electrical energy to cause severe burns or even death if they are discharged so that the flow of charge can pass through the heart. Under the proper conditions, however, they can be used to sustain life by stopping cardiac fibrillation in heart attack victims. When fibrillation occurs, the heart produces a rapid, irregular pattern of beats. A fast discharge of electrical energy through the heart can return the organ to its normal beat pattern. Emergency medical teams use portable defibrillators that contain batteries capable of charging a capacitor to a high voltage. (The circuitry actually permits the capacitor to be charged to a much higher voltage than the battery.) In this case and others (camera flash units and lasers used for fusion experiments), capacitors serve as

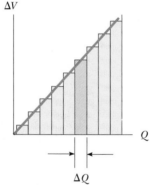

FIGURE 16.21 A plot of voltage versus charge for a capacitor is a straight line with the slope $1/C$. The work required to move a charge of ΔQ through a potential difference of ΔV across the capacitor plates is $\Delta W = \Delta V \Delta Q$, which equals the area of the blue rectangle. The *total work* required to charge the capacitor to a final charge of Q is the area under the straight line, which equals $Q\Delta V/2$.

DEFIBRILLATORS

energy reservoirs that can be slowly charged and then discharged quickly to provide large amounts of energy in a short pulse. The stored electrical energy is released through the heart by conducting electrodes, called paddles, that are placed on both sides of the victim's chest. The paramedics must wait between applications of the electrical energy due to the time necessary for the capacitors to become fully charged. The high voltage on the capacitor can be obtained from a low-voltage battery in a portable machine with the phenomenon of *electromagnetic induction,* to be studied in Chapter 20.

Webnote 16.2

Researchers report evidence suggesting that the heart's erratic shivering during cardiac fibrillation is a form of chaos. Read more in

http://www.sciencenews.org/sn_arc97/ 1_25_97/fob1.htm

Example 16.8 Typical Voltage, Energy, and Discharge Time for a Defibrillator

A fully charged defibrillator contains 1.2 kJ of energy stored in a 110-μF capacitor. In a discharge through a patient, 600 J of electrical energy are delivered in 2.5 ms. (This is impossible for a battery, which cannot supply such a large amount of energy in such a short time.)

A Find the voltage needed to store 1.2 kJ in the unit.

Solution Because we know the energy stored and the capacitance, we can use Equation 16.18 to find the required voltage:

$$\text{Energy stored} = \tfrac{1}{2}C\,\Delta V^2$$

$$\Delta V = \sqrt{\frac{2 \times (\text{energy stored})}{C}} = \sqrt{\frac{2(1.2 \times 10^3\,\text{J})}{1.1 \times 10^{-4}\,\text{F}}} = \boxed{4.7 \times 10^3\,\text{V}}$$

B What power is delivered to the patient in watts?

Solution Because 600 J of energy are delivered in 2.5 ms, we can find the power delivered to the patient as follows:

$$\mathcal{P} = \frac{\text{Energy delivered}}{\Delta t} = \frac{600\,\text{J}}{2.5 \times 10^{-3}\,\text{s}} = \boxed{2.4 \times 10^5\,\text{W}}$$

APPLYING PHYSICS 16.2

You have three capacitors and two batteries. How should you connect the batteries to all three capacitors so that the capacitors will store the maximum possible energy?

Explanation The energy stored in the capacitor is proportional to the capacitance and the square of the potential difference. Thus, we would like to maximize each of these quantities. We can do this by connecting the three capacitors in parallel (so that the capacitances add) across the two batteries in series (so that the potential differences add).

Quick Quiz 16.5 You charge a parallel-plate capacitor, remove it from the battery, and prevent the wires connected to the plates from touching each other. When you pull the plates farther apart, do the following quantities increase, decrease, or stay the same? (a) C, (b) Q, (c) E between the plates, (d) ΔV, (e) energy stored in the capacitor.

16.10 CAPACITORS WITH DIELECTRICS

A **dielectric** is an insulating material, such as rubber, plastic, or waxed paper. When a dielectric is inserted between the plates of a capacitor, the capacitance increases. If the dielectric completely fills the space between the plates, the capacitance is multiplied by the factor κ, called the **dielectric constant.**

The following experiment illustrates the effect of a dielectric in a capacitor. Consider a parallel-plate capacitor of charge Q_0 and capacitance C_0 in the absence of a dielectric. The potential difference across the capacitor plates can be measured, and it is given by $\Delta V_0 = Q_0/C_0$ (Fig. 16.22a). Because the capacitor is not connected to an external circuit, there is no pathway for charge to leave or be added to the plates. If a dielectric is now inserted between the plates as in Figure 16.22b, it is found that the voltage across the plates is *reduced* by the factor κ to the value ΔV, where

$$\Delta V = \frac{\Delta V_0}{\kappa}$$

Because $\kappa > 1$, ΔV is less than ΔV_0. Because the charge Q_0 on the capacitor does not change, we conclude that the capacitance C in the presence of the dielectric must change to the value

$$C = \frac{Q_0}{\Delta V} = \frac{Q_0}{\Delta V_0/\kappa} = \frac{\kappa Q_0}{\Delta V_0}$$

or

$$C = \kappa C_0 \qquad [16.19]$$

According to this result, the capacitance is *multiplied* by the factor κ when the dielectric completely fills the region between the plates. For a parallel-plate capacitor, where the capacitance in the absence of a dielectric is $C_0 = \epsilon_0 A/d$, we can express the capacitance in the presence of a dielectric as

$$C = \kappa \epsilon_0 \frac{A}{d} \qquad [16.20]$$

From this result it appears that the capacitance could be made very large by decreasing d, the plate separation. In practice, the lowest value of d is limited by the electric discharge that can occur through the dielectric material separating the plates. For any given plate separation, there is a maximum electric field that can be produced in the dielectric before it breaks down and begins to conduct. This maximum electric field is called the **dielectric strength,** and for air its value is

FIGURE 16.22 (a) With air between the plates, the voltage across the capacitor is ΔV_0, the capacitance is C_0, and the charge is Q_0. (b) With a dielectric between the plates, the charge remains at Q_0, but the voltage and capacitance both change.

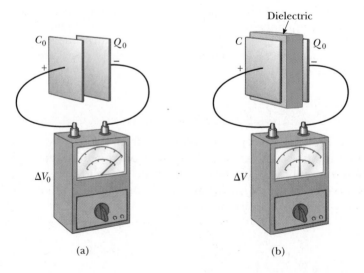

(a)   (b)

TABLE 16.1	Dielectric Constants and Dielectric Strengths of Various Materials at Room Temperature	
Material	Dielectric Constant, κ	Dielectric Strength (V/m)
Vacuum	1.000 00	—
Air	1.000 59	3×10^6
Bakelite	4.9	24×10^6
Fused quartz	3.78	8×10^6
Pyrex glass	5.6	14×10^6
Polystyrene	2.56	24×10^6
Teflon	2.1	60×10^6
Neoprene rubber	6.7	12×10^6
Nylon	3.4	14×10^6
Paper	3.7	16×10^6
Strontium titanate	233	8×10^6
Water	80	—
Silicone oil	2.5	15×10^6

Dielectric breakdown in air. Sparks are produced when a large alternating voltage is applied across the wires using a high-voltage induction coil power supply. *(© Loren Winters/Visuals Unlimited)*

about 3×10^6 V/m. Most insulating materials have dielectric strengths greater than that of air, as indicated by the values in Table 16.1.

Commercial capacitors are often made using metal foil interlaced with thin sheets of paraffin-impregnated paper or mylar, which serves as the dielectric material. These alternating layers of metal foil and dielectric are then rolled into a small cylinder (Fig. 16.23a). A high-voltage capacitor commonly consists of a number of interwoven metal plates immersed in silicone oil (Fig. 16.23b). Small capacitors are often constructed from ceramic materials. Variable capacitors (typically 10–500 pF) usually consist of two interwoven sets of metal plates, one fixed and the other movable, with air as the dielectric.

An electrolytic capacitor (Fig. 16.23c) is often used to store large amounts of charge at relatively low voltages. It consists of a metal foil in contact with an electrolyte—a solution that conducts charge by virtue of the motion of the ions contained in it. When a voltage is applied between the foil and the electrolyte, a thin layer of metal oxide (an insulator) is formed on the foil, and this layer serves as the dielectric. Enormous capacitances can be attained because the dielectric layer is very thin.

When electrolytic capacitors are used in circuits, the polarity (the plus and minus signs on the device) must be observed. If the polarity of the applied voltage is opposite that intended, the oxide layer will be removed and the capacitor will conduct rather than store charge. Furthermore, reversing the polarity can result in such a large current that the capacitor may either burn or produce steam and explode.

Metal foil

Plates

Case

Electrolyte

Contacts

Metallic foil + oxide layer

Oil

Paper

(a)   (b)   (c)

FIGURE 16.23 Three commercial capacitor designs. (a) A tubular capacitor whose plates are separated by paper and then rolled into a cylinder, (b) a high-voltage capacitor consisting of many parallel plates separated by oil, and (c) an electrolytic capacitor.

A collection of capacitors used in a variety of applications. *(Paul Silverman/Fundamental Photographs)*

APPLICATION

STUD FINDERS

APPLYING PHYSICS 16.3

If you have ever tried to hang a picture on a wall securely, you know that it can be difficult to locate a wooden stud in which to anchor your nail or screw. The principles discussed in this section can be used to detect a stud electronically. The primary element of an electronic stud finder is a capacitor with its plates arranged side by side instead of facing one another, as in Figure 16.24. How does this device work?

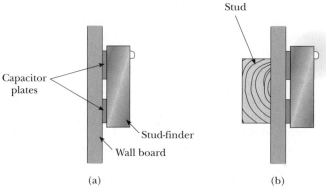

FIGURE 16.24 (Applying Physics 16.3) A stud finder. (a) The materials between the plates of the capacitor are the drywall and the air behind it. (b) The materials become drywall and wood when the detector moves across a stud in the wall. The change in the dielectric constant causes a signal light to illuminate.

Explanation As the detector is moved along a wall, its capacitance changes when it passes across a stud because the dielectric constant of the material "between" the plates changes. The change in capacitance can be used to cause a light to come on, signaling the presence of the stud.

Webnote 16.3

At home you might have a touch-sensitive lamp (turns on when you touch it). This device elegantly makes use of capacitance. Visit *http://www.howstuffworks.com/question42.htm*

Quick Quiz 16.6

A fully charged parallel-plate capacitor remains connected to a battery while you slide a dielectric between the plates. Do the following quantities increase, decrease, or stay the same? (a) C, (b) Q, (c) E between the plates, (d) ΔV, (e) energy stored in the capacitor.

Example 16.9 A Paper-Filled Capacitor

A parallel-plate capacitor has plates 2.0 cm by 3.0 cm. The plates are separated by a 1.0-mm thickness of paper.

A Find the capacitance of this device.

Solution Because $\kappa = 3.7$ for paper (Table 16.1), we get

$$C = \kappa \epsilon_0 \frac{A}{d} = 3.7 \left(8.85 \times 10^{-12} \frac{C^2}{N \cdot m^2} \right) \left(\frac{6.0 \times 10^{-4} \, m^2}{1.0 \times 10^{-3} \, m} \right)$$

$$= 20 \times 10^{-12} \, F = \boxed{20 \, pF}$$

B Find the maximum charge that can be placed on the capacitor.

Solution From Table 16.1 we see that the dielectric strength of paper is equal to 16×10^6 V/m. Because the paper thickness is 1.0 mm, the maximum voltage that can be applied before electrical breakdown occurs can be calculated using Equation 16.4:

$$\Delta V_{max} = E_{max} d = (16 \times 10^6 \, V/m)(1.0 \times 10^{-3} \, m) = 16 \times 10^3 \, V$$

Hence, the maximum charge that can be placed on the capacitor is

$$Q_{max} = C \Delta V_{max} = (20 \times 10^{-12} \, F)(16 \times 10^3 \, V) = \boxed{0.32 \, \mu C}$$

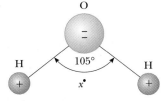

FIGURE 16.25 The water molecule, H_2O, has a permanent polarization resulting from its bent geometry. The point labeled x is the center of positive charge.

AN ATOMIC DESCRIPTION OF DIELECTRICS

The explanation of why a dielectric increases the capacitance of a capacitor is based on an atomic description of the material, which in turn involves a property of some molecules called **polarization.** A molecule is said to be polarized when there is a separation between the "centers of gravity" of its negative charge and its positive charge. In some molecules, such as water, this condition is always present. To see why, consider the geometry of a water molecule (Fig. 16.25).

The molecule is arranged so that the negative oxygen atom is bonded to the positively charged hydrogen atoms with a 105° angle between the two bonds. The center of negative charge is at the oxygen atom, and the center of positive charge lies at a point midway along the line joining the hydrogen atoms (point x in the diagram). Materials composed of molecules that are permanently polarized in this fashion have large dielectric constants, and indeed Table 16.1 shows that the dielectric constant of water is quite large ($\kappa = 80$).

A symmetric molecule (Fig. 16.26a) can have no permanent polarization, but a polarization can be induced by an external electric field. A field directed to the left, as in Figure 16.26b, would cause the center of positive charge to shift to the left from its initial position, and the center of negative charge to shift to the right. This *induced polarization* is the effect that predominates in most materials used as dielectrics in capacitors.

To understand why the polarization of a dielectric can affect capacitance, consider Figure 16.27, which shows a slab of dielectric placed between the plates of a parallel-plate capacitor. The dielectric becomes polarized as shown because it is in the electric field that exists between the metal plates. Notice that a net positive charge appears on the dielectric surface adjacent to the negatively charged metal plate. The presence of this positive charge on the dielectric effectively reduces some of the negative charge on the metal, allowing more negative charge to be stored on the capacitor plates for a given applied voltage. From the definition of capacitance, $C = Q \Delta V$, we see that, because the plates can store more charge for a given voltage, the capacitance must increase.

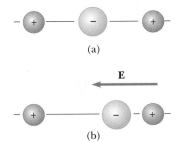

FIGURE 16.26 (a) A symmetric molecule has no permanent polarization. (b) An external electric field induces a polarization in the molecule.

FIGURE 16.27 When a dielectric is placed between the plates of a charged parallel-plate capacitor, the dielectric becomes polarized. This creates a net positive induced charge on the left side of the dielectric and a net negative induced charge on the right side. As a result, the capacitance of the device is multiplied by the factor κ.

APPLYING **PHYSICS** **16.4**

Consider a parallel-plate capacitor with a dielectric material between the plates. Is the capacitance higher on a cold day or a hot day?

Explanation The polarization of the molecules in the dielectric increases the capacitance when the dielectric is added. As the temperature increases, there is more vibrational motion of the polarized molecules. This disturbs the orderly arrangement of the polarized molecules, and the net polarization decreases. Thus, the capacitance must decrease as the temperature increases.

SUMMARY

The **difference in electric potential ΔV** between two points A and B is

$$\Delta V \equiv V_B - V_A = \frac{\Delta PE}{q} \qquad [16.2]$$

where ΔPE is the *change* in electrical potential energy experienced by a charge q as it moves between A and B. The units of potential difference are joules per coulomb, or **volts;** $1\,\text{J/C} = 1\,\text{V}$.

The **electric potential difference** between two points A and B in a uniform electric field $\mathbf{E}$ is

$$V_B - V_A = -Ed \qquad [16.4]$$

where d is the distance between A and B and E is the strength of the electric field in that region.

The **electric potential** due to a point charge q at distance r from the point charge is

$$V = k_e \frac{q}{r} \qquad [16.5]$$

The **electrical potential energy** of a pair of point charges separated by distance r is

$$PE = k_e \frac{q_1 q_2}{r} \qquad [16.6]$$

Every point on the surface of a charged conductor in electrostatic equilibrium is at the same potential. Furthermore, the potential is constant everywhere inside the conductor and equals its value on the surface.

The **electron volt** is defined as the energy that an electron (or proton) gains when accelerated through a potential difference of 1 V. The conversion between electron volts and joules is

$$1\,\text{eV} = 1.60 \times 10^{-19}\,\text{C}\cdot\text{V} = 1.60 \times 10^{-19}\,\text{J} \qquad [16.8]$$

A **capacitor** consists of two metal plates with charges that are equal in magnitude but opposite in sign. The capacitance (C) of any capacitor is the ratio of the magnitude of the charge Q on either plate to the potential difference ΔV between them:

$$C \equiv \frac{Q}{\Delta V} \qquad [16.9]$$

Capacitance has the units coulombs per volt, or farads; $1\,\text{C/V} = 1\,\text{F}$.

The capacitance of two parallel metal plates of area A separated by distance d is

$$C = \epsilon_0 \frac{A}{d} \qquad [16.10]$$

where ϵ_0 is a constant called the **permittivity of free space,** with the value $\epsilon_0 = 8.85 \times 10^{-12}\,\text{C}^2/\text{N}\cdot\text{m}^2$.

The **equivalent capacitance of a parallel combination** of capacitors is

$$C_{eq} = C_1 + C_2 + C_3 + \cdots \qquad \textbf{[16.13]}$$

If two or more capacitors are connected in series, the **equivalent capacitance of the series combination** is

$$\frac{1}{C_{eq}} = \frac{1}{C_1} + \frac{1}{C_2} + \frac{1}{C_3} + \cdots \qquad \textbf{[16.16]}$$

Three equivalent expressions for calculating the **energy stored** in a charged capacitor are

$$\text{Energy stored} = \tfrac{1}{2}Q\,\Delta V = \tfrac{1}{2}C(\Delta V)^2 = \frac{Q^2}{2C} \qquad \textbf{[16.18]}$$

When a nonconducting material, called a **dielectric,** is placed between the plates of a capacitor, the capacitance is multiplied by the factor κ, which is called the **dielectric constant** and is a property of the dielectric material. The capacitance of a parallel-plate capacitor filled with a dielectric is

$$C = \kappa \epsilon_0 \frac{A}{d} \qquad \textbf{[16.20]}$$

CONCEPTUAL QUESTIONS

1. Criticize this statement: "There is no potential in a region where the electric field is uniform."
2. "The equipotentials of a uniformly charged conducting sphere are spheres." Is this statement true? Discuss.
3. A parallel-plate capacitor is charged by a battery, and the battery is then disconnected from the capacitor. Because the charges on the capacitor plates are opposite in sign, they attract each other. Hence, it takes positive work to increase the plate separation. Show that the external work done when the plate separation is increased leads to an increase in the energy stored in the capacitor.
4. Distinguish between electric potential and electrical potential energy.
5. Suppose you are sitting in a car and a 20-kV power line drops across the car. Should you stay in the car or get out? The power line potential is 20 kV compared to the potential of the ground.
6. Why is it important to avoid sharp edges or points on conductors used in high-voltage equipment?
7. Explain why, under static conditions, all points in a conductor must be at the same electric potential.
8. If you are given three different capacitors C_1, C_2, and C_3, how many different combinations of capacitance can you produce, using all capacitors in your circuits?
9. Why is it dangerous to touch the terminals of a high-voltage capacitor even after the voltage source that charged the battery is disconnected from the capacitor? What can be done to make the capacitor safe to handle after the voltage source has been removed?
10. The plates of a capacitor are connected to a battery. What happens to the charge on the plates if the connecting wires are removed from the battery? What happens to the charge if the wires are removed from the battery and connected to each other?
11. Can electric field lines ever cross? Why or why not? Can equipotentials ever cross? Why or why not?
12. What happens to the energy stored in a capacitor (a) if the potential difference across the plates is doubled? (b) if the plate separation is doubled while the capacitor is connected to a battery?
13. (a) What happens to the *charge* stored in a capacitor if the potential difference across the plates is doubled? (b) What happens to the *energy* stored in a capacitor if the plate separation is doubled while the charge on the capacitor remains constant (as it would after being disconnected from a battery)?
14. Is it always possible to reduce a combination of capacitors to one equivalent capacitor with the rules developed in this chapter? Explain.
15. If you were asked to design a capacitor where small size and large capacitance were required, what factors would be important in your design?
16. Explain why a dielectric increases the maximum operating voltage of a capacitor although the physical size of the capacitor does not change.
17. What happens to the energy stored in a capacitor (a) if a dielectric is inserted into the capacitor while it is connected to a battery? (b) if a dielectric is inserted into the capacitor while the charge on the capacitor remains constant (as it would after being disconnected from a battery)?

PROBLEMS

1, 2, 3 = straightforward, intermediate, challenging   ☐ = full solution available in Student Solutions Manual/Study Guide
web = solution posted at **http://info.brookscole.com/serway**   = biomedical application

Section 16.1   Potential Difference and Electric Potential

1. A proton moves 2.00 cm parallel to a uniform electric field with $E = 200$ N/C. (a) How much work is done by the field on the proton? (b) What change occurs in the potential energy of the proton? (c) Through what potential difference did the proton move?

2. A uniform electric field of magnitude 250 V/m is directed in the positive x direction. A positive 12-μC charge moves from the origin to the point $(x, y) = (20$ cm, 50 cm). (a) What is the change in the potential energy of this charge? (b) Through what potential difference does the charge move?

3. A potential difference of 90 mV exists between the inner and outer surfaces of a cell membrane. The inner surface is negative relative to the outer surface. How much work is required to eject a positive sodium ion (Na^+) from the interior of the cell?

4. An ion accelerated through a potential difference of 115 V experiences an increase in kinetic energy of 7.37×10^{-17} J. Calculate the charge on the ion.

5. The difference in potential between the accelerating plates of a TV set is about 25 kV. If the distance between these plates is 1.5 cm, find the magnitude of the uniform electric field in this region.

6. To recharge a 12-V battery, a battery charger must move 3.6×10^5 C of charge from the negative terminal to the positive terminal. How much work is done by the battery charger? Express your answer in joules.

7. A pair of oppositely charged parallel plates are separated
web by 5.33 mm. A potential difference of 600 V exists between the plates. (a) What is the magnitude of the electric field between the plates? (b) What is the magnitude of the force on an electron between the plates? (c) How much work must be done on the electron to move it to the negative plate if it is initially positioned 2.90 mm from the positive plate?

8. Suppose an electron is released from rest in a uniform electric field whose strength is 5.90×10^3 V/m. (a) Through what potential difference will it have passed after moving 1.00 cm? (b) How fast will the electron be moving after it has traveled 1.00 cm?

9. A 4.00-kg block carrying a charge $Q = 50.0$ μC is connected to a spring for which $k = 100$ N/m. The block lies on a frictionless horizontal track, and the system is immersed in a uniform electric field of magnitude $E = 5.00 \times 10^5$ V/m directed as in Figure P16.9. (a) If the block is released at rest when the spring is unstretched (at

FIGURE P16.9

$x = 0$), by what maximum amount does the spring expand? (b) What is the equilibrium position of the block?

10. On planet Tehar, the free-fall acceleration is the same as that on Earth, but there is also a strong downward electric field that is uniform close to the planet's surface. A 2.00-kg ball having a charge of 5.00 μC is thrown upward at a speed of 20.1 m/s, and it hits the ground after an interval of 4.10 s. What is the potential difference between the starting point and the top point of the trajectory?

Section 16.2   Electric Potential and Potential Energy Due to Point Charges

Section 16.3   Potentials and Charged Conductors

Section 16.4   Equipotential Surfaces

11. (a) Find the potential 1.00 cm from a proton. (b) What is the potential difference between two points that are 1.00 cm and 2.00 cm from a proton?

12. Find the potential at point P for the rectangular grouping of charges shown in Figure P16.12.

FIGURE P16.12

13. (a) Find the electric potential, taking zero at infinity, at the upper-right corner (the corner without a charge) of the rectangle in Figure P16.13. (b) Repeat if the 2.00-μC charge is replaced with a charge of -2.00 μC.

FIGURE P16.13   (Problems 13 and 14)

14. Three charges are situated at corners of a rectangle as in Figure P16.13. How much energy would be expended in moving the 8.00-μC charge to infinity?

15. Two point charges $Q_1 = +5.00$ nC and $Q_2 = -3.00$ nC are separated by 35.0 cm. (a) What is the electric potential at a point midway between the charges? (b) What is the potential energy of the pair of charges? What is the significance of the algebraic sign of your answer?

16. Calculate the speed of (a) an electron that has a kinetic energy of 1.00 eV and (b) a proton that has a kinetic energy of 1.00 eV.

17. The three charges in Figure P16.17 are at the vertices of an isosceles triangle. Calculate the electric potential at the midpoint of the base, taking $q = 7.00$ nC.

FIGURE P16.17

18. An electron starts from rest 3.00 cm from the center of a uniformly charged sphere of radius 2.00 cm. If the sphere carries a total charge of 1.00×10^{-9} C, how fast will the electron be moving when it reaches the surface of the sphere?

19. In Rutherford's famous scattering experiments (which led **web** to the planetary model of the atom), alpha particles (having charges of $+2e$ and masses of 6.64×10^{-27} kg) were fired toward a gold nucleus with charge $+79e$. An alpha particle, initially very far from the gold nucleus, is fired at 2.00×10^{7} m/s directly toward the gold nucleus as in Figure P16.19. How close does the alpha particle get to the gold nucleus before turning around? Assume the gold nucleus remains stationary.

FIGURE P16.19

20. Starting with the definition of work, prove that at every point on an equipotential surface the surface must be perpendicular to the local electric field.

21. A small spherical object carries a charge of 8.00 nC. At what distance from the center of the object is the potential equal to 100 V? 50.0 V? 25.0 V? Is the spacing of the equipotentials proportional to the change in potential?

Section 16.6 Capacitance

Section 16.7 The Parallel-Plate Capacitor

22. (a) How much charge is on each plate of a 4.00-μF capacitor when it is connected to a 12.0-V battery? (b) If this same capacitor is connected to a 1.50-V battery, what charge is stored?

23. Consider Earth and a cloud layer 800 m above Earth to be the plates of a parallel-plate capacitor. (a) If the cloud layer has an area of 1.0 km^2 $= 1.0 \times 10^6$ m^2, what is the capacitance? (b) If an electric field strength greater than 3.0×10^6 N/C causes the air to break down and conduct charge (lightning), what is the maximum charge the cloud can hold?

24. The potential difference between a pair of oppositely charged parallel plates is 400 V. (a) If the spacing between the plates is doubled without altering the charge on the plates, what is the new potential difference between the plates? (b) If the plate spacing is doubled while the potential difference between the plates is kept constant, what is the ratio of the final charge on one of the plates to the original charge?

25. An air-filled capacitor consists of two parallel plates, each with an area of 7.60 cm^2, separated by a distance of 1.80 mm. If a 20.0-V potential difference is applied to these plates, calculate (a) the electric field between the plates, (b) the capacitance, and (c) the charge on each plate.

26. A 1-megabit computer memory chip contains many 60.0×10^{-15} F capacitors. Each capacitor has a plate area of 21.0×10^{-12} m^2. Determine the plate separation of such a capacitor (assume a parallel-plate configuration). The diameter of an atom is on the order of 10^{-10} m $= 1$ Å. Express the plate separation in angstroms.

27. The plates of a parallel-plate capacitor are separated by 0.100 mm. If the material between the plates is air, what plate area is required to provide a capacitance of 2.00 pF?

28. A small object with a mass of 350 mg carries a charge of 30.0 nC and is suspended by a thread between the vertical plates of a parallel-plate capacitor. The plates are separated by 4.00 cm. If the thread makes an angle of 15.0° with the vertical, what is the potential difference between the plates?

Section 16.8 Combinations of Capacitors

29. A series circuit consists of a 0.050-μF capacitor, a 0.100-μF capacitor, and a 400-V battery. Find the charge (a) on each of the capacitors and (b) on each of the capacitors if they are reconnected in parallel across the battery.

30. Three capacitors $C_1 = 5.00$ μF, $C_2 = 4.00$ μF, and $C_3 = 9.00$ μF are connected together. Find the effective capacitance of the group (a) if they are all in parallel, and (b) if they are all in series.

31. (a) Find the equivalent capacitance of the group of capacitors in Figure P16.31. (b) Find the charge on and the potential difference across each.

FIGURE P16.31

32. Two capacitors when connected in parallel give an equivalent capacitance of 9.00 pF and give an equivalent capacitance of 2.00 pF when connected in series. What is the capacitance of each capacitor?

33. Four capacitors are connected as shown in Figure P16.33. (a) Find the equivalent capacitance between points *a* and *b*. (b) Calculate the charge on each capacitor if a 15.0-V battery is connected across points *a* and *b*.

FIGURE P16.33

34. Consider the combination of capacitors in Figure P16.34. (a) What is the equivalent capacitance of the group? (b) Determine the charge on each capacitor.

FIGURE P16.34

35. Find the charge on each of the capacitors in Figure P16.35.

FIGURE P16.35

36. To repair a power supply for a stereo amplifier, an electronics technician needs a 100-μF capacitor capable of withstanding a potential difference of 90 V between the plates. The only available supply is a box of five 100-μF capacitors, each having a maximum voltage capability of 50 V. Can the technician substitute a combination of these capacitors that has the proper electrical characteristics, and if so, what will be the maximum voltage across any of the capacitors used? (*Hint:* The technician may not have to use all the capacitors in the box.)

37. A 25.0-μF capacitor and a 40.0-μF capacitor are charged
web by being connected across separate 50.0-V batteries. (a) Determine the resulting charge on each capacitor. (b) The capacitors are then disconnected from their batteries and connected to each other, with each negative plate connected to the other positive plate. What is the final charge of each capacitor, and what is the final potential difference across the 40.0-μF capacitor?

38. A 10.0-μF capacitor is fully charged across a 12.0-V battery. The capacitor is then disconnected from the battery and connected across an initially uncharged capacitor, *C*. The resulting voltage across each capacitor is 3.00 V. What is the capacitance *C*?

39. A 1.00-μF capacitor is first charged by being connected across a 10.0-V battery. It is then disconnected from the battery and connected across an uncharged 2.00-μF capacitor. Determine the resulting charge on each capacitor.

40. Find the equivalent capacitance between points *a* and *b* for the group of capacitors connected as shown in Figure P16.40 if $C_1 = 5.00$ μF, $C_2 = 10.0$ μF, and $C_3 = 2.00$ μF.

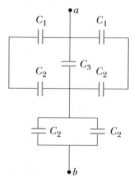

FIGURE P16.40   (Problems 40 and 41)

41. For the network described in the previous problem if the potential between points *a* and *b* is 60.0 V, what charge is stored on C_3?

42. Find the equivalent capacitance between points *a* and *b* in the combination of capacitors shown in Figure P16.42.

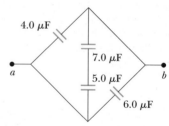

FIGURE P16.42

Section 16.9   Energy Stored in a Charged Capacitor

43. A parallel-plate capacitor has 2.00-cm^2 plates that are separated by 5.00 mm with air between them. If a 12.0-V battery is connected to this capacitor, how much energy does it store?

44. Two capacitors $C_1 = 25.0$ μF and $C_2 = 5.00$ μF are connected in parallel and charged with a 100-V power supply. (a) Calculate the total energy stored in the two capacitors. (b) What potential difference would be required across the same two capacitors connected in *series* in order that the combination store the same energy as in (a)?

45. Consider the parallel-plate capacitor formed by Earth and a cloud layer as described in Problem 16.23. Assume this capacitor will discharge (that is, lightning occurs) when the electric field strength between the plates reaches 3.0×10^6 N/C. What is the energy released if the capacitor discharges completely during a lightning strike?

46. A certain storm cloud has a potential difference of 1.00×10^8 V relative to a tree. If, during a lightning storm, 50.0 C of charge is transferred through this potential difference and 1.00% of the energy is absorbed by the tree, how much water (sap in the tree) initially at 30.0°C can be boiled away? Water has a specific heat of 4 186 J/kg·°C, a boiling point of 100°C, and a heat of vaporization of 2.26×10^6 J/kg.

Section 16.10 Capacitors with Dielectrics

47. A capacitor with air between its plates is charged to 100 V and then disconnected from the battery. When a piece of glass is placed between the plates, the voltage across the capacitor drops to 25 V. What is the dielectric constant of this glass? (Assume the glass completely fills the space between the plates.)

48. Two parallel plates, each of area 2.00 cm^2, are separated by 2.00 mm with purified nonconducting water between them. A voltage of 6.00 V is applied between the plates. Calculate (a) the magnitude of the electric field between the plates, (b) the charge stored on each plate, and (c) the charge stored on each plate if the water is removed and replaced with air.

49. Determine (a) the capacitance and (b) the maximum voltage that can be applied to a Teflon-filled parallel-plate capacitor having a plate area of 175 cm^2 and insulation thickness of 0.040 0 mm.

50. A commercial capacitor is constructed as in Figure 16.23a. This particular capacitor is made from two strips of aluminum separated by a strip of paraffin-coated paper. Each strip of foil and paper is 7.00 cm wide. The foil is 0.004 00 mm thick, and the paper is 0.025 0 mm thick and has a dielectric constant of 3.70. What length should the strips be if a capacitance of 9.50×10^{-8} F is desired before the capacitor is rolled up? (Use the parallel-plate formula. Adding a second strip of paper and rolling up the capacitor doubles its capacitance by allowing both surfaces of each strip of foil to store charge.)

51. A model of a red blood cell portrays the cell as a spherical capacitor—a positively charged liquid sphere of surface area A, separated by a membrane of thickness t from the surrounding negatively charged fluid. Tiny electrodes introduced into the interior of the cell show a potential difference of 100 mV across the membrane. The membrane's thickness is estimated to be 100 nm and its dielectric constant to be 5.00. (a) If an average red blood cell has a mass of 1.00×10^{-12} kg, estimate the volume of the cell and thus find its surface area. The density of blood is 1 100 kg/m^3. (b) Estimate the capacitance of the cell. (c) Calculate the charge on the surface of the membrane. How many electronic charges does this represent?

ADDITIONAL PROBLEMS

52. Three parallel-plate capacitors are constructed, each having the same plate spacing d, and with C_1 having plate area A_1, C_2 having area A_2, and C_3 having area A_3. Show that the total capacitance C of these three capacitors connected in parallel is the same as a capacitor having plate spacing d and plate area $A = A_1 + A_2 + A_3$.

53. Three parallel-plate capacitors are constructed, each having the same plate area A, and with C_1 having plate
web

spacing d_1, C_2 having plate spacing d_2, and C_3 having plate spacing d_3. Show that the total capacitance C of these three capacitors connected in series is the same as a capacitor of plate area A and with plate spacing $d = d_1 + d_2 + d_3$.

54. Charges of equal magnitude 1.00×10^{-15} C and opposite sign are distributed over the inner and outer surfaces of the cell wall in Figure P16.54. Find the force on the potassium ion (K$^+$) if the ion is (a) 2.70 μm from the center of the cell, (b) 2.92 μm from the center, and (c) 4.00 μm from the center.

FIGURE P16.54

55. A virus rests on the bottom plate of oppositely charged parallel plates in the vacuum chamber of an electron microscope. The electric field strength between the plates is 2.00×10^5 N/C, and the bottom plate is negative. If the virus has a mass of 1.00×10^{-15} kg and suddenly acquires a charge of -1.60×10^{-19} C, what are its velocity and position 75.0 ms later? Do not disregard gravity.

56. A plastic pellet has a mass of 7.50×10^{-10} kg and carries a charge of 2.00×10^{-9} C. The pellet is fired horizontally between a pair of oppositely charged parallel plates, as shown in Figure P16.56. If the electric field strength between the plates is 9.20×10^4 N/C and the pellet enters the plates with an initial speed of 150 m/s, determine the deflection angle δ. (*Hint:* This exercise is similar to those involving the motion of a projectile under gravity.)

FIGURE P16.56

57. Find the equivalent capacitance of the group of capacitors in Figure P16.57.

58. A spherical capacitor consists of a spherical conducting shell of radius b and charge $-Q$ concentric with a smaller conducting sphere of radius a and charge Q. (a) Find the capacitance of this device. (b) Show that as the radius b of

FIGURE P16.57

the outer sphere approaches infinity, the capacitance approaches the value of $a/k_e = 4\pi\epsilon_0 a$.

59. The immediate cause of many deaths is ventricular fibrillation, uncoordinated quivering of the heart as opposed to proper beating. An electric shock to the chest can cause momentary paralysis of the heart muscle, after which the heart will sometimes start organized beating again. A *defibrillator* is a device that applies a strong electric shock to the chest over a time of a few milliseconds. The device contains a capacitor of several microfarads, charged to several thousand volts. Electrodes called paddles, about 8 cm across and coated with conducting paste, are held against the chest on both sides of the heart. Their handles are insulated to prevent injury to the operator, who calls, "Clear!" and pushes a button on one paddle to discharge the capacitor through the patient's chest. Assume that an energy of 300 W · s is to be delivered from a 30.0 μF capacitor. To what potential difference must it be charged?

60. When a certain air-filled parallel-plate capacitor is connected across a battery, it acquires a charge (on each plate) of 150 μC. While the battery connection is maintained, a dielectric slab is inserted into and fills the region between the plates. This results in the accumulation of an additional charge of 200 μC on each plate. What is the dielectric constant of the dielectric slab?

61. Capacitors $C_1 = 6.0$ μF and $C_2 = 2.0$ μF are charged as a parallel combination across a 250-V battery. The capacitors are disconnected from the battery and from each other. They are then connected positive plate to negative plate and negative plate to positive plate. Calculate the resulting charge on each capacitor.

62. Capacitors $C_1 = 4.0$ μF and $C_2 = 2.0$ μF are charged as a series combination across a 100-V battery. The two capacitors are disconnected from the battery and from each other. They are then connected positive plate to positive plate and negative plate to negative plate. Calculate the resulting charge on each capacitor.

63. The charge distribution shown in Figure P16.63 is referred to as a linear quadrupole. (a) Show that the electric poten-

tial at a point on the *x* axis where $x > d$ is

$$V = \frac{2k_e Qd^2}{x^3 - xd^2}$$

(b) Show that the expression obtained in (a) when $x \gg d$ reduces to

$$V = \frac{2k_e Qd^2}{x^3}$$

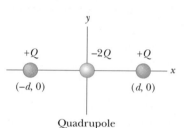

Quadrupole
FIGURE P16.63

64. The energy stored in a 52.0-μF capacitor is used to melt a 6.00-mg sample of lead. To what voltage must the capacitor be initially charged, assuming that the initial temperature of the lead is 20.0°C? (Lead has a specific heat of 128 J/kg · °C, a melting point of 327.3°C, and a latent heat of fusion of 24.5 kJ/kg.)

65. Consider a parallel-plate capacitor with charge Q and area A, filled with dielectric material having dielectric constant κ. It can be shown that the magnitude of the attractive force exerted by each plate on the other is given by $F = Q^2/(2\kappa\epsilon_0 A)$. When a potential difference of 100 V exists between the plates of an air-filled 20-μF parallel-plate capacitor, what force does each plate exert on the other if they are separated by 2.0 mm?

66. An electron is fired at a speed of $v_0 = 5.6 \times 10^6$ m/s and at an angle of $\theta_0 = -45°$ between two parallel conducting plates that are $D = 2.0$ mm apart, as in Figure P16.66. If the potential difference between the plates is $\Delta V = 100$ V, determine (a) how close d the electron will get to the bottom plate and (b) where the electron will strike the top plate.

FIGURE P16.66

GROUP ACTIVITIES

G.1 It takes an electric field of about 30 kV/cm to cause a spark in dry air. Shuffle across a rug and reach toward a doorknob. By estimating the length of the spark, determine the electric potential difference between your finger and the doorknob that existed just before you touched the knob. Try this experiment again on a very humid day and you will find that the spark is much shorter or is imperceptible. Why?

G.2 Suppose you are given a battery, a capacitor, two switches, a lightbulb and several pieces of connecting wire. On a sheet of paper, design a circuit that will do the following. (1) When switch 1 is closed and switch 2 is open, the capacitor charges but no current moves through the lightbulb. (2) Then when switch 1 is opened and switch 2 closed, the lightbulb is connected to the capacitor but not to the battery. Describe the motion of charge in the circuit when switch 1 is closed and switch 2 is open. Is energy being stored in the capacitor? What measurements would you have to make to determine how much energy is stored if any? What happens to the lightbulb when switch 1 is opened after the capacitor has charged and switch 2 then closed? Will the bulb light and stay lit? What happens to the charge on the capacitor when switch 2 is closed in this way?

G.3 (a) Under what conditions can the equation $V_B - V_A = -Ed$ be used? (b) Can this equation be used to find the difference in potential between two points in an electric field set up by a point charge? Defend your answer. (c) Can this equation be used to find the difference in potential between two points in the electric field between the plates of a charged, parallel-plate capacitor? Defend your answer. (d) What is the strength of the electric field between two parallel metal plates attached to a battery of 12 V when the plates are separated by 5 mm?

G.4 Two charges of 1.0 μC and -2.0 μC are 0.5 m apart at two vertices of an equilateral triangle. (a) What is the direction of the potential set up by the 1.0-μC charge at the third vertex of the triangle? Be careful, and defend your answer. (b) What is the direction of the potential set up by the charge of -2.0 μC at this third vertex? Again be very careful before you defend your answer too strongly. (c) Find the electric potential at this point.

G.5 (a) A group of capacitors connected in parallel have the same (i) charge on them, (ii) potential difference across them, or (iii) neither of the above.

(b) A group of capacitors connected in series have the same (i) charge on them, (ii) potential difference across them, or (iii) neither of the above.

(c) The equivalent capacitance for a group of capacitors connected in parallel is (i) greater than any of the capacitors in the group, (ii) less than any of the capacitors in the group, or (iii) neither of the above.

(d) The equivalent capacitance for a group of capacitors connected in series is (i) greater than any of the capacitors in the group, (ii) less than any of the capacitors in the group, or (iii) neither of the above.

(e) Defend your answers by examining a circuit consisting of a 3.0-μF and a 5.0-μF capacitor used with a 12-V battery. Be sure that the answers you find are consistent with your answers to parts (a) through (d).

G.6 Figure GA16.6 shows a contour map of a hilly island. The outer part of the figure is at sea level (marked 0). Each contour line from the region marked zero shows a level 10 m higher than the previous. The maximum height is 70 m and is shown by the number 70. Answer the following questions by giving the pair of grid markers (a letter and a number) closest to the point being requested.

(a) Where is there a steep cliff?

(b) Where is there a pass between two hills?

(c) Where is the easiest slope to climb?

FIGURE GA16.6

Group Activity 6 is courtesy of Edward F. Redish. For more problems of this type, see
http://www.physics.umd.edu/perg/

17

Current and Resistance

Chapter Outline

This lightbulb emits light when its tungsten filaments carry sufficient current. Tungsten is an ideal filament material because it is strong and has a high melting point (about 3 410°C). The interior of the bulb contains an inert gas, such as a mixture of argon and nitrogen. Air must be avoided because the oxygen it contains would corrode the tungsten filaments. *(Tek Image/Science Photo Library/Photo Researchers, Inc.)*

M any practical applications and devices are based on the principles of static electricity, but electricity was destined to become an inseparable part of our daily lives when scientists learned how to produce a continuous flow of charge for relatively long periods of time by using the battery. The battery or voltaic cell was invented in 1800 by the Italian physicist Alessandro Volta. The battery is a device that supplies a continuous flow of charge at low potential, as opposed to the previously invented electrostatic machines, which produced a tiny flow of charge at high potential for brief periods. This steady source of electric current allowed scientists to perform experiments to learn how to control the flow of electric charges in circuits. Today, electric currents power our lights, radios, television sets, air conditioners, and refrigerators; they ignite the gasoline in automobile engines, travel through miniature components making up the chips of microcomputers, and perform countless other invaluable tasks.

In this chapter we define current and discuss some of the factors that contribute to the resistance to flow of charge in conductors. We also discuss energy transformations in electric circuits. These topics will be the foundation for additional work with circuits in later chapters.

17.1 ELECTRIC CURRENT

Whenever electric charges of like signs move, a *current* is said to exist. To define current more precisely, suppose the charges are moving perpendicularly to a surface of area A, as in Figure 17.1. (This area could be the cross-sectional area of a wire, for example.) **The current is the rate at which charge flows through this surface.** If ΔQ is the amount of charge that passes through this area in a time interval of Δt, the current I is equal to the ratio of the charge to the time interval:

$$I \equiv \frac{\Delta Q}{\Delta t} \qquad [17.1]$$

The SI unit of current is the **ampere (A):**

$$1\,A = 1\,C/s \qquad [17.2]$$

Thus, 1 A of current is equivalent to 1 C of charge passing through the cross-sectional area in a time interval of 1 s.

When charges flow through a surface as in Figure 17.1, they can be positive, negative, or both. **The direction of conventional current used in this book is the direction positive charge would flow.** (This historical convention originated about 200 years ago when the ideas of positive and negative charge were introduced.) In a common conductor, such as copper, the current is due to the motion of negatively charged electrons. Therefore, when we speak of current in such a conductor, the direction of the current is opposite the direction of motion of electrons. On the other hand, if one considers a beam of positively charged protons in an accelerator, the current is in the direction of motion of the protons. In some cases—gases and electrolytes, for example—the current is the result of the flows of both positive and negative charges. It is common to refer to a moving charge (whether it is positive or negative) as a *charge carrier.* In a metal, for example, the charge carriers are electrons.

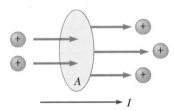

FIGURE 17.1 Charges in motion through an area A. The time rate of flow of charge through the area is defined as the current I. The direction of the current is the direction of flow of positive charges.

◀ Direction of current

Tip 17.1 CURRENT FLOW IS REDUNDANT

The phrases *flow* of *current* or *current flow* are commonly used but are incorrect because current *is* a flow (of charge). We shall avoid these redundant phrases and speak of *flow of charge* or *charge flow.*

Example 17.1 The Current in a Lightbulb

The amount of charge that passes through the filament of a certain lightbulb in 2.00 s is 1.67 C. Find (a) the current in the lightbulb and (b) the number of electrons that pass through the filament in 1 s.

Solution

A From Equation 17.1 we have

$$I = \frac{\Delta Q}{\Delta t} = \frac{1.67 \text{ C}}{2.00 \text{ s}} = \boxed{0.835 \text{ A}}$$

B In 1 s, 0.835 C of charge must pass the cross-sectional area of the filament. This total charge per second is equal to the number of electrons, N, times the charge on a single electron.

$$Nq = N(1.60 \times 10^{-19} \text{ C/electron}) = 0.835 \text{ C}$$

$$N = \boxed{5.22 \times 10^{18}} \text{ electrons}$$

Quick Quiz 17.1 Consider positive and negative charges moving horizontally through the four regions in Figure 17.2. Rank the currents in these four regions, from lowest to highest.

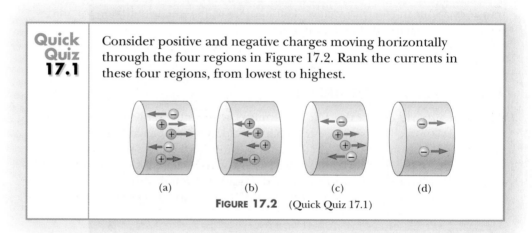

(a)   (b)   (c)   (d)

FIGURE 17.2   (Quick Quiz 17.1)

17.2   A MICROSCOPIC VIEW: CURRENT AND DRIFT SPEED

It is instructive to develop a model that will allow us to relate the macroscopic current to the motion of the microscopic charged particles that make up the current. We will show that current depends on the average speed of the charge carriers in the direction of the current, the number of charge carriers per unit volume, and the size of the charge carried by each.

Consider identically charged particles moving in a conductor of cross-sectional area A (Fig. 17.3). The volume of an element of length Δx of the conductor is $A \, \Delta x$. If n represents the number of mobile charge carriers per unit volume, then the number of carriers in the volume element is $nA \, \Delta x$. The mobile charge ΔQ in this element is therefore

$$\Delta Q = \text{number of carriers} \times \text{charge per carrier} = (nA \, \Delta x)q$$

where q is the charge on each carrier. If the carriers move with a constant average speed v_d, called the **drift speed,** the distance they move in time interval Δt is $\Delta x = v_d \, \Delta t$. Therefore, we can write ΔQ as

$$\Delta Q = (nAv_d \, \Delta t)q$$

If we divide both sides of this equation by Δt, we see that the current in the conductor is

$$I = \frac{\Delta Q}{\Delta t} = nqv_dA \qquad \text{[17.3]}$$

FIGURE 17.3   A section of a uniform conductor of cross-sectional area A. The charge carriers move with a speed v_d, and the distance they travel in the time Δt is given by $\Delta x = v_d \, \Delta t$. The number of mobile charge carriers in the section of length Δx is given by $nAv_d \, \Delta t$, where n is the number of mobile carriers per unit volume.

Current is proportional to the drift speed ▶

To understand the meaning of drift speed, consider a conductor in which the charge carriers are free electrons. If the conductor is isolated, these electrons undergo random motion similar to that of gas molecules. Note that the drift speed is normally much smaller than the free electrons' average speed between collisions with the fixed atoms of the conductor. When a potential difference is applied between the ends of the conductor (say, by means of a battery), an electric field is set up in the conductor, creating an electric force on the electrons and hence a current. In reality, the electrons do not simply move in straight lines along the conductor. Instead, they undergo repeated collisions with the metal atoms, and the result is a complicated zigzag motion with only a small average drift speed along the wire (Fig. 17.4). The energy transferred from the electrons to the metal atoms during collision increases the vibrational energy of the atoms and causes a corresponding increase in the temperature of the conductor. However, despite the collisions, the electrons move slowly along the conductor in a direction opposite $\mathbf{E}$ with the drift velocity $\mathbf{v}_d$.

FIGURE 17.4 A schematic representation of the zigzag motion of a charge carrier in a conductor. The sharp changes in direction are due to collisions with atoms in the conductor. Note that the net motion of electrons is opposite the direction of the electric field.

Example 17.2 Just How Slow Is the Drift Speed of Electrons in a Copper Wire?

A copper wire of cross-sectional area $3.00 \times 10^{-6}\ \mathrm{m}^2$ carries a current of 10.0 A. (a) Assuming that each copper atom contributes one free electron to the metal, find the drift speed of the electrons in this wire. The density of copper is $8.95\ \mathrm{g/cm^3}$. (b) Compare the drift speed to the random rms speed an electron would have at 20°C using the ideal gas model.

Reasoning All the variables in Equation 17.3 are known except n, the number of free charge carriers per unit volume. We can find n by recalling that one mole of copper contains Avogadro's number (6.02×10^{23}) of atoms, and each atom contributes one charge carrier to the metal. The volume of one mole can be found from copper's known density and its atomic mass.

Solution

A From the periodic table of the elements (inside back cover), we find that the mass of one mole of copper is 63.5 g. Knowing the density of copper enables us to calculate the volume occupied by 63.5 g of copper:

$$V = \frac{m}{\rho} = \frac{63.5\ \mathrm{g}}{8.95\ \mathrm{g/cm^3}} = 7.09\ \mathrm{cm^3}$$

If we now assume that each copper atom contributes one free electron to the body of the material, we have

$$n = \frac{6.02 \times 10^{23}\ \mathrm{electrons}}{7.09\ \mathrm{cm^3}} = 8.48 \times 10^{22}\ \mathrm{electrons/cm^3}$$

$$= \left(8.48 \times 10^{22}\ \frac{\mathrm{electrons}}{\mathrm{cm^3}}\right)\left(10^6\ \frac{\mathrm{cm^3}}{\mathrm{m^3}}\right) = 8.48 \times 10^{28}\ \mathrm{electrons/m^3}$$

From Equation 17.3, we find that the drift speed is

$$v_d = \frac{I}{nqA} = \frac{10.0\ \mathrm{C/s}}{(8.48 \times 10^{28}\ \mathrm{electrons/m^3})(1.60 \times 10^{-19}\ \mathrm{C})(3.00 \times 10^{-6}\ \mathrm{m^2})}$$

$$= 2.46 \times 10^{-4}\ \mathrm{m/s}$$

B Equation 10.18 states that the randomly directed rms speed of an ideal gas particle of mass m at absolute temperature T is given by

$$v_{\mathrm{rms}} = \sqrt{\frac{3k_{\mathrm{B}}T}{m}}$$

Because the free electrons in a metal behave approximately as an ideal gas, we have for the rms speed of an electron

$$v_{rms} = \sqrt{\frac{3(1.38 \times 10^{-23} \text{ J/K})(293 \text{ K})}{9.11 \times 10^{-31} \text{ kg}}} = 1.15 \times 10^5 \text{ m/s}$$

Thus, the drift speed of an electron in a wire is only about one billionth of its random thermal speed.

Example 17.2 shows that drift speeds are typically very small. In fact, the drift speed is much smaller than the average speed between collisions; for instance, electrons traveling at 2.46×10^{-4} m/s would take about 68 min to travel 1 m! In view of this low speed, you might wonder why a light turns on almost instantaneously when a switch is thrown. Think of the flow of water through a pipe. If a drop of water is forced into one end of a pipe that is already filled with water, a drop must be pushed out the other end of the pipe. Although it may take an individual drop a long time to make it through the pipe, a flow initiated at one end produces a similar flow at the other end very quickly. Another familiar analogy is the motion of a bicycle chain. When the sprocket moves one link, the other links all move more or less immediately, even though it takes a given link some time to make a complete rotation. In a conductor, the electric field that drives the free electrons travels with a speed close to that of light. Thus, when you flip a light switch, the message for the electrons to start moving through the wire (the electric field) reaches them at a speed on the order of 10^8 m/s.

Tip 17.2   ELECTRONS ARE EVERYWHERE IN THE CIRCUIT

Electrons do not have to travel from the light switch to the light for the light to operate. Electrons already in the filament of the lightbulb move in response to the electric field set up by a battery (or other source of potential difference). Also, the battery does *not* provide electrons to the circuit.

APPLYING PHYSICS 17.1

Suppose a current-carrying wire has a cross-sectional area that gradually becomes smaller along the wire, so that the wire has the shape of a very long cone. How does the drift velocity vary along the wire?

Explanation   Every portion of the wire is carrying the same current because charge cannot build up at any point on the wire. Thus, as the cross-sectional area decreases, the drift velocity must increase to maintain the constant value of the current. This increased drift velocity is a result of the electric field lines in the wire being compressed into a smaller area, thus increasing the strength of the field.

APPLYING PHYSICS 17.2

We have seen that an electric field must exist inside a conductor that carries a current. How is this possible in view of the fact that in electrostatics we concluded that the electric field is zero inside a conductor?

Explanation   In the electrostatic case in which charges are stationary, the internal electric field must be zero because a nonzero field would produce a current (by interacting with the free electrons in the conductor), which would violate the condition of static equilibrium. In this chapter we deal with conductors that carry current, a nonelectrostatic situation. The current arises because of a potential difference applied between the ends of the conductor, which produces an internal electric field. So there is no paradox.

17.3   CURRENT AND VOLTAGE MEASUREMENTS IN CIRCUITS

So far we have tried to keep things simple and concentrate on concepts like current without mentioning how one measures current or even sets up a circuit to make charge flow. In order to understand better the concepts covered later in this chapter, let us digress briefly to discuss what a circuit is and how you measure current in and voltage across some circuit element.

Look at the circuit shown in Figure 17.5a. This is the actual circuit that would have to be set up to measure the current in Example 17.1. Figure 17.5b shows a stylized figure called a **circuit diagram** (as first introduced in Chapter 16) that represents the actual circuit of Figure 17.5a. This circuit consists of only a battery and a lightbulb. The word *circuit* means that current circulates around a closed loop of some sort. The battery pumps charge through the bulb and around the loop. No charge would flow if you did not have a complete conducting path from the positive terminal of the battery into one side of the bulb, out the other, and through the copper conducting wires back to the negative terminal of the battery. The most important quantities that characterize how the bulb will work in different situations are the current I in the bulb and the potential difference ΔV *across* the bulb. To measure the current in the bulb, we place an ammeter, the device for measuring current, in line with the bulb so that there is no path for the current to sneak around the meter; all of the charge passing through the bulb must also pass through the ammeter. The voltmeter measures the potential difference or voltage between the two ends of the bulb's filament. If we use two meters simultaneously as in Figure 17.5a, we can remove the voltmeter and see if its presence effects the current reading. Figure 17.5c shows a digital **multimeter,** a convenient device with a digital readout that can be used to measure voltage, current, or resistance. An advantage of using a digital multimeter as a voltmeter is that it usually will not affect the current, because a digital meter has enormous resistance to the flow of charge in the voltmeter mode.

At this point, you can measure the current as a function of voltage (the I-ΔV curve) of various devices in the lab. All you need is a variable voltage supply (an adjustable battery) capable of supplying potential differences from about -5 V to $+5$ V, a bulb, a resistor, some wires and alligator clips, and a couple of multimeters. Always start your measurements using the highest multimeter scales, that is, 10 A and 1 000 V, and increase the sensitivity one scale at a time to obtain the highest accuracy without overloading the meters. (To increase the sensitivity means that we lower the maximum current or voltage that the scale reads.) Also note that the meters must be connected with the proper polarity with respect to the voltage supply, as shown in Figure 17.5b. Finally, follow your instructor's directions carefully to avoid damaged meters and a soaring lab fee.

FIGURE 17.5 (a) A sketch of an actual circuit used to measure the current in a flashlight bulb and the potential difference across it. (b) A schematic diagram of the circuit shown in part (a). (c) A digital multimeter that can be used to measure both currents and potential differences. Here, the meter is measuring the potential difference across the terminals of a battery. *(c, Michael Dalton, Fundamental Photographs)*

(a)   (b)   (c)   (d)

FIGURE 17.6 (Quick Quiz 17.2)

> **Quick Quiz 17.2**
>
> Look at the four "circuits" shown in Figure 17.6 and select those that will light the bulb.

17.4 RESISTANCE AND OHM'S LAW

When a voltage (potential difference) ΔV is applied across the ends of a metallic conductor as in Figure 17.7, the current in the conductor is found to be proportional to the applied voltage; that is, $I \propto \Delta V$. If the proportionality holds, we can write $\Delta V = IR$, where the proportionality constant R is called the resistance of the conductor. In fact, we define the **resistance** as the ratio of the voltage across the conductor to the current it carries:

  Resistance

$$R \equiv \frac{\Delta V}{I} \qquad [17.4]$$

Resistance has the SI units volts per ampere, called **ohms** (Ω). Thus, if a potential difference of 1 V across a conductor produces a current of 1 A, the resistance of the conductor is 1 Ω. For example, if an electrical appliance connected to a 120-V source carries a current of 6 A, its resistance is 20 Ω.

It is useful to compare the concepts of electric current, voltage, and resistance with the flow of water in a river. As water flows downhill in a river of constant width and depth, the flow rate (water current) depends on the steepness of descent of the river and the effects of rocks, the river bank, and other obstructions. Based on this analogy, it seems reasonable that increasing the voltage applied to a circuit should increase the current in the circuit just as increasing the steepness of descent increases the water current. Also, increasing the obstructions in the river's path will reduce the water current just as increasing the resistance in a circuit will lower the electric current. Resistance in a circuit arises due to collisions between the electrons carrying the current with fixed atoms inside the conductor. These collisions inhibit the movement of charges in much the same way as would a force of friction. For many materials, including most metals, experiments show that **the resistance remains constant over a wide range of applied voltages or currents.** This statement is known as **Ohm's law** after Georg Simon Ohm (1789–1854), who was the first to conduct a systematic study of electrical resistance.

It is common practice to express Ohm's law as

FIGURE 17.7 A uniform conductor of length l and cross-sectional area A. The current I in the conductor is proportional to the applied voltage $\Delta V = V_b - V_a$. The electric field **E** set up in the conductor is also proportional to the current.

Ohm's law

$$\Delta V = IR \qquad [17.5]$$

where R is understood to be independent of ΔV, the potential drop across the resistor, and I, the current in the resistor. We shall continue to use this traditional form of Ohm's law when discussing electrical circuits. A **resistor** is a conductor that provides a specified resistance in an electric circuit. The symbol for a resistor in circuit diagrams is a zigzag line, ——⋀⋀——.

Ohm's law is an empirical relationship that is valid only for certain materials. Materials that obey Ohm's law, and hence have a constant resistance over a wide

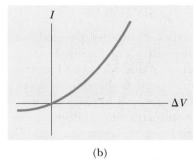

(a)   (b)

FIGURE 17.8 (a) The current-voltage curve for an ohmic material. The curve is linear, and the slope gives the resistance of the conductor. (b) A nonlinear current-voltage curve for a semiconducting diode. This device does not obey Ohm's law.

GEORG SIMON OHM (1787–1854)

Ohm, a high school teacher in Cologne and later a professor at Munich, formulated the concept of resistance and discovered the proportionalities expressed in Equation 17.6. But Henry Cavendish had earlier determined the proportionality of current to voltage that we call Ohm's law. *(© Bettmann/CORBIS)*

range of voltages, are said to be **ohmic.** Materials whose resistance changes with voltage or current are **nonohmic.** Ohmic materials have a linear current-voltage relationship over a large range of applied voltages (Fig. 17.8a). Nonohmic materials have a nonlinear current-voltage relationship (Fig. 17.8b). One common semiconducting device that is nonohmic is the **diode,** a circuit element that acts like a one-way valve for current (see Problem 62). Its resistance is small for currents in one direction (positive ΔV) and large for currents in the reverse direction (negative ΔV). Most modern electronic devices, such as transistors, have nonlinear current-voltage relationships; their operation depends on the particular ways in which they violate Ohm's law.

Quick Quiz 17.3

In Figure 17.8b, does the resistance of the diode (a) increase or (b) decrease as the positive voltage ΔV increases?

Example 17.3 The Resistance of a Steam Iron

All electric devices are required to have identifying plates that specify their electrical characteristics. The plate on a certain steam iron states that the iron carries a current of 6.4 A when connected to a 120-V source. What is the resistance of the steam iron?

Solution From Ohm's law, we find the resistance to be

$$R = \frac{\Delta V}{I} = \frac{120 \text{ V}}{6.4 \text{ A}} = \boxed{19 \ \Omega}$$

EXERCISE The resistance of a hot plate is 48 Ω. How much current does the plate carry when connected to a 120-V source?

ANSWER 2.5 A

An assortment of resistors used for a variety of applications in electronic circuits. *(Courtesy of Henry Leap and Jim Lehman)*

17.5   RESISTIVITY

In an earlier section we pointed out that electrons do not move in straight-line paths through a conductor. Instead, they undergo repeated collisions with the metal atoms. Consider a conductor with a voltage applied across its ends. An electron gains speed as the electric force associated with the internal electric field accelerates it, giving it a velocity in the direction opposite that of the electric field. A collision with an atom randomizes the electron's velocity, thus reducing its velocity in the direction opposite the field. The process then repeats itself. Together these collisions affect the electron somewhat as a force of internal friction would. This is the origin of a material's resistance. The resistance of an ohmic conductor is proportional to its length l and inversely proportional to its cross-sectional area A. That is,

$$R = \rho \frac{l}{A}$$ [17.6]

where the constant of proportionality ρ is called the **resistivity** of the material.[1] Every material has a characteristic resistivity that depends on its electronic structure and on temperature. Good electric conductors have very low resistivities, and good insulators have very high resistivities. Table 17.1 lists the resistivities of a variety of materials at 20°C. Because resistance values are in ohms, resistivity values must be in ohm-meters.

Equation 17.6 states that the resistance of a cylindrical conductor is proportional to its length and inversely proportional to its cross-sectional area. This may be understood by analogy to the flow of liquid through a pipe. As the length of the pipe is increased, the resistance to liquid flow increases because of a gain in friction between the fluid and the walls of the pipe. As its cross-sectional area is in-

TABLE 17.1	Resistivities and Temperature Coefficients of Resistivity for Various Materials (at 20°C)	
Material	**Resistivity ($\Omega \cdot$ m)**	**Temperature Coefficient of Resistivity [(°C)$^{-1}$]**
Silver	1.59×10^{-8}	3.8×10^{-3}
Copper	1.7×10^{-8}	3.9×10^{-3}
Gold	2.44×10^{-8}	3.4×10^{-3}
Aluminum	2.82×10^{-8}	3.9×10^{-3}
Tungsten	5.6×10^{-8}	4.5×10^{-3}
Iron	10.0×10^{-8}	5.0×10^{-3}
Platinum	11×10^{-8}	3.92×10^{-3}
Lead	22×10^{-8}	3.9×10^{-3}
Nichrome*	150×10^{-8}	0.4×10^{-3}
Carbon	3.5×10^{5}	-0.5×10^{-3}
Germanium	0.46	-48×10^{-3}
Silicon	640	-75×10^{-3}
Glass	$10^{10} - 10^{14}$	
Hard rubber	$\approx 10^{13}$	
Sulfur	10^{15}	
Quartz (fused)	75×10^{16}	

* A nickel-chromium alloy commonly used in heating elements.

[1] The symbol ρ used for resistivity should not be confused with the same symbol used earlier in the book for density. Very often, a single symbol is used to represent different quantities.

creased, the pipe can transport more fluid in a given time interval, so its resistance drops.

APPLYING **PHYSICS** *17.3*

It is a common observation that as a lightbulb ages, it gives off less light than when new. Why?

APPLICATION

DIMMING OF AGING LIGHTBULBS

Explanation There are two reasons for this, one electrical and one optical, but both are related to the same phenomenon occurring within the bulb. The filament of a lightbulb is made of a tungsten wire that, in an old lightbulb, has been kept at a high temperature for many hours. These high temperatures cause tungsten to be evaporated from the filament, decreasing its radius. From $R = \rho\, l/A$, we see that a decreased cross-sectional area leads to an increase in resistance of the filament. This increasing resistance with age means that the filament will carry less current for the same applied voltage. With less current in the filament, there is less light output, and the filament glows more dimly.

At the high operating temperature of the filament, tungsten atoms leave the surface of the filament, much as water molecules evaporate from a puddle of water. These atoms are carried away by convection currents in the gas in the bulb and are deposited on the inner surface of the glass. In time, the glass becomes less transparent because of this tungsten coating, which decreases the amount of light that passes through the glass.

Example 17.4 The Resistance of Nichrome Wire

A Calculate the resistance per unit length of a 22-gauge nichrome wire of radius 0.321 mm.

Solution The cross-sectional area of this wire is

$$A = \pi r^2 = \pi (0.321 \times 10^{-3}\,\text{m})^2 = 3.24 \times 10^{-7}\,\text{m}^2$$

The resistivity of nichrome is $1.5 \times 10^{-6}\,\Omega \cdot \text{m}$ (Table 17.1). Thus, we can use Equation 17.6 to find the resistance per unit length:

$$\frac{R}{l} = \frac{\rho}{A} = \frac{1.5 \times 10^{-6}\,\Omega \cdot \text{m}}{3.24 \times 10^{-7}\,\text{m}^2} = \boxed{4.6\,\Omega/\text{m}}$$

B If a potential difference of 10.0 V is maintained across a 1.0-m length of the nichrome wire, what is the current in the wire?

Solution Because a 1.0-m length of this wire has a resistance of 4.6 Ω, Ohm's law gives

$$I = \frac{\Delta V}{R} = \frac{10.0\,\text{V}}{4.6\,\Omega} = \boxed{2.2\,\text{A}}$$

Note from Table 17.1 that the resistivity of nichrome is about 100 times that of copper, a typical good conductor. Therefore, a copper wire of the same radius would have a resistance per unit length of only 0.052 Ω/m, and a 1.0-m length of copper wire of the same radius would carry the same current (2.2 A) with an applied voltage of only 0.11 V.

Because of its resistance to oxidation, nichrome is often used for heating elements in toasters, irons, and electric heaters.

EXERCISE What is the resistance of a 6.0-m length of 22-gauge nichrome wire? How much current does it carry when connected to a 120-V source?

ANSWER 28 Ω; 4.3 A

17.6   TEMPERATURE VARIATION OF RESISTANCE

The resistivity, and hence the resistance, of a conductor depends on a number of factors. One of the most important is the temperature of the metal. For most metals, resistivity increases with increasing temperature. This correlation can be understood as follows. As the temperature of the material increases, its constituent atoms vibrate with increasingly greater amplitudes. Just as it is more difficult to weave one's way through a crowded room when the people are in motion than when they are standing still, so do the electrons find it more difficult to pass atoms vibrating with large amplitudes. The increased electron scattering with increasing temperature results in increased resistivity.

For most metals, resistivity increases approximately linearly with temperature over a limited temperature range, according to the expression

$$\rho = \rho_0[1 + \alpha(T - T_0)] \qquad \text{[17.7]}$$

where ρ is the resistivity at some temperature T (in Celsius degrees), ρ_0 is the resistivity at some reference temperature T_0 (usually taken to be 20°C), and α is a parameter called the **temperature coefficient of resistivity.** The temperature coefficients for various materials are provided in Table 17.1. The interesting negative values of α for semiconductors arise because these materials possess weakly bound charge carriers that become free to move and contribute to the current as the temperature is raised.

Because the resistance of a conductor with uniform cross section is proportional to the resistivity, according to Equation 17.6 ($R = \rho l/A$), the temperature variation of resistance can be written

$$R = R_0[1 + \alpha(T - T_0)] \qquad \text{[17.8]}$$

Precise temperature measurements are often made using this property, as shown by the following example.

An old-fashioned carbon-filament incandescent lamp. The resistance of such a lamp is typically 10 Ω, but it changes with temperature. *(Courtesy of Central Scientific Company)*

Example 17.5   A Platinum Resistance Thermometer

A resistance thermometer, which measures temperature by measuring the change in resistance of a conductor, is made of platinum and has a resistance of 50.0 Ω at 20.0°C. When the device is immersed in a vessel containing melting indium, its resistance increases to 76.8 Ω. From this information, find the melting point of indium.

Solution If we solve Equation 17.8 for $T - T_0$ and get α for platinum from Table 17.1, we obtain

$$T - T_0 = \frac{R - R_0}{\alpha R_0} = \frac{76.8 \ \Omega - 50.0 \ \Omega}{[3.92 \times 10^{-3} \ (°C)^{-1}][50.0 \ \Omega]} = 137°C$$

Because $T_0 = 20.0°C$, we find that the melting point of indium is

$$T = \boxed{157°C}$$

$R(\Omega)$

FIGURE 17.9 Resistance versus temperature for a sample of mercury. The graph follows that of a normal metal above the critical temperature T_c. The resistance drops to zero at the critical temperature, which is 4.1 K for mercury, and remains at zero for lower temperatures.

17.7 SUPERCONDUCTORS

There is a class of metals and compounds whose resistances fall virtually to *zero* below a certain temperature T_c called the *critical temperature*. These materials are known as **superconductors.** The resistance-temperature graph for a superconductor follows that of a normal metal at temperatures above T_c (Fig. 17.9). When the temperature is at or below T_c, the resistance suddenly drops to zero. This phenomenon was discovered in 1911 by the Dutch physicist H. Kamerlingh Onnes as he and a graduate student worked with mercury, which is a superconductor below 4.1 K. Recent measurements have shown that the resistivities of superconductors below T_c are less than $4 \times 10^{-25} \ \Omega \cdot$m—around 10^{17} times smaller than the resistivity of copper and in practice considered to be zero.

Today, thousands of superconductors are known, including such common metals as aluminum, tin, lead, zinc, and indium. Table 17.2 lists the critical temperatures of several superconductors. The value of T_c is sensitive to chemical composition, pressure, and crystalline structure. Interestingly, copper, silver, and gold, which are excellent conductors, do not exhibit superconductivity.

One of the truly remarkable features of superconductors is the fact that, once a current is set up in them, it persists *without any applied voltage* (because $R = 0$). In fact, steady currents in superconducting loops have been observed to persist for many years with no apparent decay!

An important development in physics that created much excitement in the scientific community was the discovery of high-temperature copper-oxide-based superconductors. The excitement began with a 1986 publication by J. Georg Bednorz and K. Alex Müller, scientists at the IBM Zurich Research Laboratory in Switzerland, in which they reported evidence for superconductivity at a temperature near 30 K in an oxide of barium, lanthanum, and copper. Bednorz and Müller were awarded the Nobel prize for physics in 1987 for their discovery. This discovery was remarkable in view of the fact that the critical temperature was significantly higher than that of any previously known superconductor. Shortly thereafter, a new family of compounds was investigated, and research activity in the field of superconductivity proceeded vigorously. In early 1987, groups at the University of Alabama at Huntsville and the University of Houston announced the discovery of superconductivity at about 92 K in an oxide of yttrium, barium, and copper ($YBa_2Cu_3O_7$), shown as the gray disk in Figure 17.10. Late in 1987, teams of scientists from Japan and the United States reported superconductivity at 105 K in an oxide of bismuth, strontium, calcium, and copper. More recently, scientists have reported superconductivity at temperatures as high as 150 K in an oxide containing mercury. At this point one cannot rule out the possibility of room-temperature superconductivity, and the search for novel superconducting materials continues. It is an important search both for scientific reasons and because practical applications become more probable and widespread as the critical temperature is raised and the expense of cooling materials decreases.

An important and useful application is superconducting magnets in which the magnetic field intensities are about ten times greater than those of the best normal electromagnets. Such magnets are being considered as a means of storing

TABLE 17.2

Critical Temperatures for Various Superconductors

Material	T_c (K)
Zn	0.88
Al	1.19
Sn	3.72
Hg	4.15
Pb	7.18
Nb	9.46
Nb_3Sn	18.05
Nb_3Ge	23.2
$YBa_2Cu_3O_7$	90
Bi-Sr-Ca-Cu-O	105
Tl-Ba-Ca-Cu-O	125
$HgBa_2Ca_2Cu_3O_8$	134

Webnote 17.1

The Department of Energy has a good Web site on superconductivity at *http://www.eren.doe.gov/ superconductivity/*

FIGURE 17.10 A small permanent magnet floats freely above a ceramic disk of the superconductor $YBa_2Cu_3O_7$ cooled by liquid nitrogen at 77 K. The superconductor has zero electric resistance at temperatures below 92 K and expels any applied magnetic field. *(Courtesy of IBM Research Laboratory)*

energy. The idea of using superconducting power lines for transmitting power efficiently is also receiving serious consideration. Modern superconducting electronic devices consisting of two thin-film superconductors separated by a thin insulator have been constructed. They include magnetometers (magnetic-field measuring devices) and various microwave devices.

17.8 ELECTRICAL ENERGY AND POWER

If a battery is used to establish an electric current in a conductor, chemical energy stored in the battery is continuously transformed into kinetic energy of the charge carriers. This kinetic energy is quickly lost as a result of collisions between the charge carriers and fixed atoms in the conductor, causing an increase in the temperature of the conductor. Thus, the chemical energy stored in the battery is continuously transformed into internal energy.

In order to understand the process of energy transfer in a simple circuit, consider a battery whose terminals are connected to a resistor (Fig. 17.11). (Remember that the positive terminal of the battery is always at the higher potential.) Now imagine following a quantity of positive charge ΔQ around the circuit from point A through the battery and resistor and back to A. Point A is a reference point that is grounded (the ground symbol is ⏚), and its potential is taken to be zero. As the charge moves from A to B through the battery, whose potential difference is ΔV, the electrical potential energy of the system increases by the amount $\Delta Q \, \Delta V$, and the chemical potential energy in the battery decreases by the same amount. (Recall from Chapter 16 that $\Delta PE = q \, \Delta V$.) However, as the charge moves from C to D through the resistor, it loses this electrical potential energy during collisions with atoms in the resistor. In this process, the energy is transformed to internal energy corresponding to increased vibrational motion of the atoms in the resistor. Because we have ignored the resistance of the interconnecting wires, no energy transformation occurs for paths BC and DA. When the charge returns to point A, the net result is that some of the chemical energy in the battery has been delivered to the resistor and causes its temperature to rise. Feeling the resistor, we would often notice that it was warmer than its surroundings.

The charge ΔQ loses energy $\Delta Q \, \Delta V$ as it passes through the resistor. If Δt is the time it takes the charge to pass through the resistor, then the rate at which it loses electrical potential energy is

$$\frac{\Delta Q}{\Delta t} \Delta V = I \Delta V$$

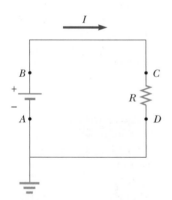

FIGURE 17.11 A circuit consisting of a battery and resistance R. Positive charge flows clockwise from the positive to the negative terminal of the battery. Point A is grounded.

where I is the current in the resistor and ΔV is the potential difference across it. Of course, the charge regains this energy when it passes through the battery, at the expense of chemical energy in the battery. The rate at which the system loses potential energy as the charge passes through the resistor is equal to the rate at which the system gains internal energy in the resistor. Thus, the power $\mathcal{P}$ representing the rate at which energy is delivered to the resistor is

$$\mathcal{P} = I \Delta V \qquad [17.9]$$

We have developed this result by considering a battery delivering energy to a resistor. However, Equation 17.9 can be used to determine the power transferred from a voltage source to *any* device carrying a current I and having a potential difference ΔV between its terminals.

Using Equation 17.9 and the fact that $\Delta V = IR$ for a resistor, we can express the power delivered to the resistor in the alternative forms

$$\mathcal{P} = I^2 R = \frac{(\Delta V)^2}{R} \qquad [17.10]$$

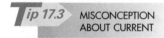

Tip 17.3 MISCONCEPTION ABOUT CURRENT

Note that current is *not* "used up" in the resistor. The current is the same *everywhere* in the circuit. Charges flow in the same rotational sense at *all* points in the circuit.

▶ Power delivered to a resistor

When I is in amperes, ΔV in volts, and R in ohms, the SI unit of power is the watt (introduced in Chapter 5). The power delivered to a conductor of resistance R is often referred to as an I^2R *loss*. Note that Equation 17.10 applies only to resistors and not to nonohmic devices like lightbulbs and diodes.

Regardless of the ways in which you use electrical energy in your home, you ultimately must pay for it or risk having your power turned off. The unit of energy used by electric companies to calculate consumption, the **kilowatt-hour,** is defined in terms of the unit of power and the amount of time it is supplied. One kilowatt-hour (kWh) is the energy converted or consumed in 1 h at the constant rate of 1 kW. It has the numerical value

$$1 \text{ kWh} = (10^3 \text{ W})(3\ 600 \text{ s}) = 3.60 \times 10^6 \text{ J} \qquad \text{[17.11]}$$

◀ The kilowatt-hour is a unit of energy

On an electric bill, the amount of electricity used in a given period is usually stated in multiples of kilowatt-hours.

APPLYING PHYSICS 17.4

When is more power delivered to a lightbulb—just after it is turned on and the glow of the filament is increasing or after it has been on for a few seconds and the glow is steady?

Explanation Once the switch is closed, the line voltage is applied across the lightbulb. As the voltage is applied across the cold filament when first turned on, the resistance of the filament is low, the current is high, and a relatively large amount of power is delivered to the bulb. As the filament warms, its resistance rises and the current decreases. As a result, the power delivered to the bulb decreases. The large current spike at the beginning of operation is the reason that lightbulbs often fail just after they are turned on.

Quick Quiz 17.6   For the two resistors shown in Figure 17.12, rank the currents at points a through f, from largest to smallest.

Quick Quiz 17.7   Two resistors, A and B, are connected across the same potential difference. The resistance of A is twice that of B. (a) Which resistor dissipates more power? (b) Which carries the greater current?

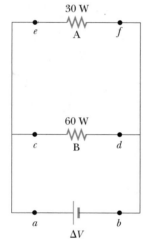

FIGURE 17.12   (Quick Quiz 17.6)

Example 17.6   The Power Converted by an Electric Heater

An electric heater is operated by applying a potential difference of 50.0 V to a nichrome wire of total resistance 8.00 Ω. Find the current carried by the wire and the power rating of the heater.

Solution   Because $\Delta V = IR$, we have

$$I = \frac{\Delta V}{R} = \frac{50.0 \text{ V}}{8.00 \text{ }\Omega} = \boxed{6.25 \text{ A}}$$

We can find the power rating using $\mathcal{P} = I^2R$:

$$\mathcal{P} = I^2R = (6.25 \text{ A})^2(8.00 \text{ }\Omega) = \boxed{313 \text{ W}}$$

EXERCISE   If we doubled the applied voltage to the heater, what would happen to the current and power?

ANSWER   The current would double, and the power would quadruple.

PHYSICS *IN ACTION*

The photo shows electrical workers at dusk, trying to restore power to the eastern Ontario town of St. Isadore that was without power for several days in January 1998 because of a severe ice storm. When power transmission lines fall because of lightning, ice storms, or earthquakes, it is very dangerous to touch them due to the high electric potential, possibly hundreds of thousands of volts relative to the ground. When transporting electrical energy through power lines, utility companies seek to minimize the power transformed into internal energy in the lines and maximize the energy delivered to the consumer. Because $\mathcal{P} = I\,\Delta V$, the same amount of power can be transported either at high currents and low potential differences or at low currents and high potential differences. Utility companies choose to transport electrical energy at low currents and high potential differences primarily for economic reasons. Copper wire is very expensive, and so it is cheaper to use high-resistance wire (that is, wire having a small cross-sectional area; see Eq. 17.6). In the expression for power delivered to a resistor, $\mathcal{P} = I^2 R$, the resistance of the transmission wire is fixed at a relatively high value for economic considerations. The $I^2 R$ loss can be reduced by keeping the current I as low as possible.

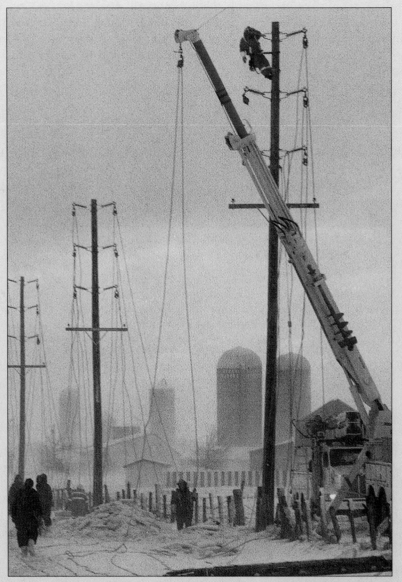

(AP/Wide World Photos/Fred Chartrand)

Example 17.7 Lighting Up Your Life

A circuit provides a current of 20.0 A at an operating voltage of 120 V. How many 75-W bulbs can operate with this voltage source?

Solution The total power that can be delivered by the voltage source is given by Equation 17.9:

$$\mathcal{P}_{\text{total}} = I\,\Delta V = (20.0\ \text{A})(120\ \text{V}) = 2.40 \times 10^3\ \text{W}$$

Because each bulb requires a power of 75 W, the number of bulbs that can operate from this voltage source is

$$\text{Number of bulbs} = \frac{\mathcal{P}_{\text{total}}}{\mathcal{P}_{\text{bulb}}} = \frac{2.40 \times 10^3\ \text{W}}{75\ \text{W}} = \boxed{32}$$

Example 17.8 The Cost of Operating a Lightbulb

How much does it cost to burn a 100-W lightbulb for 24 h if electric energy costs 12 cents or $0.12 per kilowatt-hour?

Solution A 100-W lightbulb is equivalent to a 0.10-kW bulb. Because energy consumed equals power × time, the amount of energy you must pay for, expressed in kilowatt-hours, is

$$\text{Energy} = (0.10 \text{ kW})(24 \text{ h}) = 2.4 \text{ kWh}$$

If energy is purchased at 12 cents or $0.12 per kilowatt-hour, the 24-h cost is

$$\text{Cost} = (2.4 \text{ kWh})(\$0.12/\text{kWh}) = \boxed{\$0.29}$$

This is a small amount of money, but when larger and more complex electric devices are used, the cost goes up rapidly.

EXERCISE If electric energy costs $0.12/kWh, what does it cost to operate an electric oven, which operates at 20.0 A and 220 V, for 5.0 h?

ANSWER $2.70

17.9 ELECTRICAL ACTIVITY IN THE HEART

ELECTROCARDIOGRAMS

Every action involving the body's muscles is initiated by electrical activity. The voltages produced by muscular action in the heart are particularly important to physicians. Voltage pulses cause the heart to beat, and the waves of electrical excitation that sweep across the heart associated with the heartbeat are conducted through the body via the body fluids. These voltage pulses are large enough to be detected by suitable monitoring equipment attached to the skin. A sensitive voltmeter making good electrical contact with the skin by means of contacts attached with conducting paste can be used to measure heart pulses that are typically of the order of 1 mV at the surface of the body. The voltage pulses can be recorded on an instrument called an **electrocardiograph,** and the pattern recorded by this instrument is called an **electrocardiogram** (EKG). In order to understand the information contained in an EKG pattern, it is necessary first to describe the underlying principles concerning electrical activity in the heart.

The right atrium of the heart contains a specialized set of muscle fibers called the SA (sinoatrial) node, which initiate the heartbeat (Fig. 17.13). Electric impulses that originate in these fibers gradually spread from cell to cell throughout the right and left atrial muscles, causing them to contract. The pulse that passes through the muscle cells is often called a *depolarization wave* because of its effect on individual cells. If an individual muscle cell were examined in its resting state, a double-layered electric charge distribution would be found on its surface, as shown in Figure 17.14a. The impulse generated by the SA node momentarily and locally allows positive charge on the outside of the cell to flow in and neutralize the negative charge on the inside layer. This changes the cell's charge distribution to that shown in Figure 17.14b. Once the depolarization wave has passed through a cell, an individual heart muscle cell recovers the resting state charge distribution (positive out, negative in) shown in Figure 17.14a in about 250 ms. When the impulse reaches the AV (atrioventricular) node (Fig. 17.13), the muscles of the atria begin to relax, and the pulse is directed by the AV node to the ventricular muscles. The muscles of the ventricles contract as the depolarization wave spreads through the ventricles along a group of fibers called the *Purkinje fibers*. The ventricles then

APPLICATION

ELECTROCARDIOGRAMS

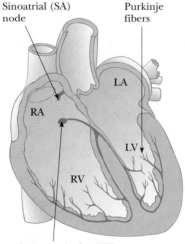

FIGURE 17.13 The electrical conduction system of the human heart. (RA: right atrium; LA: left atrium; RV: right ventricle; LV: left ventricle.)

(a)   (b)

Depolarization wave front

FIGURE 17.14 (a) Charge distribution of a muscle cell in the atrium before a depolarization wave has passed through the cell. (b) Charge distribution as the wave passes.

relax after the pulse has passed through. At this point, the SA node is again triggered and the cycle is repeated.

A sketch of the electrical activity registered on an EKG for one beat of a normal heart is shown in Figure 17.15. The pulse indicated by *P* occurs just before the atria begin to contract. The *QRS* pulse occurs in the ventricles just before they contract, and the *T* pulse occurs when the cells in the ventricles begin to recover. EKGs for an abnormal heart are shown in Figure 17.16. The *QRS* portion of the pattern shown in Figure 17.16a is wider than normal. This indicates that the patient may have an enlarged heart. (Why?) Figure 17.16b indicates that there is no constant relationship between the *P* pulse and the *QRS* pulse. This suggests a blockage in the electrical conduction path between the SA and AV nodes, which results in the atria and ventricles beating independently, leading to inefficient heart pumping. Finally, Figure 17.16c shows a situation in which there is no *P* pulse and an irregular spacing between the *QRS* pulses. This is symptomatic of irregular atrial contraction, which is called *fibrillation*. In this situation, the atrial and ventricular contractions are irregular.

As noted previously, the sinoatrial node directs the heart to beat at the appropriate rate, usually about 72 beats per minute. However, disease or the aging process can damage the heart and slow its beating, and a medical assist may be

FIGURE 17.15 An EKG response for a normal heart.

Webnote 17.2

Electrocardiograms have unique signatures associated with malfunctions. To see and learn about these signatures, visit the home page of the following EKG library: *http://www.ecglibrary.com/ ecghome.html*

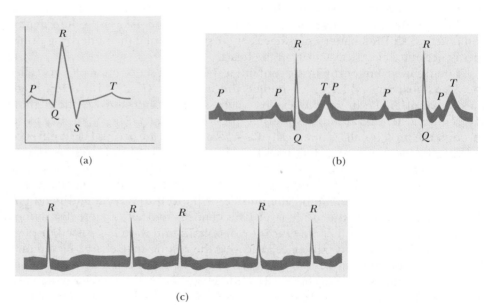

(a)   (b)

(c)

FIGURE 17.16 Abnormal EKGs.

necessary in the form of a *cardiac pacemaker* attached to the heart. This matchbox-sized electrical device implanted under the skin has a lead that is connected to the wall of the right ventricle. Pulses from this lead stimulate the heart to maintain its proper rhythm. In general, a pacemaker is designed to produce pulses at a rate of about 60 per minute, slightly slower than the normal beats per minute but sufficient to maintain life. The circuitry basically consists of a capacitor charging up from a lithium battery to a certain voltage and then discharging. The design of the circuit is such that if the heart is beating normally the capacitor is never allowed to charge completely and send pulses to the heart.

APPLICATION

CARDIAC PACEMAKERS

An Emergency Room in Your Chest

The operation in June 2001 on Vice President Dick Cheney focused attention on the progress in treating heart problems with tiny implanted electrical devices. Aptly termed "an emergency room in your chest" by Cheney's attending physician, current devices called Implanted Cardioverter Defibrillators **(ICDs)** can monitor, record, and logically process heart signals and then supply different corrective signals to hearts beating too slowly, too rapidly, or irregularly, even monitoring and sending independent signals to atria and ventricles! Figure 17.17a shows a sketch of an ICD with conducting leads that are implanted in the heart. Figure 17.17b shows an actual titanium-encapsulated dual chamber ICD.

APPLICATION

IMPLANTED CARDIOVERTER DEFIBRILLATORS

The latest ICDs are quite sophisticated devices capable of

1. monitoring both atrial and ventricular chambers to differentiate between atrial and potentially fatal ventricular arrhythmias, which require prompt regulation;
2. storing about $\frac{1}{2}$ hour of heart signals, which can easily be read out by a physician;
3. being easily reprogrammed with an external magnetic wand;
4. performing complicated signal analysis and comparison;

(a)   (b)

FIGURE 17.17 (a) A dual chamber ICD with leads in the heart. One lead monitors/stimulates the right atrium and the other monitors/stimulates the right ventricle. (b) Medtronic Dual Chamber ICD. *(Courtesy of Medtronic, Inc.)*

TABLE 17.3	Properties of Implanted Cardioverter Defibrillators[a]
Physical Specifications	
Mass (g)	85
Size (cm)	7.3 × 6.2 × 1.3 (about 5 stacked silver dollars)
Antitachycardia Pacing	ICD delivers a burst of critically timed, low-energy pulses
Number of bursts	1–15
Burst cycle length (ms)	200–552
Number of pulses per burst	2–20
Pulse amplitude (V)	7.5 or 10
Pulsewidth (ms)	1.0 or 1.9
High-Voltage Defibrillation	
Pulse energy (J)	37 stored/33 delivered
Pulse amplitude (V)	801
Bradycardia Pacing	A dual chamber ICD can steadily deliver repetitive pulses to both the atrium and ventricle
Base frequency (beats/min)	40–100
Pulse amplitude (V)	0.25–7.5
Pulsewidth (ms)	0.05, 0.1–1.5, 1.9

[a] For more information, see www.photonicd.com/specs.html

5. supplying 0.25- to 10-V repetitive pacing signals to speed up or slow down a malfunctioning heart, or a high-voltage pulse of about 800 V to halt the potentially fatal condition of ventricular fibrillation, in which the heart rapidly quivers rather than beats (people who have experienced such a high-voltage jolt say that it feels like a kick or a bomb going off in the chest);

6. automatically adjusting the number of pacing pulses per minute to match the patient's activity.

ICDs are powered by lithium batteries and have implanted lifetimes of 4–6 years. Some basic properties of these adjustable ICDs are given in Table 17.3. In this table, *tachycardia* means rapid heartbeat and *bradycardia* means slow heartbeat. A key factor in developing tiny electrical implants that serve as defibrillators is the development of capacitors with relatively large capacitance (125 μF) and small physical size.

SUMMARY

The **electric current** I in a conductor is defined as

$$I \equiv \frac{\Delta Q}{\Delta t} \qquad [17.1]$$

where ΔQ is the charge that passes through a cross section of the conductor in time Δt. The SI unit of current is the **ampere** (A); 1 A = 1 C/s. By convention, the direction of current is in the direction of flow of positive charge.

The current in a conductor is related to the motion of the charge carriers by

$$I = nqv_dA \qquad [17.3]$$

where n is the number of mobile charge carriers per unit volume, q is the charge on each carrier, v_d is the drift speed of the charges, and A is the cross-sectional area of the conductor.

The **resistance** R of a conductor is defined as the ratio of the potential difference across the conductor to the current:

$$R \equiv \frac{\Delta V}{I} \qquad \text{[17.4]}$$

The SI units of resistance are volts per ampere, or **ohms** (Ω); $1\ \Omega = 1\ \text{V/A}$.

Ohm's law describes many conductors for which the applied voltage is directly proportional to the current it causes. The proportionality constant is the resistance:

$$\Delta V = IR \qquad \text{[17.5]}$$

If a conductor has length l and cross-sectional area A, its **resistance** is

$$R = \rho \frac{l}{A} \qquad \text{[17.6]}$$

where ρ is an intrinsic property of the conductor called the **resistivity.** The SI unit of resistivity is the **ohm-meter** ($\Omega \cdot \text{m}$).

The resistivity of a conductor varies with temperature over a limited temperature range, according to the expression

$$\rho = \rho_0[1 + \alpha(T - T_0)] \qquad \text{[17.7]}$$

where α is the **temperature coefficient of resistivity** and ρ_0 is the resistivity at some reference temperature T_0 (usually taken to be 20°C).

The resistance of a conductor varies with temperature according to the expression

$$R = R_0[1 + \alpha(T - T_0)] \qquad \text{[17.8]}$$

If a potential difference ΔV is maintained across an electrical device, the **power,** or rate at which energy is supplied to the device, is

$$\mathcal{P} = I \Delta V \qquad \text{[17.9]}$$

Because the potential difference across a resistor is $\Delta V = IR$, the **power delivered to a resistor** can be expressed as

$$\mathcal{P} = I^2 R = \frac{(\Delta V)^2}{R} \qquad \text{[17.10]}$$

A **kilowatt-hour** is the amount of energy converted or consumed in 1 h by a device supplied with power at the rate of 1 kW. This is equivalent to

$$1\ \text{kWh} = 3.60 \times 10^6\ \text{J} \qquad \text{[17.11]}$$

CONCEPTUAL QUESTIONS

1. A coulomb is a very large unit of charge, and yet currents of several amperes are quite common. How is this possible?
2. Edison's original lightbulb contained a carbon filament. How should carbon be described electrically?
3. Why don't the free electrons in a metal fall to the bottom of the metal due to gravity? And charges in a conductor are supposed to reside on the surface—why don't the free electrons all go to the surface?
4. In an analogy between traffic flow and electrical current, what would correspond to the charge Q? What would correspond to the current I?
5. Newspaper articles often have statements such as, "10 000 volts of electricity surged *through* the victim's body." What is wrong with this statement?
6. "A microvoltmeter measured the resistance of the sciatic nerve." Criticize this statement. How might it be corrected?
7. When the voltage across a certain conductor is doubled, the current is observed to triple. What can you conclude about the conductor?
8. There is an old admonition given to experimenters to "keep one hand in the pocket" when working around high voltages. Why might this be a good idea?
9. What factors affect the resistance of a conductor?
10. Some homes have light dimmers that are operated by rotation of a knob. What is being changed in the electric circuit when the knob is rotated?
11. Two wires A and B with circular cross section are made of the same metal and have equal lengths, but the resistance of wire A is three times greater than that of wire B. What is the ratio of their cross-sectional areas? How do the radii compare?
12. What single experimental requirement makes superconducting devices expensive to operate? In principle, can this limitation be overcome?

13. What could happen to the drift velocity of the electrons in a wire and to the current in the wire if the electrons could move freely without resistance through this metallic conductor?

14. Use the atomic theory of matter to explain why the resistance of a material should increase as its temperature increases.

PROBLEMS

1, 2, 3 = straightforward, intermediate, challenging   ☐ = full solution available in Student Solutions Manual/Study Guide

web = solution posted at **http://info.brookscole.com/serway**   🔬 = biomedical application

Section 17.1 Electric Current

Section 17.2 A Microscopic View: Current and Drift Speed

1. If a current of 80.0 mA exists in a metal wire, how many electrons flow past a given cross section of the wire in 10.0 min? Sketch the directions of the current and the electrons' motion.

2. A small sphere that carries a charge q is whirled in a circle at the end of an insulating string. The angular frequency of rotation is ω. What average current does this rotating charge represent?

3. A total charge of 6.0 mC passes through a cross-sectional area of a wire in 2.0 s. What is the current in the wire?

4. In a particular television picture tube, the measured beam current is 60.0 μA. How many electrons strike the screen every second?

5. In the Bohr model of the hydrogen atom, an electron in the lowest energy state moves at a speed of 2.19×10^6 m/s in a circular path having a radius of 5.29×10^{-11} m. What is the effective current associated with this orbiting electron?

6. If 3.25×10^{-3} kg of gold is deposited on the negative electrode of an electrolytic cell in a period of 2.78 h, what is the current through the cell in this period? Assume that the gold ions carry one elementary unit of positive charge.

7. A 200-km-long high-voltage transmission line 2.0 cm in diameter carries a steady current of 1 000 A. If the conductor is copper with a free charge density of 8.5×10^{28} electrons per cubic meter, how long (in years) does it take one electron to travel the full length of the cable?

8. An aluminum wire with a cross-sectional area of **web** 4.0×10^{-6} m^2 carries a current of 5.0 A. Find the drift speed of the electrons in the wire. The density of aluminum is 2.7 g/cm^3. (Assume that one electron is supplied by each atom.)

9. If the current carried by a conductor is doubled, what happens to the (a) charge carrier density and (b) electron drift velocity?

Section 17.4 Resistance and Ohm's Law

Section 17.5 Resistivity

10. A lightbulb has a resistance of 240 Ω when operating at a voltage of 120 V. What is the current through the lightbulb?

11. A person notices a mild shock if the current along a path 🔬 through the thumb and index finger exceeds 80 μA. Compare the maximum allowable voltage without shock across the thumb and index finger with a dry-skin resistance of 4.0×10^5 Ω and a wet-skin resistance of 2 000 Ω.

12. Suppose that you wish to fabricate a uniform wire out of 1.00 g of copper. If the wire is to have a resistance of $R = 0.500$ Ω, and if all of the copper is to be used, what will be (a) the length and (b) the diameter of this wire?

13. Calculate the diameter of a 2.0-cm length of tungsten filament in a small lightbulb if its resistance is 0.050 Ω.

14. Eighteen-gauge wire has a diameter of 1.024 mm. Calculate the resistance of 15 m of 18-gauge copper wire at 20°C.

15. A potential difference of 12 V is found to produce a current of 0.40 A in a 3.2-m length of wire with a uniform radius of 0.40 cm. What is (a) the resistance of the wire and (b) the resistivity of the wire?

16. A length L_0 of copper wire has a resistance R_0. The wire is cut into three pieces of equal length. The pieces are then connected as parallel lengths between points A and B. What resistance will this new "wire" of length $L_0/3$ have between points A and B?

17. A wire 50.0 m long and 2.00 mm in diameter is connected to a source with a potential difference of 9.11 V, and the current is found to be 36.0 A. Assume a temperature of 20°C and, using Table 17.1, identify the metal of the wire.

18. A rectangular block of copper has sides of length 10 cm, 20 cm, and 40 cm. If the block is connected to a 6.0-V source across opposite faces of the rectangular block, what are (a) the maximum current and (b) minimum current that can be carried?

19. The breathing monitor shown in Figure P17.19 girds the 🔬 patient with a mercury-filled rubber tube and measures the variation of the tube resistance. The tube has an unstretched length of 1.25 m and an inside diameter of 2.51 mm. The monitor is connected to a 100-mV power supply, and the total resistance of the circuit is that due to

FIGURE P17.19

the mercury *plus* 1.00 Ω (an internal resistance of the power supply). Determine the change of current through the monitor as a patient draws in a breath and stretches the hose by 10.0 cm. Take $\rho_{Hg} = 9.40 \times 10^{-7}\ \Omega \cdot m$.

Section 17.6 Temperature Variation of Resistance

20. A certain lightbulb has a tungsten filament with a resistance of 19 Ω when cold and 140 Ω when hot. Assume that Equation 17.8 can be used over the large temperature range involved here, and find the temperature of the filament when it is hot. Assume an initial temperature of 20°C.

21. While taking photographs in Death Valley on a day when the temperature is 58.0°C, Bill Hiker finds that a certain voltage applied to a copper wire produces a current of 1.000 A. Bill then travels to Antarctica and applies the same voltage to the same wire. What current does he register there if the temperature is −88.0°C? Assume that no change occurs in the wire's shape and size.

22. If a silver wire has a resistance of 10.0 Ω at 20.0°C, what resistance does it have at 40.0°C? Neglect any change in length or cross-sectional area resulting from the change in temperature.

23. At 20°C the carbon resistor in an electric circuit, connected to a 5.0-V battery, has a resistance of 200 Ω. What is the current in the circuit when the temperature of the carbon rises to 80°C?

24. At 40.0°C, the resistance of a segment of gold wire is 100.0 Ω. When the wire is placed in a liquid bath, the resistance decreases to 97.0 Ω. What is the temperature of the bath? (*Hint:* First determine the resistance of the gold wire at room temperature, 20°C.)

25. The copper wire used in a house has a cross-sectional area of 3.00 mm². If 10.0 m of this wire is used to wire a circuit in the house at 20.0°C, find the resistance of the wire at temperatures of (a) 30.0°C and (b) 10.0°C.

26. An aluminum rod has a resistance of 1.234 Ω at 20.0°C. Calculate the resistance of the rod at 120°C by accounting for the changes in both the resistivity and the dimensions of the rod.

27. (a) A 34.5-m length of copper wire at 20.0°C has a radius of 0.25 mm. If a potential difference of 9.0 V is applied across the length of the wire, determine the current in the wire. (b) If the wire is heated to 30.0°C while the 9.0-V potential difference is maintained, what is the resulting current in the wire?

28. A toaster rated at 1 050 W operates on a 120-V household circuit and has a 4.00-m length of nichrome wire as its heating element. The operating temperature of this element is 320°C. What is the cross-sectional area of the wire?

29. In one form of plethysmograph (a device for measuring volume), a rubber capillary tube with an inside diameter of 1.00 mm is filled with mercury at 20°C. The resistance of the mercury is measured with the aid of electrodes sealed into the ends of the tube. If 100.00 cm of the tube is wound in a spiral around a patient's upper arm, the blood flow during a heartbeat causes the arm to expand, stretching the tube to a length of 100.04 cm. From this observation (assuming cylindrical symmetry) you can find the change in volume of the arm, which gives an indication of blood flow. (a) Calculate the resistance of the mercury. (b) Calculate the fractional change in resistance during the heartbeat. (*Hint:* The fraction by which the cross-sectional area of the mercury thread decreases is the fraction by which the length increases, since the volume of mercury is constant.) Take $\rho_{Hg} = 9.4 \times 10^{-7}\ \Omega \cdot m$.

30. A platinum resistance thermometer has resistances of 200.0 Ω when placed in a 0°C ice bath and 253.8 Ω when immersed in a crucible containing melting potassium. What is the melting point of potassium? (*Hint:* First determine the resistance of the platinum resistance thermometer at room temperature, 20°C.)

Section 17.8 Electrical Energy and Power

31. A toaster is rated at 600 W when connected to a 120-V source. What current does the toaster carry, and what is its resistance?

32. The output power of the Sun is 4.0×10^{26} W. Calculate at eight cents per kilowatt-hour the cost of running the Sun for one second.

33. How many 100-W lightbulbs can you use in a 120-V circuit without tripping a 15-A circuit breaker? (The bulbs are connected in parallel, which means that the potential difference across each lightbulb is 120 V.)

34. A high-voltage transmission line with a resistance of 0.31 Ω/km carries a current of 1 000 A. The line is at a potential of 700 kV at the power station and carries the current to a city located 160 km from the power station. (a) What is the power loss due to resistance in the line? (b) What fraction of the transmitted power does this loss represent?

35. The heating element of a coffee maker operates at 120 V and carries a current of 2.00 A. Assuming that the water absorbs all of the energy converted by the resistor, calculate how long it takes to heat 0.500 kg of water from room temperature (23.0°C) to the boiling point.

36. The power supplied to a typical black-and-white television set is 90 W when the set is connected to 120 V. (a) How much electric energy does this set consume in one hour? (b) A color television set draws about 2.5 A when connected to 120 V. How much time is required for it to consume the same energy as the black-and-white model consumes in one hour?

37. What is the required resistance of an immersion heater that will increase the temperature of 1.50 kg of water from 10.0°C to 50.0°C in 10.0 min while operating at 120 V?

38. A certain toaster has a heating element made of nichrome resistance wire. When the toaster is first connected to a 120-V source of potential difference (and the wire is at a temperature of 20.0°C) the initial current is 1.80 A. However, the current begins to decrease as the resistive element warms up. When the toaster has reached its final operating temperature, the current has dropped to 1.53 A. (a) Find the power the toaster converts when it is at its operating temperature. (b) What is the final temperature of the heating element?

39. A copper cable is designed to carry a current of 300 A with a power loss of 2.00 W/m. What is the required radius of this cable?

40. A small motor draws a current of 1.75 A from a 120-V line. The output power of the motor is 0.20 hp. (a) At a rate of $0.060/kWh, what is the cost of operating the motor for 4.0 h? (b) What is the efficiency of the motor?

41. We estimate that there are 270 million plug-in electric clocks in the United States, approximately one clock for each person. The clocks convert energy at the average rate of 2.50 W. To supply this energy, how many metric tons of coal are burned per hour in coal-fired electric-generating plants that are, on average, 25.0% efficient? The heat of combustion for coal is 33.0 MJ/kg.

42. The cost of electricity varies widely throughout the United States; $0.120/kWh is one typical value. At this unit price, calculate the cost of (a) leaving a 40.0-W porch light on for two weeks while you are on vacation, (b) making a piece of dark toast in 3.00 min with a 970-W toaster, and (c) drying a load of clothes in 40.0 min in a 5 200-W dryer.

43. How much does it cost to watch a complete 21-hour-long World Series on a 180-W television set? Assume that electricity costs $0.070/kWh.

44. A house is heated by a 24-kW electric furnace using resistance heating. The rate for electrical energy is $0.080/kWh. If the heating bill for January is $200, how long must the furnace have been running on an average January day?

45. An 11-W energy-efficient fluorescent lamp is designed to produce the same illumination as a conventional 40-W lamp. How much does the energy-efficient lamp save during 100 hours of use? Assume a cost of $0.080/kWh for electrical energy.

46. An electric resistance heater is to deliver 1 500 kcal/h to a room using 110-V electricity. If fuses come in 10-A, 20-A, and 30-A sizes, what is the smallest fuse that can safely be used in the heater circuit?

47. The heating coil of a hot water heater has a resistance of
web 20 Ω and operates at 210 V. If electrical energy costs $0.080/kWh, what does it cost to raise the 200 kg of water in the tank from 15°C to 80°C? (See Chapter 11.)

ADDITIONAL PROBLEMS

48. One lightbulb is marked "25 W 120 V" and another "100 W 120 V"; this means that each converts its respective power when plugged into a constant 120-V potential difference. (a) Find the resistance of each bulb. (b) How long does it take for 1.00 C to pass through the dim bulb? How is this charge different upon its exit versus its entry into the bulb? (c) How long does it take for 1.00 J to pass through the dim bulb? How is this energy different upon its exit versus its entry into the bulb? (d) Find the cost of running the dim bulb continuously for 30.0 days if the electric company sells its product at $0.070 0 per kWh. What physical quantity *does* the electric company sell? What is its price for one SI unit of this quantity?

49. A particular wire has a resistivity of 3.0×10^{-8} $\Omega \cdot$ m and a cross-sectional area of 4.0×10^{-6} m². A length of this wire is to be used as a resistor that will develop 48 W of power when connected across a 20-V battery. What length of wire is required?

50. A steam iron draws 6.0 A from a 120-V line. (a) How many joules of internal energy are produced in 20 min? (b) How much does it cost, at $0.080/kWh, to run the steam iron for 20 min?

51. An experiment is conducted to measure the electrical resistivity of nichrome in the form of wires with different lengths and cross-sectional areas. For one set of measurements, a student uses 30-gauge wire, which has a cross-sectional area of 7.30×10^{-8} m². The student measures the potential difference across the wire and the current in the wire with a voltmeter and an ammeter, respectively. For each of the measurements given in the table taken on wires of three different lengths, calculate the resistance of the wires and the corresponding values of the resistivity. What is the average value of the resistivity, and how does this value compare with the value given in Table 17.1?

L (m)	ΔV (V)	I (A)	R (Ω)	ρ (Ω·m)
0.540	5.22	0.500		
1.028	5.82	0.276		
1.543	5.94	0.187		

52. Birds resting on high-voltage power lines are a common sight. The copper wire on which a bird stands is 2.2 cm in diameter and carries a current of 50 A. If the bird's feet are 4.0 cm apart, calculate the potential difference across its body.

53. A small sphere that carries a charge of 8.00 nC is whirled in a circle at the end of an insulating string. The angular speed is 100π rad/s. What average current does this rotating charge represent?

54. The current in a conductor varies in time as shown in Figure P17.54. (a) How many coulombs of charge pass through a cross section of the conductor in the interval $t = 0$ to $t = 5.0$ s? (b) What constant current would transport the same total charge during the 5.0-s interval as does the actual current?

FIGURE P17.54

55. An electric car is designed to run off a bank of 12.0-V batteries with a total energy storage of 2.00×10^7 J. (a) If the electric motor draws 8.00 kW, what is the current delivered to the motor? (b) If the electric motor draws 8.00 kW as the car moves at a steady speed of 20.0 m/s, how far will the car travel before it is "out of juice"?

56. (a) A 115-g mass of aluminum is formed into a right circular cylinder, shaped so that its diameter equals its height. Calculate the resistance between the top and bottom faces of the cylinder at 20°C. (b) Calculate the resistance between opposite faces if the same mass of aluminum is formed into a cube.

57. A length of metal wire has a radius of 5.00×10^{-3} m and a resistance of 0.100 Ω. When the potential difference across the wire is 15.0 V, the electron drift speed is found to be 3.17×10^{-4} m/s. Based on these data, calculate the density of free electrons in the wire.

58. A carbon wire and a nichrome wire are connected one after the other. If the combination has total resistance of 10.0 kΩ at 20°C, what is the resistance of each wire at 20°C so that the resistance of the combination does not change with temperature?

59. (a) Determine the resistance of a lightbulb marked 100 W @ 120 V. (b) Assuming that the filament is tungsten and has a cross-sectional area of 0.010 mm², determine the length of the wire inside the bulb when the bulb is operating. (c) Why do you think the wire inside the bulb is tightly coiled? (d) If the temperature of the tungsten wire is 2 600°C when the bulb is operating, what is the length of the wire after the bulb is turned off and has cooled to 20°C? (See Chapter 10, and use $4.5 \times 10^{-6}/°C$ as the coefficient of linear expansion for tungsten.)

60. In a certain stereo system, each speaker has a resistance of 4.00 Ω. The system is rated at 60.0 W in each channel. Each speaker circuit includes a fuse rated at a maximum current of 4.00 A. Is this system adequately protected against overload?

61. A resistor is constructed by forming a material of resistivity 3.5×10^5 $\Omega \cdot$m into the shape of a hollow cylinder of length 4.0 cm and inner and outer radii of 0.50 cm and 1.2 cm, respectively. In use, a potential difference is applied between the ends of the cylinder, producing a current parallel to the length of the cylinder. Find the resistance of the cylinder.

62. The graph in Figure P17.62a shows the current I in a diode as a function of potential difference ΔV across the diode. Figure P17.62b shows the circuit used to make the measurements. The symbol ⟶▶⊢ represents the diode. (a) Using Equation 17.4, make a table of the resis-

tance of the diode for different values of ΔV in the range from -1.5 V to $+1.0$ V. (b) Based on your results, what amazing electrical property does a diode possess?

63. An x-ray tube used for cancer therapy operates at 4.0 MV, with a beam current of 25 mA striking the metal target. Nearly all the power in this beam is transferred to a stream of water flowing through holes drilled in the target. What rate of flow, in kilograms per second, is needed if the temperature rise (ΔT) of the water is not to exceed 50°C?

64. A 50.0-g sample of a conducting material is all that is available. The resistivity of the material is measured to be 11×10^{-8} $\Omega \cdot$m, and the density is 7.86 g/cm³. The material is to be shaped into a solid cylindrical wire that has a total resistance of 1.5 Ω. (a) What length is required? (b) What must be the diameter of the wire?

65. (a) A sheet of copper ($\rho = 1.7 \times 10^{-8}$ $\Omega \cdot$m) is 2.0 mm thick and has surface dimensions of 8.0 cm $\times$ 24 cm. If the long edges are joined to form a tube 24 cm in length, what is the resistance between the ends? (b) What mass of copper is required to manufacture a 1 500-m-long spool of copper cable with a total resistance of 4.5 Ω?

66. When a straight wire is heated, its resistance changes according to the equation

$$R = R_0[1 + \alpha(T - T_0)]$$

where α is the temperature coefficient of resistivity. (a) Show that a more precise result, which includes the fact that the length and area of a wire change when it is heated, is

$$R = \frac{R_0[1 + \alpha(T - T_0)][1 + \alpha'(T - T_0)]}{[1 + 2\alpha'(T - T_0)]}$$

where α' is the coefficient of linear expansion (see Chapter 10). (b) Compare these two results for a 2.00-m-long copper wire of radius 0.100 mm, starting at 20.0°C and heated to 100.0°C.

(a)

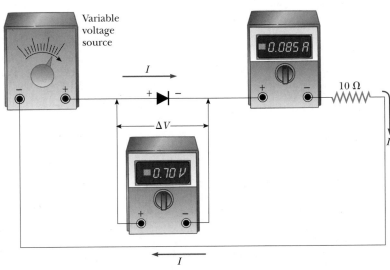

(b)

FIGURE P17.62

GROUP ACTIVITIES

G.1 Connect one terminal of a D-cell battery to the base of a flashlight bulb using insulated wire, tape a second wire to the other battery terminal, and tape a third wire to the center conductor of the bulb as in Figure GA17.1. Make sure to remove about 1 cm of insulation from the ends of all wires before making connections. Now bridge the gap between the open wires with different objects, such as a plastic pen, an aluminum can, a penny, a rubber band, and a spoon. Which objects make the bulb light up? Explain your observations.

Touch objects with these wires

FIGURE GA17.1

G.2 When the lightbulbs in your home are used, they are always connected across the same potential difference. Which do you believe has a filament with the highest resistance when cool, a 60-W bulb or a 100-W bulb? To check your prediction, ask your instructor to lend you a device called an ohmmeter and to instruct you in its use. A resistor must always be disconnected from a circuit when its resistance is measured with an ohmmeter.

G.3 Examine the label on several household appliances such as a toaster, television, lamp, stereo, air conditioner, and clock. From the label, determine the power rating of the device in watts. Check the billing statement from your electric utility company to find the cost of electrical energy per kilowatt-hour. (Prices usually range from about a nickel to twenty cents.) Calculate the cost of running each appliance for 1 h. Estimate how many hours per day each appliance is used. Then calculate the monthly cost of using each appliance based on your daily estimate.

G.4 A piece of material is connected to a battery. All of the following factors affect the current in the wire. Explain why these influence the current and discuss whether the current is directly proportional or inversely proportional to each factor, or has another pattern of dependence. (a) The cross-sectional area of the material, the length of the material, the temperature of the material, the number of free charge carriers in the material. (b) "When I turn on a wall switch, the current in a room seems to come on instantly. Thus, I know that the drift velocity of the electrons in a wire is extremely fast." Argue for or against this statement. (c) Assume a certain material has 9.00×10^{28} free electrons per cubic meter, a cross-sectional area of 3.00×10^{-6} m^2, and a current of 0.300 A. Find the drift speed of the electrons in the material.

G.5 (a) Explain why temperature affects the resistance of a material. (b) Explain why length affects the resistance of a material. (c) Explain why cross-sectional area affects the resistance of a material. (d) A tungsten-filament lightbulb has a resistance of 240 Ω when connected to a 120-V outlet, and it is found that the temperature of the filament is about 2 000°C under these conditions. What is the current in the bulb when it is connected to a 2.00-V battery?

G.6 Two identical, parallel copper wires used as cable TV lines are placed underground between two points 1.00 mile apart. These wires are usually not connected, but a construction accident shorts the wires together at some point. Problem: Where is that point? To try to isolate the spot so that repair can be initiated, a technician goes to end A of the lines and finds that a 12.0-V battery connected across the wires at that end produces a current of 1.00 A. Doing the same at the other end, end B, produces a current of 0.20 A. (a) Is the break closer to end A or to end B? (b) How far from the 0.20-A end should digging begin?

G.7 You are cooking breakfast for yourself and a friend using a 1 200-W waffle iron and a 500-W coffee pot. Usually, you operate these appliances from a 110-V outlet for 0.50 h each day. (a) At eight cents per kWh, how much do you spend to cook breakfast each month? (b) You find yourself addicted to waffles and would like to upgrade to a 2 400-W waffle iron that will enable you to cook twice as many waffles during a half-hour period, but you know that the circuit breaker in your kitchen is a 20-A breaker. Can you do the upgrade?

Direct Current Circuits

18

This versatile circuit enables the experimenter to examine the properties of circuit elements such as capacitors and resistors and their effects on circuit behavior. *(Courtesy of Central Scientific Company)*

Chapter Outline

T his chapter analyzes some simple circuits whose elements include batteries, resistors, and capacitors in varied combinations. Such analysis is simplified by the use of two rules known as Kirchhoff's rules, which follow from the principle of conservation of energy and the law of conservation of charge. Most of the circuits are assumed to be in *steady state*, which means that the currents are constant in magnitude and direction. We close the chapter with a discussion of circuits containing resistors and capacitors, in which current varies with time.

18.1   SOURCES OF emf

The source that maintains the current in a closed circuit is called a source of emf.[1] Any devices (such as batteries and generators) that increase the potential energy of charges circulating in circuits are sources of emf. One can think of such a source as a "charge pump" that forces electrons to move in a direction opposite the electrostatic field inside the source. The emf $\mathcal{E}$ of a source is the work done per unit charge, and hence the SI unit of emf is the volt.

Consider the circuit in Figure 18.1a, consisting of a battery connected to a resistor. We assume that the connecting wires have no resistance. If we neglect the internal resistance of the battery, the potential difference across the battery (the terminal voltage) equals the emf of the battery. However, because a real battery always has some internal resistance r, the terminal voltage is not equal to the emf. The circuit of Figure 18.1a can be described schematically by the diagram in Figure 18.1b. The battery, represented by the dashed rectangle, consists of a source of emf $\mathcal{E}$ in series with an internal resistance r. Now imagine a positive charge moving from a to b in Figure 18.1b. As the charge passes from the negative to the positive terminal of the battery, the potential of the charge increases by $\mathcal{E}$. However, as the charge moves through the resistance r its potential decreases by the amount Ir, where I is the current in the circuit. Thus, the terminal voltage of the battery, $\Delta V = V_b - V_a$, is

$$\Delta V = \mathcal{E} - Ir \qquad [18.1]$$

Note from this expression that $\mathcal{E}$ **is equal to the terminal voltage when the current is zero,** called the **open-circuit voltage.** By inspecting Figure 18.1b, we see that the terminal voltage ΔV must also equal the potential difference across the external resistance R, often called the **load resistance;** that is, $\Delta V = IR$. Combining this with Equation 18.1, we see that

$$\mathcal{E} = IR + Ir \qquad [18.2]$$

Battery

Resistor

(a)

(b)

FIGURE 18.1   (a) A circuit consisting of a resistor connected to the terminals of a battery. (b) A circuit diagram of a source of emf $\mathcal{E}$ having internal resistance r connected to an external resistor R.

An assortment of batteries. *(George Semple)*

---

[1] The term was originally an abbreviation for electromotive force, but emf is not really a force, so the long form is discouraged.

Solving for the current gives

$$I = \frac{\mathcal{E}}{R + r}$$

This shows that the current in this simple circuit depends on both the resistance external to the battery and the internal resistance. If R is much greater than r, we can neglect r in our analysis and we do, for many circuits. If we multiply Equation 18.2 by the current I, we get

$$I\mathcal{E} = I^2R + I^2r$$

This equation tells us that the total power output $I\mathcal{E}$ of the source of emf is converted to the rate I^2R at which energy is delivered to the load resistance *plus* the rate I^2r at which energy is delivered to the internal resistance. Again, if $r \ll R$, most of the power delivered by the battery is transferred to the load resistance.

Unless otherwise stated, we will assume in our examples and end-of-chapter problems that the internal resistance of a battery in a circuit is negligible.

18.2 RESISTORS IN SERIES

When two or more resistors are connected end-to-end as in Figure 18.2, they are said to be in *series*. The resistors could be simple devices such as lightbulbs or heating elements. When the two resistors R_1 and R_2 are connected to a battery as in Figure 18.2, **the current is the same in the two resistors because any charge that flows through R_1 must also flow through R_2.** This is analogous to water flowing through a pipe with two constrictions, corresponding to R_1 and R_2. Whatever volume of water flows in one end in a given time interval must exit the opposite end in the same time interval.

Because the potential difference between a and b in Figure 18.2b equals IR_1 and the potential difference between b and c equals IR_2, the potential difference between a and c is

$$\Delta V = IR_1 + IR_2 = I(R_1 + R_2)$$

Regardless of how many resistors we have in series, the sum of the potential differences across the resistors is equal to the total potential difference across the combination. As we will show later, this is a consequence of the conservation of energy. Figure 18.2c shows an equivalent resistor R_{eq} that can replace the two resistors of the original circuit. The equivalent resistance has the same effect on the circuit as

Tip 18.1 WHAT IS CONSTANT IN A BATTERY?

Equation 18.2 shows that the current in a circuit depends on the resistance of the battery. Note that the battery is not a source of constant current. Furthermore, the terminal voltage of a battery given by Equation 18.1 is also not constant. A battery is a source of constant emf.

◀ For a series connection of resistors, the current is the same in all the resistors.

(a)   (b)   (c)

FIGURE 18.2 A series connection of two resistors R_1 and R_2. The currents in the resistors are the same, and the equivalent resistance of the combination is given by $R_{eq} = R_1 + R_2$.

the combination of resistors because it results in the same current in the circuit. Applying Ohm's law to this resistor, we have

$$\Delta V = I R_{eq}$$

Equating the preceding two expressions, we have

$$I R_{eq} = I(R_1 + R_2)$$

or

$$R_{eq} = R_1 + R_2 \qquad [18.3]$$

An extension of this analysis shows that the equivalent resistance of three or more resistors connected in series is

$$R_{eq} = R_1 + R_2 + R_3 + \cdots \qquad [18.4]$$

▶ Equivalent resistance of a series combination of resistors.

Therefore, **the equivalent resistance of a series combination of resistors is the algebraic sum of the individual resistances and is always greater than any individual resistance.**

Note that if the filament of one lightbulb in Figure 18.2 were to fail, the circuit would no longer be complete (an open-circuit condition would exist) and the second bulb would also go out.

APPLYING **PHYSICS** 18.1

APPLICATION

CHRISTMAS LIGHTS IN SERIES

A new design for Christmas tree lights allows them to be connected in series. One might expect that a failed bulb in such a string would result in an open circuit and all of the bulbs would go out. How can the bulbs be designed to prevent this from happening?

Explanation If the string of lights contained normal bulbs, a failed bulb would be hard to locate. Each bulb would have to be replaced with a good bulb, one by one, until the failed bulb is found. If there happened to be two or more failed bulbs in the string of lights, finding them would be a formidable task.

Christmas lights use specially designed bulbs that have an insulated loop of wire (jumper) across the conducting supports to the bulb filaments (Fig. 18.3). If the filament breaks and the bulb fails, the resistance of this bulb increases dramatically. As a result, most of the applied voltage appears across the loop of wire. This voltage causes the insulation around the loop of wire to melt, causing the metal wire to make electrical contact with the supports. This produces a conducting path through the bulb, enabling the other bulbs to remain lit.

Webnote 18.1

Holiday lights are a tradition at the end of the year. Learn how these colorful and festive decorations are constructed at *http://www.howstuffworks.com/ christmas-lights.htm*

Filament

Jumper

Glass insulator

FIGURE 18.3 (Applying Physics 18.1) Diagram of a modern miniature holiday lightbulb, with a jumper connection to provide a current if the filament breaks.

Quick Quiz 18.1

When a piece of wire is used to connect points b and c in Figure 18.2b, the brightness of bulb R_1 (a) increases, (b) decreases, or (c) stays the same. The brightness of bulb R_2 (a) increases, (b) decreases, or (c) stays the same.

Quick Quiz 18.2

With the switch in the circuit of Figure 18.4a closed, no current exists in R_2 because the current has an alternate zero-resistance path through the switch. Current does exist in R_1 and this current is measured with the ammeter at the right side of the circuit. If the switch is opened (Fig. 18.4b), current exists in R_2. After the switch is opened, the reading on the ammeter (a) increases, (b) decreases, (c) does not change.

FIGURE 18.4 (Quick Quiz 18.2)

(a)   (b)

Example 18.1 Four Resistors in Series

Four resistors are arranged as shown in Figure 18.5a. Find (a) the equivalent resistance and (b) the current in the circuit if the emf of the battery is 6.0 V.

2.0 Ω 4.0 Ω 5.0 Ω 7.0 Ω
R_1 R_2 R_3 R_4

18.0 Ω

(a)   6.0 V   (b)   6.0 V

FIGURE 18.5 (Example 18.1) (a) Four resistors connected in series. (b) The equivalent resistance of the circuit in (a).

Solution

A The equivalent resistance is found from Equation 18.4:

$$R_{eq} = R_1 + R_2 + R_3 + R_4 = 2.0\ \Omega + 4.0\ \Omega + 5.0\ \Omega + 7.0\ \Omega = \boxed{18.0\ \Omega}$$

B If we apply Ohm's law to the equivalent resistor in Figure 18.5b, we find the current in the circuit to be

$$I = \frac{\Delta V}{R_{eq}} = \frac{6.0\ \text{V}}{18.0\ \Omega} = \boxed{\tfrac{1}{3}\ \text{A}}$$

Exercise Because the current in the equivalent resistor is $\tfrac{1}{3}$ A, this must also be the current in each resistor of the original circuit. Find the voltage drop across each resistor.

Answer $\Delta V_{2\Omega} = \tfrac{2}{3}$ V; $\Delta V_{4\Omega} = \tfrac{4}{3}$ V; $\Delta V_{5\Omega} = \tfrac{5}{3}$ V; $\Delta V_{7\Omega} = \tfrac{7}{3}$ V

18.3 RESISTORS IN PARALLEL

Now consider two resistors connected in parallel, as in Figure 18.6. In this case, **the potential differences across the resistors are the same because each is connected directly across the battery terminals.** The currents are generally not the same. When the charge reaches point a (called a junction) in Figure 18.6b, the current splits into two parts, I_1 going through R_1 and I_2 going through R_2. If R_1 is greater than R_2, then I_1 is less than I_2. That is, the charge tends to follow the path of least resistance. **Because charge is conserved, the current I that enters point a must equal the total current leaving that point, $I_1 + I_2$.** That is,

$$I = I_1 + I_2$$

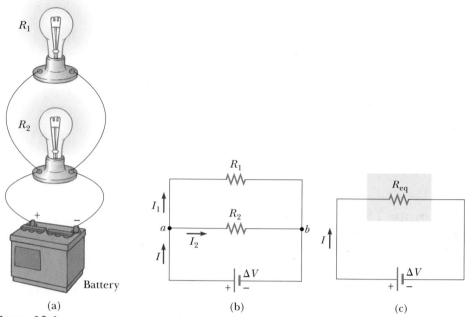

FIGURE 18.6   (a) A parallel connection of two resistors. (b) A circuit diagram for the parallel combination. (c) The voltages across the resistors are the same, and the equivalent resistance of the combination is given by the reciprocal relationship $1/R_{eq} = 1/R_1 + 1/R_2$.

The potential drop must be the same for the two resistors and must also equal the potential drop across the battery. Ohm's law applied to each resistor gives

$$I_1 = \frac{\Delta V}{R_1} \qquad I_2 = \frac{\Delta V}{R_2}$$

Ohm's law applied to the equivalent resistor in Figure 18.6c gives

$$I = \frac{\Delta V}{R_{eq}}$$

When these expressions for the current are substituted into the equation $I = I_1 + I_2$, and ΔV is cancelled, we obtain

$$\frac{1}{R_{eq}} = \frac{1}{R_1} + \frac{1}{R_2} \qquad \text{(parallel combination)} \qquad \text{[18.5]}$$

An extension of this analysis to three or more resistors in parallel produces the following general expression for the equivalent resistance:

▶ Equivalent resistance of a parallel combination of resistors.

$$\frac{1}{R_{eq}} = \frac{1}{R_1} + \frac{1}{R_2} + \frac{1}{R_3} + \cdots \qquad \text{[18.6]}$$

From this expression, it can be seen that **the inverse of the equivalent resistance of two or more resistors connected in parallel is the algebraic sum of the inverses of the individual resistances. The equivalent resistance is always less than the smallest resistance in the group.**

Household circuits are always wired so that the electrical devices are connected in parallel, as in Figure 18.6a. In this manner, each device operates independently of the others, so that if one is switched off, the others remain on. For example, if one of the lightbulbs in Figure 18.6 were removed from its socket, the other would continue to operate. Equally important, each device operates on the same voltage. If the devices were connected in series, the voltage across any one device would depend on how many devices were in the combination and on their individual resistances.

In many household circuits, circuit breakers are used in series with other circuit elements for safety purposes. A circuit breaker is designed to switch off and open the circuit at some maximum current (typically 15 A or 20 A) whose value depends on the nature of the circuit. If a circuit breaker were not used, excessive currents caused by operating several devices simultaneously could result in excessive wire temperatures and perhaps cause a fire. In older home construction, fuses were used in place of circuit breakers. When the current in a circuit exceeds some value, the conductor in a fuse melts and opens the circuit. The disadvantage of fuses is that they are destroyed in the process of opening the circuit, whereas circuit breakers can be reset.

◄ **APPLICATION**

CIRCUIT BREAKERS

PROBLEM-SOLVING STRATEGY   **Resistors**

1. When two or more unequal resistors are connected in *series*, they carry the same current, but the potential differences across them are not the same. The resistors add directly to give the equivalent resistance of the series combination.

2. When two or more unequal resistors are connected in *parallel*, the potential differences across them are the same. Because the current is inversely proportional to the resistance, the currents through them are not the same. The equivalent resistance of a parallel combination of resistors is found through reciprocal addition, and the equivalent resistance is always *less* than the smallest individual resistor in the combination.

3. A complicated circuit consisting of several resistors and batteries can often be reduced to a simple circuit with only one resistor. To do so, examine the initial circuit and replace any resistors in series or any in parallel using the procedures outlined in Steps 1 and 2. Sketch the new circuit after these changes have been made. Examine the new circuit and replace any series or parallel combinations. Continue this process until a single equivalent resistance is found.

4. If the current in or the potential difference across a resistor in the complicated circuit is to be identified, start with the final circuit found in Step 3 and gradually work back through the circuits, using $\Delta V = IR$ and the procedures of Steps 1 and 2.

APPLYING **PHYSICS** **18.2**

Compare the brightness of the four identical bulbs shown in Figure 18.7. What happens if bulb A fails, so that it cannot conduct current? What if C fails? What if D fails?

Explanation   Bulbs A and B are connected in series across the emf of the battery, whereas bulb C is connected by itself across the battery. Thus, the emf is split between bulbs A and B. As a result, bulb C will be brighter than bulbs A and B, which should be equally as bright as each other. Bulb D has a wire connected across it. Thus, there is no potential difference across D, and it does not glow at all. If bulb A fails, B goes out, but C stays lit. If C fails, there is no effect on the other bulbs. If D fails, the event is undetectable, because D was not glowing initially.

FIGURE 18.7 (Applying Physics 18.2)

FIGURE 18.8 (Applying Physics 18.3)

APPLYING PHYSICS 18.3

Figure 18.8 illustrates how a three-way lightbulb is constructed to provide three levels of light intensity. The socket of the lamp is equipped with a three-way switch for selecting different light intensities. The bulb contains two filaments. Why are the filaments connected in parallel? Explain how the two filaments are used to provide three different light intensities.

Explanation If the filaments were connected in series and one of them were to fail, there would be no current in the bulb and the bulb would give no illumination, regardless of the switch position. However, when the filaments are connected in parallel and one of them (say the 75-W filament) fails, the bulb will still operate in one of the switch positions because there is current in the other (100-W) filament. The three light intensities are made possible by selecting one of three values of filament resistance using a single value of 120 V for the applied voltage. The 75-W filament offers one value of resistance, the 100-W filament offers a second value, and the third resistance is obtained by combining the two filaments in parallel. When switch S_1 is closed and switch S_2 is opened, only the 75-W filament carries current. When switch S_1 is open and switch S_2 is closed, only the 100-W filament carries current. When both switches are closed, both filaments carry current and a total illumination corresponding to 175 W is obtained.

Quick Quiz 18.3

With the switch in the circuit of Figure 18.9a open, there is no current in R_2. There is current in R_1 and this current is measured with the ammeter at the right side of the circuit. If the switch is closed (Fig. 18.9b), there is current in R_2. When the switch is closed, the reading on the ammeter (a) increases, (b) decreases, or (c) remains the same.

(a)   (b)

FIGURE 18.9 (Quick Quiz 18.3)

Quick Quiz 18.4

You have a large supply of lightbulbs and a battery. You start with one lightbulb connected to the battery and notice its brightness. You then add one lightbulb at a time, each new bulb being added in parallel to the previous bulbs. As the lightbulbs are added, what happens (a) to the brightness of the bulbs, (b) to the current in the bulbs, (c) to the power delivered by the battery, (d) to the lifetime of the battery, (e) to the terminal voltage of the battery? *Hint:* Do not ignore the internal resistance of the battery.

Example 18.2 Three Resistors in Parallel

Three resistors are connected in parallel as in Figure 18.10. A potential difference of 18 V is maintained between points a and b.

A Find the current in each resistor.

Solution Because the resistors are in parallel, the potential difference across each is 18 V. Let us apply $\Delta V = IR$ to find the current in each resistor:

$$I_1 = \frac{\Delta V}{R_1} = \frac{18\ \text{V}}{3.0\ \Omega} = \boxed{6.0\ \text{A}}$$

$$I_2 = \frac{\Delta V}{R_2} = \frac{18\ \text{V}}{6.0\ \Omega} = \boxed{3.0\ \text{A}}$$

$$I_3 = \frac{\Delta V}{R_3} = \frac{18\ \text{V}}{9.0\ \Omega} = \boxed{2.0\ \text{A}}$$

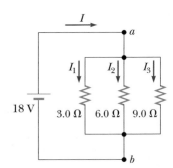

FIGURE 18.10 (Example 18.2) Three resistors connected in parallel. The voltage across each resistor is 18 V.

There is a saying that holds that current, like politicians, follows the path of least resistance. This is only partially correct, as our example demonstrates. The smallest (3.0-Ω) resistor carries the largest current, while the other larger resistors of 6.0 Ω and 9.0 Ω carry smaller currents.

B Calculate the power delivered to each resistor and the total power delivered to the three resistors.

Solution Applying $\mathcal{P} = I^2 R$ to each resistor gives

$$3\ \Omega: \quad \mathcal{P}_1 = I_1{}^2 R_1 = (6.0\ \text{A})^2 (3.0\ \Omega) = \boxed{110\ \text{W}}$$

$$6\ \Omega: \quad \mathcal{P}_2 = I_2{}^2 R_2 = (3.0\ \text{A})^2 (6.0\ \Omega) = \boxed{54\ \text{W}}$$

$$9\ \Omega: \quad \mathcal{P}_3 = I_3{}^2 R_3 = (2.0\ \text{A})^2 (9.0\ \Omega) = \boxed{36\ \text{W}}$$

(Note that you can also use $\mathcal{P} = (\Delta V)^2 / R$ to find the power dissipated by each resistor.) Summing the three quantities gives a total power of 200 W.

EXERCISE Calculate the equivalent resistance of the three resistors, and from this result find the total power dissipated.

ANSWER $\frac{18}{11}\ \Omega$; 200 W

Example 18.3 Equivalent Resistance

Four resistors are connected as shown in Figure 18.11a.

A Find the equivalent resistance between points a and c.

Solution The circuit can be reduced in steps, as shown in Figures 18.11b and 18.11c. The 8.0-Ω and 4.0-Ω resistors are in series, and so the equivalent resistance between a and b is 12 Ω (Eq. 18.4). The 6.0-Ω and 3.0-Ω resistors are in parallel, and so from Equation 18.6 we find that the equivalent resistance from b to c is 2.0 Ω. Hence, the equivalent resistance from a to c is 14 Ω.

B What is the current in each resistor if the terminals of a 42-V battery are connected between a and c?

Solution The current I is the same in the 8.0-Ω and 4.0-Ω resistors because they are in series. Using Ohm's law and the results of (a), we get

$$I = \frac{\Delta V_{ac}}{R_{eq}} = \frac{42\ \text{V}}{14\ \Omega} = \boxed{3.0\ \text{A}}$$

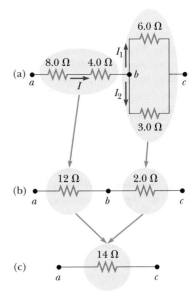

FIGURE 18.11 (Example 18.3) The four resistors shown in (a) can be reduced in steps to an equivalent 14-Ω resistor.

When this current enters the junction at b, it splits; part of it passes through the 6.0-Ω resistor (I_1), and part passes through the 3.0-Ω resistor (I_2). Because the potential difference ΔV_{bc} across these resistors is the *same* (they are in parallel), $(6\ \Omega)I_1 = (3\ \Omega)I_2$, or $I_2 = 2I_1$. Using this result and the fact that $I_1 + I_2 = 3.0$ A, we find that $I_1 = \boxed{1.0\ \text{A}}$ and $I_2 = \boxed{2.0\ \text{A}}$. We could have guessed this from the start by noting that the current in the 3.0-Ω resistor has to be twice the current in the 6.0-Ω resistor in view of their relative resistances and the fact that the same voltage is applied to both.

As a final check, note that $\Delta V_{bc} = (6\ \Omega)I_1 = (3\ \Omega)I_2 = 6.0$ V and $\Delta V_{ab} = (12\ \Omega)I_1 = 36$ V; therefore, $\Delta V_{ac} = \Delta V_{ab} + \Delta V_{bc} = 42$ V, as expected.

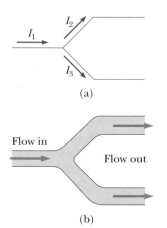

(a)

(b)

FIGURE 18.12 (a) A schematic diagram illustrating Kirchhoff's junction rule. Conservation of charge requires that whatever current enters a junction must leave that junction. Therefore, in this case, $I_1 = I_2 + I_3$. (b) A mechanical analog of the junction rule: The net flow in must equal the net flow out.

18.4 KIRCHHOFF'S RULES AND COMPLEX DC CIRCUITS

As demonstrated in the preceding section, we can analyze simple circuits using Ohm's law and the rules for series and parallel combinations of resistors. However, there are many ways in which resistors can be connected so that the circuits formed cannot be reduced to a single equivalent resistor. The procedure for analyzing more complex circuits is greatly simplified by the use of two simple rules called **Kirchhoff's rules:**

> 1. **The sum of the currents entering any junction must equal the sum of the currents leaving that junction. (This rule is often referred to as the junction rule.)**
> 2. **The sum of the potential differences across all the elements around any closed-circuit loop must be zero. (This rule is usually called the loop rule.)**

The junction rule is a statement of *conservation of charge.* Because charge must be conserved, whatever current enters a point in a circuit must equal the total current leaving that point. If we apply this rule to the junction in Figure 18.12a, we get

$$I_1 = I_2 + I_3$$

Figure 18.12b represents a mechanical analog to this situation, in which water flows through a branched pipe with no leaks. The flow rate into the pipe equals the total flow rate out of the two branches.

The loop rule is equivalent to the principle of *conservation of energy.* Any charge that moves around any closed loop in a circuit (starting and ending at the same point) must gain as much energy as it loses. It gains energy as it is pumped through a source of emf. Its energy may decrease in the form of a potential drop, $-IR$, across a resistor or as a result of flowing backward through a source of emf— that is, from the positive to the negative terminal inside the battery. In the latter case, electrical energy is converted to chemical energy as the battery is charged.

When applying Kirchhoff's rules, you must make two decisions at the beginning of a problem.

1. Assign symbols and directions to the currents in all branches of the circuit. Do not be alarmed that you might guess the direction of a current incorrectly; the resulting answer will be negative, but *its magnitude will be correct.*
2. When applying the loop rule, you must choose a direction for traversing the loop and be consistent in going either clockwise or counterclockwise. As you traverse the loop, record voltage drops and rises according to the rules stated next. They are summarized in Figure 18.13, where it is assumed that the traversal is from point a toward point b:

(a) $\Delta V = V_b - V_a = -IR$

(b) $\Delta V = V_b - V_a = +IR$

(c) ε
$\Delta V = V_b - V_a = +\varepsilon$

(d) ε
$\Delta V = V_b - V_a = -\varepsilon$

FIGURE 18.13 Rules for determining the potential changes across a resistor and a battery, assuming the battery has no internal resistance.

(a) If a resistor is traversed in the direction of the current, the change in electric potential across the resistor is $-IR$ (Fig. 18.13a).

(b) If a resistor is traversed in the direction opposite the current, the change in electric potential across the resistor is $+IR$ (Fig. 18.13b).

(c) If a source of emf is traversed in the direction of the emf (from $-$ to $+$ on the terminals), the change in electric potential is $+\mathcal{E}$ (Fig. 18.13c).

(d) If a source of emf is traversed in the direction opposite the emf (from $+$ to $-$ on the terminals), the change in electric potential is $-\mathcal{E}$ (Fig. 18.13d).

There are limits to the numbers of times the junction rule and the loop rule can be used. You can use the junction rule as often as needed so long as, each time you write an equation, you include in it a current that has not been used in a previous junction-rule equation. (If this procedure is not followed, a new equation will not be produced.) In general, the number of times the junction rule can be used is one fewer than the number of junction points in the circuit. The loop rule can be used as often as needed so long as a new circuit element (resistor or battery) or a new current appears in each new equation. In general, **to solve a particular circuit problem, you need as many independent equations as you have unknowns.**

GUSTAV KIRCHHOFF, GERMAN PHYSICIST (1824–1887)

Kirchhoff, a professor at Heidelberg, together with Robert Bunsen invented the spectroscopy that we study in Chapter 28. He also formulated another rule that states "a cool substance will absorb light of the same wavelengths that it emits when hot." *(AIP ESVA, W. F. Meggers Collection)*

PROBLEM-SOLVING STRATEGY | **Kirchhoff's Rules**

1. First, draw the circuit diagram and assign labels and symbols to all the known and unknown quantities. Assign *directions* to the currents in each part of the circuit. Although the assignment of current directions is arbitrary, you must stick with them throughout as you apply Kirchhoff's rules.

2. Apply the junction rule to any junction in the circuit. The junction rule may be applied as many times as a new current (one not used in a previous application) appears in the resulting equation.

3. Now apply Kirchhoff's loop rule to as many loops in the circuit as are needed to solve for the unknowns. In order to apply this rule, you must correctly identify the change in electric potential as you cross each element in traversing the closed loop. Watch out for signs! (Use the conventions in Fig. 18.13.)

4. Solve the equations simultaneously for the unknown quantities. Be careful in your algebraic steps, and check your numerical answers for consistency.

Example 18.4 Applying Kirchhoff's Rules

Find the currents in the circuit shown in Figure 18.14.

Reasoning There are three unknown currents in this circuit, and so we must obtain three independent equations. We can find the equations with one application of the junction rule and two applications of the loop rule.

Solution The first step is to assign a current to each branch of the circuit; these are our unknowns and are labeled I_1, I_2, and I_3 in Figure 18.14. It is also necessary to guess directions for the currents. Your experience with circuits such as this should tell you that the directions of all three have been chosen correctly. (However, recall that if a current direction is chosen incorrectly, the numerical answer will turn out negative, but the magnitude will be correct. This point will be demonstrated in the next example.)

We now apply Kirchhoff's rules. First we can apply the junction rule using either c or d, the only two junctions in the circuit. Let us choose junction c. The net current into

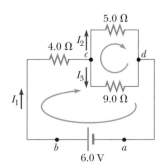

FIGURE 18.14 (Example 18.4) A multiloop circuit.

this junction is I_1, and the net current leaving it is $I_2 + I_3$. Thus, the junction rule applied to c gives

$$I_1 = I_2 + I_3$$

Recall that you may apply the junction rule over and over until you reach a situation in which no new currents appear in an equation. In this example we have reached that point with one application. If we apply the junction rule at d, we find that $I_1 = I_2 + I_3$, exactly the same equation.

We have three unknowns in our problem, I_1, I_2, and I_3; thus, we need two more independent equations before we can find a solution. We obtain these equations by applying the loop rule to the two loops indicated in the figure. Note that there are actually three loops in the circuit, but these two are sufficient to complete the problem. (Where is the loop that we do not use?)

When applying the loop rule, we must first choose the loops to be traversed, and then the directions in which to traverse them. We have selected the two loops indicated in the figure and have decided to traverse both of them clockwise. Other choices could be made, but the final result would be the same.

Starting at point a and moving clockwise around the large loop, we encounter the following voltage changes (see Fig. 18.13 for the basic sign conventions):

- From a to b, we encounter a voltage change of 6.0 V.
- From b to c through the 4.0-Ω resistor, we encounter a voltage change of $-(4.0\ \Omega)I_1$.
- From c to d through the 9.0-Ω resistor, we encounter a voltage change of $-(9.0\ \Omega)I_3$.

No voltage change occurs from d back to a. Now that we have made a complete traversal of the loop, we can equate the sum of the voltage changes to zero:

$$6.0\ \text{V} - (4.0\ \Omega)I_1 - (9.0\ \Omega)I_3 = 0$$

Moving clockwise around the small loop from point c, we encounter the following:

- From c to d through the 5.0-Ω resistor, a voltage change of $-(5.0\ \Omega)I_2$
- From d to c through the 9.0-Ω resistor, a voltage change of $+(9.0\ \Omega)I_3$

We find that

$$-(5.0\ \Omega)I_2 + (9.0\ \Omega)I_3 = 0$$

Thus, we have the following three equations to be solved for the three unknowns:

$$I_1 - I_2 - I_3 = 0$$

$$6.0 - 4.0I_1 - 9.0I_3 = 0$$

$$-5.0I_2 + 9.0I_3 = 0$$

where we have dropped units of volts and ohms and realize that the currents will have units of amperes.

If you need help in solving three equations with three unknowns, see Example 18.5. You should be able to obtain the following answers:

$$I_1 = \boxed{0.83\ \text{A}} \qquad I_2 = \boxed{0.53\ \text{A}} \qquad I_3 = \boxed{0.30\ \text{A}}$$

EXERCISE Solve this same problem by using the methods learned earlier for series and parallel combinations of resistors. First find the equivalent resistance of the circuit, which you can then use to obtain I_1.

Example 18.5 Another Application of Kirchhoff's Rules

Find I_1, I_2, and I_3 in Figure 18.15.

Reasoning To find the three unknown currents, we apply the junction rule once and the loop rule twice.

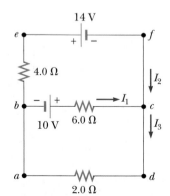

FIGURE 18.15 (Example 18.5) A circuit containing three loops.

Tip 18.2 CURRENT DOES NOT TAKE THE PATH OF LEAST RESISTANCE

The statement "current takes the path of least resistance" is incorrect in reference to a parallel combination of current paths. In this context, the current takes all paths. Those paths with lower resistance have larger currents, whereas those with higher resistance have lower currents.

Solution   We choose the directions of the currents as shown in the figure. Applying Kirchhoff's first rule to junction c gives

$$I_1 + I_2 = I_3 \tag{1}$$

The circuit has three loops: *abcda, befcb,* and *aefda.* We need only two loop equations to determine the unknown currents. The third loop equation would give no new information. Applying Kirchhoff's second rule to loops *abcda* and *befcb* and traversing these loops clockwise, we obtain the following expressions:

Loop *abcda:* $\qquad\qquad$ $10\text{ V} - (6.0\ \Omega)I_1 - (2.0\ \Omega)I_3 = 0 \tag{2}$

Loop *befcb:* $\qquad$ $-14\text{ V} + (6.0\ \Omega)I_1 - 10\text{ V} - (4.0\ \Omega)I_2 = 0 \tag{3}$

Note that in loop *befcb,* a positive sign is obtained when the 6.0-Ω resistor is traversed, because the direction of the path is opposite the direction of the current I_1. A third loop equation for *aefda* gives $-14\text{ V} - (2.0\ \Omega)I_3 - (4.0\ \Omega)I_2 = 0$, which is just the sum of (2) and (3). Expressions (1), (2), and (3) represent three linear, independent equations with three unknowns.

We can solve the problem as follows: substitution of (1) into (2) gives, with units ignored for the moment,

$$10 - 6.0I_1 - 2.0(I_1 + I_2) = 0$$

$$10 = 8.0I_1 + 2.0I_2 \tag{4}$$

Dividing each term in (3) by 2 and rearranging the equation gives

$$-12 = -3.0I_1 + 2.0I_2 \tag{5}$$

Subtracting (5) from (4) eliminates I_2, giving

$$22 = 11I_1$$

$$I_1 = 2.0\text{ A}$$

Using this value of I_1 in (5) yields a value for I_2:

$$2.0I_2 = 3.0I_1 - 12 = 3.0(2.0) - 12 = -6.0$$

$$I_2 = -3.0\text{ A}$$

Finally, $I_3 = I_1 + I_2 = -1.0$ A. Hence, the currents have the values

$$I_1 = \boxed{2.0\text{ A}} \qquad I_2 = \boxed{-3.0\text{ A}} \qquad I_3 = \boxed{-1.0\text{ A}}$$

The fact that I_2 and I_3 are both negative merely indicates that we chose the wrong directions for these currents. The numerical values are correct.

EXERCISE   Find the potential difference between junctions b and c.

ANSWER $V_b - V_c = 2.0$ V

18.5   *RC* CIRCUITS

So far, we have been concerned with circuits with constant currents. We now consider direct-current circuits containing capacitors, in which the currents vary with time. Consider the series circuit in Figure 18.16a. Let us assume that the capacitor is initially uncharged with the switch opened. After the switch is closed, the battery begins to charge the plates of the capacitor and a current passes through the resistor. The charging process continues until the capacitor is charged to its maximum equilibrium value $Q = C\mathcal{E}$, where $\mathcal{E}$ is the maximum voltage across the capacitor. Once the capacitor is fully charged, the current in the circuit is zero. If we assume that the capacitor is uncharged before the switch is closed, and the switch is closed

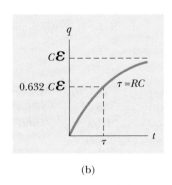

FIGURE 18.16 (a) A capacitor in series with a resistor, a battery, and a switch. (b) A plot of the charge on the capacitor versus time after the switch for the circuit is closed. After one time constant τ, the charge is 63% of the maximum value $C\mathcal{E}$. The charge approaches its maximum value as t approaches infinity.

at $t = 0$, we find that the charge on the capacitor varies with time according to the expression

$$q = Q(1 - e^{-t/RC}) \qquad [18.7]$$

where $e = 2.718 \ldots$ is Euler's constant, the base of the natural logarithms. Figure 18.16b is a graph of this expression. Note that the charge is zero at $t = 0$ and approaches its maximum value Q as t approaches infinity. The voltage ΔV across the capacitor at any time is obtained by dividing the charge by the capacitance. That is, $\Delta V = q/C$.

As you can see from Equation 18.7, it takes an infinite amount of time for the capacitor to become fully charged. The term RC that appears in Equation 18.7, is called the **time constant,** τ (Greek letter tau), so

$$\tau = RC \qquad [18.8]$$

The time constant represents the time required for the charge to increase from zero to 63.2% of its maximum equilibrium value. That is, in one time constant, the charge on the capacitor increases from zero to $0.632Q$. This can be seen by substituting $t = \tau = RC$ in Equation 18.7 and solving for q. (Note that $1/e = 0.632$.) It is important to note that a capacitor charges very slowly in a circuit with a long time constant, whereas it charges very rapidly in a circuit with a short time constant.

Now consider the circuit in Figure 18.17a, consisting of a capacitor with an initial charge Q, a resistor, and a switch. Before the switch is closed, the potential difference across the charged capacitor is Q/C. Once the switch is closed, the charge begins to flow through the resistor from one capacitor plate to the other until the capacitor is fully discharged. If the switch is closed at $t = 0$, it can be shown that the charge q on the capacitor varies with time according to the expression

$$q = Qe^{-t/RC} \qquad [18.9]$$

Webnote 18.2

You can see the effects of charging and discharging the capacitor in an *RC* circuit by studying the following java applet:
http://www.phy.ntnu.edu.tw/java/rc/rc.html

That is, the charge decreases exponentially with time as shown in Figure 18.17b. In the interval $t = \tau = RC$, the charge decreases from its initial value Q to $0.368Q$. In other words, in one time constant, the capacitor loses 63.2% of its initial charge. Because $\Delta V = q/C$, we see that the voltage across the capacitor also decreases ex-

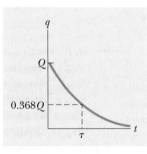

FIGURE 18.17 (a) A charged capacitor connected to a resistor and a switch. (b) A graph of the charge on the capacitor versus time after the switch is closed.

(a) (b)

ponentially with time according to the expression $\Delta V = \mathcal{E}e^{-t/RC}$, where $\mathcal{E}$ (which equals Q/C) is the initial voltage across the fully charged capacitor.

APPLYING PHYSICS 18.4

Many automobiles are equipped with windshield wipers that can be used intermittently during a light rainfall. How does the operation of this feature depend on the charging and discharging of a capacitor?

Explanation The wipers are part of an *RC* circuit whose time constant can be varied by selecting different values of *R* through a multiposition switch. The brief time that the wipers remain on and the time they are off are determined by the value of the time constant of the circuit.

APPLICATION

TIMED WINDSHIELD WIPERS

APPLYING PHYSICS 18.5

In biological applications concerned with population growth, an equation is used that is similar to the exponential equations encountered in the analysis of *RC* circuits. It is

$$N_f = N_i 2^n$$

where N_f is the number of bacteria present after n doubling times, N_i is the number present initially, and n is the number of growth cycles or doubling times. Doubling times vary according to the organism. The doubling time for the bacteria responsible for leprosy is about 30 days, and that for the salmonella bacteria responsible for food poisoning is about 20 minutes. For those of you with a biology inclination, consider that only 10 salmonella bacteria find their way onto a turkey leg after your Thanksgiving meal. Four hours later you come back for a midnight snack. How many bacteria are present now?

Explanation The number of doubling times is 240 min/20 min = 12. Thus, we have

$$N_f = N_i 2^n = (10 \text{ bacteria})(2^{12}) = 40\,960 \text{ bacteria}$$

So, your system will have to deal with an invading host of about 41 000 bacteria, which are going to continue to double in a very promising environment.

APPLICATION

BACTERIAL GROWTH

FIGURE 18.18 (Applying Physics 18.6)

APPLYING PHYSICS 18.6

Many roadway construction sites have flashing yellow lights to warn motorists of possible dangers. What causes the lights to flash?

Explanation A typical circuit for such a flasher is shown in Figure 18.18. The lamp L is a gas-filled lamp that acts as an open circuit until a large potential difference causes a discharge, which gives off a bright light. During this discharge, charge flows through the gas between the electrodes of the lamp. When the switch is closed, the battery charges up the capacitor. At the beginning, the current is high and the charge on the capacitor is low, so most of the potential difference appears across the resistance *R*. As the capacitor charges, more potential difference appears across it, reflecting the lower current and, thus, lower potential difference across the resistor. Eventually, the potential difference across the capacitor reaches a value at which the lamp will conduct, causing a flash. This discharges the capacitor through the lamp and the process of charging begins again. The period between flashes can be adjusted by changing the time constant of the *RC* circuit.

APPLICATION

ROADWAY FLASHERS

Example 18.6 Charging a Capacitor in an *RC* Circuit

An uncharged capacitor and a resistor are connected in series to a battery, as in Figure 18.16a. If $\mathcal{E} = 12$ V, $C = 5.0$ μF, and $R = 8.0 \times 10^5$ Ω, find the time constant of the circuit, the maximum charge on the capacitor, and the charge on the capacitor after one time constant—that is, when $t = RC$.

Solution The time constant of the circuit is

$$\tau = RC = (8.0 \times 10^5 \ \Omega)(5.0 \times 10^{-6} \ \text{F}) = \boxed{4.0 \ \text{s}}$$

The maximum charge on the capacitor is

$$Q = C\mathcal{E} = (5.0 \times 10^{-6} \ \text{F})(12 \ \text{V}) = \boxed{60 \ \mu\text{C}}$$

After one time constant, the charge on the capacitor is 63.2% of its maximum value:

$$q = 0.632Q = 0.632(60 \times 10^{-6} \ \text{C}) = \boxed{38 \ \mu\text{C}}$$

EXERCISE Find the charge on the capacitor and the voltage across the capacitor after time t has elapsed.

ANSWER $q = (60 \ \mu\text{C})(1 - e^{-t/4})$; $\Delta V = (12 \ \text{V})(1 - e^{-t/4})$

Example 18.7 Discharging a Capacitor in an *RC* Circuit

Consider a capacitor C being discharged through a resistor R as in Figure 18.17a. After how many time constants does the charge on the capacitor drop to one fourth of its initial value?

Solution The charge on the capacitor varies with time according to Equation 18.9:

$$q = Qe^{-t/RC}$$

where Q is the initial charge on the capacitor. To find the time it takes the charge q to drop to one fourth of its initial value, we substitute $q = Q/4$ into this expression and solve for t:

$$\tfrac{1}{4}Q = Qe^{-t/RC}$$

$$\tfrac{1}{4} = e^{-t/RC}$$

Taking logarithms of both sides, we find that

$$\ln(\tfrac{1}{4}) = -\ln 4 = -\frac{t}{RC}$$

$$t = RC \ln 4 = \boxed{1.39RC}$$

EXERCISE If $R = 8.0 \times 10^5$ Ω, $C = 5.0$ μF, and the initial voltage across the capacitor is 6.0 V, what is the voltage across the capacitor after time t has elapsed?

ANSWER $\Delta V = (6.0 \ \text{V})e^{-t/4}$

FIGURE 18.19 A wiring diagram for a household circuit. The resistances R_1 and R_2 represent appliances or other electrical devices that operate at an applied voltage of 120 V.

18.6 HOUSEHOLD CIRCUITS

Household circuits are a very practical application of some of the ideas presented in this chapter. In a typical installation, the utility company distributes electric power to individual houses with a pair of wires, or power lines. Electrical devices in a house are then connected in parallel to these lines, as shown in Figure 18.19.

The potential drop between the two wires is about 120 V. (These are actually alternating currents and voltages, but for the present discussion we shall assume that they are direct, constant currents and voltages.) One of the wires is connected to ground, and the other wire, sometimes called the "hot" wire, is at a potential of 120 V. A meter and a circuit breaker or fuse are connected in series with the wire entering the house, as indicated in Figure 18.19.

In modern homes, circuit breakers are used in place of fuses. When the current in a circuit exceeds some value (typically 20 A), the circuit breaker acts as a switch and opens the circuit. Figure 18.20 shows one design for a circuit breaker. Current passes through a bimetallic strip, the top of which bends to the left when excessive current heats it. If the strip bends far enough to the left, it settles into a groove in the spring-loaded metal bar. When this occurs, the bar drops enough to open the circuit at the contact point. The bar also flips a switch that indicates that the circuit breaker has interrupted the current. (After the overload is removed, the switch can be flipped back on.) Circuit breakers based on this design have the disadvantage that some time is required for the heating of the strip, and thus the circuit is not opened rapidly enough when it is overloaded. As a consequence, many circuit breakers are now designed to use electromagnets, which we shall discuss in Chapter 19.

The wire and circuit breaker are carefully selected to meet the current demands of a circuit. If the circuit is to carry currents as large as 30 A, a heavy-duty wire and appropriate circuit breaker must be used. Household circuits that are normally used to power lamps and small appliances often require only 20 A. Each circuit has its own circuit breaker to accommodate its maximum safe current.

As an example, consider a circuit that powers a toaster, a microwave oven, and a heater (represented by R_1, R_2, . . . in Fig. 18.19). We can calculate the current carried by each appliance using the equation $\mathcal{P} = I\,\Delta V$. The toaster, rated at 1 000 W, draws a current of $1\,000/120 = 8.33$ A. The microwave oven, rated at 800 W, draws a current of 6.67 A, and the heater, rated at 1 300 W, draws a current of 10.8 A. If the three appliances are operated simultaneously, they draw a total current of 25.8 A. Therefore, the breaker should be able to handle at least this much current, or else it will be tripped. As an alternative, one could operate the toaster and microwave oven on one 20-A circuit and the heater on a separate 20-A circuit.

Many heavy-duty appliances, such as electric ranges and clothes dryers, require 240 V to operate. The power company supplies this voltage by providing, in addition to a live wire that is 120 V above ground potential, a wire, also considered live, that is 120 V below ground potential (Fig. 18.21). Therefore, the potential drop across the two live wires is 240 V. An appliance operating from a 240-V line requires half the current of one operating from a 120-V line; therefore, smaller wires can be used in the higher-voltage circuit without becoming overheated.

▶ APPLICATION

FUSES AND CIRCUIT BREAKERS

FIGURE 18.20 A circuit breaker that uses a bimetallic strip for its operation.

Webnote 18.3

At home you probably have at least one light that is controlled by two switches. Flipping either switch changes the light from off to on, or from on to off. Discover how this works at *http://www.howstuffworks.com/three-way.htm*

FIGURE 18.21 Power connections for a 240-V appliance. *(b, George Semple)*

18.7 ELECTRICAL SAFETY

A person can be electrocuted by touching a live wire (which commonly is live because of a frayed cord and exposed conductors) while the person is in contact with ground. The ground contact might be made by touching a water pipe (which is normally at ground potential) or by standing on the ground with wet feet, because impure water is a good conductor. Obviously, such situations should be avoided at all costs.

Electric shock can result in fatal burns, or it can cause the muscles of vital organs, such as the heart, to malfunction. The degree of damage to the body depends on the magnitude of the current, the length of time it acts, and the part of the body through which it passes. Currents of 5 mA or less can cause a sensation of shock but ordinarily do little or no damage. If the current is larger than about 10 mA, the hand muscles contract and the person may be unable to let go of the live wire. If a current of about 100 mA passes through the body for just a few seconds, it can be fatal. Such large currents paralyze the respiratory muscles. In some cases, currents of about 1 A through the body produce serious burns.

APPLICATION

THIRD WIRE ON CONSUMER APPLIANCES

As an additional safety feature for consumers, electrical equipment manufacturers now use electrical cords that have a third wire, called a ground. To understand how this works, consider the drill being used in Figure 18.22. A two-wire device has one wire, called the "hot" wire, connected to the high-potential (120-V)

(a)

(b)

FIGURE 18.22   The "hot" wire, at 120 V, always includes a circuit breaker for safety. (a) When the drill is operated with two wires, the normal current path is from the "hot" wire through the motor connections and back to ground through the "neutral" wire. However, here the high-voltage wire has come in contact with the drill case so that the person holding the drill receives an electrical shock. (b) Shock can be prevented by a third wire running from the drill case to the ground.

side of the input power line, and the second wire is connected to ground (0 V). If the high-voltage wire comes in contact with the case of the drill (Fig. 18.22a), a "short circuit" occurs. In this undesirable circumstance, the pathway for the current is from the high-voltage wire through the person holding the drill and to Earth—a pathway that can kill. Protection is provided by a third wire, connected to the case of the drill (Fig. 18.22b). In this situation, if a short occurs, the path of least resistance for the current is from the high-voltage wire through the case and back to ground through the third wire. The resulting high current produced will trip a circuit breaker before the person is injured.

Special power outlets called ground-fault interrupters (GFIs) are now being used in kitchens, bathrooms, basements, and other hazardous areas of new homes. They are designed to protect people from electrical shock by sensing small currents—approximately 5 mA and greater—leaking to ground. When current above this level is detected, the device shuts off (interrupts) the current in less than a millisecond. Ground-fault interrupters will be discussed further in Chapter 19.

Webnote 18.4

Read more about the electrical safety features of three-prong cords at *http://www.howstuffworks.com/question110.htm*

18.8 CONDUCTION OF ELECTRICAL SIGNALS BY NEURONS[2]

The most remarkable use of electrical phenomena in living organisms is found in the nervous system of animals. Specialized cells in the body called **neurons** form a complex network that receives, processes, and transmits information from one part of the body to another. The center of this network is located in the brain, which has the ability to store and analyze information. Based on this information, the nervous system controls parts of the body.

The nervous system is very complex and consists of about 10^{10} interconnected neurons. Neurons are the basic units of the nervous system. Some aspects of the nervous system are well known. Over the past 45 years, the method of signal propagation through the nervous system has been established. The messages are voltage pulses called *action potentials* transmitted by neurons. When a neuron receives a strong enough stimulus, it produces repetitive voltage pulses that are actively propagated along its structure. The strength of the stimulus is conveyed by the number of pulses produced. When the pulses reach the end of the neuron, they activate either muscle cells or other neurons. It is interesting that there is a "firing threshold" for neurons. That is, action potentials will propagate along a neuron only if the stimulus is sufficiently strong.

Neurons can be divided into three classes: sensory neurons, motor neurons, and interneurons. The sensory neurons receive stimuli from sensory organs that monitor the external and internal environment of the body. Depending on their specialized functions, the sensory neurons convey messages about factors such as light, temperature, pressure, muscle tension, and odor to higher centers in the nervous system. The motor neurons carry messages that control the muscle cells. These messages are based on the information provided by the sensory neurons and by the brain. The interneurons transmit information from one neuron to another.

Each neuron consists of a cell body to which are attached input ends called **dendrites** and a long tail called the **axon,** which transmits the signal away from the cell (Fig. 18.23). The far end of the axon branches into nerve endings that transmit the signal across small gaps to other neurons or to muscle cells. A simple sensory-motor neuron circuit is shown in Figure 18.24. A stimulus from a muscle produces nerve impulses that travel to the spine. Here the signal is transmitted to a motor neuron, which in turn sends impulses to control the muscle. Figure 18.25 shows an electron microscope image of neurons in the brain.

---

[2] This section is based upon an essay by Paul Davidovits of Boston College.

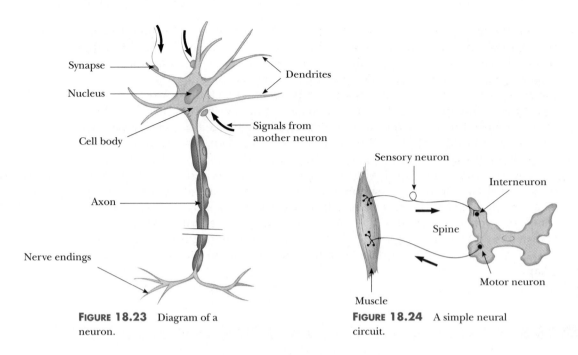

FIGURE 18.23 Diagram of a neuron.

FIGURE 18.24 A simple neural circuit.

FIGURE 18.25 Stellate neuron from human cortex. (*Juergen Berger, Max-Planck Institute/Science Photo Library/Photo Researchers, Inc.*)

The axon, which is an extension of the neuron cell, conducts the electric impulses away from the cell body. Some axons are extremely long. In humans, for example, the axons connecting the spine with the fingers and toes are more than 1 m long. The neuron can transmit messages because of the special active electrical characteristics of the axon. (The axon acts as an *active* source of energy like a battery rather than like a *passive* stretch of resistive wire.) Much of the information about the electrical and chemical properties of the axon is obtained by inserting small needle-like probes into the axon. Figure 18.26 shows an experimental setup. Note that the outside of the axon is grounded so that all measured voltages are with respect to a zero potential on the outside of the axon. With such probes it is possible to inject current into the axon, measure the resulting action potential as a function of time at a fixed point, and sample the cell's chemical composition. Such experiments are usually difficult to run because the diameter of most axons is very small. Even the largest axons in the human nervous system have a diameter of only about 20×10^{-4} cm. The giant squid, however, has an axon with a diameter of about 0.5 mm, which is large enough for the convenient insertion of probes. Much of the information about signal transmission in the nervous system has come from experiments with the squid axon.

In the aqueous environment of the body, salts and other molecules dissociate into positive and negative ions. As a result, body fluids are relatively good conductors of electricity. The inside of the axon is filled with an ionic fluid that is sepa-

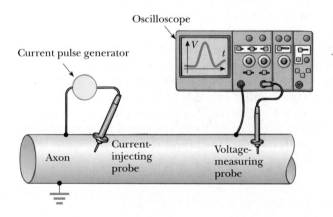

FIGURE 18.26 An axon stimulated electrically. The left probe injects a short current pulse and the right probe measures the resulting action potential as a function of time.

rated from the surrounding body fluid by a thin membrane that is only about 5 to 10 nm thick. The resistivities of the internal and external fluids are about the same, but their chemical compositions are substantially different. The external fluid is similar to sea water. Its ionic solutes are mostly positive sodium ions and negative chloride ions. Inside the axon, the positive ions are mostly potassium ions and the negative ions are mostly large organic ions.

Because there is a large concentration of sodium ions outside the axon and a large concentration of potassium ions inside, one might wonder why the concentrations are not equalized by diffusion. In other words, why don't the sodium ions leak into the axon and the potassium ions leak out of it? The answer lies in the properties of the axon membrane, which, as part of a living cell with an energy supply, can change its permeability on the time scale of milliseconds!

When the axon is not conducting an electric pulse, the axon membrane is highly permeable to potassium ions, slightly permeable to sodium ions, and impermeable to large organic ions. Thus, although sodium ions cannot easily enter the axon, potassium ions can leave it. As the positive potassium ions leave the axon, however, they leave behind large negative organic ions, which cannot follow them through the membrane. As a result, a negative potential builds up inside the axon with respect to the outside. The final negative potential reached, which has been measured at about -70 mV, holds back the outflow of potassium ions so that at equilibrium, the concentration of ions is as we have stated.

The mechanism for the production of an electric signal by the neuron is conceptually simple but was experimentally difficult to sort out. When a neuron changes its resting potential because of an appropriate stimulus, the properties of its membrane change locally. As a result, there is a sudden flow of sodium ions into the cell that lasts for about 2 ms. This produces the $+30$-mV peak in the action potential shown in Figure 18.27a. Immediately after, there is an increase in potassium ion flow out of the cell, which restores the resting action potential of -70 mV in an additional 3 ms. Both the Na^+ and K^+ ion flows have been

(a)

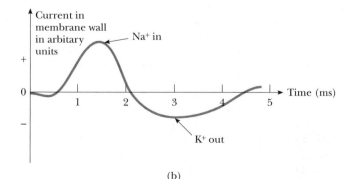

(b)

FIGURE 18.27 (a) Typical action potential as a function of time. (b) Current in the axon membrane wall as a function of time.

measured by using radioactive Na and K tracers. The nerve signal has been measured to travel along the axon at speeds of 50 m/s to about 150 m/s. This flow of charged particles (or signal transmission) in a nerve axon is *unlike* signal transmission in a metal wire. In an axon, charges move in a direction perpendicular to the direction of travel of the nerve signal, and the nerve signal moves much slower than a voltage pulse traveling along a metallic wire.

Although the axon is a complex structure, and much of how ion channels selectively pass Na^+ and K^+ is not understood,[3] standard electric circuit concepts of current and capacitance can be used to analyze axons. It is left as a problem to show that the axon, having equal and opposite charges separated by a thin dielectric membrane, acts like a capacitor (see Problem 18.41).

SUMMARY

A **source of emf** is any device that transforms nonelectrical energy into electrical energy.
The **equivalent resistance** of a set of resistors connected in **series** is

$$R_{eq} = R_1 + R_2 + R_3 + \cdots \qquad \text{[18.4]}$$

The **equivalent resistance** of a set of resistors connected in **parallel** is

$$\frac{1}{R_{eq}} = \frac{1}{R_1} + \frac{1}{R_2} + \frac{1}{R_3} + \cdots \qquad \text{[18.6]}$$

Complex circuits are conveniently analyzed by using **Kirchhoff's rules:**

1. The sum of the currents entering any junction must equal the sum of the currents leaving that junction.
2. The sum of the potential differences across all the elements around any closed circuit loop must be zero.

The first rule is a statement of **conservation of charge.** The second is a statement of **conservation of energy.**

As a capacitor is charged by a battery through a resistor, the charge increases from zero to some maximum value. The **time constant** $\tau = RC$ represents the time it takes the charge on the capacitor to increase from zero to 63% of its maximum value.

CONCEPTUAL QUESTIONS

1. Is the direction of current through a battery always from the negative terminal to the positive one? Explain.
2. Given three lightbulbs and a battery, sketch as many different circuits as you can.
3. If the energy transferred to a dead battery during charging is *W*, is the total energy transferred out of the battery to an external circuit during use in which it completely discharges also *W*?
4. How would you connect resistors so that the equivalent resistance is larger than the individual resistances? Give an example involving two or three resistors.
5. If you have your headlights on while you start your car, why do they dim while the car is starting?
6. How would you connect resistors so that the equivalent resistance is smaller than the individual resistances? Give an example involving two or three resistors.

7. Electrical devices are often rated with a voltage and a current—for example, 120 V, 5 A. Batteries, however, are only rated with a voltage—for example, 1.5 V. Why?
8. A "short circuit" is a circuit containing a path of very low resistance in parallel with some other part of the circuit. Discuss the effect of a short circuit on the portion of the circuit it parallels. Use a lamp with a frayed cord as an example.
9. Connecting batteries in series increases the emf applied to a circuit. What advantage might there be to connecting them in parallel?
10. If electrical power is transmitted over long distances, the resistance of the wires becomes significant. Why? Which mode of transmission would result in less energy loss—high current and low voltage or low current and high voltage? Discuss.

[3] For recent developments on this topic, see the November 1, 2001 issue of *Nature*.

11. In Figure Q18.11, describe what happens to the lightbulb after the switch is closed. Assume the capacitor has a large capacitance and is initially uncharged, and assume that the light illuminates when connected directly across the battery terminals.

Switch   Battery

FIGURE Q18.11

12. Two sets of Christmas tree lights are available. For set A, when one bulb is removed, the remaining bulbs remain illuminated. For set B, when one bulb is removed, the remaining bulbs do not operate. Explain the difference in wiring for the two sets.

13. Why is it possible for a bird to sit on a high-voltage wire without being electrocuted?

FIGURE Q18.13   Birds on a high-voltage wire. *(Superstock)*

14. Are the two headlights on a car wired in series or in parallel? How can you tell?

15. Embodied in Kirchhoff's rules are two conservation laws. What are they?

16. A ski resort consists of a few chairlifts and several interconnected downhill runs on the side of a mountain, with a lodge at the bottom. The lifts are analogous to batteries and the runs are analogous to resistors. Describe how two runs can be in series. Describe how three runs can be in parallel. Sketch a junction of one lift and two runs. One of the skiers is carrying an altimeter. State Kirchhoff's junction rule and Kirchhoff's loop rule for ski resorts.

17. Suppose you are flying a kite when it strikes a high-voltage wire (a very dangerous situation). What factors determine how great a shock you will receive?

18. Why is it dangerous to turn on a light when you are in a bathtub?

19. Suppose a parachutist lands on a high-voltage wire and grabs the wire as she prepares to be rescued. Will she be electrocuted? If the wire then breaks, should she continue to hold onto the wire as she falls to the ground?

20. Would a fuse or circuit breaker work successfully if it were placed in parallel with the device it was supposed to protect?

21. A series circuit consists of three identical lamps connected to a battery as in Figure Q18.21. When the switch S is closed, what happens (a) to the intensities of lamps A and B, (b) to the intensity of lamp C, (c) to the current in the circuit, and (d) to the voltage drop across the three lamps? (e) Does the power dissipated in the circuit increase, decrease, or remain the same?

A   B   C

ε S

FIGURE Q18.21

22. Figure Q18.22 shows a series connection of three lamps, all rated at 120 V, with power ratings of 60 W, 75 W, and 200 W. Why do the intensities of the lamps differ? Which lamp has the greatest resistance? How would their intensities differ if they were connected in parallel?

FIGURE Q18.22   *(Courtesy of Henry Leap and Jim Lehman)*

PROBLEMS

1, 2, 3 = straightforward, intermediate, challenging ☐ = full solution available in Student Solutions Manual/Study Guide

web = solution posted at **http://info.brookscole.com/serway** 🐾 = biomedical application

Section 18.1   Sources of emf

Section 18.2   Resistors in Series

Section 18.3   Resistors in Parallel

1. A battery having an emf of 9.00 V delivers 117 mA when connected to a 72.0-Ω load. Determine the internal resistance of the battery.

2. A 4.0-Ω resistor, an 8.0-Ω resistor, and a 12-Ω resistor are connected in series with a 24-V battery. What are (a) the equivalent resistance and (b) the current in each resistor? (c) Repeat for the case in which all three resistors are connected in parallel across the battery.

3. A lightbulb marked "75 W [at] 120 V" is screwed into a socket at one end of a long extension cord in which each of the two conductors has a resistance of 0.800 Ω. The other end of the extension cord is plugged into a 120-V outlet. Draw a circuit diagram, and find the actual power of the bulb in this circuit.

4. A 9.0-Ω resistor and a 6.0-Ω resistor are connected in series with a power supply. (a) The voltage drop across the 6.0-Ω resistor is measured to be 12 V. Find the voltage output of the power supply. (b) The two resistors are connected in parallel across a power supply, and the current through the 9.0-Ω resistor is found to be 0.25 A. Find the voltage setting of the power supply.

5. (a) Find the equivalent resistance between points *a* and *b* in Figure P18.5. (b) Calculate the current in each resistor if a potential difference of 34.0 V is applied between points *a* and *b*.

FIGURE P18.5

6. Find the equivalent resistance of the circuit in Figure P18.6.

FIGURE P18.6

7. Find the equivalent resistance of the circuit in Figure P18.7.

FIGURE P18.7

8. (a) Find the equivalent resistance of the circuit in Figure P18.8. (b) If the total power supplied to the circuit is 4.00 W, find the emf of the battery.

FIGURE P18.8

9. Consider the circuit shown in Figure P18.9. Find (a) the current in the 20.0-Ω resistor and (b) the potential difference between points *a* and *b*.

FIGURE P18.9

10. Two resistors, *A* and *B*, are connected in parallel across a 6.0-V battery. The current through *B* is found to be 2.0 A. When the two resistors are connected in series to the 6.0-V battery, a voltmeter connected across resistor *A* measures a voltage of 4.0 V. Find the resistances of *A* and *B*.

11. The resistance between terminals *a* and *b* in Figure P18.11 is 75 Ω. If the resistors labeled *R* have the same value, determine *R*.

12. Three 100-Ω resistors are connected as shown in Figure P18.12. The maximum power that can safely be delivered to any one resistor is 25.0 W. (a) What is the maximum voltage that can be applied to the terminals *a* and *b*?

FIGURE P18.11

(b) For the voltage determined in part (a), what is the power delivered to each resistor? What is the total power delivered?

FIGURE P18.12

13. Find the current in the 12-Ω resistor in Figure P18.13.
web

FIGURE P18.13

14. Calculate the power delivered to each resistor in the circuit shown in Figure P18.14.

FIGURE P18.14

15. (a) You need a 45-Ω resistor, but the stockroom has only 20-Ω and 50-Ω resistors. How can the desired resistance be achieved under these circumstances? (b) What can you do if you need a 35-Ω resistor?

Section 18.4 Kirchhoff's Rules and Complex DC Circuits

Note: The currents are not necessarily in the direction shown for some circuits.

16. The ammeter shown in Figure P18.16 reads 2.00 A. Find I_1, I_2, and $\mathcal{E}$.

FIGURE P18.16

17. Determine the current in each branch of the circuit shown in Figure P18.17.

FIGURE P18.17 (Problems 17 and 18)

18. In Figure P18.17, show how to add just enough ammeters to measure every different current in the circuit. Show how to add just enough voltmeters to measure the potential difference across each resistor and across each battery.

19. Figure P18.19 shows a circuit diagram. Determine (a) the current, (b) the potential of wire A relative to ground, and (c) the voltage drop across the 1 500-Ω resistor.

FIGURE P18.19

20. In the circuit of Figure P18.20, the current I_1 is 3.0 A and the values of $\mathcal{E}$ and R are unknown. What are the currents I_2 and I_3?

21. What is the emf $\mathcal{E}$ of the battery in the circuit of Figure P18.21?

22. Find the current in each of the three resistors of Figure P18.22 (a) by the rules for resistors in series and parallel and (b) by the use of Kirchhoff's rules.

FIGURE P18.20

FIGURE P18.21

FIGURE P18.22

23. (a) Using Kirchhoff's rules, find the current in each resistor shown in Figure P18.23 and (b) find the potential difference between points *c* and *f*.

FIGURE P18.23

24. Two 1.50-V batteries—with their positive terminals in the same direction—are inserted in series into the barrel of a flashlight. One battery has an internal resistance of 0.255 Ω, the other an internal resistance of 0.153 Ω. When the switch is closed, a current of 0.600 A passes through the lamp. (a) What is the lamp's resistance? (b) What fraction of the power dissipated is dissipated in the batteries?

25. Calculate each of the unknown currents I_1, I_2, and I_3 for the circuit of Figure P18.25.

FIGURE P18.25

26. A car with a dead battery is being given a jumpstart by connecting the live battery of another car to the dead battery with jumper cables (Fig. P18.26). Determine the current in the starter and in the dead battery.

FIGURE P18.26

27. Find the current in each resistor in Figure P18.27.

FIGURE P18.27

28. (a) Determine the potential difference ΔV_{ab} for the circuit in Figure P18.28. Note that each battery has an internal resistance as indicated in the figure. (b) If points *a* and *b* are connected by a 7.0-Ω resistor, what is the current through this resistor?

29. Find the potential difference across each resistor in Figure P18.29.

Section 18.5   RC Circuits

30. Show that $\tau = RC$ has units of time.

31. Consider a series *RC* circuit for which $C = 6.0$ μF, $R = 2.0 \times 10^6$ Ω, and $\mathcal{E} = 20$ V. Find (a) the time con-

$\mathcal{E} = 18$ V
$r = 2.0$ Ω
8.0 Ω

$\mathcal{E} = 12$ V 3.0 Ω
6.0 Ω $r = 5.0$ Ω
a b

$\mathcal{E} = 36$ V
$r = 4.0$ Ω
1.0 Ω

FIGURE P18.28

12.0 V   3.00 V   18.0 V

5.00 Ω   4.00 Ω   2.00 Ω

3.00 Ω

FIGURE P18.29

stant of the circuit and (b) the maximum charge on the capacitor after a switch in the circuit is closed.

32. An uncharged capacitor and a resistor are connected in series to a source of emf. If $\mathcal{E} = 9.00$ V, $C = 20.0$ μF, and $R = 100$ Ω, find (a) the time constant of the circuit, (b) the maximum charge on the capacitor, and (c) the charge on the capacitor after one time constant.

33. Consider a series RC circuit for which $R = 1.0$ MΩ, **web** $C = 5.0$ μF, and $\mathcal{E} = 30$ V. The capacitor is initially uncharged when the switch is open. Find the charge on the capacitor 10 s after the switch is closed.

34. A series combination of a 12-kΩ resistor and an unknown capacitor is connected to a 12-V battery. One second after the circuit is completed, the voltage across the capacitor is 10 V. Determine the capacitance.

35. A capacitor in an RC circuit is charged to 60.0% of its maximum value in 0.900 s. What is the time constant of the circuit?

36. A series RC circuit has a time constant of 0.960 s. The battery has an emf of 48.0 V, and the maximum current in the circuit is 0.500 mA. What are (a) the value of the capacitance and (b) the charge stored in the capacitor 1.92 s after the switch is closed?

Section 18.6   Household Circuits

37. An electric heater is rated at 1 300 W, a toaster is rated at 1 000 W, and an electric grill is rated at 1 500 W. The three appliances are connected in parallel to a common 120-V circuit. (a) How much current does each appliance draw? (b) Is a 30.0-A circuit breaker sufficient in this situation? Explain.

38. A lamp ($R = 150$ Ω), an electric heater ($R = 25$ Ω), and a fan ($R = 50$ Ω) are connected in parallel across a 120-V line. (a) What total current is supplied to the circuit?

(b) What is the voltage across the fan? (c) What is the current in the lamp? (d) What power is expended in the heater?

39. A heating element in a stove is designed to dissipate 3 000 W when connected to 240 V. (a) Assuming that the resistance is constant, calculate the current in this element if it is connected to 120 V. (b) Calculate the power it dissipates at this voltage.

40. Your toaster oven and coffeemaker each dissipate 1 200 W of power. Can you operate them together if the 120-V line that feeds them has a circuit breaker rated at 15 A? Explain.

Section 18.8   Conduction of Electrical Signals by Neurons

41. Assume that a length of axon membrane of about 10 cm is excited by an action potential. (Length excited = nerve speed × pulse duration = 50 m/s × 2.0 ms = 10 cm.) In the resting state, the outer surface of the axon wall is charged positively with K^+ ions and the inner wall has an equal and opposite charge of negative organic ions as shown in Figure P18.41. Model the axon as a parallel plate capacitor and use $C = \kappa \epsilon_0 A/d$ and $Q = C\Delta V$ to investigate the charge as follows. Use typical values for a cylindrical axon of cell wall thickness $d = 1.0 \times 10^{-8}$ m, axon radius $r = 10$ μm, and cell wall dielectric constant $\kappa = 3.0$. (a) Calculate the positive charge on the outside of a 10-cm piece of axon when it is not conducting an electric pulse. How many K^+ ions are on the outside of the axon? Is this a large charge per unit area? [*Hint:* Calculate the charge per unit area in terms of the number of square angstroms (Å^2) per electronic charge. An atom has a cross section of about 1 Å^2 (1 Å = 10^{-10} m).] (b) How much positive charge must flow through the cell membrane to reach the excited state of + 30 mV from the resting state of − 70 mV? How many sodium ions is this? (c) If it takes 2.0 ms for the Na^+ ions to enter the axon, what is the average current in the axon wall in this process? (d) How much energy does it take to raise the potential of the inner axon wall to + 30 mV starting from the resting potential of − 70 mV?

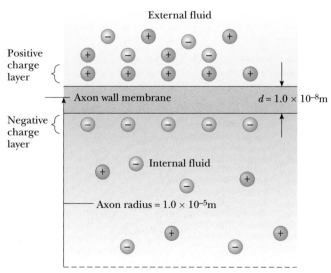

FIGURE P18.41   (Problems 41 and 42)

42. Continuing with the model of the axon as a capacitor from Problem 41 and Figure P18.41, (a) how much energy does it take to restore the inner wall of the axon to -70 mV starting from $+30$ mV? (b) Find the average current in the axon wall during this process, if it takes 3.0 ms.

43. Using Figure 18.27b and the results of Problem 18.41d and Problem 18.42a, find the average power supplied by the axon during firing and recovery.

ADDITIONAL PROBLEMS

44. Consider an *RC* circuit in which the capacitor is being charged by a battery connected in the circuit. After a time equal to two time constants, what percent of the *final* charge is present on the capacitor?

45. Find the equivalent resistance between points *a* and *b* in Figure P18.45.

FIGURE P18.45

46. For the circuit in Figure P18.46, calculate (a) the equivalent resistance of the circuit and (b) the power dissipated by the entire circuit. (c) Find the current in the 5.0-Ω resistor.

FIGURE P18.46

47. Find (a) the equivalent resistance of the circuit in Figure P18.47, (b) each current in the circuit, (c) the potential difference across each resistor, and (d) the power dissipated by each resistor.

FIGURE P18.47

48. Three 60.0-W, 120-V lightbulbs are connected across a 120-V power source, as shown in Figure P18.48. Find

(a) the total power delivered to the three bulbs and (b) the potential difference across each. Assume that the resistance of each bulb is constant (even though in reality the resistance increases markedly with current).

FIGURE P18.48

49. An automobile battery has an emf of 12.6 V and an internal resistance of 0.080 Ω. The headlights have total resistance of 5.00 Ω (assumed constant). What is the potential difference across the headlight bulbs (a) when they are the only load on the battery and (b) when the starter motor is operated, taking an additional 35.0 A from the battery?

50. In Figure P18.50, suppose that the switch has been closed for a length of time sufficiently long for the capacitor to become fully charged. (a) Find the steady-state current in each resistor and (b) find the charge on the capacitor.

FIGURE P18.50

51. Find the values of I_1, I_2, and I_3 for the circuit in Figure P18.51.

FIGURE P18.51

52. The resistance between points *a* and *b* in Figure P18.52 drops to one half its original value when switch S is closed. Determine the value of *R*.

53. A generator has a terminal voltage of 110 V when it delivers 10.0 A, and 106 V when it delivers 30.0 A. Calculate the emf and the internal resistance of the generator.

54. An emf of 10 V is connected to a series *RC* circuit consisting of a resistor of 2.0×10^6 Ω and a capacitor of 3.0 μF. Find the time required for the charge on the capacitor to reach 90% of its final value.

FIGURE P18.52

55. The student engineer of a campus radio station wishes to verify the effectiveness of the lightning rod on the antenna mast (Fig. P18.55). The unknown resistance R_x is between points C and E. Point E is a "true ground" but is inaccessible for direct measurement since this stratum is several meters below the Earth's surface. Two identical rods are driven into the ground at A and B, introducing an unknown resistance R_y. The procedure is as follows: measure resistance R_1 between points A and B, then connect A and B with a heavy conducting wire and measure resistance R_2 between points A and C. (a) Derive a formula for R_x in terms of the observable resistances R_1 and R_2. (b) A satisfactory ground resistance would be $R_x < 2.0$ Ω. Is the grounding of the station adequate if measurements give $R_1 = 13$ Ω and $R_2 = 6.0$ Ω?

FIGURE P18.55

56. The resistor R in Figure P18.56 dissipates 20 W of power. Determine the value of R.

FIGURE P18.56

57. A voltage ΔV is applied to a series configuration of n resis-
web tors, each of value R. The circuit components are recon-

nected in a parallel configuration, and voltage ΔV is again applied. Show that the power consumed by the series configuration is $1/n^2$ times the power consumed by the parallel configuration.

58. For the network in Figure P18.58, show that the resistance between points a and b is $R_{ab} = \frac{27}{17}$ Ω. (*Hint:* Connect a battery with emf $\mathcal{E}$ across points a and b and determine $\mathcal{E}/I$, where I is the current in the battery.)

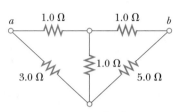

FIGURE P18.58

59. A battery with an internal resistance of 10.0 Ω produces an open-circuit voltage of 12.0 V. A variable load resistance with a range of 0 to 30.0 Ω is connected across the battery. (*Note:* A battery has a resistance that depends on the condition of its chemicals and increases as the battery ages. This internal resistance can be represented in a simple circuit diagram as a resistor in series with the battery.) (a) Graph the power dissipated in the load resistor as a function of the load resistance. (b) With your graph, demonstrate the following important theorem: *the power delivered to a load is a maximum if the load resistance equals the internal resistance of the source.*

60. The circuit in Figure P18.60 contains two resistors, $R_1 = 2.0$ kΩ and $R_2 = 3.0$ kΩ, and two capacitors, $C_1 = 2.0$ μF and $C_2 = 3.0$ μF, connected to a battery with emf $\mathcal{E} = 120$ V. If there are no charges on the capacitors before switch S is closed, determine as functions of time the charges q_1 and q_2 on capacitors C_1 and C_2, respectively, after the switch is closed. (*Hint:* First reconstruct the circuit so that it becomes a simple RC circuit containing a single resistor and single capacitor in series, connected to the battery, and then determine the total charge q stored in the circuit.)

FIGURE P18.60

GROUP ACTIVITIES

G.1 Connect one terminal of a D-cell battery to the base of a flashlight bulb using insulated wire, tape a second wire to the other battery terminal, and tape a third wire to the center conductor of the bulb as shown in the Figure

GA18.1a. Make sure to remove about 1 cm of insulation from the ends of all wires before making connections. Connect the two open wires together to complete the circuit, and note the illumination of the bulb. Now add a second D-cell battery to the circuit as in Figure GA18.1b to give a total voltage of 3.0 V, connect the two open wires together to complete the circuit, and note the illumination of the bulb. Why does the bulb grow brighter in this case?

1.5 V

Touch
wires

(a)

3.0 V

Touch
wires

(b)

FIGURE GA18.1

G.2 Use the basic equipment of activity 1 plus a few more items to test some additional features of circuits. First, note the brightness of a single bulb connected to the battery. Now connect two bulbs in series with each other and the battery. Predict whether the bulbs will be dimmer or brighter than when operated separately. Try it and see. Continue for three bulbs in series.

G.3 Repeat activity 2, but in this activity you must connect the bulbs in parallel with each other. Predict how the brightness will change as you add more bulbs in parallel. Why?

G.4 Continue your experimentation with circuits by connecting the battery to one bulb followed in the same circuit by two bulbs in parallel then back to the battery. Predict the brightness of each bulb in this situation before you connect the circuit. Explain the results you obtain.

G.5 (a) A group of resistors connected in parallel have the same (i) current in them, (ii) potential difference across them, or (iii) neither of the above. (b) A group of resistors connected in series have the same (i) current in them, (ii) potential difference across them, or (iii) neither of the above. (c) The equivalent resistance for a group of resistors connected in parallel is (i) greater than any of the resistors in the group, (ii) less than any of the resistors in the group, or (iii) neither of the above. (d) The equivalent resistance of a group of resistors connected in series is (i) greater than any of the resistors in the group, (ii) less than any of the resistors in

the group, or (iii) neither of the above. (e) Defend your answers by examining a circuit consisting of a 3.0-Ω resistor and a 5.0-Ω resistor connected across a 12-V battery. Be sure that the answers you find are consistent with your answers above.

G.6 (a) Two resistors are connected in series. The power converted by each resistor is (i) the same, or (ii) not necessarily the same. (b) Two resistors are connected in parallel. The power converted by each resistor is (i) the same, or (ii) not necessarily the same. (c) Suppose two resistors of 100 Ω and 200 Ω are connected to a 10.0-V power supply. Find the power converted by the two resistors in the configurations of parts (a) and (b). Are your answers consistent with your selections in parts (a) and (b)?

G.7 In this activity, select the correct answer from within the parentheses. (a) You traverse a 20-Ω resistor in a circuit such that you are moving in the same direction as a current of 2.0 A. In Kirchhoff's loop rule, the potential difference is (− 40 V, 40 V). (b) You traverse the same resistor moving against the current. The potential difference encountered is (− 40 V, 40 V). (c) You traverse a 12-V battery such that you are moving from the negative terminal to the positive terminal. In Kirchhoff's loop rule the potential difference is considered to be (12 V, − 12 V). (d) You traverse the same battery moving from the positive terminal to the negative terminal. The potential difference encountered is (12 V, − 12 V). (e) When working a Kirchhoff's rules problem, if you make an incorrect guess for the direction of the current in a branch, the final answer found for that current will be (positive, negative, meaningless).

G.8 Consider the two arrangements of batteries and bulbs shown in Figure GA18.8. All four bulbs are identical and have resistance R, and the two batteries are identical with output voltage ΔV. Answer the following questions and explain your answer. (a) In case 1, which bulb is brighter? (b) In case 1, what is the current in each bulb and the power supplied to each bulb? (c) In case 2, which bulb is brighter? (d) In case 2, what is the current in each bulb, the voltage drop across each bulb, and the power delivered to each bulb? (e) Which bulbs are brighter, those in case 1 or those in case 2? (f) If one bulb in each case fails, will the other go out as well? If the other bulb doesn't fail, will it get brighter or stay the same?

Case 1   Case 2

FIGURE GA18.8

G.9 In Figure GA18.9, some bulbs and batteries are connected together in different ways. The bulbs are all identical. For each case, give the letter that tells which of the bulbs will be brighter. If two or more will be equal in brightness, give the letters of all the bulbs that will be of the greatest brightness.

G.10 (a) All of the bulbs in Figure GA18.10a have the same resistance R. If bulb B is removed from the circuit, what happens to the current in bulb A, bulb D, and the battery? Indicate whether it increases, decreases or remains the same. For each case explain your reasoning. (b) A wire is added to the circuit as in Figure GA18.10b. What happens to the current in bulb A, bulb D, and the battery? Indicate whether it increases, decreases, or remains the same. For each case, explain your reasoning. (c) What is the equivalent resistance of the network in Figure GA18.10c? Again, all the bulbs have the same resistance R. If the current in the bulb on the right is I, what is the current in each bulb and in the battery?

(Problems 8, 9, and 10 are courtesy of E. F. Redish. For other problems of this type, visit http://www.physics.umd.edu/perg/)

Case 1   Case 2   Case 3

FIGURE GA18.9

(a)   (b)   (c)

FIGURE GA18.10

19

Magnetism

Chapter Outline

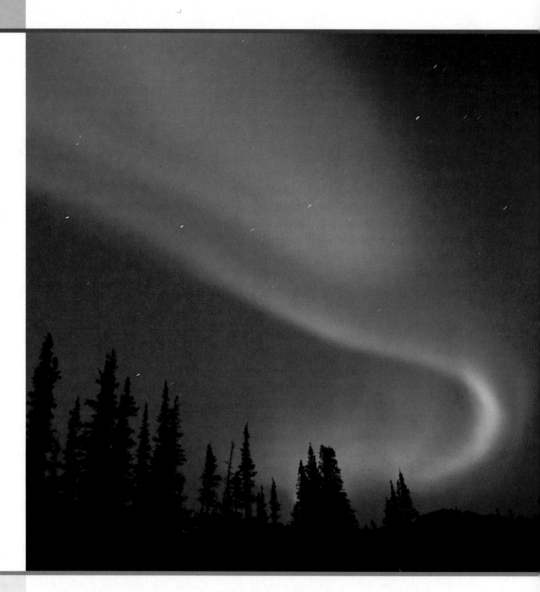

Aurora borealis, the Northern Lights. These displays are caused by cosmic ray particles trapped in the magnetic field of Earth. When the particles collide with atoms in the atmosphere, they cause the atoms to emit visible light. (*Johnny Johnson/Stone/Getty*)

586

The list of important technological applications of magnetism is very long. For instance, large electromagnets are used to pick up heavy loads. Magnets are used in such devices as meters, motors, and loudspeakers. Magnetic tapes are routinely used in sound and video recording equipment and for computer data storage. Intense magnetic fields are currently being used in magnetic resonance imaging devices (MRI) to explore the human body with better resolution and greater safety than x-rays.

As we investigate magnetism in this chapter, we shall find that this subject cannot be divorced from electricity. For example, magnetic fields affect moving charges and moving charges produce magnetic fields. The ultimate source of all magnetic fields is electric current.

19.1 MAGNETS

Most people have had experience with some form of magnet. You are most likely familiar with the common iron horseshoe magnet that can pick up iron-containing objects such as paper clips and nails. Several commercially available magnets are shown in Figure 19.1. In the discussion that follows, we shall assume that the magnet has the shape of a bar. Iron objects are most strongly attracted to the ends of such a bar magnet, called its **poles.** One end is called the **north pole** and the other the **south pole.** The names come from the behavior of a magnet in the presence of Earth's magnetic field. If a bar magnet is suspended from its midpoint by a piece of string so that it can swing freely in a horizontal plane, it will rotate until its north pole points to the north and its south pole points to the south. The same idea is used to construct a simple compass. Magnetic poles also exert attractive or repulsive forces on each other similar to the electrical forces between charged objects. In fact, simple experiments with two bar magnets show that **like poles repel each other and unlike poles attract each other.** Although the force between two magnetic poles is similar to the force between two electric charges, there is an important difference between electric and magnetic phenomena. Electric charges can be isolated (witness the proton and the electron), but magnetic poles cannot be isolated. In fact, no matter how many times a permanent magnet is cut, each piece always has a north pole and a south pole. Thus, magnetic poles always occur in north-south pairs. There is some theoretical basis for the speculation that magnetic monopoles (isolated north or south poles) may exist in nature, and attempts to detect them are currently an active experimental field of investigation. However, none of these attempts has yet proven successful.

Another similarity between electric and magnetic effects concerns methods for making magnets. In Chapter 15 we learned that when two materials such as rubber and wool are rubbed together, each becomes charged, one positively and the other negatively. In a somewhat analogous fashion, an unmagnetized piece of iron can be magnetized by stroking it with a magnet. Magnetism can also be induced in iron (and other materials) by other means. For example, if a piece of unmagnetized iron is placed near a strong permanent magnet, the piece of iron eventually becomes magnetized. The process can be accelerated by heating and cooling the iron. Naturally occurring magnetic materials, such as magnetite, achieve their magnetism in this manner, because they have been subjected to Earth's magnetic field over very long periods of time. The extent to which a piece of material retains its magnetism depends on whether it is classified as being magnetically hard or soft. **Soft** magnetic materials, such as iron, are easily magnetized but also tend to lose their magnetism easily. In contrast, **hard** magnetic materials such as cobalt and nickel are difficult to magnetize but tend to retain their magnetism.

FIGURE 19.1 An assortment of commercially available magnets. The four red magnets and the large black magnet on the left are made of an alloy of iron, aluminum, and cobalt. The six horseshoe magnets on the right are made of different nickel alloy steels. The rectangular magnets on the lower right are ceramics made of iron, nickel, and beryllium oxides. *(Courtesy of Central Scientific Company)*

(a)   (b)

FIGURE 19.2 (a) Tracing the magnetic field of a bar magnet. (b) Several magnetic field lines of a bar magnet.

In earlier chapters we described the interaction between charged objects in terms of electric fields. Recall that an electric field surrounds any stationary electric charge. The region of space surrounding a *moving* charge also includes a magnetic field. A magnetic field also surrounds any magnetic material.

To describe any type of vector field, we must define its magnitude, or strength, and its direction. The direction of the magnetic field **B** at any location is the direction in which the north pole of a compass needle points at that location. Figure 19.2a shows how the magnetic field of a bar magnet can be traced with the aid of a compass, defining a **magnetic field line.** Several magnetic field lines of a bar magnet traced out in this manner appear in the two-dimensional representation in Figure 19.2b. Magnetic field patterns can be displayed by small iron filings placed in the vicinity of a magnet, as in Figure 19.3.

Forensic scientists use a technique similar to that shown in Figure 19.3 to find fingerprints at a crime scene. One way to find latent, or invisible, prints is by sprinkling a powder of iron dust on a surface. The iron adheres to perspiration or body oils present and can be spread around on the surface with a magnetic brush that never comes into contact with the powder or the surface.

APPLICATION

DUSTING FOR FINGERPRINTS

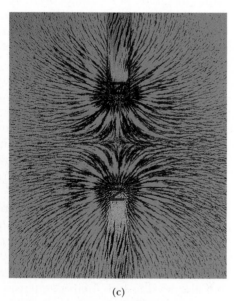

(a)   (b)   (c)

FIGURE 19.3 (a) The magnetic field pattern of a bar magnet, displayed by iron filings on a sheet of paper. (b) The magnetic field pattern between *unlike* poles of two bar magnets, displayed by iron filings. (c) The magnetic field pattern between two *like* poles. *(Courtesy of Henry Leap and Jim Lehman)*

19.2 EARTH'S MAGNETIC FIELD

When we speak of a small bar magnet as having north and south poles, we should more properly say that it has a "north-seeking" pole and a "south-seeking" pole. By this we mean that if such a magnet is used as a compass, one end will seek, or point to, the geographic North Pole of the Earth. Thus, we conclude that **Earth's geographic North Pole corresponds to a magnetic south pole, and its geographic South Pole corresponds to a magnetic north pole.** In fact, the configuration of Earth's magnetic field, pictured in Figure 19.4, very much resembles what would be achieved by burying a huge bar magnet deep in Earth's interior.

If a compass needle is suspended in bearings that allow it to rotate in the vertical plane as well as in the horizontal plane, the needle is horizontal with respect to Earth's surface only near the Equator. As the device is moved northward, the needle rotates so that it points more and more toward Earth's surface. The angle between the direction of the magnetic field and the horizontal is called the **dip angle.** Finally, at a point just north of Hudson Bay in Canada, the north pole of the needle points directly downward, and the dip angle is 90°. This site, first found in 1832, is considered to be the location of Earth's south magnetic pole. It is approximately 1 300 mi from Earth's geographic North Pole and its position varies with time. Similarly, Earth's magnetic north pole is about 1 200 miles from its geographic South Pole. Thus, it is only approximately correct to say that a compass needle points north. The difference between true north, defined as the geographic North Pole, and north indicated by a compass varies from point to point on Earth, and the difference is referred to as *magnetic declination.* For example, along a line through South Carolina and the Great Lakes, a compass indicates true north, whereas in Washington state it aligns 25° east of true north (Fig. 19.5).

Although Earth's magnetic field pattern is similar to the pattern that would be set up by a bar magnet placed at its center, we can easily explain why the source of Earth's field cannot be large masses of permanently magnetized material. Earth does have large deposits of iron ore deep beneath its surface, but the high temperatures in its core prevent the iron from retaining any permanent magnetization. The most likely source of Earth's magnetic field is believed to be electric current in the liquid part of its core. This current is not well understood but is thought to be driven by an interaction between Earth's rotation and convective motion in the hot liquid core. There is some evidence that the strength of a planet's magnetic field is related to the planet's rate of rotation. For example, Jupiter rotates faster

Tip 19.1 EARTH'S NORTH POLE IS A MAGNETIC SOUTH POLE

Don't be confused because the north pole of a magnet points to the geographic North Pole of Earth. Earth's geographic North Pole is really a magnetic south pole.

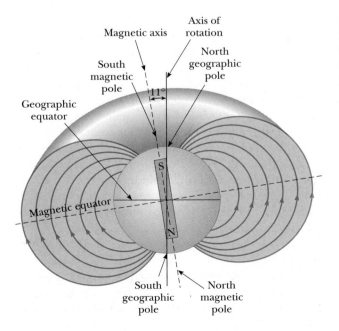

FIGURE 19.4 Earth's magnetic field lines. Note that magnetic south is at the north geographic pole, and magnetic north is at the south geographic pole.

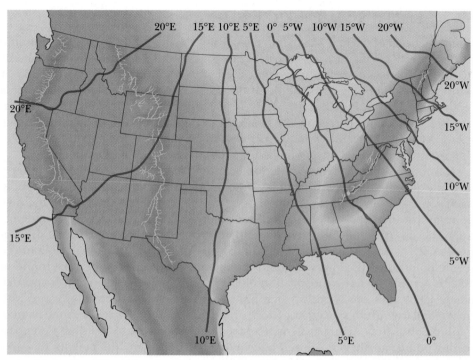

FIGURE 19.5 A map of the lower forty-eight United States showing the declination of a compass from true north.

Webnote 19.1

Bacteria that have their own built-in directional magnetic sensors are described as magnetotactic. Learn more about these creatures that existed long before humans navigated Earth with compasses.
http://www.wpi.edu/~ppham/bacteria. html
http://www.calpoly.edu/~rfrankel/ mtbcalpoly.html
Tips for collecting your own sample of magnetotactic bacteria can be found at
http://www.home.duq.edu/~stolz/ magbachunt.html

APPLICATION

MAGNETIC BACTERIA

APPLICATION

LABELING AIRPORT RUNWAYS

than Earth, and recent space probes indicate that Jupiter's magnetic field is stronger than ours even though Jupiter lacks an iron core. Venus, on the other hand, rotates more slowly than Earth, and its magnetic field is found to be weaker. Investigation into the cause of Earth's magnetism continues.

An interesting sidelight concerning Earth's magnetic field is that the field direction reverses every few million years. Evidence for this is provided by basalt (an iron-containing rock) that is sometimes spewed forth by volcanic activity on the ocean floor. As the lava cools, it solidifies and retains a picture of Earth's magnetic field direction. When the basalt deposits are dated, they provide evidence for periodic reversals of the magnetic field. The origin of these field reversals is still not understood.

It has long been speculated that some animals, such as birds, use Earth's magnetic field to guide their migrations. Studies have shown that a type of anaerobic bacterium that lives in swamps has a magnetized chain of magnetite as part of its internal structure. (The term *anaerobic* means that these bacteria live and grow without oxygen; in fact, oxygen is toxic to them.) The magnetized chain acts as a compass needle that enables the bacteria to align with Earth's magnetic field. When they find themselves out of the mud on the bottom of the swamp, they return to their oxygen-free environment by following the magnetic field lines. Further evidence for their magnetic sensing ability is the fact that bacteria found in the Northern Hemisphere have internal magnetite chains that are opposite in polarity to those of similar bacteria in the Southern Hemisphere. This is consistent with the fact that in the Northern Hemisphere Earth's field has a downward component, whereas in the Southern Hemisphere it has an upward component. Recently, a meteorite originating on Mars has been found to contain a chain of magnetite. NASA scientists believe it may be a fossil of ancient Martian bacterial life.

Earth's magnetic field is used to label runways at airports according to their direction. A large number is painted on the end of the runway so that it can be read by the pilot of an incoming airplane. This number describes the direction in which the airplane is traveling, expressed as the magnetic heading, in degrees measured clockwise from magnetic north divided by 10. Thus, a runway marked 9 would be directed toward the east (90° divided by 10), and one marked 18 would be directed toward the magnetic south.

APPLYING **PHYSICS** *19.1*

On a business trip to Australia, you take along your American-made compass that you use on camping trips. Does this compass work correctly in Australia?

Explanation You will have no problem using the compass in Australia. The north pole of the magnet in the compass is attracted to the south magnetic pole near the geographic North Pole, just as it is in the United States. The only difference in the magnetic field lines is that they have an upward component in Australia, whereas they have a downward component in the United States. Your compass held in a horizontal plane cannot detect this, however—it only displays the direction of the horizontal component of the magnetic field.

19.3 MAGNETIC FIELDS

Experiments show that a stationary charged particle does not interact with a static magnetic field. However, **when moving through a magnetic field, a charged particle experiences a magnetic force.** This force has its maximum value when the charge moves perpendicularly to the magnetic field lines, decreases in value at other angles, and becomes zero when the particle moves along the field lines. We can use these observations to describe the magnetic field.

In our discussion of electricity, we defined the electric field at some point in space as the electric force per unit charge acting on some test charge placed at that point. In a similar manner, we can describe the properties of the magnetic field **B** at some point in terms of the magnetic force exerted on a test charge at that point. Our test object is a charge q moving with velocity **v**. It is found experimentally that the strength of the magnetic force on the particle is proportional to the magnitude of the charge q, the magnitude of the velocity **v**, the strength of the external magnetic field **B**, and the sine of the angle θ between the direction of **v** and the direction of **B**.

These observations can be summarized by writing the magnitude of the magnetic force as

$$F = qvB \sin \theta \qquad [19.1]$$

This expression is used to define the magnitude of the magnetic field as

$$B \equiv \frac{F}{qv \sin \theta} \qquad [19.2]$$

If F is in newtons, q in coulombs, and v in meters per second, the SI unit of magnetic field is the **tesla** (T), also called the **weber** (Wb) **per square meter** (that is, $1\text{ T} = 1\text{ Wb/m}^2$). Thus, if a 1-C charge moves through a magnetic field of magnitude 1 T with a speed of 1 m/s perpendicularly to the field (so that $\sin \theta = 1$), the magnetic force exerted on the charge is 1 N. We can express the units of **B** as

$$[B] = T = \frac{Wb}{m^2} = \frac{N}{C \cdot m/s} = \frac{N}{A \cdot m} \qquad [19.3]$$

In practice, the cgs unit for magnetic field, the **gauss** (G), is often used. The gauss is related to the tesla through the conversion

$$1\text{ T} = 10^4\text{ G}$$

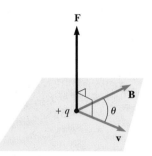

FIGURE 19.6   The direction of the magnetic force on a positively charged particle moving with a velocity **v** in the presence of a magnetic field. When **v** is at an angle θ with **B**, the magnetic force is perpendicular to both **v** and **B**.

FIGURE 19.7   Right-hand rule #1 for determining the direction of the magnetic force on a positive charge moving with a velocity **v** in a magnetic field **B**. With your thumb in the direction of **v** and your four fingers in the direction of **B**, the force is directed out of the palm of your hand.

Conventional laboratory magnets can produce magnetic fields as large as about 25 000 G, or 2.5 T. Superconducting magnets that can generate magnetic fields as great as 3×10^5 G, or 30 T, have been constructed. These values can be compared with the value of Earth's magnetic field near its surface, which is about 0.5 G, or 0.5×10^{-4} T.

From Equation 19.1 we see that the force on a charged particle moving in a magnetic field has its maximum value when the particle moves *perpendicularly* to the magnetic field, corresponding to $\theta = 90°$, so that $\sin \theta = 1$. The magnitude of this maximum force has the value

$$F_{max} = qvB \qquad \text{[19.4]}$$

Also, note from Equation 19.1 that F is zero when **v** is parallel to **B** (corresponding to $\theta = 0°$ or $180°$). Thus, no magnetic force is exerted on a charged particle when it moves in the direction of the magnetic field or opposite the field.

Experiments show that the direction of the magnetic force is always perpendicular to both **v** and **B**, as shown in Figure 19.6 for a positively charged particle. To determine the direction of the force, we employ **right-hand rule #1:**

> **Hold your right hand open as illustrated in Figure 19.7, and then place your fingers in the direction of B with your thumb pointing in the direction of v. The force F on a positive charge is directed *out* of the palm of your hand.**

If the charge is negative rather than positive, the force is directed *opposite* that shown in Figures 19.6 and 19.7. That is, if q is negative, simply use the right-hand rule to find the direction of **F** for positive q, and then reverse this direction for the negative charge.

Quick Quiz 19.1	A charged particle moves in a straight line through a certain region of space. The magnetic field in that region (a) has a magnitude of zero, (b) has a zero component perpendicular to the particle's velocity, and (c) has a zero component parallel to the particle's velocity.
Quick Quiz 19.2	The north-pole end of a bar magnet is held near a stationary positively charged piece of plastic. Is the plastic (a) attracted, (b) repelled, or (c) unaffected by the magnet?

Example 19.1   A Proton Traveling in Earth's Magnetic Field

A proton moves with a speed of 1.0×10^5 m/s through Earth's magnetic field, which has a value of 55 μT at a particular location. When the proton moves eastward, the magnetic force acting on it is a maximum, and when it moves northward, no magnetic force acts on it. What is the strength of the magnetic force, and what is the direction of the magnetic field?

Solution   The magnitude of the force can be found from Equation 19.4:

$$F_{max} = qvB = (1.6 \times 10^{-19} \text{ C})(1.0 \times 10^5 \text{ m/s})(55 \times 10^{-6} \text{ T})$$
$$= 8.8 \times 10^{-19} \text{ N}$$

The direction of the magnetic field cannot be determined precisely from the information given in the problem. Because no magnetic force acts on a charged particle when it is moving parallel to the field, all that we can say for sure is that the magnetic field is directed either northward or southward.

EXERCISE   Calculate the gravitational force on the proton and compare it with the magnetic force. Note that the mass of the proton is 1.67×10^{-27} kg.

ANSWER 1.6×10^{-26} N; $F_{grav}/F_{max} = 1.9 \times 10^{-8}$

Example 19.2   A Proton Moving in a Magnetic Field

A proton moves at 8.0×10^6 m/s along the x axis. It enters a region in which there is a magnetic field of magnitude 2.5 T, directed at an angle of 60° with the x axis and lying in the xy plane (Fig. 19.8). Calculate the initial force on and acceleration of the proton.

Solution   From Equation 19.1 we get

$$F = qvB \sin \theta = (1.6 \times 10^{-19} \text{ C})(8.0 \times 10^6 \text{ m/s})(2.5 \text{ T})(\sin 60°)$$
$$= \boxed{2.8 \times 10^{-12} \text{ N}}$$

Use right-hand rule #1, noting that the charge is positive, to see that the force is in the positive z direction. Verify that the units of F in the calculation reduce to newtons.

Because the mass of the proton is 1.67×10^{-27} kg, its initial acceleration is

$$a = \frac{F}{m} = \frac{2.8 \times 10^{-12} \text{ N}}{1.67 \times 10^{-27} \text{ kg}} = \boxed{1.7 \times 10^{15} \text{ m/s}^2}$$

in the positive z direction.

EXERCISE   Calculate the acceleration of an electron that moves through the same magnetic field at the same speed as the proton. The mass of an electron is 9.11×10^{-31} kg.

ANSWER 3.0×10^{18} m/s² in the negative z direction

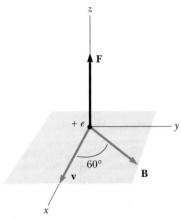

FIGURE 19.8   (Example 19.2) The magnetic force **F** on a proton is in the positive z direction when **v** and **B** lie in the xy plane.

19.4   MAGNETIC FORCE ON A CURRENT-CARRYING CONDUCTOR

If a force is exerted on a single charged particle when it moves through a magnetic field, it should be no surprise that a current-carrying wire also experiences a force when placed in a magnetic field (see Figure 19.9). This follows from the fact that the current is a collection of many charged particles in motion; hence, the resultant force on the wire is due to the sum of the individual forces on the charged particles. The force on the particles is transmitted to the "bulk" of the wire through collisions with the atoms making up the wire.

Before we continue, some explanation is in order concerning notation in many of the figures. To indicate the direction of **B**, we use the following convention.

> **If B is directed *into* the page, as in Figure 19.10, we use a series of blue crosses, representing the tails of arrows. If B is directed *out* of the page, we use a series of blue dots, representing the tips of arrows. If B lies in the plane of the page, we use a series of blue field lines with arrowheads.**

The force on a current-carrying conductor can be demonstrated by hanging a wire between the poles of a magnet, as in Figure 19.10. In this figure, the magnetic

FIGURE 19.9   This apparatus demonstrates the force on a current-carrying conductor in an external magnetic field. Why does the bar swing *away* from the magnet after the switch is closed? *(Courtesy of Henry Leap and Jim Lehman)*

FIGURE 19.10   A segment of a flexible vertical wire partially stretched between the poles of a magnet, with the field (blue crosses) directed into the page. (a) When there is no current in the wire, it remains vertical. (b) When the current is upward, the wire deflects to the left. (c) When the current is downward, the wire deflects to the right.

(a)   (b)   (c)

FIGURE 19.11   A section of a wire containing moving charges in an external magnetic field **B**.

ip 19.2   THE ORIGIN OF THE MAGNETIC FORCE ON A WIRE

When a magnetic field is applied at some angle to a wire carrying a current, a magnetic force is exerted on each moving charge in the wire. The total magnetic force on the wire is the sum of all the magnetic forces on the individual charges producing the current.

FIGURE 19.12   A wire carrying a current I in the presence of an external magnetic field **B** that makes an angle θ with the wire.

APPLICATION

LOUDSPEAKER OPERATION

field is directed into the page and covers the region within the shaded circle. The wire deflects to the right or left when a current is passed through it.

Let us quantify this discussion by considering a straight segment of wire of length ℓ and cross-sectional area A carrying current I in a uniform external magnetic field **B**, as in Figure 19.11. We assume that the magnetic field is perpendicular to the wire and is directed into the page. Each charge carrier in the wire experiences a force of magnitude $F_{max} = qv_dB$, where v_d is the drift velocity of the charge. To find the total force on the wire, we multiply the force on one charge carrier by the number of carriers in the segment. Because the volume of the segment is $A\ell$, the number of carriers is $nA\ell$, where n is the number of carriers per unit volume. Hence, the magnitude of the total magnetic force on the wire of length ℓ is

Total force = force on each charge carrier × total number of carriers

$$F_{max} = (qv_dB)(nA\ell)$$

From Chapter 17, however, we know that the current in the wire is given by $I = nqv_dA$. Therefore, F_{max} can be expressed as

$$F_{max} = BI\ell \qquad \text{[19.5]}$$

This equation can be used *only* when the current and the magnetic field are at right angles to each other.

If the wire is not perpendicular to the field but is at some arbitrary angle, as in Figure 19.12, the magnitude of the magnetic force on the wire is

$$F = BI\ell \sin \theta \qquad \text{[19.6]}$$

where θ is the angle between **B** and the direction of the current. The direction of this force can be obtained by using right-hand rule #1. However, in this case you must place your thumb in the direction of the current rather than in the direction of **v**. In Figure 19.12, the direction of the magnetic force on the wire is out of the page.

Finally, when the current is either in the direction of the field or opposite the direction of the field, the magnetic force on the wire is zero.

The fact that a magnetic force acts on a current-carrying wire in a magnetic field is the operating principle of most speakers in sound systems. One speaker design, shown in Figure 19.13, consists of a coil of wire (called the voice coil), a flexible paper cone that acts as the speaker, and a permanent magnet. The coil of wire surrounding the north pole of the magnet is shaped so that the magnetic field lines are directed radially outward from the coil's axis. When an electrical signal is sent to the coil, producing a current in the coil as in Figure 19.13, a magnetic force to the left acts on the coil. (This can be seen by applying right-hand rule #1 to each turn of wire.) When the current reverses direction, as it would for a sinusoidally varying current, the magnetic force on the coil also reverses direction, and the cone, which is attached to the coil, accelerates to the right. An alternating current through the coil causes an alternating force on the coil, which results in vibrations of the cone. The vibrating cone creates sound waves as it pushes and pulls on the air in front of it. In this way a 1-kHz electrical signal is converted to a 1-kHz sound wave.

An unusual application of the force on a current-carrying conductor is illustrated by the electromagnetic pump shown in Figure 19.14. Artificial hearts require a pump to keep the blood flowing, and kidney dialysis machines also require a pump to assist the heart in pumping blood that is to be cleansed. Ordinary mechanical pumps create difficulties because they damage the blood cells as cells move through the pump. The mechanism shown in Figure 19.14 has demonstrated some promise in such applications. A magnetic field is established across a segment of the tube containing the blood, flowing in the direction shown by the velocity **v**. An electric current passing through the fluid perpendicular to the blood velocity as shown has a magnetic force acting on it in the direction of **v**, as application of right-hand rule #1 shows. (Place your right thumb in the direction of the *current!*) This force of magnetic origin helps to keep the blood in motion.

FIGURE 19.13 Diagram of a loudspeaker.

APPLICATION

ELECTROMAGNETIC PUMPS FOR ARTIFICIAL HEARTS AND KIDNEYS

FIGURE 19.14 A simple electromagnetic pump has no moving parts to damage a conducting fluid, such as blood, passing through. Application of right-hand rule #1 (fingers along **B**, thumb along *I* in this case) shows that the force on the current-carrying segment of the fluid is in the direction of the velocity.

APPLYING PHYSICS 19.2

In a lightning strike, there is a rapid movement of negative charge from a cloud to the ground. In what direction is a lightning strike deflected by Earth's magnetic field?

Explanation The downward flow of negative charge in a lightning stroke is equivalent to an upward-moving current. Thus, we have an upward-moving current in a northward-directed magnetic field. According to right-hand rule #1, the lightning strike deflects toward the west.

Example 19.3 A Current-Carrying Wire in Earth's Magnetic Field

A wire carries a current of 22 A from east to west. Assume that at this location Earth's magnetic field is horizontal and directed from south to north, and that it has a magnitude of 0.50×10^{-4} T. Find the magnetic force on a 36-m length of wire. How does the force change if the current runs west to east?

Solution Because the directions of the current and magnetic field are at right angles, we can use Equation 19.5. The magnitude of the magnetic force is

$$F_{max} = BI\ell = (0.50 \times 10^{-4}\ \text{T})(22\ \text{A})(36\ \text{m}) = \boxed{4.0 \times 10^{-2}\ \text{N}}$$

Right-hand rule #1 shows that the force on the wire is directed toward Earth.

If the current is directed from west to east, the force has the same magnitude but its direction is upward, away from Earth.

EXERCISE If the current is directed north to south, what is the magnetic force on the wire?

ANSWER Zero

19.5 TORQUE ON A CURRENT LOOP AND ELECTRIC MOTORS

In the preceding section we showed how a force is exerted on a current-carrying conductor when the conductor is placed in an external magnetic field. With this as a starting point, we now show that a torque is exerted on a current loop placed in

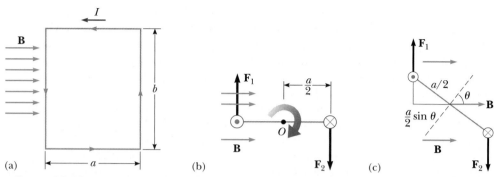

FIGURE 19.15 (a) Top view of a rectangular loop in a uniform magnetic field **B**. No magnetic forces act on the sides of length *a* parallel to **B**, but forces do act on the sides of length *b*. (b) Side view of the rectangular loop shows that the forces **F**$_1$ and **F**$_2$ on the sides of length *b* create a torque that tends to twist the loop clockwise. (c) If **B** is at an angle θ with a line perpendicular to the plane of the loop, the torque is given by *BIA* sin θ.

a magnetic field. The results of this analysis will be of great practical value when we discuss generators and motors, here and in Chapter 20.

Consider a rectangular loop carrying current *I* in the presence of an external uniform magnetic field in the plane of the loop, as shown in Figure 19.15a. The forces on the sides of length *a* are zero because these wires are parallel to the field. The magnitude of the magnetic forces on the sides of length *b*, however, is

$$F_1 = F_2 = BIb$$

The direction of **F**$_1$, the force on the left side of the loop, is out of the page, and that of **F**$_2$, the force on the right side of the loop, is into the page. If we view the loop from the side, as in Figure 19.15b, the forces are directed as shown. If we assume that the loop is pivoted so that it can rotate about point *O*, we see that these two forces produce a torque about *O* that rotates the loop clockwise. The magnitude τ_{max} of this torque is

$$\tau_{\text{max}} = F_1 \frac{a}{2} + F_2 \frac{a}{2} = (BIb) \frac{a}{2} + (BIb) \frac{a}{2} = BIab$$

where the moment arm about *O* is *a/2* for both forces. Because the area of the loop is *A = ab*, the torque can be expressed as

$$\tau_{\text{max}} = BIA \qquad \text{[19.7]}$$

Note that this result is valid only when the magnetic field is parallel to the plane of the loop, as in the view in Figure 19.15b. If the field makes an angle θ with a line perpendicular to the plane of the loop, as in Figure 19.15c, the moment arm for each force is given by $(a/2) \sin \theta$. An analysis such as the one just used produces, for the magnitude of the torque,

$$\tau = BIA \sin \theta \qquad \text{[19.8]}$$

This result shows that the torque has the *maximum* value *BIA* when the field is parallel to the plane of the loop ($\theta = 90°$) and is *zero* when the field is perpendicular to the plane of the loop ($\theta = 0$). As seen in Figure 19.15c, the loop tends to rotate to smaller values of θ (so that the normal to the plane of the loop rotates toward the direction of the magnetic field).

Although this analysis is for a rectangular loop, a more general derivation would indicate that Equation 19.8 applies regardless of the shape of the loop. Furthermore, the torque on a coil with *N* turns is

$$\tau = NBIA \sin \theta \qquad \text{[19.9]}$$

FIGURE 19.16 (Example 19.4) An edge view of a circular current loop in an external magnetic field **B**.

Example 19.4 The Torque on a Circular Loop in a Magnetic Field

A circular wire loop of radius 50.0 cm is placed in a magnetic field of magnitude 0.50 T. The normal to the plane of the loop makes an angle of 30.0° with the magnetic field as shown in an edge view in Figure 19.16. The current in the loop is 2.0 A in the direction shown. Find the magnitude of the torque at this instant.

Solution Regardless of the shape of the loop, Equation 19.8 is valid:

$$\tau = BIA \sin \theta = (0.50 \text{ T})(2.0 \text{ A})[\pi(0.50 \text{ m})^2](\sin 30.0°) = \boxed{0.39 \text{ N} \cdot \text{m}}$$

EXERCISE Find the torque on the loop if it has three turns rather than one.

ANSWER The torque is 3 times that on the one-turn loop, or 1.2 N·m.

ELECTRIC MOTORS

It is difficult to imagine life in the 21st century without electric motors. Among the numerous appliances that contain motors are computer disk drives, CD players, VCR and DVD players, food processors and blenders, car starters, furnaces, and air conditioners. The motors convert electrical energy to kinetic energy of rotation, and consist basically of a rigid current-carrying loop that rotates when placed in the field of a magnet.

As we have just seen (Figure 19.15), the torque on such a loop rotates the loop to smaller values of θ until the torque becomes zero when the magnetic field is perpendicular to the plane of the loop and θ = 0. If the loop turns past this angle, and the current remains in the direction shown in Figure 19.15, the torque reverses direction and turns the loop in the opposite direction, that is, counterclockwise. To overcome this difficulty and provide continuous rotation in one direction, the current in the loop must periodically reverse direction. In alternating current (ac) motors, this occurs naturally as the current reverses direction 120 times each second. In direct current (dc) motors, the current reversal is accomplished mechanically with split-ring contacts (commutators) and brushes as shown in Figure 19.17.

Although actual motors contain many current loops and commutators, for simplicity Figure 19.17 shows only a single loop and a single set of split-ring

> **APPLICATION**
>
> ELECTRIC MOTORS

FIGURE 19.17 Simplified sketch of a dc electric motor.

FIGURE 19.18 The Honda Insight combines a three-cylinder gasoline automobile engine with a thin electric motor for improved efficiency and added power when needed. The electric motor (circled) also acts as a generator during braking or coasting downhill to recharge the batteries, with the result that they never need to be recharged by the owner. *(Courtesy of America Honda Motor Co., Inc.)*

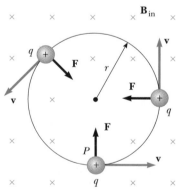

The bending of an electron beam in an external magnetic field. The tube contains gas at very low pressure, and the beam is made visible as the electrons collide with the gas atoms, which in turn emit visible light. The beam is projected to the right and deflects downward in the presence of a magnetic field produced by a pair of current-carrying coils. What is the direction of the magnetic field? *(Courtesy of Central Scientific Company)*

contacts rigidly attached to and rotating with the loop. A set of electrical stationary contacts called *brushes* are maintained in electrical contact with the rotating split ring. These brushes are usually made of graphite because graphite is a good electrical conductor as well as a good lubricant. Just as the loop becomes perpendicular to the magnetic field and the torque becomes zero, inertia carries the loop forward in the clockwise direction and the brushes cross the gaps in the ring, causing the loop current to reverse its direction. This provides another pulse of torque in the clockwise direction for another 180°, the current reverses, and the process repeats itself. Figure 19.18 shows a modern motor used to provide added power to a hybrid gasoline-electric car.

19.6 MOTION OF A CHARGED PARTICLE IN A MAGNETIC FIELD

FIGURE 19.19 When the velocity of a charged particle is perpendicular to a uniform magnetic field, the particle moves in a circle whose plane is perpendicular to **B**, which is directed into the page. (The crosses represent the tails of the magnetic field vectors.) The magnetic force **F** on the charge is always directed toward the center of the circle.

Consider the case of a positively charged particle moving in a uniform magnetic field so that the direction of the particle's velocity is *perpendicular to the field,* as in Figure 19.19. The label $\mathbf{B}_{in}$ and the blue crosses indicate that **B** is directed into the page. Application of right-hand rule #1 at point *P* shows that the direction of the magnetic force **F** at this location is upward. This causes the particle to alter its direction of travel and to follow a curved path. Application of right-hand rule #1 at any point shows that **the magnetic force is always *toward* the center of the circular path;** therefore, the magnetic force causes a centripetal acceleration, which changes only the direction of **v** and not its magnitude. Because **F** produces the centripetal acceleration, we can equate its magnitude, qvB in this case, to the mass of the particle multiplied by the centripetal acceleration v^2/r. From Newton's second law, we find that

$$F = qvB = \frac{mv^2}{r}$$

which gives

$$r = \frac{mv}{qB} \qquad [19.10]$$

This says that the radius of the path is proportional to the momentum mv of the particle and is inversely proportional to the magnetic field.

If the initial direction of the velocity of the charged particle is not perpendicular to the magnetic field but instead is directed at an angle to the field, as shown in Figure 19.20, the path followed by the particle is a spiral (called a helix) along the magnetic field lines.

Webnote 19.2

See an animation of positive and negative charges being deflected by a magnetic field at *http://www.lightlink.com/sergey/java/ java/partmagn/index.html*

APPLYING PHYSICS 19.3

Suppose a uniform magnetic field exists in a finite region of space. Can you inject a charged particle into this region from the outside and have it stay trapped in the region by the magnetic force?

Explanation Let us consider separately the components of the particle velocity parallel and perpendicular to the field lines in the region. For the component parallel to the field lines, there is no force on the particle—it continues to move with the parallel component of the velocity. Now consider the component of velocity perpendicular to the field lines. This component results in a magnetic force that is perpendicular to both the field lines and the velocity component. The path of a particle for which the force is always perpendicular to the velocity is a circle. Thus, the particle follows a circular arc and exits the field on the other side of the circle, as shown in Figure 19.21 for a particle with constant kinetic energy. On the other hand, a particle can become trapped if it loses some kinetic energy in a collision after entering the field, as in Figure 19.20.

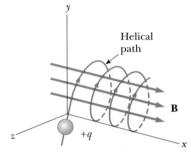

FIGURE 19.20 A charged particle having a velocity directed at an angle with a uniform magnetic field moves in a helical path.

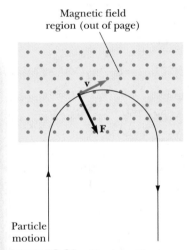

FIGURE 19.21 (Applying Physics 19.3)

Quick Quiz 19.3
As a charged particle moves freely in a circular path in the presence of a constant magnetic field applied perpendicular to the particle's velocity, its kinetic energy (a) remains constant, (b) increases, or (c) decreases.

Quick Quiz 19.4
Two charged particles are projected into a region in which a magnetic field is perpendicular to their velocities. After they enter the magnetic field, you can conclude that (a) the charges are deflected in opposite directions, (b) the charges continue to move in a straight line, (c) the charges move in circular paths, or (d) the charges move in circular paths but in opposite directions.

Example 19.5 A Proton Moving Perpendicularly to a Uniform Magnetic Field

A proton is moving in a circular orbit of radius 14 cm in a uniform magnetic field of magnitude 0.35 T, directed perpendicularly to the velocity of the proton. Find the orbital speed of the proton.

Solution From Equation 19.10, we get

$$v = \frac{qBr}{m} = \frac{(1.6 \times 10^{-19}\ \text{C})(0.35\ \text{T})(14 \times 10^{-2}\ \text{m})}{1.67 \times 10^{-27}\ \text{kg}} = \boxed{4.7 \times 10^6\ \text{m/s}}$$

EXERCISE If an electron moves perpendicularly to the same magnetic field with this speed, what is the radius of its circular orbit?

ANSWER 7.6×10^{-5} m

Webnote 19.3

To see how mass spectrometry is used to screen newborns for genetic disorders, visit *http://www.duke.edu/~mdfeezor/NSHome/index.html*

APPLICATION

MASS SPECTROMETERS

FIGURE 19.22   (Example 19.6) Two isotopes leave the slit at point *S* and travel in different circular paths before striking a photographic plate at *P*.

Example 19.6   The Mass Spectrometer

Two singly ionized atoms move out of a slit at point *S* in Figure 19.22 and into a magnetic field of 0.10 T. Each has a speed of 1.0×10^6 m/s. The nucleus of the first atom contains one proton and has a mass of 1.67×10^{-27} kg, and the nucleus of the second atom contains a proton and a neutron and has a mass of 3.34×10^{-27} kg. Atoms with the same number of protons in the nucleus but different masses are called isotopes. The two isotopes here are hydrogen and deuterium. Find their distance of separation when they strike a photographic plate at *P*.

Solution   The radius of the circular path followed by the lighter isotope, hydrogen, is

$$r_1 = \frac{m_1 v}{qB} = \frac{(1.67 \times 10^{-27}\ \text{kg})(1.0 \times 10^6\ \text{m/s})}{(1.6 \times 10^{-19}\ \text{C})(0.10\ \text{T})} = 0.10\ \text{m}$$

The radius of the path of the heavier isotope, deuterium, is

$$r_2 = \frac{m_2 v}{qB} = \frac{(3.34 \times 10^{-27}\ \text{kg})(1.0 \times 10^6\ \text{m/s})}{(1.6 \times 10^{-19}\ \text{C})(0.10\ \text{T})} = 0.21\ \text{m}$$

The distance of separation is

$$x = 2r_2 - 2r_1 = \boxed{0.21\ \text{m}}$$

The concepts used in this example underlie the operation of a device called a **mass spectrometer,** which is sometimes used to separate isotopes according to their mass-to-charge ratios, but more often is used to measure masses.

19.7   MAGNETIC FIELD OF A LONG, STRAIGHT WIRE AND AMPÈRE'S LAW

During a lecture demonstration in 1819, the Danish scientist Hans Oersted (1777–1851) found that an electric current in a wire deflected a nearby compass needle. This momentous discovery, linking a magnetic field with an electric current for the first time, was the beginning of our understanding of the origin of magnetism.

A simple experiment first carried out by Oersted in 1820 clearly demonstrates that a current-carrying conductor produces a magnetic field. In this experiment, several compass needles are placed in a horizontal plane near a long, vertical wire, as in Figure 19.23a. When there is no current in the wire, all needles point in the same direction (that of Earth's field), as we would expect. However, when the wire

HANS CHRISTIAN OERSTED (1777–1851), DANISH PHYSICIST AND CHEMIST.

Oersted is best known for observing that a compass needle deflects when placed near a wire carrying a current. This important discovery was the first evidence of the connection between electric and magnetic phenomena. Oersted was also the first to prepare pure aluminum. *(North Wind Picture Archives)*

FIGURE 19.23   (a) When there is no current in the vertical wire, all compass needles point in the same direction. (b) When the wire carries a strong current, the compass needles deflect in directions tangent to the circle, pointing in the direction of **B** due to the current.

(a)   (b)

FIGURE 19.24   (a) Right-hand rule #2 for determining the direction of the magnetic field due to a long, straight wire carrying a current. Note that the magnetic field lines form circles around the wire. (b) Circular magnetic field lines surrounding a current-carrying wire, displayed by iron filings. (© *Richard Megna, Fundamental Photographs*)

ANDRÉ-MARIE AMPÈRE (1775–1836)

Ampère, a Frenchman, is credited with the discovery of electromagnetism—the relationship between electric currents and magnetic fields. Ampère's genius, particularly in mathematics, became evident by the age of 12, but his personal life was filled with tragedy. His father, a wealthy city official, was guillotined during the French Revolution, and his wife died young, in 1803. Ampère died at the age of 61 of pneumonia. His judgment of his life is clear from the epitaph he chose for his gravestone: *Tandem felix* (Happy at last). (*Leonard de Selva/CORBIS*)

carries a strong, steady current, the needles all deflect in directions tangent to the circle, as in Figure 19.23b. These observations show that the direction of **B** is consistent with the following convenient rule, **right-hand rule #2:**

> **If the wire is grasped in the right hand with the thumb in the direction of the current, as in Figure 19.24a, the fingers will curl in the direction of B.**

When the current is reversed, the filings in Figure 19.24b also reverse.

Because the filings point in the direction of **B**, we conclude that the lines of **B** form circles about the wire. By symmetry, the magnitude of **B** is the same everywhere on a circular path centered on the wire and lying in a plane perpendicular to the wire. By varying the current and distance from the wire, we find that **B** is proportional to the current and inversely proportional to the distance from the wire.

Shortly after Oersted's discovery, scientists arrived at an expression for the strength of the magnetic field due to the current in a long, straight wire. The magnetic field strength at distance r from a wire carrying current I is

$$B = \frac{\mu_0 I}{2\pi r}$$ [19.11]

◀ Magnetic field due to a long, straight wire

This result shows that the magnitude of the magnetic field is proportional to the current and decreases as the distance from the wire increases, as we might intuitively expect. The proportionality constant μ_0, called the **permeability of free space,** has the value

$$\mu_0 \equiv 4\pi \times 10^{-7}\ \mathrm{T \cdot m/A}$$ [19.12]

Tip 19.3   USE YOUR RIGHT HAND, NOT THE LEFT

Note that we have introduced two right-hand rules in this chapter. Be sure to use *only* your right hand when applying these rules.

AMPÈRE'S LAW AND A LONG, STRAIGHT WIRE

Equation 19.11 enables us to calculate the magnetic field due to a long, straight wire carrying a current. A general procedure for deriving such equations was proposed by the French scientist André-Marie Ampère (1775–1836); it provides a relation between the current in an arbitrarily shaped wire and the magnetic field produced by the wire.

Consider an arbitrary closed path surrounding a current as in Figure 19.25. The path consists of many short segments, each of length $\Delta \ell$. Let us now multiply one of these lengths by the component of the magnetic field parallel to that segment, where the product is labeled $B_\parallel \Delta \ell$. According to Ampère, the sum of all such products over the closed path is equal to μ_0 times the net current I that passes through the surface bounded by the closed path. This statement, known as **Ampère's circuital law,** can be written

Ampère's circuital law

$$\sum B_\parallel \, \Delta\ell = \mu_0 I \qquad \text{[19.13]}$$

where $B_\parallel$ is the component of **B** parallel to the segment of length $\Delta\ell$ and $\sum B_\parallel \, \Delta\ell$ means that we take the sum over all the products $B_\parallel \, \Delta\ell$ around the closed path. Ampère's law is the fundamental law describing how electric currents create magnetic fields in the surrounding empty space.

We can use Ampère's circuital law to derive the magnetic field due to a long, straight wire carrying a current I. As discussed earlier, the magnetic field lines of this configuration form circles with the wire at their centers. The magnetic field is tangent to this circle at every point and has the same value B over the entire circumference of a circle of radius r, so that $B_\parallel = B$, as shown in Figure 19.26. We now calculate the sum $\sum B_\parallel \, \Delta\ell$ over the circular path and note that $B_\parallel$ can be removed from the sum (because it has the same value B for each element on the circle). Equation 19.13 then gives

$$\sum B_\parallel \, \Delta\ell = B_\parallel \sum \Delta\ell = B(2\pi r) = \mu_0 I$$

Dividing both sides by the circumference $2\pi r$, we obtain

$$B = \frac{\mu_0 I}{2\pi r}$$

This is identical to Equation 19.11, which is the magnetic field of a long, straight current.

Ampère's circuital law provides an elegant and simple method for calculating the magnetic fields of highly symmetric current configurations. However, it cannot be easily used to calculate magnetic fields for complex current configurations that lack symmetry. Furthermore, Ampère's circuital law is valid only when the currents and fields do not change with time.

FIGURE 19.25 An arbitrary closed path around a current is used to calculate the magnetic field due to the current by use of Ampère's rule.

FIGURE 19.26 A closed circular path of radius r around a long, straight current-carrying wire is used to calculate the magnetic field set up by the wire.

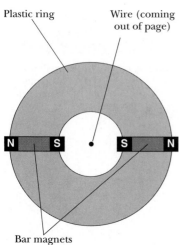

Plastic ring   Wire (coming out of page)

Bar magnets

FIGURE 19.27 (Applying Physics 19.4)

APPLYING PHYSICS 19.4

Consider a plastic ring encircling a long, straight wire, which is coming out of the page in Figure 19.27. On the plastic ring are fastened two bar magnets as shown. If a current flows in the wire in a direction out of the page, is there a net torque on the ring-magnet combination? If so, in which direction?

Explanation Each of the poles on the bar magnets experiences a magnetic force due to the circular magnetic field of the wire. The magnetic field falls off inversely with distance from the wire. Thus, the force on the north poles is smaller than the force on the south poles. The torque, however, increases linearly with distance from the wire. Thus, the effects of the decrease in field strength and the increase in torque with distance cancel, and there is no net torque on the ring-magnet combination.

Example 19.7   The Magnetic Field of a Long Wire

A long, straight wire carries a current of 5.00 A. At one instant, a proton, 4.00 mm from the wire, travels at 1.50×10^3 m/s parallel to the wire and in the same direction as the current (Fig. 19.28). Find the magnitude and direction of the magnetic force that is acting on the proton because of the magnetic field produced by the wire.

Solution From Equation 19.11, the magnitude of the magnetic field produced by the current at a point 4.00 mm from the wire is

$$B = \frac{\mu_0 I}{2\pi r} = \frac{(4\pi \times 10^{-7} \text{ T} \cdot \text{m/A})(5.00 \text{ A})}{2\pi(4.00 \times 10^{-3} \text{ m})} = 2.50 \times 10^{-4} \text{ T}$$

This field is directed into the page at the location of the proton, as shown by right-hand rule #2 for a long, straight wire (see Fig. 19.24a).

The magnitude of the magnetic force on the proton is

$$F = qvB = (1.60 \times 10^{-19} \text{ C})(1.50 \times 10^3 \text{ m/s})(2.50 \times 10^{-4} \text{ T})$$
$$= \boxed{6.00 \times 10^{-20} \text{ N}}$$

The force is directed toward the wire, as shown by right-hand rule #1 for the force on a moving charge (see Fig. 19.7).

FIGURE 19.28 (Example 19.7) The magnetic field due to the current is into the page at the location of the proton, and the magnetic force on the proton is to the left.

19.8 MAGNETIC FORCE BETWEEN TWO PARALLEL CONDUCTORS

As we have seen, a magnetic force acts on a current-carrying conductor when the conductor is placed in an external magnetic field. Because a conductor carrying a current creates a magnetic field around itself, it is easy to understand that two current-carrying wires placed close together exert magnetic forces on each other. Consider two long, straight, parallel wires separated by the distance d and carrying currents I_1 and I_2 in the same direction, as shown in Figure 19.29. Let us determine the magnetic force on one wire due to a magnetic field set up by the other wire.

Wire 2, which carries current I_2, sets up magnetic field $\mathbf{B}_2$ at wire 1. The direction of $\mathbf{B}_2$ is perpendicular to the wire, as shown in the figure. Using Equation 19.11, we see that the magnitude of this magnetic field is

$$B_2 = \frac{\mu_0 I_2}{2\pi d}$$

According to Equation 19.5, the magnitude of the magnetic force on wire 1 in the presence of field $\mathbf{B}_2$ due to I_2 is

$$F_1 = B_2 I_1 \ell = \left(\frac{\mu_0 I_2}{2\pi d}\right) I_1 \ell = \frac{\mu_0 I_1 I_2 \ell}{2\pi d}$$

We can rewrite this in terms of the force per unit length:

$$\frac{F_1}{\ell} = \frac{\mu_0 I_1 I_2}{2\pi d} \qquad \text{[19.14]}$$

FIGURE 19.29 Two parallel wires, each carrying a steady current, exert forces on each other. The field $\mathbf{B}_2$ at wire 1 due to wire 2 produces a force on wire 1 given by $F_1 = B_2 I_1 \ell$. The force is attractive if the currents have the same direction, as shown, and repulsive if the two currents have opposite directions.

The direction of $\mathbf{F}_1$ is downward, toward wire 2, as indicated by right-hand rule #1. If we consider the field set up at wire 2 due to wire 1, the force $\mathbf{F}_2$ on wire 2 is found to be equal in magnitude and opposite in direction to $\mathbf{F}_1$. This is what we would expect from Newton's third law of action-reaction.

We have shown that parallel conductors carrying currents in the same direction *attract* each other. You should use the approach indicated by Figure 19.29 and the steps leading to Equation 19.14 to show that parallel conductors carrying currents in opposite directions *repel* each other.

The force between two parallel wires carrying a current is used to define the SI unit of current, the **ampere** (A), as follows:

◀ Definition of the ampere

> If two long, parallel wires 1 m apart carry the same current, and the magnetic force per unit length on each wire is 2×10^{-7} N/m, then the current is defined to be 1 A.

Definition of the coulomb ▶

The SI unit of charge, the **coulomb** (C), can now be defined in terms of the ampere as follows:

> **If a conductor carries a steady current of 1 A, then the quantity of charge that flows through any cross section in 1 s is 1 C.**

Quick Quiz 19.5

If $I_1 = 2$ A and $I_2 = 6$ A in Figure 19.29, which of the following is true: (a) $F_1 = 3F_2$, (b) $F_1 = F_2$, or (c) $F_1 = F_2/3$?

Example 19.8   Levitating a Wire

Two wires, each having a weight per unit length of 1.0×10^{-4} N/m, are strung parallel to one another above Earth's surface, one directly above the other. The wires are aligned in a north-south direction so that Earth's magnetic field will not affect them. When their distance of separation is 0.10 m, what must be the current in each in order for the lower wire to levitate the upper wire? Assume that the wires carry the same currents, traveling in opposite directions.

Solution   If the upper wire is to float, it must be in equilibrium under the action of two forces: the force of gravity and magnetic repulsion. The weight per unit length — here 1.0×10^{-4} N/m — must be equal and opposite the magnetic force per unit length given in Equation 19.14. Because the currents are the same, we have

$$\frac{F_1}{\ell} = \frac{mg}{\ell} = \frac{\mu_0 I^2}{2\pi d}$$

$$1.0 \times 10^{-4}\,\text{N/m} = \frac{(4\pi \times 10^{-7}\,\text{T} \cdot \text{m/A})\,(I^2)}{(2\pi)(0.10\,\text{m})}$$

We solve for the current to find

$$I = \boxed{7.1\ \text{A}}$$

EXERCISE   If the current in each wire is doubled, what is the equilibrium separation of the two wires?

ANSWER   0.40 m

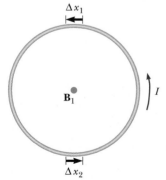

FIGURE 19.30 All segments of the current loop produce a magnetic field at the center of the loop, directed *out of the page.*

19.9   MAGNETIC FIELD OF A CURRENT LOOP

The strength of the magnetic field set up by a piece of wire carrying a current can be enhanced at a specific location if the wire is formed into a loop. You can understand this by considering the effect of several small segments of the current loop, as in Figure 19.30. The small segment at the top of the loop, labeled Δx_1, produces at the loop's center a magnetic field of magnitude B_1, directed out of the page. The direction of **B** can be verified using right-hand rule #2 for a long, straight wire. Imagine holding the wire with your right hand, with your thumb pointing in the direction of the current. Your fingers curl around in the direction of **B**.

A segment at the bottom of the loop of length Δx_2 also contributes to the field at the center, thus increasing its strength. The field produced at the center of the

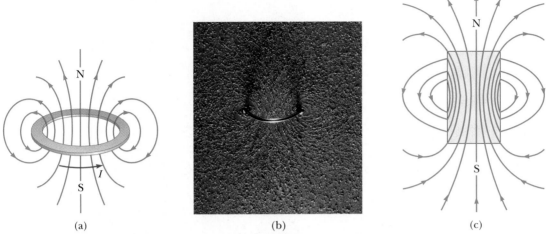

(a)   (b)   (c)

FIGURE 19.31 (a) Magnetic field lines for a current loop. Note that the magnetic field lines of the current loop resemble those of a bar magnet. (b) Field lines of a current loop, displayed by iron filings. *(© Richard Megna, Fundamental Photographs)* (c) The magnetic field of a bar magnet is similar to that of a current loop.

current loop by the segment Δx_2 has the same magnitude as B_1 and is also directed out of the page. Similarly, all other such segments of the current loop contribute to the field. The net effect is a magnetic field for the current loop as pictured in Figure 19.31a.

Notice in Figure 19.31a that the magnetic field lines enter at the bottom of the current loop and exit at the top. Thus, one side of the loop acts as though it were the north pole of a magnet, and the other acts as a south pole. The fact that the field set up by such a current loop bears a striking resemblance to the field of a bar magnet (Fig. 19.31c) will be of interest to us in a future section.

APPLYING PHYSICS 19.5

In electrical circuits, it is often the case that wires carrying currents in opposite directions are twisted together. What is the advantage of doing this?

Explanation If the wires are not twisted together, the combination of the two wires forms a current loop, which produces a relatively strong magnetic field. In fact, the magnetic field generated by the loop could be strong enough to affect adjacent circuits or components.

19.10 MAGNETIC FIELD OF A SOLENOID

If a long, straight wire is bent into a coil of several closely spaced loops, the resulting device is a **solenoid,** often called an **electromagnet.** This device is important in many applications because it acts as a magnet only when it carries a current. As we shall see, the magnetic field inside a solenoid increases with the current and is proportional to the number of coils per unit length.

Figure 19.32 shows the magnetic field lines of a loosely wound solenoid of length ℓ and total number of turns N. Note that the field lines inside the solenoid are nearly parallel, uniformly spaced, and close together. This indicates that the field inside the solenoid is nearly uniform and strong. The exterior field at the sides of the solenoid is nonuniform, is much weaker than the interior field, and is in the *opposite direction* to the field inside the solenoid.

If the turns are closely spaced, the field lines are as shown in Figure 19.33a, entering at one end of the solenoid and emerging at the other. This means that

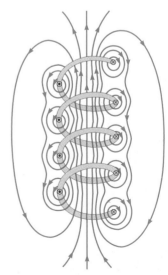

FIGURE 19.32 The magnetic field lines for a loosely wound solenoid.

FIGURE 19.33 (a) Magnetic field lines for a tightly wound solenoid of finite length carrying a steady current. The field inside the solenoid is nearly uniform and strong. Note that the field lines resemble those of a bar magnet, so the solenoid effectively has north and south poles. (b) The magnetic field pattern of a bar magnet, displayed by small iron filings on a sheet of paper. Compare with Figure 19.31c. *(Courtesy of Henry Leap and Jim Lehman)*

(a)

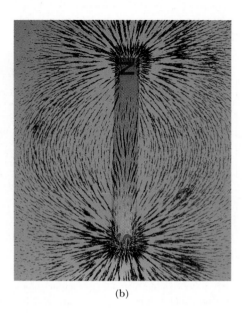

(b)

one end of the solenoid acts as a north pole and the other end acts as a south pole. If the length of the solenoid is much greater than its radius, the lines that leave the north end of the solenoid spread out over a wide region before returning to enter the south end. Hence, as you can see in Figure 19.33a, the magnetic field lines outside are widely separated, indicative of a weak field. This is in contrast to a much stronger field inside the solenoid, where the lines are close together. Also, the field inside the solenoid has a constant magnitude at all points far from its ends. The expression for the magnetic field inside the solenoid is

▶ The magnetic field inside a solenoid

$$B = \mu_0 nI \qquad \text{[19.15]}$$

where $n = N/\ell$ is the number of turns per unit length of the solenoid.

Example 19.9 The Magnetic Field Inside a Solenoid

A certain solenoid consists of 100 turns of wire and has a length of 10.0 cm.

A Find the magnetic field inside the solenoid when it carries a current of 0.500 A.

Solution The number of turns per unit length is

$$n = \frac{N}{\ell} = \frac{100 \text{ turns}}{0.10 \text{ m}} = 1\,000 \text{ turns/m}$$

so

$$B = \mu_0 nI = (4\pi \times 10^{-7} \text{ T} \cdot \text{m/A})(1\,000 \text{ turns/m})(0.500 \text{ A})$$
$$= 6.28 \times 10^{-4} \text{ T}$$

By comparison, this is about 10 times stronger than Earth's magnetic field.

B Assume that the field has one half of this value just outside the solenoid, at the point labeled N in Figure 19.33a. Find the magnitude and direction of the magnetic force acting on an electron that is moving from right to left in the figure, through point N, at 375 m/s.

Solution The magnitude of the magnetic force on the electron is

$$F = qvB = (1.60 \times 10^{-19} \text{ C})(375 \text{ m/s})(3.14 \times 10^{-4} \text{ T}) = 1.88 \times 10^{-20} \text{ N}$$

By use of right-hand rule #1 as in Figure 19.7, the direction of this force is found to be out of the page. (Remember to change the direction of the magnetic force for the negatively charged electron.) This force will deflect the electron from its original direction of motion.

EXERCISE   How many turns should the solenoid have (assuming it carries the same current) if the field inside is to be 5 times as great?

ANSWER   500 turns

So-called steering magnets placed along the neck of the picture tube in a television set, as in Figure 19.34, are used to make the electron beam move to the desired locations on the screen, thus tracing out the images of your favorite program.

> **APPLICATION**
>
> CONTROLLING THE ELECTRON BEAM IN A TELEVISION SET

AMPÈRE'S LAW APPLIED TO A SOLENOID

We can use Ampère's law to obtain the expression for the magnetic field inside a solenoid carrying a current I. A cross section taken along the length of part of our solenoid is shown in Fig. 19.35. **B** inside the solenoid is uniform and parallel to the axis, and **B** outside is approximately zero. Consider a rectangular path of length L and width w as shown in Figure 19.35. We can apply Ampère's law to this path by evaluating the sum of $B_\parallel \, \Delta\ell$ over each side of the rectangle. The contribution along side 3 is clearly zero, because **B** = 0 in this region. The contributions from sides 2 and 4 are both zero, because **B** is perpendicular to $\Delta\ell$ along these paths. Side 1 of length L gives a contribution of BL to the sum, because **B** along this path is uniform and parallel to $\Delta\ell$. Therefore, the sum over the closed rectangular path has the value

$$\sum B_\parallel \, \Delta\ell = BL$$

The right side of Ampère's law involves the total current that passes through the area bounded by the path chosen. In our case, the total current through the rectangular path equals the current through each turn of the solenoid multiplied by the number of turns. If N is the number of turns in the length L, then the total current through the rectangular path equals NI. Therefore, Ampère's law applied to this path gives

$$\sum B_\parallel \, \Delta\ell = BL = \mu_0 NI$$

$$B = \mu_0 \frac{N}{L} I = \mu_0 nI$$

where $n = N/L$ is the number of turns per unit length.

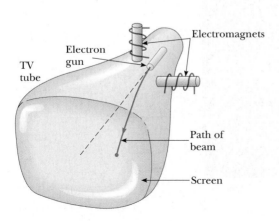

FIGURE 19.34   Electromagnets are used to deflect electrons to desired positions on the screen of a television tube.

Electromagnets

Electron gun

TV tube

Path of beam

Screen

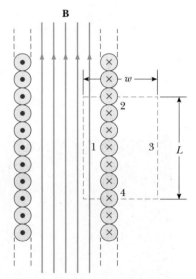

FIGURE 19.35   A cross-sectional view of a tightly wound solenoid. If the solenoid is long relative to its radius, we can assume that the magnetic field inside is uniform and the field outside is zero. Ampère's law applied to the red dashed rectangular path can then be used to calculate the field inside the solenoid.

19.11 MAGNETIC DOMAINS

The magnetic field produced by a current in a coil of wire gives us a hint as to what might cause certain materials to exhibit strong magnetic properties. A single coil like that in Figure 19.31a has a north pole and a south pole, but if this is true for a coil of wire, it should also be true for any current confined to a circular path. *In particular, an individual atom should act as a magnet because of the motion of the electrons about the nucleus.* Each electron, with its charge of 1.6×10^{-19} C, circles the atom once in about 10^{-16} s. If we divide the electronic charge by this time interval, we see that the orbiting electron is equivalent to a current of 1.6×10^{-3} A. Such a current produces a magnetic field on the order of 20 T at the center of the circular path. From this we see that a very strong magnetic field would be produced if several of these atomic magnets could be aligned inside a material. This does not occur, however, because the simple model we have described is not the complete story. A thorough analysis of atomic structure shows that the magnetic field produced by one electron in an atom is often canceled by an oppositely revolving electron in the same atom. The net result is that **the magnetic effect produced by the electrons orbiting the nucleus is either zero or very small for most materials.**

The magnetic properties of many materials are explained by the fact that an electron not only circles in an orbit but also spins on its axis like a top (Fig. 19.36). (This classical description should not be taken too literally. The property of *spin*

FIGURE 19.36 Classical model of a spinning electron.

PHYSICS *IN ACTION*

The Motion of Charged Particles in Magnetic Fields

The white arc in the photograph on the left indicates the circular path followed by an electron beam in a magnetic field. The vessel contains gas at very low pressure, and the beam is made visible as the electrons collide with the gas atoms, which emit visible light. The magnetic field is produced by two coils (not shown). The apparatus can be used to measure the ratio of e/m for the electron.

In the photograph on the right, oxygen, a paramagnetic substance, is attracted to a magnetic field. The liquid oxygen in this photograph is suspended between the poles of a permanent magnet. Paramagnetic substances contain atoms (or ions) that behave as tiny bar magnets. These magnetic atoms interact weakly with each other and are randomly oriented in the absence of an external magnetic field. When the substance is placed in an external magnetic field, the tiny bar magnets associated with the atoms of the substance tend to line up with the field.

(Courtesy of Central Scientific Company)

(Courtesy of Leon Lewandowski)

can be understood only with the methods of quantum mechanics, which we shall not discuss here.) The spinning electron represents a charge in motion that produces a magnetic field. The field due to the spinning is generally stronger than the field due to the orbital motion. In atoms containing many electrons, the electrons usually pair up with their spins opposite each other, so that their fields cancel each other. That is why most substances are not magnets. However, in certain strongly magnetic materials such as iron, cobalt, and nickel, the magnetic fields produced by the electron spins do not cancel completely. Such materials are said to be **ferromagnetic.** In ferromagnetic materials, strong coupling occurs between neighboring atoms, forming large groups of atoms called **domains** in which the spins are aligned. Typically, the sizes of these domains range from about 10^{-4} cm to 0.1 cm. In an unmagnetized substance the domains are randomly oriented, as shown in Figure 19.37a. When an external field is applied, domains that are already aligned with the field tend to grow at the expense of the others (Fig. 19.37b and c). This causes the material to become magnetized.

In what are called hard magnetic materials, domain alignment persists after the external field is removed; the result is a **permanent magnet.** In soft magnetic materials, such as iron, once the external field is removed, thermal agitation produces domain motion and the material quickly returns to an unmagnetized state.

The alignment of domains explains why the strength of an electromagnet is increased dramatically by the insertion of an iron core into the magnet's center. The magnetic field produced by the current in the loops causes alignment of the domains, thus producing a large net external field. The use of iron as a core is also advantageous because it is a soft magnetic material and loses its magnetism almost instantaneously after the current in the coils is turned off.

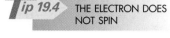

Tip 19.4 THE ELECTRON DOES NOT SPIN

Even though we use the word *spin,* the electron is not physically spinning. The electron has an intrinsic angular momentum *as if it were spinning,* but the concept of spin angular momentum is actually a relativistic effect.

Webnote 19.4

To see actual photographs of magnetic domains in ferromagnetic materials and the equipment used to record them, visit
http://www.physics.brown.edu/ Studies/Demo/em/demo/5g2020.htm

Webnote 19.5

Try this experiment! Grapes are repelled by a magnet. Find out why.
http://www.exploratorium.edu/snacks/ diamagnetism_www/index.html

---

APPLYING PHYSICS 19.6

Suppose a cardboard tube is filled with very fine iron filings. Each filing is in the form of a sliver—long and thin. If the tube of filings is shaken, what is the likelihood that all of them will wind up aligned in the same direction? Suppose the shaking is repeated while the tube is in a strong magnetic field. What is the likelihood now that the filings will be aligned? Is this a macroscopic analog to anything discussed in this section?

Explanation When the tube is shaken outside of a magnetic field, the orientation of the filings will be random—it is extremely unlikely that they will all be aligned in the same direction. If the shaking occurs in a magnetic field, the filings, which become magnetized, can align with the field while they are momentarily airborne and free to rotate. Thus, it is likely that a large number of filings will be aligned with the field after the shaking. The filings are an analog to magnetic domains. When the material is placed in a magnetic field, more domains end up aligned with the field than misaligned, similar to the iron filings.

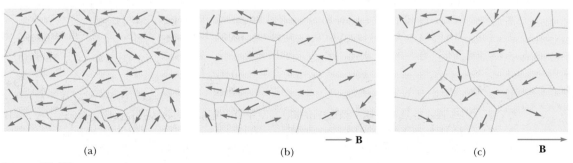

(a)	(b)	(c)
	$\longrightarrow$ **B**	**B**

FIGURE 19.37 (a) Random orientation of domains in an unmagnetized substance. (b) When an external magnetic field **B** is applied, the domains aligned with **B** grow larger. (c) As the field is made even stronger, the domains not aligned with the external field become very small.

SUMMARY

The **magnetic force** that acts on a charge q moving with velocity **v** in a magnetic field **B** has the magnitude

$$F = qvB \sin \theta \qquad [19.1]$$

where θ is the angle between **v** and **B**.

To find the direction of this force, you can use **right-hand rule #1:** Place the fingers of your open right hand in the direction of **B** and point your thumb in the direction of the velocity **v**. The force **F** on a positive charge is directed out of the palm of your hand. If the charge is *negative* rather than positive, the force is directed opposite the force given by the right-hand rule.

The SI unit of magnetic field is the **tesla** (T), or weber per square meter (Wb/m^2). An additional commonly used unit for magnetic field is the **gauss** (G); 1 T = 10^4 G.

If a straight conductor of length ℓ carries current I, the magnetic force on that conductor when it is placed in a uniform external magnetic field **B** is

$$F = BI\ell \sin \theta \qquad [19.6]$$

Right-hand rule #1 also gives the direction of the magnetic force on the conductor. In this case, however, you must place your thumb in the direction of the current rather than in the direction of **v**.

The torque τ on a current-carrying loop of wire in a magnetic field, **B**, has the magnitude

$$\tau = BIA \sin \theta \qquad [19.8]$$

where I is the current in the loop and A is its cross-sectional area. The angle between **B** and a line drawn perpendicularly to the plane of the loop is θ.

If a charged particle moves in a uniform magnetic field so that its initial velocity is perpendicular to the field, it will move in a circular path whose plane is perpendicular to the magnetic field. The radius r of the circular path is

$$r = \frac{mv}{qB} \qquad [19.10]$$

where m is the mass of the particle and q is its charge.

The magnetic field at distance r from a **long, straight wire** carrying current I has the magnitude

$$B = \frac{\mu_0 I}{2\pi r} \qquad [19.11]$$

where $\mu_0 = 4\pi \times 10^{-7}$ T·m/A is the **permeability of free space.** The magnetic field lines around a long, straight wire are circles concentric with the wire.

Ampère's law can be used to find the magnetic field around certain simple current-carrying conductors. It can be written

$$\sum B_\parallel \Delta\ell = \mu_0 I \qquad [19.13]$$

where $B_\parallel$ is the component of **B** tangent to a small current element of length $\Delta\ell$ that is part of a closed path, and I is the total current that penetrates the closed path.

The force per unit length on each of two parallel wires separated by the distance d and carrying currents I_1 and I_2 has the magnitude

$$\frac{F}{\ell} = \frac{\mu_0 I_1 I_2}{2\pi d} \qquad [19.14]$$

The forces are attractive if the currents are in the same direction and repulsive if they are in opposite directions.

The magnetic field inside a solenoid has the magnitude

$$B = \mu_0 n I \qquad [19.15]$$

where n is the number of turns of wire per unit length, $n = N/\ell$.

CONCEPTUAL QUESTIONS

1. Is a net force exerted on a bar magnet in a uniform magnetic field?

2. A coil with a current alternating in direction demagnetizes a watch or tape head. Explain why an alternating current is necessary to achieve this effect.

3. A proton moving horizontally enters a region where a uniform magnetic field is directed perpendicular to the proton's velocity, as shown in Figure Q19.3. Describe the subsequent motion of the proton. How would an electron behave under the same circumstances?

FIGURE Q19.3

4. A current-carrying conductor experiences no magnetic force when placed in a certain manner in a uniform magnetic field. Explain.

5. How can the motion of a charged particle be used to distinguish between a magnetic field and an electric field in a certain region?

6. Which way would a compass point if you were at Earth's north magnetic pole?

7. Why does the picture on a television screen become distorted when a magnet is brought near the screen as in Figure Q19.7? (*Caution:* You should not do this at home on a color television set, because it may permanently affect the television picture quality.)

8. A magnet attracts a piece of iron. The iron can then attract another piece of iron. On the basis of domain alignment, explain what happens in each piece of iron.

9. A Hindu ruler once suggested that he be entombed in a magnetic coffin with the polarity arranged so that he could be forever suspended between heaven and Earth. Is such magnetic levitation possible? Discuss.

10. Will a nail be attracted to either pole of a magnet? Explain what is happening inside the nail when it is placed near the magnet.

FIGURE Q19.7 (© *Loren Winters/Visuals Unlimited*)

11. Suppose you move along a wire at the same speed as the drift speed of the electrons in the wire. Do you now measure a magnetic field of zero?

12. Why does hitting a magnet with a hammer cause the magnetism to be reduced?

13. Can you use a compass to detect the currents in wires in the walls near light switches in your home?

14. It is found that charged particles from outer space, called cosmic rays, strike Earth more frequently at the poles than at the Equator. Why?

15. Two wires carry currents in opposite directions and are oriented parallel, with one above the other. The wires repel each other. Is the upper wire in a stable levitation over the lower wire? Suppose the current in one wire is reversed, so that the wires now attract. Is the lower wire hanging in a stable attraction to the upper wire?

16. How can a current loop be used to determine the presence of a magnetic field in a given region of space?

17. A hanging Slinky toy is attached to a powerful battery and a switch. When the switch is closed so that current suddenly flows through the Slinky, does the Slinky compress or expand?

18. Is it possible to orient a current loop in a uniform magnetic field such that the loop will not tend to rotate?

19. Parallel wires exert magnetic forces on each other. What about perpendicular wires? Imagine two wires oriented perpendicular to each other and almost touching. Each wire carries a current. Is there a force between the wires?

20. Can a constant magnetic field set into motion an electron at rest? Explain your answer.

PROBLEMS

1, 2, 3 = straightforward, intermediate, challenging   ☐ = full solution available in Student Solutions Manual/Study Guide

web = solution posted at **http://info.brookscole.com/serway**   🔬 = biomedical application

Section 19.3 Magnetic Fields

1. An electron gun fires electrons into a magnetic field that is directed straight downward. Find the direction of the force exerted by the field on an electron for each of the following directions of the electron's velocity: (a) horizontal and due north; (b) horizontal and 30° west of north; (c) due north, but at 30° below the horizontal; (d) straight upward. (Remember that an electron has a negative charge.)

2. (a) Find the direction of the force on a proton (a positively charged particle) moving through the magnetic fields in Figure P19.2, as shown. (b) Repeat part (a), assuming the moving particle is an electron.

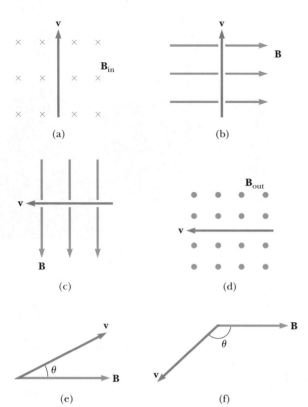

(a)

(b)

(c)

(d)

(e)

(f)

FIGURE P19.2 (Problems 2 and 13) For Problem 13, replace the velocity vector with a current in that direction.

3. Find the direction of the magnetic field acting on the positively charged particle moving in the various situations shown in Figure P19.3, if the direction of the magnetic force acting on it is as indicated.

(a)

(b)

(c)

FIGURE P19.3 (Problems 3 and 12) For Problem 12, replace the velocity vector with a current in that direction.

4. Determine the initial direction of the deflection of charged particles as they enter the magnetic fields shown in Figure P19.4.

5. At the Equator near Earth's surface, the magnetic field is approximately 50.0 μT northward and the electric field is about 100 N/C downward in fair weather. Find the gravitational, electric, and magnetic forces on an electron with an instantaneous velocity of 6.00×10^6 m/s directed to the east in this environment.

6. A proton travels with a speed of 3.0×10^6 m/s at an angle of 37° with the direction of a magnetic field of 0.30 T in the $+y$ direction. What are (a) the magnitude of the magnetic force on the proton and (b) the proton's acceleration?

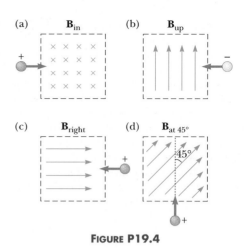

FIGURE P19.4

7. What velocity would a proton need to circle Earth
web 1 000 km above the magnetic equator, where Earth's magnetic field is directed horizontally north and has a magnitude of 4.00×10^{-8} T?

8. An electron is accelerated through 2 400 V from rest and then enters a region where there is a uniform 1.70-T magnetic field. What are the (a) maximum and (b) minimum magnitudes of the magnetic force this charge can experience?

9. A proton moves perpendicularly to a uniform magnetic field **B** at 1.0×10^7 m/s and experiences an acceleration of 2.0×10^{13} m/s² in the $+x$ direction when its velocity is in the $+z$ direction. Determine the magnitude and direction of the field.

10. Sodium ions (Na⁺) move at 0.851 m/s through a bloodstream in the arm of a person standing near a large magnet. The magnetic field has a strength of 0.254 T and makes an angle of 51.0° with the motion of the sodium ions. The arm contains 100 cm³ of blood with 3.00×10^{20} Na⁺ ions per cubic centimeter. If no other ions were present in the arm, what would be the magnetic force on the arm?

Section 19.4 Magnetic Force on a Current-Carrying Conductor

11. A current $I = 15$ A is directed along the positive x axis and perpendicularly to a magnetic field. The conductor experiences a magnetic force per unit length of 0.12 N/m in the negative y direction. Calculate the magnitude and direction of the magnetic field in the region through which the current passes.

12. In Figure P19.3, assume that in each case the velocity vector shown is replaced with a wire carrying a current in the direction of the velocity vector. For each case, find the direction of the magnetic field that will produce the magnetic force shown.

13. In Figure P19.2, assume that in each case the velocity vector shown is replaced with a wire carrying a current in the direction of the velocity vector. For each case, find the direction of the magnetic force acting on the wire.

14. A wire carries a steady current of 2.40 A. A straight section of the wire is 0.750 m long and lies along the x axis within a uniform magnetic field of magnitude 1.60 T in the posi-

tive z direction. If the current is in the $+x$ direction, what is the magnetic force on the section of wire?

15. A wire carries a current of 10.0 A in a direction that makes an angle of 30.0° with the direction of a magnetic field of strength 0.300 T. Find the magnetic force on a 5.00-m length of the wire.

16. At a certain location, Earth has a magnetic field of 0.60×10^{-4} T pointing 75° below the horizontal in a north-south plane. A 10.0-m-long straight wire carries a 15-A current. (a) If the current is directed horizontally toward the east, what are the magnitude and direction of the magnetic force on the wire? (b) What are the magnitude and direction of the force if the current is directed vertically upward?

17. A wire with a mass per unit length of 1.00 g/cm is placed on a horizontal surface with a coefficient of friction of 0.200. The wire carries a current of 1.50 A eastward and moves horizontally to the north. What are the magnitude and the direction of the *smallest* vertical magnetic field that enables the wire to move in this fashion?

18. A conductor suspended by two flexible wires as shown in Figure P19.18 has a mass per unit length of 0.040 0 kg/m. What current must exist in the conductor for the tension in the supporting wires to be zero when the magnetic field is 3.60 T into the page? What is the required direction for the current?

FIGURE P19.18

19. An unusual message delivery system is pictured in Figure
web P19.19. A 15-cm length of conductor that is free to move is held in place between two thin conductors. When a 5.0-A current is directed as shown in the figure, the wire segment moves upward at a constant velocity. If the mass of the wire is 15 g, find the magnitude and direction of the minimum magnetic field that is required to move the wire. (The wire slides without friction on the two vertical conductors.)

FIGURE P19.19

20. A thin, horizontal copper rod is 1.00 m long and has a mass of 50.0 g. What is the minimum current in the rod that can cause it to float in a horizontal magnetic field of 2.00 T?

21. In Figure P19.21, the cube is 40.0 cm on each edge. Four straight segments of wire—*ab*, *bc*, *cd*, and *da*—form a closed loop that carries a current $I = 5.00$ A, in the direction shown. A uniform magnetic field of magnitude $B = 0.020\ 0$ T is in the positive y direction. Determine the magnitude and direction of the magnetic force on each segment.

FIGURE P19.21

Section 19.5 Torque on a Current Loop and Electric Motors

22. A current of 17.0 mA is maintained in a single circular loop with a circumference of 2.00 m. A magnetic field of 0.800 T is directed parallel to the plane of the loop. What is the magnitude of the torque exerted by the magnetic field on the loop?

23. An 8-turn coil encloses an elliptical area having a major axis of 40.0 cm and a minor axis of 30.0 cm (Fig. P19.23). The coil lies in the plane of the page and has a 6.00-A current flowing clockwise around it. If the coil is in a uniform magnetic field of 2.00×10^{-4} T, directed toward the left of the page, what is the magnitude of the torque on the coil? (*Hint:* The area of an ellipse is $A = \pi ab$, where a and b are the semi-major and semi-minor axes of the ellipse.)

FIGURE P19.23

24. A rectangular loop consists of 100 closely wrapped turns and has dimensions 0.40 m by 0.30 m. The loop is hinged along the y axis, and the plane of the coil makes an angle of 30.0° with the x axis (Fig. P19.24). What is the magnitude of the torque exerted on the loop by a uniform magnetic field of 0.80 T directed along the x axis, when the current in the windings has a value of 1.2 A in the direction shown? What is the expected direction of rotation of the loop?

25. A long piece of wire with a mass of 0.100 kg and a total length of 4.00 m is used to make a square coil with a side

FIGURE P19.24

of 0.100 m. The coil is hinged along a horizontal side, carries a 3.40-A current, and is placed in a vertical magnetic field with a magnitude of 0.010 0 T. (a) Determine the angle that the plane of the coil makes with the vertical when the coil is in equilibrium. (b) Find the torque acting on the coil due to the magnetic force at equilibrium.

26. A copper wire is 8.00 m long, and has a cross-sectional area of 1.00×10^{-4} m^2. This wire forms a 1-turn loop in the shape of a square and is then connected to a battery that applies a potential difference of 0.100 V. If the loop is placed in a uniform magnetic field of magnitude 0.400 T, what is the maximum torque that can act on it? The resistivity of copper is 1.70×10^{-8} $\Omega \cdot$ m.

Section 19.6 Motion of a Charged Particle in a Magnetic Field

27. A particle with a $+2.0$ μC charge and a kinetic energy of 0.090 J is fired into a uniform magnetic field of magnitude 0.10 T. If the particle moves in a circular path of radius 3.0 m, determine its mass.

28. A cosmic-ray proton in interstellar space has an energy of 10.0 MeV and executes a circular orbit having a radius equal to that of Mercury's orbit around the Sun $(5.80 \times 10^{10}$ m). What is the magnetic field in that region of space?

29. Figure P19.29a is a diagram of a device called a velocity selector, in which particles of a specific velocity pass through undeflected but those with greater or lesser velocities are deflected either upward or downward. An electric field is directed perpendicularly to a magnetic field. This produces on the charged particle an electric force and a magnetic force that can be equal in magnitude and opposite in direction (Fig. P19.29b), and hence cancel. Show that particles with a speed of $v = E/B$ will pass through undeflected.

(a)

(b)

FIGURE P19.29

30. Consider the mass spectrometer shown schematically in Figure P19.30. The electric field between the plates of the velocity selector is 950 V/m, and the magnetic fields in both the velocity selector and the deflection chamber have magnitudes of 0.930 T. Calculate the radius of the path in the system for a singly charged ion with mass $m = 2.18 \times 10^{-26}$ kg. (*Hint:* See Problem 29.)

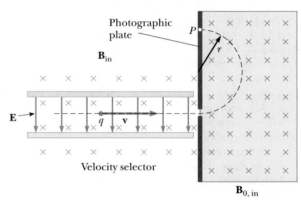

FIGURE P19.30 A mass spectrometer. Charged particles are first sent through a velocity selector. They then enter a region where a magnetic field $\mathbf{B}_0$ (inward) causes positive ions to move in a semicircular path and strike a photographic film at *P*.

31. A singly charged positive ion has a mass of 2.50×10^{-26} kg. After being accelerated through a potential difference of 250 V, the ion enters a magnetic field of 0.500 T, in a direction perpendicular to the field. Calculate the radius of the ion's path in the field.

32. A mass spectrometer is used to examine the isotopes of uranium. Ions in the beam emerge from the velocity selector at a speed of 3.00×10^5 m/s and enter a uniform magnetic field of 0.600 T directed perpendicularly to the velocity of the ions. What is the distance between the impact points formed on the photographic plate by singly charged ions of ^{235}U and ^{238}U?

33. An electron moves in a circular path perpendicular to a constant magnetic field with a magnitude of 1.00 mT. If the angular momentum of the electron about the center of the circle is 4.00×10^{-25} J$\cdot$s, determine (a) the radius of the circular path and (b) the speed of the electron.

Section 19.7 Magnetic Field of a Long, Straight Wire and Ampère's Law

34. Find the direction of the current in the wire in Figure P19.34 that would produce a magnetic field directed as shown, in each case.

35. A lightning bolt may carry a current of 1.00×10^4 A for a short period of time. What is the resulting magnetic field 100 m from the bolt? Suppose that the bolt extends far above and below the point of observation.

36. In 1962, measurements of the magnetic field of a large tornado were made at the Geophysical Observatory in Tulsa, Oklahoma. If the tornado's field was $B = 1.50 \times 10^{-8}$ T pointing north when the tornado was 9.00 km east of the observatory, what current was carried up or down the funnel of the tornado? Model the vortex as a long straight wire carrying a current.

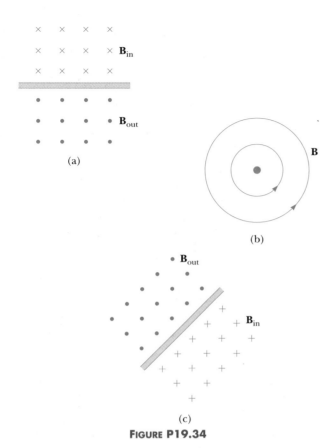

(a)

(b)

(c)

FIGURE P19.34

37. At what distance from a long, straight wire carrying a current of 5.0 A is the magnetic field due to the wire equal to the strength of Earth's field, approximately 5.0×10^{-5} T?

38. The two wires shown in Figure P19.38 carry currents of 5.00 A in opposite directions and are separated by 10.0 cm. Find the direction and magnitude of the net magnetic field (a) at a point midway between the wires, (b) at point P_1 (10.0 cm to the right of the wire on the right), and (c) at point P_2 (20.0 cm to the left of the wire on the left).

FIGURE P19.38

39. Four long, parallel conductors carry equal currents of $I = 5.00$ A. Figure P19.39 is an end view of the conductors. The current direction is into the page at points A and B (indicated by the crosses) and out of the page at C and D (indicated by the dots). Calculate the magnitude and direction of the magnetic field at point P, located at the center of the square of edge length 0.200 m.

40. The two wires in Figure P19.40 carry currents of 3.00 A and 5.00 A in the direction indicated. (a) Find the direction and magnitude of the magnetic field at a point mid-

FIGURE P19.39

way between the wires. (b) Find the magnitude and direction of the magnetic field at point P, located 20.0 cm above the wire carrying the 5.00-A current.

FIGURE P19.40

41. A wire carries a 7.00-A current along the x axis and another wire carries a 6.00-A current along the y axis as shown in Figure P19.41. What is the magnetic field at point P located at $x = 4.00$ m, $y = 3.00$ m?

FIGURE P19.41

42. A long, straight wire lies on a horizontal table and carries a current of 1.20 μA. In a vacuum, a proton moves parallel to the wire (opposite the current) with a constant velocity of 2.30×10^4 m/s at a constant distance d above the wire. Determine the value of d. You may ignore the magnetic field due to Earth.

43. The magnetic field 40.0 cm away from a long, straight wire carrying current 2.00 A is 1.00 μT. (a) At what distance is it 0.100 μT? (b) At one instant, the two conductors in a long household extension cord carry equal 2.00-A currents in opposite directions. The two wires are 3.00 mm apart. Find the magnetic field 40.0 cm away from the middle of the straight cord, in the plane of the two wires. (c) At what distance is it one tenth as large? (d) The center wire in a coaxial cable carries current 2.00 A in one direction, and the sheath around it carries current 2.00 A in the opposite direction. What magnetic field does the cable create at points outside?

Section 19.8 Magnetic Force Between Two Parallel Conductors

44. Two parallel wires are 10.0 cm apart, and each carries a current of 10.0 A. (a) If the currents are in the same direction, find the force per unit length exerted by one of the wires on the other. Are the wires attracted or repelled? (b) Repeat the problem with the currents in opposite directions.

45. A wire with a weight per unit length of 0.080 N/m is suspended directly above a second wire. The top wire carries a current of 30.0 A, and the bottom wire carries a current of 60.0 A. Find the distance of separation between the wires so that the top wire will be held in place by magnetic repulsion.

46. In Figure P19.46, the current in the long, straight wire is $I_1 = 5.00$ A, and the wire lies in the plane of the rectangular loop, which carries 10.0 A. The dimensions are $c = 0.100$ m, $a = 0.150$ m, and $\ell = 0.450$ m. Find the magnitude and direction of the net force exerted by the magnetic field due to the straight wire on the loop.

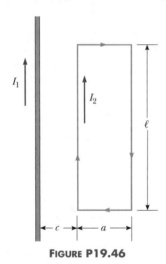

FIGURE P19.46

Section 19.10 Magnetic Field of a Solenoid

47. What current is required in the windings of a long solenoid that has 1 000 turns uniformly distributed over a length of 0.400 m in order to produce a magnetic field of magnitude 1.00×10^{-4} T at the center of the solenoid?

48. It is desired to construct a solenoid that has a resistance of 5.00 Ω (at 20°C) and that produces a magnetic field at its center of 4.00×10^{-2} T when it carries a current of 4.00 A. The solenoid is to be constructed from copper wire having a diameter of 0.500 mm. If the radius of the solenoid is to be 1.00 cm, determine (a) the number of turns of wire needed and (b) the length the solenoid should have.

49. A single-turn square loop of wire, 2.00 cm on a side, carries a counterclockwise current of 0.200 A. The loop is inside a solenoid, with the plane of the loop perpendicular to the magnetic field of the solenoid. The solenoid has 30 turns per centimeter and carries a counterclockwise current of 15.0 A. Find the force on each side of the loop and the torque acting on it.

50. An electron moves at a speed of 1.0×10^4 m/s in a circular path of radius of 2.0 cm inside a solenoid. The magnetic field of the solenoid is perpendicular to the plane of the electron's path. Find (a) the strength of the magnetic field inside the solenoid and (b) the current in the solenoid if it has 25 turns per centimeter.

ADDITIONAL PROBLEMS

51. A circular coil consisting of a single loop of wire has a radius of 30.0 cm and carries a current of 25 A. It is placed in an external magnetic field of 0.30 T. Find the torque on the wire when the plane of the coil makes an angle of 35° with the direction of the field.

52. An electron enters a region of magnetic field of magnitude 0.010 0 T, traveling perpendicular to the linear boundary of the region. The direction of the field is perpendicular to the velocity of the electron. (a) Determine the time it takes for the electron to leave the "field-filled" region, noting that its path is a semicircle. (b) Find the kinetic energy of the electron if the radius of its semicircular path is 2.00 cm.

53. Two long, straight wires cross each other at right angles, as shown in Figure P19.53. (a) Find the direction and magnitude of the magnetic field at point P, which is in the same plane as the two wires. (b) Find the magnetic field at a point 30.0 cm above the point of intersection (30.0 cm out of the page, toward you).

FIGURE P19.53

54. A 0.200-kg metal rod carrying a current of 10.0 A glides on two horizontal rails 0.500 m apart. What vertical magnetic field is required to keep the rod moving at a constant speed if the coefficient of kinetic friction between the rod and rails is 0.100?

55. Two species of singly charged positive ions of masses 20.0×10^{-27} kg and 23.4×10^{-27} kg enter a magnetic field at the same location with a speed of 1.00×10^5 m/s. If the strength of the field is 0.200 T, and the ions move perpendicularly to the field, find their distance of separation after they complete one half of their circular path.

56. Two parallel conductors carry currents in opposite directions, as shown in Figure P19.56. One conductor carries a current of 10.0 A. Point A is the midpoint between the wires, and point C is 5.00 cm to the right of the 10.0-A current. I is adjusted so that the magnetic field at C is zero. Find (a) the value of the current I and (b) the value of the magnetic field at A.

FIGURE P19.56

57. A heart surgeon monitors the flow rate of blood through an artery using an electromagnetic flowmeter (shown schematically in Fig. P19.57). Electrodes A and B make contact with the outer surface of the blood vessel, which has interior diameter 3.00 mm. (a) For a magnetic field magnitude of 0.040 0 T, a potential difference of 160 μV appears between the electrodes. Calculate the speed of the blood. (b) Verify that electrode A is positive, as shown. Does the sign of the emf depend on whether the mobile ions in the blood are predominantly positively or negatively charged? Explain.

FIGURE P19.57

58. Two circular loops are parallel, coaxial, and almost in contact, 1.00 mm apart (Fig. P19.58). Each loop is 10.0 cm in radius. The top loop carries a clockwise current of 140 A. The bottom loop carries a counterclockwise current of 140 A. (a) Calculate the magnetic force that the bottom loop exerts on the top loop. (b) The upper loop has a mass of 0.021 0 kg. Calculate its acceleration, assuming that the only forces acting on it are the force in part (a) and its weight. (*Hint:* The distance between the loops is

FIGURE P19.58

small in comparison to the radius of curvature, so the loops may be treated as long, straight, parallel wires.)

59. A 1.00-kg ball having net charge $Q = 5.00$ μC is thrown out of a window horizontally at a speed $v = 20.0$ m/s. The window is at a height $h = 20.0$ m above the ground. A uniform horizontal magnetic field of magnitude $B = 0.010\ 0$ T is perpendicular to the plane of the ball's trajectory. Find the magnitude of the magnetic force acting on the ball just before it hits the ground. (*Hint:* Ignore magnetic forces in finding the ball's final velocity.)

60. At the Fermilab accelerator in Batavia, Illinois, protons having momentum 4.80×10^{-16} kg·m/s are held in a circular orbit of radius 1.00 km by an upward magnetic field. What is the magnitude of this field?

61. Two long, parallel conductors carry currents $I_1 = 3.00$ A and $I_2 = 3.00$ A, both directed into the page in Figure P19.61. Determine the magnitude and direction of the resultant magnetic field at P.

FIGURE P19.61

62. A uniform horizontal wire with a linear mass density of 0.50 g/m carries a 2.0-A current. It is placed in a constant magnetic field, with a strength of 4.0×10^{-3} T, that is horizontal and perpendicular to the wire. As the wire moves upward starting from rest, (a) what is its acceleration and (b) how long does it take to rise 50 cm? Neglect the magnetic field of Earth.

63. Three long, parallel conductors carry currents of $I = 2.0$ A. Figure P19.63 is an end view of the conductors, with each current coming out of the page. Given that $a = 1.0$ cm, determine the magnitude and direction of the magnetic field at points (a) A, (b) B, and (c) C.

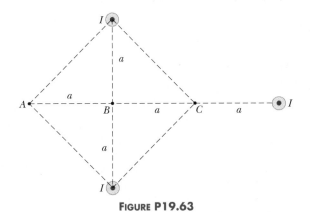

FIGURE P19.63

64. Two long, parallel wires, each with a mass per unit length of 40 g/m, are supported in a horizontal plane by 6.0-cm-long strings, as shown in Figure P19.64. Each wire carries

the same current *I*, causing the wires to repel each other so that the angle θ between the supporting strings is 16°. (a) Are the currents in the same or opposite directions? (b) Determine the magnitude of each current.

FIGURE P19.64

65. Protons having a kinetic energy of 5.00 MeV are moving in the positive *x* direction and enter a magnetic field of 0.050 0 T in the *z* direction, out of the plane of the page, and extending from *x* = 0 to *x* = 1.00 m as in Figure P19.65. (a) Calculate the *y* component of the protons' momentum as they leave the magnetic field. (b) Find the angle α between the initial velocity vector of the proton beam and the velocity vector after the beam emerges from the field. (*Hint:* Neglect relativistic effects and note that 1 eV = 1.60×10^{-19} J.)

FIGURE P19.65

66. A straight wire of mass 10.0 g and length 5.0 cm is suspended from two identical springs that, in turn, form a closed circuit (Fig. P19.66). The springs stretch a distance of 0.50 cm under the weight of the wire. The circuit has a total resistance of 12 Ω. When a magnetic field is turned on, directed out of the page (indicated by the dots in Fig. P19.66), the springs are observed to stretch an additional 0.30 cm. What is the strength of the magnetic field? (The upper portion of the circuit is fixed.)

FIGURE P19.66

GROUP ACTIVITIES

G.1 For this experiment, you will need a small bar magnet, a small plastic container, and a bowl of water. Tape the magnet to the bottom of the small container, and float the container and magnet on the surface of the bowl as in Figure GA19.1. The magnet and small container should rotate and come to equilibrium with the magnet pointing along a north-south line. The compass you have constructed is similar to the type used in early sailing vessels. How can you determine which direction is north and which is south?

FIGURE GA19.1

G.2 In the northern hemisphere the direction of Earth's magnetic field becomes more and more nearly vertical the farther north one goes. To find the variation from the horizontal of the magnetic field in your locale, try the following. Press an unmagnetized needle through a Ping-Pong ball and balance the combination between two drinking glasses that are lined up along an east-west line. Now press a magnetized needle through the ball at right angles to the unmagnetized needle so that the needle points north. The magnetized needle can now rotate in the vertical direction and will point in the direction of Earth's magnetic field, which is at some angle below the horizontal. Take several measurements of this dip angle and obtain an average value.

G.3 Construct an electromagnet by wrapping about 1 m of small-diameter insulated wire around a steel nail. Tape the ends of the wires to a D-cell battery as in Figure GA19.3. How many staples or paper clips can you pick up with your electromagnet? How would you increase the magnetic field set up by the nail? Disconnect the wires from the battery and test to see how much magnetism is retained by the nail by seeing how many staples it can pick up. A convenient way to test the strength of a magnet is to attach a paper clip to a rubber band. Note how far the rubber band is stretched before the clip comes free of the magnet. Test your electromagnet in this way. Where is the magnetic field of the electromagnet strongest, at the ends of the nail or near its center? When you have your nail magnetized, bang it against a table or the floor and then check its magnetism. Why does the nail lose its magnetism by this procedure?

FIGURE GA19.3

G.4 You can trace out the field pattern of a magnet with iron filings. Any machine shop will supply the filings, which should be soaked in a soap solution to remove grit and oil, and then dried. Scatter them lightly over the surface of a paper covering the magnet, and then lightly tap the paper to jar the filings into alignment. Explain why the filings form their pattern. Examine the field pattern set up in the following situations. (a) Arrange two bar magnets about 4 cm apart, aligned with opposite poles facing each other. (b) Use two bar magnets about 4 cm apart, aligned with like poles facing each other. (c) Examine the pattern set up by a horseshoe magnet.

G.5 In your home television set, a beam of electrons moves from the back of the picture tube to the screen where they strike a fluorescent dot that glows with a particular color when hit. The electrons are accelerated through a voltage of about 20 000 V at the back of the tube. (a) Does the orientation of the television set in Earth's magnetic field affect the beam? (b) In which direction(s) should the set be oriented so that the beam is affected the most? (c) What is the force on an individual electron in direction and magnitude when the beam is moving from east to west? (Assume the magnetic field is horizontal at the location of the television set and has a value of 0.50×10^{-4} T.)

G.6 A beam of alpha particles ($m = 6.64 \times 10^{-27}$ kg, $q = 3.2 \times 10^{-19}$ C) is to be used to produce some radioactive nuclei in a hospital setting where the nuclei are needed as cancer-fighting agents. The beam must enter through one wall of a room directly at the center of the wall, turn in a 90° angle and exit directly at the center of the adjacent wall. The room is 2.5 m by 2.5 m, and it is known that the alpha particles travel at a speed of 2.00×10^5 m/s. (a) Sketch a design of the room and the direction of the magnetic field in the room that will allow this movement to take place. (b) If a beam of electrons of the same speed is directed along the same initial path into the same magnetic field used for the alpha particles, how will they move? Show this trajectory on your sketch. Is the radius of curvature larger or smaller? (c) What is the magnitude of the magnetic field you need to cause the alpha particles to follow the desired path?

20

Induced Voltages and Inductance

The vibrating strings induce a voltage in pickup coils that detect and amplify the musical sounds being produced. The details of how this works are discussed in this chapter. (*Photo Disc/Getty Images*)

*I*n 1819 Hans Christian Oersted discovered that a magnetic compass experiences a force in the vicinity of an electric current. Although there had long been speculation that such a relationship existed, this was the first evidence of a link between electricity and magnetism. Because nature is often symmetric, the discovery that electric currents produce magnetic fields led scientists to suspect that magnetic fields could produce electric currents. Indeed, experiments conducted by Michael Faraday in England and independently by Joseph Henry in the United States in 1831 showed that a changing magnetic field could induce an electric current in a circuit. The results of these experiments led to a basic and important law known as Faraday's law. In this chapter we discuss Faraday's law and several practical applications, one of which is the production of electrical energy in power generation plants throughout the world.

MICHAEL FARADAY, BRITISH PHYSICIST AND CHEMIST (1791–1867)

Faraday is often regarded as the greatest experimental scientist of the 1800s. His many contributions to the study of electricity include the invention of the electric motor, electric generator, and transformer, as well as the discovery of electromagnetic induction and the laws of electrolysis. Greatly influenced by religion, he refused to work on military poison gas for the British government. *(By kind permission of the President and Council of the Royal Society)*

20.1 INDUCED emf AND MAGNETIC FLUX

We begin this chapter by describing an experiment, first conducted by Faraday, demonstrating that a current can be produced by a changing magnetic field. The apparatus shown in Figure 20.1 consists of a coil connected to a switch and a battery. We refer to this coil as the *primary coil* and to the corresponding circuit as the primary circuit. The coil is wrapped around an iron ring to intensify the magnetic field produced by the current in the coil. A second coil, at the right, is wrapped around the iron ring and is connected to an ammeter. We refer to this as the *secondary coil* and to the corresponding circuit as the secondary circuit. Note that there is no battery in the secondary circuit. The only purpose of this circuit is to detect any current that might be produced by a magnetic field.

At first glance, you might guess that no current would ever be detected in the secondary circuit. However, when the switch in the primary circuit is suddenly closed or opened, something quite amazing happens. Just after the switch is closed, the ammeter in the secondary circuit deflects in one direction and then returns to zero. When the switch is opened, the ammeter deflects in the opposite direction and again returns to zero. Finally, when there is a steady current in the primary circuit, the ammeter reads zero.

From observations such as these, Faraday concluded that an electric current can be produced by a *changing* magnetic field. (A steady magnetic field cannot produce a current.) The current produced in the secondary circuit occurs only for an instant while the magnetic field through the secondary coil is changing. In effect, the secondary circuit behaves as though a source of emf were connected to it

FIGURE 20.1 Faraday's experiment. When the switch in the primary circuit at the left is closed, the ammeter in the secondary circuit at the right deflects momentarily. The emf in the secondary circuit is induced by the changing magnetic field through the coil in this circuit.

FIGURE 20.2 (a) A uniform magnetic field **B** making an angle θ with the normal to the plane of a wire loop of area A. (b) An edge view of the loop.

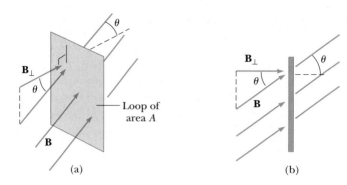

for a short time. It is customary to say that **an induced emf is produced in the secondary circuit by the changing magnetic field.**

MAGNETIC FLUX

In order to evaluate induced emfs quantitatively, it is first necessary to understand what factors affect the phenomenon. The emf is actually induced by a change in a quantity called the magnetic flux rather than simply by a change in the magnetic field. Magnetic flux is defined in a similar manner to the electric flux (Section 15.10) and is proportional to both the strength of the magnetic field passing through the plane of a loop of wire and the area of the loop.

Consider a loop of wire in the presence of a uniform magnetic field **B**. If the loop has an area A, the **magnetic flux Φ_B** through the loop is defined as

 Magnetic flux

$$\Phi_B = B_\perp A = BA \cos \theta \qquad [20.1]$$

where $B_\perp$ is the component of **B** perpendicular to the plane of the loop as in Figure 20.2a, and θ is the angle between **B** and the normal (perpendicular) to the plane of the loop. Figure 20.2b is an edge view of the loop and the penetrating magnetic field lines. When the field is perpendicular to the plane of the loop as in Figure 20.3a, $\theta = 0$ and Φ_B has a maximum value, $\Phi_{B,\max} = BA$. When the plane of the loop is parallel to **B** as in Figure 20.3b, $\theta = 90°$ and $\Phi_B = 0$. The flux can also be negative. For example, when $\theta = 180°$, the flux is equal to $-BA$. Because B is in SI units of teslas, or webers per square meter, the units of flux are $T \cdot m^2$, or webers (Wb).

We can emphasize the qualitative meaning of Equation 20.1 by first drawing magnetic field lines, as in Figure 20.3. The number of lines per unit area increases as the field strength increases. **The value of the magnetic flux is proportional to the total number of lines passing through the loop.** Thus, we see that the most lines pass through the loop when its plane is perpendicular to the field as in Figure 20.3a, and so the flux has its maximum value. As Figure 20.3b shows, no lines pass through the loop when its plane is parallel to the field, and so in this case $\Phi_B = 0$.

FIGURE 20.3 An edge view of a loop in a uniform magnetic field. (a) When the field lines are perpendicular to the plane of the loop, the magnetic flux through the loop is a maximum and equal to $\Phi_B = BA$. (b) When the field lines are parallel to the plane of the loop, the magnetic flux through the loop is zero.

APPLYING PHYSICS 20.1

Argentina has more land area (2.8×10^6 km^2) than Greenland (2.2×10^6 km^2). Yet the magnetic flux of Earth's magnetic field is larger through Greenland than through Argentina. Why?

Explanation   There are two reasons for the larger magnetic flux through Greenland. Greenland (latitude 60°N to 80°N) is closer to a magnetic pole than is Argentina (latitude 20°S to 50°S). As a result, the angle that the magnetic field lines make with the vertical is smaller in Greenland than in Argentina. Thus, more field lines penetrate the surface in Greenland. The second reason is also a result of Greenland's proximity to a magnetic pole. The field strength is larger in the vicinity of a magnetic pole than it is farther from the pole. Therefore, the magnitude of **B** in the definition of flux is larger for Greenland than for Argentina. These two influences dominate over the slightly larger area of Argentina.

20.2 FARADAY'S LAW OF INDUCTION

The usefulness of the concept of magnetic flux is made obvious by another simple experiment that demonstrates the basic idea of electromagnetic induction. Consider a wire loop connected to an ammeter as in Figure 20.4. If a magnet is moved toward the loop, the ammeter needle deflects in one direction as in Figure 20.4a. When the magnet is held stationary as in Figure 20.4b, the needle is not deflected. If the magnet is moved away from the loop, the needle deflects in the opposite direction as in Figure 20.4c. If the magnet is held stationary and the loop is moved

(a)

(b)

(c)

FIGURE 20.4   (a) When a magnet is moved toward a wire loop connected to an ammeter, the ammeter deflects as shown, indicating that a current I is induced in the loop. (b) When the magnet is held stationary, no current is induced in the loop, even when the magnet is inside the loop. (c) When the magnet is moved away from the loop, the ammeter deflects in the opposite direction, indicating that the induced current is opposite that shown in part (a).

either toward or away from the magnet, the needle also deflects. From these observations, it can be concluded that **a current is set up in the circuit as long as there is relative motion between the magnet and the loop.** Note that the same experimental results are found whether the loop moves or the magnet moves. We call such a current an **induced current** because it is produced by an **induced emf.**

This experiment has something in common with the Faraday experiment discussed in Section 20.1. In each case, an emf is induced in a circuit when the magnetic flux through the circuit changes with time. In fact, we can make the following general summary of such experiments involving an induced emf and a changing magnetic flux:

Tip 20.1 INDUCED CURRENT REQUIRES A CHANGE IN MAGNETIC FLUX

Note that the existence of magnetic flux through an area is not sufficient to create an induced emf. A *change* in the magnetic flux over some time interval Δt must occur for an emf to be induced.

> **The instantaneous emf induced in a circuit equals the time rate of change of magnetic flux through the circuit.**

If a circuit contains N tightly wound loops and the flux through each loop changes by the amount $\Delta\Phi_B$ during the interval Δt, the average emf induced in the circuit during time Δt is

Faraday's law ▶

$$\varepsilon = -N\frac{\Delta\Phi_B}{\Delta t}$$ [20.2]

This is a statement of **Faraday's law of magnetic induction.** Because $\Phi_B = BA \cos \theta$, note that a change with time of any of the factors B, A, or θ produces an emf. We explore the effect of a change of each of these factors in the following sections. The minus sign in Equation 20.2 is included to indicate the polarity of the induced emf, which can be found by use of **Lenz's law:**

Lenz's law ▶

> **The polarity of the induced emf is such that it produces a current whose magnetic field opposes the change in magnetic flux through the loop. That is, the induced current tends to maintain the original flux through the circuit.**

APPLICATION

GROUND FAULT INTERRUPTERS

We consider several applications of Lenz's law in Section 20.4.

The ground fault interrupter (GFI) is an interesting safety device that protects people against electric shock when they touch appliances and power tools. Its operation makes use of Faraday's law. Figure 20.5 shows the essential parts of a ground fault interrupter. Wire 1 leads from the wall outlet to the appliance to be protected, and wire 2 leads from the appliance back to the wall outlet. An iron ring surrounds the two wires to confine the magnetic field set up by each wire. A sensing coil, which can activate a circuit breaker when changes in magnetic flux occur, is wrapped around part of the iron ring. Because the currents in the wires are in opposite directions, the net magnetic field through the sensing coil due to

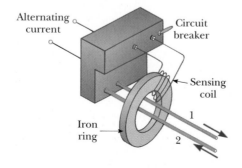

FIGURE 20.5 Essential components of a ground fault interrupter (contents of the gray box in Fig. 20.6a). In newer homes, such devices are built directly into wall outlets. The purpose of the sensing coil and circuit breaker is to cut off the current before damage is done.

(a)

(b)

FIGURE 20.6 (a) This hair dryer has been plugged into a ground fault interrupter that is in turn plugged into an unprotected wall outlet. (b) You have likely seen such a ground fault interrupter in a hotel bathroom, where hair dryers and electric shavers will be used by people just out of the shower or who might touch a water pipe, providing a ready path to ground in the event of a short circuit. *(b, Richard Megna, Fundamental Photographs, NYC)*

the currents is zero. However, if a short circuit occurs in the appliance so that there is no returning current, the net magnetic field through the sensing coil is no longer zero. This can happen, for example, if one of the wires loses its insulation, providing a path through you to ground if you happen to be touching the appliance and are grounded as in Figure 18.22a. Because the current is alternating, the magnetic flux through the sensing coil changes with time, producing an induced voltage in the coil. This induced voltage is used to trigger a circuit breaker, stopping the current quickly in about a millisecond before it reaches a level that might be harmful to the person using the appliance. A ground fault interrupter provides faster and more complete protection than even the case-ground-and-circuit-breaker combination shown in Figure 18.22b. For this reason, ground fault interrupters are commonly found in bathrooms, where electricity poses a hazard to people (see Figure 20.6).

Another interesting application of Faraday's law is the production of sound in an electric guitar. A vibrating string induces an emf in a coil (Fig. 20.7). The

(a)

(b)

FIGURE 20.7 (a) In an electric guitar, a vibrating string induces a voltage in the pickup coil. (b) Several pickups allow the vibration to be detected from different portions of the string. *(b, Charles D. Winters)*

FIGURE 20.8 This infant is wearing a monitor designed to alert caregivers if breathing stops. Note the two wires attached to opposite sides of the chest. *(Courtesy of PedsLink Pediatric Healthcare Resources, Newport Beach, CA)*

APPLICATION

ELECTRIC GUITAR PICKUPS

pickup coil is placed near the vibrating guitar string, which is made of a metal that can be magnetized. The permanent magnet inside the coil magnetizes the portion of the string nearest the coil. When the guitar string vibrates at some frequency, its magnetized segment produces a changing magnetic flux through the pickup coil. The changing flux induces a voltage in the coil; the voltage is fed to an amplifier. The output of the amplifier is sent to the loudspeakers, producing the sound waves that we hear.

APPLICATION

APNEA MONITORS

Sudden Infant Death Syndrome (SIDS) is a devastating affliction in which a baby suddenly stops breathing during sleep without an apparent cause. One type of monitoring device sometimes used to alert caregivers of the cessation of breathing uses induced currents, as shown in Figure 20.8. A coil of wire attached to one side of the chest carries an alternating current. The varying magnetic flux produced by this current passes through a pickup coil attached to the opposite side of the chest. Expansion and contraction of the chest caused by breathing or movement changes the strength of the voltage induced in the pickup coil. However, if breathing stops, the pattern of the induced voltage stabilizes, and external circuits monitoring the voltage sound an alarm to the caregivers after a momentary pause to ensure that a problem actually does exist.

Quick Quiz 20.1

Figure 20.9 is a graph of the magnitude *B* versus time *t* for a magnetic field that passes through a fixed loop and is oriented perpendicular to the plane of the loop. Rank the magnitudes of the emf generated in the loop at the three instants indicated (*a*, *b*, *c*), from largest to smallest.

FIGURE 20.9 (Quick Quiz 20.1)

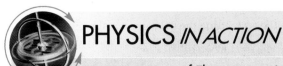

PHYSICS *IN ACTION*

Demonstrations of Electromagnetic Induction

On the left is a simple demonstration involving a permanent magnet, a conducting coil, and a *galvanometer*, which is a device used in the construction of analog ammeters and voltmeters. The galvanometer needle can deflect to either side of the zero mark, indicating the direction of a current. When a strong magnet is moved toward or away from the coil attached to the galvanometer, a current is induced in the coil, indicated by the momentary deflection of the galvanometer during the movement of the magnet. What is the cause of this induced current? Would the galvanometer needle deflect if the coil were moved toward a stationary magnet?

On the right, to demonstrate electromagnetic induction, an ac voltage is applied to the lower coil in the apparatus. A voltage is induced in the upper coil, as indicated by the illuminated flashlight bulb connected to this coil. What do you think happens to the bulb's intensity as the upper coil is moved over the vertical tube? To answer this question, note that the magnetic field associated with the lower coil varies along the axis of the tube.

(Richard Megna, Fundamental Photographs)

(Courtesy of Central Scientific Company)

Example 20.1   Application of Faraday's Law (Change of *B* with time)

A coil with 25 turns of wire is wrapped on a square frame 1.80 cm on a side. Each turn has the same area, equal to that of the frame, and the total resistance of the coil is 0.35 Ω. A uniform magnetic field is applied perpendicularly to the plane of the coil. If the field changes uniformly from 0 to 0.500 T in 0.800 s, find the magnitude of the induced emf in the coil while the field is changing.

Reasoning   The magnitude of the induced emf can be found from Faraday's law of induction, $\mathcal{E} = -N(\Delta\Phi_B/\Delta t)$. In this equation, $\Delta\Phi_B$ is the difference between the final and initial fluxes, where $\Phi_B = BA \cos\theta = BA \cos 0° = BA$.

Solution   The area of the coil is $(0.018\,0\text{ m})^2 = 3.24 \times 10^{-4}\text{ m}^2$. The magnetic flux through the coil at $t = 0$ is zero because $B = 0$. At $t = 0.800$ s, the magnetic flux through the coil is

$$\Phi_{B,f} = BA = (0.500\text{ T})(3.24 \times 10^{-4}\text{ m}^2) = 1.62 \times 10^{-4}\text{ T} \cdot \text{m}^2$$

Therefore, the *change* in flux through the coil during the 0.800-s interval is

$$\Delta\Phi_B = \Phi_{B,f} - \Phi_{B,i} = 1.62 \times 10^{-4} \, \text{T} \cdot \text{m}^2$$

Faraday's law of induction enables us to find the magnitude of the induced emf:

$$|\boldsymbol{\mathcal{E}}| = N\frac{\Delta\Phi_B}{\Delta t} = (25 \text{ turns})\left(\frac{1.62 \times 10^{-4} \, \text{T} \cdot \text{m}^2}{0.800 \, \text{s}}\right) = \boxed{5.1 \text{ mV}}$$

Note that $1 \, \text{T} \cdot \text{m}^2/\text{s} = 1(\text{N} \cdot \text{m/C} \cdot \text{s})(\text{m}^2/\text{s}) = 1 \, \text{N} \cdot \text{m/C} = 1 \, \text{J/C} = 1 \, \text{V}$.

EXERCISE Find the magnitude of the induced current in the coil while the field is changing.

ANSWER 14 mA

20.3 MOTIONAL emf

In Section 20.2, we considered a situation in which an emf is induced in a circuit when the magnetic field changes with time. In this section we describe a particular application of Faraday's law in which a so-called **motional emf** is produced. This is the emf induced in a conductor moving through a magnetic field.

First consider a straight conductor of length ℓ moving with constant velocity through a uniform magnetic field directed into the paper, as in Figure 20.10. For simplicity, we assume that the conductor is moving perpendicularly to the field. The electrons in the conductor experience a force of magnitude $F = qvB$ directed downward along the conductor. Because of this magnetic force, the free electrons move to the lower end and accumulate there, leaving a net positive charge at the upper end. As a result of this charge separation, an electric field is produced in the conductor. The charge at the ends builds up until the downward magnetic force qvB is balanced by the upward electric force qE. At this point, charge stops flowing and the condition for equilibrium requires that

$$qE = qvB \qquad \text{or} \qquad E = vB$$

Because the electric field is uniform, the field produced in the conductor is related to the potential difference across the ends by $\Delta V = E\ell$. Thus,

$$\Delta V = E\ell = B\ell v \qquad\qquad \text{[20.3]}$$

Because there is an excess of positive charge at the upper end of the conductor and an excess of negative charge at the lower end, the upper end is at a higher potential than the lower end. Thus,

> **A potential difference is maintained across the conductor as long as there is motion through the field. If the motion is reversed, the polarity of the potential difference is also reversed.**

A more interesting situation occurs if the moving conductor is part of a closed conducting path. This situation is particularly useful for illustrating how a changing loop area induces a current in a closed circuit described by Faraday's law. Consider a circuit consisting of a conducting bar of length ℓ, sliding along two fixed parallel conducting rails, as in Figure 20.11a. For simplicity, we assume that the moving bar has zero resistance and that the stationary part of the circuit has constant resistance R. A uniform and constant magnetic field **B** is applied perpendicularly to the plane of the circuit. As the bar is pulled to the right with velocity **v** under the influence of an applied force $\mathbf{F}_{\text{app}}$, the free charges in the bar experience a magnetic force along the length of the bar. This force in turn sets up an induced current because the charges are free to move in a closed conducting path. In this

FIGURE 20.10 A straight conductor of length ℓ moving with velocity **v** through a uniform magnetic field **B** directed perpendicularly to **v**. The vector **F** is the force on an electron in the conductor. An emf of $B\ell v$ is induced between the ends of the bar.

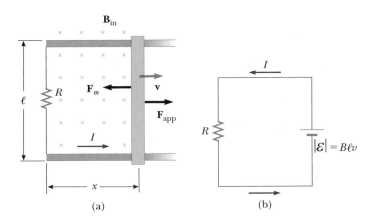

FIGURE 20.11 (a) A conducting bar sliding with velocity **v** along two conducting rails under the action of an applied force **F**$_{app}$. The magnetic force **F**$_m$ opposes the motion, and a counterclockwise current is induced in the loop. (b) The equivalent circuit of (a).

case, the changing magnetic flux through the loop and the corresponding induced emf across the moving bar arise from the *change in area of the loop* as the bar moves through the magnetic field.

Let us assume that the bar moves a distance Δx in time Δt, as shown in Figure 20.12. The increase in flux $\Delta\Phi_B$ through the loop in that time is the amount of flux that now passes through the portion of the circuit that has area $\ell\,\Delta x$:

$$\Delta\Phi_B = BA = B\ell\,\Delta x$$

Using Faraday's law and noting that there is one loop ($N = 1$), we find that the magnitude of the induced emf is

$$\left|\mathcal{E}\right| = \frac{\Delta\Phi_B}{\Delta t} = B\ell\,\frac{\Delta x}{\Delta t} = B\ell v \qquad \text{[20.4]}$$

This induced emf is often called a **motional emf** because it arises from the motion of a conductor through a magnetic field.

Furthermore, if the resistance of the circuit is R, the magnitude of the induced current in the circuit is

$$I = \frac{\left|\mathcal{E}\right|}{R} = \frac{B\ell v}{R} \qquad \text{[20.5]}$$

Figure 20.11b is the equivalent circuit diagram for this example.

Webnote 20.1

Change the shape of a coil in a magnetic field and watch as a current is induced in the coil. Visit *http://www.lightlink.com/sergey/java/java/indcur/index.html*

FIGURE 20.12 As the bar moves to the right, the area of the loop increases by the amount $\ell\,\Delta x$ and the magnetic flux through the loop increases by $B\ell\,\Delta x$.

APPLYING **PHYSICS 20.2**

We have discussed applying a force on the bar, which results in an induced emf in the circuit shown in Figure 20.11. Suppose we remove the external magnetic field in the diagram and replace the resistor with a high-voltage source and a switch, as in Figure 20.13. What happens when the switch is closed? Will the bar move, and does it matter which way we connect the high-voltage source?

Explanation Suppose the source is capable of establishing high current. The two horizontal conducting rods will create a strong magnetic field in the area between them, directed into the page. (The movable bar also creates a magnetic field, but this field cannot exert force on the bar itself.) As the current passes downward through the movable bar, it experiences a magnetic force to the right. Hence, it accelerates along the rails away from the power supply. If the polarity of the power were reversed, the magnetic field would be out of the page, the current in the bar would be upward, and the force on the bar would still be to the right. (This is the essence of a railgun.) The $BI\ell$ force exerted by a magnetic field per Equation 19.6 causes the bar to accelerate away from the voltage source. Studies have shown that it is possible to launch payloads into space with this technology. Very large accelerations can be obtained with currently available technology, with payloads being accelerated to a speed of several kilometers per second in a fraction of a second. This is a larger acceleration than humans can withstand.

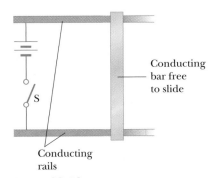

Conducting bar free to slide

Conducting rails

FIGURE 20.13 (Applying Physics 20.2)

A PPLICATION

RAILGUNS

Quick Quiz 20.2 As an airplane flies due north from Los Angeles to Seattle, it cuts through Earth's magnetic field. As a result, an emf is developed between the wing tips. Which wing tip is positively charged?

Quick Quiz 20.3 You wish to move a rectangular loop of wire into a region of uniform magnetic field at a given speed so as to induce an emf in the loop. The plane of the loop must remain perpendicular to the magnetic field lines. In which orientation should you hold the loop while you move it into the region of magnetic field in order to generate the largest emf? (a) With the long dimension of the loop parallel to the velocity vector. (b) With the short dimension of the loop parallel to the velocity vector. (c) Either way—the emf is the same regardless of orientation.

Example 20.2   The Electrified Airplane Wing

An airplane with a wing span of 30.0 m flies parallel to Earth's surface at a location at which the downward component of Earth's magnetic field is 0.60×10^{-4} T. Find the difference in potential between the wing tips when the speed of the plane is 250 m/s.

Solution   Because the plane is flying horizontally, we do not have to concern ourselves with the horizontal component of Earth's magnetic field. Thus, we find that

$$\mathcal{E} = B\ell v = (0.60 \times 10^{-4}\ \text{T})(30.0\ \text{m})(250\ \text{m/s}) = \boxed{0.45\ \text{V}}$$

Example 20.3   Where Is the Energy Source?

A The sliding bar in Figure 20.11a has a length of 0.50 m and moves at 2.0 m/s in a magnetic field of magnitude 0.25 T. Find the induced voltage in the moving rod.

Solution   We use Equation 20.3 and find that

$$\mathcal{E} = B\ell v = (0.25\ \text{T})(0.50\ \text{m})(2.0\ \text{m/s}) = \boxed{0.25\ \text{V}}$$

B If the resistance in the circuit is 0.50 Ω, find the current in the circuit.

Solution   The current is found from Ohm's law to be

$$I = \frac{\mathcal{E}}{R} = \frac{0.25\ \text{V}}{0.50\ \Omega} = \boxed{0.50\ \text{A}}$$

C Find the amount of energy delivered to the 0.50-Ω resistor in 1 s.

Solution   The power delivered to the resistor is

$$\mathcal{P} = I\,\Delta V = (0.50\ \text{A})(0.25\ \text{V}) = 0.13\ \text{W}$$

Because power is defined as the rate at which energy is converted in a device, the energy *W* delivered to the resistor in 1 s is

$$W = \mathcal{P}t = (0.125\ \text{W})(1.0\ \text{s}) = \boxed{0.13\ \text{J}}$$

D The source of the energy calculated in part C is some external agent that keeps the bar moving at a constant speed of 2.0 m/s by exerting an applied force F_{app}. Find the value of F_{app}.

Solution   From part C, we know that the work done by the applied force in 1 s is 0.13 J. In 1 s, the bar moves a distance of

$$d = vt = (2.0 \text{ m/s})(1.0 \text{ s}) = 2.0 \text{ m}$$

Thus, from the definition of work, we find that $W = F_{app}d$, or

$$F_{app} = \frac{W}{d} = \frac{0.13 \text{ J}}{2.0 \text{ m}} = \boxed{0.063 \text{ N}}$$

EXERCISE   If the rod is to move at constant speed, the applied force must be equal in magnitude to the retarding magnetic force, $BI\ell$. Show that this approach also gives $F_{app} = 0.063$ N, as found in part (d).

20.4 LENZ'S LAW REVISITED (THE MINUS SIGN IN FARADAY'S LAW)

To attain a better understanding of Lenz's law, let us return to the example of a bar moving to the right on two parallel rails in the presence of a uniform magnetic field directed into the paper (Fig. 20.14a). As the bar moves to the right, the magnetic flux through the circuit increases with time because the area of the loop increases. Lenz's law says that the induced current must be in a direction such that the flux *it* produces opposes the change in the external magnetic flux. Because the flux due to the external field is increasing *into* the paper, the induced current, to oppose the change, must produce a flux *out* of the paper. Hence, the induced current must be counterclockwise when the bar moves to the right. (Use right-hand rule #2 from Chapter 19 to verify this direction.) On the other hand, if the bar is moving to the left, as in Figure 20.14b, the magnetic flux through the loop decreases with time. Because the flux is into the paper, the induced current has to be clockwise to produce its own flux into the paper (which opposes the decrease in the external flux). In either case, the induced current tends to maintain the original flux through the circuit.

Now let us examine this situation from the viewpoint of energy conservation. Suppose that the bar is given a slight push to the right. In the preceding analysis, we found that this motion led to a counterclockwise current in the loop. What would happen if we assume that the current is clockwise, that is, opposite the direction required by Lenz's law? For a clockwise current I, the direction of the magnetic force $BI\ell$ on the sliding bar would be to the right. This force would accelerate the rod and increase its velocity. This, in turn, would cause the area of the loop to increase more rapidly, thereby increasing the induced current, which would increase the force, which would increase the current, which would. . . . In effect, the system would acquire energy with zero input energy. This is clearly inconsistent both with our experience and with the law of conservation of energy. We are forced to conclude that the current must be counterclockwise.

Consider another situation. A bar magnet is moved to the right toward a stationary loop of wire, as in Figure 20.15a. As the magnet moves, the magnetic flux through the loop increases with time. To counteract this, the induced current produces a flux to the left, as in Figure 20.15b; hence, the induced current is in the direction shown. Note that the magnetic field lines associated with the induced current oppose the motion of the magnet. Therefore, the left face of the current loop is a north pole and the right face is a south pole.

On the other hand, if the magnet moves to the left, as in Figure 20.15c, its flux through the loop, which is toward the right, decreases in time. Under these circumstances, the induced current in the loop is in a direction to set up a field directed from left to right through the loop, in an effort to maintain a constant

(a)

(b)

FIGURE 20.14   (a) As the conducting bar slides on the two fixed conducting rails, the magnetic flux through the loop increases in time. By Lenz's law, the induced current must be *counterclockwise* so as to produce a counteracting flux *out of the paper*. (b) When the bar moves to the left, the induced current must be *clockwise*. Why?

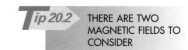

Tip 20.2   THERE ARE TWO MAGNETIC FIELDS TO CONSIDER

When applying Lenz's law, note that there are *two* magnetic fields to consider. The first is the external changing magnetic field that induces the current in a conducting loop. The second is the magnetic field produced *by* the induced current in the loop.

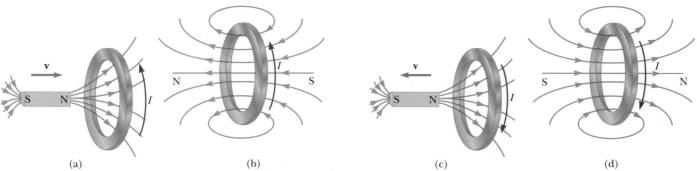

(a)   (b)   (c)   (d)

FIGURE 20.15 (a) When the magnet is moved toward the stationary conducting loop, a current is induced in the direction shown. (b) This induced current produces its own flux to the left to counteract the increasing external flux to the right. (c) When the magnet is moved away from the stationary conducting loop, a current is induced in the direction shown. (d) This induced current produces its own flux to the right to counteract the decreasing external flux to the right.

number of flux lines. Hence, the induced current in the loop is as shown in Figure 20.15d. In this case, the left face of the loop is a south pole and the right face is a north pole.

Quick Quiz 20.4 A bar magnet is falling through a loop of wire with constant velocity with the north pole entering first. Viewed from the same side of the loop as the magnet, as the north pole approaches the loop, the induced current will be in what direction? (a) clockwise (b) zero (c) counterclockwise (d) along the length of the magnet

Example 20.4 Application of Lenz's Law

A coil of wire is placed near an electromagnet, as in Figure 20.16a. Find the direction of the induced current in the coil (a) at the instant the switch is closed, (b) after the switch has been closed for several seconds, and (c) when the switch is opened.

Reasoning and Solution

A When the switch is closed, the situation changes from a condition in which no lines of flux pass through the coil to one in which lines of flux pass through in the direction

FIGURE 20.16 (Example 20.4)

shown in Figure 20.16b. To counteract this change in the number of lines, the coil must set up a field from left to right in the figure. This requires a current directed as shown in Figure 20.16b.

B After the switch has been closed for several seconds, there is no change in the number of lines through the loop; hence, the induced current is zero.

C Opening the switch causes the magnetic field to change from a condition in which flux lines thread through the coil from right to left to a condition of zero flux. The induced current must then be as shown in Figure 20.16c, so as to set up its own field from right to left.

TAPE RECORDERS

One common practical use of induced currents and emfs is in the tape recorder. Many different types of tape recorders are made, but the basic principles are the same for all. A magnetic tape moves past a recording and playback head, as in Figure 20.17a. The tape is a plastic ribbon coated with iron oxide or chromium oxide.

The recording process uses the fact that a current in an electromagnet produces a magnetic field. Figure 20.17b illustrates the steps in the process. A sound wave sent into a microphone is transformed into an electric current, amplified, and allowed to pass through a wire coiled around a doughnut-shaped piece of iron, which functions as the recording head. The iron ring and the wire constitute an electromagnet, in which the lines of the magnetic field are contained completely inside the iron except at the point where a slot is cut in the ring. Here the magnetic field fringes out of the iron and magnetizes the small pieces of iron oxide embedded in the tape. Thus, as the tape moves past the slot, it becomes magnetized in a pattern that reproduces both the frequency and the intensity of the sound signal entering the microphone.

To reconstruct the sound signal, the previously magnetized tape is allowed to pass through a recorder head operating in the playback mode. A second wirewound doughnut-shaped piece of iron with a slot in it passes close to the tape, so that the varying magnetic fields on the tape produce changing field lines through the wire coil. The changing flux induces a current in the coil that corresponds to the current in the recording head that originally produced the tape. This changing electric current can be amplified and used to drive a speaker. Playback is thus an example of induction of a current by a moving magnet.

> ◄ **APPLICATION**
>
> MAGNETIC TAPE RECORDERS

Webnote 20.2

Take a look at audiotapes, tape decks, and tape recorders by visiting
http://www.howstuffworks.com/cassette.htm

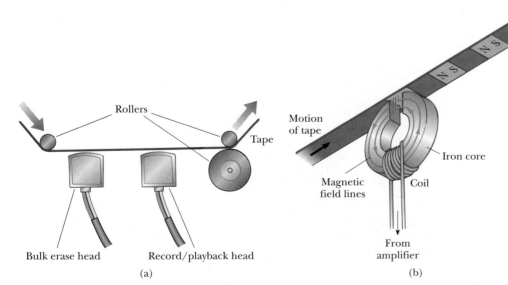

FIGURE 20.17 (a) Major parts of a magnetic tape recorder. If a new recording is to be made, the bulk erase head wipes the tape clean of signals before recording. (b) The fringing magnetic field magnetizes the tape during recording.

20.5   GENERATORS

APPLICATION

ALTERNATING CURRENT
GENERATORS

Generators and motors are important practical devices that operate on the principle of electromagnetic induction. First, let us consider the **alternating current (AC) generator,** a device that converts mechanical energy to electrical energy. In its simplest form, the AC generator consists of a wire loop rotated in a magnetic field by some external means (Fig. 20.18a). In commercial power plants, the energy required to rotate the loop can be derived from a variety of sources. For example, in a hydroelectric plant, falling water directed against the blades of a turbine produces the rotary motion; in a coal-fired plant, heat produced by burning coal is used to convert water to steam, and this steam is directed against the turbine blades. As the loop rotates, the magnetic flux through it changes with time, inducing an emf and a current in an external circuit. The ends of the loop are connected to slip rings that rotate with the loop. Connections to the external circuit are made by stationary brushes in contact with the slip rings.

We can derive an expression for the emf generated in the rotating loop by making use of the expression for motional emf, $\mathcal{E} = B\ell v$. Figure 20.19a shows a loop of wire rotating clockwise in a uniform magnetic field directed to the right. The magnetic force (qvB) on the charges in wires AB and CD is not along the lengths of the wires. (The force on the electrons in these wires is perpendicular to the wires.) Hence, an emf is generated only in wires BC and AD. At any instant, wire BC has velocity **v** at an angle θ with the magnetic field, as shown in Figure 20.19b. (Note that the component of velocity parallel to the field has no effect on the charges in the wire, whereas the component of velocity perpendicular to the field produces a magnetic force on the charges that moves electrons from C to B.) The emf generated in wire BC equals $B\ell v_\perp$, where ℓ is the length of the wire and $v_\perp$ is the component of velocity perpendicular to the field. An emf of $B\ell v_\perp$ is also generated in wire DA, and the sense of this emf is the same as in wire BC. Because $v_\perp = v \sin \theta$, the total induced emf is

$$\mathcal{E} = 2B\ell v_\perp = 2B\ell v \sin \theta \qquad [20.6]$$

If the loop rotates with a constant angular speed ω, we can use the relation $\theta = \omega t$ in Equation 20.6. Furthermore, because every point on wires BC and DA rotates in a circle about the axis of rotation with the same angular speed ω, we have $v = r\omega = (a/2)\omega$, where a is the length of sides AB and CD. Therefore, Equation 20.6 reduces to

$$\mathcal{E} = 2B\ell \left(\frac{a}{2}\right) \omega \sin \omega t = B\ell a\omega \sin \omega t$$

If a coil has N turns, the emf is N times as large because each loop has the same

FIGURE 20.18 (a) Diagram of an AC generator. An emf is induced in a coil, which rotates by some external means in a magnetic field. (b) Plot of the alternating emf induced in the loop versus time.

(a)

(b)

(a) (b)

emf induced in it. Furthermore, because the area of the loop is $A = \ell a$, the total emf is

$$\mathcal{E} = NBA\omega \sin \omega t \qquad \text{[20.7]}$$

This result shows that the emf varies sinusoidally with time, as plotted in Figure 20.18b. Note that the maximum emf has the value

$$\mathcal{E}_{max} = NBA\omega \qquad \text{[20.8]}$$

which occurs when $\omega t = 90°$ or $270°$. In other words, $\mathcal{E} = \mathcal{E}_{max}$ when the plane of the loop is parallel to the magnetic field. Furthermore, the emf is zero when $\omega t = 0$ or $180°$, that is, when the magnetic field is perpendicular to the plane of the loop. In the United States and Canada, the frequency of rotation for commercial generators is 60 Hz, whereas in some European countries 50 Hz is used. (Recall that $\omega = 2\pi f$, where f is the frequency in hertz.)

The **direct current (DC) generator** is illustrated in Figure 20.20a. The components are essentially the same as those of the AC generator, except that the contacts to the rotating loop are made by a split ring, or commutator. In this design, the output voltage always has the same polarity and the current is a pulsating direct current, as in Figure 20.20b. This can be understood by noting that the contacts to the split ring reverse their roles every half cycle. At the same time, the polarity of the induced emf reverses. Hence, the polarity of the split ring remains the same.

A pulsating DC current is not suitable for most applications. To produce a steady DC current, commercial DC generators use many loops and commutators distributed around the axis of rotation so that the sinusoidal pulses from the loops overlap in phase. When these pulses are superimposed, the DC output is almost free of fluctuations.

Turbines turn electric generators at a hydroelectric power plant. *(Luis Castaneda/The IMAGE Bank)*

APPLICATION

DIRECT CURRENT GENERATORS

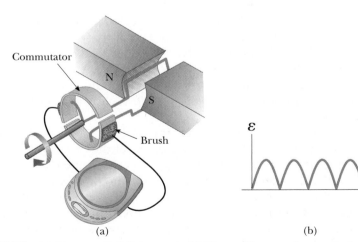

(a) (b)

FIGURE 20.20 (a) Diagram of a DC generator. (b) The emf fluctuates in magnitude but always has the same polarity.

Example 20.5   Emf Induced in an AC Generator

An AC generator consists of eight turns of wire of area $A = 0.090\ 0\ \text{m}^2$ with a total resistance of 12.0 Ω. The loop rotates in a magnetic field of 0.500 T at a constant frequency of 60.0 Hz.

A Find the maximum induced emf.

Solution   First note that $\omega = 2\pi f = 2\pi(60.0\ \text{Hz}) = 377\ \text{rad/s}$. When we substitute the appropriate numerical values into Equation 20.8, we obtain

$$\mathcal{E}_{\text{max}} = NAB\omega = 8(0.90\ 0\ \text{m}^2)(0.500\ \text{T})(377\ \text{rad/s}) = \boxed{136\ \text{V}}$$

B What is the maximum induced current?

Solution   From Ohm's law and the result of part A, we find that

$$I_{\text{max}} = \frac{\mathcal{E}_{\text{max}}}{R} = \frac{136\ \text{V}}{12.0\ \Omega} = \boxed{11.3\ \text{A}}$$

C Determine the induced emf as a function of time.

Solution   We can use Equation 20.7 to obtain the time variation of $\mathcal{E}$:

$$\mathcal{E} = \mathcal{E}_{\text{max}} \sin \omega t = \boxed{(136\ \text{V}) \sin 377t}$$

where *t* is in seconds.

Exercise   Determine the time variation of the induced current.

Answer $I = (11.3\ \text{A}) \sin 377t$

APPLICATION

MOTORS

MOTORS AND BACK emf

Motors are devices that convert electrical energy to mechanical energy. Essentially, **a motor is a generator run in reverse.** Instead of a current being generated by a rotating loop, a current is supplied to the loop by a source of emf, and the magnetic torque on the current-carrying loop causes it to rotate.

A motor can perform useful mechanical work when a shaft connected to its rotating coil is attached to some external device. As the coil in the motor rotates, however, the changing magnetic flux through it induces an emf, which acts to reduce the current in the coil. If it increased the current, Lenz's law would be violated. The phrase **back emf** is used for an emf that tends to reduce the applied current. The back emf increases in magnitude as the rotational speed of the coil increases. We can picture this state of affairs as the equivalent circuit in Figure 20.21. For illustrative purposes, assume that the external power source attempting to supply current in the coil of the motor has a voltage of 120 V, that the coil has a resistance of 10 Ω, and that the back emf induced in the coil at this instant is 70 V. Thus, the voltage available to supply current equals the difference between the applied voltage and the back emf, 50 V in this case. It is clear that the current is reduced by the back emf.

When a motor is turned on, there is no back emf initially, and the current is very large because it is limited only by the resistance of the coil. As the coil begins to rotate, the induced back emf opposes the applied voltage and the current in the coil is reduced. If the mechanical load increases, the motor slows down, which decreases the back emf. This reduction in the back emf increases the current in the coil and therefore also increases the power needed from the external voltage source. As a consequence, the power requirements for starting a motor and for running it under heavy loads are greater than those for running the motor under average loads. If the motor is allowed to run under no mechanical load, the back

10 Ω coil   70 V
resistance   back emf

120 V external source

FIGURE 20.21   A motor can be represented as a resistance plus a back emf.

emf reduces the current to a value just large enough to balance energy losses by heat and friction.

Example 20.6 The Induced Current in a Motor

A motor has coils with a resistance of 10 Ω and is supplied by a voltage of 120 V. When the motor is running at its maximum speed, the back emf is 70 V. Find the current in the coils (a) when the motor is first turned on and (b) when the motor has reached maximum speed.

Solution

A When the motor is first turned on, the back emf is zero. (The coils are motionless.) Thus, the current in the coils is a maximum and is

$$I = \frac{\mathcal{E}}{R} = \frac{120 \text{ V}}{10 \text{ Ω}} = \boxed{12 \text{ A}}$$

B At the maximum speed, the back emf has its maximum value. Thus, the effective supply voltage is now that of the external source minus the back emf, and the current is reduced to

$$I = \frac{\mathcal{E} - \mathcal{E}_{\text{back}}}{R} = \frac{120 \text{ V} - 70 \text{ V}}{10 \text{ Ω}} = \frac{50 \text{ V}}{10 \text{ Ω}} = \boxed{5.0 \text{ A}}$$

EXERCISE If the current in the motor is 8.0 A at some instant, what is the back emf at this time?

ANSWER 40 V

20.6 SELF-INDUCTANCE

Consider a circuit consisting of a switch, a resistor, and a source of emf, as in Figure 20.22. When the switch is closed, the current does not immediately change from zero to its maximum value, $\mathcal{E}/R$. The law of electromagnetic induction, Faraday's law, prevents this. What happens instead is the following. As the current increases with time, the magnetic flux through the loop due to this current also increases. The increasing flux induces an emf in the circuit that opposes the change in magnetic flux. By Lenz's law, the induced emf is in the direction indicated by the dashed battery in Figure 20.22. The net potential difference across the resistor is the emf of the battery minus the opposing induced emf. As the magnitude of the current increases, the *rate* of increase lessens and hence the induced emf decreases. This opposing emf results in a gradual increase in the current. For the same reason, when the switch is opened, the current does not immediately decrease to zero. This effect is called **self-induction** because the changing flux through the circuit arises from the circuit itself. The emf that is set up in this case is called a **self-induced emf.**

As a second example of self-inductance, consider Figure 20.23, which shows a coil wound on a cylindrical iron core. (A practical device would have several hundred turns.) Assume that the current changes with time. When the current is in the direction shown, a magnetic field is set up inside the coil, directed from right to left. As a result, some lines of magnetic flux pass through the cross-sectional area of the coil. As the current changes with time, the flux through the coil changes and induces an emf in the coil. Application of Lenz's law shows that this induced emf has a direction so as to oppose the change in the current. That is, if the current is increasing, the induced emf is as pictured in Figure 20.23b, and if the current is decreasing, the induced emf is as shown in Figure 20.23c.

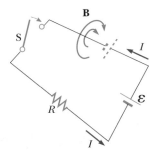

FIGURE 20.22 After the switch in the circuit is closed, the current produces its own magnetic flux through the loop. As the current increases toward its equilibrium value, the flux changes in time and induces an emf in the loop. The battery drawn with dashed lines is a symbol for the self-induced emf.

FIGURE 20.23 (a) A current in the coil produces a magnetic field directed to the left. (b) If the current increases, the coil acts as a source of emf directed as shown by the dashed battery. (c) The induced emf in the coil changes its polarity if the current decreases.

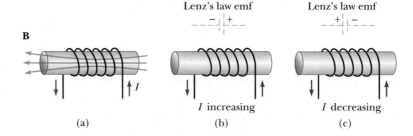

To evaluate self-inductance quantitatively, first note that, according to Faraday's law, the induced emf is given by Equation 20.2:

$$\varepsilon = -N\frac{\Delta\Phi_B}{\Delta t}$$

The magnetic flux is proportional to the magnetic field, which is proportional to the current in the coil. Thus, **the self-induced emf must be proportional to the time rate of change of the current:**

Self-induced emf ▶

$$\varepsilon \equiv -L\frac{\Delta I}{\Delta t} \qquad \text{[20.9]}$$

where L is a proportionality constant called the **inductance** of the device. The negative sign indicates that a changing current induces an emf in opposition to that change. This means that if the current is increasing (ΔI positive), the induced emf is negative, to indicate opposition to the increase in current. Likewise, if the current is decreasing (ΔI negative), the sign of the induced emf is positive to indicate that the emf is acting to oppose the decrease.

The inductance of a coil depends on the cross-sectional area of the coil and other quantities, all of which can be grouped under the general heading of geometric factors. The SI unit of inductance is the **henry** (H), which, from Equation 20.9, is equal to 1 volt-second per ampere:

$$1\,\text{H} = 1\,\text{V}\cdot\text{s/A}$$

Examples 20.7 and 20.8 discuss simple situations for which self-inductances can be calculated. In the process, it is often convenient to equate Equations 20.2 and 20.9 to find an expression for L:

$$N\frac{\Delta\Phi_B}{\Delta t} = L\frac{\Delta I}{\Delta t}$$

Inductance ▶

$$L = N\frac{\Delta\Phi_B}{\Delta I} = \frac{N\Phi_B}{I} \qquad \text{[20.10]}$$

APPLYING **PHYSICS** **20.3**

In some circuits, a spark occurs between the poles of a switch when the switch is opened. Yet, when the switch for this circuit is closed, there is no spark. Why is there this difference?

Explanation According to Lenz's law, induced emfs are in a direction such as to attempt to maintain the original magnetic flux when a change occurs. When the switch is opened, the sudden drop in the magnetic field in the circuit induces an emf in a direction that attempts to maintain the original current. This can cause a spark as the current bridges the air gap between the poles of the switch. The spark does not occur when the switch is closed, because the original current is zero, and the induced emf attempts to maintain it at zero.

Example 20.7   Inductance of a Solenoid

Find the inductance of a uniformly wound solenoid with N turns and length ℓ. Assume that ℓ is large compared with the radius and that the core of the solenoid is air.

Reasoning   The inductance can be found from $L = N\Phi_B/I$. The flux through each turn is $\Phi_B = BA$, and $B = \mu_0 nI$.

Solution   We take the interior field to be uniform and given by Equation 19.15:

$$B = \mu_0 nI = \mu_0 \frac{N}{\ell} I$$

where $n = N/\ell$ is the number of turns per unit length. The flux through each turn is

$$\Phi_B = BA = \mu_0 \frac{N}{\ell} AI$$

where A is the cross-sectional area of the solenoid. From this expression and Equation 20.10, we find that

$$L = \frac{N\Phi_B}{I} = \frac{\mu_0 N^2 A}{\ell} \qquad (1)$$

This shows that L depends on the geometric factors ℓ and A and on μ_0 and is proportional to the square of the number of turns. Because $N = n\ell$ we can also express the result in the form

$$L = \mu_0 \frac{(n\ell)^2}{\ell} A = \mu_0 n^2 A\ell = \mu_0 n^2 V \qquad (2)$$

where $V = A\ell$ is the volume of the solenoid.

JOSEPH HENRY, AMERICAN PHYSICIST (1797–1878)

Henry became the first director of the Smithsonian Institution and first president of the Academy of Natural Science. He was the first to produce an electric current with a magnetic field, but he failed to publish his results as early as Faraday because of his heavy teaching duties at the Albany Academy in New York state. He improved the design of the electromagnet and constructed one of the first motors. He also discovered the phenomena of self-induction. The unit of inductance, the henry, is named in his honor. (*North Wind Picture Archives*)

Example 20.8   Calculating Inductance and Self-Induced emf

A Calculate the inductance of a solenoid containing 300 turns if the length of the solenoid is 25.0 cm and its cross-sectional area is 4.00 cm^2 = 4.00×10^{-4} m^2.

Solution   Using the results to Example 20.7, we get

$$L = \frac{\mu_0 N^2 A}{\ell} = (4\pi \times 10^{-7}\,\text{T}\cdot\text{m/A}) \frac{(300)^2(4.00 \times 10^{-4}\,\text{m}^2)}{25.0 \times 10^{-2}\,\text{m}}$$

$$= 1.81 \times 10^{-4}\,\text{T}\cdot\text{m}^2/\text{A} = \boxed{0.181\,\text{mH}}$$

B Calculate the self-induced emf in the solenoid described in part A if the current through it is decreasing at the rate of 50.0 A/s.

Solution   Equation 20.9 can be combined with $\Delta I/\Delta t = -50.0$ A/s to give

$$\varepsilon = -L\frac{\Delta I}{\Delta t} = -(1.81 \times 10^{-4}\,\text{H})(-50.0\,\text{A/s}) = \boxed{9.05\,\text{mV}}$$

20.7   *RL* CIRCUITS

A circuit element that has a large inductance, such as a closely wrapped coil of many turns, is called an **inductor.** The circuit symbol for an inductor is ⎍⎍⎍ . We shall always assume that the self-inductance of the remainder of the circuit is negligible compared with that of the inductor in the circuit.

FIGURE 20.24   A comparison of the effect of a resistor to that of an inductor in a simple circuit.

FIGURE 20.25   A series *RL* circuit. As the current increases toward its maximum value, the inductor produces an emf that opposes the increasing current.

FIGURE 20.26   Plot of current versus time for the *RL* circuit shown in Figure 20.25. The switch is closed at $t = 0$, and the current increases toward its maximum value $\mathcal{E}/R$. The time constant τ is the time it takes the current to reach 63.2% of its maximum value.

To gain some insight into the effect of an inductor in a circuit, consider the two circuits in Figure 20.24. Figure 20.24a shows a resistor connected to the terminals of a battery. For this circuit, Kirchhoff's loop rule is $\mathcal{E} - IR = 0$. The voltage drop across the resistor is

$$\Delta V_R = -IR \qquad [20.11]$$

In this case, we interpret **resistance as a measure of opposition to the current.** Now consider the circuit in Figure 20.24b, consisting of an inductor connected to the terminals of a battery. At the instant the switch in this circuit is closed, because $IR = 0$, the emf of the battery equals the back emf generated in the coil. Thus, we have

$$\mathcal{E}_L = -L\frac{\Delta I}{\Delta t} \qquad [20.12]$$

From this expression, **we can interpret L as a measure of opposition to the rate of change in current.**

Figure 20.25 shows a circuit consisting of a resistor, inductor, and battery. Suppose the switch is closed at $t = 0$. The current begins to increase, but the inductor produces an emf that opposes the increasing current. Thus, the current is unable to change from zero to its maximum value of $\mathcal{E}/R$ instantaneously. Equation 20.12 shows that the induced emf is a maximum when the current is changing most rapidly, which occurs when the switch is first closed. As the current approaches its steady-state value, the back emf of the coil falls off because the current is changing more slowly. Finally, when the current reaches its steady-state value, the rate of change is zero and the back emf is also zero. Figure 20.26 plots current in the circuit as a function of time.[1] This plot is similar to that of the charge on a capacitor as a function of time, discussed in Chapter 18. In that case, we found it convenient to introduce a quantity called the *time constant of the circuit,* which told us something about the time required for the capacitor to approach its steady-state charge. In the same fashion, time constants are defined for circuits containing resistors and inductors. The **time constant** τ for an *RL* circuit is the time required for the current in the circuit to reach 63.2% of its final value $\mathcal{E}/R$; the time constant of an *RL* circuit is given by

Time constant for an *RL* circuit ▶

$$\tau = \frac{L}{R} \qquad [20.13]$$

---

[1] The equation for the current in the circuit as a function of time may be obtained from calculus and is

$$I = \frac{\mathcal{E}}{R}(1 - e^{-Rt/L})$$

Quick Quiz 20.5

The switch in the circuit shown in Figure 20.27 is closed and the lightbulb glows steadily. The inductor is a simple air-core solenoid. An iron rod is inserted into the interior of the solenoid, which increases the magnitude of the magnetic field in the solenoid. As the rod is inserted into the solenoid, the brightness of the lightbulb (a) increases, (b) decreases, or (c) remains the same.

FIGURE 20.27   (Quick Quiz 20.5)

Example 20.9   The Time Constant for an *RL* Circuit

The circuit shown in Figure 20.25 consists of a 30-mH inductor, a 6.0-Ω resistor, and a 12-V battery. The switch is closed at $t = 0$.

A Find the time constant of the circuit.

Solution   The time constant is given by Equation 20.13:

$$\tau = \frac{L}{R} = \frac{30 \times 10^{-3}\ \text{H}}{6.0\ \Omega} = \boxed{5.0\ \text{ms}}$$

B Find the current after one time constant has elapsed.

Solution   After one time constant, the current in the circuit has risen to 63.2% of its final value. Thus, the current is

$$I = 0.632\,\frac{\mathcal{E}}{R} = (0.632)\left(\frac{12\ \text{V}}{6.0\ \Omega}\right) = \boxed{1.3\ \text{A}}$$

EXERCISE   What is the voltage drop across the resistor (a) at $t = 0$? (b) after one time constant?

ANSWER   (a) 0, (b) 7.6 V

20.8   ENERGY STORED IN A MAGNETIC FIELD

The emf induced by an inductor prevents a battery from establishing an instantaneous current in a circuit. The battery has to do work to produce a current. We can think of this needed work as energy stored by the inductor in its magnetic field. In a manner quite similar to that used in Section 16.9 to find the energy stored by a capacitor, we find that the energy stored by an inductor is

Energy stored in an inductor ▶

$$PE_L = \tfrac{1}{2}LI^2 \qquad\qquad \textbf{[20.14]}$$

Note that the result is similar in form to the expression for the energy stored in a charged capacitor (Eq. 16.18):

Energy stored in a capacitor ▶

$$PE_C = \tfrac{1}{2}C(\Delta V)^2$$

SUMMARY

The magnetic flux Φ_B through a closed loop is defined as

$$\Phi_B \equiv BA \cos \theta \qquad\qquad \textbf{[20.1]}$$

where B is the strength of the uniform magnetic field, A is the cross-sectional area of the loop, and θ is the angle between **B** and the direction perpendicular to the plane of the loop.

Faraday's law of induction states that the instantaneous emf induced in a circuit equals the rate of change of magnetic flux through the circuit:

$$\varepsilon = -N\frac{\Delta\Phi_B}{\Delta t} \qquad\qquad \textbf{[20.2]}$$

where N is the number of loops in the circuit.

Lenz's law states that the polarity of the induced emf is such that it produces a current whose magnetic field opposes the *change* in magnetic flux through a circuit.

If a conducting bar of length ℓ moves through a magnetic field with a speed v so that **B** is perpendicular to the bar, the emf induced in the bar, often called a **motional emf,** is

$$|\varepsilon| = B\ell v \qquad\qquad \textbf{[20.4]}$$

When a coil of wire with N turns, each of area A, rotates with constant angular speed ω in a uniform magnetic field **B** as in Figure 20.19, the emf induced in the coil is

$$\varepsilon = NAB\omega \sin \omega t \qquad\qquad \textbf{[20.7]}$$

When the current in a coil changes with time, an emf is induced in the coil according to Faraday's law. This **self-induced emf** is defined by the expression

$$\varepsilon \equiv -L\frac{\Delta I}{\Delta t} \qquad\qquad \textbf{[20.9]}$$

where L is the inductance of the coil. The SI unit for inductance is the henry (H); $1\,\text{H} = 1\,\text{V}\cdot\text{s/A}$.

The **inductance** of a coil can be found from the expression

$$L = \frac{N\Phi_B}{I} \qquad\qquad \textbf{[20.10]}$$

where N is the number of turns on the coil, I is the current in the coil, and Φ_B is the magnetic flux through the coil produced by that current.

If a resistor and inductor are connected in series to a battery and a switch is closed at $t = 0$, the current in the circuit does not rise instantly to its maximum value. After one **time constant** $\tau = L/R$ the current in the circuit is 63.2% of its final value, ε/R.

The **energy stored** in the magnetic field of an inductor carrying current I is

$$PE_L = \tfrac{1}{2}LI^2 \qquad\qquad \textbf{[20.14]}$$

CONCEPTUAL QUESTIONS

1. A circular loop is located in a uniform and constant magnetic field. Describe how an emf can be induced in the loop in this situation.

2. Does dropping a magnet down a copper tube produce a current in the tube? Explain.

3. A spacecraft orbiting Earth has a coil of wire in it. An astronaut measures a small current in the coil, although no battery is connected to it and there are no magnets in the spacecraft. What is causing the current?

4. A loop of wire is placed in a uniform magnetic field. For what orientation of the loop is the magnetic flux a maximum? For what orientation is the flux zero?

5. As the conducting bar in Figure Q20.5 moves to the right, an electric field is set up directed downward. If the bar were moving to the left, explain why the electric field would be upward.

FIGURE Q20.5   (Conceptual Questions 5 and 6)

6. As the bar in Figure Q20.5 moves perpendicular to the field, is an external force required to keep it moving with constant speed?

7. Wearing a metal bracelet in a region of strong magnetic field can be hazardous. Discuss.

8. How is electrical energy produced in dams (that is, how is the energy of motion of the water converted to ac electricity)?

9. Eddy currents are induced currents set up in a piece of metal when it moves through a nonuniform magnetic field. For example, consider the flat metal plate swinging at the end of a bar as a pendulum as shown in Figure Q20.9. At position 1, the pendulum is moving from a region where there is no magnetic field into a region where the field $\mathbf{B}_{in}$ is directed into the paper. Show that at position 1, the direction of the eddy current is counterclockwise. Also, at position 2

the pendulum is moving out of the field into a region of zero field. Show that the direction of the eddy current is clockwise in this case. Use right-hand rule #2 to show that these eddy currents lead to a magnetic force on the plate directed as shown in the figure. Because the induced eddy current always produces a retarding force when the plate enters or leaves the field, the swinging plate quickly comes to rest.

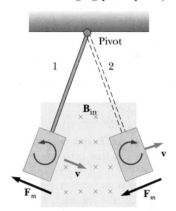

FIGURE Q20.9

10. Suppose you would like to steal power for your home from the electric company by placing a loop of wire near a transmission cable in order to induce an emf in the loop (an illegal procedure). Should you locate the loop so that the transmission cable passes through your loop or simply place your loop near the transmission cable? Does the orientation of the loop matter?

11. A piece of aluminum is dropped vertically downward between the poles of an electromagnet. Does the magnetic field affect the velocity of the aluminum? (*Hint:* See Conceptual Question 9.)

12. A bar magnet is dropped toward a conducting ring lying on the floor. As the magnet falls toward the ring, does it move as a freely falling object?

13. If the current in an inductor is doubled, by what factor does the stored energy change?

14. Is it possible to induce a constant emf for an infinite amount of time?

15. Why is the induced emf that appears in an inductor called a back (counter) emf?

16. A magneto is used to cause the spark in a spark plug in many lawn mowers today. A magneto consists of a permanent magnet mounted on the flywheel so that it spins past a fixed coil. Explain how this arrangement generates a large enough potential difference to cause the spark.

PROBLEMS

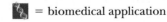

1, 2, 3 = straightforward, intermediate, challenging   ☐ = full solution available in Student Solutions Manual/Study Guide

web = solution posted at **http://info.brookscole.com/serway**   🔬 = biomedical application

Section 20.1   Induced emf and Magnetic Flux

1. A magnetic field of strength 0.30 T is directed perpendicular to a plane circular loop of wire of radius 25 cm.

Find the magnetic flux through the area enclosed by this loop.

2. A circular loop with a radius of 0.200 m is placed in a uniform magnetic field of magnitude 0.850 T. The normal to

the loop makes an angle of 30.0° with respect to the direction of **B**. If the field increases to 0.950 T, what is the increase in magnetic flux through the loop?

3. Consider a place where Earth's magnetic field has strength of 0.520×10^{-4} T and makes an angle of 62.0° with the horizontal. At this location a magnetic compass points toward true north. What is the magnetic flux in a rectangular loop of wire that is 15.0 by 25.0 cm and is lying on a table? What is the magnetic flux in this loop if it is mounted vertically on a north wall? On an east wall?

4. A long, straight wire carrying a current of 2.00 A is placed along the axis of a cylinder of radius 0.500 m and a length of 3.00 m. Determine the total magnetic flux through the cylinder.

5. A long, straight wire lies in the plane of a circular coil with a radius of 0.010 m. The wire carries a current of 2.0 A and is placed along a diameter of the coil. (a) What is the net flux through the coil? (b) If the wire passes through the center of the coil and is perpendicular to the plane of the coil, find the net flux through the coil.

6. A solenoid 4.00 cm in diameter and 20.0 cm long has 250 turns and carries a current of 15.0 A. Calculate the magnetic flux through the circular cross-sectional area of the solenoid.

7. A cube of edge length $\ell = 2.5$ cm is positioned as shown in Figure P20.7. There is a uniform magnetic field throughout the region with components of $B_x = +5.0$ T, $B_y = +4.0$ T, and $B_z = +3.0$ T. (a) Calculate the flux through the shaded face of the cube. (b) What is the total flux emerging from the volume enclosed by the cube (that is, total flux through all six faces)?

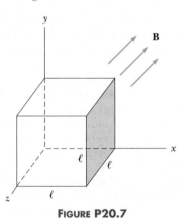

FIGURE P20.7

Section 20.2   Faraday's Law of Induction

8. A circular coil of radius 20 cm is placed in an external magnetic field of strength 0.20 T so that the plane of the coil is perpendicular to the field. The coil is pulled out of the field in 0.30 s. Find the average induced emf during this interval.

9. A 25-turn circular coil of wire has a diameter of 1.00 m. It is placed with its axis along the direction of Earth's magnetic field of 50.0 μT, and then in 0.200 s it is flipped 180°. An average emf of what magnitude is generated in the coil?

10. The flexible loop in Figure P20.10 has a radius of 12 cm and is in a magnetic field of strength 0.15 T. The loop is grasped at points A and B and stretched until its area is

nearly zero. If it takes 0.20 s to close the loop, find the magnitude of the average induced emf in it during this time.

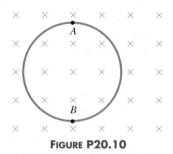

FIGURE P20.10

11. A strong electromagnet produces a uniform field of 1.60 T over a cross-sectional area of 0.200 m². We place a coil having 200 turns and a total resistance of 20.0 Ω around the electromagnet. We then smoothly decrease the current in the electromagnet until it reaches zero in 20.0 ms. What is the current induced in the coil?

12. A 500-turn circular-loop coil 15.0 cm in diameter is initially aligned so that its axis is parallel to Earth's magnetic field. In 2.77 ms, the coil is flipped so that its axis is perpendicular to Earth's magnetic field. If an average voltage of 0.166 V is thereby induced in the coil, what is the value of Earth's magnetic field at that location?

13. The plane of a rectangular coil, 5.0 cm by 8.0 cm, is perpendicular to the direction of a magnetic field **B**. If the coil has 75 turns and a total resistance of 8.0 Ω, at what rate must the magnitude of **B** change to induce a current of 0.10 A in the windings of the coil?

14. A square, single-turn wire loop 1.00 cm on a side is placed inside a solenoid that has a circular cross section of radius 3.00 cm, as shown in Figure P20.14. The solenoid is 20.0 cm long and wound with 100 turns of wire. (a) If the current in the solenoid is 3.00 A, find the flux through the loop. (b) If the current in the solenoid is reduced to zero in 3.00 s, find the magnitude of the average induced emf in the loop.

FIGURE P20.14

15. A 300-turn solenoid with a length of 20 cm and a radius of 1.5 cm carries a current of 2.0 A. A second coil of four turns is wrapped tightly about this solenoid so that it can be considered to have the same radius as the solenoid. Find (a) the change in the magnetic flux through the coil and (b) the magnitude of the average induced emf in the coil when the current in the solenoid increases to 5.0 A in a period of 0.90 s.

16. A circular coil, enclosing an area of 100 cm², is made of 200 turns of copper wire. The wire making up the coil has resistance of 5.0 Ω, and the ends of the wire are connected

to form a closed loop. Initially, a 1.1-T uniform magnetic field points perpendicularly upward through the plane of the coil. The direction of the field then reverses so that the final magnetic field has a magnitude of 1.1 T and points downward through the coil. If the time required for the field to reverse directions is 0.10 s, what average current flows through the coil during this time?

17. The person in Figure P20.17 is girded about the chest with a breathing monitor, which is a 100-turn coil. As a breath is inhaled, the area of the coil varies from 0.120 m² to 0.124 m². Earth's magnetic field is 50.0 μT and makes an angle of 22.5° with respect to the normal to the loop. If the patient inhales a breath in 1.59 s, what is the average voltage induced in the coil during the inhalation?

Coil

FIGURE P20.17

Section 20.3 Motional emf

18. Consider the arrangement shown in Figure P20.18. Assume that $R = 6.00\ \Omega$ and $\ell = 1.20$ m, and that a uniform 2.50-T magnetic field is directed *into* the page. At what speed should the bar be moved to produce a current of 0.500 A in the resistor?

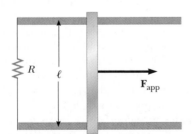

R ℓ $\mathbf{F}_{app}$

FIGURE P20.18   (Problems 18 and 57)

19. A Boeing-747 jet with a wing span of 60.0 m is flying horizontally at a speed of 300 m/s over Phoenix, Arizona, at a location where Earth's magnetic field is 50.0 μT at 58.0° below the horizontal. What voltage is generated between the wingtips?

20. Over a region where the *vertical* component of Earth's magnetic field is 40.0 μT directed downward, a 5.00-m length of wire is held in an east-west direction and moved horizontally to the north with a speed of 10.0 m/s. Calculate the potential difference between the ends of the wire, and determine which end is positive.

21. An automobile has a vertical radio antenna 1.20 m long. The automobile travels at 65.0 km/h on a horizontal road where Earth's magnetic field is 50.0 μT directed toward the north and downward at an angle of 65.0° below the horizontal. (a) Specify the direction that the automobile should move in order to generate the maximum motional emf in the antenna, with the top of the antenna positive relative to the bottom. (b) Calculate the magnitude of this induced emf.

22. A helicopter has blades of length 3.0 m, rotating at 2.0 rev/s about a central hub. If the vertical component of Earth's magnetic field is 5.0×10^{-5} T, what is the emf induced between the blade tip and the central hub?

Section 20.4 Lenz's Law Revisited (The Minus Sign in Faraday's Law)

23. A bar magnet is positioned near a coil of wire as shown in Figure P20.23. What is the direction of the current through the resistor when the magnet is moved (a) to the left? (b) to the right?

R

FIGURE P20.23

24. A bar magnet is held above the center of a wire loop in a horizontal plane, as shown in Figure P20.24. The south end of the magnet is toward the loop. The magnet is dropped. Find the direction of the current through the resistor (a) while the magnet is falling toward the loop and (b) after the magnet has passed through the loop and moves away from it.

R

FIGURE P20.24

25. What is the direction of the current induced in the resistor when the current in the long, straight wire in Figure P20.25 decreases rapidly to zero?

R

I

FIGURE P20.25

26. In Figure P20.26, what is the direction of the current induced in the resistor at the instant the switch is closed?

FIGURE P20.26

27. A copper bar is moved to the right while its axis is maintained in a direction perpendicular to a magnetic field, as shown in Figure P20.27. If the top of the bar becomes positive relative to the bottom, what is the direction of the magnetic field?

FIGURE P20.27

28. Find the direction of the current through the resistor in Figure P20.28, (a) at the instant the switch is closed, (b) after the switch has been closed for several minutes, and (c) at the instant the switch is opened.

FIGURE P20.28

29. Find the direction of the current in resistor R in Figure P20.29 after each of the following steps (taken in the order given). (a) The switch is closed. (b) The variable resistance in series with the battery is decreased. (c) The circuit containing resistor R is moved to the left. (d) The switch is opened.

FIGURE P20.29

Section 20.5   Generators

30. A 100-turn square wire coil of area 0.040 m^2 rotates about a vertical axis at 1 500 rev/min, as indicated in Figure P20.30. The horizontal component of Earth's magnetic field at the location of the loop is 2.0×10^{-5} T. Calculate the maximum emf induced in the coil by Earth's field.

FIGURE P20.30

31. An automobile generator produces 12.0 V when turning at 500 rev/min. What potential difference will it produce at 1 200 rev/min?

32. A motor has coils with a resistance of 30 Ω and operates from a voltage of 240 V. When the motor is operating at its maximum speed, the back emf is 145 V. Find the current in the coils (a) when the motor is first turned on and (b) when the motor has reached maximum speed. (c) If the current in the motor is 6.0 A at some instant, what is the back emf at that time?

33. When the coil of a motor is rotating at maximum speed, the current in the windings is 4.0 A. When the motor is first turned on, the current in the windings is 11 A. If the motor is operated at 120 V, find (a) the resistance of the windings and (b) the back emf in the coil at maximum speed.

34. A loop of area 0.10 m^2 is rotating at 60 rev/s with its axis of rotation perpendicular to a 0.20-T magnetic field. (a) If there are 1 000 turns on the loop, what is the maximum voltage induced in the loop? (b) When the maximum induced voltage occurs, what is the orientation of the loop with respect to the magnetic field?

35. In a model ac generator, a 500-turn rectangular coil, web 8.0 cm by 20 cm, rotates at 120 rev/min in a uniform magnetic field of 0.60 T. (a) What is the maximum emf induced in the coil? (b) What is the instantaneous value of the emf in the coil at $t = (\pi/32)$ s? Assume that the emf is zero at $t = 0$. (c) What is the smallest value of t for which the emf will have its maximum value?

Section 20.6   Self-Inductance

36. A coiled telephone cord forms a spiral with 70.0 turns, a diameter of 1.30 cm, and an unstretched length of 60.0 cm. Determine the self-inductance of one conductor in the unstretched cord.

37. A 2.00-H inductor carries a steady current of 0.500 A. When the switch in the circuit is thrown open, the current is effectively zero after 10.0 ms. What is the average induced emf in the inductor during this time?

38. Show that the two expressions for inductance given by

$$L = \frac{N\Phi_B}{I} \quad \text{and} \quad L = \frac{-\mathcal{E}}{\Delta I / \Delta t}$$

have the same units.

39. A solenoid of radius 2.5 cm has 400 turns and a length of 20 cm. Find (a) its inductance and (b) the rate at which current must change through it to produce an emf of 75 mV.

40. An emf of 24.0 mV is induced in a 500-turn coil when the current is changing at a rate of 10.0 A/s. What is the magnetic flux through each turn of the coil at an instant when the current is 4.00 A?

Section 20.7 RL Circuits

41. Show that the SI units for the inductive time constant $\tau = L/R$ are seconds.

42. An RL circuit with $L = 3.00$ H and an RC circuit with $C = 3.00$ μF have the same time constant. If the two circuits have the same resistance R, (a) what is the value of R, and (b) what is this common time constant?

43. A 6.0-V battery is connected in series with a resistor and an inductor. The series circuit has a time constant of 600 μs, and the maximum current is 300 mA. What is the value of the inductance?

44. A 25-mH inductor, an 8.0-Ω resistor, and a 6.0-V battery are connected in series. The switch is closed at $t = 0$. Find the voltage drop across the resistor (a) at $t = 0$ and (b) after one time constant has passed. Also, find the voltage drop across the inductor (c) at $t = 0$ and (d) after one time constant has elapsed.

45. Calculate the resistance in an RL circuit in which
web $L = 2.50$ H and the current increases to 90.0% of its final value in 3.00 s.

46. Consider the circuit in Figure P20.46, taking $\mathcal{E} = 6.00$ V, $L = 8.00$ mH, and $R = 4.00$ Ω. (a) What is the inductive time constant of the circuit? (b) Calculate the current in the circuit 250 μs after the switch is closed. (c) What is the value of the final steady-state current? (d) How long does it take the current to reach 80.0% of its maximum value?

FIGURE P20.46

20.8 Energy Stored in a Magnetic Field

47. How much energy is stored in a 70.0-mH inductor at an instant when the current is 2.00 A?

48. An air-core solenoid with 68 turns is 8.00 cm long and has a diameter of 1.20 cm. How much energy is stored in its magnetic field when it carries a current of 0.770 A?

49. A 24-V battery is connected in series with a resistor and an inductor, where $R = 8.0$ Ω and $L = 4.0$ H. Find the energy stored in the inductor (a) when the current reaches its

maximum value and (b) one time constant after the switch is closed.

ADDITIONAL PROBLEMS

50. What is the time constant for (a) the circuit shown in Figure P20.50a? (b) the circuit shown in Figure P20.50b?

FIGURE P20.50

51. In Figure P20.51, the bar magnet is being moved toward the loop. Is $(V_a - V_b)$ positive, negative, or zero during this motion? Explain.

FIGURE P20.51

52. Your physics teacher asks you to help her set up a demonstration of Faraday's law for the class. The apparatus consists of a strong permanent magnet that has a field of 0.10 T, a small 10-turn coil of radius 2.0 cm cemented on a wood frame with a handle, some flexible connecting wires, and an ammeter as in Figure P20.52. The idea is to pull the coil out of the center of the magnetic field as quickly as possible and read the average current registered on the meter. The combined resistance of the coil, leads, and meter is 2.0 Ω and you must flip the coil out of the field in about 0.20 s. The ammeter you must use has a full-scale sensitivity of 1 000 μA. Will this meter be sensitive enough to show clearly the induced current?

53. An 820-turn wire coil of resistance 24.0 Ω is placed on top
web of a 12 500-turn, 7.00-cm-long solenoid, as in Figure P20.53. Both coil and solenoid have cross-sectional areas of 1.00×10^{-4} m^2. (a) How long does it take the solenoid current to reach 0.632 times its maximum value? (b) Determine the average back emf caused by the self-inductance of the solenoid during this interval. The magnetic field produced by the solenoid at the location of the coil is one-half as strong as the field at the center of the

FIGURE P20.52

FIGURE P20.54

solenoid. (c) Determine the average rate of change in magnetic flux through each turn of the coil during this interval. (d) Find the magnitude of the average induced current in the coil.

FIGURE P20.53

54. Figure P20.54 is a graph of induced emf versus time for a coil of N turns rotating with angular speed ω in a uniform magnetic field directed perpendicularly to the axis of rotation of the coil. Copy this sketch (increasing the scale), and on the same set of axes show the graph of emf versus t when (a) the number of turns in the coil is doubled, (b) the angular speed is doubled, and (c) the angular speed is doubled while the number of turns in the coil is halved.

55. The plane of a square loop of wire with edge length $a = 0.200$ m is perpendicular to Earth's magnetic field at a point where $B = 15.0 \ \mu T$, as in Figure P20.55. The total resistance of the loop and the wires connecting it to the ammeter is 0.500 Ω. If the loop is suddenly collapsed by horizontal forces as shown, what total charge passes through the ammeter?

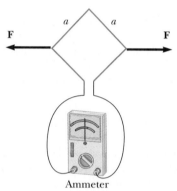

FIGURE P20.55

56. A novel method of storing electrical energy has been proposed. A huge underground superconducting coil, 1.00 km in diameter, would be fabricated. It would carry a maximum current of 50.0 kA through each winding of a 150-turn Nb_3Sn solenoid. (a) If the inductance of this huge coil is 50.0 H, what is the total energy stored? (b) What is the compressive force per meter length acting between two adjacent windings 0.250 m apart? (*Hint:* Because the radius of the coil is so large, the magnetic field created by one winding and acting on an adjacent turn can be considered to be that of a long, straight wire.)

57. A conducting rod of length ℓ moves on two horizontal frictionless rails, as in Figure P20.18. A constant force of magnitude 1.00 N moves the bar at a uniform speed of 2.00 m/s through a magnetic field **B** that is into the page. (a) What is the current in an 8.00-Ω resistor R? (b) What is the rate of energy dissipation in the resistor? (c) What is the mechanical power delivered by the constant force?

58. The square loop in Figure P20.58 is made of wires with total series resistance 10.0 Ω. It is placed in a uniform 0.100-T magnetic field directed perpendicular into the plane of the paper. The loop, which is hinged at each corner, is pulled as shown until the separation between points A and B is 3.00 m. If this process takes 0.100 s, what is the average current generated in the loop? What is the direction of the current?

59. The bolt of lightning depicted in Figure P20.59 passes 200 m from a 100-turn coil oriented as shown. If the cur-

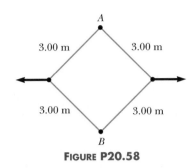

FIGURE P20.58

rent in the lightning bolt falls from 6.02×10^6 A to zero in 10.5 μs, what is the average voltage induced in the coil? Assume that the distance to the center of the coil determines the average magnetic field at the coil's position. Treat the lightning bolt as a long, vertical wire.

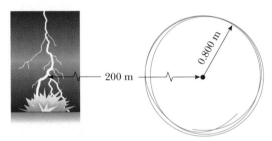

FIGURE P20.59

60. The wire shown in Figure P20.60 is bent in the shape of a "tent" with $\theta = 60°$ and $L = 1.5$ m, and is placed in a uniform magnetic field of 0.30 T perpendicular to the tabletop. The wire is "hinged" at points a and b. If the tent is flattened out on the table in 0.10 s, what is the average induced emf in the wire during this time?

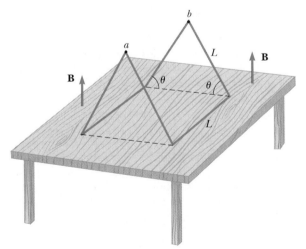

FIGURE P20.60

61. The magnetic field shown in Figure P20.61 has a uniform magnitude of 25.0 mT directed into the paper. The initial diameter of the kink is 2.00 cm. (a) The wire is quickly pulled taut, and the kink shrinks to a diameter of zero in 50.0 ms. Determine the average voltage induced between endpoints A and B. Include the polarity. (b) Suppose the kink is undisturbed, but the magnetic field increases to

100 mT in 4.00×10^{-3} s. Determine the average voltage across terminals A and B, including polarity, during this period.

FIGURE P20.61

62. An aluminum ring of radius 5.00 cm and resistance 3.00×10^{-4} Ω is placed around the top of a long air-core solenoid with 1 000 turns per meter and smaller radius 3.00 cm as in Figure P20.62. If the current in the solenoid is increasing at a constant rate of 270 A/s, what is the induced current in the ring? Assume that the magnetic field produced by the solenoid over the area at the end of the solenoid is one-half as strong as the field at the center of the solenoid. Assume that the solenoid produces a negligible field outside its cross-sectional area.

FIGURE P20.62

63. In Figure P20.63, the rolling axle, 1.50 m long, is pushed along horizontal rails at a constant speed $v = 3.00$ m/s. A resistor $R = 0.400$ Ω is connected to the rails at points a and b, directly opposite each other. (The wheels make good electrical contact with the rails, and so the axle, rails, and R form a closed-loop circuit. The only significant resistance in the circuit is R.) There is a uniform magnetic field $B = 0.800$ T vertically downward. (a) Find the induced current I in the resistor. (b) What horizontal force **F** is required to keep the axle rolling at constant speed? (c) Which end of the resistor, a or b, is at the higher electric potential? (d) After the axle rolls past the resistor, does the current in R reverse direction?

64. In 1832 Faraday proposed that the apparatus shown in Figure P20.64 could be used to generate electric current

FIGURE P20.63

FIGURE P20.64

from the flowing water in the Thames River.[2] Two conducting plates of lengths a and widths b are placed facing one another on opposite sides of the river, a distance w apart and immersed entirely. The flow velocity of the river is **v**, and the vertical component of Earth's magnetic field is B. Show that the current in the load resistor R is

$$I = \frac{abvB}{\rho + abR/w}$$

where ρ is the resistivity of the water. (b) Calculate the short-circuit current ($R = 0$) if $a = 100$ m, $b = 5.00$ m, $v = 3.00$ m/s, $B = 50.0$ μT, and $\rho = 100$ Ω m.

GROUP ACTIVITIES

G.1 Experimenting with induced currents is not easy. For small magnets and small coils of wire, the resulting induced currents are so small that they are difficult to detect. Thus, you may have to try this exercise several times before you are satisfied with your results. Wind a coil of wire on a cardboard mailing tube. Use insulated wire with as small a diameter as possible because you need as many turns as possible on the coil. Connect the coil to a flashlight bulb and see if you can get it to light by moving a bar magnet into and out of the coil in rapid succession. Why does the speed of movement make a difference? If you are unsuccessful, place two magnets side by side and repeat.

After you have experimented with the bulb, ask your instructor to let you use a galvanometer as a current detector. These devices are capable of measuring very small currents and they have the added advantage of detecting the direction of the current in the circuit.

Use your equipment to observe or test the following. (a) Does the magnitude of the induced current depend on the speed of movement of the magnet? (b) Can you induce a current by holding the magnet still and moving the coil over it? (c) Does the direction of the current depend on whether the magnet is pushed in or pulled out of the coil? (d) Does the direction of the current depend on whether the inserted pole of the magnet is north or south? (e) Can you predict the direction of the current by using Lenz's law? (f) Replace your bar magnet with the electromagnet you constructed in the last chapter and repeat the observations above.

G.2 As explained in the text, a cassette tape is made up of tiny particles of metal oxide attached to a long plastic strip. Pull a tape out of a cassette that you do not mind destroy-

ing and see if it is repelled or attracted by a refrigerator magnet. Also, try this with an expendable floppy computer disk.

G.3 This experiment takes steady hands, a dime, and a strong magnet. After verifying that a dime is not attracted to the magnet, carefully balance the coin on its edge. (This will not work with other coins because they require too much force to topple them.) Hold one pole of the magnet within a millimeter of the face of the dime, but do not make contact with it. Now very rapidly pull the magnet straight back away from the coin. Which way does the dime tip? Does the coin fall the same way most of the time? Explain what is going on in terms of Lenz's law.

G.4 A ramp runs from the bed of a truck down to the level ground. The ramp holds two parallel conducting rails connected at its base. A metal bar slides on the rails without friction. A magnet supplies an external magnetic field

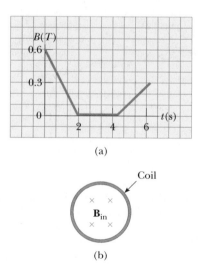

(a)

(b)

FIGURE GA20.5

[2] The idea for this problem and Figure P20.64 is from Oleg D. Jefimenko, *Electricity and Magnetism: An Introduction to the Theory of Electric and Magnetic Fields,* Star City, WV, Electret Scientific Co., 1989.

directed toward the ground. It is found that the bar slides down the ramp at a constant speed. (a) In a sketch, indicate the direction of the induced current in the bar. (b) Find the direction of the magnetic force acting on the bar. (c) What is the relationship between the forces that must be acting on the bar if it is to move down the ramp at constant speed? (d) How are the directions of the current and the force altered if the magnetic field is upward, away from the ground? Can the bar still slide at constant speed down the ramp in this case?

G.5 A one-turn coil of wire of area 0.20 m^2 and resistance 0.25 Ω is in a magnetic field that varies with time as shown in Figure GA20.5a. The magnetic flux through the coil at $t = 0$ is as shown in Figure GA20.5b. (a) When is the induced current the largest? (b) When is it zero? (c) Is the induced current always in the same direction? (d) Find the direction (clockwise or counterclockwise) and magnitude of the current between $t = 0$ and $t = 2.0$ s, between $t = 2.0$ s and $t = 4.0$ s, and between $t = 4.0$ s and $t = 6.0$ s.

21

Alternating Current Circuits and Electromagnetic Waves

Radiowaves from distant galaxies are detected by this array of radio telescopes. Everything that we know about celestial objects is determined by studying electromagnetic waves emitted by the objects. In this chapter, we will discuss the various kinds of electromagnetic waves that surround us. *(The Image Bank/Getty Images)*

*I*t is important to understand the basic principles of alternating-current (AC) circuits because they are so much a part of our everyday life. Every time we turn on a television set or a stereo, or any of a multitude of other electric appliances, we are calling on alternating currents to provide the power to operate them. We begin our study of AC circuits by examining the characteristics of a circuit containing a source of emf and one other circuit element: either a resistor, a capacitor, or an inductor. Then we examine what happens when these elements are connected in combination with each other. Our discussion is limited to situations in which the elements are arranged in simple series configurations.

We conclude this chapter with a discussion of **electromagnetic waves,** which are composed of fluctuating electric and magnetic fields. Electromagnetic waves in the form of visible light enable us to view the world around us; infrared waves warm our environment; radio-frequency waves carry our favorite television and radio programs; the list goes on and on.

21.1 RESISTORS IN AN AC CIRCUIT

An AC circuit consists of combinations of circuit elements and an AC generator or an AC source, which provides the alternating current. We have seen that the output of an AC generator is sinusoidal and varies with time according to

$$\Delta v = \Delta V_{max} \sin 2\pi ft \qquad [21.1]$$

where Δv is the instantaneous voltage, ΔV_{max} is the maximum voltage of the AC generator, and f is the frequency at which the voltage changes, measured in hertz (Hz). (Compare Eqs. 20.7 and 20.8 with Eq. 21.1.) We first consider a simple circuit consisting of a resistor and an AC source (designated by the symbol

——(∼)——) as in Figure 21.1. The current and the voltage across the resistor are shown in Figure 21.2.

Let us try to explain the concept of alternating current by briefly discussing the current-versus-time curve in Figure 21.2. At point *a* on the curve, the current has a maximum value in one direction, arbitrarily called the *positive direction*. Between points *a* and *b*, the current is decreasing in magnitude but is still in the positive direction. At point *b*, the current is momentarily zero; it then begins to increase in the opposite (negative) direction between points *b* and *c*. At point *c*, the current has reached its maximum value in the negative direction.

Note that the current and voltage are in step with each other because they vary identically with time. **Because the current and the voltage reach their maximum values at the same time, they are said to be in phase.** Also note that **the average value of the current over one cycle is zero.** That is, the current is maintained in one direction (the positive direction) for the same amount of time and at the same magnitude as it is in the opposite direction (the negative direction). However, the direction of the current has no effect on the behavior of the resistor in the circuit. This can be understood by realizing that collisions between electrons and the fixed atoms of the resistor result in an increase in the temperature of the resistor. Although this temperature increase depends on the magnitude of the current, it is independent of its direction.

We can quantify this discussion by recalling that the rate at which electrical energy is dissipated in a resistor, which is the power $\mathscr{P}$, is

$$\mathscr{P} = i^2 R$$

where i is the *instantaneous* current in the resistor. Because the heating effect of a current is proportional to the *square* of the current, it makes no difference whether the current is direct or alternating—that is, whether the sign associated with the current is positive or negative. However, the heating effect produced by an

$$\Delta v = \Delta V_{max} \sin 2\pi ft$$

FIGURE 21.1 A series circuit consisting of a resistor R connected to an AC generator.

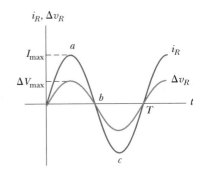

FIGURE 21.2 A plot of current and voltage across a resistor versus time.

FIGURE 21.3 (a) Plot of the current in a resistor as a function of time. (b) Plot of the current squared in a resistor as a function of time. Notice that the gray shaded regions *under* the curve and *above* the dashed lines for $I_{max}^2/2$ have the same area as the gray shaded regions *above* the curve and *below* the dashed line for $I_{max}^2/2$. Thus, the average value of i^2 is $I_{max}^2/2$.

alternating current with a maximum value of I_{max} *is not the same* as that produced by a direct current of the same value. This is because the alternating current has this maximum value for only an instant of time during a cycle. What is important in an AC circuit is a special kind of average value of current, referred to as the rms current. The **rms current** is the direct current that would dissipate the same amount of energy in a resistor as is dissipated by the actual alternating current. To find the rms current, first we square the current, next find its average value, and then take the square root of this average value. Hence, the rms current is the square *root* of the average (*mean*) of the *square* of the current. Because i^2 varies as $\sin^2 2\pi ft$, one can show that the average value of i^2 is $\frac{1}{2}I_{max}^2$ (Fig. 21.3).[1] Therefore, the rms current I_{rms} is related to the maximum value of the alternating current I_{max} by

rms current ▶

$$I_{rms} = \frac{I_{max}}{\sqrt{2}} = 0.707 I_{max} \qquad [21.2]$$

This equation says that an alternating current with a maximum value of 3 A produces the same heating effect in a resistor as a direct current of $(3/\sqrt{2})$ A. Thus, we can say that the average power dissipated in a resistor that carries alternating current I is $\mathscr{P}_{av} = I_{rms}^2 R$.

Alternating voltages are also best discussed in terms of rms voltages, with the relationship being identical to the preceding one; that is, the rms voltage ΔV_{rms} is related to the maximum value of the alternating voltage ΔV_{max} by

rms voltage ▶

$$\Delta V_{rms} = \frac{\Delta V_{max}}{\sqrt{2}} = 0.707 \Delta V_{max} \qquad [21.3]$$

[1] The fact that $(i^2)_{av} = I_{max}^2/2$ can be shown as follows. The current in the circuit varies with time according to the expression $i = I_{max} \sin 2\pi ft$, and so $i^2 = I_{max}^2 \sin^2 2\pi ft$. Therefore, we can find the average value of i^2 by calculating the average value of $\sin^2 2\pi ft$. Note that a graph of $\cos^2 2\pi ft$ versus time is identical to a graph of $\sin^2 2\pi ft$ versus time, except that the points are shifted on the time axis. Thus, the time average of $\sin^2 2\pi ft$ is equal to the time average of $\cos^2 2\pi ft$ when taken over one or more cycles. That is,

$$(\sin^2 2\pi ft)_{av} = (\cos^2 2\pi ft)_{av}$$

With this fact and the trigonometric identity $\sin^2 \theta + \cos^2 \theta = 1$, we get

$$(\sin^2 2\pi ft)_{av} + (\cos^2 2\pi ft)_{av} = 2(\sin^2 2\pi ft)_{av} = 1$$

$$(\sin^2 2\pi ft)_{av} = \frac{1}{2}$$

When this result is substituted into the expression $i^2 = I_{max}^2 \sin^2 2\pi ft$, we get $(i^2)_{av} = I_{rms}^2 = I_{max}^2/2$, or $I_{rms} = I_{max}/\sqrt{2}$, where I_{rms} is the rms current.

TABLE 21.1	Notation Used in This Chapter	
	Voltage	**Current**
Instantaneous value	Δv	i
Maximum value	ΔV_{max}	I_{max}
rms value	ΔV_{rms}	I_{rms}

When we speak of measuring an AC voltage of 120 V from an electric outlet, we really mean an rms voltage of 120 V. A quick calculation using Equation 21.3 shows that such an AC voltage actually has a peak value of about 170 V. In this chapter we use rms values when discussing alternating currents and voltages. One reason is that AC ammeters and voltmeters are designed to read rms values. Furthermore, if we use rms values, many of the equations for alternating current will have the same form as those used in the study of direct-current (DC) circuits. Table 21.1 summarizes the notations used in this chapter.

Consider the series circuit in Figure 21.1, consisting of a resistor connected to an AC generator. A resistor impedes the current in an AC circuit just as it does in a DC circuit. Therefore, Ohm's law is valid for an AC circuit, and we have

$$\Delta V_{R,rms} = I_{rms}R \qquad [21.4]$$

That is, **the rms voltage across a resistor is equal to the rms current in the circuit times the resistance.** This equation also applies if maximum values of current and voltage are used. That is, the maximum voltage drop across a resistor equals the maximum current in the resistor times the resistance.

Quick Quiz 21.1   Which of the following statements might be true for a resistor connected to an AC generator? (a) $\mathcal{P}_{av} = 0$ and $i_{av} = 0$, (b) $\mathcal{P}_{av} = 0$ and $i_{av} > 0$, (c) $\mathcal{P}_{av} > 0$ and $i_{av} = 0$, (d) $\mathcal{P}_{av} > 0$ and $i_{av} > 0$.

Example 21.1   What Is the rms Current?

An AC voltage source has an output of $\Delta v = (200 \text{ V}) \sin 2\pi ft$. This source is connected to a 100-Ω resistor as in Figure 21.1. Find the rms current in the resistor.

Reasoning   Compare the expression for the voltage output just given with the general form, $\Delta v = \Delta V_{max} \sin 2\pi ft$.

Solution   By comparison, we see that the maximum output voltage of the device is 200 V. Thus, the rms voltage output of the source is

$$\Delta V_{rms} = \frac{\Delta V_{max}}{\sqrt{2}} = \frac{200 \text{ V}}{\sqrt{2}} = 141 \text{ V}$$

Ohm's law can be used in purely resistive AC circuits as well as in DC circuits. The calculated rms voltage can be used with Ohm's law to find the rms current in the circuit:

$$I_{rms} = \frac{\Delta V_{rms}}{R} = \frac{141 \text{ V}}{100 \text{ }\Omega} = \boxed{1.41 \text{ A}}$$

EXERCISE   Find the maximum current in the circuit and the power delivered to the circuit.

ANSWER   2.00 A; 199 W

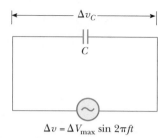

$$\Delta v = \Delta V_{max} \sin 2\pi ft$$

FIGURE 21.4 A series circuit consisting of a capacitor C connected to an AC generator.

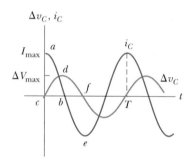

FIGURE 21.5 Plots of current and voltage across a capacitor versus time in an AC circuit. The voltage lags the current by 90°.

The voltage across a capacitor ▶ lags behind the current by 90°.

Capacitive reactance ▶

21.2 CAPACITORS IN AN AC CIRCUIT

To understand the effect of a capacitor on the behavior of a circuit containing an AC voltage source, let us first review what happens when a capacitor is placed in a circuit containing a DC source, such as a battery. At the instant a switch is closed in a series circuit containing a battery, a resistor, and a capacitor, there is zero charge on the plates of the capacitor. Therefore, the motion of charge through the circuit is relatively free, and initially there is a large current in the circuit. As more charge accumulates on the capacitor, the voltage across it increases, opposing the current. After some time interval, which depends on the time constant RC, the current approaches zero. From this, we see that a capacitor in a DC circuit limits or impedes the current so that it approaches zero after a brief time.

Now consider the simple series circuit in Figure 21.4, consisting of a capacitor connected to an AC generator. We sketch curves of current versus time and voltage versus time and then attempt to make the graphs seem reasonable. The curves are shown in Figure 21.5. First, note that the segment of the current curve from a to b indicates that the current starts out at a rather large value. This can be understood by recognizing that there is no charge on the capacitor at $t = 0$; as a consequence, there is nothing in the circuit except the resistance of the wires to hinder the flow of charge at this instant. However, the current decreases as the voltage across the capacitor increases from c to d on the voltage curve. When the voltage is at point d, the current reverses and begins to increase in the opposite direction (from b to e on the current curve). During this time, the voltage across the capacitor decreases from d to f because the plates are now losing the charge they accumulated earlier. The remainder of the cycle for both voltage and current is a repeat of what happened during the first half of the cycle. The current reaches a maximum value in the opposite direction at point e on the current curve and then decreases as the voltage across the capacitor builds up.

Note that the current and voltage are not in step with each other, as they are in a purely resistive circuit. The curves of Figure 21.5 indicate that when an alternating voltage is applied across a capacitor, the voltage reaches its maximum value one quarter of a cycle after the current reaches its maximum value. In this situation, it is common to say that **the voltage across a capacitor always lags behind the current by 90°.**

The impeding effect of a capacitor on the current in an AC circuit is expressed in terms of a factor called the **capacitive reactance,** X_C, defined as

$$X_C \equiv \frac{1}{2\pi fC}$$ [21.5]

You will be asked in Problem 7 at the end of the chapter to show that when C is in farads and f is in hertz, the unit of X_C is the ohm.

Equation 21.5 indicates that as the frequency f of the voltage source increases, the capacitive reactance X_C, or the impeding effect of the capacitor, decreases and therefore the current increases. This may be understood by noting that the voltage across the capacitor controls the current in the circuit. At high frequency, there is less time available to charge the capacitor, and less charge and voltage accumulate on the capacitor resulting in less opposition to charge flow and higher current. The analogy between capacitive reactance and resistance allows us to write an equation of the same form as Ohm's law to describe AC circuits containing capacitors. This equation relates the rms voltage and rms current in the circuit to the reactance as

$$\Delta V_{C,rms} = I_{rms}X_C$$ [21.6]

Example 21.2   A Purely Capacitive AC Circuit

An 8.00-μF capacitor is connected to the terminals of an AC generator with an rms voltage of 150 V and a frequency of 60.0 Hz. Find the capacitive reactance and the rms current in the circuit.

Solution   From Equation 21.5 and the fact that $2\pi f = 377$ s^{-1}, we have

$$X_C = \frac{1}{2\pi fC} = \frac{1}{(377 \text{ s}^{-1})(8.00 \times 10^{-6} \text{ F})} = \boxed{332 \; \Omega}$$

If we substitute this result into Equation 21.6, we find that

$$I_{\text{rms}} = \frac{\Delta V_{C,\text{rms}}}{X_C} = \frac{150 \text{ V}}{332 \; \Omega} = \boxed{0.452 \text{ A}}$$

EXERCISE   If the frequency is doubled, what happens to the capacitive reactance and the current?

ANSWER X_C is halved, and I_{rms} is doubled.

21.3   INDUCTORS IN AN AC CIRCUIT

Now consider an AC circuit consisting only of an inductor connected to the terminals of an AC source, as in Figure 21.6. (In any real circuit, there is some resistance in the wire forming the inductive coil, but we ignore this for now.) The changing current output of the generator produces a back emf in the coil of magnitude

$$\Delta v_L = L \frac{\Delta I}{\Delta t} \qquad \text{[21.7]}$$

Thus, the current in the circuit is impeded by the back emf of the inductor. The effective resistance of the coil in an AC circuit is measured by a quantity called the **inductive reactance,** X_L:

$$X_L \equiv 2\pi fL \qquad \text{[21.8]}$$

You will be asked in Problem 13 at the end of the chapter to show that when f is in hertz and L is in henries, the unit of X_L is the ohm. Note that the inductive reactance increases with increasing frequency and increasing inductance.

To understand the meaning of inductive reactance, let us compare this equation for X_L with Equation 21.7. First, note from Equation 21.8 that the inductive reactance depends on the inductance L. This seems reasonable because the back emf (Eq. 21.7) is large for large values of L. Second, note that the inductive reactance depends on the frequency f. This, too, seems reasonable because the back emf depends on $\Delta I/\Delta t$, a quantity that is large when the current changes rapidly, as it would for high frequencies.

With inductive reactance defined in this manner, we can write an equation of the same form as Ohm's law for the voltage across the coil or inductor:

$$\Delta V_{L,\text{rms}} = I_{\text{rms}} X_L \qquad \text{[21.9]}$$

where $\Delta V_{L,\text{rms}}$ is the rms voltage across the coil and I_{rms} is the rms current in the coil.

Figure 21.7 shows the instantaneous voltage and instantaneous current across the coil as functions of time. When a sinusoidal voltage is applied across an

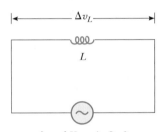

$\Delta v = \Delta V_{\text{max}} \sin 2\pi ft$

FIGURE 21.6   A series circuit consisting of an inductor L connected to an AC generator.

◀ Inductive reactance

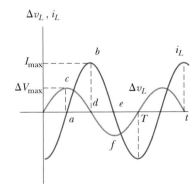

FIGURE 21.7   Plots of current and voltage across an inductor versus time in an AC circuit. The voltage leads the current by 90°.

The voltage across an inductor ▶ leads the current by 90°.

inductor, the voltage reaches its maximum value one quarter of an oscillation period before the current reaches its maximum value. In this situation, we say that **the voltage across an inductor always leads the current by 90°.**

To see why there is a phase relationship between voltage and current, let us examine a few points on the curves of Figure 21.7. Note that at point *a* on the current curve, the current is beginning to increase in the positive direction. At this instant, the rate of change of current, $\Delta I/\Delta t$ (the slope of the current curve), is at a maximum, and we see from Equation 21.7 that the voltage across the inductor is consequently also at a maximum at this time. As the current rises between points *a* and *b* on the curve, $\Delta I/\Delta t$ gradually decreases until it reaches zero at point *b*. As a result, the voltage across the inductor is decreasing during this same time interval, as the segment between *c* and *d* on the voltage curve indicates. Immediately after point *b*, the current begins to decrease, although it still has the same direction it had during the previous quarter cycle. As the current decreases to zero (from *b* to *e* on the curve), a voltage is again induced in the coil (*d* to *f*), but the sense of this voltage is opposite the sense of the voltage induced between *c* and *d*. This occurs because back emfs are always directed to oppose the change in the current.

We could continue to examine other segments of the curves, but no new information would be gained because the current and voltage variations are repetitive.

Example 21.3   A Purely Inductive AC Circuit

In a purely inductive AC circuit (see Fig. 21.6), $L = 25.0$ mH and the rms voltage is 150 V. Find the inductive reactance and rms current in the circuit if the frequency is 60.0 Hz.

Solution   First, note that $2\pi f = 2\pi(60.0) = 377$ s^{-1}. Equation 21.8 then gives

$$X_L = 2\pi fL = (377 \text{ s}^{-1})(25.0 \times 10^{-3} \text{ H}) = \boxed{9.43 \ \Omega}$$

Substituting this result into Equation 21.9 gives

$$I_{\text{rms}} = \frac{\Delta V_{L,\text{rms}}}{X_L} = \frac{150 \text{ V}}{9.43 \ \Omega} = \boxed{15.9 \text{ A}}$$

EXERCISE   Calculate the inductive reactance and rms current in the circuit if the frequency is 6 kHz.

ANSWER $X_L = 943 \ \Omega$; $I = 0.159$ A

21.4   THE *RLC* SERIES CIRCUIT

In the foregoing sections, we examined the effects of an inductor, a capacitor, and a resistor when they are connected separately across an AC voltage source. We now consider what happens when these devices are combined.

Figure 21.8 shows a circuit containing a resistor, an inductor, and a capacitor connected in series across an AC source that supplies a total voltage Δv at some instant. The current in the circuit is the same at all points in the circuit at any instant and varies sinusoidally with time, as indicated in Figure 21.9a. Thus,

$$i = I_{\text{max}} \sin 2\pi ft$$

Earlier we learned that the voltage across each element may or may not be in phase with the current. The instantaneous voltages across the three elements, shown in Figure 21.9, have the following phase relations to the instantaneous current:

1. The instantaneous voltage Δv_R across the resistor is *in phase* with the instantaneous current. (See Fig. 21.9b.)

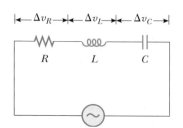

FIGURE 21.8   A series circuit consisting of a resistor, an inductor, and a capacitor connected to an AC generator.

2. The instantaneous voltage Δv_L across the inductor *leads* the current by 90°. (See Fig. 21.9c.)

3. The instantaneous voltage Δv_C across the capacitor *lags behind* the current by 90°. (See Fig. 21.9d.)

The net instantaneous voltage Δv supplied by the AC source equals the sum of the instantaneous voltages across the separate elements; that is, $\Delta v = \Delta v_R + \Delta v_C + \Delta v_L$. This does not mean, however, that the voltages measured with an AC voltmeter across R, C, and L sum up to the measured source voltage! The measured voltages do not sum to the measured source voltage because the voltages across R, C, and L all have different phases.

To account for the different phases of the voltage drops, we use a technique involving vectors. We represent the voltage across each element with a rotating vector, as in Figure 21.10. The rotating vectors are referred to as **phasors,** and the diagram is called a **phasor diagram.** This particular diagram represents the circuit voltage given by the expression $\Delta v = \Delta V_{max} \sin(2\pi ft + \phi)$, where ΔV_{max} is the maximum voltage (the magnitude or length of the rotating vector or phasor) and ϕ is the angle between the phasor and the $+x$ axis when $t = 0$. The phasor can be viewed as a vector of magnitude ΔV_{max} rotating at a constant frequency f so that its projection along the y axis is the instantaneous voltage in the circuit. Because ϕ is the phase angle between the voltage and current in the circuit, the phasor for the current (not shown in Fig. 21.10) lies along the $+x$ axis when $t = 0$ and is expressed by the relation $i = I_{max} \sin(2\pi ft)$.

The phasor diagrams in Figure 21.11 are useful for analyzing the *series RLC* circuit. Voltages in phase with the current are represented by vectors along the $+x$ axis, and voltages out of phase with the current lie along other directions. Thus, ΔV_R is horizontal and to the right because it is in phase with the current. Likewise, ΔV_L is represented by a phasor along the $+y$ axis because it leads the current by 90°. Finally, ΔV_C is along the $-y$ axis because it lags behind the current by 90°.[2] If the phasors are added as vector quantities in order to account for the different phases of the voltages across R, L, and C, Figure 21.11a shows that the only x component for the voltages is ΔV_R, and the net y component is $\Delta V_L - \Delta V_C$. We now add the phasors vectorially to find the phasor ΔV_{max} (Fig. 21.11b), where ΔV_{max} represents the maximum voltage. The right triangle in Figure 21.11b gives the

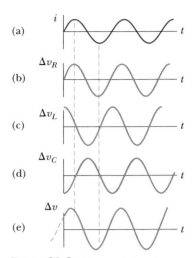

FIGURE 21.9 Phase relations in the series *RLC* circuit shown in Figure 21.8.

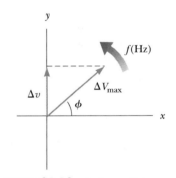

FIGURE 21.10 A phasor diagram for the voltage in an AC circuit, where ϕ is the phase angle between the voltage and current and Δv is the instantaneous voltage.

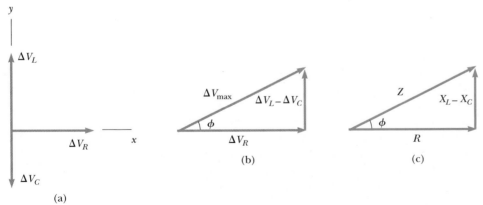

FIGURE 21.11 (a) A phasor diagram for the *RLC* circuit. (b) Addition of the phasors as vectors gives $\Delta V_{max} = \sqrt{\Delta V_R^2 + (\Delta V_L - \Delta V_C)^2}$. (c) The reactance triangle that gives the impedance relation, $Z = \sqrt{R^2 + (X_L - X_C)^2}$.

---

[2] A mnemonic to help you remember the phase relationships in *RLC* circuits is "*ELI* the *ICE* man." *E* represents the voltage, *I* the current, *L* the inductance, and *C* the capacitance. Thus, the name *ELI* means that in an inductive circuit, the voltage *E* leads the current *I*. In a capacitive circuit, *ICE* means that the current leads the voltage.

following equations for the maximum voltage and the phase angle ϕ between the maximum voltage and the current:

$$\Delta V_{max} = \sqrt{\Delta V_R^2 + (\Delta V_L - \Delta V_C)^2} \qquad \text{[21.10]}$$

$$\tan \phi = \frac{\Delta V_L - \Delta V_C}{\Delta V_R} \qquad \text{[21.11]}$$

where all voltages are maximum values. Note that although we choose to use maximum voltages in our analysis, the preceding equations apply equally well to rms voltages, because the two quantities are related to each other by the same factor for all circuit elements. The result for the maximum voltage ΔV_{max} as given by Equation 21.10 reinforces the fact that **the voltages across the resistor, capacitor, and inductor are not in phase, so one cannot simply add them to get the voltage across the combination of elements or the source voltage.**

Quick Quiz 21.2 For the circuit of Figure 21.8, is the voltage of the source equal to (a) the sum of the maximum voltages across the elements, (b) the sum of the instantaneous voltages across the elements, or (c) the sum of the rms voltages across the elements?

We can write Equation 21.10 in the form of Ohm's law, using the relations $\Delta V_R = I_{max}R$, $\Delta V_L = I_{max}X_L$, and $\Delta V_C = I_{max}X_C$, where I_{max} is the maximum current in the circuit:

$$\Delta V_{max} = I_{max}\sqrt{R^2 + (X_L - X_C)^2} \qquad \text{[21.12]}$$

It is convenient to define a parameter called the **impedance,** Z, of the circuit as

Impedance

$$Z \equiv \sqrt{R^2 + (X_L - X_C)^2} \qquad \text{[21.13]}$$

so that Equation 21.12 becomes

$$\Delta V_{max} = I_{max}Z \qquad \text{[21.14]}$$

Note that Equation 21.14 is in the form of Ohm's law, $\Delta V = IR$, where R is replaced by the impedance in ohms. Equation 21.14 can be regarded as a generalized form of Ohm's law applied to a series AC circuit. Note that the impedance and therefore the current in an AC circuit depend on the resistance, the inductance, the capacitance, *and* the frequency (because the reactances are frequency dependent).

It is useful to represent the impedance Z with a vector diagram such as the one depicted in Figure 21.11c. A right triangle is constructed; the right side is the quantity $X_L - X_C$, the base is R, and the hypotenuse is Z. Applying the Pythagorean theorem to this triangle, we see that

$$Z = \sqrt{R^2 + (X_L - X_C)^2}$$

which is consistent with Equation 21.13. Furthermore, we see from the vector diagram in Figure 21.11c that the phase angle ϕ between the current and the voltage obeys the relationship

Phase angle ϕ

$$\tan \phi = \frac{X_L - X_C}{R} \qquad \text{[21.15]}$$

The physical significance of the phase angle will become apparent in Section 21.5.

TABLE 21.2	Impedance Values and Phase Angles for Various Combinations of Circuit Elements[a]	
Circuit Elements	**Impedance Z**	**Phase Angle ϕ**
R	R	$0°$
C	X_C	$-90°$
L	X_L	$+90°$
R C	$\sqrt{R^2 + X_C^2}$	Negative, between $-90°$ and $0°$
R L	$\sqrt{R^2 + X_L^2}$	Positive, between $0°$ and $90°$
R L C	$\sqrt{R^2 + (X_L - X_C)^2}$	Negative if $X_C > X_L$ Positive if $X_C < X_L$

[a] In each case, an AC voltage (not shown) is applied across the combination of elements (that is, across the dots).

Table 21.2 provides impedance values and phase angles for some series circuits containing different combinations of circuit elements.

Parallel alternating-current circuits are also useful in everyday applications. We shall not discuss them here, however, because their analysis is beyond the scope of this book.

Webnote 21.1

Take a look at a real-time Java applet of voltage across each element of an *RLC* series circuit. Go to
http://www.phy.ntnu.edu.tw/java/rlc/rlc.html

Quick Quiz 21.3

The switch in the circuit shown in Figure 21.12 is closed and the lightbulb glows steadily. The inductor is a simple air-core solenoid. As an iron rod is being inserted into the interior of the solenoid, the brightness of the lightbulb (a) increases, (b) decreases, or (c) remains the same.

FIGURE 21.12 (Quick Quiz 21.3)

PROBLEM-SOLVING STRATEGY   **Alternating Current**

The following procedures are recommended for solving alternating-current problems:

1. The first step in analyzing alternating-current circuits is to calculate as many of the unknown quantities, such as X_L and X_C, as possible. (When you calculate X_C, express the capacitance in farads rather than, say, microfarads.)
2. Apply the equation $\Delta V_{max} = I_{max} Z$ to the portion of the circuit that is of interest. For example, if you want to know the voltage drop across the combination of an inductor and a resistor, the equation for the voltage drop reduces to $\Delta V_{max} = I_{max} \sqrt{R^2 + X_L{}^2}$.

Example 21.4 Analyzing a Series *RLC* AC Circuit

Analyze a series *RLC* AC circuit for which $R = 250\ \Omega$, $L = 0.600$ H, $C = 3.50\ \mu$F, $f = 60$ Hz, and $\Delta V_{\text{max}} = 150$ V.

Solution The reactances are given by $X_L = 2\pi fL = 226\ \Omega$ and $X_C = 1/2\pi fC = 758\ \Omega$. Therefore, the impedance is

$$Z = \sqrt{R^2 + (X_L - X_C)^2} = \sqrt{(250\ \Omega)^2 + (226\ \Omega - 758\ \Omega)^2} = 588\ \Omega$$

The maximum current is

$$I_{\text{max}} = \frac{\Delta V_{\text{max}}}{Z} = \frac{150\text{ V}}{588\ \Omega} = 0.255\text{ A}$$

The phase angle between the current and voltage is

$$\phi = \tan^{-1}\left(\frac{X_L - X_C}{R}\right) = \tan^{-1}\left(\frac{226\ \Omega - 758\ \Omega}{250\ \Omega}\right) = -64.8°$$

Because the circuit is more capacitive than inductive (that is, $X_C > X_L$), ϕ is negative. A negative phase angle means that the current leads the applied voltage.

The maximum voltages across the elements are

$$\Delta V_R = I_{\text{max}}R = (0.255\text{ A})(250\ \Omega) = 63.8\text{ V}$$

$$\Delta V_L = I_{\text{max}}X_L = (0.255\text{ A})(226\ \Omega) = 57.6\text{ V}$$

$$\Delta V_C = I_{\text{max}}X_C = (0.255\text{ A})(758\ \Omega) = 193\text{ V}$$

Note that the sum of the maximum voltages across the elements is $\Delta V_R + \Delta V_L + \Delta V_C = 314$ V, which is much greater than the maximum voltage of the generator, 150 V. As we saw in Quick Quiz 21.2, the sum of the maximum voltages is a meaningless quantity because when alternating voltages are added, *both their amplitudes and their phases* must be taken into account. We know that the maximum voltages across the various elements occur at different times. That is, the voltages must be added in a way that takes account of the different phases. When this is done, Equation 21.12 is satisfied. You should verify this result.

21.5 POWER IN AN AC CIRCUIT

No power losses are associated with capacitors and pure inductors in an AC circuit. (A pure inductor is defined as one with no resistance or capacitance.) Let us begin by analyzing the power dissipated in an AC circuit that contains only a generator and a capacitor.

When the current begins to increase in one direction in an AC circuit, charge begins to accumulate on the capacitor and a voltage drop appears across it. When this voltage reaches its maximum value, the energy stored in the capacitor is

$$PE_C = \tfrac{1}{2}C(\Delta V_{\text{max}})^2$$

However, this energy storage is only momentary. When the current reverses direction, the charge leaves the capacitor plates and returns to the voltage source. Thus, during one half of each cycle the capacitor is being charged, and during the other half the charge is being returned to the voltage source. Therefore, the average power supplied by the source is zero. In other words, **no power losses occur in a capacitor in an AC circuit.**

Similarly, the source must do work against the back emf of an inductor that is carrying a current. When the current reaches its maximum value, the energy stored in the inductor is a maximum and is given by

$$PE_L = \tfrac{1}{2}LI_{\text{max}}^2$$

When the current begins to decrease in the circuit, this stored energy is returned

NIKOLA TESLA (1856–1943)
Tesla was born in Croatia but spent most of his professional life as an inventor in the United States. He was a key figure in the development of alternating-current electricity, high-voltage transformers, and the transport of electrical power using AC transmission lines. Tesla's viewpoint was at odds with the ideas of Edison, who committed himself to the use of direct current in power transmission. Tesla's AC approach won out. *(CORBIS/Bettmann)*

to the source as the inductor attempts to maintain the current in the circuit. The average power delivered to a resistor in an *RLC* circuit is

$$\mathcal{P}_{av} = I_{rms}^2 R \qquad [21.16]$$

Note **that the average power delivered by the generator is converted to internal energy in the resistor. No power loss occurs in an ideal capacitor or inductor.**

An alternative equation for the average power loss in an AC circuit can be found by substituting (from Ohm's law) $R = \Delta V_R/I_{rms}$ into Equation 21.16:

$$\mathcal{P}_{av} = I_{rms}\,\Delta V_R$$

It is convenient to refer to a voltage triangle that shows the relationship among ΔV_{rms}, ΔV_R, and $\Delta V_L - \Delta V_C$, such as Figure 21.11b. (Remember that Figure 21.11 was valid for *both* maximum and rms voltages.) From this figure, we see that the voltage drop across a resistor can be written in terms of the voltage of the source, ΔV_{rms}:

$$\Delta V_R = \Delta V_{rms} \cos\phi$$

Hence, the average power delivered by a generator in an AC circuit is

$$\mathcal{P}_{av} = I_{rms}\,\Delta V_{rms} \cos\phi \qquad [21.17]$$

◀ Average power

where the quantity $\cos\phi$ is called the **power factor.**

Equation 21.17 shows that the power delivered by an AC source to any circuit depends on the phase difference between the source voltage and the resulting current. This fact has many interesting applications. For example, factories often use large motors in devices such as machines, generators, and transformers that have a large inductive load due to all the windings. To deliver greater power to such devices without using excessively high voltages, factory technicians introduce capacitance in the circuits to shift the phase.

◀ **APPLICATION**

SHIFTING PHASE TO DELIVER MORE POWER

Example 21.5   Average Power in a *RLC* Series Circuit

Calculate the average power delivered to the series *RLC* circuit described in Example 21.4.

Solution   First, let us calculate the rms voltage and rms current using the value of Z calculated in Example 21.4:

$$\Delta V_{rms} = \frac{\Delta V_{max}}{\sqrt{2}} = \frac{150\text{ V}}{\sqrt{2}} = 106\text{ V}$$

$$I_{rms} = \frac{I_{max}}{\sqrt{2}} = \frac{\Delta V_{max}/Z}{\sqrt{2}} = \frac{0.255\text{ A}}{\sqrt{2}} = 0.180\text{ A}$$

Because $\phi = -64.8°$, the power factor, $\cos\phi$, is 0.426, and hence the average power using Equation 21.17 is

$$\mathcal{P}_{av} = I_{rms}\Delta V_{rms} \cos\phi = (0.180\text{ A})(106\text{ V})(0.426) = \boxed{8.12\text{ W}}$$

The same result can be obtained using Equation 21.16.

21.6   RESONANCE IN A SERIES *RLC* CIRCUIT

In general, the rms current in a series *RLC* circuit can be written

$$I_{rms} = \frac{\Delta V_{rms}}{Z} = \frac{\Delta V_{rms}}{\sqrt{R^2 + (X_L - X_C)^2}} \qquad [21.18]$$

Resonance frequency ▶

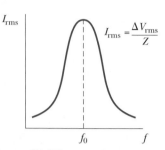

FIGURE 21.13 A plot of current amplitude in a series *RLC* circuit versus frequency of the generator voltage. Note that the current reaches its maximum value at the resonance frequency f_0.

From this we see that if the frequency is varied, the current has its *maximum* value when the impedance has its *minimum* value. This occurs when $X_L = X_C$. In such a circumstance, the impedance of the circuit reduces to $Z = R$. The frequency f_0 at which this happens is called the **resonance frequency** of the circuit. To find f_0, we set $X_L = X_C$, which gives, from Equations 21.5 and 21.8,

$$2\pi f_0 L = \frac{1}{2\pi f_0 C}$$

$$f_0 = \frac{1}{2\pi\sqrt{LC}} \qquad [21.19]$$

Figure 21.13 is a plot of current as a function of frequency for a circuit containing a fixed value for both the capacitance and inductance. From Equation 21.18 it must be concluded that the current would become infinite at resonance when $R = 0$. Although Equation 21.18 predicts this result, real circuits always have some resistance, which limits the value of the current.

The tuning circuit of a radio is an important application of a series resonance circuit. The radio is tuned to a particular station (which transmits a specific radio-frequency signal) by varying a capacitor, which changes the resonance frequency of the tuning circuit. When this resonance frequency matches that of the incoming radio wave, the current in the tuning circuit increases.

APPLICATION ▶

TUNING YOUR RADIO

APPLICATION ▶

METAL DETECTORS IN AIRPORTS

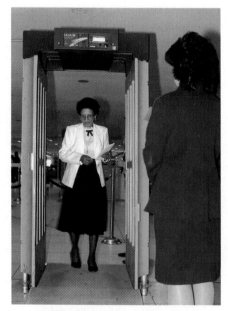

FIGURE 21.14 (Applying Physics 21.1) An airport metal detector. *(Ryan Williams/International Stock Photography)*

APPLYING **PHYSICS** **21.1**

When you walk through the doorway of an airport metal detector as in Figure 21.14, you are really walking through a coil of many turns. How might this work?

Explanation The metal detector is essentially a resonant circuit. The portal you step through is an inductor (a large loop of conducting wire) that is part of the circuit. The frequency of the circuit is tuned to the resonant frequency of the circuit when there is no metal in the inductor. When you walk through with metal in your pocket, you change the effective inductance of the resonance circuit, resulting in a change in the current in the circuit. This change in current is detected, and an electronic circuit causes a sound to be emitted as an alarm.

Example 21.6 **The Capacitance of a Circuit in Resonance**

Consider a series *RLC* circuit for which $R = 150\ \Omega$, $L = 20$ mH, $\Delta V_{rms} = 20$ V, and $2\pi f = 5.0 \times 10^3\ \mathrm{s}^{-1}$. Determine the value of the capacitance for which the rms current is a maximum.

Reasoning The current is a maximum at the resonance frequency f_0, which should be made to match the driving frequency, $5.0 \times 10^3\ \mathrm{s}^{-1}$.

Solution In this problem,

$$2\pi f_0 = 5.0 \times 10^3\ \mathrm{s}^{-1} = \frac{1}{\sqrt{LC}}$$

$$C = \frac{1}{(25 \times 10^6\ \mathrm{s}^{-2})L} = \frac{1}{(25 \times 10^6\ \mathrm{s}^{-2})(20.0 \times 10^{-3}\ \mathrm{H})} = 2.0\ \mu F$$

EXERCISE Calculate the maximum rms current in the circuit.

ANSWER 0.13 A

21.7 THE TRANSFORMER

In many situations it is desirable or necessary to change a small AC voltage to a larger one or vice versa. Before we examine a few such cases, let us consider the device that makes these conversions possible, the transformer.

In its simplest form, the **AC transformer** consists of two coils of wire wound around a core of soft iron, as in Figure 21.15. The coil on the left, which is connected to the input AC voltage source and has N_1 turns, is called the primary winding, or the *primary*. The coil on the right, which is connected to a resistor R and consists of N_2 turns, is the *secondary*. The purpose of the common iron core is to increase the magnetic flux and to provide a medium in which nearly all the flux through one coil passes through the other.

When an input AC voltage ΔV_1 is applied to the primary, the induced voltage across it is given by

$$\Delta V_1 = -N_1 \frac{\Delta \Phi_B}{\Delta t} \qquad [21.20]$$

where Φ_B is the magnetic flux through each turn. If we assume that no flux leaks from the iron core, then the flux through each turn of the primary equals the flux through each turn of the secondary. Hence, the voltage across the secondary coil is

$$\Delta V_2 = -N_2 \frac{\Delta \Phi_B}{\Delta t} \qquad [21.21]$$

The term $\Delta \Phi_B / \Delta t$ is common to Equations 21.20 and 21.21. Therefore, we see that

$$\Delta V_2 = \frac{N_2}{N_1} \Delta V_1 \qquad [21.22]$$

When N_2 is greater than N_1, and thus ΔV_2 exceeds ΔV_1, the transformer is referred to as a *step-up transformer*. When N_2 is less than N_1, making ΔV_2 less than ΔV_1, we speak of a *step-down transformer*.

It should be clear that a voltage is generated across the secondary only when there is a *change* in the number of flux lines passing through the secondary. Thus, the input current in the primary must change with time, which is what happens when an alternating current is used. However, when the input at the primary is a direct current, a voltage output occurs at the secondary only at the instant a switch in the primary circuit is opened or closed. Once the current in the primary reaches a steady value, the output voltage at the secondary is zero.

It may seem that a transformer is a device in which it is possible to get something for nothing. For example, a step-up transformer can change an input voltage from, say, 10 V to 100 V. This means that each 1 coulomb of charge leaving the secondary has 100 J of energy, whereas each coulomb of charge entering the primary has only 10 J of energy. However, this is not an example of a breakdown in the principle of conservation of energy, because **the power input to the primary equals the power output at the secondary;** that is,

$$I_1 \Delta V_1 = I_2 \Delta V_2 \qquad [21.23]$$

Thus, if the voltage at the secondary is ten times that at the primary, the current at the secondary is reduced by a factor of 10. Equation 21.23 assumes an **ideal transformer,** in which there are no power losses between the primary and the secondary. Real transformers typically have power efficiencies ranging from 90% to 99%. Power losses occur because of such factors as eddy currents induced in the iron core of the transformer, which dissipate energy in the form of I^2R losses.

When electric power is transmitted over large distances, it is economical to use a high voltage and a low current because the power lost via resistive heating in the transmission lines varies as I^2R. This means that if a utility company can reduce

FIGURE 21.15 An ideal transformer consists of two coils wound on the same soft iron core. An AC voltage ΔV_1 is applied to the primary coil, and the output voltage ΔV_2 is observed across the load resistance R.

◀ In an ideal transformer, the input power equals the output power.

APPLICATION

LONG-DISTANCE ELECTRIC POWER TRANSMISSION

the current by a factor of 10, for example, the power loss is reduced by a factor of 100. In practice, the voltage is stepped up to around 230 000 V at the generating station, then stepped down to around 20 000 V at a distribution station, and finally stepped down to 120 V at the customer's utility pole.

Example 21.7 Distributing Power to a City

A generator at a utility company produces 100 A of current at 4 000 V. The voltage is stepped up to 240 000 V by a transformer before it is sent on a high-voltage transmission line across a rural area to a city. Assume that the effective resistance of the power line is 30.0 Ω.

A Determine the percentage of power lost.

Solution From Equation 21.23, the current in the transmission line is

$$I_2 = \frac{I_1 \Delta V_1}{\Delta V_2} = \frac{(100\ \text{A})(4.00 \times 10^3\ \text{V})}{2.40 \times 10^5\ \text{V}} = 1.67\ \text{A}$$

and the power lost in the transmission line is

$$\mathcal{P}_{\text{lost}} = I_2^2 R = (1.67\ \text{A})^2 (30.0\ \Omega) = 83.7\ \text{W}$$

The power output of the generator is

$$\mathcal{P} = I \Delta V = (100\ \text{A})(4\ 000\ \text{V}) = 4.00 \times 10^5\ \text{W}$$

From this we can find the percentage of power lost as

$$\%\ \text{power lost} = \left(\frac{83.7\ \text{W}}{4.00 \times 10^5\ \text{W}} \right) \times 100 = \boxed{0.0209\%}$$

B What percentage of the original power would be lost in the transmission line if the voltage were not stepped up?

Solution If the voltage were not stepped up, the current in the transmission line would be 100 A and the power lost in the line would be

$$\mathcal{P}_{\text{lost}} = I^2 R = (100\ \text{A})^2 (30.0\ \Omega) = 3.00 \times 10^5\ \text{W}$$

In this case, the percentage of power lost would be

$$\%\ \text{power lost} = \left(\frac{3.00 \times 10^5\ \text{W}}{4.00 \times 10^5\ \text{W}} \right) \times 100 = \boxed{75\%}$$

This example illustrates the advantage of high-voltage transmission lines. At the city, a transformer at a substation steps the voltage back down to about 4 000 V, and this voltage is maintained across utility lines throughout the city. When the power is to be used at a home or business, a transformer on a utility pole near the establishment reduces the voltage to 240 V or 120 V.

EXERCISE If the transmission line is cooled so that the resistance is reduced to 5.0 Ω, how much power is lost in the line if it carries a current of 0.89 A?

ANSWER 4.0 W

This cylindrical step-down transformer drops the voltage from 4 000 V to 220 V for delivery to a group of residences. *(George Semple)*

21.8 MAXWELL'S PREDICTIONS

During the early stages of their study and development, electric and magnetic phenomena were thought to be unrelated. In 1865, however, James Clerk Maxwell (1831–1879) provided a mathematical theory that showed a close relationship between all electric and magnetic phenomena. In addition to unifying the formerly separate fields of electricity and magnetism, his brilliant theory predicted that electric and magnetic fields can move through space as waves. The theory he developed is based on the following four pieces of information:

1. Electric field lines originate on positive charges and terminate on negative charges.
2. Magnetic field lines always form closed loops—that is, they do not begin or end anywhere.
3. A varying magnetic field induces an emf and hence an electric field. This is a statement of Faraday's law (Chapter 20).
4. Magnetic fields are generated by moving charges (or currents), as summarized in Ampère's law (Chapter 19).

Let us examine these statements further in order to understand their significance and Maxwell's great contribution to the theory of electromagnetism. The first statement is a consequence of the nature of the electrostatic force between charged particles, given by Coulomb's law. It embodies the fact that **free charges (electric monopoles) exist in nature.**

The second statement—that magnetic fields form continuous loops—is exemplified by the magnetic field lines around a long, straight wire, which are closed circles, and the magnetic field lines of a bar magnet, which form closed loops.

The third statement is equivalent to Faraday's law of induction, and the fourth statement is equivalent to Ampère's law.

In one of the greatest theoretical developments of the 19th century, Maxwell used these four statements within a corresponding mathematical framework to prove that electric and magnetic fields play symmetric roles in nature. It was already known from experiments that a changing magnetic field produced an electric field according to Faraday's law. Maxwell believed that nature was symmetric, and he therefore hypothesized that a changing electric field should produce a magnetic field. This hypothesis could not be proven experimentally at the time it was developed, because the magnetic fields generated by changing electric fields are generally very weak and therefore difficult to detect.

To justify his hypothesis, Maxwell searched for other phenomena that might be explained by it. He turned his attention to the motion of rapidly oscillating (accelerating) charges, such as those in a conducting rod connected to an alternating voltage. Such charges experience accelerations and, according to Maxwell's predictions, generate changing electric and magnetic fields. The changing fields cause electromagnetic disturbances that travel through space as waves, similar to the spreading water waves created by a pebble thrown into a pool. The waves sent out by the oscillating charges are fluctuating electric and magnetic fields, and so they are called *electromagnetic waves*. From Faraday's law and from his generalization of Ampère's law, Maxwell calculated their speed to be equal to the speed of light, $c = 3 \times 10^8$ m/s. He concluded that visible light and other electromagnetic waves consist of fluctuating electric and magnetic fields traveling through empty space, with each varying field inducing the other! This was truly one of the greatest discoveries of science on a par with Newton's discovery of the laws of motion. Like Newton's laws, it had a profound influence on later scientific developments.

JAMES CLERK MAXWELL, SCOTTISH THEORETICAL PHYSICIST (1831–1879)

Maxwell developed the electromagnetic theory of light and the kinetic theory of gases and explained the nature of Saturn's rings and color vision. Maxwell's successful interpretation of the electromagnetic field resulted in the equations that bear his name. Formidable mathematical ability combined with great insight enabled him to lead the way in the study of electromagnetism and kinetic theory. Although it is difficult to know what influences and characteristics shape a great scientist, Maxwell already showed remarkable curiosity at the age of three. He was described in a letter by his mother as follows: "He is a very happy man, and has improved much since the weather got moderate; he has great work with doors, locks, keys, etc., and 'Show me how it doos' is never out of his mouth. He also investigates the hidden course of streams and bellwires, the way the water gets from the pond through the wall . . . and down a drain into Water Orr (River Urr) . . . As to the bells, they will not rust; he stands sentry in the kitchen . . . or he rings, and sends Bessy to see and shout to let him know, and he drags papa all over to show him the holes where the wires go through." (From *Faraday, Maxwell, and Kelvin*, D. K. C. MacDonald, Doubleday-Anchor Books, 1964.) *(North Wind Picture Archives)*

21.9 HERTZ'S CONFIRMATION OF MAXWELL'S PREDICTIONS

In 1887, after Maxwell's death, Heinrich Hertz (1857–1894) was the first to generate and detect electromagnetic waves in a laboratory setting. To appreciate the details of his experiment, let us re-examine the properties of an *LC* circuit. In such a circuit, a charged capacitor is connected to an inductor, as in Figure 21.16. When the switch is closed, oscillations occur in the current in the circuit and in the charge on the capacitor. If the resistance of the circuit is neglected, no energy is dissipated, and the oscillations continue.

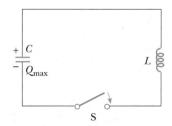

FIGURE 21.16 A simple *LC* circuit. The capacitor has an initial charge of Q_{max} and the switch is closed at $t = 0$.

HEINRICH RUDOLF HERTZ, GERMAN PHYSICIST (1857–1894)

Hertz made his most important discovery of radio waves in 1887. After finding that the speed of a radio wave was the same as that of light, Hertz showed that radio waves, like light waves, could be reflected, refracted, and diffracted. Hertz died of blood poisoning at the age of 36. During his short life, he made many contributions to science. The hertz, equal to one complete vibration or cycle per second, is named after him. *(Hulton-Deutsch Collection/CORBIS)*

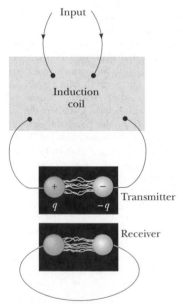

FIGURE 21.17 A schematic diagram of Hertz's apparatus for generating and detecting electromagnetic waves. The transmitter consists of two spherical electrodes connected to an induction coil, which provides short voltage surges to the spheres, setting up oscillations in the discharge. The receiver is a nearby single loop of wire containing a second spark gap.

In the following analysis, we shall neglect the resistance in the circuit. Let us assume that the capacitor has an initial charge of Q_{max} and that the switch is closed at $t = 0$. It is convenient to describe what ensues from an energy viewpoint. When the capacitor is fully charged, the total energy in the circuit is stored in the electric field of the capacitor and is equal to $Q_{max}^2/2C$. At this time, the current is zero and so no energy is stored in the inductor. As the capacitor begins to discharge, the energy stored in its electric field decreases. At the same time, the current increases and energy equal to $LI^2/2$ is now stored in the magnetic field of the inductor. Thus, energy is transferred from the electric field of the capacitor to the magnetic field of the inductor. When the capacitor is fully discharged, it stores no energy. At this time, the current reaches its maximum value, and all of the energy is stored in the inductor. The process then repeats in the reverse direction. The energy continues to transfer between the inductor and the capacitor, corresponding to oscillations in the current and charge.

As we saw in Section 21.6, the frequency of oscillation of an *LC* circuit is called the *resonance frequency* of the circuit and is given by

$$f_0 = \frac{1}{2\pi\sqrt{LC}}$$

The circuit Hertz used in his investigations of electromagnetic waves is similar to that just discussed and is shown schematically in Figure 21.17. An induction coil (a large coil of wire) is connected to two metal spheres with a narrow gap between them to form a capacitor. Oscillations are initiated in the circuit by short voltage pulses sent via the coil to the spheres, charging one positive, the other negative. Because *L* and *C* are quite small in this circuit, the frequency of oscillation is quite high, $f \approx 100$ MHz. This circuit is called a *transmitter* because it produces electromagnetic waves.

Several meters from the transmitter circuit, Hertz placed a second circuit, the receiver, which consisted of a single loop of wire connected to two spheres. It had its own effective inductance, capacitance, and natural frequency of oscillation. Hertz found that energy was being sent from the transmitter to the receiver when the resonance frequency of the receiver was adjusted to match that of the transmitter. The energy transfer was detected when the voltage across the spheres in the receiver circuit became high enough to produce ionization in the air, which caused sparks to appear in the air gap separating the spheres. Hertz's experiment is analogous to the mechanical phenomenon in which a tuning fork picks up the vibrations from another, identical tuning fork.

Hertz hypothesized that the energy transferred from the transmitter to the receiver is carried in the form of waves, which are now known to be electromagnetic waves. In a series of experiments, he also showed that the radiation generated by the transmitter exhibits wave properties: interference, diffraction, reflection, refraction, and polarization. As you will see shortly, all of these properties are exhibited by light. Thus, it became evident that these electromagnetic waves had the same known properties of light waves (already known at the time) and differed only in frequency and wavelength. Hertz effectively confirmed Maxwell's theory by showing that Maxwell's mysterious electromagnetic waves existed and had all the properties of light waves.

Perhaps the most convincing experiment Hertz performed was the measurement of the speed of waves from the transmitter, accomplished as follows. Waves of known frequency from the transmitter were reflected from a metal sheet so that an interference pattern was set up, much like the standing wave pattern on a stretched string. As we saw in our discussion of standing waves, the distance between nodes is $\lambda/2$, so Hertz was able to determine the wavelength λ. Using the relationship $v = \lambda f$, he found that v was close to 3×10^8 m/s, the known speed of visible light. Hertz's experiments thus provided the first evidence in support of Maxwell's theory.

21.10 PRODUCTION OF ELECTROMAGNETIC WAVES BY AN ANTENNA

In the previous section, we found that the energy stored in an *LC* circuit is continually transferred between the electric field of the capacitor and the magnetic field of the inductor. However, this energy transfer continues for prolonged periods of time only when the changes occur slowly. If the current alternates rapidly, the circuit loses some of its energy in the form of electromagnetic waves. In fact, electromagnetic waves are radiated by *any* circuit carrying an alternating current. The fundamental mechanism responsible for this radiation is the acceleration of a charged particle. Whenever a charged particle undergoes an acceleration, it must radiate energy.

An alternating voltage applied to the wires of an antenna forces an electric charge in the antenna to oscillate. This is a common technique for accelerating charged particles and is the source of the radio waves emitted by the broadcast antenna of a radio station.

Figure 21.18 illustrates the production of an electromagnetic wave by oscillating electric charges in an antenna. Two metal rods are connected to an AC source, which causes charges to oscillate between the two rods. The output voltage of the generator is sinusoidal. At $t = 0$, the upper rod is given a maximum positive charge and the bottom rod an equal negative charge, as in Figure 21.18a. The electric field near the antenna at this instant is also shown in Figure 21.18a. As the charges oscillate, the rods become less charged, the field near the rods decreases in strength, and the downward-directed maximum electric field produced at $t = 0$ moves away from the rod. When the charges are neutralized, as in Figure 21.18b, the electric field has dropped to zero. This occurs after an interval equal to one quarter of the period of oscillation. Continuing in this fashion, the upper rod soon obtains a maximum negative charge and the lower rod becomes positive, as in Figure 21.19c, resulting in an electric field directed upward. This occurs after an interval equal to one-half the period of oscillation. The oscillations continue as indicated in Figure 21.18d. Note that the electric field near the antenna oscillates in phase with the charge distribution. That is, the field points down when the upper rod is positive and up when the upper rod is negative. Furthermore, the magnitude of the field at any instant depends on the amount of charge on the rods at that instant.

As the charges continue to oscillate (and accelerate) between the rods, the electric field set up by the charges moves away from the antenna in all directions at the speed of light. Figure 21.18 shows the electric field pattern on one side of the antenna at certain times during the oscillation cycle. As you can see, one cycle of charge oscillation produces one full wavelength in the electric field pattern.

Because the oscillating charges create a current in the rods, a magnetic field is also generated when the current in the rods is upward, as shown in Figure 21.19. The magnetic field lines circle the antenna (right-hand rule #2) and are perpendicular to the electric field at all points. As the current changes with time, the

▶ An accelerating charge radiates energy.

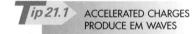

APPLICATION

RADIO-WAVE TRANSMISSION

Tip 21.1 ACCELERATED CHARGES PRODUCE EM WAVES

Stationary charges produce only electric fields, while charges in uniform motion (i.e., constant velocity) produce electric and magnetic fields, but no electromagnetic waves. In contrast, accelerated charges produce electromagnetic waves as well as electric and magnetic fields.

(a) $t = 0$ (b) $t = \frac{T}{4}$ (c) $t = \frac{T}{2}$ (d) $t = T$

FIGURE 21.18 An electric field set up by oscillating charges in an antenna. The field moves away from the antenna at the speed of light.

FIGURE 21.19 Magnetic field lines around an antenna carrying a changing current.

FIGURE 21.20 An electromagnetic wave sent out by oscillating charges in an antenna, represented at one instant of time and far from the antenna. Note that the electric field is perpendicular to the magnetic field, and both are perpendicular to the direction of wave propagation.

magnetic field lines spread out from the antenna. At great distances from the antenna, the strengths of the electric and magnetic fields become very weak. However, at these distances it is necessary to take into account the facts that (1) a changing magnetic field produces an electric field and (2) a changing electric field produces a magnetic field, as predicted by Maxwell. These induced electric and magnetic fields are in phase: At any point, the two fields reach their maximum values at the same instant. This is illustrated at one instant of time in Figure 21.20. Note that (1) **E** and **B** fields are perpendicular to each other, and (2) both fields are perpendicular to the direction of motion of the wave. This second property is characteristic of transverse waves. Hence, we see that **an electromagnetic wave is a transverse wave.**

Webnote 21.2

Watch the propagation of an electromagnetic wave through the real-time Java applet at *http://www.phy.ntnu.edu.tw/~hwang/ emWave/emWave.html*

21.11 PROPERTIES OF ELECTROMAGNETIC WAVES

We have seen that Maxwell's detailed analysis predicted the existence and properties of electromagnetic waves. We have already examined some of those properties. In this section we summarize what we know about electromagnetic waves thus far and consider some additional properties. In our discussion here and in future sections, we shall often make reference to a type of wave called a **plane wave.** A plane electromagnetic wave is a wave traveling from a very distant source. Figure 21.20 pictures such a wave at a given instant of time. In this case, the oscillations of the electric and magnetic fields take place in planes perpendicular to the *x* axis and thus to the direction of travel for the wave. Because the electric and magnetic fields are perpendicular to the direction of travel of the wave, electromagnetic waves are transverse waves. In Figure 21.20, the electric field **E** is in the *y* direction and the magnetic field **B** is in the *z* direction.

Electromagnetic waves travel with the speed of light. In fact, it can be shown that the speed of an electromagnetic wave is related to the permeability and permittivity of the medium through which it travels. Maxwell found this relationship for free space to be

Speed of light ▶

$$c = \frac{1}{\sqrt{\mu_0 \epsilon_0}}$$

[21.24]

where *c* is the speed of light, $\mu_0 = 4\pi \times 10^{-7}$ N·s²/C² is the permeability constant of vacuum, and $\epsilon_0 = 8.854\ 19 \times 10^{-12}\ C^2/N \cdot m^2$ is the permittivity of free space. Substituting these values into Equation 21.24, we find that

$$c = 2.997\ 92 \times 10^8\ \text{m/s} \qquad \text{[21.25]}$$

Because electromagnetic waves travel at a speed that is precisely the same as the speed of light in vacuum, one is led to believe (correctly) that **light is an electromagnetic wave.**

Maxwell also proved that the ratio of the electric to the magnetic field in an electromagnetic wave equals the speed of light. That is,

$$\frac{E}{B} = c \qquad \text{[21.26]}$$

Electromagnetic waves carry energy as they travel through space, and this energy can be transferred to objects placed in their paths. The average rate at which energy passes through an area perpendicular to the direction of travel of a wave, or the average power per unit area, is given by

$$\text{Average power per unit area} = \frac{E_{max}B_{max}}{2\mu_0} \qquad \text{[21.27]}$$

where E_{max} and B_{max} are the *maximum* values of E and B. As in Chapter 14, we call this quantity the *intensity* of the wave. From Equation 21.26, we see that $E_{max} = cB_{max} = B_{max}/\sqrt{\mu_0\epsilon_0}$. Therefore, Equation 21.27 can also be expressed as

$$\text{Average power per unit area} = \frac{E_{max}^2}{2\mu_0 c} = \frac{c}{2\mu_0} B_{max}^2 \qquad \text{[21.28]}$$

Note that in these expressions we use the *average* power per unit area. A detailed analysis would show that the energy carried by an electromagnetic wave is shared equally by the electric and magnetic fields.

Electromagnetic waves transport linear momentum as well as energy. Hence it follows that pressure is exerted on a surface when an electromagnetic wave impinges on it. In what follows, we assume that the electromagnetic wave transports a total energy U to a surface in a time t. If the surface absorbs all the incident energy U in this time, Maxwell showed that the total momentum **p** delivered to this surface has a magnitude

$$p = \frac{U}{c} \qquad \text{(complete absorption)} \qquad \text{[21.29]}$$

If the surface is a perfect reflector, then the momentum delivered in a time t for normal incidence is twice that given by Equation 21.29. That is, a momentum U/c is delivered first by the incident wave and then again by the reflected wave, in analogy with a ball colliding elastically with a wall. Therefore,

$$p = \frac{2U}{c} \qquad \text{(complete reflection)} \qquad \text{[21.30]}$$

Although radiation pressures are very small (about $5 \times 10^{-6}\ N/m^2$ for direct sunlight), they have been measured with a device such as the one shown in Figure 21.21. Light is allowed to strike a mirror and a black disk that are connected to each other by a horizontal bar suspended from a fine fiber. Light striking the black disk is completely absorbed, so *all* of the momentum of the light is transferred to the disk. Light striking the mirror head on is totally reflected; hence, the momentum transfer to the mirror is twice that transmitted to the disk. As a result, the horizontal bar supporting the disks twists counterclockwise as seen from above. The bar comes to equilibrium at some angle under the action of the torques caused by radiation pressure and the twisting of the fiber. The radiation pressure can be determined by measuring the angle at which equilibrium occurs. The apparatus

◀ Light is an electromagnetic wave and transports energy and momentum.

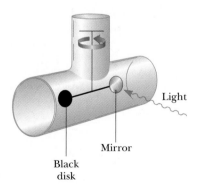

FIGURE 21.21 An apparatus for measuring the radiation pressure of light. In practice, the system is contained in a high vacuum.

must be placed in a high vacuum to eliminate the effects of air currents. It is interesting to note that similar experiments demonstrate that electromagnetic waves carry angular momentum as well.

In summary, electromagnetic waves traveling through free space have the following properties:

1. Electromagnetic waves travel at the speed of light.
2. Electromagnetic waves are transverse waves, because the electric and magnetic fields are perpendicular to the direction of propagation of the wave and to each other.
3. The ratio of the electric field to the magnetic field in an electromagnetic wave equals the speed of light.
4. Electromagnetic waves carry both energy and momentum, which can be delivered to a surface.

APPLYING PHYSICS 21.2

In the interplanetary space in the Solar System, there is a large amount of dust. Although interplanetary dust can in theory have a variety of sizes—from molecular size upward—there are very few dust particles smaller than about 0.2 μm in our Solar System. Why? (*Hint:* The Solar System originally contained dust particles of all sizes.)

Explanation Dust particles in the solar system are subject to two forces—the gravitational force toward the Sun and the force from radiation pressure, which is away from the Sun. The gravitational force is proportional to the cube of the radius of a spherical dust particle, because it is proportional to the mass (ρV) of the particle. The radiation pressure is proportional to the square of the radius, because it depends on the cross-sectional area of the particle. For large particles, the gravitational force is larger than the force of radiation pressure and the weak attraction to the Sun causes such particles to move slowly toward the Sun. For small particles, less than about 0.2 μm, the larger force from radiation pressure sweeps these particles out of the Solar System.

Quick Quiz 21.4

In an apparatus such as that in Figure 21.21, suppose the black disk is replaced by one with half the radius. Which of the following are different after the disk is replaced: (a) radiation pressure on the disk, (b) radiation force on the disk, (c) radiation momentum delivered to the disk in a given time interval?

Example 21.8 Solar Energy

Assume that the Sun delivers an average power per unit area of about 1 000 W/m^2 to Earth's surface. Calculate the total power incident on a roof 8.00 m by 20.0 m. Assume that the radiation is incident *normal* to the roof (the Sun is directly overhead).

Solution The power per unit area, or light intensity, is 1 000 W/m^2. For normal incidence, the power supplied to the roof is

$$\mathcal{P} = (1\,000 \text{ W/m}^2)(8.00 \times 20.0 \text{ m}^2) = \boxed{1.60 \times 10^5 \text{ W}}$$

Note that if this power could *all* be converted to electric power, it would be more than enough for the average home. Unfortunately, solar energy is not easily harnessed, and the prospects for large-scale conversion are not as bright as they may appear from this

(Example 21.8) A solar home in Oregon. (*John Neal/Photo Researchers, Inc.*)

simple calculation. For example, the conversion efficiency from solar to electrical energy is far less than 100%; 10% is typical for photovoltaic cells. Roof systems for using solar energy to raise the temperature of water with efficiencies of around 50% have been built. However, other practical problems must be considered, such as overcast days, geographic location, and energy storage.

EXERCISE How much solar energy (in joules) is incident on the roof in 1.00 h?

ANSWER 5.76×10^8 J

Example 21.9 A High-Intensity Laser Beam

The intensity of light ranges from about 1 W/m² for a candle to about 30 MW/m² for a modest-size laser. (A laser is a high-intensity light source that concentrates several watts of power into a very narrow beam producing MW/m² intensity. See Chapter 28.) A particular laser produces a power of 4.0 W in a beam 0.40 mm in diameter. (a) What is its average intensity? (b) Find the peak electric field of the laser light. (c) Find the peak magnetic field of the laser light. (d) If the laser beam is aimed upward to levitate a 20-μm-diameter sphere (this has actually been done), what is the maximum mass and density of this sphere? Assume the sphere is perfectly reflecting.

Solution

A Intensity is power per unit area carried by the beam:

$$\text{Power per unit area} = \frac{\mathcal{P}}{\pi r^2} = \frac{4.0 \text{ W}}{(3.14)(0.20 \times 10^{-3} \text{ m})^2} = \boxed{32 \text{ MW/m}^2}$$

B Since the average power per unit area $= E_{max}^2 / 2\mu_0 c$,

$$\begin{aligned} E_{max} &= \sqrt{(2\mu_0 c)(\text{average power per unit area})} \\ &= \sqrt{(8\pi \times 10^{-7} \text{ Ns}^2/C^2)(3.0 \times 10^8 \text{ m/s})(32 \times 10^6 \text{ W/m}^2)} \\ &= \boxed{1.6 \times 10^5 \text{ V/m}} \end{aligned}$$

C E_{max} and B_{max} are not independent in an electromagnetic wave, but are related by $B_{max} = E_{max}/c$:

$$B_{max} = \frac{E_{max}}{c} = \frac{1.6 \times 10^5 \text{ V/m}}{3.0 \times 10^8 \text{ m/s}} = \boxed{5.3 \times 10^{-4} \text{ T}}$$

D To suspend a 20-μm-diameter sphere, the upward electromagnetic force F_{EM} must balance the sphere's weight.

$$F_{EM} = mg \quad \text{or} \quad m = \frac{F_{EM}}{g}$$

Because $F_{EM} = \Delta p / \Delta t$, and Equation 21.30 states that the momentum p delivered to a perfectly reflecting object is given by $p = 2U/c$ where U is the electromagnetic energy incident on the object, $\Delta p / \Delta t = 2U/c\Delta t$. Finally, we find for the mass

$$m = \frac{2U/\Delta t}{gc}$$

$U/\Delta t$ is the energy per unit time incident on the sphere and is given by the product of the laser intensity and the cross-sectional area of the sphere ($A = \pi r^2$).

$$\frac{U}{\Delta t} = \text{intensity} \cdot \pi r^2 = (3.2 \times 10^7 \text{ W/m}^2)(3.14)(1.0 \times 10^{-10} \text{ m}^2) = 0.010 \text{ W}$$

$$= 0.010 \text{ J/s}$$

The maximum mass that can be supported by this laser beam is

$$m = \frac{2U/\Delta t}{gc} = \frac{0.020 \text{ J/s}}{(9.8 \text{ m/s}^2)(3.0 \times 10^8 \text{ m/s})} = \boxed{6.8 \times 10^{-12} \text{ kg}}$$

The maximum density is given by

$$\rho_{\max} = \frac{\text{mass}}{\text{volume}} = \frac{m}{(4/3)\pi r^3} = \frac{6.8 \times 10^{-12} \text{ kg}}{(4/3)(3.14)(1.0 \times 10^{-5} \text{ m})^3} = \boxed{1.6 \times 10^3 \text{ kg/m}^3}$$

which is about the density of bone.

21.12   THE SPECTRUM OF ELECTROMAGNETIC WAVES

We have seen that all electromagnetic waves travel in vacuum with the speed of light, c. These waves transport energy and momentum from some source to a receiver. In 1887, Hertz successfully generated and detected the radio-frequency electromagnetic waves predicted by Maxwell. Maxwell himself had recognized as electromagnetic waves both visible light and the infrared radiation discovered in 1800 by William Herschel. It is now known that other forms of electromagnetic waves exist that are distinguished by their frequencies and wavelengths.

Because all electromagnetic waves travel through vacuum with a speed c, their frequency f and wavelength λ are related by the important expression

$$c = f\lambda \qquad \text{[21.31]}$$

The types of electromagnetic waves are presented in Figure 21.22. Note the wide and overlapping range of frequencies and wavelengths. For instance, an AM radio wave with a frequency of 5.00 MHz (a typical value) has a wavelength of

$$\lambda = \frac{c}{f} = \frac{3.00 \times 10^8 \text{ m/s}}{5.00 \times 10^6 \text{ s}^{-1}} = 60.0 \text{ m}$$

The following abbreviations are often used to designate short wavelengths and distances:

$$1 \text{ micrometer } (\mu\text{m}) = 10^{-6} \text{ m}$$

$$1 \text{ nanometer } (\text{nm}) = 10^{-9} \text{ m}$$

$$1 \text{ angstrom } (\text{Å}) = 10^{-10} \text{ m}$$

The wavelengths of visible light, for example, range from 0.4 μm to 0.7 μm, or 400 nm to 700 nm, or 4 000 Å to 7 000 Å.

Brief descriptions of these wave types follow, in order of decreasing wavelength. There is no sharp division between one kind of electromagnetic wave and the next. Note that all forms of radiation are produced by accelerating charges.

Frequency, Hz   Wavelength

FIGURE 21.22 The electromagnetic spectrum. Note the overlap between one type of wave and the next. There is no sharp division between the types of waves.

Radio waves, which were discussed in Section 21.10, are the result of charges accelerating through conducting wires. They are, of course, used in radio and television communication systems.

Microwaves (short-wavelength radio waves) have wavelengths ranging between about 1 mm and 30 cm and are generated by electronic devices. Their short wavelengths make them well suited for the radar systems used in aircraft navigation and for the study of atomic and molecular properties of matter. Microwave ovens are an interesting domestic application of these waves. It has been suggested that solar energy might be harnessed by beaming microwaves to Earth from a solar collector in space.

Infrared waves (sometimes and incorrectly called "heat waves"), produced by hot objects and molecules, have wavelengths ranging from about 1 mm to the longest wavelength of visible light, 7×10^{-7} m. They are readily absorbed by most materials. The infrared energy absorbed by a substance causes it to get warmer. This is because the energy agitates the atoms of the object, increasing their vibrational or translational motion, and the result is a temperature rise. Infrared radiation has many practical and scientific applications, including physical therapy, infrared photography, and the study of the vibrations of atoms.

Visible light, the most familiar form of electromagnetic waves, may be defined as the part of the spectrum that is detected by the human eye. Light is produced by the rearrangement of electrons in atoms and molecules. The wavelengths of visible light are classified as colors ranging from violet ($\lambda \approx 4 \times 10^{-7}$ m) to red ($\lambda \approx 7 \times 10^{-7}$ m). The eye's sensitivity is a function of wavelength and is greatest at a wavelength of about 5.6×10^{-7} m (yellow-green).

Ultraviolet (UV) light covers wavelengths ranging from about 4×10^{-7} m (400 nm) down to 6×10^{-10} m (0.6 nm). The Sun is an important source of ultraviolet light (which is the main cause of suntans). Most of the ultraviolet light from

Webnote 21.3

For much more detail on the radio spectrum and the electromagnetic spectrum in general, go to *http://www.howstuffworks.com/ radio-spectrum.htm*

Webnote 21.4

Microwave ovens produce microwaves at a frequency of 2.45 GHz. How can they heat food from the inside out? Find out at *http://www.howstuffworks.com/ microwave.htm*

(a)   (b)   (c)   (d)

FIGURE 21.23   Observations in different parts of the electromagnetic spectrum show different features of the Crab Nebula. (a) X-ray image *(NASA/CXC/SAO)* (b) Optical image *(Palomar Observatory)* (c) Infrared image *(WM Keck Observatory)* (d) Radio image *(VLA/NRAO)*.

the Sun is absorbed by atoms in the upper atmosphere, or stratosphere. This is fortunate, because UV light in large quantities has harmful effects on humans. One important constituent of the stratosphere is ozone (O_3) from reactions of oxygen with ultraviolet radiation. This ozone shield causes lethal high-energy ultraviolet radiation to warm the stratosphere.

X-rays are electromagnetic waves with wavelengths from about 10^{-8} m (10 nm) down to 10^{-13} m (10^{-4} nm). The most common source of x-rays is the acceleration of high-energy electrons bombarding a metal target. X-rays are used as a diagnostic tool in medicine and as a treatment for certain forms of cancer. Because x-rays easily penetrate and damage or destroy living tissues and organisms, care must be taken to avoid unnecessary exposure and overexposure.

Gamma rays, electromagnetic waves emitted by radioactive nuclei, have wavelengths ranging from about 10^{-10} m to less than 10^{-14} m. They are highly penetrating and cause serious damage when absorbed by living tissues. As a consequence, those working near such radiation must be protected by garments containing heavily absorbing materials, such as layers of lead.

When astronomers observe the same celestial object using detectors sensitive to different regions of the electromagnetic spectrum, striking variations in the object's features can be seen. Figure 21.23 shows images of the Crab Nebula made in four different wavelength ranges. The Crab Nebula is the remnant of a supernova explosion that was seen from Earth in 1054 A.D. (Compare Fig. 8.28).

APPLYING **PHYSICS** 21.3

The center of sensitivity of our eyes coincides with the center of the wavelength distribution of the Sun. Is this an amazing coincidence?

Explanation   This is not a coincidence—it is the result of biological evolution. Humans have evolved with vision most sensitive to wavelengths that are strongest from the Sun. It is an interesting conjecture to imagine aliens from another planet, with a Sun of a different temperature, arriving at Earth. Their eyes would have a center of sensitivity at different wavelengths than ours. How would their vision of the Earth compare to ours?

21.13 THE DOPPLER EFFECT FOR ELECTROMAGNETIC WAVES

As we saw in Section 14.6, sound waves exhibit the Doppler effect when the observer, the source, or both are moving relative to the medium of propagation. Recall that in the Doppler effect, the observed frequency of the wave is larger or smaller than the frequency emitted by the source of the wave.

A Doppler effect also occurs for electromagnetic waves, but it differs from the Doppler effect for sound waves in two ways. First, in the Doppler effect for sound waves, motion relative to the medium is most important because sound waves require a medium in which to propagate. In contrast, the medium of propagation plays no role in the Doppler effect for electromagnetic waves because the waves require no medium in which to propagate. Second, the speed of sound that appears in the equation for the Doppler effect for sound depends on the reference frame in which it is measured. In contrast, as we shall see in Chapter 26, the speed of electromagnetic waves has the same value in all coordinate systems that are either at rest or moving at constant velocity with respect to one another.

The single equation that describes the Doppler effect for electromagnetic waves is given by the approximate expression

$$f' = f \left(1 \pm \frac{u}{c} \right) \qquad \text{if } u \ll c \qquad \text{[21.32]}$$

where f' is the observed frequency, f is the frequency emitted by the source, c is the speed of light in vacuum, and u is the *relative* speed of the observer and source. Note that Equation 21.32 is valid only if u is much smaller than c. Furthermore, this result can also be used for sound as long as the relative velocity of the source and observer is much less than the velocity of sound. The positive sign in Equation 21.32 must be used when the source and observer are moving toward one another, and the negative sign must be used when they are moving away from each other. Thus, we anticipate an increase in the observed frequency if the source and observer are approaching each other, and a decrease in observed frequency if the source and observer recede from each other.

Astronomers have made important discoveries using Doppler observations on light reaching Earth from distant stars and galaxies. Such measurements have shown that most distant galaxies are moving away from the Earth. Thus, the Universe is expanding. This is called a *red shift* because the observed wavelengths are shifted toward the red portion (longest wavelength) of the visible spectrum. Furthermore, measurements show that the speed of a galaxy increases with increasing distance from the Earth. More recent Doppler effect measurements made with the Hubble Space Telescope have shown that a galaxy labeled M87 is rotating—one edge moving toward us and the other moving away. Its measured speed of rotation was used to identify a supermassive black hole located at the center of this galaxy.

SUMMARY

If an AC circuit consists of a generator and a resistor, the current in the circuit is in phase with the voltage. That is, the current and voltage reach their maximum values at the same time.

In discussions of voltages and currents in AC circuits, **rms values** of voltages are usually used. One reason is that AC ammeters and voltmeters are designed to read rms values. The rms values of currents and voltage (I_{rms} and ΔV_{rms}) are related to the maximum values of these quantities (I_{max} and ΔV_{max}) as follows:

$$I_{rms} = \frac{I_{max}}{\sqrt{2}} \qquad \Delta V_{rms} = \frac{\Delta V_{max}}{\sqrt{2}} \qquad \text{[21.2, 21.3]}$$

The rms voltage across a resistor is related to the rms current in the resistor by **Ohm's law:**

$$\Delta V_{R,rms} = I_{rms} R \qquad \text{[21.4]}$$

If an AC circuit consists of a generator and a capacitor, the voltage lags behind the current by 90°. That is, the voltage reaches its maximum value one quarter of a period after the current reaches its maximum value.

The impeding effect of a capacitor on current in an AC circuit is given by the **capacitive reactance** X_C, defined as

$$X_C \equiv \frac{1}{2\pi f C} \qquad \text{[21.5]}$$

where f is the frequency of the AC voltage source.

The rms voltage across and the rms current in a capacitor are related by

$$\Delta V_{C,\text{rms}} = I_{\text{rms}} X_C \qquad \text{[21.6]}$$

If an AC circuit consists of a generator and an inductor, the voltage leads the current by 90°. That is, the voltage reaches its maximum value one quarter of a period before the current reaches its maximum value.

The effective impedance of a coil in an AC circuit is measured by a quantity called the **inductive reactance** X_L, defined as

$$X_L \equiv 2\pi f L \qquad \text{[21.8]}$$

The rms voltage across a coil is related to the rms current in the coil by

$$\Delta V_{L,\text{rms}} = I_{\text{rms}} X_L \qquad \text{[21.9]}$$

In an *RLC* series AC circuit, the maximum applied voltage ΔV is related to the maximum voltages across the resistor (ΔV_R), capacitor (ΔV_C), and inductor (ΔV_L) by

$$\Delta V_{\text{max}} = \sqrt{\Delta V_R^2 + (\Delta V_L - \Delta V_C)^2} \qquad \text{[21.10]}$$

If an AC circuit contains a resistor, an inductor, and a capacitor connected in series, the limit they place on the current is given by the **impedance Z** of the circuit, defined as

$$Z \equiv \sqrt{R^2 + (X_L - X_C)^2} \qquad \text{[21.13]}$$

The relationship between the maximum voltage supplied to an *RLC* circuit and the maximum current in the circuit that is the same in every element is

$$\Delta V_{\text{max}} = I_{\text{max}} Z \qquad \text{[21.14]}$$

In an *RLC* series AC circuit, the applied rms voltage and current are out of phase. The **phase angle** ϕ between the current and voltage is given by

$$\tan \phi = \frac{X_L - X_C}{R} \qquad \text{[21.15]}$$

The **average power** delivered by the voltage source in an *RLC* series AC circuit is

$$\mathcal{P}_{\text{av}} = I_{\text{rms}} \Delta V_{\text{rms}} \cos \phi \qquad \text{[21.17]}$$

where the constant $\cos \phi$ is called the **power factor.**

Electromagnetic waves were predicted by James Clerk Maxwell and later experimentally confirmed by Heinrich Hertz. These waves have the following properties:

1. Electromagnetic waves are transverse waves, because the electric and magnetic fields are perpendicular to the direction of travel.
2. Electromagnetic waves travel with the speed of light.
3. The ratio of the electric field to the magnetic field at a given point in an electromagnetic wave equals the speed of light. That is,

$$\frac{E}{B} = c \qquad \text{[21.26]}$$

4. Electromagnetic waves carry energy as they travel through space. The average power per unit area is

$$\frac{E_{\text{max}} B_{\text{max}}}{2\mu_0} = \frac{E_{\text{max}}^2}{2\mu_0 c} = \frac{c}{2\mu_0} B_{\text{max}}^2 \qquad \text{[21.27, 21.28]}$$

where E_{max} and B_{max} are the maximum values of the electric and magnetic fields.

5. Electromagnetic waves transport linear and angular momentum as well as energy. The speed *c*, frequency *f*, and wavelength λ of an electromagnetic wave are related by

$$c = f\lambda \qquad\qquad [21.31]$$

The **electromagnetic spectrum** includes waves covering a broad range of frequencies and wavelengths. These waves have a variety of applications and characteristics, depending on their frequencies or wavelengths.

CONCEPTUAL QUESTIONS

1. Before the advent of cable television and satellite dishes, homeowners either mounted a television antenna on the roof or used "rabbit ears" atop their sets (see Fig. Q21.1). Certain orientations of the receiving antenna on a television set gave better reception than others. Furthermore, the best orientation varied from station to station. Explain.

FIGURE Q21.1   *(George Semple)*

2. What is the impedance of an *RLC* circuit at the resonance frequency?

3. When a DC voltage is applied to a transformer, the primary coil sometimes will overheat and burn. Why?

4. Why are the primary and secondary coils of a transformer wrapped on an iron core that passes through both coils?

5. Receiving radio antennas can be in the form of conducting lines or loops. What should the orientation of each of these antennas be relative to a broadcasting antenna that is vertical?

6. If the fundamental source of a sound wave is a vibrating object, what is the fundamental source of an electromagnetic wave?

7. In radio transmission, a radio wave serves as a carrier wave and the sound signal is superimposed on the carrier wave. In amplitude modulation (AM) radio, the amplitude of the carrier wave varies according to the sound wave. The Navy sometimes uses flashing lights to send Morse code between neighboring ships, a process that has similarities to radio broadcasting. Is this AM or FM? What is the carrier frequency? What is the signal frequency? What is the broadcasting antenna? What is the receiving antenna?

8. When light (or other electromagnetic radiation) travels across a given region, what is it that moves?

9. In space sailing, which is a proposed alternative method to reach the planets, a spacecraft carries a very large sail that experiences a force due to radiation pressure from the Sun. Should the sail be absorptive or reflective to be most effective?

10. How can the average value of an alternating current be zero yet the square root of the average squared value not be zero?

11. Suppose a creature from another planet had eyes that were sensitive to infrared radiation. Describe what it would see if it looked around the room that you are now in. That is, what would be bright and what would be dim?

12. Why should an infrared photograph of a person look different from a photograph taken with visible light?

13. Radio stations often advertise "instant news." If what they mean is that you hear the news at the instant they speak it, is their claim true? About how long would it take for a message to travel across this country by radio waves, assuming that these waves could travel this great distance and still be detected?

14. Would an inductor and a capacitor used together in an AC circuit dissipate any energy?

15. Does a wire connected to a battery emit an electromagnetic wave?

16. Do all current-carrying conductors emit electromagnetic waves? Explain.

17. If the resistance in an *RLC* circuit is doubled but the capacitance and inductance are unchanged, how will the resonance frequency change?

18. If the resistance in an *RLC* circuit remains the same but the capacitance and inductance are each doubled, how will the resonance frequency change?

PROBLEMS

1, 2, 3 = straightforward, intermediate, challenging   □ = full solution available in Student Solutions Manual/Study Guide

web = solution posted at **http://info.brookscole.com/serway**   🐾 = biomedical application

Section 21.1 Resistors in an AC Circuit

1. An rms voltage of 100 V is applied to a purely resistive load of 5.00 Ω. Find (a) the maximum voltage applied, (b) the rms current supplied, (c) the maximum current supplied, and (d) the power dissipated.

2. (a) What is the resistance of a lightbulb that uses an average power of 75.0 W when connected to a 60-Hz power source with a peak voltage of 170 V? (b) What is the resistance of a 100-W bulb?

3. An AC power supply produces a maximum voltage of $\Delta V_{max} = 100$ V. This power supply is connected to a 24.0-Ω resistor, and the current and resistor voltage are measured with an ideal AC ammeter and an ideal AC voltmeter, as shown in Figure P21.3. What does each meter read? Recall that an ideal ammeter has zero resistance and an ideal voltmeter has infinite resistance.

$\Delta V_{max} = 100$ V

$R = 24$ Ω

FIGURE P21.3

4. Figure P21.4 shows three lamps connected to a 120-V AC (rms) household supply voltage. Lamps 1 and 2 have 150-W bulbs; lamp 3 has a 100-W bulb. Find the rms current and resistance of each bulb.

120 V   1   2   3

FIGURE P21.4

5. An audio amplifier, represented by the AC source and the resistor R in Figure P21.5, delivers alternating voltages at audio frequencies to the speaker. If the source puts out an alternating voltage of 15.0 V (rms), resistance R is 8.20 Ω, and the speaker is equivalent to a resistance of 10.4 Ω, what is the time-averaged power delivered to the speaker?

R

Speaker

FIGURE P21.5

6. An AC voltage source has an output of $\Delta v = 150 \sin 377t$. Find (a) the rms voltage output, (b) the frequency of the source, and (c) the voltage at $t = 1/120$ s. (d) Find the maximum current in the circuit when the generator is connected to a 50.0-Ω resistor.

Section 21.2 Capacitors in an AC Circuit

7. Show that the SI unit of capacitive reactance, X_C, is the ohm.

8. What maximum current is delivered by a 2.20-μF capacitor when connected across (a) a North American outlet having $\Delta V_{rms} = 120$ V, $f = 60.0$ Hz; and (b) a European outlet having $\Delta V_{rms} = 240$ V, $f = 50.0$ Hz?

9. When a 4.0-μF capacitor is connected to a generator whose rms output is 30 V, the current in the circuit is observed to be 0.30 A. What is the frequency of the source?

10. What maximum current is delivered by an AC generator with $\Delta V_{max} = 48.0$ V and $f = 90.0$ Hz when connected across a 3.70-μF capacitor?

11. What value of capacitor must be inserted in a 60-Hz circuit **web** in series with a generator of 170 V maximum output voltage to produce an rms current output of 0.75 A?

12. The generator in a purely capacitive AC circuit has an angular frequency of 120π rad/s. If $\Delta V_{max} = 140$ V and $C = 6.00$ μF, what is the rms current in the circuit?

Section 21.3 Inductors in an AC Circuit

13. Show that the inductive reactance X_L has SI units of ohms.

14. In a purely inductive AC circuit, as in Figure 21.6, $\Delta V_{max} = 100$ V. (a) If the maximum current is 7.50 A at 50.0 Hz, calculate the inductance L. (b) At what angular frequency ω is the maximum current 2.50 A?

15. An inductor has a 54.0-Ω reactance at 60.0 Hz. What will be the *maximum* current if this inductor is connected to a 50.0-Hz source that produces a 100-V rms voltage?

16. A 2.40-μF capacitor is connected across an alternating voltage with an rms value of 9.00 V. The rms current in the circuit is 25.0 mA. (a) What is the source frequency? (b) If the capacitor is replaced by an ideal coil with an inductance of 0.160 H, what is the rms current in the coil?

17. Determine the maximum magnetic flux through an inductor connected to a standard outlet ($\Delta V_{rms} = 120$ V, $f = 60.0$ Hz).

Section 21.4 The RLC Series Circuit

18. An inductor ($L = 400$ mH), a capacitor ($C = 4.43$ μF), and a resistor ($R = 500$ Ω) are connected in series. A 50.0-Hz AC generator connected in series to these elements produces a maximum current of 250 mA in the circuit. (a) Calculate the required maximum voltage ΔV_{max}. (b) Determine the phase angle by which the current leads or lags the applied voltage.

19. A 10.0-μF capacitor and a 2.00-H inductor are connected in series with a 60.0-Hz source whose rms output is 100 V. Find (a) the rms current in the circuit, (b) the voltage drop across the inductor, (c) the voltage drop across the capacitor, and (d) the phase angle for the circuit. (e) Sketch the phasor diagram for this circuit.

20. A series AC circuit contains the following components: $R = 150\ \Omega$, $L = 250$ mH, $C = 2.00\ \mu F$, and a generator with $\Delta V_{max} = 210$ V operating at 50.0 Hz. Calculate the (a) inductive reactance, (b) capacitive reactance, (c) impedance, (d) maximum current, and (e) phase angle between the current and the generator voltage.

21. A resistor ($R = 900\ \Omega$), a capacitor ($C = 0.25\ \mu F$), and an inductor ($L = 2.5$ H) are connected in series across a 240-Hz AC source for which $\Delta V_{max} = 140$ V. Calculate the (a) impedance of the circuit, (b) maximum current delivered by the source, and (c) phase angle between the current and voltage. (d) Is the current leading or lagging behind the voltage?

22. A sinusoidal voltage $\Delta v(t) = (40.0\ \text{V})\sin(100t)$ is applied to a series RLC circuit with $L = 160$ mH, $C = 99.0\ \mu F$, and $R = 68.0\ \Omega$. (a) What is the impedance of the circuit? (b) What is the maximum current?

23. A 60.0-Ω resistor, a 3.00-μF capacitor, and a 0.400-H inductor are connected in series to a 90.0-V, 60.0-Hz source. Find (a) the voltage drop across the LC combination and (b) the voltage drop across the RC combination.

24. A 50.0-Ω resistor is connected in series with a 15.0-μF capacitor and a 60.0-Hz, 120-V source. (a) Find the current in the circuit. (b) What is the value of the inductor that must be inserted in the circuit to reduce the current to one-half that found in (a)?

25. A person is working near the secondary of a transformer, as shown in Figure P21.25. The primary voltage is 120 V (rms) at 60.0 Hz. The capacitance C_s, which is the stray capacitance between the hand and the secondary winding, is 20.0 pF. Assuming the person has a body resistance to ground of $R_b = 50.0$ kΩ, determine the rms voltage across the body. (*Hint:* Redraw the circuit with the secondary of the transformer as a simple AC source.)

FIGURE P21.25

26. A coil of resistance 35.0 Ω and inductance 20.5 H is in series with a capacitor and a 200-V (rms), 100-Hz source. The rms current in the circuit is 4.00 A. (a) Calculate the capacitance in the circuit. (b) What is ΔV_{rms} across the coil?

27. An AC source with a maximum voltage of 150 V and
web $f = 50.0$ Hz is connected between points a and d in Figure P21.27. Calculate the rms voltages between points (a) a and b, (b) b and c, (c) c and d, (d) b and d.

FIGURE P21.27

Section 21.5 Power in an AC Circuit

28. A 50.0-Ω resistor is connected to a 30.0-μF capacitor and to a 60.0-Hz, 100-V (rms) source. (a) Find the power factor and the average power delivered to the circuit. (b) Repeat part (a) when the capacitor is replaced with a 0.300-H inductor.

29. A multimeter in an RL circuit records an rms current of 0.500 A and a 60.0-Hz rms generator voltage of 104 V. A wattmeter shows that the average power delivered to the resistor is 10.0 W. Determine (a) the impedance in the circuit, (b) the resistance R, and (c) the inductance L.

30. In a certain RLC circuit, the rms current is 6.0 A, the rms voltage is 240 V, and the current leads the voltage by 53°. (a) What is the total resistance of the circuit? (b) Calculate the total reactance $X_L - X_C$. (c) Find the average power delivered to the circuit.

31. An inductor and a resistor are connected in series. When connected to a 60-Hz, 90-V source, the voltage drop across the resistor is found to be 50 V and the power delivered to the circuit is 14 W. Find (a) the value of the resistance and (b) the value of the inductance.

32. An AC voltage of the form $\Delta v = (100\ \text{V})\sin(1\ 000t)$ is applied to a series RLC circuit. If $R = 400\ \Omega$, $C = 5.00\ \mu F$, and $L = 0.500$ H, find the average power delivered to the circuit.

Section 21.6 Resonance in a Series RLC Circuit

33. An RLC circuit is used to tune a radio to an FM station broadcasting at 88.9 MHz. The resistance in the circuit is 12.0 Ω and the capacitance is 1.40 pF. What inductance should be present in the circuit?

34. A resonant circuit in a radio receiver is tuned to a certain station when the inductor has a value of 0.200 mH and the capacitor has a value of 30.0 pF. Find the frequency of the radio station and the wavelength sent out by the station.

35. The AM band extends from approximately 500 kHz to 1 600 kHz. If a 2.0-μH inductor is used in a tuning circuit for a radio, what are the extremes that a capacitor must reach in order to cover the complete band of frequencies?

36. A series circuit contains a 3.00-H inductor, a 3.00-μF capacitor, and a 30.0-Ω resistor connected to a 120-V (rms) source of variable frequency. Find the power delivered to the circuit when the frequency of the source is (a) the resonance frequency, (b) one-half the resonance frequency, (c) one-fourth the resonance frequency, (d) two times the resonance frequency, (e) four times the resonance frequency. From your calculations, can you draw a conclusion about the frequency at which the maximum power is delivered to the circuit?

37. A 10.0-Ω resistor, 10.0-mH inductor, and a 100-μF capacitor are connected in series to a 50.0-V (rms) source having variable frequency. Find the energy delivered to the circuit during one period if the operating frequency is twice the resonance frequency.

Section 21.7 The Transformer

38. A step-down transformer is used for recharging the batteries of portable devices such as tape players. The turns ratio inside the transformer is 13:1 and it is used with 120 V (rms) household service. If a particular ideal transformer draws 0.350 A from the house outlet, what are (a) the voltage and (b) the current supplied to a tape player from the transformer? (c) How much power is delivered?

39. A step-up transformer is designed to have an output voltage of 2 200 V (rms) when the primary is connected across a 110-V (rms) source. (a) If there are 80 turns on the primary winding, how many turns are required on the secondary? (b) If a load resistor across the secondary draws a current of 1.5 A, what is the current in the primary, assuming ideal conditions?

40. A transformer has $N_1 = 350$ turns and $N_2 = 2\,000$ turns. If the input voltage is $\Delta v(t) = (170\text{ V})\cos\omega t$, what rms voltage is developed across the secondary coil?

41. A transformer on a pole near a factory steps the voltage down from 3 600 V to 120 V. The transformer is to deliver 1 000 kW to the factory at 90% efficiency. Find (a) the power delivered to the primary, (b) the current in the primary, and (c) the current in the secondary.

42. A transmission line that has a resistance per unit length of $4.50 \times 10^{-4}\ \Omega/\text{m}$ is to be used to transmit 5.00 MW over 400 miles (6.44×10^5 m). The output voltage of the generator is 4.50 kV. (a) What is the line loss if a transformer is used to step up the voltage to 500 kV? (b) What fraction of the input power is lost to the line under these circumstances? (c) What difficulties would be encountered on attempting to transmit the 5.00 MW at the generator voltage of 4.50 kV?

Section 21.10 Production of Electromagnetic Waves by an Antenna

Section 21.11 Properties of Electromagnetic Waves

43. The U.S. Navy has long proposed the construction of extremely low-frequency (ELF) communications systems; such waves could penetrate the oceans to reach distant submarines. Calculate the length of a quarter-wavelength antenna for a transmitter generating ELF waves of frequency 75 Hz. How practical is this?

44. Experimenters at the National Institute of Standards and Technology have made precise measurements of the speed of light using the fact that, in vacuum, the speed of electromagnetic waves is $c = 1/\sqrt{\mu_0\epsilon_0}$, where the constants $\mu_0 = 4\pi \times 10^{-7}\ \text{N}\cdot\text{s}^2/\text{C}^2$ and $\epsilon_0 = 8.854 \times 10^{-12}\ \text{C}^2/\text{N}\cdot\text{m}^2$. What value (to four significant figures) does this give for the speed of light in vacuum?

45. An electromagnetic wave in vacuum has an electric field amplitude of 220 V/m. Calculate the amplitude of the corresponding magnetic field.

46. A particular electromagnetic wave traveling in vacuum has a magnetic field amplitude of 1.5×10^{-7} T. Find (a) the electric field amplitude and (b) the average power per unit area associated with the wave.

47. A microwave oven is powered by an electron tube called a magnetron that generates electromagnetic waves of frequency 2.45 GHz. The microwaves enter the oven and are reflected by the walls. The standing wave pattern produced in the oven can cook food unevenly, with hot spots in the food at antinodes and cool spots at nodes, so a turntable is often used to rotate the food and distribute the energy. If a microwave oven is used with a cooking dish in a fixed position, the antinodes can appear as burn marks on foods such as carrot strips or cheese. The separation distance between the burns is measured to be 6.00 cm. Calculate the speed of the microwaves from these data.

48. Assume that the solar radiation incident on Earth is 1 340 W/m² (at the top of Earth's atmosphere). Calculate the total power radiated by the Sun, taking the average separation between Earth and the Sun to be 1.49×10^{11} m.

49. The Sun delivers an average power of 1 340 W/m² to the top of Earth's atmosphere. Find the magnitudes of $\mathbf{E}_{\text{max}}$ and $\mathbf{B}_{\text{max}}$ for the electromagnetic waves at the top of the atmosphere.

Section 21.12 The Spectrum of Electromagnetic Waves

50. The eye is most sensitive to light of wavelength 5.50×10^{-7} m, which is in the green-yellow region of the visible electromagnetic spectrum. What is the frequency of this light?

51. What are the wavelength ranges in the (a) AM radio band (540–1 600 kHz) and (b) the FM radio band (88–108 MHz)?

52. A diathermy machine, used in physiotherapy, generates electromagnetic radiation that gives the effect of "deep heat" when absorbed in tissue. One assigned frequency for diathermy is 27.33 MHz. What is the wavelength of this radiation?

53. A radar pulse returns to the receiver after a total travel time of 4.00×10^{-4} s. How far away is the object that reflected the wave?

54. An important news announcement is transmitted by radio waves to people who are 100 km away, sitting next to their radios, and by sound waves to people sitting across the newsroom, 3.0 m from the newscaster. Who receives the news first? Explain. Take the speed of sound in air to be 343 m/s.

ADDITIONAL PROBLEMS

55. An AC adapter for a telephone answering unit uses a transformer to reduce the line voltage of 120 V (rms) to a voltage of 9.0 V. The rms current delivered to the answering system is 400 mA. (a) If the primary (input) coil in the transformer in the adapter has 240 turns, how many turns are there on the secondary (output) coil? (b) What is the rms power delivered to the transformer? Assume an ideal transformer.

56. The primary coil of a certain transformer has an inductance of 2.50 H and a resistance of 80.0 Ω. (a) If the primary coil is connected to an AC source with a frequency of 60.0 Hz and a voltage of 110 V (rms), what is the rms current in the primary? (b) If the primary is connected to 110 V DC, what is the current in the primary? (c) In each case, compare the power delivered to the coil.

57. As a way of determining the inductance of a coil used in a research project, a student first connects the coil to a 12.0-V battery and measures a current of 0.630 A. The student then connects the coil to a 24.0-V (rms), 60.0-Hz generator and measures an rms current of 0.570 A. What is the inductance?

58. The intensity of solar radiation at the top of Earth's atmosphere is 1 340 W/m³. Assuming that 60% of the incoming solar energy reaches Earth's surface and assuming that you absorb 50% of the incident energy, make an order-of-magnitude estimate of the amount of solar energy you absorb in a 60-minute sunbath.

59. A 200-Ω resistor is connected in series with a 5.0-μF capacitor and a 60-Hz, 120-V rms line. If electrical energy costs \$0.080/kWh, how much does it cost to leave this circuit connected for 24 h?

60. A series RLC circuit has a resonance frequency of $2\,000/\pi$ Hz. When it is operating at a frequency of

$\omega > \omega_0$, $X_L = 12 \ \Omega$ and $X_C = 8.0 \ \Omega$. Calculate the values of L and C for the circuit.

61. Two connections allow contact with two circuit elements in
web series inside a box, but it is not known whether the circuit elements are R, L, or C. In an attempt to find what is inside the box, you make some measurements, with the following results. When a 3.0-V DC power supply is connected across the terminals, there is a direct current of 300 mA in the circuit. When a 3.0-V, 60-Hz source is connected, the current becomes 200 mA. (a) What are the two elements in the box? (b) What are their values of R, L, or C?

62. (a) What capacitance will resonate with a one-turn loop of inductance 400 pH to give a radar wave of wavelength 3.0 cm? (b) If the capacitor has square parallel plates separated by 1.0 mm of air, what should the edge length of the plates be? (c) What is the common reactance of the loop and capacitor at resonance?

63. A dish antenna with a diameter of 20.0 m receives (at normal incidence) a radio signal from a distant source, as shown in Figure P21.63. The radio signal is a continuous sinusoidal wave with amplitude $E_{max} = 0.20 \ \mu\text{V/m}$. Assume the antenna absorbs all the radiation that falls on the dish. (a) What is the amplitude of the magnetic field in this wave? (b) What is the intensity of the radiation received by this antenna? (c) What is the power received by the antenna?

64. A particular inductor has appreciable resistance. When the inductor is connected to a 12-V battery, the current in the inductor is 3.0 A. When it is connected to an AC source with an rms output of 12 V and a frequency of 60 Hz, the current drops to 2.0 A. What are (a) the impedance at 60 Hz and (b) the inductance of the inductor?

FIGURE P21.63

65. A possible means of space flight is to place a perfectly reflecting aluminized sheet into Earth's orbit and to use the light from the Sun to push this solar sail. Suppose such a sail, of area $6.00 \times 10^4 \ \text{m}^2$ and mass 6 000 kg, is placed in orbit facing the Sun. (a) What force is exerted on the sail? (b) What is the sail's acceleration? (c) How long does it take for this sail to reach the Moon, 3.84×10^8 m away? Ignore all gravitational effects, and assume a solar intensity of 1 340 W/m^2. [*Hint:* The radiation pressure by a reflected wave is given by 2(average power per unit area)/c.]

66. Suppose you wish to use a transformer as an impedance-matching device between an audio amplifier that has an output impedance of 8.0 kΩ and a speaker that has an input impedance of 8.0 Ω. What should be the ratio of primary to secondary turns on the transformer?

67. Compute the average energy content of a liter of sunlight as it reaches the top of Earth's atmosphere, where its intensity is 1 340 W/m^2.

GROUP ACTIVITIES

G.1 For this observation, you will need some items that can be found at many electronics stores. You will need a bicolored LED (light-emitting diode), a resistor of about 100 Ω, 2 m of flexible wire, and a step-down transformer with an output of 3 to 6 V. Use the wire to connect the LED and the resistor in series with the transformer. A bi-colored LED is designed such that it emits a red color when the current in the LED is in one direction and green when the current reverses. When connected to an AC source, the LED is yellow. Why?

Hold the wires and whirl the LED in a circular path. In a darkened room, you will see red and green bars at equally spaced intervals along the path of the LED. Why?

As you continue to whirl the LED in a circular path, have your partner count the number of green bars in the circle then measure the time it takes for the LED to travel ten times around the circular path. Based on this information, determine the time it takes for the color of the LED to change from green to red to green. You should obtain an answer of (1/60) s. Why?

G.2 Rotate a portable radio (with a telescoping antenna) about a horizontal axis while it is tuned to a weak station. Such an antenna detects the varying electric field produced by the station. What can you determine about the direction of the electric field produced by the transmitter?

Now turn on your radio to a nearby station and experiment with shielding the radio from incoming waves. Is the reception affected by surrounding the radio by aluminum foil? By plastic wrap? Use any other material you have available. What kinds of material block the signal? Why?

G.3 An *RLC* circuit is brought to resonance. (a) What is the impedance of the circuit at resonance? (b) When will the current be greatest—at resonance, at 10% below the resonant frequency, or at 10% above the resonant frequency? (c) Check your guesses by examining a circuit with $\Delta V = 12$ V, $R = 15 \ \Omega$, $C = 75 \ \mu\text{F}$, and $L = 200$ mH.

G.4 A transformer is to be used to provide power for a computer disk drive that needs 6.0 V instead of the 120 V from the wall outlet. (a) Should the power supply have more turns in the primary as compared to the secondary, or the reverse? (b) Will the larger current be in the primary or the secondary? (c) Assume the number of turns in the primary is 400 and that it delivers 4.0 A at the 6.0-V output. Do the calculations to confirm your answers.

G.5 Consider an *RLC* circuit with $R = 25 \ \Omega$, $L = 6.0$ mH, and $C = 25 \ \mu\text{F}$. This circuit is connected to a 10-V, 600-Hz AC source. (a) Is the sum of the voltage drops across R, L, and C equal to 10 V? (b) Which is greatest, the power delivered to the resistor, the capacitor, or the inductor? (c) Do the calculations to verify your predictions.

Light and Optics

Scientists have long been intrigued by the nature of light, and philosophers have argued endlessly concerning the proper definition and perception of light. It is important to understand the nature of this basic ingredient of life on Earth. Plants convert light energy from the Sun to chemical energy through photosynthesis. Light is the means by which we transmit and receive information from objects around us and throughout the Universe.

The ancient Greeks believed that vision worked by means of "fire" emanating from the eye, which then interacted with objects and returned to the eye carrying object properties with it. Around 1 000 A.D., it was proposed that light consisted of tiny particles that were emitted by a light source. These particles stimulated the perception of vision on striking the observer's eye. Newton used this particle model to explain the reflection and refraction of light. In 1670 one of Newton's contemporaries, the Dutch scientist Christian Huygens, succeeded in explaining many properties of light by proposing that light was wave-like. In 1801 Thomas Young gave strong support to the wave theory by showing that light beams can interfere with one another. In 1865 Maxwell developed a brilliant theory that electromagnetic waves travel with the speed of light (Chapter 21). By that time, the wave theory of light seemed to be on firm ground.

However, at the beginning of the 20th century, Max Planck introduced the notion that vibrating electrons in the wall of a glowing solid had quantized (discrete) energies in order to explain the light spectrum emitted by hot objects. Albert Einstein extended this idea to a full-fledged particle theory of light to explain the photoelectric effect (electrons emitted by a metal exposed to light). We shall discuss those and other modern topics in the last part of this book.

Today, scientists view light as having a dual nature. Experiments can be devised that will display either its particle-like or its wave-like nature. All we can say about light is that whenever we design an experiment to determine its wave-like nature, we will find it, and whenever we look for its particle-like nature, we will find that. Nature prevents us from testing both qualities at the same time.

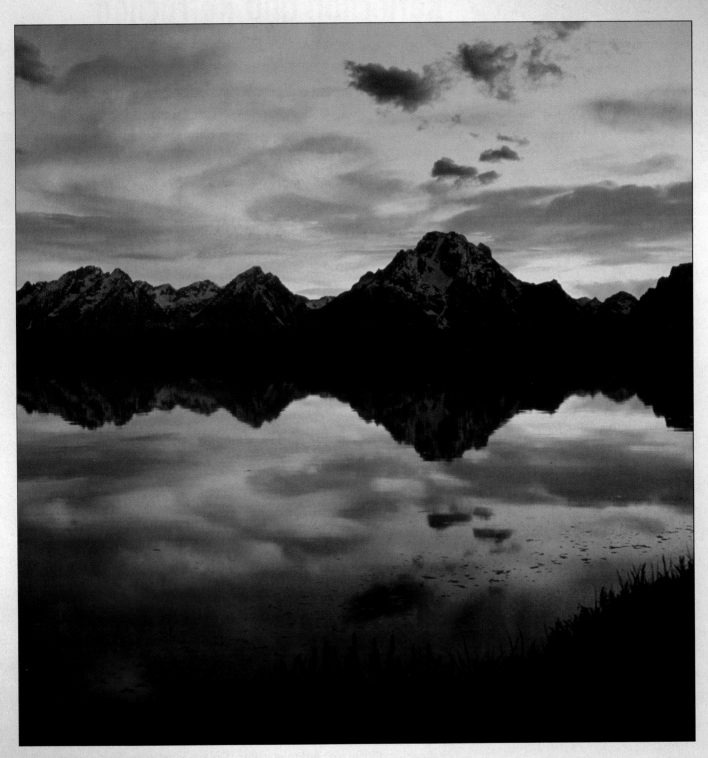

This striking photograph shows the Grand Tetons in Wyoming reflected in a lake which acts like a mirror. Note that the image of the mountains has the same size as the mountains.

(David Muench/CORBIS)

22

Reflection and Refraction of Light

The prism in this photograph demonstrates both refraction and reflection. A beam of white light at the top left is refracted and strikes the right face of the prism. At this face, some of the light is refracted once more, exiting the prism as a spectrum of light with different colors. The spectrum arises from the fact that the incident white light contains different wavelengths that refract at different angles. Part of the light is reflected at the right face and exits the prism at the bottom left as white light. *(David Parker/Science Photo Library/Photo Researchers, Inc.)*

*I*n this part of the book, we concentrate on the aspects of light that are best understood through the wave model. First, we discuss the reflection of light at the boundary between two media and the refraction (bending) of light as it travels from one medium into another. We use these ideas to study the refraction of light as it passes through lenses and the reflection of light from mirrored surfaces. Finally, we describe how lenses and mirrors can be used to view objects with telescopes and microscopes and how lenses are used in photography.

22.1 THE NATURE OF LIGHT

Until the beginning of the 19th century, light was considered to be a stream of particles, emitted by a light source, that stimulated the sense of sight on entering the eye. The chief architect of the particle theory of light was Newton. With this theory he provided simple explanations of some known experimental facts concerning the nature of light—namely, the laws of reflection and refraction.

Most scientists accepted Newton's particle theory of light. However, during Newton's lifetime another theory was proposed. In 1678 a Dutch physicist and astronomer, Christian Huygens (1629–1695), showed that a wave theory of light could also explain the laws of reflection and refraction. The wave theory did not receive immediate acceptance for several reasons. All the waves known at the time (sound, water, and so on) traveled through some sort of medium, but light from the Sun could travel to the Earth through empty space. Furthermore, it was argued that if light were some form of wave, it would bend around obstacles; hence, we should be able to see around corners. It is now known that light does indeed bend around the edges of objects. This phenomenon, known as *diffraction*, is not easy to observe because light waves have such short wavelengths. Even though experimental evidence for the diffraction of light was discovered by Francesco Grimaldi (1618–1663) around 1660, for more than a century most scientists rejected the wave theory and adhered to Newton's particle theory. This was also due to Newton's great reputation as a scientist.

The first clear demonstration of the wave nature of light was provided in 1801 by Thomas Young (1773–1829), who showed that under appropriate conditions, light exhibits interference behavior. That is, light waves emitted by a single source and traveling along two different paths can arrive at some point, combine, and cancel each other by destructive interference. Such behavior could not be explained at that time by a particle model, because scientists could not imagine how two or more particles could come together and cancel one another.

The most important development concerning the theory of light was the work of Maxwell, who in 1865 predicted that light was a form of high-frequency electromagnetic wave (Chapter 21). His theory predicted that these waves should have a speed of 3×10^8 m/s, in agreement with the measured value.

Although the classical theory of electricity and magnetism explained most known properties of light, some subsequent experiments could not be explained by the assumption that light was a wave. The most striking of these was the *photoelectric effect* (which we shall examine more closely in Chapter 27), discovered by Hertz. Hertz found that clean metal surfaces emit charges when exposed to ultraviolet light.

In 1905 Einstein published a paper that formulated the theory of light quanta ("particles") and explained the photoelectric effect. He reached the conclusion that light is composed of corpuscles, or discontinuous quanta of energy. These corpuscles or quanta are now called *photons* to emphasize their particle-like nature. According to Einstein's theory, the energy of a photon is proportional to the frequency of the electromagnetic wave:

$$E = hf \qquad [22.1]$$

◄ Energy of a photon

CHRISTIAN HUYGENS, DUTCH PHYSICIST AND ASTRONOMER (1629–1695)

Huygens is best known for his contributions to the fields of optics and dynamics. To Huygens, light was a vibratory motion in the ether, spreading out and producing the sensation of light when impinging on the eye. On the basis of this theory, he deduced the laws of reflection and refraction and explained the phenomenon of double refraction. *(Courtesy of Rijksmuseum voor de Geschiedenis der Natuurwetenschappen. Courtesy AIP Niels Bohr Library)* where $h = 6.63 \times 10^{-34}$ J·s is *Planck's constant*. This theory retains some features of both the wave and particle theories of light. As we shall discuss later, the photoelectric effect is the result of energy transfer from a single photon to an electron in the metal. That is, the electron interacts with one photon of light as if the electron had been struck by a particle. Yet the photon has wave-like characteristics, as implied by the fact that a frequency is used in its definition.

In view of these developments, light must be regarded as having a *dual nature*. That is, **in some cases light acts as a wave and in others it acts as a particle.** Classical electromagnetic wave theory provides adequate explanations of light propagation and of the effects of interference, whereas the photoelectric effect and other experiments involving the interaction of light with matter are best explained by assuming that light is a particle. However, the question "Is light a wave or a particle?" is inappropriate; sometimes it acts as one, sometimes as the other. Fortunately, it never acts as both in the same experiment.

22.2 THE RAY APPROXIMATION IN GEOMETRIC OPTICS

In studying geometric optics here and in future chapters, we shall make use of an important property of light that can be understood based on common experience. **Light travels in a straight-line path in a homogeneous medium until it encounters a boundary between two different materials.** As we shall see, when light strikes a boundary it either is reflected from that boundary or passes into the material on the other side of the boundary or partially does both.

Based on this observation, we will use what is called the **ray approximation** to represent beams of light. As shown in Figure 22.1, a ray of light is an imaginary line drawn along the direction of travel of the light beam. For example, a beam of sunlight passing through a darkened room traces out the path of a light ray. We will also have occasion to refer to wave fronts of light. A **wave front** is a surface passing through the points of a wave that have the same phase and amplitude. For instance, the wave fronts in Figure 22.1 could be surfaces passing through the crests of waves. You should note that the rays, corresponding to the direction of wave motion, are straight lines perpendicular to the wave fronts. Also, note that when light rays travel in parallel paths, the wave fronts are planes perpendicular to the rays.

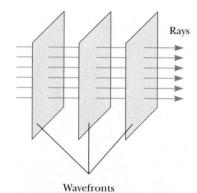

FIGURE 22.1 A plane wave traveling to the right. Note that the rays, corresponding to the direction of wave motion, are straight lines perpendicular to the wave fronts.

22.3 REFLECTION AND REFRACTION

REFLECTION OF LIGHT

When a light ray traveling in a transparent medium encounters a boundary leading into a second medium, part of the incident ray is reflected back into the first medium. Figure 22.2a shows several rays of a beam of light incident on a smooth, mirror-like, reflecting surface. The reflected rays are parallel to each other, as indicated in the figure. Reflection of light from such a smooth surface is called **specular reflection.** On the other hand, if the reflecting surface is rough, as in Figure 22.2b, the surface reflects the rays in a variety of directions. Reflection from any rough surface is known as **diffuse reflection.** A surface behaves as a smooth surface as long as the surface variations are small compared with the wavelength of the incident light. Figures 22.2c and 22.2d are photographs of specular and diffuse reflection of laser light.

For instance, consider the two types of reflection from a road surface that one sees while driving at night. When the road is dry, light from oncoming vehicles is scattered off the road in different directions (diffuse reflection) and the road is quite visible. On a rainy night, when the road is wet, the road irregularities are filled with water. Because the wet surface is quite smooth, the light undergoes

APPLICATION

SEEING THE ROAD ON A RAINY NIGHT

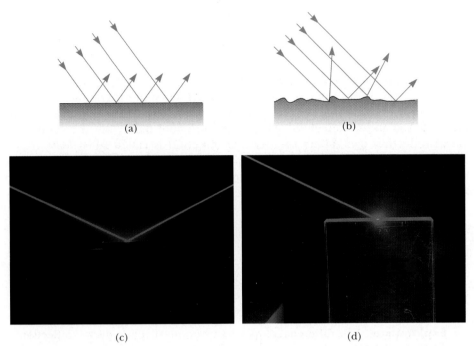

(c)   (d)

FIGURE 22.2   A schematic representation of (a) specular reflection, where the reflected rays are all parallel to each other, and (b) diffuse reflection, where the reflected rays travel in random directions. (c, d) Photographs of specular and diffuse reflection, using laser light. *(Photographs courtesy of Henry Leap and Jim Lehman)*

(a)   (b)

FIGURE 22.3   (Quick Quiz 22.1) *(Photos by Charles D. Winters)*

specular reflection. This means that the light is reflected straight ahead, and the driver of a car sees only what is directly in front. Light from the side never reaches the driver's eye. In this book we concern ourselves only with specular reflection, and we use the term *reflection* to mean specular reflection.

Quick Quiz 22.1	Which part of Figure 22.3, (a) or (b), shows specular reflection of light from the roadway?

Consider a light ray traveling in air and incident at some angle on a flat, smooth surface, as in Figure 22.4. The incident and reflected rays make angles θ_1 and θ_1', respectively, **with a line perpendicular to the surface** at the point where the incident ray strikes the surface. We call this line the *normal* to the surface. Experiments

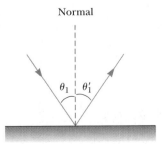

FIGURE 22.4   According to the law of reflection, $\theta_1 = \theta_1'$.

show that the angle of reflection equals *the angle of incidence;* that is,

$$\theta_1' = \theta_1 \qquad \text{[22.2]}$$

You may have noticed a common occurrence in photographs of individuals—their eyes appear to be glowing red. This occurs when a photographic flash device is used and the flash unit is very close to the camera lens. Light from the flash unit enters the eye and is reflected back along its original path from the retina. This type of reflection back along the original direction is called *retroreflection.* If the flash unit and lens are close together, this retroreflected light can enter the lens. Most of the light reflected from the retina is red, due to the blood vessels at the back of the eye, giving the red-eye effect in the photograph.

APPLYING PHYSICS 22.1

An observer on the west-facing beach of a large lake is watching the beginning of a sunset. The water is very smooth except for some areas with small ripples. The observer notices that some areas of the water are blue and some are pink. Why does the water appear to be different colors in different areas?

Explanation The different colors arise from specular and diffuse reflection. The smooth areas of the water will specularly reflect the light from the west, which is the pink light from the sunset. The areas with small ripples will reflect the light diffusely. Thus, light from all parts of the sky will be reflected into the observer's eyes. Because most of the sky is still blue at the beginning of the sunset, these areas will appear to be blue.

APPLYING PHYSICS 22.2

When looking outdoors through a glass window at night, you sometimes see a double image of yourself. Why?

Explanation Reflection occurs whenever there is an interface between two different media. For the glass in the window, there are two such surfaces. The first is the inner surface of the glass, and the second is the outer surface. Each of these interfaces results in an image.

Example 22.1   The Double-Reflecting Light Ray

Two mirrors make an angle of 120° with each other, as in Figure 22.5. A ray is incident on mirror M_1 at an angle of 65° to the normal. Find the angle the ray makes with the normal to M_2 after it is reflected from both mirrors.

Reasoning and Solution From the law of reflection, we see that the first reflected ray also makes an angle of 65° with the normal to M_1. It follows that this same ray makes an angle of 90° − 65°, or 25°, with the horizontal. From the triangle made by the first reflected ray and the two mirrors, we see that the first reflected ray makes an angle of 35° with M_2 (because the sum of the interior angles of any triangle is 180°). This means that this ray makes an angle of 55° with the normal to M_2. From the law of reflection, it follows that the second reflected ray makes an angle of 55° with the normal to M_2.

FIGURE 22.5 (Example 22.1) Mirrors M_1 and M_2 make an angle of 120° with each other.

REFRACTION OF LIGHT

When a ray of light traveling through a transparent medium encounters a boundary leading into another transparent medium, as in Figure 22.6a, part of the ray is reflected and part enters the second medium. The ray that enters the second

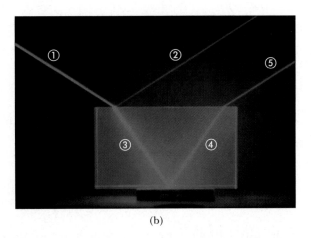

FIGURE 22.6 (a) A ray obliquely incident on an air-glass interface. The refracted ray is bent toward the normal because $v_2 < v_1$. (b) Light incident on the Lucite block bends both when it enters the block and when it leaves the block. *(Courtesy of Henry Leap and Jim Lehman)*

medium is bent at the boundary and is said to be *refracted*. The incident ray, the reflected ray, the refracted ray, and the normal at the point of incidence all lie in the same plane. The **angle of refraction,** θ_2, in Figure 22.6a depends on the properties of the two media and on the angle of incidence, through the relationship

$$\frac{\sin \theta_2}{\sin \theta_1} = \frac{v_2}{v_1} = \text{constant} \qquad \text{[22.3]}$$

where v_1 is the speed of light in medium 1 and v_2 is the speed of light in medium 2. Note that the angle of refraction is also measured with respect to the normal. In Section 22.7 we shall derive the laws of reflection and refraction using Huygens's principle.

Experiment shows that **the path of a light ray through a refracting surface is reversible.** For example, the ray in Figure 22.6a travels from point *A* to point *B*. If the ray originated at *B*, it would follow the same path to reach point *A*, but the reflected ray would be in the glass.

Quick Quiz 22.2 If beam 1 is the incoming beam in Figure 22.6b, which of the other four beams are reflected and which are refracted?

When light moves from a material in which its speed is high to a material in which its speed is lower, the angle of refraction θ_2 is less than the angle of incidence. The refracted ray therefore bends toward the normal, as shown in Figure 22.7a. If the ray moves from a material in which it travels slowly to a material in

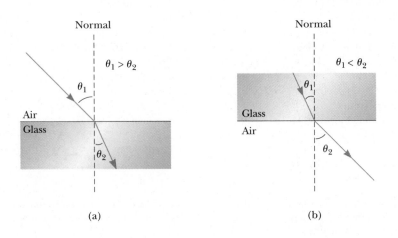

(a) (b)

FIGURE 22.7 (a) When the light beam moves from air into glass, its path is bent toward the normal. (b) When the beam moves from glass into air, its path is bent away from the normal.

which it travels more rapidly, θ_2 is greater than θ_1, and so the ray bends away from the normal, as shown in Figure 22.7b.

22.4 THE LAW OF REFRACTION

When light passes from one transparent medium to another, it is refracted because the speed of light is different in the two media.[1] It is convenient to define the **index of refraction,** n, of a medium as the ratio

Index of refraction

$$n \equiv \frac{\text{speed of light in vacuum}}{\text{speed of light in a medium}} = \frac{c}{v} \qquad [22.4]$$

From this definition, we see that the index of refraction is a dimensionless number that is greater than or equal to unity because v is always less than c. Furthermore, n is equal to unity for vacuum. Table 22.1 lists the indices of refraction for various substances.

As light travels from one medium to another, its frequency does not change. To see why, consider Figure 22.8. Wave fronts pass an observer at point A in medium 1 with a certain frequency and are incident on the boundary between medium 1 and medium 2. The frequency at which the wave fronts pass an observer at point B in medium 2 must equal the frequency at which they arrive at point A. If this were not the case, the wave fronts would either pile up at the boundary or be destroyed or created at the boundary. Because this does not occur, the frequency must remain the same as a light ray passes from one medium into another.

Therefore, because the relation $v = f\lambda$ must be valid in both media and because $f_1 = f_2 = f$, we see that

$$v_1 = f\lambda_1 \qquad \text{and} \qquad v_2 = f\lambda_2$$

Because $v_1 \neq v_2$, it follows that $\lambda_1 \neq \lambda_2$. A relationship between index of refraction and wavelength can be obtained by dividing these two equations and making use of the definition of index of refraction given by Equation 22.4:

$$\frac{\lambda_1}{\lambda_2} = \frac{v_1}{v_2} = \frac{c/n_1}{c/n_2} = \frac{n_2}{n_1} \qquad [22.5]$$

Tip 22.1 AN INVERSE RELATIONSHIP

The index of refraction is *inversely* proportional to the wave speed. Therefore, as the wave speed v decreases, the index of refraction n *increases*.

Tip 22.2 THE FREQUENCY REMAINS THE SAME

Note that the *frequency* of a wave does *not* change as the wave passes from one medium to another. Both the wave speed and wavelength *do* change, but the frequency remains constant.

FIGURE 22.8 As the wave moves from medium 1 to medium 2, its wavelength changes but its frequency remains constant.

TABLE 22.1	Indices of Refraction for Various Substances, Measured with Light of Vacuum Wavelength $\lambda_0 = 589$ nm		
Substance	**Index of Refraction**	**Substance**	**Index of Refraction**
Solids at 20°C		**Liquids at 20°C**	
Diamond (C)	2.419	Benzene	1.501
Fluorite (CaF_2)	1.434	Carbon disulfide	1.628
Fused quartz (SiO_2)	1.458	Carbon tetrachloride	1.461
Glass, crown	1.52	Ethyl alcohol	1.361
Glass, flint	1.66	Glycerine	1.473
Ice (H_2O) (at 0°C)	1.309	Water	1.333
Polystyrene	1.49		
Sodium chloride (NaCl)	1.544	**Gases at 0°C, 1 atm**	
Zircon	1.923	Air	1.000 293
		Carbon dioxide	1.000 45

[1] The speed of light varies between media because the time lags caused by absorption and reemission of light as it travels from atom to atom depends on the particular electronic structure of the atoms constituting each material.

which gives

$$\lambda_1 n_1 = \lambda_2 n_2 \qquad [22.6]$$

Let medium 1 be the vacuum so that $n_1 = 1$. It follows from Equation 22.6 that the index of refraction of any medium can be expressed as the ratio

$$n = \frac{\lambda_0}{\lambda_n} \qquad [22.7]$$

where λ_0 is the wavelength of light in vacuum and λ_n is the wavelength in the medium whose index of refraction is n. Figure 22.9 is a schematic representation of this reduction in wavelength when light passes from vacuum into a transparent medium.

We are now in a position to express Equation 22.3 in an alternative form. If we substitute Equation 22.5 into Equation 22.3, we get

$$n_1 \sin \theta_1 = n_2 \sin \theta_2 \qquad [22.8]$$

The experimental discovery of this relationship is usually credited to Willebord Snell (1591–1627) and is therefore known as **Snell's law of refraction.**

$n_2 > n_1$
$\lambda_2 < \lambda_1$

FIGURE 22.9 A schematic diagram of the *reduction* in wavelength when light travels from a medium with a low index of refraction to one with a higher index of refraction.

◀ Snell's law of refraction

Quick Quiz 22.3

A material has an index of refraction that increases continuously from top to bottom. Of the three paths shown in Figure 22.10, which path will a light ray follow as it passes through the material.

(a)   (b)   (c)

FIGURE 22.10 (Quick Quiz 22.3)

Quick Quiz 22.4

As light travels from vacuum ($n = 1$) to a medium such as glass ($n > 1$), which of the following properties remains the same: (a) wavelength, (b) wave speed, or (c) frequency?

Example 22.2   Angle of Refraction for Glass

A light ray of wavelength 589 nm (produced by a sodium lamp) traveling through air is incident on a smooth, flat slab of crown glass at an angle of 30.0° to the normal, as sketched in Figure 22.11. Find the angle of refraction, θ_2.

Solution   Snell's law (Eq. 22.8) can be rearranged as

$$\sin \theta_2 = \frac{n_1}{n_2} \sin \theta_1$$

From Table 22.1, we find that $n_1 = 1.00$ for air and $n_2 = 1.52$ for crown glass. Therefore, the unknown refraction angle is determined by

$$\sin \theta_2 = \left(\frac{1.00}{1.52}\right)(\sin 30.0°) = 0.329$$

$$\theta_2 = \sin^{-1}(0.329) = \boxed{19.2°}$$

We see that the ray is bent *toward* the normal, as expected.

FIGURE 22.11 (Example 22.2) Refraction of light by glass.

EXERCISE   If the light ray moves from inside the glass toward the glass-air interface at an angle of 30.0° to the normal, determine the angle of refraction.

ANSWER   49.5° *away* from the normal

Example 22.3   The Speed of Light in Fused Quartz

Light of wavelength 589 nm in vacuum passes through a piece of fused quartz of index of refraction $n = 1.458$.

A Find the speed of light in fused quartz.

Solution   The speed of light in fused quartz can be obtained from Equation 22.4:

$$v = \frac{c}{n} = \frac{3.00 \times 10^8 \text{ m/s}}{1.458} = \boxed{2.06 \times 10^8 \text{ m/s}}$$

It is interesting to note that the speed of light in vacuum, 3.00×10^8 m/s, is an upper limit for the speed of material objects. In our treatment of relativity in Chapter 26, we shall find that this upper limit is consistent with experimental observations. However, it is possible for a particle moving in a medium to have a speed that exceeds the speed of light in that medium. For example, it is theoretically possible for a particle to travel through fused quartz at a speed greater than 2.06×10^8 m/s, but it must have a speed less than 3.00×10^8 m/s.

B What is the wavelength of this light in fused quartz?

Solution   We can use $\lambda_n = \lambda_0/n$ (Eq. 22.7) to calculate the wavelength in fused quartz, noting that we are given $\lambda_0 = 589$ nm $= 589 \times 10^{-9}$ m:

$$\lambda_n = \frac{\lambda_0}{n} = \frac{589 \text{ nm}}{1.458} = \boxed{404 \text{ nm}}$$

EXERCISE   Find the frequency of the light passing through the fused quartz.

ANSWER 5.09×10^{14} Hz

A transparent block has refracted three incident light beams, which exit parallel to each other but are off-set from their original direction of travel. Note the partial reflection at both interfaces. (© *Leonard Lessin/Peter Arnold, Inc.*)

Webnote 22.1

Scientists have been able to slow down the speed of light to as low as 38 mph! See how this was achieved at
http://ajanta.sci.ccny.cuny.edu/~jupiter/ pub/sciinfo/slowlight.html

Example 22.4   Light Passing Through a Slab

A light beam traveling through a transparent medium of index of refraction n_1 passes through a thick transparent slab with parallel faces and index of refraction n_2 (Fig. 22.12). Show that the emerging beam is parallel to the incident beam.

Reasoning   To solve this problem, it is necessary to apply Snell's law twice, once at the upper surface and once at the lower surface. The two equations will be related because the angle of refraction at the upper surface equals the angle of incidence at the lower surface. The ray passing through the slab makes equal angles with the normal at the entry and exit points. This procedure will enable us to compare angles θ_1 and θ_3.

Solution   First, let us apply Snell's law to the upper surface:

$$\sin \theta_2 = \frac{n_1}{n_2} \sin \theta_1 \tag{1}$$

Applying Snell's law to the lower surface gives

$$\sin \theta_3 = \frac{n_2}{n_1} \sin \theta_2 \tag{2}$$

FIGURE 22.12   (Example 22.4) When light passes through a flat slab of material, the emerging beam is parallel to the incident beam, and therefore $\theta_1 = \theta_3$.

Substituting (1) into (2) gives

$$\sin \theta_3 = \frac{n_2}{n_1} \left(\frac{n_1}{n_2} \sin \theta_1 \right) = \sin \theta_1$$

Thus, $\theta_3 = \theta_1$, and so the slab does not alter the direction of the beam as shown by the dashed line in Figure 22.12. It does, however, produce a lateral displacement of the beam.

Example 22.5 Refraction of Laser Light in a DVD (Digital Video Disk)

A DVD is a video recording consisting of a spiral track of digital information about 1.0 μm wide (see Fig. 22.13a). The digital information consists of a series of pits that are "read" by a laser beam sharply focused on a track in the information layer. If the width of the laser beam a at the information layer must equal 1.0 μm to distinguish individual tracks, and the width of the beam w as it enters the plastic is 0.70 mm, find the angle θ_1 at which the conical beam should enter the plastic (see Fig. 22.13b). Assume the plastic has a thickness $t = 1.2$ mm and an index of refraction $n = 1.55$. Note that this system is relatively immune to small dust particles degrading the video quality, because particles would have to be as large as 0.70 mm to obscure the beam at the point where it enters the plastic.

APPLICATION

READING DATA ON A DVD

(a)

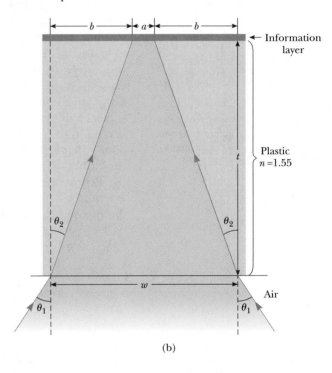

(b)

FIGURE 22.13 (a) A micrograph of a DVD surface showing tracks and pits along each track. *(Courtesy of Sony Disc Manufacturing)* (b) Cross section of a cone-shaped laser beam used to read a DVD.

Solution From Figure 22.13b we see that

$$w = 2b + a$$

$$b = \frac{w - a}{2} = \frac{699 \ \mu m}{2} = 349.5 \ \mu m$$

Also,

$$\tan \theta_2 = \frac{b}{t} = \frac{349.5 \ \mu m}{1200 \ \mu m}; \qquad \theta_2 = 16.2°$$

Finally we can use Snell's law to find θ_1.

$$n_1 \sin \theta_1 = n_2 \sin \theta_2$$

$$\sin \theta_1 = \frac{n_2 \sin \theta_2}{n_1} = \frac{1.55 \sin 16.2°}{1.00} = 0.433$$

$$\theta_1 = \boxed{26°}$$

22.5 DISPERSION AND PRISMS

In Table 22.1, we presented the index of refraction values for various materials. If we make careful measurements, however, we find that the index of refraction in anything but vacuum depends on the wavelength of light. The dependence of the index of refraction on wavelength is called **dispersion.** Figure 22.14 is a graphical representation of this variation in index of refraction with wavelength. Because n is a function of wavelength, Snell's law indicates that **the angle of refraction when light enters a material depends on the wavelength of the light.** As seen in Figure 22.14, the index of refraction for a material usually decreases with increasing wavelength. This means that violet light ($\lambda \cong 400$ nm) refracts more than red light ($\lambda \cong 650$ nm) when passing from air into a material.

To understand the effects of dispersion on light, consider what happens when light strikes a prism, as in Figure 22.15a. A ray of light of a single wavelength that is incident on the prism from the left emerges bent away from its original direction of travel by an angle δ, called the **angle of deviation.** Now suppose a beam of white light (a combination of all visible wavelengths) is incident on a prism. Because of dispersion, the different colors refract through different angles of deviation, and the rays that emerge from the second face of the prism spread out in a series of colors known as a visible **spectrum,** as shown in Figure 22.16. These colors, in order of decreasing wavelength, are red, orange, yellow, green, blue, and violet. Violet light deviates the most, red light the least, and the remaining colors in the visible spectrum fall between these extremes.

Prisms are often used in an instrument known as a **prism spectrometer,** the essential elements of which are shown in Figure 22.17a. This instrument is commonly used to study the wavelengths emitted by a light source, such as a sodium

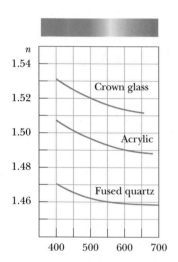

FIGURE 22.14 Variations of index of refraction in the visible spectrum with respect to vacuum wavelength for three materials.

(a)

(b)

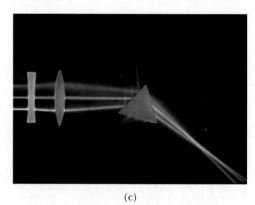

(c)

FIGURE 22.15 (a) A prism refracts a light ray and deviates the light through the angle δ. (b) When light is incident on a prism, the blue light is bent more than the red. (c) Light of different colors passes through a prism and two lenses. Note that as the light passes through the prism, different wavelengths are refracted at different angles. (*David Parker/SPL/Photo Researchers*)

FIGURE 22.16 (a) Dispersion of white light by a prism. Since *n* varies with wavelength, the prism disperses the white light into its various spectral components. (b) Different colors of light that pass through a prism are refracted at different angles because the index of refraction of the glass depends on wavelength. The violet light bends the most; red light bends the least. *(b, courtesy of Bausch & Lomb)*

vapor lamp. Light from the source is sent through a narrow, adjustable slit and lens to produce a parallel, or collimated, beam. The light then passes through the prism and is dispersed into a spectrum. The refracted light is observed through a telescope. The experimenter sees different colored images of the slit through the eyepiece of the telescope. The telescope can be moved or the prism can be rotated in order to view the various wavelengths, which have different angles of deviation. Figure 22.17b shows one type of prism spectrometer used in undergraduate laboratories.

All hot, low-pressure gases emit their own characteristic spectra. Thus, one use of a prism spectrometer is to identify gases. For example, sodium emits only two wavelengths in the visible spectrum: two closely spaced yellow lines. (The bright line-like images of the slit seen in a spectroscope are called *spectral lines*.) A gas emitting these and only these colors can thus be identified as sodium. Likewise, mercury vapor has its own characteristic spectrum, consisting of four prominent wavelengths—orange, green, blue, and violet lines—along with some wavelengths of lower intensity. The particular wavelengths emitted by a gas serve as "fingerprints" of that gas. Spectral analysis, which is the measurement of the wavelengths emitted or absorbed by a substance, has proven to be a powerful general tool in many areas: Chemists and biologists use infrared spectroscopy to identify molecules, astronomers use visible light spectroscopy to identify elements on distant stars, and geologists use spectral analysis to identify minerals.

◄ **APPLICATION**

IDENTIFYING GASES USING A
SPECTROMETER

FIGURE 22.17 (a) A diagram of a prism spectrometer. The colors in the spectrum are viewed through a telescope. (b) A prism spectrometer with interchangeable components. *(Courtesy of PASCO Scientific)*

APPLYING **PHYSICS** **22.3**

When a beam of light enters a glass prism, which has nonparallel sides, the rainbow of color exiting the prism is a testimonial to the dispersion occurring in the glass. Suppose a beam of light enters a slab of material with parallel sides. When the beam exits the other side, traveling in the same direction as the original beam, is there any evidence of dispersion?

Explanation Due to dispersion, light at the violet end of the spectrum will exhibit a larger angle of refraction on entering the glass than light at the red end. All colors of light will return to the original direction of propagation on refracting back out into the air. Thus, the outgoing beam will be white. But the net shift in the position of the violet light along the edge of the slab will be larger than that of the red light. Thus, one edge of the outgoing beam will have a bluish tinge (it will appear blue rather than violet, because the eye is not very sensitive to violet light), whereas the other edge will have a reddish tinge. This effect is indicated in Figure 22.18. The colored edges of the outgoing beam of white light are evidence of dispersion.

FIGURE 22.18 (Applying Physics 22.3)

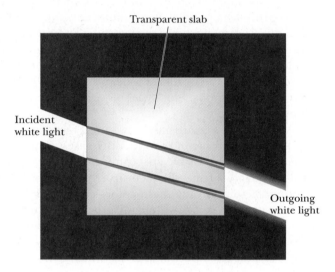

Transparent slab

Incident white light

Outgoing white light

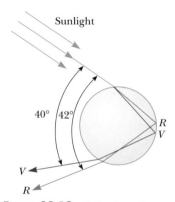

Sunlight

40° 42°

R
V

V

R

FIGURE 22.19 Refraction of sunlight by a spherical raindrop.

Webnote 22.2

See what makes the bow in a rainbow. Go to
http://www.unidata.ucar.edu/staff/blynds/rnbw.html

22.6 THE RAINBOW

The dispersion of light into a spectrum is demonstrated most vividly in nature through the formation of a rainbow, often seen by an observer positioned between the Sun and a rain shower. To understand how a rainbow is formed, consider Figure 22.19. A ray of light passing overhead strikes a drop of water in the atmosphere and is refracted and reflected as follows. It is first refracted at the front surface of the drop, with the violet light deviating the most and the red light the least. At the back surface of the drop, the light is reflected and returns to the front surface, where it again undergoes refraction as it moves from water into air. The rays leave the drop so that the angle between the incident white light and the returning violet ray is 40°, and the angle between the white light and the returning red ray is 42° (Fig. 22.19). This small angular difference between the returning rays causes us to see the bow as explained in the next paragraph.

Now consider an observer viewing a rainbow, as in Figure 22.20a. If a raindrop high in the sky is being observed, the red light returning from the drop can reach the observer because it is deviated the most, but the violet light passes over the observer because it is deviated the least. Hence, the observer sees this drop as being red. Similarly, a drop lower in the sky would direct violet light toward the observer and appear to be violet. (The red light from this drop would strike the ground

(a)

(b)

FIGURE 22.20 (a) The formation of a rainbow. (b) This photograph of a rainbow shows a distinct secondary rainbow with the colors reversed. *(Mark D. Phillips/Photo Researchers, Inc.)*

and not be seen.) The remaining colors of the spectrum would reach the observer from raindrops lying between these two extreme positions. Figure 22.20b shows a beautiful rainbow.

22.7 HUYGENS'S PRINCIPLE

The laws of reflection and refraction can be deduced using a geometric method proposed by Huygens in 1678. Huygens assumed that light is a form of wave motion rather than a stream of particles. He had no knowledge of the nature of light or of its electromagnetic character. Nevertheless, his simplified wave model is adequate for understanding many practical aspects of the propagation of light.

Huygens's principle is a geometric construction for determining at some instant the position of a new wave front from the knowledge of the wave front that preceded it. (A wave front is a surface passing through those points of a wave that have the same phase and amplitude. For instance, a wave front could be a surface passing through the crests of waves.) In Huygens's construction, **all points on a**

FIGURE 22.21 Huygens's constructions for (a) a plane wave propagating to the right and (b) a spherical wave.

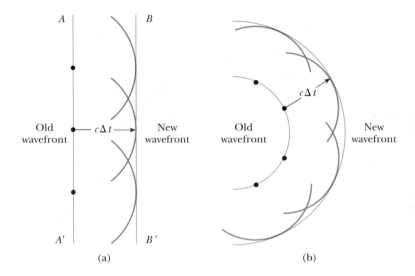

(a)

(b)

Huygens's principle ▶ given wave front are taken as point sources for the production of spherical secondary waves, called wavelets, which propagate in the forward direction with speeds characteristic of waves in that medium. After some time has elapsed, the new position of the wave front is the surface tangent to the wavelets.

Figure 22.21 illustrates two simple examples of Huygens's construction. First, consider a plane wave moving through free space, as in Figure 22.21a. At $t = 0$, the wave front is indicated by the plane labeled AA'. In Huygens's construction, each point on this wave front is considered a point source. For clarity, only a few points on AA' are shown. With these points as sources for the wavelets, we draw circles each of radius $c \Delta t$, where c is the speed of light in vacuum and Δt is the period of propagation from one wave front to the next. The surface drawn tangent to these wavelets is the plane BB', which is parallel to AA'. In a similar manner, Figure 22.21b shows Huygens's construction for an outgoing spherical wave.

Figure 22.22 shows a convincing demonstration of Huygens's principle. Plane waves coming from far off shore emerge from the openings between the barriers as two-dimensional circular waves propagating outward.

FIGURE 22.22 This photograph of the beach at Tel Aviv, Israel, shows Huygens wavelets radiating from each opening between breakwalls. Note how the beach has been shaped by the wave action. *(Courtesy of Sabina Zigman/Benjamin Cardozo High School, and by permission of PHYSICS TEACHER, vol. 37, January 1999, p. 55)*

FIGURE 22.23 (a) Huygens's construction for proving the law of reflection. (b) Triangle *ADC* is identical to triangle *AA'C*.

(a)

(b)

HUYGENS'S PRINCIPLE APPLIED TO REFLECTION AND REFRACTION

The laws of reflection and refraction were stated earlier in this chapter without proof. We shall now derive these laws using Huygens's principle. For the law of reflection, refer to Figure 22.23a. The line AA' represents a wave front of the incident light. As ray 3 travels from A' to C, ray 1 reflects from A and produces a spherical wavelet of radius AD. (Recall that the radius of a Huygens wavelet is $v\Delta t$.) Because the two wavelets having radii $A'C$ and AD are in the same medium, they have the same speed, v, and thus $AD = A'C$. Meanwhile, the spherical wavelet centered at B has spread only half as far as the one centered at A, because ray 2 strikes the surface later than ray 1.

From Huygens's principle, we find that the reflected wave front is CD, a line tangent to all the outgoing spherical wavelets. The remainder of our analysis depends on geometry, as summarized in Figure 22.23b. Note that the right triangles ADC and $AA'C$ are congruent because they have the same hypotenuse, AC, and because $AD = A'C$. From Figure 22.23b we have

$$\sin \theta_1 = \frac{A'C}{AC} \quad \text{and} \quad \sin \theta_1' = \frac{AD}{AC}$$

Thus,

$$\sin \theta = \sin \theta_1'$$

$$\theta_1 = \theta_1'$$

which is the law of reflection.

Now let us use Huygens's principle and Figure 22.24a to derive Snell's law of refraction. Note that in the time interval Δt, ray 1 moves from A to B and ray 2 moves from A' to C. The radius of the outgoing spherical wavelet centered at A is equal to $v_2 \Delta t$. The distance $A'C$ is equal to $v_1 \Delta t$. Geometric considerations show

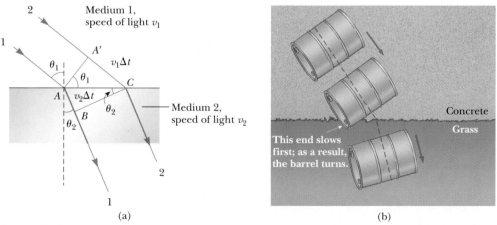

(a)

(b)

FIGURE 22.24 (a) Huygens's construction for proving the law of refraction. (b) Overhead view of a barrel rolling from concrete onto grass.

that angle $A'AC$ equals θ_1 and angle ACB equals θ_2. From triangles $AA'C$ and ACB we find that

$$\sin \theta_1 = \frac{v_1 \Delta t}{AC} \quad \text{and} \quad \sin \theta_2 = \frac{v_2 \Delta t}{AC}$$

If we divide these two equations, we get

$$\frac{\sin \theta_1}{\sin \theta_2} = \frac{v_1}{v_2}$$

But from Equation 22.4 we know that $v_1 = c/n_1$ and $v_2 = c/n_2$. Therefore,

$$\frac{\sin \theta_1}{\sin \theta_2} = \frac{c/n_1}{c/n_2} = \frac{n_2}{n_1}$$

$$n_1 \sin \theta_1 = n_2 \sin \theta_2$$

which is the law of refraction.

A mechanical analog of refraction is shown in Figure 22.24b. When the left end of the rolling barrel reaches the grass, it slows down, while the right end remains on the concrete and moves at its original speed. This difference in speeds causes the barrel to pivot, and this changes the direction of travel.

22.8 TOTAL INTERNAL REFLECTION

An interesting effect called *total internal reflection* can occur when light attempts to move from a medium with a *high* index of refraction to one with a *lower* index of refraction. Consider a light beam traveling in medium 1 and meeting the boundary between medium 1 and medium 2, where n_1 is greater than n_2 (Fig. 22.25). Possible directions of the beam are indicated by rays 1 through 5. Note that the refracted rays are bent away from the normal because n_1 is greater than n_2. At some particular angle of incidence θ_c, called the **critical angle,** the refracted light ray moves parallel to the boundary so that $\theta_2 = 90°$ (Fig. 22.25b). *For angles of incidence greater than* θ_c, the beam is entirely reflected at the boundary, as is ray 5 in Figure 22.25a. This ray is reflected at the boundary as though it had struck a perfectly reflecting surface. It and all rays like it obey the law of reflection; that is, the angle of incidence equals the angle of reflection.

We can use Snell's law to find the critical angle. When $\theta_1 = \theta_c$, $\theta_2 = 90°$ and Snell's law (Eq. 22.8) gives

$$n_1 \sin \theta_c = n_2 \sin 90° = n_2$$

$$\sin \theta_c = \frac{n_2}{n_1} \quad \text{for } n_1 > n_2 \qquad \text{[22.9]}$$

This photograph shows nonparallel light rays entering a glass prism. The bottom two rays undergo total internal reflection at the longest side of the prism. The top three rays are refracted at the longest side as they leave the prism. *(Courtesy of Henry Leap and Jim Lehman)*

FIGURE 22.25 (a) Rays from a medium with index of refraction n_1 travel to a medium with index of refraction n_2, where $n_1 > n_2$. As the angle of incidence increases, the angle of refraction θ_2 increases until θ_2 is 90° (ray 4). For even larger angles of incidence, total internal reflection occurs (ray 5). (b) The angle of incidence producing a 90° angle of refraction is often called the *critical angle, θ_c.*

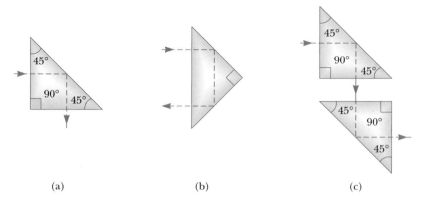

FIGURE 22.26 Internal reflection in a prism. (a) The ray is deviated by 90°. (b) The direction of the ray is reversed. (c) Two prisms used as a periscope.

(a) (b) (c)

Note that this equation can be used only when n_1 is greater than n_2. That is, **total internal reflection occurs only when light attempts to move from a medium of high index of refraction to a medium of lower index of refraction.** If n_1 were less than n_2, Equation 22.9 would give sin $\theta_c > 1$, which is an absurd result because the sine of an angle can never be greater than unity.

When medium 2 is air, the critical angle is small for substances with large indices of refraction, such as diamond, where $n = 2.42$ and $\theta_c = 24.0°$. By comparison, for crown glass, $n = 1.52$ and $\theta_c = 41.0°$. This property, combined with proper faceting, causes a diamond to sparkle brilliantly.

One can use a prism and the phenomenon of total internal reflection to alter the direction of travel of a light beam. Figure 22.26 illustrates two such possibilities. In one case the light beam is deflected by 90.0° (Fig. 22.26a), and in the second case the path of the beam is reversed (Fig. 22.26b). A common application of total internal reflection is a submarine periscope. In this device, two prisms are arranged, as in Figure 22.26c so that an incident beam of light follows the path shown and the user can "see around corners."

◀ APPLICATION

SUBMARINE PERISCOPES

APPLYING **PHYSICS** 22.4

A beam of white light is incident on the curved edge of a semicircular piece of glass, as shown in Figure 22.27. The light enters the curved surface along the normal, so it shows no refraction. It encounters the straight side at the center of curvature of the curved side and refracts into the air. The incoming beam is moved clockwise (so that the angle θ increases) such that the beam always enters along the normal to the curved side and encounters the straight side at the center of curvature of the curved side. As the refracted beam approaches a direction parallel to the straight side, it becomes redder. Why?

Explanation When the outgoing beam approaches the direction parallel to the straight side, the incident angle is approaching the critical angle for total internal reflection. Dispersion occurs as the light passes out of the glass. The index of refraction for light at the violet end of the visible spectrum is larger than at the red end. Thus, as the outgoing beam approaches the straight side, the violet light experiences total internal reflection first, followed by the other colors. The red light is the last to experience total internal reflection, so just before the outgoing light disappears, it is composed of light from the red end of the visible spectrum.

FIGURE 22.27 (Applying Physics 22.4)

Example 22.6   A View from the Fish's Eye

A Find the critical angle for a water-air boundary if the index of refraction of water is 1.33.

Webnote 22.3
See what total internal reflection would look like beneath the surface of a pond. Go to
http://www.lightlink.com/sergey/java/java/totintrefl/index.html

This finger tip is being illuminated by a bundle of optical fibers. The light source is not shown in the photograph. (*© Novastock/Photo Researchers, Inc.*)

> **APPLICATION**
>
> FIBER OPTICS IN MEDICAL DIAGNOSIS AND SURGERY

> **APPLICATION**
>
> FIBER OPTICS IN TELECOMMUNICATIONS

FIGURE 22.28 Light travels in a curved transparent rod by multiple internal reflections.

Solution   Applying Equation 22.9, we find the critical angle to be

$$\sin \theta_c = \frac{n_2}{n_1} = \frac{1.00}{1.33} = 0.752$$

$$\theta_c = 48.8°$$

B Use the results of (part A) to predict what a fish will see if it looks up toward the water surface at angles of 40.0°, 48.8°, and 60.0°.

Reasoning and Solution   Because the path of a light ray is reversible, the fish can see out of the water if it looks toward the surface at an angle less than the critical angle. Thus, at 40.0°, the fish can see into the air above the water. At an angle of 48.8°, the critical angle for water, the light that reaches the fish has to skim along the water surface before being refracted to the fish's eye. At angles greater than the critical angle, the light reaching the fish comes via internal reflection at the surface. Thus, beyond 60.0°, the fish sees a reflection of some object on the bottom of the pool.

FIBER OPTICS

Another interesting application of total internal reflection is the use of solid glass or transparent plastic rods to "pipe" light from one place to another. As indicated in Figure 22.28, light is confined to traveling within the rods, even around gentle curves, as a result of successive internal reflections. Such a light pipe can be quite flexible if thin fibers are used rather than thick rods. If a bundle of parallel fibers is used to construct an optical transmission line, images can be transferred from one point to another.

Very little light intensity is lost in these fibers as a result of reflections on the sides. Any loss of intensity is due essentially to reflections from the two ends and absorption by the fiber material. Fiber-optic devices are particularly useful for viewing images produced at inaccessible locations. Physicians often use fiber-optic cables to aid in the diagnosis and correction of certain medical problems without the intrusion of major surgery. For example, a fiber-optic cable can be threaded through the esophagus and into the stomach to look for ulcers. In this application, the cable actually consists of two fiber-optic lines, one to transmit a beam of light into the stomach for illumination and the other to allow this light to be transmitted out of the stomach. The resulting image can, in some cases, be viewed directly by the physician but most often is displayed on a television monitor or captured on film. In a similar fashion, the cables can be used to examine the colon or to do repair work without the need for large incisions.

Most important, the field of fiber optics has revolutionized the entire communications industry. Billions of kilometers of optical fiber have been installed in the United States to carry high-speed Internet traffic, radio and television signals, and telephone calls. The fibers can carry much higher volumes of telephone calls and other forms of communication than electrical wires because of the higher frequency of the infrared light used to carry the information on optical fibers. Optical fibers are also preferable to copper wires because they are insulators and do not pick up stray electric and magnetic fields or electronic "noise."

> **APPLYING PHYSICS 22.5**
>
> An optical fiber consists of a transparent core, surrounded by cladding, which is a material with a lower index of refraction than the core. There is a cone of angles, called the *acceptance cone*, at the entrance to the fiber. Incoming light at angles within this cone will be transmitted through the fiber, whereas light entering

the core from angles outside the cone will not be transmitted. Figure 22.29 shows a light ray entering the fiber just within the acceptance cone and experiencing total internal reflection at the interface between the core and the cladding. If it is technologically difficult to produce light entering the fiber from a small range of angles, how could you adjust the indices of refraction of the core and cladding to increase the size of the acceptance cone—would you design them to be farther apart or closer together?

Explanation The acceptance cone would become larger if the critical angle (θ_c in the diagram) could be made smaller. This can be done by making the index of refraction of the cladding material smaller, so that the indices of refraction of the core and cladding material would be farther apart.

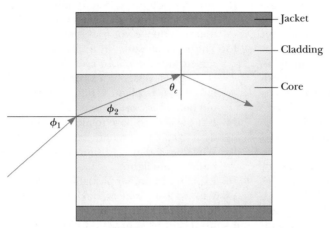

FIGURE 22.29 (Applying Physics 22.5)

SUMMARY

The **index of refraction** of a material, n, is defined as

$$n \equiv \frac{c}{v} \tag{22.4}$$

where c is the speed of light in a vacuum and v is the speed of light in the material. The index of refraction of a material is also

$$n = \frac{\lambda_0}{\lambda_n} \tag{22.7}$$

where λ_0 is the wavelength of the light in vacuum and λ_n is its wavelength in the material.

The **law of reflection** states that a wave reflects from a surface so that the *angle of reflection* θ_1' equals the *angle of incidence* θ_1.

The **law of refraction,** or **Snell's law,** states that

$$n_1 \sin \theta_1 = n_2 \sin \theta_2 \tag{22.8}$$

where n_1 and n_2 are the indices of refraction in the two media. The incident ray, the reflected ray, the refracted ray, and the normal to the surface all lie in the same plane.

Huygens's principle states that all points on a wave front are point sources for the production of spherical secondary waves called wavelets. These wavelets propagate forward at a speed characteristic of waves in a particular medium. After some time has elapsed, the new position of the wave front is the surface tangent to the wavelets.

Total internal reflection can occur when light attempts to move from a material with a high index of refraction to one with a lower index of refraction. The *maximum angle of incidence* θ_c for which light can move from a medium with index n_1 into a medium with index n_2, where n_1 is greater than n_2, is called the **critical angle** and is given by

$$\sin \theta_c = \frac{n_2}{n_1} \qquad \text{for } n_1 > n_2 \tag{22.9}$$

CONCEPTUAL QUESTIONS

1. Under certain circumstances, sound can be heard from extremely far away. This frequently happens over a body of water, where the air near the water surface is cooler than the air at higher altitudes. Explain how the refraction of sound waves could increase the distance over which sound can be heard.

2. What are some reasons that most ceilings are made of white textured material?

3. The color of an object is said to depend on wavelength. So, if you view colored objects under water, in which the wavelength of the light will be different, does the color change?

4. How is it possible that a complete circle of a rainbow can sometimes be seen from an airplane?

5. Pass white light through a prism to produce a multicolored spectrum. Let that spectrum fall on a piece of cardboard that has a slit in it such that only green light passes through. Now let this green light fall on a second prism. Describe the light that comes out of the second prism.

6. Why does the arc of a rainbow appear with red on top and violet on the bottom?

7. A scientific supply catalog advertises a material having an index of refraction of 0.85. Is this a good product to buy? Why or why not?

8. Under what conditions is a mirage formed? On a hot day, what are we seeing when we observe a mirage water puddle on the road?

9. In dispersive materials, the angle of refraction for a light ray depends on the wavelength of the light. Does the angle of reflection from the surface of the material depend on the wavelength? Why or why not?

10. A type of mirage called a *pingo* is observed often in Alaska. These occur when the light from a small hill passes to an observer by a path that takes the light over a body of water warmer than the air. What is seen is the hill and an inverted image directly below it. Explain how these mirages are formed.

11. Explain why a diamond loses most of its sparkle when submerged in carbon disulfide.

12. Suppose you are told that only two colors of light (X and Y) are sent through a glass prism and that X is bent more than Y. Which color travels more slowly in the prism?

13. The level of water in a clear, colorless glass is easily observed with the naked eye. The level of liquid helium in a clear glass vessel is extremely difficult to see with the naked eye. Explain. (*Hint:* The index of refraction of liquid helium is close to that of air.)

14. Is it possible to have total internal reflection for light incident from air on water? Explain.

15. Why does a diamond show flashes of color when observed under white light?

16. Explain why an oar partially in water appears to be bent.

17. Why do astronomers looking at distant galaxies talk about looking backward in time?

18. If a beam of light with a given cross section enters a new medium, the cross section of the refracted beam is different from the incident beam. Is it larger or smaller, or is there no definite direction to the change?

PROBLEMS

1, 2, 3 = straightforward, intermediate, challenging   ☐ = full solution available in Student Solutions Manual/Study Guide

web = solution posted at **http://info.brookscole.com/serway**   🎵 = biomedical application

Section 22.1   The Nature of Light

1. During the Apollo XI Moon landing, a retroreflecting panel was erected on the Moon's surface. The speed of light may be found by measuring the time it takes a laser beam to travel from Earth, reflect from the panel, and return to Earth. If this interval is measured to be 2.51 s, what is the measured speed of light? Take the center-to-center distance from Earth to the Moon to be 3.84×10^8 m. Assume the Moon is directly overhead and do not neglect the sizes of Earth and the Moon.

2. Figure P22.2 shows the apparatus used by Armand H. L. Fizeau (1819–1896) to measure the speed of light. The basic idea is to measure the total time it takes light to travel from some point to a distant mirror and back. If *d* is the distance between the light source and the mirror and if the transit time for one round trip is *t*, then the speed of light is $c = 2d/t$. To measure the transit time, Fizeau used a rotating toothed wheel, which converts an otherwise continuous beam of light to a series of light pulses. The rotation of the wheel controls what an observer at the light source sees. For example, assume that the toothed wheel of the Fizeau experiment has 360 teeth and is rotating at a

speed of 27.5 rev/s when the light from the source is extinguished—that is, when a burst of light passing through opening *A* in Figure P22.2 is blocked by tooth *B* on return. If the distance to the mirror is 7 500 m, find the speed of light.

FIGURE P22.2   (Problems 2 and 3)

3. In an experiment to measure the speed of light using the apparatus of Fizeau described in the preceding problem, the distance between light source and mirror was 11.45 km and the wheel had 720 notches. The experimen-

tally determined value of *c* was 2.998×10^8 m/s. Calculate the minimum angular speed of the wheel for this experiment.

4. Albert A. Michelson very carefully measured the speed of light using an alternative version of the technique developed by Fizeau. (See Problem 2.) Figure P22.4 shows the approach he used. Light was reflected from one face of a rotating eight-sided mirror toward a stationary mirror 35.0 km away. At certain rates of rotation, the returning beam of light was directed toward the eye of an observer as shown. (a) What minimum angular speed must the rotating mirror have in order that side A will have rotated to position B, causing the light to be reflected to the eye? (b) What is the next-higher angular velocity that will enable the source of light to be seen?

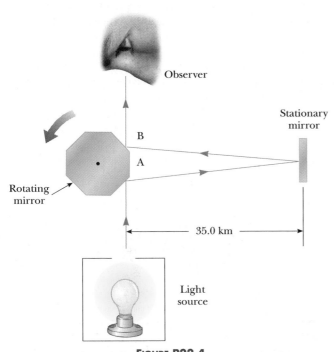

FIGURE P22.4

5. Figure P22.5 shows an apparatus used to measure the speed distribution of gas molecules. It consists of two slotted rotating disks separated by a distance *d*, with the slots displaced by the angle θ. Suppose the speed of light is measured by sending a light beam toward the left disk of this apparatus. (a) Show that a light beam will be seen in the detector (that is, will make it through both slots) only

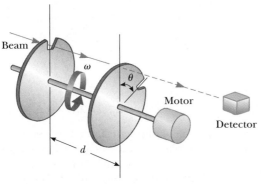

FIGURE P22.5

if its speed is given by $c = \omega d / \theta$, where ω is the angular speed of the disks and θ is measured in radians. (b) What is the measured speed of light if the distance between the two slotted rotating disks is 2.500 m, the slot in the second disk is displaced 1/60 of one degree from the slot in the first disk, and the disks are rotating at 5 555 rev/s?

Section 22.3 Reflection and Refraction

Section 22.4 The Law of Refraction

6. The two mirrors in Figure P22.6 meet at a right angle. The beam of light in the vertical plane *P* strikes mirror 1 as shown. (a) Determine the distance the reflected light beam travels before striking mirror 2. (b) In what direction does the light beam travel after being reflected from mirror 2?

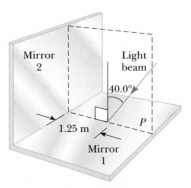

FIGURE P22.6

7. An underwater scuba diver sees the Sun at an apparent angle of 45.0° from the vertical. What is the actual direction of the Sun?

8. Light is incident normally on a 1.00-cm layer of water that lies on top of a flat Lucite plate with a thickness of 0.500 cm. How much more time is required for light to pass through this double layer than is required to traverse the same distance in air ($n_{\text{Lucite}} = 1.59$)?

9. A laser beam is incident at an angle of 30.0° to the vertical onto a solution of corn syrup in water. If the beam is refracted to 19.24° to the vertical, (a) what is the index of refraction of the syrup solution? Suppose the light is red, with vacuum wavelength 632.8 nm. Find its (b) wavelength, (c) frequency, and (d) speed in the solution.

10. Find the speeds of light in (a) flint glass, (b) water, and (c) zircon.

11. Light of wavelength λ_0 in vacuum has a wavelength of 438 nm in water and a wavelength of 390 nm in benzene. (a) What is the wavelength λ_0 of this light in vacuum? (b) Using only the given wavelengths, determine the ratio of the index of refraction of benzene to that of water.

12. Light of wavelength 436 nm in air enters a fishbowl filled with water, then exits through the crown-glass wall of the container. Find the wavelengths of the light (a) in the water and (b) in the glass.

13. A ray of light is incident on the surface of a block of clear ice at an angle of 40.0° with the normal. Part of the light is reflected and part is refracted. Find the angle between the reflected and refracted light.

14. A narrow beam of sodium yellow light ($\lambda_0 = 589$ nm) is incident from air on a smooth surface of water at an angle

of $\theta_1 = 35.0°$. Determine the angle of refraction θ_2 and the wavelength of the light in water.

15. A beam of light, traveling in air, strikes the surface of mineral oil at an angle of 23.1° with the normal to the surface. If the light travels at 2.17×10^8 m/s through the oil, what is the angle of refraction?

16. A flashlight on the bottom of a 4.00-m-deep swimming pool sends a ray upward and at an angle so that the ray strikes the surface of the water 2.00 m from the point directly above the flashlight. What angle (in air) does the emerging ray make with the water's surface?

17. How many times will the incident beam shown in Figure P22.17 be reflected by each of the parallel mirrors?

FIGURE P22.17

18. A ray of light strikes a flat, 2.00-cm-thick block of glass ($n = 1.50$) at an angle of 30.0° with the normal (Fig. P22.18). Trace the light beam through the glass and find the angles of incidence and refraction at each surface.

FIGURE P22.18 (Problems 18, 19, and 20)

19. When the light ray in Problem 18 passes through the glass **web** block, it is shifted laterally by a distance d (Fig. P22.18). Find the value of d.

20. Find the time required for the light to pass through the glass block described in Problem 19.

21. The light beam shown in Figure P22.21 makes an angle of 20.0° with the normal line NN' in the linseed oil. Determine the angles θ and θ'. (The refractive index for linseed oil is 1.48.)

22. A submarine is 300 m horizontally out from the shore and 100 m beneath the surface of the water. A laser beam is sent from the sub so that it strikes the surface of the water at a point 210 m from the shore. If the beam just strikes the top of a building standing directly at the water's edge, find the height of the building.

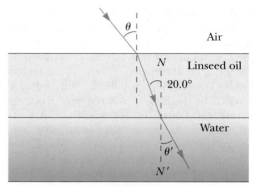

FIGURE P22.21

23. Two light pulses are emitted simultaneously from a source. The pulses take parallel paths to a detector 6.20 m away, but one moves through air and the other through a block of ice. Determine the difference in the pulses' times of arrival at the detector.

24. A narrow beam of ultrasonic waves reflects off the liver tumor in Figure P22.24. If the speed of the wave is 10.0% less in the liver than in the surrounding medium, determine the depth of the tumor.

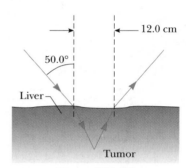

FIGURE P22.24

25. A beam of light both reflects and refracts at the surface between air and glass as shown in Figure P22.25. If the index of refraction of the glass is n_g, find the angle of incidence θ_1 in the air that would result in the reflected ray and the refracted ray being perpendicular to each other. [*Hint:* Remember the identity $\sin(90° - \theta) = \cos \theta$.]

FIGURE P22.25

26. Three sheets of plastic have unknown indices of refraction. Sheet 1 is placed on top of sheet 2, and a laser beam is directed onto the sheets from above so that it strikes the interface at an angle of 26.5° with the normal. The refracted beam in sheet 2 makes an angle of 31.7° with the

normal. The experiment is repeated with sheet 3 on top of sheet 2 and, with the same angle of incidence, the refracted beam makes an angle of 36.7° with the normal. If the experiment is repeated again with sheet 1 on top of sheet 3, what is the expected angle of refraction in sheet 3? Assume the same angle of incidence.

27. An opaque cylindrical tank with an open top has a diameter of 3.00 m and is completely filled with water. When the afternoon Sun reaches an angle of 28.0° above the horizon, sunlight ceases to illuminate the bottom of the tank. How deep is the tank?

28. A cylindrical cistern, constructed below ground level, is 3.0 m in diameter and 2.0 m deep and is filled to the brim with a liquid whose index of refraction is 1.5. A small object rests on the bottom of the cistern at its center. How far from the edge of the cistern can a girl whose eyes are 1.2 m from the ground stand and still see the object?

Section 22.5 Dispersion and Prisms

29. The index of refraction for red light in water is 1.331, and that for blue light is 1.340. If a ray of white light enters the water at an angle of incidence of 83.00°, what are the underwater angles of refraction for the blue and red components of the light?

30. A certain kind of glass has an index of refraction of 1.650 for blue light of wavelength 430 nm and an index of 1.615 for red light of wavelength 680 nm. If a beam containing these two colors is incident at an angle of 30.00° on a piece of this glass, what is the angle between the two beams inside the glass?

31. A ray of light strikes the midpoint of one face of an equiangular (60°-60°-60°) glass prism ($n = 1.5$) at an angle of incidence of 30°. (a) Trace the path of the light ray through the glass, and find the angles of incidence and refraction at each surface. (b) If a small fraction of light is also reflected at each surface, find the angles of reflection at these surfaces.

32. The index of refraction for violet light in silica flint glass is 1.66 and that for red light is 1.62. What is the angular dispersion of visible light passing through an equilateral prism of apex angle 60.0° if the angle of incidence is 50.0°? (See Fig. P22.32.)

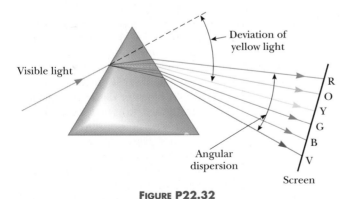

FIGURE P22.32

Section 22.8 Total Internal Reflection

33. Calculate the critical angles for the following materials when surrounded by air: (a) zircon, (b) fluorite, (c) ice. Assume that $\lambda = 589$ nm.

34. For 589-nm light, calculate the critical angle for the following materials surrounded by air: (a) diamond and (b) flint glass.

35. Repeat Problem 34 when the materials are surrounded by water.

36. A beam of light is incident from air on the surface of a liquid. If the angle of incidence is 30.0° and the angle of refraction is 22.0°, find the critical angle for the liquid when surrounded by air.

37. A light pipe consists of a central strand of material surrounded by an outer coating. The interior portion of the pipe has an index of refraction of 1.60. If all rays striking the interior walls of the pipe with incident angles greater than 59.5° are subject to total internal reflection, what is the index of refraction of the coating?

38. Determine the maximum angle θ for which the light rays incident on the end of the light pipe in Figure P22.38 are subject to total internal reflection along the walls of the pipe. Assume that the light pipe has an index of refraction of 1.36 and that the outside medium is air.

FIGURE P22.38

39. Consider a common mirage formed by super-heated air just above a roadway. A truck driver whose eyes are 2.00 m above the road, where $n = 1.000\ 3$, looks forward. She has the illusion of seeing a patch of water ahead on the road, where her line of sight makes an angle of 1.20° below the horizontal. Find the index of refraction of the air just above the road surface. (*Hint:* Treat this as a problem in total internal reflection.)

40. A jewel thief hides a diamond by placing it on the bottom of a public swimming pool. He places a circular raft on the surface of the water directly above and centered on the diamond, as shown in Figure P22.40. If the surface of the water is calm and the pool is 2.00 m deep, find the minimum diameter of the raft that would prevent the diamond from being seen.

FIGURE P22.40

41. A room contains air in which the speed of sound is 343 m/s. The walls of the room are made of concrete, in which the speed of sound is 1 850 m/s. (a) Find the critical angle for total internal reflection of sound at the

concrete-air boundary. (b) In which medium must the sound be traveling in order to undergo total internal reflection? (c) "A bare concrete wall is a highly efficient mirror for sound." Give evidence for or against this statement.

42. A light ray is incident normally to the long face (the hypotenuse) of a 45°-45°-90° prism surrounded by air, as shown in Figure 22.26b. Calculate the minimum index of refraction of the prism for which the ray will follow the path shown.

43. The light beam in Figure P22.43 strikes surface 2 at the critical angle. Determine the angle of incidence θ_i.

FIGURE P22.46

FIGURE P22.43

FIGURE P22.47

ADDITIONAL PROBLEMS

44. (a) Consider a horizontal interface between air above and glass with an index of 1.55 below. Draw a light ray incident from the air at an angle of incidence of 30.0°. Determine the angles of the reflected and refracted rays and show them on the diagram. (b) Suppose instead that the light ray is incident from the glass at an angle of incidence of 30.0°. Determine the angles of the reflected and refracted rays and show all three rays on a new diagram. (c) For rays incident from the air onto the air-glass surface, determine and tabulate the angles of reflection and refraction for all the angles of incidence at 10.0° intervals from 0° to 90.0°. (d) Do the same for light rays traveling up to the interface through the glass.

45. A layer of ice, having parallel sides, floats on water. If light
web is incident on the upper surface of the ice at an angle of incidence of 30.0°, what is the angle of refraction in the water?

46. A light ray of wavelength 589 nm is incident at an angle θ on the top surface of a block of polystyrene surrounded by air, as shown in Figure P22.46. (a) Find the maximum value of θ for which the refracted ray will undergo total internal reflection at the left vertical face of the block. (b) Repeat the calculation for the case in which the polystyrene block is immersed in water. (c) What happens if the block is immersed in carbon disulfide?

47. Figure P22.47 shows the path of a beam of light through several layers of different indices of refraction. (a) If $\theta_1 = 30.0°$, what is the angle θ_2 of the emerging beam? (b) What must the incident angle θ_1 be in order to have total internal reflection at the surface between the $n = 1.20$ medium and the $n = 1.00$ medium?

48. The walls of a prison cell are perpendicular to the four cardinal compass directions. On the first day of spring, light from the rising Sun enters a rectangular window in the eastern wall. The light traverses 2.37 m horizontally to shine perpendicularly on the wall opposite the window. A prisoner observes the patch of light moving across this western wall and for the first time forms his own understanding of the rotation of the Earth. (a) With what speed does the illuminated rectangle move? (b) The prisoner holds a small square mirror flat against the wall at one corner of the rectangle of light. The mirror reflects light back to a spot on the eastern wall close beside the window. How fast does the smaller square of light move across that wall? (c) Seen from a latitude of 40.0° north, the rising Sun moves through the sky along a line making a 50.0° angle with the southeastern horizon. In what direction does the rectangular patch of light on the western wall of the prisoner's cell move? (d) In what direction does the smaller square of light on the eastern wall move?

49. As shown in Figure P22.49, a light ray is incident normally on one face of a 30°-60°-90° block of dense flint glass (a prism) that is immersed in water. (a) Determine the exit angle θ_4 of the ray. (b) A substance is dissolved in the water to increase the index of refraction. At what value of n_2 does total internal reflection cease at point P?

50. A narrow beam of light is incident from air onto a glass surface with index of refraction 1.56. Find the angle of incidence for which the corresponding angle of refraction is one half the angle of incidence. (*Hint:* You might want to use the trigonometric identity $\sin 2\theta = 2 \sin \theta \cos \theta$.)

51. One technique to measure the angle of a prism is shown in Figure P22.51. A parallel beam of light is directed on

FIGURE P22.49

FIGURE P22.55

FIGURE P22.56

the apex of the prism so that the beam reflects from opposite faces of the prism. Show that the angular separation of the two reflected beams is given by $B = 2A$.

FIGURE P22.51

52. A 4.00-m-long pole stands vertically in a lake having a depth of 2.00 m. When the Sun is 40.0° above the horizontal, determine the length of the pole's shadow on the bottom of the lake. Take the index of refraction for water to be 1.33.

53. A piece of wire is bent through an angle θ. The bent wire is partially submerged in benzene (index of refraction = 1.50) so that to a person looking along the dry part, the wire appears to be straight and makes an angle of 30.0° with the horizontal. Determine the value of θ.

54. When you look through a window, by how much time is the light you see delayed by having to go through glass instead of air? Make an order-of-magnitude estimate on the basis of data you specify. By how many wavelengths is it delayed?

55. A transparent cylinder of radius $R = 2.00$ m has a mir-**web** rored surface on its right half, as shown in Figure P22.55. A light ray traveling in air is incident on the left side of the cylinder. The incident light ray and the exiting light ray are parallel and $d = 2.00$ m. Determine the index of refraction of the material.

56. A laser beam strikes one end of a slab of material, as in Figure P22.56. The index of refraction of the slab is 1.48. Determine the number of internal reflections of the beam before it emerges from the opposite end of the slab.

57. For this problem, refer to Figure 22.15. For various angles of incidence, it can be shown that the deviation angle δ is a minimum when the ray passes through the glass so that the interior ray is parallel to the base of the prism. A measurement of this minimum angle of deviation enables one to find the index of refraction of the prism material. Show that n is given by the expression

$$n = \frac{\sin[\frac{1}{2}(A + \delta_{\min})]}{\sin\left(\frac{A}{2}\right)}$$

where A is the apex angle of the prism.

58. A hiker stands on a mountain peak near sunset and observes a rainbow caused by water droplets in the air about 8.00 km away. The valley is 2.00 km below the mountain peak and entirely flat. What fraction of the complete circular arc of the rainbow is visible to the hiker?

59. A light ray is incident on a prism and refracted at the first surface, as shown in Figure P22.59. Let ϕ represent the apex angle of the prism and n its index of refraction. Find, in terms of n and ϕ, the smallest allowed value of the angle of incidence at the first surface for which the refracted ray will not undergo internal reflection at the second surface.

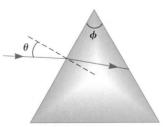

FIGURE P22.59

GROUP ACTIVITIES

G.1 Tape a coin to the bottom of a large opaque bowl, as shown in Figure GA22.1a. Stand over the bowl so that you are looking at the coin and then move backwards away from the bowl until you can no longer see the coin over the bowl's rim. Remain at that position and have your co-investigator fill the bowl with water, as shown in Figure GA22.1b. You can now see the coin again because the light is refracted at the water-air interface.

(a)

(b)

FIGURE GA22.1

G.2 Tape a piece of black paper to the end of a flashlight and cut a narrow slit in the middle of the paper, as shown in Figure GA22.2. Lean a flat mirror against one end of a tray partially filled with water. Shine your flashlight on that part of the mirror that is under water, and hold a sheet of white paper such that the reflected light shines on the paper. You should observe a spectrum of colors on the paper as the light is dispersed when it travels from air into water, and then from water into air. According to your observations, which color is bent the most? Which is bent the least?

G.3 Create an artificial rainbow by standing with your back to the Sun and spraying water into the air with a hose. You should cover the end of the hose slightly with a finger or use a nozzle so that the water is broken up into tiny

FIGURE GA22.2

droplets. You should be able to form a rainbow in this way. Which color is on the outside of the arc? The inside?

If the droplets are close to you in the experiment above, you may be able to see two rainbows, one for each eye. Close one eye and only one bow is seen.

With the Sun high in the sky, stand on a ladder and spray the hose toward the ground. In this case, you should be able to form a complete circle rainbow.

G.4 You can demonstrate that light rays travel in straight-line paths by observing the shadows formed by large and small light sources. Figure GA22.4 shows the shadows formed on a screen by a baseball when the light from a lightbulb (a large source) falls on it. Use the figure to explain why the shadow is dark at location A and less dark at B. Replace the light source with a small source such as a high-intensity lamp with a small clear bulb. (A substitute for a high-intensity lamp is an ordinary lightbulb with a piece of cardboard close in front of it with a hole about the size of a penny punched in the cardboard. The hole serves as the small source of light.) What kind of shadow is formed in this case? Why?

G.5 A ray of light is moving from a material having a high index of refraction into a material with a lower index of refraction. (a) Is the ray bent toward the normal or away from it? (b) If the wavelength is 600 nm in the high-index-of-refraction material, is it greater, smaller, or the

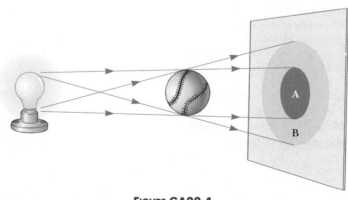

FIGURE GA22.4

same in the lower-index material? (c) How does the frequency change as the light moves between the two materials? Does it increase, decrease, or remain the same? (d) Assume the high-index material has $n = 1.55$, the lower index is 1.33, and the angle of incidence in the higher material is, say, 30°. Do the calculations to see if they agree with your predictions.

G.6 Light attempts to move from cubic zirconia ($n = 1.61$) to air. In a nearby experiment, light is attempting to move from cubic zirconia to water. Finally, in a third trial, light is attempting to move from cubic zirconia to a material with an index of refraction of 1.62. (a) Without doing any calculations, in which case is the angle of refraction the largest? (b) Without calculation, can there be situations in the preceding cases in which the light will not emerge from the cubic zirconia? Explain. (c) Verify your predictions by doing the calculations with an angle of incidence of 40°.

G.7 Light containing three wavelengths—400 nm, 500 nm, and 600 nm—is incident from air on a block of crown glass at an angle of 30°. Are all colors refracted alike or is one color bent more than the others? If you think one is bent more than the others, which one is it? Explain your answer. Calculate the angle of refraction in each case to verify your answer.

23

Mirrors and Lenses

Chapter Outline

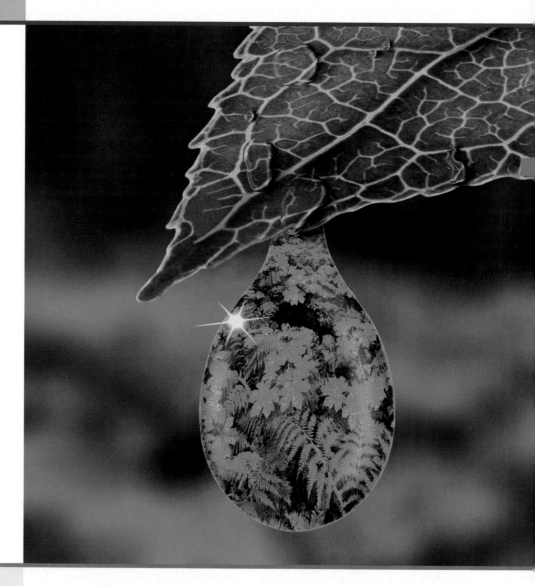

This beautiful photograph shows a raindrop suspended from a leaf. The raindrop acts as a lens. It refracts light twice to produce a real, inverted image of the foliage beyond. *(Don Hammond/CORBIS)*

his chapter is concerned with the formation of images when plane and spherical light waves fall on plane and spherical surfaces. Images can be formed by either reflection or refraction. Mirrors and lenses form images in both ways. In our study of mirrors and lenses, we continue to use the ray approximation and to assume that light travels in straight lines (in other words, we ignore diffraction).

23.1 FLAT MIRRORS

We begin our investigation by examining the simplest possible mirror, the flat mirror. Consider a point source of light placed at O in Figure 23.1, a distance p in front of a flat mirror. The distance p is called the **object distance.** Light rays leave the source and are reflected from the mirror. After reflection, the rays diverge (spread apart), but they appear to the viewer to come from a point I behind the mirror. Point I is called the **image** of the object at O. Regardless of the system under study, **images are formed at the point at which rays of light actually intersect or at which they appear to originate.** Because the rays in Figure 23.1 appear to originate at I, which is a distance q behind the mirror, this is the location of the image. The distance q is called the **image distance.**

Images are classified as real or virtual. A *real image* is one in which light actually passes through the image point; a *virtual image* is one in which the light does not pass through the image point but appears to come (diverge) from that point. The image formed by the flat mirror in Figure 23.1 is a virtual image. In fact, the images seen in flat mirrors are always virtual (for real objects). Real images can be displayed on a screen (as at a movie), but virtual images cannot.

We shall examine some of the properties of the images formed by flat mirrors by using the simple geometric techniques shown in Figure 23.2. To find out where an image is formed, it is always necessary to follow at least two rays of light as they reflect from the mirror. One of those rays starts at P, follows a horizontal path PQ to the mirror, and reflects back on itself. The second ray follows the oblique path PR and reflects as shown. An observer to the left of the mirror would trace the two reflected rays back to the point from which they appear to have originated—that is, point P'. A continuation of this process for points other than P on the object would result in a virtual image (drawn as a yellow arrow) to the right of the mirror. Because triangles PQR and $P'QR$ are identical, $PQ = P'Q$. Hence, we conclude that **the image formed by an object placed in front of a flat mirror is as far behind the mirror as the object is in front of the mirror.** Geometry also shows that the object height h equals the image height h'. **Lateral magnification** M is defined as follows:

$$M \equiv \frac{\text{image height}}{\text{object height}} = \frac{h'}{h} \qquad [23.1]$$

This is a general definition of the lateral magnification of any type of mirror. For a flat mirror, $M = 1$ because $h' = h$.

In summary, the image formed by a flat mirror has the following properties:

1. **The image is as far behind the mirror as the object is in front.**
2. **The image is unmagnified, virtual, and upright. (By *upright* we mean that if the object arrow points upward, as in Figure 23.2, so does the image arrow. The opposite of an upright image is an inverted image.)**

Finally, note that a flat mirror produces an image having an *apparent* left-right reversal. This reversal can be seen by standing in front of a mirror and raising your

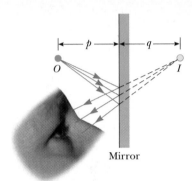

FIGURE 23.1 An image formed by reflection from a flat mirror. The image point I is behind the mirror at distance q, which is equal in magnitude to the object distance p.

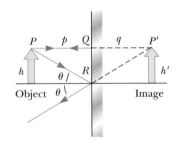

FIGURE 23.2 A geometric construction to locate the image of an object placed in front of a flat mirror. Because the triangles PQR and $P'QR$ are identical, $p = |q|$ and $h = h'$.

Tip 23.1 MAGNIFICATION ≠ ENLARGEMENT

Note that the word *magnification* used in optics does not always mean *enlargement* because the image could be smaller than the object.

right hand. The image you see raises its left hand. Likewise, your hair appears to be parted on the opposite side and a mole on your right cheek appears to be on your left cheek.

Quick Quiz 23.1

In the overhead view of Figure 23.3, the image of the stone seen by observer 1 is at *C*. Where does observer 2 see the image—at *A*, at *B*, at *C*, at *E*, or not at all?

A   *B*   *C*   *D*   *E*

2   1

FIGURE 23.3   (Quick Quiz 23.1)

Example 23.1   "I Can See Myself!"

A 1.80-m tall man stands in front of a mirror in hopes of seeing his full height, no more and no less. If his eyes are 0.10 m from the top of his head, what is the minimum height of the mirror?

Reasoning and Solution   Figure 23.4 shows two rays of light originating at the extremes of the body, reflecting from the mirror, and entering the eye of the viewer. The ray from the feet just strikes the bottom of the mirror, so if the mirror were longer, it would be too long; if shorter, the ray would not be reflected. The angle of incidence and the angle of reflection are equal, labeled θ. The two triangles, *ABD* and *DBC* are identical because they are right triangles with a common side (*DB*) and two identical angles (θ). Thus, we have

$$AD = DC = \tfrac{1}{2}AC = \tfrac{1}{2}(1.80 \text{ m} - 0.10 \text{ m}) = 0.85 \text{ m}$$

Furthermore, because $\tfrac{1}{2}CF = \tfrac{1}{2}(0.10 \text{ m}) = 0.05$ m, we find that

$$d = FA - AD - \tfrac{1}{2}CF = 1.80 \text{ m} - 0.85 \text{ m} - 0.05 \text{ m} = \boxed{0.90 \text{ m}}$$

Thus, the mirror must be exactly equal to half the height of the man in order for him to see only his full height and nothing more or less.

EXERCISE   Is the result of this problem still valid if the man moves farther away from the mirror?

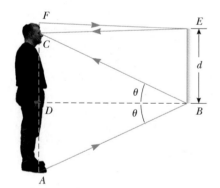

FIGURE 23.4   (Example 23.1)

APPLICATION

DAY AND NIGHT SETTINGS FOR REARVIEW MIRRORS

Most rearview mirrors in cars have a day setting and a night setting. The night setting greatly diminishes the intensity of the image so that lights from trailing cars will not blind the driver. To understand how such a mirror works, consider Figure 23.5. The mirror is a wedge of glass with a reflecting metallic coating on the back side. When the mirror is in the day setting, as in Figure 23.5a, light from an object

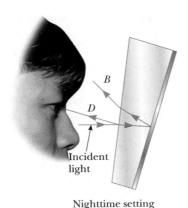

Reflecting
side of mirror

Incident
light

Daytime setting

(a)

Incident
light

Nighttime setting

(b)

FIGURE 23.5 Cross-sectional views of a rearview mirror. (a) With the day setting, the silvered back surface of the mirror reflects a bright ray B into the driver's eyes. (b) With the night setting, the glass of the unsilvered front surface of the mirror reflects a dim ray D into the driver's eyes.

behind the car strikes the mirror at point 1. Most of the light enters the wedge, is refracted, and reflects from the back of the mirror to return to the front surface, where it is refracted again as it re-enters the air as ray B (for *bright*). In addition, a small portion of the light is reflected at the front surface, as indicated by ray D (for *dim*). This dim reflected light is responsible for the image observed when the mirror is in the night setting, as in Figure 23.5b. In this case, the wedge is rotated so that the path followed by the bright light (ray B) does not lead to the eye. Instead, the dim light reflected from the front surface travels to the eye, and the brightness of trailing headlights does not become a hazard.

APPLYING PHYSICS 23.1

The professor in the box shown in Figure 23.6 appears to be balancing himself on a few fingers with both of his feet elevated from the floor. The professor can maintain this position for a long time, and he appears to defy gravity. How do you suppose this illusion was created?

Explanation This is one example of an optical illusion, used by magicians, that makes use of a mirror. The box that the professor is standing in is a cubical frame that contains a flat vertical mirror through a diagonal plane. The professor straddles the mirror so that the foot you see is in front of the mirror and the other foot is behind the mirror where you cannot see it. When he raises the foot that you see in front of the mirror, the reflection of this foot also rises, so he appears to float in air.

FIGURE 23.6 (Applying Physics 23.1) *(Courtesy of Henry Leap and Jim Lehman)*

23.2 IMAGES FORMED BY SPHERICAL MIRRORS

CONCAVE MIRRORS

A **spherical mirror,** as its name implies, has the shape of a segment of a sphere. Figure 23.7 shows a spherical mirror with light reflecting from its silvered inner, concave surface; this is called a **concave mirror.** The mirror has radius of curvature R, and its center of curvature is at point C. Point V is the center of the spherical segment, and a line drawn from C to V is called the **principal axis** of the mirror.

Now consider a point source of light placed at point O in Figure 23.7, on the principal axis and outside point C. Several diverging rays originating at O are shown. After reflecting from the mirror, these rays converge to meet at I, called the **image point.** The rays then continue and diverge from I as if there were an

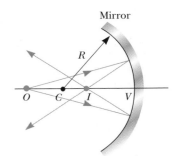

FIGURE 23.7 A point object placed at O, outside the center of curvature of a concave spherical mirror, forms a real image at I as shown. If the rays diverge from O at small angles, they all reflect through the same image point.

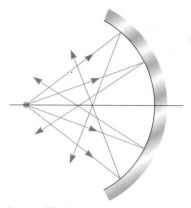

FIGURE 23.8 Rays at large angles from the horizontal axis reflect from a spherical concave mirror to intersect the principal axis at different points, resulting in a blurred image. This is called *spherical aberration.*

object there. As a result, a real image is formed. **Whenever reflected light actually passes through a point, the image formed there is real.**

We often assume that all rays that diverge from the object make small angles with the principal axis. All such rays reflect through the image point, as in Figure 23.7. Rays that make a large angle with the principal axis, as in Figure 23.8, converge to other points on the principal axis, producing a blurred image. This effect, called **spherical aberration,** is present to some extent with any spherical mirror and will be discussed in Section 23.7.

We can use the geometry shown in Figure 23.9 to calculate the image distance q from the object distance p and radius of curvature R. By convention, these distances are measured from point V. Figure 23.9 shows two rays of light leaving the tip of the object. One ray passes through the center of curvature C of the mirror, hitting the mirror head on (perpendicularly to the mirror surface) and reflecting back on itself. The second ray strikes the mirror at point V and reflects as shown, obeying the law of reflection. The image of the tip of the arrow is at the point at which the two rays intersect. From the largest triangle in Figure 23.9, we see that $\tan \theta = h/p$; the light blue triangle gives $\tan \theta = -h'/q$. The negative sign has been introduced to satisfy our convention that h' is negative when the image is inverted with respect to the object, as it is here. Thus, from Equation 23.1 and these results, we find that the magnification of the mirror is

$$M = \frac{h'}{h} = -\frac{q}{p} \qquad [23.2]$$

We also note, from two other triangles in the figure, that

$$\tan \alpha = \frac{h}{p - R} \qquad \text{and} \qquad \tan \alpha = -\frac{h'}{R - q}$$

from which we find that

$$\frac{h'}{h} = -\frac{R - q}{p - R} \qquad [23.3]$$

If we compare Equation 23.2 to Equation 23.3, we see that

$$\frac{R - q}{p - R} = \frac{q}{p}$$

Simple algebra reduces this to

Mirror equation ▶

$$\frac{1}{p} + \frac{1}{q} = \frac{2}{R} \qquad [23.4]$$

This expression is called the **mirror equation.**

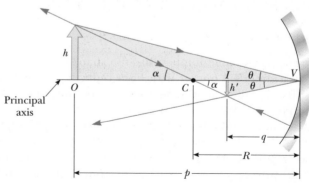

FIGURE 23.9 The image formed by a spherical concave mirror, where the object at O lies outside the center of curvature, C.

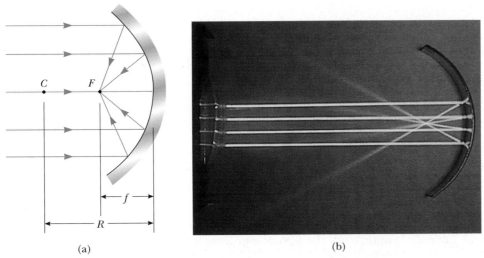

(a) (b)

FIGURE 23.10 (a) Light rays from a distant object ($p = \infty$) reflect from a concave mirror through the focal point F. In this case, the image distance $q = R/2 = f$, where f is the focal length of the mirror. (b) A photograph of the reflection of parallel rays from a concave mirror. *(Courtesy of Jim Lehman, James Madison University)*

If the object is very far from the mirror—that is, if the object distance p is great enough compared with R that p can be said to approach infinity—then $1/p \approx 0$, and we see from Equation 23.4 that $q \approx R/2$. In other words, when the object is very far from the mirror, **the image point is halfway between the center of curvature and the center of the mirror,** as in Figure 23.10a. The incoming rays are essentially parallel in this figure because the source is assumed to be very far from the mirror. In this special case we call the image point the **focal point,** F, and the image distance the **focal length,** f, where

$$f = \frac{R}{2} \qquad\qquad \textbf{[23.5]}$$

 Focal length

The mirror equation can therefore be expressed in terms of the focal length:

$$\frac{1}{p} + \frac{1}{q} = \frac{1}{f} \qquad\qquad \textbf{[23.6]}$$

Note that rays from objects at infinity are always focused at the focal point.

Tip 23.2 FOCAL POINT $\neq$ FOCUS POINT

The focal point *is not* the point at which light rays focus to form an image. The focal point of a mirror is determined *solely* by its curvature—it does not depend on the location of any object.

23.3 CONVEX MIRRORS AND SIGN CONVENTIONS

Figure 23.11 shows the formation of an image by a **convex mirror,** which is silvered so that light is reflected from the outer, convex surface. This is sometimes called a **diverging mirror** because the rays from any point on the object diverge after reflection as though they were coming from some point behind the mirror. The image in Figure 23.11 is virtual rather than real because it lies behind the mirror at the point at which the reflected rays appear to originate. In general, as shown in the figure, the image formed by a convex mirror is upright, virtual, and smaller than the object.

We shall not derive any equations for convex spherical mirrors. If we did, we would find that the equations developed for concave mirrors can be used with convex mirrors if particular sign conventions are used. Let us call the region in which light rays move the *front side* of the mirror, and the other side, where virtual images are formed, the *back side.* For example, in Figures 23.9 and 23.11, the side to the left of the mirror is the front side, and the side to the right is the back side.

FIGURE 23.11 Formation of an image by a spherical convex mirror. Note that the image is virtual and upright.

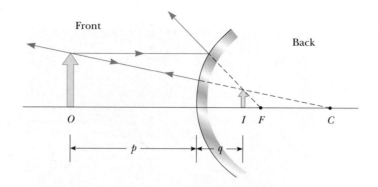

Front

Back

O I F C

p

q

Front, or real, side	Back, or virtual, side
p and q positive	p and q negative

→ Incident light

← Reflected light | No light

Convex or concave mirror

FIGURE 23.12 A diagram describing the signs of p and q for convex and concave mirrors.

Figure 23.12 is helpful for understanding the rules for object and image distances, and Table 23.1 summarizes the sign conventions for all the necessary quantities.

RAY DIAGRAMS FOR MIRRORS

We can determine conveniently the positions and sizes of images formed by mirrors by constructing *ray diagrams* similar to the ones we have been using. This kind of graphical construction tells us the overall nature of the image and can be used to check parameters calculated from the mirror and magnification equations. To make a ray diagram, one needs to know the position of the object and the location of the center of curvature. To locate the image, three rays are constructed (rather than just the two we have been constructing so far), as shown by the examples in Figure 23.13. All three rays start from the same object point; for these examples the tip of the arrow was chosen. For the concave mirrors in Figure 23.13a and b, the rays are drawn as follows:

1. **Ray 1 is drawn parallel to the principal axis and is reflected back through the focal point *F*.**
2. **Ray 2 is drawn through the focal point. Thus, it is reflected parallel to the principal axis.**
3. **Ray 3 is drawn through the center of curvature *C* and is reflected back on itself.**

Note that rays actually go in all directions from the object; we choose to follow those moving in a direction that simplifies our drawing.

The intersection of any *two* of these rays at a point locates the image. The third ray serves as a check of construction. The image point obtained in this fashion must always agree with the value of q calculated from the mirror formula.

In the case of a concave mirror, note what happens as the object is moved closer to the mirror. The real, inverted image in Figure 23.13a moves to the left as the object approaches the focal point. When the object is at the focal point, the

TABLE 23.1	**Sign Conventions for Mirrors**	
Quantity	**Positive When**	**Negative When**
Object location (p)	Object is in front of mirror	Object is behind mirror
Image location (q)	Image is behind mirror	Image is in front of mirror
Image height (h')	Image is upright	Image is inverted
Focal length (f) and radius (R)	Mirror is concave	Mirror is convex
Magnification (M)	Image is upright	Image is inverted

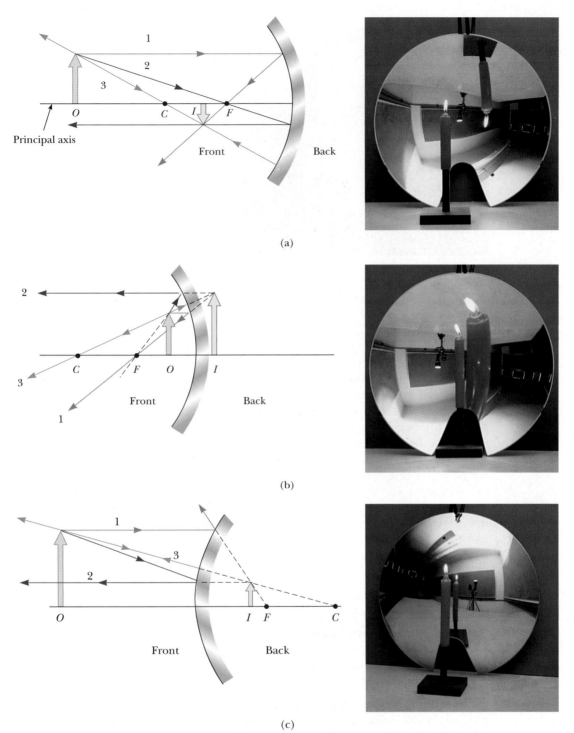

(a)

(b)

(c)

FIGURE 23.13 Ray diagrams for spherical mirrors, and corresponding photographs of the images of candles. (a) When an object is outside the center of curvature of a concave mirror, the image is real, inverted, and reduced in size. (b) When an object is between a concave mirror and the focal point, the image is virtual, upright, and magnified. (c) When an object is in front of a convex mirror, the image is virtual, upright, and reduced in size. *(Photos courtesy of David Rogers)*

image is infinitely far to the left. However, when the object lies between the focal point and the mirror surface, as in Figure 23.13b, the image is virtual and upright.

With the convex mirror shown in Figure 23.13c, the image of a real object is always virtual and upright. As the object distance increases, the virtual image

FIGURE 23.14 A convex sideview mirror on a vehicle produces an upright image that is smaller than the object. (© *Junebug Clark 1988/Photo Researchers, Inc.*)

shrinks and approaches the focal point as p approaches infinity. You should construct a ray diagram to verify this.

The image-forming characteristics of curved mirrors obviously determine their uses. For example, suppose you want to design a mirror that will help people shave or apply cosmetics, such as the one in Figure 23.13b. That is, you need a concave mirror that puts the user inside the focal point. In such a situation, the image is upright and greatly enlarged. In contrast, suppose that the primary purpose of a mirror is to observe a large field of view, in which case you need a convex mirror such as the one in Figure 23.13c. The diminished size of the image means that a fairly large field of view is seen in the mirror. Mirrors such as this are often placed in stores to help employees watch for shoplifting. A second use is as a sideview mirror on a car (Fig. 23.14). This kind of mirror is usually placed on the passenger side of the car and carries the warning "Objects are closer than they appear." Without this warning, a driver might think she is looking into a flat mirror, which does not alter the size of the image. Thus, she could be fooled into believing that a truck is far away because it looks small, when it is actually a large semi very close behind her but diminished in size because of the image formation characteristics of the convex mirror.

APPLYING **PHYSICS** 23.2

For a concave mirror, a virtual image can be anywhere behind the mirror. For a convex mirror, however, there is a maximum distance at which the image can exist behind the mirror. Why?

Explanation Let us consider the concave mirror first and imagine two different light rays leaving a tiny object and striking the mirror. If the object is at the focal point, the light rays reflecting from the mirror will be parallel to the mirror axis. They can be interpreted as forming a virtual image infinitely far away behind the mirror. As the object is brought closer to the mirror, the reflected rays will diverge through larger and larger angles, resulting in their extensions converging closer and closer to the back of the mirror. When the object is brought right up to the mirror, the image is right behind the mirror. When the object is much closer to the mirror than the focal length, the mirror acts like a flat mirror, and the image is just as far behind the mirror as the object is in front of it. Thus, the image can be anywhere from infinitely far away to right at the surface of the mirror. For the convex mirror, an object at infinity produces a virtual image at the focal point. As the object is brought closer, the reflected rays diverge more sharply and the image moves closer to the mirror. Thus, the virtual image is restricted to the region between the mirror and the focal point.

APPLYING PHYSICS 23.3

Large trucks often have a sign on the back saying, "If you can't see my mirror, I can't see you." Explain this sign.

Explanation　The trucking companies are making use of the principle of reversibility of light rays. In order for an image of you to be formed in the driver's mirror, there must be a pathway for rays of light to reach the mirror, allowing the driver to see your image. If you can't see the mirror, obviously there is no such pathway.

Example 23.2　Images Formed by a Concave Mirror

Assume that a certain concave spherical mirror has a focal length of 10.0 cm. Locate the images for object distances of (a) 25.0 cm, (b) 10.0 cm, and (c) 5.00 cm. Describe the image in each case.

Solution

A　For an object distance of 25.0 cm, we find the image distance using the mirror equation:

$$\frac{1}{p} + \frac{1}{q} = \frac{1}{f}$$

$$\frac{1}{25.0 \text{ cm}} + \frac{1}{q} = \frac{1}{10.0 \text{ cm}}$$

$$q = \boxed{16.7 \text{ cm}}$$

The magnification is given by Equation 23.2:

$$M = -\frac{q}{p} = -\frac{16.7 \text{ cm}}{25.0 \text{ cm}} = -0.667$$

Thus, the image is smaller than the object. Furthermore, the image is inverted because M is negative. Finally, because q is positive, the image is on the front side of the mirror and is real. This situation is pictured in Figure 23.13a.

B　When the object distance is 10.0 cm, the object is at the focal point. Substituting the values $p = 10.0$ cm and $f = 10.0$ cm into the mirror equation, we find that

$$\frac{1}{10.0 \text{ cm}} + \frac{1}{q} = \frac{1}{10.0 \text{ cm}}$$

$$q = \boxed{\infty}$$

Thus, we see that rays of light originating from an object at the focal point of a concave mirror are reflected so that the image is formed an infinite distance from the mirror—that is, the rays travel parallel to one another after reflection. Furthermore, because the image distance is infinite, the magnification M is infinite. In other words, light diverging from a point of zero height forms an image of height $2R$ and so M is infinite.

C　When the object is at 5.00 cm, inside the focal point of the mirror, the mirror equation gives

$$\frac{1}{5.00 \text{ cm}} + \frac{1}{q} = \frac{1}{10.0 \text{ cm}}$$

$$q = \boxed{-10.0 \text{ cm}}$$

That is, the image is virtual because it is behind the mirror. The magnification is

$$M = -\frac{q}{p} = -\left(\frac{-10.0 \text{ cm}}{5.00 \text{ cm}}\right) = 2.00$$

We see that the image height is magnified by a factor of 2, and the positive sign indicates that the image is upright (Fig. 23.13b).

Note the characteristics of an image formed by a concave spherical mirror. When the object is outside the focal point, the image is inverted and real; at the focal point, the image is formed at infinity; inside the focal point, the image is upright and virtual.

EXERCISE   If the object distance is 20.0 cm, find the image distance and the magnification of the mirror.

ANSWER $q = 20.0$ cm, $M = -1.00$

Example 23.3   Images Formed by a Convex Mirror

An object 3.00 cm high is placed 20.0 cm from a convex mirror with a focal length of 8.00 cm. Find (a) the position of the final image and (b) the magnification of the mirror.

Solution

A Because the mirror is convex, its focal length is negative. To find the image position, we use the mirror equation:

$$\frac{1}{p} + \frac{1}{q} = \frac{1}{f}$$

$$\frac{1}{20.0 \text{ cm}} + \frac{1}{q} = \frac{1}{-8.00 \text{ cm}}$$

$$q = \boxed{-5.71 \text{ cm}}$$

The negative value of q indicates that the image is virtual, or behind the mirror, as in Figure 23.13c.

B The magnification of the mirror is

$$M = -\frac{q}{p} = -\left(\frac{-5.71 \text{ cm}}{20.0 \text{ cm}}\right) = \boxed{0.286}$$

The image is upright because M is positive.

EXERCISE   Find the height of the image.

ANSWER   0.857 cm

Webnote 23.1

To see how objects are formed into virtual images for a convex (diverging) mirror, go to *http://www.lightlink.com/sergey/java/java/dmirr/index.html*

Example 23.4   An Enlarged Image

When a woman stands with her face 40.0 cm from a cosmetic mirror, the upright image is twice as tall as her face. What is the focal length of the mirror?

Reasoning   Most of the problems we have encountered so far have been simple applications of the mirror equation. However, to find f in this example, we must first find q, the image distance. Because the problem states that the image is upright, the magnification must be positive (in this case, $M = +2$), and because $M = -q/p$, we can determine q.

Solution   The magnification equation gives us a relationship between the object and image distances:

$$M = -\frac{q}{p} = 2$$

$$q = -2p = -2(40.0 \text{ cm}) = -80.0 \text{ cm}$$

First, note that a virtual image is formed because the woman is able to see her upright image in the mirror. This explains why the image distance is negative. Substitute $q = -80.0$ cm into the mirror equation to obtain

$$\frac{1}{40.0 \text{ cm}} - \frac{1}{80.0 \text{ cm}} = \frac{1}{f}$$

$$f = \boxed{80.0 \text{ cm}}$$

The positive sign for the focal length indicates that the mirror is concave, a fact that we already knew because the mirror magnified the object. (A convex mirror would have produced a diminished image.)

23.4 IMAGES FORMED BY REFRACTION

In this section we describe how images are formed by refraction at a spherical surface. Consider two transparent media with indices of refraction n_1 and n_2, where the boundary between the two media is a spherical surface of radius R (Fig. 23.15). Let us assume that the medium to the right has a higher index of refraction than the one to the left; that is, $n_2 > n_1$. This would be the case for light entering a curved piece of glass from air or for light entering the water in a fishbowl from air. The rays originating at the object location O are refracted at the spherical surface and then converge to the image point I. We can begin with Snell's law of refraction and use simple geometric techniques to show that the object distance, image distance, and radius of curvature are related by the equation

$$\frac{n_1}{p} + \frac{n_2}{q} = \frac{n_2 - n_1}{R} \qquad \text{[23.7]}$$

Furthermore, the magnification of a refracting surface is

$$M = \frac{h'}{h} = -\frac{n_1 q}{n_2 p} \qquad \text{[23.8]}$$

As with mirrors, we must use a sign convention if we are to apply these equations to a variety of circumstances. First note that real images are formed on the side of the surface *opposite* the side from which the light comes. This is in contrast with mirrors, where real images are formed on the *same* side of the reflecting surface. We define the side of the surface in which light rays originate as the front side. The other side is called the back side. Because of the difference in location of real images, the refraction sign conventions for q and R are the opposite of those for reflection. For example, p, q, and R are all positive in Figure 23.15. The sign convention for spherical refracting surfaces is summarized in Table 23.2.

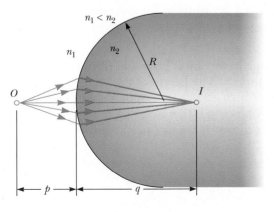

FIGURE 23.15 An image formed by refraction at a spherical surface. Rays making small angles with the principal axis diverge from a point object at O and pass through the image point I.

TABLE 23.2	Sign Conventions for Refracting Surfaces	
Quantity	**Positive When**	**Negative When**
Object location (p)	Object is in front of surface	Object is in back of surface
Image location (q)	Image is in back of surface	Image is in front of surface
Image height (h')	Image is upright	Image is inverted
Radius (R)	Center of curvature is in back of surface	Center of curvature is in front of surface

APPLYING PHYSICS 23.4

Why does a person with normal vision see a blurry image if the eyes are opened underwater without goggles or a diving mask?

Explanation   The eye presents a spherical refraction surface. The eye normally functions so that light entering from the air is refracted to form an image in the retina located at the back of the eyeball. The difference in index of refraction between water and the eye is smaller than the difference in index of refraction between air and the eye. Thus, light entering the eye from the water does not experience as much refraction as does light entering from the air, and the image is formed behind the retina. A diving mask or swimming goggles have no optical action of their own; they are simply flat pieces of glass or plastic in a rubber mount. However, they provide a region of air adjacent to the eyes, so that the correct refraction relationship is established, and images will be in focus.

APPLICATION

OPENING YOUR EYES UNDERWATER

FLAT REFRACTING SURFACES

If the refracting surface is flat, then R approaches infinity and Equation 23.7 reduces to

$$\frac{n_1}{p} = -\frac{n_2}{q}$$

$$q = -\frac{n_2}{n_1}p \qquad \text{[23.9]}$$

From Equation 23.9 we see that the sign of q is opposite that of p. Thus, **the image formed by a flat refracting surface is on the same side of the surface as the object.** This is illustrated in Figure 23.16 for the situation in which n_1 is greater than n_2, where a virtual image is formed between the object and the surface. Note that the refracted ray bends *away* from the normal in this case, because $n_1 > n_2$.

$n_1 > n_2$

FIGURE 23.16   The image formed by a flat refracting surface is virtual; that is, it forms to the left of the refracting surface. Note that if the light rays are reversed in direction, we have the situation described in Example 22.6.

Quick Quiz 23.2   A person spear fishing from a boat sees a fish located 3 m from the boat at an apparent depth of 1 m. To spear the fish, should the person aim (a) at, (b) above, or (c) below the image of the fish?

Quick Quiz 23.3   True or false? (a) The image of an object placed in front of a concave mirror is always upright. (b) The height of the image of an object placed in front of a concave mirror must be smaller than or equal to the height of the object. (c) The image of an object placed in front of a convex mirror is always upright and smaller than the object.

Example 23.5 Gaze into the Crystal Ball

A coin 2.00 cm in diameter is embedded in a solid glass ball of radius 30.0 cm (Fig. 23.17). The index of refraction of the ball is 1.50, and the coin is 20.0 cm from the surface. Find the position and height of the image of the coin.

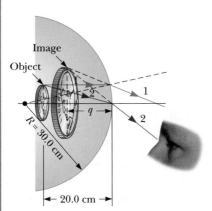

FIGURE 23.17 (Example 23.5) A coin embedded in a glass ball forms a virtual image between the coin and the glass surface.

Solution Because they are moving from a medium of high index of refraction to a medium of lower index of refraction, the rays originating at the coin are refracted away from the normal at the surface and diverge outward. The image is formed in the glass and is virtual. Applying Equation 23.7 and taking $n_1 = 1.50$, $n_2 = 1.00$, $p = 20.0$ cm, and $R = -30.0$ cm, we get

$$\frac{n_1}{p} + \frac{n_2}{q} = \frac{n_2 - n_1}{R}$$

$$\frac{1.50}{20.0 \text{ cm}} + \frac{1.00}{q} = \frac{1.00 - 1.50}{-30.0 \text{ cm}}$$

$$q = \boxed{-17.1 \text{ cm}}$$

The negative sign indicates that the image is in the same medium as the object (the side of incident light), in agreement with our ray diagram, and therefore must be virtual.

To find the image height, we first use Equation 23.8 for the magnification:

$$M = -\frac{n_1 q}{n_2 p} = -\frac{1.50(-17.1 \text{ cm})}{1.00(20.0 \text{ cm})} = \frac{h'}{h}$$

Therefore,

$$h' = 1.28h = (1.28)(2.00 \text{ cm}) = \boxed{2.56 \text{ cm}}$$

The positive value for M indicates an upright image.

Example 23.6 The One That Got Away

A small fish is swimming at a depth of d below the surface of a pond (Fig. 23.18). What is the *apparent depth* of the fish as viewed from directly overhead?

Reasoning In this example, the refracting surface is flat, and so R is infinite. Hence, we can use Equation 23.9 to determine the location of the image, which is the apparent location of the fish.

Solution The facts that $n_1 = 1.33$ for water and $p = d$ give us

$$q = -\frac{n_2}{n_1} p = -\frac{1}{1.33} d = \boxed{-0.752d}$$

FIGURE 23.18 (Example 23.6) The apparent depth q of the fish is less than the true depth d.

Again, because q is negative, the image is virtual, as indicated in Figure 23.18. The apparent depth is three-fourths the actual depth. For instance, if d = 4.0 m, q = − 3.0 m.

EXERCISE   If the fish is 12 cm long, how long is its image?

ANSWER   12 cm

23.5   ATMOSPHERIC REFRACTION

Images formed by refraction in our atmosphere lead to some interesting results. In this section we look at two examples. A situation that occurs daily is the visibility of the Sun at dusk even though it has passed below the horizon. Figure 23.19 shows why this occurs. Rays of light from the Sun strike the Earth's atmosphere (represented by the shaded area around the Earth) and are bent as they pass into a medium that has an index of refraction different from that of the almost empty space in which they have been traveling. The bending in this situation differs somewhat from the bending we have considered previously in that it is gradual and continuous as the light moves through the atmosphere toward an observer at point *O*. This is because the light moves through layers of air that have a continuously changing index of refraction. When the rays reach the observer, the eye follows them back along the direction from which they appear to have come (indicated by the dashed path in the figure). The end result is that the Sun is seen to be above the horizon even after it has fallen below it.

The **mirage** is another phenomenon of nature produced by refraction in the atmosphere. A mirage can be observed when the ground is so hot that the air directly above it is warmer than the air at higher elevations. The desert is a region in which such circumstances prevail, but mirages are also seen on heated roadways during the summer. The layers of air at different heights above the Earth have different densities and different refractive indices. The effect this can have is pictured in Figure 23.20a. In this situation the observer sees the sky and a tree in two different ways. One group of light rays reaches the observer by the straight-line path *A*, and the eye traces these rays back to see the tree in the normal fashion. In addition, a second group of rays travels along the curved path *B*. These rays are directed toward the ground and are then bent as a result of refraction. As a consequence, the observer also sees an inverted image of the tree and sky background as he traces these rays back to the point at which they appear to have originated. Because an upright image and an inverted image are seen when the image of a tree is observed in a reflecting pool of water, the observer unconsciously calls on this past experience and concludes that the sky background is reflected by a pool of water in front of the tree.

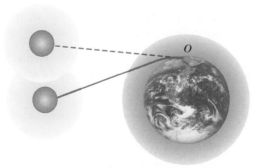

FIGURE 23.19   Because light is refracted by the Earth's atmosphere, an observer at *O* sees the Sun even though it has fallen below the horizon.

(a)

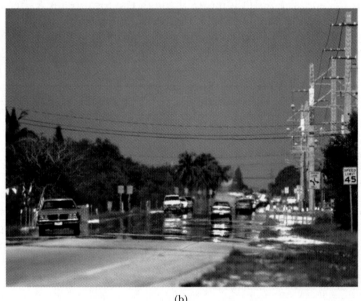

(b)

FIGURE 23.20 (a) A mirage is produced by the bending of light rays in the atmosphere when there are large temperature differences between the ground and the air. (b) (*John M. Dunay IV, Fundamental Photographs, NYC*)

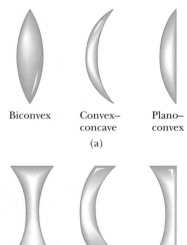

Biconvex   Convex–   Plano–
            concave   convex

(a)

Biconcave   Convex–   Plano–
             concave   concave

FIGURE 23.21 Lens shapes.
(a) Converging lenses have positive focal lengths and are thickest at the middle. (b) Diverging lenses have negative focal lengths and are thickest at the edges.

23.6 THIN LENSES

A typical **thin lens** consists of a piece of glass or plastic, ground so that each of its two refracting surfaces is a segment of either a sphere or a plane. Lenses are commonly used to form images by refraction in optical instruments, such as cameras, telescopes, and microscopes. The equation that relates object and image distances for a lens is virtually identical to the mirror equation derived earlier, and the method used to derive it is also similar.

Figure 23.21 shows some representative shapes of lenses. Notice that we have placed these lenses in two groups. Those in Figure 23.21a are thicker at the center than at the rim, and those in Figure 23.21b are thinner at the center than at the rim. The lenses in the first group are examples of **converging lenses,** and those in the second group are **diverging lenses.** The reason for these names will become apparent shortly.

As we did for mirrors, it is convenient to define a point called the **focal point** for a lens. For example, in Figure 23.22a, a group of rays parallel to the axis passes through the focal point *F* after being converged by the lens. The distance from the focal point to the lens is called the **focal length,** *f*. **The focal length is the image distance that corresponds to an infinite object distance.** Recall that we are considering the lens to be very thin. As a result, it makes no difference whether we take the focal length to be the distance from the focal point to the surface of the lens or the distance from the focal point to the center of the lens, because the difference between these two lengths is negligible. A thin lens has *two* focal points, as illustrated in Figure 23.22, corresponding to parallel rays traveling from the left and from the right.

Rays parallel to the axis diverge after passing through a lens of biconcave shape in Figure 23.22b. In this case, the focal point is defined to be the point at which the diverged rays appear to originate, labeled *F* in the figure. Figures 23.22a and 23.22b indicate why the names *converging* and *diverging* are applied to these lenses.

Consider a ray of light passing through the center of a lens, labeled ray 1 in Figure 23.23. For a thin lens, a ray passing through the center is undeviated. Ray 2 in Figure 23.23 is parallel to the principal axis of the lens (the horizontal axis passing through *O*), and as a result it passes through the focal point *F* after refraction. Rays 1 and 2 intersect at the point which is the tip of the image arrow.

We first note that the tangent of the angle α can be found by using the blue and gold shaded triangles in Figure 23.23:

$$\tan \alpha = \frac{h}{p} \quad \text{or} \quad \tan \alpha = -\frac{h'}{q}$$

FIGURE 23.22   (*Left*) Photographs of the effects of converging and diverging lenses on parallel rays. (*Courtesy of Jim Lehman, James Madison University*) (*Right*) The focal points of (a) the biconvex lens and (b) the biconcave lens.

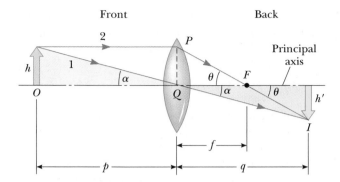

Front   Back

FIGURE 23.23 A geometric construction for developing the thin-lens equation.

From this we find that

$$M = \frac{h'}{h} = -\frac{q}{p} \qquad [23.10]$$

Thus, the equation for magnification by a lens is the same as the equation for magnification by a mirror. We also note from Figure 23.23 that the tangent of θ is

$$\tan \theta = \frac{PQ}{f} \qquad \text{or} \qquad \tan \theta = -\frac{h'}{q-f}$$

However, the height PQ used in the first of these equations is the same as h, the height of the object. Therefore,

$$\frac{h}{f} = -\frac{h'}{q-f}$$

$$\frac{h'}{h} = -\frac{q-f}{f}$$

Using this in combination with Equation 23.10 gives

$$\frac{q}{p} = \frac{q-f}{f}$$

which reduces to

$$\frac{1}{p} + \frac{1}{q} = \frac{1}{f} \qquad [23.11]$$

◀ Thin-lens equation

This equation, called the **thin-lens equation,** can be used with both converging and diverging lenses if we adhere to a set of sign conventions. Figure 23.24 is useful for obtaining the signs of p and q, and Table 23.3 gives the complete sign conventions for lenses. Note that **a converging lens has a positive focal length** under this convention, and **a diverging lens has a negative focal length.** Hence the names *positive* and *negative* are often given to these lenses.

FIGURE 23.24 A diagram for obtaining the signs of p and q for a thin lens or a refracting surface.

TABLE 23.3	**Sign Conventions for Thin Lenses**	
Quantity	**Positive When**	**Negative When**
Object location (p)	Object is in front of lens	Object is in back of lens
Image location (q)	Image is in back of lens	Image is in front of lens
Image height (h')	Image is upright	Image is inverted
R_1 and R_2	Center of curvature is in back of lens	Center of curvature is in front of lens
Focal length (f)	Converging lens	Diverging lens

The focal length for a lens in air is related to the curvatures of its front and back surfaces and to the index of refraction n of the lens material by

 Lens maker's equation ▶

$$\frac{1}{f} = (n - 1)\left(\frac{1}{R_1} - \frac{1}{R_2}\right)$$ [23.12]

where R_1 is the radius of curvature of the front surface of the lens and R_2 is the radius of curvature of the back surface. (As with mirrors, we arbitrarily call the side from which the light approaches the *front* of the lens.) Table 23.3 gives the sign conventions for R_1 and R_2. Equation 23.12 enables us to calculate the focal length from the known properties of the lens. It is called the **lens maker's equation.**

RAY DIAGRAMS FOR THIN LENSES

Ray diagrams are essential for understanding the overall image formation by a thin lens or a system of lenses. They should also help clarify the sign conventions we have already discussed. Figure 23.25 illustrates this method for three single-lens situations. To locate the image formed by a converging lens (Fig. 23.25a and b), the following three rays are drawn from the top of the object:

1. The first ray is drawn parallel to the principal axis. After being refracted by the lens, this ray passes through (or appears to come from) one of the focal points.
2. The second ray is drawn through the center of the lens. This ray continues in a straight line.
3. The third ray is drawn through the other focal point and emerges from the lens parallel to the principal axis.

A similar construction is used to locate the image formed by a diverging lens, as shown in Figure 23.25c. The point of intersection of *any two* of the rays in these diagrams can be used to locate the image. The third ray serves as a check on construction.

For the converging lens in Figure 23.25a, where the object is *outside* the front focal point ($p > f$), the ray diagram shows the image is real and inverted. When the real object is *inside* the front focal point ($p < f$), as in Figure 23.25b, the image is virtual and upright. For the diverging lens of Figure 23.25c, the image is virtual and upright.

Tip 23.3 WE CHOOSE ONLY A FEW RAYS

Note that although our ray diagrams in Figure 23.25 only show three rays leaving an object, there are an infinite number of rays that can be drawn between the object and its image.

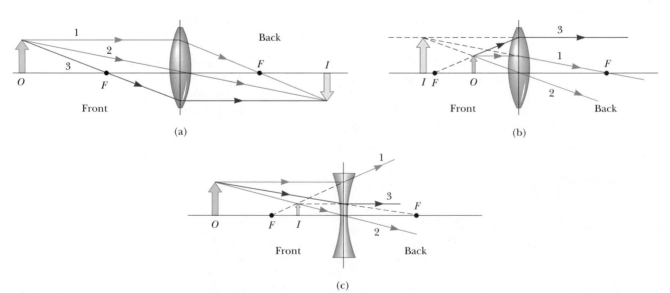

FIGURE 23.25 Ray diagrams for locating the image of an object. (a) The object is outside the focal point of a converging lens. (b) The object is inside the focal point of a converging lens. (c) The object is outside the focal point of a diverging lens.

PROBLEM-SOLVING STRATEGY   **Lenses and Mirrors**

Your success or failure in working lens and mirror problems will be determined largely by whether or not you make sign errors when substituting into the lens and mirror equations. The only way to ensure that you don't make sign errors is to become adept at using the sign conventions. The best way to do this is to work a multitude of problems on your own and to make confirming ray diagrams. Watching an instructor or reading the example problems is no substitute for practice.

APPLYING PHYSICS 23.5

Diving masks often have a lens built into the glass for divers who do not have perfect vision. This allows the individual to dive without the necessity of glasses, because the lenses in the faceplate perform the necessary refraction to produce clear vision. Normal glasses have lenses that are curved on both the front and rear surfaces. The lenses in a diving mask faceplate often only have curved surfaces on the inside of the glass. Why is this design desirable?

Explanation   The main reason for curving only the inner surface of the lenses in the diving mask faceplate is so that the diver can see clearly while underwater and in the air. If there were curved surfaces on both the front and the back of the diving lens, there would be two refractions. The lens could be designed so that these two refractions would give clear vision while the diver is in air. When the diver goes underwater, however, the refraction between the water and the glass at the first interface is now different, because the index of refraction of water is different from that of air. Thus, the vision will not be clear underwater.

> **APPLICATION**
>
> VISION AND DIVING MASKS

APPLYING PHYSICS 23.6

Consider a glass plano-convex lens, one that is flat on one side and convex on the other. You project three laser beams through it, as shown in Figure 23.26a, and measure the focal length f—that is, the distance from the lens to the point at which the three beams cross. Now you hold the flat side of the lens against the glass of an aquarium filled with water, as in Figure 23.26b. When you shine the

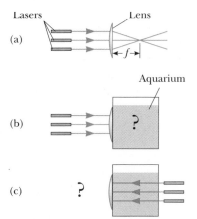

FIGURE 23.26 (Applying Physics 23.6)

laser beams through the lens from the outside of the aquarium, will the beams cross at a point closer to the lens than before, farther away, or at the same distance? What if you direct the laser beams through the lens from the other side of the aquarium, as in Figure 23.26c? What happens in this case?

Explanation   The laser beam going through the center of the lens is undeviated by the lens in all three cases. We need to look at what happens to the outer beams. In Figure 23.26a, these two laser beams will refract toward the normal as they pass from the air into the glass of the lens. Thus, they will be deviated *toward* the central beam. As they reach the other side, and pass from the glass back into the air, they will deviate away from the normal, which causes additional deviation toward the central beam. As a result, all three beams cross at the focal point. When the flat side of the lens is held against the glass side of the aquarium as in Figure 23.26b, the outgoing laser beams pass through the glass of the aquarium without additional refraction, but then enter *water*. The change in index of refraction in going from glass to water is much smaller than in the previous case, and the deviation toward the central beam is less. As a result, the crossing point for the three beams is farther from the lens. When the laser beams are directed through the water and then through the lens, as in Figure 23.26c, we must consider two factors. First, the focal length of a lens is independent of which way the light passes through it. Second, in the situation shown, the laser beams experience no refraction on entering the flat side of the lens, because they strike the glass at normal incidence. Thus, all of the refraction occurs at the interface of the air with the curved side of the lens. This is exactly the same situation as if the laser beams had struck the flat side of the lens while it was in air. Thus, there is no effect of the water in the aquarium. The laser beams cross at the same distance from the lens as they did in the first diagram. This third situation is very similar to the curvature of a lens on the inside of a diving mask, discussed previously.

Example 23.7   Images Formed by a Converging Lens

A converging lens of focal length 10.0 cm forms images of objects placed (a) 30.0 cm, (b) 10.0 cm, and (c) 5.00 cm from the lens. In each case, construct a ray diagram, find the image distance, and describe the image.

Solution

A First we construct a ray diagram as shown in Figure 23.27a. The diagram shows that a real, inverted, smaller image is formed on the far side of the lens. The thin-lens equation, Equation 23.11, can be used to find the image distance:

$$\frac{1}{p} + \frac{1}{q} = \frac{1}{f}$$

$$\frac{1}{30.0 \text{ cm}} + \frac{1}{q} = \frac{1}{10.0 \text{ cm}}$$

$$q = \boxed{+15.0 \text{ cm}}$$

The positive sign for the image distance tells us that the image is real and on the back side of the lens. The magnification of the lens is

$$M = -\frac{q}{p} = -\frac{15.0 \text{ cm}}{30.0 \text{ cm}} = \boxed{-0.500}$$

Thus, the image is reduced in height by one half, and the negative sign for M tells us that the image is inverted.

B No calculation is necessary for this case because we know that when the object is placed at the focal point, the image is formed at infinity. This is readily verified by substituting $p = 10.0$ cm into the lens equation.

FIGURE 23.27   (Example 23.7)

[C] We now move inside the focal point. The ray diagram in Figure 23.27b immediately shows that in this case the lens is being used as a magnifying glass; that is, the image is magnified, erect, on the same side as the object, and virtual. Because the object distance is 5.00 cm, the lens equation gives

$$\frac{1}{5.00 \text{ cm}} + \frac{1}{q} = \frac{1}{10.0 \text{ cm}}$$

$$q = \boxed{-10.0 \text{ cm}}$$

$$M = -\frac{q}{p} = -\left(\frac{-10.0 \text{ cm}}{5.00 \text{ cm}}\right) = \boxed{+2.00}$$

The negative image distance tells us that the image is virtual and formed on the side of the lens from which the light is incident, the front side. The image is enlarged, and the positive sign for M tells us that the image is upright.

Example 23.8   The Case of a Diverging Lens

Repeat the problem of Example 23.7 for a *diverging* lens of focal length 10.0 cm.

Solution

[A] We begin by constructing a ray diagram as in Figure 23.28a taking the object distance to be 30.0 cm. The diagram shows that the image is virtual, smaller than the object, and upright. Let us now apply the lens equation with $p = 30.0$ cm:

$$\frac{1}{p} + \frac{1}{q} = \frac{1}{f}$$

$$\frac{1}{30.0 \text{ cm}} + \frac{1}{q} = -\frac{1}{10.0 \text{ cm}}$$

$$q = \boxed{-7.50 \text{ cm}}$$

FIGURE 23.28   (Example 23.8)

The magnification is

$$M = -\frac{q}{p} = -\left(\frac{-7.50 \text{ cm}}{30.0 \text{ cm}}\right) = \boxed{+0.250}$$

This result confirms that the image is virtual, smaller than the object, and upright.

B When the object is at the focal point, $p = 10.0$ cm, we have

$$\frac{1}{10.0 \text{ cm}} + \frac{1}{q} = -\frac{1}{10.0 \text{ cm}}$$

$$q = \boxed{-5.00 \text{ cm}}$$

$$M = -\frac{q}{p} = -\left(\frac{-5.00 \text{ cm}}{10.0 \text{ cm}}\right) = \boxed{+0.500}$$

C When the object is inside the focal point, at $p = 5.00$ cm, the ray diagram in Figure 23.28b shows that we have a virtual image that is smaller than the object and upright. In this case, the lens equation gives

$$\frac{1}{5.00 \text{ cm}} + \frac{1}{q} = -\frac{1}{10.0 \text{ cm}}$$

$$q = \boxed{-3.33 \text{ cm}}$$

$$M = -\left(\frac{-3.33 \text{ cm}}{5.00 \text{ cm}}\right) = \boxed{+0.667}$$

This confirms that the image is virtual, smaller than the object, and upright.

COMBINATIONS OF THIN LENSES

If two thin lenses are used to form an image, the system can be treated in the following manner. First, the image produced by the first lens is calculated as though the second lens were not present. The light then approaches the second lens *as if* it had come from the image formed by the first lens. Hence, **the image formed by the first lens is treated as the object for the second lens.** The image formed by the second lens is the final image of the system. If the image formed by the first lens lies on the back side of the second lens, then the image is treated as a virtual object for the second lens (that is, p is negative). The same procedure can be extended to a system of three or more lenses. The overall magnification of a system of thin lenses is the *product* of the magnifications of the separate lenses.

Webnote 23.2

For a more advanced version of lens and mirror combinations, go to
http://webphysics.ph.msstate.edu/ javamirror/ntnujava/Lens/lens_e.html

Example 23.9   Two Lenses in a Row

Two converging lenses are placed 20.0 cm apart, as shown in Figure 23.29a. If the first lens has a focal length of 10.0 cm and the second has a focal length of 20.0 cm, locate the final image formed of an object 30.0 cm in front of the first lens. Find the magnification of the system.

Reasoning   We apply the thin-lens equation to both lenses. The image formed by the first lens is treated as the object for the second lens. Also, we use the fact that the total magnification of the system is the product of the magnifications produced by the separate lenses.

Solution   First we make ray diagrams roughly to scale to see where the image from the first lens falls and how it acts as the object for the second lens. (See Fig. 23.29b.) The location of the image formed by the first lens is found via the thin-lens equation:

$$\frac{1}{30.0 \text{ cm}} + \frac{1}{q} = \frac{1}{10.0 \text{ cm}}$$

$$q = \boxed{+15.0 \text{ cm}}$$

The magnification of this lens is

$$M_1 = -\frac{q}{p} = -\frac{15.0 \text{ cm}}{30.0 \text{ cm}} = \boxed{-0.500}$$

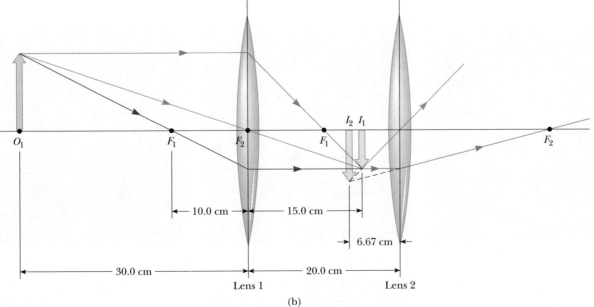

FIGURE 23.29   (Example 23.9)

The image formed by this lens becomes the object for the second lens. Thus, the object distance for the second lens is 20.0 cm − 15.0 cm = 5.00 cm. We again apply the thin-lens equation to find the location of the final image.

$$\frac{1}{5.00 \text{ cm}} + \frac{1}{q} = \frac{1}{20.0 \text{ cm}}$$

$$q = \boxed{-6.67 \text{ cm}}$$

The magnification of the second lens is

$$M_2 = -\frac{q}{p} = -\frac{(-6.67 \text{ cm})}{5.00 \text{ cm}} = \boxed{+1.33}$$

Thus, the final image is 6.67 cm to the left of the second lens, and the overall magnification of the system is

$$M = M_1 M_2 = (-0.500)(1.33) = \boxed{-0.667}$$

The negative sign indicates that the final image is inverted with respect to the initial object.

EXERCISE If the two lenses in Figure 23.29 are separated by 10.0 cm, locate the final image and find the magnification of the system.

ANSWER 4.00 cm behind the second lens; $M = -0.400$

23.7 LENS AND MIRROR ABERRATIONS

One of the basic problems of systems containing mirrors and lenses is the imperfect quality of the images, which is largely the result of defects in shape and form. The simple theory of mirrors and lenses assumes that rays make small angles with the principal axis and that all rays reaching the lens or mirror from a point source are focused at a single point, producing a sharp image. Clearly, this is not always true in the real world. Where the approximations used in this theory do not hold, imperfect images are formed.

(a)

(b)

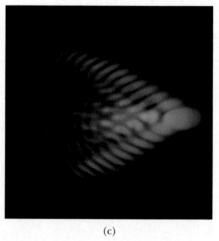
(c)

FIGURE 23.30 Lenses can produce varied forms of aberrations, as shown by these blurred photographic images of a point source. (a) Spherical aberration occurs when light passing through the lens at different distances from the principal axis is focused at different points. (b) Astigmatism is an aberration that occurs when the object is not on the principal axis of the lens. (c) Coma. This aberration occurs when light passing through the lens far from the principal axis focuses at a different part of the focal plane from light passing near the center of the lens. *(Photos by Norman Goldberg)*

If one wishes to analyze image formation precisely, it is necessary to trace each ray, using Snell's law, at each refracting surface. This procedure shows that there is no single-point image; instead, the image is blurred. The departures of real (imperfect) images from the ideal predicted by the simple theory are called **aberrations.** Two common types of aberrations are spherical aberration and chromatic aberration. Photographs of three forms of lens aberrations are shown in Figure 23.30.

SPHERICAL ABERRATION

Spherical aberration results from the fact that the focal points of light rays far from the principal axis of a spherical lens (or mirror) are different from the focal points of rays with the same wavelength passing near the axis. Figure 23.31 illustrates spherical aberration for parallel rays passing through a converging lens. Rays near the middle of the lens are imaged farther from the lens than rays at the edges. Hence, there is no single focal length for a spherical lens.

Most cameras are equipped with an adjustable aperture to control the light intensity and, when possible, reduce spherical aberration. (An aperture is an opening that controls the amount of light transmitted through the lens.) As the aperture size is reduced, sharper images are produced, because only the central portion of the lens is exposed to the incident light when the aperture is very small. At the same time, however, progressively less light is imaged. To compensate for this loss, a longer exposure time is used. An example of the results obtained with small apertures is the sharp image produced by a "pinhole" camera, with an aperture size of approximately 0.1 mm.

In the case of mirrors used for very distant objects, one can eliminate, or at least minimize, spherical aberration by employing a parabolic rather than spherical surface. Parabolic surfaces are not used in many applications, however, because they are very expensive to make with high-quality optics. Parallel light rays incident on such a surface focus at a common point. Parabolic reflecting surfaces are used in many astronomical telescopes to enhance the image quality. They are also used in flashlights, in which a nearly parallel light beam is produced from a small lamp placed at the focus of the reflecting surface.

CHROMATIC ABERRATION

The fact that different wavelengths of light refracted by a lens focus at different points gives rise to chromatic aberration. In Chapter 22 we described how the index of refraction of a material varies with wavelength. When white light passes through a lens, one finds, for example, that violet light rays are refracted more than red light rays (Fig. 23.32); thus, the focal length for red light is greater than that for violet light. Other wavelengths (not shown in Fig. 23.32) would have intermediate focal points. The chromatic aberration for a diverging lens is opposite that for a converging lens. Chromatic aberration can be greatly reduced by the use of a combination of converging and diverging lenses.

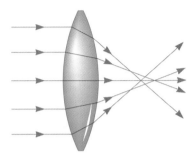

FIGURE 23.31 Spherical aberration produced by a converging lens. Does a diverging lens produce spherical aberration? (Angles are greatly exaggerated for clarity.)

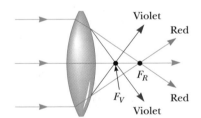

FIGURE 23.32 Chromatic aberration produced by a converging lens. Rays of different wavelengths focus at different points. (Angles are greatly exaggerated for clarity.)

Webnote 23.3

To learn more about optical aberrations, visit
http://nikon.topica.ne.jp/bi_e/encyclo/ad.htm

SUMMARY

Images are formed where rays of light intersect or where they appear to originate. A **real image** is one in which light intersects, or passes through, an image point. A **virtual image** is one in which the light does not pass through the image point but appears to diverge from that point.

The image formed by a flat mirror has the following properties:

1. The image is as far behind the mirror as the object is in front.
2. The image is unmagnified, virtual, and upright.

The **magnification** M of a mirror is defined as the ratio of **image height** h' to **object height** h, which is the negative of the ratio of image distance q to object distance p:

$$M = \frac{h'}{h} = -\frac{q}{p} \qquad [23.2]$$

The **object distance** and **image distance** for a spherical mirror of radius R are related by the **mirror equation:**

$$\frac{1}{p} + \frac{1}{q} = \frac{1}{f} \qquad [23.6]$$

where $f = R/2$ is the **focal length** of the mirror.

An image can be formed by refraction at a spherical surface of radius R. The object and image distances for refraction from such a surface are related by

$$\frac{n_1}{p} + \frac{n_2}{q} = \frac{n_2 - n_1}{R} \qquad [23.7]$$

The **magnification of a refracting surface** is

$$M = \frac{h'}{h} = -\frac{n_1 q}{n_2 p} \qquad [23.8]$$

where the object is located in the medium with index of refraction n_1 and the image is formed in the medium with index of refraction n_2.

The **magnification for a thin lens** is

$$M = \frac{h'}{h} = -\frac{q}{p} \qquad [23.10]$$

and the object and image distances are related by the **thin-lens equation:**

$$\frac{1}{p} + \frac{1}{q} = \frac{1}{f} \qquad [23.11]$$

Aberrations are responsible for the formation of imperfect images by lenses and mirrors. **Spherical aberration** results from the fact that the focal points of light rays far from the principal axis of a spherical lens or mirror are different from those of rays passing through the center. **Chromatic aberration** arises from the fact that light rays of different wavelengths focus at different points when refracted by a lens.

CONCEPTUAL QUESTIONS

1. Tape a picture of yourself on a bathroom mirror. Stand several centimeters away from the mirror. Can you focus your eyes on *both* the picture taped to the mirror *and* your image in the mirror *at the same time?* So, where is the image of yourself?

2. One method for determining the position of an image, either real or virtual, is by means of *parallax.* If a finger or another object is placed at the position of the image, as shown in Figure Q23.2, and the finger and the image are viewed simultaneously (the image is viewed through the lens if it is virtual), the finger and image have the same parallax; that is, if the image is viewed from different positions, it will appear to move along with the finger. Use this method to locate the image formed by a lens. Explain why this method works.

3. A flat mirror creates a virtual image of your face. Suppose the flat mirror is combined with another optical element. Can a flat mirror form a real image in such a combination?

4. Explain why a mirror cannot give rise to chromatic aberration.

5. You are taking a picture of yourself with a camera that uses an ultrasonic range finder to measure the distance to the object. When you take a picture of yourself in a mirror with this camera, your image is out of focus. Why?

6. A solar furnace can be constructed by using a concave mirror to reflect and focus sunlight into a furnace enclosure. What factors in the design of the reflecting mirror will guarantee that very high temperatures can be achieved?

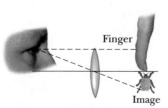

FIGURE Q23.2

7. A virtual image is often described as one through which the light rays do not actually travel, as they do for a real image. Can a virtual image be photographed?

8. What is wrong with the caption of the cartoon shown in Figure Q23.8?

FIGURE Q23.8 "Most mirrors reverse left and right. This one reverses top and bottom."

9. Suppose you want to use a converging lens to project the image of two trees onto a screen. One tree is distance *x* from the lens; the other is at 2*x*, as in Figure Q23.9. You adjust the screen so that the near tree is in focus. If you now want the far tree to be in focus, do you move the screen toward or away from the lens?

FIGURE Q23.9

10. Why does a clear stream always appear to be shallower than it actually is?

11. Can a converging lens be made to diverge light if placed in a liquid? How about a converging mirror?

12. A common mirage is formed when the air gets gradually cooler as the height above the ground increases. What might happen if the air grows gradually warmer as the height is increased? This often happens over bodies of water or snow-covered ground; the effect is called *looming*.

13. In a Jules Verne novel, a piece of ice is shaped into a magnifying lens to focus sunlight to start a fire. Is this possible?

14. Lenses used in eyeglasses, whether converging or diverging, are always designed such that the middle of the lens curves away from the eye. Why?

15. Why does the focal length of a mirror not depend on the mirror material whereas the focal length of a lens does depend on the lens material?

16. If a cylinder of solid glass or clear plastic is placed above the words LEAD OXIDE and viewed from the side, as shown in Figure Q23.16, the word LEAD appears inverted but the word OXIDE does not. Explain.

FIGURE Q23.16 *(Richard Megna/Fundamental Photographs, NYC)*

17. What is the focal length of a flat mirror? Is that value consistent with the mirror equation?

PROBLEMS

1, 2, 3 = straightforward, intermediate, challenging ☐ = full solution available in Student Solutions Manual/Study Guide

web = solution posted at **http://info.brookscole.com/serway** 🔬 = biomedical application

Section 23.1 Flat Mirrors

1. Does your bathroom mirror show you older or younger than your actual age? Compute an order-of-magnitude estimate for the age difference, based on data that you specify.

2. Use Figure 23.2 to give a geometric proof that the virtual image formed by a plane mirror is the same distance behind the mirror as the object is in front of it.

3. A person walks into a room that has, on opposite walls, two plane mirrors producing multiple images. When the person is 5.00 ft from the mirror on the left wall and 10.0 ft

from the mirror on the right wall, find the distances from the person to the first three images seen in the left-hand mirror.

4. In a church choir loft, two parallel walls are 5.30 m apart. The singers stand against the north wall. The organist faces the south wall, sitting 0.800 m away from it. So that she can see the choir, a flat mirror 0.600 m wide is mounted on the south wall, straight in front of the organist. What width of the north wall can she see? (*Hint:* Draw a top-view diagram to justify your answer.)

Section 23.2 Images Formed by Spherical Mirrors

Section 23.3 Convex Mirrors and Sign Conventions

In the following problems, algebraic signs are not given. We leave it to you to determine the correct sign to use with each quantity, based on an analysis of the problem and the sign conventions in Table 23.1.

5. At an intersection of hospital hallways, a convex mirror is mounted high on a wall to help people avoid collisions. The mirror has a radius of curvature of 0.550 m. Locate and describe the image of a patient 10.0 m from the mirror. Determine the magnification.

6. A spherical Christmas tree ornament is 6.00 cm in diameter. What is the magnification of an object placed 10.0 cm away from the ornament?

7. A concave spherical mirror has a radius of curvature of 20.0 cm. Locate the images for object distances of (a) 40.0 cm, (b) 20.0 cm, and (c) 10.0 cm. In each case, state whether the image is real or virtual and upright or inverted, and find the magnification.

8. A dentist uses a mirror to examine a tooth. The tooth is 1.00 cm in front of the mirror, and the image is formed 10.0 cm behind the mirror. Determine (a) the mirror's radius of curvature and (b) the magnification of the image.

9. A large church has a niche in one wall. On the floor plan it appears as a semicircular indentation of radius 2.50 m. A worshiper stands on the center line of the niche, 2.00 m out from its deepest point, and whispers a prayer. Where is the sound concentrated after reflection from the back wall of the niche?

10. A convex mirror has a focal length of 20.0 cm. Determine the object location for which the image will be one-half as tall as the object.

11. A 2.00-cm-high object is placed 3.00 cm in front of a concave mirror. If the image is 5.00 cm high and virtual, what is the focal length of the mirror?

12. A 2.00-cm-high object is placed 10.0 cm in front of a mirror. What type of mirror and what radius of curvature are needed to create an upright image that is 4.00 cm high?

13. A concave makeup mirror is designed so that a person 25 cm in front of it sees an upright image magnified by a factor of 2. What is the radius of curvature of the mirror?

14. A concave mirror has a focal length of 40.0 cm. Determine the object position for which the resulting image is upright and four times the size of the object.

15. A man standing 1.52 m in front of a shaving mirror produces an inverted image 18.0 cm in front of it. How close to the mirror should he stand if he wants to form an upright image of his chin that is twice the chin's actual size?

16. A convex spherical mirror with a radius of curvature of 10.0 cm produces a virtual image one-third the size of the real object. Where is the object?

17. A child holds a candy bar 10.0 cm in front of a convex mirror and notices that the image is only one-half the size of the candy bar. What is the radius of curvature of the mirror?

18. It is observed that the size of a *real* image formed by a concave mirror is four times the size of the object when the object is 30.0 cm in front of the mirror. What is the radius of curvature of this mirror?

19. A spherical mirror is to be used to form an image, five times as tall as an object, on a screen positioned 5.0 m from the mirror. (a) Describe the type of mirror required. (b) Where should the mirror be positioned relative to the object?

20. A ball is dropped from rest 3.00 m directly above the vertex of a concave mirror having a radius of 1.00 m and lying in a horizontal plane. (a) Describe the motion of the ball's image in the mirror. (b) At what time do the ball and its image coincide?

Section 23.4 Images Formed by Refraction

21. A cubical block of ice 50.0 cm on an edge is placed on a level floor over a speck of dust. Locate the image of the speck, when viewed from directly above, if the index of refraction of ice is 1.309.

22. The top of a swimming pool is at ground level. If the pool is 2 m deep, how far below ground level does the bottom of the pool appear to be located when (a) the pool is completely filled with water? (b) the pool is filled halfway with water?

23. A paperweight is made of a solid glass hemisphere of index of refraction 1.50. The radius of the circular cross section is 4.0 cm. The hemisphere is placed on its flat surface with the center directly over a 2.5-mm-long line drawn on a sheet of paper. What length of line is seen by someone looking vertically down on the hemisphere?

24. A flint glass plate ($n = 1.66$) rests on the bottom of an aquarium tank. The plate is 8.00 cm thick (vertical dimension) and covered with water ($n = 1.33$) to a depth of 12.0 cm. Calculate the apparent thickness of the plate as viewed from above the water. (Assume nearly normal incidence.)

25. One end of a long glass rod ($n = 1.50$) is formed into the shape of a convex surface of radius 8.00 cm. An object is positioned in air along the axis of the rod. Find the image position that corresponds to each of the following object positions: (a) 20.0 cm, (b) 8.00 cm, (c) 4.00 cm, (d) 2.00 cm.

26. A goldfish is swimming at 2.00 cm/s toward the front wall of a rectangular aquarium. What is the apparent speed of the fish measured by an observer looking in from outside the front wall of the tank? The index of refraction of water is 1.333.

Section 23.6 Thin Lenses

27. A contact lens is made of plastic with an index of refraction of 1.50. The lens has an outer radius of curvature of + 2.00 cm and an inner radius of curvature of + 2.50 cm. What is the focal length of the lens?

28. The left face of a biconvex lens has a radius of curvature of 12.0 cm, and the right face has a radius of curvature of 18.0 cm. The index of refraction of the glass is 1.44. (a) Calculate the focal length of the lens. (b) Calculate the focal length if the radii of curvature of the two faces are interchanged.

29. A converging lens has a focal length of 20.0 cm. Locate the images for object distances of (a) 40.0 cm, (b) 20.0 cm, and (c) 10.0 cm. For each case, state whether the image is real or virtual and upright or inverted, and find the magnification.

30. Where must an object be placed to have unit magnification ($|M| = 1.00$) (a) for a converging lens of focal length 12.0 cm? (b) for a diverging lens of focal length 12.0 cm?

31. A diverging lens has a focal length of 20.0 cm. Locate the images for object distances of (a) 40.0 cm, (b) 20.0 cm, and (c) 10.0 cm. For each case, state whether the image is real or virtual and upright or inverted, and find the magnification.

32. A convex lens of focal length 15.0 cm is used as a magnifying glass. At what distance from a postage stamp should you hold this lens to get a magnification of + 2.00?

33. A transparent photographic slide is placed in front of a converging lens with a focal length of 2.44 cm. The lens forms an image of the slide 12.9 cm from the slide. How far is the lens from the slide if the image is (a) real? (b) virtual?

34. The nickel's image in Figure P23.34 has twice the diameter of the nickel when the lens is 2.84 cm from the nickel. Determine the focal length of the lens.

FIGURE P23.34

35. A certain LCD projector contains a single thin lens. An object 24.0 mm high is to be projected so that its image fills a screen 1.80 m high. The object-to-screen distance is 3.00 m. (a) Determine the focal length of the projection lens. (b) How far from the object should the lens of the projector be placed in order to form the image on the screen?

36. An object's distance from a converging lens is ten times the focal length. How far is the image from the focal point? Express the answer as a fraction of the focal length.

37. A diverging lens is to be used to produce a virtual image one-third as tall as the object. Where should the object be placed?

web

38. An object is 5.00 m to the left of a flat screen. A converging lens for which the focal length is $f = 0.800$ m is placed between object and screen. (a) Show that there are two lens positions that form an image on the screen, and determine how far these positions are from the object. (b) How do the two images differ from each other?

39. A converging lens is placed 30.0 cm to the right of a diverging lens of focal length 10.0 cm. A beam of parallel light enters the diverging lens from the left, and the beam is again parallel when it emerges from the converging lens. Calculate the focal length of the converging lens.

40. An object is placed 20.0 cm to the left of a converging lens of focal length 25.0 cm. A diverging lens of focal length 10.0 cm is 25.0 cm to the right of the converging lens. Find the position and magnification of the final image.

41. Two converging lenses, each of focal length 15.0 cm, are placed 40.0 cm apart, and an object is placed 30.0 cm in front of the first. Where is the final image formed, and what is the magnification of the system?

42. Object O_1 is 15.0 cm to the left of a converging lens of 10.0-cm focal length. A second lens is positioned 10.0 cm to the right of the first lens and is observed to form a final image at the position of the original object O_1. (a) What is the focal length of the second lens? (b) What is the overall magnification of this system? (c) What is the nature (i.e., real or virtual, upright or inverted) of the final image?

43. A 1.00-cm-high object is placed 4.00 cm to the left of a converging lens of focal length 8.00 cm. A diverging lens of focal length −16.00 cm is 6.00 cm to the right of the converging lens. Find the position and height of the final image. Is the image inverted or upright? Real or virtual?

44. Two converging lenses having focal lengths of 10.0 cm and 20.0 cm are placed 50.0 cm apart, as shown in Figure P23.44. The final image is to be located between the lenses, at the position indicated. (a) How far to the left of the first lens should the object be positioned? (b) What is the overall magnification? (c) Is the final image upright or inverted?

FIGURE P23.44

45. Lens L_1 in Figure P23.45 has a focal length of 15.0 cm and is located a fixed distance in front of the film plane of a camera. Lens L_2 has a focal length of 13.0 cm, and its distance d from the film plane can be varied from 5.00 cm to 10.0 cm. Determine the range of distances for which objects can be focused on the film.

FIGURE P23.45

46. Consider two thin lenses, one of focal length f_1 and the other of focal length f_2, placed in contact with each other as shown in Figure P23.46. Apply the thin-lens equation to each of these lenses and combine these results to show that this combination of lenses behaves like a thin lens having a focal length f given by $1/f = 1/f_1 + 1/f_2$. Assume that the thicknesses of the lenses can be ignored in comparison to the other distances involved.

FIGURE P23.46

ADDITIONAL PROBLEMS

47. An object placed 10.0 cm from a concave spherical mirror produces a real image 8.00 cm from the mirror. If the object is moved to a new position 20.0 cm from the mirror, what is the position of the image? Is the final image real or virtual?

48. An object is placed 12 cm to the left of a diverging lens of focal length -6.0 cm. A converging lens of focal length 12 cm is placed a distance of d to the right of the diverging lens. Find the distance d that places the final image at infinity.

49. The distance between an object and its upright image is 20.0 cm. If the magnification is 0.500, what is the focal length of the lens being used to form the image?

50. The object in Figure P23.50 is midway between the lens and the mirror. The mirror's radius of curvature is 20.0 cm, and the lens has a focal length of -16.7 cm. Considering only the light that leaves the object and travels first toward the mirror, locate the final image formed by this system. Is this image real or virtual? Is it upright or inverted? What is the overall magnification?

FIGURE P23.50

51. The lens and mirror in Figure P23.51 are separated by
web 1.00 m and have focal lengths of $+80.0$ cm and -50.0 cm, respectively. If an object is placed 1.00 m to the left of the lens, locate the final image. State whether the image is upright or inverted, and determine the overall magnification.

52. A diverging lens ($n = 1.50$) is shaped like that in Figure 23.25c. The radius of the first surface is 15.0 cm and that of the second surface is 10.0 cm. (a) Find the focal length of the lens. Determine the positions of the images for object distances of (b) infinity, (c) $3|f|$, (d) $|f|$, and (e) $|f|/2$.

FIGURE P23.51

53. A parallel beam of light enters a glass hemisphere perpendicular to the flat face, as shown in Figure P23.53. The radius is $R = 6.00$ cm, and the index of refraction is $n = 1.560$. Determine the point at which the beam is focused. (Assume paraxial rays; that is, all rays are located close to the principal axis.)

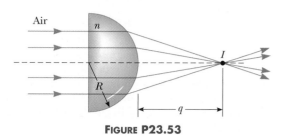

FIGURE P23.53

54. A converging lens of focal length 20.0 cm is separated by 50.0 cm from a converging lens of focal length 5.00 cm. (a) Find the position of the final image of an object placed 40.0 cm in front of the first lens. (b) If the height of the object is 2.00 cm, what is the height of the final image? Is the image real or virtual? (c) If the two lenses are now placed in contact with each other and the object is 5.00 cm in front of this combination, where will the image be located? (See Problem 46.)

55. To work this problem, use the fact that the image formed by the first surface becomes the object for the second surface. Figure P23.55 shows a piece of glass with index of refraction of 1.50. The ends are hemispheres with radii 2.00 cm and 4.00 cm, and the centers of the hemispherical ends are separated by a distance of 8.00 cm. A point object is in air, 1.00 cm from the left end of the glass. Locate the image of the object due to refraction at the two spherical surfaces.

FIGURE P23.55

56. In a darkened room, a burning candle is placed 1.50 m from a white wall. A lens is placed between candle and wall at a location that causes a larger, inverted image to form on the wall. When the lens is moved 90.0 cm toward the wall, another image of the candle is formed. Find (a) the two object distances that produce the images and (b) the focal length of the lens. (c) Characterize the second image.

57. An object 2.00 cm high is placed 40.0 cm to the left of a converging lens having a focal length of 30.0 cm. A diverging lens having a focal length of -20.0 cm is placed 110 cm to the right of the converging lens. (a) Determine the final position and magnification of the final image. (b) Is the image upright or inverted? (c) Repeat parts (a) and (b) for the case where the second lens is a converging lens having a focal length of $+20.0$ cm.

58. A "floating strawberry" illusion consists of two parabolic mirrors, each with a focal length of 7.5 cm, facing each other so that their centers are 7.5 cm apart (Fig. P23.58). If a strawberry is placed on the lower mirror, an image of the strawberry forms at the small opening at the center of the top mirror. Show that the final image forms at that location, and describe its characteristics. (*Note:* A flashlight beam shone on these *images* has a very startling effect. Even at a glancing angle, the incoming light beam is seemingly reflected off the *images* of the strawberry! Do you understand why?)

FIGURE P23.58　(*Photo © Michael Levin/Opti-Gone Associates*)

59. Figure P23.59 shows a converging lens with radii $R_1 = 9.00$ cm and $R_2 = -11.00$ cm in front of a concave spherical mirror of radius $R = 8.00$ cm. The focal points (F_1 and F_2) for the thin lens and the center of curvature (C) of the mirror are also shown. (a) If the focal points F_1 and F_2 are 5.00 cm from the vertex of the thin lens, determine the index of refraction for the lens. (b) If the lens and mirror are 20.0 cm apart, and an object is placed 8.00 cm to the left of the lens, determine the position of the final image and its magnification as seen by the eye in the figure. (c) Is the final image inverted or upright? Explain.

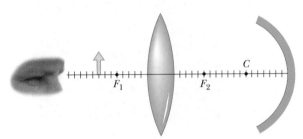

FIGURE P23.59

60. Find the object distances (in terms of f) for a thin converging lens of focal length f if (a) the image is real and the image distance is four times the focal length; (b) the image is virtual and the image distance is three times the focal length. (c) Calculate the magnification of the lens for case (a) and (b).

61. The lens maker's equation for a lens with index n_1 immersed in a medium with index n_2 takes the form

$$\frac{1}{f} = \left(\frac{n_1}{n_2} - 1\right)\left(\frac{1}{R_1} - \frac{1}{R_2}\right)$$

A thin diverging glass lens (index = 1.50) with $R_1 = -3.00$ m and $R_2 = -6.00$ m is surrounded by air. An arrow is placed 10.0 m to the left of the lens. (a) Determine the position of the image. Repeat part (a) with the arrow and lens immersed in (b) water (index = 1.33); (c) a medium with an index of refraction of 2.00. (d) How can a lens that is diverging in air be changed into a converging lens?

GROUP ACTIVITIES

G.1 This experiment will enable you to examine some of the properties of images formed by flat mirrors. As shown in Figure GA23.1, place a clear plastic sheet between two small candles of the same height as in the figure. Light one candle placed about 6 inches from the sheet and observe the reflected flame from the front side. Move the unlit candle until it also appears to be lit. The candles should be approximately equidistant from the sheet at this point. Why? When you place your finger on the wick of the unlit candle, you will get the illusion that your finger is burning. Explain your observation.

   A similar experiment can be performed with a flat mirror and two pencils. Hold one of the pencils in front of the mirror at a distance of about 10 inches from the

mirror and at a position such that about half of the image appears in the mirror. Move the second pencil back and forth behind the mirror until it appears to align perfectly with the image in the mirror. When alignment is achieved, measure the distance from the mirror to the location of the second pencil. This pencil is located at the apparent position of the virtual image. Why? The two pencils should be at equal distances from the mirror. Why?

G.2 Fill a clear glass tumbler with water and place a pencil or straw into the tumbler. Now observe the pencil from the side at an angle of about 45° to the surface, and note that the line of the portion of the pencil under water is not parallel with the line of the portion in air. That is, the

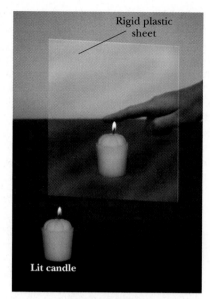

FIGURE GA23.1

pencil appears to be bent at the point where it enters the water. Use the techniques of Section 23.4 to explain this observation.

FIGURE GA23.2

G.3 View yourself in a full-length mirror. Standing close to the mirror, place one piece of tape at the top of the image of your head and another piece of tape at the very bottom of the image of your feet. Now step back a few meters and observe your image. How big is it relative to the original size? How does the distance between the pieces of tape compare with your actual height?

Move to a position in front of the mirror such that you can see a full image of yourself with the top of your head just level with the top of the mirror. Have your co-worker gradually block off the lower portion of the mirror with a sheet or newspaper page until you can see your complete image but no more. Measure the length of the mirror and compare this measurement to your height. How do the two compare?

G.4 Compare the images formed of your face when you look first at the front side and then at the back side of a shiny soupspoon. For each side observe the change in the image of your face as you move closer to and farther away from the spoon.

G.5 An object is placed 25 cm in front of a concave makeup mirror that has a focal length of 40 cm. (a) The sign of the focal length is _____ (plus or minus). (b) The sign of the object distance is _____. (c) The sign of the image distance will be _____. (d) Use the mirror equation to find the image distance. (e) Repeat for the case in which the object is 5.0 m away.

G.6 Describe in words what happens to the following rays of light when they strike a converging lens: (a) a ray parallel to the optic axis, (b) a ray passing through the focal point before striking the lens, (c) a ray passing through the center of curvature of the lens. (d) Repeat for the case of a diverging lens.

G.7 An object is placed 25 cm in front of a converging lens that has a focal length of 10 cm. (a) The sign of the focal length to be used in the lens equation is _____. (b) The sign of the object distance is _____. (c) The sign of the image distance will be _____. (d) Verify your answer to (c) by a calculation. (d) Repeat for the case when the lens is diverging with a focal length of 15 cm.

G.8 Draw a sketch of you and your image in a plane mirror when you are walking away from it with a velocity **v**. Indicate on your diagram the velocity vectors for you and your image. Keep the relative lengths of the two velocities as you believe they will be. Explain your answer. (b) Repeat for when you are walking toward the mirror. Explain why you have the vectors drawn as you do.

Wave Optics

Peacock feathers appear to be brilliant in color, especially in the blues and greens. However, these colors are not caused by pigments in the feathers. This chapter describes diffraction, the optical process responsible for this bird's magnificent display. *(Terry Qing/FPG International, Inc.)*

T hus far our discussion of light has been concerned with what happens when light passes through a lens or reflects from a mirror. Because explanations of such phenomena rely on a geometric analysis of light rays, that part of optics is often called geometric optics. We now expand our study of light into an area called *wave optics*. The three primary topics we examine in this chapter are interference, diffraction, and polarization. These phenomena cannot be adequately explained with ray optics, but the wave theory leads us to satisfying descriptions.

24.1   CONDITIONS FOR INTERFERENCE

When you blew bubbles as a child, did you notice the brilliant colors playing on the surface of the bubbles? Have you ever wondered what causes the vivid rainbow of colors reflected from the oil films floating in parking lot puddles on rainy days? Or, have you noticed that good-quality lenses used in cameras have a violet hue? All of these effects are caused by interference in thin films, a subject we shall discuss in this chapter. However, before we can examine interference, we must pause to consider the conditions necessary for it to occur.

In our discussion of interference of mechanical waves in Chapter 13, we found that two waves can add together either constructively or destructively. In constructive interference, the amplitude of the resultant wave is greater than that of either of the individual waves, whereas in destructive interference, the resultant amplitude is less than that of either individual wave. Light waves also interfere with each other. Fundamentally, all interference associated with light waves arises when the electromagnetic fields that constitute the individual waves combine.

Interference effects in light waves are not easy to observe because of the short wavelengths involved (about 4×10^{-7} m to about 7×10^{-7} m). For sustained interference between two sources of light to be observed, the following conditions must be met:

Conditions for interference

1. The sources must be **coherent**—that is, they must maintain a constant phase with respect to each other.
2. The waves must have identical wavelengths.

Let us examine the characteristics of coherent sources. Two sources (producing two traveling waves) are needed to create interference. To produce a stable interference pattern, the individual waves must maintain a constant phase with one another. When this situation prevails, the sources are said to be coherent. The sound waves emitted by two side-by-side loudspeakers driven by a single amplifier can produce interference because the two speakers respond to the amplifier in the same way at the same time—that is, they are in phase.

If two light sources are placed side by side, however, no interference effects are observed, because the light waves from one source are emitted independently of the waves from the other source. Hence, the emissions from the two sources do not maintain a constant phase relationship with each other during the time of observation. An ordinary light source undergoes random changes about once every 10^{-8} s. Therefore, the conditions for constructive interference, destructive interference, and intermediate states have durations on the order of 10^{-8} s. The result is that no interference effects are observed, because the eye cannot follow such short-term changes. Such light sources are said to be **incoherent.**

An older method for producing two coherent light sources is to pass light from a single wavelength (monochromatic) source through a narrow slit and then allow this light to fall on a screen containing two narrow slits. The first slit is needed to insure that light comes from a tiny region of the light source in which excited atoms emit coherently. The light emerging from the two slits is coherent because a single source produces the original light beam and the slits serve only to separate the original beam into two parts (which is exactly what was done to the sound signal just mentioned). Any random change in the light emitted by the source will occur in the two separate beams at the same time, and interference effects can be observed.

Currently it is much more common to use a laser as a coherent source to demonstrate interference. A laser produces an intense, coherent, monochromatic beam over a width of several millimeters. This means that the laser can be used to illuminate multiple slits directly, and that interference effects can be observed easily in a fully lighted room. The principles of operation of lasers are explained in Chapter 28.

24.2 YOUNG'S DOUBLE-SLIT INTERFERENCE

Thomas Young first demonstrated interference in light waves from two sources in 1801. Figure 24.1a is a diagram of the apparatus used in this experiment. (Young used pinholes rather than slits in his original experiments.) Light is incident on a screen in which there is a narrow slit S_0. The light waves emerging from this slit arrive at a second screen that contains two narrow, parallel slits S_1 and S_2. These slits serve as a pair of coherent light sources because waves emerging from them originate from the same wave front and therefore are always in phase. The light from the two slits produces a visible pattern on screen C consisting of a series of bright and dark parallel bands called **fringes** (Fig. 24.1b). When the light from slits S_1 and S_2 arrives at a point on the screen so that constructive interference occurs at that location, a bright fringe appears. When the light from the two slits combines destructively at any location on the screen, a dark fringe results.

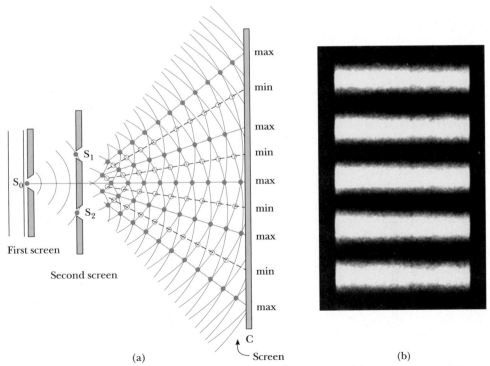

FIGURE 24.1 (a) A diagram of Young's double-slit experiment. The narrow slits act as sources of waves. Slits S_1 and S_2 behave as coherent sources that produce an interference pattern on screen C. (This drawing is not to scale.) (b) The fringe pattern formed on screen C could look like this.

Figure 24.2 is a photograph of an interference pattern produced by two coherent vibrating sources in a water tank.

Figure 24.3 is a diagram of some of the ways in which the two waves can combine at screen C. In Figure 24.3a, the two waves, which leave the two slits in phase, strike the screen at the central point P. Because these waves travel equal distances, they arrive in phase at P, and as a result constructive interference occurs there and a bright fringe is observed. In Figure 24.3b, the two light waves again start in phase, but the upper wave has to travel one wavelength farther to reach point Q on the screen. Because the upper wave falls behind the lower one by exactly one wavelength, the two waves still arrive in phase at Q, and so a second bright fringe appears at that location. Now consider point R, midway between P and Q in Figure 24.3c. At that point, the upper wave has fallen half a wavelength behind the lower wave. This means that the trough of the bottom wave overlaps the crest of the upper wave, giving rise to destructive interference at R. As a consequence, a dark fringe can be observed at R.

We can describe Young's experiment quantitatively with the help of Figure 24.4. Consider point P on the viewing screen; the screen is positioned a perpendicular distance L from the screen containing slits S_1 and S_2, which are separated by distance d, and r_1 and r_2 are the distances the secondary waves travel from slit to screen. Let us assume the waves emerging from S_1 and S_2 have the same constant frequency, the same amplitude, and start out in phase. The light intensity on the screen at P is the resultant of the light from both slits. Note that a wave from the lower slit travels farther than a wave from the upper slit by the amount $d \sin \theta$. This distance is called the **path difference** δ (lowercase Greek delta), where

▶ Path difference

$$\delta = r_2 - r_1 = d \sin \theta \qquad [24.1]$$

This equation assumes that the two waves travel in parallel lines, which is approximately true, because L is much greater than d. As noted earlier, the value of this path difference determines whether the two waves are in phase when they arrive at P. If the path difference is either zero or some integral multiple of the wavelength, the two waves are in phase at P and constructive interference results. Therefore, the condition for bright fringes, or **constructive interference,** at P is

▶ Condition for constructive interference (two slits)

$$\delta = d \sin \theta_{\text{bright}} = m\lambda \qquad m = 0, \pm 1, \pm 2, \ldots \qquad [24.2]$$

The number m is called the **order number.** The central bright fringe at $\theta_{\text{bright}} = 0$ ($m = 0$) is called the *zeroth-order maximum.* The first maximum on either side, where $m = \pm 1$, is called the *first-order maximum,* and so forth.

When δ is an odd multiple of $\lambda/2$, the two waves arriving at P are 180° out of phase and give rise to **destructive interference.** Therefore, the condition for

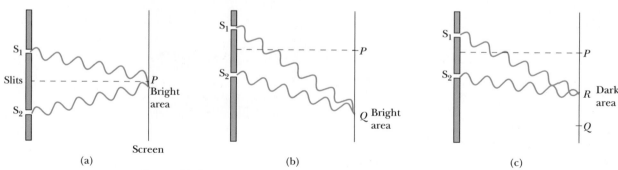

FIGURE 24.3 (a) Constructive interference occurs at P when the waves combine. (b) Constructive interference also occurs at Q. (c) Destructive interference occurs at R when the wave from the upper slit falls half a wavelength behind the wave from the lower slit. (These figures are not drawn to scale.)

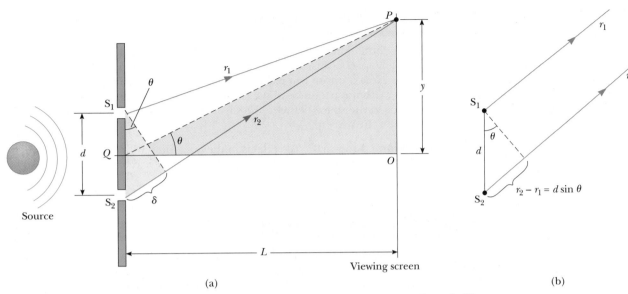

FIGURE 24.4 A geometric construction that describes Young's double-slit experiment. The path difference between the two rays is $r_2 - r_1 = d \sin \theta$. (This figure is not drawn to scale.)

dark fringes at P is

$$\delta = d \sin \theta_{dark} = (m + \tfrac{1}{2})\lambda \qquad m = 0, \pm 1, \pm 2, \ldots \qquad \text{[24.3]}$$

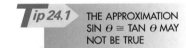

◄ Condition for destructive interference (two slits)

If $m = 0$ in this equation, the path difference is $\delta = \lambda/2$, which is the condition for the location of the first dark fringe on either side of the central (bright) maximum. Likewise, if $m = 1$, $\delta = 3\lambda/2$, which is the condition for the second dark fringe on each side, and so forth.

It is useful to obtain expressions for the positions of the bright and dark fringes measured vertically from O to P. In addition to our assumption that $L \gg d$, we assume that $d \gg \lambda$. These can be valid assumptions because in practice, L is often of the order of 1 m, d a fraction of a millimeter, and λ a fraction of a micrometer for visible light. Under these conditions, θ is small; we can thus use the approximation $\sin \theta \cong \tan \theta$. Then, from triangle OPQ in Figure 24.4, we see that

$$y = L \tan \theta \approx L \sin \theta \qquad \text{[24.4]}$$

Solving Equation 24.2 for $\sin \theta$ and substituting the result into Equation 24.4, we see that the positions of the *bright fringes*, measured from O are

$$y_{bright} = \frac{\lambda L}{d} m \qquad m = 0, \pm 1, \pm 2, \ldots \qquad \text{[24.5]}$$

Using Equations 24.3 and 24.4, we find that the *dark fringes* are located at

$$y_{dark} = \frac{\lambda L}{d}(m + \tfrac{1}{2}) \qquad m = 0, \pm 1, \pm 2, \ldots \qquad \text{[24.6]}$$

As we shall demonstrate in Example 24.1, Young's double-slit experiment provides a method for measuring the wavelength of light. In fact, Young used this technique to do just that. In addition, his experiment gave the wave model of light a great deal of credibility. It was inconceivable that particles of light coming through the slits could cancel each other in a way that would explain the dark fringes.

Tip 24.1 THE APPROXIMATION SIN $\theta \cong$ TAN θ MAY NOT BE TRUE

The small-angle approximation $\sin \theta \cong \tan \theta$ is true to three-digit precision only for angles less than about 4°.

As the waves pass through a narrow gap, they spread out (diffract), and the interference of two waveforms is manifested in cross-patterned areas. *(John S. Shelton)*

Webnote 24.1

See the Young's two-slit diffraction pattern by visiting *http://vsg.tripod.com/interfer.htm*

APPLICATION

TELEVISION SIGNAL INTERFERENCE

APPLYING **PHYSICS** 24.1

Consider a double-slit experiment in which a laser beam is passed through a pair of very closely spaced slits, and a clear interference pattern is displayed on a distant screen. Now suppose you place smoke particles between the double slit and the screen. With the presence of the smoke particles, will you see the effects of interference in the space between the slits and the screen, or will you only see the effects on the screen?

Explanation   You will see the effects in the area filled with smoke. There will be bright lines directed toward the bright areas on the screen and dark lines directed toward the dark areas on the screen.

APPLYING **PHYSICS** 24.2

Suppose you are watching television by means of an antenna rather than a cable system. If an airplane flies near your location, you will sometimes notice wavering ghost images in the television picture. What might cause this?

Explanation   Your television antenna receives two signals—the direct signal from the transmitting antenna and a signal reflected from the surface of the airplane. As the airplane changes position, there are times when these two signals are in phase and other times when they are out of phase. As a result, the intensity of the combined signal received at your antenna varies. The wavering of the ghost images of the picture is evidence for this variation.

Example 24.1   **Measuring the Wavelength of a Light Source**

A screen is separated from a double-slit source by 1.2 m. The distance between the two slits is 0.030 mm. The second-order bright fringe ($m = 2$) is measured to be 4.5 cm from the centerline. Determine (a) the wavelength of the light and (b) the distance between adjacent bright fringes.

Reasoning and Solution

A   We can use Equation 24.5 with $m = 2$, $y_2 = 4.5 \times 10^{-2}$ m, $L = 1.2$ m, and $d = 3.0 \times 10^{-5}$ m:

$$\lambda = \frac{y_2 d}{mL} = \frac{(4.5 \times 10^{-2}\ \text{m})(3.0 \times 10^{-5}\ \text{m})}{2(1.2\ \text{m})} = 5.6 \times 10^{-7}\ \text{m} = \boxed{560\ \text{nm}}$$

B   Because the positions of the bright fringes are given by Equation 24.5, we see that the distance between *any* adjacent bright fringes (say, those characterized by m and $m + 1$) is

$$\Delta y = y_{m+1} - y_m = \frac{\lambda L}{d}(m + 1) - \frac{\lambda L}{d}m = \frac{\lambda L}{d}$$

$$= \frac{(5.6 \times 10^{-7}\ \text{m})(1.2\ \text{m})}{3.0 \times 10^{-5}\ \text{m}} = \boxed{2.2\ \text{cm}}$$

FIGURE 24.5   Lloyd's mirror. An interference pattern is produced on a screen at *P* as a result of the combination of the direct ray (blue) and the reflected ray (brown). The reflected ray undergoes a phase change of 180°.

24.3   CHANGE OF PHASE DUE TO REFLECTION

Young's method of producing two coherent light sources involves illuminating a pair of slits with a single source. Another simple, although ingenious, arrangement for producing an interference pattern with a single light source is known as *Lloyd's mirror*. A light source is placed at point *S*, close to a mirror, as illustrated in Figure 24.5. Waves can reach the viewing point *P* either by the direct path *SP* or by the

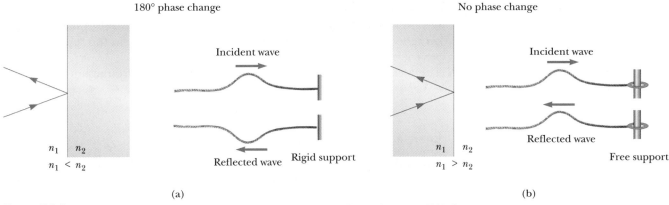

FIGURE 24.6 (a) A ray reflecting from a medium of higher refractive index undergoes a 180° phase change. The right side shows the analogy with a reflected pulse on a string. (b) A ray reflecting from a medium of lower refractive index undergoes no phase change.

path involving reflection from the mirror. The reflected ray can be treated as a ray originating at the source S' behind the mirror. Source S', which is the image of S, can be considered a virtual source.

At points far from the source, we would expect an interference pattern due to waves from S and S', just as is observed for two real coherent sources. An interference pattern is indeed observed. However, the positions of the dark and bright fringes are *reversed* relative to the pattern of two real coherent sources (Young's experiment). This is because the coherent sources S and S' differ in phase by 180°. This 180° phase change is produced by reflection.

To illustrate this further, consider point P', at which the mirror meets the screen. This point is equidistant from S and S'. If path difference alone were responsible for the phase difference, we would expect to see a bright fringe at P' (because the path difference is zero for this point), corresponding to the central fringe of the two-slit interference pattern. Instead, we observe a *dark* fringe at P' because of the 180° phase change produced by reflection. In general, an electromagnetic wave undergoes a phase change of 180° upon reflection from a medium of higher index of refraction than the one in which it was traveling.

It is useful to draw an analogy between reflected light waves and the reflections of a transverse wave on a stretched string when the wave meets a boundary, as in Figure 24.6. The reflected pulse on a string undergoes a phase change of 180° when it is reflected from the boundary of a denser string, or from a rigid barrier, and no phase change when it is reflected from the boundary of a less dense string. Similarly, an electromagnetic wave undergoes a 180° phase change when reflected from the boundary of a medium of higher index of refraction than the one in which it has been traveling. There is no phase change when the wave is reflected from a boundary leading to a medium of lower index of refraction. The transmitted wave that crosses the boundary also undergoes no phase change.

24.4 INTERFERENCE IN THIN FILMS

Interference effects are commonly observed in thin films, such as soap bubbles and thin layers of oil on water. The varied colors observed with incoherent white light result from the interference of waves reflected from the opposite surfaces of the film.

Consider a film of uniform thickness t and index of refraction n, as in Figure 24.7. Let us assume that the light rays traveling in air are nearly normal to the two surfaces of the film. To determine whether the reflected rays interfere constructively or destructively, we must first note the following facts:

FIGURE 24.7 Interference observed in light reflected from a thin film is due to a combination of rays reflected from the upper and lower surfaces.

1. An electromagnetic wave traveling from a medium of index of refraction n_1 toward a medium of index of refraction n_2 undergoes a 180° phase change on reflection when $n_2 > n_1$. There is no phase change in the reflected wave if $n_2 < n_1$.

2. The wavelength of light λ_n in a medium with index of refraction n is

$$\lambda_n = \frac{\lambda}{n} \qquad \text{[24.7]}$$

where λ is the wavelength of light in vacuum.

We apply these rules to the film of Figure 24.7. According to the first rule, ray 1, which is reflected from the upper surface A, undergoes a phase change of 180° with respect to the incident wave. Ray 2, which is reflected from the lower surface B, undergoes no phase change with respect to the incident wave. Therefore, ray 1 is 180° out of phase with respect to ray 2, a situation that is equivalent to a path difference of $\lambda_n/2$. However, we must also consider the fact that ray 2 travels an extra distance of $2t$ before the waves recombine. For example, if $2t = \lambda_n/2$, rays 1 and 2 recombine in phase and constructive interference results. In general, the condition for *constructive interference* is

$$2t = (m + \tfrac{1}{2})\lambda_n \qquad m = 0, 1, 2, \ldots \qquad \text{[24.8]}$$

This condition takes into account two factors: (a) the difference in optical path length for the two rays (the term $m\lambda_n$) and (b) the 180° phase change upon reflec-

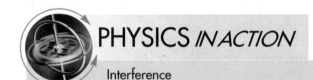

PHYSICS *IN ACTION*

Interference

The image on the left shows a layer of soap-film bubbles on water. The colors, produced just before the bubbles burst, are due to interference between light rays reflected from the front and back of the thin film of soap making up the bubble. The color depends on the thickness of the film, ranging from black where the film is at its thinnest to magenta where it is thickest.

On the right, a thin film of oil on water displays interference, evidenced by the pattern of colors when white light is incident on the film. Variations in the film's thickness in the vicinity of the blade produces the intersecting color pattern.

(Dr. Jeremy Burgess/Science Photo Library)

(Peter Aprahamian/Science Photo Library/Photo Researchers, Inc.)

tion (the term $\lambda_n/2$). Because $\lambda_n = \lambda/n$, we can write Equation 24.8 in the form

$$2nt = (m + \tfrac{1}{2})\lambda \qquad m = 0, 1, 2, \ldots \qquad \text{[24.9]}$$

◀ Condition for constructive interference (thin film)

If the extra distance $2t$ traveled by ray 2 is a multiple of λ_n, the two waves combine out of phase and destructive interference results. The general equation for *destructive interference* is

$$2nt = m\lambda \qquad m = 0, 1, 2, \ldots \qquad \text{[24.10]}$$

◀ Condition for destructive interference (thin film)

It is important to realize that two factors influence interference: (1) possible phase reversals on reflection and (2) differences in travel distance. The foregoing conditions for constructive and destructive interference are valid when the medium above the top surface of the film is the same as the medium below the bottom surface. The surrounding medium may have a refractive index less than or greater than that of the film. In either case, the rays reflected from the two surfaces will be out of phase by 180°. If the film is placed between two *different* media, one of lower refractive index than the film and one of higher refractive index, the conditions for constructive and destructive interference are reversed. In this case, either there is a phase change of 180° for both ray 1 reflecting from surface A and ray 2 reflecting from surface B, or there is no phase change for either ray; hence, the net change in relative phase due to the reflections is *zero*.

Tip 24.2   BE CAREFUL WITH THIN FILMS

Be sure to include *both* effects—path length and phase change—when analyzing an interference pattern from a thin film.

NEWTON'S RINGS

Another method for observing interference of light waves is to place a planoconvex lens on top of a flat glass surface, as in Figure 24.8a. With this arrangement, the air film between the glass surfaces varies in thickness from zero at the point of contact to some value t at P. If the radius of curvature R of the lens is very large compared with the distance r, and if the system is viewed from above using light of wavelength λ, a pattern of light and dark rings is observed (Fig. 24.8b). These circular fringes are called **Newton's rings** after their discoverer. Newton's particle model of light could not explain the origin of the rings.

The interference is due to the combination of ray 1, reflected from the plate, with ray 2, reflected from the lower surface of the lens. Ray 1 undergoes a phase change of 180° on reflection, because it is reflected from a boundary leading into a medium of higher refractive index, whereas ray 2 undergoes no phase change. Hence, the conditions for constructive and destructive interference are given by

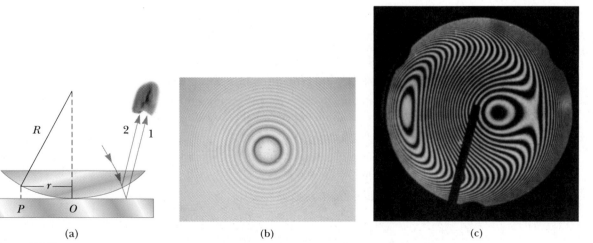

(a)   (b)   (c)

FIGURE 24.8   (a) The combination of rays reflected from the glass plate and the curved surface of the lens gives rise to an interference pattern known as Newton's rings. (b) A photograph of Newton's rings. *(Richard Megna, Fundamental Photographs, NYC)* (c) This asymmetric interference pattern indicates imperfections in the lens. *(From Physical Science Study Committee, College Physics, Lexington, Mass., D. C. Heath and Co., 1968)*

Equations 24.9 and 24.10, respectively, with $n = 1$, because the "film" is air. Here again, we might guess that the contact point O would be bright, corresponding to constructive interference. Instead, it is dark, as seen in Figure 24.8b, because ray 1, reflected from the plate, undergoes a 180° phase change with respect to ray 2. Using the geometry shown in Figure 24.8a, we can obtain expressions for the radii of the bright and dark bands in terms of the radius of curvature R, and vacuum wavelength λ. For example, the dark rings have radii of $r \approx \sqrt{m\lambda R/n}$. In Problem 64 at the end of the chapter, you will be asked to supply the details.

One of the important uses of Newton's rings is in the testing of optical lenses. A circular pattern like that in Figure 24.8b is achieved only when the lens is ground to a perfectly spherical curvature. Variations from such symmetry might produce a pattern like that in Figure 24.8c. These variations give an indication of how the lens must be ground and polished to remove the imperfections.

APPLICATION

CHECKING FOR
IMPERFECTIONS IN OPTICAL
LENSES

PROBLEM-SOLVING STRATEGY | **Thin-Film Interference**

The following features should be kept in mind when you work thin-film interference problems:

1. Identify the thin film causing the interference.
2. The type of interference—constructive or destructive—that occurs is determined by the phase relationship between the portion of the wave reflected at the upper surface of the film and the portion reflected at the lower surface.
3. Phase differences between the two portions of the wave have two causes: (a) differences in the distances traveled by the two portions and (b) phase changes occurring on reflection. *Both* causes must be considered when you are determining whether constructive or destructive interference occurs.
4. The interference is constructive if the path difference between the two waves is an integral multiple of λ, and destructive if the equivalent path difference is $\lambda/2$, $3\lambda/2$, $5\lambda/2$, and so forth. However, the conditions for constructive and destructive interference are reversed if one of the waves undergoes a phase change on reflection.

Example 24.2 Interference in a Soap Film

Calculate the minimum thickness of a soap-bubble film ($n = 1.33$) that will result in constructive interference in the reflected light if the film is illuminated by light with a wavelength in free space of 602 nm.

Reasoning The minimum film thickness for constructive interference corresponds to $m = 0$ in Equation 24.9. This gives $2nt = \lambda/2$.

Solution Because $2nt = \lambda/2$, we have

$$t = \frac{\lambda}{4n} = \frac{602 \text{ nm}}{(4)(1.33)} = \boxed{113 \text{ nm}}$$

EXERCISE What other film thicknesses will produce constructive interference?

ANSWER 338 nm, 564 nm, 789 nm, and so on

Example 24.3 Nonreflective Coatings for Solar Cells

Semiconductors such as silicon are used to fabricate solar cells—devices that generate electric energy when exposed to sunlight. Solar cells are often coated with a transparent thin film, such as silicon monoxide (SiO; $n = 1.45$) to minimize reflective losses (Fig. 24.9). A silicon solar cell ($n = 3.50$) is coated with a thin film of silicon monoxide for this purpose. Assuming normal incidence, determine the minimum thickness of the film that will produce the least reflection at a wavelength of 552 nm.

Reasoning Reflection is least when rays 1 and 2 in Figure 24.9 meet the condition of destructive interference. Note that both rays undergo 180° phase changes on reflection. Hence, the net change in phase due to reflection is zero, and the condition for a reflection *minimum* is a path difference of $\lambda_n/2$; therefore, $2t = \lambda_n/2 = \lambda/2n$.

Solution Because $2t = \lambda/2n$, the required thickness is

$$t = \frac{\lambda}{4n} = \frac{552 \text{ nm}}{(4)(1.45)} = \boxed{95.2 \text{ nm}}$$

Typically, such coatings reduce the reflective loss from 30% (with no coating) to 10% (with coating), thereby increasing the cell's efficiency, because more light is available to create charge carriers in the cell. In reality, the coating is never perfectly nonreflecting, because the required thickness is wavelength-dependent and the incident light covers a wide range of wavelengths.

Glass lenses used in cameras and other optical instruments are usually coated with a transparent thin film, such as magnesium fluoride (MgF_2), to reduce or eliminate unwanted reflection. More important, such coatings enhance the transmission of light through the lenses.

FIGURE 24.9 (Example 24.3) Reflective losses from a silicon solar cell are minimized by coating it with a thin film of silicon monoxide, SiO.

Example 24.4 Interference in a Wedge-Shaped Film

A thin, wedge-shaped film of refractive index n is illuminated with monochromatic light of wavelength λ, as illustrated in Figure 24.10a. Describe the interference pattern observed in this case.

Reasoning and Solution The interference pattern is that of a thin film of variable thickness consisting of air. Hence, the pattern is a series of alternating bright and dark

FIGURE 24.10 (Example 24.4) (a) Interference bands in reflected light can be observed by illuminating a wedge-shaped film with monochromatic light. The dark areas correspond to positions of destructive interference. (b) Interference in a vertical film of variable thickness. The top of the film appears darkest where the film is thinnest. *(Richard Megna, Fundamental Photographs)*

(a)   (b)

parallel bands. A dark band corresponding to destructive interference appears at point O, the apex, because the upper reflected ray undergoes a $180°$ phase change and the lower one does not. According to Equation 24.10, other dark bands appear when $2nt = m\lambda$, so that $t_1 = \lambda/2n$, $t_2 = \lambda/n$, $t_3 = 3\lambda/2n$, and so on. Similarly, bright bands are observed when the thickness satisfies the condition $2nt = (m + \frac{1}{2})\lambda$, corresponding to thicknesses of $\lambda/4n$, $3\lambda/4n$, $5\lambda/4n$, and so on. If white light is used, bands of different colors are observed at different points, corresponding to the different wavelengths of light present. This is shown in the soap film in Figure 24.10b.

24.5 USING INTERFERENCE TO READ CDs AND DVDs

Compact disks (CDs) and digital video disks (DVDs) have revolutionized the computer and entertainment industries by providing fast access; high-density storage of text, graphics, and movies; and high-quality sound recordings. The data on these disks are stored digitally—that is, as a series of zeros and ones—and these zeros and ones are read by laser light reflected from the disk. Strong reflections (constructive interference) from the disk are chosen to represent zeros and weak reflections (destructive interference) represent ones.

To see in more detail how thin-film interference plays a crucial role in reading CDs, consider Figure 24.11. This shows a photomicrograph of several CD tracks, which consist of a sequence of pits (when viewed from the top or label side of the disk) of varying length formed in a reflecting-metal information layer. A cross-sectional view of a CD as shown in Figure 24.12 reveals that the pits appear as bumps to the laser beam, which shines on the metallic layer through a clear plastic coating from below. As the disk rotates, the laser reflects off the sequence of bumps and lower areas into a photodetector, which converts the fluctuating reflected light intensity into an electrical string of zeros and ones. To make the light fluctuations more pronounced and easier to detect, the pit depth t is made equal to one quarter of a wavelength of the laser light in the plastic. When the beam hits a rising or falling bump edge, part of the beam reflects from the top of the bump and part from the lower adjacent area, ensuring destructive interference and very low intensity when the reflected beams combine at the detector. Thus bump edges are read as ones, and flat bump tops and intervening flat plains are read as zeros.

Example 24.5 determines the pit depth for a standard CD using an infrared laser of wavelength 780 nm. DVDs use shorter-wavelength lasers of 635 nm, and the track separation, pit depth, and minimum pit length are all smaller. This allows a DVD to store about 30 times more information than a CD.

FIGURE 24.11 A photomicrograph of adjacent tracks on a compact disc (CD). The information encoded in these pits and smooth areas is read by a laser beam. *(Courtesy of Sony Disc Manufacturing)*

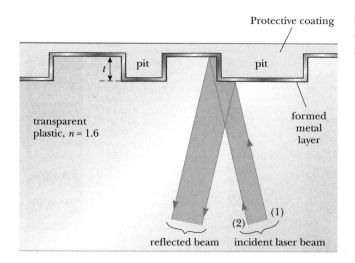

Protective coating

pit   pit

t

transparent
plastic, $n = 1.6$

formed
metal
layer

(1)

(2)

reflected beam   incident laser beam

FIGURE 24.12 Cross section of a CD showing metallic pits of depth t and a laser beam detecting a pit edge.

Example 24.5 Pit Depth in a CD

Find the pit depth in a CD, which has a plastic transparent layer with index of refraction of 1.6 and is designed for use in a CD player, using a laser that has a wavelength of 780 nm in air.

Reasoning and Solution Rays 1 and 2 undergo a 180° phase change on reflection from the metal layer, which acts like a mirror. (See Fig. 24.12.) Thus, there is no phase difference between rays 1 and 2 because of reflection, but there is a phase difference caused by the extra distance $2t$ traveled by ray 2. The wavelength of light in a substance with index of refraction n is λ/n, where λ is the wavelength in air. Hence, for destructive interference of rays 1 and 2, we know that $2t = \lambda/2n$, or

$$t = \frac{\lambda}{4n} = \frac{780 \text{ nm}}{(4)(1.60)} = \boxed{1.2 \times 10^2 \text{ nm}}$$

24.6 DIFFRACTION

Suppose a light beam is incident on two slits, as in Young's double-slit experiment. If the light truly traveled in straight-line paths after passing through the slits, as in Figure 24.13a, the waves would not overlap and no interference pattern would be seen. Instead, Huygens's principle requires that the waves spread out from the slits, as shown in Figure 24.13b. In other words, the light deviates from a straight-line path and enters the region that would otherwise be shadowed. This spreading out of light from its initial line of travel is called **diffraction.**

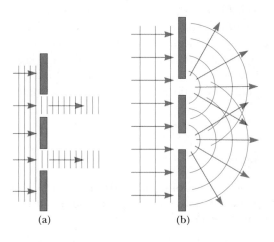

(a)   (b)

FIGURE 24.13 (a) If light did not spread out after passing through the slits, no interference would occur. (b) The light from the two slits overlaps as it spreads out, filling the expected shadowed regions with light and producing interference fringes.

FIGURE 24.14 The diffraction pattern that appears on a screen when light passes through a narrow, vertical slit. The pattern consists of a broad central band and a series of less intense and narrower side bands.

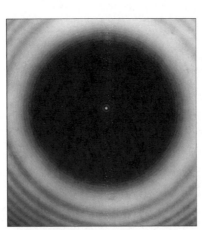

FIGURE 24.15 The diffraction pattern of a penny placed midway between the screen and the source. *(Courtesy of P. M. Rinard, from* Am. J. Phys., *44:70, 1976)*

FIGURE 24.16 (a) The Fraunhofer diffraction pattern of a single slit. The parallel rays are brought into focus on the screen with a converging lens. The pattern consists of a central bright region flanked by much weaker maxima. (This drawing is not to scale.) (b) A photograph of a single-slit Fraunhofer diffraction pattern. *(From M. Cagnet, M. Francon, and J. C. Thierr,* Atlas of Optical Phenomena, *Berlin, Springer-Verlag, 1962, plate 18)*

In general, diffraction occurs when waves pass through small openings, around obstacles, or by sharp edges. For example, when a single narrow slit is placed between a distant light source (or a laser beam) and a screen, the light produces a diffraction pattern like that in Figure 24.14. The pattern consists of a broad, intense central band flanked by a series of narrower, less intense secondary bands (called **secondary maxima**) and a series of dark bands, or **minima.** This cannot be explained within the framework of geometric optics, which says that light rays traveling in straight lines should cast a sharp image of the slit on the screen.

Figure 24.15 shows the diffraction pattern and shadow of a penny. The pattern consists of the shadow, a bright spot at its center, and a series of bright and dark circular bands of light near the edge of the shadow. The bright spot at the center (called the *Fresnel bright spot*) is explained by Augustin Fresnel's wave theory of light, which predicts constructive interference at this point for certain locations of the penny. In contrast, from the viewpoint of geometric optics, the center of the pattern would be completely screened by the penny, and so we would never observe a central bright spot.

One type of diffraction, called **Fraunhofer diffraction,** occurs when the rays leave the diffracting object in parallel directions. This can be achieved experimentally either by placing the observing screen far from the slit or by using a converging lens to focus the parallel rays on a nearby screen, as in Figure 24.16a. A bright fringe is observed along the axis at $\theta = 0$, with alternating dark and bright fringes on each side of the central bright fringe. Figure 24.16b is a photograph of a single-slit Fraunhofer diffraction pattern.

24.7   SINGLE-SLIT DIFFRACTION

Until now we have assumed that slits are line sources of light. In this section we determine how their finite widths are the basis for understanding the nature of the Fraunhofer diffraction pattern produced by a single slit.

We can deduce some important features of this problem by examining waves coming from various portions of the slit, as shown in Figure 24.17. According to Huygens's principle, **each portion of the slit acts as a source of waves. Hence, light from one portion of the slit can interfere with light from another portion,** and the resultant intensity on the screen depends on the direction θ.

To analyze the diffraction pattern, it is convenient to divide the slit into halves, as in Figure 24.17. All the waves that originate at the slit are in phase. Consider waves 1 and 3, which originate at the bottom and center of the slit, respectively. Wave 1 travels farther than wave 3 by an amount equal to the path difference $(a/2)\sin \theta$, where a is the width of the slit. Similarly, the path difference between waves 3 and 5 is also $(a/2)\sin \theta$. If this path difference is exactly half of a wavelength (corresponding to a phase difference of 180°), the two waves cancel each other and destructive interference results. This is true, in fact, for any two waves that originate at points separated by half the slit width, because the phase difference between two such points is 180°. Therefore, waves from the upper half of the slit interfere *destructively* with waves from the lower half of the slit when

$$\frac{a}{2} \sin \theta = \frac{\lambda}{2} \qquad \text{or when} \qquad \sin \theta = \frac{\lambda}{a}$$

If we divide the slit into four parts rather than two, and use similar reasoning, we find that the screen is also dark when

$$\sin \theta = \frac{2\lambda}{a}$$

Likewise, we can divide the slit into six parts and show that darkness occurs on the screen when

$$\sin \theta = \frac{3\lambda}{a}$$

Therefore, the general condition for **destructive interference** for a single slit of width a is

$$\sin \theta_{\text{dark}} = m \frac{\lambda}{a} \qquad m = \pm 1, \pm 2, \pm 3, \ldots \qquad \text{[24.11]}$$

Equation 24.11 gives the values of θ for which the diffraction pattern has zero intensity—that is, a dark fringe is formed. However, this equation tells us nothing about the variation in intensity along the screen. The general features of the intensity distribution along the screen are shown in Figure 24.18. A broad central bright fringe is observed, flanked by much weaker bright fringes alternating with dark fringes. The various dark fringes (points of zero intensity) occur at the values of θ that satisfy Equation 24.11. The points of constructive interference lie approximately halfway between the dark fringes. Note that the central bright fringe is twice as wide as the weaker maxima having $m > 1$.

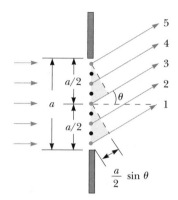

FIGURE 24.17 Diffraction of light by a narrow slit of width a. Each portion of the slit acts as a point source of waves. The path difference between rays 1 and 3 or between rays 2 and 4 is equal to $(a/2)\sin \theta$. (This drawing is not to scale, and the rays are assumed to converge at a distant point.)

Webnote 24.2

To see the diffraction pattern of a single slit as a function of the incident wavelength, explore *http://www.lightlink.com/sergey/java/ java/slitdiffr/index.html*

◀ Condition for destructive interference (single slit)

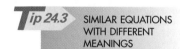

Tip 24.3 SIMILAR EQUATIONS WITH DIFFERENT MEANINGS

Although Equations 24.2 and 24.11 have the same form, they have different meanings. Equation 24.2 describes the *bright* regions in a two-slit interference pattern, whereas Equation 24.11 describes the *dark* regions in a single-slit interference pattern.

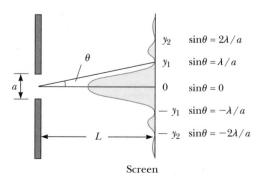

FIGURE 24.18 Positions of the minima for the Fraunhofer diffraction pattern of a single slit of width a. (This is not to scale.)

$y_2 \quad \sin\theta = 2\lambda/a$

$y_1 \quad \sin\theta = \lambda/a$

$0 \quad \sin\theta = 0$

$-y_1 \quad \sin\theta = -\lambda/a$

$-y_2 \quad \sin\theta = -2\lambda/a$

Screen

Quick Quiz 24.1 In a single-slit diffraction experiment, as the width of the slit is made smaller, the width of the central maximum of the diffraction pattern becomes (a) smaller, (b) larger, or (c) remains the same.

APPLYING **PHYSICS** 24.3

If a classroom door is open even just a small amount, you can hear sounds coming from the hallway. Yet you cannot see what is going on in the hallway. Why is there this difference?

Explanation The space between the slightly open door and the wall is acting as a single slit for waves. Sound waves have wavelengths larger than the slit width, so sound is effectively diffracted by the opening and the central maximum spreads throughout the room. Light wavelengths are much smaller than the slit width, so there is virtually no diffraction for the light. You must have a direct line of sight to detect the light waves.

Example 24.6 Where Are the Dark Fringes?

Light of wavelength 580 nm is incident on a slit of width 0.30 mm. The observing screen is placed 2.0 m from the slit. Find the positions of the first dark fringes and the width of the central bright fringe.

Solution The first dark fringes that flank the central bright fringe correspond to $m = \pm 1$ in Equation 24.11:

$$\sin \theta = \pm \frac{\lambda}{a} = \pm \frac{5.8 \times 10^{-7} \text{ m}}{0.30 \times 10^{-3} \text{ m}} = \pm 1.9 \times 10^{-3}$$

From the triangle in Figure 24.16, we see that $\tan \theta = y_1/L$. Because θ is very small, we can use the approximation $\sin \theta \approx \tan \theta$, so that $\sin \theta \approx y_1/L$. Therefore, the positions of the first minima, measured from central axis, are

$$y_1 \approx L \sin \theta = \pm L \frac{\lambda}{a} = \boxed{\pm 3.9 \times 10^{-3} \text{ m}}$$

The positive and negative signs correspond to the first dark fringes on either side of the central bright fringe. Hence, the width of the central bright fringe is given by $2|y_1| = 7.8 \times 10^{-3}$ m = $\boxed{7.8 \text{ mm}}$. Note that this value is much greater than the width of the slit. However, as the width of the slit is *increased*, the diffraction pattern *narrows*, corresponding to smaller values of θ. In fact, for large values of a, the maxima and minima are so closely spaced that the only observable pattern is a large central bright area resembling the geometric image of the slit. Because the width of the geometric image increases as the slit width increases, the narrowest image occurs when the geometric and diffraction widths are equal.

EXERCISE Determine the width of the first-order bright fringe.

ANSWER 3.9 mm

24.8 THE DIFFRACTION GRATING

The diffraction grating, a very useful device for analyzing light sources, consists of many equally spaced parallel slits. A grating can be made by scratching parallel lines on a glass plate with a precision machining technique. The scratches act to scatter the light and so behave as slits. A typical grating contains several thousand lines per centimeter. For example, a grating ruled with 5 000 lines/cm has a slit spacing d equal to the inverse of this number; hence, $d = (1/5\,000)$ cm $= 2 \times 10^{-4}$ cm.

Figure 24.19 is a diagram of a section of a plane diffraction grating. A plane wave is incident from the left, normal to the plane of the grating. The intensity of the pattern on the screen is the result of the combined effects of interference and diffraction. Each slit produces diffraction, and the diffracted beams in turn interfere with one another to produce the pattern. Moreover, each slit acts as a source of waves, and all waves start at the slits in phase. However, for some arbitrary direction θ measured from the horizontal, the waves must travel *different* path lengths before reaching a particular point P on the screen. From Figure 24.19, note that the path difference between waves from any two adjacent slits is $d \sin \theta$. If this path difference equals one wavelength or some integral multiple of a wavelength, waves from all slits will be in phase at P and a bright line will be observed at this point. Therefore, the condition for **maxima** in the interference pattern at the angle θ is

$$d \sin \theta_{\text{bright}} = m\lambda \qquad m = 0, 1, 2, \ldots \qquad \text{[24.12]}$$

◀ Condition for maxima in the interference pattern of a diffraction grating

Light emerging from a slit at an angle other than that for a maximum interferes with nearly complete destruction with light from some other slit on the grating. All such pairs will result in little or no transmission in that direction, as illustrated in Figure 24.20.

Equation 24.12 can be used to calculate the wavelength from the grating spacing and the angle of deviation, θ. The integer m is the **order number** of the diffraction pattern. If the incident radiation contains several wavelengths, each wavelength deviates through a specific angle, which can be found from Equation 24.12. All wavelengths are focused at $\theta = 0$, corresponding to $m = 0$. This is called the *zeroth-order maximum*. The *first-order maximum*, corresponding to $m = 1$, is observed at an angle that satisfies the relationship $\sin \theta = \lambda / d$; the *second-order maximum*, corresponding to $m = 2$, is observed at a larger angle θ, and so on. Figure 24.20 is a sketch of the intensity distribution for some of the orders produced by a diffrac-

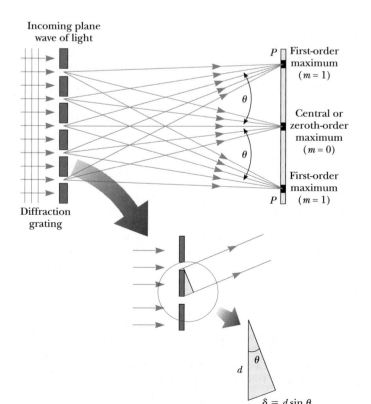

Incoming plane wave of light

P — First-order maximum ($m = 1$)

Central or zeroth-order maximum ($m = 0$)

First-order maximum ($m = 1$)

P

Diffraction grating

d

θ

$\delta = d \sin \theta$

FIGURE 24.19 A side view of a diffraction grating. The slit separation is d, and the path difference between adjacent slits is $d \sin \theta$.

FIGURE 24.20 Intensity versus sin θ for the diffraction grating. The zeroth-, first-, and second-order principal maxima are shown.

FIGURE 24.21 A diagram of a diffraction grating spectrometer. The collimated beam incident on the grating is diffracted into the various orders at the angles θ that satisfy the equation $d \sin \theta = m\lambda$, where $m = 0, 1, 2, \ldots$.

tion grating. Note the sharpness of the principal maxima and the broad range of the dark areas. This is in contrast to the broad, bright fringes characteristic of the two-slit interference pattern.

A simple arrangement that can be used to measure the angles in a diffraction pattern is shown in Figure 24.21. This is a form of diffraction-grating spectrometer. The light to be analyzed passes through a slit and is formed into a parallel beam by a lens. The light then strikes the grating at a 90° angle. The diffracted light leaves the grating at angles that satisfy Equation 24.12. A telescope is used to view the image of the slit. The wavelength can be determined by measuring the angles at which the images of the slit appear for the various orders.

Quick Quiz 24.2 If laser light is reflected from a phonograph record or a compact disc, a diffraction pattern appears. This occurs because both devices contain parallel tracks of information that act as a reflection diffraction grating. Which device, record or compact disc, results in diffraction maxima that are farther apart?

APPLYING PHYSICS 24.4

White light enters through an opening in an opaque box, exits through an opening on the other side of the box, and a spectrum of colors appears on the wall. How would you determine whether the box *contains* a prism or a diffraction grating?

Explanation The determination could be made by noticing the order of the colors in the spectrum relative to the direction of the original beam of white light. For a prism, in which the separation of light is a result of dispersion, the violet light will be refracted more than the red light. Hence, the order of the spectrum from a prism will be from red, closest to the original direction, to violet. For a diffraction grating, the angle of diffraction increases with wavelength. Thus, the spectrum from the diffraction grating will have colors in the order of violet, closest to the original direction, to red. Furthermore, the diffraction grating will produce *two* first-order spectra on either side of the grating, whereas the prism will only produce a single spectrum.

APPLYING **PHYSICS** 24.5

White light reflected from the surface of a compact disc has a multicolored appearance, as shown in Figure 24.22. Furthermore, the observation depends on the orientation of the disc relative to the eye and the position of the light source. Explain how this works.

Explanation   The surface of a compact disc has a spiral-shaped track (with a spacing of approximately 1 μm) that acts as a reflection grating. The light scattered by these closely spaced parallel tracks interferes constructively in certain directions that depend on the wavelength and on the direction of the incident light. Any one section of the disc serves as a diffraction grating for white light, sending beams of constructive interference for different colors in different directions. The different colors you see when viewing one section of the disc change as the light source, the disc, or you move to change the angles of incidence or diffraction.

FIGURE 24.22   (Applying Physics 24.5) Compact discs act as diffraction gratings when observed under white light. *(© Kristen Brochmann/Fundamental Photographs)*

USE OF DIFFRACTION GRATING IN CD TRACKING

If a CD player is to reproduce sound faithfully, the laser beam must follow the spiral track of information perfectly. However, sometimes the laser beam can drift off track, and without a feedback procedure to let the player know this is happening, the fidelity of the music would be greatly reduced.

Figure 24.23 shows how a diffraction grating is used in a three-beam method to keep the beam on track. The central maximum of the diffraction pattern is

APPLICATION

TRACKING INFORMATION ON A CD

FIGURE 24.23   The laser beam in a CD player is able to follow the spiral track by using three beams produced with a diffraction grating.

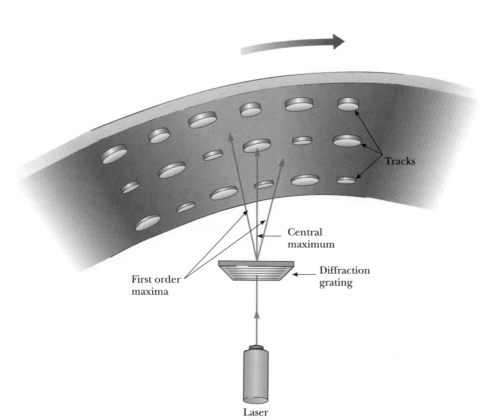

Tracks

Central maximum

Diffraction grating

First order maxima

Laser

used to read the information on the CD track, and the two first-order maxima are used for steering. The grating is designed so that the first-order maxima fall on the smooth surfaces on either side of the information track. Both of these reflected beams have their own detectors, and because both beams are reflected from smooth surfaces, they should both have the same strong intensity when they are detected. If the central beam wanders off the track however, one of the steering beams will begin to strike bumps on the information track and the amount of light reflected will decrease. This information is then used by electronic circuits to drive the main beam back to its desired location.

Example 24.7   The Orders of a Diffraction Grating

Monochromatic light from a helium-neon laser ($\lambda = 632.8$ nm) is incident normally on a diffraction grating containing 6 000 lines/cm. Find the angles at which we would observe the first-order maximum, the second-order maximum, and so forth.

Solution   First we must calculate the slit separation, which is the inverse of the number of lines per centimeter:

$$d = \frac{1}{6\ 000}\text{ cm} = 1.667 \times 10^{-4}\text{ cm} = 1\ 667\text{ nm}$$

For the first-order maximum ($m = 1$), we get

$$\sin \theta_1 = \frac{\lambda}{d} = \frac{632.8\text{ nm}}{1\ 667\text{ nm}} = 0.379\ 6$$

$$\theta_1 = \boxed{22.31°}$$

For $m = 2$, we find that

$$\sin \theta_2 = \frac{2\lambda}{d} = \frac{2(632.8\text{ nm})}{1\ 667\text{ nm}} = 0.759\ 2$$

$$\theta_2 = \boxed{49.39°}$$

However, for $m = 3$, we find that $\sin \theta_3 = 1.139$. Because $\sin \theta$ cannot exceed unity, this is not a realistic solution. Hence, only zeroth-, first-, and second-order maxima would be observed in this situation.

24.9   POLARIZATION OF LIGHT WAVES

In Chapter 21 we described the transverse nature of electromagnetic waves. Figure 24.24 shows that the electric and magnetic field vectors associated with an electromagnetic wave are at right angles to each other and also to the direction of wave propagation. The phenomenon of polarization, described in this section, is firm evidence of the transverse nature of electromagnetic waves.

An ordinary beam of light consists of a large number of electromagnetic waves emitted by the atoms or molecules of the light source. The vibrating charges asso-

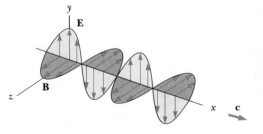

FIGURE 24.24   A diagram of a polarized electromagnetic wave propagating in the *x* direction. The electric field vector **E** vibrates in the *xy* plane, and the magnetic field vector **B** vibrates in the *xz* plane.

ciated with the atoms act as tiny antenna. Each atom produces a wave with its own orientation of **E**, as in Figure 24.24, corresponding to the direction of atomic vibration. However, because all directions of vibration are possible, the resultant electromagnetic wave is a superposition of waves produced by the individual atomic sources. The result is an **unpolarized** light wave, represented schematically in Figure 24.25a. The direction of wave propagation in this figure is perpendicular to the page. Note that *all* directions of the electric field vector are equally probable and lie in a plane (such as the plane of this page) perpendicular to the direction of propagation.

A wave is said to be **linearly polarized** if the resultant electric field **E** vibrates in the same direction *at all times* at a particular point, as in Figure 24.25b. (Sometimes such a wave is described as *plane-polarized* or simply *polarized*.) The wave in Figure 24.24 is an example of a wave linearly polarized in the *y* direction. As the wave propagates in the *x* direction, **E** is always in the *y* direction. The plane formed by **E** and the direction of propagation is called the *plane of polarization* of the wave. In Figure 24.24, the plane of polarization is the *xy* plane.

It is possible to obtain a linearly polarized beam from an unpolarized beam by removing all waves from the beam except those whose electric field vectors oscillate in a single plane. We now discuss three processes for doing this: (1) selective absorption, (2) reflection, and (3) scattering.

POLARIZATION BY SELECTIVE ABSORPTION

The most common technique for polarizing light is to use a material that transmits waves whose electric field vectors vibrate in a plane parallel to a certain direction and absorbs those waves whose electric field vectors vibrate in directions perpendicular to that direction.

In 1932, E. H. Land discovered a material, which he called **polaroid,** that polarizes light through selective absorption by oriented molecules. This material is fabricated in thin sheets of long-chain hydrocarbons, which are stretched during manufacture so that the molecules align. After a sheet is dipped into a solution containing iodine, the molecules become good electrical conductors. However, the conduction takes place primarily along the hydrocarbon chains, because the valence electrons of the molecules can move easily only along the chains. (Recall that valence electrons are "free" electrons that can readily move through the conductor.) As a result, the molecules readily *absorb* light whose electric field vector is parallel to their lengths, and *transmit* light whose electric field vector is perpendicular to their lengths. It is common to refer to the direction perpendicular to the molecular chains as the **transmission axis.** In an ideal polarizer, all light with **E** parallel to the transmission axis is transmitted, and all light with **E** perpendicular to the transmission axis is absorbed.

Let us now describe the intensity of light that passes through a polarizing material. In Figure 24.26, an unpolarized light beam is incident on the first polarizing sheet, called the **polarizer,** where the transmission axis is as indicated. The light that is passing through this sheet is polarized vertically, and the transmitted elec-

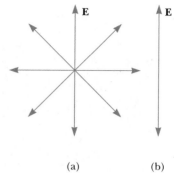

(a)   (b)

FIGURE 24.25   (a) An unpolarized light beam viewed along the direction of propagation (perpendicular to the page). The transverse electric field vector can vibrate in any direction with equal probability. (b) A linearly polarized light beam with the electric field vector vibrating in the vertical direction.

Webnote 24.3

Polarized light has helped dermatologic surgeons estimate the size (area) of skin cancers. Explore *http://omlc.ogi.edu/news/feb98/ polarization/index.html*

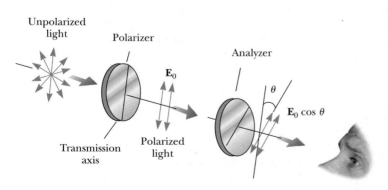

FIGURE 24.26   Two polarizing sheets whose transmission axes make an angle *θ* with each other. Only a fraction of the polarized light incident on the analyzer is transmitted.

(a)

(b)

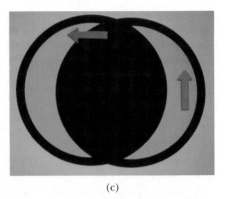
(c)

FIGURE 24.27 The intensity of light transmitted through two polarizers depends on the relative orientations of their transmission axes. (a) The transmitted light has *maximum* intensity when the transmission axes are *aligned* with each other. (b) The transmitted light intensity diminishes when the transmission axes are at an angle of 45° with each other. (c) The transmitted light intensity is a *minimum* when the transmission axes are at *right angles* to each other. *(Photos courtesy of Henry Leap)*

tric field vector is $\mathbf{E}_0$. A second polarizing sheet, called the **analyzer,** intercepts this beam with its transmission axis at an angle of θ to the axis of the polarizer. The component of $\mathbf{E}_0$ that is perpendicular to the axis of the analyzer is completely absorbed. The component of $\mathbf{E}_0$ parallel to the analyzer axis, $E_0 \cos \theta$, is allowed to pass through the analyzer. Because the intensity of the transmitted beam varies as the *square* of its amplitude E, we conclude that the intensity of the (polarized) beam transmitted through the analyzer varies as

$$I = I_0 \cos^2 \theta \qquad [24.13]$$

where I_0 is the intensity of the polarized wave incident on the analyzer. This expression, known as **Malus's law,** applies to any two polarizing materials whose transmission axes are at an angle of θ to each other. From this expression, note that the transmitted intensity is a maximum when the transmission axes are parallel ($\theta = 0$ or $180°$) and zero (complete absorption by the analyzer) when the transmission axes are perpendicular to each other. This variation in transmitted intensity through a pair of polarizing sheets is illustrated in Figure 24.27.

When unpolarized light of intensity I_0 is sent through a single ideal polarizer, the transmitted linearly polarized light has intensity $I_0/2$. This fact follows from Malus's law because the average value of $\cos^2 \theta$ is one half.

APPLYING **PHYSICS** 24.6

APPLICATION

POLARIZING MICROWAVES

A polarizer for microwaves can be made as a grid of parallel metal wires about a centimeter apart. Is the electric field vector for microwaves transmitted through this polarizer parallel to, or perpendicular to, the metal wires?

Explanation Electric field vectors parallel to the metal wires cause electrons in the metal to oscillate parallel to the wires. Thus, the energy from the waves with these electric field vectors is transferred to the metal by accelerating these electrons and is eventually transformed to internal energy through the metal's resistance. Waves with electric field vectors perpendicular to the metal wires are not able to accelerate electrons and therefore pass through. Thus, the electric field polarization is perpendicular to the metal wires.

POLARIZATION BY REFLECTION

When an unpolarized light beam is reflected from a surface, the reflected light is completely polarized, partially polarized, or unpolarized, depending on the angle of incidence. If the angle of incidence is either 0° or 90° (a normal or grazing an-

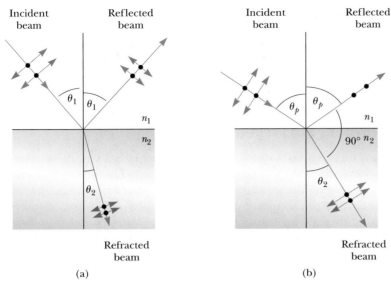

Incident beam   Reflected beam

θ_1 θ_1 n_1 n_2 θ_2

Refracted beam

(a)

Incident beam   Reflected beam

θ_p θ_p n_1 $90°$ n_2 θ_2

Refracted beam

(b)

FIGURE 24.28 (a) When unpolarized light is incident on a reflecting surface, the reflected and refracted beams are partially polarized. (b) The reflected beam is completely polarized when the angle of incidence equals the polarizing angle θ_p, satisfying the equation $n = \tan \theta_p$.

gle), the reflected beam is unpolarized. However, for angles of incidence between 0° and 90°, the reflected light is polarized to some extent. For one particular angle of incidence, the reflected beam is completely polarized. Let us now investigate that special angle.

Suppose an unpolarized light beam is incident on a surface, as in Figure 24.28a. The beam can be described by two electric field components, one parallel to the surface (represented by dots) and the other perpendicular to the first component and to the direction of propagation (represented by brown arrows). It is found that the parallel component reflects more strongly than the other components, and this results in a partially polarized beam. Furthermore, the refracted beam is also partially polarized.

Now suppose that the angle of incidence θ_1 is varied until the angle between the reflected and refracted beams is 90° (Fig. 24.28b). At this particular angle of incidence, the reflected beam is completely polarized, with its electric field vector parallel to the surface, whereas the refracted beam is partially polarized. The angle of incidence at which this occurs is called the **polarizing angle** θ_p.

An expression relating the polarizing angle to the index of refraction of the reflecting surface can be obtained by use of Figure 24.28b. From this figure we see that, at the polarizing angle, $\theta_p + 90° + \theta_2 = 180°$, so that $\theta_2 = 90° - \theta_p$. Using Snell's law and taking $n_1 = n_{air} = 1.00$ and $n_2 = n$,

$$n = \frac{\sin \theta_1}{\sin \theta_2} = \frac{\sin \theta_p}{\sin \theta_2}$$

Because $\sin \theta_2 = \sin(90° - \theta_p) = \cos \theta_p$, the expression for n can be written

$$n = \frac{\sin \theta_p}{\cos \theta_p} = \tan \theta_p \qquad [24.14]$$

◀ Brewster's law

This expression is called **Brewster's law,** and the polarizing angle θ_p is sometimes called **Brewster's angle** after its discoverer, Sir David Brewster (1781–1868). For example, Brewster's angle for crown glass ($n = 1.52$) has the value $\theta_p = \tan^{-1}(1.52) = 56.7°$. Because n varies with wavelength for a given substance, Brewster's angle is also a function of wavelength.

APPLICATION

POLAROID SUNGLASSES

Polarization by reflection is a common phenomenon. Sunlight reflected from water, glass, or snow is partially polarized. If the surface is horizontal, the electric

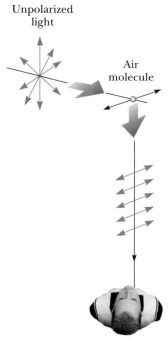

Unpolarized
light

Air
molecule

FIGURE 24.29 The scattering of unpolarized sunlight by air molecules. The light observed at right angles is linearly polarized because the vibrating molecule has a horizontal component of vibration.

field vector of the reflected light has a strong horizontal component. Sunglasses made of polarizing material reduce the glare, which *is* the reflected light. The transmission axes of the lenses are oriented vertically to absorb the strong horizontal component of the reflected light. Because the reflected light is mostly polarized, most of the glare can be eliminated without removing most of the normal light.

POLARIZATION BY SCATTERING

When light is incident on a system of particles, such as a gas, the electrons in the medium can absorb and reradiate part of the light. The absorption and reradiation of light by the medium, called **scattering,** is what causes sunlight reaching an observer on Earth from straight overhead to be polarized. You can observe this effect by looking directly up through a pair of sunglasses made of polarizing glass. Less light passes through at certain orientations of the lenses than at others.

Figure 24.29 illustrates how the sunlight becomes polarized. The left side of the figure shows an incident unpolarized beam of sunlight on the verge of striking an air molecule. When the beam strikes the air molecule, it sets the electrons of the molecule into vibration. These vibrating charges act like those in an antenna except that they vibrate in a complicated pattern. The horizontal part of the electric field vector in the incident wave causes the charges to vibrate horizontally, and the vertical part of the vector simultaneously causes them to vibrate vertically. A horizontally polarized wave is emitted by the electrons as a result of their horizontal motion, and a vertically polarized wave is emitted parallel to Earth as a result of their vertical motion.

Scientists have found that bees and homing pigeons use the polarization of sunlight as a navigational aid.

OPTICAL ACTIVITY

Many important practical applications of polarized light involve the use of certain materials that display the property of **optical activity.** A substance is said to be optically active if it rotates the plane of polarization of transmitted light. Suppose unpolarized light is incident on a polarizer from the left, as in Figure 24.30a. The transmitted light is polarized vertically, as shown. If this light is then incident on an analyzer with its axis perpendicular to that of the polarizer, no light emerges from it. If an optically active material is placed between the polarizer and analyzer, as in Figure 24.30b, the material causes the direction of the polarized beam to ro-

FIGURE 24.30 (a) When crossed polarizers are used, none of the polarized light can pass through the analyzer. (b) An optically active material rotates the direction of polarization through the angle θ, enabling some of this light to pass through the analyzer.

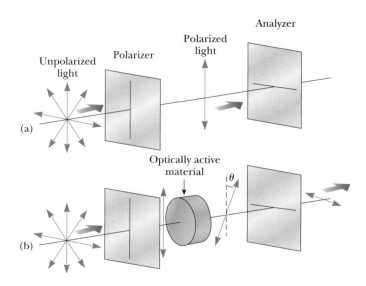

Analyzer

Polarized
light

Polarizer

Unpolarized
light

(a)

Optically active
material

θ

(b)

(a)   (b)

Figure 24.31 (a) Strain distribution in a plastic model of a hip replacement used in a medical research laboratory. The pattern is produced when the plastic model is placed between a polarizer and analyzer oriented perpendicular to each other. *(Sepp Seitz 1981)* (b) A plastic model of an arch structure under load conditions observed between perpendicular polarizers. Such patterns are useful in the optimum design of architectural components. *(Peter Aprahamian/Science Photo Library)*

tate through the angle θ. As a result, some light is able to pass through the analyzer. The angle through which the light is rotated by the material can be found by rotating the polarizer until the light is again extinguished. It is found that the angle of rotation depends on the length of the sample and, if the substance is in solution, on the concentration. One optically active material is a solution of common sugar, dextrose. A standard method for determining the concentration of sugar solutions is to measure the rotation produced by a fixed length of the solution.

Optical activity occurs in a material because of an asymmetry in the shape of its constituent molecules. For example, some proteins are optically active because of their spiral shapes. Other materials, such as glass and plastic, become optically active when placed under stress. If polarized light is passed through an unstressed piece of plastic and then through an analyzer with an axis perpendicular to that of the polarizer, none of the polarized light is transmitted. However, if the plastic is placed under stress, the regions of greatest stress produce the largest angles of rotation of polarized light. Hence, we observe a series of light and dark bands in the transmitted light. Engineers often use this procedure in the design of structures ranging from bridges to small tools. A plastic model is built and analyzed under different load conditions to determine positions of potential weakness and failure under stress. If the design is poor, patterns of light and dark bands will indicate the points of greatest weakness, and the design can be corrected at an early stage. Figure 24.31 shows examples of stress patterns in plastic.

> ◄ **APPLICATION**
>
> FINDING THE CONCENTRATIONS OF SOLUTIONS USING THEIR OPTICAL ACTIVITY

LIQUID CRYSTALS

An effect similar to rotation of the plane of polarization is used to create the familiar displays on pocket calculators, wristwatches, laptop computers, and so forth. The properties of a unique substance called a liquid crystal make these displays (called LCDs for *liquid crystal displays*) possible. As its name implies, a **liquid crystal** is a substance with properties intermediate between those of a crystalline solid and those of a liquid—that is, the molecules of the substance are more orderly than those in a liquid but less than those in a pure crystalline solid. The forces that hold the molecules together in such a state are just barely strong enough to enable the substance to maintain a definite shape, so it is reasonable to call it a solid. However, small inputs of mechanical or electrical energy can disrupt these weak bonds and make the substance flow, rotate, or twist.

To see how liquid crystals are used to create a display, consider Figure 24.32a. The liquid crystal is placed between two glass plates in the pattern shown, and electrical contacts, indicated by the thin lines, are made to the liquid crystal. When a voltage is applied across any segment in the display, that segment turns dark. In this fashion, any number between 0 and 9 can be formed by the pattern, depending on the voltages applied to the seven segments.

> ◄ **APPLICATION**
>
> LIQUID CRYSTAL DISPLAYS (LCDS)

FIGURE 24.32 (a) The light-segment pattern of a liquid crystal display. (b) Rotation of a polarized light beam by a liquid crystal when the applied voltage is zero. (c) Molecules of the liquid crystal align with the electric field when a voltage is applied.

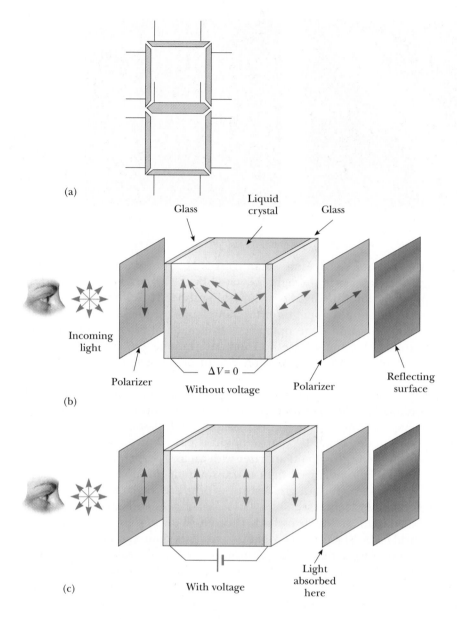

To see why a segment can be changed from dark to light by the application of a voltage, consider Figure 24.32b, which shows the basic construction of a portion of the display. The liquid crystal is placed between two glass substrates that are packaged between two pieces of Polaroid material with their transmission axes perpendicular. A reflecting surface is placed behind one of the pieces of Polaroid. First consider what happens when light falls on this package and no voltages are applied to the liquid crystal, as shown in Figure 24.32b. Incoming light is polarized by the polarizer on the left and then falls on the liquid crystal. As the light passes through the crystal, its plane of polarization is rotated by 90°, allowing it to pass through the polarizer on the right. It reflects from the reflecting surface and retraces its path through the crystal. Thus, an observer to the left of the crystal sees the segment as being bright. When a voltage is applied as in Figure 24.32c, the molecules of the liquid crystal do not rotate the plane of polarization of the light. In this case, the light is absorbed by the polarizer on the right and none is reflected back to the observer to the left of the crystal. Thus, the observer sees this segment as black. Changing the applied voltage to the crystal in a precise pattern and at precise times can make the pattern tick off the seconds on a watch, display a letter on a computer display, and so forth.

SUMMARY

Interference occurs when two or more light waves overlap at a given point. A sustained interference pattern is observed if (1) the sources are coherent (that is, they maintain a constant phase relationship with one another); (2) the sources have identical wavelengths; and (3) the superposition principle is applicable.

In **Young's double-slit experiment,** two slits separated by distance d are illuminated by a single-wavelength light source. An interference pattern consisting of bright and dark fringes is observed on a screen a distance L from the slits. The condition for **bright fringes** (constructive interference) is

$$d \sin \theta_{\text{bright}} = m\lambda \qquad m = 0, \pm 1, \pm 2, \ldots \qquad \text{[24.2]}$$

The number m is called the **order number** of the fringe. The condition for **dark fringes** (destructive interference) is

$$d \sin \theta_{\text{dark}} = (m + \tfrac{1}{2})\lambda \qquad m = 0, \pm 1, \pm 2, \ldots \qquad \text{[24.3]}$$

An electromagnetic wave undergoes a phase change of 180° on reflection from a medium with an index of refraction higher than that of the medium in which the wave is traveling.

The wavelength λ_n of light in a medium with index of refraction n is

$$\lambda_n = \frac{\lambda}{n} \qquad \text{[24.7]}$$

where λ is the wavelength of the light in free space.

The **diffraction pattern** produced by a single slit on a distant screen consists of a central bright maximum flanked by less-bright fringes alternating with dark regions. The angles θ at which the diffraction pattern has zero intensity (regions of destructive interference) are described by

$$\sin \theta_{\text{dark}} = m\frac{\lambda}{a} \qquad m = \pm 1, \pm 2, \pm 3, \ldots \qquad \text{[24.11]}$$

where a is the width of the slit and λ is the wavelength of the light incident on the slit.

A **diffraction grating** consists of many equally spaced, identical slits. The condition for **maximum intensity** in the interference pattern of a diffraction grating is

$$d \sin \theta_{\text{bright}} = m\lambda \qquad m = 0, 1, 2, \ldots \qquad \text{[24.12]}$$

where d is the spacing between adjacent slits and m is the order number of the diffraction pattern.

Unpolarized light can be polarized by selective absorption, reflection, and scattering.

In general, light reflected from an amorphous material, such as glass, is partially polarized. Reflected light is completely polarized with its electric field parallel to the surface when the angle of incidence produces a 90° angle between the reflected and refracted beams. This angle of incidence, called the **polarizing angle** θ_p, satisfies **Brewster's law,** given by

$$n = \tan \theta_p \qquad \text{[24.14]}$$

where n is the index of refraction of the reflecting medium.

CONCEPTUAL QUESTIONS

1. Your automobile has two headlights. What sort of interference pattern do you expect to see from them? Why?

2. Holding your hand at arm's length, you can readily block sunlight from your eyes. Why can you not block sound from your ears this way?

3. Consider a dark fringe in an interference pattern, at which almost no light energy is arriving. Light from both slits is arriving at this point, but the waves are canceling. Where does the energy go?

4. If Young's double-slit experiment were performed under water, how would the observed interference pattern be affected?

5. In a laboratory accident, you spill two liquids onto water, neither of which mixes with the water. They both form

thin films on the water surface. You notice, as the films become very thin as they spread, that one film becomes bright and the other black in reflected light. Why might this be?

6. If white light is used in Young's double-slit experiment rather than monochromatic light, how does the interference pattern change?

7. In our discussion of thin-film interference, we looked at light *reflecting* from a thin film. Consider one light ray, the direct ray, that transmits through the film without reflecting. Consider a second ray, the reflected ray, that transmits through the first surface, reflects back to the second, reflects again from the first, and then transmits out into the air, parallel to the direct ray. For normal incidence, how thick must the film be, in terms of the wavelength of the light, for the outgoing rays to interfere destructively? Is it the same thickness as for reflected destructive interference?

8. What is the necessary condition on path-length difference between two waves that interfere (a) constructively and (b) destructively? Assume that the wave sources are coherent.

9. A lens with outer radius of curvature R and index of refraction n rests on a flat glass plate, and the combination is illuminated from white light from above. Is there a dark spot or a light spot at the center of the lens? What does it mean if the observed rings are noncircular?

10. Often fingerprints left on a piece of glass such as a window pane show colored spectra like that from a diffraction grating. Why?

11. In our everyday experience, radio waves are polarized, but light is not. Why?

12. Suppose reflected white light is used to observe a thin, transparent coating on glass as the coating material is gradually deposited by evaporation in a vacuum. Describe possible color changes that might occur during the process of building up the thickness of the coating.

13. Would it be possible to place a nonreflective coating on an airplane to cancel radar waves of wavelength 3 cm?

14. Certain sunglasses use a polarizing material to reduce the intensity of light reflected from shiny surfaces, such as water or the hood of a car. What orientation of the transmission axis should the material have to be most effective?

15. Why is it so much easier to perform interference experiments with a laser than with an ordinary light source?

16. A simple way to observe an interference pattern is to look at a distant light source through a stretched handkerchief or an open umbrella. Explain how this works.

17. Although we can hear around corners, we cannot see around corners. How can you explain this in view of the fact that sound and light are both waves?

18. Can a sound wave be polarized? Explain.

19. When you receive a chest x-ray at a hospital, the x-rays pass through a series of parallel ribs in your chest. Do the ribs act as a diffraction grating for x-rays?

PROBLEMS

1, 2, 3 = straightforward, intermediate, challenging   ☐ = full solution available in Student Solutions Manual/Study Guide

web = solution posted at **http://info.brookscole.com/serway**   🧬 = biomedical application

Section 24.2   Young's Double-Slit Interference

1. A laser beam ($\lambda = 632.8$ nm) is incident on two slits 0.200 mm apart. How far apart are the bright interference fringes on a screen 5.00 m away from the double slits?

2. In a Young's double-slit experiment, a set of parallel slits with a separation of 0.100 mm is illuminated by light having a wavelength of 589 nm and the interference pattern observed on a screen 4.00 m from the slits. (a) What is the difference in path lengths from each of the slits to the screen location of a third-order bright fringe? (b) What is the difference in path lengths from the two slits to the screen location of the third dark fringe away from the center of the pattern?

3. A pair of narrow, parallel slits separated by 0.250 mm are illuminated by the green component from a mercury vapor lamp ($\lambda = 546.1$ nm). The interference pattern is observed on a screen 1.20 m from the plane of the parallel slits. Calculate the distance (a) from the central maximum to the first bright region on either side of the central maximum and (b) between the first and second dark bands in the interference pattern.

4. Light of wavelength 460 nm falls on two slits spaced 0.300 mm apart. What is the required distance from the slit to a screen if the spacing between the first and second dark fringes is to be 4.00 mm?

5. In a location where the speed of sound is 354 m/s, a 2 000-Hz sound wave impinges on two slits 30.0 cm apart. (a) At what angle is the first maximum located? (b) If the sound wave is replaced by 3.00-cm microwaves, what slit separation gives the same angle for the first maximum? (c) If the slit separation is 1.00 μm, what frequency light gives the same first maximum angle?

6. White light spans the wavelength range between about 400 nm and 700 nm. If white light passes through two slits 0.30 mm apart and falls on a screen 1.5 m from the slits, find the distance between the first-order violet and the first-order red fringes.

7. Two radio antennas separated by 300 m, as shown in Figure P24.7, simultaneously transmit identical signals of the same wavelength. A radio in a car traveling due north receives the signals. (a) If the car is at the position of the second maximum, what is the wavelength of the signals? (b) How much farther must the car travel to encounter the next minimum in reception? (*Hint:* Determine the path difference between the two signals at the two locations of the car.)

8. Light of wavelength 575 nm falls on a double slit, and the first bright fringe is seen at an angle of 16.5°. Find the distance between the two slits.

9. Waves from a radio station have a wavelength of 300 m.

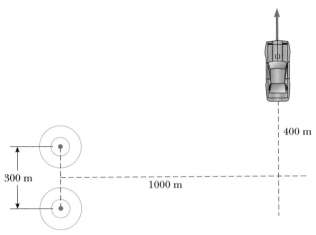

FIGURE P24.7

They travel by two paths to a home receiver 20.0 km from the transmitter. One path is a direct path, and the second is by reflection from a mountain directly behind the home receiver. What is the minimum distance from the mountain to the receiver that produces destructive interference at the receiver? (Assume that no phase change occurs on reflection from the mountain.)

10. A Young's interference experiment is performed with blue-green argon laser light. The separation between the slits is 0.500 mm, and the interference pattern on a screen 3.30 m away shows the first maximum 3.40 mm from the center of the pattern. What is the wavelength of argon laser light?

11. A riverside warehouse has two open doors, as in Figure P24.11. Its interior is lined with a sound-absorbing material. A boat on the river sounds its horn. To person A the sound is loud and clear. To person B the sound is barely audible. The principal wavelength of the sound waves is 3.00 m. Assuming person B is at the position of the first minimum, determine the distance between the doors, center to center.

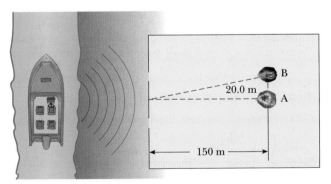

FIGURE P24.11

12. The waves from a radio station can reach a home receiver **web** by two different paths. One is a straight-line path from the transmitter to the home, a distance of 30.0 km. The second path is by reflection from a storm cloud. Assume that this reflection takes place at a point midway between receiver and transmitter. If the wavelength broadcast by the radio station is 400 m, find the minimum height of the storm cloud that will produce destructive interference be-

tween the direct and reflected beams. (Assume no phase changes on reflection.)

13. Radio waves from a star, of wavelength 250 m, reach a radio telescope by two separate paths as shown in Figure P24.13. One is a direct path to the receiver, which is situated on the edge of a cliff by the ocean. The second is by reflection off the water. The first minimum of destructive interference occurs when the star is 25.0° above the horizon. Find the height of the cliff. (Assume no phase change on reflection.)

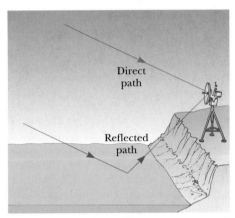

FIGURE P24.13

Section 24.3 Change of Phase Due to Reflection

Section 24.4 Interference in Thin Films

14. A soap bubble ($n = 1.33$) is floating in air. If the thickness of the bubble wall is 115 nm, what is the wavelength of the visible light that is most strongly reflected?

15. Suppose the film shown in Figure 24.7 has an index of refraction of 1.36 and is surrounded by air on both sides. Find the minimum thickness that will produce constructive interference in the reflected light when the film is illuminated by light of wavelength 500 nm.

16. A thin film of oil ($n = 1.25$) is located on a smooth, wet pavement. When viewed perpendicular to the pavement, the film appears to be predominantly red (640 nm) and has no blue color (512 nm). How thick is the oil film?

17. A thin layer of liquid methylene iodide ($n = 1.756$) is sandwiched between two flat parallel plates of glass ($n = 1.50$). What must be the thickness of the liquid layer if normally incident light with $\lambda = 600$ nm in air is to be strongly reflected?

18. A transparent oil of index of refraction 1.29 spills on the surface of water (index of refraction 1.33), producing a maximum of reflection with normally incident orange light (wavelength 600 nm in air). Assuming the maximum occurs in the first order, determine the thickness of the oil slick.

19. A possible means for making an airplane invisible to radar is to coat the plane with an antireflective polymer. If radar waves have a wavelength of 3.00 cm and the index of refraction of the polymer is $n = 1.50$, how thick would you make the coating?

20. A beam of light of wavelength 580 nm passes through two closely spaced glass plates, as shown in Figure P24.20. For what minimum nonzero value of the plate separation d

will the transmitted light be bright? This arrangement is often used to measure the wavelength of light and is called a Fabry-Perot interferometer.

FIGURE P24.20

21. Astronomers observe the chromosphere of the Sun with a filter that passes the red hydrogen spectral line of wavelength 656.3 nm, called the H_α line. The filter consists of a transparent dielectric of thickness d held between two partially aluminized glass plates. The filter is held at a constant temperature. (a) Find the minimum value of d that will produce maximum transmission of perpendicular H_α light, if the dielectric has index of refraction 1.378. (b) If the temperature of the filter increases above the normal value, thereby increasing its thickness, what happens to the transmitted wavelength? (c) The dielectric will also pass what near-visible wavelength? One of the glass plates is colored red to absorb this light.

22. Two rectangular optically flat plates ($n = 1.52$) are in contact along one end and are separated along the other end by a 2.00-μm-thick spacer (Fig. P24.22). The top plate is illuminated by monochromatic light of wavelength 546.1 nm. Calculate the number of dark parallel bands crossing the top plate (including the dark band at zero thickness along the edge of contact between the two plates).

FIGURE P24.22   (Problems 22 and 23)

23. An air wedge is formed between two glass plates separated at one edge by a very fine wire as in Figure P24.22. When the wedge is illuminated from above by 600-nm light, 30 dark fringes are observed. Calculate the radius of the wire.

24. A planoconvex lens with radius of curvature $R = 3.0$ m is in contact with a flat plate of glass. A light source and the observer's eye are both close to the normal, as shown in Figure 24.8. The radius of the 50th bright Newton's ring is found to be 9.8 mm. What is the wavelength of the light produced by the source?

25. A planoconvex lens rests with its curved side on a flat glass surface and is illuminated from above by light of wavelength 500 nm (see Fig. 24.8). A dark spot is observed at the center, surrounded by 19 concentric dark rings (with bright rings in between). How much thicker is the air wedge at the position of the 19th dark ring than at the center?

26. Nonreflective coatings on camera lenses reduce the loss of light at the surfaces of multi-lens systems and prevent internal reflections that might mar the image. Find the minimum thickness of a layer of magnesium fluoride ($n = 1.38$) on flint glass ($n = 1.66$) that will cause destructive interference of reflected light of wavelength 550 nm near the middle of the visible spectrum.

27. A thin film of MgF_2 ($n = 1.38$) with thickness 1.00×10^{-5} cm is used to coat a camera lens. Are any wavelengths in the visible spectrum intensified in the reflected light?

28. A flat piece of glass is supported horizontally above the flat end of a 10.0-cm-long metal rod that has its lower end rigidly fixed. The thin film of air between the rod and glass is observed to be bright when illuminated by light of wavelength 500 nm. As the temperature is slowly increased by 25.0°C, the film changes from bright to dark and back to bright 200 times. What is the coefficient of linear expansion of the metal?

Section 24.7   Single-Slit Diffraction

29. Helium-neon laser light ($\lambda = 632.8$ nm) is sent through a 0.300-mm-wide single slit. What is the width of the central maximum on a screen 1.00 m from the slit?

30. Light of wavelength 600 nm falls on a 0.40-mm-wide slit and forms a diffraction pattern on a screen 1.5 m away. (a) Find the position of the first dark band on each side of the central maximum. (b) Find the width of the central maximum.

31. Light of wavelength 587.5 nm illuminates a single 0.75-mm-wide slit. (a) At what distance from the slit should a screen be placed if the first minimum in the diffraction pattern is to be 0.85 mm from the central maximum? (b) Calculate the width of the central maximum.

32. Microwaves of wavelength 5.00 cm enter a long, narrow window in a building that is otherwise essentially opaque to the microwaves. If the window is 36.0 cm wide, what is the distance from the central maximum to the first-order minimum along a wall 6.50 m from the window?

33. A slit of width 0.50 mm is illuminated with light of wavelength 500 nm, and a screen is placed 120 cm in front of the slit. Find the widths of the first and second maxima on each side of the central maximum.

34. A screen is placed 50.0 cm from a single slit, which is illuminated with light of wavelength 680 nm. If the distance between the first and third minima in the diffraction pattern is 3.00 mm, what is the width of the slit?

Section 24.8   The Diffraction Grating

35. Three discrete spectral lines occur at angles of 10.1°, 13.7°, and 14.8° in the first-order spectrum of a diffraction grating spectrometer. (a) If the grating has 3 660 slits/cm, what are the wavelengths of the light? (b) At what angles are these lines found in the second-order spectra?

36. Intense white light is incident on a diffraction grating that has 600 lines/mm. (a) What is the highest order in which the complete visible spectrum can be seen using this grating? (b) What is the angular separation between the violet edge (400 nm) and the red edge (700 nm) of the first-order spectrum produced by this grating?

37. The hydrogen spectrum has a red line at 656 nm and a violet line at 434 nm. What is the angular separation between these two spectral lines obtained with a diffraction grating that has 4 500 lines/cm?

38. A grating with 1 500 slits per centimeter is illuminated with light of wavelength 500 nm. (a) What is the highest-order number that can be observed with this grating? (b) Repeat for a grating of 15 000 slits per centimeter.

39. A light source emits two major spectral lines, an orange line of wavelength 610 nm and a blue-green line of wavelength 480 nm. If the spectrum is resolved by a diffraction grating having 5 000 lines/cm and viewed on a screen 2.00 m from the grating, what is the distance (in centimeters) between the two spectral lines in the second-order spectrum?

40. White light is spread out into its spectral components by a diffraction grating. If the grating has 2 000 lines per centimeter, at what angle does red light of wavelength 640 nm appear in the first-order spectrum?

41. Sunlight is incident on a diffraction grating that has 2 750 lines/cm. The second-order spectrum over the visible range (400–700 nm) is to be limited to 1.75 cm along a screen that is distance L from the grating. What is the required value of L?

42. Light containing two different wavelengths passes through a diffraction grating with 1 200 slits/cm. On a screen 15.0 cm from the grating, the third-order maximum of the shorter wavelength falls midway between the central maximum and the first side maximum for the longer wavelength. If the neighboring maxima of the longer wavelength are 8.44 mm apart on the screen, what are the wavelengths in the light? (*Hint:* Use the small-angle approximation.)

43. A beam of 541-nm light is incident on a diffraction grating
web that has 400 lines/mm. (a) Determine the angle of the second-order ray. (b) If the entire apparatus is immersed in water, determine the new second-order angle of diffraction. (c) Show that the two diffracted rays of parts (a) and (b) are related through the law of refraction.

44. Light from an helium-neon laser ($\lambda = 632.8$ nm) is incident on a single slit. What is the maximum width for which no diffraction minima are observed? (*Hint:* Values of $\sin \theta > 1$ are not possible.)

Section 24.9 Polarization of Light Waves

45. The angle of incidence of a light beam in air onto a reflecting surface is continuously variable. The reflected ray is found to be completely polarized when the angle of incidence is 48.0°. (a) What is the index of refraction of the reflecting material? (b) If some of the incident light (at an angle of 48°) passes into the material below the surface, what is the angle of refraction?

46. Unpolarized light passes through two polaroid sheets. The axis of the first is vertical, and that of the second is at 30.0° to the vertical. What fraction of the initial light is transmitted?

47. The index of refraction of a glass plate is 1.52. What is the Brewster's angle when the plate is (a) in air? (b) in water? (See Problem 51.)

48. At what angle above the horizon is the Sun if light from it is completely polarized upon reflection from water?

49. A light beam is incident on heavy flint glass ($n = 1.65$) at the polarizing angle. Calculate the angle of refraction for the transmitted ray.

50. The critical angle for total internal reflection for sapphire surrounded by air is 34.4°. Calculate the Brewster's angle for sapphire if the light is incident from the air.

51. Equation 24.14 assumes that the incident light is in air. If the light is incident from a medium of index n_1 on a medium of index n_2, follow the procedure used to derive Equation 24.14 to show that $\tan \theta_p = n_2/n_1$.

52. Plane-polarized light is incident on a single polarizing disk with the direction of E_0 parallel to the direction of the transmission axis. Through what angle should the disk be rotated so that the intensity in the transmitted beam is reduced by a factor of (a) 3.00, (b) 5.00, (c) 10.0?

53. Three polarizing plates whose planes are parallel are centered on a common axis. The directions of the transmission axes relative to the common vertical direction are shown in Figure P24.53. A linearly polarized beam of light with the plane of polarization parallel to the vertical reference direction is incident from the left on the first disk with intensity $I_i = 10.0$ units (arbitrary). Calculate the transmitted intensity I_f when $\theta_1 = 20.0°$, $\theta_2 = 40.0°$, and $\theta_3 = 60.0°$. (*Hint:* Make repeated use of Malus's law.)

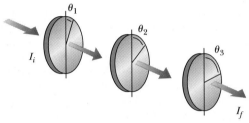

FIGURE P24.53 (Problems 53 and 62)

54. Light of intensity I_0 and polarized parallel to the transmis-
web sion axis of a polarizer, is incident on an analyzer. (a) If the transmission axis of the analyzer makes an angle of 45° with the axis of the polarizer, what is the intensity of the transmitted light? (b) What should the angle between the transmission axes be to make $I/I_0 = 1/3$?

55. Light with a wavelength in vacuum of 546.1 nm falls perpendicularly on a biological specimen that is 1.000 μm thick. The light splits into two beams polarized at right angles, for which the indices of refraction are 1.320 and 1.333. (a) Calculate the wavelength of each component of the light while it is traversing the specimen. (b) Calculate the phase difference between the two beams when they emerge from the specimen.

ADDITIONAL PROBLEMS

56. A beam containing light of wavelengths λ_1 and λ_2 is incident on a set of parallel slits. In the interference pattern, the fourth bright line of the λ_1 light occurs at the same position as the fifth bright line of the λ_2 light. If λ_1 is known to be 540 nm, what is the value of λ_2?

57. Light of wavelength 546 nm (the intense green line from a mercury source) produces a Young's interference pattern in which the second minimum from the central maximum is along a direction that makes an angle of 18.0 min of arc with the axis through the central maximum. What is the distance between the parallel slits?

58. The two speakers of a boom box are 35.0 cm apart. A single oscillator makes the speakers vibrate in phase at a frequency of 2.00 kHz. At what angles, measured from the perpendicular bisector of the line joining the speakers, would a distant observer hear maximum sound intensity?

minimum sound intensity? (Take the speed of sound as 340 m/s.)

59. Interference effects are produced at point *P* on a screen as a result of direct rays from a 500-nm source and reflected rays off the mirror, as in Figure P24.59. If the source is 100 m to the left of the screen, and 1.00 cm above the mirror, find the distance *y* (in millimeters) to the first dark band above the mirror.

FIGURE P24.59

60. Many cells are transparent and colorless. Structures of great interest in biology and medicine can be practically invisible to ordinary microscopy. An *interference microscope* reveals a difference in refractive index as a shift in interference fringes, to indicate the size and shape of cell structures. The idea is exemplified in the following problem: An air wedge is formed between two glass plates in contact along one edge and slightly separated at the opposite edge. When the plates are illuminated with monochromatic light from above, the reflected light has 85 dark fringes. Calculate the number of dark fringes that appear if water (*n* = 1.33) replaces the air between the plates.

61. When a monochromatic beam of light is incident from air at an angle of 37.0° with the normal on the surface of a glass block, it is observed that the refracted ray is directed at 22.0° with the normal. What angle of incidence from air would result in total polarization of the reflected beam?

62. Three polarizers, centered on a common axis and with their planes parallel to each other, have transmission axes oriented at angles of θ_1, θ_2, and θ_3 from the vertical as shown in Figure P24.53. Light of intensity I_i, polarized with its plane of polarization oriented vertically, is incident from the left on the first polarizer. What is the ratio I_f/I_i of the final transmitted intensity to the incident intensity if (a) $\theta_1 = 45°$, $\theta_2 = 90°$, and $\theta_3 = 0°$? (b) $\theta_1 = 0°$, $\theta_2 = 45°$, and $\theta_3 = 90°$?

63. Figure P24.63 shows a radio wave transmitter and a receiver, both *h* = 50.0 m above the ground and *d* = 600 m

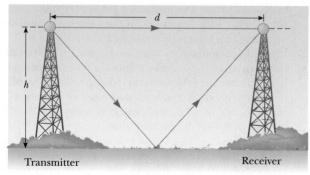

FIGURE P24.63

apart. The receiver can receive signals both directly, from the transmitter, and indirectly, from signals that bounce off the ground. If the ground is level between the transmitter and receiver and a $\lambda/2$ phase shift occurs upon reflection, determine the longest wavelengths that interfere (a) constructively and (b) destructively.

64. A planoconvex lens (flat on one side, convex on the other) with index of refraction *n* rests with its curved side (radius of curvature *R*) on a flat glass surface of the same index of refraction with a film of index n_{film} between them. The lens is illuminated from above by light of wavelength λ. Show that the dark Newton rings have radii of

$$r \approx \sqrt{m\lambda R / n_{film}}$$

where *m* is an integer.

65. When a liquid is introduced into the air space between the lens and the plate in a Newton's rings apparatus, the diameter of the tenth ring changes from 1.50 to 1.31 cm. Find the index of refraction of the liquid. (See Problem 64.)

66. (a) If light is incident at an angle θ from a medium of index n_1 on a medium of index n_2 so that the angle between the reflected ray and refracted ray is β, show that

$$\tan \theta = \frac{n_2 \sin \beta}{n_1 - n_2 \cos \beta}$$

Hint: Use the following identity.

$$\sin(A + B) = \sin A \cos B + \cos A \sin B$$

(b) Show that the foregoing equation for tan θ reduces to Brewster's law when $\beta = 90°$, $n_1 = 1$, and $n_2 = n$.

67. A diffraction pattern is produced on a screen 140 cm from a single slit, using monochromatic light of wavelength 500 nm. The distance from the center of the central maximum to the first-order maximum is 3.00 mm. Calculate the slit width. (*Hint:* Assume that the first-order maximum is halfway between the first- and second-order minima.)

68. A glass plate (*n* = 1.61) is covered with a thin, uniform layer of oil (*n* = 1.20). A light beam of variable wavelength is normally incident from air onto the oil surface. Observation of the reflected beam shows destructive interference at 500 nm and constructive interference at 750 nm. From this information, calculate the thickness of the oil film.

69. The condition for constructive interference by reflection from a thin film in air, as developed in Section 24.4, assumes nearly normal incidence. (a) Show that for large angles of incidence, the condition for constructive interference of light reflecting from a thin film of thickness *t*, index of refraction *n*, and surrounded by air may be written as

$$2nt \cos \theta_2 = (m + \tfrac{1}{2})\lambda$$

where θ_2 is the angle of refraction. (b) Calculate the minimum thickness for constructive interference if sodium light ($\lambda = 590$ nm) is incident at an angle of 30.0° on a film with index of refraction 1.38.

70. Figure P24.70 illustrates the formation of an interference pattern by the Lloyd's mirror method. Light from source *S* reaches the screen via two different pathways. One is a direct path, and the second is by reflection from a horizontal mirror. The interference is as though light from two

different sources S and S' had interfered as in the Young's double-slit arrangement. Assume that the actual source S and the virtual source S' are in a plane 25 cm to the left of the mirror, and the screen is a distance of $L = 120$ cm to the right of this plane. Source S is a distance of $h = 2.5$ mm above the top surface of the mirror, and the light is monochromatic with $\lambda = 620$ nm. Determine the distance of the first bright fringe above the surface of the mirror.

FIGURE P24.70

GROUP ACTIVITIES

G.1 A simple way to observe interference effects is to look at a distant light source through a stretched handkerchief or an opened umbrella. You will observe the light passing through the threads of the cloth spreads out to form streaks. Explain why the streaks look as they do.

G.2 Place a clear dish or plate on a black surface such as a sheet of black construction paper. Now add a thin layer of water to the glass and place a few drops of kerosene or light machine oil on the water. Darken the room and shine a flashlight from an angle, as in Figure GA24.2. Note the interference pattern of various colors you observe under the white light. How does the pattern change if you cover the flashlight with a sheet of red, blue, or green cellophane, which acts as a filter?

As an extension of the above, observe the colors that appear to swirl on the surface of a soap bubble. What color do you see just before a bubble bursts?

Oil

Water

FIGURE GA24.2

G.3 Make a V with your index and middle fingers. Hold your hand up very close to your eye so that you are looking between your two fingers toward a distant light source. Now bring the fingers together until there is only a very tiny slit between them. You should be able to see a series of parallel lines. Although the lines appear to be located in the narrow space between your fingers, what you are actually seeing is a diffraction pattern cast upon your retina.

G.4 Stand a couple of meters from a lightbulb. Facing away from the light, hold a compact disc about 10 cm from your eye and tilt it until the reflection of the bulb is located in the hole at the disc's center. You should see spectra radiating out from the center, with violet on the inside and red on the outside. Now move the disc away

from your eye until the violet band is at the outer edge. Carefully measure the distance from your eye to the center of the disc and also determine the radius of the disc. Use this information to find the angle θ to the first-order maximum for violet light. Now use $d \sin \theta = m\lambda$ to determine the spacing between the grooves of the disc. The industry standard is 1.6 μm. How close did you come? While you are observing the spectrum from a CD, note that the color of the light from a given point changes with the viewing angle. Explain this in terms of changes in the path length. It is of interest that the blues and blue-greens in peacock feathers and butterflies are caused by diffraction off finely aligned structures in feathers and wings (see chapter opener photo).

G.5 (a) Devise a way to use a protractor, desklamp, and polarizing sunglasses to measure Brewster's angle for the glass in a window. From this, determine the index of refraction of the glass. (b) Put on a pair of polarizing sunglasses and close one eye. Hold up a lens of a second pair of polarizing glasses in front of your open eye so that light must pass through a lens of each pair before entering your eye. Now rotate the second pair of glasses around. You will note that the light reaching your eye is considerably reduced at some orientations and will pass freely at others. (c) On a sunny day, rotate your polarizing sunglasses in front of your eye and observe light reflection from a window or the surface of water. Note the change in the amount of light entering your eye for various orientations of the glasses. (d) For a final observation concerning polarized light, rotate a pair of polarizing sunglasses as you observe various areas of the sky. From what direction do you find the light to be most highly polarized?

G.6 (a) In case 1, light from a laser passes through a double slit that has a separation d. The light then forms an interference pattern on a screen a distance L away. In case 2, the slit is replaced by a slit with a separation $2d$. In which case is the distance from the central maximum to the next maximum the greatest? (b) Check your logic by considering light of wavelength $\lambda = 694.3$ nm, the distance $d = 0.100$ mm, and $L = 2.50$ m.

G.7 Light of a certain wavelength passes through a double-slit arrangement that has a separation d and forms a pattern on a screen a distance L away. (a) If light having a wavelength 10% smaller now shines on the slits, is the distance from the central maximum to the next maximum larger or smaller? (b) Based on your logic for case (a), what will happen when visible light containing all wavelengths falls on the slits? (c) Let the first light have a wavelength of $\lambda = 694.3$ nm, the second wavelength to be 10% smaller, $d = 0.100$ mm and $L = 2.50$ m to check your reasoning.

G.8 Light of wavelength λ in air shines on an oil film of index n_o floating on water where $n_w > n_o$. (a) What is the wavelength of the light in the oil? (b) Is there a phase change upon reflection at the oil-water boundary? (c) Use $\lambda = 450$ nm and $n_o = 1.25$ to find the minimum thickness for constructive interference to occur at the top of the oil film.

G.9 Light in air that is reflected from a water surface is found to be completely polarized at an angle θ. (a) If the light is instead reflected from a glass coffee table, will the new angle for complete polarization be larger or smaller? (b) Use $n_g = 1.54$ for glass and $n_w = 1.33$ for water to check your prediction.

G.10 Light of a certain wavelength passes through a diffraction grating that has a slit separation d and forms a pattern on a screen. (a) If light having a wavelength 10% smaller now shines on the grating, are the bright lines separated by larger, smaller, or the same angles? (b) Based on your logic for case (a), what will happen when visible light containing all wavelengths falls on the slits? (c) Let the first light have a wavelength $\lambda = 694.3$ nm, take the second wavelength to be 10% smaller, assume $d = 1\,500$ nm, and find the diffraction angles.

Optical Instruments

A normal lens for a camera might have a focal length of 50 mm. A telephoto lens might have a focal length of 200 mm, resulting in a larger image on the film.

(Courtesy of Canon, USA)

We use devices made from lenses, mirrors, or other optical components every time we put on a pair of eyeglasses or contact lenses, take a photograph, look at the sky through a telescope, and so on. In this chapter we examine how these and other optical instruments work. For the most part, our analyses will involve the laws of reflection and refraction and the procedures of geometric optics. However, to explain certain phenomena we must use the wave nature of light.

25.1   THE CAMERA

The single-lens photographic **camera** is a simple optical instrument whose essential features are shown in Figure 25.1. It consists of a light-tight box, a converging lens that produces a real image, and a film behind the lens to receive the image. Focusing is accomplished by varying the distance between lens and film—with an adjustable bellows in antique cameras and with some other mechanical arrangements in contemporary models. For proper focusing, which leads to sharp images, the lens-to-film distance will depend on the object distance as well as on the focal length of the lens. The shutter, located behind the lens, is a mechanical device that is opened for selected time intervals. With this arrangement, moving objects can be photographed by using short exposure times, and dark scenes (low light levels) by using long exposure times. If this adjustment were not available, it would be impossible to take stop-action photographs. For example, a rapidly moving vehicle could move far enough while the shutter was open to produce a blurred image. Another major cause of blurred images is movement of the camera while the shutter is open. To prevent such movement, you should mount the camera on a tripod or use short exposure times. Typical shutter speeds (that is, exposure times) are 1/30, 1/60, 1/125, and 1/250 s. Stationary objects are often shot with a shutter speed of 1/60 s.

Most cameras also have an aperture of adjustable diameter to further control the intensity of the light reaching the film. When an aperture of small diameter is used, only light from the central portion of the lens reaches the film, and so spherical aberration is reduced.

The intensity I of the light reaching the film is proportional to the area of the lens. Because this area is proportional to the square of the lens diameter D, the intensity is also proportional to D^2. Light intensity is a measure of the rate at which energy is received by the film per unit area of the image. Because the area of the image is proportional to q^2, and $q \approx f$ (when $p \gg f$, so that p can be approximated as infinite), we conclude that the intensity is also proportional to $1/f^2$, so that $I \propto D^2/f^2$. The brightness of the image formed on the film depends on the light intensity, and so we see that the image brightness depends on both the focal length and diameter D of the lens.

The ratio f/D is called the ***f*-number** of a lens:

$$f\text{-number} \equiv \frac{f}{D} \qquad\qquad \textbf{[25.1]}$$

The *f*-number is often given as a description of the lens "speed." The lens with a low *f*-number is a "fast" lens. Extremely fast lenses, which have an *f*-number as low as approximately 1.2, are expensive because it is very difficult to keep aberrations acceptably small with light rays passing through a large area of the lens. Camera lenses are often marked with a range of *f*-numbers such as 1.4, 2, 2.8, 4, 5.6, 8, 11, ... Any one of these settings can be selected by adjusting the aperture, which changes the value of D. Increasing the setting from one *f*-number to the next higher value (for example, from 2.8 to 4) decreases the area of the aperture by a factor of 2. The lowest *f*-number setting on a camera corresponds to the aperture wide open and the maximum possible lens area in use.

FIGURE 25.1   A cross-sectional view of a simple camera.

Simple cameras usually have a fixed focal length and fixed aperture size, with an *f*-number of about 11. This high value for the *f*-number allows for a large **depth of field.** This means that objects at a wide range of distances from the lens form reasonably sharp images on the film. In other words, the camera does not have to be focused. Most cameras with variable *f*-numbers adjust them automatically.

25.2 THE EYE

Like a camera, a normal eye focuses light and produces a sharp image. However, the mechanisms by which the eye controls the amount of light admitted and adjusts to produce correctly focused images are far more complex, intricate, and effective than those in even the most sophisticated camera. In all respects, the eye is a physiological wonder.

Figure 25.2 shows the essential parts of the eye. Light entering the eye passes through a transparent structure called the *cornea,* behind which are a clear liquid (the *aqueous humor*), a variable aperture (the *pupil,* which is an opening in the *iris*), and the *crystalline lens.* Most of the refraction occurs at the outer surface of the eye, at which the cornea is covered with a film of tears. Relatively little refraction occurs in the crystalline lens because the aqueous humor in contact with the lens has an average index of refraction close to that of the lens. The iris, which is the colored portion of the eye, is a muscular diaphragm that controls pupil size. The iris regulates the amount of light entering the eye by dilating the pupil in low-light conditions and contracting the pupil in high-light conditions. The *f*-number range of the eye is from about 2.8 to 16.

The cornea-lens system focuses light onto the back surface of the eye, the *retina,* which consists of millions of sensitive receptors called *rods* and *cones.* When stimulated by light, these structures send impulses via the optic nerve to the brain, which converts these impulses into our conscious view of the world. The process by which the brain performs this conversion is not well understood and is the subject of much speculation and research. Unlike film in a camera, the rods and cones chemically adjust their sensitivity according to the prevailing light conditions. This adjustment, which takes about 15 minutes, is responsible for the experience of

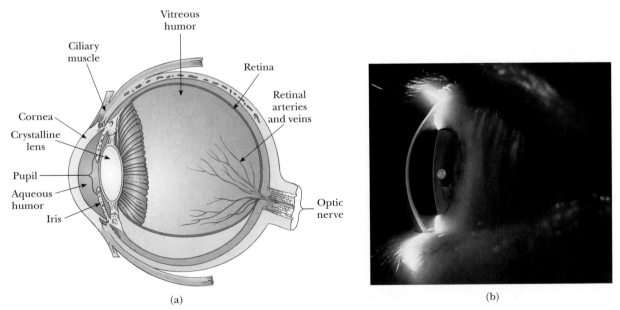

(a)   (b)

FIGURE 25.2 (a) Essential parts of the eye. Can you correlate the essential parts of the eye with those of the simple camera in Figure 25.1? (b) Close-up photograph of the human cornea. *(From Lennart Nilson, in collaboration with Jan Lindberg,* Behold Man: A Photographic Journey of Discovery Inside the Body, *Boston, Little, Brown & Co., 1974)* "getting used to the dark" in such places as movie theaters. Iris aperture control, which takes less than 1 second, helps protect the retina from overload in the adjustment process.

The eye focuses on an object by varying the shape of the pliable crystalline lens through an amazing process called **accommodation.** An important component in accommodation is the *ciliary muscle,* which is situated in a circle around the rim of the lens. Thin filaments, called *zonules,* run from this muscle to the edge of the lens. When the eye is focused on a distant object, the ciliary muscle is relaxed, tightening the zonules that attach the ciliary muscle to the edge of the lens. The force of the zonules causes the lens to flatten, increasing its focal length. For an object distance of infinity, the focal length of the eye is equal to the fixed distance between lens and retina, about 1.7 cm. The eye focuses on nearby objects by tensing the ciliary muscle, which relaxes the zonules. This action allows the lens to bulge a bit and its focal length decreases, resulting in the image being focused on the retina. All these lens adjustments take place so swiftly that we are not even aware of the change. In this respect, even the finest electronic camera is a toy compared with the eye.

There is a limit to accommodation because objects that are very close to the eye produce blurred images. The **near point** is the closest distance for which the lens can accommodate to focus light on the retina. This distance usually increases with age and has an average value of 25 cm. Typically, at age 10 the near point of the eye is about 18 cm. This increases to about 25 cm at age 20, to 50 cm at age 40, and to 500 cm or greater at age 60. The **far point** of the eye represents the largest distance for which the lens of the relaxed eye can focus light on the retina. A person with normal vision is able to see very distant objects, such as the Moon, and thus has a far point at infinity.

CONDITIONS OF THE EYE

When the eye suffers a mismatch between the focusing power of the lens-cornea system and the length of the eye such that light rays reach the retina before they converge to form an image, as in Figure 25.3a, the condition is known as **farsight-**

FIGURE 25.3 (a) A farsighted eye is slightly shorter than normal; hence, the image of a nearby object focuses *behind* the retina. (b) The condition can be corrected with a converging lens. (The object is assumed to be very small in these figures.)

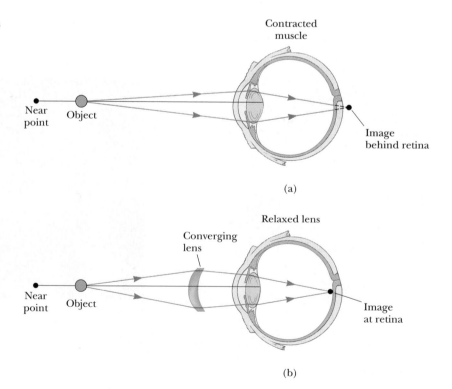

edness (or *hyperopia*). A farsighted person can usually see faraway objects clearly but not nearby objects. Although the near point of a normal eye is approximately 25 cm, the near point of a farsighted person is much farther than this. The eye of a farsighted person tries to focus by accommodation, that is, by shortening its focal length. This works for distant objects, but because the focal length of the far-sighted eye is longer than normal, the light from nearby objects cannot be brought to a sharp focus before it reaches the retina, causing a blurred image. There is insufficient refracting power in the cornea and lens to satisfactorily focus the light from all but distant objects. The condition can be corrected by placing a converging lens in front of the eye, as in Figure 25.3b. The lens refracts the incoming rays more toward the principal axis before entering the eye, allowing them to converge and focus on the retina.

Nearsightedness (or *myopia*) is another mismatch condition in which a person is able to focus on nearby objects but not faraway objects. In the case of *axial myopia,* nearsightedness is caused by the lens being too far from the retina. It is also possible to have *refractive myopia,* in which case the lens-cornea system is too powerful for the normal length of the eye. The far point of the nearsighted eye is not at infinity and may be less than a meter. The maximum focal length of the nearsighted eye is insufficient to produce a sharp image on the retina, and rays from a distant object converge to a focus in front of the retina. They then continue past that point, diverging before they finally reach the retina and causing blurred vision (Fig. 25.4a). Nearsightedness can be corrected with a diverging lens, as shown in Figure 25.4b. The lens refracts the rays away from the principal axis before they enter the eye, a modification that allows them to focus on the retina.

Beginning with middle age, most people lose some of their accommodation ability as the ciliary muscle weakens and the lens hardens. Unlike farsightedness, which is a mismatch of focusing power and eye length, **presbyopia** (literally, "old-age vision") is due to a reduction in accommodation ability. The cornea and lens do not have sufficient focusing power to bring nearby objects into focus on the retina. The symptoms are the same as with farsightedness, and the condition can be corrected with converging lenses.

APPLICATION

CORRECTING FOR DEFECTS WITH OPTICAL LENSES

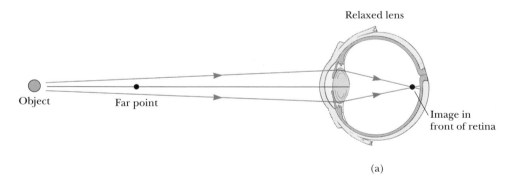

Relaxed lens

Object   Far point

Image in front of retina

(a)

FIGURE 25.4 (a) A nearsighted eye is slightly longer than normal; hence, the image of a distant object focuses *in front* of the retina. (b) The condition can be corrected with a diverging lens. (The object is assumed to be very small in these figures.)

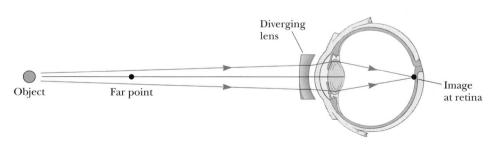

Diverging lens

Object   Far point

Image at retina

(b)

In the eye defect known as **astigmatism,** light from a point source produces a line image on the retina. This condition arises when either the cornea or the lens or both are not perfectly symmetric. Astigmatism can be corrected with lenses having different curvatures in two mutually perpendicular directions.

Optometrists and ophthalmologists usually prescribe lenses measured in **diopters:**

> **The power $\mathscr{P}$ of a lens in diopters equals the inverse of the focal length in meters: $\mathscr{P} = 1/f$.**

For example, a converging lens of focal length $+20$ cm has a power of $+5.0$ diopters, and a diverging lens of focal length -40 cm has a power of -2.5 diopters.

APPLYING **PHYSICS** 25.1

A classic science fiction story, *The Invisible Man,* tells of a person who becomes invisible by changing the index of refraction of his body to that of air. This story has been criticized by students who know how the eye works; they claim the invisible man would be unable to see. On the basis of your knowledge of the eye, could he see or not?

Explanation   He would not be able to see. In order for the eye to "see" an object, incoming light must be refracted at the cornea and lens to form an image on the retina. If the cornea and lens have the same index of refraction as air, refraction cannot occur, and an image would not be formed.

Example 25.1   Prescribing a Lens

The near point of an eye is 50.0 cm.

A What focal length must a corrective lens have to enable the eye to see clearly an object 25.0 cm away?

Reasoning   The thin-lens equation (Eq. 23.11) enables us to solve this problem. We have placed an object at 25.0 cm, and we want the lens to form an image at the closest point that the eye can see clearly. This corresponds to the near point, 50.0 cm.

Solution   Applying the thin-lens equation, we have

$$\frac{1}{25.0 \text{ cm}} + \frac{1}{(-50.0 \text{ cm})} = \frac{1}{f}$$

$$f = \boxed{50.0 \text{ cm}}$$

Why did we use a negative sign for the image distance? Notice that the focal length is positive, indicating the need for a converging lens to correct farsightedness such as this.

B What is the power of this lens?

Solution   The power is the reciprocal of the focal length in meters:

$$\mathscr{P} = \frac{1}{f} = \frac{1}{0.500 \text{ m}} = \boxed{2.00 \text{ diopters}}$$

Example 25.2   A Case of Nearsightedness

A particular nearsighted person cannot see objects clearly when they are beyond 25 cm (the far point of the eye). What focal length should the prescribed lens have to correct this problem?

Solution   The purpose of the lens in this instance is to "move" an object from infinity to a distance where it can be seen clearly. This is accomplished by having the lens produce an image at the far point. From the thin-lens equation, we have

$$\frac{1}{p} + \frac{1}{q} = \frac{1}{\infty} + \frac{1}{(-25\ \text{cm})} = \frac{1}{f}$$

$$f = \boxed{-25\ \text{cm}}$$

Why did we use a negative sign for the image distance? As you should have suspected, the lens must be diverging (one with a negative focal length) to correct nearsightedness.

EXERCISE   What is the power of this lens?

ANSWER -0.40 diopters

Webnote 25.1

To study how corrective laser eye surgery is related to the structures of the eye, visit
http://www.medscape.com/ viewarticle/200310_4

25.3   THE SIMPLE MAGNIFIER

The **simple magnifier** is one of the most basic of all optical instruments because it consists only of a single converging lens. As the name implies, this device is used to increase the apparent size of an object. Suppose an object is viewed at some distance p from the eye, as in Figure 25.5. Clearly, the size of the image formed at the retina depends on the angle θ subtended by the object at the eye. As the object moves closer to the eye, θ increases and a larger image is observed. However, a normal eye cannot focus on an object closer than about 25 cm, the near point (Fig. 25.6a). (Try it!) Therefore, θ is maximum at the near point.

To further increase the apparent angular size of an object, a converging lens can be placed in front of the eye with the object positioned at point O, just inside the focal point of the lens, as in Figure 25.6b. At this location, the lens forms a virtual, upright, and enlarged image, as shown. The lens allows the object to be

FIGURE 25.5   The size of the image formed on the retina depends on the angle θ subtended at the eye.

(a)

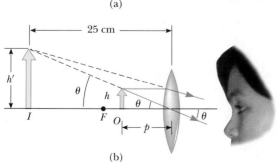

(b)

FIGURE 25.6   (a) An object placed at the near point ($p = 25$ cm) subtends an angle of $\theta_0 \approx h/25$ at the eye. (b) An object placed near the focal point of a converging lens produces a magnified image, which subtends an angle of $\theta \approx h'/25$ at the eye. Note that in this situation, $q = -25$ cm.

viewed closer to the eye than is possible without the lens. We define the **angular magnification** m as the ratio of the angle subtended by the object when the lens is in use (angle θ in Fig. 25.6b) to the angle subtended by the object placed at the near point with no lens in use (angle θ_0 in Fig. 25.6a):

Angular magnification with the object at the near point

$$m \equiv \frac{\theta}{\theta_0} \qquad\qquad \text{[25.2]}$$

The angular magnification is a maximum when the image formed by the lens is at the near point of the eye—that is, when $q = -25$ cm (see Fig. 25.6b). The object distance corresponding to this image distance can be calculated from the thin-lens equation:

$$\frac{1}{p} + \frac{1}{(-25 \text{ cm})} = \frac{1}{f}$$

$$\qquad\qquad \text{[25.3]}$$

$$p = \frac{25f}{25 + f}$$

where f is the focal length of the magnifier in centimeters. From Figures 25.6a and 25.6b, the small-angle approximation gives

$$\tan \theta_0 \approx \theta_0 \approx \frac{h}{25} \qquad \text{and} \qquad \tan \theta \approx \theta \approx \frac{h}{p} \qquad\qquad \text{[25.4]}$$

Thus, Equation 25.2 becomes

$$m_{\text{max}} = \frac{\theta}{\theta_0} = \frac{h/p}{h/25} = \frac{25}{p} = \frac{25}{25f/(25 + f)}$$

$$= 1 + \frac{25 \text{ cm}}{f} \qquad\qquad \text{[25.5]}$$

The maximum angular magnification given by Equation 25.5 is the ratio of the angular size seen with the lens to the angular size seen without the lens, with the object at the near point of the eye. Although the normal eye can focus on an image formed anywhere between the near point and infinity, it is most relaxed when the image is at infinity (Sec. 25.2). For the image formed by the magnifying lens to appear at infinity, the object must be placed at the focal point of the lens— that is, $p = f$. In this case, Equation 25.4 becomes

$$\theta_0 \approx \frac{h}{25} \qquad \text{and} \qquad \theta \approx \frac{h}{f}$$

and the angular magnification is

$$m = \frac{\theta}{\theta_0} = \frac{25 \text{ cm}}{f} \qquad\qquad \text{[25.6]}$$

With a single lens, it is possible to achieve angular magnifications up to about 4 without serious aberrations. Magnifications up to about 20 can be achieved by using one or two additional lenses to correct for aberrations.

Example 25.3   Maximum Angular Magnification of a Lens

What is the maximum angular magnification of a lens with a focal length of 10.0 cm, and what is the angular magnification of this lens when the eye is relaxed?

Reasoning   The maximum angular magnification occurs when the image formed by the lens is at the near point of the eye. Under these circumstances, Equation 25.5 gives us the maximum angular magnification.

Solution Using Equation 25.5, we have

$$m_{max} = 1 + \frac{25\ cm}{f} = 1 + \frac{25\ cm}{10.0\ cm} = \boxed{3.5}$$

When the eye is relaxed, the image is at infinity. In this case, we use Equation 25.6:

$$m = \frac{25\ cm}{f} = \frac{25\ cm}{10.0\ cm} = \boxed{2.5}$$

25.4 THE COMPOUND MICROSCOPE

A simple magnifier provides only limited assistance with inspection of the minute details of an object. Greater magnification can be achieved by combining two lenses in a device called a compound microscope, a diagram of which is shown in Figure 25.7a. It consists of two lenses: an objective with a very short focal length f_o (where $f_o < 1$ cm), and an ocular lens, or eyepiece, with a focal length f_e of a few centimeters. The two lenses are separated by distance L which is much greater than either f_o or f_e.

As you read the discussion that follows, note that the basic approach used to analyze the image formation properties of a microscope is that of two lenses in a row—that is, the image formed by the first becomes the object for the second. The object O placed just outside the focal length of the objective, forms a real, inverted image at I_1, which is at or just inside the focal point of the eyepiece. This image is real and much enlarged. (For clarity, the enlargement of I_1 is not shown in Fig. 25.7a.) The eyepiece, which serves as a simple magnifier, uses the image at I_1 as its object and produces an image at I_2. The image seen by the eye at I_2 is virtual, inverted, and very much enlarged.

The lateral magnification M_1 of the first image is $-q_1/p_1$. Note in Figure 25.7a that q_1 is approximately equal to L. This occurs because the object is placed close to the focal point of the objective lens, which ensures that the image formed will be far from the objective lens. Furthermore, because the object is very close to the focal point of the objective lens, $p_1 \approx f_o$. This gives a lateral magnification of

$$M_1 = -\frac{q_1}{p_1} \approx -\frac{L}{f_o}$$

for the objective. The angular magnification of the eyepiece for an object (corresponding to the image at I_1) placed at the focal point is found from Equation 25.6 to be

$$m_e = \frac{25\ cm}{f_e}$$

(a)

(b)

FIGURE 25.7 (a) A diagram of a compound microscope, which consists of an objective and an eyepiece, or ocular lens. (b) A compound microscope. The three-objective turret allows the user to switch to several different powers of magnification. Combinations of eyepieces with different focal lengths and different objectives can produce a wide range of magnifications. *(Henry Leap and Jim Lehman)*

The overall magnification of the compound microscope is defined as the product of the lateral and angular magnifications:

Magnification of a microscope ▶

$$m = M_1 m_e = -\frac{L}{f_o}\left(\frac{25\text{ cm}}{f_e}\right) \qquad \text{[25.7]}$$

The negative sign indicates that the image is inverted with respect to the object.

The microscope has extended our vision into the previously unknown realm of incredibly small objects, and the capabilities of this instrument have increased steadily with improved techniques in precision grinding of lenses. A question that is often asked about microscopes is, "With extreme patience and care, would it be possible to construct a microscope that would enable us to see an atom?" The answer to this question is no, as long as visible light is used to illuminate the object. The reason is that, in order to be seen, the object under a microscope must be at least as large as a wavelength of light. An atom is many times smaller than the wavelength of visible light, and so its mysteries must be probed via other techniques.

The wavelength dependence of the "seeing" ability of a wave can be illustrated by water waves set up in a bathtub in the following manner. Imagine that you vibrate your hand in the water until waves with a wavelength of about 6 in. are moving along the surface. If you fix a small object, such as a toothpick, in the path of the waves, you will find that the waves are not appreciably disturbed by the toothpick but continue along their path. Now suppose you fix a larger object, such as a toy sailboat, in the path of the waves. In this case, the waves are considerably "disturbed" by the object. The toothpick was much smaller than the wavelength of the waves, and as a result the waves did not "see" it. The toy sailboat, in contrast, is about the same size as the wavelength of the waves and hence creates a disturbance. Light waves behave in this same general way. The ability of an optical microscope to view an object depends on the size of the object relative to the wavelength of the light used to observe it. Hence, it will never be possible to observe atoms or molecules with such a microscope, because their dimensions are so small (≈ 0.1 nm) relative to the wavelength of the light (≈ 500 nm).

Webnote 25.2

How do you effectively light and use a compound microscope? This web site will help you learn these skills. Visit
*http://www.microscopy-uk.org.uk/
mag/art98/incid1.html*

Example 25.4 Magnifications of a Microscope

A certain microscope has two interchangeable objectives. One has a focal length of 20.0 mm, and the other has a focal length of 2.0 mm. Also available are two eyepieces of focal lengths 2.5 cm and 5.0 cm. If the length of the microscope is 18 cm, what magnifications are possible?

Reasoning and Solution The solution consists of applying Equation 25.7 to four different combinations of lenses. For the combination of the two long focal lengths, we have

$$m = -\frac{L}{f_o}\left(\frac{25\text{ cm}}{f_e}\right) = -\frac{18}{2.0}\left(\frac{25}{5.0}\right) = \boxed{-45}$$

The combination of the 20.0-mm objective and the 2.5-cm eyepiece gives

$$m = -\frac{18}{2.0}\left(\frac{25}{2.5}\right) = \boxed{-90}$$

The 2.0-mm and 5.0-cm combination produces

$$m = -\frac{18}{0.20}\left(\frac{25}{5.0}\right) = \boxed{-450}$$

Finally, the two short focal lengths give

$$m = -\frac{18}{0.20}\left(\frac{25}{2.5}\right) = \boxed{-900}$$

25.5 THE TELESCOPE

There are two fundamentally different types of telescopes, both designed to help us view distant objects such as the planets in our Solar System. These two types are (1) the **refracting telescope,** which uses a combination of lenses to form an image, and (2) the **reflecting telescope,** which uses a curved mirror and a lens to form an image. Once again, we will be able to analyze the telescope by considering it to be a system of two optical elements in a row. The basic technique followed is that the image formed by the first element becomes the object for the second.

Let us first consider the refracting telescope. In this device, two lenses are arranged so that the objective forms a real, inverted image of the distant object very near the focal point of the eyepiece (Fig. 25.8a). Furthermore, the image at I_1 is formed at the focal point of the objective because the object is essentially at infinity. Hence, the two lenses are separated by the distance $f_o + f_e$, which corresponds to the length of the telescope's tube. The eyepiece finally forms, at I_2, an enlarged, inverted image of the image at I_1.

The angular magnification of the telescope is given by θ/θ_o, where θ_o is the angle subtended by the object at the objective and θ is the angle subtended by the final image. From the triangles in Figure 25.8a, and for small angles, we have

$$\theta \approx \frac{h'}{f_e} \quad \text{and} \quad \theta_o \approx \frac{h'}{f_o}$$

Therefore, the angular magnification of the telescope can be expressed as

$$m = \frac{\theta}{\theta_o} = \frac{h'/f_e}{h'/f_o} = \frac{f_o}{f_e} \tag{25.8}$$

This says that the angular magnification of a telescope equals the ratio of the objective focal length to the eyepiece focal length. Here again, the angular magnification is the ratio of the angular size seen with the telescope to the angular size seen with the unaided eye.

In some applications—for instance, the observation of relatively nearby objects such as the Sun, Moon, or planets—angular magnification is important.

The Hubble Space telescope now enables us to see both further into space and back in time than possible before its launch. *(NASA)*

◀ Angular magnification of a telescope

Webnote 25.3

How does the Hubble Space Telescope operate? Take a fascinating look to see how it is built. Visit *http://www.howstuffworks.com/hubble.htm*
Another good Hubble site is *http://hubble.stsci.edu/gallery*

(a)   (b)

FIGURE 25.8   (a) A diagram of a refracting telescope, with the object at infinity. (b) A photograph of a refracting telescope. *(Photo courtesy of Henry Leap and Jim Lehman)*

A •

M

Eyepiece

Parabolic
mirror

FIGURE 25.9 A reflecting telescope with a Newtonian focus.

Stars, in contrast, are so far away that they always appear as small points of light, regardless of how much angular magnification is used. The large research telescopes used to study very distant objects must have great diameters to gather as much light as possible. It is difficult and expensive to manufacture such large lenses for refracting telescopes. In addition, the heaviness of large lenses leads to sagging, which is another source of aberration.

These problems can be partially overcome by replacing the objective lens with a reflecting, concave mirror, which usually has a parabolic shape to avoid spherical aberration. Figure 25.9 shows the design for a typical reflecting telescope. Incoming light rays pass down the barrel of the telescope and are reflected by a parabolic mirror at the base. These rays converge toward point A in the figure, where an image would be formed on a photographic plate or another detector. However, before this image is formed, a small flat mirror at point M reflects the light toward an opening in the side of the tube that passes into an eyepiece. This design is said to have a *Newtonian focus*, after its developer. Note that in the reflecting telescope the light never passes through glass (except the small eyepiece). As a result, problems associated with chromatic aberration are virtually eliminated.

The largest optical telescopes in the world are the two 10-m-diameter Keck reflectors on Mauna Kea in Hawaii. The largest single-mirrored reflecting telescope in the United States is the 5-m-diameter instrument on Mount Palomar in California (see Figure 25.10). In contrast, the largest refracting telescope in the world, at the Yerkes Observatory in Williams Bay, Wisconsin, has a diameter of only 1 m.

Example 25.5 Angular Magnification of a Reflecting Telescope

A reflecting telescope has an 8-in.-diameter objective mirror with a focal length of 1 500 mm. What is the angular magnification of this telescope when an eyepiece with an 18-mm focal length is used?

Solution The equation for finding the angular magnification of a reflector is the same as that for a refractor. Thus, Equation 25.8 gives

$$m = \frac{f_o}{f_e} = \frac{1\ 500\ \text{mm}}{18\ \text{mm}} = \boxed{83}$$

FIGURE 25.10 The Hale telescope at Mount Palomar Observatory. Just before taking the elevator up to the prime-focus cage, a first-time observer is always told, "Good viewing! And, if you should fall, try to miss the mirror." *(Courtesy of Palomar Observatory/ California Institute of Technology)*

25.6 RESOLUTION OF SINGLE-SLIT AND CIRCULAR APERTURES

The ability of an optical system such as the eye, a microscope, or telescope to distinguish between closely spaced objects is limited because of the wave nature of light. To understand this difficulty, consider Figure 25.11, which shows two light sources far from a narrow slit of width *a*. The sources can be taken as two point sources S_1 and S_2 that are *not* coherent. For example, they could be two distant stars. If no diffraction occurred, we would observe two distinct bright spots (or images) on the screen at the right in the figure. However, because of diffraction, each source is imaged as a bright central region flanked by weaker bright and dark rings. What is observed on the screen is the sum of two diffraction patterns, one from S_1 and the other from S_2.

If the two sources are separated so that their central maxima do not overlap, as in Figure 25.11a, their images can be distinguished and are said to be *resolved*. If the sources are close together, however, as in Figure 25.11b, the two central maxima may overlap and the images are *not resolved*. To decide whether two images are resolved, the following condition is often applied to their diffraction patterns:

> **When the central maximum of one image falls on the first minimum of another image, the images are said to be just resolved. This limiting condition of resolution is known as Rayleigh's criterion.**

◀ Rayleigh's criterion

Figure 25.12 shows diffraction patterns in three situations. The images are just resolved when their angular separation satisfies Rayleigh's criterion (Fig. 25.12a). As the objects are brought closer together, their images are barely resolved (Fig. 25.12b). Finally, when the sources are very close to each other, their images are not resolved (Fig. 25.12c).

From Rayleigh's criterion, we can determine the minimum angular separation θ_{min} subtended by the source at the slit so that the images will be just resolved. In Chapter 24 we found that the first minimum in a single-slit diffraction pattern occurs at the angle that satisfies the relationship

$$\sin \theta = \frac{\lambda}{a}$$

Slit   Screen

(a)

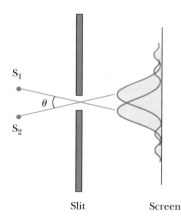

Slit   Screen

(b)

FIGURE 25.11 Each of two point sources at some distance from a small aperture produces a diffraction pattern. (a) The angle subtended by the sources at the aperture is large enough so that the diffraction patterns are distinguishable. (b) The angle subtended by the sources is so small that the diffraction patterns are not distinguishable. (Note that the angles are greatly exaggerated.)

FIGURE 25.12 The diffraction patterns of two point sources (solid curves) and the resultant pattern (dashed curve) for three angular separations of the sources. (a) The sources are separated such that their patterns are just resolved. (b) The sources are closer together, and their patterns are barely resolved. (c) The sources are so close together that their patterns are not resolved. *(From M. Cagnet, M. Francon, and J. C. Thierr, Atlas of Optical Phenomena, Berlin, Springer-Verlag, 1962, plate 16)*

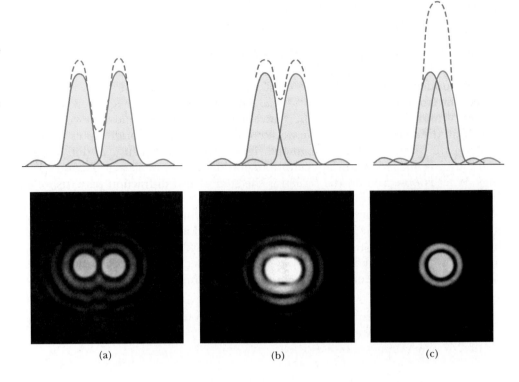

(a)   (b)   (c)

where *a* is the width of the slit. According to Rayleigh's criterion, this expression gives the smallest angular separation for which the two images can be resolved. Because $\lambda \ll a$ in most situations, sin θ is small and we can use the approximation sin $\theta \approx \theta$. Therefore, the limiting angle of resolution for a slit of width *a* is

Limiting angle for a slit ▶

$$\theta_{min} \approx \frac{\lambda}{a} \qquad \text{[25.9]}$$

where θ_{min} is in radians. Hence, the angle subtended by the two sources at the slit must be *greater* than λ/a if the images are to be resolved.

Many optical systems use circular apertures rather than slits. The diffraction pattern of a circular aperture (Fig. 25.13) consists of a central, circular bright region surrounded by progressively fainter rings. Analysis shows that the limiting angle of resolution of the circular aperture is

Limiting angle for a circular aperture ▶

$$\theta_{min} = 1.22 \frac{\lambda}{D} \qquad \text{[25.10]}$$

where *D* is the diameter of the aperture. Note that Equation 25.10 is similar to Equation 25.9 except for the factor 1.22, which arises from a complex mathematical analysis of diffraction from a circular aperture.

FIGURE 25.13 The diffraction pattern of a circular aperture consists of a central bright disk surrounded by concentric bright and dark rings. *(From M. Cagnet, M. Francon, and J. C. Thierr, Atlas of Optical Phenomena, Berlin, Springer-Verlag, 1962, plate 34)*

Quick Quiz 25.1 Suppose you are observing a binary star with a telescope and are having difficulty resolving the two stars. You decide to use a colored filter to help you. Should you choose a blue filter or a red filter?

APPLYING PHYSICS 25.2

Cats' eyes have vertical pupils when in dim light. Which would cats be most successful at resolving at night—headlights on a distant car or vertically separated running lights on a distant boat's mast having the same separation?

Explanation The effective slit width in the vertical direction of the cat's eye is larger than that in the horizontal direction. Thus, it has more resolving power for lights separated in the vertical direction and would be more effective at resolving the mast lights on the boat.

Example 25.6 Limiting Resolution of a Microscope

Sodium light of wavelength 589 nm is used to view an object under a microscope. The aperture of the objective has a diameter of 0.90 cm.

A Find the limiting angle of resolution.

Solution From Equation 25.10, we find the limiting angle of resolution to be

$$\theta_{min} = 1.22 \left(\frac{589 \times 10^{-9} \text{ m}}{0.90 \times 10^{-2} \text{ m}} \right) = \boxed{8.0 \times 10^{-5} \text{ rad}}$$

This means that any two points on the object subtending an angle of less than 8.0×10^{-5} rad at the objective cannot be distinguished in the image.

B Using visible light of any wavelength you desire, what is the maximum limit of resolution for this microscope?

Solution To obtain the maximum resolution, we have to use the shortest wavelength available in the visible spectrum. Violet light of wavelength 400 nm gives us a limiting angle of resolution of

$$\theta_{min} = 1.22 \left(\frac{400 \times 10^{-9} \text{ m}}{0.90 \times 10^{-2} \text{ m}} \right) = \boxed{5.4 \times 10^{-5} \text{ rad}}$$

C Suppose water of index of refraction 1.33 filled the space between the object and the objective. What effect would this have on the resolving power of the microscope?

Solution In this case, the wavelength of the sodium light in the water is found by $\lambda_w = \lambda_a/n$ (Chapter 22). Thus, we have

$$\lambda_w = \frac{\lambda_a}{n} = \frac{589 \text{ nm}}{1.33} = 443 \text{ nm}$$

The limiting angle of resolution at this wavelength is

$$\theta_{min} = 1.22 \left(\frac{443 \times 10^{-9} \text{ m}}{0.90 \times 10^{-2} \text{ m}} \right) = \boxed{6.0 \times 10^{-5} \text{ rad}}$$

Example 25.7 Resolution of a Telescope

The Hale telescope at Mount Palomar has a diameter of 200 in. What is its limiting angle of resolution at a wavelength of 600 nm?

Solution Because $D = 200$ in. $= 5.08$ m and $\lambda = 6.00 \times 10^{-7}$ m, Equation 25.10 gives

$$\theta_{min} = 1.22 \frac{\lambda}{D} = 1.22 \left(\frac{6.00 \times 10^{-7} \text{ m}}{5.08 \text{ m}} \right) = \boxed{1.44 \times 10^{-7} \text{ rad}}$$

Therefore, any two stars that subtend an angle greater than or equal to this value could be resolved if the air above the telescope were perfectly steady.

It is interesting to compare this value with the resolution of a large radio telescope, such as the system at Arecibo, Puerto Rico, which has a diameter of 1 000 ft (305 m). This telescope detects radio waves at a wavelength of 0.75 m. The corresponding minimum angle of resolution is calculated to be 3.0×10^{-3} rad (10 min 19 s of arc), which is more than 10 000 times larger than the calculated minimum angle for the Hale telescope.

Example 25.8 Comparing Two Telescopes

Two telescopes have the following properties:

Telescope	Diameter of Objective (in.)	Focal Length of Objective (mm)	Focal Length of Eyepiece (mm)
A	6.00	1 000	6.00
B	8.00	1 250	25.00

A Which has the better resolving power?

Solution The telescope with the larger objective has the greater ability to discriminate between objects that are close together, and this is telescope B.

B Which has the greater light-gathering ability?

Solution The telescope with the larger objective can collect more light. Hence, B is again the choice.

C Which produces a greater magnification?

Solution The magnification of telescope A is

$$m_{\mathrm{A}} = \frac{f_o}{f_e} = \frac{1\ 000\ \mathrm{mm}}{6.00\ \mathrm{mm}} = 167$$

That of telescope B is

$$m_{\mathrm{B}} = \frac{1\ 250\ \mathrm{mm}}{25.0\ \mathrm{mm}} = 50.0$$

Thus, telescope A has the greater magnification.

RESOLVING POWER OF THE DIFFRACTION GRATING

The diffraction grating, which we studied in Chapter 24, is most useful for making accurate wavelength measurements. Like the prism, it can be used to disperse a spectrum into its components. Of the two devices, the grating is more precise if we want to distinguish between two closely spaced wavelengths. We say that the grating spectrometer has a higher *resolution* than the prism spectrometer. If λ_1 and λ_2 are two nearly equal wavelengths between which the spectrometer can just barely distinguish, the **resolving power** R of the grating is defined as

$$R \equiv \frac{\lambda}{\lambda_2 - \lambda_1} = \frac{\lambda}{\Delta\lambda} \qquad \text{[25.11]}$$

where $\lambda \approx \lambda_1 \approx \lambda_2$ and $\Delta\lambda = \lambda_2 - \lambda_1$. Thus, we see that a grating with a high resolving power can distinguish small differences in wavelength. Furthermore, if N lines of the grating are illuminated, it can be shown that the resolving power in the mth-order diffraction equals the product Nm:

$$R = Nm \qquad [25.12]$$

◀ Resolving power of a grating

Thus, the resolving power increases with order number. Furthermore, R is large for a grating with a great number of illuminated slits. Note that for $m = 0$, $R = 0$, which signifies that *all wavelengths are indistinguishable* for the zeroth-order maximum (all wavelengths fall at the same point on the screen). However, consider the second-order diffraction pattern of a grating that has 5 000 rulings illuminated by the light source. The resolving power of such a grating in second order is $R = 5\,000 \times 2 = 10\,000$. Therefore, the *minimum* wavelength separation between two spectral lines that can be just resolved, assuming a mean wavelength of 600 nm, is calculated from Equation 25.12 to be $\Delta\lambda = \lambda/R = 6 \times 10^{-2}$ nm. For the third-order principal maximum, $R = 15\,000$ and $\Delta\lambda = 4 \times 10^{-2}$ nm, and so on.

Example 25.9 Resolving the Sodium Spectral Lines

Two bright lines in the spectrum of sodium have wavelengths of 589.00 nm and 589.59 nm.

A What must the resolving power of a grating be in order to distinguish these wavelengths?

Solution From Equation 25.11, we find that

$$R = \frac{\lambda}{\Delta\lambda} = \frac{589 \text{ nm}}{589.59 \text{ nm} - 589.00 \text{ nm}} = \frac{589}{0.59} = \boxed{998}$$

B To resolve these lines in the second-order spectrum, how many lines of the grating must be illuminated?

Solution From Equation 25.12 and the result of Part A, we find that

$$N = \frac{R}{m} = \frac{998}{2} = \boxed{499 \text{ lines}}$$

25.7 THE MICHELSON INTERFEROMETER

The camera and the telescope are examples of commonly used optical instruments. In contrast, the Michelson interferometer is an optical instrument that is unfamiliar to most people. It has great scientific importance, however. Invented by the American physicist A. A. Michelson (1852–1931), it is an ingenious device that splits a light beam into two parts and then recombines them to form an interference pattern. The interferometer is used to make accurate length measurements.

Figure 25.14 is a diagram of an interferometer. A beam of light provided by a monochromatic source is split into two rays by a partially silvered mirror, M, inclined at an angle of 45° relative to the incident light beam. One ray is reflected vertically upward to mirror M_1, and the other ray is transmitted horizontally through mirror M to mirror M_2. Hence, the two rays travel separate paths, L_1 and L_2. After reflecting from mirrors M_1 and M_2, the two rays eventually recombine to produce an interference pattern, which can be viewed through a telescope. The glass plate P equal in thickness to mirror M is placed in the path of the horizontal ray to ensure that the two rays travel the same distance through glass.

The interference pattern for the two rays is determined by the difference in their path lengths. When the two rays are viewed as shown, the image of M_2 is at

FIGURE 25.14 A diagram of the Michelson interferometer. A single beam is split into two rays by the half-silvered mirror, M. The path difference between the two rays is varied with the adjustable mirror, M_1.

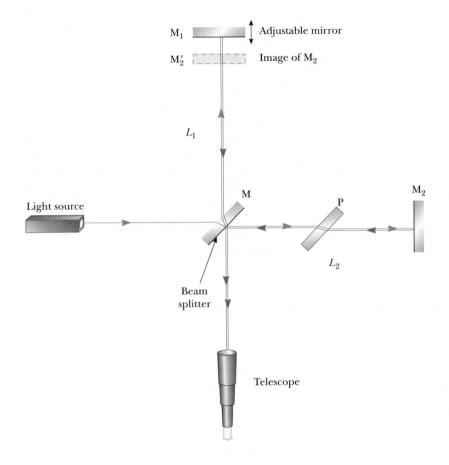

M_2' parallel to M_1. Hence, the space between M_2' and M_1 forms the equivalent of a parallel air film. The effective thickness of the air film is varied by using a finely threaded screw to move mirror M_1 in the direction indicated by the arrows in Figure 25.14. If one of the mirrors is tipped slightly with respect to the other, the thin film between the two is wedge-shaped, and an interference pattern consisting of parallel fringes is set up, as described in Example 24.4.

Now suppose we focus on one of the dark lines with the crosshairs of a telescope. As the mirror M_1 is moved to lengthen the path L_1, the thickness of the wedge increases. When the thickness increases by $\lambda/4$, the destructive interference that initially produced the dark fringe has changed to constructive interference, and we observe a bright fringe at the location of the crosshairs. The term *fringe shift* is used to describe the change in a fringe from dark to light or light to dark. Thus, successive light and dark fringes are formed each time M_1 is moved a distance of $\lambda/4$. The wavelength of light can be measured by counting the number of fringe shifts for a measured displacement of M_1. Conversely, if the wavelength is accurately known (as with a laser beam), the mirror displacement can be determined to within a fraction of the wavelength. Because the interferometer can measure displacements precisely, it is often used to make highly accurate measurements of the dimensions of mechanical components.

If the mirrors are perfectly aligned, rather than tipped with respect to one another, the path difference differs slightly for different angles of view. This results in an interference pattern that resembles Newton's rings. The pattern can be used in a fashion similar to that for tipped mirrors. One mirror concentrates on the center spot in the interference pattern. For example, suppose the spot is initially dark, indicating that destructive interference is occurring. If M_1 is now moved by a distance of $\lambda/4$, this central spot changes to a light region, corresponding to a fringe shift.

SUMMARY

The light-concentrating power of a lens of focal length f and diameter D is determined by the ***f*-number,** defined as

$$f\text{-number} \equiv \frac{f}{D} \qquad \textbf{[25.1]}$$

The smaller the f-number of a lens, the brighter the image formed.

Hyperopia (farsightedness) is a defect of the eye that occurs either when the eyeball is too short or when the ciliary muscle cannot change the shape of the lens enough to form a properly focused image. **Myopia** (nearsightedness) occurs either when the eye is longer than normal or when the maximum focal length of the lens is insufficient to produce a clearly focused image on the retina.

The **power** of a lens in **diopters** is the inverse of the focal length in meters.

The **angular magnification of a lens** is defined as

$$m \equiv \frac{\theta}{\theta_0} \qquad \textbf{[25.2]}$$

where θ is the angle subtended by an object at the eye with a lens in use and θ_0 is the angle subtended by the object when it is placed at the near point of the eye and no lens is used. The **maximum angular magnification of a lens** is

$$m_{\max} = 1 + \frac{25 \text{ cm}}{f} \qquad \textbf{[25.5]}$$

When the eye is relaxed, the angular magnification is

$$m = \frac{25 \text{ cm}}{f} \qquad \textbf{[25.6]}$$

The overall **magnification of a compound microscope** of length L is the product of the magnification produced by the objective, of focal length f_o, and the magnification produced by the eyepiece, of focal length f_e:

$$m = -\frac{L}{f_o}\left(\frac{25 \text{ cm}}{f_e}\right) \qquad \textbf{[25.7]}$$

The **angular magnification of a telescope** is

$$m = \frac{f_o}{f_e} \qquad \textbf{[25.8]}$$

where f_o is the focal length of the objective and f_e is the focal length of the eyepiece.

Two images are said to be **just resolved** when the central maximum of the diffraction pattern for one image falls on the first minimum of the other image. This limiting condition of resolution is known as **Rayleigh's criterion.** The limiting angle of resolution for a **slit** of width a is

$$\theta_{\min} \approx \frac{\lambda}{a} \qquad \textbf{[25.9]}$$

The limiting angle of resolution of a **circular aperture** is

$$\theta_{\min} = 1.22 \frac{\lambda}{D} \qquad \textbf{[25.10]}$$

where D is the diameter of the aperture.

The **resolving power** of a diffraction grating in the mth order is

$$R = Nm \qquad \textbf{[25.12]}$$

where N is the number of illuminated rulings on the grating.

CONCEPTUAL QUESTIONS

1. A lens is used to examine an object across the room. Is the lens probably being used as a simple magnifier?

2. Why is it difficult or impossible to focus a microscope on an object across the room?

3. The optic nerve and the brain invert the image formed on the retina. Why do we not see everything upsidedown?

4. Why is it difficult or impossible to focus a telescope on an object very close to you?

5. Suppose you are observing the interference pattern formed by a Michelson interferometer in a laboratory, and a joking colleague holds a lit match in the light path of one arm of the interferometer. Will this have an effect on the interference pattern?

6. Compare and contrast the eye and a camera. What parts of the camera correspond to the iris, the retina, and the cornea of the eye?

7. Large telescopes are usually reflecting rather than refracting. List some reasons for this choice.

8. If you want to use a converging lens to set fire to a piece of paper, why should the light source be farther from the lens than its focal point?

9. Explain why it is theoretically impossible to see an object as small as an atom regardless of the quality of the light microscope being used.

10. Which is most important in the use of a camera photoflash unit, the intensity of the light (the energy per unit area per unit time), or the product of the intensity and the time of the flash, assuming the time is less than the shutter speed?

PROBLEMS

1, 2, 3 = straightforward, intermediate, challenging   ☐ = full solution available in Student Solutions Manual/Study Guide

web = solution posted at **http://info.brookscole.com/serway**   🔬 = biomedical application

Section 25.1   The Camera

1. A camera used by a professional photographer to shoot portraits has a focal length of 25.0 cm. The photographer takes a portrait of a person 1.50 m in front of the camera. Where is the image formed, and what is the lateral magnification?

2. The lens of a certain 35-mm camera (where 35-mm is the width of the film strip) has a focal length of 55 mm and a speed (an f-number) of f/1.8. Determine the diameter of the lens.

3. The image area of a typical 35-mm slide is 23.5 mm by 35.0 mm. If a camera has a 55.0-mm focal length lens, will the full image this camera forms of the constellation Orion, which is 20° across, fit on a 35-mm slide?

4. The full Moon is photographed using a camera with a 120-mm focal length lens. Determine the diameter of the Moon's image on the film. (*Note:* The radius of the Moon is 1.74×10^6 m, and the Earth-Moon distance is 3.84×10^8 m.)

5. A camera is being used with the correct exposure at $f/4$ **web** and a shutter speed of 1/32 s. In order to "stop" a fast-moving subject, the shutter speed is changed to 1/256 s. Find the new *f*-stop that should be used to maintain satisfactory exposure, assuming no change in lighting conditions.

6. (a) Use conceptual arguments to show that the intensity of light (energy per unit area per unit time) reaching the film in a camera is proportional to the square of the reciprocal of the *f*-number, as

$$I \propto \frac{1}{(f/D)^2}$$

(b) The correct exposure time for a camera set to $f/1.8$ is (1/500) s. Calculate the correct exposure time if the *f*-number is changed to $f/4$ under the same lighting conditions.

7. A certain type of film requires an exposure time of 0.010 s with an $f/11$ lens setting. Another type of film requires twice the light energy to produce the same level of expo-

sure. What *f*-stop does the second type of film need with the 0.010-s exposure time?

8. Assume that the camera in Figure 25.1 has a fixed focal length of 65.0 mm and is adjusted to properly focus the image of a distant object. How far and in what direction must the lens be moved to focus the image of an object that is 2.00 m away?

Section 25.2   The Eye

9. A retired bank president can easily read the fine print of the financial page when the newspaper is held no closer than arm's length, 60.0 cm from the eye. What should be the focal length of an eyeglass lens that allows her to read at the more comfortable distance of 24.0 cm?

10. A person has far points 84.4 cm from the right eye and 122 cm from the left eye. Write a prescription for the powers of the corrective lenses.

11. The accommodation limits for Nearsighted Nick's eyes are 18.0 cm and 80.0 cm. When he wears his glasses, he is able to see faraway objects clearly. At what minimum distance is he able to see objects clearly?

12. The near point of an eye is 100 cm. A corrective lens is to be used to allow this eye to clearly focus on objects 25.0 cm in front of it. (a) What should be the focal length of this lens? (b) What is the power of the needed corrective lens?

13. An individual is nearsighted; his near point is 13.0 cm and his far point is 50.0 cm. (a) What lens power is needed to correct his nearsightedness? (b) When the lenses are in use, what is this person's near point?

14. A person sees clearly wearing eyeglasses that have a power of −4.00 diopters and sit 2.00 cm in front of the eyes. If the person wants to switch to contact lenses, which are placed directly on the eyes, what lens power should be prescribed?

15. An artificial lens is implanted in a person's eye to replace a diseased lens. The distance between the artificial lens and the retina is 2.80 cm. In the absence of the lens, an image

of a distant object (Formed by refraction at the cornea) falls 2.53 cm behind the retina. The lens is designed to put the image of the distant object on the retina. What is the power of the implanted lens? (*Hint:* Consider the image formed by the cornea as a virtual object.)

16. A person is to be fitted with bifocals. She can see clearly when the object is between 30 cm and 1.5 m from the eye. (a) The upper portions of the bifocals (Fig. P25.16) should be designed to enable her to see distant objects clearly. What power should they have? (b) The lower portions of the bifocals should enable her to see objects comfortably at 25 cm. What power should they have?

Far vision

Near vision

FIGURE P25.16

Section 25.3 The Simple Magnifier

17. A philatelist examines the printing detail on a stamp using a biconvex lens of focal length 10.0 cm as a simple magnifier. The lens is held close to the eye, and the lens-to-object distance is adjusted so that the virtual image is formed at the normal near point (25.0 cm). Calculate the magnification.

18. A lens having a focal length of 25 cm is used as a simple magnifier. (a) What is the angular magnification obtained when the image is formed at the normal near point ($q = -25$ cm)? (b) What is the angular magnification produced by this lens when the eye is relaxed?

19. A biology student uses a simple magnifier to examine the structural features of the wing of an insect. The wing is held 3.50 cm in front of the lens, and the image is formed 25.0 cm from the eye. (a) What is the focal length of the lens? (b) What angular magnification is achieved?

20. A lens that has a focal length of 5.00 cm is used as a magnifying glass. (a) To obtain maximum magnification, where should the object be placed? (b) What is the magnification?

21. A leaf of length h is positioned 71.0 cm in front of a converging lens with a focal length of 39.0 cm. An observer views the image of the leaf from a position 1.26 m behind the lens, as shown in Figure P25.21. (a) What is the magnitude of the lateral magnification (ratio of image size to object size) produced by the lens? (b) What angular magnifi-

cation is achieved by viewing the image of the leaf rather than viewing the leaf directly?

Section 25.4 The Compound Microscope

Section 25.5 The Telescope

22. The distance between eyepiece and objective lens in a certain compound microscope is 23.0 cm. The focal length of the eyepiece is 2.50 cm, and that of the objective is 0.400 cm. What is the overall magnification of the microscope?

23. The desired overall magnification of a compound microscope is 140×. The objective alone produces a lateral magnification of 12×. Determine the required focal length of the eyepiece.

24. A microscope has an objective lens with a focal length of 16.22 mm and an eyepiece with a focal length of 9.50 mm. With the length of the barrel set at 29.0 cm, the diameter of a red blood cell's image subtends an angle of 1.43 mrad with the eye. If the final image distance is 29.0 cm from the eyepiece, what is the actual diameter of the red blood cell?

25. The length of a microscope tube is 15.0 cm. The focal web length of the objective is 1.00 cm, and the focal length of the eyepiece is 2.50 cm. What is the magnification of the microscope, assuming it is adjusted so that the eye is relaxed?

26. The Yerkes refracting telescope has a 1.00-m-diameter objective lens of focal length 20.0 m. Assume it is used with an eyepiece of focal length 2.50 cm. (a) Determine the angular magnification of the planet Mars as seen through this telescope. (b) Are the Martian polar caps right side up or upside down?

27. The lenses of an astronomical telescope are 92 cm apart when adjusted for viewing a distant object with minimum eyestrain. The angular magnification produced by the telescope is 45. Compute the focal length of each lens.

28. An elderly sailor is shipwrecked on a desert island but manages to save his eyeglasses. The lens for one eye has a power of +1.20 diopters, and the other lens has a power of +9.00 diopters. (a) What is the magnifying power of the telescope he can construct with these lenses? (b) How far apart are the lenses when the telescope is adjusted for minimum eyestrain?

29. Astronomers often take photographs with the objective lens or mirror of a telescope alone, without an eyepiece. (a) Show that the image size h' for this telescope is given by $h' = fh/(f - p)$ where h is the object size, f the objective focal length, and p the object distance. (b) Simplify the expression in part (a) if the object distance is much greater than objective focal length. (c) The "wingspan" of the International Space Station is 108.6 m, the overall width of its solar panel configuration. When it is orbiting at an altitude of 407 km, find the width of the image formed by a telescope objective of focal length 4.00 m.

30. Galileo devised a simple terrestrial telescope that produces an upright image. It consists of a converging objective lens and a diverging eyepiece at opposite ends of the telescope tube. For distant objects, the tube length is the objective focal length less the absolute value of the eyepiece focal length. (a) Does the user of the telescope see a real or virtual image? (b) Where is the final image? (c) If a telescope is to be constructed with a tube of length 10.0 cm and a magnification of 3.00, what are the focal lengths of the objective and eyepiece?

Leaf

71.0 cm   1.26 m

FIGURE P25.21

31. A person decides to use an old pair of eyeglasses to make some optical instruments. He knows that the near point in his left eye is 50.0 cm and the near point in his right eye is 100 cm. (a) What is the maximum angular magnification he can produce in a telescope? (b) If he places the lenses 10.0 cm apart, what is the maximum overall magnification he can produce in a microscope? (Go back to basics and use the thin-lens equation to solve part (b).)

Section 25.6 Resolution of Single-Slit and Circular Apertures

32. If the distance from Earth to the Moon is 3.8×10^8 m, what diameter would be required for a telescope objective to resolve a Moon crater 300 m in diameter? Assume a wavelength of 500 nm.

33. A converging lens with a diameter of 30.0 cm forms an image of a satellite passing overhead. The satellite has two green lights (wavelength 500 nm) spaced 1.00 m apart. If the lights can just be resolved according to the Rayleigh criterion, what is the altitude of the satellite?

34. The pupil of a cat's eye narrows to a vertical slit of width 0.500 mm in daylight. What is the angular resolution for a pair of horizontally separated mice? (Use 500-nm light in your calculation.)

35. To increase the resolving power of a microscope, the object and the objective are immersed in oil ($n = 1.5$). If the limiting angle of resolution without the oil is 0.60 μrad, what is the limiting angle of resolution with the oil? (*Hint:* The oil changes the wavelength of the light.)

36. Two motorcycles, separated laterally by 2.00 m, are approaching an observer holding an infrared detector that is sensitive to radiation of wavelength 885 nm. What aperture diameter is required in the detector if the two headlights are to be resolved at a distance of 10.0 km?

37. A helium-neon laser emits light that has a wavelength of 632.8 nm. The circular aperture through which the beam emerges has a diameter of 0.500 cm. Estimate the diameter of the beam 10.0 km from the laser.

38. A spy satellite circles Earth at an altitude of 200 km and carries out surveillance with a special high-resolution telescopic camera having a lens diameter of 35 cm. If the angular resolution of this camera is limited by diffraction, estimate the separation of two small objects on Earth's surface that are just resolved in yellow-green light ($\lambda = 550$ nm).

39. Suppose a 5.00-m-diameter telescope is constructed on the Moon, where the absence of atmospheric distortion permits excellent viewing. If observations are made using 500-nm light, what minimum separation between two objects could just be resolved on Mars at closest approach (when Mars is 8.0×10^7 km from the Moon)?

Section 25.7 The Michelson Interferometer

40. Light of wavelength 550 nm is used to calibrate a Michelson interferometer. By use of a micrometer screw, the platform on which one mirror is mounted is moved 0.180 mm. How many fringe shifts are counted?

41. An interferometer is used to measure the length of a bacterium. The wavelength of the light used is 650 nm. As one arm of the interferometer is moved from one end of the cell to the other, 310 fringe shifts are counted. How long is the bacterium?

42. Mirror M_1 in Figure 25.14 is displaced a distance ΔL. During this displacement, 250 fringe shifts are counted. The light being used has a wavelength of 632.8 nm. Calculate the displacement ΔL.

43. Monochromatic light is beamed into a Michelson interferometer. The movable mirror is displaced 0.382 mm, causing the central spot in the interferometer pattern to change from bright to dark and back to bright 1 700 times. Determine the wavelength of the light. What color is it?

44. The Michelson interferometer can be used to measure the index of refraction of a gas by placing an evacuated transparent tube in the light path along one arm of the device. Fringe shifts occur as the gas is slowly added to the tube. Assume that 600-nm light is used, the tube is 5.00 cm long, and that 160 fringe shifts occur as the pressure of the gas in the tube increases to atmospheric pressure. What is the index of refraction of the gas? (*Hint:* The fringe shifts occur because the wavelength of the light changes inside the gas-filled tube.)

45. The light path in one arm of a Michelson interferometer includes a transparent cell that is 5.00 cm long. How many fringe shifts would be observed if all the air were evacuated from the cell? The wavelength of the light source is 590 nm and the refractive index of air is 1.000 29. (See the Hint in Problem 44.)

46. The H_α line in hydrogen has a wavelength of 656.20 nm. This line differs in wavelength from the corresponding spectral line in deuterium (the heavy stable isotope of hydrogen) by 0.18 nm. (a) Determine the minimum number of lines a grating must have to resolve these two wavelengths in the first order. (b) Repeat part (a) for the second order.

47. A 15.0-cm-long grating has 6 000 slits per centimeter. Can two lines of wavelengths 600.000 nm and 600.003 nm be separated using this grating? Explain.

ADDITIONAL PROBLEMS

48. A person with a nearsighted eye has near and far points of 16 cm and 25 cm, respectively. (a) Assuming a lens is placed 2.0 cm from the eye, what power must the lens have to correct this condition? (b) Contact lenses placed directly on the cornea are used to correct the eye in this example. What is the power of the lens required in this case, and what is the new near point? (*Hint:* The contact lens and the eyeglass lens require slightly different powers because they are at different distances from the eye.)

49. The near point of an eye is 75.0 cm. (a) What should be the power of a corrective lens prescribed to enable the eye to see an object clearly at 25.0 cm? (b) If, using the corrective lens, the user can see an object clearly at 26.0 cm but not 25.0 cm, by how many diopters did the lens grinder miss the prescription?

50. A compound microscope has an objective of focal length 0.300 cm and an eyepiece of focal length 2.50 cm. If an object is 3.40 mm from the objective, what is the magnification? (*Hint:* Use the lens equation for the objective.)

51. A cataract-impaired lens in an eye can be surgically removed and replaced by a manufactured lens. The focal length required for the new lens is determined by the lens-to-retina distance, which is measured by a sonar-like device, and by the requirement that the implant provide for correct distant vision. (a) If the distance from lens to retina is 22.4 mm, calculate the power of the implanted lens in diopters. (b) Since there is no accommodation and the implant allows for correct distant vision, a corrective lens for close work or reading must be used. Assume a

reading distance of 33.0 cm and calculate the power of the lens in the reading glasses.

52. Estimate the minimum angle subtended at the eye of a hawk flying at an altitude of 50 m to recognize a mouse on the ground.

53. The wavelengths of the sodium spectrum are $\lambda_1 = 589.00$ nm and $\lambda_2 = 589.59$ nm. Determine the minimum number of lines in a grating that will allow resolution of the sodium spectrum in (a) the first order and (b) the third order.

54. The text discusses the astronomical telescope. Another type is the Galilean telescope, in which an objective lens gathers light (Fig. P25.54), and tends to form an image at point A. An eyepiece, consisting of a diverging lens, intercepts the light before it comes to a focus and forms a virtual image at point B. When adjusted for minimum eyestrain, point B is an infinite distance in front of the lens and parallel rays emerge from the lens, as in Figure P25.54b. An opera glass, which is a Galilean telescope, is used to view a 30.0-cm-tall singer's head that is 40.0 m from the objective lens. The focal length of the objective is + 8.00 cm, and that of the eyepiece is − 2.00 cm. The telescope is adjusted so parallel rays enter the eye. Compute (a) the size of the real image that would have been formed by the objective, (b) the virtual object distance for the diverging lens, (c) the distance between the lenses, and (d) the overall angular magnification.

55. A laboratory (astronomical) telescope is used to view a scale that is 300 cm from the objective, which has a focal length of 20.0 cm; the eyepiece has a focal length of

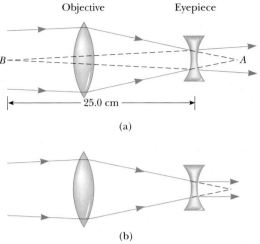

FIGURE P25.54

2.00 cm. Calculate the angular magnification when the telescope is adjusted for minimum eyestrain. (*Note:* The object is not at infinity, and so the simple expression $m = f_o/f_e$ is not sufficiently accurate for this problem. Also, assume small angles so that $\tan \theta \approx \theta$.)

56. If the aqueous humor of the eye has an index of refraction of 1.34 and the distance from the vertex of the cornea to the retina is 2.00 cm, what is the radius of curvature of the cornea for which distant objects will be focused on the retina? (For simplicity, assume that all refraction occurs in the aqueous humor.)

GROUP ACTIVITIES

G.1 (a) Move this book toward your face until the letters just begin to blur. The distance from the book to your eye is your near point. (b) On a sheet of paper make a dot near the center and about 3 inches to the left and right place an x. With one eye shut and while looking at the dot, slowly move the paper toward your eye. You will notice at some distance from your eye, one of the x's will disappear. This is the location of the blind spot of your eye, the point where the optic nerve enters the eye. (c) Stand before a mirror in a darkened room for a few minutes. Then, turn on a light in the room and observe your pupils in the mirror as they change size. Such dark adaptation also takes place at the rods and cones as they chemically adjust their sensitivity. This adjustment takes 15–30 minutes as you have noted when you enter a darkened movie theater. Iris aperture control takes less than a second and helps protect the retina from overload.

G.2 On a sunny day, hold a magnifying glass above a nonflammable surface, such as a sidewalk, so that the image of the Sun forms a round spot of light on the surface. Note where the spot formed by the lens is most distinct, or smallest. Use a ruler to measure the distance between the glass and the image. The distance is equal to the focal length of the lens.

G.3 Hold a pair of prescription glasses about 12 cm from your eye, and look at different objects through the lenses. Try this with different types of glasses, such as those for farsightedness and nearsightedness, and describe what effects the differences have on the image you see. If you

have bifocals, how do the images produced by the top and bottom portions of the bifocal lens compare?

G.4 If you have never experimented with a 35-mm camera with adjustable f-numbers and shutter speeds, use up a couple of rolls of film to see what happens. Take several shots of the same object with different settings for these two variables. (You should record your f-numbers and shutter speeds for each photograph.) Explain any differences you see in the final images in terms of the settings used.

G.5 (a) A patient has a near point of 1.25 m. Is she nearsighted or farsighted? (b) Should the corrective lens be converging or diverging? (c) For a comfortable reading distance of 25.0 cm from her eye, what diopter lens should be prescribed? (d) Does the power of the lens change if you assume that the lens is always placed 2.00 cm from the eye?

G.6 A lens with a certain power is used as a simple magnifier. (a) If the power of the lens is doubled, does the angular magnification increase or decrease? (b) What assumptions are made when you use the equation $m = 25$ cm$/f$? (c) What is the angular magnification of a lens having a focal length of 5.0 cm?

G.7 A telescope uses an objective lens of focal length 40 cm and an eyepiece of 2.0 cm. (a) Without doing a calculation, how will the view through the telescope change if the eyepiece is changed to one having a focal length of 4.0 cm? (b) Do the calculations to verify your prediction. (c) Does your answer depend on whether you are using a refracting or a reflecting telescope?

Modern Physics

At the end of the 19th century, scientists believed that they had learned most of what there was to know about physics. Newton's laws of motion and his universal theory of gravitation, Maxwell's theoretical work in unifying electricity and magnetism, and the laws of thermodynamics and kinetic theory were highly successful in explaining a wide variety of phenomena.

However, at the turn of the 20th century, a major revolution shook the world of physics. In 1900 Planck provided the basic ideas that led to the formulation of the quantum theory, and in 1905 Einstein formulated his brilliant special theory of relativity. The excitement of the times is captured in Einstein's own words: "It was a marvelous time to be alive." Both ideas were to have a profound effect on our understanding of nature. Within a few decades, these two theories inspired new developments in the fields of atomic physics, nuclear physics, and condensed-matter physics.

Our discussion of modern physics will begin with a treatment of the special theory of relativity in Chapter 26. The theory provides us with a new and deeper view of physical laws. In Chapter 27 we shall discuss various developments in quantum theory, which provides us with a successful model for understanding electrons, atoms, and molecules. The last three chapters of the text are concerned with applications of quantum theory. Chapter 28 discusses the structure and properties of atoms using concepts from quantum mechanics. Chapter 29 is concerned with the structure and properties of the atomic nucleus. Chapter 30 discusses practical applications of nuclear physics and concludes with a discussion of elementary particles.

Keep in mind that although modern physics was developed during the 20th century and has led to a multitude of important technological achievements, the story is still incomplete as we begin this 21st century. Surprising discoveries will continue to be made during our lifetimes, and many of these discoveries will deepen or refine our understanding of nature and the world around us. It is still "a marvelous time to be alive."

The 20th century saw the development of large accelerators capable of producing high-energy collisions between atoms and the subatomic particles within them. Charged particles resulting from these collisions follow curved tracks similar to the ones shown above when the detector is placed in a strong magnetic field. *(Science VU/FNL/DOE)*

26 Relativity

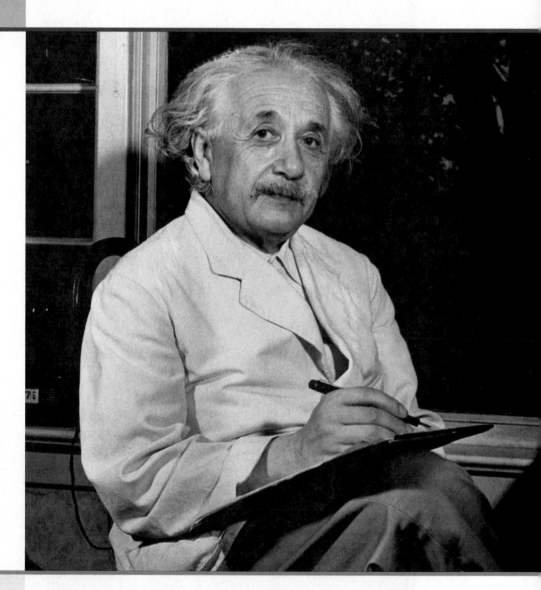

Albert Einstein, one of the greatest theoretical physicists of all times, revolutionized the world of modern physics. He is quoted as saying, "I have no particular talent. I am merely inquisitive." Although he is recognized primarily for his monumental intellectual achievements, he was also deeply concerned with the social impact of scientific discovery. (© *Bettmann/CORBIS*)

M ost of our everyday experiences and observations have to do with objects that move at speeds much less than the speed of light. Newtonian mechanics was formulated to describe the motion of such objects, and this formalism is quite successful in describing a wide range of phenomena that occur at low speeds. It fails, however, when applied to particles whose speeds approach that of light.

This chapter is concerned with the theory of special relativity. Although the concepts of special relativity often violate our common sense, the theory correctly predicts the results of experiments involving speeds near the speed of light.

26.1 INTRODUCTION

Experimentally, the predictions of Newtonian theory can be tested at high speeds by accelerating electrons or other charged particles through a large electric potential difference. For example, it is possible to accelerate an electron to a speed of $0.99c$ (where c is the speed of light) by using a potential difference of several million volts. According to Newtonian mechanics, if the potential difference is increased by a factor of 4, the electron's kinetic energy is four times greater and its speed should double to $1.98c$. However, experiments show that the speed of the electron—as well as the speed of any other particle in the Universe—always remains *less* than the speed of light, regardless of the size of the accelerating voltage. Because it places no upper limit on the speed, Newtonian mechanics is contrary to modern experimental results and is clearly a limited theory.

In 1905, at the age of 26, Einstein published his special theory of relativity. Regarding the theory, Einstein wrote,

> The relativity theory arose from necessity, from serious and deep contradictions in the old theory from which there seemed no escape. The strength of the new theory lies in the consistency and simplicity with which it solves all these difficulties, using only a few very convincing assumptions.[1]

Although Einstein made many other important contributions to science, his theory of relativity alone represents one of the greatest intellectual achievements of all time. With this theory, experimental observations can be correctly predicted over the range of speeds from $v = 0$ to speeds approaching the speed of light. Newtonian mechanics, which was accepted for more than 200 years, is in fact a specialized case of Einstein's theory. This chapter introduces the special theory of relativity, with emphasis on some of the consequences of the theory. A brief discussion of general relativity and some of its consequences is presented in Section 26.11.

As we shall see, the special theory of relativity is based on two postulates:

1. **The laws of physics are the same in all coordinate systems either at rest or moving at constant velocity with respect to one another.**
2. **The speed of light in a vacuum has the same value, 3×10^8 m/s, regardless of the velocity of the observer or the velocity of the source emitting the light.**

Special relativity covers such phenomena as the slowing down of moving clocks and the contraction of lengths in reference frames moving with constant velocity relative to an observer. We also discuss the relativistic forms of momentum and energy, as well as some consequences of the famous mass-energy formula, $E = mc^2$.

---

[1] A. Einstein and L. Infeld, *The Evolution of Physics*, New York, Simon and Schuster, 1961.

APPLYING **PHYSICS** *26.1*

Imagine a very powerful lighthouse with a rotating beacon. Imagine also drawing a horizontal circle around the lighthouse, with the lighthouse at the center. Along the circumference of the circle, the light beam lights up a portion of the circle and the lit portion of the circle moves around the circle at a certain tangential speed. If we now imagine a circle twice as big in radius, the tangential speed of the lit portion is faster, because it must travel a larger circumference in the time of one rotation of the light source. Imagine that we continue to make the circle larger and larger, eventually moving it out into space. The tangential speed of the lit portion will keep increasing. Is it possible that the tangential speed could become larger than the speed of light? Would this violate a principle of special relativity?

Explanation   For a large-enough circle, it is possible that the tangential speed of the lit portion of the circle could be larger than the speed of light. This does not violate a principle of special relativity, however, because no matter or information is traveling faster than the speed of light.

26.2   THE PRINCIPLE OF GALILEAN RELATIVITY

In order to describe a physical event, it is necessary to choose a *frame of reference*. For example, when you perform an experiment in a laboratory, you select a coordinate system, or frame of reference, that is at rest with respect to the laboratory. However, suppose an observer in a passing car moving at a constant velocity with respect to the lab were to observe your experiment. Would the observations made by the moving observer differ dramatically from yours? That is, if you found Newton's first law to be valid in your frame of reference, would the moving observer agree with you? According to the principle of Galilean relativity, **the laws of mechanics must be the same in all inertial frames of reference.** Inertial frames of reference are those reference frames in which Newton's laws are valid. Practically, such frames are those in which objects subjected to no forces move in straight lines at constant speed—thus the name *inertial frame* because objects observed from these frames obey Newton's first law, the law of inertia. For the situation just described, the laboratory coordinate system and the coordinate system of the moving car are both inertial frames of reference. As a consequence, if the laws of mechanics are found to be true in the lab, the person in the car must also observe the same laws.[2]

Let us describe a hypothetical observation to illustrate the equivalence of the laws of mechanics in different inertial frames. Consider an airplane in flight, moving with a constant velocity, as in Figure 26.1a. If a passenger in the airplane throws a ball straight up in the air, the passenger observes that the ball moves in a vertical path. The motion of the ball is precisely the same as it would be if the ball were thrown while at rest on Earth. The law of gravity and the equations of motion under constant acceleration are obeyed whether the airplane is at rest or in uniform motion. Now consider the same experiment when viewed by another observer at rest on Earth. This stationary observer views the path of the ball in the plane to be a parabola, as in Figure 26.1b. Furthermore, according to this observer, the ball has a velocity to the right equal to the velocity of the plane. Although the two observers disagree on the shape of the ball's path, both agree that the motion of the ball obeys the law of gravity and Newton's laws of motion and even agree on how long the ball is in the air. Thus, we draw the following important conclusion: **There is no preferred frame of reference for describing the laws of mechanics.**

---

[2] What is an example of a noninertial frame? A frame undergoing translational acceleration or a frame rapidly rotating with respect to the two inertial frames mentioned above.

(a)

(b)

FIGURE 26.1 (a) The observer on the airplane sees the ball move in a vertical path when thrown upward. (b) The Earth observer views the path of the ball to be a parabola.

26.3 THE SPEED OF LIGHT

It is quite natural to ask whether the concept of Galilean relativity in mechanics also applies to experiments in electricity, magnetism, optics, and other areas. Experiments indicate that the answer is no. For example, if we assume that the laws of electricity and magnetism are the same in all inertial frames, a paradox concerning the speed of light immediately arises. This can be understood by recalling that according to electromagnetic theory, the speed of light always has the fixed value of $2.997\ 924\ 58 \times 10^8$ m/s in free space. But this is in direct contradiction to common sense. For example, suppose a light pulse is sent out by an observer in a boxcar moving with a velocity **v** (Fig. 26.2). The light pulse has a velocity **c** relative to observer S′ in the boxcar. According to Galilean relativity, the velocity of the pulse relative to the stationary observer S outside the boxcar should be **c + v.** This obviously contradicts Einstein's theory, which postulates that the velocity of the light pulse is the same for all observers.

In order to resolve this paradox, we must conclude either that (1) the addition law for velocities is incorrect or that (2) the laws of electricity and magnetism are not the same in all inertial frames. Assume that the second conclusion is true; then a preferred reference frame must exist in which the speed of light has the value c, but in any other reference frame the speed of light must have a value that is greater or less than c. It is useful to draw an analogy with sound waves, which propagate through a medium such as air. The speed of sound in air is about 330 m/s when measured in a reference frame in which the air is stationary. However, the speed of sound is greater or less than this value when measured from a reference frame that is moving with respect to the air.

In the case of light signals (electromagnetic waves), recall that electromagnetic theory predicted that such waves must propagate through free space with a

FIGURE 26.2 A pulse of light is sent out by a person in a moving boxcar. According to Newtonian relativity, the speed of the pulse should be **c + v** relative to a stationary observer.

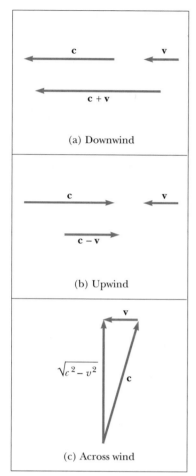

FIGURE 26.3 If the speed of the ether wind relative to Earth is v, and c is the speed of light relative to the ether, the speed of light relative to Earth is (a) $c + v$ in the downwind direction, (b) $c - v$ in the upwind direction, and (c) $(c^2 - v^2)^{1/2}$ in the direction perpendicular to the wind.

speed equal to the speed of light. However, the theory does not require the presence of a medium for wave propagation. This is in contrast to other types of waves that we have studied, such as water and sound waves, that do require a medium to support the disturbances. In the 19th century, physicists thought that electromagnetic waves also required a medium in order to propagate. They proposed that such a medium existed, and they gave it the wonderful name **luminiferous ether.** The ether was assumed to be present everywhere, even in empty space, and light waves were viewed as ether oscillations. Furthermore, the ether would have to be a massless but rigid medium with no effect on the motion of planets or other objects. These are strange concepts indeed. In addition, it was found that the troublesome laws of electricity and magnetism would take on their simplest forms in a special frame of reference at *rest* with respect to the ether. This frame was called the *absolute frame.* The laws of electricity and magnetism would be valid in this absolute frame, but they would have to be modified in any reference frame moving with respect to the absolute frame.

As a result of the importance attached to the ether and this absolute frame, it became of considerable interest in physics to prove by experiment that they existed. Since it was considered likely that Earth was in motion through the ether, from the view of an experimenter on Earth, there was an "ether wind" blowing through his lab. A direct method for detecting the ether wind would use an apparatus fixed to Earth to measure the wind's influence on the speed of light. If v is the speed of the ether relative to Earth, then the speed of light should have its maximum value $c + v$ when propagating downwind, as shown in Figure 26.3a. Likewise, the speed of light should have its minimum value $c - v$ when propagating upwind, as in Figure 26.3b, and an intermediate value $(c^2 - v^2)^{1/2}$ in the direction perpendicular to the ether wind, as in Figure 26.3c. If the Sun is assumed to be at rest in the ether, then the velocity of the ether wind would be equal to the orbital velocity of Earth around the Sun, which has a magnitude of approximately 3×10^4 m/s. Because $c = 3 \times 10^8$ m/s, it should be possible to detect a change in speed of about 1 part in 10^4 for measurements in the upwind or downwind directions. However, as we shall see in the next section, all attempts to detect such changes and establish the existence of the ether (and hence the absolute frame) proved futile!

In conclusion, we see that the second hypothesis in our introduction to this section is false—and we now believe that **the laws of electricity and magnetism are the same in all inertial frames.** It is the simple classical addition laws for velocities that are incorrect and must be modified, as shown in Section 26.8.

26.4 THE MICHELSON–MORLEY EXPERIMENT

The most famous experiment designed to detect small changes in the speed of light was first performed in 1881 by Michelson and later repeated under various conditions by Michelson and Edward W. Morley (1838–1923). We state at the outset that the outcome of the experiment contradicted the ether hypothesis.

The experiment was designed to determine the velocity of Earth relative to the hypothetical ether. The experimental tool used was the Michelson interferometer, which was discussed in Section 25.7 and is shown again in Figure 26.4. Arm 2 is aligned along the direction of Earth's motion through space. Earth moving through the ether at speed v is equivalent to the ether flowing past Earth in the opposite direction with speed v. This ether wind blowing in the direction opposite the direction of Earth's motion should cause the speed of light measured in Earth's frame to be $c - v$ as the light approaches mirror M_2 and $c + v$ after reflection, where c is the speed of light in the ether frame.

The two beams reflected from M_1 and M_2 recombine, and an interference pattern consisting of alternating dark and bright fringes is formed. The interference pattern was observed while the interferometer was rotated through an angle of 90°. This rotation supposedly would change the speed of the ether wind along the direction of arm 1. The effect of this rotation should have been to cause the fringe pattern to shift slightly but measurably, but measurements failed to show any change in the interference pattern! The Michelson–Morley experiment was repeated at different times of the year when the ether wind was expected to change direction, but the results were always the same: **No fringe shift of the magnitude required was ever observed.**

The negative results of the Michelson–Morley experiment not only contradicted the ether hypothesis but also showed that it was impossible to measure the absolute velocity of Earth with respect to the ether frame. However, as we shall see in the next section, Einstein suggested a postulate in the special theory of relativity that places quite a different interpretation on these negative results. In later years, when more was known about the nature of light, the idea of an ether that permeates all of space was relegated to the ash heap of worn-out concepts. **Light is now understood to be an electromagnetic wave, which requires no medium for its propagation.** As a result, the idea of an ether in which these waves could travel became unnecessary.

FIGURE 26.4 According to the ether wind theory, the speed of light should be $c - v$ as the beam approaches mirror M_2 and $c + v$ after reflection.

DETAILS OF THE MICHELSON–MORLEY EXPERIMENT

As we mentioned earlier, the Michelson–Morley experiment was designed to detect the motion of Earth with respect to the ether. Before we examine the details of this historical experiment, it is instructive to first consider a race between two airplanes, as shown in Figure 26.5a. One airplane flies from point O to point A perpendicular to the direction of the wind, and the second airplane flies from point O to point B parallel to the wind. We shall assume that they start at O at the same time, travel the same distance L with the same cruising speed c with respect to the wind, and return to O. Which airplane will win the race? In order to answer this question, we calculate the time of flight for both airplanes.

First, consider the airplane that moves along path I parallel to the wind. As it moves to the right, its speed is enhanced by the wind, and its speed with respect to Earth is $c + v$. As it moves to the left on its return journey, it must fly opposite the wind; hence its speed with respect to Earth is $c - v$. Thus, the times of flight to the right and to the left are, respectively,

$$t_R = \frac{L}{c + v} \quad \text{and} \quad t_L = \frac{L}{c - v}$$

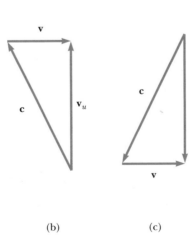

| (a) | (b) | (c) |

FIGURE 26.5 (a) If an airplane wishes to travel from O to A with a wind blowing to the right, it must head into the wind at some angle. (b) Vector diagram for determining the airplane's direction for the trip from O to A. (c) Vector diagram for determining its direction for the trip from A to O.

and the total time of flight for the airplane moving along path I is

$$t_1 = t_R + t_L = \frac{L}{c+v} + \frac{L}{c-v} = \frac{2Lc}{c^2 - v^2} \qquad \text{[26.1]}$$

$$= \frac{2L}{c\left(1 - \dfrac{v^2}{c^2}\right)}$$

Now consider the airplane flying along path II. If the pilot aims the airplane directly toward point *A*, it will be blown off course by the wind and will not reach its destination. To compensate for the wind, the pilot must point the airplane into the wind at some angle, as shown in Figure 26.5a. This angle must be selected so that the vector sum of **c** and **v** leads to a velocity vector pointed directly toward *A*. The resultant vector diagram is shown in Figure 26.5b, where $\mathbf{v}_u$ is the velocity of the airplane with respect to the ground as it moves from *O* to *A*. From the Pythagorean theorem, the magnitude of the vector $\mathbf{v}_u$ is

$$v_u = \sqrt{c^2 - v^2} = c\sqrt{1 - \frac{v^2}{c^2}}$$

Likewise, on the return trip from *A* to *O*, the pilot must again head into the wind so that the airplane's velocity $\mathbf{v}_d$ with respect to Earth will be directed toward *O*, as shown in Figure 26.5c. From this figure, we see that

$$v_d = \sqrt{c^2 - v^2} = c\sqrt{1 - \frac{v^2}{c^2}}$$

Thus, the total time of flight for the trip along path II is

$$t_2 = \frac{L}{v_u} + \frac{L}{v_d} = \frac{L}{c\sqrt{1 - \dfrac{v^2}{c^2}}} + \frac{L}{c\sqrt{1 - \dfrac{v^2}{c^2}}} \qquad \text{[26.2]}$$

$$= \frac{2L}{c\sqrt{1 - \dfrac{v^2}{c^2}}}$$

Comparing Equations 26.1 and 26.2, we see that the airplane flying along path II wins the race. The difference in flight times is given by

$$\Delta t = t_1 - t_2 = \frac{2L}{c}\left[\frac{1}{\left(1 - \dfrac{v^2}{c^2}\right)} - \frac{1}{\sqrt{1 - \dfrac{v^2}{c^2}}}\right]$$

This expression can be simplified by noting that the ratio v/c of wind speed to plane speed is usually much smaller than 1 and by using the following binomial expansions in v/c after dropping all terms higher than second order:

$$\left(1 - \frac{v^2}{c^2}\right)^{-1} \approx 1 + \frac{v^2}{c^2}$$

and

$$\left(1 - \frac{v^2}{c^2}\right)^{-1/2} \approx 1 + \frac{1}{2}\frac{v^2}{c^2}$$

Therefore, the difference in flight times is

$$\Delta t \approx \frac{Lv^2}{c^3} \qquad \text{for} \qquad v/c \ll 1 \qquad \text{[26.3]}$$

The analogy between this airplane race and the Michelson–Morley experiment is shown in Figure 26.6a. Two beams of light travel along two arms of an

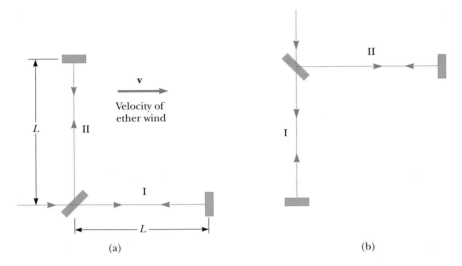

FIGURE 26.6 (a) Top view of the Michelson–Morley interferometer, where **v** is the velocity of the ether and L is the length of each arm. (b) When the interferometer is rotated by 90°, the role of each arm is reversed.

interferometer. In this case, the "wind" is the ether blowing across Earth from left to right as Earth moves through the ether from right to left. Because the speed of Earth in its orbital path is approximately 3×10^4 m/s, it is reasonable to use this value for the speed of the ether wind. Also notice that in this case $v/c \approx 1 \times 10^{-4} \ll 1$. The two light beams start out in phase and return to form an interference pattern. Let us assume that the interferometer is adjusted for parallel fringes and that a telescope is focused on one of these fringes. The time difference between the two light beams gives rise to a phase difference between the beams, producing an interference pattern when they combine at the position of the telescope. The difference in the pattern is detected by rotating the interferometer through 90° in a horizontal plane so that the two beams exchange roles (Fig. 26.6b). This results in a net time shift of twice the time difference given by Equation 26.3. Thus, the net time difference is

$$\Delta t_{\text{net}} = 2\,\Delta t = \frac{2Lv^2}{c^3} \qquad \qquad [26.4]$$

The corresponding path difference is

$$\Delta d = c\,\Delta t_{\text{net}} = \frac{2Lv^2}{c^2} \qquad \qquad [26.5]$$

In the first experiments by Michelson and Morley, each light beam was reflected by the mirrors many times to give an increased effective path length L of about 11 meters. Using this value and taking v to be equal to 3×10^4 m/s gives a path difference of

$$\Delta d = \frac{2(11 \text{ m})(3.0 \times 10^4 \text{ m/s})^2}{(3.0 \times 10^8 \text{ m/s})^2} = 2.2 \times 10^{-7} \text{ m}$$

This extra travel distance should produce a noticeable shift in the fringe pattern. Specifically, calculations show that if the pattern is viewed while the interferometer is rotated through 90°, a shift of about 0.4 fringes should be observed. The instrument used by Michelson and Morley was capable of detecting a shift in the fringe pattern as small as 0.01 fringes. However, *it detected no shift whatsoever in the fringe pattern.* Since then, the experiment has been repeated many times by different scientists under a wide variety of conditions and no fringe shift has ever been detected. Thus, it was concluded that the motion of Earth with respect to the ether cannot be detected.

Many efforts were made to explain the null results of the Michelson–Morley experiment and to save the ether frame concept and the Galilean addition law for the velocity of light. All proposals resulting from these efforts have been shown to be wrong. No experiment in the history of physics has been the subject of such valiant efforts to explain the absence of an expected result as was the Michelson–Morley experiment. The stage was set for Einstein, who solved the problem in 1905 with his special theory of relativity.

26.5   EINSTEIN'S PRINCIPLE OF RELATIVITY

In the previous section we noted the serious contradiction between the Galilean addition law for velocities and the fact that the speed of light is the same for all observers. In 1905 Albert Einstein proposed a theory that resolved this contradiction but at the same time completely altered our notions of space and time. He based his special theory of relativity on two postulates:

Postulates of relativity ▶

1. **The principle of relativity: All the laws of physics are the same in all inertial frames.**
2. **The constancy of the speed of light: The speed of light in a vacuum has the same value, $c = 2.997\ 924\ 58 \times 10^8$ m/s, in all inertial reference frames, regardless of the velocity of the observer or the velocity of the source emitting the light.**

The first postulate asserts that *all* the laws of physics are the same in all reference frames moving with constant velocity relative to each other. This postulate is a sweeping generalization of the principle of Galilean relativity, which refers only to the laws of mechanics. From an experimental point of view, Einstein's principle of relativity means that any kind of experiment—mechanical, thermal, optical, or electrical—performed in a laboratory at rest must give the same result when performed in a laboratory moving at a constant speed past the first one. Hence no preferred inertial reference frame exists, and it is impossible to detect absolute motion.

Postulate 2 was a brilliant theoretical insight on Einstein's part in 1905, and it has since been confirmed experimentally in many ways. Perhaps the most direct demonstration involves measuring the speed of photons emitted by particles traveling at 99.99% of the speed of light. The measured photon speed in this case agrees to five significant figures with the speed of light in empty space.

The null result of the Michelson–Morley experiment can be readily understood within the framework of Einstein's theory. According to his principle of relativity, the premises of the Michelson–Morley experiment were incorrect. In the process of trying to explain the expected results, we stated that when light traveled against the ether wind its speed was $c - v$. However, if the state of motion of the observer or of the source has no influence on the value found for the speed of light, one always measures the value to be c. Likewise, the light makes the return trip after reflection from the mirror at a speed of c, not at a speed of $c + v$. Thus, the motion of Earth does not influence the fringe pattern observed in the Michelson–Morley experiment and a null result should be expected.

If we accept Einstein's theory of relativity, we must conclude that uniform relative motion is unimportant when measuring the speed of light. At the same time, we must alter our common-sense notions of space and time and be prepared for some rather bizarre consequences.

26.6   CONSEQUENCES OF SPECIAL RELATIVITY

Almost everyone who has dabbled even superficially in science is aware of some of the startling predictions that arise because of Einstein's approach to relative motion. As we examine some of the consequences of relativity in this section, we shall find that they conflict with some of our basic notions of space and time. We shall restrict our discussion to the concepts of length, time, and simultaneity, which are

ALBERT EINSTEIN, GERMAN AMERICAN PHYSICIST (1879–1955)

Einstein, one of the greatest physicists of all times, was born in Ulm, Germany. In 1905, at the age of 26, he published four scientific papers that revolutionized physics. Two of these papers were concerned with what is now considered his most important contribution, the special theory of relativity. In 1916 Einstein published his work on the general theory of relativity. The most dramatic prediction of this theory is the degree to which light is deflected by a gravitational field. Measurements made by astronomers on bright stars in the vicinity of the eclipsed Sun in 1919 confirmed Einstein's prediction, and as a result Einstein became a world celebrity. Einstein was deeply disturbed by the development of quantum mechanics in the 1920s despite his own role as a scientific revolutionary. In particular, he could never accept the probabilistic view of events in nature that is a central feature of quantum theory. The last few decades of his life were devoted to an unsuccessful search for a unified theory that would combine gravitation and electromagnetism. *(AIP Niels Bohr Library)* quite different in relativistic mechanics from what they are in Newtonian mechanics. For example, in relativistic mechanics, the distance between two points and the time interval between two events depend on the frame of reference in which they are measured. That is, **in relativistic mechanics, there is no such thing as absolute length or absolute time.** Furthermore, **events at different locations that are observed to occur simultaneously in one frame are not observed to be simultaneous in another frame moving uniformly past the first.**

◀ Absolute length and absolute time intervals are meaningless in relativity.

SIMULTANEITY AND THE RELATIVITY OF TIME

A basic premise of Newtonian mechanics is that a universal time scale exists that is the same for all observers. In fact, Newton wrote, "Absolute, true, and mathematical time, of itself, and from its own nature, flows equably without relation to anything external." Newton and his followers simply took simultaneity for granted. In his special theory of relativity, Einstein abandoned this assumption.

Einstein devised the following thought experiment to illustrate this point. A boxcar moves with uniform velocity, and two lightning bolts strike its ends, as in Figure 26.7a, leaving marks on the boxcar and the ground. The marks on the boxcar are labeled A' and B', and those on the ground are labeled A and B. An observer at O' moving with the boxcar is midway between A' and B', and an observer on the ground at O is midway between A and B. The events recorded by the observers are the striking of the boxcar by the two lightning bolts.

The light signals recording the instant at which the two bolts struck reach observer O at the same time, as indicated in Figure 26.7b. This observer realizes that the signals have traveled at the same speed over equal distances, and so rightly concludes that the events at A and B occurred simultaneously. Now consider the same events as viewed by observer O'. By the time the signals have reached observer O, observer O' has moved, as indicated in Figure 26.7b. Thus, the signal from B' has already swept past O', but the signal from A' has not yet reached O'. In other words, O' sees the signal from B' before seeing the signal from A'. According to Einstein, *the two observers must find that light travels at the same speed.* Therefore, observer O' concludes that the lightning struck the front of the boxcar before it struck the back.

This thought experiment clearly demonstrates that the two events that appear to be simultaneous to observer O do not appear to be simultaneous to observer O'. In other words,

two events that are simultaneous in one reference frame are in general not simultaneous in a second frame moving relative to the first. That is, simultaneity is not an absolute concept but rather one that depends on the state of motion of the observer.

Tip 26.1 WHO'S RIGHT?

Which person is correct concerning the simultaneity of the two events? Both are correct, because the principle of relativity states that no inertial frame of reference is preferred. Although the two observers may reach different conclusions, both are correct in their own reference frame. Any uniformly moving frame of reference can be used to describe events and do physics.

FIGURE 26.7 Two lightning bolts strike the ends of a moving boxcar. (a) The events appear to be simultaneous to the stationary observer at O, who is midway between A and B. (b) The events do not appear to be simultaneous to the observer at O', who claims that the front of the train is struck *before* the rear.

At this point, you might wonder which observer is right concerning the two events. The answer is that *both* are correct because the principle of relativity states that **there is no preferred inertial frame of reference.** Although the two observers reach different conclusions, both are correct in their own reference frames because the concept of simultaneity is not absolute. In fact, this is the central point of relativity. Any inertial frame of reference can be used to describe events and do physics.

TIME DILATION

We can illustrate the fact that observers in different inertial frames may measure different time intervals between a pair of events by considering a vehicle moving to the right with a speed v as in Figure 26.8a. A mirror is fixed to the ceiling of the vehicle, and an observer O' at rest in this system holds a laser a distance d below the mirror. At some instant, the laser emits a pulse of light directed toward the mirror (event 1), and at some later time after reflecting from the mirror, the pulse arrives back at the laser (event 2). Observer O' carries a clock and uses it to measure the time interval, Δt_p, between these two events which she views to occur at the same place. (The subscript p stands for *proper*, as we shall see in a moment.) Because the light pulse has a speed c, the time it takes it to travel from point A to the mirror and back to point A is

$$\Delta t_p = \frac{\text{distance traveled}}{\text{speed}} = \frac{2d}{c} \qquad \text{[26.6]}$$

The time interval Δt_p measured by O' requires only a single clock located at the same place as the laser in this frame.

Now consider the same set of events as viewed by O in a second frame, as shown in Fig. 26.8b. According to this observer, the mirror and laser are moving to the right with a speed v, and as a result the sequence of events appears different. By the time the light from the laser reaches the mirror, the mirror has moved to the right a distance $v \, \Delta t/2$, where Δt is the time it takes the light pulse to travel from point A to the mirror and back to point A as measured by O. In other words, O concludes that, because of the motion of the vehicle, if the light is to hit the mirror, it must leave the laser at an angle with respect to the vertical direction. Comparing Figures 26.8a and 26.8b, we see that the light must travel farther in (b) than in (a). (Note that neither observer "knows" that he or she is moving; each is at rest in his or her own inertial frame.)

According to the second postulate of the special theory of relativity, both observers must measure c for the speed of light. Because the light travels farther in

(a)

(b)   (c)

FIGURE 26.8 (a) A mirror is fixed to a moving vehicle, and a light pulse leaves O' at rest in the vehicle. (b) Relative to a stationary observer on Earth, the mirror and O' move with a speed v. Note that the distance the pulse travels is greater than $2d$ as measured by the stationary observer. (c) The right triangle for calculating the relationship between Δt and Δt_p.

the frame of O, it follows that the time interval Δt measured by O is longer than the time interval Δt_p measured by O'. To obtain a relationship between these two time intervals, it is convenient to use the right triangle shown in Figure 26.8c. The Pythagorean theorem gives

$$\left(\frac{c\,\Delta t}{2}\right)^2 = \left(\frac{v\,\Delta t}{2}\right)^2 + d^2$$

Solving for Δt gives

$$\Delta t = \frac{2d}{\sqrt{c^2 - v^2}} = \frac{2d}{c\sqrt{1 - v^2/c^2}}$$

Because $\Delta t_p = 2d/c$, we can express this result as

$$\Delta t = \frac{\Delta t_p}{\sqrt{1 - v^2/c^2}} = \gamma\,\Delta t_p \qquad\qquad \text{[26.7]}$$

◀ Time dilation

where

$$\gamma = \frac{1}{\sqrt{1 - v^2/c^2}} \qquad\qquad \text{[26.8]}$$

Because γ is always greater than unity, Equation 26.7 says that **the time interval Δt between two events measured by an observer moving with respect to a clock[3] is longer than the time interval Δt_p between the same two events measured by an observer at rest with respect to the clock.** Thus, $\Delta t > \Delta t_p$ and the proper time interval is expanded or dilated by the factor γ. Hence this effort is known as **time dilation.**

For example, suppose the observer who is at rest with respect to the clock measures the time required for the light flash to leave the laser and return. Let us assume that the measured time interval Δt_p in this frame of reference is 1 second. (This would require a very tall vehicle.) Now let us find the time interval as measured by observer O moving with respect to the same clock. If observer O is traveling at half the speed of light ($v = 0.500c$), then $\gamma = 1.15$, and according to Equation 26.7 $\Delta t = \gamma\,\Delta t_p = 1.15(1.00 \text{ s}) = 1.15 \text{ s}$. Thus, when observer O' claims that 1.00 s has passed, observer O claims that 1.15 s has passed. Observer O considers the clock of O' to be reading too low a value for the elapsed time between events and says that the clock of O' is "runn ing slow." From this we may conclude that

A clock moving past an observer at speed v runs more slowly than an identical clock at rest with respect to the observer by a factor of γ^{-1}.

◀ A clock in motion runs more slowly than an identical stationary clock.

The time interval Δt_p in Equations 26.6 and 26.7 is called the **proper time.** In general, **proper time is the time interval between two events as measured by an observer who sees two events occur at the same position.**

Although you may have realized it by now, it is important to spell out that relativity is a scientific democracy—the view of O' that O is really the one moving with speed v to the left and that O's clock is running more slowly is just as valid. The principle of relativity requires that the views of two observers in uniform relative motion must be equally valid and capable of being checked experimentally.

We have seen that moving clocks run slow by a factor of γ^{-1}. This is true for ordinary mechanical clocks as well as for the light clock just described. In fact, we can generalize these results by stating that all physical processes, including chemical and biological ones, slow down relative to a clock when those processes occur

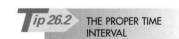

Tip 26.2 THE PROPER TIME INTERVAL

You must be able to correctly identify the observer who measures the proper time interval. The proper time interval between two events is the time interval measured by an observer for whom the two events take place at the same position.

[3] Actually Figure 26.8 shows the clock moving and not the observer, but this is equivalent to the observer O moving to the left with velocity v with respect to the clock.

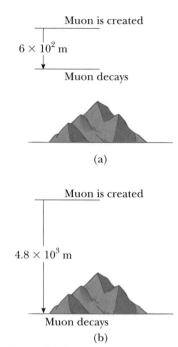

Muon is created

6×10^2 m

Muon decays

(a)

Muon is created

4.8×10^3 m

Muon decays

(b)

FIGURE 26.9 (a) The muons travel only about 600 m as measured in their reference frame, in which their lifetime is about 2.2 μs. (b) Because of time dilation, the muons' lifetime is longer as measured by the observer on Earth. Muons traveling with a speed of $0.99c$ travel a distance of about 4 800 m as measured by an observer on Earth.

in a frame moving with respect to the clock. For example, the heartbeat of an astronaut moving through space would keep time with a clock inside the spaceship. Both the astronaut's clock and heartbeat would be slowed down relative to a clock back on Earth (although the astronaut would have no sensation of life slowing down in the spaceship).

Time dilation is a very real phenomenon that has been verified by various experiments involving the ticking of natural clocks. An interesting example of time dilation involves the observation of *muons,* unstable elementary particles that have a charge equal to that of the electron and a mass 207 times that of the electron. Muons can be produced by the collision of cosmic radiation with atoms high in the atmosphere. These particles have a lifetime of $\Delta t_p = 2.2$ μs when measured in a reference frame at rest with respect to them. If we take 2.2 μs as the average lifetime of a muon and assume that their speed is close to the speed of light, we find that these particles can travel only about 600 m before they decay (Fig. 26.9a). Hence, they could never reach Earth from the upper atmosphere where they are produced. However, experiments show that a large number of muons *do* reach Earth, and the phenomenon of time dilation explains how. Relative to an observer on Earth, the muons have a lifetime equal to $\gamma \Delta t_p$, where $\Delta t_p = 2.2$ μs is the lifetime in a frame of reference traveling with the muons (the "proper" lifetime). For example, for $v = 0.99c$, $\gamma \approx 7.1$ and $\gamma \Delta t_p \approx 16$ μs. Hence, the average distance traveled as measured by an observer on Earth is $\gamma v \Delta t_p \approx 4\,800$ m, as indicated in Figure 26.9b. Hence, muons can reach Earth's surface.

In 1976 experiments with muons were conducted at the laboratory of the European Council for Nuclear Research (CERN) in Geneva. Muons were injected into a large storage ring, reaching speeds of about $0.9994c$. Electrons produced by the decaying muons were detected by counters around the ring, enabling scientists to measure the decay rate and hence the lifetime of the muons. The lifetime of the moving muons was measured to be about 30 times as long as that of stationary muons to within two parts in a thousand, in agreement with the prediction of relativity.

Quick Quiz 26.1

Imagine that you are an astronaut who is being paid according to the time spent traveling in space as measured by a clock on Earth. You take a long voyage traveling at a speed near that of light. Upon your return to Earth, your paycheck will be (a) smaller than if you had remained on Earth, (b) larger than if you had remained on Earth, or (c) the same as if you had remained on Earth.

APPLYING **PHYSICS** 26.2

Suppose a student explains time dilation with the following argument: "If you start running at $0.99c$ away from a clock at 12:00, you would not see the time change, because the light from the clock representing 12:01 would never reach you." What is the flaw in this argument?

Explanation This argument infers that the velocity of light relative to the runner is approximately zero—"the light . . . would never reach you." This is a Galilean relativity point of view, in which the relative velocity is a simple subtraction of running velocity from the light velocity. From the point of view of special relativity, one of the fundamental postulates is that the speed of light is the same for all observers, including one running away from the light source at the speed of light. Thus, the light from 12:01 will move toward the runner at the speed of light. Section 26.8 gives the remarkable relativistic formulas for velocity addition that always give c for the speed of light.

Example 26.1   What Is the Period of the Pendulum?

The period of a pendulum is measured to be 3.0 s in the inertial frame of the pendulum. What is the period when measured by an observer moving at a speed of $0.95c$ with respect to the pendulum?

Reasoning and Solution   In this case, the proper time is 3.0 s. We can use Equation 26.7 to calculate the period measured by the moving observer:

$$T = \gamma\, T_p = \frac{1}{\sqrt{1 - \dfrac{(0.95c)^2}{c^2}}}\, T_p = (3.2)(3.0\ \text{s}) = \boxed{9.6\ \text{s}}$$

That is, the observer moving at a speed of $0.95c$ observes that the pendulum slows down.

THE TWIN PARADOX

An intriguing consequence of time dilation is the so-called twin paradox (Fig. 26.10). Consider an experiment involving a set of twins named Speedo and Goslo. When they are 20 years old, Speedo, the more adventuresome of the two, sets out on an epic journey to Planet X, located 20 lightyears from Earth. Furthermore, his spaceship is capable of reaching a speed of $0.95c$ relative to the inertial frame of his twin brother back home. After reaching Planet X, Speedo becomes homesick and immediately returns to Earth at the same speed, $0.95c$. Upon his return, Speedo is shocked to discover that Goslo has aged $2D/v = 2(20y)/(0.95y/y)$ 42 years and is now 62 years old. Speedo, on the other hand, has aged only 13 years.

At this point, it is fair to raise the following question—which twin is the traveler and which is really younger as a result of this experiment? From Goslo's frame of reference, he was at rest while his brother traveled at a high speed. From Speedo's perspective, it is he who was at rest while Goslo was on the high-speed space journey. According to Speedo, he himself remained stationary while Goslo and Earth raced away from him on a 6.5-year journey and then headed back for another 6.5 years. This leads to an apparent contradiction. Which twin has aged the most?

To resolve this apparent paradox, recall that the special theory of relativity deals with inertial frames of reference moving relative to each other at uniform speed. However, the trip in our current problem is not symmetrical. Speedo, the space traveler, must experience a series of accelerations during his journey. As a re-

▶ The space traveler ages more slowly than his twin who remains on Earth.

      (a)   (b)

FIGURE 26.10   (a) As the twins depart, they are the same age. (b) When Speedo returns from his journey to Planet X, he is younger than his twin Goslo who remained on Earth.

Tip 26.3   THE PROPER LENGTH

You must be able to correctly identify the observer who measures the proper length. The proper length between two points in space is the length measured by an observer at rest with respect to the length. Very often, the proper time interval and the proper length are not measured by the same observer.

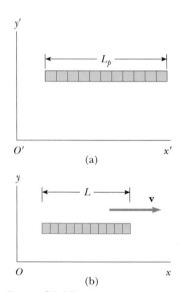

FIGURE 26.11   A meter stick moves to the right with a speed v. (a) The meter stick as viewed by an observer at rest with respect to the meter stick. (b) The meter stick as seen by an observer moving with a speed v with respect to the meter stick. The moving meter stick is always measured to be *shorter* than in its own rest frame by a factor of $\sqrt{1 - v^2/c^2}$.

Length contraction ▶

sult, his speed is not always uniform and consequently he is not in an inertial frame at all times. He cannot be regarded as always being at rest while Goslo is in uniform motion because to do so would be an incorrect application of the special theory of relativity. Therefore there is no paradox. During each passing year noted by Goslo, slightly less than four months elapsed for Speedo.

The conclusion that Speedo is in a noninertial frame is inescapable. Each twin observes the other as accelerating, but it is Speedo who actually undergoes dynamical acceleration due to the real forces acting on him. The time required to accelerate and decelerate Speedo's spaceship may be made very small by using large rockets, so Speedo can claim that he spends most of his time traveling to Planet X at $0.95c$ in an inertial frame. However, Speedo must slow down, reverse his motion, and return to Earth in an altogether different inertial frame. At the very best, Speedo is in two different inertial frames during his journey. Only Goslo, who is in a single inertial frame, can apply the simple time-dilation formula to Speedo's trip. Thus, Goslo finds that instead of aging 42 years, Speedo ages only $(1 - v^2/c^2)^{1/2}(42 \text{ years}) = 13 \text{ years}$. Of this 13 years, Speedo spends 6.5 years traveling to Planet X and 6.5 years returning, for a total travel time of 13 years, in agreement with our earlier statement.

LENGTH CONTRACTION

The measured distance between two points depends on the frame of reference of the observer. The **proper length,** L_p, of an object is **the length of the object measured by someone at rest relative to the object.** The length of an object measured in a reference frame that is moving with respect to the object is always less than the proper length. This effect is known as **length contraction.**

To understand length contraction quantitatively, consider a spaceship traveling with a speed v from one star to another. There are two observers: one on Earth and the other in the spaceship. The observer at rest on Earth (and also assumed to be at rest with respect to the two stars) measures the distance between the stars to be L_p. According to this observer, the time it takes the spaceship to complete the voyage is $\Delta t = L_p/v$. Because of time dilation, the space traveler using the spaceship clock measures a smaller time of travel: $\Delta t_p = \Delta t/\gamma$. The space traveler claims to be at rest and sees the destination star moving toward the spaceship with speed v. Because the spaceship reaches the star in the time Δt_p, the traveler concludes that the distance L between the stars is shorter than L_p. The distance measured by the space traveler is

$$L = v \Delta t_p = v \frac{\Delta t}{\gamma}$$

Because $L_p = v \Delta t$, we see that

$$L = \frac{L_p}{\gamma} = L_p \sqrt{1 - v^2/c^2} \qquad \text{[26.9]}$$

According to this result, illustrated in Figure 26.11, if an observer at rest with respect to an object measures its length to be L_p, an observer moving at a speed v relative to the object will find it to be shorter than its proper length by the factor $\sqrt{1 - v^2/c^2}$. Note that **length contraction takes place only along the direction of motion.**

Time dilation and length contraction effects have interesting applications for future space travel to distant stars. In order for the star to be reached in a fraction of human lifetime, the trip must be taken at very high speeds. According to an Earth-bound observer, the time for a spacecraft to reach the destination star will be dilated compared to the time interval measured by the travelers. Thus, as was discussed in the treatment of the twin paradox, the travelers will be younger than their twins when they return to Earth. Thus, by the time the travelers reach the star, they have aged by some number of years, while their partners back on Earth will have aged a larger number of years, the exact ratio depending on the speed of the spacecraft. At a spacecraft speed of $0.94c$, this ratio is about $3:1$.

Quick Quiz 26.2 You are packing for a trip to another star, to which you will be traveling at $0.99c$. Should you buy smaller sizes of your clothing, because you will be skinnier on the trip? Can you sleep in a smaller cabin than usual, because you will be shorter when you lie down? Explain your answers.

Quick Quiz 26.3 You are observing a rocket moving away from you. Compared to its length when it was at rest on the ground, you will measure its length to be (a) shorter, (b) longer, or (c) the same. Now you see a clock through a window on the rocket. Compared to the passage of time measured by the watch on your wrist, you observe that the passage of time on the rocket's clock is (d) faster, (e) slower, or (f) the same. Answer the same questions if the rocket turns around and comes toward you.

Example 26.2   The Contraction of a Spaceship

A spaceship is measured to be 120 m long while it is at rest with respect to an observer. If this spaceship now flies past the observer with a speed of $0.99c$, what length will the observer measure for the spaceship?

Solution   From Equation 26.9, the length measured by the observer is

$$L = L_p \sqrt{1 - v^2/c^2} = (120 \text{ m}) \sqrt{1 - \frac{(0.99c)^2}{c^2}} = \boxed{17 \text{ m}}$$

EXERCISE   If the ship moves past the observer with a speed of $0.01000c$, what length will the observer measure?

ANSWER   119.994 m

Example 26.3   How High Is the Spaceship?

An observer on Earth sees a spaceship at an altitude of 435 m moving downward toward Earth with a speed of $0.970c$. What is the distance from the spaceship to the ground as measured by an observer in the spaceship?

Solution   The moving observer in the spaceship finds the altitude to be

$$L = L_p \sqrt{1 - v^2/c^2} = (435 \text{ m}) \sqrt{1 - \frac{(0.970c)^2}{c^2}} = \boxed{106 \text{ m}}$$

Example 26.4   The Triangular Spaceship

A spaceship in the form of a triangle flies by an observer with a speed of $0.95c$. When the spaceship is at rest (Fig. 26.12a), the distances x and y are found to be 52 m and 25 m, respectively. What is the shape of the spaceship as seen by an observer at rest when the spaceship is in motion along the direction shown in Figure 26.12b?

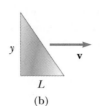

FIGURE 26.12 (Example 26.4) (a) When the spaceship is at rest, its shape is as shown. (b) The spaceship appears to look like this when it moves to the right with a speed *v*. Note that only its *x* dimension is contracted in this case.

Solution The observer sees the horizontal length of the spaceship to be contracted to a length of

$$L = L_p \sqrt{1 - v^2/c^2} = (52 \text{ m}) \sqrt{1 - \frac{(0.95c)^2}{c^2}} = 16 \text{ m}$$

The 25-m vertical height is unchanged because it is perpendicular to the direction of relative motion between the observer and the spaceship. Figure 26.12b represents the shape of the spaceship as seen by the observer at rest.

26.7   RELATIVISTIC MOMENTUM

In order to describe properly the motion of particles within the framework of special relativity, we must generalize Newton's laws of motion and the definitions of momentum and energy. As we shall see, these generalized definitions reduce to the classical (nonrelativistic) definitions when *v* is much less than *c*.

First, recall that conservation of momentum states that when two objects collide, the total momentum of the system remains constant, assuming that the objects are isolated (that is, they interact only with each other). However, analyzing such collisions from rapidly moving inertial frames, it is found that momentum is not conserved if the classical definition of momentum, $p = mv$, is used. In order to have momentum conservation in all inertial frames, even those moving at an appreciable fraction of *c*, the definition of momentum must be modified as follows:

Momentum

$$p \equiv \frac{mv}{\sqrt{1 - v^2/c^2}} = \gamma mv \qquad \qquad \text{[26.10]}$$

where *v* is the speed of the particle and *m* is its mass as measured by an observer at rest with respect to the mass. Note that when *v* is much less than *c*, the denominator of Equation 26.10 approaches unity, so that *p* approaches *mv*. Therefore, the relativistic equation for momentum reduces to the classical expression when *v* is small compared with *c*.

Webnote 26.1

The Earth's upper atmosphere is a very violent setting of high-energy cosmic rays colliding with atoms. See what takes place at
http://www.sciam.com/specialissues/0398cosmos/0398croninbox3.html

Webnote 26.2

AMANDA is a detector being constructed at the South Pole whose purpose is to observe high-energy neutrinos from astrophysical point sources. Visit the project home page and take a look at photos of the science measurement site at
http://amanda.berkeley.edu

Example 26.5   The Relativistic Momentum of an Electron

An electron, which has a mass of 9.11×10^{-31} kg, moves with a speed of $0.75c$. Find its relativistic momentum and compare this value to the momentum calculated from the classical expression.

Solution From Equation 26.10, with $v = 0.75c$, we have

$$p = \frac{mv}{\sqrt{1 - v^2/c^2}}$$

$$= \frac{(9.11 \times 10^{-31} \text{ kg})(0.75 \times 3 \times 10^8 \text{ m/s})}{\sqrt{1 - (0.75c)^2/c^2}} = \boxed{3.1 \times 10^{-22} \text{ kg} \cdot \text{m/s}}$$

The classical expression gives

$$\text{Momentum} = mv = 2.1 \times 10^{-22} \text{ kg} \cdot \text{m/s}$$

The (correct) relativistic result is 50% greater than the classical result!

26.8 RELATIVISTIC ADDITION OF VELOCITIES

Imagine a motorcycle rider moving with a speed of $0.80c$ past a stationary observer, as shown in Figure 26.13. If the rider tosses a ball in the forward direction with a speed of $0.70c$ relative to himself, what is the speed of the ball as seen by the stationary observer at the side of the road? Common sense and the ideas of Newtonian relativity say that the speed should be the sum of the two speeds, or $1.50c$. This answer must be incorrect because it contradicts the assertion that no material object can travel faster than the speed of light.

Einstein resolved this dilemma by deriving an equation for the relativistic addition of velocities. For one-dimensional motion, this equation is

$$v_{ab} = \frac{v_{ad} + v_{db}}{1 + \dfrac{v_{ad}v_{db}}{c^2}} \qquad [26.11]$$

◀ Velocity addition

where v_{ad} is the speed of some object a with respect to a moving frame d, and v_{db} is the speed of the moving frame d with respect to frame b. The left side of this equation and the numerator on the right are like the equations of Galilean relativity discussed in Chapter 3, and the evaluation of subscripts is applied in the same fashion as discussed in Section 3.6. The denominator of Equation 26.11 is a correction to Galilean relativity based on length contraction and time dilation. Let us apply this equation to the case of the speedy motorcycle rider and the stationary observer.

We are given

v_{bm} = the velocity of the ball with respect to the motorcycle = $0.70c$

v_{mo} = the velocity of the motorcycle with respect to the stationary observer
     = $0.80c$,

and we want to find

v_{bo} = the velocity of the ball with respect to the stationary observer.

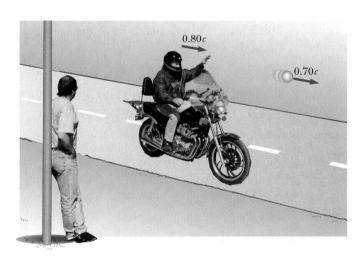

0.80c

0.70c

FIGURE 26.13 A motorcycle moves past a stationary observer with a speed of $0.80c$; the motorcyclist throws a ball in the direction of motion with a speed of $0.70c$ relative to himself.

The speed of light is the speed limit of the Universe.

Thus,

$$v_{bo} = \frac{v_{bm} + v_{mo}}{1 + \dfrac{v_{bm}v_{mo}}{c^2}} = \frac{0.70c + 0.80c}{1 + \dfrac{(0.70c)(0.80c)}{c^2}} = 0.96c$$

Example 26.6 Measuring the Speed of a Light Beam

Suppose that the motorcyclist moving with a speed of $0.80c$ turns on a beam of light that moves away from the motorcycle with a speed of c in the same direction as the moving motorcycle. What speed would the stationary observer measure for the beam of light?

Solution   In this case, we have

v_{lm} = the velocity of the light with respect to the motorcycle = c

v_{mo} = the velocity of the motorcycle with respect to the stationary observer = $0.80c$

and we want

v_{lo} = the velocity of the light with respect to the stationary observer

Thus,

$$v_{lo} = \frac{v_{lm} + v_{mo}}{1 + \dfrac{v_{lm}v_{mo}}{c^2}} = \frac{c + 0.80c}{1 + \dfrac{(c)(0.80c)}{c^2}} = c$$

This is consistent with the statement made earlier that **all observers measure the speed of light to be c regardless of the motion of the source of light.**

26.9 RELATIVISTIC ENERGY AND THE EQUIVALENCE OF MASS AND ENERGY

We have seen that the definition of momentum required generalization to make it compatible with the principle of relativity. Likewise, the definition of kinetic energy requires modification in relativistic mechanics. Einstein found that the correct expression for the **kinetic energy** of an object is

Kinetic energy ▶

$$KE = \gamma mc^2 - mc^2 \tag{26.12}$$

The constant term mc^2 in Equation 26.12, which is independent of the speed of the object, is called the **rest energy,** E_R, of the object.

Rest energy ▶

$$E_R = mc^2 \tag{26.13}$$

The term γmc^2 in Equation 26.12 depends on the object's speed and is the sum of the kinetic and rest energies. We define γmc^2 to be the **total energy,** E; that is,

Total energy = kinetic energy + rest energy

or using Equation 26.12, gives

$$E = KE + mc^2 = \gamma mc^2 \tag{26.14}$$

Because $\gamma = (1 - v^2/c^2)^{-1/2}$, we can also express E as

Total energy ▶

$$E = \frac{mc^2}{\sqrt{1 - v^2/c^2}} \tag{26.15}$$

This is Einstein's famous mass-energy equivalence equation.[4] The relation $E = \gamma mc^2 = KE + mc^2$ shows the amazing result that **a particle has energy by virtue of its mass alone; that is, a stationary particle with zero kinetic energy has an energy proportional to its mass.** Furthermore, this result shows that a small mass corresponds to an enormous amount of energy because the proportionality constant between mass and energy is a large quantity— $c^2 = 9 \times 10^{16} \text{ m}^2/\text{s}^2$. The equation $E_R = mc^2$, as Einstein first suggested, indicates that the mass of a particle may be completely convertible to energy, and the reverse—pure energy, say electromagnetic energy, may be converted to particles—is also true. This is indeed the case as has been shown in the laboratory many times. For example, the coming together of a slow-moving electron and its antiparticle, the positron, a particle with the same mass m_e as the electron but opposite charge, results in the disappearance of both particles and the appearance of a burst of electromagnetic energy in the amount $2m_e c^2$. The reverse process is also fairly easily observed in the lab: A high-energy pulse of electromagnetic energy, a gamma ray, disappears near an atom and an electron-positron pair is created with nearly 100% conversion of the gamma ray's energy into mass. Such a pair-production process is shown in the bubble-chamber photo of Figure 26.14. We shall discuss pair production and annihilation in more detail in Section 26.10.

On a larger scale, nuclear power plants produce energy by the fission of uranium, which involves the conversion of a small amount of the mass of the uranium into energy. The Sun, too, converts mass into energy and continually loses mass in pouring out a tremendous amount of electromagnetic energy in all directions.

It is extremely interesting that while we have been talking about the interconversion of mass and energy for particles, the expression $E = mc^2$ is universal and applies to all objects, processes, and systems: A hot object has slightly more mass and is slightly more difficult to accelerate than an identical cold object because it has more thermal energy; a stretched spring has more elastic potential energy and more mass than an identical unstretched spring. A key point, however, is that these mass changes are often far too small to measure. Our best bet for measuring mass changes is in nuclear transformations, where a measurable fraction of the mass is converted to energy.

FIGURE 26.14 Bubble-chamber photograph of electron (green) and positron (red) tracks produced by energetic gamma rays. The highly curved tracks at the top are due to the electron and positron in an electron-positron pair bending in opposite directions in the magnetic field. *(Lawrence Berkeley Laboratory/Science Photo Library/Photo Researchers, Inc.)*

ENERGY AND RELATIVISTIC MOMENTUM

In many situations, the momentum or energy of a particle is measured rather than its speed. It is therefore useful to have an expression relating the total energy E to the relativistic momentum p. This is accomplished by using the expressions $E = \gamma mc^2$ and $p = \gamma mv$. By squaring these equations and subtracting, we can eliminate v. The result, after some algebra, is

$$E^2 = p^2 c^2 + (mc^2)^2 \qquad \text{[26.16]}$$

When the particle is at rest, $p = 0$ and so $E = E_R = mc^2$. That is, the total energy equals the rest energy. For the case of particles that have zero mass, such as photons (massless, chargeless particles of light), we set $m = 0$ in Equation 26.16, and we see that

$$E = pc \qquad \text{[26.17]}$$

This equation is an exact expression relating energy and momentum for photons, which always travel at the speed of light.

When dealing with subatomic particles, it is convenient to express their energy in electron volts (eV), because the particles are given energy by acceleration

---

[4] Although this doesn't look exactly like the famous equation $E = mc^2$, it used to be common to write $m = \gamma m_0$ (Einstein himself wrote it that way), where m is the effective mass of an object moving at speed v and m_0 is the mass of that object measured by an observer at rest with respect to the object. Then our $E = \gamma mc^2$ becomes the familiar $E = mc^2$. It is currently unfashionable to use $m = \gamma m_0$.

through an electrostatic potential difference. The conversion factor is

$$1 \text{ eV} = 1.60 \times 10^{-19} \text{ J}$$

For example, the mass of an electron is 9.11×10^{-31} kg. Hence, the rest energy of the electron is

$$m_e c^2 = (9.11 \times 10^{-31} \text{ kg})(3.00 \times 10^8 \text{ m/s})^2 = 8.20 \times 10^{-14} \text{ J}$$

Converting this to eV, we have

$$m_e c^2 = (8.20 \times 10^{-14} \text{ J})(1 \text{ eV}/1.60 \times 10^{-19} \text{ J}) = 0.511 \text{ MeV}$$

Because we frequently use the expression $E = \gamma mc^2$ in nuclear physics and m is usually in atomic mass units, u, it is useful to have the conversion factor $1 \text{ u} = 939.494 \text{ MeV}/c^2$. For example, this makes it easy to find the rest energy in MeV of a nucleus with mass of 235.043 924 u:

$$E_R = mc^2 = (235.043\,924 \text{ u})(931.494 \text{ MeV/u}c^2)(c^2) = 2.189\,42 \times 10^5 \text{ MeV}$$

Quick Quiz 26.4   A photon is reflected from a mirror. **True or false:** (a) Because a photon has a zero mass, it does not exert a force on the mirror. (b) Although the photon has energy, it cannot transfer any energy to the surface because it has zero mass. (c) The photon carries momentum, and when it reflects off the mirror, it undergoes a change in momentum and exerts a force on the mirror. (d) Although the photon carries momentum, its change in momentum is zero when it reflects from the mirror, so it cannot exert a force on the mirror.

Example 26.7   The Energy Contained in a Baseball

If a 0.50-kg baseball could be converted completely to energy of forms other than mass, how much energy of other forms would be released?

Solution   The energy equivalent of the baseball is found from Equation 26.12 (with $KE = 0$):

$$E_R = mc^2 = (0.50 \text{ kg})(3.0 \times 10^8 \text{ m/s})^2 = 4.5 \times 10^{16} \text{ J}$$

This is enough energy to keep a 100-W lightbulb burning for approximately 10 million years. However, it is generally impossible to achieve complete conversion from mass to energy of other forms. For example, mass is converted to energy in nuclear power plants, but only a small fraction of the mass actually undergoes conversion.

Example 26.8   The Energy of a Speedy Electron

An electron moves with a speed of $v = 0.850c$. Find its total energy and kinetic energy in electron volts.

Solution   The fact that the rest energy of an electron is 0.511 MeV, along with Equation 26.15, gives

$$E = \frac{m_e c^2}{\sqrt{1 - v^2/c^2}} = \frac{0.511 \text{ MeV}}{\sqrt{1 - \dfrac{(0.850c)^2}{c^2}}}$$

$$= 1.90(0.511 \text{ MeV}) = 0.970 \text{ MeV}$$

The kinetic energy is obtained by subtracting the rest energy from the total energy:

$$KE = E - m_e c^2 = 0.970 \text{ MeV} - 0.511 \text{ MeV} = 0.459 \text{ MeV}$$

Example 26.9 The Conversion of Mass to Kinetic Energy in Uranium Fission

The fission or splitting of uranium was discovered in 1938 by Otto Hahn and is a process in which about 0.1% of the mass is converted to kinetic energy of the fission fragments. The fission of $^{235}_{92}\text{U}$ begins with the absorption of a slow-moving neutron that produces an unstable nucleus of ^{236}U. The ^{236}U nucleus then quickly decays into two heavy fragments, moving at high speed, as well as several neutrons. Most of the kinetic energy released in such a fission is carried off by the two large fragments. (a) For the typical fission process

$$^{1}_{0}\text{n} + {}^{235}_{92}\text{U} \rightarrow {}^{141}_{56}\text{Ba} + {}^{92}_{36}\text{Kr} + 3{}^{1}_{0}\text{n}$$

calculate the kinetic energy in MeV carried off by the fission fragments. (b) Assuming the barium and krypton fragments split the kinetic energy equally between themselves leaving nothing for the three neutrons, find the speed of each fragment. The atomic masses involved are given next in atomic mass units.

$$^{1}_{0}\text{n} = 1.008\,665 \qquad {}^{235}_{92}\text{U} = 235.043\,924 \qquad {}^{141}_{56}\text{Ba} = 140.903\,496 \qquad {}^{92}_{36}\text{Kr} = 91.907\,720$$

Solution

A We use the conservation of relativistic energy to find the kinetic energy released:

$$(KE + mc^2)_{\text{initial}} = (KE + mc^2)_{\text{final}}$$

$$0 + m_n c^2 + m_U c^2 = m_{\text{Ba}} c^2 + m_{\text{Kr}} c^2 + 3m_n c^2 + KE_{\text{final}}$$

where we assumed that the initial kinetic energy was zero. Solving for KE_{final},

$$KE_{\text{final}} = [(m_n + m_U) - (m_{\text{Ba}} + m_{\text{Kr}} + 3m_n)] c^2$$

$$KE_{\text{final}} = [(1.008\,665 \text{ u} + 235.043\,924 \text{ u}) - (140.903\,496 \text{ u} + 91.907\,936 \text{ u}$$

$$+ 3(1.008\,665 \text{ u}))] c^2$$

$$= (0.215\,162 \text{ u}) c^2 = (0.215\,162 \text{ u})(931.494 \text{ MeV/u} \cdot c^2)(c^2) = \boxed{200.422 \text{ MeV}}$$

This means that a little over 0.2 u of the original mass is converted to kinetic energy in this fission.

B According to relativity, the total energy E of a particle of mass m moving at velocity v is given by

$$E = \frac{mc^2}{\sqrt{1 - v^2/c^2}}$$

Solving for v/c, we find

$$\frac{v}{c} = \sqrt{1 - (mc^2)^2/E^2}$$

Applying this expression for v/c to the barium fragment,

$$m_{\text{Ba}} c^2 = (140.903\,496 \text{ u})(931.49 \text{ MeV/u} \cdot c^2)(c^2) = 1.312\,5 \times 10^5 \text{ MeV}$$

$$E_{\text{Ba}} = KE_{\text{Ba}} + m_{\text{Ba}} c^2 = 200.42 \text{ MeV}/2 + 1.312\,5 \times 10^5 \text{ MeV} = 1.313\,5 \times 10^5 \text{ MeV}$$

$$\frac{v_{\text{Ba}}}{c} = \sqrt{1 - (1.312\,5 \times 10^5 \text{ MeV})^2/(1.313\,5 \times 10^5 \text{ MeV})^2} = 0.039\,013$$

Thus $v_{\text{Ba}} = 0.039\,013c \approx \boxed{1.2 \times 10^7 \text{ m/s}}$. In a similar fashion we find

$$v_{\text{Kr}} = 0.0483\,39c \approx \boxed{1.5 \times 10^7 \text{ m/s}}.$$

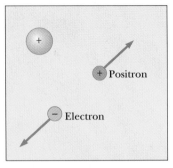

FIGURE 26.15 Representation of the process of pair production.

FIGURE 26.16 Representation of the process of pair annihilation.

26.10 PAIR PRODUCTION AND ANNIHILATION

We shall now describe a process in which the energy of a photon is converted completely into mass. This is a striking verification of the equivalence of mass and other forms of energy as predicted by Einstein's theory of relativity.

A common process in which a photon creates matter is called **pair production,** illustrated in Figure 26.15. In this process, an electron and a positron are simultaneously produced, while the photon disappears. (Note that the positron is a positively charged particle having the same mass as an electron. The positron is often called the *antiparticle* of the electron.) In order for pair production to occur, energy, momentum, and charge must all be conserved during the process. It is impossible for a photon to produce a single electron because the photon has zero charge, and charge would not be conserved in the process.

As we explain in more detail in Chapter 27, the energy of a photon having a frequency f is given by $E = hf$, where h is Planck's constant. The *minimum* energy that a photon must have to produce an electron-positron pair can be found using conservation of energy by equating the photon energy, hf_{min}, to the total rest energy of the pair. That is,

$$hf_{min} = 2m_ec^2 \qquad \text{[26.18]}$$

Because the energy of an electron is $m_ec^2 = 0.51$ MeV, the minimum energy required for pair production is 1.02 MeV.

Pair production cannot occur in a vacuum, but can only take place in the presence of a massive particle such as an atomic nucleus. The massive particle must participate in the interaction in order that energy and momentum be conserved simultaneously.

Pair annihilation is a process in which an electron-positron pair produces two photons, the inverse of pair production. Figure 26.16 is one example of pair annihilation in which an electron and positron initially at rest combine with each other, disappear, and create two photons. Because the initial momentum of the pair is zero, it is impossible to produce a single photon. Momentum can be conserved only if two photons moving in opposite directions, both with the same energy and magnitude of momentum, are produced. We shall discuss particles and their antiparticles further in Chapter 30.

26.11 GENERAL RELATIVITY

Up to this point, we have sidestepped a curious puzzle. Mass has two seemingly different properties: a *gravitational attraction* for other masses and an *inertial* property that resists acceleration. To designate these two attributes, we use the subscripts g and i and write

Gravitational property $\qquad F_g = G\dfrac{m_g m_g'}{r^2}$

Inertial property $\qquad F_i = m_i a$

The value for the gravitational constant G was chosen to make the magnitudes of m_g and m_i numerically equal. Regardless of how G is chosen, however, the strict proportionality of m_g and m_i has been established experimentally to an extremely high degree: a few parts in 10^{12}. Thus, it appears that gravitational mass and inertial mass may indeed be exactly proportional.

But why? They seem to involve two entirely different concepts: a force of mutual gravitational attraction between two masses and the resistance of a single mass to being accelerated. This question, which puzzled Newton and many other physicists over the years, was answered when Einstein published his theory of gravita-

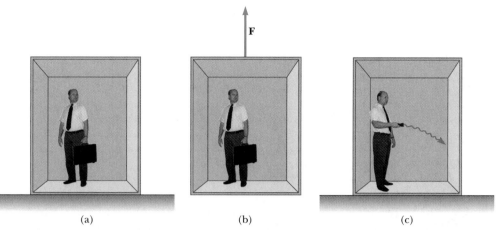

FIGURE 26.17 (a) The observer is at rest in a uniform gravitational field *g*. (b) The observer is in a region in which gravity is negligible, but the frame of reference is accelerated by an external force **F** that produces an acceleration *g*. According to Einstein, the frames of reference in parts (a) and (b) are equivalent in every way. No local experiment could distinguish any difference between the two frames. (c) If parts (a) and (b) are truly equivalent, as Einstein proposed, then a ray of light would bend in a gravitational field.

tion, known as *general relativity,* in 1916. Because it is a mathematically complex theory, we merely offer a hint of its elegance and insight.

In Einstein's view, the remarkable coincidence that m_g and m_i were exactly proportional was evidence for a very intimate and basic connection between the two concepts. He pointed out that no mechanical experiment (such as dropping a mass) could distinguish between the two situations illustrated in Figures 26.17a and 26.17b. In each case, a mass released by the observer undergoes a downward acceleration of *g* relative to the floor.

Einstein carried this idea further and proposed that *no* experiment, mechanical or otherwise, could distinguish between the two cases. This extension to include all phenomena (not just mechanical ones) has interesting consequences. For example, suppose that a light pulse is sent horizontally across the box, as in Figure 26.17c. The trajectory of the light pulse bends downward as the box accelerates upward to meet it. Einstein proposed that a beam of light should also be bent downward by a gravitational field. (No such bending is predicted in Newton's theory of gravitation.)

The two postulates of Einstein's **general relativity** are as follows:

1. All the laws of nature have the same form for observers in any frame of reference, whether accelerated or not.
2. In the vicinity of any given point, a gravitational field is equivalent to an accelerated frame of reference without a gravitational field. (This is the *principle of equivalence.*)

The second postulate implies that gravitational mass and inertial mass are completely equivalent, not just proportional. What were thought to be two different types of mass are actually identical.

One interesting effect predicted by general relativity is that time scales are altered by gravity. A clock in the presence of gravity runs more slowly than one where gravity is negligible. As a consequence, the frequencies of radiation emitted by atoms in the presence of a strong gravitational field are shifted to lower frequencies when compared with the same emissions in a weak field. This gravitational shift has been detected in spectral lines emitted by atoms in massive stars. It has also been verified on Earth by comparing the frequencies of gamma rays emitted from nuclei separated vertically by about 20 m.

FIGURE 26.18 Deflection of starlight passing near the Sun. Because of this effect, the Sun and other remote objects can act as a *gravitational lens*. In his general theory of relativity, Einstein calculated that starlight just grazing the Sun's surface should be deflected by an angle of 1.75″.

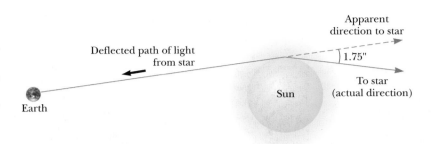

Deflected path of light from star

Apparent direction to star

1.75″

Sun

To star (actual direction)

Earth

Quick Quiz 26.5	Two identical clocks are in the same house, one upstairs in a bedroom and the other downstairs in the kitchen. Which statement is correct? (a) The clock in the kitchen runs more slowly than the clock in the bedroom. (b) The clock in the bedroom runs more slowly than the clock in the kitchen. (c) Both clocks keep the same time.

The second postulate suggests that a gravitational field may be "transformed away" at any point if we choose an appropriate accelerated frame of reference—a freely falling one. Einstein developed an ingenious method of describing the acceleration necessary to make the gravitational field "disappear." He specified a certain quantity, the *curvature of space-time,* that describes the gravitational effect at every point. In fact, the curvature of space-time completely replaces Newton's gravitational theory. According to Einstein, there is no such thing as a gravitational force. Rather, the presence of a mass causes a curvature of space-time in the vicinity of the mass, and this curvature dictates the space-time path that all freely moving objects must follow. In 1979, John Wheeler summarized Einstein's general theory of relativity in a single sentence: "Mass one tells space-time how to curve; curved space-time tells mass two how to move."

One important test of the general theory of relativity is the prediction that a light ray passing near the Sun should be deflected by the curved space-time created by the Sun's mass. This prediction was confirmed by astronomers during a total solar eclipse shortly following World War I when they detected starlight being bent near the Sun (Fig. 26.18). When this discovery was announced, Einstein became an international celebrity.

If the concentration of mass becomes very great, as is believed to occur when a large star exhausts its nuclear fuel and collapses to a very small volume, a **black hole** may form. Here the curvature of space-time is so extreme that, within a certain distance from the center of the black hole, all matter and light become trapped.

Webnote 26.3

Have you ever wondered what it would look like to travel to a black hole? Learn more about the fascinating topic of black holes by visiting
http://antwrp.gsfc.nasa.gov/htmltest/rjn_bht.html

APPLYING PHYSICS 26.3

Atomic clocks are extremely accurate; in fact, an error of 1 second in 3 million years is typical. This error can be described as about one part in 10^{14}. On the other hand, the atomic clock in Boulder, Colorado, is often 15 ns faster than the one in Washington after only one day. This is an error of about one part in 6×10^{12}, which is about 17 times larger than the previously expressed error. If atomic clocks are so accurate, why does a clock in Boulder not remain in synchronism with one in Washington? (*Hint:* Denver, near Boulder, is known as the Mile High City.)

> **Explanation**   According to the general theory of relativity, a clock's rate depends on gravity—clocks run more slowly in strong gravitational fields. Washington is at an elevation very close to sea level, whereas Boulder is about a mile higher in altitude. This will result in a weaker gravitational field at Boulder than at Washington. As a result, an atomic clock runs more rapidly in Boulder than in Washington.

SUMMARY

The two basic postulates of the **special theory of relativity** are as follows:

1. The laws of physics are the same in all inertial frames of reference.
2. The speed of light is the same for all inertial observers, independent of their motion or of the motion of the source of light.

Some of the consequences of the special theory of relativity are as follows:

1. Clocks in motion relative to an observer slow down. This is known as **time dilation.** The relationship between time intervals in the moving and at-rest systems is

$$\Delta t = \gamma \, \Delta t_p \qquad\qquad \textbf{[26.7]}$$

where Δt is the time interval measured in the system in relative motion with respect to the clock,

$$\gamma = 1/\sqrt{1 - v^2/c^2} \qquad\qquad \textbf{[26.8]}$$

and Δt_p is the proper time interval measured in the system moving with the clock.

2. The length of an object in motion is *contracted* in the direction of motion. The equation for **length contraction** is

$$L = L_p \sqrt{1 - v^2/c^2} \qquad\qquad \textbf{[26.9]}$$

where L is the length measured by an observer in motion relative to the object and L_p is the proper length measured by an observer for whom the object is at rest.

3. Events that are simultaneous for one observer are not simultaneous for another observer in motion relative to the first.

The relativistic expression for the **momentum** of a particle moving with a velocity v is

$$p \equiv \frac{mv}{\sqrt{1 - v^2/c^2}} = \gamma mv \qquad\qquad \textbf{[26.10]}$$

The relativistic expression for the addition of velocities is

$$v_{ab} = \frac{v_{ad} + v_{db}}{1 + \dfrac{v_{ad} v_{db}}{c^2}} \qquad\qquad \textbf{[26.11]}$$

where v_{ab} is the velocity of object a with respect to object b, v_{ad} is the velocity of object a with respect to object d, and so forth.

The relativistic expression for the **kinetic energy** of an object is

$$KE = \gamma mc^2 - mc^2 \qquad\qquad \textbf{[26.12]}$$

where mc^2 is the **rest energy,** E_R, of the object.

The **total energy** of a particle is

$$E = \frac{mc^2}{\sqrt{1 - v^2/c^2}} \qquad\qquad \textbf{[26.15]}$$

This is Einstein's famous mass-energy equivalence equation.

The relativistic momentum is related to the total energy through the equation

$$E^2 = p^2 c^2 + (mc^2)^2 \qquad\qquad \textbf{[26.16]}$$

Pair production is a process in which the energy of a photon is converted into mass. In this process, the photon disappears as an electron-positron pair is created. Likewise, the energy of an electron-positron pair can be converted into electromagnetic radiation by the process of **pair annihilation.**

CONCEPTUAL QUESTIONS

1. Why must you be careful with such phrases as *at the same time* and *right now* when dealing with very high speeds?

2. Without looking outside the laboratory, is it possible to conduct a laboratory experiment to tell if, relative to the stars, you are at rest? Moving at constant velocity? Moving at constant acceleration?

3. You are in a speedboat on a lake. You see ahead of you a wave front, caused by the previous passage of another boat, moving away from you. You accelerate, catch up with, and pass the wave front. Is this scenario possible if you are in a rocket and you detect a wave front of light ahead of you?

4. What two speed measurements will two observers in relative motion always agree on?

5. In the *Principia*, Newton suggested that the "fixed" stars constitute an absolute frame of reference. Criticize this suggestion.

6. A rocket is moving with half the speed of light. Discuss the value of the speed of a light pulse emitted from the nose of the rocket as measured by (a) the rocket captain, (b) an observer who sees the rocket approaching, (c) an observer who sees the rocket receding.

7. Some distant star-like objects, called quasars, are receding from us at half the speed of light (or greater). What is the speed of the light we receive from these quasars?

8. It is said that Einstein, in his teenage years, asked the question, "What would I see in a mirror if I carried it in my hands and ran at a speed near that of light?" How would you answer this question?

9. List some ways our day-to-day lives would change if the speed of light were only 50 m/s.

10. Two identically constructed clocks are synchronized. One is put in orbit around Earth while the other remains on Earth. Which clock runs more slowly? When the moving clock returns to Earth, will the two clocks still be synchronized?

11. As discussed in Chapter 22, the speed of light in a transparent material medium is less than the speed of light in a vacuum. Is this a violation of relativity? Discuss.

12. Imagine an astronaut on a trip to Sirius, which lies 8 lightyears from Earth. Upon arrival at Sirius, the astronaut finds that the trip lasted 6 years. If the trip was made at a constant speed of $0.8c$, how can the 8-lightyear distance be reconciled with the 6-year duration?

13. Explain why it is necessary, when defining length, to specify that the positions of the ends of a rod are to be measured simultaneously.

14. The equation $E = mc^2$ is often given in popular descriptions of Einstein's theory of relativity. Is this expression strictly correct? For example, does it accurately account for the kinetic energy of a moving mass?

PROBLEMS

1, 2, 3 = straightforward, intermediate, challenging ☐ = full solution available in Student Solutions Manual/Study Guide

web = solution posted at **http://info.brookscole.com/serway** 🦠 = biomedical application

Section 26.4 The Michelson–Morley Experiment

1. Two airplanes fly paths I and II as specified in Figure 26.5a. Both planes have airspeeds of 100 m/s and fly a distance $L = 200$ km. The wind blows at 20.0 m/s in the direction shown in the figure. Find (a) the time of flight to each city, (b) the time to return, and (c) the difference in total flight times.

2. In one version of the Michelson–Morley experiment, the lengths L in Figure 26.6 were 28 m. Take v to be 3.0×10^4 m/s and find the time difference caused by rotation of the interferometer and (b) the expected fringe shift, assuming that the light used has a wavelength of 550 nm.

Section 26.6 Consequences of Special Relativity

3. A deep-space probe moves away from Earth with a speed of $0.80c$. An antenna on the probe requires 3.0 s, probe time, to rotate through 1.0 rev. How much time is required for 1.0 rev according to an observer on Earth?

4. How fast must a meter stick be moving if its length is observed to shrink to 0.500 m?

5. At what speed does a clock have to move in order to be seen to run at a rate that is one-half the rate of a clock at rest?

6. An astronaut at rest on Earth has a heartbeat rate of 70 beats/min. When the astronaut is traveling in a spaceship at $0.90c$, what will this rate be as measured by (a) an observer also in the ship and (b) an observer at rest on Earth?

7. The average lifetime of a pi meson in its own frame of reference (i.e., the proper lifetime) is 2.6×10^{-8} s. If the meson moves with a speed of $0.98c$, what is (a) its mean lifetime as measured by an observer on Earth and (b) the average distance it travels before decaying as measured by an observer on Earth? (c) What distance would it travel if time dilation did not occur?

8. An astronaut is traveling in a space vehicle that has a speed of $0.500c$ relative to Earth. The astronaut measures his pulse rate at 75.0 per minute. Signals generated by the

astronaut's pulse are radioed to Earth when the vehicle is moving perpendicularly to a line that connects the vehicle with an Earth observer. What pulse rate does Earth observer measure? What would be the pulse rate if the speed of the space vehicle were increased to $0.990c$?

9. A muon formed high in Earth's atmosphere travels at a
web speed $v = 0.99c$ for a distance of 4.6 km before it decays into an electron, a neutrino, and an antineutrino ($\mu^- \rightarrow e^- + \nu + \overline{\nu}$). (a) How long does the muon live, as measured in its reference frame? (b) How far does the muon travel, as measured in its frame?

10. A box is cubical with sides of proper lengths $L_1 = L_2 = L_3$, as shown in Figure P26.10, when viewed in its own rest frame. If this block moves parallel to one of its edges with a speed of $0.80c$ past an observer, (a) what shape does it appear to have to this observer, and (b) what is the length of each side as measured by this observer?

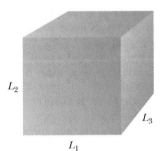

L_2

L_3

L_1

FIGURE P26.10

11. The proper length of one spaceship is three times that of another. The two spaceships are traveling in the same direction and, while both are passing overhead, an Earth observer measures the two spaceships to be the same length. If the slower spaceship is moving with a speed of $0.350c$, determine the speed of the faster spaceship.

12. An atomic clock moves at 1 000 km/h for one hour as measured by an identical clock on Earth. How many nanoseconds slow will the moving clock be at the end of the one-hour interval? (*Hint:* The following approximation will prove helpful: $\sqrt{1 - x} \approx 1 - (x/2)$ for $x \ll 1$.)

13. A supertrain of proper length 100 m travels at a speed of $0.95c$ as it passes through a tunnel having proper length 50 m. As seen by a trackside observer, is the train ever completely within the tunnel? If so, by how much?

14. An observer moving at a speed of $0.995c$ relative to a rod (Fig. P26.14) measures its length to be 2.00 m and sees its length to be oriented at $30.0°$ with respect to the direction of motion. What is the proper length of the rod? (b) What is the orientation angle in a reference frame moving with the rod?

2.00 m

$30.0°$

Motion of observer

FIGURE P26.14   View of rod as seen by an observer moving to the right.

15. In 1963 when Mercury astronaut Gordon Cooper orbited Earth 22 times, the press stated that for each orbit he aged 2 millionths of a second less than if he had remained on Earth. (a) Assuming that he was 160 km above Earth in a circular orbit, determine the time difference between someone on Earth and the orbiting astronaut for the 22 orbits. You will need to use the approximation

$$\frac{1}{\sqrt{1 - x}} \approx 1 + \frac{x}{2} \qquad \text{for } x \ll 1$$

(b) Did the press report accurate information? Explain.

16. An interstellar space probe is launched from Earth. After a brief period of acceleration, it moves with a constant velocity, 70.0% of the speed of light. Its nuclear-powered batteries supply the energy to keep its data transmitter active continuously. The batteries have a lifetime of 15.0 years as measured in a rest frame. (a) How long do the batteries on the space probe last as measured by mission control on Earth? (b) How far is the probe from Earth when its batteries fail, as measured by mission control? (c) How far is the probe from Earth, as measured by its built-in trip odometer, when its batteries fail? (d) For what total time after launch is data received from the probe by mission control? Note that radio waves travel at the speed of light and fill the space between the probe and Earth at the time of battery failure.

Section 26.7 Relativistic Momentum

17. An electron has a speed of $v = 0.90c$. At what speed will a proton have a momentum equal to that of the electron?

18. Calculate the momentum of an electron moving with a speed of (a) $0.010c$, (b) $0.50c$, (c) $0.90c$.

19. An unstable particle at rest breaks up into two fragments of *unequal mass*. The mass of the lighter fragment is 2.50×10^{-28} kg, and that of the heavier fragment is 1.67×10^{-27} kg. If the lighter fragment has a speed of $0.893c$ after the breakup, what is the speed of the heavier fragment?

20. The nonrelativistic expression for the momentum of a particle, $p = mv$, can be used if $v \ll c$. For what speed does the use of this formula give an error in the momentum of (a) 1.00% and (b) 10.0%?

Section 26.8 Relativistic Addition of Velocities

21. An electron moves to the right with a speed of $0.90c$ relative to the laboratory frame. A proton moves to the left with a speed of $0.70c$ relative to the electron. Find the speed of the proton relative to the laboratory frame.

22. Spaceship R is moving to the right at a speed of $0.70c$ with respect to Earth. A second spaceship, L, moves to the left at the same speed with respect to Earth. What is the speed of L with respect to R?

23. A Klingon space ship moves away from Earth at a speed of $0.800c$ (Fig. P26.23). The Starship *Enterprise* pursues at a speed of $0.900c$ relative to Earth. Observers on Earth see the *Enterprise* overtaking the Klingon ship at a relative speed of $0.100c$. With what speed is the *Enterprise* overtaking the Klingon ship as seen by the crew of the *Enterprise*?

24. A spaceship travels at $0.750c$ relative to Earth. If the spaceship fires a small rocket in the forward direction, how fast

FIGURE P26.23

(relative to the ship) must it be fired for it to travel at 0.950*c* relative to Earth?

25. A rocket moves with a velocity of 0.92*c* to the right with respect to a stationary observer A. An observer B moving relative to observer A finds that the rocket is moving with a velocity of 0.95*c* to the left. What is the velocity of observer B relative to observer A? (*Hint:* Consider observer B's velocity in the frame of reference of the rocket.)

26. A pulsar is a stellar object that emits light in short bursts. Suppose a pulsar with a speed of 0.950*c* approaches Earth, and a rocket with a speed of 0.995*c* heads toward the pulsar (both speeds measured in Earth's frame of reference). If the pulsar emits 10.0 pulses per second in its own frame of reference, at what rate are the pulses emitted in the rocket's frame of reference?

27. Spaceship I, which contains students taking a physics exam, approaches Earth with a speed of 0.60*c*, while spaceship II, which contains instructors proctoring the exam, moves away from Earth at 0.28*c*, as in Figure P26.27. If the instructors in spaceship II stop the exam after 50 min have passed on *their clock*, how long does the exam last as measured by (a) the students? (b) an observer on Earth?

FIGURE P26.27

Section 26.9 Relativistic Energy
and the Equivalence of Mass and Energy

28. A proton moves with a speed of 0.950*c*. Calculate its (a) rest energy, (b) total energy, and (c) kinetic energy.

29. What is the speed of a particle whose kinetic energy is equal to its own rest energy?

30. A proton in a high-energy accelerator is given a kinetic energy of 50.0 GeV. Determine (a) the momentum and (b) the speed of the proton.

31. In a color television tube, electrons are accelerated through a potential difference of 20 000 V. With what speed do the electrons strike the screen?

32. Determine the energy required to accelerate an electron from (a) 0.500*c* to 0.900*c* and (b) 0.900*c* to 0.990*c*.

33. A cube of steel has a volume of 1.00 cm³ and a mass of 8.00 g when at rest on Earth. If this cube is now given a speed of *v* = 0.900*c*, what is its density as measured by a stationary observer? Note that relativistic density is E_R/c^2V.

34. An unstable particle with a mass equal to 3.34×10^{-27} kg is initially at rest. The particle decays into two fragments that fly off with velocities of 0.987*c* and −0.868*c*. Find the

masses of the fragments. (*Hint:* Conserve both mass-energy and momentum.)

Section 26.10 Pair Production and Annihilation

35. How much total kinetic energy will an electron-positron pair have if produced by a photon of energy 3.00 MeV?

36. If an electron-positron pair with a total kinetic energy of 2.50 MeV is produced, find (a) the energy of the photon that produced the pair and (b) its frequency.

37. Two photons are produced when a proton and an antiproton annihilate each other. What is the minimum frequency and corresponding wavelength of each photon?

38. An electron moving at a speed of 0.60*c* collides head on with a positron also moving at 0.60*c*. Determine the energy and momentum of each photon produced in the process.

ADDITIONAL PROBLEMS

39. What is the speed of a proton that has been accelerated from rest through a difference of potential of (a) 500 V and (b) 5.00×10^8 V?

40. An electron has a total energy equal to 5 times its rest energy. (a) What is its momentum? (b) Repeat for a proton.

41. What is the momentum (in units of MeV/*c*) of an electron with a kinetic energy of 1.00 MeV?

42. An astronomer on Earth observes a meteoroid in the southern sky approaching at a speed of 0.800*c*. At the time of its discovery, the meteoroid is 20.0 lightyears from Earth. Calculate (a) the time required for the meteoroid to reach Earth as measured by the Earthbound astronomer, (b) this time as measured by a tourist on the meteoroid, and (c) the distance to Earth as measured by the tourist.

43. Ted and Mary are playing a game of catch in frame S′, which is moving with a speed of 0.60*c*; Jim in frame S is watching (Fig. P26.43). Ted throws the ball to Mary with a speed of 0.80*c* (according to Ted) and their separation (measured in S′) is 1.8×10^{12} m. (a) According to Mary, how fast is the ball moving? (b) According to Mary, how long will it take the ball to reach her? (c) According to Jim, how far apart are Ted and Mary, and how fast is the ball moving?

FIGURE P26.43

44. An alarm clock is set to sound in 10 h. At *t* = 0, the clock is placed in a spaceship moving with a speed of 0.75*c* (relative to Earth). What distance, as determined by an Earth observer, does the spaceship travel before the alarm clock sounds?

45. A radioactive nucleus moves with a speed of v relative to a laboratory observer. The nucleus emits an electron in the positive x direction with a speed of $0.70c$ relative to the decaying nucleus, and a speed of $0.85c$ in the $+x$ direction relative to the laboratory observer. What is the value of v?

46. A certain quasar recedes from Earth at $v = 0.870c$. A jet of material ejected from the quasar back toward Earth moves at $0.550c$ relative to the quasar. Find the speed of the ejected material relative to Earth.

47. An astronaut wishes to visit the Andromeda galaxy, making a one-way trip that will take 30.0 y in the spaceship's frame of reference. Assume that the galaxy is 2.00 million lightyears away and that the speed is constant. (a) How fast must the astronaut travel relative to Earth? (b) What will be the kinetic energy of the spacecraft, whose mass is 1.00×10^6 kg? (c) What is the cost of this energy if it is purchased at a typical consumer price for electric energy, 13.0 cents per kWh? The following approximation will prove useful:

$$\frac{1}{\sqrt{1 + x}} \approx 1 - \frac{x}{2} \quad \text{for } x \ll 1$$

48. The cosmic rays of highest energy are protons that have kinetic energy on the order of 10^{13} MeV. (a) How long would it take a proton of this energy to travel across the Milky Way galaxy, having a diameter $\sim 10^5$ lightyears, as measured in the proton's frame? (b) From the point of view of the proton, how many kilometers across is the galaxy?

49. A spaceship of proper length 300 m takes 0.75 μs to pass an Earth observer. Determine the speed of this spaceship as measured by Earth observer.

50. Find the kinetic energy of a 78.0-kg spacecraft launched out of the solar system with a speed of 106 km/s by using (a) the classical equation $KE = \frac{1}{2}mv^2$ and (b) the relativistic equation. You will need to use the approximation

$$\frac{1}{\sqrt{1 - x}} \approx 1 + \frac{x}{2} \quad \text{for } x \ll 1$$

51. An alien spaceship traveling at $0.600c$ toward Earth **web** launches a landing craft with an advance guard of purchasing agents. The lander travels in the same direction with a velocity $0.800c$ relative to the spaceship. As observed on Earth, the spaceship is 0.200 lightyears from Earth when the lander is launched. (a) With what velocity is the lander observed to be approaching by observers on Earth? (b) What is the distance to Earth at the time of lander launch, as observed by the aliens on the mother ship? (c) How long does it take the lander to reach Earth as observed by the aliens on the mother ship? (d) If the lander has a mass of 4.00×10^5 kg, what is its kinetic energy as observed in Earth's reference frame?

52. (a) Show that a potential difference of 1.02×10^6 V would be sufficient to give an electron a speed equal to twice the speed of light if Newtonian mechanics remained valid at high speeds. (b) What speed would an electron actually acquire in falling through a potential difference of 1.02×10^6 V?

53. The muon is an unstable particle that spontaneously decays into an electron and two neutrinos. In a reference frame in which the muons are stationary, if the number of muons at $t = 0$ is N_0, the number at time t is given by $N = N_0 e^{-t/\tau}$, where τ is the mean lifetime, equal to 2.2 μs. Suppose that the muons move at a speed of $0.95c$ and that there are 5.0×10^4 muons at $t = 0$. (a) What is the observed lifetime of the muons? (b) How many muons remain after traveling a distance of 3.0 km?

54. An observer in a rocket moves toward a mirror at speed v relative to the reference frame labeled by S in Figure P26.54. The mirror is stationary with respect to S. A light pulse emitted by the rocket travels toward the mirror and is reflected back to the rocket. The front of the rocket is a distance d from the mirror (as measured by observers in S) at the moment the light pulse leaves the rocket. What is the total travel time of the pulse as measured by observers in (a) the S frame and (b) the front of the rocket?

FIGURE P26.54

55. A physics professor on Earth gives an exam to her students who are on a rocket ship traveling at a speed of v with respect to Earth. The moment the ship passes the professor, she signals the start of the exam. If she wishes her students to have T_0 (rocket time) to complete the exam, show that she should wait a time of

$$T = T_0 \sqrt{\frac{1 - v/c}{1 + v/c}}$$

(Earth time) before sending a light signal telling them to stop. (*Hint:* Remember that it takes some time for the second light signal to travel from the professor to the students.)

56. Imagine that the entire Sun collapses to a sphere of radius R_g such that the work required to remove a small mass m from the surface would be equal to its rest energy, mc^2. This radius is called the *gravitational radius* for the Sun. Find R_g. (It is believed that the ultimate fate of very massive stars is to collapse beyond their gravitational radii into black holes.)

57. A rod of length L_0 moves with a speed of v along the horizontal direction. The rod makes an angle of θ_0 with respect to the axis of a coordinate system moving with the rod. (a) Show that the length of the rod as measured by a stationary observer is given by

$$L = L_0 \left[1 - \left(\frac{v^2}{c^2} \right) \cos^2 \theta_0 \right]^{1/2}$$

(b) Show that the angle the rod makes with the axis as seen by the stationary observer is given by the expression $\tan \theta = \gamma \tan \theta_0$. These results show that the rod is both contracted and rotated. (Take the lower end of the rod to be at the origin of the moving coordinate system.)

GROUP ACTIVITIES

G.1 How would your life change if the speed of light were 55 mi/h? As you and your co-worker go about your daily activities, make notes about how events or occurrences would be changed. See which of the two of you can find the largest number of ways in which your day-to-day life would be affected.

G.2 Consider two travelers on the surface of Earth walking directly toward the North Pole but from different starting locations. How is the fact that they are getting closer together as they go north similar to the curvature of space discussed in the general theory of relativity?

G.3 Find a comfortable location on campus to observe life going on about you. Take note of all the motion going on about you and describe how this motion would look in a system moving at a constant velocity with respect to you.

G.4 Two rockets are of length L and each carries an antenna also of length L that is supported in a direction perpendicular to the length of the rocket. Two pilots, A and B, are in the rockets, which are initially at rest with respect to one another. Rocket A takes off and travels at a speed of $0.400c$ away from B. Answer the following questions for observers A and B either with greater than L, less than L, or equal to L.
(a) The length of A as seen by pilot A is _____.
(b) The length of A as seen by pilot B is _____.
(c) The length of B as seen by pilot B is _____.
(d) The length of B as seen by pilot A is _____.
(e) The length of the antenna of A as seen by pilot B is _____.
(f) The length of the antenna of B as seen by pilot A is _____.
(g) Take L to be 100 m to find precise values for all of the above answers.

G.5 Two rockets carry identical clocks. When they are at rest with respect to each other, pilot A on ship A and pilot B on ship B agree that the second hand on each other's watch takes precisely time ΔT to go around the dial of the watch once. Rocket A takes off and travels at a speed of $0.500c$ away from B. Answer the following questions for observers A and B either with greater than ΔT, less than ΔT, or equal to ΔT.
(a) The interval on A as seen by pilot A is _____.
(b) The interval on A as seen by pilot B is _____.
(c) The interval on B as seen by pilot B is _____.
(d) The interval on B as seen by pilot A is _____.
(e) Find the time intervals when ΔT is equal to 1 min.

G.6 An alien spacecraft fires torpedoes forward with a speed of $0.75c$ relative to the spacecraft. The spacecraft has a speed of $0.30c$ relative to Earth. In such a case, will the torpedoes travel at $1.05c$ with respect to Earth? If you do not agree with the speed of $1.05c$, find the speed at which the torpedoes do travel with respect to Earth.

Quantum Physics

Color-enhanced scanning electron micrograph of the storage mite *Lepidoglyphus destructor*. **These common mites grow up to 0.75 mm and feed on molds, flour and rice. They thrive at 25°C and high humidity, and can trigger allergies.** (*© Eye of Science/Science Source/Photo Researchers, Inc.*)

Chapter Outline

lthough many problems were resolved by the theory of relativity in the early part of the 20th century, many other problems remained unsolved. Attempts to explain the behavior of matter on the atomic level with the laws of classical physics were consistently unsuccessful. Various phenomena, such as the electromagnetic radiation emitted by a heated object (blackbody radiation), the emission of electrons by illuminated metals (photoelectric effect), and the emission of sharp spectral lines by gas atoms in an electric discharge tube, could not be understood within the framework of classical physics. Between 1900 and 1930, however, a modern version of mechanics called *quantum mechanics* or *wave mechanics* was highly successful in explaining the behavior of atoms, molecules, and nuclei. Moreover, quantum mechanics reduces to classical physics when applied to macroscopic systems. As with relativity, the quantum theory requires a modification of our everyday ideas concerning the physical world.

The earliest and most basic ideas of quantum theory were introduced by Planck, and most of the subsequent mathematical developments, interpretations, and improvements were made by a number of distinguished physicists, including Einstein, Bohr, Schrödinger, de Broglie, Heisenberg, Born, and Dirac. An extensive study of quantum theory is certainly beyond the scope of this book. This chapter is simply an introduction to the underlying ideas of quantum theory and the wave-particle nature of matter. We also discuss some simple applications of quantum theory, including the photoelectric effect, the Compton effect, and x-rays.

27.1   BLACKBODY RADIATION AND PLANCK'S HYPOTHESIS

An object at any temperature is known to emit electromagnetic radiation that is sometimes referred to as **thermal radiation.** Stefan's law, which we discussed in Section 11.7, describes the total power radiated. The spectrum of the radiation depends on the temperature and properties of the object. At low temperatures, the wavelengths of the thermal radiation are mainly in the infrared region and hence not observable by the eye. As the temperature of an object increases, the object eventually begins to glow red. At sufficiently high temperatures, it appears to be white, as in the glow of the hot tungsten filament of a lightbulb. A careful study of thermal radiation shows that it consists of a continuous distribution of wavelengths from the infrared, visible, and ultraviolet portions of the spectrum.

From a classical viewpoint, thermal radiation originates from accelerated charged particles near the surface of an object; such charges emit radiation much as small antennas do. The thermally agitated charges can have a distribution of frequencies, which accounts for the continuous spectrum of radiation emitted by the object. By the end of the 19th century, it had become apparent that the classical theory of thermal radiation was inadequate. The basic problem was in understanding the observed distribution of energy as a function of wavelength in the radiation emitted by a blackbody. By definition, a blackbody is an ideal system that absorbs *all* radiation incident on it. A good approximation of a blackbody is a small hole leading to the inside of a hollow object, as shown in Figure 27.1. The nature of the radiation emitted through the small hole leading to the cavity depends *only on the temperature* of the cavity walls, and not at all on the material composition of the object, its shape, or other factors.

Experimental data for the distribution of energy in blackbody radiation at three temperatures are shown in Figure 27.2. The radiated energy varies with wavelength and temperature. As the temperature of the blackbody increases, the total amount of energy (area under the curve) it emits increases. Also, with increasing temperature, the peak of the distribution shifts to shorter wavelengths. This shift was found to obey the following relationship, called **Wien's displacement law:**

$$\lambda_{max} T = 0.2898 \times 10^{-2} \text{ m} \cdot \text{K} \qquad [27.1]$$

FIGURE 27.1 The opening in the cavity of a body is a good approximation of a blackbody. As light enters the cavity through the small opening, part is reflected and part is absorbed on each reflection from the interior walls. After many reflections, essentially all of the incident energy is absorbed.

Wavelength (μm)

FIGURE 27.2 Intensity of blackbody radiation versus wavelength at three different temperatures. Note that the total radiation emitted (the area under the curve) increases with increasing temperature.

where λ_{max} is the wavelength at which the curve peaks and T is the absolute temperature of the object emitting the radiation.

APPLYING **PHYSICS** *27.1*

If you look carefully at stars in the night sky, you can distinguish three main colors—red, white, and blue. What is the reason for these colors?

Explanation   These colors are a result of the thermal radiation from the surfaces of the stars. A relatively cool star, with a surface temperature of 3 000 K, has a radiation curve like the middle curve in Figure 27.2. Notice that the peak in this curve is above the visible wavelengths, 0.4 μm–0.7 μm. Thus, significantly more radiation is emitted within the visible range at the red end than the blue end. As a result, the star appears to be reddish in color, similar to the red glow from the burner of an electric stove.

A hotter star has a radiation curve more like the upper curve in Figure 27.2. In this case, the star emits significant radiation throughout the visible range and the combination of all colors causes the star to look white. This is the case with our own Sun, with a surface temperature of 5 800 K. For very hot stars, the peak can be shifted so far below the visible range that significantly more blue radiation is emitted than red, so that the star appears bluish in color.

> ◀ **APPLICATION**
>
> COLOR OF THE STARS

Attempts to use classical ideas to explain the shape of the curves shown in Figure 27.2 failed. Figure 27.3 shows an experimental plot of the blackbody radiation spectrum (red curve) together with the theoretical picture of what this curve should look like based on classical theories (blue curve). At long wavelengths, classical theory is in good agreement with the experimental data. At short wavelengths, however, major disagreement exists between classical theory and experiment. As λ approaches zero, classical theory predicts that the amount of energy being radiated should increase. In fact, theory predicts that the intensity should be infinite. This is contrary to the experimental data, which show that as λ approaches zero, the amount of energy carried by short-wavelength radiation also approaches zero. This contradiction is called the **ultraviolet catastrophe** because theory and experiment disagree strongly in the short-wavelength ultraviolet region of the spectrum.

In 1900 Planck developed a formula for blackbody radiation that was in complete agreement with experiments at all wavelengths, yielding an intensity versus wavelength distribution that matches the red line in Figure 27.3. Planck's original theoretical approach is rather abstract in that it involves arguments based on entropy and thermodynamics. We shall present arguments that are easier to visualize physically, while attempting to convey the spirit and revolutionary impact of Planck's original work.

Planck hypothesized that blackbody radiation was produced by submicroscopic charged oscillators, which he called *resonators*. He assumed that the walls of a glowing cavity were composed of literally billions of these resonators whose exact nature was unknown. In his theory, Planck made a bold and controversial assumption concerning the nature of the oscillating molecular charges in the blackbody:

The resonators were allowed to have only certain discrete energies, E_n, given by

$$E_n = nhf \qquad \text{[27.2]}$$

where n is a positive integer called a **quantum number,** f is the frequency of vibra-

FIGURE 27.3 Comparison of experimental data with the classical theory of blackbody radiation. Planck's theory perfectly matches the experimental data.

Webnote 27.1

Physics 2000 from the University of Colorado at Boulder is an interactive journey through modern physics using applets, animations, and cartoons. Go to *http://www.colorado.edu/physics/2000/index.pl*

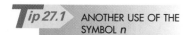
tion of the resonator, and *h* is a constant, known as **Planck's constant,** which has the value

$$h = 6.626 \times 10^{-34} \, \text{J} \cdot \text{s} \qquad [27.3]$$

Because the energy of each resonator can have only discrete values given by Equation 27.2, we say the energy is *quantized*. Each discrete energy value represents a different *quantum state*, with each value of *n* representing a specific quantum state. (When the resonator is in the *n* = 1 quantum state, its energy is *hf*; when it is in the *n* = 2 quantum state, its energy is 2*hf*; and so on.)

The key point in Planck's theory is the radical assumption of quantized energy states. This development marked the birth of the quantum theory. When Planck presented his theory, most scientists (including Planck!) did not consider the quantum concept to be realistic since it violated several centuries of belief that energy was a continuous quantity. Hence, Planck and others continued to search for a classical explanation of blackbody radiation. However, subsequent developments showed that a theory based on the quantum concept (rather than on classical concepts) had to be used to explain a number of other phenomena at the atomic level.

MAX PLANCK, GERMAN PHYSICIST (1858–1947)

Planck was born in Germany. His first son was killed in World War I, and his second son was executed by the Nazis for conspiring to assassinate Hitler. During an air raid on Berlin, he lost his home and valuable library. Planck introduced the concept of a "quantum of action" (Planck's constant, *h*) in an attempt to explain the spectral distribution of blackbody radiation, which laid the foundations for quantum theory. In 1918 he was awarded the Nobel prize for this discovery of the quantized nature of energy. The work leading to the "lucky" blackbody radiation formula was described by Planck in his Nobel prize acceptance speech (1920): "But even if the radiation formula proved to be perfectly correct, it would after all have been only an interpolation formula found by lucky guesswork and, thus, would have left us rather unsatisfied." (© *Bettmann/CORBIS*)

Example 27.1 Thermal Radiation from the Human Body

The temperature of the skin is approximately 35°C. At what wavelength does the radiation emitted from the skin reach its peak?

Solution From Wien's displacement law, Equation 27.1, we have

$$\lambda_{\text{max}} T = 0.2898 \times 10^{-2} \, \text{m} \cdot \text{K}$$

Solving for λ_{max}, noting that 35°C corresponds to an absolute temperature of 308 K, we find

$$\lambda_{\text{max}} = \frac{0.2898 \times 10^{-2} \, \text{m} \cdot \text{K}}{308 \, \text{K}} = \boxed{940 \, \mu\text{m}}$$

This radiation is in the infrared region of the spectrum.

EXERCISE (a) Find the wavelength corresponding to the peak of the radiation curve for the heating element of an electric oven at a temperature of 1.20×10^3 K. (Note that although this radiation peak lies in the infrared, there is enough visible radiation at this temperature to give the element a red glow.) (b) Calculate the wavelength corresponding to the peak of the radiation curve for an object whose temperature is 5.00×10^3 K, an approximate temperature for the surface of the Sun.

ANSWER (a) 2.42 μm; (b) 580 nm in the visible region

Example 27.2 The Quantized Macroscopic Oscillator

A 2.0-kg object is attached to a massless spring of force constant *k* = 25 N/m. The spring is stretched 0.40 m from its equilibrium position and released.

A Find the total energy and frequency of oscillation according to classical calculations.

Solution The total energy of a simple harmonic oscillator having an amplitude *A* is $kA^2/2$. Therefore,

$$E = \tfrac{1}{2}kA^2 = \tfrac{1}{2}(25 \, \text{N/m})(0.40 \, \text{m})^2 = \boxed{2.0 \, \text{J}}$$

The frequency of oscillation is

$$f = \frac{1}{2\pi}\sqrt{\frac{k}{m}} = \frac{1}{2\pi}\sqrt{\frac{25 \text{ N/m}}{2.0 \text{ kg}}} = \boxed{0.56 \text{ Hz}}$$

B Assume that Planck's law of energy quantization applies to any oscillator (atomic or large-scale) and find the quantum number n for this system.

Solution If the energy is quantized, we have $E_n = nhf$, and from the result of Part A, we have

$$E_n = nhf = n(6.63 \times 10^{-34} \text{ J} \cdot \text{s})(0.56 \text{ Hz}) = 2.0 \text{ J}$$

Therefore,

$$n = \boxed{5.4 \times 10^{33}}$$

and we see that macroscopic systems have huge quantum numbers.

C How much energy would be carried away in a one-quantum change?

Solution The energy carried away in a one-quantum change of energy is $\Delta E = E_{n+1} - E_n = hf$, so

$$\Delta E = hf = (6.63 \times 10^{-34} \text{ J} \cdot \text{s})(0.56 \text{ Hz}) = \boxed{3.7 \times 10^{-34} \text{ J}}$$

The energy carried away by a one-quantum change is such a small fraction of the total energy of the oscillator that we could not expect to measure it. Thus, the energy of an object-spring system decreases by such small quantum transitions that the decrease in energy of this macroscopic oscillator appears to be continuous.

27.2 THE PHOTOELECTRIC EFFECT AND THE PARTICLE THEORY OF LIGHT

In the latter part of the 19th century, experiments showed that when light is incident on certain metallic surfaces, electrons are emitted from the surfaces. This phenomenon is known as the **photoelectric effect,** and the emitted electrons are called **photoelectrons.** The first discovery of this phenomenon was made by Hertz, who was also the first to produce the electromagnetic waves predicted by Maxwell.

Figure 27.4 is a schematic diagram of a photoelectric effect apparatus. An evacuated glass tube known as a photocell contains a metal plate E connected to the negative terminal of a variable power supply. Another metal plate C is maintained at a positive potential by the power supply. When the tube is kept in the dark, the ammeter reads zero, indicating that there is no current in the circuit. However, when plate E is illuminated by light having a wavelength shorter than some particular wavelength that depends on the material used to make plate E, a current is detected by the ammeter, indicating a flow of charges across the gap between E and C. This current arises from photoelectrons emitted from the negative plate (the emitter) and collected at the positive plate (the collector).

Figure 27.5 is a plot of the photoelectric current versus the potential difference (voltage) ΔV between E and C for two light intensities. At large values of ΔV, the current reaches a maximum value. In addition, the current increases as the incident light intensity increases, as you might expect. Finally, when ΔV is negative—that is, when the power supply in the circuit is reversed to make E positive and C negative—the current drops to a low value because most of the emitted photoelectrons are repelled by the now negative plate C. In this situation, only those electrons having a kinetic energy greater than the magnitude of $e\,\Delta V$ reach C, where e is the charge on the electron.

When ΔV is equal to or more negative than $-\Delta V_s$, the **stopping potential,** no electrons reach C and the current is zero. The stopping potential is *independent* of the radiation intensity. The maximum kinetic energy of the photoelectrons is

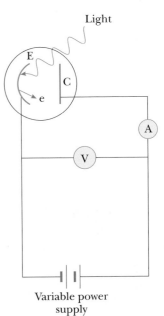

FIGURE 27.4 Circuit diagram for observing the photoelectric effect. When light strikes plate E, photoelectrons are ejected from the plate. Electrons collected at C and passing through the ammeter constitute a current in the circuit.

FIGURE 27.5 Photoelectric current versus applied voltage for two light intensities. The current increases with intensity but reaches a saturation level for large values of ΔV. At voltages equal to or less than $-\Delta V_s$, the current is zero.

related to the stopping potential through the relationship

$$KE_{max} = e\,\Delta V_s \qquad [27.4]$$

Several features of the photoelectric effect cannot be explained with classical physics or with the wave theory of light:

- No electrons are emitted if the incident light frequency falls below some **cutoff frequency,** f_c, that is characteristic of the material being illuminated. This is inconsistent with the wave theory, which predicts that the photoelectric effect should occur at any frequency, provided the light intensity is sufficiently high.
- The maximum kinetic energy of the photoelectrons is independent of light intensity. According to wave theory, light of higher intensity should carry more energy into the metal per unit time and therefore eject photoelectrons having higher kinetic energies.
- The maximum kinetic energy of the photoelectrons increases with increasing light frequency. The wave theory predicts no relationship between photoelectron energy and incident light frequency.
- Electrons are emitted from the surface almost instantaneously (less than 10^{-9} s after the surface is illuminated), even at low light intensities. Classically, we expect the photoelectrons to require some time to absorb the incident radiation before they acquire enough kinetic energy to escape from the metal.

A successful explanation of the photoelectric effect was given by Einstein in 1905, the same year he published his special theory of relativity. As part of a general paper on electromagnetic radiation, for which he received the Nobel prize in 1921, Einstein extended Planck's concept of quantization to electromagnetic waves. He suggested that a tiny packet of light energy or **photon** would be emitted when a quantized oscillator made a jump from an energy state $E_n = nhf$ to the next lower state $E_{n-1} = (n-1)hf$. Conservation of energy would require the decrease in oscillator energy, hf, to be equal to the photon's energy, E, so that

Energy of a photon ▶

$$E = hf \qquad [27.5]$$

h is Planck's constant and f is the frequency of the light which is equal to the frequency of the Planckian oscillator.

The key point here is that the light energy lost by the emitter, hf, stays sharply localized in a tiny packet or particle called a photon. In Einstein's model, a photon is so localized that it can give *all* its energy hf to a single electron in the metal. According to Einstein, the maximum kinetic energy for these liberated photoelectrons is

Photoelectric effect equation ▶

$$KE_{max} = hf - \phi \qquad [27.6]$$

where ϕ is called the **work function** of the metal. The work function represents the minimum energy with which an electron is bound in the metal and is on the order of a few electron volts. Table 27.1 lists work functions for various metals.

With the photon theory of light, we can explain the previously mentioned features of the photoelectric effect that cannot be understood using concepts of classical physics:

- That the effect is not observed below a certain cutoff frequency follows from the fact that the photon energy must be greater than or equal to ϕ. If the energy of the incoming photon does not satisfy this condition, the electrons are never ejected from the surface, regardless of light intensity.
- That KE_{max} is independent of the light intensity can be understood with the following argument. If the light intensity is doubled, the number of photons is doubled, which doubles the number of photoelectrons emitted. However, their maximum kinetic energy, which equals $hf - \phi$, depends only on the light frequency and the work function, not on the light intensity.
- That KE_{max} increases with increasing frequency is easily understood with Equation 27.6.

TABLE 27.1

Work Functions of Selected Metals

Metal	ϕ (eV)
Na	2.46
Al	4.08
Cu	4.70
Zn	4.31
Ag	4.73
Pt	6.35
Pb	4.14
Fe	4.50

- That the electrons are emitted almost instantaneously is consistent with the particle theory of light, in which the incident energy arrives at the surface in small spatial packets and there is a one-to-one interaction between photons and photoelectrons. In this interaction, the photon's energy is imparted to an electron that then has enough energy to leave the metal. This is in contrast to the wave theory, in which the incident energy is distributed uniformly over a large area of the metal surface.

Experimental observation of a linear relationship between f and KE_{max} would be a final confirmation of Einstein's theory. Indeed, such a linear relationship is observed, as sketched in Figure 27.6. The intercept on the horizontal axis gives the cutoff frequency below which no photoelectrons are emitted, regardless of light intensity. The cutoff frequency is related to the work function through the relationship $f_c = \phi/h$. This cutoff frequency corresponds to a **cutoff wavelength** of

$$\lambda_c = \frac{c}{f_c} = \frac{c}{\phi/h} = \frac{hc}{\phi} \qquad [27.7]$$

where c is the speed of light. Wavelengths *greater* than λ_c incident on a material with a work function of ϕ do not result in the emission of photoelectrons.

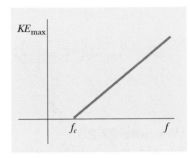

FIGURE 27.6 A sketch of KE_{max} versus frequency of incident light for photoelectrons in a typical photoelectric effect experiment. Photons with frequency less than f_c do not have sufficient energy to eject an electron from the metal.

Example 27.3 The Energy of a "Yellow" Photon

Yellow light with a frequency of approximately 6.0×10^{14} Hz is the predominant frequency in sunlight. What is the energy carried by a photon (a quantum of light) having this frequency?

Solution The energy carried by a photon having this frequency is given by Equation 27.5:

$$E = hf = (6.63 \times 10^{-34}\,\text{J} \cdot \text{s})(6.0 \times 10^{14}\,\text{Hz}) = 4.0 \times 10^{-19}\,\text{J} = \boxed{2.5\ \text{eV}}$$

Example 27.4 The Photoelectric Effect for Sodium

A sodium surface is illuminated with light of wavelength $0.300\ \mu$m. The work function for sodium is 2.46 eV. Find (a) the maximum kinetic energy of the ejected photoelectrons and (b) the cutoff wavelength for sodium.

Solution

A The energy of each photon of the illuminating light beam is

$$E = hf = \frac{hc}{\lambda} = \frac{(6.63 \times 10^{-34}\,\text{J} \cdot \text{s})(3.00 \times 10^8\,\text{m/s})}{3.00 \times 10^{-7}\,\text{m}}$$

$$= 6.63 \times 10^{-19}\,\text{J} = \frac{6.63 \times 10^{-19}\,\text{J}}{1.60 \times 10^{-19}\,\text{J/eV}} = 4.14\ \text{eV}$$

where we have used the conversion $1\ \text{eV} = 1.6 \times 10^{-19}$ J. Using Equation 27.6 gives

$$KE_{max} = hf - \phi = 4.14\ \text{eV} - 2.46\ \text{eV} = \boxed{1.68\ \text{eV}}$$

B The cutoff wavelength can be calculated from Equation 27.7 after we convert ϕ from electron volts to joules:

$$\phi = 2.46\ \text{eV} = (2.46\ \text{eV})(1.6 \times 10^{-19}\,\text{J/eV}) = 3.94 \times 10^{-19}\,\text{J}$$

$$\lambda_c = \frac{hc}{\phi} = \frac{(6.63 \times 10^{-34}\,\text{J} \cdot \text{s})(3.00 \times 10^8\,\text{m/s})}{3.94 \times 10^{-19}\,\text{J}}$$

$$= 5.05 \times 10^{-7}\,\text{m} = \boxed{505\ \text{nm}}$$

This wavelength is in the yellow-green region of the visible spectrum.

PHOTOCELLS

Webnote 27.2

To see an applet of the photoelectric effect, go to
http://home.a-city.de/walter.fendt/ physengl/photoeffect.htm

The photoelectric effect has many interesting applications using a device called the *photocell*. The photocell shown in Figure 27.4 acts much like a switch in an electric circuit with associated active elements in that it produces a current in the circuit when light of sufficiently high frequency falls on the cell, but it does not allow a current in the dark. Many practical devices in our everyday lives make use of photocells. For example, a use familiar to everyone is that of turning streetlights on at night and off in the morning. A photoelectric control unit in the base of the light activates a switch to turn off the streetlight when ambient light strikes it. Many garage-door systems and elevators use a light beam and a photocell as a safety feature in their design. When the light beam strikes the photocell, the electric current generated is sufficiently large to maintain a closed circuit. When an object or person blocks the light beam, the current is interrupted, which signals the door to open.

27.3 X-RAYS

In 1895 at the University of Wurzburg, Wilhelm Roentgen (1845–1923) was studying electrical discharges in low-pressure gases when he noted that a fluorescent screen glowed even when placed several meters from the gas discharge tube and even when black cardboard was placed between the tube and the screen. He concluded that the effect was caused by a mysterious type of radiation, which he called **x-rays** because of their unknown nature. Subsequent study showed that these rays traveled at or near the speed of light and that they could not be deflected by either electric or magnetic fields. This last fact indicated that x-rays did not consist of beams of charged particles, although the possibility that they were beams of uncharged particles remained.

In 1912 Max von Laue (1879–1960) suggested that if x-rays were electromagnetic waves with very short wavelengths, one should be able to diffract x-rays by using the regular atomic spacings of a crystal lattice as a diffraction grating, just as visible light is diffracted by a ruled grating. Shortly thereafter, researchers demonstrated that such a diffraction pattern could be observed, similar to that shown in Figure 27.7 for NaCl. The wavelengths of the x-rays were then determined from the diffraction data and the known values of the spacing between atoms in the crystal. X-ray diffraction has proved to be an invaluable technique for understanding the structure of matter. We shall discuss this subject in more detail in the next section.

FIGURE 27.7 X-ray diffraction pattern of NaCl.

Typical x-ray wavelengths are about 0.1 nm, which is on the order of the atomic spacing in a solid. We now know that x-rays are a part of the electromagnetic spectrum, characterized by frequencies higher than those of ultraviolet radiation and having the ability to penetrate most materials with relative ease.

X-rays are produced when high-speed electrons are suddenly slowed down, for example, when a metal target is struck by electrons that have been accelerated through a potential difference of several thousand volts. Figure 27.8a shows a schematic diagram of an x-ray tube. A current in the filament causes electrons to be emitted, and these freed electrons are accelerated toward a dense metal target, such as tungsten, which is held at a higher potential than the filament.

Figure 27.9 represents a plot of x-ray intensity versus wavelength for the spectrum of radiation emitted by an x-ray tube. Note that the spectrum has two distinct components. One component is a continuous broad spectrum that depends on the voltage applied to the tube. Superimposed on this component is a series of sharp, intense lines that depend on the nature of the target material. The accelerating voltage must exceed a certain value, called the **threshold voltage,** in order to observe these sharp lines, which represent radiation emitted by the target atoms as their electrons undergo rearrangements. We shall discuss this further in Chap-

FIGURE 27.9 The x-ray spectrum of a metal target consists of a broad continuous spectrum plus a number of sharp lines, which are due to *characteristic x-rays*. The data shown were obtained when 35-keV electrons bombarded a molybdenum target. Note that 1 pm = 10^{-12} m = 10^{-3} nm.

(a)   (b)

FIGURE 27.8 (a) Diagram of an x-ray tube. (b) Photograph of an x-ray tube. *(Courtesy of GE Medical Systems)*

ter 28. The continuous radiation is sometimes called **bremsstrahlung,** a German word meaning "braking radiation." The term arises from the nature of the mechanism responsible for the radiation. That is, electrons emit radiation when they undergo an acceleration inside the target.

Figure 27.10 illustrates how x-rays are produced when an electron passes near a charged target nucleus. As an electron passes close to a positively charged nucleus contained in the target material, it is deflected from its path because of its electrical attraction to the nucleus, and hence experiences an acceleration. An analysis from classical physics shows that any charged particle will emit electromagnetic radiation when it is accelerated. (An example of this is the production of electromagnetic waves by accelerated charges in a radio antenna, as described in Chapter 21.) According to quantum theory, this radiation must appear in the form of photons. Because the radiated photon shown in Figure 27.10 carries energy, the electron must lose kinetic energy because of its encounter with the target nucleus. Let us consider an extreme example in which the electron loses all of its energy in a single collision. In this case, the initial energy of the electron ($e \Delta V$) is transformed completely into the energy of the photon (hf_{max}). In equation form, we have

$$e \Delta V = hf_{max} = \frac{hc}{\lambda_{min}} \qquad [27.8]$$

where $e \Delta V$ is the energy of the electron after it has been accelerated through a potential difference of ΔV volts, and e is the charge on the electron. This says that the shortest-wavelength radiation that can be produced is

$$\lambda_{min} = \frac{hc}{e \Delta V} \qquad [27.9]$$

The reason that all the radiation produced does not have this particular wavelength is because many of the electrons are not stopped in a single collision. This results in the production of the continuous spectrum of wavelengths.

Interesting insights into the process of painting and revising a masterpiece are being revealed by x-rays. Long-wavelength x-rays are absorbed in varying degrees by some paints, such as those having lead, cadmium, chromium, or cobalt as a base. The x-ray interactions with the paints give contrast because the different elements in the paints have different electron densities. Also, thicker layers will absorb more than thin layers. To examine a painting by an old master, a film is placed behind it while it is x-rayed from the front. Ghost outlines of earlier paintings and earlier forms of the final masterpiece are sometimes revealed when the film is developed.

FIGURE 27.10 An electron passing near a charged target atom experiences an acceleration, and a photon is emitted in the process.

> **APPLICATION**

USING X-RAYS TO STUDY THE WORK OF MASTER PAINTERS

Example 27.5   **The Minimum X-Ray Wavelength**

Calculate the minimum wavelength produced when electrons are accelerated through a potential difference of 100 000 V, a not-uncommon voltage for an x-ray tube.

Solution   From Equation 27.9, we have

$$\lambda_{\min} = \frac{(6.63 \times 10^{-34}\,\text{J} \cdot \text{s})(3.00 \times 10^{8}\,\text{m/s})}{(1.60 \times 10^{-19}\,\text{C})(10^{5}\,\text{V})} = \boxed{1.24 \times 10^{-11}\,\text{m}}$$

27.4   DIFFRACTION OF X-RAYS BY CRYSTALS

In Chapter 24 we described how a diffraction grating can be used to measure the wavelength of light. In principle, the wavelength of *any* electromagnetic wave can be measured if a grating having a suitable line spacing can be found. The spacing between lines must be approximately equal to the wavelength of the radiation to be measured. X-rays are electromagnetic waves with wavelengths on the order of 0.1 nm. It would be impossible to construct a grating with such a small spacing. However, as noted in the previous section, Max von Laue suggested that the regular array of atoms in a crystal could act as a three-dimensional grating for observing the diffraction of x-rays.

One experimental arrangement for observing x-ray diffraction is shown in Figure 27.11. A narrow beam of x-rays with a continuous wavelength range is incident on a crystal such as sodium chloride. The diffracted radiation is very intense in certain directions, corresponding to constructive interference from waves reflected from layers of atoms in the crystal. The diffracted radiation is detected by a photographic film and forms an array of spots known as a Laue pattern. The crystal structure is determined by analyzing the positions and intensities of the various spots in the pattern.

The arrangement of atoms in a crystal of NaCl is shown in Figure 27.12. The smaller red spheres represent Na^+ ions and the larger blue spheres represent Cl^- ions. The spacing between successive Na^+ (or Cl^-) ions in this cubic structure, denoted by the symbol *a* in Figure 27.12, is approximately 0.563 nm.

A careful examination of the NaCl structure shows that the ions lie in various planes. The shaded areas in Figure 27.12 represent one example in which the atoms lie in equally spaced planes. Now suppose an x-ray beam is incident at grazing angle θ on one of the planes, as in Figure 27.13. The beam can be reflected from both the upper and lower plane of atoms. However, the geometric construction in Figure 27.13 shows that the beam reflected from the lower surface travels farther than the beam reflected from the upper surface by a distance of $2d \sin \theta$. The two portions of the reflected beam will combine to produce constructive in-

FIGURE 27.11 Schematic diagram of the technique used to observe the diffraction of x-rays by a single crystal. The array of spots formed on the film by the diffracted beams is called a Laue pattern. (See Figure 27.7.)

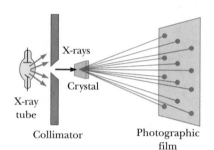

FIGURE 27.12 A model of the cubic crystalline structure of sodium chloride. The blue spheres represent the Cl^- ions, and the red spheres represent the Na^+ ions. The length of the cube edge is $a = 0.563$ nm.

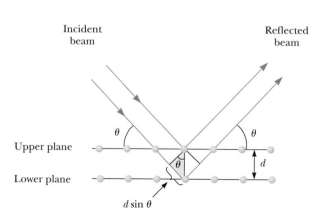

FIGURE 27.13 A two-dimensional description of the reflection of an x-ray beam from two parallel crystalline planes separated by a distance *d*. The beam reflected from the lower plane travels farther than the one reflected from the upper plane by an amount equal to $2d \sin \theta$.

terference when this path difference equals some integral multiple of the wavelength λ. The condition for constructive interference is given by

$$2d \sin \theta = m\lambda \qquad (m = 1, 2, 3, \ldots) \qquad \text{[27.10]}$$

 Bragg's law

This condition is known as **Bragg's law** after W. L. Bragg (1890–1971), who first derived the relationship. If the wavelength and diffraction angle are measured, Equation 27.10 can be used to calculate the spacing between atomic planes.

The method of x-ray diffraction to determine crystalline structures was thoroughly developed in England by W. H. Bragg and his son W. L. Bragg, who shared a Nobel prize in 1915 for their work. Since then, thousands of crystalline structures have been investigated. Most important, the technique of x-ray diffraction has been and is being used to determine the atomic arrangement of complex organic molecules such as proteins. Proteins are large molecules containing thousands of atoms that help to regulate and speed up chemical life processes in cells. Some proteins are amazing catalysts, speeding up the slow room-temperature reactions in cells by 17 orders of magnitude. In order to understand this incredible biochemical reactivity, it is important to determine the structure of these intricate molecules.

The main technique used to determine the molecular structure of proteins, DNA, and RNA is x-ray diffraction using x-rays of wavelength of about 1.0 A. This allows the experimenter to "see" individual atoms that are separated by about this distance in molecules. Because the biochemical x-ray diffraction sample is prepared in crystal form, the *geometry* (position of the bright spots in space) of the diffraction pattern is determined by the regular three-dimensional crystal lattice arrangement of molecules in the sample. The *intensities* of the bright diffraction spots are determined by the atoms and their electronic distributions in the fundamental building block of the crystal, the unit cell. Using complicated computational techniques, investigators can essentially deduce the molecular structure by matching the observed intensities of diffracted beams with a series of assumed atomic positions, resulting in the atomic structure and electron density of the molecule. Figure 27.14 shows a classic x-ray diffraction image of DNA made by Rosalind Franklin in 1952. This and similar x-ray diffraction photos played an important role in the determination of the double-helix structure of DNA by F. H. C. Crick and J. D. Watson in 1953. A model of the famous DNA double helix is shown in Figure 27.15.

FIGURE 27.14 An x-ray diffraction photograph of DNA taken by Rosalind Franklin. The cross pattern of spots was a clue that DNA has a helical structure. *(Science Source/Photo Researchers, Inc.)*

FIGURE 27.15 The double-helix structure of DNA.

Example 27.6 X-Ray Diffraction from Calcite

If the spacing between certain planes in a crystal of calcite ($CaCO_3$) is 0.314 nm, find the grazing angles at which first- and third-order interference will occur for x-rays of wavelength 0.070 nm.

Solution For first-order interference, the value of m in Equation 27.10 is 1. Thus, the grazing angle corresponding to this order of interference is found as follows:

$$\sin \theta = \frac{m\lambda}{2d} = \frac{(0.0700 \text{ nm})}{2(0.314 \text{ nm})} = 0.111$$

$$\theta = 6.37°$$

In third-order interference, $m = 3$, and we find

$$\sin \theta = \frac{m\lambda}{2d} = \frac{3(0.0700 \text{ nm})}{2(0.314 \text{ nm})} = 0.334$$

$$\theta = 19.5°$$

Webnote 27.3

To learn much more about x-ray crystallography, its techniques, and the equipment used for measurement, explore *http://www-structure.llnl.gov/xray/101index.html*

27.5   THE COMPTON EFFECT

Further justification for the photon nature of light came from an experiment conducted by Arthur H. Compton in 1923. In his experiment, Compton directed an x-ray beam of wavelength λ_0 toward a block of graphite. He found that the scattered x-rays had a slightly longer wavelength λ than the incident x-rays, and hence the energies of the scattered rays were lower. The amount of energy reduction depended on the angle at which the x-rays were scattered. The change in wavelength, $\Delta\lambda$, between a scattered x-ray and an incident x-ray is called the **Compton shift.**

In order to explain this effect, Compton assumed that if a photon behaves like a particle, its collision with other particles is similar to that between two billiard balls. Hence, the x-ray photon carries both measurable *energy and momentum* and these two quantities must be conserved in a collision. If the incident photon collides with an electron initially at rest, as in Figure 27.16, the photon transfers some of its energy and momentum to the electron. As a consequence, the energy and frequency of the scattered photon are lowered and its wavelength increases. Applying relativistic energy and momentum conservation to the collision described in Figure 27.16, the shift in wavelength of the scattered photon is given by

The Compton shift formula

$$\Delta\lambda = \lambda - \lambda_0 = \frac{h}{m_e c}(1 - \cos\theta) \qquad [27.11]$$

where m_e is the mass of the electron and θ is the angle between the directions of the scattered and incident photons. The quantity $h/m_e c$ is called the **Compton wavelength** and has a value of $h/m_e c = 0.002\ 43$ nm. Note that the Compton wavelength is very small relative to the wavelengths of visible light, and hence the shift in wavelength would be difficult to detect if visible light were used. Furthermore, note that the Compton shift depends on the scattering angle θ and not on the wavelength. Experimental results for x-rays scattered from various targets obey Equation 27.11 and strongly support the photon concept.

ARTHUR HOLLY COMPTON, AMERICAN PHYSICIST (1892–1962)

Compton was born in Wooster, Ohio, and attended Wooster College and Princeton University. He became director of the laboratory at the University of Chicago where experimental work concerned with sustained chain reactions was conducted. This work was of central importance to the construction of the first atomic bomb. His discovery of the Compton effect and his work with cosmic rays led to his sharing the 1927 Nobel prize in physics with Charles Wilson. *(Courtesy of AIP Niels Bohr Library)*

> **Quick Quiz 27.1** An x-ray photon is scattered by an electron. The frequency of the scattered photon relative to that of the incident photon (a) increases, (b) decreases, or (c) remains the same.

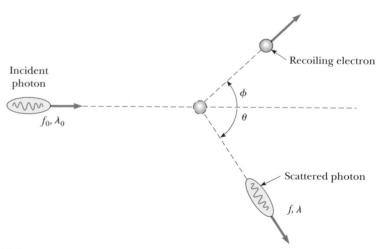

FIGURE 27.16   Diagram representing Compton scattering of a photon by an electron. The scattered photon has less energy (or longer wavelength) than the incident photon.

Quick Quiz 27.2

A photon of energy E_0 strikes a free electron, with the scattered photon of energy E moving in the direction opposite that of the incident photon. In this Compton effect interaction, the resulting kinetic energy of the electron is (a) E_0, (b) E, (c) $E_0 - E$, (d) $E_0 + E$, (e) none of the above.

Quick Quiz 27.3

A photon of energy E_0 strikes a free electron with the scattered photon of energy E moving in the direction opposite that of the incident photon. In this Compton effect interaction, the resulting momentum of the electron is (a) E_0/c, (b) $< E_0/c$, (c) $> E_0/c$, (d) $(E_0 - E)/c$, (e) $(E - E_0)/c$.

APPLYING PHYSICS 27.2

The Compton effect involves a change in wavelength as photons are scattered through different angles. Suppose we illuminate a piece of material with a beam of light and then view the material from different angles relative to the beam of light. Will we see a color change corresponding to the change in wavelength of the scattered light?

Explanation There will be a wavelength change for visible light scattered by the material, but the change will be far too small to detect as a color change. The largest possible wavelength change, at 180° scattering, will be twice the Compton wavelength, about 0.005 nm. This represents a change of less than 0.001% of the wavelength of red light. The Compton effect is only detectable for wavelengths that are very short to begin with, so that the Compton wavelength is an appreciable fraction of the incident wavelength. As a result, the usual radiation for observing the Compton effect is in the x-ray range of the electromagnetic spectrum.

Example 27.7 Compton Scattering at 45°

X-rays of wavelength $\lambda_0 = 0.200\,000$ nm are scattered from a block of material. The scattered x-rays are observed at an angle of 45.0° to the incident beam. Calculate the wavelength of the x-rays scattered at this angle.

Solution The shift in wavelength of the scattered x-rays is given by Equation 27.11. Taking $\theta = 45.0°$, we find that

$$\Delta\lambda = \frac{h}{m_e c}(1 - \cos\theta)$$

$$= \frac{6.626 \times 10^{-34}\,\text{J}\cdot\text{s}}{(9.11 \times 10^{-31}\,\text{kg})(3.00 \times 10^8\,\text{m/s})}(1 - \cos 45.0°)$$

$$= 7.10 \times 10^{-13}\,\text{m} = 0.000\,710\,\text{nm}$$

Hence, the wavelength of the scattered x-ray at this angle is

$$\lambda = \Delta\lambda + \lambda_0 = \boxed{0.200\,710\,\text{nm}}$$

EXERCISE Find the fraction of energy lost by the photon in this collision.

ANSWER Fraction $= \Delta E/E = 0.003\,54$

27.6 PHOTONS AND ELECTROMAGNETIC WAVES

Phenomena such as the photoelectric effect and the Compton effect offer iron-clad evidence that when light (or other forms of electromagnetic radiation) and matter interact, the light behaves as if it were composed of particles having energy *hf* and momentum *h/λ*. An obvious question that arises at this point is, "How can light be considered a photon (in other words, a particle) when we know it is a wave?" On the one hand, we describe light in terms of photons having energy and momentum. On the other hand, we must also recognize that light and other electromagnetic waves exhibit interference and diffraction effects that are consistent only with a wave interpretation.

Which model is correct? Is light a wave or a particle? The answer depends on the phenomenon being observed. Some experiments can be better explained with the photon concept, whereas others are best described with a wave model. The end result is that we must use both models and admit that the true nature of light is not describable in terms of a single classical picture. We can say that

Light has a dual nature ▶

Light has a dual nature. It exhibits both wave and particle characteristics.

To understand why photons are compatible with electromagnetic waves, consider 2.5-MHz radio waves as an example. The energy of a photon having this frequency is only about 10^{-8} eV, too small to allow the photon to be detected. A sensitive radio receiver might require as many as 10^{10} of these photons to produce a detectable signal. Such a large number of photons would appear, on the average, as a continuous wave. With so many photons reaching the detector every second, it is unlikely that any graininess would appear in the detected signal. That is, with 2.5-MHz waves, we would not be able to detect the individual photons striking the antenna.

Now consider what happens as we go to higher frequencies. In the visible region, it is possible to observe both the particle characteristics and the wave characteristics of light. As we mentioned earlier, a light beam shows interference phenomena (thus, it is a wave) and at the same time can produce photoelectrons (thus, it is a particle). At even higher frequencies, the momentum and energy of the photons increase. Consequently, the particle nature of light becomes more evident than its wave nature. For example, absorption of an x-ray photon is easily detected as a single event, but wave effects are difficult to observe.

LOUIS DE BROGLIE, FRENCH PHYSICIST (1892–1987)

De Broglie was born in Dieppe, France. At the Sorbonne in Paris, he studied history in preparation for what he hoped to be a career in the diplomatic service. The world of science is lucky that he changed his career path to become a theoretical physicist. De Broglie was awarded the Nobel prize in 1929 for his discovery of the wave nature of electrons. "It would seem that the basic idea of quantum theory is the impossibility of imaging an isolated quantity of energy without associating with it a certain frequency." *(AIP Niels Bohr Library)*

27.7 THE WAVE PROPERTIES OF PARTICLES

The "magic" duality of light mentioned in the last section is not the only duality we have encountered in this course. The mass-energy duality mentioned in Chapter 26 is probably about as easy or difficult to accept as the duality of light. All these dualities should make it easy to accept that *matter* has a dual nature as well!

In 1924, in his doctoral dissertation, Louis de Broglie postulated that **because photons have wave and particle characteristics, perhaps all forms of matter have both properties.** This was a highly revolutionary idea with no experimental confirmation at that time. According to de Broglie, electrons, just like light, have a dual particle-wave nature. Let us follow de Broglie's argument for the wavelength of a material particle.

In Chapter 26 we found that the relationship between energy and momentum for a photon, which has a rest energy of zero, is $p = E/c$. We also know from Equation 27.5 that the energy of a photon is

Energy of a photon ▶

$$E = hf = \frac{hc}{\lambda}$$

[27.12]

Thus, the momentum of a photon can be expressed as

$$p = \frac{E}{c} = \frac{hc}{c\lambda} = \frac{h}{\lambda}$$

[27.13] ◀ Momentum of a photon

From this equation we see that the photon wavelength can be specified by its momentum, or $\lambda = h/p$. De Broglie suggested that all material particles of momentum p should have a characteristic wavelength $\lambda = h/p$. Because the momentum of a particle of mass m and speed v is $mv = p$, the **de Broglie wavelength** of a particle is

$$\lambda = \frac{h}{p} = \frac{h}{mv}$$

[27.14] ◀ De Broglie's hypothesis

Furthermore, in analogy with photons, de Broglie postulated that the frequencies of matter waves (that is, waves associated with particles having nonzero rest energy) obey the Einstein relationship $E = hf$, so that

$$f = \frac{E}{h}$$

[27.15] ◀ Frequency of matter waves

The dual nature of matter is quite apparent in Equations 27.14 and 27.15 because each contains both particle concepts (mv and E) and wave concepts (λ and f). The fact that these relationships had been established experimentally for photons made the de Broglie hypothesis that much easier to accept.

THE DAVISSON-GERMER EXPERIMENT

De Broglie's proposal in 1923 that matter exhibits both wave and particle properties was first regarded as pure speculation. If particles such as electrons had wave-like properties, then under the correct conditions they should exhibit diffraction effects. In 1927, three years after de Broglie published his work, C. J. Davisson (1881–1958) and L. H. Germer (1896–1971) of the United States succeeded in measuring the wavelength of electrons. Their important discovery provided the first experimental confirmation of the matter waves proposed by de Broglie.

Interestingly, the intent of the initial Davisson–Germer experiment was not to confirm the de Broglie hypothesis. In fact, their discovery was made by accident (as is often the case). The experiment involved the scattering of low-energy electrons (about 54 eV) from a nickel target in a vacuum. During one experiment, the nickel surface was badly oxidized because of an accidental break in the vacuum system. After the nickel target was heated in a flowing stream of hydrogen to remove the oxide coating, electrons scattered by it exhibited intensity maxima and minima at specific angles. The experimenters finally realized that the nickel had formed large crystalline regions upon heating and that the regularly spaced planes of atoms in the crystalline regions served as a diffraction grating for electron matter waves (see Section 27.4).

Shortly thereafter, Davisson and Germer performed more extensive diffraction measurements on electrons scattered from single-crystal targets. Their results showed conclusively the wave nature of electrons and confirmed the de Broglie relation $\lambda = h/p$. In the same year, G. P. Thomson (1892–1975) of Scotland also observed electron diffraction patterns by passing electrons through very thin gold foils. Diffraction patterns have since been observed for helium atoms, hydrogen atoms, and neutrons. Hence, the universal nature of matter waves has been established in various ways.

A nonrelativistic electron and a nonrelativistic proton are moving and have the same de Broglie wavelength. Which of the following are also the same for the two particles: (a) speed, (b) kinetic energy, (c) momentum, (d) frequency?

We have seen two wavelengths assigned to the electron, the Compton wavelength and the de Broglie wavelength. Which is an actual *physical* wavelength associated with the electron: (a) the Compton wavelength, (b) the de Broglie wavelength, (c) both wavelengths, (d) neither wavelength?

Example 27.8 The Wavelength of an Electron

Calculate the de Broglie wavelength for an electron ($m_e = 9.11 \times 10^{-31}$ kg) moving with a speed of 1.00×10^7 m/s.

Solution Equation 27.14 gives

$$\lambda = \frac{h}{m_e v} = \frac{6.63 \times 10^{-34}\,\text{J} \cdot \text{s}}{(9.11 \times 10^{-31}\,\text{kg})(1.00 \times 10^7\,\text{m/s})} = \boxed{7.28 \times 10^{-11}\,\text{m}}$$

This wavelength corresponds to that of x-rays in the electromagnetic spectrum.

EXERCISE Find the de Broglie wavelength of a proton ($m_p = 1.67 \times 10^{-27}$ kg) moving with a speed of 1.00×10^7 m/s.

ANSWER 3.97×10^{-14} m

Example 27.9 The Wavelength of a Baseball

A baseball of mass 0.145 kg is thrown with a speed of 40.0 m/s. What is the de Broglie wavelength of the ball?

Solution From Equation 27.14, we have

$$\lambda = \frac{h}{mv} = \frac{6.63 \times 10^{-34}\,\text{J} \cdot \text{s}}{(0.145\,\text{kg})(40.0\,\text{m/s})} = \boxed{1.14 \times 10^{-34}\,\text{m}}$$

This wavelength is much smaller than any aperture through which the baseball could possibly pass. This means that we could not observe diffraction effects, and as a result the wave properties of large-scale objects cannot be observed.

APPLICATION: THE ELECTRON MICROSCOPE

APPLICATION

ELECTRON MICROSCOPES

A practical device that relies on the wave characteristics of electrons is the **electron microscope.** A *transmission* electron microscope, used for viewing flat, thin samples, is shown in Figure 27.17. In many respects it is similar to an optical microscope, but the electron microscope has a much greater resolving power because it can accelerate electrons to very high kinetic energies, giving them very short wavelengths. No microscope can resolve details that are significantly smaller than the wavelength of the radiation used to illuminate the object. Typically, the wavelengths of electrons are about 100 times smaller than those of the visible light used

(a)   (b)

FIGURE 27.17 (a) Diagram of a transmission electron microscope for viewing a thin, sectioned sample. The "lenses" that control the electron beam are magnetic deflection coils. (b) An electron microscope. *(W. Ormerod/Visuals Unlimited)*

in optical microscopes. (Radiation of the same wavelength as the electrons in an electron microscope is in the x-ray region of the spectrum.)

The electron beam in an electron microscope is controlled by electrostatic or magnetic deflection, which acts on the electrons to focus the beam to an image. Rather than examining the image through an eyepiece as in an optical microscope, the viewer looks at an image formed on a fluorescent screen. (The viewing screen must be fluorescent because otherwise the image produced would not be visible.)

APPLYING PHYSICS 27.3

Electron microscopes (Fig. 27.17) take advantage of the wave nature of particles. Electrons are accelerated to high speeds, giving them a short de Broglie wavelength. Imagine an electron microscope using electrons with a de Broglie wavelength of 0.2 nm. Why don't we design a microscope using 0.2-nm *photons* to do the same thing?

Explanation Because electrons are charged particles, they interact electrically with the sample in the microscope and scatter according to the shape and density of various portions of the sample, providing a means of viewing the sample. Photons of wavelength 0.2 nm are uncharged and in the x-ray region of the spectrum. They tend to simply pass through the thin sample without interacting.

Webnote 27.4

To see how an electron microscope works, visit
http://www.mos.org/sln/sem

ERWIN SCHRÖDINGER, AUSTRIAN THEORETICAL PHYSICIST (1887–1961)

Schrödinger is best known as the creator of wave mechanics. He demonstrated the mathematical equivalence between wave mechanics and the more abstract matrix mechanics developed by Heisenberg. In 1933 Schrödinger left Germany and eventually settled at the Dublin Institute of Advanced Study, where he spent 17 happy, creative years working on problems in general relativity, cosmology, and the application of quantum physics to biology. In 1956 he returned home to Austria and to his beloved Tirolean mountains, where he died in 1961. (*AIP Emilio Segrè Visual Archives*)

27.8 THE WAVE FUNCTION

De Broglie's revolutionary idea that particles should have a wave nature soon moved out of the realm of skepticism to the point where it was viewed as a necessary concept in understanding the subatomic world. In 1926 the Austrian-German physicist Erwin Schrödinger proposed a wave equation that described the manner in which matter waves change in space and time. The Schrödinger wave equation represents a key element in the theory of quantum mechanics. It is as important in quantum mechanics as Newton's laws in classical mechanics. Schrödinger's equation has been successfully applied to the hydrogen atom and to many other microscopic systems. Its importance in most aspects of modern physics cannot be overemphasized.

We shall not go through a mathematical derivation of Schrödinger's wave equation, nor shall we even state the equation here because it involves mathematical operations beyond the scope of this textbook, but we do wish to discuss the general features of the solution to this equation. When we attempt to solve the Schrödinger equation, the basic entity we seek to determine is a quantity Ψ, called the **wave function.** Each particle is represented by a wave function Ψ that depends both on its position and on time. Once Ψ is found, what information about the particle does it give? To answer this question, let us consider an analogy with light.

In Chapter 24 we discussed Young's double-slit experiment and explained experimental observations of the interference pattern solely in terms of the wave nature of light. Let us now discuss this same experiment in terms of both the wave and particle nature of light.

First, recall from Chapter 21 that the intensity of a light beam is proportional to the square of the electric field strength E associated with the beam. That is, $I \propto E^2$. According to the wave model of light, there are certain points on the viewing screen where the net electric field is zero as a result of destructive interference of waves from the two slits. Because E is zero at these points, the intensity is also zero, and the screen is dark at these locations. Likewise, at points on the screen at which constructive interference occurs, E is large, as is the intensity; hence these locations are bright.

Now consider the same experiment when light is viewed as having a particle nature. The number of photons reaching a point on the screen per second increases as the intensity (brightness) increases. Thus, the number of photons that strikes a unit area on the screen each second is proportional to the square of the electric field, or $N \propto E^2$. Now let us consider the behavior of a single photon. What will be the fate of the photon as it moves through the slits in Young's experiment? From a probabilistic point of view, a photon has a high probability of striking the screen at a point at which the intensity (and E^2) is high and a low probability of striking the screen where the intensity is low.

When describing particles rather than photons, Ψ rather than E plays the role of the amplitude. Using an analogy with the description of light, we make the following interpretation of Ψ for particles: If Ψ is a wave function used to describe a single particle, the value of Ψ^2 at some location at a given time is proportional to the probability of finding the particle at that location at that time.

27.9 THE UNCERTAINTY PRINCIPLE

If you were to measure the position and speed of a particle at any instant, you would always be faced with experimental uncertainties in your measurements. According to classical mechanics, no fundamental barrier to an ultimate refinement of the apparatus or experimental procedures exists. In other words, it is possible, in principle, to make such measurements with arbitrarily small uncertainty. Quantum theory predicts, however, that such a barrier does exist. In 1927 Werner

Heisenberg (1901–1976) introduced this notion, which is now known as the **uncertainty principle:**

> If a measurement of position of a particle is made with precision Δx and a simultaneous measurement of linear momentum is made with precision Δp_x, then the product of the two uncertainties can never be smaller than $h/4\pi$:

$$\Delta x \, \Delta p_x \geq \frac{h}{4\pi}$$ [27.16]   ◀ Uncertainty principle

In other words, **it is physically impossible to measure simultaneously the exact position and exact linear momentum of a particle.** If Δx is very small, then Δp_x is large, and vice versa.

To understand the physical origin of the uncertainty principle, consider the following thought experiment introduced by Heisenberg. Suppose you wish to measure the position and linear momentum of an electron as accurately as possible. You might be able to do this by viewing the electron with a powerful light microscope. For you to see the electron and thus determine its location, at least one photon of light must bounce off the electron, as shown in Figure 27.18a, and pass through the microscope into your eye, as shown in Figure 27.18b. When it strikes the electron, however, the photon transfers some unknown amount of its momentum to the electron. Thus, in the process of locating the electron very accurately (that is, by making Δx very small), the light that enables you to succeed in your measurement changes the electron's momentum to some undeterminable extent (making Δp_x very large.)

Let us analyze the collision by first noting that the incoming photon has a momentum of h/λ. As a result of the collision, the photon transfers part or all of its momentum along the x axis to the electron. Thus, the *uncertainty* in the electron's momentum after the collision is as great as the momentum of the incoming photon: $\Delta p_x = h/\lambda$. Furthermore, because the photon also has wave properties, we expect to be able to determine the electron's position to within one wavelength of the light being used to view it, so $\Delta x = \lambda$. Multiplying these two uncertainties gives

$$\Delta x \, \Delta p_x = \lambda \left(\frac{h}{\lambda} \right) = h$$

WERNER HEISENBERG, GERMAN THEORETICAL PHYSICIST (1901–1976)

Heisenberg obtained his Ph.D. in 1923 at the University of Munich, where he studied under Arnold Sommerfeld. While physicists such as de Broglie and Schrödinger tried to develop physical models of the atom, Heisenberg developed an abstract mathematical model called *matrix mechanics* to explain the wavelengths of spectral lines. The more successful wave mechanics of Schrödinger, announced a few months after Heisenberg's model, was shown to be equivalent to Heisenberg's approach. Heisenberg made many other significant contributions to physics, including his famous uncertainty principle, for which he received the Nobel prize in 1932; the prediction of two forms of molecular hydrogen; and theoretical models of the nucleus. *(Courtesy of the University of Hamburg)*

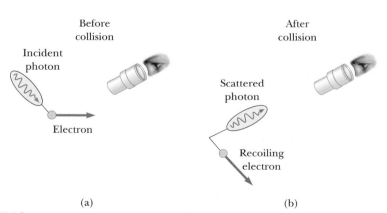

(a)   (b)

FIGURE 27.18 A thought experiment for viewing an electron with a powerful microscope. (a) The electron is viewed before colliding with the photon. (b) The electron recoils (is disturbed) as the result of the collision with the photon.

The value h represents the minimum in the products of the uncertainties. Because the uncertainty can always be greater than this minimum, we have

$$\Delta x \, \Delta p_x \geq h$$

Apart from the numerical factor $1/4\pi$ introduced by Heisenberg's more precise analysis, this agrees with Equation 27.16.

Another form of the uncertainty relationship sets a limit on the accuracy with which the energy of a system, ΔE, can be measured in a finite time interval Δt:

$$\Delta E \, \Delta t \geq \frac{h}{4\pi} \qquad\qquad \text{[27.17]}$$

It can be inferred from this relationship that the energy of a particle cannot be measured with complete precision in a very short interval of time. Thus, when an electron is viewed as a particle, the uncertainty principle tells us that (a) its position and velocity cannot both be known precisely at the same time and (b) its energy can be uncertain for a period of time given by $\Delta t = h/(4\pi \, \Delta E)$.

APPLYING PHYSICS 27.4

A common, but erroneous, description of the absolute zero of temperature is "that temperature at which all molecular motion ceases." How can the uncertainty principle be used to argue against this description?

Explanation   Let us imagine molecules in a piece of material. The molecules are confined within the material, so there is a fixed uncertainty in their position along one axis, Δx, which is the size of the piece of material along this axis. If the molecular motion were to cease at absolute zero, we would be claiming that the molecule velocity is zero, with no uncertainty, $\Delta v = 0$. The product of zero uncertainty in velocity and a nonzero uncertainty in position will be zero, violating the uncertainty principle. Thus, even at absolute zero, there must be some molecular motion.

Example 27.10   Locating an Electron

The speed of an electron is measured to be 5.00×10^3 m/s to an accuracy of 0.003 00%. Find the uncertainty in determining the position of this electron.

Solution   The momentum of the electron is

$$p = m_e v = (9.11 \times 10^{-31} \text{ kg})(5.00 \times 10^3 \text{ m/s}) = 4.56 \times 10^{-27} \text{ kg} \cdot \text{m/s}$$

Because the uncertainty in p is 0.00 300% of this value, we get

$$\Delta p = 0.000\,0300p = (0.000\,0300)(4.56 \times 10^{-27} \text{ kg} \cdot \text{m/s})$$
$$= 1.37 \times 10^{-31} \text{ kg} \cdot \text{m/s}$$

The uncertainty in position can now be calculated by using this value of Δp and Equation 27.16:

$$\Delta x \, \Delta p_x \geq \frac{h}{4\pi}$$

$$\Delta x \geq \frac{h}{4\pi \, \Delta p_x} = \frac{6.626 \times 10^{-34} \text{ J} \cdot \text{s}}{4\pi (1.37 \times 10^{-31} \text{ kg} \cdot \text{m/s})}$$
$$= 0.384 \times 10^{-3} \text{ m} = \boxed{0.384 \text{ mm}}$$

Webnote 27.5

See an applet that shows you the conjugate effect of defining position and momentum. Visit *http://www2.adnc.com/~topquark/ quantum/heisenbergmain.html*

Example 27.11   Excited States of Atoms

As we shall see in the next chapter, electrons in atoms can be found in higher states of energy called **excited states** for short periods of time. If the average time that an electron exists in one of these states is 1.00×10^{-8} s, what is the minimum uncertainty in energy of the excited state?

Solution   From the uncertainty principle in the form of Equation 27.17, we find that the minimum uncertainty in energy is

$$\Delta E = \frac{h}{4\pi \Delta t} = \frac{(6.63 \times 10^{-34}\,\text{J} \cdot \text{s})}{4\pi(1.00 \times 10^{-8}\,\text{s})} = 5.28 \times 10^{-27}\,\text{J} = \boxed{3.30 \times 10^{-8}\,\text{eV}}$$

27.10   THE SCANNING TUNNELING MICROSCOPE[1]

One of the basic phenomena of quantum mechanics—tunneling—is at the heart of a very practical device, the scanning tunneling microscope, or STM, which enables us to get highly detailed images of surfaces with resolution comparable to the size of a single atom.

Figure 27.19 is an image of a ring of 48 iron atoms located on a copper surface. Note the high quality of the STM image and the recognizable ring of iron atoms. What makes this image so remarkable is that at its resolution the size of the smallest detail that can be discerned is about 0.2 nm. For an ordinary microscope, the resolution is limited by the wavelength of the waves used to make the image. Thus, an optical microscope has a resolution no better than 200 nm, about half the wavelength of visible light, and so could never show the detail displayed in Figure 27.19. Electron microscopes can have a resolution of 0.2 nm by using electron waves of this wavelength, given by the de Broglie formula $\lambda = h/p$. The electron momentum p required to give this wavelength is 10 000 eV/c, corresponding to an electron speed of 2% of the speed of light. Electrons traveling at this speed would penetrate into the interior of the sample in Figure 27.19 and so could not give us information about individual surface atoms.

> **◄ APPLICATION**
>
> SCANNING TUNNELING
> MICROSCOPES

FIGURE 27.19   This is a photograph of a "quantum-corral" consisting of a ring of 48 iron atoms located on a copper surface. The diameter of the ring is 143 nm, and the photograph was obtained using a low-temperature scanning tunneling microscope (STM). *(IBM Corporation Research Division)*

[1] This section was written by Roger A. Freedman and Paul K. Hansma, University of California, Santa Barbara.

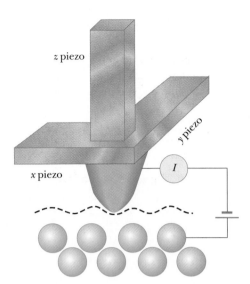

FIGURE 27.20 A schematic view of an STM. The tip, shown as a rounded cone, is mounted on a piezoelectric *x, y, z* scanner. A scan of the tip over the sample can reveal contours of the surface down to the atomic level. An STM image is composed of a series of scans displaced laterally from each other. *(Based on a drawing from P. K. Hansma, V. B. Elings, O. Marti, and C. Bracker,* Science 242:209, 1988, *Copyright 1988 by the AAAS.)*

The STM achieves its very fine resolution by using the basic idea shown in Figure 27.20. A conducting probe with a sharp tip is brought near the surface to be studied. Because it is attracted to the positive ions in the surface, an electron in the surface has a lower total energy than an electron in the empty space between surface and tip. The same thing is true for an electron in the probe tip, which is attracted to the positive ions in the tip. In Newtonian mechanics, this means that electrons cannot move between surface and tip because they lack the energy to escape either material. Because the electrons obey quantum mechanics, however, they can "tunnel" across the barrier of empty space. By applying a voltage between surface and tip, the electrons can be made to tunnel preferentially from surface to tip. In this way the tip samples the distribution of electrons just above the surface.

Because of the nature of tunneling, the STM is very sensitive to the distance *z* from tip to surface. The reason is that in the empty space between tip and surface, the electron wave function falls off exponentially with a decay length of order 0.1 nm—that is, the wave function decreases by $1/e$ over that distance. For distances *z* greater than 1 nm (that is, beyond a few atomic diameters), essentially no tunneling takes place. This exponential behavior causes the current of electrons tunneling from surface to tip to depend very strongly on *z*. This sensitivity is the basis of the operation of the STM: By monitoring the tunneling current as the tip is scanned over the surface, scientists obtain a sensitive measure of the topography of the electron distribution on the surface. The result of this scan is used to make images like that in Figure 27.19. In this way, the STM can measure the height of surface features to within 0.001 nm, approximately 1/100 of an atomic diameter!

The STM has, however, one serious limitation: It depends on electrical conductivity of the sample and the tip. Unfortunately, most materials are not electrically conductive at their surface. Even metals such as aluminum are covered with nonconductive oxides. A newer microscope, the atomic force microscope, or AFM, overcomes this limitation. It measures the force between a tip and the sample rather than an electrical current. This force depends very strongly on the tip-sample separation, just as the electron tunneling current does for the STM. Thus the AFM has comparable sensitivity for measuring topography and has become widely used for technological applications.

Perhaps the most remarkable thing about the STM is that its operation is based on a quantum-mechanical phenomenon—tunneling—that was well understood in the 1920s, even though the first STM was not built until the 1980s. What other applications of quantum mechanics may yet be waiting to be discovered?

Webnote 27.6

Take a fascinating look at an exquisite gallery of STM photographs. Jump to *http://www.almaden.ibm.com/vis/ stm/gallery.html*

Webnote 27.7

Learn more about an atomic force microscope. Explore *http://stm2.nrl.navy.mil/how-afm/ how-afm.html*

Webnote 27.8

See some exciting visuals from an atomic force microscope. Go to *http://www.physics.purdue.edu/ nanophys/afm.html*

SUMMARY

The characteristics of **blackbody radiation** cannot be explained using classical concepts. The peak of a blackbody radiation curve is given by **Wien's displacement law:**

$$\lambda_{max} T = 0.2898 \times 10^{-2} \, \text{m} \cdot \text{K} \qquad \textbf{[27.1]}$$

where λ_{max} is the wavelength at which the curve peaks and T is the absolute temperature of the object emitting the radiation.

Planck first introduced the quantum concept when he assumed that the subatomic oscillators responsible for blackbody radiation could have only discrete amounts of energy given by

$$E_n = nhf \qquad \textbf{[27.2]}$$

where n is a positive integer called a **quantum number** and f is the frequency of vibration of the resonator.

The **photoelectric effect** is a process whereby electrons are ejected from a metal surface when light is incident on that surface. Einstein provided a successful explanation of this effect by extending Planck's quantum hypothesis to electromagnetic waves. In this model, light is viewed as a stream of particles called photons, each with energy $E = hf$, where f is the light frequency and h is **Planck's constant.** The maximum kinetic energy of the ejected photoelectrons is

$$KE_{max} = hf - \phi \qquad \textbf{[27.6]}$$

where ϕ is the **work function** of the metal.

X-rays are produced when high-speed electrons are suddenly decelerated. When electrons have been accelerated through a voltage V, the shortest-wavelength radiation that can be produced is

$$\lambda_{min} = \frac{hc}{e \, \Delta V} \qquad \textbf{[27.9]}$$

The regular array of atoms in a crystal can act as a diffraction grating for x-rays and for electrons. The condition for constructive interference of the diffracted rays is given by **Bragg's law:**

$$2d \sin \theta = m\lambda \qquad (m = 1, 2, 3, \ldots) \qquad \textbf{[27.10]}$$

X-rays from an incident beam are scattered at various angles by electrons in a target such as carbon. In such a scattering event, a shift in wavelength is observed for the scattered x-rays. This phenomenon is known as the **Compton shift.** Conservation of momentum and energy applied to a photon-electron collision yields the following expression for the shift in wavelength of the scattered x-rays:

$$\Delta\lambda = \lambda - \lambda_0 = \frac{h}{m_e c} (1 - \cos \theta) \qquad \textbf{[27.11]}$$

where m_e is the mass of the electron, c is the speed of light, and θ is the scattering angle.

De Broglie proposed that all matter has both a particle and a wave nature. The **de Broglie wavelength** of any particle of mass m and speed v is

$$\lambda = \frac{h}{p} = \frac{h}{mv} \qquad \textbf{[27.14]}$$

De Broglie also proposed that the frequencies of the waves associated with particles obey the Einstein relationship, $E = hf$.

In the theory of **quantum mechanics,** each particle is described by a quantity Ψ called the **wave function.** The probability of finding the particle at a particular point at some instant is proportional to Ψ^2. Quantum mechanics is very successful in describing the behavior of atomic and molecular systems.

According to Heisenberg's **uncertainty principle,** it is impossible to measure simultaneously the exact position and exact momentum of a particle. If Δx is the uncertainty in the measured position and Δp_x the uncertainty in the momentum, the product $\Delta x \, \Delta p_x$ is given by

$$\Delta x \, \Delta p_x \geq \frac{h}{4\pi} \qquad \text{[27.16]}$$

Also,

$$\Delta E \, \Delta t \geq \frac{h}{4\pi} \qquad \text{[27.17]}$$

where ΔE is the uncertainty in the energy of the particle and Δt is the uncertainty in the time it takes to measure the energy.

CONCEPTUAL QUESTIONS

1. If you observe objects inside a very hot kiln, it is difficult to discern the shapes of the objects. Why?
2. Why is an electron microscope more suitable than an optical microscope for "seeing" objects of an atomic size?
3. Are blackbodies black?
4. Why is it impossible to simultaneously measure the position and velocity of a particle with infinite accuracy?
5. All objects radiate energy. Why, then, are we not able to see all objects in a dark room?
6. Is light a wave or a particle? Support your answer by citing specific experimental evidence.
7. A student claims that he is going to eject electrons from a piece of metal by placing a radio transmitter antenna adjacent to the metal and sending a strong AM radio signal into the antenna. The work function of a metal is typically a few electron volts. Will this work?
8. In view of Heisenberg's uncertainty principle, how can we predict such things as maximum height and range when describing something like a baseball?
9. In the photoelectric effect, explain why the stopping potential depends on the frequency of the light but not on the intensity.
10. Which has more energy, a photon of ultraviolet radiation or a photon of yellow light?
11. Why does the existence of a cutoff frequency in the photoelectric effect favor a particle theory of light rather than a wave theory?
12. What effect, if any, would you expect the temperature of a material to have on the ease with which electrons can be ejected from it in the photoelectric effect?
13. If photons that cause photoelectrons to be released all have the same energy, how can the photoelectrons have different energies?
14. The brightest star in the constellation Lyra is the bluish star Vega, whereas the brightest star in Bootes is the reddish star Arcturus. How do you account for the difference in color of the two stars?
15. If the photoelectric effect is observed for one metal, can you conclude that the effect will also be observed for another metal under the same conditions? Explain.
16. Discuss whether the behavior of an electron is mainly wave-like or particle-like in each of the following situations: (a) traversing a circular orbit in a magnetic field; (b) absorbing a photon and being photoelectrically ejected from the surface of a metal; (c) forming an interference pattern.
17. If a photon is deflected via the Compton effect, can its wavelength ever become shorter?
18. In an adult, 13% of the bone marrow is in the head. Discuss this in relation to dental x-rays.

PROBLEMS

1, 2, 3 = straightforward, intermediate, challenging ☐ = full solution available in Student Solutions Manual/Study Guide
web = solution posted at **http://info.brookscole.com/serway** 🦠 = biomedical application

Section 27.1 Blackbody Radiation and Planck's Hypothesis

Section 27.2 The Photoelectric Effect and the Particle Theory of Light

1. (a) What is the surface temperature of Betelgeuse, a red giant star in the constellation of Orion, which radiates with a peak wavelength of about 970 nm? (b) Rigel, a bluish-white star in Orion, radiates with a peak wavelength of 145 nm. Find the temperature of Rigel's surface.
2. (a) Lightning produces a maximum air temperature on the order of 10^4 K, while (b) a nuclear explosion produces a temperature on the order of 10^7 K. Use Wien's displacement law to find the order of magnitude of the wavelength of the thermally produced photons radiated with greatest intensity by each of these sources. Name the part of the electromagnetic spectrum where you would expect each to radiate most strongly.
3. (a) Assuming that the tungsten filament of a lightbulb is a blackbody, determine its peak wavelength if its temperature is 2 900 K. (b) Why does your answer to part (a) suggest that more energy from a lightbulb goes into heat than into light?
4. Calculate the energy, in electron volts, of a photon whose frequency is (a) 620 THz, (b) 3.10 GHz, (c) 46.0 MHz. (d) Determine the corresponding wavelengths for these

photons and state the classification of each on the electromagnetic spectrum.

5. Calculate the energy in electron volts of a photon having a wavelength in (a) the microwave range, 5.00 cm; (b) the visible light range, 500 nm; and (c) the x-ray range, 5.00 nm.

6. A sodium-vapor lamp has a power output of 1 000 W. Using 589.3 nm as the average wavelength of this source, calculate the number of photons emitted per second.

7. An FM radio transmitter has a power output of 150 kW and operates at a frequency of 99.7 MHz. How many photons per second does the transmitter emit?

8. The threshold of dark-adapted (scotopic) vision is 4.0×10^{-11} W/m^2 at a central wavelength of 500 nm. If light with this intensity and wavelength enters the eye when the pupil is open to its maximum diameter of 8.5 mm, how many photons per second enter the eye?

9. A 1.5-kg mass vibrates at an amplitude of 3.0 cm on the end of a spring of spring constant 20 N/m. (a) If the energy of the spring is quantized, find its quantum number. (b) If n changes by 1, find the fractional change in energy of the spring.

10. A 0.50-kg mass falls from a height of 3.0 m. If all of the energy of this mass could be converted to visible light of wavelength 5.0×10^{-7} m, how many photons would be produced?

11. When light of wavelength 350 nm falls on a potassium surface, electrons are emitted that have a maximum kinetic energy of 1.31 eV. Find (a) the work function of potassium, (b) the cutoff wavelength, and (c) the frequency corresponding to the cutoff wavelength.

12. Electrons are ejected from a metallic surface with speeds ranging up to 4.6×10^5 m/s when light with a wavelength of $\lambda = 625$ nm is used. (a) What is the work function of the surface? (b) What is the cutoff frequency for this surface?

13. Molybdenum has a work function of 4.20 eV. (a) Find the cutoff wavelength and threshold frequency for the photoelectric effect. (b) Calculate the stopping potential if the incident light has a wavelength of 180 nm.

14. Lithium, beryllium, and mercury have work functions of 2.30 eV, 3.90 eV, and 4.50 eV, respectively. If 400-nm light is incident on each of these metals, determine (a) which metals exhibit the photoelectric effect and (b) the maximum kinetic energy for the photoelectrons in each case.

15. From the scattering of sunlight, Thomson calculated that the classical radius of the electron has a value of 2.82×10^{-15} m. If sunlight having an intensity of 500 W/m^2 falls on a disk with this radius, find the time required to accumulate 1.00 eV of energy. Assume that light is a classical wave and that the light striking the disk is completely absorbed. How does your value compare with the observation that photoelectrons are promptly (within 10^{-9} s) emitted?

16. An isolated copper sphere of radius 5.00 cm, initially uncharged, is illuminated by ultraviolet light of wavelength 200 nm. What charge will the photoelectric effect induce on the sphere? The work function for copper is 4.70 eV.

17. When light of wavelength 254 nm falls on cesium, the required stopping potential is 3.00 V. If light of wavelength 436 nm is used, the stopping potential is 0.900 V. Use this information to plot a graph like that shown in Figure 27.6, and from the graph determine the cutoff frequency for cesium and its work function.

18. Ultraviolet light is incident normally on the surface of a certain substance. The binding energy of the electrons in this substance is 3.44 eV. The incident light has an intensity of 0.055 W/m^2. The electrons are photoelectrically emitted with a maximum speed of 4.2×10^5 m/s. How many electrons are emitted from a square centimeter of the surface each second? Assume that the absorption of every photon ejects an electron.

Section 27.3 X-Rays

19. The extremes of the x-ray portion of the electromagnetic spectrum range from approximately 1.0×10^{-8} m to 1.0×10^{-13} m. Find the minimum accelerating voltages required to produce wavelengths at these two extremes.

20. Calculate the minimum wavelength x-ray that can be produced when a target is struck by an electron that has been accelerated through a potential difference of (a) 15.0 kV, (b) 100 kV.

21. What minimum accelerating voltage would be required to produce an x-ray with a wavelength of 0.0300 nm?

Section 27.4 Diffraction of X-Rays by Crystals

22. A monochromatic x-ray beam is incident on a NaCl crystal surface where $d = 0.353$ nm. The second-order maximum in the reflected beam is found when the angle between the incident beam and the surface is 20.5°. Determine the wavelength of the x-rays.

23. Potassium iodide has an interplanar spacing of $d = 0.296$ nm. A monochromatic x-ray beam shows a first-order diffraction maximum when the grazing angle is 7.6°. Calculate the x-ray wavelength.

24. The spacing between certain planes in a crystal is known to be 0.30 nm. Find the smallest grazing angle at which constructive interference will occur for wavelength 0.070 nm.

25. X-rays of wavelength 0.140 nm are reflected from a certain crystal, and the first-order maximum occurs at an angle of 14.4°. What value does this give for the interplanar spacing of this crystal?

Section 27.5 The Compton Effect

26. X-rays are scattered from electrons in a carbon target. The measured wavelength shift is 1.50×10^{-3} nm. Calculate the scattering angle.

27. Calculate the energy and momentum of a photon of wavelength 700 nm.

28. A beam of 0.68-nm photons undergoes Compton scattering from free electrons. What are the energy and momentum of the photons that emerge at a 45° angle with respect to the incident beam?

29. A 0.001 6-nm photon scatters from a free electron. For what (photon) scattering angle will the recoiling electron and scattered photon have the same kinetic energy?

30. X-rays with an energy of 300 keV undergo Compton scattering from a target. If the scattered rays are deflected at 37.0° relative to the direction of the incident rays, find (a) the Compton shift at this angle, (b) the energy of the scattered x-ray, and (c) the kinetic energy of the recoiling electron.

31. A 0.110-nm photon collides with a stationary electron. After the collision, the electron moves forward and the photon recoils backward. Find the momentum and kinetic energy of the electron.

32. After a 0.800 nm x-ray photon scatters from a free electron, the electron recoils with a speed equal to 1.40×10^6 m/s. (a) What was the Compton shift in the photon's wavelength? (b) Through what angle was the photon scattered?

33. A 0.45-nm x-ray photon is deflected through a 23° angle after scattering from a free electron. (a) What is the kinetic energy of the recoiling electron? (b) What is its speed?

Section 27.7   The Wave Properties of Particles

34. Calculate the de Broglie wavelength for an electron that has kinetic energy (a) 50.0 eV and (b) 50.0 keV (ignore relativistic effects).

35. (a) If the wavelength of an electron is equal to 5.00×10^{-7} m, how fast is it moving? (b) If the electron has a speed of 1.00×10^7 m/s, what is its wavelength?

36. Through what potential difference would an electron have to be accelerated from rest to give it a de Broglie wavelength of 1.0×10^{-10} m?

37. The nucleus of an atom is on the order of 10^{-14} m in diameter. For an electron to be confined to a nucleus, its de Broglie wavelength would have to be of this order of magnitude or smaller. (a) What would be the kinetic energy of an electron confined to this region? (b) On the basis of this result, would you expect to find an electron in a nucleus? Explain.

38. After learning about de Broglie's hypothesis that particles of momentum p have wave characteristics with wavelength $\lambda = h/p$, an 80-kg student has grown concerned about being diffracted when passing through a 75-cm-wide doorway. Assume that significant diffraction occurs when the width of the diffraction aperture is less than 10 times the wavelength of the wave being diffracted. (a) Determine the order of magnitude of the maximum speed at which the student can pass through the doorway in order to be significantly diffracted. (b) With that speed, how long will it take the student to pass through a doorway in a wall 15 cm thick? Compare your order-of-magnitude result to the currently accepted age of the Universe, which is 4×10^{17} s. (c) Should this student worry about being diffracted?

39. De Broglie postulated that the relationship $\lambda = h/p$ is
web valid for relativistic particles. What is the de Broglie wavelength for a (relativistic) electron whose kinetic energy is 3.00 MeV?

40. At what speed must an electron move so that its de Broglie wavelength equals its Compton wavelength? (*Hint:* This electron is relativistic.)

41. The resolving power of a microscope is proportional to the wavelength used. A resolution of approximately 1.0×10^{-11} m (0.010 nm) would be required in order to "see" an atom. (a) If electrons were used (electron microscope), what minimum kinetic energy would be required for the electrons? (b) If photons were used, what minimum photon energy would be needed to obtain 1.0×10^{-11} m resolution?

Section 27.9   The Uncertainty Principle

42. A 50.0-g ball moves at 30.0 m/s. If its speed is measured to an accuracy of 0.10%, what is the minimum uncertainty in its position?

43. A 0.50-kg block rests on the icy surface of a frozen pond, which we can assume to be frictionless. If the location of the block is measured to a precision of 0.50 cm, what speed must the block acquire because of the measurement process?

44. Suppose Fuzzy, a quantum-mechanical duck, lives in a world in which $h = 2\pi$ J·s. Fuzzy has a mass of 2.00 kg and is initially known to be within a pond 1.00 m wide. (a) What is the minimum uncertainty in his speed? (b) Assuming this uncertainty in speed to prevail for 5.00 s, determine the uncertainty in position after this time.

45. Suppose optical radiation ($\lambda = 5.00 \times 10^{-7}$ m) is used to determine the position of an electron to within the wavelength of the light. What will be the resulting uncertainty in the electron's velocity?

46. (a) Show that the kinetic energy of a nonrelativistic particle can be written in terms of its momentum as $KE = p^2/2m$. (b) Use the results of (a) to find the minimum kinetic energy of a proton confined within a nucleus having a diameter of 1.0×10^{-15} m.

ADDITIONAL PROBLEMS

47. Figure P27.47 shows the spectrum of light emitted by a firefly. Determine the temperature of a blackbody that would emit radiation peaked at the same frequency. Based on your result, would you say firefly radiation is blackbody radiation?

FIGURE P27.47

48. An x-ray tube is operated at 50 000 V. (a) Find the minimum wavelength of the radiation emitted by this tube. (b) If this radiation is directed at a crystal, the first-order maximum in the reflected radiation occurs when the grazing angle is 2.50°. What is the spacing between reflecting planes in the crystal?

49. The spacing between planes of nickel atoms in a nickel crystal is 0.352 nm. At what angle does a second-order Bragg reflection occur in nickel for 11.3-keV x-rays?

50. Johnny Jumper's favorite trick is to step out of his 16th story window and fall 50.0 m into a pool. A news reporter takes a picture of 75.0-kg Johnny just before he makes a splash, using an exposure time of 5.00 ms. Find (a) Johnny's de Broglie wavelength at this moment, (b) the uncertainty of his kinetic energy measurement during such a period of time, and (c) the percent error caused by such an uncertainty.

51. Photons of wavelength 450 nm are incident on a metal. The most energetic electrons ejected from the metal are bent into a circular arc of radius 20.0 cm by a magnetic field with a magnitude of 2.00×10^{-5} T. What is the work function of the metal?

52. A 200-MeV photon is scattered at 40.0° by a free proton initially at rest. Find the energy (in MeV) of the scattered photon.

53. A light source of wavelength λ illuminates a metal and **web** ejects photoelectrons with a maximum kinetic energy of 1.00 eV. A second light source of wavelength $\lambda/2$ ejects photoelectrons with a maximum kinetic energy of 4.00 eV. What is the work function of the metal?

54. Red light of wavelength 670 nm produces photoelectrons from a certain photoemissive material. Green light of wavelength 520 nm produces photoelectrons from the same material with 1.50 times the maximum kinetic energy. What is the material's work function?

55. How fast must an electron be moving if all its kinetic energy is lost to a single x-ray photon (a) at the long-wavelength end of the x-ray electromagnetic spectrum with a wavelength of 1.00×10^{-8} m; (b) at the short-wavelength end of the x-ray electromagnetic spectrum with a wavelength of 1.00×10^{-13} m?

56. Show that if an electron were confined inside an atomic nucleus of diameter 2.0×10^{-15} m, it would have to be moving relativistically, while a proton confined to the same nucleus can be moving at less than one-tenth the speed of light.

57. A photon strikes a metal with a work function of ϕ and produces a photoelectron with a de Broglie wavelength equal to the wavelength of the original photon. (a) Show that the energy of this photon must have been given by

$$E = \frac{\phi(m_e c^2 - \phi/2)}{m_e c^2 - \phi}$$

where m_e is the mass of the electron. (*Hint:* Begin with conservation of energy, $E + m_e c^2 = \phi + \sqrt{(pc)^2 + (m_e c^2)^2}$.) (b) If one of these photons strikes platinum ($\phi = 6.35$ eV), determine the resulting maximum speed of the photoelectron.

58. In a Compton scattering event, the scattered photon has an energy of 120.0 keV and the recoiling electron has a kinetic energy of 40.0 keV. Find (a) the wavelength of the incident photon, (b) the angle θ at which the photon is scattered, and (c) the recoil angle of the electron. (*Hint:* Conserve both mass-energy and relativistic momentum.)

59. A woman on a ladder drops small pellets toward a spot on the floor. (a) Show that according to the uncertainty principle, the average miss distance must be at least

$$\Delta x = \left(\frac{h}{2\pi m}\right)^{1/2}\left(\frac{H}{2g}\right)^{1/4}$$

where H is the initial height of each pellet above the floor and m is the mass of each pellet. (b) If $H = 2.00$ m and $m = 0.500$ g, what is Δx?

60. Show that the speed of a particle having de Broglie wavelength λ and Compton wavelength $\lambda_C = h/(mc)$ is

$$v = \frac{c}{\sqrt{1 + (\lambda/\lambda_C)^2}}$$

GROUP ACTIVITIES

G.1 Use a black marker or pieces of dark electrical tape to make a very dark area on the outside of a shoebox. Poke a hole in the center of the dark area with a pencil. Now put a lid on the box and compare the blackness of the hole with the blackness of the surrounding dark area. Based on your observation explain why the radiation emitted from the hole is like that emitted from a blackbody.

G.2 On a clear night, go outdoors far from city lights and find the constellation Orion. Your instructor should be able to furnish you with a star chart to assist you in locating this grouping of stars. Look very carefully at the color of the two stars Betelgeuse and Rigel. Can you tell which star is hotter? Orion is visible from November through April in the evening sky. Thus, if Orion is not visible, compare two of the brightest stars you can see, such as Vega in the constellation Lyra and Arcturus in Bootes.

G.3 (a) If a pulse of blue light and a pulse of red light both carry energy E, which pulse contains more photons? (b) If the wavelength of the red light is 690 nm, the wavelength of the blue light is 420 nm, and the energy E is 1 000 eV, find the number of photons in each beam to verify your prediction of part (a).

G.4 (a) The threshold frequency for the photoelectric effect for a material is f_0. Are electrons emitted from the material when light incident on the material is of frequency (i) greater than f_0? (ii) less than f_0? (iii) equal to f_0? (b) The work function for zinc is 4.31 eV. What is the lowest fre-

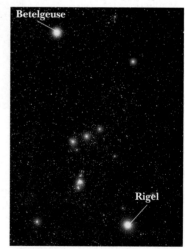

FIGURE GA27.2 *(John Chumack/Photo Researchers, Inc.)*

quency of light that releases photoelectrons when incident on the material?

G.5 Light sometimes acts like a wave and sometimes like a particle. For the following situations, which best describes the behavior of light? Defend your answers. (a) The photoelectric effect. (b) The Compton effect. (c) Young's double-slit experiment.

28

Atomic Physics

"Neon lights", commonly used in advertising signs, consist of thin glass tubes filled with various gases such as neon and helium. The gas atoms are excited to higher energy levels by electric discharge through the tube. When the electrons in these excited levels return to lower energy levels, the atoms emit light having a wavelength (color) that depends on the type of gas in the tube. For example, a tube filled with neon produces a red-orange color, while helium produces pink.

(Terry Gleason, Visuals Unlimited)

A large portion of this chapter is concerned with the study of the hydrogen atom. Although the hydrogen atom is the simplest atomic system, it is especially important for several reasons:

- The quantum numbers used to characterize the allowed states of hydrogen can also be used to describe (approximately) the allowed states of more complex atoms. This enables us to understand the periodic table of the elements, one of the great triumphs of quantum mechanics.
- The hydrogen atom is an ideal system for performing precise comparisons of theory and experiment and for improving our overall understanding of atomic structure.
- Much of what we know about the hydrogen atom with its single electron can be extended to such single-electron ions as He^+ and Li^{2+}.

In this chapter we first discuss the Bohr model of hydrogen, which helps us understand many features of hydrogen but fails to explain the finer details of atomic structure. Next we examine the hydrogen atom from the viewpoint of quantum mechanics and the quantum numbers used to characterize various atomic states. In addition, we examine the physical significance of the quantum numbers and the effect of a magnetic field on certain quantum states. The Pauli exclusion principle is also presented. This physical principle is extremely important in understanding the properties of complex atoms and the arrangement of elements in the periodic table. Finally, we apply our knowledge of atomic structure to describe the mechanisms involved in the production of x-rays, the operation of lasers, and the behavior of solid-state devices such as diodes and transistors.

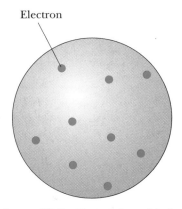

FIGURE 28.1 Thomson's model of the atom, with the electrons embedded inside the positive charge like seeds in a watermelon.

28.1 EARLY MODELS OF THE ATOM

The model of the atom in the days of Newton was a tiny, hard, indestructible sphere. Although this model was a good basis for the kinetic theory of gases, new models had to be devised when later experiments revealed the electronic nature of atoms. J. J. Thomson (1856–1940) suggested a model of the atom as a volume of positive charge with electrons embedded throughout the volume, much like the seeds in a watermelon (Fig. 28.1).

In 1911 Ernest Rutherford (1871–1937) and his students Hans Geiger and Ernest Marsden performed a critical experiment showing that Thomson's model could not be correct. In this experiment, a beam of positively charged **alpha particles** was projected against a thin metal foil, as in Figure 28.2a. The results of the experiment astounded scientists. Most of the alpha particles passed through the foil as if it were empty space, but a few particles deflected from their original direc-

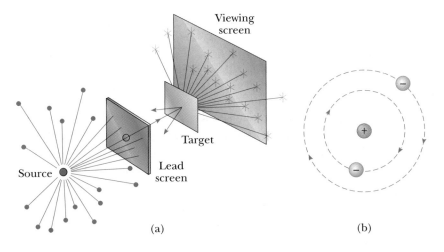

(a)   (b)

FIGURE 28.2 (a) Geiger and Marsden's technique for observing the scattering of alpha particles from a thin foil target. The source is a naturally occurring radioactive substance, such as radium. (b) Rutherford's planetary model of the atom.

SIR JOSEPH JOHN THOMSON, ENGLISH PHYSICIST (1856–1940)

Thomson, usually considered the discoverer of the electron, opened up the field of subatomic particle physics with his extensive work on the deflection of cathode rays (electrons) in an electric field. Thomson received the 1906 Nobel prize for his discovery of the electron. *(Stock Montage, Inc.)*

tion of travel were scattered through large angles. Some particles even deflected backward, reversing their direction of travel. When Geiger informed Rutherford of these results, Rutherford wrote, "It was quite the most incredible event that has ever happened to me in my life. It was almost as incredible as if you fired a 15-inch [artillery] shell at a piece of tissue paper and it came back and hit you."

Such large deflections were not expected on the basis of Thomson's model. According to this model, a positively charged alpha particle would never come close enough to a large positive charge to cause large-angle deflections. Rutherford explained these astounding results by assuming that the positive charge in an atom was concentrated in a region that was small relative to the size of the atom. He called this concentration of positive charge the **nucleus** of the atom. Any electrons belonging to the atom were assumed to be in the relatively large volume outside the nucleus. In order to explain why electrons in this outer region of the atom were not pulled into the nucleus, Rutherford viewed them as moving in orbits about the positively charged nucleus in the same manner as the planets orbit the Sun, as shown in Figure 28.2b. Alpha particles themselves were later identified as the nuclei of helium atoms.

There are two basic difficulties with Rutherford's planetary model. As we shall see in the next section, an atom emits certain discrete characteristic frequencies of electromagnetic radiation and no others; the Rutherford model is unable to explain this phenomenon. A second difficulty is that Rutherford's electrons are undergoing a centripetal acceleration. According to Maxwell's theory of electromagnetism, centripetally accelerated charges revolving with frequency f should radiate electromagnetic waves of the same frequency. Unfortunately, this classical model falls apart when applied to the atom. As the electron radiates energy, the radius of its orbit steadily decreases and its frequency of revolution increases. This leads to an ever-increasing frequency of emitted radiation and a rapid collapse of the atom as the electron spirals into the nucleus.

28.2 ATOMIC SPECTRA

As you may have already learned in chemistry, the hydrogen atom is the simplest atomic system, and an especially important one to understand. Much of what we know about the hydrogen atom (which consists of one proton and one electron) can be extended directly to other single-electron ions such as He^+ and Li^{2+}. Furthermore, a thorough understanding of the physics underlying the hydrogen atom can then be used to describe more complex atoms and the periodic table of the elements.

Suppose an evacuated glass tube is filled with hydrogen (or some other gas) at low pressure. If a voltage applied between metal electrodes in the tube is great enough to produce an electric current in the gas, the tube emits light whose color is characteristic of the gas in the tube. (This is how a neon sign works.) When the emitted light is analyzed with a spectrometer, a series of discrete bright lines is observed, each line having a different wavelength, or color. Such a series of spectral lines is commonly referred to as an **emission spectrum.** The wavelengths contained in a given line spectrum are characteristic of the element emitting the light (Fig. 28.3a). Because no two elements emit the same line spectrum, this phenomenon represents a marvelous and reliable technique for identifying elements in a gaseous substance.

The emission spectrum of hydrogen shown in Figure 28.4 includes four prominent lines that occur at wavelengths of 656.3 nm, 486.1 nm, 434.1 nm, and 410.2 nm. In 1885 Johann Balmer (1825–1898) found that the wavelengths of these and less prominent lines can be described by the simple empirical equation:

Balmer series ▶

$$\frac{1}{\lambda} = R_H\left(\frac{1}{2^2} - \frac{1}{n^2}\right)$$ [28.1]

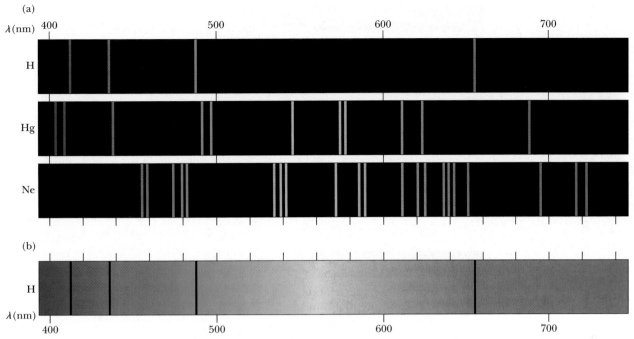

FIGURE 28.3 Visible spectra. (a) Line spectra produced by emission in the visible range for the elements hydrogen, mercury, and neon. (b) The absorption spectrum for hydrogen. The dark absorption lines occur at the same wavelengths as the emission lines for hydrogen shown in (a).

where n has integral values of 3, 4, 5, . . . , and R_H is a constant, called the **Rydberg constant.** If the wavelength is in meters, R_H has the value

$$R_H = 1.097\ 373\ 2 \times 10^7\,\text{m}^{-1} \qquad\qquad [28.2]$$

◀ Rydberg constant

The first line in the Balmer series, at 656.3 nm, corresponds to $n = 3$ in Equation 28.1; the line at 486.1 nm corresponds to $n = 4$; and so on. In addition to this Balmer series of spectral lines, a Lyman series was subsequently discovered in the far ultraviolet, with the radiated wavelengths described by a similar equation.

In addition to emitting light at specific wavelengths, an element can also absorb light at specific wavelengths. The spectral lines corresponding to this process form what is known as an **absorption spectrum.** An absorption spectrum can be obtained by passing a continuous radiation spectrum (one containing all wavelengths) through a vapor of the element being analyzed. The absorption spectrum consists of a series of dark lines superimposed on the otherwise bright continuous spectrum. Each line in the absorption spectrum of a given element coincides with a line in the emission spectrum of the element. That is, if hydrogen is the absorbing vapor, dark lines will appear at the visible wavelengths 656.3 nm, 486.1 nm, 434.1 nm, and 410.2 nm, as shown in Figures 28.3b and 28.4.

The absorption spectrum of an element has many practical applications. For example, the continuous spectrum of radiation emitted by the Sun must pass through the cooler gases of the solar atmosphere before reaching Earth. The various absorption lines observed in the solar spectrum have been used to identify elements in the solar atmosphere. It is interesting to note that when the solar spectrum was first studied, some lines were found that did not correspond to any known element. A new element had been discovered! Because the Greek word for Sun is *helios,* the new element was named *helium.* It was later identified in underground gases on Earth. Scientists are able to examine the light from stars other than our Sun in this fashion, but elements other than those present on Earth have never been detected.

FIGURE 28.4 A series of spectral lines for atomic hydrogen. The prominent labeled lines are part of the Balmer series.

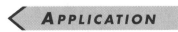

APPLICATION

DISCOVERY OF HELIUM

You are observing a yellow candle flame, and your laboratory partner claims that the light from the flame originates from excited sodium atoms in the flame. You disagree, claiming that the candle flame is hot, so the radiation must be thermal in origin. Before your disagreement leads to fisticuffs, how can you determine who is correct?

Explanation A simple determination can be made by observing the light from the candle flame through a spectrometer, which is a slit and diffraction grating combination discussed in Chapter 25. If the spectrum of the light is continuous, then it is most likely thermal in origin. If the spectrum shows discrete lines, it is atomic in origin. The results of the experiment show that the light is indeed thermal in origin and originates from random molecular motion in the candle flame.

At extreme northern latitudes, the aurora borealis provides a beautiful and colorful display in the nighttime sky. A similar display occurs near the southern polar region and is called the aurora australis. What is the origin of the various colors of radiation seen in the auroras?

Explanation The auroras are due to high-speed particles interacting with Earth's magnetic field and entering the atmosphere, as described on the back cover of this book. When these particles collide with molecules in the atmosphere, they excite the molecules in a way similar to the voltage in the spectrum tubes discussed earlier in this section. In response, the molecules emit colors of light according to the characteristic spectrum of their atomic constituents. For our atmosphere, the primary constituents are nitrogen and oxygen, which provide the red, blue, and green colors of the auroras.

APPLICATION ▷

THE NORTHERN AND SOUTHERN LIGHTS

28.3 THE BOHR THEORY OF HYDROGEN

At the beginning of the 20th century, scientists were perplexed by the failure of classical physics to explain the characteristics of spectra. Why did atoms of a given element emit only certain lines? Furthermore, why did the atoms absorb only those wavelengths that they emitted? In 1913 Bohr provided an explanation of atomic spectra that includes some features of the currently accepted theory. Using the simplest atom, hydrogen, Bohr developed a model of what he thought must be the atom's structure in an attempt to explain why the atom was stable. His model of the hydrogen atom contains some classical features as well as some revolutionary postulates that could not be justified within the framework of classical physics. The basic assumptions of the Bohr theory as it applies to the hydrogen atom are as follows:

1. The electron moves in circular orbits about the proton under the influence of the Coulomb force of attraction, as in Figure 28.5. In this case, the Coulomb force produces the electron's centripetal acceleration.
2. Only certain electron orbits are stable. These are orbits in which the hydrogen atom does not emit energy in the form of electromagnetic radiation. Hence, the total energy of the atom remains constant, and classical mechanics can be used to describe the electron's motion.
3. Radiation is emitted by the hydrogen atom when the electron "jumps" from a more energetic initial state to a lower state. The "jump" cannot be visualized or treated classically. In particular, the frequency, f, of the radiation emitted in the jump is related to the change in the atom's energy and is

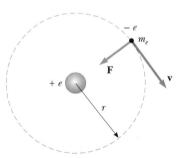

FIGURE 28.5 Bohr's model of the hydrogen atom. In this model, the orbiting electron is allowed only in specific orbits of discrete radius.

generally not the same as the frequency of the electron's orbital motion. The frequency of the emitted radiation is given by

$$E_i - E_f = hf \qquad \text{[28.3]}$$

where E_i is the energy of the initial state, E_f is the energy of the final state, h is Planck's constant, and $E_i > E_f$.

4. The size of the allowed electron orbits is determined by a condition imposed on the electron's orbital angular momentum: The allowed orbits are those for which the electron's orbital angular momentum about the nucleus is an integral multiple of $\hbar$ (pronounced "h bar"), where $\hbar = h/2\pi$.

$$m_e v r = n\hbar \qquad n = 1, 2, 3, \ldots \qquad \text{[28.4]}$$

With these assumptions, we can calculate the allowed energies and emission wavelengths of the hydrogen atom. We shall use the model pictured in Figure 28.5, in which the electron travels in a circular orbit of radius r with an orbit speed v.

The electrical potential energy of the atom is

$$PE = k_e \frac{q_1 q_2}{r} = k_e \frac{(-e)(e)}{r} = -k_e \frac{e^2}{r}$$

where k_e is the Coulomb constant. Assuming the nucleus is at rest the total energy E of the atom is the sum of the kinetic and potential energy:

$$E = KE + PE = \frac{1}{2} m_e v^2 - k_e \frac{e^2}{r} \qquad \text{[28.5]}$$

Let us apply Newton's second law to the electron. We know that the electric force of attraction on the electron, $k_e e^2/r^2$, must equal $m_e a_r$, where $a_r = v^2/r$ is the centripetal acceleration of the electron. Thus,

$$k_e \frac{e^2}{r^2} = m_e \frac{v^2}{r} \qquad \text{[28.6]}$$

From this equation, we see that the kinetic energy of the electron is

$$\frac{1}{2} m v^2 = \frac{k_e e^2}{2r} \qquad \text{[28.7]}$$

We can combine this result with Equation 28.5 and express the energy of the atom as

$$E = -\frac{k_e e^2}{2r} \qquad \text{[28.8]}$$

where the negative value of the energy indicates that the electron is bound to the proton.

An expression for r is obtained by solving Equations 28.4 and 28.6 for v and equating the results:

$$v^2 = \frac{n^2 \hbar^2}{m_e^2 r^2} = \frac{k_e e^2}{m_e r}$$

$$r_n = \frac{n^2 \hbar^2}{m_e k_e e^2} \qquad n = 1, 2, 3, \ldots \qquad \text{[28.9]}$$

This equation is based on the assumption that the **electron can exist only in certain allowed orbits determined by the integer n.**

The orbit with the smallest radius, called the **Bohr radius,** a_0, corresponds to $n = 1$ and has the value

$$a_0 = \frac{\hbar^2}{m_e k_e e^2} = 0.052\ 9\ \text{nm} \qquad \text{[28.10]}$$

NIELS BOHR, DANISH PHYSICIST (1885–1962)

Bohr was an active participant in the early development of quantum mechanics and provided much of its philosophical framework. During the 1920s and 1930s, Bohr headed the Institute for Advanced Studies in Copenhagen. The institute was a magnet for many of the world's best physicists and provided a forum for the exchange of ideas. When Bohr visited the United States in 1939 to attend a scientific conference, he brought news that the fission of uranium had been observed by Hahn and Strassman in Berlin. The results were the foundations of the atomic bomb developed in the United States during World War II. Bohr was awarded the 1922 Nobel prize for his investigation of the structure of atoms and of the radiation emanating from them. *(Princeton University/Courtesy of AIP Emilio Segre Visual Archives)*

◀ Energy of the hydrogen atom

◀ The radii of the Bohr orbits are quantized

A general expression for the radius of any orbit in the hydrogen atom is obtained by substituting Equation 28.10 into Equation 28.9:

$$r_n = n^2 a_0 = n^2(0.052\ 9\ \text{nm}) \qquad \text{[28.11]}$$

The first three Bohr orbits for hydrogen are shown in Figure 28.6.

Equation 28.9 may be substituted into Equation 28.8 to give the following expression for the energies of the quantum states:

$$E_n = -\frac{m_e k_e^2 e^4}{2\hbar^2}\left(\frac{1}{n^2}\right) \qquad n = 1, 2, 3, \ldots \qquad \text{[28.12]}$$

If we insert numerical values into Equation 28.12, we find

$$E_n = -\frac{13.6}{n^2}\ \text{eV} \qquad \text{[28.13]}$$

▷ **Allowed energies of the hydrogen atom**

The lowest energy state, or **ground state,** corresponds to $n = 1$ and has an energy $E_1 = -m_e k_e^2 e^4/2\hbar^2 = -13.6$ eV. The next state, corresponding to $n = 2$, has an energy $E_2 = E_1/4 = -3.40$ eV, and so on. An energy level diagram showing the energies (horizontal lines) of these stationary states and the corresponding quantum numbers is shown in Figure 28.7. The uppermost level shown, corresponding to $E = 0$ and $n \to \infty$, represents the state for which the electron is completely removed from the atom. In this case, both the electron's *KE* and *PE* are zero, which means that the electron is at rest infinitely far away from the proton. The minimum energy required to ionize the atom, that is, to completely remove the electron, is called the **ionization energy.** The ionization energy for hydrogen is 13.6 eV.

ip 28.1   ENERGY DEPENDS ON *N* ONLY FOR HYDROGEN

According to Equation 28.13, the energy depends only on the quantum number *n*. Note that this is only true for the hydrogen atom. For more complicated atoms, the energy levels depend primarily on *n*, but also on other quantum numbers.

Equations 28.3 and 28.12 and the third Bohr postulate show that if the electron jumps from one orbit, whose quantum number is n_i, to a second orbit, whose quantum number is n_f, it emits a photon of frequency *f*, given by

$$f = \frac{E_i - E_f}{h} = \frac{m_e k_e^2 e^4}{4\pi\hbar^3}\left(\frac{1}{n_f^2} - \frac{1}{n_i^2}\right) \qquad \text{[28.14]}$$

where $n_f < n_i$.

Finally, to compare this result with the empirical formulas for the various spectral series, we use the fact that for light, $\lambda f = c$ and Equation 28.14 to get

$$\frac{1}{\lambda} = \frac{f}{c} = \frac{m_e k_e^2 e^4}{4\pi c\hbar^3}\left(\frac{1}{n_f^2} - \frac{1}{n_i^2}\right) \qquad \text{[28.15]}$$

A comparison of this result with Equation 28.1 gives the following expression for the Rydberg constant:

$$R_H = \frac{m_e k_e^2 e^4}{4\pi c\hbar^3} \qquad \text{[28.16]}$$

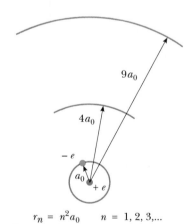

$$r_n = n^2 a_0 \qquad n = 1, 2, 3, \ldots$$

FIGURE 28.6   The first three circular orbits predicted by the Bohr model of the hydrogen atom.

If we insert the known values of $m_e, k_e, e, c,$ and $\hbar$ into this expression, the resulting theoretical value for R_H is found to be in excellent agreement with the value determined experimentally for the Rydberg constant. When Bohr demonstrated this agreement, it was recognized as a major accomplishment of his theory.

In order to compare Equation 28.15 with spectroscopic data, it is convenient to express it in the form

$$\frac{1}{\lambda} = R_H\left(\frac{1}{n_f^2} - \frac{1}{n_i^2}\right) \qquad \text{[28.17]}$$

We can use this expression to evaluate the wavelengths for the various series in the hydrogen spectrum. For example, in the Balmer series, $n_f = 2$ and $n_i = 3, 4, 5, \ldots$

(Eq. 28.1). For the Lyman series, we take $n_f = 1$ and $n_i = 2, 3, 4, \ldots$. The energy level diagram for hydrogen, shown in Figure 28.7, indicates the origin of the spectral lines described previously. The transitions between levels are represented by vertical arrows. Note that whenever a transition occurs between a state designated by n_i to one designated by n_f (where $n_i > n_f$), a photon with a frequency of $(E_i - E_f)/h$ is emitted. This can be interpreted as follows. The lines in the visible part of the hydrogen spectrum arise when the electron jumps from the third, fourth, or even higher orbit to the second orbit. Likewise, the lines of the Lyman series (in the ultraviolet) arise when the electron jumps from the second, third, or even higher orbit to the innermost ($n_f = 1$) orbit. Hence, the Bohr theory successfully predicts the wavelengths of all observed spectral lines of hydrogen.

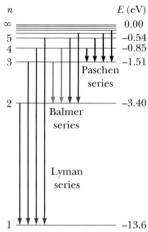

FIGURE 28.7 An energy level diagram for hydrogen. In such diagrams the discrete allowed energies are plotted on the vertical axis. Nothing is plotted on the horizontal axis, but the horizontal extent of the diagram is made large enough to show allowed transitions. Note that the quantum numbers are given on the left and the energies (in eV) are on the right.

APPLYING PHYSICS 28.3

According to the Bohr model of the hydrogen atom, the electron in the ground state moves in a circular orbit of radius 0.529×10^{-10} m, and the speed of the electron in this state is 2.2×10^6 m/s. How could Bohr's model be so successful initially when its specific orbital radii seem to contradict the "fuzziness" demanded by the Heisenberg uncertainty principle?

Explanation We have no answer. The Bohr theory works much better than it should. The theory pictures an electron as having definite position and momentum, ignoring quantum uncertainty. The theory proceeds from Newton's second law, but Schrödinger's equation describes the wave motion of the electron. Scattering experiments show that the electron does not lie on a flat circle, but fills a sphere around the nucleus.

Example 28.1 An Electronic Transition in Hydrogen

The electron in the hydrogen atom makes a transition from the $n = 2$ energy state to the ground state (corresponding to $n = 1$). Find the wavelength and frequency of the emitted photon.

Solution We can use Equation 28.17 directly to obtain λ, with $n_i = 2$ and $n_f = 1$:

$$\frac{1}{\lambda} = R_H \left(\frac{1}{n_f^2} - \frac{1}{n_i^2} \right)$$

$$\frac{1}{\lambda} = R_H \left(\frac{1}{1^2} - \frac{1}{2^2} \right) = \frac{3R_H}{4}$$

$$\lambda = \frac{4}{3R_H} = \frac{4}{3(1.097 \times 10^7 \text{ m}^{-1})} = 1.215 \times 10^{-7} \text{ m} = \boxed{121.5 \text{ nm}}$$

This wavelength lies in the ultraviolet region.

Because $c = f\lambda$, the frequency of the photon is

$$f = \frac{c}{\lambda} = \frac{3.00 \times 10^8 \text{ m/s}}{1.215 \times 10^{-7} \text{ m}} = \boxed{2.47 \times 10^{15} \text{ Hz}}$$

EXERCISE What is the wavelength of the photon emitted by hydrogen when the electron makes a transition from the $n = 3$ state to the $n = 1$ state?

ANSWER $\dfrac{9}{8R_H} = 102.6$ nm

$$
\begin{array}{cc}
n & E\ (\text{eV}) \\
\infty & 0.00 \\
6 & -0.38 \\
5 & -0.54 \\
4 & -0.85 \\
3 & -1.51 \\
2 & -3.40 \\
\end{array}
$$

Balmer
series

FIGURE 28.8 (Example 28.2) Transitions responsible for the Balmer series for the hydrogen atom. All transitions terminate at the $n = 2$ level.

Example 28.2 The Balmer Series for Hydrogen

The Balmer series for the hydrogen atom corresponds to electronic transitions that terminate in the state of quantum number $n = 2$, as shown in Figure 28.8.

A Find the longest-wavelength photon emitted and determine its energy.

Solution The longest-wavelength photon in the Balmer series results from the transition from $n = 3$ to $n = 2$. Using Equation 28.17 gives

$$\frac{1}{\lambda} = R_H \left(\frac{1}{n_f^2} - \frac{1}{n_i^2} \right)$$

$$\frac{1}{\lambda_{\text{max}}} = R_H \left(\frac{1}{2^2} - \frac{1}{3^2} \right) = \frac{5}{36} R_H$$

$$\lambda_{\text{max}} = \frac{36}{5R_H} = \frac{36}{5(1.097 \times 10^7 \text{ m}^{-1})} = \boxed{656.3 \text{ nm}}$$

This wavelength is in the red region of the visible spectrum.

The energy of this photon is

$$
\begin{aligned}
E_{\text{photon}} = hf &= \frac{hc}{\lambda_{\text{max}}} \\
&= \frac{(6.626 \times 10^{-34} \text{ J} \cdot \text{s})(3.00 \times 10^8 \text{ m/s})}{656.3 \times 10^{-9} \text{ m}} \\
&= 3.03 \times 10^{-19} \text{ J} = \boxed{1.89 \text{ eV}}
\end{aligned}
$$

We could also obtain the energy of the photon by using Equation 28.3 in the form $hf = E_3 - E_2$, where E_2 and E_3 are the energy levels of the hydrogen atom, calculated from Equation 28.13. Note that this is the lowest-energy photon in this series, because it involves the smallest energy change.

B Find the shortest-wavelength photon emitted in the Balmer series.

Solution The shortest-wavelength photon in the Balmer series is emitted when the electron makes a transition from $n = \infty$ to $n = 2$. Therefore,

$$\frac{1}{\lambda_{\text{min}}} = R_H \left(\frac{1}{2^2} - \frac{1}{\infty} \right) = \frac{R_H}{4}$$

$$\lambda_{\text{min}} = \frac{4}{R_H} = \frac{4}{1.097 \times 10^7 \text{ m}^{-1}} = \boxed{364.6 \text{ nm}}$$

This wavelength is in the ultraviolet region and corresponds to the series limit.

BOHR'S CORRESPONDENCE PRINCIPLE

In our study of relativity in Chapter 26, we found that Newtonian mechanics cannot be used to describe phenomena that occur at speeds approaching the speed of light. Newtonian mechanics is a special case of relativistic mechanics and is usable only when v is much smaller than c. Similarly, **quantum mechanics is in agreement with classical physics when the energy differences between quantized levels are very small.** This principle, first set forth by Bohr, is called the **correspondence principle.**

For example, consider the hydrogen atom with $n > 10\ 000$. For such large values of n, the energy differences between adjacent levels approach zero and the levels are nearly continuous as Equation 28.13 shows. As a consequence, the classical model is reasonably accurate in describing the system for large values of n. According to the classical model, the frequency of the light emitted by the atom is equal to the frequency of revolution of the electron in its orbit about the nucleus. Calculations show that for $n > 10\,000$, the frequency of the emitted light is the same as the electron's frequency of revolution to within 0.015%.

28.4 MODIFICATION OF THE BOHR THEORY

The Bohr theory of the hydrogen atom was a tremendous success in certain areas because it explains several features of the hydrogen spectrum that previously defied explanation. It accounts for the Balmer series and other series; it predicts a value for the Rydberg constant that is in excellent agreement with the experimental value; it gives an expression for the radius of the atom; and it predicts the energy levels of hydrogen. Although these successes are important to scientists, it is perhaps even more significant that the Bohr theory gives us a model of what the atom looks like and how it behaves. Once a basic model is constructed, refinements and modifications can be made to enlarge on the concept and to explain finer details.

The analysis used in the Bohr theory is also successful when applied to *hydrogen-like* atoms. An atom is said to be hydrogen-like when it contains only one electron. Examples are singly ionized helium, doubly ionized lithium, triply ionized beryllium, and so forth. The results of the Bohr theory for hydrogen can be extended to hydrogen-like atoms by substituting Ze^2 for e^2 in the hydrogen equations, where Z is the atomic number of the element. For example, Equations 28.12 and 28.15 become

$$E_n = -\frac{m_e k_e^2 Z^2 e^4}{2\hbar^2}\left(\frac{1}{n^2}\right) \qquad n = 1, 2, 3, \ldots \qquad \text{[28.18]}$$

and

$$\frac{1}{\lambda} = \frac{m_e k_e^2 Z^2 e^4}{4\pi c\hbar^3}\left(\frac{1}{n_f^2} - \frac{1}{n_i^2}\right) \qquad \text{[28.19]}$$

Although many attempts have been made to extend the Bohr theory to more complex multi-electron atoms, the results have thus far been unsuccessful. Even today, only approximate methods are available for treating multi-electron atoms.

Within a few months following the publication of Bohr's theory, Arnold Sommerfeld (1868–1951) extended the results to include elliptical orbits. We shall examine his model briefly because much of the nomenclature used in this treatment is still in use today. Bohr's concept of quantization of angular momentum led to the **principal quantum number** n, which determines the energy of the allowed states of hydrogen. Sommerfeld's theory retained n, but also introduced a new quantum number ℓ called the **orbital quantum number,** where the value of ℓ ranges from 0 to $n - 1$ in integer steps. According to this model, an electron in any one of the allowed energy states of a hydrogen atom may move in any one of a number of orbits corresponding to different ℓ values. For each value of n there are n possible orbits corresponding to different ℓ values. Because $n = 1$ and $\ell = 0$ for the first energy level (ground state), there is only one possible orbit for this state. The second energy level, with $n = 2$, has two possible orbits corresponding to $\ell = 0$ and $\ell = 1$. The third energy level, with $n = 3$, has three possible orbits corresponding to $\ell = 0$, $\ell = 1$, and $\ell = 2$.

For historical reasons, **all states with the same principal quantum number n are said to form a shell.** Shells are identified by the letters K, L, M, . . . , which designate the states for which $n = 1, 2, 3, \ldots$. Likewise, **the states with given values of n and ℓ are said to form a subshell.** The letters s, p, d, f, g, . . . are used to designate the states for which $\ell = 0, 1, 2, 3, 4, \ldots$. These notations are summarized in Table 28.1.

States that violate the rules given in Table 28.1 cannot exist. For instance, the $2d$ state, which would have $n = 2$ and $\ell = 2$, cannot exist because the highest al-

TABLE 28.1		**Shell and Subshell Notations**	
n	Shell Symbol	ℓ	Subshell Symbol
1	K	0	s
2	L	1	p
3	M	2	d
4	N	3	f
5	O	4	g
6	P	5	h
⋮		⋮	

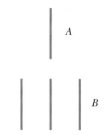

FIGURE 28.9 A single line (*A*) can split into three separate lines (*B*) in a magnetic field.

lowed value of ℓ is $n - 1$, or 1 in this case. Thus, for $n = 2$, $2s$ and $2p$ are allowed subshells but $2d$, $2f$, . . . are not. For $n = 3$, the allowed states are $3s$, $3p$, and $3d$. The maximum number of electrons allowed in any given subshell is $2(2\ell + 1)$. For example, the p subshell ($\ell = 1$) is filled when it contains six electrons. This fact will be important to us later when we discuss the *Pauli exclusion principle*.

Another modification of the Bohr theory arose when it was discovered that the spectral lines of a gas are split into several closely spaced lines when the gas is placed in a strong magnetic field. (This is called the *Zeeman effect*, after its discoverer.) Figure 28.9 shows a single spectral line being split into three closely spaced lines. This observation indicates that the energy of an electron is slightly modified when the atom is immersed in a magnetic field. In order to explain this observation, a new quantum number m_ℓ called the **orbital magnetic quantum number,** was introduced. The theory is in accord with experimental results when m_ℓ is restricted to values ranging from $-\ell$ to $+\ell$, in integer steps.

Finally, very high resolution spectrometers revealed that spectral lines of gases are in fact two very closely spaced lines even in the absence of an external magnetic field. This splitting was referred to as **fine structure.** In 1925 Samuel Goudsmit and George Uhlenbeck introduced the idea of an electron spinning about its own axis to explain the origin of fine structure. The results of their work introduced yet another quantum number m_s called the **spin magnetic quantum number.** We shall save further discussion of this quantum number for a later section. It is interesting to note that each of the new concepts introduced into the original Bohr theory added new quantum numbers and improved the original model. However, the most profound step forward in our understanding of atomic structure came with the development of quantum mechanics.

Example 28.3   Singly Ionized Helium

Singly ionized helium, He$^+$, a hydrogen-like system, has one electron in the $1s$ orbit when the atom is in its ground state.

A Find the energy of the system in the ground state.

Solution   From Equation 28.18, the energy of a hydrogen-like system whose principal quantum number is n is given by

$$E_n = -\frac{m_e k_e^2 Z^2 e^4}{2\hbar^2}\left(\frac{1}{n^2}\right)$$

This can be expressed in eV units as

$$E_n = -\frac{Z^2(13.6)}{n^2} \text{ eV}$$

Because $Z = 2$ for helium, and $n = 1$ in the ground state, we have

$$E_1 = -4(13.6) \text{ eV} = \boxed{-54.4 \text{ eV}}$$

B Find the radius of the ground-state orbit.

Solution   The radius of the ground-state orbit can be found with the help of Equation 28.9. This equation must be modified in the case of a hydrogen-like atom by substituting Ze^2 for e^2 to obtain

$$r_n = \frac{n^2\hbar^2}{m_e k_e Z e^2} = \frac{n^2}{Z}(0.052\ 9 \text{ nm})$$

For our case, $n = 1$ and $Z = 2$, and the result is

$$r_1 = \boxed{0.026\ 5 \text{ nm}}$$

28.5 DE BROGLIE WAVES AND THE HYDROGEN ATOM

One of the postulates made by Bohr in his theory of the hydrogen atom was that angular momentum of the electron is quantized in units of $\hbar$ or

$$m_e v r = n\hbar$$

For more than a decade following Bohr's publication, no one was able to explain why the angular momentum of the electron was restricted to these discrete values. Finally, de Broglie gave a direct physical way of interpreting this condition. He assumed that an electron orbit would be stable (allowed) only if it contained an integral number of electron wavelengths. Figure 28.10a demonstrates this point when three complete wavelengths are contained in one circumference of the orbit. Similar patterns can be drawn for orbits containing one wavelength, two wavelengths, four wavelengths, five wavelengths, and so forth. This situation is analogous to that of standing waves on a string, discussed in Chapter 14. There we found that strings have preferred (resonant) frequencies of vibration. Figure 28.10b shows a standing-wave pattern containing three wavelengths for a string fixed at each end. Now imagine that the vibrating string is removed from its supports at A and B and bent into a circular shape that brings points A and B together. The end result is a pattern such as the one shown in Figure 28.10a.

In general, the condition for a de Broglie standing wave in an electron orbit is that the circumference must contain an integral number of electron wavelengths. We can express this condition as

$$2\pi r = n\lambda \qquad n = 1, 2, 3, \ldots$$

Because the de Broglie wavelength of an electron is $\lambda = h/m_e v$, we can write the preceding equation as $2\pi r = nh/m_e v$, or

$$m_e v r = n\hbar$$

This is precisely the quantization of angular momentum condition imposed by Bohr in his original theory of hydrogen.

The electron orbit shown in Figure 28.10a contains three complete wavelengths and corresponds to the case in which the principal quantum number $n = 3$. The orbit with one complete wavelength in its circumference corresponds to the first Bohr orbit, $n = 1$; the orbit with two complete wavelengths corresponds to the second Bohr orbit, $n = 2$, and so forth.

By applying the wave theory of matter to electrons in atoms, de Broglie was able to explain the appearance of integers in the Bohr theory as a natural consequence of standing-wave patterns. This was the first convincing argument that the wave nature of matter was at the heart of the behavior of atomic systems. Although the analysis provided by de Broglie was a promising first step, gigantic strides were made subsequently with the development of Schrödinger's wave equation and its application to atomic systems.

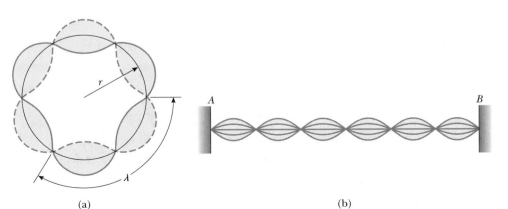

(a)   (b)

FIGURE 28.10 (a) Standing-wave pattern for an electron wave in a stable orbit of hydrogen. There are three full wavelengths in this orbit. (b) Standing-wave pattern for a vibrating stretched string fixed at its ends. This pattern has three full wavelengths.

In an analysis relating Bohr's theory to the de Broglie wavelength of electrons, when an electron moves from the $n = 1$ level to the $n = 3$ level, the circumference of its orbit becomes 9 times greater. This occurs because (a) there are 3 times as many wavelengths in the new orbit, (b) there are 3 times as many wavelengths and each wavelength is 3 times as long, (c) the wavelength of the electron becomes 9 times as long, or (d) the electron is moving 9 times as fast.

28.6   QUANTUM MECHANICS AND THE HYDROGEN ATOM

One of the first great achievements of quantum mechanics was the solution of the wave equation for the hydrogen atom. We shall not attempt to carry out this solution. Rather, we will simply describe its properties and some of its implications with regard to atomic structure.

According to quantum mechanics, the energies of the allowed states are in exact agreement with the values obtained by the Bohr theory (Eq. 28.12), when the allowed energies depend only on the principal quantum number n.

In addition to the principal quantum number, two other quantum numbers emerge from the solution of the wave equation, ℓ and m_ℓ. The quantum number ℓ is called the **orbital quantum number,** and m_ℓ is called the **orbital magnetic quantum number.** As pointed out in Section 28.4, these quantum numbers had already appeared in modifications made to the Bohr theory. The significance of quantum mechanics is that these quantum numbers and the restrictions placed on their values arise directly from mathematics and not from any ad hoc assumptions to make the theory consistent with experimental observation. Because we shall need to make use of the various quantum numbers in the next sections, we repeat the allowed ranges of their values here.

> The values of n can range from 1 to ∞ in integer steps. The values of ℓ can range from 0 to $n - 1$ in integer steps. The values of m_ℓ can range from $-\ell$ to ℓ in integer steps.

For example, if $n = 1$, only $\ell = 0$ and $m_\ell = 0$ are permitted. If $n = 2$, the value of ℓ may be 0 or 1; if $\ell = 0$, then $m_\ell = 0$, but if $\ell = 1$, then m_ℓ may be 1, 0, or -1. Table 28.2 summarizes the rules for determining the allowed values of ℓ and m_ℓ for a given value of n.

States that violate the rules given in Table 28.2 cannot exist. For instance, one state that cannot exist is the $2d$ state, which would have $n = 2$ and $\ell = 2$. This state

TABLE 28.2   **Three Quantum Numbers for the Hydrogen Atom**

Quantum Number	Name	Allowed Values	Number of Allowed States
n	Principal quantum number	$1, 2, 3, \ldots$	Any number
ℓ	Orbital quantum number	$0, 1, 2, \ldots, n - 1$	n
m_ℓ	Orbital magnetic quantum number	$-\ell, -\ell + 1, \ldots,$ $0, \ldots, \ell - 1, \ell$	$2\ell + 1$

is not allowed because the highest allowed value of ℓ is $n - 1$, or 1 in this case. Thus, for $n = 2$, $2s$ and $2p$ are allowed states but $2d$, $2f$, . . . are not. For $n = 3$, the allowed states are $3s$, $3p$, and $3d$.

Example 28.4   The *n* = 2 Level of Hydrogen

Determine the number of states in the hydrogen atom for $n = 2$ and calculate the energies of these states.

Solution   For $n = 2$, ℓ can have the values 0 and 1. For $\ell = 0$, m_ℓ can only be 0; for $\ell = 1$, m_ℓ can be -1, 0, or 1. Hence we have one state designated as the $2s$ state associated with the quantum numbers $n = 2$, $\ell = 0$, and $m_\ell = 0$, and three states designated as $2p$ states, for which the quantum numbers are $n = 2$, $\ell = 1$, $m_\ell = -1$; $n = 2$, $\ell = 1$, $m_\ell = 0$; and $n = 2$, $\ell = 1$, $m_\ell = 1$.

Because all these states have the same principal quantum number, $n = 2$, they also have the same energy, which can be calculated using Equation 28.13, $E_n = -(13.6/n^2)\text{eV}$. For $n = 2$, this gives

$$E_2 = -\frac{13.6}{2^2}\text{ eV} = \boxed{-3.40\text{ eV}}$$

EXERCISE   How many possible states are there for the $n = 3$ level of hydrogen? For the $n = 4$ level?

ANSWER   9 states with different values of ℓ or m_ℓ for $n = 3$, and 16 states for $n = 4$

> **Quick Quiz 28.2**   How many possible orbital states are there for (a) the $n = 3$ level of hydrogen? (b) the $n = 4$ level?
>
> **Quick Quiz 28.3**   When the principal quantum number is $n = 5$, how many different values of (a) ℓ and (b) m_ℓ are possible?

28.7   THE SPIN MAGNETIC QUANTUM NUMBER

Example 28.4 was presented to give you some practice in manipulating quantum numbers, but as we shall see in this section, there actually are *eight* states corresponding to $n = 2$ for hydrogen, not four. This happens because another quantum number, m_s, the **spin magnetic quantum number,** has to be introduced to explain the splitting of each level into two.

As pointed out in Section 28.4, the need for this new quantum number first came about because of an unusual feature in the spectra of certain gases, such as sodium vapor. Close examination of one of the prominent lines of sodium shows that it is, in fact, two very closely spaced lines. The wavelengths of these lines occur in the yellow region at 589.0 nm and 589.6 nm. In 1925, when this doublet was first noticed, atomic theory could not explain it. To resolve the dilemma, Samuel Goudsmit and George Uhlenbeck, following a suggestion by the Austrian physicist Wolfgang Pauli, proposed that a fourth quantum number, called the *spin quantum number,* be introduced to describe any atomic level.

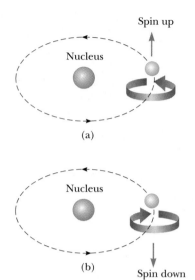

Spin up

Nucleus

(a)

Nucleus

(b)

Spin down

FIGURE 28.11   As an electron moves in its orbit about the nucleus, its spin can be either (a) up or (b) down.

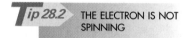

Tip 28.2   THE ELECTRON IS NOT SPINNING

The electron is *not* physically spinning. Electron spin is a purely quantum effect that gives the electron an angular momentum as if it were physically spinning.

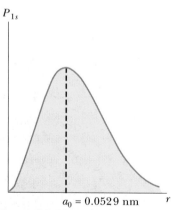

P_{1s}

$a_0 = 0.0529$ nm r

FIGURE 28.12   The probability of finding the electron versus distance from the nucleus for the hydrogen atom in the 1*s* (ground) state. Note that the probability has its maximum value when *r* equals the first Bohr radius, a_0.

In order to describe the spin quantum number, it is convenient (but technically incorrect) to think of the electron as spinning on its axis as it orbits the nucleus, just as Earth spins on its axis as it orbits the Sun. Strangely, there are only two ways in which the electron can spin as it orbits the nucleus, as shown in Figure 28.11. If the direction of spin is as shown in Figure 28.11a, the electron is said to have "spin up." If the direction of spin is reversed, as in Figure 28.11b, the electron is said to have "spin down." The energy of the electron is slightly different for the two spin directions, and this energy difference accounts for the sodium doublet. The quantum numbers associated with electron spin are $m_s = \frac{1}{2}$ for the spin-up state and $m_s = -\frac{1}{2}$ for the spin-down state. As we shall see in Example 28.5, this new quantum number doubles the number of allowed states specified by the quantum numbers n, ℓ, and m_ℓ.

Any classical description of electron spin is technically incorrect because quantum mechanics tells us that, because the electron cannot be precisely located in space, it cannot be considered to be a spinning solid object, as pictured in Figure 28.11. In spite of this conceptual difficulty, all experimental evidence supports the fact that an electron does have some intrinsic property that can be described by the spin magnetic quantum number.

Example 28.5   The Quantum Numbers for the 2*p* Subshell

List the quantum numbers for electrons in the 2*p* subshell.

Solution   For this subshell, $n = 2$ and $\ell = 1$. The magnetic quantum number can have the values -1, 0, 1, and the spin quantum number is always $+\frac{1}{2}$ or $-\frac{1}{2}$. Thus, the six possibilities are as follows:

n	ℓ	m_ℓ	m_s
2	1	-1	$-\frac{1}{2}$
2	1	-1	$\frac{1}{2}$
2	1	0	$-\frac{1}{2}$
2	1	0	$\frac{1}{2}$
2	1	1	$-\frac{1}{2}$
2	1	1	$\frac{1}{2}$

28.8   ELECTRON CLOUDS

The solution of the wave equation, discussed in Section 27.10, yields a wave function Ψ that depends on the quantum numbers n, ℓ, and m_ℓ. Let us assume that we have found a wave function Ψ and see what it may tell us about the hydrogen atom. We choose a value of $n = 1$ for the principal quantum number, which corresponds to the lowest energy state for hydrogen. For $n = 1$, the restrictions placed on the remaining quantum numbers are that $\ell = 0$ and $m_\ell = 0$.

The quantity Ψ^2 has great physical significance because it is proportional to the probability of finding the electron at a given position. Figure 28.12 gives the probability of finding the electron at various radial distances from the nucleus in the 1*s* state of hydrogen. Some useful and surprising information can be extracted from this curve. First, the curve peaks at a value of $r = 0.052\ 9$ nm, the Bohr value of the radius of the first electron orbit in hydrogen. This means that there is a maximum probability of finding the electron in a small interval centered at this distance from the nucleus. However, as the curve indicates, there is also a probability of finding the electron in a small interval centered at any other distance from the nucleus. In other words, the electron is not confined to a particular orbital distance from the nucleus, as assumed in the Bohr model. The electron may be found at various distances from the nucleus, but **the probability of finding it at a distance corresponding to the first Bohr orbit is a maximum.** Quantum mechanics also predicts that the wave function for the hydrogen atom in the ground state is spherically symmetric; hence, the electron can be found in a spherical region surrounding the nucleus. This is in contrast to the Bohr theory, which confines the position of the electron to points in a plane. The quantum mechanical result is often interpreted by viewing the electron as a cloud surrounding the nucleus. An attempt at picturing this cloud-like behavior is shown in Figure 28.13. The densest regions of the cloud represent those locations where the electron is most likely to be found.

If a similar analysis is carried out for the $n = 2$, $\ell = 0$ state of hydrogen, a peak of the probability curve is found at $4a_0$. Likewise, for the $n = 3$, $\ell = 0$ state, the curve peaks at $9a_0$. Thus, quantum mechanics predicts a most-probable electron distance to the nucleus that is in agreement with the location predicted by the Bohr theory.

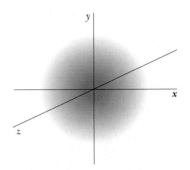

FIGURE 28.13 The spherical electron cloud for the hydrogen atom in its $1s$ state.

28.9 THE EXCLUSION PRINCIPLE AND THE PERIODIC TABLE

Earlier we found that the state of an electron in an atom is specified by four quantum numbers: n, ℓ, m_ℓ, and m_s. For example, an electron in the ground state of hydrogen could have quantum numbers of $n = 1$, $\ell = 0$, $m_\ell = 0$, $m_s = \frac{1}{2}$. As it turns out, the state of an electron in any other atom may also be specified by this same set of quantum numbers. In fact, these four quantum numbers can be used to describe all the electronic states of an atom regardless of the number of electrons in its structure.

An obvious question that arises here is, How many electrons in an atom can have a particular set of quantum numbers? This important question was answered by Pauli in 1925 in a powerful statement known as the **Pauli exclusion principle:**

> **No two electrons in an atom can ever be in the same quantum state; that is, no two electrons in the same atom can have exactly the same values for the set of quantum numbers n, ℓ, m_ℓ, and m_s.**

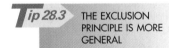

Tip 28.3 THE EXCLUSION PRINCIPLE IS MORE GENERAL

The exclusion principle stated here is a limited form of the more general exclusion principle, which states that no two *fermions* (particles with spin $\frac{1}{2}$, $\frac{3}{2}$, ...) can be in the same quantum state.

◀ The Pauli exclusion principle

The Pauli exclusion principle explains the electronic structure of complex atoms as a succession of filled levels with different quantum numbers increasing in energy, where the outermost electrons are primarily responsible for the chemical properties of the element. It is interesting to note that if this principle were not valid, every electron would end up in the lowest energy state of the atom and the chemical behavior of the elements would be grossly different. Nature as we know it would not exist—and *we* would not exist to wonder about it!

As a general rule, the order that electrons fill an atom's subshell is as follows. Once one subshell is filled, the next electron goes into the vacant subshell that is lowest in energy. We can understand this principle by recognizing that if the atom were not in the lowest energy state available to it, it would radiate energy until it reached this state. A subshell is filled when it contains $2(2\ell + 1)$ electrons. This rule is based on the analysis of quantum numbers to be described later. Following this rule, shells and subshells can contain a number of electrons according to the pattern given in Table 28.3.

The exclusion principle can be illustrated by examining the electronic arrangement in a few of the lighter atoms.

Hydrogen has only one electron, which, in its ground state, can be described by either of two sets of quantum numbers: $1, 0, 0, \frac{1}{2}$ or $1, 0, 0, -\frac{1}{2}$. The electronic

Wolfgang Pauli and Niels Bohr watch a spinning top. *(Courtesy of AIP Niels Bohr Library, Margarethe Bohr Collection)*

TABLE 28.3 Number of Electrons in Filled Subshells and Shells

Shell	Subshell	Number of Electrons in Filled Subshell	Number of Electrons in Filled Shell
K ($n = 1$)	$s(\ell = 0)$	2	2
L ($n = 2$)	$s(\ell = 0)$ $p(\ell = 1)$	2 ⎱ 6 ⎰	8
M($n = 3$)	$s(\ell = 0)$ $p(\ell = 1)$ $d(\ell = 2)$	2 ⎫ 6 ⎬ 10 ⎭	18
N ($n = 4$)	$s(\ell = 0)$ $p(\ell = 1)$ $d(\ell = 2)$ $f(\ell = 3)$	2 ⎫ 6 ⎪ 10 ⎬ 14 ⎭	32

configuration of this atom is often designated as $1s^1$. The notation $1s$ refers to a state for which $n = 1$ and $\ell = 0$, and the superscript indicates that one electron is present in this level.

Neutral *helium* has two electrons. In the ground state, the quantum numbers for these two electrons are 1, 0, 0, $\frac{1}{2}$ and 1, 0, 0, $-\frac{1}{2}$. No other possible combinations of quantum numbers exist for this level, and we say that the K shell is filled. The helium electronic configuration is designated as $1s^2$.

Neutral *lithium* has three electrons. In the ground state, two of these are in the $1s$ subshell and the third is in the $2s$ subshell because it is lower in energy than the $2p$ subshell. Hence, the electronic configuration for lithium is $1s^2 2s^1$.

A list of electronic ground-state configurations for a number of atoms is provided in Table 28.4. In 1871 Dmitri Mendeleev (1834–1907), a Russian chemist, arranged the elements known at that time in a table according to their atomic masses and chemical similarities. The first table Mendeleev proposed contained many blank spaces, and he boldly stated that the gaps were there only because those elements had not yet been discovered. By noting the column in which these missing elements should be located, he was able to make rough predictions about their chemical properties. Within 20 years of this announcement, these elements were indeed discovered.

The elements in our current version of the periodic table are still arranged so that all those in a vertical column have similar chemical properties. For example, consider the elements in the last column: He (helium), Ne (neon), Ar (argon), Kr (krypton), Xe (xenon), and Rn (radon). The outstanding characteristic of these elements is that they do not normally take part in chemical reactions; that is, they do not join with other atoms to form molecules, and are therefore classified as inert. Because of this aloofness, they are referred to as the *noble gases*. We can partially understand their behavior by looking at the electronic configurations shown in Table 28.4. The element helium has the electronic configuration $1s^2$. In other words, one shell is filled. The electrons in this filled shell are considerably separated in energy from the next available level, the $2s$ level.

The electronic configuration for neon is $1s^2 2s^2 2p^6$. Again, the outer shell is filled and the difference in energy between the $2p$ level and the $3s$ level is large. Argon has the configuration $1s^2 2s^2 2p^6 3s^2 3p^6$. Here, the $3p$ subshell is filled and a wide gap in energy exists between the $3p$ subshell and the $3d$ subshell. Through all the noble gases, the pattern remains the same. A noble gas is formed when either a shell or a subshell is filled and there is a large gap in energy before the next possible level is encountered.

TABLE 28.4		Electronic Configuration of Some Elements						
Z	Symbol	Ground-State Configuration	Ionization Energy (eV)	Z	Symbol	Ground-State Configuration		Ionization Energy (eV)
1	H	$1s^1$	13.595	19	K	[Ar]	$4s^1$	4.339
2	He	$1s^2$	24.581	20	Ca		$4s^2$	6.111
				21	Sc		$3d4s^2$	6.54
3	Li	[He] $2s^1$	5.390	22	Ti		$3d^24s^2$	6.83
4	Be	$2s^2$	9.320	23	V		$3d^34s^2$	6.74
5	B	$2s^22p^1$	8.296	24	Cr		$3d^54s^1$	6.76
6	C	$2s^22p^2$	11.256	25	Mn		$3d^54s^2$	7.432
7	N	$2s^22p^3$	14.545	26	Fe		$3d^64s^2$	7.87
8	O	$2s^22p^4$	13.614	27	Co		$3d^74s^2$	7.86
9	F	$2s^22p^5$	17.418	28	Ni		$3d^84s^2$	7.633
10	Ne	$2s^22p^6$	21.559	29	Cu		$3d^{10}4s^1$	7.724
				30	Zn		$3d^{10}4s^2$	9.391
11	Na	[Ne] $3s^1$	5.138	31	Ga		$3d^{10}4s^24p^1$	6.00
12	Mg	$3s^2$	7.644	32	Ge		$3d^{10}4s^24p^2$	7.88
13	Al	$3s^23p^1$	5.984	33	As		$3d^{10}4s^24p^3$	9.81
14	Si	$3s^23p^2$	8.149	34	Se		$3d^{10}4s^24p^4$	9.75
15	P	$3s^23p^3$	10.484	35	Br		$3d^{10}4s^24p^5$	11.84
16	S	$3s^23p^4$	10.357	36	Kr		$3d^{10}4s^24p^6$	13.996
17	Cl	$3s^23p^5$	13.01					
18	Ar	$3s^23p^6$	15.755					

Note: The bracket notation is used as a shorthand method to avoid repetition in indicating inner-shell electrons. Thus, [He] represents $1s^2$, [Ne] represents $1s^22s^22p^6$, [Ar] represents $1s^22s^22p^63s^23p^6$, and so on.

The elements in the first column of the periodic table are called the *alkali metals* and are very active chemically. Referring to Table 28.4, we can understand why these elements interact so strongly with other elements. All of these alkali metals have a single outer electron in an *s* subshell. This electron is shielded from the nucleus by all the electrons in the inner shells. Thus, it is only loosely bound to the atom and can readily be accepted by other atoms that bind it more tightly to form molecules.

All the elements in the seventh column of the periodic table (called the *halogens*) are also very active chemically. Note that all these elements lack one electron in a subshell. As a consequence, they readily accept electrons from other atoms to form molecules.

Quick Quiz 28.4

Krypton (atomic number 36) has how many electrons in its next to outer shell ($n = 3$)?
(a) 2   (b) 4   (c) 8   (d) 18

APPLYING PHYSICS 28.4

As we move from left to right across one row of the periodic table, the effective size of the atoms first decreases and then increases. What would cause this behavior?

Explanation   If we begin at the left side of the periodic table and move toward the middle, the nuclear charge increases. As a result, the Coulomb attraction between the nucleus and the electrons increases, and the electrons are pulled into an average position that is closer to the nucleus. From the middle of the row to the right side, the increasing number of electrons being placed in proximity to each other results in a mutual repulsion that increases the average distance from the nucleus and causes the atomic size to grow.

28.10 CHARACTERISTIC X-RAYS

When a metal target is bombarded with high-energy electrons, x-rays are emitted. The x-ray spectrum typically consists of a broad continuous band and a series of intense sharp lines that are dependent on the type of metal used for the target, as shown in Figure 28.14. These discrete lines, called **characteristic x-rays,** were discovered in 1908, but their origin remained unexplained until the details of atomic structure were developed.

The first step in the production of characteristic x-rays occurs when a bombarding electron collides with an electron in an inner shell of a target atom with sufficient energy to remove the electron from the atom. The vacancy created in the shell is filled when an electron in a higher level drops down into the lower energy level containing the vacancy. The time it takes for this to happen is very short, less than 10^{-9} s. This transition is accompanied by the emission of a photon whose energy equals the difference in energy between the two levels. Typically, the energy of such transitions is greater than 1 000 eV, and the emitted x-ray photons have wavelengths in the range of 0.01–1 nm.

Let us assume that the incoming electron has dislodged an atomic electron from the innermost shell, the K shell. If the vacancy is filled by an electron dropping from the next higher shell, the L shell, the photon emitted in the process is referred to as the K_α line on the curve of Figure 28.14. If the vacancy is filled by an electron dropping from the M shell, the line produced is called the K_β line.

Other characteristic x-ray lines are formed when electrons drop from upper levels to vacancies other than those in the K shell. For example, L lines are produced when vacancies in the L shell are filled by electrons dropping from higher shells. An L_α line is produced as an electron drops from the M shell to the L shell, and an L_β line is produced by a transition from the N shell to the L shell.

We can estimate the energy of the emitted x-rays as follows. Consider two electrons in the K shell of an atom whose atomic number is Z. Each electron partially shields the other from the charge of the nucleus, Ze, and so each is subject to an effective nuclear charge of $Z_{eff} = (Z - 1)e$. We can now use a modified form of Equation 28.18 to estimate the energy of either electron in the K shell (with $n = 1$):

$$E_K = -m_e Z_{eff}^2 \frac{k_e^2 e^4}{2\hbar^2} = -Z_{eff}^2 E_0$$

where E_0 is the ground-state energy. Substituting $Z_{eff} = Z - 1$ gives

$$E_K = -(Z - 1)^2 (13.6 \text{ eV}) \qquad \text{[28.20]}$$

As Example 28.6 shows, we can estimate the energy of an electron in an L or M shell in a similar fashion. Taking the energy difference between these two levels, we can then calculate the energy and wavelength of the emitted photon.

In 1914 Henry G. J. Moseley plotted the Z values for a number of elements versus $\sqrt{1/\lambda}$, where λ is the wavelength of the K_α line for each element. He found that such a plot produced a straight line, as in Figure 28.15. This is consistent with our rough calculations of the energy levels based on Equation 28.20. From this plot, Moseley was able to determine the Z values of other elements, providing a periodic chart in excellent agreement with the known chemical properties of the elements.

FIGURE 28.14 The x-ray spectrum of a metal target consists of broad continuous spectrum plus a number of sharp lines that are due to *characteristic x-rays*. The data shown were obtained when 35-keV electrons bombarded a molybdenum target. (*Note:* 1 pm = 10^{-12} m = 0.001 nm.)

FIGURE 28.15 A Moseley plot. A straight line is obtained when $\sqrt{1/\lambda}$ is plotted versus Z for the K_α x-ray lines of a number of elements.

Example 28.6 **Estimating the Energy of an x-Ray**

Estimate the energy of the characteristic x-ray emitted from a tungsten target when an electron drops from an M shell ($n = 3$ state) to a vacancy in the K shell ($n = 1$ state).

Solution The atomic number for tungsten is $Z = 74$. Using Equation 28.20, we see that the energy of the electron in the K shell state is approximately

$$E_K = -(74 - 1)^2(13.6 \text{ eV}) = -72\,500 \text{ eV}$$

The electron in the M shell ($n = 3$) is subject to an effective nuclear charge that depends on the number of electrons in the $n = 1$ and $n = 2$ states, which shield the nucleus. Because there are eight electrons in the $n = 2$ state and one electron in the $n = 1$ state, roughly nine electrons shield the nucleus, and so $Z_{\text{eff}} = Z - 9$. Hence, the energy of an electron in the M shell ($n = 3$), following Equation 28.20, is equal to

$$E_M = -Z_{\text{eff}}^2 E_3 = -(Z - 9)^2 \frac{E_0}{3^2} = -(74 - 9)^2 \frac{(13.6 \text{ eV})}{9} = -6\,380 \text{ eV}$$

where E_3 is the energy of an electron in the $n = 3$ level of the hydrogen atom. Therefore, the emitted x-ray has an energy equal to $E_M - E_K = -6\,380 \text{ eV} - (-72\,500 \text{ eV}) = $ 66 100 eV . Note that this energy difference is also equal to hf, where $hf = hc/\lambda$, and where λ is the wavelength of the emitted x-ray.

EXERCISE Calculate the wavelength of the emitted x-ray for this transition.

ANSWER 0.018 8 nm

28.11 ATOMIC TRANSITIONS

We have seen that an atom emits radiation only at certain frequencies that correspond to the energy separation between the various allowed states. Consider an atom with many allowed energy states, labeled E_1, E_2, E_3, . . . , as in Figure 28.16. When light is incident on the atom, only those photons whose energy, hf, matches the energy separation, ΔE, between two levels can be absorbed by the atom. A diagram representing this **stimulated absorption process** is shown in Figure 28.17. At ordinary temperatures, most of the atoms in a sample are in the ground state. If a vessel containing many atoms of a gas is illuminated with a light beam containing all possible photon frequencies (that is, a continuous spectrum), only those photons of energies $E_2 - E_1$, $E_3 - E_1$, $E_4 - E_1$, and so on, can be ab-

FIGURE 28.16 Energy level diagram of an atom with various allowed states. The lowest energy state, E_1, is the ground state. All others are excited states.

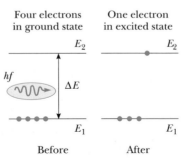

FIGURE 28.17 *Stimulated absorption* of a photon by an atom. The blue dots represent electrons in the various states. One electron is transferred from the ground state to the excited state when the atom absorbs one photon whose energy is $hf = E_2 - E_1$.

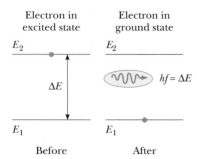

Before   After

FIGURE 28.18 *Spontaneous emission* of a photon by an atom that is initially in the excited state E_2. When the electron falls to the ground state, the atom emits a photon whose energy is $hf = E_2 - E_1$.

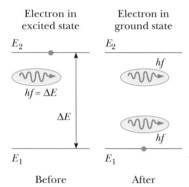

Before   After

FIGURE 28.19 *Stimulated emission* of a photon by an incoming photon of energy hf. Initially, the atom is in the excited state. The incoming photon stimulates the atom to emit a second photon of energy $hf = E_2 - E_1$, which is in phase with the first.

Webnote 28.1

Visualize how a photon interacts with an atom. Go to http://www.lightlink.com/sergey/java/java/atomphoton/index.html

sorbed. As a result of this absorption, some atoms are raised to various allowed higher energy levels, called **excited states.**

Once an atom is in an excited state, there is a constant probability that it will jump back to a lower level by emitting a photon, as shown in Figure 28.18. This process is known as **spontaneous emission.** Typically, an atom remains in an excited state for only about 10^{-8} s.

A third process that is important in lasers, **stimulated emission,** was predicted by Einstein in 1917. Suppose an atom is in the excited state E_2, as in Figure 28.19, and a photon with energy $hf = E_2 - E_1$ is incident on it. The incoming photon increases the probability that the excited atom will return to the ground state and thereby emit a second photon having the same energy hf. Note that two identical photons result from stimulated emission—the incident photon and the emitted photon. *The emitted photon is exactly in phase with the incident photon.* These photons can stimulate other atoms to emit photons in a chain of similar processes. The many photons produced in this fashion are the source of the intense, coherent (in phase) light in a laser.

APPLYING PHYSICS 28.5

A physics student is watching a meteor shower in the early morning hours. She notices that the streaks of light from the meteoroids entering the very high regions of the atmosphere last for as long as 2 or 3 seconds before fading. She also notices a lightning storm off in the distance. The streaks of light from the lightning fade away almost immediately after the flash, certainly in much less than 1 second. Both lightning and meteors cause the air to turn into a plasma because of the very high temperatures generated. The light is given off when the stripped electrons in the plasma recombine with the ionized atoms. Why would this light last longer for meteors than for lightning?

Explanation To answer this question, let us examine the phrase—". . . the streaks of light from the meteoroids entering the very high regions of the atmosphere." In the very high regions of the atmosphere, the pressure is very low. Thus, the density is very low, and the atoms of the gas are relatively far apart. Thus, after the air is ionized by the passing meteoroid, the probability of freed electrons finding an ionized atom with which to recombine is relatively low. As a result, the recombination process occurs over a relatively long time, measured in seconds. However, lightning occurs in the lower regions of the atmosphere (the troposphere) in which the pressure and density are relatively high. After the ionization by the lightning flash, the electrons and ionized atoms are much closer together than in the upper atmosphere. The probability of a recombination is much higher, and the time for the recombination to occur is much shorter.

28.12 LASERS AND HOLOGRAPHY

We have described how an incident photon can cause atomic transitions either upward (stimulated absorption) or downward (stimulated emission). The two processes are equally probable. When light is incident on a system of atoms, there is usually a net absorption of energy because, when the system is in thermal equilibrium, many more atoms are in the ground state than in excited states. However, if the situation can be inverted so that more atoms are in an excited state than in the ground state, a net emission of photons can result. Such a condition is called **population inversion.** This is the fundamental principle involved in the operation of a laser, an acronym for *l*ight *a*mplification by *s*timulated *e*mission of *r*adiation. The amplification corresponds to a buildup of photons in the system as the result of a chain reaction of events. To achieve laser action, the following three conditions must be satisfied:

   1. The system must be in a state of population inversion (that is, more atoms in an excited state than in the ground state).

2. The excited state of the system must be a *metastable state,* which means its lifetime must be long compared with the usually short lifetimes of excited states. When that is the case, stimulated emission will occur before spontaneous emission.

3. The emitted photons must be confined in the system long enough to allow them to stimulate further emission from other excited atoms. This is achieved by using reflecting mirrors at the ends of the system. One end is totally reflecting, and the other is slightly transparent to allow the laser beam to escape.

One device that exhibits stimulated emission of radiation is the helium-neon gas laser. Figure 28.20 is an energy level diagram for the neon atom in this system. The mixture of helium and neon is confined to a glass tube sealed at the ends by mirrors. A high voltage applied to the tube causes electrons to sweep through the tube, colliding with the atoms of the gas and raising them into excited states. Neon atoms are excited to state E_3^* through this process and also as a result of collisions with excited helium atoms. When a neon atom makes a transition to state E_2, it stimulates emission by neighboring excited atoms. This results in the production of coherent light at a wavelength of 632.8 nm. Figure 28.21a summarizes the steps in the production of a laser beam.

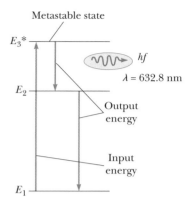

FIGURE 28.20 Energy level diagram for the neon atom, which emits photons at a wavelength of 632.8 nm through stimulated emission. The photon at this wavelength arises from the transition $E_3^* \rightarrow E_2$. This is the source of coherent light in the helium-neon gas laser.

(a)

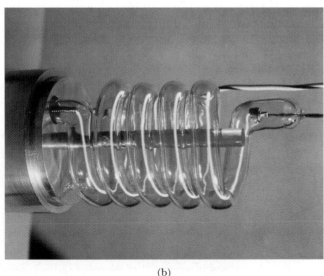

(b)

FIGURE 28.21 (a) Steps in the production of a laser beam. The tube contains atoms, which represent the active medium. An external source of energy (optical, electrical, etc.) is needed to "pump" the atoms to excited energy states. The parallel end mirrors provide the feedback of the stimulating wave. (b) Photograph of the first ruby laser showing the flash lamp surrounding the ruby rod. *(Courtesy of HRL Laboratories LLC, Malibu, CA)*

Scientist checking the performance of an experimental laser-cutting device mounted on a robot arm. The laser is being used to cut through a metal plate. *(Philippe Plailly/Photo Researchers)*

APPLICATION

LASER TECHNOLOGY

Since the development of the first laser in 1960, laser technology has experienced tremendous growth. Lasers that cover wavelengths in the infrared, visible, and ultraviolet regions are now available. Applications include surgical "welding" of detached retinas, "lasik" surgery, precision surveying and length measurement, a potential source for inducing nuclear fusion reactions, precision cutting of metals and other materials, and telephone communication along optical fibers. These and other applications are possible because of the unique characteristics of laser light. In addition to being highly monochromatic and coherent, laser light is also highly directional and can be sharply focused to produce regions of extremely intense light energy.

APPLICATION

HOLOGRAPHY

HOLOGRAPHY

Webnote 28.2

You can learn more about holograms, and your instructor can produce holograms on a realistic budget by viewing *http://members.aol.com/gakall/ holopg.html*

One interesting application of the laser is holography, the production of three-dimensional images of objects. Figure 28.22a shows how a hologram is made. Light from the laser is split into two parts by a half-silvered mirror at *B*. One part of the beam reflects off the object to be photographed and strikes an ordinary photographic film. The other half of the beam is diverged by lens L_2, reflects from mirrors M_1 and M_2, and finally strikes the film. The two beams overlap to form an extremely complicated interference pattern on the film, one that can be produced only if the phase relationship of the two waves is constant throughout the exposure of the film. This condition is met through the use of light from a laser because such light is coherent. The hologram records not only the intensity of the light scattered from the object (as in a conventional photograph) but also the phase difference between the reference beam and the beam scattered from the object. Because of this phase difference, an interference pattern is formed that produces an image with full three-dimensional perspective.

Webnote 28.3

Take a look at the popular *Lasik* procedure. Go to *http://www.lasik.md/lasiksection/ normaleye.htm*

A hologram is best viewed by allowing coherent light to pass through the developed film while looking back along the direction from which the beam comes. Figure 28.22b is a photograph of a hologram made using a cylindrical film.

28.13   ENERGY BANDS IN SOLIDS

In this section we trace the changes that occur in the discrete energy levels of isolated atoms when atoms group together and form a solid. We find that in solids, the discrete levels of isolated atoms broaden into allowed energy bands separated by forbidden gaps. The separation and electron population of the highest bands determines whether a given solid is a conductor, insulator, or semiconductor.

(a)

(b)

FIGURE 28.22   (a) Experimental arrangement for producing a hologram. (b) Photograph of a hologram made using a cylindrical film. Note the detail of the Volkswagen image. *(Courtesy of Central Scientific Company)*

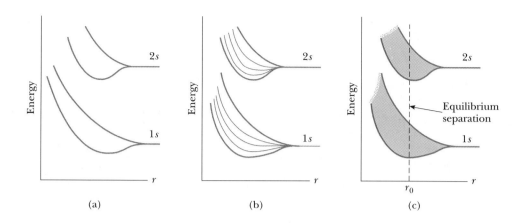

FIGURE 28.23 (a) Splitting of the 1*s* and 2*s* states when two atoms are brought together. (b) Splitting of the 1*s* and 2*s* states when five atoms are brought close together. (c) Formation of energy bands when a large number of atoms are assembled to form a solid.

Consider two identical atoms, initially widely separated, that are brought closer and closer together. If two identical atoms are very far apart, they do not interact, and their electronic energy levels can be considered to be those of isolated atoms. Hence, the energy levels are exactly the same. As the atoms come close together, they essentially become one quantum system, and the Pauli exclusion principle demands that the two electrons be in different quantum states for this single system. The exclusion principle manifests itself as a changing or splitting of electron energy levels that were identical in the widely separated atoms as shown in Figure 28.23a. Figure 28.23b shows that with five atoms, each energy level in the isolated atom splits into five different, more closely spaced levels.

If we extend this argument to the large number of atoms found in solids (of the order of 10^{23} atoms/cm^3), we obtain a large number of levels so closely spaced that they may be regarded as a continuous **band** of energy levels, as in Figure 28.23c. An electron can have any energy within an allowed energy band, but cannot have an energy in the **band gap** or the region between allowed bands. In practice we are only interested in the band structure of a solid at some equilibrium separation of its atoms r_0, and so we remove the distance scale on the r axis and simply plot the allowed energy bands of a solid as a series of horizontal bands as shown in Figure 28.24 for sodium.

CONDUCTORS AND INSULATORS

Figure 28.24 shows that the band structure of a particular solid is quite complicated with individual atomic levels broadening by varying amounts and some levels (3*s* and 3*p*) broadening so much that they overlap. Nevertheless, it is possible to gain a qualitative understanding of whether a solid is a conductor, insulator, or semiconductor by considering only the structure of the upper one or two energy bands, and whether they are occupied by electrons.

Deciding whether an energy band is empty (unoccupied by electrons), partially filled, or full is carried out in basically the same way as we decided on the energy level population of atoms: We distribute the total number of electrons from the lowest energy levels up in a way consistent with the exclusion principle. Although we omit the details of this process here, one important case is shown in Figure 28.25a, where the highest-energy occupied band is only partially full. The other important case, where the highest occupied band is completely full, is shown in Figure 28.25b. Notice that this figure also shows that the highest filled band is the **valence band** and the next higher empty band is the **conduction band.** The energy band gap, which varies with the type of material, is also indicated as the energy difference E_g between the top of the valence band and the bottom of the conduction band.

With these ideas and definitions we are now in a position to understand what determines, quantum mechanically, whether a solid will be a conductor or an insu-

FIGURE 28.24 Energy bands of sodium. Note the energy gaps (white regions) between the allowed bands; electrons cannot occupy states that lie in these forbidden gaps. Blue represents energy bands occupied by the sodium electrons when the atom is in its ground state. Gold represents energy bands that are empty. Note that the 3*s* and 3*p* levels broaden so much that they overlap.

FIGURE 28.25 (a) Half-filled band of a metal, an electrical conductor. (b) An electrical insulator at $T = 0$ K has a filled valence band and an empty conduction band. (c) Band structure of a semiconductor at ordinary temperatures ($T \approx 300$ K). The energy gap is much smaller than in an insulator, and many electrons occupy states in the conduction band.

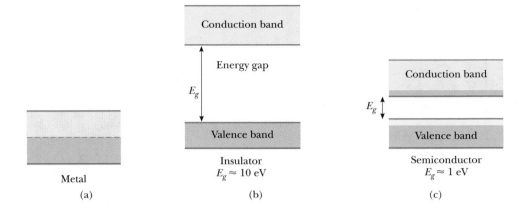

lator. When a modest voltage is applied to a good conductor, the electrons accelerate and gain energy. In quantum terms, electron energies increase *if there are higher unoccupied energy levels for electrons to jump to.* For example, electrons near the top of the partially filled band in sodium need to gain very little energy from the applied voltage to reach one of the nearby, closely spaced, empty states. Thus, it is easy for a small voltage to kick electrons into higher energy states, and charge flows easily in sodium, an excellent conductor.

Now consider the case of a material where the highest occupied band is completely full of electrons and a band gap separates this filled valence band from the vacant conduction band, as in Figure 28.25b. A typical case might be that of diamond (carbon) where the band gap is about 10 eV. When a voltage is applied, electrons cannot easily gain energy because there are no vacant states nearby in energy to which electrons can make transitions. Because the only empty band is the conduction band, an electron must gain an amount of energy at least equal to the band gap for it to move through the solid. This large amount of energy can't be supplied by a modest applied voltage, and so no charge flows and diamond is a good insulator. In summary then, a conductor has a highest-energy occupied band that is *partially filled,* and an insulator has a highest-energy occupied band that is *completely filled* with a large energy gap between its valence and conduction bands.

SEMICONDUCTORS

Thus far, we have completely ignored the influence of temperature on the electronic populations of energy bands. Recalling that the average thermal energy of a particle at temperature T is $3k_B T/2$, we find that an electron at room temperature has an average energy of about 0.04 eV. Because this energy is about 100 times smaller than the band gap in a typical insulator, very few electrons have enough random thermal energy to jump the energy gap in an insulator and contribute to conduction. However, things are different for a semiconductor. As we see in Figure 28.25c, a **semiconductor** is a material with a small band gap of about 1 eV whose conductivity results from appreciable thermal excitation of electrons across the band gap into the conduction band at room temperature. The most commonly used semiconductors are silicon and gallium arsenide with band gaps of 1.14 eV and 1.43 eV, respectively, at 300 K. As you might expect, the resistivity of semiconductors usually decreases with increasing temperature, because $k_B T$ becomes a larger fraction of the band gap energy.

It is interesting that the electrons in the conduction band of a semiconductor do not carry the entire current when a voltage is applied, as Figure 28.25c shows. (We are tempted to say that conduction electrons do not constitute the "whole" story.) The missing electrons in the valence band, shown as a narrow white band in the figure, provide a few empty states called **holes** for valence band electrons to fill; so some electrons in the valence band gain energy and move toward a positive electrode and thus also carry the current. Since the valence band electrons that fill holes leave behind other holes, it is equally valid and more common to view the conduction process in the valence band as a flow of positive holes toward the negative electrode applied to a semiconductor. Thus, a pure semiconductor, such as silicon, can be viewed in a symmetric way: Silicon has equal numbers of mobile electrons in the conduction band and holes in the valence band. Furthermore, when an external voltage is applied to the semiconductor, electrons move toward the positive electrode, and holes move toward the negative electrode. This symmetric current process in a semiconductor is summarized in Figure 28.26. In the next section we shall look at the concepts of electron and hole in a simpler, more graphic way as the presence or absence of an outer-shell electron at a particular location in a crystal lattice.

When small amounts of impurities are added to a semiconductor (about one impurity atom per 10^7 silicon atoms), both the band structure of the semiconductor and its resistivity are modified. The process of adding impurities, called **doping,** is important in making devices that have well-defined regions of different resistivity. For example, when an atom containing five outer-shell electrons, such as arsenic, is added to a semiconductor such as silicon, four of the arsenic electrons form shared bonds with atoms of the semiconductor and one is left over. This extra electron is nearly free of its parent atom and has an energy level that lies in the energy gap, just below the conduction band. Such a pentavalent atom in effect donates an electron to the structure and hence is referred to as a **donor atom.** Because the spacing between the energy level of the electron of the donor atom and the bottom of the conduction band is very small (typically, about 0.05 eV), only a small amount of thermal energy is needed to cause this electron to move into the conduction band. (Recall that the average thermal energy of an electron at room temperature is $3k_B T/2 \approx 0.04$ eV). Semiconductors doped with donor atoms are called ***n*-type semiconductors** because the charge carriers are electrons, the charge of which is **n**egative.

If a semiconductor is doped with atoms containing three outer-shell electrons, such as aluminum, the three electrons form shared bonds with neighboring semiconductor atoms, leaving an electron deficiency—a hole—where the fourth bond would be if an impurity-atom electron was available to form it. The energy level of this hole lies in the energy gap, just above the valence band. An electron from the valence band has enough energy at room temperature to fill this impurity level, leaving behind a hole in the valence band. Because a trivalent atom in effect accepts an electron from the valence band, such impurities are referred to as **acceptor atoms.** A semiconductor doped with acceptor impurities is known as a ***p*-type semiconductor** because the majority of charge carriers are **p**ositively charged holes.

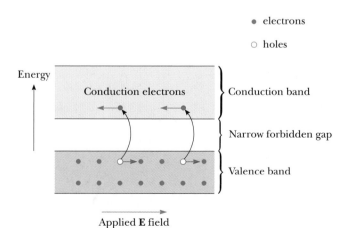

FIGURE 28.26 Movement of charges (holes and electrons) in a semiconductor. The electrons move in the direction opposite the direction of the external electric field, and the holes move in the direction of the field.

28.14 SEMICONDUCTOR DEVICES

THE *P-N* JUNCTION

Now let us consider what happens when a *p*-semiconductor is joined to an *n*-semiconductor to form a *p-n* junction. The junction consists of the three distinct regions shown in Figure 28.27a: a *p* region, a depletion region, and an *n* region.

The depletion region, which extends several micrometers to either side of the center of the junction, may be visualized as arising when the two halves of the junction are brought together. Mobile donor electrons from the *n* side nearest the junction (blue area in Figure 28.27a) diffuse to the *p* side, leaving behind immobile positive ions. At the same time, holes from the *p* side nearest the junction diffuse to the *n* side and leave behind a region (red area in Figure 28.27a) of fixed negative ions. Thus, the depletion region is so named because it is depleted of mobile charge carriers.

The depletion region contains an internal electric field (arising from the charges of the fixed ions) on the order of 10^4–10^6 V/cm. This field sweeps mobile charge out of the depletion region and keeps it truly depleted. This internal electric field creates an internal potential difference ΔV_0 that prevents further diffusion of holes and electrons across the junction and thereby ensures zero current through the junction when no external potential difference is applied.

Perhaps the most notable feature of the *p-n* junction is its ability to pass current in only one direction. Such *diode* action is easiest to understand in terms of the potential difference graph shown in Figure 28.27c. If an external voltage ΔV is applied to the junction such that the *p* side is connected to the positive terminal of a voltage source as in Figure 28.27a, the internal potential difference ΔV_0 across the junction is decreased, resulting in a current that increases exponentially with increasing forward voltage, or *forward bias*. For *reverse bias* (where the *n* side of the junction is connected to the positive terminal of a voltage source), the internal potential difference ΔV_0 increases with increasing reverse bias. This results in a very small reverse current that quickly reaches a saturation value I_0. The current-

FIGURE 28.27 (a) Physical arrangement of a *p-n* junction. (b) Internal electric field versus *x* for the *p-n* junction. (c) Internal electric potential ΔV versus *x* for the *p-n* junction. ΔV_0 represents the potential difference across the junction in the absence of an applied electric field.

(a)

Forward bias

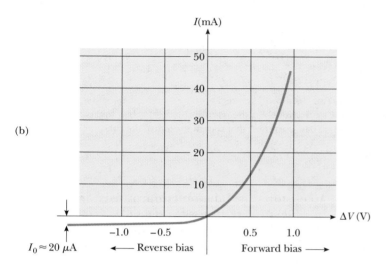

(b)

$I_0 \approx 20\ \mu A$ ← Reverse bias   Forward bias →

FIGURE 28.28 (a) Schematic of a
p-n junction under forward bias.
(b) The characteristic curve for a real
p-n junction.

voltage relationship for an ideal diode is

$$I = I_0(e^{q\Delta V/k_B T} - 1)$$ **[28.21]**

where e is Euler's number, q is the electron charge, k_B is Boltzmann's constant, and
T is the temperature in kelvins. Figure 28.28 shows an I-ΔV plot characteristic of a
real *p-n* junction along with a schematic of such a device under forward bias.

The most common use of the semiconductor diode is as a rectifier, a device
that changes 120-V AC voltage supplied by the power company to, say the 12-V DC
voltage needed by your music keyboard. We can understand how a diode rectifies
a current by considering Figure 28.29a, which shows a diode connected in series
with a resistor and an AC source. Because appreciable current can pass through
the diode only in one direction, the alternating current in the resistor is reduced
to the form shown in Figure 28.29b. The diode is said to act as a **half-wave rectifier** because there is current in the circuit during only half of each cycle.

Figure 28.30a shows a circuit that lowers the AC voltage to 12 V with a step-
down transformer and then rectifies both halves of the 12 V AC. Such a rectifier is
called a **full-wave rectifier** and when combined with a step-down transformer is
the most common DC power supply around the home today. A capacitor added in
parallel with the load will yield an even steadier DC voltage.

THE JUNCTION TRANSISTOR

The invention of the transistor by John Bardeen (1908–1991), Walter Brattain
(1902–1987), and William Shockley (1910–1989) in 1948 totally revolutionized
the world of electronics. For this work, these three men shared a Nobel prize in
1956. By 1960, the transistor had replaced the vacuum tube in many electronic applications. The advent of the transistor created a multitrillion dollar industry that
produced such popular devices as pocket radios, handheld calculators, computers,
television receivers, and electronic games. In this section we explain how a transistor acts as an amplifier to boost the tiny voltages and currents generated in a microphone to the ear-splitting levels required to drive a speaker.

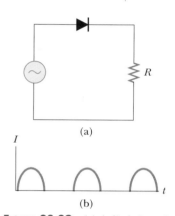

(a)

(b)

FIGURE 28.29 (a) A diode in series
with a resistor allows current to pass
in only one direction. (b) The current versus time for the circuit in (a).

(a)

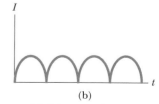

(b)

FIGURE 28.30 (a) A full-wave rectifier circuit. (b) The current versus
time in the resistor *R*.

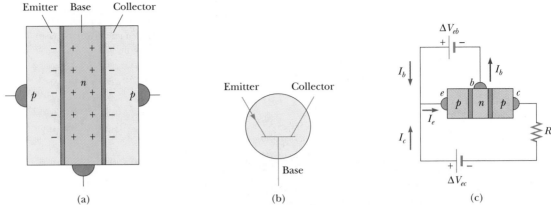

FIGURE 28.31 (a) The *pnp* transistor consists of an *n* region (base) sandwiched between two *p* regions (emitter and collector). (b) Circuit symbol for the *pnp* transistor. (c) A bias voltage ΔV_{eb} applied to the base as shown produces a small base current I_b that is used to control the collector current I_c in a *pnp* transistor.

One simple form of the transistor, called the **junction transistor,** consists of a semiconducting material in which a very narrow *n* region is sandwiched between two *p* regions. This configuration is called a **pnp transistor.** Another configuration is the **npn transistor,** which consists of a *p* region sandwiched between two *n* regions. Because the operation of the two transistors is essentially the same, we describe only the *pnp* transistor. The structure of the *pnp* transistor, together with its circuit symbol, is shown in Figure 28.31. The outer regions are called the **emitter** and **collector,** and the narrow central region is called the **base.** The configuration contains two junctions: the emitter-base interface and the collector-base interface.

Suppose a voltage is applied to the transistor so that the emitter is at a higher electric potential than the collector. (This is accomplished with the battery labeled ΔV_{ec} in Figure 28.31c.) If we think of the transistor as two diodes back to back, we see that the emitter-base junction is forward-biased and the base-collector junction is reverse-biased. The emitter is heavily doped relative to the base, and as a result nearly all the current consists of holes moving across the emitter-base junction. Most of these holes do not recombine with electrons in the base because the base is very narrow. Instead, they are accelerated across the reverse-biased base-collector junction, producing the collector current I_c in Figure 28.31c.

Although only a small percentage of holes recombine in the base, those that do limit the emitter current to a small value because positive charge carriers accumulating in the base prevent holes from flowing in. In order to prevent this current limitation, some of the positive charge on the base must be drawn off; this is accomplished by connecting the base to the battery labeled ΔV_{eb} in Figure 28.31c. Those positive charges that are not swept across the base-collector junction leave the base through this added pathway. **This base current I_b is very small, but a small change in it can significantly change the collector current I_c.** If the transistor is properly biased, the collector (output) current is directly proportional to the base (input) current, and the transistor acts as a current amplifier. This condition may be written

$$I_c = \beta I_b$$

where β, the *current gain* factor, is typically in the range of 10–100. Thus, the transistor may be used to amplify a small signal. The small voltage to be amplified is placed in series with the battery ΔV_{eb}. The input signal produces a small variation in the base current, resulting in a large change in the collector current and hence a large change in the voltage across the output resistor.

(a)   (b)

FIGURE 28.32   (a) Jack Kilby's first integrated circuit tested on September 12, 1958. *(Courtesy of Texas Instruments, Inc.)* (b) Integrated circuits continue to shrink in size and price while simultaneously growing in capability. *(Courtesy of Intel Corporation)*

THE INTEGRATED CIRCUIT

Invented independently by Jack Kilby at Texas Instruments in late 1958 and by Robert Noyce at Fairchild Camera and Instrument in early 1959, the integrated circuit has been justly called "the most remarkable technology ever to hit mankind." Kilby's first device is shown in Figure 28.32a. Integrated circuits have indeed started a "second industrial revolution" and are at the heart of computers, watches, cameras, automobiles, aircraft, robots, space vehicles, and all sorts of communication and switching networks.

In simplest terms, an **integrated circuit** is a collection of interconnected transistors, diodes, resistors, and capacitors fabricated on a single piece of silicon, known as a chip. State-of-the-art chips easily contain several million components in a 1-cm^2 area, with the number of components per square inch having doubled every year since the integrated circuit was invented.

Integrated circuits were invented partly to solve the interconnection problem spawned by the transistor. In the era of vacuum tubes, power and size considerations of individual components set significant limits on the number of components that could be interconnected in a given circuit. With the advent of the tiny, low-power, highly reliable transistor, design limits on the number of components disappeared and were replaced by the problem of wiring together hundreds of thousands of components. The magnitude of this problem can be appreciated when we consider that second-generation computers (consisting of discrete transistors rather than integrated circuits) contained several hundred thousand components requiring more than a million hand-soldered joints to be made and tested.

In addition to solving the interconnection problem, integrated circuits possess the advantages of miniaturization and fast response, two attributes critical for high-speed computers. The fast response results from the miniaturization and close packing of components, because the response time of a circuit depends on the time it takes for electrical signals traveling at about the speed of light to pass from one component to another. This time is clearly reduced by packing components closely.

SUMMARY

The **Bohr model** of the atom is successful in describing the spectra of atomic hydrogen and hydrogen-like ions. One of the basic assumptions of the model is that the electron can exist only in certain orbits such that its angular momentum, mvr, is an integral multiple of $\hbar$, where $\hbar$ is Planck's constant divided by 2π. Assuming circular orbits and a Coulomb force of attraction between electron and proton, the energies of the quantum states for hydrogen are

$$E_n = -\frac{m_e k_e^2 e^4}{2\hbar^2}\left(\frac{1}{n^2}\right) \qquad n = 1, 2, 3, \ldots \qquad \textbf{[28.12]}$$

where k_e is the Coulomb constant, e is the charge on the electron, and n is an integer called a **quantum number.**

If the electron in the hydrogen atom jumps from an orbit whose quantum number is n_i to an orbit whose quantum number is n_f, it emits a photon of frequency f, given by

$$f = \frac{m_e k_e^2 e^4}{4\pi\hbar^3}\left(\frac{1}{n_f^2} - \frac{1}{n_i^2}\right) \qquad \textbf{[28.14]}$$

Bohr's **correspondence principle** states that quantum mechanics is in agreement with classical physics when the quantum numbers for a system are very large.

One of the many successes of quantum mechanics is that the quantum numbers n, ℓ, and m_ℓ associated with atomic structure arise directly from the mathematics of the theory. The quantum number n is called the **principal quantum number,** ℓ is the **orbital quantum number,** and m_ℓ is the **orbital magnetic quantum number.** In addition, a fourth quantum number, called the **spin magnetic quantum number,** m_s, is needed to explain certain features of atomic structure.

An understanding of the periodic table of the elements became possible when Pauli formulated the **exclusion principle,** which states that no two electrons in an atom can ever be in the same quantum state; that is, no two electrons in the same atom can have the same values for the set of quantum numbers, n, ℓ, m_ℓ, and m_s.

Characteristic x-rays are produced when a bombarding electron collides with an electron in an inner shell of an atom with sufficient energy to remove the electron from the atom. The vacancy thus created is filled when an electron from a higher level drops down into the level containing the vacancy.

Lasers are monochromatic, coherent light sources that work on the principle of **stimulated emission** of radiation from a system of atoms.

In solids, the discrete energy levels of isolated atoms broaden into **bands,** and the structure and electron population of the bands determine whether a substance is a conductor, insulator, or semiconductor. By **doping** semiconductors, specific regions with known resistivity and positive or negative majority charge carriers are created, and devices such as diodes, transistors, and integrated circuits may be fabricated. These solid state devices have revolutionized life in the 20th century by their reliability, low power consumption, speed of response, and their virtually limitless interconnection capability in chips.

CONCEPTUAL QUESTIONS

1. In the hydrogen atom, the quantum number n can increase without limit. Because of this, does the frequency of possible spectral lines from hydrogen also increase without limit?

2. Does the light emitted by a neon sign constitute a continuous spectrum or only a few colors? Defend your answer.

3. In an x-ray tube, if the energy with which the electrons strike the metal target is increased, the wavelengths of the characteristic x-rays do not change. Why not?

4. Must an atom first be ionized before it can emit light? Discuss.

5. Is it possible for a spectrum from an x-ray tube to show the continuous spectrum of x-rays without the presence of the characteristic x-rays?

6. Suppose that the electron in the hydrogen atom obeyed classical mechanics rather than quantum mechanics. Why should such a hypothetical atom emit a continuous spectrum rather than the observed line spectrum?

7. When a hologram is produced, the system (including light source, object, beam splitter, and so on) must be held motionless within a quarter of the light's wavelength. Why?

8. If matter has a wave nature, why is this not observable in our daily experiences?

9. Discuss some consequences of the exclusion principle.

10. Can the electron in the ground state of hydrogen absorb a photon of energy less than 13.6 eV? Can it absorb a photon of energy greater than 13.6 eV? Explain.

11. Why do lithium, potassium, and sodium exhibit similar chemical properties?

12. List some ways in which quantum mechanics altered our view of the atom pictured by the Bohr theory.

13. It is easy to understand how two electrons (one spin-up, one spin-down) can fill the $1s$ shell for a helium atom.

How is it possible that eight more electrons can fit into the $2s$, $2p$ level to complete the $1s^2 2s^2 2p^6$ shell for a neon atom?

14. The ionization energies for Li, Na, K, Rb, and Cs are 5.390, 5.138, 4.339, 4.176, and 3.893 eV, respectively. Explain why this pattern of decrease should be expected in terms of the atomic structures.

15. How can the total energy of an atom be negative? For that matter, how can any energy be negative?

16. A 0.1-mW laser might damage your eye if you look directly at it, yet there is no harm in looking at a bare 100-W lightbulb. Why?

17. Why is stimulated emission so important in the operation of a laser?

PROBLEMS

1, 2, 3 = straightforward, intermediate, challenging ☐ = full solution available in Student Solutions Manual/Study Guide

web = solution posted at **http://info.brookscole.com/serway** 🎵 = biomedical application

Section 28.1 Early Models of the Atom

Section 28.2 Atomic Spectra

1. Use Equation 28.1 to calculate the wavelength of the first three lines in the Balmer series for hydrogen.

2. Show that the wavelengths for the Balmer series satisfy the equation

$$\lambda = \frac{364.5 n^2}{n^2 - 4} \text{ nm} \qquad \text{where } n = 3, 4, 5, \ldots$$

3. The "size" of the *atom* in Rutherford's model is about 1.0×10^{-10} m. (a) Determine the attractive electrical force between an electron and a proton separated by this distance. (b) Determine (in eV) the electrical potential energy of the atom.

4. The "size" of the *nucleus* in Rutherford's model of the atom is about 1.0 fm = 1.0×10^{-15} m. (a) Determine the repulsive electrical force between two protons separated by this distance. (b) Determine (in MeV) the electrical potential energy of the pair of protons.

5. The "size" of the atom in Rutherford's model is about
web 1.0×10^{-10} m. (a) Determine the speed of an electron moving in a circle around the proton using the attractive electrical force between an electron and a proton separated by this distance. (b) Does this speed suggest that Einsteinian relativity must be considered when studying the atom? (c) Compute the de Broglie wavelength of the electron as it moves about the proton. (d) Does this wavelength suggest that wave effects, such as diffraction and interference, must be considered when studying the atom?

6. In a Rutherford scattering experiment, an α-particle (charge = $+2e$) heads directly toward a gold nucleus (charge = $+79e$). The α-particle has a kinetic energy of 5.0 MeV when very far ($r \rightarrow \infty$) from the nucleus. Assuming the gold nucleus to be fixed in space, determine the distance of closest approach. (*Hint:* Use conservation of energy with $PE = k_e q_1 q_2 / r$.)

Section 28.3 The Bohr Theory of Hydrogen

7. A hydrogen atom is in its first excited state ($n = 2$). Using the Bohr theory of the atom, calculate (a) the radius of the orbit, (b) the linear momentum of the electron, (c) the angular momentum of the electron, (d) the kinetic energy, (e) the potential energy, and (f) the total energy.

8. For a hydrogen atom in its ground state, use the Bohr model to compute (a) the orbital speed of the electron, (b) the kinetic energy of the electron, and (c) the electrical potential energy of the atom.

9. Show that the speed of the electron in the nth Bohr orbit in hydrogen is given by

$$v_n = \frac{k_e e^2}{n \hbar}$$

10. A photon is emitted as a hydrogen atom undergoes a transition from the $n = 6$ state to the $n = 2$ state. Calculate (a) the energy, (b) the wavelength, and (c) the frequency of the emitted photon.

11. Calculate the Coulomb force of attraction on the electron when it is in the ground state of the hydrogen atom.

12. Four possible transitions for a hydrogen atom are listed below:

I. $n_i = 2$; $n_f = 5$ II. $n_i = 5$; $n_f = 3$
III. $n_i = 7$; $n_f = 4$ IV. $n_i = 4$; $n_f = 7$

(a) Which transition will emit the shortest-wavelength photon? (b) For which transition will the atom gain the most energy? (c) For which transition(s) does the atom lose energy?

13. What is the energy of the photon that, when absorbed by a hydrogen atom, could cause (a) an electronic transition from the $n = 3$ state to the $n = 5$ state and (b) an electronic transition from the $n = 5$ state to the $n = 7$ state?

14. A hydrogen atom initially in its ground state ($n = 1$) absorbs a photon and ends up in the state for which $n = 3$. (a) What is the energy of the absorbed photon? (b) If the

atom eventually returns to the ground state, what photon energies could the atom emit?

15. Determine both the longest and the shortest wavelengths in (a) the Lyman series ($n_f = 1$) and (b) the Paschen series ($n_f = 3$) of hydrogen.

16. How much energy is required to ionize hydrogen when it is in (a) the ground state and (b) the state for which $n = 3$?

17. A monochromatic beam of light is absorbed by a collection of ground-state hydrogen atoms in such a way that six different wavelengths are observed when the hydrogen relaxes back to the ground state. What is the wavelength of the incident beam?

18. A particle of charge q and mass m, moving with a constant speed v, perpendicular to a constant magnetic field, B, follows a circular path. If the angular momentum about the center of this circle is quantized so that $mvr = n\hbar$, show that the allowed radii for the particle are

$$r_n = \sqrt{\frac{n\hbar}{qB}} \qquad \text{where } n = 1, 2, 3, \ldots$$

19. (a) If an electron makes a transition from the $n = 4$ Bohr orbit to the $n = 2$ orbit, determine the wavelength of the photon created in the process. (b) Assuming that the atom was initially at rest, determine the recoil speed of the hydrogen atom when this photon is emitted.

20. An electron is in the first Bohr orbit of hydrogen. Find (a) the speed of the electron, (b) the time required for the electron to circle the nucleus, and (c) the current in amperes corresponding to the motion of the electron.

21. Analyze the Earth-Sun system following the Bohr model, where the gravitational force between Earth (mass m) and the Sun (mass M) replaces the Coulomb force between the electron and proton (so that $F = GMm/r^2$ and $PE = -GMm/r$). Show that (a) Earth's total energy in an orbit of radius r is given by $E = -GMm/2r$, (b) the radius of the nth orbit is given by $r_n = r_0 n^2$, where $r_0 = \hbar^2/GMm^2 = 2.32 \times 10^{-138}$ m, and (c) the energy of the nth orbit is given by $E_n = -E_0/n^2$, where $E_0 = G^2M^2m^3/2\hbar^2 = 1.71 \times 10^{182}$ J. (d) Using the Earth-Sun orbit radius of $r = 1.49 \times 10^{11}$ m, determine the value of the quantum number n. (e) Should you expect to observe quantum effects in the Earth-Sun system?

22. An electron is in the nth Bohr orbit of the hydrogen atom. (a) Show that the period of the electron is $T = t_0 n^3$, and determine the numerical value of t_0. (b) On the average, an electron remains in the $n = 2$ orbit for about 10 μs before it jumps down to the $n = 1$ (ground-state) orbit. How many revolutions does the electron make before it jumps to the ground state? (c) If one revolution of the electron is defined as an "electron-year" (analogous to an Earth-year being one revolution of Earth around the Sun), does the electron in the $n = 2$ orbit "live" very long? Explain. (d) How does the above calculation support the "electron cloud" concept?

23. Consider a hydrogen atom. (a) Calculate the frequency f of the $n = 2 \rightarrow n = 1$ transition and compare with the frequency f_{orb} of the electron orbital motion in the $n = 2$ state. (b) Make the same calculation for the $n = 10\,000 \rightarrow n = 9\,999$ transition. Comment on the results.

24. Two hydrogen atoms collide head-on and end up with zero kinetic energy. Each then emits a 121.6-nm photon

($n = 2$ to $n = 1$ transition). At what speed were the atoms moving before the collision?

25. Two hydrogen atoms, both initially in the ground state, undergo a head-on collision. If both atoms are to be excited to the $n = 2$ level in this collision, what is the minimum speed each atom can have before the collision?

26. (a) Calculate the angular momentum of the Moon due to its orbital motion about Earth. In your calculation, use 3.84×10^8 m as the average Earth-Moon distance and 2.36×10^6 s as the period of the Moon in its orbit. (b) If the angular momentum of the moon obeys Bohr's quantization rule ($L = n\hbar$), determine the value of the quantum number, n. (c) By what fraction would the Earth-Moon radius have to be increased to increase the quantum number by 1?

Section 28.4   Modification of the Bohr Theory

Section 28.5   de Broglie Waves and the Hydrogen Atom

27. (a) Find the energy of the electron in the ground state of doubly ionized lithium, which has an atomic number $Z = 3$. (b) Find the radius of its ground-state orbit.

28. (a) Construct an energy level diagram for the He^+ ion, for which $Z = 2$. (b) What is the ionization energy for He^+?

29. The orbital radii of a hydrogen-like atom is given by the equation $r = n^2\hbar^2/Zm_e k_e e^2$. What is the radius of the first Bohr orbit in (a) He^+, (b) Li^{2+}, and (c) Be^{3+}?

30. Construct an energy level diagram like that in Figure 28.7 for doubly ionized lithium (Li^{2+}), for which $Z = 3$.

31. Determine the wavelength of an electron in the third excited orbit of the hydrogen atom, with $n = 4$.

32. Using the concept of standing waves, de Broglie was able to derive Bohr's stationary orbit postulate. He assumed that a confined electron could exist only in states where its de Broglie waves form standing-wave patterns, as in Figure 28.10a. Consider a particle confined in a box of length L to be equivalent to a string of length L and fixed at both ends. Apply de Broglie's concept to show that (a) the linear momentum of this particle is quantized with $p = mv = nh/2L$ *and* (b) the allowed states correspond to particle energies of $E_n = n^2E_0$ where $E_0 = h^2/(8mL^2)$.

Section 28.6   Quantum Mechanics and the Hydrogen Atom

Section 28.7   The Spin Magnetic Quantum Number

33. List the possible sets of quantum numbers for electrons in the $3p$ subshell.

34. List the possible sets of quantum numbers for electrons in the $3d$ subshell.

35. The ρ-meson has a charge of $-e$, a spin quantum number of 1, and a mass 1 507 times that of the electron. If the electrons in atoms were replaced by ρ-mesons, list the possible sets of quantum numbers for ρ-mesons in the $3d$ subshell.

Section 28.9   The Exclusion Principle and the Periodic Table

36. (a) Write out the electronic configuration of the ground state for oxygen ($Z = 8$). (b) Write out values for the set of

quantum numbers n, ℓ, m_ℓ, and m_s for each of the electrons in oxygen.

37. Two electrons in the same atom both have $n = 3$ and $\ell = 1$. List the quantum numbers for the possible states of the atom. (a) How many states are possible? (b) How many states would be possible if the exclusion principle did not apply to the atom?

38. How many different sets of quantum numbers are possible for an electron for which (a) $n = 1$, (b) $n = 2$, (c) $n = 3$, (d) $n = 4$, and (e) $n = 5$? Check your results to show that they agree with the general rule that the number of different sets of quantum numbers is equal to $2n^2$.

39. Zirconium ($Z = 40$) has two electrons in an incomplete d
web subshell. (a) What are the values of n and ℓ for each electron? (b) What are all possible values of m_ℓ and m_s? (c) What is the electron configuration in the ground state of zirconium?

Section 28.10 Characteristic X-Rays

40. The K shell ionization energy of copper is 8 979 eV. The L shell ionization energy is 951 eV. Determine the wavelength of the K_α emission line of copper. What must the minimum voltage be on an x-ray tube with a copper target in order to see the K_α line?

41. The K_α x-ray is emitted when an electron undergoes a transition from the L shell ($n = 2$) to the K shell ($n = 1$). Use the method illustrated in Example 28.7 to calculate the wavelength of the K_α x-ray from a nickel target ($Z = 28$).

42. When an electron drops from the M shell ($n = 3$) to a vacancy in the K shell ($n = 1$), the measured wavelength of the emitted x-ray is found to be 0.101 nm. Identify the element.

43. The K series of the discrete spectrum of tungsten contains wavelengths of 0.018 5 nm, 0.020 9 nm, and 0.021 5 nm. The K shell ionization energy is 69.5 keV. Determine the ionization energies of the L, M, and N shells. Sketch the transitions that produce the above wavelengths.

ADDITIONAL PROBLEMS

44. In a hydrogen atom, what is the principal quantum number of the electron orbit with a radius closest to 1.0 μm?

45. (a) How much energy is required to cause an electron in hydrogen to move from the $n = 1$ state to the $n = 2$ state? (b) If the electrons gain this energy by collision between hydrogen atoms in a high-temperature gas, find the minimum temperature of the heated hydrogen gas. The thermal energy of the heated atoms is given by $3k_B T/2$, where k_B is the Boltzmann constant.

46. A pulsed ruby laser emits light at 694.3 nm. For a 14.0-ps pulse containing 3.00 J of energy, find (a) the physical length of the pulse as it travels through space and (b) the number of photons in it. (c) If the beam has a circular cross section of 0.600-cm diameter, find the number of photons per cubic millimeter.

47. The Lyman series for a (new!?) one-electron atom is observed in a distant galaxy. The wavelengths of the first four lines and the short-wavelength limit of this Lyman series are given by the energy level diagram in Figure P28.47. Based on this information, calculate (a) the energies of the ground state and first four excited states for this one-electron atom, and (b) the longest-wavelength (alpha) lines and the short-wavelength series limit in the Balmer series for this atom.

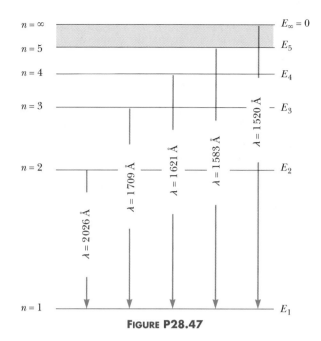

FIGURE P28.47

48. A dimensionless number that often appears in atomic physics is the fine-structure constant $\alpha = k_e e^2/\hbar c$, where k_e is the Coulomb constant. (a) Obtain a numerical value for $1/\alpha$. (b) In terms of α, what is the ratio of the Bohr radius a_0 to the Compton wavelength $\lambda_C = h/m_e c$? (c) In terms of α, what is the ratio of the reciprocal of the Rydberg constant $1/R_H$ to the Bohr radius?

49. Mercury's ionization energy is 10.39 eV. The three longest wavelengths of the absorption spectrum of mercury are 253.7 nm, 185.0 nm, and 158.5 nm. (a) Construct an energy level diagram for mercury. (b) Indicate all emission lines that can occur when an electron is raised to the third level above the ground state. (c) Disregarding recoil of the mercury atom, determine the minimum speed an electron must have in order to make an inelastic collision with a mercury atom in its ground state.

50. Suppose the ionization energy of an atom is 4.100 eV. In this same atom, we observe emission lines with wavelengths 310.0 nm, 400.0 nm, and 1 378 nm. Use this information to construct the energy level diagram with the least number of levels. Assume the higher energy levels are closer together.

51. A laser used in eye surgery emits a 3.00-mJ pulse in 1.00 ns, focused to a spot 30.0 μm in diameter on the retina. (a) Find (in SI units) the power per unit area at the retina. (This quantity is called the *irradiance*.) (b) What energy is delivered per pulse to an area of molecular size—say, a circular area 0.600 nm in diameter?

52. An electron has a de Broglie wavelength equal to the diameter of a hydrogen atom in its ground state. (a) What is the kinetic energy of the electron? (b) How does this energy compare with the ground-state energy of the hydrogen atom?

53. Use Bohr's model of the hydrogen atom to show that when the atom makes a transition from the state n to the state $n - 1$, the frequency of the emitted light is given by

$$f = \frac{2\pi^2 m k_e^2 e^4}{h^3}\left[\frac{2n-1}{(n-1)^2 n^2}\right]$$

54. Calculate the classical frequency for the light emitted by an atom. To do so, note that the frequency of revolution is $v/2\pi r$, where r is the Bohr radius. Show that as n approaches infinity in the equation of the preceding problem, the expression given there varies as $1/n^3$ and reduces to the classical frequency. (This is an example of the correspondence principle, which requires that the classical and quantum models agree for large values of n.)

55. A π-meson (π^-) of charge $-e$ and mass 273 times greater
web than that of the electron is captured by a helium nucleus ($Z = +2$) as shown in Figure P28.55. (a) Draw an energy level diagram (in units of eV) for this "Bohr-type" atom up

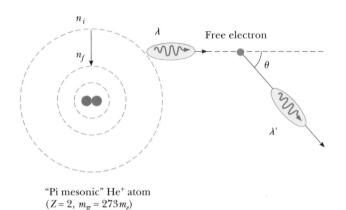

"Pi mesonic" He$^+$ atom
($Z = 2$, $m_\pi = 273\,m_e$)

FIGURE P28.55

to the first six energy levels. (b) When the π-meson makes a transition between two orbits, a photon is emitted that Compton scatters off a free electron initially at rest, producing a scattered photon of wavelength $\lambda' = 0.089\,929\,3$ nm at an angle of $\theta = 42.68°$, as shown on the right-hand side of Figure P28.55. Between which two orbits did the π-meson make a transition?

56. When a muon with charge $-e$ is captured by a proton, the resulting bound system forms a "muonic atom," which is the same as hydrogen except with a muon (of mass 207 times the mass of an electron) replacing the electron. For this "muonic atom" determine: (a) the Bohr radius and (b) the three lowest energy levels.

57. In this problem you will estimate the classical lifetime of the hydrogen atom. An accelerating charge loses electromagnetic energy at a rate given by $\mathcal{P} = -2k_e q^2 a^2/(3c^3)$, where k_e is the Coulomb constant, q is the charge of the particle, a is its acceleration, and c is the speed of light in a vacuum. Assuming that the electron is one Bohr radius (0.052 9 nm) from the center of the hydrogen atom, (a) determine its acceleration. (b) Show that $\mathcal{P}$ has units of energy per unit time and determine the rate of energy loss. (c) Calculate the kinetic energy of the electron and determine how long it will take for all of this energy to be converted into electromagnetic waves, assuming that the rate calculated in part (b) remains constant throughout the electron's motion.

GROUP ACTIVITIES

G.1 With your partner not looking, use modeling clay to build one or more mounds on top of a table. Place a piece of cardboard over your mound(s), and assign your partner the task of determining the size, shape, and number of mounds without looking. He is to do this by rolling marbles at the unseen mounds and observing how they emerge. This experiment models the Rutherford scattering experiment.

G.2 Your instructor can probably lend you a small plastic diffraction grating that will enable you to examine the spectra of different light sources. You can use this grating to examine a source by holding the grating very close to your eye with the grooves vertical and noting the spectrum appearing off to the side when you look at a small or distant light source. You should look at light sources such as sodium vapor lights and mercury vapor lights used in many parking lots, neon lights used in many signs, black lights, ordinary incandescent light bulbs, and so forth.

G.3 Assume that a quantum mechanical oscillator behaves in a fashion similar to a simple harmonic motion oscillator in our large-scale world. That is, its acceleration and ve-

locity vary in the same way. Assume the oscillation takes place along a segment of the x axis and sketch a graph of the probability of finding the oscillator versus distance from the origin. Where does the probability distribution peak? Where would it have the smallest amplitude?

G.4 Photons are emitted when an electron drops from the $n = 3$ and $n = 4$ levels to the $n = 2$ level in the hydrogen atom. (a) In which case is the wavelength of the emitted photon greater? (b) In which case is the energy of the emitted photon greater? (c) Use the Bohr theory to justify your answers.

G.5 What are the quantum numbers that lead to the minimum and maximum wavelengths in the Balmer series for hydrogen? (b) Use the Bohr theory to find these wavelengths. Do your results agree with your predictions?

G.6 Consider the following quantum numbers, n, ℓ, m_ℓ, and m_s. (a) Which of these are integers and which are fractional? (b) Which of the above are always positive and which can be negative? (c) If $n = 2$, the largest value of ℓ is _____. (d) If $\ell = 1$, the values of m_ℓ range from _____ to _____.

Nuclear Physics

29

Chapter Outline

Aerial view of a nuclear power plant that generates electrical power. Energy is generated in such plants from the process of nuclear fission, in which a heavy nucleus such as ^{235}U splits into two smaller nuclei. The fission process and nuclear reactors are discussed in more detail in Chapter 30. *(Courtesy of Public Service Electric and Gas Company)*

ERNEST RUTHERFORD, NEW ZEALAND PHYSICIST (1871–1937)

Rutherford was awarded the Nobel Prize in 1908 for discovering that atoms can be broken apart by alpha rays, and for studying radioactivity. "On consideration, I realized that this scattering backward must be the result of a single collision, and when I made calculations I saw that it was impossible to get anything of that order of magnitude unless you took a system in which the greater part of the mass of the atom was concentrated in a minute nucleus. It was then that I had the idea of an atom with a minute massive center carrying a charge." *(North Wind Picture Archives)*

Webnote 29.1

See a timeline of the atomic and nuclear age from 1942 to 1998. Go to
http://www.pbs.org/wgbh/amex/three/timeline/index.html

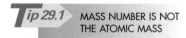

MASS NUMBER IS NOT THE ATOMIC MASS

Do not confuse the mass number A with the atomic mass. Mass number is an integer that specifies an isotope and has no units—it is simply equal to the number of nucleons. Atomic mass is generally not an integer. It is an average of the masses of the isotopes of a given element and has units of u.

*I*n 1896, the year that marks the birth of nuclear physics, Henri Becquerel (1852–1908) discovered radioactivity in uranium compounds. A great deal of activity followed this discovery as researchers attempted to understand and characterize the radiation that we now know to be emitted by radioactive nuclei. Pioneering work by Rutherford showed that the radiation was of three types, which he called *alpha, beta,* and *gamma rays.* These types are classified according to the nature of their electric charge and according to their ability to penetrate matter. Later experiments showed that alpha rays are helium nuclei, beta rays are electrons, and gamma rays are high-energy photons.

In 1911 Rutherford and his students Geiger and Marsden performed a number of important scattering experiments involving alpha particles. These experiments established that the nucleus of an atom can be regarded as essentially a point mass and point charge and that most of the atomic mass is contained in the nucleus. Furthermore, such studies demonstrated a wholly new type of force, the *nuclear force,* which is predominant at distances of less than about 10^{-14} m and zero at greater distances.

Other milestones in the development of nuclear physics include

- the first observations of nuclear reactions by Rutherford and coworkers in 1919 in which naturally occurring alpha particles bombarded nitrogen nuclei to produce oxygen,
- the first use of artificially accelerated protons to produce nuclear reactions by Cockcroft and Walton in 1932,
- the discovery of the neutron by Chadwick in 1932,
- the discovery of artificial radioactivity by Joliot and Irene Curie in 1933,
- the discovery of nuclear fission by Hahn and Strassman in 1938,
- the development of the first controlled fission reactor by Fermi and his collaborators in 1942.

In this chapter we discuss the properties and structure of the atomic nucleus. We start by describing the basic properties of nuclei and follow with a discussion of the phenomenon of radioactivity. Finally, we explore nuclear reactions and the various processes by which nuclei decay.

29.1 SOME PROPERTIES OF NUCLEI

All nuclei are composed of two types of particles: protons and neutrons. The only exception is the ordinary hydrogen nucleus, which is a single proton. In describing some of the properties of nuclei, such as their charge, mass, and radius, we make use of the following quantities:

- the **atomic number** Z, which equals the number of protons in the nucleus,
- the **neutron number** N, which equals the number of neutrons in the nucleus,
- the **mass number** A, which equals the number of nucleons in the nucleus. (*Nucleon* is a generic term used to refer to either a proton or a neutron.)

We use the symbol $^{A}_{Z}X$ to represent nuclei, where X represents the chemical symbol for the element. For example, $^{27}_{13}Al$ has the mass number 27 and the atomic number 13; therefore, it contains 13 protons and 14 neutrons. When no confusion is likely to arise, we often omit the subscript Z because the chemical symbol can always be used to determine Z.

The nuclei of all atoms of a particular element must contain the same number of protons, but they may contain different numbers of neutrons. Nuclei that are related in this way are called **isotopes. The isotopes of an element have the same Z value but different N and A values.** The natural abundances of isotopes can differ substantially. For example, $^{11}_{6}C$, $^{12}_{6}C$, $^{13}_{6}C$, and $^{14}_{6}C$ are four isotopes of carbon. The natural abundance of the $^{12}_{6}C$ isotope is about 98.9%, whereas that of

TABLE 29.1	Mass of the Proton, Neutron, and Electron in Various Units		
		Mass	
Particle	**kg**	**u**	**MeV/c^2**
Proton	$1.672\ 6 \times 10^{-27}$	$1.007\ 276$	938.28
Neutron	$1.675\ 0 \times 10^{-27}$	$1.008\ 665$	939.57
Electron	9.109×10^{-31}	5.486×10^{-4}	0.511

the $^{13}_{6}$C isotope is only about 1.1%. Some isotopes do not occur naturally but can be produced in the laboratory through nuclear reactions. Even the simplest element, hydrogen, has isotopes: $^{1}_{1}$H, hydrogen; $^{2}_{1}$H, deuterium; and $^{3}_{1}$H, tritium.

CHARGE AND MASS

The proton carries a single positive charge, $+e$, the electron carries a single negative charge, $-e$, where $e = 1.602\ 177\ 33 \times 10^{-19}$ C, and the neutron is electrically neutral. Because the neutron has no charge, it is difficult to detect. The proton is about 1 836 times as massive as the electron, and the masses of the proton and the neutron are almost equal (Table 29.1).

It is convenient to define, for atomic masses, the **unified mass unit** u in such a way that the mass of one atom of the isotope ^{12}C is exactly 12 u, where 1 u $= 1.660\ 559 \times 10^{-27}$ kg. The proton and neutron each have a mass of about 1 u, and the electron has a mass that is only a small fraction of an atomic mass unit.

▷ Definition of the unified mass unit, u

Because the rest energy of a particle is given by $E_R = mc^2$, it is often convenient to express the particle's mass in terms of its energy equivalent. For one atomic mass unit, we have an energy equivalent of

$$E_R = mc^2 = (1.660\ 559 \times 10^{-27}\ \text{kg})(2.997\ 92 \times 10^8\ \text{m/s})^2$$
$$= 1.492\ 431 \times 10^{-10}\ \text{J} = 931.494\ \text{MeV}$$

Nuclear physicists often express *mass* in terms of the unit MeV/c^2 in calculations, where

$$1\ \text{u} = 931.494\ \text{MeV}/c^2$$

THE SIZE OF NUCLEI

The size and structure of nuclei were first investigated in the scattering experiments of Rutherford, discussed in Section 28.1. Using the principle of conservation of energy, Rutherford found an expression for how close an alpha particle moving directly toward the nucleus can come to the nucleus before being turned around by Coulomb repulsion.

In such a head-on collision, the kinetic energy of the incoming alpha particle must be converted completely to electrical potential energy when the particle stops at the point of closest approach and turns around (Fig. 29.1). If we equate the initial kinetic energy of the alpha particle to the maximum electrical potential energy of the system (alpha particle + target nucleus), we have

$$\tfrac{1}{2}mv^2 = k_e \frac{q_1 q_2}{r} = k_e \frac{(2e)(Ze)}{d}$$

where d is the distance of closest approach. Solving for d, we get

$$d = \frac{4k_e Ze^2}{mv^2}$$

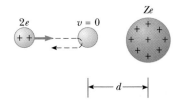

FIGURE 29.1 An alpha particle on a head-on collision course with a nucleus of charge Ze. Because of the Coulomb repulsion between the like charges, the alpha particle will stop instantaneously at a distance d from the nucleus.

From this expression, Rutherford found that alpha particles approached to within 3.2×10^{-14} m of a nucleus when the foil was made of gold. Thus, the radius of the gold nucleus must be less than this value. For silver atoms, the distance of closest approach was 2×10^{-14} m. From these results, Rutherford concluded that the positive charge in an atom is concentrated in a small sphere, which he called the nucleus, whose radius is no greater than about 10^{-14} m. Because such small lengths are common in nuclear physics, a convenient unit of length is the *femtometer* (fm), sometimes called the **fermi**, defined as

$$1 \text{ fm} \equiv 10^{-15} \text{ m}$$

Since the time of Rutherford's scattering experiments, a multitude of other experiments have shown that most nuclei are approximately spherical and have an average radius given by

$$r = r_0 A^{1/3} \qquad \text{[29.1]}$$

where A is the total number of nucleons and r_0 is a constant equal to 1.2×10^{-15} m. Because the volume of a sphere is proportional to the cube of its radius, it follows from Equation 29.1 that the volume of a nucleus (assumed to be spherical) is directly proportional to A, the total number of nucleons. This suggests that **all nuclei have nearly the same density.** Nucleons combine to form a nucleus *as though* they were tightly packed spheres (Fig. 29.2).

FIGURE 29.2 A nucleus can be visualized as a cluster of tightly packed spheres in which each sphere is a nucleon.

Example 29.1   Nuclear Volume and Density

A Find an approximate expression for the mass of a nucleus of mass number A.

Solution   The mass of the proton is approximately equal to that of the neutron. Thus, if the mass of one of these particles is m, the mass of the nucleus is approximately Am.

B Find an expression for the volume of this nucleus in terms of the mass number.

Solution   Assuming the nucleus is spherical and using Equation 29.1, we find that the volume is

$$V = \tfrac{4}{3}\pi r^3 = \tfrac{4}{3}\pi r_0^3 A$$

C Find a numerical value for its density.

Solution   The nuclear density can be found as follows:

$$\rho_n = \frac{\text{mass}}{\text{volume}} = \frac{Am}{\tfrac{4}{3}\pi r_0^3 A} = \frac{3m}{4\pi r_0^3}$$

The fact that A cancels out of our expression for nuclear density proves our statement above that all nuclei have roughly the same density.

Taking $r_0 = 1.2 \times 10^{-15}$ m and $m = 1.67 \times 10^{-27}$ kg, we find that

$$\rho_n = \frac{3(1.67 \times 10^{-27} \text{ kg})}{4\pi(1.2 \times 10^{-15} \text{ m})^3} = 2.3 \times 10^{17} \text{ kg/m}^3$$

Note that the nuclear density is about 2.3×10^{14} times greater than the density of water $(1 \times 10^3 \text{ kg/m}^3)$!

NUCLEAR STABILITY

Given that the nucleus consists of a closely packed collection of protons and neutrons, you might be surprised that it can exist. The very large repulsive electrostatic forces between protons should cause the nucleus to fly apart. However, nuclei are stable because of the presence of another, short-range (about 2 fm) force, the

FIGURE 29.3 A plot of the neutron number N versus the proton number Z for the stable nuclei (solid points). They are centered on the so-called line of stability. The dashed straight line corresponds to the condition $N = Z$. The shaded area shows radioactive (unstable) nuclei.

MARIA GOEPPERT-MAYER, GERMAN PHYSICIST (1906–1972)

Goeppert-Mayer was born and educated in Germany. She is best known for her devlopment of the shell model of the nucleus, published in 1950. A similar model was simultaneously developed by Hans Jensen, a German scientist. Maria Goeppert-Mayer and Hans Jensen were awarded the Nobel prize in physics in 1963 for their extraordinary work in understanding the structure of the nucleus. *(Courtesy of Louise Barker/AIP Niels Bohr Library)*

nuclear force. This is an attractive force that acts between all nuclear particles. The protons attract each other via the nuclear force, and at the same time they repel each other through the Coulomb force. The attractive nuclear force also acts between pairs of neutrons and between neutrons and protons.

The nuclear attractive force is stronger than the Coulomb repulsive force within the nucleus (at short ranges). If this were not the case, stable nuclei would not exist. Moreover, the strong nuclear force is nearly independent of charge. In other words, the nuclear forces associated with the proton-proton, proton-neutron, and neutron-neutron interactions are approximately the same.

There are about 260 stable nuclei; hundreds of others have been observed but are unstable. A plot of N versus Z for a number of stable nuclei is given in Figure 29.3. Note that light nuclei are most stable if they contain equal numbers of protons and neutrons—that is, if $N = Z$—but heavy nuclei are more stable if $N > Z$. This can be partially understood by recognizing that, as the number of protons increases, the strength of the Coulomb force increases, which tends to break the nucleus apart. As a result, more neutrons are needed to keep the nucleus stable, because neutrons experience only the attractive nuclear forces. In effect, the additional neutrons "dilute" the nuclear charge density. Eventually, when $Z = 83$, the repulsive forces between protons cannot be compensated by the addition of more neutrons. Elements that contain more than 83 protons do not have stable nuclei, but decay or disintegrate into other particles in various amounts of time. The masses of several stable particles are given in Table 29.1. The masses and some other properties of selected isotopes are provided in Appendix B.

Webnote 29.2

A detailed, "clickable" chart of the nuclide's extending information in Figure 29.3 can be found at *http://www2.bnl.gov/ton*

29.2 BINDING ENERGY

The total mass of a nucleus is always less than the sum of the masses of its nucleons. Because mass is another manifestation of energy, **the total energy of the bound system (the nucleus) is less than the combined energy of the sepa-**

rated nucleons. This difference in energy is called the **binding energy** of the nucleus and can be thought of as the energy that must be added to a nucleus to break it apart into its separated neutrons and protons.

Example 29.2 The Binding Energy of the Deuteron

The nucleus of the deuterium atom, called the deuteron, consists of a proton and a neutron. Calculate the deuteron's binding energy, given that its atomic mass, that is, *the mass of a deuterium nucleus plus an electron,* is measured to be 2.014 102 u.

Solution We know that the proton and neutron masses are

$$m_p = 1.007\ 825\ \text{u} \qquad m_n = 1.008\ 665\ \text{u}$$

Note that the masses used for the proton and deuteron in this example are actually those of the neutral atoms, as are all the atomic masses found in Appendix B. We are able to use atomic masses for these calculations because the electron masses cancel. That is, when m_d containing one electron mass is subtracted from $(m_p + m_n)$ containing one electron mass in the line below, the electron masses cancel. Therefore,

$$m_p + m_n = 2.016\ 490\ \text{u}$$

To calculate the mass difference, we subtract the deuteron mass from this value:

$$\Delta m = (m_p + m_n) - m_d$$
$$= 2.016\ 490\ \text{u} - 2.014\ 102\ \text{u} = 0.002\ 388\ \text{u}$$

Because 1 u corresponds to an equivalent energy of 931.494 MeV (that is, $1\ \text{u} \cdot c^2 = 931.494$ MeV), the mass difference corresponds to the binding energy

$$E_b = (0.002\ 388\ \text{u})(931.494\ \text{MeV/u}) = \boxed{2.224\ \text{MeV}}$$

This result tells us that, to separate a deuteron into a proton and a neutron, it is necessary to add 2.224 MeV of energy to the deuteron to overcome the attractive nuclear force between proton and neutron. One way of supplying the deuteron with this energy is by bombarding it with energetic particles.

If the binding energy of a nucleus were zero, the nucleus would separate into its constituent protons and neutrons without the addition of any energy—that is, it would spontaneously break apart.

FIGURE 29.4 A plot of the binding energy per nucleon versus the mass number A for nuclei that are along the line of stability shown in Figure 29.3.

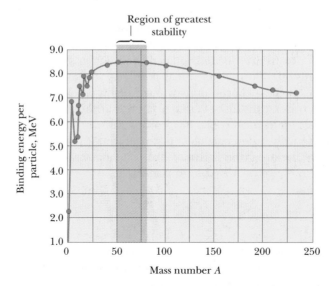

It is interesting to examine a plot of binding energy per nucleon, E_b/A, as a function of mass number for various stable nuclei (Fig. 29.4). Except for the lighter nuclei, the average binding energy per nucleon is about 8 MeV. Note that the curve peaks in the vicinity of $A = 60$. That is, nuclei with mass numbers greater or less than 60 are not as strongly bound as those near the middle of the periodic table. As we shall see later, this fact allows energy to be released in fission and fusion reactions. The curve is slowly varying for $A > 40$, which suggests that the nuclear force saturates. In other words, a particular nucleon can interact with only a limited number of other nucleons, which can be viewed as the "nearest neighbors" in the close-packed structure illustrated in Figure 29.2.

APPLYING PHYSICS 29.1

Figure 29.4 shows a graph of the amount of energy required to remove a nucleon from the nucleus. The figure indicates that an approximately constant amount of energy is necessary to remove a nucleon (above $A = 40$), whereas we saw in Chapter 28 that widely varying amounts of energy are required to remove an electron from the atom. Why does this difference occur?

Explanation In the case of Figure 29.4, the approximately constant value of the nuclear binding energy is a result of the short-range nature of the nuclear force. A given nucleon interacts only with its few nearest neighbors, rather than with all of the nucleons in the nucleus. Thus, no matter how many nucleons are present in the nucleus, pulling any one nucleon out involves separating it only from its nearest neighbors. The energy to do this, therefore, is approximately independent of how many nucleons are present. For clearest comparison with the electron, think of averaging the energies required to strip all of the electrons out of a particular atom, from the outermost valence electron to the innermost K shell electron. This average increases steeply with increasing atomic number. The electrical force binding the electrons to the nucleus in an atom is a long-range force. An electron in the atom interacts with all the protons in the nucleus. As the nuclear charge increases, the attraction between the nucleus and the electrons becomes stronger. Therefore, as the nuclear charge increases, more energy is necessary to remove an average electron.

MARIE CURIE, POLISH SCIENTIST (1867–1934)

In 1903 Marie Curie shared the Nobel prize in physics with her husband, Pierre, and with Becquerel for their studies of radioactive substances. In 1911 she was awarded a second Nobel prize in chemistry for the discovery of radium and polonium. Marie Curie died of leukemia caused by years of exposure to radioactive substances. "I persist in believing that the ideas that then guided us are the only ones which can lead to the true social progress. We cannot hope to build a better world without improving the individual. Toward this end, each of us must work toward his own highest development, accepting at the same time his share of responsibility in the general life of humanity." *(FPG/Getty)*

29.3 RADIOACTIVITY

In 1896 Becquerel accidentally discovered that uranium salt crystals emit an invisible radiation that can darken a photographic plate even if the plate is covered to exclude light. After several such observations under controlled conditions, he concluded that the radiation emitted by the crystals was of a new type, one that required no external stimulation. This spontaneous emission of radiation was soon called **radioactivity.** Subsequent experiments by other scientists showed that other substances were also radioactive.

The most significant investigations of this type were conducted by Marie and Pierre Curie. After several years of careful and laborious chemical separation processes on tons of pitchblende, a radioactive ore, the Curies reported the discovery of two previously unknown elements, both of which were radioactive. These were named polonium and radium. Subsequent experiments, including Rutherford's famous work on alpha-particle scattering, suggested that radioactivity was the result of the decay, or disintegration, of unstable nuclei.

Three types of radiation can be emitted by a radioactive substance: alpha (α) particles, in which the emitted particles are ${}^4_2\text{He}$ nuclei; beta (β) particles, in which the emitted particles are either electrons or positrons; and gamma (γ) rays, in which the emitted "rays" are high-energy photons. A **positron** is a particle similar to the electron in all respects except that it has a charge of $+ e$. (The positron is

The hands and numbers of this luminous watch contain minute amounts of radium salt. The radioactive decay of radium causes the phosphors to glow in the dark. (© *Richard Megna 1990, Fundamental Photographs*)

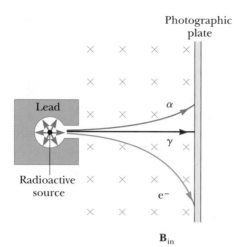

FIGURE 29.5 The radiation from a radioactive source, such as radium, can be separated into three components using a magnetic field to deflect the charged particles. The photographic plate at the right records the events.

said to be the **antiparticle** of the electron.) The symbol e^- is used to designate an electron and e^+ designates a positron.

It is possible to distinguish these three forms of radiation using the scheme shown in Figure 29.5. The radiation from a radioactive sample is directed into a region with a magnetic field, and the beam splits into three components, two bending in opposite directions and the third not changing direction. From this simple observation, it can be concluded that the radiation of the undeflected beam carries no charge (the gamma ray), the component deflected upward contains positively charged particles (alpha particles), and the component deflected downward contains negatively charged particles (e^-). If the beam includes a positron (e^+), it is deflected upward.

The three types of radiation have quite different penetrating powers. Alpha particles barely penetrate a sheet of paper, beta particles can penetrate a few millimeters of aluminum, and gamma rays can penetrate several centimeters of lead.

THE DECAY CONSTANT AND HALF-LIFE

If a radioactive sample contains N radioactive nuclei at some instant, it is found that the number of nuclei, ΔN, that decay in a small time interval Δt is proportional to N:

$$\frac{\Delta N}{\Delta t} \propto N$$

or

$$\Delta N = -\lambda N \Delta t \qquad [29.2]$$

where λ is a constant called the **decay constant.** The negative sign signifies that N decreases with time; that is, ΔN is negative. The value of λ for any isotope determines the rate at which that isotope will decay. **The decay rate, or activity, R, of a sample is defined as the number of decays** per second. From Equation 29.2, we see that the decay rate is

Decay rate ▷

$$R = \left| \frac{\Delta N}{\Delta t} \right| = \lambda N \qquad [29.3]$$

Thus, we see that isotopes with a large λ value decay at a rapid rate and those with a small λ value decay slowly.

A general decay curve for a radioactive sample is shown in Figure 29.6. It can be shown from Equation 29.2 (using calculus) that the number of nuclei present varies with time according to the expression

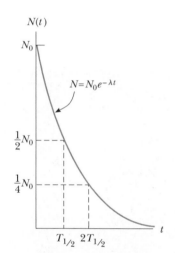

FIGURE 29.6 Plot of the exponential decay law for radioactive nuclei. The vertical axis represents the number of radioactive nuclei present at any time t and the horizontal axis is time. The parameter $T_{1/2}$ is the half-life of the sample.

Exponential decay ▷

$$N = N_0 e^{-\lambda t} \qquad [29.4]$$

where N is the number of radioactive nuclei present at time t, N_0 is the number present at time $t = 0$, and $e = 2.718 \ldots$ is Euler's constant. Processes that obey Equation 29.4 are sometimes said to undergo exponential decay.[1]

Another parameter that is useful for characterizing radioactive decay is the **half-life, $T_{1/2}$. The half-life of a radioactive substance is the time it takes for half of a given number of radioactive nuclei to decay.** Setting $N = N_0/2$ and $t = T_{1/2}$ in Equation 29.4 gives

$$\frac{N_0}{2} = N_0 e^{-\lambda T_{1/2}}$$

Writing this in the form $e^{\lambda T_{1/2}} = 2$ and taking the natural logarithm of both sides, we get

$$T_{1/2} = \frac{\ln 2}{\lambda} = \frac{0.693}{\lambda} \qquad \text{[29.5]}$$

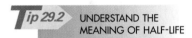 ◀ Half-life

This is a convenient expression relating the half-life to the decay constant. Note that after an elapsed time of one half-life, $N_0/2$ radioactive nuclei remain (by definition); after two half-lives, half of these will have decayed and $N_0/4$ radioactive nuclei will be left; after three half-lives, $N_0/8$ will be left; and so on.

The unit of activity R is the **curie** (Ci), defined as

$$1 \text{ Ci} \equiv 3.7 \times 10^{10} \text{ decays/s} \qquad \text{[29.6]}$$

This unit was selected as the original activity unit because it is the approximate activity of 1 g of radium. The SI unit of activity is the **becquerel** (Bq):

$$1 \text{ Bq} = 1 \text{ decay/s} \qquad \text{[29.7]}$$

Therefore, 1 Ci = 3.7×10^{10} Bq. The most commonly used units of activity are the millicurie (10^{-3} Ci) and the microcurie (10^{-6} Ci).

Tip 29.2 UNDERSTAND THE MEANING OF HALF-LIFE

Be sure that you understand the concept of the half-life of a substance. It is simply the time it takes half of a given number of nuclei to decay. Note that not all of the nuclei have decayed after two half-lives.

Quick Quiz 29.1 What fraction of a radioactive sample has decayed after two half-lives have elapsed? (a) $\frac{1}{4}$ (b) $\frac{1}{2}$ (c) $\frac{3}{4}$ (d) not enough information to say

APPLYING PHYSICS 29.2

The isotope $^{14}_{6}$C is radioactive and has a half-life of 5 730 years. If you start with a sample of 1 000 carbon-14 nuclei, how many will still be around in 17 190 years?

Explanation In 5 730 years, half the sample will have decayed, leaving 500 radioactive $^{14}_{6}$C nuclei. In another 5 730 years (for a total elapsed time of 11 460 years), the number will be reduced to 250 nuclei. After another 5 730 years (total time 17 190 years), 125 nuclei remain.

These numbers represent ideal circumstances. Radioactive decay is an averaging process over a very large number of atoms, and the actual outcome depends on statistics. Our original sample in this example contained only 1 000 nuclei, certainly not a very large number. Thus, if we were actually to count the number remaining after one half-life for this small sample, it probably would not be exactly 500.

---

[1] Other examples of exponential decays were discussed in Chapter 18 in connection with RC circuits, and in Chapter 20 in connection with RL circuits.

Example 29.3 The Activity of Radium

The half-life of the radioactive nucleus $^{226}_{86}\text{Ra}$ is 1.6×10^3 yr. If a sample contains 3.0×10^{16} such nuclei, determine the activity at this time.

Solution First, let us convert the half-life to seconds:

$$T_{1/2} = (1.6 \times 10^3 \text{ yr})(3.16 \times 10^7 \text{ s/yr}) = 5.0 \times 10^{10} \text{ s}$$

Now we can use this value in Equation 29.5 to get the decay constant:

$$\lambda = \frac{0.693}{T_{1/2}} = \frac{0.693}{5.0 \times 10^{10} \text{ s}} = 1.4 \times 10^{-11} \text{ s}^{-1}$$

We can calculate the activity of the sample at $t = 0$ using $R_0 = \lambda N_0$, where R_0 is the decay rate at $t = 0$ and N_0 is the number of radioactive nuclei present at $t = 0$:

$$R_0 = \lambda N_0 = (1.4 \times 10^{-11} \text{ s}^{-1})(3.0 \times 10^{16}) = 4.1 \times 10^5 \text{ decays/s}$$

Because 1 Ci $= 3.7 \times 10^{10}$ decays/s, the activity, or decay rate, at $t = 0$ is

$$R_0 = \boxed{11.1 \ \mu\text{Ci}}$$

Example 29.4 The Activity of Radon Gas

Radon, $^{222}_{86}\text{Rn}$, is a radioactive gas that can be trapped in the basement of homes, and its presence in high concentrations is a known health hazard. Radon has a half-life of 3.83 days. A gas sample contains 4.0×10^8 radon atoms initially.

A How many atoms will remain after 12 days have passed if no more radon leaks in?

Reasoning First note that 12 days corresponds to about 3.1 half-lives. In three half-lives, the number of radon atoms is reduced by a factor $2^3 = 8$. Therefore, the number of remaining radon atoms is approximately $4.0 \times 10^8/8 = 5.0 \times 10^7$.

Solution A more precise answer is obtained by first finding the decay constant from Equation 29.5:

$$\lambda = \frac{0.693}{T_{1/2}} = \frac{0.693}{3.83 \text{ days}} = 0.181 \text{ days}^{-1}$$

We now use Equation 29.4, taking $N_0 = 4.0 \times 10^8$ and the value of λ just found to obtain the number N remaining after 12 days:

$$N = N_0 e^{-\lambda t} = (4.0 \times 10^8 \text{ atoms}) e^{-(0.181 \text{ days}^{-1})(12 \text{ days})} = \boxed{4.6 \times 10^7 \text{ atoms}}$$

This is very close to our original estimate of 5.0×10^7 atoms.

B What is the initial activity of the radon sample?

Solution First, we must express the decay constant in units of s^{-1}.

$$\lambda = \frac{0.693}{(3.83 \text{ days})(8.64 \times 10^4 \text{ s/day})} = 2.09 \times 10^{-6} \text{ s}^{-1}$$

From Equation 29.3 and the above value of λ, we find that the initial activity is

$$R = \lambda N_0 = (2.09 \times 10^{-6} \text{ s}^{-1})(4.0 \times 10^8) = 840 \text{ decays/s} = \boxed{840 \text{ Bq}}$$

EXERCISE Find the activity of the radon sample after 12 days have elapsed.

ANSWER 95 Bq

Quick Quiz 29.2 The activity of a newly discovered radioactive isotope reduces to 96% of its original value in an interval of 2 hours. What is its half-life? (a) 10.2 h (b) 34.0 h (c) 44.0 h (d) 68.6 h

29.4 THE DECAY PROCESSES

As stated in the previous section, radioactive nuclei spontaneously decay via alpha, beta, and gamma decay. Let us discuss these processes in more detail.

ALPHA DECAY

If a nucleus emits an alpha particle ($^4_2$He), it loses two protons and two neutrons. Therefore, the neutron number N of a single nucleus decreases by 2, Z decreases by 2, and A decreases by 4. The decay can be written symbolically as

$$^A_Z X \longrightarrow ^{A-4}_{Z-2} Y + ^4_2 He \qquad [29.8]$$

where X is called the **parent nucleus** and Y the **daughter nucleus.** As examples, ^{238}U and ^{226}Ra are both alpha emitters and decay according to the schemes

$$^{238}_{92} U \longrightarrow ^{234}_{90} Th + ^4_2 He \qquad [29.9]$$

$$^{226}_{88} Ra \longrightarrow ^{222}_{86} Rn + ^4_2 He \qquad [29.10]$$

The half-life for ^{238}U decay is 4.47×10^9 yr, and the half-life for ^{226}Ra decay is 1.60×10^3 yr. In both cases, note that the A of the daughter nucleus is 4 less than that of the parent nucleus. Likewise, Z is reduced by 2.

The decay of ^{226}Ra is shown in Figure 29.7. When one element changes into another, as happens in alpha decay, the process is called **spontaneous decay** or transmutation. As a general rule, (1) the sum of the mass numbers A must be the same on both sides of the equation, and (2) the sum of the atomic numbers Z must be the same on both sides of the equation.

In order for alpha emission to occur, the mass of the parent must be greater than the combined mass of the daughter and the alpha particle. In the decay process, this excess mass is converted into energy of other forms and appears in the form of kinetic energy in the daughter nucleus and the alpha particle. Most of the kinetic energy is carried away by the alpha particle because it is much less massive than the daughter nucleus. This can be understood by first noting that a particle's kinetic energy and momentum, p, are related as follows:

$$KE = \frac{p^2}{2m}$$

Because momentum is conserved, the two particles emitted in the decay of a nucleus at rest must have equal, but oppositely directed, momenta. Thus, the lighter particle has more kinetic energy than the more massive particle.

FIGURE 29.7 Alpha decay of radium. The radium nucleus is initially at rest. After the decay, the radon nucleus has kinetic energy KE_{Rn}, and momentum $\mathbf{p}_{Rn}$, and the alpha particle has kinetic energy KE_α and momentum $\mathbf{p}_\alpha$.

Quick Quiz 29.3 If a nucleus such as ^{226}Ra that is initially at rest undergoes alpha decay, which of the following statements is true? (a) The alpha particle has more kinetic energy than the daughter nucleus. (b) The daughter nucleus has more kinetic energy than the alpha particle. (c) The daughter nucleus and the alpha particle have the same kinetic energy.

APPLYING **PHYSICS** 29.3

In comparing alpha decay energies from a number of radioactive nuclides, it is found that the half-life of the decay goes down as the energy of the decay goes up. Why is this?

Explanation It should seem reasonable that the higher the energy of the alpha particle, the more likely it is to escape the confines of the nucleus. The higher probability of escape translates to a faster rate of decay, which appears as a shorter half-life.

Example 29.5 The Energy Liberated When Radium Decays

We showed that the $^{226}_{88}$Ra nucleus undergoes alpha decay to $^{222}_{86}$Rn (Eq. 29.10). Calculate the amount of energy liberated in this decay. Take the mass of $^{226}_{88}$Ra to be 226.025 402 u, that of $^{222}_{86}$Rn to be 222.017 571 u and that of $^{4}_{2}$He to be 4.002 602 u as found in Appendix B.

Solution After decay, the mass of the daughter m_d plus the mass of the alpha particle m_α is

$$m_d + m_\alpha = 222.017\ 571\ \text{u} + 4.002\ 602\ \text{u} = 226.020\ 173\ \text{u}$$

Thus, calling the mass of the parent nucleus M_p, we find that the mass lost during decay is

$$\Delta m = M_p - (m_d + m_\alpha) = 226.025\ 402\ \text{u} - 226.020\ 173\ \text{u} = 0.005\ 229\ \text{u}$$

Using the relationship that 1 u corresponds to 931.494 MeV, we find that the energy liberated is

$$E = (0.005\ 229\ \text{u})(931.494\ \text{MeV/u}) = \boxed{4.87\ \text{MeV}}$$

ENRICO FERMI, ITALIAN PHYSICIST (1901–1954)

Fermi was awarded the Nobel prize in 1938 for producing the transuranic elements by neutron irradiation and for his discovery of nuclear reactions brought about by slow neutrons. He made many other outstanding contributions to physics, including his theory of beta decay, the free electron theory of metals, and the development of the world's first fission reactor in 1942. Fermi was truly a gifted theoretical and experimental physicist. He was also well known for his ability to present physics in a clear and exciting manner. "Whatever Nature has in store for mankind, unpleasant as it may be, men must accept, for ignorance is never better than knowledge." *(National Accelerator Laboratory)*

BETA DECAY

When a radioactive nucleus undergoes beta decay, the daughter nucleus has the same number of nucleons as the parent nucleus, but the atomic number is changed by 1:

$$^{A}_{Z}\text{X} \longrightarrow\ ^{A}_{Z+1}\text{Y} + \text{e}^{-} \qquad \text{[29.11]}$$

$$^{A}_{Z}\text{X} \longrightarrow\ ^{A}_{Z-1}\text{Y} + \text{e}^{+} \qquad \text{[29.12]}$$

Again, note that the nucleon number and total charge are both conserved in these decays. However, as we shall see shortly, these processes are not described completely by these expressions. A typical beta decay event is

$$^{14}_{6}\text{C} \longrightarrow\ ^{14}_{7}\text{N} + \text{e}^{-} \qquad \text{[29.13]}$$

The emission of electrons from a *nucleus* is surprising because, in all our previous discussions, we stated that the nucleus is composed of protons and neutrons only. This apparent discrepancy can be explained by noting that the emitted electron is created in the nucleus by a process in which a neutron is transformed into a proton. This can be represented by the equation

$$^{1}_{0}\text{n} \longrightarrow\ ^{1}_{1}\text{p} + \text{e}^{-} \qquad \text{[29.14]}$$

Let us consider the energy of the system of Equation 29.13 before and after decay. As with alpha decay, energy must be conserved in beta decay. The following example illustrates how to calculate the amount of energy released in the beta decay of $^{14}_{6}$C.

Example 29.6   The Beta Decay of Carbon-14

Find the energy liberated in the beta decay of $^{14}_{6}\text{C}$ to $^{14}_{7}\text{N}$ as represented by Equation 29.13. Equation 29.13 refers to nuclei, whereas Appendix B shows masses of neutral atoms. Adding six electrons to both sides of Equation 29.13 gives

$$^{14}_{6}\text{C atom} \longrightarrow \ ^{14}_{7}\text{N atom}$$

Solution   We find from Appendix B that $^{14}_{6}\text{C}$ has a mass of 14.003 242 u and $^{14}_{7}\text{N}$ has a mass of 14.003 074 u. Here, the mass difference between the initial and final states is

$$\Delta m = 14.003\ 242\ \text{u} - 14.003\ 074\ \text{u} = 0.000\ 168\ \text{u}$$

This corresponds to an energy release of

$$E = (0.000\ 168\ \text{u})(931.494\ \text{MeV/u}) = \boxed{0.156\ \text{MeV}}$$

From Example 29.6, we see that the energy released in the beta decay of ^{14}C is approximately 0.16 MeV. As with alpha decay, we expect the electron to carry away virtually all of this as kinetic energy because apparently it is the lightest particle produced in the decay. However, as Figure 29.8 shows, only a small number of electrons have this maximum kinetic energy, represented as KE_{max} on the graph; most of the electrons emitted have kinetic energies lower than this predicted value. If the daughter nucleus and the electron are not carrying away this liberated energy, then the energy conservation requirement leads to the question, What accounts for the missing energy? As an additional complication, further analysis of beta decay shows that the principles of conservation of both angular momentum and linear momentum appear to be violated!

In 1930 Pauli proposed that a third particle must be present to carry away the "missing" energy and to conserve momentum. Enrico Fermi later developed a complete theory of beta decay and named this particle the **neutrino** ("little neutral one") because it had to be electrically neutral and have little or no mass. Although it eluded detection for many years, the neutrino (ν) was finally detected experimentally in 1956. The neutrino has the following properties:

- Zero electric charge
- A mass much smaller than that of the electron, but probably not zero (Recent experiments suggest that the neutrino definitely has mass, but the value is uncertain—perhaps less than 1 eV/c^2.)
- A spin of $\frac{1}{2}$
- Very weak interaction with matter, making it quite difficult to detect

Thus, with the introduction of the neutrino, we are now able to represent the beta decay process of Equation 29.13 in its correct form:

$$^{14}_{6}\text{C} \longrightarrow \ ^{14}_{7}\text{N} + \text{e}^{-} + \bar{\nu} \qquad \textbf{[29.15]}$$

where the bar in the symbol $\bar{\nu}$ indicates an **antineutrino**. To explain what an antineutrino is, let us first consider the following decay:

$$^{12}_{7}\text{N} \longrightarrow \ ^{12}_{6}\text{C} + \text{e}^{+} + \nu \qquad \textbf{[29.16]}$$

Here we see that when ^{12}N decays into ^{12}C, a particle is produced that is identical to the electron except that it has a positive charge of $+ e$. This particle is called a **positron.** Because it is like the electron in all respects except charge, the positron is said to be the **antiparticle** to the electron. We shall discuss antiparticles further in Chapter 30. For now, suffice it to say that **in beta decay, an electron and an antineutrino are emitted or a positron and a neutrino are emitted.**

FIGURE 29.8   A typical beta-decay spectrum.

 Properties of the neutrino

Webnote 29.3

Visit a neutrino measurement facility in Japan. Check out *http://neutrino.kek.jp*

 MASS NUMBER OF THE ELECTRON

Another notation that is sometimes used for an electron is $_{-1}^{0}\text{e}$. This notation does not imply that the electron has zero rest energy. The mass of the electron is much smaller than that of the lightest nucleon, so we can approximate it as zero when we study nuclear decays and reactions.

GAMMA DECAY

Very often a nucleus that undergoes radioactive decay is left in an excited energy state. The nucleus can then undergo a second decay to a lower energy state, perhaps to the ground state, by emitting one or more high-energy photons. The process is very similar to the emission of light by an atom. An atom emits radiation to release some extra energy when an electron "jumps" from a state of high energy to a state of lower energy. Likewise, the nucleus uses essentially the same method to release any extra energy it may have following a decay or some other nuclear event. In nuclear de-excitation, the "jumps" that release energy are made by protons or neutrons in the nucleus as they move from a higher energy level to a lower level. The photons emitted in such a de-excitation process are called **gamma rays,** which have very high energy relative to the energy of visible light.

A nucleus may reach an excited state as the result of a violent collision with another particle. However, it is more common for a nucleus to be in an excited state as a result of alpha or beta decay. The following sequence of events represents a typical situation in which gamma decay occurs:

$$\,^{12}_{5}\text{B} \longrightarrow \,^{12}_{6}\text{C*} + \text{e}^- + \overline{\nu} \qquad [29.17]$$

$$\,^{12}_{6}\text{C*} \longrightarrow \,^{12}_{6}\text{C} + \gamma \qquad [29.18]$$

Equation 29.17 represents a beta decay in which ^{12}B decays to $^{12}\text{C*}$ where the asterisk indicates that the carbon nucleus is left in an excited state following the decay. The excited carbon nucleus then decays to the ground state by emitting a gamma ray, as indicated by Equation 29.18. Note that gamma emission does not result in any change in either Z or A.

PRACTICAL USES OF RADIOACTIVITY

Carbon Dating

The beta decay of ^{14}C given by Equation 29.15 is commonly used to date organic samples. Cosmic rays (high-energy particles from outer space) in the upper atmosphere cause nuclear reactions that create ^{14}C from ^{14}N. In fact, the ratio of ^{14}C to ^{12}C (by numbers of nuclei) in the carbon dioxide molecules of our atmosphere has a constant value of about 1.3×10^{-12} as determined by measuring carbon ratios in tree rings. All living organisms have the same ratio of ^{14}C to ^{12}C because they continuously exchange carbon dioxide with their surroundings. When an organism dies, however, it no longer absorbs ^{14}C from the atmosphere, and so the ratio of ^{14}C to ^{12}C decreases as the result of the beta decay of ^{14}C. It is therefore possible to determine the age of a material by measuring its activity per unit mass as a result of the decay of ^{14}C. Using carbon dating, samples of wood, charcoal, bone, and shell have been identified as having lived from 1 000 to 25 000 years ago. This knowledge has helped researchers to reconstruct the history of living organisms—including humans—during this time span.

A particularly interesting example is the dating of the Dead Sea Scrolls. This group of manuscripts was first discovered by a young Bedouin boy in a cave at Qumran near the Dead Sea in 1947. Translation showed them to be religious documents, including most of the books of the Old Testament. Because of their historical and religious significance, scholars wanted to know their age. Carbon dating applied to fragments of the scrolls and to the material in which they were wrapped established that they were about 1 950 years old. The scrolls are now stored at the Israel Museum in Jerusalem.

Smoke Detectors Smoke detectors are frequently used in homes and industry for fire protection. Most of the common ones are the ionization type that use radioactive materials (see Fig. 29.9). A smoke detector consists of an ionization chamber, a sensitive current detector, and an alarm. A weak radioactive source ionizes the

FIGURE 29.9 An ionization-type smoke detector. Smoke entering the chamber reduces the detected current, causing the alarm to sound.

Webnote 29.4

Learn a lot more about ^{14}C dating by exploring
http://abcnews.go.com/sections/science/DailyNews/carbon0220.html

> **APPLICATION**

CARBON DATING OF THE DEAD SEA SCROLLS

> **APPLICATION**

SMOKE DETECTORS

air in the chamber of the detector, which creates charged particles. A voltage is maintained between the plates inside the chamber, setting up a small but detectable current in the external circuit. As long as the current is maintained, the alarm is deactivated. However, if smoke drifts into the chamber, the ions become attached to the smoke particles. These heavier particles do not drift as readily as do the lighter ions, which causes a decrease in the detector current. The external circuit senses this decrease in current and sets off the alarm.

Radon Detecting Radioactivity can also affect our daily lives in harmful ways. Soon after the discovery of radium by the Curies, it was found that the air in contact with radium compounds becomes radioactive. It was shown that this radioactivity came from the radium itself, and the product was therefore called "radium emanation." Rutherford and Soddy succeeded in condensing this "emanation," confirming that it is a real substance—the inert, gaseous element now called **radon,** Rn. We now know that the air in uranium mines is radioactive because of the presence of radon gas. The mines must therefore be well ventilated to help protect miners. The fear of radon pollution has now moved from uranium mines into our own homes (see Example 29.4). Because certain types of rock, soil, brick, and concrete contain small quantities of radium, some of the resulting radon gas finds its way into our homes and other buildings. The most serious problems arise from leakage of radon from the ground into the structure. One practical remedy is to exhaust the air through a pipe just above the underlying soil or gravel directly to the outdoors by means of a small fan or blower.

> **◀ APPLICATION**
>
> RADON POLLUTION

> **Webnote 29.5**
>
> Read more on the risks associated with radon gas. Go to *http://www.howstuffworks.com/radon.htm*

APPLYING **PHYSICS** 29.4

In 1991 a German tourist discovered the well-preserved remains of the Iceman trapped in a glacier in the Italian Alps (see Fig. 29.10). Radioactive dating of a sample of Iceman revealed an age of 5 300 years. Why did scientists date the sample using the isotope ^{14}C, rather than ^{11}C, a beta emitter with a half-life of 20.4 min?

Explanation Carbon-14 has a long half-life of 5 730 years, so the fraction of ^{14}C nuclei remaining after one half-life is high enough to measure accurate changes in the sample's activity. The ^{11}C isotope, which has a very short half-life, is not useful because its activity decreases to a vanishingly small value over the age of the sample, making it impossible to detect.

> **◀ APPLICATION**
>
> RADIOACTIVE DATING OF ICEMAN

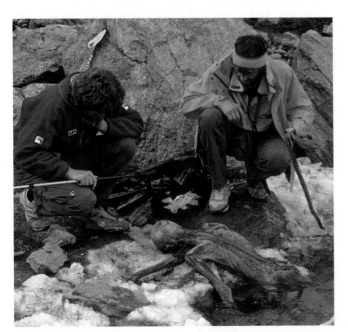

FIGURE 29.10 (Applying Physics 29.4) The body of an ancient man (dubbed the Iceman) was exposed by a melting glacier in the Alps. *(Hanny Paul/Gamma Liaison)*

If a sample to be dated is not very old, say about 50 years, then you should select the isotope of some other element whose half-life is comparable with the age of the sample. For example, if the sample contained hydrogen, you could measure the activity of ^{3}H (tritium), a beta emitter of half-life 12.3 years. As a general rule, the expected age of the sample should be long enough to measure a change in activity, but not so long that its activity cannot be detected.

APPLYING **PHYSICS** 29.5

A wooden coffin is found that contains a skeleton holding a gold statue. Which of the three objects (coffin, skeleton, and gold statue) can be carbon dated to find out how old it is?

Explanation   Only the coffin and the skeleton can be carbon dated. These two items exchanged air with the environment during their lifetimes, resulting in a fixed ratio of ^{14}C to ^{12}C. Once the human and the tree died, the amount of ^{14}C began to decrease due to radioactive decay. The gold in the statue was never alive and did not have an uptake of carbon; therefore, there is no ^{14}C to detect.

Example 29.7   Should We Report This to Homicide?

A 50.0-g sample of carbon is taken from the pelvis bone of a skeleton and is found to have a ^{14}C decay rate of 200.0 decays/min. It is known that carbon from a living organism has a decay rate of 15.0 decays/min · g and that ^{14}C has a half-life of 5 730 yr = 3.01×10^9 min. Find the age of the skeleton.

Solution   Let us start with Equation 29.4

$$N = N_0 e^{-\lambda t}$$

and multiply both sides by λ to get

$$\lambda N = \lambda N_0 e^{-\lambda t}$$

But from Equation 29.3 we see that this is equivalent to

$$R = R_0 e^{-\lambda t}$$

or

$$\frac{R}{R_0} = e^{-\lambda t}$$

where R is the present activity and R_0 was the activity when the skeleton was a part of a living organism. We can solve for the time by taking the natural log of both sides of this equation.

$$\ln\left(\frac{R}{R_0}\right) = \ln(e^{-\lambda t}) = -\lambda t$$

$$t = -\frac{\ln\left(\frac{R}{R_0}\right)}{\lambda}$$

Because we are given the decay rate and mass of the sample, we can find R_0 as

$$R_0 = \left(15.0 \, \frac{\text{decays}}{\text{min} \cdot \text{g}}\right)(50.0 \text{ g}) = 750 \, \frac{\text{decays}}{\text{min}}$$

The decay constant is found from Equation 29.5 as

$$\lambda = \frac{0.693}{T_{1/2}} = \frac{0.693}{3.01 \times 10^9 \text{ min}} = 2.30 \times 10^{-10} \text{ min}^{-1}$$

Thus, we make the following substitutions:

$$t = -\frac{\ln\left(\dfrac{R}{R_0}\right)}{\lambda} = -\frac{\ln\left(\dfrac{200.0 \text{ decays/min}}{750 \text{ decays/min}}\right)}{2.30 \times 10^{-10} \text{ min}^{-1}} = \frac{1.32}{2.30 \times 10^{-1}\text{min}^{-1}}$$

$$= 5.74 \times 10^9 \text{ min} = \boxed{10\ 900 \text{ yr}}$$

CARBON-14 AND THE SHROUD OF TURIN

Since the Middle Ages, many people have marveled at a 14-foot-long, yellowing piece of linen found in Turin, Italy, purported to be the burial shroud of Jesus Christ (Fig. 29.11). The cloth bears a remarkable, full-size likeness of a crucified body, with wounds on the head that could have been caused by a crown of thorns, and another in the side that could have been the cause of death. Skepticism over the authenticity of the shroud has existed since its first public showing in 1354; in fact, a French bishop declared it to be a fraud at the time. Because of its controversial nature, religious bodies have taken a neutral stance on its authenticity.

In 1978 the bishop of Turin allowed the cloth to be subjected to scientific analysis, but notably missing from these tests was a ^{14}C dating. The reason for this omission was that, at the time, carbon-dating techniques required a piece of cloth about the size of a handkerchief. By 1988 the process had been refined to the point that pieces as small as 1 in^2 were sufficient, and at that time permission was granted to allow the dating to proceed. Three labs were selected for the testing, and each was given four pieces of material. One of these was a piece of the shroud, and the other three pieces were control pieces similar in appearance to the shroud.

The testing procedure consisted of burning the cloth to produce carbon dioxide, which was then converted chemically to graphite. The graphite sample was subjected to ^{14}C analysis, and in the end all three labs agreed amazingly well on the age of the shroud. The average of their results gave a date for the cloth of A.D. 1320 ± 60 years, with an assurance that the cloth could not be older than A.D. 1200. Carbon-14 dating has thus unraveled the most important mystery concerning the shroud, but others remain. For example, investigators have not yet been able to explain how the image was imprinted.

APPLICATION

^{14}C DATING OF THE SHROUD OF TURIN

FIGURE 29.11 The Shroud of Turin as it appears in a photographic negative image. *(Santi Visali/The IMAGE Bank)*

29.5 NATURAL RADIOACTIVITY

Radioactive nuclei are generally classified into two groups: (1) unstable nuclei found in nature, which give rise to what is called **natural radioactivity,** and (2) nuclei produced in the laboratory through nuclear reactions, which exhibit **artificial radioactivity.**

Three series of naturally occurring radioactive nuclei exist (Table 29.2). Each series starts with a specific long-lived radioactive isotope whose half-life exceeds

TABLE 29.2	The Four Radioactive Series		
Series	**Starting Isotope**	**Half-life (yr)**	**Stable End Product**
Uranium	$^{238}_{92}\text{U}$	4.47×10^9	$^{206}_{82}\text{Pb}$
Actinium	$^{235}_{92}\text{U}$	7.04×10^8	$^{207}_{82}\text{Pb}$
Thorium	$^{232}_{90}\text{Th}$	1.41×10^{10}	$^{208}_{82}\text{Pb}$
Neptunium	$^{237}_{93}\text{Np}$	2.14×10^6	$^{209}_{83}\text{Bi}$

FIGURE 29.12 Decay series beginning with ^{232}Th.

that of any of its descendants. The fourth series in Table 29.2 begins with ^{237}Np, a transuranic element (one having an atomic number greater than that of uranium) not found in nature. This element has a half-life of "only" 2.14×10^6 yr.

The two uranium series are somewhat more complex than the ^{232}Th series (Fig. 29.12). Also, several other naturally occurring radioactive isotopes, such as ^{14}C and ^{40}K, are not part of either decay series.

Natural radioactivity constantly supplies our environment with radioactive elements that would otherwise have disappeared long ago. For example, because the Solar System is about 5×10^9 yr old, the supply of ^{226}Ra (whose half-life is only 1 600 yr) would have been depleted by radioactive decay long ago if it were not for the decay series that starts with ^{238}U, with a half-life of 4.47×10^9 yr.

29.6 NUCLEAR REACTIONS

It is possible to change the structure of nuclei by bombarding them with energetic particles. Such changes are called **nuclear reactions.** Rutherford was the first to observe nuclear reactions, using naturally occurring radioactive sources for the bombarding particles. He found that protons were released when alpha particles were allowed to collide with nitrogen atoms. The process can be represented symbolically as

$$\text{$_2^4$He} + \text{$_7^{14}$N} \longrightarrow X + \text{$_1^1$H} \qquad \text{[29.19]}$$

This equation says that an alpha particle ($_2^4$He) strikes a nitrogen nucleus and produces an unknown product nucleus (X) and a proton ($_1^1$H). Balancing atomic numbers and mass numbers, as we did for radioactive decay, enables us to conclude that the unknown is characterized as $_8^{17}$X. Because the element with atomic number 8 is oxygen, we see that the reaction is

$$\text{$_2^4$He} + \text{$_7^{14}$N} \longrightarrow \text{$_8^{17}$O} + \text{$_1^1$H} \qquad \text{[29.20]}$$

This nuclear reaction starts with two stable isotopes, helium and nitrogen, and produces two different stable isotopes, hydrogen and oxygen.

Since the time of Rutherford, thousands of nuclear reactions have been observed, particularly following the development of charged-particle accelerators in the 1930s. With today's advanced technology in particle accelerators and particle accelerators, it is possible to achieve particle energies of at least 1 000 GeV = 1 TeV. These high-energy particles are used to create new particles whose properties are helping us solve the mysteries of the nucleus.

Quick Quiz 29.4

Which of the following are possible reactions?
(a) $_0^1$n + $_{92}^{235}$U $\rightarrow$ $_{54}^{140}$Xe + $_{38}^{94}$Sr + 2 ($_0^1$n)
(b) $_0^1$n + $_{92}^{235}$U $\rightarrow$ $_{50}^{132}$Sn + $_{42}^{101}$Mo + 3 ($_0^1$n)
(c) $_0^1$n + $_{94}^{239}$Pu $\rightarrow$ $_{53}^{127}$I + $_{41}^{93}$Nb + 3 ($_0^1$n)

Example 29.8 The Discovery of the Neutron

A nuclear reaction of significant note occurred in 1932 when Chadwick, in England, bombarded a beryllium target with alpha particles. Analysis of the experiment indicated that the following reaction occurred:

$$\text{$_2^4$He} + \text{$_4^9$Be} \longrightarrow \text{$_6^{12}$C} + X$$

What is X in this reaction?

Solution Balancing mass numbers and atomic numbers, we see that the unknown particle must be represented as ^1_0X, that is, a particle having a mass of 1 and zero charge.

Hence, the particle X is the neutron, ^1_0n. This experiment was the first to provide positive proof of the existence of neutrons.

Example 29.9 Synthetic Elements

A A beam of neutrons is directed at a target of $^{238}_{92}U$. The reaction products are a gamma ray and another isotope. What is the isotope?

Solution
Balancing input with output gives

$$^1_0n + {}^{238}_{92}U \longrightarrow {}^{239}_{92}U + \gamma$$

B Isotope ^{239}U is radioactive and undergoes beta decay. Write the equation symbolizing this decay and identify the resulting isotope.

Solution The decay of ^{239}U by beta emission is

$$^{239}_{92}U \longrightarrow {}^{239}_{93}Np + e^- + \bar{\nu}$$

C The isotope ^{239}Np is also radioactive and decays by beta emission. What is the end product?

Solution The decay of ^{239}Np by beta emission gives

$$^{239}_{93}Np \longrightarrow {}^{239}_{94}Pu + e^- + \bar{\nu}$$

D What is the significance of these reactions?

Solution The interesting feature of these reactions is the fact that uranium is the element with the greatest number of protons, 92, that exists in nature in any appreciable amount. The reactions in parts A, B, and C do occur occasionally in nature; hence, minute traces of neptunium and plutonium are present. In 1940, however, researchers bombarded uranium with neutrons to produce plutonium and neptunium by the steps given previously. These two elements were thus the first elements made in the laboratory, and by bombarding them with neutrons and other particles, the list of synthetic elements has been extended to include those up to atomic number 112.

Q VALUES

We have just examined some nuclear reactions for which mass numbers and atomic numbers must be balanced in the equations. We shall now consider the energy involved in these reactions, because energy is another important quantity that must be conserved.

Let us illustrate this procedure by analyzing the following nuclear reaction:

$$^2_1H + {}^{14}_7N \longrightarrow {}^{12}_6C + {}^4_2He \qquad \text{[29.21]}$$

The total mass on the left side of the equation is the sum of the mass of ^2_1H (2.014 102 u) and the mass of $^{14}_7N$ (14.003 074 u), which equals 16.017 176 u. Similarly, the mass on the right side of the equation is the sum of the mass of $^{12}_6C$ (12.000 000 u) plus the mass of ^4_2He (4.002 602 u), for a total of 16.002 602 u. Thus, the total mass before the reaction is greater than the total mass after the reaction. The mass difference in this reaction is equal to 16.017 176 u − 16.002 602 u = 0.014 574 u. This "lost" mass is converted to the kinetic energy of the nuclei present after the reaction. In energy units, 0.014 574 u is equivalent to 13.576 MeV of kinetic energy carried away by the carbon and helium nuclei.

The energy required to balance the equation is called the *Q* value of the reaction. In Equation 29.21 the *Q* value is 13.576 MeV. Nuclear reactions in which there is a release of energy—that is, positive *Q* values—are said to be **exothermic reactions.**

The energy balance sheet is not complete, however. We must also consider the kinetic energy of the incident particle before the collision. As an example, let us assume that the deuteron in Equation 29.21 has a kinetic energy of 5 MeV. Adding this to our *Q* value, we find that the carbon and helium nuclei have a total kinetic energy of 18.576 MeV following the reaction.

Now consider the reaction

$$\,^4_2\mathrm{He} + \,^{17}_7\mathrm{N} \longrightarrow \,^{17}_8\mathrm{O} + \,^1_1\mathrm{H} \qquad \text{[29.22]}$$

Before the reaction, the total mass is the sum of the masses of the alpha particle and the nitrogen nucleus: 4.002 602 u + 14.003 074 u = 18.005 676 u. After the reaction, the total mass is the sum of the masses of the oxygen nucleus and the proton: 16.999 133 u + 1.007 825 u = 18.006 958 u. In this case, the total mass after the reaction is *greater* than the total mass before the reaction. The mass deficit is 0.001 282 u, equivalent to an energy deficit of 1.194 MeV. This deficit is expressed by the negative *Q* value of the reaction, − 1.194 MeV. Reactions with negative *Q* values are called **endothermic reactions.** Such reactions will not take place unless the incoming particle has at least enough kinetic energy to overcome the energy deficit.

At first it might appear that the reaction in Equation 29.22 could take place if the incoming alpha particle had a kinetic energy of 1.194 MeV. In practice, however, the alpha particle must have more energy than this. If it had an energy of only 1.194 MeV, energy would be conserved but careful analysis would show that momentum was not. This can easily be understood by recognizing that the incoming alpha particle has some momentum before the reaction. However, if its kinetic energy were only 1.194 MeV, the products (oxygen and a proton) would be created with zero kinetic energy and, thus, zero momentum. It can be shown that, in order to conserve both energy and momentum, the incoming particle must have a minimum kinetic energy given by

$$KE_{\min} = \left(1 + \frac{m}{M}\right)|Q| \qquad \text{[29.23]}$$

where *m* is the mass of the incident particle, *M* is the mass of the target, and the absolute value of the *Q* value is used. For the reaction given by Equation 29.22, we find

$$KE_{\min} = \left(1 + \frac{4.002\ 602}{14.003\ 074}\right)|-1.194\ \mathrm{MeV}| = 1.535\ \mathrm{MeV}$$

This minimum value of the kinetic energy of the incoming particle is called the **threshold energy.** The nuclear reaction shown in Equation 29.22 will not occur if the incoming alpha particle has a kinetic energy of less than 1.535 MeV, but can occur if its kinetic energy is equal to or greater than 1.535 MeV.

Webnote 29.6

Take a step-by-step tour of an atom smasher. Start by going to *http://www.howstuffworks.com/ atom-smasher.htm*

Quick Quiz 29.5   If the *Q* value of an endothermic reaction is − 2.17 MeV, the minimum kinetic energy needed in the reactant nuclei if the reaction is to occur must be (a) equal to 2.17 MeV, (b) greater than 2.17 MeV, (c) less than 2.17 MeV, or (d) precisely half of 2.17 MeV.

29.7 MEDICAL APPLICATIONS OF RADIATION

RADIATION DAMAGE IN MATTER

Radiation absorbed by matter can cause severe damage. The degree and type of damage depend on several factors, including the type and energy of the radiation and the properties of the absorbing material. Radiation damage in biological organisms is primarily due to ionization effects in cells. The normal function of a cell may be disrupted when highly reactive ions or radicals are formed as the result of ionizing radiation. For example, hydrogen and hydroxyl radicals produced from water molecules can induce chemical reactions that may break bonds in proteins and other vital molecules. Large acute doses of radiation are especially dangerous because damage to a great number of molecules in a cell can cause death of the cell. Also, cells that do survive the radiation may become defective, which can lead to cancer.

In biological systems, it is common to separate radiation damage into two categories: somatic damage and genetic damage. **Somatic damage** is radiation damage to any cells except the reproductive cells. Such damage can lead to cancer at high radiation levels or seriously alter the characteristics of specific organisms. **Genetic damage** affects only reproductive cells. Damage to the genes in reproductive cells can lead to defective offspring. Clearly, we must be concerned about the effect of diagnostic treatments, such as x-rays and other forms of radiation exposure.

Several units are used to quantify radiation exposure and dose. The **roentgen** (R) is defined as **that amount of ionizing radiation that will produce 2.08×10^9 ion pairs in 1 cm^3 of air under standard conditions.** Equivalently, the roentgen is **that amount of radiation that deposits 8.76×10^{-3} J of energy into 1 kg of air.**

For most applications, the roentgen has been replaced by the **rad** (which is an acronym for *r*adiation *a*bsorbed *d*ose), defined as follows: **1 rad is that amount of radiation that deposits 10^{-2} J of energy into 1 kg of absorbing material.**

Although the rad is a perfectly good physical unit, it is not the best unit for measuring the degree of biological damage produced by radiation. This is because the degree of biological damage depends not only on the dose but also on the type of radiation. For example, a given dose of alpha particles causes about 10 times more biological damage than an equal dose of x-rays. The **RBE** (*r*elative *bio*logical *e*ffectiveness) factor is defined as **the number of rad of x-radiation or gamma radiation that produces the same biological damage as 1 rad of the radiation being used.** The RBE factors for different types of radiation are given in Table 29.3. Note that the values are only approximate because they vary with particle energy and form of damage.

Finally, the **rem** (*r*oentgen *e*quivalent in *m*an) is defined as the product of the dose in rad and the RBE factor:

$$\text{Dose in rem} = \text{dose in rad} \times \text{RBE}$$

TABLE 29.3	RBE Factors for Several Types of Radiation	
Radiation	**RBE Factor**
x-Rays and gamma rays	1.0
Beta particles	1.0–1.7
Alpha particles	10–20
Slow neutrons	4–5
Fast neutrons and protons	10
Heavy ions	20

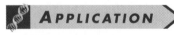

According to this definition, 1 rem of any two radiations produces the same amount of biological damage. From Table 29.3, we see that a dose of 1 rad of fast neutrons represents an effective dose of 10 rem and that 1 rad of x-radiation is equivalent to a dose of 1 rem.

Low-level radiation from natural sources, such as cosmic rays and radioactive rocks and soil, delivers to each of us a dose of about 0.13 rem/yr. The upper limit of radiation dose recommended by the U.S. government (apart from background radiation and exposure related to medical procedures) is 0.5 rem/yr. Many occupations involve higher levels of radiation exposure, and for individuals in these occupations an upper limit of 5 rem/yr has been set for whole-body exposure. Higher upper limits are permissible for certain parts of the body, such as the hands and forearms. An acute whole-body dose of 400–500 rem results in a mortality rate of about 50%. The most dangerous form of exposure is ingestion or inhalation of radioactive isotopes, especially those elements the body retains and concentrates, such as ^{90}Sr. In some cases, a dose of 1 000 rem can result from ingesting 1 mCi of radioactive material.

Sterilizing objects by exposing them to radiation has been going on for at least 25 years, but in recent years, the methods used have become safer to use and more economical. Most bacteria, worms, and insects are easily destroyed by exposure to gamma radiation from radioactive cobalt. There is no intake of radioactive nuclei by an organism in such sterilizing processes, as there is in the use of radioactive tracers. The process is very effective in destroying Trichinella worms in pork, salmonella bacteria in chickens, insect eggs in wheat, and surface bacteria on fruit and vegetables that can lead to rapid spoilage. Recently, this procedure has been expanded to include the sterilization of medical equipment while in its protective covering. Surgical gloves, sponges, sutures, and so forth are irradiated while packaged. Also, bone, cartilage, and skin used for grafting is often irradiated to reduce the chance for infection.

TRACING

Radioactive particles can be used to trace chemicals participating in various reactions. One of the most valuable uses of radioactive tracers is in medicine. For example, ^{131}I is an artificially produced isotope of iodine (the natural, nonradioactive isotope is ^{127}I). Iodine, a necessary nutrient for our bodies, is obtained largely through the intake of seafood and iodized salt. The thyroid gland plays a major role in the distribution of iodine throughout the body. In order to evaluate the performance of the thyroid, the patient drinks a small amount of radioactive sodium iodide. Two hours later, the amount of iodine in the thyroid gland is determined by measuring the radiation intensity at the neck area.

A medical application of the use of radioactive tracers occurring in emergency situations is that of locating a hemorrhage inside the body. Often the location of the site cannot easily be determined, but radioactive chromium can identify the location with a high degree of precision. Chromium is taken up by red blood cells and carried uniformly throughout the body. However, the blood will be dumped at a hemorrhage site, and the radioactivity of that region will increase markedly.

The tracer technique is also useful in agricultural research. Suppose the best method of fertilizing a plant is to be determined. A certain material in the fertilizer, such as nitrogen, can be tagged with one of its radioactive isotopes. The fertilizer is then sprayed on one group of plants, sprinkled on the ground for a second group, and raked into the soil for a third. A Geiger counter is then used to track the nitrogen through the three types of plants.

Tracing techniques are as wide-ranging as human ingenuity can devise. Present applications range from checking the absorption of fluorine by teeth to checking contamination of food-processing equipment by cleansers to monitoring deterioration inside an automobile engine. In the latter case, a radioactive material is used in the manufacture of the pistons, and the oil is checked for radioactivity to determine the amount of wear on the pistons.

COMPUTED AXIAL TOMOGRAPHY (CAT SCANS)

The normal x-ray of a human body has two primary disadvantages when used as a source of clinical diagnosis. First, it is difficult to distinguish between various types of tissue in the body because they all have similar x-ray absorption properties. Second, a conventional x-ray absorption picture is indicative of the average amount of absorption along a particular direction in the body, leading to somewhat obscured pictures. To overcome these problems, a device called a CAT scanner was developed in England in 1973; it is capable of producing pictures of much greater clarity and detail than were previously obtainable.

The operation of a CAT scanner can be understood by considering the following hypothetical experiment. Suppose a box consists of four compartments, labeled A, B, C, and D, as in Figure 29.13a. Each compartment has a different amount of absorbing material from any other compartment. What set of experimental procedures will enable us to determine the relative amounts of material in each compartment? The following steps outline one method that will provide this information. First, a beam of x-rays is passed through compartments A and C, as Figure 29.13b. The intensity of the exiting radiation is reduced by absorption by some number that we assign as 8. (The number 8 could mean, for example, that the intensity of the exiting beam is reduced by eight tenths of 1% from its initial value.) Because we do not know which of the compartments, A or C, was responsible for this reduction in intensity, half the loss is assigned to each compartment, as in Figure 29.13c. Next, a beam of x-rays is passed through compartments B and D, as in Figure 29.13b. The reduction in intensity for this beam is 10, and again we assign half the loss to each compartment. We now redirect the x-ray source so that it sends one beam through compartments A and B and another through compartments C and D, as in Figure 29.13d, and again measure the absorption. Suppose the absorption through compartments A and B in this experiment is measured to be 7 units. On the basis of our first experiment, we would have guessed it would be 9 units, 4 by compartment A, and 5 by compartment B. Thus, we have reduced the guessed absorption for each compartment by 1 unit so that the sum is 7 rather than 9, to give the numbers shown in Figure 29.13e. Likewise, when the beam is passed through compartments C and D, as in Figure 29.13d, we may find the total absorption to be 11 as compared to our first experiment of 9. In this case, we add

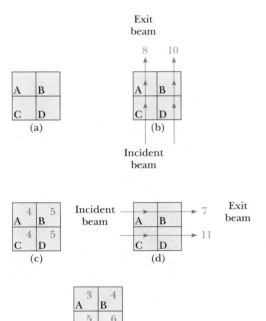

FIGURE 29.13 An experimental procedure for determining the relative amounts of x-ray absorption by four different compartments in a box.

1 unit of absorption to each compartment to give a sum of 11 as in Figure 29.13e. This somewhat crude procedure could be improved by measuring the absorption along other paths. However, these simple measurements are sufficient to enable us to conclude that compartment D contains the most absorbing material and A the least. A visual representation of these results can be obtained by assigning to each compartment a shade of gray corresponding to the particular number associated with the absorption. In our example, compartment D would be very dark and compartment A would be very light.

The steps outlined above are representative of how a CAT scanner produces images of the human body. A thin slice of the body is subdivided into perhaps 10 000 compartments, rather than 4 compartments as in our simple example. The function of the CAT scanner is to determine the relative absorption in each of these 10 000 compartments and to display a picture of its calculations in various shades of gray. Note that CAT stands for **computed axial tomography.** The term *axial* is used because the slice of the body to be analyzed corresponds to a plane perpendicular to the head-to-toe axis. *Tomos* is the Greek word for slice and *graph* is the Greek word for picture. In a typical diagnosis, the patient is placed in the position shown in Figure 29.14 and a narrow beam of x-rays is sent through the plane of interest. The emerging x-rays are detected and measured by photomultiplier tubes behind the patient. The x-ray tube is then rotated a few degrees, and the intensity is recorded again. An extensive amount of information is obtained by rotating the beam through 180° at intervals of about 1° per measurement, resulting in a set of numbers assigned to each of the 10 000 "compartments" in the slice. These numbers are then converted by the computer to a photograph in various shades of gray for this segment of the body.

A brain scan of a patient can now be made in about 2 s, and a full-body scan requires about 6 s. The final result is a picture containing much greater quantitative information and clarity than a conventional x-ray photograph. Because CAT scanners use x-rays, which are an ionizing form of radiation, the technique presents a modest health risk to the patient being diagnosed.

MAGNETIC RESONANCE IMAGING (MRI)

At the heart of magnetic resonance imaging (MRI) is the fact that when a nucleus having a magnetic moment is placed in an external magnetic field, its moment precesses about the magnetic field with a frequency that is proportional to the field. For example, a proton, whose spin is $\frac{1}{2}$, can occupy one of two energy states when placed in an external magnetic field. The lower-energy state corresponds to

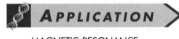

APPLICATION

CAT SCANS

APPLICATION

MAGNETIC RESONANCE
IMAGING (MRI)

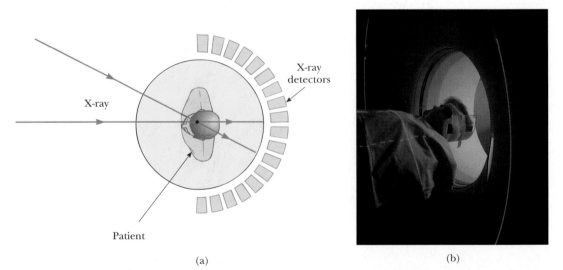

(a)   (b)

FIGURE 29.14   (a) CAT scanner detector assembly. (b) Photograph of a patient undergoing a CAT scan in a hospital. (*Jay Freis/The Image Bank*)

(a)   (b)

FIGURE 29.15 Computer-enhanced MRI images of (a) a normal human brain with the pituitary gland highlighted and (b) a human brain with a glioma tumor. *(Scott Camazine/Science Source/Photo Researchers, Inc.)*

the case in which the spin is aligned with the field, where as the higher-energy state corresponds to the case in which the spin is opposite the field. Transitions between these two states can be observed using a technique known as **nuclear magnetic resonance.** A DC magnetic field is applied to align the magnetic moments, and a second, weak oscillating magnetic field is applied perpendicular to the DC field. When the frequency of the oscillating field is adjusted to match the precessional frequency of the magnetic moments, the nuclei "flip" between the two spin states. These transitions result in a net absorption of energy by the spin system, which can be detected electronically.

In MRI, image reconstruction is obtained using spatially varying magnetic fields and a procedure for encoding each point in the sample being imaged. Some MRI images taken on a human head are shown in Figure 29.15. In practice, a computer-controlled pulse sequencing technique is used to produce signals that are captured by a suitable processing device. This signal is then subjected to appropriate mathematical manipulations to provide data for the final image. The main advantage of MRI over other imaging techniques in medical diagnostics is that it causes minimal damage to cellular structures. Photons associated with the radio frequency signals used in MRI have energies of only about 10^{-7} eV. Because molecular bond strengths are much larger (of the order of 1 eV), the rf photons cause little cellular damage. In comparison, x-rays or gamma-rays have energies ranging from 10^4 to 10^6 eV and can cause considerable cellular damage.

Webnote 29.7

Learn about other medical applications of nuclear physics by going to *http://www.howstuffworks.com/nuclear-medicine.htm*

29.8 RADIATION DETECTORS

Although most medical applications of radiation require instruments to make quantitative measurements of radioactive intensity, we have not yet explained how such instruments operate. Various devices have been developed to detect the energetic particles emitted when a radioactive nucleus decays. The **Geiger counter** (Fig. 29.16) is perhaps the most common form of device used to detect radioactivity. It can be considered the prototype of all counters that use the ionization of a medium as the basic detection process. It consists of a thin wire electrode aligned along the central axis of a cylindrical metallic tube filled with a gas at low pressure. The wire is maintained at a high positive voltage of about 1 000 V relative to the tube. When an energetic charged particle or gamma-ray photon enters the tube through a thin window at one end, some of the gas atoms are ionized. The electrons removed from these atoms are attracted toward the wire electrode, and in the process they ionize other atoms in their path. This sequential ionization results in an *avalanche* of electrons that produces a current pulse. After the pulse has been amplified, it can either be used to trigger an electronic counter or delivered

FIGURE 29.16 (a) Diagram of a Geiger counter. The voltage between the wire electrode and the metallic tube is usually about 1 000 V. (b) A Geiger counter. *(David Rogers)*

to a loudspeaker that clicks each time a particle is detected. Although a Geiger counter reliably detects the presence and quantity of radiation, it cannot be used to measure the energy of the detected radiation.

A **semiconductor diode detector** is essentially a reverse-biased *p-n* junction. As an energetic particle passes through the junction, it produces electron-hole pairs that are separated by the internal electric field. This movement of electrons and holes creates a brief pulse of current that is measured with an electronic counter. In a typical device, the duration of the pulse is 10^{-8} s.

A **scintillation counter** typically uses a solid or liquid material whose atoms are easily excited by radiation. The excited atoms then emit visible-light photons when they return to their ground state. Common materials used as scintillators are transparent crystals of sodium iodide and certain plastics. If the scintillator material is attached to one end of a device called a **photomultiplier** (PM) tube as shown in Figure 29.17, the photons emitted by the scintillator can be converted to an electrical signal. The PM tube consists of numerous electrodes, called *dynodes,* whose electric potentials increase in succession along the length of the tube. Between the top of the tube and the scintillator material is a plate called a photocathode. When photons leaving the scintillator hit this plate, electrons are emitted because of the photoelectric effect. As one of these emitted electrons strikes the first

FIGURE 29.17 (a) Diagram of a scintillation counter connected to a photomultiplier tube. (b) The sodium iodide in these scintillation crystals flashes when an energetic particle passes through it, something like the way the atmosphere flashes when a meteor passes through it.

dynode, the electron has sufficient kinetic energy to eject several other electrons from the dynode surface. When these electrons are accelerated to the second dynode, many more electrons are ejected, and thus a multiplication process occurs. The end result is 1 million or more electrons striking the last dynode. Hence, one particle striking the scintillator produces a sizable electrical pulse at the PM output, and this pulse is sent to an electronic counter.

Both the scintillator and the semiconductor diode detector are much more sensitive than a Geiger counter, mainly because of the higher mass density of the detecting medium. Both can also be used to measure particle energy from the height of the pulses produced.

Track detectors are various devices used to view the tracks or paths of charged particles directly. High-energy particles produced in particle accelerators may have energies ranging from 10^9 to 10^{12} eV. Thus, they cannot be stopped and cannot have their energy measured with the small detectors already mentioned. Instead, the energy and momentum of these energetic particles are found from the curvature of their path in a magnetic field of known magnitude and direction.

A **photographic emulsion** is the simplest example of a track detector. A charged particle ionizes the atoms in an emulsion layer. The path of the particle corresponds to a family of points at which chemical changes have occurred in the emulsion. When the emulsion is developed, the particle's track becomes visible.

A **cloud chamber** contains a gas that has been supercooled to just below its usual condensation point. An energetic charged particle passing through ionizes the gas along its path. The ions serve as centers for condensation of the supercooled gas. The track can be seen with the naked eye and can be photographed. A magnetic field can be applied to determine the charges of the radioactive particles, as well as their momentum and energy.

A device called a **bubble chamber,** invented in 1952 by D. Glaser, uses a liquid (usually liquid hydrogen) maintained near its boiling point. Ions produced by incoming charged particles leave bubble tracks, which can be photographed (Fig. 29.18). Because the density of the liquid in a bubble chamber is much higher than the density of the gas in a cloud chamber, the bubble chamber has a much higher sensitivity.

A **wire chamber** consists of thousands of closely spaced parallel wires that collect the electrons created by a passing ionizing particle. A second grid, with wires

(a)   (b)

FIGURE 29.18 (a) Artificially colored bubble-chamber photograph showing tracks of particles that have passed through the chamber. *(Photo Researchers, Inc./Science Photo Library)* (b) This research scientist is studying a photograph of particle tracks made in a bubble chamber at Fermilab.

Webnote 29.8

Nuclear aircraft? Nuclear automobiles? Nuclear excavation? These were some of the ideas that were proposed following World War II. To learn why one of them failed, go to *http://www.megazone.org/ANP*

perpendicular to the first, allows the *x,y* position of the particle in the plane of the two sets of wires to be determined. Finally several such *x,y* grids arranged parallel to each other along the *z* axis can be used to determine the particle's track in three dimensions. Wire chambers form a part of most detectors used at high-energy accelerator labs, and provide electronic readout to a computer for rapid track reconstruction and display.

SUMMARY

Nuclei are represented symbolically as $^A_Z X$, where X represents the chemical symbol for the element. The quantity *A* is the **mass number,** which equals the total number of nucleons (neutrons plus protons) in the nucleus. The quantity *Z* is the **atomic number,** which equals the number of protons in the nucleus. Nuclei that contain the same number of protons but different numbers of neutrons are called **isotopes.** In other words, isotopes have the same *Z* value but different *A* values.

Most nuclei are approximately spherical, with an average radius given by

$$r = r_0 A^{1/3} \qquad \text{[29.1]}$$

where *A* is the mass number and r_0 is a constant equal to 1.2×10^{-15} m.

The total mass of a nucleus is always less than the sum of the masses of its individual nucleons. This mass difference, Δm, multiplied by c^2 gives the **binding energy** of the nucleus.

The spontaneous emission of radiation by certain nuclei is called **radioactivity.** There are three processes by which a radioactive substance can decay: alpha (α) decay, in which the emitted particles are $^4_2 He$ nuclei; beta (β) decay, in which the emitted particles are electrons or positrons; and gamma (γ) decay, in which the emitted particles are high-energy photons.

The **decay rate,** or **activity,** *R*, of a sample is given by

$$R = \left| \frac{\Delta N}{\Delta t} \right| = \lambda N \qquad \text{[29.3]}$$

where *N* is the number of radioactive nuclei at some instant and λ is a constant for a given substance called the **decay constant.**

Nuclei in a radioactive substance decay in such a way that the number of nuclei present varies with time according to the expression

$$N = N_0 e^{-\lambda t} \qquad \text{[29.4]}$$

where *N* is the number of radioactive nuclei present at time *t*, N_0 is the number at time $t = 0$, and $e = 2.718. \ldots$

The **half-life,** $T_{1/2}$, of a radioactive substance is the time required for half of a given number of radioactive nuclei to decay. The half-life is related to the decay constant as

$$T_{1/2} = \frac{0.693}{\lambda} \qquad \text{[29.5]}$$

If a nucleus decays by alpha emission, it loses two protons and two neutrons. A typical alpha decay is

$$^{238}_{92}U \longrightarrow \, ^{234}_{90}Th + \, ^4_2 He \qquad \text{[29.9]}$$

Note that in this decay, as in all radioactive decay processes, the sum of the *Z* values on the left equals the sum of the *Z* values on the right; the same is true for the *A* values.

A typical beta decay is

$$^{14}_6 C \longrightarrow \, ^{14}_7 N + e^- + \bar{\nu} \qquad \text{[29.15]}$$

When a nucleus beta decays, an **antineutrino** is emitted along with an electron, or a **neutrino** along with a position. A neutrino has zero electric charge and a small mass (which may be zero) and interacts weakly with matter.

Nuclei are often in an excited state following radioactive decay, and release their extra energy by emitting a high-energy photon called a **gamma ray** (γ). A typical gamma-ray emission is

$$^{12}_6 C^* \longrightarrow \, ^{12}_6 C + \gamma \qquad \text{[29.18]}$$

where the asterisk indicates that the carbon nucleus was in an excited state before gamma emission.

Nuclear reactions can occur when a bombarding particle strikes another nucleus. A typical nuclear reaction is

$$\,^4_2\text{He} + \,^{14}_7\text{N} \longrightarrow \,^{17}_8\text{O} + \,^1_1\text{H} \qquad\qquad \textbf{[29.20]}$$

In this reaction an alpha particle strikes a nitrogen nucleus, producing an oxygen nucleus and a proton. As in radioactive decay, atomic numbers and mass numbers balance on the two sides of the arrow.

Nuclear reactions in which energy is released are said to be **exothermic reactions** and are characterized by positive Q values. Reactions with negative Q values, called **endothermic reactions,** cannot occur unless the incoming particle has at least enough kinetic energy to overcome the energy deficit. In order to conserve both energy and momentum, the incoming particle must have a minimum kinetic energy, called the **threshold energy,** given by

$$KE_{\text{min}} = \left(1 + \frac{m}{M}\right)|Q| \qquad\qquad \textbf{[29.23]}$$

where m is the mass of the incident particle and M is the mass of the target atom.

CONCEPTUAL QUESTIONS

1. Isotopes of a given element have many different properties, such as mass, but the same chemical properties. Why is this?
2. Physicists have attempted to measure the solar neutrinos associated with nuclear processes in the Sun's interior. These studies involve placing a large volume of detecting material in a deep mine shaft to reduce the background from cosmic rays. Explain why neutrinos are not similarly reduced.
3. A student claims that a heavy form of hydrogen decays by alpha emission. How do you respond?
4. Explain the main differences between alpha, beta, and gamma rays.
5. In beta decay, the energy of the electron or positron emitted from the nucleus lies somewhere in a relatively large range of possibilities. In alpha decay, however, the alpha-particle energy can only have discrete values. Why is there this difference?
6. If film is kept in a box, alpha particles from a radioactive source outside the box cannot expose the film but beta particles can. Explain.
7. In positron decay, a proton in the nucleus becomes a neutron, and the positive charge is carried away by the positron. But a neutron has a larger rest energy than a proton. How is this possible?
8. An alpha particle has twice the charge of a beta particle. Why does the former deflect less than the latter when passing between electrically charged plates, assuming they both have the same speed?
9. Can carbon-14 dating be used to measure the age of a stone?
10. Pick any beta-decay process and show that the neutrino must have zero charge.
11. Why do heavier elements require more neutrons in order to maintain stability?
12. Suppose it could be shown that cosmic-ray intensity was much greater 10 000 years ago. How would this affect the ages we assign to ancient samples of once-living matter?
13. If no more people were to be born, the law of population growth would strongly resemble the radioactive decay law. Discuss this statement.
14. Why is carbon dating unable to provide accurate estimates of very old materials?
15. Two samples of the same radioactive nuclide are prepared. Sample A has twice the intial activity of sample B. How does the half-life of A compare with the half-life of B? After each has passed through five half-lives, what is the ratio of their activities?
16. Which number determines the chemical characteristics of an atom: A, N, or Z?
17. Compare and contrast a photon and a neutrino.
18. Why do you think neutrons are more effective at starting a nuclear reaction than alpha particles or protons?

PROBLEMS

1, 2, 3 = straightforward, intermediate, challenging ☐ = full solution available in Student Solutions Manual/Study Guide

web = solution posted at **http://info.brookscole.com/serway** 🔥 = biomedical application

Note: Table 29.4 will be useful for many of these problems. A more complete list of atomic masses is given in Appendix B.

Section 29.1 Some Properties of Nuclei

1. Compare the nuclear radii of the following nuclides: ^2_1H, $^{60}_{27}\text{Co}$ $^{197}_{79}\text{Au}$ $^{239}_{94}\text{Pu}$

2. What is the order of magnitude of the number of protons in your body? Of the number of neutrons? Of the number of electrons?
3. Using the result of Example 29.1, find the radius of a sphere of nuclear matter that would have a mass equal to that of Earth. (Earth has a mass of 5.98×10^{24} kg and equatorial radius of 6.38×10^6 m.)

TABLE 29.4	Some Atomic Masses
Element	**Atomic Mass (u)**
$(_{-1}^{0}e)$	0.000 549
$(_{1}^{0}n)$	1.008 665
$_{1}^{1}H$	1.007 825
$_{1}^{2}H$	2.014 102
$_{2}^{4}He$	4.002 602
$_{3}^{7}Li$	7.016 003
$_{4}^{9}Be$	9.012 174
$_{5}^{10}B$	10.012 936
$_{6}^{12}C$	12.000 000
$_{6}^{13}C$	13.003 355
$_{7}^{14}N$	14.003 074
$_{7}^{15}N$	15.000 108
$_{8}^{15}O$	15.003 065
$_{8}^{17}O$	16.999 131
$_{8}^{18}O$	17.999 160
$_{9}^{18}F$	18.000 937
$_{10}^{20}Ne$	19.992 435
$_{11}^{23}Na$	22.989 770
$_{12}^{23}Mg$	22.994 127
$_{13}^{27}Al$	26.981 538
$_{15}^{30}P$	29.978 310
$_{20}^{40}Ca$	39.962 591
$_{20}^{42}Ca$	41.958 63
$_{20}^{43}Ca$	42.958 770
$_{26}^{56}Fe$	55.934 940
$_{30}^{64}Zn$	63.929 144
$_{29}^{64}Cu$	63.929 599
$_{41}^{93}Nb$	92.906 376 8
$_{79}^{197}Au$	196.966 543
$_{80}^{202}Hg$	201.970 617
$_{84}^{216}Po$	216.001 790
$_{86}^{220}Rn$	220.011 401
$_{90}^{234}Th$	234.043 583
$_{92}^{238}U$	238.050 784

4. Consider the hydrogen atom to be a sphere of radius equal to the Bohr radius, 0.53×10^{-10} m, and calculate the approximate value of the ratio of the nuclear density to the atomic density.

5. An alpha particle ($Z = 2$, mass 6.64×10^{-27} kg) approaches to within 1.00×10^{-14} m of a carbon nucleus ($Z = 6$). What are (a) the maximum Coulomb force on the alpha particle, (b) the acceleration of the alpha particle at this point, and (c) the potential energy of the alpha particle at this point?

6. Singly ionized carbon is accelerated through 1 000 V and passed into a mass spectrometer to determine the isotopes present (see Chapter 19). The magnetic field strength in

the spectrometer is 0.200 T. (a) Determine the orbit radii for the ^{12}C and ^{13}C isotopes as they pass through the field. (b) Show that the ratio of radii may be written in the form

$$\frac{r_1}{r_2} = \sqrt{\frac{m_1}{m_2}}$$

and verify that your radii in part (a) agree with this.

7. (a) Find the speed an alpha particle requires to come **web** within 3.2×10^{-14} m of a gold nucleus. (b) Find the energy of the alpha particle in MeV.

8. Find the nucleus that has a radius approximately equal to one half of the radius of uranium $_{92}^{238}U$.

Section 29.2 Binding Energy

9. Calculate the average binding energy per nucleon of $_{41}^{93}Nb$ and $^{197}_{79}Au$.

10. Calculate the binding energy per nucleon for (a) ^{2}H, (b) ^{4}He, (c) ^{56}Fe, and (d) ^{238}U.

11. A pair of nuclei for which $Z_1 = N_2$ and $Z_2 = N_1$ are called *mirror isobars* (the atomic and neutron numbers are interchangeable). Binding energy measurements on such pairs can be used to obtain evidence of the charge independence of nuclear forces. Charge independence means that the proton-proton, proton-neutron, and neutron-neutron nuclear forces are approximately equal. Calculate the difference in binding energy for the two mirror nuclei, $^{15}_{8}O$ and $^{15}_{7}N$.

12. The peak of the stability curve occurs at ^{56}Fe. Elements up to iron are produced in the cores of massive stars by exothermic fusion reactions. This is the fundamental reason that iron and lighter elements are much more common in the Universe than elements with higher mass numbers. Show that ^{56}Fe has a higher binding energy per nucleon than its neighbors ^{55}Mn and ^{59}Co. Compare your results with Figure 29.4.

13. Two nuclei having the same mass number are known as *isobars*. Calculate the difference in binding energy per nucleon for the isobars $_{11}^{23}Na$ and $_{12}^{23}Mg$. How do you account for the difference?

14. Compare the average binding energy per nucleon of $_{12}^{24}Mg$ and $_{37}^{85}Rb$.

Section 29.3 Radioactivity

15. The half-life of an isotope of phosphorus is 14 days. If a sample contains 3.0×10^{16} such nuclei, determine its activity. Express your answer in curies.

16. A drug tagged with $_{43}^{99}Tc$ (half-life = 6.05 h) is prepared for a patient. If the original activity of the sample was 1.1×10^4 Bq, what is its activity after it has sat on the shelf for 2.0 h?

17. The half-life of ^{131}I is 8.04 days. (a) Calculate the decay constant for this isotope. (b) Find the number of ^{131}I nuclei necessary to produce a sample with an activity of 0.50 μCi.

18. The half-life of ^{131}I is 8.04 days. On a certain day, the activity of an ^{131}I sample is 6.40 mCi. What is its activity 40.2 days later?

19. A radioactive sample contains 3.50 μg of pure ^{11}C, which has a half-life of 20.4 min. (a) How many moles of ^{11}C is present initially? (b) Determine the number of nuclei present initially. What is the activity of the sample (c) initially and (d) after 8.00 h?

20. How much time elapses before 90.0% of the radioactivity of a sample of $^{72}_{33}$As disappears, as measured by its activity? The half-life of $^{72}_{33}$As is 26 h.

21. Many smoke detectors use small quantities of the isotope ^{241}Am in their operation. The half-life of ^{241}Am is 432 yr. How long will it take for the activity of this material to decrease to 1.00×10^{-3} of the original activity?

22. After a plant or animal dies, its ^{14}C content decreases with a half-life of 5 730 yr. If an archaeologist finds an ancient firepit containing partially consumed firewood, and the ^{14}C content of the wood is only 12.5% that of an equal carbon sample from a present-day tree, what is the age of the ancient site?

23. A freshly prepared sample of a certain radioactive isotope has an activity of 10.0 mCi. After 4.00 h, the activity is 8.00 mCi. (a) Find the decay constant and half-life of the isotope. (b) How many atoms of the isotope were contained in the freshly prepared sample? (c) What is the sample's activity 30 h after it is prepared?

24. A building has become accidentally contaminated with radioactivity. The longest-lived material in the building is strontium-90 (the atomic mass of $^{90}_{38}$Sr is 89.907 7). If the building initially contained 5.0 kg of this substance and the safe level is less than 10.0 counts/min, how long will the building be unsafe?

Section 29.4 The Decay Processes

25. Complete the following radioactive decay formulas:

$$^{212}_{83}\text{Bi} \longrightarrow ? + ^4_2\text{He}$$
$$^{95}_{36}\text{Kr} \longrightarrow ? + e^-$$
$$? \longrightarrow ^4_2\text{He} + ^{140}_{58}\text{Ce}$$

26. Complete the following radioactive decay formulas:
    (a) $^{12}_5\text{B} \rightarrow ? + e^-$
    (b) $^{234}_{90}\text{Th} \rightarrow ^{230}_{88}\text{Ra} + ?$
    (c) $? \rightarrow ^{14}_7\text{N} + e^-$

27. Complete the following radioactive decay formulas:
    (a) $? \rightarrow e^+ + ^{40}_{19}\text{K}$
    (b) $^{94}_{44}\text{Ru} \rightarrow ^4_2\text{He} + ?$
    (c) $^{144}_{60}\text{Nd} \rightarrow ? + ^{140}_{58}\text{Ce}$

28. Figure P29.28 shows the steps by which $^{235}_{92}\text{U}$ decays to $^{207}_{82}\text{Pb}$. Enter the correct isotope symbol in each square.

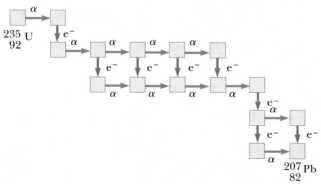

FIGURE P29.28

29. The mass of ^{56}Fe is 55.934 9 u and the mass of ^{56}Co is 55.939 9 u. Which isotope decays into the other, and by what process?

30. Find the energy released in the alpha decay of $^{238}_{92}\text{U}$. The following mass value will be useful: $^{234}_{90}\text{Th}$ has a mass of 234.043 583 u.

31. A nucleus of mass 228 u, initially at rest, undergoes alpha decay. If the alpha particle emitted has a kinetic energy of 4.00 MeV, what is the kinetic energy of the recoiling daughter nucleus?

32. $^{66}_{28}\text{Ni}$ (mass = 65.929 1 u) undergoes beta decay to $^{66}_{29}\text{Cu}$ (mass = 65.928 9 u). (a) Write the complete decay formula for this process. (b) Find the maximum kinetic energy of the emerging electrons.

33. An ^{3}H nucleus beta-decays into ^{3}He by creating an electron and an antineutrino according to the reaction

$$^3_1\text{H} \longrightarrow ^3_2\text{He} + e^- + \bar{\nu}$$

Use Appendix B to determine the total energy released in this reaction.

34. A piece of charcoal used for cooking is found at the remains of an ancient campsite. A 1.00-kg sample of carbon from the wood has an activity of 2.00×10^3 decays per minute. Find the age of the charcoal. (*Hint:* Living material has an activity of 15.0 decays/minute per gram of carbon present.)

35. A wooden artifact is found in an ancient tomb. Its ^{14}C activity is measured to be 60.0% of that in a fresh sample of wood from the same region. Assuming the same amount of ^{14}C was initially present in the wood from which the artifact was made, determine the age of the artifact.

36. A living specimen in equilibrium with the atmosphere contains one atom of ^{14}C (half-life = 5 730 yr) for every 7.70×10^{11} stable carbon atoms. An archeological sample of wood (cellulose, $C_{12}H_{22}O_{11}$) contains 21.0 mg of carbon. When the sample is placed inside a shielded beta counter with 88.0% counting efficiency, 837 counts are accumulated in one week. Assuming that the cosmic-ray flux and Earth's atmosphere have not changed appreciably since the sample was formed, find the age of the sample.

Section 29.6 Nuclear Reactions

37. The first known reaction in which the product nucleus was radioactive (achieved in 1934) was one in which $^{27}_{13}\text{Al}$ was bombarded with alpha particles. Produced in the reaction were a neutron and a product nucleus. (a) What was the product nucleus? (b) Find the Q value of the reaction.

38. Complete the following nuclear reactions:
    (a) $? + ^{14}_7\text{N} \rightarrow ^1_1\text{H} + ^{17}_8\text{O}$
    (b) $^7_3\text{Li} + ^1_1\text{H} \rightarrow ^4_2\text{He} + ?$

39. Identify the unknown particles X and X′ in the following nuclear reactions:
    (a) $X + ^4_2\text{He} \rightarrow ^{24}_{12}\text{Mg} + ^1_0\text{n}$
    (b) $^{235}_{92}\text{U} + ^1_0\text{n} \rightarrow ^{90}_{38}\text{Sr} + X + 2 ^1_0\text{n}$
    (c) $2 ^1_1\text{H} \rightarrow ^2_1\text{H} + X + X'$

40. The first nuclear reaction utilizing particle accelerators was performed by Cockcroft and Walton. Accelerated protons were used to bombard lithium nuclei, producing the following reaction:

$$^1_1\text{H} + ^7_3\text{Li} \longrightarrow ^4_2\text{He} + ^4_2\text{He}$$

Since the masses of the particles involved in the reaction were well known, these results were used to obtain an early proof of the Einstein mass-energy relation. Calculate the Q value of the reaction.

41. (a) Suppose $^{10}_{5}$B is struck by an alpha particle, releasing a proton and a product nucleus in the reaction. What is the product nucleus? (b) An alpha particle and a product nucleus are produced when $^{13}_{6}$C is struck by a proton. What is the product nucleus?

42. (a) Determine the product of the reaction $^{7}_{3}$Li $+ ^{4}_{2}$He $\rightarrow$? $+$ n. (b) What is the Q value of the reaction?

43. A beam of 6.61-MeV protons is incident on a target of $^{27}_{13}$Al. Those that collide produce the reaction

$$p + ^{27}_{13}\text{Al} \longrightarrow ^{27}_{14}\text{Si} + n$$

($^{27}_{14}$Si has a mass of 26.986 721 u.) Neglect any recoil of the product nucleus and determine the kinetic energy of the emerging neutrons.

44. Find the threshold energy that the incident neutron must have to produce the reaction:

$$^{1}_{0}\text{n} + ^{4}_{2}\text{He} \longrightarrow ^{2}_{1}\text{H} + ^{3}_{1}\text{H}$$

45. When ^{18}O is struck by a proton, ^{18}F and another particle are produced. (a) What is the other particle? (b) This reaction has a Q value of -2.453 MeV, and the atomic mass of ^{18}O is 17.999 160 u. What is the atomic mass of ^{18}F?

Section 29.7 Medical Applications of Radiation

46. In terms of biological damage, how many rad of heavy ions are equivalent to 100 rad of x-rays?

47. A person whose mass is 75.0 kg is exposed to a whole-body dose of 25.0 rad. How many joules of energy are deposited in the person's body?

48. A 200-rad dose of radiation is administered to a patient in an effort to combat a cancerous growth. Assuming all of the energy deposited is absorbed by the growth, (a) calculate the amount of energy delivered per unit mass. (b) Assuming the growth has a mass of 0.25 kg and a specific heat equal to that of water, calculate its temperature rise.

49. A "clever" technician decides to heat some water for his coffee with an x-ray machine. If the machine produces 10 rad/s, how long will it take to raise the temperature of a cup of water by 50°C? Ignore heat losses during this time.

50. An x-ray technician works 5 days per week, 50 weeks per year. Assume that the technician takes an average of eight x-rays per day and receives a dose of 5.0 rem/yr as a result. (a) Estimate the dose in rem per x-ray taken. (b) How does this result compare with the amount of low-level background radiation the technician is exposed to?

51. A patient swallows a radiopharmaceutical tagged with phosphorus-32 ($^{32}_{15}$P), a β^- emitter with a half-life of 14.3 days. The average kinetic energy of the emitted electrons is 700 keV. If the initial activity of the sample is 1.31 MBq, determine (a) the number of electrons emitted in a 10-day period, (b) the total energy deposited in the body during the 10 days, and (c) the absorbed dose if the electrons are completely absorbed in 100 g of tissue.

52. A particular radioactive source produces 100 mrad of 2-MeV gamma rays per hour at a distance of 1.0 m. (a) How long could a person stand at this distance before accumulating an intolerable dose of 1 rem? (b) Assuming the gamma radiation is emitted uniformly in all directions, at what distance would a person receive a dose of 10 mrad/h from this source?

ADDITIONAL PROBLEMS

53. A 200.0-mCi sample of a radioactive isotope is purchased by a medical supply house. If the sample has a half-life of 14.0 days, how long will it keep before its activity is reduced to 20.0 mCi?

54. One method for producing neutrons for experimental use is to bombard $^{7}_{3}$Li with protons. The neutrons are emitted according to the following reaction:

$$^{1}_{1}\text{H} + ^{7}_{3}\text{Li} \longrightarrow ^{7}_{4}\text{Be} + ^{1}_{0}\text{n}$$

What is the minimum kinetic energy the incident proton must have if this reaction is to occur?

55. Deuterons that have been accelerated are used to bombard other deuterium nuclei, resulting in the reaction

$$^{2}_{1}\text{H} + ^{2}_{1}\text{H} \longrightarrow ^{3}_{2}\text{He} + ^{1}_{0}\text{n}$$

Does this reaction require a threshold energy? If so, what is its value?

56. Consider a radioactive sample. Determine the ratio of the number of atoms decaying during the first half of its half-life to the number of atoms decaying during the second half of its half-life.

57. A by-product of some fission reactors is the isotope $^{239}_{94}$Pu, an alpha emitter having a half-life of 24 120 yr:

$$^{239}_{94}\text{Pu} \longrightarrow ^{235}_{92}\text{U} + \alpha$$

Consider a sample of 1.00 kg of pure $^{239}_{94}$Pu at $t = 0$. Calculate (a) the number of $^{239}_{94}$Pu nuclei present at $t = 0$ and (b) the initial activity in the sample. (c) How long does the sample have to be stored if a "safe" activity level is 0.100 Bq?

58. (a) Find the radius of the $^{12}_{6}$C nucleus. (b) Find the force of repulsion between a proton at the surface of a $^{12}_{6}$C nucleus and the remaining five protons. (c) How much work (in MeV) has to be done to overcome this electrostatic repulsion in order to put the last proton into the nucleus? (d) Repeat (a), (b), and (c) for $^{238}_{92}$U.

59. In a piece of rock from the Moon, the ^{87}Rb content is assayed to be 1.82×10^{10} atoms per gram of material, and the ^{87}Sr content is found to be 1.07×10^{9} atoms per gram. (The relevant decay is ^{87}Rb $\rightarrow$ ^{87}Sr $+$ e$^-$. The half-life of the decay is 4.8×10^{10} yr.) (a) Determine the age of the rock. (b) Could the material in the rock actually be much older? What assumption is implicit in using the radioactive dating method?

60. Many radioisotopes have important industrial, medical, and research applications. One of these is ^{60}Co, which has a half-life of 5.2 yr and decays by the emission of a beta particle (energy 0.31 MeV) and two gamma photons (energies 1.17 MeV and 1.33 MeV). A scientist wishes to prepare a ^{60}Co sealed source that will have an activity of at least 10 Ci after 30 months of use. What is the minimum initial mass of ^{60}Co required?

61. A medical laboratory stock solution is prepared with an initial activity due to ^{24}Na of 2.5 mCi/mL, and 10.0 mL of the stock solution is diluted at $t_0 = 0$ to a working solution whose total volume is 250 mL. After 48 h, a 5.0-mL sample of the working solution is monitored with a counter. What is the measured activity? (Note that 1 mL = 1 milliliter.)

62. A theory of nuclear astrophysics is that all the heavy elements like uranium are formed in supernova explosions of massive stars, which immediately release the elements into space. If we assume that at the time of explosion there were equal amounts of ^{235}U and ^{238}U, how long ago were the elements that formed our Earth released, given that the present ^{235}U/^{238}U ratio is 0.007? (The half-lives of ^{235}U and ^{238}U are 0.70×10^9 yr and 4.47×10^9 yr, respectively.)

63. A fission reactor is hit by a nuclear weapon, causing 5.0×10^6 Ci of ^{90}Sr ($T_{1/2} = 28.7$ yr) to evaporate into the air. The ^{90}Sr falls out over an area of 10^4 km^2. How long will it take the activity of the ^{90}Sr to reach the agriculturally "safe" level of 2.0 μCi/m^2?

64. After the sudden release of radioactivity from the Chernobyl nuclear reactor accident in 1986, the radioactivity of milk in Poland rose to 2 000 Bq/L due to iodine-131, with a half-life of 8.04 days. Radioactive iodine is particularly hazardous, because the thyroid gland concentrates iodine. The Chernobyl accident caused a measurable increase in thyroid cancers among children in Belarus. (a) For comparison, find the activity of milk due to potassium. Assume that 1 L of milk contains 2.00 g of potassium, of which 0.011 7% is the isotope ^{40}K which has a half-life of 1.28×10^9 yr. (b) After what time would the activity due to iodine fall below that due to potassium?

65. During the manufacture of a steel engine component, radioactive iron (^{59}Fe) is included in the total mass of 0.20 kg. The component is placed in a test engine when the activity due to the isotope is 20.0 μCi. After a 1 000-h test period, oil is removed from the engine and found to con-

tain enough ^{59}Fe to produce 800 disintegrations/min per liter of oil. The total volume of oil in the engine is 6.5 L. Calculate the total mass worn from the engine component per hour of operation. (The half-life for ^{59}Fe is 45.1 days.)

66. After determining that the Sun has existed for hundreds of millions of years, but before the discovery of nuclear physics, scientists could not explain why the Sun has continued to burn for such a long time. For example, if it were a coal fire, it would have burned up in about 3 000 yr. Assume that the Sun, whose mass is 1.99 10^{30} kg, originally consisted entirely of hydrogen and that its total power output is 3.76×10^{26} W. (a) If the energy-generating mechanism of the Sun is the transforming of hydrogen into helium via the net reaction

$$4\,^1_1\text{H} + 2e^- \longrightarrow\ ^4_2\text{He} + 2\nu + \gamma$$

calculate the energy (in joules) given off by this reaction. (b) Determine how many hydrogen atoms constitute the Sun. Take the mass of one hydrogen atom to be 1.67×10^{-27} kg. (c) Assuming that the total power output remains constant, after what time will all the hydrogen be converted into helium, making the Sun die? The actual projected lifetime of the Sun is about 10 billion years, because only the hydrogen in a relatively small core is available as a fuel. Only in the core are temperatures and densities high enough for the fusion reaction to be self-sustaining.

GROUP ACTIVITIES

G.1 This experiment will take a little longer to do than most that we have suggested, but the time spent is worthwhile to help you understand the concept of half-life. Obtain a box of sugar cubes and with a pencil make a mark on one side of each of about 200 cubes. Each of these cubes will represent the nucleus of a radioactive substance. Thus, at $t = 0$, you have 200 undecayed nuclei. Now, put the 200 marked cubes in a box and roll them out on a table, just as you would roll dice. Next, count and remove any cubes that have landed marked-side up. These cubes represent nuclei that emitted radiation during the roll. They are no longer radioactive and thus do not participate in the rest of the action. Record the number of undecayed cubes remaining as the number of undecayed nuclei at $t = 1$ roll.

Continue rolling, counting, and removing until you have completed 12 to 15 rolls. By then, you should have only a few cubes remaining. Plot a graph of undecayed cubes versus the roll number and from this determine the "half-roll" of the cubes.

G.2 Use a nail to punch a hole in the bottom of a large tin can. Hold the can beneath a faucet and adjust the water flow from the faucet to a fine constant stream. Although water flows from the hole at the bottom, you will note that the level of the water in the can rises. As it does so, however, the flow of water leaving the can increases due to increased water pressure caused by the greater depth of water. Unless the flow of water is too great, an equilibrium point will be reached at which the amount of water flowing out of the can each second exactly equals the amount flowing in each second. When this happens, the level of water in the can is constant. As noted in the text, carbon-

14 is continually being produced in the atmosphere and is also continually disappearing as it decays into nitrogen. What is the analogy between water entering the can, remaining in the can, and flowing out of the can and the behavior of carbon-14 in the atmosphere?

G.3 (a) Describe what happens to the number of protons and neutrons in a nucleus when the nucleus undergoes alpha decay. (b) Repeat for a beta decay. (c) In alpha decay how does the mass of the parent nuclei compare to the sum of the masses of the daughter plus alpha? Is the parent mass greater, less, equal to, or no relation to the sum of the daughter plus alpha? (d) Use the decay relationship $^{230}_{90}\text{Th} \rightarrow\ ^{226}_{98}\text{Ra} +\ ^4_2\text{He}$ to check your answer to part (c). The mass of ^{230}Th is 230.033 131 u and the mass of ^{226}Ra is 226.025 406 u.

G.4 (a) Assume the energy released in a beta decay is 0.150 MeV. If the daughter nucleus is found to have an energy of 0.100 MeV, then for energy conservation, is it true that the beta must have an energy of 0.050 MeV? (b) If you find that the beta actually has only an energy of 0.025 MeV, what happens to the remaining energy? (c) Correct the following beta decay reaction to agree with your answers to the above: $^{14}_6\text{C} \rightarrow\ ^{14}_7\text{N} + e^-$.

G.5 (a) In a radioactive decay process, which of the following are reduced by one half after one half-life, (i) the number of nuclei, (ii) the activity, or (iii) the decay constant? (b) If the half-life of a given material is 14 days, what is the ratio of the number of particles remaining after 10 days to the original number? What is the ratio of the activity after 10 days to the initial activity?

30

Nuclear Energy and Elementary Particles

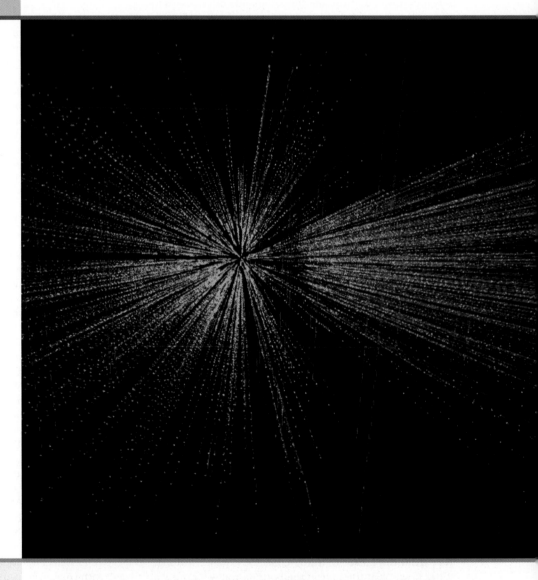

In this image from the NA49 experiment at CERN, hundreds of subatomic particles are created in the collision of high-energy lead nuclei with a lead target. The aim of the experiment was to create extreme densities in matter to break down the strong force that normally locks quarks within protons and neutrons.

(Courtesy of CERN)

932

*I*n this concluding chapter we discuss the two means by which energy can be derived from nuclear reactions. These two techniques are fission, in which a nucleus of large mass number splits, or fissions, into two smaller nuclei, and fusion, in which two light nuclei fuse to form a heavier nucleus. In either case, large amounts of energy are released that can be used destructively, through bombs, or constructively, through the production of electric power.

We end our study of physics by examining the known subatomic particles and the fundamental interactions that govern their behavior. We also discuss the current theory of elementary particles, which states that all matter in nature is constructed from only two families of particles, quarks and leptons. Finally, we describe how clarifications of such models might help us understand the evolution of the Universe.

30.1   NUCLEAR FISSION

Nuclear fission occurs when a heavy nucleus, such as ^{235}U, splits, or fissions, into two smaller nuclei. In such a reaction, **the total mass of the products is less than the original mass of the heavy nucleus.**

Nuclear fission was first observed in 1939 by Otto Hahn and Fritz Strassman, following some basic studies by Fermi. After bombarding uranium ($Z = 92$) with neutrons, Hahn and Strassman discovered among the reaction products two medium-mass elements, barium and lanthanum. Shortly thereafter, Lisa Meitner and Otto Frisch explained what had happened. The uranium nucleus had split into two nearly equal fragments after absorbing a neutron. Such an occurrence was of considerable interest to physicists attempting to understand the nucleus, but it was to have even more far-reaching consequences. Measurements showed that about 200 MeV of energy is released in each fission event, and this fact was to affect the course of human history.

The fission of ^{235}U by slow (low energy) neutrons can be represented by the equation

$$^{1}_{0}n + ^{235}_{92}U \longrightarrow ^{236}_{92}U^* \longrightarrow X + Y + neutrons \qquad [30.1]$$

where $^{236}U^*$ is an intermediate state that lasts only for about 10^{-12} s before splitting into X and Y. The resulting nuclei, X and Y, are called **fission fragments.** Many combinations of X and Y satisfy the requirements of conservation of energy and charge. In the fission of uranium, there are about 90 different daughter nuclei that can be formed. The process also results in the production of several (typically two or three) neutrons per fission event. On the average, 2.47 neutrons are released per event.

A typical reaction of this type is

$$^{1}_{0}n + ^{235}_{92}U \longrightarrow ^{141}_{56}Ba + ^{92}_{36}Kr + 3 ^{1}_{0}n \qquad [30.2]$$

The fission fragments, barium and krypton, and the released neutrons have a great deal of kinetic energy following the fission event.

The breakup of the uranium nucleus can be likened to what happens to a drop of water when excess energy is added to it. All of the atoms in the drop have energy, but not enough to break up the drop. However, if enough energy is added to set the drop vibrating, it will undergo elongation and compression until the amplitude of vibration becomes large enough to cause the drop to break apart. In the uranium nucleus, a similar process occurs (Fig. 30.1). The sequence of events is as follows:

1. The ^{235}U nucleus captures a thermal (slow-moving) neutron.
2. This capture results in the formation of $^{236}U^*$, and the excess energy of this nucleus causes it to undergo violent oscillations.
3. The $^{236}U^*$ nucleus becomes highly elongated, and the force of repulsion

◀ Sequence of events in a nuclear fission process

FIGURE 30.1 The stages involved in a nuclear fission event as described by the liquid-drop model of the nucleus.

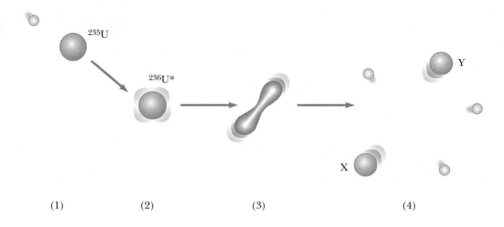

^{235}U

^{236}U*

Y

X

(1)　　　　　(2)　　　　　(3)　　　　　(4)

between protons in the two halves of the dumbbell shape tends to increase the distortion.

4. The nucleus splits into two fragments, emitting several neutrons in the process.

Let us estimate the disintegration energy, Q, released in a typical fission process. From Figure 29.4 we see that the binding energy per nucleon is about 7.2 MeV for heavy nuclei (those having a mass number of approximately 240) and about 8.2 MeV for nuclei of intermediate mass. This means that the nucleons in the fission fragments are more tightly bound and therefore have less mass than the nucleons in the original heavy nucleus. This decrease in mass per nucleon appears as released energy when fission occurs. The amount of energy released is (8.2–7.2) MeV per nucleon. Assuming a total of 240 nucleons, we find that the energy released per fission event is

$$Q = (240 \text{ nucleons})(8.2 \text{ MeV/nucleon} - 7.2 \text{ MeV/nucleon}) = 240 \text{ MeV}$$

This is indeed a very large amount of energy relative to the amount released in chemical processes. For example, the energy released in the combustion of one molecule of the octane used in gasoline engines is about one hundred-millionth the energy released in a single fission event!

APPLYING **PHYSICS** **30.1**

If a heavy nucleus were to fission into just two product nuclei, they would be very unstable. Why is this?

Explanation According to Figure 29.3, the ratio of the number of neutrons to the number of protons increases with *Z*. As a result, when a heavy nucleus splits in a fission reaction to two lighter nuclei, the lighter nuclei tend to have too many neutrons. This leads to instability, as the nucleus returns to the curve in Figure 29.3 by decay processes that reduce the number of neutrons.

Example 30.1 **The Fission of Uranium**

Two other possible ways by which ^{235}U can undergo fission when bombarded with a neutron are (1) by the release of ^{140}Xe and ^{94}Sr as fission fragments and (2) by the release of ^{132}Sn and ^{101}Mo as fission fragments. In each case, neutrons are also released. Find the number of neutrons released in each of these events.

Solution By balancing mass numbers and atomic numbers, we find that these reactions can be written

$$\,^{1}_{0}n + \,^{235}_{92}U \longrightarrow \,^{140}_{54}Xe + \,^{94}_{38}Sr + 2\,^{1}_{0}n$$

$$\,^{1}_{0}n + \,^{235}_{92}U \longrightarrow \,^{132}_{50}Sn + \,^{101}_{42}Mo + 3\,^{1}_{0}n$$

Thus, two neutrons are released in the first event and three in the second.

Quick Quiz 30.1 In the first atomic bomb, the energy released was equivalent to about 30 kilotons of TNT, where a ton of TNT releases an energy of 4.0×10^9 J. The amount of mass converted into energy in this event is nearest to (a) 1 μg, (b) 1 mg, (c) 1 g, (d) 1 kg, (e) 20 kilotons

Example 30.2 The Energy Released in the Fission of ^{235}U

Calculate the total energy released if 1.00 kg of ^{235}U undergoes fission, taking the disintegration energy per event to be $Q = 208$ MeV (a more accurate value than the estimate given previously).

Solution We need to know the number of nuclei in 1.00 kg of uranium. Because $A = 235$, the number of nuclei is

$$N = \left(\frac{6.02 \times 10^{23} \text{ nuclei/mol}}{235 \text{ g/mol}}\right)(1.00 \times 10^3 \text{ g}) = 2.56 \times 10^{24} \text{ nuclei}$$

Hence the disintegration energy is

$$E = NQ = (2.56 \times 10^{24} \text{ nuclei})\left(208 \, \frac{\text{MeV}}{\text{nucleus}}\right) = 5.32 \times 10^{26} \text{ MeV}$$

Because 1 MeV is equivalent to 4.45×10^{-20} kWh, $E = 2.37 \times 10^7$ kWh. This is enough energy to keep a 100-W lightbulb burning for about 30 000 years.

30.2 NUCLEAR REACTORS

We have seen that neutrons are emitted when ^{235}U undergoes fission. These neutrons can in turn trigger other nuclei to undergo fission, with the possibility of a chain reaction (Fig. 30.2). Calculations show that if the chain reaction is not controlled (that is, if it does not proceed slowly), it can result in a violent explosion, with the release of an enormous amount of energy, even from only 1 g of ^{235}U. If the energy in 1 kg of ^{235}U were released, it would equal that released by the detonation of about 20 000 tons of TNT! This, of course, is the principle behind the first nuclear bomb, an uncontrolled fission reaction.

A nuclear reactor is a system designed to maintain what is called a **self-sustained chain reaction.** This important process was first achieved in 1942 by a group led by Fermi at the University of Chicago, with natural uranium as the fuel. Most reactors in operation today also use uranium as fuel. Natural uranium contains only about 0.7% of the ^{235}U isotope, with the remaining 99.3% being the

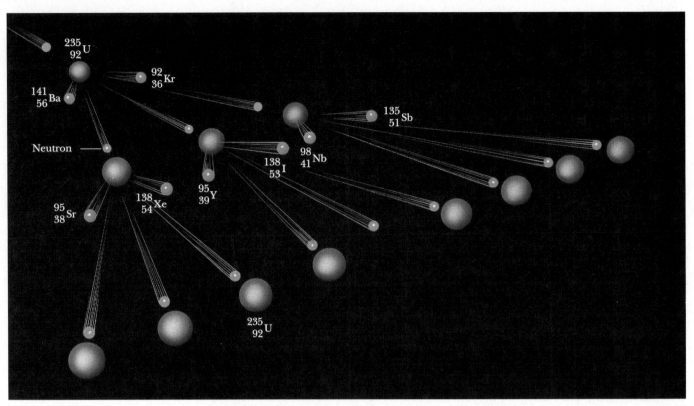

FIGURE 30.2 A nuclear chain reaction initiated by capture of a neutron.

^{238}U isotope. This is important to the operation of a reactor because ^{238}U almost never undergoes fission. Instead, it tends to absorb neutrons, producing neptunium and plutonium. For this reason, reactor fuels must be artificially enriched so that they contain several percent of the ^{235}U isotope.

Earlier, we mentioned that an average of about 2.5 neutrons are emitted in each fission event of ^{235}U. In order to achieve a self-sustained chain reaction, one of these neutrons must be captured by another ^{235}U nucleus and cause it to un-

Painting of the world's first reactor. Because of wartime secrecy, there are no photographs of the completed reactor. The reactor was composed of layers of graphite interspersed with uranium. A self-sustained chain reaction was first achieved on December 2, 1942. Word of the success was telephoned immediately to Washington with this message: "The Italian navigator has landed in the New World and found the natives very friendly." The historic event took place in an improvised laboratory in the racquet court under the west stands of the University of Chicago's Stagg Field; the Italian navigator was Fermi. *(Courtesy of Chicago Historical Society)*

dergo fission. A useful parameter for describing the level of reactor operation is the **reproduction constant *K*, defined as the average number of neutrons from each fission event that will cause another event.** As we have seen, *K* can have a maximum value of 2.5 in the fission of uranium. However, in practice *K* is less than this because of several factors, which we shall discuss in a moment.

A self-sustained chain reaction is achieved when *K* = 1. Under this condition, the reactor is said to be **critical.** When *K* is less than unity, the reactor is subcritical and the reaction dies out. When *K* is greater than unity, the reactor is said to be supercritical, and a runaway reaction occurs. In a nuclear reactor used to furnish power to a utility company, it is necessary to maintain a *K* value close to unity.

The basic design of a nuclear reactor is shown in Figure 30.3. The fuel elements consist of enriched uranium. Now let us look at the function of the remaining parts of the reactor and some aspects of its design.

FIGURE 30.3 Cross section of a reactor core surrounded by a radiation shield.

NEUTRON LEAKAGE

In any reactor, a fraction of the neutrons produced in fission will leak out of the core before inducing other fission events. If the fraction leaking out is too large, the reactor will not operate. The percentage lost is large if the reactor is very small because leakage is a function of the ratio of surface area to volume. Therefore, a critical requirement of reactor design is choosing the correct surface-area-to-volume ratio so that a sustained reaction can be achieved.

REGULATING NEUTRON ENERGIES

The neutrons released in fission events are very energetic, with kinetic energies of about 2 MeV. It is found that slow neutrons are far more likely than fast neutrons to produce fission events in ^{235}U. Furthermore, ^{238}U does not absorb slow neutrons. Therefore, in order for the chain reaction to continue, the neutrons must be slowed down. This is accomplished by surrounding the fuel with a **moderator** substance.

To understand how neutrons are slowed down, consider a collision between a light object and a very massive one. In such an event, the light object rebounds from the collision with most of its original kinetic energy. However, if the collision is between objects with nearly the same masses, the incoming projectile transfers a large percentage of its kinetic energy to the target. In the first nuclear reactor ever constructed, Fermi placed bricks of graphite (carbon) between the fuel elements. Carbon nuclei are about 12 times more massive than neutrons, but after about 100 collisions with carbon nuclei, a neutron is slowed sufficiently to increase its likelihood of fission with ^{235}U. In this design the carbon is the moderator; most modern reactors use heavy water (D_2O) as the moderator.

NEUTRON CAPTURE

In the process of being slowed down, the neutrons may be captured by nuclei that do not undergo fission. The most common event of this type is neutron capture by ^{238}U. The probability of neutron capture by ^{238}U is very high when the neutrons have high kinetic energies and very low when they have low kinetic energies. Thus, the slowing down of the neutrons by the moderator serves the dual purpose of making them available for reaction with ^{235}U and decreasing their chances of being captured by ^{238}U.

CONTROL OF POWER LEVEL

It is possible for a reactor to reach the critical stage (*K* = 1) after all neutron losses described previously are minimized. However, a method of control is needed to adjust *K* to a value near unity. If *K* were to rise above this value, the heat produced

◄ APPLICATION

NUCLEAR REACTOR DESIGN

Control rod

Uranium
fuel rod

Steam

Nuclear
reactor

Molten sodium
or liquid water
under high
pressure (carries
heat to steam
generator)

Steam turbine and
electric generator

Condenser (steam from turbine
is condensed by cold water)

Heat
exchanger

Pump

Cold water   Warm water

FIGURE 30.4 Main components of a pressurized-water reactor.

in the runaway reaction would melt the reactor. To control the power level, control rods are inserted into the reactor core (see Fig. 30.3). These rods are made of materials such as cadmium that are very efficient in absorbing neutrons. By adjusting the number and position of these control rods in the reactor core, the K value can be varied and any power level within the design range of the reactor can be achieved.

A diagram of a pressurized-water reactor is shown in Figure 30.4. This type of reactor is commonly used in electric power plants in the United States. Fission events in the reactor core supply heat to the water contained in the primary (closed) system, which is maintained at high pressure to keep it from boiling. This water also serves as the moderator. The hot water is pumped through a heat exchanger, and the heat is transferred to the water contained in the secondary system. There the hot water is converted to steam, which drives a turbine-generator to create electric power. Note that the water in the secondary system is isolated from the water in the primary system in order to prevent contamination of the secondary water and steam by radioactive nuclei from the reactor core.

REACTOR SAFETY[1]

The safety aspects of nuclear power reactors are often sensationalized by the media and misunderstood by the public. The 1979 near-disaster of Three Mile Island in Pennsylvania and the accident at the Chernobyl reactor in the Ukraine rightfully focused attention on reactor safety. Yet the safety record in the United States is enviable. The records show no fatalities attributed to commercial nuclear power generation in the history of the United States nuclear industry.

Commercial reactors achieve safety through careful design and rigid operating procedures. Radiation exposure and the potential health risks associated with such exposure are controlled by three layers of containment. The fuel and radioactive fission products are contained inside the reactor vessel. Should this vessel rupture, the reactor building acts as a second containment structure to prevent

[1] The authors are grateful to Professor Gene Skluzacek of the University of Nebraska at Omaha for rewriting this section on reactor safety.

radioactive material from contaminating the environment. Finally, the reactor facilities must be in a remote location to protect the general public from exposure should radiation escape the reactor building.

According to the Oak Ridge National Laboratory Review, "the health risk of living within 8 km (5 miles) of a nuclear reactor for 50 years is no greater than the risk of smoking 1.4 cigarettes, drinking 0.5 liters of wine, traveling 240 km by car, flying 9 600 km by jet, or having one chest x-ray in a hospital. Each of these activities is estimated to increase a person's chances of dying in any given year by one in a million."

Another potential danger in nuclear reactor operations is the possibility that the water flow could be interrupted. Even if the nuclear fission chain reaction were stopped immediately, residual heat could build up in the reactor to the point of melting the fuel elements. The molten reactor core would melt to the bottom of the reactor vessel and conceivably melt its way into the ground below—the so-called "China syndrome." Although it might appear that this deep underground burial site would be an ideal safe haven for a radioactive blob, there would be danger of a steam explosion should the molten mass encounter water. This nonnuclear explosion could spread radioactive material to the areas surrounding the power plant. To prevent such an unlikely chain of events, nuclear reactors are designed with emergency core-cooling systems, requiring no power, that automatically flood the reactor with water in the event of coolant loss. The emergency cooling water moderates heat buildup in the core, which in turn prevents the occurrence of melting.

A continuing concern in nuclear fission reactors is the safe disposal of radioactive material when the reactor core is replaced. This waste material contains long-lived, highly radioactive isotopes and must be stored over long periods of time in such a way that there is no chance of environmental contamination. At present, sealing radioactive wastes in waterproof containers and burying them in deep salt mines seems to be the most promising solution.

Transportation of reactor fuel and reactor wastes poses additional safety risks. However, neither the waste nor the fuel of nuclear power reactors can be used to construct a nuclear bomb.

Accidents during transportation of nuclear fuel could expose the public to harmful levels of radiation. The Department of Energy requires stringent crash tests on all containers used to transport nuclear materials. Container manufacturers must demonstrate that their containers will not rupture even in high-speed collisions.

The safety issues associated with nuclear power reactors are complex and often emotional. All sources of energy have associated risks. In each case, we must weigh the risks against the benefits and the availability of the energy source.

30.3 NUCLEAR FUSION

Figure 29.4 shows that the binding energy for light nuclei (those having a mass number lower than 20) is much smaller than the binding energy for heavier nuclei. This suggests a possible process that is the reverse of fission. **When two light nuclei combine to form a heavier nucleus, the process is called nuclear fusion.** Because the mass of the final nucleus is less than the masses of the original nuclei, there is a loss of mass accompanied by a release of energy. Although fusion power plants have not yet been developed, a worldwide effort is under way to harness the energy from fusion reactions in the laboratory. Later we shall discuss the possibilities and advantages of this process for generating electric power.

FUSION IN THE SUN

All stars generate their energy through fusion processes. About 90% of the stars, including the Sun, fuse hydrogen, whereas some older stars fuse helium or other heavier elements. Stars are born in regions of space containing vast clouds of dust

This photograph of the Sun, taken on December 19, 1973, during the third and final manned Skylab mission, shows one of the most spectacular solar flares ever recorded, spanning more than 588 000 km (365 000 mi) across the solar surface. The last picture, taken some 17 hours earlier, showed this feature as a large quiescent prominence on the eastern side of the Sun. The flare gives the distinct impression of a twisted sheet in the process of unwinding itself. In this photograph the solar poles are distinguished by a relative absence of granulation and a much darker tone than the central portions of the disk. Several active regions are seen on the eastern side of the disk. The photograph was taken in the light of ionized helium by the extreme ultraviolet spectroheliograph instrument of the U.S. Naval Research Laboratory. *(NASA)*

and gas. Recent mathematical models of these clouds indicate that star formation is triggered by shock waves passing through a cloud. These shock waves are similar to sonic booms and are produced by events such as the explosion of a nearby star, called a *supernova explosion.* The shock wave compresses certain regions of the cloud, causing these regions to collapse under their own gravity. As the gas falls inward toward the center, the atoms gain speed, which causes the temperature of the gas to rise. Two conditions must be met before fusion reactions in the star can sustain its energy needs: (1) The temperature must be high enough (about 10^7 K for hydrogen) to allow the kinetic energy of the positively charged hydrogen nuclei to overcome their mutual Coulomb repulsion as they collide, and (2) the density of nuclei must be high enough to ensure a high rate of collision.

When fusion reactions occur at the core of a star, the energy liberated eventually becomes sufficient to prevent further collapse of the star under its own gravity. The star then continues to live out the remainder of its life under a balance between the inward force of gravity pulling it toward collapse and the outward force due to thermal effects and radiation pressure.

The **proton-proton cycle** is a series of three nuclear reactions that are believed to be the stages in the liberation of energy in the Sun and other stars rich in hydrogen. An overall view of the proton-proton cycle is that four protons combine to form an alpha particle and two positrons, with the release of 25 MeV of energy in the process.

The specific steps in the proton-proton cycle are

$$\begin{aligned} {}^1_1\text{H} + {}^1_1\text{H} &\longrightarrow {}^2_1\text{H} + e^+ + \nu \\ {}^1_1\text{H} + {}^2_1\text{H} &\longrightarrow {}^3_2\text{He} + \gamma \end{aligned}$$

[30.3]

This second reaction is followed by either hydrogen-helium fusion or helium-helium fusion:

$$ {}^1_1\text{H} + {}^3_2\text{He} \longrightarrow {}^4_2\text{He} + e^+ + \nu $$

or

$$ {}^3_2\text{He} + {}^3_2\text{He} \longrightarrow {}^4_2\text{He} + {}^1_1\text{H} + {}^1_1\text{H} $$

The energy liberated is carried primarily by gamma rays, positrons, and neutrinos, as can be seen from the reactions. The gamma rays are soon absorbed by the dense gas, thus raising its temperature. The positrons combine with electrons to produce gamma rays, which in turn are also absorbed by the gas within a few centimeters. The neutrinos, however, almost never interact with matter; hence they escape from the star, carrying about 2% of the generated energy with them. These energy-liberating fusion reactions are called **thermonuclear fusion reactions.** The hydrogen (fusion) bomb, first exploded in 1952, is an example of an uncontrolled thermonuclear fusion reaction.

FUSION REACTORS

APPLICATION

FUSION REACTORS

The enormous amount of energy released in fusion reactions suggests the possibility of harnessing this energy for useful purposes on Earth. A great deal of effort is focused on developing a sustained and controllable thermonuclear reactor—a fusion power reactor. Controlled fusion is often called the ultimate energy source because of the availability of its fuel source: water. For example, if deuterium were used as the fuel, 0.06 g of it could be extracted from 1 gal of water at a cost of about four cents. Such rates would make the fuel costs of even an inefficient reactor almost insignificant. An additional advantage of fusion reactors is that comparatively few radioactive by-products are formed. As noted in Equation 30.3, the end product of the fusion of hydrogen nuclei is safe, nonradioactive helium. Unfortunately, a thermonuclear reactor that can deliver a net power output over a reasonable time interval is not yet a reality, and many difficulties must be solved before a successful device is constructed.

We have seen that the Sun's energy is based, in part, on a set of reactions in which ordinary hydrogen is converted to helium. Unfortunately, the proton-proton interaction is not suitable for use in a fusion reactor because the event requires very high pressures and densities. The process works in the Sun only because of the extremely high density of protons in the Sun's interior. In fact, even at the densities and temperatures that exist at the center of the Sun, the average proton takes 14 billion years to react!

The fusion reactions that appear most promising in the construction of a fusion power reactor involve deuterium and tritium, which are isotopes of hydrogen. These reactions are

$$\begin{array}{llll} {}^{2}_{1}\text{H} + {}^{2}_{1}\text{H} & \longrightarrow & {}^{3}_{2}\text{He} + {}^{1}_{0}\text{n} & \quad Q = 3.27 \text{ MeV} \\[4pt] {}^{2}_{1}\text{H} + {}^{2}_{1}\text{H} & \longrightarrow & {}^{3}_{1}\text{H} + {}^{1}_{1}\text{H} & \quad Q = 4.03 \text{ MeV} \\[4pt] {}^{2}_{1}\text{H} + {}^{3}_{1}\text{H} & \longrightarrow & {}^{4}_{2}\text{He} + {}^{1}_{0}\text{n} & \quad Q = 17.59 \text{ MeV} \end{array} \qquad \textbf{[30.4]}$$

where the Q values refer to the amount of energy released per reaction. As noted earlier, deuterium is available in almost unlimited quantities from our lakes and oceans and is very inexpensive to extract. Tritium, however, is radioactive ($T_{1/2} = 12.3$ yr) and undergoes beta decay to ${}^{3}\text{He}$. For this reason, tritium does not occur naturally to any great extent and must be artificially produced.

One of the major problems in obtaining energy from nuclear fusion is the fact that the Coulomb repulsion force between two charged nuclei must be overcome before they can fuse. The fundamental challenge is to give the two nuclei enough kinetic energy to overcome this repulsive force. This can be accomplished by heating the fuel to extremely high temperatures (about 10^8 K, far greater than the interior temperature of the Sun). As you might expect, such high temperatures are not easy to obtain in a laboratory or a power plant. At these high temperatures the atoms are ionized, and the system consists of a collection of electrons and nuclei, commonly referred to as a *plasma*.

In addition to the high temperature requirements, two other critical factors determine whether or not a thermonuclear reactor will be successful: **plasma ion density,** n, and **plasma confinement time,** τ, the time the interacting ions are maintained at a temperature equal to or greater than that required for the reaction to proceed successfully. The density and confinement time must both be large enough to ensure that more fusion energy will be released than is required to heat the plasma.

Lawson's criterion states that a net power output in a fusion reactor is possible under the following conditions:

$$\begin{array}{ll} n\tau \geq 10^{14} \text{ s/cm}^3 & \text{Deuterium-tritium interaction} \\[6pt] n\tau \geq 10^{16} \text{ s/cm}^3 & \text{Deuterium-deuterium interaction} \end{array} \qquad \textbf{[30.5]} \quad \blacktriangleleft \; \text{Lawson's criterion}$$

The problem of plasma confinement time has yet to be solved. How can a plasma be confined at a temperature of 10^8 K for times on the order of 1 s? The basic plasma-confinement technique under investigation is discussed following Example 30.3.

Example 30.3 The Deuterium-Deuterium Reaction

Find the energy released in the deuterium-deuterium reaction

$$ {}^{2}_{1}\text{H} + {}^{2}_{1}\text{H} \longrightarrow {}^{3}_{1}\text{H} + {}^{1}_{1}\text{H} $$

Solution The mass of the ${}^{2}_{1}\text{H}$ atom is 2.014 102 u. Thus, the total mass before the reaction is 4.028 204 u. After the reaction, the sum of the masses is equal to 3.016 049 u + 1.007 825 u = 4.023 874 u. Thus, the excess mass is 0.004 33 u. In energy units, this is equivalent to 4.03 MeV .

(a)

(b)

(c)

FIGURE 30.5   (a) Diagram of a tokamak used in the magnetic confinement scheme. The plasma is trapped within the spiraling magnetic field lines as shown. (b) Interior view of the Tokamak Fusion Test Reactor (TFTR) vacuum vessel located at the Princeton Plasma Physics Laboratory, Princeton University, New Jersey. *(Courtesy of Princeton Plasma Physics Laboratory)* (c) The National Spherical Torus Experiment (NSTX) that began operation in March 1999. *(Courtesy of Princeton University)*

MAGNETIC FIELD CONFINEMENT

Most fusion experiments use magnetic field confinement to contain a plasma. One device, called a **tokamak,** has a doughnut-shaped geometry (a toroid), as shown in Figure 30.5a. This device, first developed in the former Soviet Union, uses a combination of two magnetic fields to confine the plasma inside the doughnut. A strong magnetic field is produced by the current in the windings, and a weaker magnetic field is produced by the current in the toroid. The resulting magnetic field lines are helical, as in Figure 30.5a. In this configuration, the field lines spiral around the plasma and prevent it from touching the walls of the vacuum chamber.

In order for the plasma to reach ignition temperature, some form of auxiliary heating is necessary. A successful and efficient auxiliary heating technique that has been used recently is the injection of a beam of energetic neutral particles into the plasma.

When it was in operation, the Tokamak Fusion Test Reactor (TFTR) at Princeton reported central ion temperatures of 510 million degrees Celsius, more than 30 times hotter than the center of the Sun. TFTR $n\tau$ values for the D-T reaction were well above 10^{13} s/cm^3 and close to the value required by Lawson's criterion. In 1991, reaction rates of 6×10^{17} D-T fusions per second were reached in the JET tokamak at Abington, England.

One of the new generations of fusion experiments is the National Spherical Torus Experiment (NSTX) shown in Figure 30.5c. Rather than the donut-shaped

Webnote 30.1

Want to operate a tokamak on your own? Do it here with a Java applet.
http://ippex.pppl.gov/tokamak/default.htm

plasma of a tokamak, the NSTX produces a spherical plasma that has a hole through its center. The major advantage of the spherical configuration is its ability to confine the plasma at a higher pressure in a given magnetic field. This approach could lead to the development of smaller and more economical fusion reactors.

An international collaboration involving four major fusion programs is currently working on building a fusion reactor called ITER (International Thermonuclear Experimental Reactor). This facility will address the remaining technological and scientific issues concerning the feasibility of fusion power. The design is completed, and site and construction negotiations are under way. If the planned device works as expected, the Lawson number for ITER will be about six times greater than the current record holder, the JT-60U tokamak in Japan.

Webnote 30.2

One means of creating high plasma-ion density is through magnetic field confinement. Another approach is to use laser fusion. Find out how laser fusion can produce electricity at *http://www.llnl.gov/nif/library/ife.pdf* (*Note:* You will need Acrobat Reader to view this file.)

30.4 ELEMENTARY PARTICLES

The word "atom" comes from the Greek word *atomos,* meaning "indivisible." At one time atoms were thought to be the indivisible constituents of matter; that is, they were regarded as elementary particles. Discoveries in the early part of the 20th century revealed that the atom is not elementary, but has as its constituents protons, neutrons, and electrons. Until 1932 physicists viewed these three constituent particles as elementary because, with the exception of the free neutron, they are very stable. The theory soon fell apart, however, and beginning in 1937, many new particles were discovered in experiments involving high-energy collisions between known particles. These new particles are characteristically unstable and have very short half-lives, ranging between 10^{-6} s and 10^{-23} s. So far more than 300 of them have been cataloged.

Until the 1960s, physicists were bewildered by the large number and variety of subatomic particles being discovered. They wondered if the particles were like animals in a zoo or if a pattern was emerging that would provide a better understanding of the elaborate structure in the subnuclear world. In the last 30 years, physicists have made tremendous advances in our knowledge of the structure of matter by recognizing that all particles (with the exception of electrons, photons, and a few others) are made of smaller particles called *quarks.* Thus, protons and neutrons, for example, are not truly elementary but are systems of tightly bound quarks. The quark model has reduced the bewildering array of particles to a manageable number and has successfully predicted new quark combinations that were subsequently found in many experiments.

30.5 THE FUNDAMENTAL FORCES IN NATURE

The key to understanding the properties of elementary particles is to be able to describe the forces between them. All particles in nature are subject to four fundamental forces: strong, electromagnetic, weak, and gravitational.

The **strong force** is responsible for the tight binding of quarks to form neutrons and protons and for the nuclear force, a sort of residual strong force, binding neutrons and protons into nuclei. This force represents the "glue" that holds the nucleons together and is the strongest of all the fundamental forces. It is very short-ranged and is negligible for separations greater than about 10^{-15} m (the approximate size of the nucleus). The **electromagnetic force,** which is about 10^{-2} times the strength of the strong force, is responsible for the binding of atoms and molecules. It is a long-range force that decreases in strength as the inverse square of the separation between interacting particles. The **weak force** is a short-range nuclear force that tends to produce instability in certain nuclei. It is responsible for beta decay, and its strength is only about 10^{-6} times that of the strong force. (As we shall discuss later, scientists now believe that the weak and electromagnetic

TABLE 30.1	Particle Interactions		
Interaction (Force)	Relative Strength[a]	Range of Force	Mediating Field Particle
Strong	1	Short (~1 fm)	Gluon
Electromagnetic	10^{-2}	Long ($\propto 1/r^2$)	Photon
Weak	10^{-6}	Short (~ 10^{-3} fm)	$W^{\pm}$ and Z^0 bosons
Gravitational	10^{-43}	Long ($\propto 1/r^2$)	Graviton

[a] For two quarks separated by 3×10^{-17} m

forces are two manifestations of a single force called the *electroweak* force). Finally, the **gravitational force** is a long-range force with a strength of only about 10^{-43} times that of the strong force. Although this familiar interaction is the force that holds the planets, stars, and galaxies together, its effect on elementary particles is negligible. Thus, the gravitational force is the weakest of all the fundamental forces.

Modern physics often describes the forces between particles in terms of the actions of field particles or quanta. In the case of the familiar electromagnetic interaction, the field particles are photons. In the language of modern physics, it can be said that the electromagnetic force is *mediated* (carried) by photons, which are the quanta of the electromagnetic field. Likewise, the strong force is mediated by field particles called *gluons,* the weak force is mediated by particles called the W and Z *bosons,* and the gravitational force is mediated by quanta of the gravitational field called *gravitons.* All of these field quanta have been detected except for the graviton, which may never be found directly because of the weakness of the gravitational field. These interactions, their ranges, and their relative strengths are summarized in Table 30.1.

30.6 POSITRONS AND OTHER ANTIPARTICLES

In the 1920s, the theoretical physicist Paul Adrien Maurice Dirac (1902–1984) developed a version of quantum mechanics that incorporated special relativity. Dirac's theory successfully explained the origin of the electron's spin and its magnetic moment. But it had one major problem: Its relativistic wave equation required solutions corresponding to negative energy states even for free electrons. If negative energy states existed, we would expect a normal free electron in a state of positive energy to make a rapid transition to one of these lower states, emitting a photon in the process. Thus, normal electrons would not exist, and we would be left with a universe of photons and electrons locked up in negative energy states.

Dirac circumvented this difficulty by postulating that all negative energy states are normally filled. The electrons that occupy the negative energy states are said to be in the "Dirac sea" and are not directly observable when all negative energy states are filled. However, if one of these negative energy states is vacant, leaving a hole in the sea of filled states, the hole can react to external forces and therefore can be observed as the electron's positive antiparticle. The general and profound implication of Dirac's theory is that **for every particle, there is an antiparticle.** The antiparticle has the same mass as the particle, but the opposite charge. For example, the electron's antiparticle, the *positron,* has a mass of 0.511 MeV/c^2 and a positive charge of 1.6×10^{-19} C. As noted in Chapter 29, we usually designate an antiparticle with a bar over the symbol for the particle. Thus, $\overline{p}$ denotes the antiproton and $\overline{\nu}$ the antineutrino. In this book, the notation e^+ is preferred for the positron.

The positron was discovered by Carl Anderson in 1932, and in 1936 he was awarded the Nobel prize for his achievement. Anderson discovered the positron while examining tracks created by electron-like particles of positive charge in a

PAUL ADRIEN MAURICE DIRAC (1902–1984)

Winner of the Nobel prize for physics in 1933. *(Courtesy AIP Emilio Segre Visual Archives)*

Tip 30.1 ANTIPARTICLES

An antiparticle is not identified solely on the basis of opposite charge. For example, even neutral particles have antiparticles.

Webnote 30.3

In 1933, Dirac shared the Nobel Prize in Physics with Erwin Schroedinger. Read about it at *http://www.nobel.se/physics/laureates/1933*

cloud chamber. (These early experiments used cosmic rays—mostly energetic protons passing through interstellar space—to initiate high-energy reactions on the order of several GeV.) In order to discriminate between positive and negative charges, the cloud chamber was placed in a magnetic field, causing moving charges to follow curved paths. Anderson noted that some of the electron-like tracks deflected in a direction corresponding to a positively charged particle.

Since Anderson's initial discovery, the positron has been observed in a number of experiments. Perhaps the most common process for producing positrons is pair production, which was introduced in Chapter 26. In this process, a gamma ray with sufficiently high energy collides with a nucleus, creating an electron-positron pair. Because the total rest energy of the electron-positron pair is $2m_ec^2 = 1.02$ MeV, the gamma ray must have at least this much energy to create an electron-positron pair.

Practically every known elementary particle has a distinct antiparticle. Among the exceptions are the photon and the neutral pion (π^0), which are their own antiparticles. Following the construction of high-energy accelerators in the 1950s, many of these antiparticles were discovered. They included the antiproton $\bar{p}$ discovered by Emilio Segré and Owen Chamberlain in 1955 and the antineutron $\bar{n}$ discovered shortly thereafter.

The process of electron-positron annihilation is used in the medical diagnostic technique of positron emission tomography (PET). The patient is injected with glucose solution containing a radioactive substance that decays by positron emission. Examples of such substances are oxygen-15, nitrogen-13, carbon-11, and fluorine-18. The radioactive material is carried to the brain. When a decay occurs, the emitted positron annihilates with an electron in the brain tissue, resulting in two gamma-ray photons. With the assistance of a computer, we can display an image of the sites in the brain at which the glucose accumulates.

The images from a PET scan can indicate a wide variety of disorders in the brain, including Alzheimer's disease. In addition, because glucose metabolizes more rapidly in active areas of the brain, the PET scan can indicate which areas of the brain are involved in various processes such as language, music, and vision.

HIDEKI YUKAWA, JAPANESE PHYSICIST (1907–1981)
Yukawa was awarded the Nobel Prize in 1949 for predicting the existence of mesons. This photograph of Yukawa at work was taken in 1950 in his office at Columbia University. *(UPI/Corbis-Bettman)*

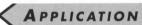

APPLICATION

POSITRON EMISSION TOMOGRAPHY (PET SCANNING)

30.7 MESONS AND THE BEGINNING OF PARTICLE PHYSICS

Physicists in the mid-1930s had a fairly simple view of the structure of matter. The building blocks were the proton, the electron, and the neutron. Three other particles were known or postulated at the time: the photon, the neutrino, and the positron. These six particles were considered the fundamental constituents of matter. Although the accepted picture of the world was marvelously simple, no one was able to provide an answer to the following important question: Because the many protons in proximity in any nucleus should strongly repel each other due to their like charges, what is the nature of the force that holds the nucleus together? Scientists recognized that this mysterious nuclear force must be much stronger than anything encountered up to that time.

The first theory to explain the nature of the nuclear force was proposed in 1935 by the Japanese physicist Hideki Yukawa (1907–1981), an effort that later earned him the Nobel prize. In order to understand Yukawa's theory, it is useful to first note that **two atoms can form a covalent chemical bond by the exchange of electrons.** Similarly, in the modern view of electromagnetic interactions, **charged particles interact by exchanging a photon.** Yukawa used this same idea to explain the nuclear force by proposing a new particle that is exchanged by nucleons in the nucleus to produce the strong force. Furthermore, he established that the range of the force is inversely proportional to the mass of this particle, and predicted that the mass would be about 200 times the mass of the electron. Because the new particle would have a mass between that of the electron and the proton, it was called a **meson** (from the Greek *meso*, meaning "middle").

Tip 30.2 THE NUCLEAR FORCE AND THE STRONG FORCE

The nuclear force discussed in Chapter 29 was originally called the *strong force*. Once the quark theory was established, however, the phrase *strong force* was reserved for the force between quarks. We shall follow this convention—the strong force is between quarks and the nuclear force is between nucleons.

FIGURE 30.6 Feynman diagram representing a photon mediating the electromagnetic force between two electrons.

Webnote 30.4

For a good summary on Feynman diagrams, take a look at *http://www2.slac.stanford.edu/vvc/theory/feynman.html*

RICHARD FEYNMAN, AMERICAN PHYSICIST (1918–1988)

With his son, Carl, in 1965. Feynman, together with Julian S. Schwinger and Shinichiro Tomonaga, won the 1965 Nobel prize for physics for fundamental work in the principles of quantum electrodynamics. His many important contributions to physics include the invention of simple diagrams to represent particle interactions graphically, the theory of the weak interaction of subatomic particles, a reformulation of quantum mechanics, the theory of superfluid helium, and his contribution to physics education through the magnificent three-volume text *The Feynman Lectures on Physics. (UPI Telephotos)*

In an effort to substantiate Yukawa's predictions, physicists began looking for the meson by studying cosmic rays that enter Earth's atmosphere. In 1937 Carl Anderson and his collaborators discovered a particle whose mass was 106 MeV/c^2, about 207 times the mass of the electron. However, subsequent experiments showed that the particle interacted very weakly with matter, and hence could not be the carrier of the nuclear force. This puzzling situation inspired several theoreticians to propose that there are two mesons with slightly different masses, an idea that was confirmed in 1947 with the discovery of the pi meson (π), or simply *pion,* by Cecil Frank Powell (1903–1969) and Guiseppe P. S. Occhialini (1907–1993). The particle discovered earlier by Anderson in 1937, the one thought to be a meson, is not really a meson. Instead, it takes part in weak and electromagnetic interactions only, and it is now called the *muon* (μ).

The pion comes in three varieties, corresponding to three charge states: π^+, π^-, and π^0. The π^+ and π^- particles have masses of 139.6 MeV/c^2, and the π^0 has a mass of 135.0 MeV/c^2. Pions and muons are very unstable particles. For example, the π^-, which has a lifetime of about 2.6×10^{-8} s, decays into a muon and an antineutrino. The muon, with a lifetime of 2.2 μs, then decays into an electron, a neutrino, and an antineutrino. The sequence of decays is

$$\pi^- \longrightarrow \mu^- + \bar{\nu}$$
$$\mu^- \longrightarrow e^- + \nu + \bar{\nu} \qquad \text{[30.6]}$$

The interaction between two particles can be understood in general with a simple diagram called a *Feynman diagram,* developed by Richard P. Feynman (1918–1988). Figure 30.6 is a Feynman diagram for the electromagnetic interaction between two electrons. In this simple case, a photon is the field particle that mediates the electromagnetic force between the electrons. The photon transfers energy and momentum from one electron to the other in this interaction. Such a photon, called a *virtual photon,* can never be detected directly because it is absorbed by the second electron very shortly after being emitted by the first electron. The existence of a virtual photon would violate the law of conservation of energy, but because of the uncertainty principle and its very short lifetime Δt, the photon's excess energy is less than the uncertainty in its energy, given by $\Delta E \approx \hbar/\Delta t$.

Now consider the pion exchange between a proton and a neutron via the nuclear force (Fig. 30.7). We can reason that the energy ΔE needed to create a pion of mass m_π is given by $\Delta E = m_\pi c^2$. Again, the existence of the pion is allowed in spite of conservation of energy if this energy is surrendered in a short enough time Δt, the time it takes the pion to transfer from one nucleon to the other. From the uncertainty principle, $\Delta E \, \Delta t \approx \hbar$, we get

$$\Delta t \approx \frac{\hbar}{\Delta E} = \frac{\hbar}{m_\pi c^2} \qquad \text{[30.7]}$$

Because the pion cannot travel faster than the speed of light, the maximum distance d it can travel in a time Δt is $c \, \Delta t$. Using Equation 30.7 and $d = c \, \Delta t$, we find this maximum distance to be

$$d \approx \frac{\hbar}{m_\pi c} \qquad \text{[30.8]}$$

The measured range of the nuclear force is about 1.5×10^{-15} m. Using this value for d in Equation 30.8, the rest energy of the pion is calculated to be

$$m_\pi c^2 \approx \frac{\hbar c}{d} = \frac{(1.05 \times 10^{-34} \, \text{J} \cdot \text{s})(3.00 \times 10^8 \, \text{m/s})}{1.5 \times 10^{-15} \, \text{m}}$$
$$= 2.1 \times 10^{-11} \, \text{J} \cong 130 \, \text{MeV}$$

This corresponds to a mass of 130 MeV/c^2 (about 250 times the mass of the electron), which is in good agreement with the observed mass of the pion.

The concept we have just described is quite revolutionary. In effect, it says that a proton can change into a proton plus a pion, as long as it returns to its original state in a very short time. High-energy physicists often say that a nucleon undergoes "fluctuations" as it emits and absorbs pions. As we have seen, these fluctuations are a consequence of a combination of quantum mechanics (through the uncertainty principle) and special relativity (through Einstein's energy-mass relation $E = mc^2$).

This section has dealt with Yukawa's early theory of particles that mediate the nuclear force, namely the pions, and the mediators of the electromagnetic force, the photons. Although Yukawa's model led to the modern view, it has been superseded by the more basic quark-gluon theory as explained in Sections 30.12 and 30.13.

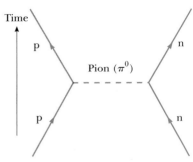

FIGURE 30.7 Feynman diagram representing a proton interacting with a neutron via the strong force. In this case, the pion mediates the nuclear force.

30.8 CLASSIFICATION OF PARTICLES

HADRONS

All particles other than photons can be classified into two broad categories, hadrons and leptons, according to their interactions. Particles that interact through the strong force are called *hadrons*. There are two classes of hadrons, known as *mesons* and *baryons*, distinguished by their masses and spins. All mesons are known to decay finally into electrons, positrons, neutrinos, and photons. The pion is the lightest of known mesons, with a mass of about 140 MeV/c^2 and a spin of 0. Another is the K meson, with a mass of about 500 MeV/c^2 and spin 0.

Baryons have masses equal to or greater than the proton mass (the name *baryon* means "heavy" in Greek), and their spin is always a noninteger value ($\frac{1}{2}$ or $\frac{3}{2}$). Protons and neutrons are baryons, as are many other particles. With the exception of the proton, all baryons decay in such a way that the end products include a proton. For example, the baryon called the Ξ hyperon first decays to a Λ^0 in about 10^{-10} s. The Λ^0 then decays to a proton and a π^- in about 3×10^{-10} s.

Today it is believed that hadrons are composed of quarks. (We shall have more to say about the quark model later.) Some of the important properties of hadrons are listed in Table 30.2.

LEPTONS

Leptons (from the Greek *leptos* meaning "small" or "light") are a group of particles that participate in the weak interaction. All leptons have a spin of $\frac{1}{2}$. Included in this group are electrons, muons, and neutrinos, which are less massive than the lightest hadron. Although hadrons have size and structure, leptons appear to be truly elementary, with no structure (that is, point-like).

Quite unlike hadrons, the number of known leptons is small. Currently, scientists believe there only are six leptons (each having an antiparticle)—the electron, the muon, the tau, and a neutrino associated with each:

$$\begin{pmatrix} e^- \\ \nu_e \end{pmatrix} \qquad \begin{pmatrix} \mu^- \\ \nu_\mu \end{pmatrix} \qquad \begin{pmatrix} \tau^- \\ \nu_\tau \end{pmatrix}$$

The tau lepton, discovered in 1975, has a mass about twice that of the proton.

Although neutrinos have masses of about zero, there is strong indirect evidence that the electron neutrino has a finite mass of about 3 eV/c^2 or a mass 1/180 000 of the electron mass. As we shall see, a firm knowledge of the neutrino's mass could have great significance in cosmological models and the future of the Universe.

Webnote 30.5

Take a look at particle physics in terms of the Big Bang Theory by visiting
http://hepwww.rl.ac.uk/Pub/Phil/ppintro/ppintro.html

Webnote 30.6

Protons showering from outer space can create muons in the atmosphere. Understand more about this effect by visiting
http://www2.slac.stanford.edu/vvc/cosmic_rays.html

TABLE 30.2   **Some Particles and Their Properties**

Category	Particle Name	Symbol	Anti-particle	Mass (MeV/c^2)	B	L_e	L_μ	L_τ	S	Lifetime(s)	Principal Decay Modes[a]
Leptons	Electron	e^-	e^+	0.511	0	+1	0	0	0	Stable	
	Electron–neutrino	ν_e	$\bar{\nu}_e$	<7 eV/c^2	0	+1	0	0	0	Stable	
	Muon	μ^-	μ^+	105.7	0	0	+1	0	0	2.20×10^{-6}	$e^-\bar{\nu}_e\nu_\mu$
	Muon–neutrino	ν_μ	$\bar{\nu}_\mu$	<0.3	0	0	+1	0	0	Stable	
	Tau	τ^-	τ^+	1 784	0	0	0	+1	0	$<4 \times 10^{-13}$	$\mu^-\bar{\nu}_\mu\nu_\tau$, $e^-\bar{\nu}_e\nu_\tau$
	Tau–neutrino	ν_τ	$\bar{\nu}_\tau$	<30	0	0	0	+1	0	Stable	
Hadrons											
Mesons	Pion	π^+	π^-	139.6	0	0	0	0	0	2.60×10^{-8}	$\mu^+\nu_\mu$
		π^0	Self	135.0	0	0	0	0	0	0.83×10^{-16}	2γ
	Kaon	K^+	K^-	493.7	0	0	0	0	+1	1.24×10^{-8}	$\mu^+\nu_\mu$, $\pi^+\pi^0$
		K_S^0	$\bar{K}_S^0$	497.7	0	0	0	0	+1	0.89×10^{-10}	$\pi^+\pi^-$, $2\pi^0$
		K_L^0	$\bar{K}_L^0$	497.7	0	0	0	0	+1	5.2×10^{-8}	$\pi^\pm e^\mp \bar{\nu}_e$, $3\pi^0$ $\pi^\pm \mu^\mp \bar{\nu}_\mu$
	Eta	η	Self	548.8	0	0	0	0	0	$<10^{-18}$	2γ, 3π
		η'	Self	958	0	0	0	0	0	2.2×10^{-21}	$\eta\pi^+\pi^-$
Baryons	Proton	p	$\bar{p}$	938.3	+1	0	0	0	0	Stable	
	Neutron	n	$\bar{n}$	939.6	+1	0	0	0	0	920	$pe^-\bar{\nu}_e$
	Lambda	Λ^0	$\bar{\Lambda}^0$	1 115.6	+1	0	0	0	-1	2.6×10^{-10}	$p\pi^-$, $n\pi^0$
	Sigma	Σ^+	$\bar{\Sigma}^-$	1 189.4	+1	0	0	0	-1	0.80×10^{-10}	$p\pi^0$, $n\pi^+$
		Σ^0	$\bar{\Sigma}^0$	1 192.5	+1	0	0	0	-1	6×10^{-20}	$\Lambda^0\gamma$
		Σ^-	$\bar{\Sigma}^+$	1 197.3	+1	0	0	0	-1	1.5×10^{-10}	$n\pi^-$
	Xi	Ξ^0	$\bar{\Xi}^0$	1 315	+1	0	0	0	-2	2.9×10^{-10}	$\Lambda^0\pi^0$
		Ξ^-	Ξ^+	1 321	+1	0	0	0	-2	1.64×10^{-10}	$\Lambda^0\pi^-$
	Omega	Ω^-	Ω^+	1 672	+1	0	0	0	-3	0.82×10^{-10}	$\Xi^0\pi^0$, $\Lambda^0 K^-$

[a] Notations in this column such as $p\pi^-$, $n\pi^0$ mean two possible decay modes. In this case, the two possible decays are $\Lambda^0 \rightarrow p + \pi^-$ and $\Lambda^0 \rightarrow n + \pi^0$.

30.9   CONSERVATION LAWS

A number of conservation laws are important in the study of elementary particles. Although the two described here have no theoretical foundation, they are supported by abundant empirical evidence.

BARYON NUMBER

The law of conservation of baryon number tells us that whenever a baryon is created in a reaction or decay, an antibaryon is also created. This can be quantified by assigning a baryon number: $B = +1$ for all baryons, $B = -1$ for all antibaryons, and $B = 0$ for all other particles. Thus, the **law of conservation of baryon number** states that whenever a nuclear reaction or decay occurs, the sum of the baryon numbers before the process must equal the sum of the baryon numbers after the process.

Conservation of baryon number ▶

Note that if baryon number is absolutely conserved, the proton must be absolutely stable. If it were not for the law of conservation of baryon number, the proton could decay into a positron and a neutral pion. However, such a decay has never been observed. At present, we can only say that the proton has a half-life of at least 10^{31} years (the estimated age of the Universe is about 10^{10} years). In one recent version of a so-called grand unified theory or (GUT), physicists have predicted that the proton is actually unstable. According to this theory, the baryon number (sometimes called the *baryonic charge*) is not absolutely conserved, whereas electric charge is always conserved.

Example 30.4 Checking Baryon Numbers

Determine whether or not each of the following reactions can occur based on the law of conservation of baryon number.

$$p + n \longrightarrow p + p + n + \bar{p} \tag{1}$$

$$p + n \longrightarrow p + p + \bar{p} \tag{2}$$

Solution Recall that $B = +1$ for baryons and $B = -1$ for antibaryons. Hence, the left side of (1) gives a total baryon number of $1 + 1 = 2$. The right side gives a total baryon number of $1 + 1 + 1 + (-1) = 2$. Thus, reaction (1) can occur provided the incoming proton has sufficient energy.

The left side of (2) gives a total baryon number of $1 + 1 = 2$. However, the right side gives $1 + 1 + (-1) = 1$. Because baryon number is not conserved, reaction (2) cannot occur.

LEPTON NUMBER

There are three conservation laws involving lepton numbers, one for each variety of lepton. The **law of conservation of electron-lepton number** states that the sum of the electron-lepton numbers before a reaction or decay must equal the sum of the electron-lepton numbers after the reaction or decay. The electron and the electron neutrino are assigned a positive electron-lepton number, $L_e = +1$; the antileptons e^+ and $\bar{\nu}_e$ are assigned the electron-lepton number $L_e = -1$; and all other particles have $L_e = 0$. For example, consider the decay of the neutron

$$n \longrightarrow p^+ + e^- + \bar{\nu}_e$$

◀ Conservation of lepton number

◀ Neutron decay

Before the decay, the electron-lepton number is $L_e = 0$; after the decay it is $0 + 1 + (-1) = 0$. Thus, the electron-lepton number is conserved. It is important to recognize that the baryon number must also be conserved. This can easily be seen by noting that before the decay $B = +1$, whereas after the decay $B = +1 + 0 + 0 = +1$.

Similarly, when a decay involves muons, the muon-lepton number L_μ is conserved. The μ^- and the ν_μ are assigned $L_\mu = +1$, the antimuons μ^+ and $\bar{\nu}_\mu$ are assigned $L_\mu = -1$, and all other particles have $L_\mu = 0$. Finally, the τ-lepton number L_τ is conserved, and similar assignments can be made for the τ lepton and its neutrino.

Example 30.5 Checking Lepton Numbers

Determine which of the following decay schemes can occur on the basis of conservation of lepton number.

$$\mu^- \longrightarrow e^- + \bar{\nu}_e + \nu_\mu \tag{1}$$

$$\pi^+ \longrightarrow \mu^+ + \nu_\mu + \nu_e \tag{2}$$

Solution Because decay 1 involves both a muon and an electron, L_μ and L_e must both be conserved. Before the decay, $L_\mu = +1$ and $L_e = 0$. After the decay, $L_\mu = 0 + 0 + 1 = +1$, and $L_e = +1 - 1 + 0 = 0$. Thus, both numbers are conserved, and on this basis the decay mode is possible.

Before decay 2 occurs, $L_\mu = 0$ and $L_e = 0$. After the decay, $L_\mu = -1 + 1 + 0 = 0$, but $L_e = +1$. Thus, the decay is not possible because the electron-lepton number is not conserved.

EXERCISE Determine whether the decay $\mu^- \rightarrow e^- + \bar{\nu}_e$ can occur.

ANSWER No. The muon-lepton number is +1 before the decay and 0 after.

Quick Quiz 30.2	Which of the following reactions cannot occur? (a) $p + p \rightarrow p + p + \overline{p}$ (b) $n \rightarrow p + e^- + \overline{\nu}_e$ (c) $\mu^- \rightarrow e^- + \overline{\nu}_e + \nu_\mu$ (d) $\pi^- \rightarrow \mu^- + \overline{\nu}_\mu$
Quick Quiz 30.3	Which of the following reactions cannot occur? (a) $p + \overline{p} \rightarrow 2\gamma$ (b) $\gamma + p \rightarrow n + \pi^0$ (c) $\pi^0 + n \rightarrow K^+ + \Sigma^-$ (d) $\pi^+ + p \rightarrow K^+ + \Sigma^+$
Quick Quiz 30.4	Suppose a claim is made that the decay of a neutron is given by $n \rightarrow p^+ + e^-$. Which of the following conservation laws are violated by this proposed decay scheme? (a) energy, (b) linear momentum, (c) spin angular momentum, (d) electric charge, (e) lepton number, (f) baryon number.

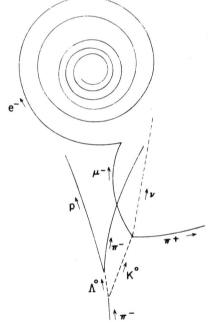

FIGURE 30.8 This drawing represents tracks of many events obtained by analyzing a bubble-chamber photograph. The strange particles Λ^0 and K^0 are formed (at the bottom) as the π^- interacts with a proton according to $\pi^- + p \rightarrow \Lambda^0 + K^0$. (Note that the neutral particles leave no tracks, as indicated by the dashed lines.) The Λ^0 and K^0 then decay according to $\Lambda^0 \rightarrow \pi^- + p$ and $K^0 \rightarrow \pi + \mu^- + \nu_\mu$. *(Courtesy Lawrence Berkeley Laboratory, University of California)*

▶ Conservation of strangeness number

30.10 STRANGE PARTICLES AND STRANGENESS

Many particles discovered in the 1950s were produced by the nuclear interaction of pions with protons and neutrons in the atmosphere. A group of these particles, namely the K, Λ, and Σ, were found to exhibit unusual properties in their production and decay, and hence were called *strange particles.*

One unusual property of strange particles is that they are always produced in pairs. For example, when a pion collides with a proton, two neutral strange particles are produced with high probability (Fig. 30.8) following the reaction

$$\pi^- + p^+ \longrightarrow K^0 + \Lambda^0$$

On the other hand, the reaction $\pi^- + p^+ \rightarrow K^0 + n$ never occurs, even though no known conservation laws are violated and the energy of the pion is sufficient to initiate the reaction.

The second peculiar feature of strange particles is that although they are produced by the strong interaction at a high rate, they do not decay into particles that interact via the strong force at a very high rate. Instead, they decay very slowly, which is characteristic of the weak interaction. Their half-lives are in the range 10^{-10} s to 10^{-8} s; most other particles that interact via the strong force have lifetimes on the order of 10^{-23} s.

To explain these unusual properties of strange particles, physicists introduced a law called *conservation of strangeness,* together with a new quantum number S called **strangeness.** The strangeness numbers for some particles are given in Table 30.2. The production of strange particles in pairs is explained by assigning $S = +1$ to one of the particles and $S = -1$ to the other. All nonstrange particles are assigned strangeness $S = 0$. The **law of conservation of strangeness** states that whenever a nuclear reaction or decay occurs, the sum of the strangeness numbers before the process must equal the sum of the strangeness numbers after the process.

We can explain the slow decay of strange particles by assuming that the strong and electromagnetic interactions obey the law of conservation of strangeness, whereas the weak interaction does not. Because the decay reaction involves the loss of one strange particle, it violates strangeness conservation and hence proceeds slowly via the weak interaction.

Webnote 30.7
Does a 600-ton bubble chamber sound big? Find out more by exploring the Imaging Cosmic and Rare Underground Signal (ICARUS) in Italy. Visit *http://www.aquila.infn.it/icarus*

APPLYING PHYSICS 30.2

A student claims to have observed a decay of an electron into two neutrinos, traveling in opposite directions. What conservation laws would be violated by this decay?

Explanation   Several conservation laws are violated. Conservation of electric charge is violated because the negative charge of the electron has disappeared. Conservation of electron-lepton number is also violated, because there is one lepton before the decay and two afterward. If both neutrinos were electron-neutrinos, electron-lepton number conservation would be violated in the final state. However, if one of the product neutrinos were other than an electron-neutrino, then another lepton conservation law would be violated, because no other leptons were in the initial state. Other conservation laws are obeyed by this decay. Energy can be conserved—the rest energy of the electron appears as the kinetic energy (and possibly some small rest energy) of the neutrinos. The opposite directions of the velocities of the two neutrinos allows for conservation of momentum. Conservation of baryon number and conservation of other lepton numbers are also upheld in this decay.

Example 30.6   Is Strangeness Conserved?

A Determine whether the following reaction occurs on the basis of conservation of strangeness.

$$\pi^0 + n \longrightarrow K^+ + \Sigma^-$$

Solution   The initial state has a total strangeness of $S = 0 + 0 = 0$. Because the strangeness of the K^+ is $S = +1$, and the strangeness of the Σ^- is $S = -1$, the total strangeness of the final state is $+1 - 1 = 0$. Thus, strangeness is conserved and the reaction is allowed.

B Show that the following reaction does not conserve strangeness.

$$\pi^- + p \longrightarrow \pi^- + \Sigma^+$$

Solution   The initial state has strangeness $S = 0 + 0 = 0$, and the final state has strangeness $S = 0 + (-1) = -1$. Thus, strangeness is not conserved.

EXERCISE   Show that the observed reaction $p^+ + \pi^- \rightarrow K^0 + \Lambda^0$ obeys the law of conservation of strangeness.

30.11   THE EIGHTFOLD WAY

As we have seen, quantities such as spin, baryon number, lepton number, and strangeness are labels we associate with particles. Many classification schemes that group particles into families based on such labels have been proposed. First, consider the first eight baryons listed in Table 30.2, all having a spin of $\frac{1}{2}$. The family consists of the proton, the neutron, and six other particles. If we plot their strangeness versus their charge using a sloping coordinate system, as in Figure 30.9a, a fascinating pattern emerges. Six of the baryons form a hexagon, and the remaining two are at the hexagon's center. (Particles with spin quantum number $\frac{1}{2}$ or $\frac{3}{2}$ are called fermions.)

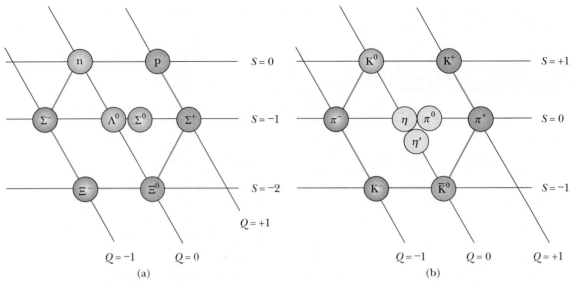

FIGURE 30.9 (a) The hexagonal eightfold-way pattern for the eight spin-$\frac{1}{2}$ baryons. This strangeness versus charge plot uses a horizontal axis for the strangeness values S, but a sloping axis for the charge number Q. (b) The eightfold-way pattern for the nine spin-0 mesons.

Now consider the family of mesons listed in Table 30.2 with spins of 0 (Particles with spin quantum number 0 or 1 are called bosons.) If we count both particles and antiparticles, there are nine such mesons. Figure 30.9b is a plot of strangeness versus charge for this family. Again, a fascinating hexagonal pattern emerges. In this case, the particles on the perimeter of the hexagon lie opposite their antiparticles, and the remaining three (which form their own antiparticles) are at its center. These and related symmetric patterns, called the **eightfold way,** were proposed independently in 1961 by Murray Gell-Mann and Yuval Ne'eman.

The groups of baryons and mesons can be displayed in many other symmetric patterns within the framework of the eightfold way. For example, the family of spin-$\frac{3}{2}$ baryons contains ten particles arranged in a pattern such as the tenpins in a bowling alley. After the pattern was proposed, one of the particles was missing—it had yet to be discovered. Gell-Mann predicted that the missing particle, which he called the *omega minus* (Ω^-), should have a spin of $\frac{3}{2}$, a charge of -1, a strangeness of -3, and a mass of about 1 680 MeV/c^2. Shortly thereafter, in 1964, scientists at the Brookhaven National Laboratory found the missing particle through careful analyses of bubble chamber photographs, and confirmed all its predicted properties.

The patterns of the eightfold way in the field of particle physics have much in common with the periodic table. Whenever a vacancy (a missing particle or element) occurs in the organized patterns, experimentalists have a guide for their investigations.

30.12 QUARKS

As we have noted, leptons appear to be truly elementary particles because they have no measurable size or internal structure, are limited in number, and do not seem to break down into smaller units. Hadrons, on the other hand, are complex particles with size and structure. Furthermore, we know that hadrons decay into other hadrons and are many in number. Table 30.2 lists only those hadrons that are stable against hadronic decay; hundreds of others have been discovered. These facts strongly suggest that hadrons cannot be truly elementary but have some substructure.

MURRAY GELL-MANN, AMERICAN PHYSICIST (B. 1929)

Gell-Mann was awarded the Nobel prize in 1969 for his theoretical studies dealing with subatomic particles. *(Photo courtesy of Michael R. Dressler)*

TABLE 30.3	Properties of Quarks and Antiquarks

Quarks

Name	Symbol	Spin	Charge	Baryon Number	Strangeness	Charm	Bottomness	Topness
Up	u	$\frac{1}{2}$	$+\frac{2}{3}e$	$\frac{1}{3}$	0	0	0	0
Down	d	$\frac{1}{2}$	$-\frac{1}{3}e$	$\frac{1}{3}$	0	0	0	0
Strange	s	$\frac{1}{2}$	$-\frac{1}{3}e$	$\frac{1}{3}$	-1	0	0	0
Charmed	c	$\frac{1}{2}$	$+\frac{2}{3}e$	$\frac{1}{3}$	0	$+1$	0	0
Bottom	b	$\frac{1}{2}$	$-\frac{1}{3}e$	$\frac{1}{3}$	0	0	$+1$	0
Top	t	$\frac{1}{2}$	$+\frac{2}{3}e$	$\frac{1}{3}$	0	0	0	$+1$

Antiquarks

Name	Symbol	Spin	Charge	Baryon Number	Strangeness	Charm	Bottomness	Topness
Anti-up	$\overline{u}$	$\frac{1}{2}$	$-\frac{2}{3}e$	$-\frac{1}{3}$	0	0	0	0
Anti-down	$\overline{d}$	$\frac{1}{2}$	$+\frac{1}{3}e$	$-\frac{1}{3}$	0	0	0	0
Anti-strange	$\overline{s}$	$\frac{1}{2}$	$+\frac{1}{3}e$	$-\frac{1}{3}$	$+1$	0	0	0
Anti-charmed	$\overline{c}$	$\frac{1}{2}$	$-\frac{2}{3}e$	$-\frac{1}{3}$	0	-1	0	0
Anti-bottom	$\overline{b}$	$\frac{1}{2}$	$+\frac{1}{3}e$	$-\frac{1}{3}$	0	0	-1	0
Anti-top	$\overline{t}$	$\frac{1}{2}$	$-\frac{2}{3}e$	$-\frac{1}{3}$	0	0	0	-1

THE ORIGINAL QUARK MODEL

In 1963 Gell-Mann and George Zweig independently proposed that hadrons have a more elementary substructure. According to their model, all hadrons are composite systems of two or three fundamental constitutents called **quarks,** which rhymes with "forks." Gell-Mann borrowed the word *quark* from the passage "Three quarks for Muster Mark" in James Joyce's book *Finnegan's Wake.* In the original model, there were three types of quarks designated by the symbols *u, d,* and *s.* These were given the arbitrary names *up, down,* and *sideways* (or, now more commonly, *strange*).

An unusual property of quarks is that they have fractional electronic charges, as shown—along with other properties—in Table 30.3. Associated with each quark is an antiquark of opposite charge, baryon number, and strangeness. The compositions of all hadrons known when Gell-Mann and Zweig presented their models could be completely specified by three simple rules:

1. Mesons consist of one quark and one antiquark, giving them a baryon number of 0, as required.
2. Baryons consist of three quarks.
3. Antibaryons consist of three antiquarks.

Table 30.4 lists the quark compositions of several mesons and baryons. Note that just two of the quarks, u and d, are contained in all hadrons encountered in ordinary matter (protons and neutrons). The third quark, s, is needed only to construct strange particles with a strangeness of either $+1$ or -1. Figure 30.10 shows the quark compositions of several particles.

TABLE 30.4

Quark Composition of Several Hadrons

Particle	Quark Composition
Mesons	
π^+	$u\overline{d}$
π^-	$\overline{u}d$
K^+	$u\overline{s}$
K^-	$\overline{u}s$
K^0	$d\overline{s}$
Baryons	
p	uud
n	udd
Λ^0	uds
Σ^+	uus
Σ^0	uds
Σ^-	dds
Ξ^0	uss
Ξ^-	dss
Ω^-	sss

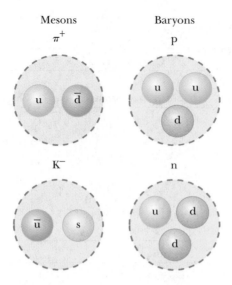

Mesons   Baryons

π^+ p

K⁻   n

FIGURE 30.10   Quark compositions of two mesons and two baryons. Note that the mesons on the left contain two quarks, and the baryons on the right contain three quarks.

APPLYING PHYSICS 30.3

We have seen a law of conservation of lepton number and a law of conservation of baryon number. Why isn't there a law of conservation of meson number?

Explanation   We can argue this from the point of view of creating particle-antiparticle pairs from available energy. If energy is converted to rest energy of a lepton-antilepton pair, then there is no net change in lepton number, because the lepton has a number of $+1$ and the antilepton has a number of -1. Energy could also be transformed into rest energy of a baryon-antibaryon pair. The baryon has baryon number $+1$, the antibaryon -1, and there is no net change in baryon number.

But now suppose energy is transformed into rest energy of a quark-antiquark pair. By definition in quark theory, a quark antiquark pair is a meson. Thus, we have created a meson from energy—there was no meson before; now there is. Thus, there is no conservation of meson number. With more energy, we can create more mesons, with no restriction from a conservation law other than that of energy.

CHARM AND OTHER RECENT DEVELOPMENTS

Although the original quark model was highly successful in classifying particles into families, there were some discrepancies between predictions of the model and certain experimental decay rates. As a consequence, a fourth quark was proposed by several physicists in 1967. The fourth quark, designated by c, was given a property called **charm.** A charmed quark would have the charge $+2e/3$, but its charm would distinguish it from the other three quarks. The new quark would have a charm of $C = +1$, its antiquark would have a charm of $C = -1$, and all other quarks would have $C = 0$, as indicated in Table 30.3. Charm, like strangeness, would be conserved in strong and electromagnetic interactions but not in weak interactions.

In 1974 a new heavy meson called the J/ψ particle (or simply ψ) was discovered independently by a group led by Burton Richter at the Stanford Linear Accelerator (SLAC) and another group led by Samuel Ting at the Brookhaven National Laboratory. Richter and Ting were awarded the Nobel prize in 1976 for this work. The J/ψ particle does not fit into the three-quark model, but has the properties of a combination of a charmed quark and its antiquark ($c\bar{c}$). It is much heavier than the other known mesons ($\sim 3\ 100\ \text{MeV}/c^2$) and its lifetime is much longer than those of other particles that decay via the strong force. In 1975 researchers at Stan-

TABLE 30.5	The Fundamental Particles and Some of Their Properties

Quarks

Particle	Rest Energy	Charge
u	360 MeV	$+\frac{2}{3}e$
d	360 MeV	$-\frac{1}{3}e$
c	1 500 MeV	$+\frac{2}{3}e$
s	540 MeV	$-\frac{1}{3}e$
t	173 GeV	$+\frac{2}{3}e$
b	5 GeV	$-\frac{1}{3}e$

Leptons

Particle	Rest Energy	Charge
e^-	511 keV	$-e$
μ^-	107 MeV	$-e$
τ^-	1 784 MeV	$-e$
ν_e	<30 eV	0
ν_μ	<0.5 MeV	0
ν_τ	<250 MeV	0

ford University reported strong evidence for the tau (τ) lepton, with a mass of 1 784 MeV/c^2. Such discoveries led to more elaborate quark models and the proposal of two new quarks, named *top* (t) and *bottom* (b). (Some physicists prefer the whimsical names *truth* and *beauty*.) To distinguish these quarks from the old ones, quantum numbers called *topness* and *bottomness* were assigned to these new particles and are included in Table 30.3. In 1977 researchers at the Fermi National Laboratory, under the direction of Leon Lederman, reported the discovery of a very massive new meson Υ whose composition is considered to be b$\bar{\text{b}}$. In March of 1995, researchers at Fermilab announced the discovery of the top quark (supposedly the last of the quarks to be found) having mass 173 GeV/c^2.

You are probably wondering whether or not such discoveries will ever end. How many "building blocks" of matter really exist? At the present, physicists believe that the fundamental particles in nature include six quarks and six leptons (together with their antiparticles). Some of the properties of these particles are given in Table 30.5.

Despite extensive experimental efforts, no isolated quark has ever been observed. Physicists now believe that quarks are permanently confined inside ordinary particles because of an exceptionally strong force that prevents them from escaping. This force, called the color force (which we discuss in Section 30.13), increases with separation distance (similar to the force of a spring). The great strength of the force between quarks has been described by one author as follows:[2]

> Quarks are slaves of their own color charge, . . . bound like prisoners of a chain gang. . . . Any locksmith can break the chain between two prisoners, but no locksmith is expert enough to break the gluon chains between quarks. Quarks remain slaves forever.

Webnote 30.8

Further investigate the importance of the search for the top quark by exploring
http://www.duke.edu/vertices/update/fall94/quark.html

---

[2] Harald Fritzsch, *Quarks, The Stuff of Matter,* London: Allen Lane, 1983.

30.13 COLORED QUARKS

Shortly after the theory of quarks was proposed, scientists recognized that certain particles had quark compositions that were in violation of the Pauli exclusion principle. Because all quarks have spins of $\frac{1}{2}$, they are expected to follow the exclusion principle. One example of a particle that violates the exclusion principle is the Ω^- (sss) baryon that contains three s quarks having parallel spins, giving it a total spin of $\frac{3}{2}$. Other examples of baryons that have identical quarks with parallel spins are the Δ^{++} (uuu) and the Δ^- (ddd). To resolve this problem, Moo-Young Han and Yoichiro Nambu suggested in 1965 that quarks possess a new property called **color or color charge.** This "charge" property is similar in many respects to electric charge except that it occurs in three varieties labeled *red, green,* and *blue!* (The antiquarks are labled *antired, antigreen,* and *antiblue.*) To satisfy the exclusion principle, all three quarks in a baryon must have different colors. Just as a combination of actual colors of light can produce the neutral color white, a combination of three quarks with different colors is also white, or colorless. A meson consists of a quark of one color and an antiquark of the corresponding anticolor. The result is that baryons and mesons are always colorless (or white).

Although the concept of color in the quark model was originally conceived to satisfy the exclusion principle, it also provided a better theory for explaining certain experimental results. For example, the modified theory correctly predicts the lifetime of the π^0 meson. The theory of how quarks interact with each other by means of color charge is called **quantum chromodynamics,** or QCD, to parallel quantum electrodynamics (the theory of interaction between electric charges). In QCD, the quark is said to carry a **color charge,** in analogy to electric charge. The strong force between quarks is often called the **color force.**

The strong force between quarks is carried by massless particles called **gluons** (analogous to photons for the electromagnetic force). According to QCD, there are eight gluons, all with color charge. When a quark emits or absorbs a gluon, its color changes. For example, a blue quark that emits a gluon may become a red quark, and the red quark that absorbs this gluon becomes a blue quark. The color force between quarks is analogous to the electric force between charges; like colors repel and opposite colors attract. Therefore, two red quarks repel each other, but a red quark will be attracted to an antired quark. The attraction between quarks of opposite color to form a meson ($q\bar{q}$) is indicated in Figure 30.11a. Differently colored quarks also attract each other, but with less intensity than opposite colors of quark and antiquark. For example, a cluster of red, blue, and green quarks all attract each other to form baryons as indicated in Figure 30.11b. Thus, every baryon contains three quarks of three different colors.

Tip 30.3   COLOR IS NOT REALLY COLOR

When we use the word color to describe a quark, it has nothing to do with visual sensation from light. It is simply a convenient name for a property analogous to electric charge.

Webnote 30.9

For more introductory remarks on quantum chromodynamics, take a look at
http://www.infoplease.com/ce6/sci/ A0840717.html

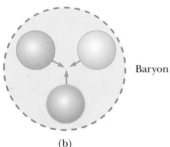

FIGURE 30.11 (a) A red quark is attracted to an antired quark. This forms a meson whose quark structure is ($q\bar{q}$). (b) Three different colored quarks attract each other to form a baryon.

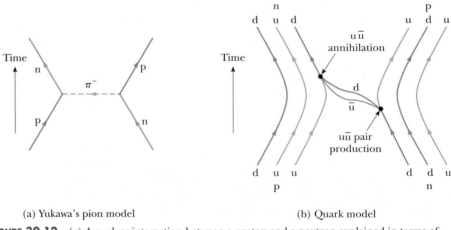

(a) Yukawa's pion model   (b) Quark model

FIGURE 30.12 (a) A nuclear interaction between a proton and a neutron explained in terms of Yukawa's pion exchange model. (b) The same interaction explained in terms of quarks and gluons. Note that the exchanged $\bar{u}d$ quark pair makes up a π^- meson.

Although the color force between two color-neutral hadrons (like a proton and a neutron) is negligible at large separations, the strong color force between their constituent quarks does not exactly cancel at small separations of about 1 fm. **This residual strong force is in fact the nuclear force that binds protons and neutrons to form nuclei.** It is similar to the residual electromagnetic force that binds neutral atoms into molecules. According to QCD, a more basic explanation of nuclear force can be given in terms of quarks and gluons, as shown in Figure 30.12, which shows contrasting Feynman diagrams of the same process. Each quark within the neutron and proton is continually emitting and absorbing virtual gluons and creating and annihilating virtual ($q\bar{q}$) pairs. When the neutron and proton approach within 1 fm of each other, these virtual gluons and quarks can be exchanged between the two nucleons, and such exchanges produce the nuclear force. Figure 30.12b depicts one likely possibility or contribution to the process shown in Figure 30.12a. A down quark emits a virtual gluon (represented by a wavy line), which creates a $u\bar{u}$ pair. Both the recoiling d quark and the $\bar{u}$ are transmitted to the proton where the $\bar{u}$ annihilates a proton u quark (with the creation of a gluon) and the d is captured.

An artist's version of a high-energy particle colliding with a nucleus. The quark structure of the nucleus is indicated by the small colored spheres inside the nucleus. *(Courtesy of Janie Martz/CEBAF)*

30.14 ELECTROWEAK THEORY AND THE STANDARD MODEL

Recall that the weak interaction is an extremely short-range force having an interaction distance of approximately 10^{-18} m (Table 30.1). Such a short-range interaction implies that the quantized particles that carry the weak field (the spin-1 W^+, W^-, and Z^0 bosons) are quite massive as is indeed the case. These amazing bosons can be thought of as structureless, point-like particles as massive as krypton atoms! The weak interaction is responsible for the decay of the c, s, b, and t quarks into lighter, more stable u and d quarks as well as the decay of the massive μ and τ leptons into (lighter) electrons. Thus, **the weak interaction is very important because it governs the stability of the basic matter particles.**

A mysterious feature of the weak interaction is its lack of symmetry, especially compared to the high degree of symmetry shown by the strong, electromagnetic, and gravitational interactions. For example, the weak interaction, unlike the strong interaction, is not symmetric under mirror reflection or charge exchange. (*Mirror reflection* means that all the quantities in a given particle reaction are exchanged as in a mirror reflection—left for right, an inward motion toward the mirror for an outward motion. *Charge exchange* means that all the electric charges in a particle reaction are converted to their opposites—all positives to negatives, and vice versa.) When we say that the weak interaction is not symmetric, we mean that the reaction with all quantities changed occurs less frequently than the direct reaction. For example, the decay of the K^0, which is governed by the weak interaction, is not symmetric under charge exchange because $K^0 \rightarrow \pi^- + e^+ + \nu_e$ occurs much more frequently than $K^0 \rightarrow \pi^+ + e^- + \bar{\nu}_e$.

In 1979, Sheldon Glashow, Abdus Salam, and Steven Weinberg won a Nobel prize for developing a theory called the **electroweak theory** that unified the electromagnetic and weak interactions. This theory postulates that the weak and electromagnetic interactions have the same strength at very high particle energies. Thus, the two interactions are viewed as two different manifestations of a single unifying electroweak interaction. The photon and the three massive bosons ($W^\pm$ and Z^0) play a key role in the electroweak theory. The theory makes many concrete predictions, but perhaps the most spectacular is the prediction of the masses of the W and Z particles at about 82 GeV/c^2 and 93 GeV/c^2, respectively. A 1984 Nobel prize was awarded to Carlo Rubbia and Simon van der Meer for their work leading to the discovery of these particles at just these energies at the CERN Laboratory in Geneva, Switzerland.

The combination of the electroweak theory and QCD for the strong interaction form what is referred to in high-energy physics as the **Standard Model.** Al-

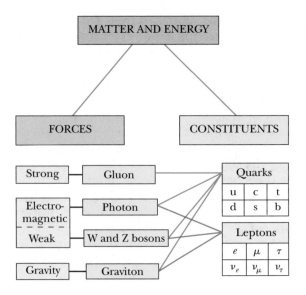

FIGURE 30.13 The Standard Model of particle physics.

A view from inside the LEP (Large Electron-Positron Collider) tunnel, which is 27 km in circumference. *(Courtesy of CERN)*

though the details of the Standard Model are complex, its essential ingredients can be summarized with the help of Figure 30.13. The strong force, mediated by gluons, holds quarks together to form composite particles such as protons, neutrons, and mesons. Leptons participate only in the electromagnetic and weak interactions. The electromagnetic force is mediated by photons, and the weak force is mediated by W and Z bosons. Note that all fundamental forces are mediated by bosons (particles with spin 1) whose properties are given, to a large extent, by symmetries involved in the theories.

However, the Standard Model does not answer all questions. A major question is why the photon has no mass, whereas the W and Z bosons do. Because of this mass difference, the electromagnetic and weak forces are quite distinct at low energies, but become similar in nature at very high energies, where the rest energies of the W and Z bosons are insignificant fractions of their total energies. This behavior as we go from high to low energies, called **symmetry breaking,** does not answer the question of the origin of particle masses. To resolve this problem, a hypothetical particle called the **Higgs boson** has been proposed that provides a mechanism for breaking the electroweak symmetry and bestowing different particle masses on different particles. The Standard Model, including the Higgs mechanism, provides a logically consistent explanation of the massive nature of the W and Z bosons. Unfortunately, the Higgs boson has not yet been found, but physicists know that its mass should be less than 1 TeV/c^2 (10^{12} eV).

In order to determine whether the Higgs boson exists, two quarks of at least 1 TeV of energy must collide, but calculations show that this requires injecting 40 TeV of energy within the volume of a proton. Scientists are convinced that because of the limited energy available in conventional accelerators using fixed targets, it is necessary to build colliding-beam accelerators called **colliders.** The concept of colliders is straightforward. Particles with equal masses and kinetic energies, traveling in opposite directions in an accelerator ring, collide head-on to produce the required reaction and the formation of new particles. Because the total momentum of the interacting particles is zero, all of their kinetic energy is available for the reaction. The Large Electron-Positron collider (LEP) at CERN near Geneva, Switzerland, and the Stanford Linear Collider in California collide both electrons and positrons. The Super Proton Synchrotron at CERN accelerates protons and antiprotons to energies of 270 GeV, and the world's highest-energy proton acclerator, the Tevatron, at the Fermi National Laboratory in Illinois produces protons at almost 1 000 GeV (or 1 TeV). CERN has started construction of the Large Hadron Collider (LHC), a proton-proton collider that will provide a center of mass energy of 14 TeV and allow an exploration of Higgs-boson physics.

The accelerator is being constructed in the same 27-km circumference tunnel as CERN's Large Electron-Positron collider, and construction is expected to be completed in 2005.

Following the success of the electroweak theory, scientists attempted to combine it with QCD in a **grand unification theory** known as GUT. In this model, the electroweak force was merged with the strong color force to form a grand unified force. One version of the theory considers leptons and quarks as members of the same family that are able to change into each other by exchanging an appropriate particle. Many GUT theories predict that protons are unstable, and will decay with a lifetime of about 10^{31} years. This is far greater than the age of the Universe, and as yet proton decays have not been observed.

APPLYING PHYSICS 30.4

Consider a car making a head-on collision with an identical car moving in the opposite direction at the same speed. Compare that collision to one in which one of the cars collides with the second car at rest. In which collision is there the larger transformation of kinetic energy to other forms during the collision? How does this relate to producing exotic particles in collisions?

Explanation In the head-on collision with both cars moving, conservation of momentum causes most, if not all, of the kinetic energy to be transformed to other forms. In the collision between a moving car and a stationary car, the cars are still moving after the collision, in the direction of the moving car, but with reduced speed. Thus, only part of the kinetic energy is transformed to other forms. This suggests the advantage of using colliding beams to produce exotic particles, as opposed to firing a beam into a stationary target. When particles moving in opposite directions collide, all of the kinetic energy is available for transformation into other forms—in this case, the creation of new particles. When a beam is fired into a stationary target, only part of the energy is available for transformation, so higher-mass particles cannot be created.

30.15 THE COSMIC CONNECTION

In this section we describe one of the most fascinating theories in all of science—the Big Bang theory of the creation of the Universe—and the experimental evidence that supports it. This theory of cosmology states that the Universe had a beginning, and that this beginning was so cataclysmic that it is impossible to look back beyond it. According to the theory, the Universe erupted from an infinitely dense singularity about 15–20 billion years ago. The first few minutes after the Big Bang saw such extremes of energy that it is believed that all four interactions of physics were unified and all matter was contained in an undifferentiated "quark soup."

The evolution of the four fundamental forces from the Big Bang to the present is shown in Figure 30.14. During the first 10^{-43} s (the ultra-hot epoch, $T \sim 10^{32}$ K), it is presumed that the strong, electroweak, and gravitational forces were joined to form a completely unified force. In the first 10^{-35} s following the Big Bang (the hot epoch, $T \sim 10^{29}$ K), gravity broke free of this unification and the strong and electroweak forces remained as one, described by a grand unification theory. This was a period when particle energies were so great ($> 10^{16}$ GeV) that very massive particles as well as quarks, leptons, and their antiparticles existed. Then, after 10^{-35} s, the Universe rapidly expanded and cooled (the warm epoch, $T \sim 10^{29}$ to 10^{15} K), the strong and electroweak forces parted company, and the grand unification scheme was broken. As the Universe continued to cool, the electroweak force split into the weak force and the electromagnetic force about 10^{-10} s after the Big Bang.

GEORGE GAMOW (1904–1968)
Gamow and two of his students, Ralph Alpher and Robert Herman, were the first to take the first half-hour of the Universe seriously. In a mostly overlooked paper published in 1948, they made truly remarkable cosmological predictions. They correctly calculated the abundances of hydrogen and helium after the first half-hour (75% H and 25% He) and predicted that radiation from the Big Bang should still be present and have an apparent temperature of about 5 K. *(Courtesy of AIP Emilio Segre Visual Archives)*

FIGURE 30.14 A brief history of the Universe from the Big Bang to the present. The four forces became distinguishable during the first microsecond. Following this, all the quarks combined to form particles that interact via the strong force. The leptons remained separate, however, and exist as individually observable particles to this day.

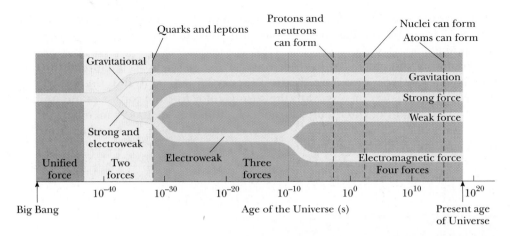

After a few minutes, protons condensed out of the hot soup. For half an hour the Universe underwent thermonuclear detonation, exploding as a hydrogen bomb and producing most of the helium nuclei now present. The Universe continued to expand, and its temperature dropped. Until about 700 000 years after the Big Bang, the Universe was dominated by radiation. Energetic radiation prevented matter from forming single hydrogen atoms because collisions would instantly ionize any atoms that might form. Photons experienced continuous Compton scattering from the vast number of free electrons, resulting in a Universe that was opaque to radiation. By the time the Universe was about 700 000 years old, it had expanded and cooled to about 3 000 K, and protons could bind to electrons to form neutral hydrogen atoms. Because the energies of the atoms were quantized, far more wavelengths of radiation were not absorbed by atoms than were, and the Universe suddenly became transparent to photons. Radiation no longer dominated the Universe, and clumps of neutral matter steadily grew—first atoms, followed by molecules, gas clouds, stars, and finally galaxies.

OBSERVATION OF RADIATION FROM THE PRIMORDIAL FIREBALL

In 1965 Arno A. Penzias (b. 1933) and Robert W. Wilson (b. 1936) of Bell Laboratories made an amazing discovery while testing a sensitive microwave receiver. A pesky signal producing a faint background hiss was interfering with their satellite communications experiments. In spite of their valiant efforts, the signal remained. Ultimately it became clear that they were observing microwave background radiation (at a wavelength of 7.35 cm) representing the leftover "glow" from the Big Bang.

The microwave horn that served as their receiving antenna is shown in Figure 30.15. The intensity of the detected signal remained unchanged as the antenna was pointed in different directions. The fact that the radiation had equal strengths in all directions suggested that the entire Universe was the source of this radiation. Evicting a flock of pigeons from the 20-foot horn and cooling the microwave detector both failed to remove the signal. Through a casual conversation, Penzias and Wilson discovered that a group at Princeton had predicted the residual radiation from the Big Bang and were planning an experiment to confirm the theory. The excitement in the scientific community was high when Penzias and Wilson announced that they had already observed an excess microwave background compatible with a 3-K blackbody source.

Because Penzias and Wilson made their measurements at a single wavelength, they did not completely confirm the radiation as 3-K blackbody radiation. Subsequent experiments by other groups added intensity data at different wavelengths as shown in Figure 30.16. The results confirm that the radiation is that of a black body at 2.9 K. This figure is perhaps the most clear-cut evidence for the Big Bang

FIGURE 30.15 Robert W. Wilson *(left)* and Arno A. Penzias *(right)* with Bell Telephone Laboratories' horn-reflector antenna. *(AT&T Bell Laboratories)*

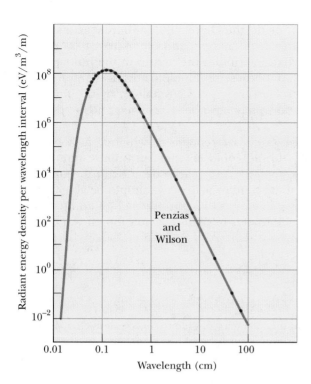

FIGURE 30.16 Theoretical blackbody (red curve) and measured radiation spectra (black points) of the Big Bang. Most of the data were collected from the Cosmic Background Explorer (COBE) satellite. The data of Wilson and Penzias are indicated.

theory. The 1978 Nobel prize in physics was awarded to Penzias and Wilson for this most important discovery.

The discovery of the cosmic background radiation produced a problem, however—the radiation was too uniform. Scientists believed there had to be slight fluctuations in this background in order for such objects as galaxies to form. In 1989 NASA launched a satellite called COBE (KOH-bee), for <u>Co</u>smic <u>B</u>ackground <u>E</u>xplorer, to study this radiation in greater detail. In 1992 George Smoot (b. 1945) at the Lawrence Berkeley Laboratory, based on the data collected, found that the background was not perfectly uniform but instead contained irregularities corresponding to temperature variations of 0.000 3 K. It is these small variations that provided nucleation sites for the formation of the galaxies and other objects we now see in the sky.

30.16 PROBLEMS AND PERSPECTIVES

While particle physicists have been exploring the realm of the very small, cosmologists have been exploring cosmic history back to the first microsecond of the Big Bang. Observation of the events that occur when two particles collide in an accelerator is essential in reconstructing the early moments in cosmic history. Perhaps the key to understanding the early Universe is to first understand the world of elementary particles. Cosmologists and particle physicists find that they have many common goals and are now joining efforts to attempt to study the physical world at its most fundamental level.

Our understanding of physics at short and long distances is far from complete. Particle physics is faced with many questions. Why is there so little antimatter in the Universe? Do neutrinos have a small mass, and if so, how much do they contribute to the "dark matter" holding the Universe together gravitationally? How can we understand the latest astronomical measurements, which show that the expansion of the Universe is accelerating and that there may be a kind of "antigravity force" acting between widely separated galaxies? Is it possible to unify the strong and electroweak theories in a logical and consistent manner? Why do quarks and leptons form three similar but distinct families? Are muons the same as electrons (apart from their different masses), or do they have subtle differences that have not been detected? Why are some particles charged and others neutral? Why do quarks carry a fractional charge? What determines the masses of the fundamental particles? The questions go on and on. Because of the rapid advances and new discoveries in the related fields of particle physics and cosmology, by the time you read this book some of these questions may have been resolved and others may have emerged.

An important and obvious question that remains is whether leptons and quarks have a substructure. If they do, we could envision an infinite number of deeper structure levels. However, if leptons and quarks are indeed the ultimate constituents of matter, as physicists today tend to believe, we should be able to construct a final theory of the structure of matter as Einstein dreamed of doing. In the view of many physicists, the end of the road is in sight, but how long it will take to reach that goal is anyone's guess.

SUMMARY

In **nuclear fission** and **nuclear fusion,** the total mass of the products is always less than the original mass of the reactants. Nuclear fission occurs when a heavy nucleus splits, or fissions, into two smaller nuclei. In nuclear fusion, two light nuclei combine to form a heavier nucleus.

A **nuclear reactor** is a system designed to maintain a self-sustaining chain reaction. Nuclear reactors using controlled fission events are currently being used to generate electric power.

Controlled fusion events offer the hope of plentiful supplies of energy in the future. The nuclear fusion reactor is considered by many scientists to be the ultimate energy source because its fuel is water. **Lawson's criterion** states that a fusion reactor will provide a net output power if the product of the plasma ion density, n, and the plasma confinement time, τ, satisfies the following relationships:

$$n\tau \geq 10^{14} \text{ s/cm}^3 \quad \text{Deuterium-tritium interaction}$$
$$n\tau \geq 10^{16} \text{ s/cm}^3 \quad \text{Deuterium-deuterium interaction}$$

[30.5]

Four fundamental forces in nature: **strong** (hadronic), **electromagnetic, weak,** and **gravitational.** The strong force is the force between nucleons that keeps the nucleus together. The weak force is responsible for beta decay. The electromagnetic and weak forces are now considered to be manifestations of a single force called the **electroweak** force.

Every fundamental interaction is said to be mediated by the exchange of field particles. The electromagnetic interaction is mediated by the photon; the weak interaction is mediated by the $W^\pm$ and Z^0 bosons; the gravitational interaction is mediated by gravitons; and the strong interaction is mediated by gluons.

An antiparticle and a particle have the same mass but opposite charge, and other properties may also have opposite values, such as lepton number and baryon number. It is possible to produce particle-antiparticle pairs in nuclear reactions if the available energy is greater than $2mc^2$, where m is the mass of the particle (or antiparticle).

Particles other than photons are classified as hadrons or leptons. **Hadrons** interact primarily through the strong force. They have size and structure and hence are not elementary particles. There are two types of hadrons, *baryons* and *mesons*. Mesons have a baryon number of 0 and have either zero or integer spin. Baryons, which generally are the most massive particles, have nonzero baryon numbers and spins of $\frac{1}{2}$ or $\frac{3}{2}$. The neutron and proton are examples of baryons.

Leptons have no structure or size and are considered truly elementary particles. Leptons interact only through the weak and electromagnetic forces. There are six leptons: the electron, e^-; the muon, μ^-; the tau, τ^-; and their associated neutrinos, ν_e, ν_μ, and ν_τ.

In all reactions and decays, quantities such as energy, linear momentum, angular momentum, electric charge, baryon number, and lepton number are strictly conserved. Certain particles have properties called **strangeness** and **charm.** These unusual properties are conserved only in those reactions and decays that occur via the strong force.

Recent theories postulate that all hadrons are composed of smaller units known as **quarks,** which have fractional electric charges and baryon numbers of $\frac{1}{3}$ and come in six "flavors": up, down, strange, charmed, top, and bottom. Each baryon contains three quarks, and each meson contains one quark and one antiquark.

According to the theory of **quantum chromodynamics,** quarks have a property called **color,** and the strong force between quarks is referred to as the **color force.**

Observation of background microwave radiation by Penzias and Wilson strongly confirmed that the Universe started with a Big Bang about 15 billion years ago. The background radiation is equivalent to that of a blackbody at a temperature of about 3 K.

CONCEPTUAL QUESTIONS

1. If high-energy electrons with the de Broglie wavelengths smaller than the size of the nucleus are scattered from nuclei, the behavior of the electrons is consistent with scattering from very dense structures much smaller in size than the nucleus—quarks. How is this similar to another classic experiment that detected small structures in atoms?

2. What factors make a fusion reaction difficult to achieve?

3. Doubly charged baryons are known to exist. Why are there no doubly charged mesons?

4. Why would a fusion reactor produce less radioactive waste than a fission reactor?

5. Atoms did not exist until hundreds of thousands of years after the Big Bang. Why?

6. Particles known as resonances have very short half-lives, of the order of 10^{-23} s. Would you guess they are hadrons or leptons?

7. Describe the quark model of hadrons, including the properties of quarks.

8. In the theory of quantum chromodynamics, quarks come in three colors. How would you justify the statement that "all baryons and mesons are colorless"?

9. Describe the properties of baryons and mesons and the important differences between them.

10. Identify the particle decays in Table 30.2 that occur by the electromagnetic interaction. Justify your answer.

11. Kaons all decay into final states that contain no protons or neutrons. What is the baryon number of kaons?

12. When an electron and a positron meet at low speeds in free space, why are *two* 0.511-MeV gamma rays produced, rather than *one* gamma ray with an energy of 1.02 MeV?

13. Two protons in a nucleus interact via the strong interaction. Are they also subject to the weak interaction?

14. Why is a neutron stable inside the nucleus? (In free space, the neutron decays in 900 s.)

15. An antibaryon interacts with a meson. Can a baryon be produced in such an interaction? Explain.

16. Why is water a better shield against neutrons than lead or steel?

17. How many quarks are there in (a) a baryon, (b) an antibaryon, (c) a meson, (d) an antimeson? How do you account for the fact that baryons have half-integral spins and mesons have spins of 0 or 1? (*Hint:* Quarks have spin $\frac{1}{2}$.)

18. Which of the four fundamental forces have effects on an electron? On a proton? On a neutron?

19. List similarities and differences between nuclear-powered electric generating stations and coal-powered ones.

PROBLEMS

1, 2, 3 = straightforward, intermediate, challenging   ☐ = full solution available in Student Solutions Manual/Study Guide

web = solution posted at **http://info.brookscole.com/serway**   🐾 = biomedical application

Section 30.1 Nuclear Fission

Section 30.2 Nuclear Reactors

1. When ^{235}U absorbs a neutron, one possible fission reaction produces ^{141}Ba and ^{92}Kr as products. Write down this reaction. How many neutrons are released?

2. Find the energy released in the fission reaction

$$n + {}^{235}_{92}\text{U} \longrightarrow {}^{98}_{40}\text{Zr} + {}^{135}_{52}\text{Te} + 3\,n$$

The atomic masses of the fission products are: $^{98}_{40}$Zr, 97.912 0 u; $^{135}_{52}$Te, 134.908 7 u.

3. Find the energy released in the following fission reaction:

$$^{1}_{0}n + {}^{235}_{92}\text{U} \longrightarrow {}^{88}_{38}\text{Sr} + {}^{136}_{54}\text{Xe} + 12\,{}^{1}_{0}n$$

4. Strontium-90 is a particularly dangerous fission product of ^{235}U because it is radioactive and it substitutes for calcium in bones. What other direct fission products would accompany it in the neutron-induced fission of ^{235}U? (*Note:* This reaction may release two, three, or four free neutrons.)

5. Assume that ordinary soil contains natural uranium in amounts of 1 part per million by mass. (a) How much uranium is in the top 1.00 meter of soil on a one acre (43 560 ft^2) plot of ground, assuming the specific gravity of soil is 4.00? (b) How much of the isotope ^{235}U, appropriate for nuclear reactor fuel, is in this soil? (*Hint:* See Appendix B for the percent abundance of $^{235}_{92}$U.)

6. A typical nuclear fission power plant produces about 1.00 GW of electrical power. Assume that the plant has an

overall efficiency of 40.0% and that each fission produces 200 MeV of thermal energy. Calculate the mass of ^{235}U consumed each day.

7. Suppose that the water exerts an average frictional drag of 1.0×10^5 N on a nuclear-powered ship. How far can the ship travel per kilogram of fuel if the fuel consists of enriched uranium containing 1.7% of the fissionable isotope ^{235}U, and the ship's engine has a efficiency of 20%? (Assume 208 MeV released per fission event.)

web

8. The first atomic bomb released an energy equivalent to 20 kilotons of TNT. If 1 ton of TNT releases about 4.0×10^9 J, how much uranium was fissioned in this bomb? (Assume 208 MeV released per fission.)

9. An all-electric home uses approximately 2 000 kWh of electric energy per month. How much uranium-235 would be required to provide this house with its energy needs for 1 year? (Assume 100% conversion efficiency and 208 MeV released per fission.)

Section 30.3 Nuclear Fusion

10. Find the energy released in the fusion reaction

$$^1_1\text{H} + ^2_1\text{H} \longrightarrow ^3_2\text{He} + \gamma$$

11. When a star has exhausted its hydrogen fuel, it may fuse other nuclear fuels. At temperatures above 1.0×10^8 K, helium fusion can occur. Write the equation for the processes described below. (a) Two alpha particles fuse to produce a nucleus A and a gamma ray. What is nucleus A? (b) Nucleus A absorbs an alpha particle to produce a nucleus B and a gamma ray. What is nucleus B? (c) Find the total energy released in the sequence of reactions given in (a) and (b).

12. Another series of nuclear reactions that can produce energy in the interior of stars is the cycle described below. This process is most efficient when the central temperature in a star is above 1.6×10^7 K. Because the temperature at the center of the Sun is only 1.5×10^7 K, the cycle below produces less than 10% of the Sun's energy. (a) A high-energy proton is absorbed by ^{12}C. Another nucleus, A, is produced in the reaction, along with a gamma ray. Identify nucleus A. (b) Nucleus A decays through positron emission to form nucleus B. Identify nucleus B. (c) Nucleus B absorbs a proton to produce nucleus C and a gamma ray. Identify nucleus C. (d) Nucleus C absorbs a proton to produce nucleus D and a gamma ray. Identify nucleus D. (e) Nucleus D decays through positron emission to produce nucleus E. Identify nucleus E. (f) Nucleus E absorbs a proton to produce nucleus F plus an alpha particle. What is nucleus F? (*Note:* If nucleus F is not ^{12}C, the nucleus you started with, you have made an error and should review the sequence of events.)

13. If an all-electric home uses approximately 2 000 kWh of electric energy per month, how many fusion events described by the reaction $^2_1\text{H} + ^3_1\text{H} \rightarrow ^4_2\text{He} + ^1_0\text{n}$ would be required to keep this house running for one year?

14. To understand why plasma containment is necessary, consider the rate at which an unconfined plasma would be lost. (a) Estimate the rms speed of deuterons in a plasma at 4.00×10^8 K. (b) Estimate the order of magnitude of the time such a plasma would remain in a 10-cm cube if no steps were taken to contain it.

15. The oceans have a volume of 317 million cubic miles and contain 1.32×10^{21} kg of water. Of all the hydrogen nuclei in this water, 0.0300% of the mass is deuterium. (a) If all of these deuterium nuclei were fused to helium via the first reaction in Equation 30.4, determine the total amount of energy that could be released. (b) Present world electric power consumption is about 7.00×10^{12} W. If consumption were 100 times greater, how many years would the energy supply calculated in (a) last?

Section 30.6 Positrons and Other Antiparticles

16. Two photons are produced when a proton and an antiproton annihilate each other. What is the minimum frequency and corresponding wavelength of each photon?

17. A photon produces a proton-antiproton pair according to the reaction $\gamma \rightarrow \text{p} + \bar{\text{p}}$. What is the minimum possible frequency of the photon? What is its wavelength?

web

18. A photon with an energy of 2.09 GeV creates a proton-antiproton pair in which the proton has a kinetic energy of 95.0 MeV. What is the kinetic energy of the antiproton?

Section 30.7 Mesons and the Beginning of Particle Physics

19. When a high-energy proton or pion traveling near the speed of light collides with a nucleus, it travels an average distance of 3.0×10^{-15} m before interacting. From this information, estimate the order of magnitude of the time for the strong interaction to occur.

20. Calculate the order of magnitude of the range of the force that might be produced by the virtual exchange of a proton.

21. One of the mediators of the weak interaction is the Z^0 boson, whose mass is 93 GeV/c^2. Use this information to find the order of magnitude of the range of the weak interaction.

22. If a π^0 at rest decays into two γ's, what is the energy of each of the γ's?

Section 30.9 Conservation Laws

Section 30.10 Strange Particles and Strangeness

23. Each of the following reactions is forbidden. Determine a conservation law that is violated for each reaction.
    (a) $\text{p} + \bar{\text{p}} \rightarrow \mu^+ + e^-$ (d) $\text{p} + \text{p} \rightarrow \text{p} + \text{p} + \text{n}$
    (b) $\pi^- + \text{p} \rightarrow \text{p} + \pi^+$ (e) $\gamma + \text{p} \rightarrow \text{n} + \pi^0$
    (c) $\text{p} + \text{p} \rightarrow \text{p} + \pi^+$

24. (a) Show that baryon number and charge are conserved in the following reactions of a pion with a proton.

$$\pi^+ + \text{p} \longrightarrow \text{K}^+ + \Sigma^+ \tag{1}$$

$$\pi^+ + \text{p} \longrightarrow \pi^+ + \Sigma^+ \tag{2}$$

(b) The first reaction is observed, but the second never occurs. Explain these observations.

25. Identify the unknown particle on the left side of the reaction

web

$$? + \text{p} \longrightarrow \text{n} + \mu^+$$

26. Determine the type of neutrino or antineutrino involved in each of the following processes:
    (a) $\pi^+ \rightarrow \pi^0 + e^+ + ?$ (c) $\Lambda^0 \rightarrow \text{p} + \mu^- + ?$
    (b) $? + \text{p} \rightarrow \mu^- + \text{p} + \pi^+$ (d) $\tau^+ \rightarrow \mu^+ + ? + ?$

27. The following reactions or decays involve one or more neutrinos. Supply the missing neutrinos.
(a) $\pi^- \to \mu^- + ?$ (d) $? + n \to p + e^-$
(b) $K^+ \to \mu^+ + ?$ (e) $? + n \to p + \mu^-$
(c) $? + p \to n + e^+$ (f) $\mu^- \to e^- + ? + ?$

28. Determine which of the reactions below can occur. For those that cannot occur, determine the conservation law (or laws) that each violates.
(a) $p \to \pi^+ + \pi^0$ (d) $\pi^+ \to \mu^+ + \nu_\mu$
(b) $p + p \to p + p + \pi^0$ (e) $n \to p + e^- + \overline{\nu}_e$
(c) $p + p \to p + \pi^+$ (f) $\pi^+ \to \pi^+ + n$

29. Which of the following processes are allowed by the strong interaction, the electromagnetic interaction, the weak interaction, or no interaction at all?
(a) $\pi^- + p \to 2\eta^0$ (d) $\Omega^- \to \Xi^- + \pi^0$
(b) $K^- + n \to \Lambda^0 + \pi^-$ (e) $\eta^0 \to 2\gamma$
(c) $K^- \to \pi^- + \pi^0$

30. A K^0 particle at rest decays into a π^+ and a π^-. What will be the speed of each of the pions? The mass of the K^0 is 497.7 MeV/c^2 and the mass of each pion is 139.6 MeV/c^2.

31. Determine whether or not strangeness is conserved in the following decays and reactions.
(a) $\Lambda^0 \to p + \pi^-$ (d) $\pi^- + p \to \pi^- + \Sigma^+$
(b) $\pi^- + p \to \Lambda^0 + K^0$ (e) $\Xi^- \to \Lambda^0 + \pi^-$
(c) $\overline{p} + p \to \overline{\Lambda}^0 + \Lambda^0$ (f) $\Xi^0 \to p + \pi^-$

32. Fill in the missing particle. Assume that (a) occurs via the strong interaction and that (b) and (c) involve the weak interaction.
(a) $K^+ + p \to \underline{\quad} + p$
(b) $\Omega^- \to \underline{\quad} + \pi^-$
(c) $K^+ \to \underline{\quad} + \mu^+ + \nu_\mu$

33. Identify the conserved quantities in the following processes:
(a) $\Xi^- \to \Lambda^0 + \mu^- + \nu_\mu$ (d) $\Sigma^0 \to \Lambda^0 + \gamma$
(b) $K^0 \to 2\pi^0$ (e) $e^+ + e^- \to \mu^+ + \mu^-$
(c) $K^- + p \to \Sigma^0 + n$ (f) $\overline{p} + n \to \overline{\Lambda}^0 + \Sigma^-$

Section 30.12 Quarks

Section 30.13 Colored Quarks

34. The quark composition of the proton is uud, whereas that of the neutron in udd. Show that the charge, baryon number, and strangeness of these particles equal the sums of these numbers for their quark constituents.

35. Find the number of electrons, and of each species of quarks, in 1 liter of water.

36. The quark compositions of the K^0 and Λ^0 particles are d$\overline{s}$ and uds, respectively. Show that the charge, baryon number, and strangeness of these particles equal the sums of these numbers for the quark constituents.

37. Identify the particles corresponding to the following quark states: (a) suu; (b) $\overline{u}$d; (c) $\overline{s}$d; (d) ssd.

38. What is the electrical charge of the baryons with the quark compositions (a) $\overline{u}\overline{u}\overline{d}$ and (b) $\overline{u}\overline{d}\overline{d}$? What are these baryons called?

39. Analyze the first three of the following reactions at the quark level and show that each conserves the net number of each type quark. In the last reaction, identify the mystery particle.
(a) $\pi^- + p \to K^0 + \Lambda^0$
(b) $\pi^+ + p \to K^+ + \Sigma^+$
(c) $K^- + p \to K^+ + K^0 + \Omega^-$
(d) $p + p \to K^0 + p + \pi^+ + ?$

40. Imagine binding energies could be neglected. Find the masses of the u and d quarks from the masses of the proton and neutron.

web

ADDITIONAL PROBLEMS

41. A Σ^0 particle traveling through matter strikes a proton. A Σ^+ and a gamma ray emerge, as well as a third particle. Use the quark model of each to determine the identity of the third particle.

42. It was stated in the text that the reaction $\pi^- + p^+ \to K^0 + \Lambda^0$ occurs with high probability, whereas the reaction $\pi^- + p^+ \to K^0 + n$ never occurs. Analyze these reactions at the quark level and show that the first conserves the net number of each type of quark while the second reaction does not.

43. Two protons approach each other with equal and opposite velocities. What is the minimum kinetic energy of each of the protons if they are to produce a π^+ meson at rest in the reaction,

$$p + p \longrightarrow p + n + \pi^+$$

44. Name at least one conservation law that prevents each of the following reactions:
(a) $\pi^- + p \to \Sigma^+ + \pi^0$
(b) $\mu^- \to \pi^- + \nu_e$
(c) $p \to \pi^+ + \pi^+ + \pi^-$

45. Find the energy released in the fusion reaction

$$^1_1H + ^3_2He \longrightarrow ^4_2He + e^+ + \nu$$

46. Occasionally, high-energy muons collide with electrons and produce two neutrinos according to the reaction $\mu^+ + e^- \to 2\nu$. What kind of neutrinos are these?

47. Each of the following decays is forbidden. For each process, determine a conservation law that is violated.
(a) $\mu^- \to e^- + \gamma$ (d) $p \to e^+ + \pi^0$
(b) $n \to p + e^- + \nu_e$ (e) $\Xi^0 \to n + \pi^0$
(c) $\Lambda^0 \to p + \pi^0$

48. Two protons approach each other head on, each with 70.4 MeV of kinetic energy, and engage in a reaction in which a proton and positive pion emerge at rest. What third particle, obviously uncharged and therefore difficult to detect, must have been created?

49. The atomic bomb dropped on Hiroshima on August 6, 1945, released 5×10^{13} J of energy (equivalent to that from 12 000 tons of TNT). Estimate (a) the number of $^{235}_{92}$U nuclei fissioned, and (b) the mass of this $^{235}_{92}$U.

50. A 2.0-MeV neutron is emitted in a fission reactor. If it loses one half of its kinetic energy in each collision with a moderator atom, how many collisions must it undergo in order to achieve thermal energy (0.039 eV)?

51. If baryon number were not conserved, then one possible mechanism by which a proton could decay would be

$$p \longrightarrow e^+ + \gamma$$

(a) Show that this reaction violates conservation of baryon number. (b) Assuming that this reaction occurs, and that the proton is initially at rest, determine the energy and momentum of the photon after the reaction. (*Hint:* Recall that energy and momentum must be conserved in the reaction.) (c) Determine the speed of the positron after the reaction.

52. Classical general relativity views the space-time manifold as a deterministic structure completely well defined down to arbitrarily small distances. On the other hand, quantum general relativity forbids distances smaller than the Planck length given by $L = (\hbar G/c^3)^{1/2}$. (a) Calculate the value of L. The answer suggests that after the Big Bang (when all the known Universe was reduced to a singularity), nothing could be observed until that singularity grew larger than the Planck length, L. Because the size of the singularity grew at the speed of light, we can infer that during the time it took for light to travel the Planck length, no observations were possible. (b) Determine this time (known as the Planck time, T) and compare it to the ultra-hot epoch discussed in the text. (c) Does this suggest that we may never know what happened between the time $t = 0$ and the time $t = T$?

GROUP ACTIVITIES

G.1 (a) A typical chemical reaction is one in which a water molecule is formed by combining hydrogen and oxygen. In such a reaction about 2.5 eV of energy is released. Compare this reaction to $^1_0 n + ^{235}_{92}U \rightarrow ^{136}_{53}I + ^{98}_{39}Y + 2^1_0 n$ (Take the mass of ^{136}I to be 135.914 650 u and the mass of ^{98}Y to be 97.922 300 u.) Would you expect the energy released in this reaction to be much greater, much less, or about the same as the chemical reaction? (b) Find the actual energy released in this reaction.

G.2 (a) Explain in a couple of sentences what is meant by the conservation of baryons. (b) Use conservation of baryons to show that a proton cannot decay into a positron and a neutral pion.

G.3 (a) Explain in a couple of sentences what is meant by conservation of strangeness. (b) Use conservation of strangeness to show that a reaction in which a negative pion strikes a proton producing a neutral kaon and a neutron cannot occur.

"QUARKS. NEUTRINOS. MESONS. ALL THOSE DAMN PARTICLES YOU CAN'T SEE. THAT'S WHAT DROVE ME TO DRINK. BUT NOW I CAN SEE THEM."

Mathematical Review

A.1 MATHEMATICAL NOTATION

Many mathematical symbols are used throughout this book. You are no doubt familiar with some, such as the symbol = to denote the equality of two quantities.

The symbol $\propto$ denotes a proportionality. For example, $y \propto x^2$ means that y is proportional to the square of x.

The symbol $<$ means *is less than*, and $>$ means *is greater than*. For example, $x > y$ means x is greater than y.

The symbol $\ll$ means *is much less than*, and $\gg$ means *is much greater than*.

The symbol $\approx$ indicates that two quantities are *approximately equal* to each other.

The symbol $\equiv$ means *is defined as*. This is a stronger statement than a simple $=$.

It is convenient to use the notation Δx (read as "delta x") to indicate the *change in the quantity x*. (Note that Δx does not mean "the product of Δ and x.") For example, suppose that a person out for a morning stroll starts measuring her distance away from home when she is 10 m from her doorway. She then moves along a straight-line path and stops strolling 50 m from the door. Her change in position during the walk is $\Delta x = 50 \text{ m} - 10 \text{ m} = 40 \text{ m}$ or, in symbolic form,

$$\Delta x = x_f - x_i$$

In this equation x_f is the *final position* and x_i is the *initial position*.

We often have occasion to add several quantities. A useful abbreviation for representing such a sum is the Greek letter Σ (capital sigma). Suppose we wish to add a set of five numbers represented by x_1, x_2, x_3, x_4, and x_5. In the abbreviated notation, we would write the sum as

$$x_1 + x_2 + x_3 + x_4 + x_5 = \sum_{i=1}^{5} x_i$$

where the subscript i on x represents any one of the numbers in the set. For example, if there are five masses in a system, m_1, m_2, m_3, m_4, and m_5, the total mass of the system $M = m_1 + m_2 + m_3 + m_4 + m_5$ could be expressed as

$$M = \sum_{i=1}^{5} m_i$$

Finally, the magnitude of a quantity x, written $|x|$, is simply the absolute value of that quantity. The sign of $|x|$ is always positive, regardless of the sign of x. For example, if $x = -5$, $|x| = 5$; if $x = 8$, $|x| = 8$.

A.2 SCIENTIFIC NOTATION

Many quantities that scientists deal with often have very large or very small values. For example, the speed of light is about 300 000 000 m/s and the ink required to make the dot over an i in this textbook has a mass of about 0.000 000 001 kg. Obviously, it is cumbersome to read, write, and keep track of numbers such as these. We avoid this problem by using a method dealing with powers of the number 10:

$$10^0 = 1$$

$$10^1 = 10$$

$$10^2 = 10 \times 10 = 100$$

$$10^3 = 10 \times 10 \times 10 = 1\,000$$

$$10^4 = 10 \times 10 \times 10 \times 10 = 10\,000$$

$$10^5 = 10 \times 10 \times 10 \times 10 \times 10 = 100\,000$$

and so on. The number of zeros corresponds to the power to which 10 is raised, called the **exponent** of 10. For example, the speed of light, 300 000 000 m/s, can be expressed as 3×10^8 m/s.

For numbers less than one, we note the following:

$$10^{-1} = \frac{1}{10} = 0.1$$

$$10^{-2} = \frac{1}{10 \times 10} = 0.01$$

$$10^{-3} = \frac{1}{10 \times 10 \times 10} = 0.001$$

$$10^{-4} = \frac{1}{10 \times 10 \times 10 \times 10} = 0.000\,1$$

$$10^{-5} = \frac{1}{10 \times 10 \times 10 \times 10 \times 10} = 0.000\,01$$

In these cases, the number of places the decimal point is to the left of the digit 1 equals the value of the (negative) exponent. Numbers that are expressed as some power of 10 multiplied by another number between 1 and 10 are said to be in **scientific notation.** For example, the scientific notation for 5 943 000 000 is 5.943×10^9 and that for 0.000 083 2 is 8.32×10^{-5}.

When numbers expressed in scientific notation are being multiplied, the following general rule is very useful:

$$10^n \times 10^m = 10^{n+m} \qquad \text{[A.1]}$$

where n and m can be *any* numbers (not necessarily integers). For example, $10^2 \times 10^5 = 10^7$. The rule also applies if one of the exponents is negative. For example, $10^3 \times 10^{-8} = 10^{-5}$.

When dividing numbers expressed in scientific notation, note that

$$\frac{10^n}{10^m} = 10^n \times 10^{-m} = 10^{n-m} \qquad \text{[A.2]}$$

Exercises

With help from the above rules, verify the answers to the following:

1. $86\,400 = 8.64 \times 10^4$
2. $9\,816\,762.5 = 9.816\,762\,5 \times 10^6$
3. $0.000\,000\,039\,8 = 3.98 \times 10^{-8}$
4. $(4.0 \times 10^8)(9.0 \times 10^9) = 3.6 \times 10^{18}$
5. $(3.0 \times 10^7)(6.0 \times 10^{-12}) = 1.8 \times 10^{-4}$
6. $\dfrac{75 \times 10^{-11}}{5.0 \times 10^{-3}} = 1.5 \times 10^{-7}$
7. $\dfrac{(3 \times 10^6)(8 \times 10^{-2})}{(2 \times 10^{17})(6 \times 10^5)} = 2 \times 10^{-18}$

A.3 ALGEBRA

A. SOME BASIC RULES

When algebraic operations are performed, the laws of arithmetic apply. Symbols such as x, y, and z are frequently used to represent quantities that are not specified, what are called the **unknowns.**

First, consider the equation

$$8x = 32$$

If we wish to solve for x, we can divide (or multiply) each side of the equation by the same factor without destroying the equality. In this case, if we divide both sides by 8, we have

$$\frac{8x}{8} = \frac{32}{8}$$

$$x = 4$$

Next consider the equation

$$x + 2 = 8$$

In this type of expression, we can add or subtract the same quantity from each side. If we subtract 2 from each side, we get

$$x + 2 - 2 = 8 - 2$$

$$x = 6$$

In general, if $x + a = b$, then $x = b - a$.

Now consider the equation

$$\frac{x}{5} = 9$$

If we multiply each side by 5, we are left with x on the left by itself and 45 on the right:

$$\left(\frac{x}{5}\right)(5) = 9 \times 5$$

$$x = 45$$

In all cases, **whatever operation is performed on the left side of the equality must also be performed on the right side.**

The following rules for multiplying, dividing, adding, and subtracting fractions should be recalled, where a, b, and c are three numbers:

	Rule	Example
Multiplying	$\left(\dfrac{a}{b}\right)\left(\dfrac{c}{d}\right) = \dfrac{ac}{bd}$	$\left(\dfrac{2}{3}\right)\left(\dfrac{4}{5}\right) = \dfrac{8}{15}$
Dividing	$\dfrac{(a/b)}{(c/d)} = \dfrac{ad}{bc}$	$\dfrac{2/3}{4/5} = \dfrac{(2)(5)}{(4)(3)} = \dfrac{10}{12}$
Adding	$\dfrac{a}{b} \pm \dfrac{c}{d} = \dfrac{ad \pm bc}{bd}$	$\dfrac{2}{3} - \dfrac{4}{5} = \dfrac{(2)(5) - (4)(3)}{(3)(5)} = -\dfrac{2}{15}$

Exercises

In the following exercises, solve for x:

Answers

1. $a = \dfrac{1}{1+x}$ $\qquad\qquad$ $x = \dfrac{1-a}{a}$

2. $3x - 5 = 13$ $\qquad\qquad$ $x = 6$

3. $ax - 5 = bx + 2$ $\qquad\quad$ $x = \dfrac{7}{a-b}$

4. $\dfrac{5}{2x+6} = \dfrac{3}{4x+8}$ $\qquad$ $x = -\dfrac{11}{7}$

B. POWERS

When powers of a given quantity x are multiplied, the following rule applies:

$$x^n x^m = x^{n+m} \qquad\qquad \text{[A.3]}$$

For example, $x^2 x^4 = x^{2+4} = x^6$.

When dividing the powers of a given quantity, note that

$$\frac{x^n}{x^m} = x^{n-m} \qquad\qquad \text{[A.4]}$$

For example, $x^8/x^2 = x^{8-2} = x^6$.

A power that is a fraction, such as $\frac{1}{3}$, corresponds to a root as follows:

$$x^{1/n} = \sqrt[n]{x} \qquad\qquad \text{[A.5]}$$

For example, $4^{1/3} = \sqrt[3]{4} = 1.587\ 4$. (A scientific calculator is useful for such calculations.)

Finally, any quantity x^n that is raised to the mth power is

$$(x^n)^m = x^{nm} \qquad\qquad \text{[A.6]}$$

Table A.1 summarizes the rules of exponents.

TABLE A.1

Rules of Exponents

$$x^0 = 1$$
$$x^1 = x$$
$$x^n x^m = x^{n+m}$$
$$x^n/x^m = x^{n-m}$$
$$x^{1/n} = \sqrt[n]{x}$$
$$(x^n)^m = x^{nm}$$

Exercises

Verify the following:

1. $3^2 \times 3^3 = 243$
2. $x^5 x^{-8} = x^{-3}$
3. $x^{10}/x^{-5} = x^{15}$
4. $5^{1/3} = 1.709\ 975$ (Use your calculator.)
5. $60^{1/4} = 2.783\ 158$ (Use your calculator.)
6. $(x^4)^3 = x^{12}$

C. FACTORING

Some useful formulas for factoring an equation are

$$ax + ay + az = a(x + y + z) \qquad \text{common factor}$$
$$a^2 + 2ab + b^2 = (a + b)^2 \qquad \text{perfect square}$$
$$a^2 - b^2 = (a + b)(a - b) \qquad \text{differences of squares}$$

D. QUADRATIC EQUATIONS

The general form of a quadratic equation is

$$ax^2 + bx + c = 0 \qquad\qquad \text{[A.7]}$$

where x is the unknown quantity and a, b, and c are numerical factors referred to as **coefficients** of the equation. This equation has two roots, given by

$$x = \frac{-b \pm \sqrt{b^2 - 4ac}}{2a} \qquad \text{[A.8]}$$

If $b^2 \geq 4ac$, the roots will be real.

Example

The equation $x^2 + 5x + 4 = 0$ has the following roots corresponding to the two signs of the square root term:

$$x = \frac{-5 \pm \sqrt{5^2 - (4)(1)(4)}}{2(1)} = \frac{-5 \pm \sqrt{9}}{2} = \frac{-5 \pm 3}{2}$$

that is,

$$x_+ = \frac{-5 + 3}{2} = -1 \qquad x_- = \frac{-5 - 3}{2} = -4$$

where x_+ refers to the root corresponding to the positive sign and x_- refers to the root corresponding to the negative sign.

Exercises

Solve the following quadratic equations:

Answers

1. $x^2 + 2x - 3 = 0$ $\qquad x_+ = 1 \qquad\qquad x_- = -3$
2. $2x^2 - 5x + 2 = 0$ $\qquad x_+ = 2 \qquad\qquad x_- = 1/2$
3. $2x^2 - 4x - 9 = 0$ $\qquad x_+ = 1 + \sqrt{22}/2 \qquad x_- = 1 - \sqrt{22}/2$

E. LINEAR EQUATIONS

A linear equation has the general form

$$y = ax + b \qquad \text{[A.9]}$$

where a and b are constants. This equation is referred to as being linear because the graph of y versus x is a straight line, as shown in Figure A.1. The constant b, called the **intercept**, represents the value of y at which the straight line intersects the y axis. The constant a is equal to the **slope** of the straight line. If any two points on the straight line are specified by the coordinates (x_1, y_1) and (x_2, y_2), as in Figure A.1, then the slope of the straight line can be expressed

$$\text{Slope} = \frac{y_2 - y_1}{x_2 - x_1} = \frac{\Delta y}{\Delta x} \qquad \text{[A.10]}$$

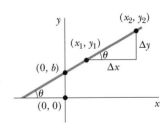

FIGURE A.1

Note that a and b can have either positive or negative values. If $a > 0$, the straight line has a positive slope, as in Figure A.1. If $a < 0$, the straight line has a *negative* slope. In Figure A.1, both a and b are positive. Three other possible situations are shown in Figure A.2: $a > 0$, $b < 0$; $a < 0$, $b > 0$; and $a < 0$, $b < 0$.

Exercises

1. Draw graphs of the following straight lines:
   (a) $y = 5x + 3$ (b) $y = -2x + 4$ (c) $y = -3x - 6$
2. Find the slopes of the straight lines described in Exercise 1.
   Answers: (a) 5   (b) -2 (c) -3

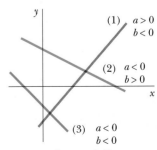

FIGURE A.2

3. Find the slopes of the straight lines that pass through the following sets of points: **(a)** $(0, -4)$ and $(4, 2)$, **(b)** $(0, 0)$ and $(2, -5)$, and **(c)** $(-5, 2)$ and $(4, -2)$

Answers: **(a)** $\frac{3}{2}$ **(b)** $-\frac{5}{2}$ **(c)** $-\frac{4}{9}$

F. SOLVING SIMULTANEOUS LINEAR EQUATIONS

Consider an equation such as $3x + 5y = 15$, which has two unknowns, x and y. Such an equation does not have a unique solution. That is, $(x = 0, y = 3)$, $(x = 5, y = 0)$ and $(x = 2, y = \frac{9}{5})$ are all solutions to this equation.

If a problem has two unknowns, a unique solution is possible only if we have *two* independent equations. In general, if a problem has n unknowns, its solution requires n independent equations. In order to solve two simultaneous equations involving two unknowns, x and y, we solve one of the equations for x in terms of y and substitute this expression into the other equation.

Example

Solve the following two simultaneous equations:

$$(1) \quad 5x + y = -8 \qquad (2) \quad 2x - 2y = 4$$

Solution From (2), we find that $x = y + 2$. Substitution of this into (1) gives

$$5(y + 2) + y = -8$$

$$6y = -18$$

$$\boxed{y = -3}$$

$$x = y + 2 = \boxed{-1}$$

Alternate Solution Multiply each term in (1) by the factor 2 and add the result to (2):

$$10x + 2y = -16$$

$$\underline{2x - 2y = 4}$$

$$12x = -12$$

$$x = \boxed{-1}$$

$$y = x - 2 = \boxed{-3}$$

Two linear equations with two unknowns can also be solved by a graphical method. If the straight lines corresponding to the two equations are plotted in a conventional coordinate system, the intersection of the two lines represents the solution. For example, consider the two equations

$$x - y = 2$$

$$x - 2y = -1$$

These are plotted in Figure A.3. The intersection of the two lines has the coordinates $x = 5$, $y = 3$. This represents the solution to the equations. You should check this solution by the analytical technique discussed above.

Exercises

Solve the following pairs of simultaneous equations involving two unknowns:

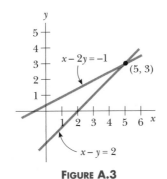

FIGURE A.3

Answers

1. $x + y = 8$ $x = 5, y = 3$
   $x - y = 2$
2. $98 - T = 10a$ $T = 65, a = 3.27$
   $T - 49 = 5a$
3. $6x + 2y = 6$ $x = 2, y = -3$
   $8x - 4y = 28$

G. LOGARITHMS

Suppose that a quantity x is expressed as a power of some quantity a:

$$x = a^y \qquad \text{[A.11]}$$

The number a is called the **base** number. The **logarithm** of x with respect to the base a is equal to the exponent to which the base must be raised in order to satisfy the expression $x = a^y$:

$$y = \log_a x \qquad \text{[A.12]}$$

Conversely, the **antilogarithm** of y is the number x:

$$x = \text{antilog}_a y \qquad \text{[A.13]}$$

In practice, the two bases most often used are base 10, called the *common* logarithm base, and base $e = 2.718 \ldots$, called the *natural* logarithm base. When common logarithms are used,

$$y = \log_{10} x \qquad (\text{or } x = 10^y) \qquad \text{[A.14]}$$

When natural logarithms are used,

$$y = \ln_e x \qquad (\text{or } x = e^y) \qquad \text{[A.15]}$$

For example, $\log_{10} 52 = 1.716$, so that $\text{antilog}_{10} 1.716 = 10^{1.716} = 52$. Likewise, $\ln_e 52 = 3.951$, so $\text{antiln}_e 3.951 = e^{3.951} = 52$.

In general, note that you can convert between base 10 and base e with the equality

$$\ln_e x = (2.302\ 585)\log_{10} x \qquad \text{[A.16]}$$

Finally, some useful properties of logarithms are

$$\log (ab) = \log a + \log b \qquad \ln e = 1$$

$$\log (a/b) = \log a - \log b \qquad \ln e^a = a$$

$$\log (a^n) = n \log a \qquad \ln\left(\frac{1}{a}\right) = -\ln a$$

A.4 GEOMETRY

Table A.2 gives the areas and volumes for several geometric shapes used throughout this text:

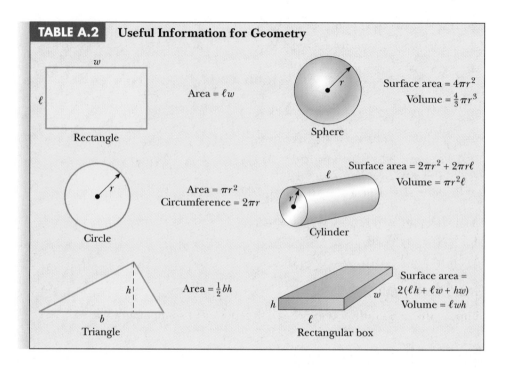

TABLE A.2	Useful Information for Geometry

Area = ℓw

Rectangle

Surface area = $4\pi r^2$
Volume = $\frac{4}{3}\pi r^3$

Sphere

Area = πr^2
Circumference = $2\pi r$

Circle

Surface area = $2\pi r^2 + 2\pi r\ell$
Volume = $\pi r^2 \ell$

Cylinder

Area = $\frac{1}{2}bh$

Triangle

Surface area = $2(\ell h + \ell w + hw)$
Volume = ℓwh

Rectangular box

A.5 TRIGONOMETRY

Some of the most basic facts concerning trigonometry are presented in Chapter 1, and we encourage you to study the material presented there if you are having trouble with this branch of mathematics. In addition to the discussion of Chapter 1, certain useful trig identities that can be of value to you follow.

$$\sin^2\theta + \cos^2\theta = 1$$

$$\sin\theta = \cos(90° - \theta)$$

$$\cos\theta = \sin(90° - \theta)$$

$$\sin 2\theta = 2\sin\theta\cos\theta$$

$$\cos 2\theta = \cos^2\theta - \sin^2\theta$$

$$\sin(\theta \pm \phi) = \sin\theta\cos\phi \pm \cos\theta\sin\phi$$

$$\cos(\theta \pm \phi) = \cos\theta\cos\phi \mp \sin\theta\sin\phi$$

The following relationships apply to *any* triangle, as shown in Figure A.4:

$$\alpha + \beta + \gamma = 180°$$

$$a^2 = b^2 + c^2 - 2bc\cos\alpha$$

Law of cosines
$$b^2 = a^2 + c^2 - 2ac\cos\beta$$

$$c^2 = a^2 + b^2 - 2ab\cos\gamma$$

Law of sines
$$\frac{a}{\sin\alpha} = \frac{b}{\sin\beta} = \frac{c}{\sin\gamma}$$

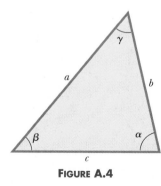

FIGURE A.4

An Abbreviated Table of Isotopes

Atomic Number, Z	Element	Symbol	Chemical Atomic Mass (u)	Mass Number (* Indicates Radioactive) A	Atomic Mass (u)[a]	Percentage Abundance	Half-Life (if Radioactive) $T_{1/2}$
0	(Neutron)	n		1*	1.008 665		10.4 min
1	Hydrogen	H	1.007 9	1	1.007 825	99.985	
	Deuterium	D		2	2.014 102	0.015	
	Tritium	T		3*	3.016 049		12.33 yr
2	Helium	He	4.002 60	3	3.016 029	0.000 14	
				4	4.002 602	99.999 86	
3	Lithium	Li	6.941	6	6.015 121	7.5	
				7	7.016 003	92.5	
4	Beryllium	Be	9.012 2	7*	7.016 928		53.3 days
				9	9.012 174	100	
5	Boron	B	10.81	10	10.012 936	19.9	
				11	11.009 305	80.1	
6	Carbon	C	12.011	11*	11.011 433		20.4 min
				12	12.000 000	98.90	
				13	13.003 355	1.10	
				14*	14.003 242		5730 yr
7	Nitrogen	N	14.006 7	13*	13.005 738		9.96 min
				14	14.003 074	99.63	
				15	15.000 108	0.37	
8	Oxygen	O	15.999 4	15*	15.003 065		122 s
				16	15.994 915	99.761	
				18	17.999 160	0.20	
9	Fluorine	F	18.998 40	19	18.998 404	100	
10	Neon	Ne	20.180	20	19.992 435	90.48	
				22	21.991 383	9.25	
11	Sodium	Na	22.989 87	22*	21.994 434		2.61 yr
				23	22.989 770	100	
				24*	23.990 961		14.96 h
12	Magnesium	Mg	24.305	24	23.985 042	78.99	
				25	24.985 838	10.00	
				26	25.982 594	11.01	
13	Aluminum	Al	26.981 54	27	26.981 538	100	
14	Silicon	Si	28.086	28	27.976 927	92.23	
15	Phosphorus	P	30.973 76	31	30.973 762	100	
				32*	31.973 908		14.26 days

(Table continues)

[a] The masses in the sixth column are atomic masses, which include the mass of Z electrons. Data are from the National Nuclear Data Center, Brookhaven National Laboratory, prepared by Jagdish K. Tuli, July 1990. The data are based on experimental results reported in *Nuclear Data Sheets* and *Nuclear Physics* and also from *Chart of the Nuclides,* 14th ed. Atomic masses are based on those by A. H. Wapstra, G. Audi, and R. Hoekstra. Isotopic abundances are based on those by N. E. Holden.

Atomic Number, Z	Element	Symbol	Chemical Atomic Mass (u)	Mass Number (* Indicates Radioactive) A	Atomic Mass (u)[a]	Percentage Abundance	Half-Life (if Radioactive) $T_{1/2}$
16	Sulfur	S	32.066	32	31.972 071	95.02	
				35*	34.969 033		87.5 days
17	Chlorine	Cl	35.453	35	34.968 853	75.77	
				37	36.965 903	24.23	
18	Argon	Ar	39.948	40	39.962 384	99.600	
19	Potassium	K	39.098 3	39	38.963 708	93.258 1	
				40*	39.964 000	0.0117	1.28×10^9 yr
20	Calcium	Ca	40.08	40	39.962 591	96.941	
21	Scandium	Sc	44.955 9	45	44.955 911	100	
22	Titanium	Ti	47.88	48	47.947 947	73.8	
23	Vanadium	V	50.941 5	51	50.943 962	99.75	
24	Chromium	Cr	51.996	52	51.940 511	83.79	
25	Manganese	Mn	54.938 05	55	54.938 048	100	
26	Iron	Fe	55.847	56	55.934 940	91.72	
27	Cobalt	Co	58.933 20	59	58.933 198	100	
				60*	59.933 820		5.27 yr
28	Nickel	Ni	58.693	58	57.935 346	68.077	
				60	59.930 789	26.223	
29	Copper	Cu	63.54	63	62.929 599	69.17	
				65	64.927 791	30.83	
30	Zinc	Zn	65.39	64	63.929 144	48.6	
				66	65.926 035	27.9	
				68	67.924 845	18.8	
31	Gallium	Ga	69.723	69	68.925 580	60.108	
				71	70.924 703	39.892	
32	Germanium	Ge	72.61	70	69.924 250	21.23	
				72	71.922 079	27.66	
				74	73.921 177	35.94	
33	Arsenic	As	74.921 6	75	74.921 594	100	
34	Selenium	Se	78.96	78	77.917 307	23.78	
				80	79.916 519	49.61	
35	Bromine	Br	79.904	79	78.918 336	50.69	
				81	80.916 287	49.31	
36	Krypton	Kr	83.80	82	81.913 481	11.6	
				83	82.914 136	11.5	
				84	83.911 508	57.0	
				86	85.910 615	17.3	
37	Rubidium	Rb	85.468	85	84.911 793	72.17	
				87*	86.909 186	27.83	4.75×10^{10} yr
38	Strontium	Sr	87.62	86	85.909 266	9.86	
				88	87.905 618	82.58	
				90*	89.907 737		29.1 yr
39	Yttrium	Y	88.905 8	89	88.905 847	100	
40	Zirconium	Zr	91.224	90	89.904 702	51.45	
				91	90.905 643	11.22	
				92	91.905 038	17.15	
				94	93.906 314	17.38	
41	Niobium	Nb	92.906 4	93	92.906 376	100	
42	Molybdenum	Mo	95.94	92	91.906 807	14.84	
				95	94.905 841	15.92	
				96	95.904 678	16.68	
				98	97.905 407	24.13	
43	Technetium	Tc		98*	97.907 215		4.2×10^6 yr
				99*	98.906 254		2.1×10^6 yr

(Table continues)

Atomic Number, Z	Element	Symbol	Chemical Atomic Mass (u)	Mass Number (* Indicates Radioactive) A	Atomic Mass (u)[a]	Percentage Abundance	Half-Life (if Radioactive) $T_{1/2}$
44	Ruthenium	Ru	101.07	99	98.905 939	12.7	
				100	99.904 219	12.6	
				101	100.905 558	17.1	
				102	101.904 348	31.6	
				104	103.905 428	18.6	
45	Rhodium	Rh	102.905 5	103	102.905 502	100	
46	Palladium	Pd	106.42	104	103.904 033	11.14	
				105	104.905 082	22.33	
				106	105.903 481	27.33	
				108	107.903 893	26.46	
				110	109.905 158	11.72	
47	Silver	Ag	107.868	107	106.905 091	51.84	
				109	108.904 754	48.16	
48	Cadmium	Cd	112.41	110	109.903 004	12.49	
				111	110.904 182	12.80	
				112	111.902 760	24.13	
				113*	112.904 401	12.22	9.3×10^{15} yr
				114	113.903 359	28.73	
49	Indium	In	114.82	115*	114.903 876	95.7	4.4×10^{14} yr
50	Tin	Sn	118.71	116	115.901 743	14.53	
				118	117.901 605	24.22	
				120	119.902 197	32.59	
51	Antimony	Sb	121.76	121	120.903 820	57.36	
				123	122.904 215	42.64	
52	Tellurium	Te	127.60	126	125.903 309	18.93	
				128*	127.904 463	31.70	$>8 \times 10^{24}$ yr
				130*	129.906 228	33.87	$\leq 1.25 \times 10^{21}$ yr
53	Iodine	I	126.904 5	127	126.904 474	100	
				129*	128.904 984		1.6×10^{7} yr
				131*	130.906 118		8.04 days
54	Xenon	Xe	131.29	129	128.904 779	26.4	
				131	130.905 069	21.2	
				132	131.904 141	26.9	
				134	133.905 394	10.4	
				136*	135.907 215	8.9	$\geq 2.36 \times 10^{21}$ yr
55	Cesium	Cs	132.905 4	133	132.905 436	100	
56	Barium	Ba	137.33	137	136.905 816	11.23	
				138	137.905 236	71.70	
				144*	143.922 673		11.9 s
57	Lanthanum	La	138.905	139	138.906 346	99.909 8	
58	Cerium	Ce	140.12	140	139.905 434	88.43	
				142*	141.909 241	11.13	$>5 \times 10^{16}$ yr
59	Praseodymium	Pr	140.907 6	141	140.907 647	100	
60	Neodymium	Nd	144.24	142	141.907 718	27.13	
				144*	143.910 082	23.80	2.3×10^{15} yr
				146	145.913 113	17.19	
61	Promethium	Pm		145*	144.912 745		17.7 yr
62	Samarium	Sm	150.36	147*	146.914 894	15.0	1.06×10^{11} yr
				149*	148.917 180	13.8	$>2 \times 10^{15}$ yr
				152	151.919 728	26.7	
				154	153.922 206	22.7	
63	Europium	Eu	151.96	151	150.919 846	47.8	
				153	152.921 226	52.2	

(Table continues)

Atomic Number, Z	Element	Symbol	Chemical Atomic Mass (u)	Mass Number (* Indicates Radioactive) A	Atomic Mass (u)[a]	Percentage Abundance	Half-Life (if Radioactive) $T_{1/2}$
64	Gadolinium	Gd	157.25	156	155.922 119	20.47	
				158	157.924 099	24.84	
				160	159.927 050	21.86	
65	Terbium	Tb	158.925 3	159	158.925 345	100	
66	Dysprosium	Dy	162.50	162	161.926 796	25.5	
				163	162.928 729	24.9	
				164	163.929 172	28.2	
67	Holmium	Ho	164.930 3	165	164.930 316	100	
68	Erbium	Er	167.26	166	165.930 292	33.6	
				167	166.932 047	22.95	
				168	167.932 369	27.8	
69	Thulium	Tm	168.934 2	169	168.934 213	100	
70	Ytterbium	Yb	173.04	172	171.936 380	21.9	
				173	172.938 209	16.12	
				174	173.938 861	31.8	
71	Lutecium	Lu	174.967	175	174.940 772	97.41	
72	Hafnium	Hf	178.49	177	176.943 218	18.606	
				178	177.943 697	27.297	
				179	178.945 813	13.629	
				180	179.946 547	35.100	
73	Tantalum	Ta	180.947 9	181	180.947 993	99.988	
74	Tungsten	W	183.85	182	181.948 202	26.3	
				183	182.950 221	14.28	
				184	183.950 929	30.7	
				186	185.954 358	28.6	
75	Rhenium	Re	186.207	185	184.952 951	37.40	
				187*	186.955 746	62.60	4.4×10^{10} yr
76	Osmium	Os	190.2	188	187.955 832	13.3	
				189	188.958 139	16.1	
				190	189.958 439	26.4	
				191*	190.960 94		15.4 days
				192	191.961 468	41.0	
77	Iridium	Ir	192.2	191	190.960 585	37.3	
				193	192.962 916	62.7	
78	Platinum	Pt	195.08	194	193.962 655	32.9	
				195	194.964 765	33.8	
				196	195.964 926	25.3	
79	Gold	Au	196.966 5	197	196.966 543	100	
80	Mercury	Hg	200.59	199	198.968 253	16.87	
				200	199.968 299	23.10	
				201	200.970 276	13.10	
				202	201.970 617	29.86	
81	Thallium	Tl	204.383	203	292.972 320	29.524	
				205	204.974 400	70.476	
		(Th C″)		208*	207.981 992		3.053 min
				210*	209.990 069		1.3 min
82	Lead	Pb	207.2	204*	203.973 020	1.4	$\geq 1.4 \times 10^{17}$ yr
				206	205.974 440	24.1	
				207	206.975 871	22.1	
				208	207.976 627	52.4	
		(Ra D)		210*	209.984 163		22.3 yr
		(Ac B)		211*	210.988 734		36.1 min
		(Th B)		212*	211.991 872		10.64 h
		(Ra B)		214*	213.999 798		26.8 min

(Table continues)

Atomic Number, Z	Element	Symbol	Chemical Atomic Mass (u)	Mass Number (* Indicates Radioactive) A	Atomic Mass (u)[a]	Percentage Abundance	Half-Life (if Radioactive) $T_{1/2}$
83	Bismuth	Bi	208.980 3	209	208.980 374	100	
		(Th C)		211*	210.987 254		2.14 min
84	Polonium	Po					
		(Ra F)		210*	209.982 848		138.38 days
		(Ra C′)		214*	213.995 177		164 μs
85	Astatine	At		218*	218.008 685		1.6 s
86	Radon	Rn		222*	222.017 571		3.823 days
87	Francium	Fr					
		(Ac K)		223*	223.019 733		22 min
88	Radium	Ra		226*	226.025 402		1600 yr
		(Ms Th$_1$)		228*	228.031 064		5.75 yr
89	Actinium	Ac		227*	227.027 749		21.77 yr
90	Thorium	Th	232.038 1				
		(Rd Th)		228*	228.028 716		1.913 yr
				232*	232.038 051	100	1.40×10^{10} yr
91	Protactinium	Pa		231*	231.035 880		32.760 yr
92	Uranium	U	238.028 9	232*	232.037 131		69 yr
				233*	233.039 630		1.59×10^5 yr
		(Ac U)		235*	235.043 924	0.720	7.04×10^8 yr
				236*	236.045 562		2.34×10^7 yr
		(UI)		238*	238.050 784	99.274 5	4.47×10^9 yr
93	Neptunium	Np		237*	237.048 168		2.14×10^6 yr
94	Plutonium	Pu		239*	239.052 157		2.412×10^3 yr
				242*	242.058 737		3.73×10^5 yr
				244*	244.064 200		8.1×10^7 yr

Appendix C

Some Useful Tables

TABLE C.1	Mathematical Symbols Used in the Text and Their Meaning

Symbol	Meaning
$=$	is equal to
$\neq$	is not equal to
$\equiv$	is defined as
$\propto$	is proportional to
$>$	is greater than
$<$	is less than
$\gg$	is much greater than
$\ll$	is much less than
$\approx$	is approximately equal to
$\sim$	is on the order of magnitude of
Δx	change in x or uncertainty in x
$\Sigma\, x_i$	sum of all quantities x_i
$|x|$	absolute value of x (always a positive quantity)

TABLE C.2	Standard Symbols for Units

Symbol	Unit	Symbol	Unit
A	ampere	kcal	kilocalorie
Å	angstrom	kg	kilogram
atm	atmosphere	km	kilometer
Bq	bequerel	kmol	kilomole
Btu	British thermal unit	L	liter
C	coulomb	lb	pound
°C	degree Celsius	ly	lightyear
cal	calorie	m	meter
cm	centimeter	min	minute
Ci	curie	mol	mole
d	day	N	newton
deg	degree (angle)	nm	nanometer
eV	electronvolt	Pa	pascal
°F	degree Fahrenheit	rad	radian
F	farad	rev	revolution
ft	foot	s	second
G	Gauss	T	tesla
g	gram	u	atomic mass unit
H	henry	V	volt
h	hour	W	watt
hp	horsepower	Wb	weber
Hz	hertz	yr	year
in.	inch	μm	micrometer
J	joule	Ω	ohm
K	kelvin		

TABLE C.3	The Greek Alphabet				
Alpha	A	α	Nu	N	ν
Beta	B	β	Xi	Ξ	ξ
Gamma	Γ	γ	Omicron	O	o
Delta	Δ	δ	Pi	Π	π
Epsilon	E	ϵ	Rho	P	ρ
Zeta	Z	ζ	Sigma	Σ	σ
Eta	H	η	Tau	T	τ
Theta	Θ	θ	Upsilon	Y	υ
Iota	I	ι	Phi	Φ	ϕ
Kappa	K	κ	Chi	X	χ
Lambda	Λ	λ	Psi	Ψ	ψ
Mu	M	μ	Omega	Ω	ω

TABLE C.4	Physical Data Often Used[a]
Average Earth-Moon distance	3.84×10^8 m
Average Earth-Sun distance	1.496×10^{11} m
Equatorial radius of Earth	6.38×10^6 m
Density of air (20°C and 1 atm)	1.20 kg/m^3
Density of water (20°C and 1 atm)	1.00×10^3 kg/m^3
Free-fall acceleration	9.80 m/s^2
Mass of Earth	5.98×10^{24} kg
Mass of Moon	7.36×10^{22} kg
Mass of Sun	1.99×10^{30} kg
Standard atmospheric pressure	1.013×10^5 Pa

[a] These are the values of the constants as used in the text.

TABLE C.5	Some Fundamental Constants[a]	
Quantity	**Symbol**	**Value**[b]
Atomic mass unit	u	$1.660\ 540\ 2(10) \times 10^{-27}$ kg
		$931.494\ 32(28)$ MeV/c^2
Avogadro's number	N_A	$6.022\ 136\ 7(36) \times 10^{23}$ (mol)$^{-1}$
Bohr radius	$a_0 = \dfrac{\hbar^2}{m_e e^2 k_e}$	$0.529\ 177\ 249(24) \times 10^{-10}$ m
Boltzmann's constant	$k_B = R/N_A$	$1.380\ 658(12) \times 10^{-23}$ J/K
Compton wavelength	$\lambda_C = \dfrac{h}{m_e c}$	$2.426\ 310\ 58(22) \times 10^{-12}$ m
Coulomb's law force constant	$k_e = \dfrac{1}{4\pi\epsilon_0}$	$8.987\ 551\ 787 \times 10^9$ N·m^2/C^2 (exact)
Electron mass	m_e	$9.109\ 389\ 7(54) \times 10^{-31}$ kg
		$5.485\ 799\ 03(13) \times 10^{-4}$ u
		$0.510\ 999\ 06(15)$ MeV/c^2
Electron volt	eV	$1.602\ 177\ 33(49) \times 10^{-19}$ J
Elementary charge	e	$1.602\ 177\ 33(49) \times 10^{-19}$ C
Gas constant	R	$8.314\ 510(70)$ J/K·mol
Gravitational constant	G	$6.672\ 59(85) \times 10^{-11}$ N·m^2/kg^2
Hydrogen ionization energy	$-E_1 = \dfrac{m_e e^4 k_e^2}{2\hbar^2} = \dfrac{e^2 k_e}{2a_0}$	$13.605\ 698(40)$ eV
Neutron mass	m_n	$1.674\ 928\ 6(10) \times 10^{-27}$ kg
		$1.008\ 664\ 904(14)$ u
		$939.565\ 63(28)$ MeV/c^2
Permeability of free space	μ_0	$4\pi \times 10^{-7}$ T·m/A (exact)
Permittivity of free space	$\epsilon_0 = 1/\mu_0 c^2$	$8.854\ 187\ 817 \times 10^{-12}$ C^2/N·m^2 (exact)
Planck's constant	h	$6.626\ 075(40) \times 10^{-34}$ J·s
	$\hbar = h/2\pi$	$1.054\ 572\ 66(63) \times 10^{-34}$ J·s
Proton mass	m_p	$1.672\ 623(10) \times 10^{-27}$ kg
		$1.007\ 276\ 470(12)$ u
		$938.272\ 3(28)$ MeV/c^2
Rydberg constant	R_H	$1.097\ 373\ 153\ 4(13) \times 10^7$ m^{-1}
Speed of light in vacuum	c	$2.997\ 924\ 58 \times 10^8$ m/s (exact)

[a] These constants are the values recommended in 1986 by CODATA, based on a least-squares adjustment of data from different measurements. For a more complete list, see Cohen, E. Richard, and Barry N. Taylor, *Rev. Mod. Phys.* **59**:1121, 1987.

[b] The numbers in parentheses for the values below represent the uncertainties in the last two digits.

Appendix D

SI Units

TABLE D.1	SI Base Units	
	SI Base Unit	
Base Quantity	**Name**	**Symbol**
Length	meter	m
Mass	kilogram	kg
Time	second	s
Electric current	ampere	A
Temperature	kelvin	K
Amount of substance	mole	mol
Luminous intensity	candela	cd

TABLE D.2	Derived SI Units			
Quantity	**Name**	**Symbol**	**Expression in Terms of Base Units**	**Expression in Terms of Other SI Units**
Plane angle	radian	rad	m/m	
Frequency	hertz	Hz	s^{-1}	
Force	newton	N	$kg \cdot m/s^2$	J/m
Pressure	pascal	Pa	$kg/m \cdot s^2$	N/m^2
Energy: work	joule	J	$kg \cdot m^2/s^2$	$N \cdot m$
Power	watt	W	$kg \cdot m^2/s^3$	J/s
Electric charge	coulomb	C	$A \cdot s$	
Electric potential (emf)	volt	V	$kg \cdot m^2/A \cdot s^3$	$W/A, J/C$
Capacitance	farad	F	$A^2 \cdot s^4/kg \cdot m^2$	C/V
Electric resistance	ohm	Ω	$kg \cdot m^2/A^2 \cdot s^3$	V/A
Magnetic flux	weber	Wb	$kg \cdot m^2/A \cdot s^2$	$V \cdot s, T \cdot m^2$
Magnetic field intensity	tesla	T	$kg/A \cdot s^2$	Wb/m^2
Inductance	henry	H	$kg \cdot m^2/A^2 \cdot s^2$	Wb/A

Answers to Quick Quizzes, Selected Questions and Problems

CHAPTER 1

Conceptual Questions

1. (a) ~ 0.1 m   (b) ~ 1 m
   (c) between 10 m and 100 m
   (d) ~ 10 m   (e) ~ 100 m
3. (a) $\sim 10^6$ beats   (b) $\sim 10^9$ beats
5. $\sim 10^9$ s
7. The length of a hand varies from person to person, so it is not a useful standard of length.
9. The ark has an approximate volume of 10^4 m^3, whereas a home has an approximate volume of 10^3 m^3.
11. $\sim \$10$/lb (assumes a cost of $\$20\,000$ and weight of $2\,000$ lb), of the same order as the cost of a good steak

Problems

1. Based on units alone, the equation might be valid.
5. m^3/(kg·s^2)
7. (a) 3 significant figures   (b) 4 significant figures
   (c) 3 significant figures   (d) 2 significant figures
9. (a) 797   (b) 1.1   (c) 17.66
11. 115.9 m
13. 228.8 cm
15. 2×10^8 fathoms
17. 0.325 g
19. 1×10^{17} ft
21. 6.71×10^8 mi/h
23. 3.08×10^4 m^3
25. 9.82 cm
27. (a) 1.0×10^3 kg
    (b) $m_{cell} = 5.2 \times 10^{-16}$ kg, $m_{kidney} = 0.27$ kg, $m_{fly} = 1.3 \times 10^{-5}$ kg
29. 1800 balls (or $\sim 10^3$ balls) (assumes 81 games per season, 9 innings per game, an average of 10 hitters per inning, and 1 ball lost for every four hitters)
31. $\sim 10^7$ rev
33. $\sim 10^6$ balls
35. (2.1 m, 1.4 m)
37. $r = 2.2$ m, $\theta = 27°$
39. (a) 6.71 m   (b) 0.894   (c) 0.746
41. 3.46 mi (or about 18 300 ft)
43. (a) 3.00   (b) 3.00   (c) 0.800   (d) 0.800   (e) 1.33
45. 5.00/7.00, the angle itself is 35.5°
47. $\sim 10^{12}$ cells
49. The value of k, a dimensionless constant, cannot be found by dimensional analysis.
51. $\sim 10^{-9}$ m
53. $\sim 10^3$ tuners (assumes 1 tuner per 10 000 residents and a population of 7.5 million)
55. (a) 3.16×10^7 s
    (b) Between 10^{10} yr and 10^{11} yr.

CHAPTER 2

Quick Quizzes

1. (a) 200 yd   (b) 0   (c) 0
2. (a) False   (b) True   (c) True
3. The velocity-time graph (a) has a constant slope, indicating a constant acceleration, which is represented by acceleration-time graph (e).
   Graph (b) represents an object that has an always-increasing speed, but the speed is not increasing at a uniform rate. Thus, the acceleration must be increasing, and the acceleration-time graph that best indicates this is (d).
   Graph (c) depicts an object that first has a velocity that increases at a constant rate, which means its acceleration is constant. The motion then changes to one at constant speed, indicating that the acceleration of the object becomes zero. Thus, the best match to this situation is graph (f).
4. (b)
5. (a) blue graph   (b) red graph
   (c) green graph
6. (c)
7. (c)
8. (a) and (f)

Conceptual Questions

1. Yes. If the velocity of the particle is nonzero, the particle is in motion. If the acceleration is zero, the velocity of the particle is unchanging, or is a constant.
3. Yes. If this occurs, the acceleration of the car is opposite to the direction of motion and the car will be slowing down.
5. The free-fall acceleration is always directed downward, which is normally indicated by a negative sign. While the object is moving upward, its velocity is positive. After it reaches it maximum height and begins to fall downward, its velocity is negative. The acceleration is negative for the entire flight.
7. Once the objects leave the hand, both are in free fall, and both experience the same downward acceleration equal to the free-fall acceleration, $-g$.
9. Yes. Yes.
11. (a) At the maximum height, the ball is momentarily at rest (that is, has zero velocity). The acceleration remains constant, with magnitude equal to the free-fall acceleration g, and directed downward. Thus, even though the velocity is momentarily zero, it continues to change and the ball will begin to gain speed in the downward direction. (b) The acceleration of the ball remains constant in magnitude and direction throughout its free flight, from the instant it leaves the hand until the instant just before it strikes the ground. It is directed downward and has a magnitude equal to the free-fall acceleration g.
13. No. If the velocity of A is greater than that of B, car A is moving faster than B at this instant. However, car B may be picking up speed (that is, accelerating) at a greater rate than A. The driver of B may have stepped very hard on the gas pedal in the recent past, but car B has not yet reached the speed of A.
15. They are the same! You can see this for yourself by solving the kinematic equations for the two cases. The ball thrown upward will take longer to reach the ground, but when it arrives, it has the same velocity as the ball that was thrown downward.

17. Ignoring air resistance, in 16 s the pebble would fall a distance of $\frac{1}{2}gt^2 = \frac{1}{2}(9.80 \text{ m/s}^2)(16.0 \text{ s})^2 = 1250 \text{ m} = 1.25 \text{ km}$. Air resistance is an important force after the first few seconds, when the pebble has gained high speed. Also, part of the 16-s interval must be occupied by the sound returning up the well. Thus, the depth of the well is less than 1.25 km, but on the order of 10^3 m.

19. If the car is moving forward, but slowing down, its acceleration is directed in the rearward direction. If the acceleration remains constant after the car comes to rest, the car will begin to develop a velocity in the rearward direction (that is, it will back up).

Problems

1. (a) 52.9 km/h   (b) 90.0 km
3. (a) Boat A wins by 60 km   (b) 0
5. (a) 180 km   (b) 63.4 km/h
7. (a) 4.0 m/s   (b) -0.50 m/s
    (c) -1.0 m/s   (d) 0
9. (a) a, e, and f (b) c
    (c) d and f (d) f
    (e) d
11. 2.80 h, 218 km
13. 274 km/h
15. (a) 5.00 m/s   (b) -2.50 m/s
    (c) 0   (d) 5.00 m/s
17. (a) 4.0 m/s   (b) -4.0 m/s
    (c) 0   (d) 2.0 m/s
19. -4.00 m/s^2; the acceleration is in the negative x direction
21. 3.7 s
23. (a) 8.0 m/s^2 (b) about 12 m/s^2
25. 2.74×10^5 m/s^2, or 2.79×10^4 times g
27. 160 ft
29. (a) 2.32 m/s^2 (b) 14.4 s
31. (a) 8.94 s   (b) 89.4 m/s
33. -3.6 m/s^2
35. 200 m
37. (a) 1.5 m/s   (b) 32 m
39. (a) 8.2 s   (b) 1.3×10^2 m
43. (a) 31.9 m   (b) 2.55 s
    (c) 2.55 s   (d) -25.0 m/s
45. 12 m/s
47. (a) 21.1 m/s   (b) 19.6 m   (c) 18.1 m/s; 19.6 m
49. (a) 308 m   (b) 8.51 s   (c) 16.4 s
51. (a) 10.0 m/s upward   (b) 4.68 m/s downward
53. 15.0 s
55. (a) -3.50×10^5 m/s^2 (b) 2.86×10^4 s
57. 0.60 s
59. (a) 3.00 s   (b) -24.5 m/s, -24.5 m/s
    (c) 23.5 m
61. (a) 5.46 s   (b) 73.0 m
    (c) $v_{\text{Kathy}} = 26.7$ m/s, $v_{\text{Stan}} = 22.6$ m/s
63. (a) 4.47 m/s upward   (b) 6.31 s after release
    (c) 6.06 m/s downward   (d) 73.0 m
    (e) 17.1 m/s
65. (a) 5.5×10^3 ft
    (b) 3.7×10^2 ft/s
    (c) The plane would only travel 0.002 ft in the time it takes the light from the bolt to reach the eye.
67. (a) 7.82 m   (b) 0.782 s

CHAPTER 3

Quick Quizzes

1. (c)
2. (b)
3. (b)

4.

Vector	x component	y component
A	$-$	$+$
B	$+$	$-$
A + B	$-$	$-$

5. (a) False   (b) False
6. (c)
7. (b)
8. The answers to all three parts are the same—the acceleration is that due to gravity, 9.8 m/s^2 downward.

Conceptual Questions

1. The components of **A** will both be negative when **A** lies in the third quadrant. The components of **A** will have opposite signs when **A** lies in either the second quadrant or the fourth quadrant.
3. Assuming the ship sails in a straight line, the wrench will strike the deck at the base of the mast. The wrench has the same horizontal velocity as the ship when it is released, and will maintain that component of velocity as it falls. Thus, the wrench is always above the same spot on the deck and will strike that spot when it hits the deck. To an observer on the ship, the wrench appears to fall straight downward. To a stationary observer on shore, the wrench is seen to follow a parabolic trajectory.
5. Let v_{x0} and v_{y0} be the initial velocity components for the projectile fired on Earth. Because the projectile fired on the Moon is given the same initial velocity, its initial velocity components have the same values v_{x0} and v_{y0} as those for the projectile on Earth. From $v_y^2 = v_{y0}^2 - 2g\,\Delta y$, the maximum altitude reached by a projectile is found to be $H = v_{y0}^2/2g$. Because the free-fall acceleration g is smaller on the Moon than on Earth, the maximum altitude will be greater for the projectile fired on the Moon. If they start from the same elevation, the projectile on the Moon will have a longer time of flight and therefore a greater range. The Apollo astronauts made several long golf drives on the Moon.
7. A vector has both magnitude and direction and can be added only to other vectors representing the same physical quantity. A scalar has only a magnitude and can be added only to other scalars representing the same physical quantity.
9. (a) At the top of its flight, the velocity is horizontal and the acceleration is downward. This is the only point at which the velocity and acceleration vectors are perpendicular.
    (b) If the object is thrown straight up or down, then the velocity and acceleration will be parallel throughout the motion. For any other kind of projectile motion, the velocity and acceleration vectors are never parallel.
11. (a) The acceleration is zero, since both the magnitude and direction of the velocity remain constant.
    (b) The particle has an acceleration since the direction of **v** changes.
13. The spacecraft will follow a parabolic path. This is equivalent to a projectile thrown off a cliff with a horizontal velocity. For the projectile, gravity provides an acceleration that is always perpendicular to the initial velocity, resulting in a parabolic path. For the spacecraft, the initial velocity plays the role of the horizontal velocity of the projectile. The acceleration provided by the leaking gas provides an acceleration that plays the role of gravity for the projectile. If the orientation of the spacecraft were to change in response to the gas leak (which is by far the more likely result), then the acceleration would change direction and the motion could become quite complicated.
15. For angles $\theta < 45°$, the projectile thrown at angle θ will be in the air for a shorter time interval. For a small angle, the vertical component of the initial velocity is smaller than for the larger angle. Thus, the projectile thrown at the smaller

angle will not go as high into the air, and will spend less time in the air before landing.

17. Yes. The projectile is a freely falling body, because gravity is the only force acting on it. The vertical acceleration will be the local free-fall acceleration *g*; the horizontal acceleration will be zero.

19. If a projectile is to have zero speed at the top of its trajectory, it must be thrown straight upward, with zero horizontal velocity. To give a projectile a nonzero speed at the top of the trajectory, the projectile should be thrown at any angle other than 90° to the horizontal.

Problems

1. Approximately 7.9 m at 4° N of W
3. Approximately 15 m at 58° S of E
5. Approximately 310 km at 57° S of W
7. (a) Approximately 5.0 units at $-53°$
   (b) Approximately 5.0 units at $+53°$
9. 8.07 m at 42.0° S of E
11. (a) 5.00 blocks at 53.1° N of E
    (b) 13.0 blocks
13. 47.2 units at 122° from positive *x* axis
15. 157 km
17. 245 km at 21.4° W of N
19. 196 cm at 14.7° below the *x* axis
21. (a) (10.0 m, 16.0 m)
23. 6.13 m
25. 12 m/s
27. 48.6 m/s
29. 25 m
31. (a) 32.5 m from the base of the cliff
    (b) 1.78 s
33. (a) 52.0 m/s horizontally   (b) 212 m
35. 390 mi/h at 7.37° N of E
37. 153 km/h at 11.3° N of W
39. 249 ft upstream
41. 18.0 s
43. 196 cm at 14.7° below positive *x* axis
45. (a) 57.7 km/h at 60.0° west of vertical
    (b) 28.9 km/h downward
47. 68.6 km/h
49. (a) 1.52×10^3 m   (b) 36.1 s
    (c) 4.05×10^3 m
51. $R_{\text{moon}} = 18$ m, $R_{\text{Mars}} = 7.9$ m
53. 2.21 m/s
55. (a) 42 m/s   (b) 3.8 s
    (c) $v_x = 34$ m/s, $v_y = -13$ m/s; $v = 37$ m/s
57. 7.5 m/s in the direction the ball was thrown
61. $R/2$
63. 7.5 min
65. 10.8 m
67. (b) $y = Ax^2$ with $A = \dfrac{g}{2v_i^2}$ where v_i is the muzzle velocity
    (c) 14.5 m/s
69. 227 paces at 165° from positive *x* axis
71. (a) 20.0° above the horizontal
    (b) 3.05 s

CHAPTER 4

Quick Quizzes

1. (a) True   (b) False
2. (a) True   (b) True   (c) False
3. False
4. Because the value of *g* is smaller on the Moon than on Earth, more mass of gold would be required to represent 1 newton of weight on the Moon. Thus, your friend on the Moon is richer, by about a factor of 6!
5. (b)
6. (c); (d)
7. (c)
8. (c)
9. (b)
10. (b)
11. (b) By exerting an upward force component on the sled, you reduce the normal force of the ground and so reduce the force of kinetic friction.

Conceptual Questions

1. (a) Two external forces act on the ball. (i) One is a downward gravitational force exerted by Earth. (ii) The second force on the ball is an upward normal force exerted by the hand. The reactions to these forces are (i) an upward gravitational force exerted by the ball on Earth, and (ii) a downward force exerted by the ball on the hand.
   (b) After the ball leaves the hand, the only external force acting on the ball is the gravitational force exerted by Earth. The reaction is an upward gravitational force exerted by the ball on Earth.

3. No. If the car moves with constant acceleration, it is necessary that a net force in the direction of the acceleration be acting on it.

5. The coefficient of static friction is larger than that of kinetic friction. To start the box moving, you must counterbalance the maximum static friction force. This exceeds the kinetic friction force that you must counterbalance to maintain constant velocity motion once it is started.

7. The inertia of the suitcase would keep it moving forward as the bus stops. There would be no tendency for the suitcase to be thrown backward toward the passenger. The case should be dismissed.

9. The force causing an automobile to move is the friction exerted by the roadway on the tires as the automobile attempts to push the roadway backward. The force driving a propeller airplane forward is the reaction force exerted by the air on the propeller as the rotating propeller pushes the air backward (the action). In a rowboat, the rower pushes the water backward with the oars (the action). The water pushes forward on the oars and hence the boat (the reaction).

11. When the bus starts moving, the mass of Claudette is accelerated by the force exerted by the back of the seat on her body. Clark is standing, however, and the only force on him is the friction between his shoes and the floor of the bus. Thus, when the bus starts moving, his feet accelerate forward, but the rest of his body experiences almost no accelerating force (only that due to his being attached to his accelerating feet!). As a consequence, his body tends to stay almost at rest, according to Newton's first law, relative to the ground. Relative to Claudette, however, he is moving toward her and falls into her lap. Both performers won Academy Awards.

13. The tension in the rope is the maximum force that occurs in *both* directions. In this case, then, since both are pulling with a force of 200 N, the tension is 200 N. If the rope does not move, then the total force on each athlete must equal zero. Therefore, each athlete exerts 200 N horizontally on the ground.

15. (a) As the man takes the step, the action is the force his foot exerts on Earth; the reaction is the force exerted by Earth on his foot. (b) Here the action is the force exerted by the snowball on the girl's back; the reaction is the force exerted by the girl's back on the snowball. (c) This action is the force exerted by the glove on the ball; the reaction is the force exerted by the ball on the glove. (d) This action is the force exerted by the air molecules on the window; the reaction is the force exerted by the window on the air molecules. In each case, we could equally well interchange the terms "action" and "reaction."

17. If the brakes lock, the car will travel farther than it would travel if the wheels continued to roll because the coefficient of kinetic friction is less than that of static friction. Hence, the force of kinetic friction is less than the maximum force of static friction.

19. **(a)** The crate accelerates because of a friction force exerted by the floor of the truck on the crate. **(b)** If the driver slams on the brakes, the crate's inertia tends to keep it moving forward.

Problems

1. **(a)** 12 N   **(b)** 3.0 m/s^2
3. 2×10^4 N
5. 3.71 N, 58.7 N, 2.27 kg
7. 9.6 N
9. 1.41×10^3 N
11. **(a)** 0.200 m/s^2 **(b)** 10.0 m
     (c) 2.00 m/s
13. 4.65 m/s^2 at 35.2° above the horizontal toward the right
15. 600 N in vertical cable, 997 N in inclined cable, 796 N in horizontal cable
17. 150 N in vertical cable, 75 N in right side cable, 130 N in left side cable
19. 101 N at 13.8° to the right of vertical and downward
21. 613 N
23. 64 N
25. **(a)** 7.50×10^3 N backward   **(b)** 50.0 m
27. **(a)** 7.0 m/s^2 horizontal and to the right
     (b) 21 N
     (c) 14 N horizontal and to the right
29. 7.90 m/s
31. 1.25 kg
33. **(a)** 2.15×10^3 N forward
     (b) 645 N forward
     (c) 645 N rearward
     (d) 1.02×10^4 N at 74.1° below horizontal and rearward
35. $\mu_s = 0.38$, $\mu_k = 0.31$
37. **(a)** 0.256   **(b)** 0.509 m/s^2
39. 7.84 m/s^2, independent of the mass
41. 32.1 N
43. **(a)** 33 m/s
     (b) No. The object will speed up to 33 m/s from any lower speed and will slow down to 33 m/s from any higher speed.
45. $\mu_k = 0.287$
47. 0.465 m/s
49. 3.30 m/s^2
51. **(a)** 0.404   **(b)** 45.8 lb
53. 113 N, horizontal and at 5.20° from the centerline between the cables
55. **(a)** 84.9 N upward
     (b) 84.9 N downward
57. 50 m
59. **(a)** friction between box and truck
     (b) 2.94 m/s^2
61. **(a)** 2.22 m
     (b) 8.74 m/s down the incline
63. **(a)** 1.78 m/s^2 **(b)** 0.368
     (c) 9.37 N   **(d)** 2.67 m/s
65. **(a)** 1.7 m/s^2, 17 N   **(b)** 0.69 m/s^2, 17 N
67. 100 N, 204 N
69. **(a)** $T_1 = 78.0$ N, $T_2 = 35.9$ N   **(b)** 0.656
71. **(a)** 30.7°   **(b)** 0.843 N
73. 5.5×10^2 N
75. **(a)** 7.1×10^2 N   **(b)** 8.1×10^2 N
     (c) 7.1×10^2 N   **(d)** 6.5×10^2 N
77. 72.0 N
79. **(a)** 0.408 m/s^2 upward   **(b)** 83.3 N

CHAPTER 5

Quick Quizzes

1. (c), (a), (d), (b)
2. All three balls have the same speed the moment before they hit the ground.
3. (c)
4. (c)

Conceptual Questions

1. Since no motion is taking place, the rope experiences no displacement and no work is done on it. For the same reason, no work is being done on the pullers or the ground. Work is only being done within the bodies of the pullers. For example, the heart of each puller is applying forces on the blood to move blood through the body.

3. When the slide is frictionless, changing the length or shape of the slide will not make any difference in the final speed of the child as long as the difference in the heights of the upper and lower ends of the slide is kept constant. If friction is considered, the path length along which the friction force does negative work will be greater when the slide is made longer or given bumps. Thus, the child will arrive at the lower end with less kinetic energy (and hence less speed).

5. If we ignore the effects due to rolling friction on the tires of the car, the same amount of work would be done in driving up the switchbacks and in driving straight up the mountain, since the weight of the car is moved upward against gravity by the same vertical distance in each case. If we include friction, there is more work done in driving the switchbacks, since the distance over which the friction force acts is much longer. So why do we use switchbacks? The answer lies in the force required, not the work. The force from the engine required to follow a gentle rise is much smaller than that required to drive straight up the hill. Roadways running straight uphill would require redesigning engines to enable them to apply much larger forces. This is similar to the relative ease with which heavy objects can be moved up ramps into trucks (compared with lifting the objects vertically).

7. **(a)** The tension in the supporting cord does no work, because the motion of the pendulum is always perpendicular to the cord, and therefore to the tension force. **(b)** The air resistance does negative work at all times, because the air resistance is always acting in a direction opposite to the motion. **(c)** The force of gravity always acts downward; therefore, the work done by gravity is positive on the downswing, and negative on the upswing.

9. Because the time periods are the same for both cars, we need only to compare the work done. The sports car is moving twice as fast as the older car at the end of the time interval, so it has four times the kinetic energy. Thus, according to the work-kinetic energy theorem, four times as much work was done, and the engine must have expended four times the power.

11. During the time that the toe is in contact with the ball, the work done on the ball is given by

$$W_{toe} = \tfrac{1}{2}m_{ball}v^2 - 0 = \tfrac{1}{2}m_{ball}v^2$$

where v is the speed of the ball as it leaves the toe. After the ball loses contact with the toe, only the gravitational force and the retarding force due to air resistance continue to do work on the ball throughout its flight.

13. Yes, the total mechanical energy of the system is conserved because the only forces acting are conservative—the force of gravity and the spring force. There are two forms of potential energy in this case, gravitational potential energy and elastic potential energy stored in the spring.

15. Let's assume you lift the book slowly. In this case there are two forces on the book that are almost equal in magnitude. They are the lifting force and the force of gravity. Thus, the positive work done by you and the negative work done by gravity cancel. There is no net work performed, and no net change in the kinetic energy—the work-kinetic energy theorem is satisfied.

17. As the satellite moves in a circular orbit about Earth, its displacement along the circular path during any small time interval is always perpendicular to the gravitational force, which always acts toward the center of Earth. Therefore, the work done by the gravitational force during any displacement is zero. (Recall that the work done by a force is defined to be $F\Delta x \cos \theta$, where θ is the angle between the force and the displacement. In this case, the angle is 90°, so the work done is zero.) Since the work-kinetic energy theorem says that the net work done on an object during any displacement is equal to the change in its kinetic energy, and the work done in this case is zero, the change in the satellite's kinetic energy is zero, hence its speed remains constant.

Problems

1. 700 J
3. 15.0 MJ
5. (a) 61.3 J   (b) -46.3 J   (c) 0
7. (a) 79.4 N   (b) 1.49 kJ   (c) -1.49 kJ
9. (a) 2.00 m/s   (b) 200 N
11. 0.265 m/s
13. (a) -5.6×10^2 J   (b) 1.2 m
15. (a) 90 J   (b) 1.8×10^2 N
17. 1.0 m/s
19. (a) 6.4 J   (b) -2.5 J
21. 1.9 J
23. 1.05 J
25. (a) 8.0×10^2 J   (b) 1.1×10^2 J
    (c) 0
27. 0.459 m
29. 6.86 m/s
31. (a) 10.9 m/s   (b) 11.6 m/s
33. (a) 544 N/m   (b) 19.7 m/s
35. 10.2 m
37. 1.9 m/s
39. 2.1 kN
41. 3.8 m/s
43. (a) 5.42 m/s   (b) 0.300   (c) 147 J
45. 289 m
47. (a) 24.5 m/s   (b) Yes   (c) 206 m
    (d) Unrealistic; the actual retarding force will vary with speed.
49. (a) 0.408 m/s   (b) 2.45 kJ
51. 5.9×10^8 W
53. (a) 2.38×10^4 W = 32.0 hp   (b) 4.77×10^4 W = 63.9 hp
55. (a) 24.0 J   (b) -3.00 J   (c) 21.0 J
57. (a) The graph is a straight line passing through the points (0 m, -16 N), (2 m, 0 N), and (3 m, 8 N).
    (b) -12.0 J
59. (a) 575 N/m   (b) 46.0 J
61. (a) 4.4 m/s   (b) 1.5×10^5 N
63. (a) 3.13 m/s   (b) 4.43 m/s   (c) 1.00 m
65. 875 W
67. (a) 423 mi/gal   (b) 776 mi/gal
69. (a) 28.0 m/s   (b) 30.0 m
    (c) 89.0 m beyond the end of the track
71. 1.68 m/s
73. (a) 6.15 m/s   (b) 9.87 m/s
75. (a) 101 J   (b) 0.410 m
    (c) 2.84 m/s   (d) -9.80 mm
    (e) 2.85 m/s
77. 914 N/m

CHAPTER 6
Quick Quizzes

1. (d)
2. (c)
3. (c) and (e)
4. (d)
5. (b)
6. (a)
7. (a) Perfectly inelastic   (b) Inelastic
   (c) Inelastic
8. (b)

Conceptual Questions

1. The passenger's body must undergo a significant momentum change as the car's velocity changes abruptly in the collision. In a front-end collision, the seat belt and shoulder harness exert the required forces on the body to give it the needed impulse (change in momentum). When a car is rammed from behind, the car and the passenger's body are given a sudden increase in forward momentum. If the head is not supported by a head rest, the head tends to continue moving at its original speed until the angle of the neck changes and the neck exerts a sufficiently large forward force on the head to give it the needed momentum change. In agreement with Newton's third law, the head exerts a rearward force on the neck. This force causes whiplash injuries.

3. If all the kinetic energy disappears, there must be no motion of either object after the collision. If nothing is moving, there is no momentum. If the final momentum of the system is zero, the initial momentum of the system must also have been zero. A situation in which this could be true would be the head-on collision of two objects, having momenta of equal magnitude but opposite direction.

5. Initially the clay has momentum directed toward the wall. When it collides and sticks to the wall, nothing appears to have any momentum. Thus, it is tempting to (wrongfully) conclude that momentum is not conserved. However, the "lost" momentum is actually imparted to the wall and Earth, causing both to move. Because of Earth's enormous mass, its recoil speed is too small to detect.

7. Before the step the momentum was zero, so afterward the net momentum must also be zero. Obviously, you have some momentum, so something must have momentum in the opposite direction. That something is Earth. Earth's enormous mass ensures that its recoil speed will be too small to detect, but if you want to make Earth move, it is as simple as taking a step.

9. As the water is forced out of the holes in the spray arm, the spray arm imparts a horizontal impulse to the water. The water exerts an equal and opposite impulse on the spray arm, causing the spray arm to rotate in the direction opposite to that of the spray.

11. By discarding empty fuel tanks and other assemblies no longer needed, the mass of the remaining system is reduced. A rocket engine is capable of providing a certain impulse— that is, the product of its thrust and the time interval that thrust can be maintained. This impulse equals the change in momentum $m(\Delta v)$ given to the rocket body. By keeping the mass m as low as possible, the increase in speed Δv provided by the engine can be maximized.

13. It will be easiest to catch the medicine ball when its speed (and kinetic energy) is lowest. The first option—throwing the medicine ball at the same velocity—will be the most difficult, because the speed will not be reduced at all. The second option, throwing the medicine ball with the same momentum, will reduce the velocity by the ratio of the masses.

Since $m_t v_t = m_m v_m$,

$$v_m = v_t \left(\frac{m_t}{m_m} \right)$$

The third option, throwing the medicine ball with the same kinetic energy, will also reduce the velocity, but only by the square root of the ratio of the masses. Since

$$\tfrac{1}{2} m_t v_t^2 = \tfrac{1}{2} m_m v_m^2, \text{ then } v_m = v_t \sqrt{\frac{m_t}{m_m}}$$

Thus, the slowest—and easiest—throw will be made when the momentum is held constant. If you wish to check this, try substituting in values of $v_t = 1$ m/s, $m_t = 1$ kg, and $m_m = 100$ kg. The same-momentum throw will be caught at 1 cm/s, while the same-energy throw will be caught at 10 cm/s.

15. The follow-through keeps the club in contact with the ball as long as possible, maximizing the impulse. Thus, the ball experiences a larger change in momentum than without the follow-through, and it leaves the club with a higher velocity and travels farther. With a short shot to the green, the primary factor is control, not distance. Thus, there is little or no follow-through, allowing for the golfer to have a better feel for how hard he or she is striking the ball.

17. It is the product mv that is the same for both the bullet and the gun. The bullet has a large velocity and a small mass, while the gun has a small velocity and a large mass. Furthermore, the bullet carries much more kinetic energy than the gun.

19. When you land on the ground, a certain impulse $\overline{F}\,\Delta t$ is required to bring your body to rest. If you land with legs relaxed and allow the knees to bend, you increase the time interval Δt over which the impulse is delivered. This reduces the average force exerted on the body. Landing with the knees locked means that the time interval will be very short and the average force very high, which is more likely to result in injury to the body.

Problems

3. (a) 8.35×10^{-21} kg·m/s   (b) 4.50 kg·m/s
   (c) 750 kg·m/s   (d) 1.78×10^{29} kg·m/s
5. (a) 31.0 m/s
   (b) the bullet, 3.38×10^3 J versus 69.7 J
7. $\sim 10^3$ N upward
9. 364 kg·m/s forward, 438 N forward
11. (a) 8.0 N·s   (b) 5.3 m/s
   (c) 3.3 m/s
13. (a) 12 N·s   (b) 8.0 N·s
   (c) 8.0 m/s, 5.3 m/s
15. (a) 9.60×10^{-2} s   (b) 3.65×10^5 N
   (c) $26.6g$
17. 800 N at 60.0° below horizontal to the right
19. 65 m/s
21. (a) 1.15 m/s
   (b) 0.346 m/s directed opposite to girl's motion
23. $KE_E/KE_b \sim 10^{-25}$
25. 2.66 m/s
27. (a) 1.80 m/s   (b) 2.16×10^4 J
29. 57 m
31. 15.6 m/s
33. 273 m/s
35. (a) -6.67 cm/s, 13.3 cm/s   (b) 0.889
37. 17.1 cm/s (25.0-g object), 22.1 cm/s (10.0-g object)
39. 7.94 cm
41. (a) 2.9 m/s at 32° N of E
   (b) 7.9×10^2 J converted into internal energy
43. 5.59 m/s north
45. (a) 2.50 m/s at $-60°$ (b) elastic collision

47. 14.8 kg·m/s in the direction of the final velocity of the ball
49. 1.78×10^3 N on truck driver, 8.89×10^3 N on car driver
51. (a) 8/3 m/s (incident particle), 32/3 m/s (target particle)
   (b) $-16/3$ m/s (incident particle), 8/3 m/s (target particle)
   (c) 7.1×10^{-3} J in case (a), 2.8×10^{-2} J in case (b). The incident particle loses more kinetic energy in case (a) where the target mass is 1.0 g.
53. 1.1×10^3 N (upward)
55. (a) 1.33 m/s   (b) 235 N
   (c) 0.681 s   (d) -160 N·s, 160 N·s
   (e) 1.82 m   (f) 0.454 m
   (g) -427 J   (h) 107 J
   (i) Equal friction forces act through different distances on man and cart to do different amounts of work on them. This is a perfectly inelastic collision in which the total work on both man and cart together is -320 J, which becomes $+320$ J of internal energy.
57. (a) -2.33 m/s, 4.67 m/s   (b) 0.277 m
   (c) 2.98 m   (d) 1.49 m
59. (a) -0.667 m/s   (b) 0.952 m
61. (a) 3.54 m/s   (b) 1.77 m
   (c) 3.54×10^4 N
   (d) No, the normal force exerted by the rail contributes upward momentum to the system.
63. (a) 0.28 or 28%
   (b) 1.1×10^{-13} J for neutron, 4.5×10^{-14} J for carbon
65. $\dfrac{2v_0^2}{9\mu_k g} - \dfrac{4d}{9}$

CHAPTER 7

Quick Quizzes

1. (c)
2. (b)
3. (b) and (d)
4. (e), (a), (b)
5. (c)
6. (b) and (c)
7. (e)
8. (a)

Conceptual Questions

1. (a) The head will tend to lean toward the right shoulder (that is, toward the outside of the curve).
   (b) When there is no strap, tension in the neck muscles must produce the centripetal acceleration.
   (c) With a strap, the tension in the strap performs this function, allowing the neck muscles to remain relaxed.
3. An object can move in a circle even if the total force on it is not perpendicular to its velocity, but then its speed will change. Resolve the total force into an inward radial component and a perpendicular tangential component. If the tangential force acts in the forward direction, the object will speed up, and if the tangential force acts backward, the object will slow down.
5. The speedometer will be inaccurate. The speedometer measures the number of tire revolutions per second, so its readings will be too low.
7. The car cannot round a turn at constant velocity because constant velocity means the direction of the velocity is not changing. The statement is correct if the word "velocity" is replaced by the word "speed."
9. The gravitational force exerted on the moon by the planet produces the centripetal acceleration. Mathematically,

$$m_{\text{moon}} \left(\frac{v_t^2}{r} \right) = G \frac{M m_{\text{moon}}}{r^2}$$

or the mass of the planet is $M = rv_t^2/G$. Both r and v_t can be determined by observing the motion of the moon, and the mass of the planet is then easily computed.

11. Consider an individual standing against the inside wall of the cylinder with her head pointed toward the axis of the cylinder. As the cylinder rotates, the person tends to move in a straight-line path tangent to the circular path followed by the cylinder wall. As a result, the person presses against the wall, and the normal force exerted on her provides the radial force required to keep her moving in a circular path. If the rotational speed is adjusted such that this normal force is equal in magnitude to her weight on Earth, she would not be able to distinguish between the artificial gravity of the colony and ordinary gravity.

13. The tendency of the water is to move in a straight-line path tangent to the circular path followed by the container. As a result, at the top of the circular path, the water presses against the bottom of the pail, and the normal force exerted by the pail on the water provides the radial force required to keep the water moving in its circular path.

15. Any object that moves such that the *direction* of its velocity changes has an acceleration. A car moving in a circular path will always have a centripetal acceleration.

17. When a pilot is pulling out of a dive, blood leaves the head because there is not a great enough radial force to cause it to follow the circular path of the airplane. The loss of blood from the brain can cause the pilot to black out.

Problems

1. (a) 3.2×10^8 rad   (b) 5.0×10^7 rev
3. 1.99×10^{-7} rad/s, 0.986 deg/day
5. (a) 821 rad/s^2 (b) 4.21×10^3 rad
7. 36.5 rev
9. Main rotor: 179 m/s $= 0.522v_{\text{sound}}$
   Tail rotor: 221 m/s $= 0.644v_{\text{sound}}$
11. $\sim 10^{-2}$ cm
13. 3.2×10^6 m
15. (a) 3.37×10^{-2} m/s^2 down   (b) 0
17. (a) 0.35 m/s^2 (b) 1.0 m/s
    (c) 0.35 m/s^2, 0.94 m/s^2, 1.0 m/s^2 at 20° forward with respect to the direction of a_c
19. (a) 1.10 kN   (b) 2.04 times her weight
21. 12 m/s
23. (a) 18.0 m/s^2 (b) 900 N
    (c) 1.84; this large coefficient is unrealistic, and he will not be able to stay on the merry-go-round.
25. (a) 9.8 N   (b) 9.8 N   (c) 6.3 m/s
27. (a) 1.58 m/s^2 (b) 455 N upward
    (c) 329 N upward
    (d) 397 N directed inward and 80.8° above horizontal
29. 321 N toward Earth
31. 1.1×10^{-10} N at 72° above the $+x$ axis
33. (a) 2.50×10^{-5} N toward the 500-kg object
    (b) Between the two objects and 0.245 m from the 500-kg object
35. (a) 9.58×10^6 m   (b) 5.57 h
37. 1.90×10^{27} kg
39. (a) 1.48 h   (b) 7.79×10^3 m/s
    (c) 6.43×10^9 J
41. (a) 7.27×10^{-5} rad/s   (b) 7.13 h
43. (a) 2.51 m/s   (b) 7.90 m/s^2
    (c) 4.00 m/s
45. (a) 7.76×10^3 m/s   (b) 89.3 min
47. (a) $n = m\left(g - \dfrac{v^2}{r}\right)$ (b) 17.1 m/s
49. (a) $F_{g,\,\text{true}} = F_{g,\,\text{apparent}} + mR_E\omega^2$
    (b) 732 N (Equator), 735 N (either pole)
51. 0.131
55. 11.8 km/s
57. 0.75 m
59. (a) $v_0 = \sqrt{g\left(R - \dfrac{2h}{3}\right)}$ (b) $h' = \dfrac{R}{2} + \dfrac{2h}{3}$

CHAPTER 8

Quick Quizzes

1. (d)
2. (b)
3. (b)
4. (a)
5. (a) Nathan is correct.
6. (c) The box.
7. (c)
8. (a)

Conceptual Questions

1. In order for you to remain in equilibrium, your center of gravity must always be over your point of support, the feet. If your heels are against a wall, your center of gravity cannot remain above your feet when you bend forward, so you lose your balance.

3. There are two major differences between torque and work. The primary difference is that the displacement in the expression for work is directed along the force, while the important distance in the torque expression is perpendicular to the force. The second difference involves whether there is motion or not—in the case of work, work is done only if the force succeeds in causing a displacement of the point of application of the force. On the other hand, force applied at a perpendicular distance from a rotation axis results in a torque whether or not there is motion.

   As far as units are concerned, the mathematical expressions for both work and torque result in the product of newtons and meters, but this product is called a joule in the case of work and remains as a newton-meter in the case of torque.

5. No. For an object to be in equilibrium, the net external force acting on it must be zero. This is not possible when only one force acts on the object, unless that force should have zero magnitude. In that case, there is really no force acting on the object.

7. As the motorcycle leaves the ground, the friction between the tire and the ground suddenly disappears. If the motorcycle driver keeps the throttle open while leaving the ground, the rear tire will increase its angular speed and, hence, its angular momentum. The airborne motorcycle is now an isolated system, and its angular momentum must be conserved. The increase in angular momentum of the tire directed, say, clockwise must be compensated by an increase in angular momentum of the entire motorcycle counterclockwise. This rotation results in the nose of the motorcycle rising and the tail dropping.

9. In general, you want the rotational kinetic energy of the system to be as small a fraction of the total energy as possible— you want translation, not rotation. You want the wheels to have as little moment of inertia as possible, so that they represent the lowest resistance to changes in rotational motion. Disk-like wheels would have lower moments of inertia than hoop-like wheels, so disks are preferable. The lower the mass of the wheels, the less is the moment of inertia, so light wheels are preferable. The smaller the radius of the wheels, the less is the moment of inertia, so smaller wheels are preferable, within limits—you want the wheels to be large enough to be able to travel relatively smoothly over irregularities in the road.

11. The angular momentum of the gas cloud is conserved. Thus, the product $I\omega$ remains constant. As the cloud shrinks in size, its moment of inertia decreases, so its angular speed ω must increase.

13. We can assume fairly accurately that the driving motor will run at a constant angular speed and at a constant torque. Therefore, as the radius of the takeup reel increases, the tension in the tape will decrease.

$$T = \tau_{\text{const}} / R_{\text{takeup}} \tag{1}$$

$$\tau_{\text{source}} = FR_{\text{source}} = \tau_{\text{const}} R_{\text{source}} / R_{\text{takeup}} \tag{2}$$

As the radius of the source reel decreases, given a decreasing tension, the torque in the source reel will decrease even faster.

This torque will be partly absorbed by friction in the feed heads (which we assume to be small); some will be absorbed by friction in the source reel. Another small amount of the torque will be absorbed by the increasing angular speed of the source reel. However, in the case of a sudden jerk on the tape, the changing angular speed of the source reel becomes important. If the source reel is full, then the moment of inertia will be large, and the tension in the tape will be large. If it is nearly empty, then the angular acceleration will be large instead. Thus, the tape will be more likely to break when the source reel is nearly full. We see the same effect in the case of paper towels; it is easier to snap a towel free when the roll is new than when it is nearly empty.

15. The initial angular momentum of the system (mouse plus turntable) is zero. As the mouse begins to walk clockwise, its angular momentum increases, so the turntable must rotate in the counterclockwise direction with an angular momentum whose magnitude equals that of the mouse. This follows from the fact that the final angular momentum of the system must equal the initial angular momentum (zero).

17. When a ladder leans against a wall, both the wall and the floor exert forces of friction on the ladder. If the floor is perfectly smooth, it can exert no frictional force in the horizontal direction to counterbalance the wall's normal force. Therefore, a ladder on a smooth floor cannot stand in equilibrium. However, a smooth wall can still exert a normal force to hold the ladder in equilibrium against horizontal motion. The counterclockwise torque of this force prevents rotation about the foot of the ladder. So you should choose a rough floor.

Problems

1. 133 N
3. -207 N·m, -145 N·m, -95.7 N·m
5. 5.1 N·m
7. $F_t = 724$ N, $F_s = 716$ N
9. 312 N
11. 492 N in position (a); 1.25 kN in position (b)
13. 10-year-old: 0.39 cm below midpoint (0.33% of height)
    20-year-old: 3.4 cm below midpoint (2.0% of height)
15. $T = 1.05$ kN, $F_c = 950$ N
17. $T = 4.2$ kN, $\mathbf{F}_c = 4.1$ kN at 0.57° below the horizontal
19. $R = 107$ N, $T = 157$ N
21. $T_{\text{left wire}} = \frac{1}{3}w$, $T_{\text{right wire}} = \frac{2}{3}w$
23. $T_1 = 501$ N, $T_2 = 672$ N, $T_3 = 384$ N
25. 6.15 m
27. 209 N
29. (a) 99.0 kg·m^2 (b) 44.0 kg·m^2
    (c) 143 kg·m^2
31. (a) 24.0 N·m   (b) 0.035 6 rad/s^2
    (c) 1.07 m/s^2
33. (a) 34 N   (b) 33 cm
35. 177 N
37. 0.524
39. (a) 500 J   (b) 250 J   (c) 750 J
41. 276 J
43. 149 rad/s
45. (a) 7.08×10^{33} J·s   (b) 2.66×10^{40} J·s
47. 8.0 rev/s
49. 6.73 rad/s
51. 5.99×10^{-2} J
53. (a) 0.360 rad/s counterclockwise
    (b) 99.9 J
55. (a) $\omega = \left(\dfrac{I_1}{I_1 + I_2}\right)\omega_0$ (b) $\dfrac{KE_f}{KE_i} = \dfrac{I_1}{I_1 + I_2} < 1$
59. 21.5 N
61. 36.9°
63. $a_{\text{sphere}} = \dfrac{g\sin\theta}{1.4}$, $a_{\text{disk}} = \dfrac{g\sin\theta}{1.5}$, $a_{\text{ring}} = \dfrac{g\sin\theta}{2.0}$
    Thus, the sphere wins and the ring comes in last.
65. (a) 1.63 m/s   (b) 54.2 rad/s
67. 1.09 m
69. (a) Mvd (b) Mv^2 (c) Mvd (d) $2v$
    (e) $4Mv^2$ (f) $3Mv^2$
71. 7.5×10^{-11} s
73. (a) $T = \dfrac{Mmg}{M + 4m}$ (b) $a_t = \dfrac{4mg}{M + 4m}$

CHAPTER 9

Quick Quizzes

1. (a)
2. (c)
3. (c)
4. (c)
5. (b)
6. (c)
7. (a)

Conceptual Questions

1. The distance of the legs below the heart increases the pumping pressure required to move blood upward from these extremities. Elastic stockings increase the pressure on the blood in the legs, helping to compensate for the needed pumping pressure.

3. She exerts enough pressure on the floor to dent or puncture the floor covering. The large pressure is caused by the fact that her weight is distributed over the very small cross-sectional area of her high heels. If you are the homeowner, you might want to suggest that she remove her high heels and put on some slippers.

5. If you think of the grain stored in the silo as a fluid, the pressure the grain exerts on the walls of the silo increases with increasing depth just as water pressure in a lake increases with increasing depth. Thus, the spacing between bands is made smaller at the lower portions to counterbalance the larger outward forces on the walls in these regions.

7. The syringe does not "draw" the blood from the vein. By moving the plunger in the syringe back, the pressure inside the syringe is lowered below the pressure inside the vein. Then, the blood pressure within the body forces blood out into the syringe, which "accepts" the blood.

9. In the ocean, the ship floats due to the buoyant force from *salt water*. Salt water is denser than fresh water. As the ship is pulled up the river, the buoyant force from the fresh water in the river is not sufficient to support the weight of the ship, and it sinks.

11. The balance will not be in equilibrium—the lead side will be lower. Despite the fact that the weights on both sides of the balance are the same, the Styrofoam, due to its larger volume, experiences a larger buoyant force from the sur-

rounding air. Thus, the net force of the weight and the buoyant force is larger, in the downward direction, for the lead than for the Styrofoam.

13. The two cans displace the same volume of water and hence experience buoyant forces of equal magnitude. The total weight of the diet cola can must be less than this buoyant force, whereas the total weight of the regular cola can is greater than this force. This is possible even though the two containers are identical and contain the same volume of liquid. Because of the difference in the quantities and densities of the sweeteners used, volume V of the diet mixture has less mass than an equal volume of the regular mixture.

15. As the truck passes, the air between your car and the truck is squeezed into the channel between you and the truck and moves at a higher speed than when your car is in the open. According to Bernoulli's principle, this high-speed air has a lower pressure than the air on the outer side of your car. The difference in pressure provides a net force on your car toward the truck.

17. Opening the windows results in a smaller pressure difference between the exterior and interior of the house, and therefore less tendency for severe damage to the structure due to the Bernoulli effect.

Problems

1. ~1 cm
3. 1.8×10^6 Pa
5. 3.5×10^8 Pa
7. 4.4 mm
9. 0.024 mm
11. 36 nm
13. (a) 1.88×10^5 Pa   (b) 2.65×10^5 Pa
15. 1.33×10^3 kg/m^3
17. 3.4×10^2 m
19. 1.33 m
21. 271 kN horizontally toward the cellar
23. 1.05×10^5 Pa
25. 2.1 N·m
27. 9.41 kN
29. 1.4
31. 2.67×10^3 kg
33. 5.57 N
35. (a) 1.46×10^{-2} m^3 (b) 2.10×10^3 kg/m^3
37. 1.07 m/s^2
39. 17.3 N (upper scale), 31.7 N (lower scale)
41. (a) 80 g/s   (b) 0.27 mm/s
43. 12.6 m/s
45. 9.00 cm
47. 1.47 cm
49. (a) 28.0 m/s   (b) 28.0 m/s
    (c) 2.11 MPa
51. (b) For any y less than $y_{max} = P_0/\rho g$
53. 8.3×10^{-2} N/m
55. 5.6×10^{-2} N/m
57. 8.6 N
59. 2.1 MPa
61. 2.8 μm
63. 90 cm
65. $RN = 4.3 \times 10^3$; turbulent flow
67. 1.8×10^{-3} kg/m^3
69. 1.4×10^{-5} N·s/m^2
71. 9.28 MPa
73. 2.5×10^7 capillaries
75. (a) 1.57 kPa, 1.55×10^{-2} atm, 11.8 mm of Hg
    (b) The fluid level in the tap should rise.
    (c) Blockage of flow of the cerebrospinal fluid
77. 4.14×10^3 m^3
79. 15 m
81. 2.25 m above the level of point B
83. 6.4 m
85. 1.3 cm
87. 17.0 cm above the floor
89. 532 cm^3

CHAPTER 10

Quick Quizzes

1. (b)
2. (c)
3. (c)
4. (c)
5. (a)
6. (b)

Conceptual Questions

1. An ordinary glass dish will usually break because of stresses that build up as the glass expands when heated. The expansion coefficient for Pyrex glass is much lower than that of ordinary glass. Thus, a Pyrex dish expands much less than the dish of ordinary glass, and does not normally develop sufficient stress to cause breakage.
3. The accurate answer is that it doesn't matter! Temperatures on the Kelvin and Celsius scales differ by only 273 degrees. This difference is insignificant for temperatures on the order of 10^7 degrees. If we imagine that the temperature is given in kelvins, and ignore any problems with significant figures, then the Celsius temperature is $1.499\,972\,7 \times 10^7$ °C.
5. Mercury must have the larger coefficient of expansion. As the temperature of a thermometer rises, both the mercury and the glass expand. If they both had the same coefficient of linear expansion, the mercury and the cavity in the glass would both expand by the same amount, and there would be no apparent movement of the end of the mercury column relative to the calibration scale on the glass. If the glass expanded more than the mercury, the reading would go down as the temperature went up! Now that we have argued this conceptually, we can look in a table and find that the coefficient for mercury is about 20 times as large as that for glass, so that the expansion of the glass can sometimes be ignored.
7. We can think of each bacterium as being a small bag of liquid containing bubbles of gas at a very high pressure. If the bacterium is raised rapidly to the surface, the ideal gas law tells us that the volume must increase dramatically. In fact, this increase in volume is sufficient to rupture the bacterium.
9. Velocity is a vector quantity, so direction must be considered. If the same number of particles move to the right along the x direction as move to the left along the $-x$ direction, the x component of velocity will be zero.
11. The chip bags contain a sealed sample of air. When the bags are taken up the mountain, the external atmospheric pressure on the bags is reduced. As a result, the difference between the pressure of the air inside the bags and the reduced pressure outside results in a net force pushing the plastic of the bag outward.
13. Additional water vaporizes into the bubble, so that n increases.

Problems

1. (a) −460°C   (b) 37.0°C   (c) −280°F
3. (a) −423°F, 20.28 K   (b) 68°F, 293 K
9. (a) 1 337 K, 2 933 K   (b) 1 596°C, 1 596 K
11. 31 cm
13. 55.0°C
15. (a) −179°C (attainable)
    (b) −376°C (below 0 K, unattainable)
17. 11.507 cm (lid), 11.506 cm (jar)

19. 2.7×10^2 N
21. 1.1 L (0.29 gal)
25. (a) increases   (b) 1.603 cm
27. 8.208×10^{-2} L · atm/mol · K
29. (a) 2.5×10^{19} molecules   (b) 4.1×10^{-21} mol
31. 5.87 atm
33. 7.1 m
35. 16.0 cm^3
37. 6.21×10^{-21} J
39. 6.64×10^{-27} kg
41. (a) 8.76×10^{-21} J
    (b) $v_{rms, He} = 1.62$ km/s, $v_{rms, Ar} = 514$ m/s
43. 16 N
45. 0.663 mm at 78.2° below the horizontal
47. 3.55 L
49. 6.57 MPa
51. The expansion of the mercury is almost 20 times that of the flask (assuming Pyrex glass).
53. shorter, by 0.061 mm
55. 2.74 m
57. 1.61 MPa
59. 0.417 L
61. (a) $\theta = \dfrac{(\alpha_2 - \alpha_1)L_0(\Delta T)}{\Delta r}$
    (c) The bar bends in the opposite direction.

CHAPTER 11

Quick Quizzes

1. (a) Water, glass, iron.   (b) Iron, glass, water.
2. (d)
3. (c)
4. (c)
5. (e)

Conceptual Questions

1. Having sweat glands only in the pads of their feet, dogs rely heavily on the evaporation of fluid on the tongue for cooling. Panting involves rapid shallow breathing that accelerates evaporation of fluid from the tongue. Dogs have a rich blood supply in their tongues, and energy produced in the dog's muscles and tissues is transported to the tongue where is it eliminated through panting.
3. When you rub the surface, you increase the temperature of the rubbed region. With the metal surface, some of this energy is transferred away from the rubbing site by conduction. Thus, the temperature in the rubbed area is not as high for the metal as it is for the wood, and it feels relatively cooler than the wood.
5. The fruit loses energy into the air by radiation and convection from its surface. Before ice crystals can form inside the fruit to rupture cell walls, all of the liquid water on the skin will have to freeze. The resulting time delay may prevent damage within the fruit throughout a frosty night. Furthermore, a surface film of ice provides some insulation to slow subsequent energy loss by conduction from within the fruit.
7. The operation of an immersion coil depends on the convection of water to maintain a safe temperature. As the water near a coil warms up, the warmed water floats to the top due to Archimedes's principle. The temperature of the coil cannot go higher than the boiling temperature of water, 100°C. If the coil is operated in air, the convection process is reduced, and the upper limit of 100°C is removed. As a result, the coil can become hot enough to be damaged. If the coil is used in an attempt to warm a thick liquid like stew, the convection process cannot occur fast enough to carry energy away from the coil, so that it again may become hot enough to be damaged.

9. One of the ways that objects transfer energy is by radiation. If we consider a mailbox, the top of the box is oriented toward the clear sky. Radiation emitted by the top of the mailbox goes upward and into space. There is little radiation coming down from space to the top of the mailbox. Radiation leaving the sides of the mailbox is absorbed by the environment. Radiation from the environment (tree, houses, cars, etc.), however, can enter the sides of the mailbox, keeping them warmer than the top. As a result, the top is the coldest portion and frost forms there first.
11. The tile is a better conductor of energy than carpet. Thus, the tile conducts energy away from your feet more rapidly than does the carpeted floor.
13. The large amount of energy stored in the concrete during the day as the sun falls on it is released at night resulting in an overall higher average temperature than the countryside. The heated air in a city rises as it is displaced by cooler air moving in from the countryside. Thus, evening breezes tend to blow from country to city.
15. The fingers are wetted to create a layer of steam between the fingers and the molten lead. The steam acts as an insulator and prevents serious burns. That said, the molten lead demonstration is still dangerous, and we do not recommend it.
17. The increase in the temperature of the ethyl alcohol will be about twice that of the water.
19. When water is heated at atmospheric pressure, the temperature of the water does not exceed its boiling temperature until all the water has evaporated. If done carefully, the rate of energy transfer to the water through the paper is sufficient to keep the temperature of the paper near that of the water, and hence at or below the boiling temperature of water. This temperature is below that required to ignite the paper.

Problems

1. 1.03 kJ
3. (a) 1.67×10^{18} J   (b) 53.1 yr
5. 0.258°C
7. 2.85 km
9. 176°C
11. 88 W
13. 23°C
15. 80 g
17. 1.8×10^3 J/kg · °C
19. 0.26 kg
21. 65°C
23. 21 g
25. 2.3 km
27. 16°C
29. Liquid lead at 805°C if the specific heat is constant
31. (a) all ice melts, $T_f = 40$°C   (b) 8.0 g melts, $T_f = 0$°C
33. (a) 0.22 kW   (b) 13 mW   (c) 56 mW
35. 402 MW
37. 14 ft^2 · °F · h/Btu
39. 9.0 cm
41. 0.11 kW
43. $\sim 10^3$ W
45. 16
47. 12 kW
49. 6.0×10^3 kg
51. 2.3 kg
53. 29°C
55. 30.3 kcal/hr
57. 109°C
59. 51.2°C
61. (a) 74 stops
    (b) Assumes no energy loss to surroundings and that all internal energy generated stays with the brakes.
63. (b) 2.7×10^3 J/kg · °C

CHAPTER 12

Quick Quizzes

1. **(a)** near 0, +, +
   (b) +, near 0, +
   (c) near 0, −, −
2. A is isovolumetric, B is adiabatic, C is isothermal, D is isobaric.
3. C, B, A
4. **(b)**
5. The number 7 is the most probable outcome. The numbers 2 and 12 are the least probable outcomes.

Conceptual Questions

1. The carbon dioxide undergoes a large increase in volume as it emerges from the container and spends considerable energy doing work on the outside air. Because the spent energy is not immediately replaced by heat, there is a large decrease in the internal energy, and hence the temperature, of the gas.
3. If there is no change in internal energy, then, according to the first law, the heat is equal to the negative of the work done on the gas (and thus equal to the work done *by* the gas). $Q = -W = W_{by\ gas}$.
5. The energy that is leaving the body by work and heat is replaced by means of biological processes that transform chemical energy in the food that the individual ate into internal energy. Thus, the temperature of the body can be maintained.
7. The statement shows a misunderstanding of the concept of heat. Heat is energy in the process of being transferred, not a form of energy that is held or contained. If you wish to speak of energy that is "contained," you speak of **internal energy,** not heat. The following are correct statements: (1) "Given any two objects in thermal contact, the one with the higher temperature will transfer energy by heat to the other." (2) "Given any two objects of equal mass, the one with the higher product of absolute temperature and specific heat contains more internal energy."
9. Although no energy is transferred into or out of the system by heat, work is done on the system as the result of the agitation. Consequently, both the temperature and the internal energy of the coffee increase.
11. A decrease in temperature indicates a decrease in internal energy. This can occur, at least for a short duration, if the system does a quantity of work that exceeds the heat input during some time interval.
13. Practically speaking, it is not possible to create a heat engine on Earth that creates no thermal pollution, because there must be both a hot heat source (energy reservoir) and a cold heat sink (low-temperature energy reservoir). The heat engine will warm the cold heat sink, and will cool down the heat source. If either of those two events is undesirable, then there will be thermal pollution.

    There are some circumstances where the thermal pollution would be negligible. For example, suppose a satellite in space were to run a heat pump between its sunny side and its dark side. The satellite would intercept some of the energy that gathered on one side and would dump it to the dark side. Since neither of those effects would be particularly undesirable, it could be said that such a heat pump produced no thermal pollution.
15. The rest of the Universe must have an entropy change of + 8.0 J/K or more.
17. The first law is a statement of conservation of energy that says that we cannot devise a cyclic process that produces more energy than we put into it. If the cyclic process takes in energy by heat and puts out work, we call the device a heat engine. In addition to the first law's limitation, the second law says that during the operation of a heat engine, some energy must be rejected as heat to the environment. As a result, it is theoretically impossible to construct a engine that will work with 100% efficiency.

Problems

3. 1.1×10^4 J
5. **(a)** − 810 J   **(b)** − 507 J   **(c)** − 203 J
7. **(a)** -6.1×10^5 J   **(b)** 4.6×10^5 J
9. **(a)** 1.09×10^3 K   **(b)** − 6.81 kJ
11. **(a)** $Q < 0$, $W = 0$, $\Delta U < 0$ **(b)** $Q > 0$, $W = 0$, $\Delta U > 0$
13. **(a)** 567 J   **(b)** 167 J
15. **(a)** − 88.5 J   **(b)** 722 J
17. 720 J from the system
19. **(a)** -9.12×10^{-3} J   **(b)** − 333 J
21. **(a)** 0.95 J   **(b)** 3.2×10^5 J
    (c) 3.2×10^5 J
23. 0.489 (or 48.9%)
25. **(a)** 0.333 (or 33.3%)   **(b)** 2/3
27. **(a)** 0.672 (or 67.2%)   **(b)** 58.8 kW
29. **(a)** 0.294 (or 29.4%)   **(b)** 500 J
    (c) 1.67 kW
31. $\frac{1}{3}$
33. 0.49°C
35. **(a)** − 1.2 kJ/K   **(b)** 1.2 kJ/K
37. 57 J/K
39. 3.27 J/K
41. **(a)**

End Result	Possible Tosses	Total Number of Same Result
All H	HHHH	1
1T, 3H	HHHT, HHTH, HTHH, THHH	4
2T, 2H	HHTT, HTHT, THHT, HTTH, THTH, TTHH	6
3T, 1H	TTTH, TTHT, THTT, HTTT	4
All T	TTTT	1

Most probable result = 2H and 2T

   (b) all H or all T   **(c)** 2H and 2T
43. The maximum efficiency possible with these reservoirs = 50%; claim is invalid.
45. **(a)** $-|Q_h|/T_h$ **(b)** $|Q_c|/T_c$
    (c) $|Q_h|/T_h - |Q_c|/T_c$ **(d)** 0
47. 2.8°C
51. **(a)** 12.2 kJ   **(b)** 4.05 kJ   **(c)** 8.15 kJ
53. **(a)** 26 J   **(b)** 9.0×10^5 J
    (c) 9.0×10^5 J
55. **(a)** 2.49 kJ   **(b)** 1.50 kJ   **(c)** − 990 J
57. **(a)** 0.554 (or 55.4%)
    (b) The Carnot efficiency is 0.749 (or 74.9%)
59. **(a)** 2.6×10^3 metric tons/day
    (b) $\$7.7 \times 10^6$/yr   **(c)** 4.1×10^4 kg/s

CHAPTER 13

Quick Quizzes

1. **(d)**
2. **(c)**
3. **(b)**
4. **(d)**
5. **(b)** decreases when elevator accelerates upward
   (c) remains the same when elevator moves with constant velocity

6. (a)
7. (b)

Conceptual Questions

1. It will increase. The speed of the wave varies inversely with the mass per unit length of the rope, so it is higher in the lighter rope. The frequency is unchanged as the wave passes from one rope to the other one. The wavelength is $\lambda = v/f$, so with the frequency constant and the speed increasing, the wavelength must increase.

3. No. Because the total energy is $E = \frac{1}{2}kA^2$, changing the mass of the object while keeping A constant has no effect on the total energy. When the object is at displacement x from equilibrium, the potential energy is $\frac{1}{2}kx^2$, independent of mass, and the kinetic energy is $KE = E - \frac{1}{2}kx^2$, also independent of mass.

5. When the spring with two objects on opposite ends is set into oscillation in space, the coil in the exact center of the spring does not move. Thus, we can imagine clamping the center coil in place without affecting the motion. If we do this, we have two separate oscillating systems, one on each side of the clamp. The half-spring on each side of the clamp has twice the spring constant of the full spring, as shown by the following argument. The force exerted by a spring is proportional to the separation of the coils as the spring is extended. Imagine that we extend a spring by a given distance and measure the distance between coils. We then cut the spring in half. If one of the half-springs is now extended by the same distance, the coils will be twice as far apart as they were for the complete spring. Thus, it takes twice as much force to stretch the half-spring, from which we conclude that the half-spring has a spring constant that is twice that of the complete spring. So our clamped system of objects on two half-springs will vibrate with a frequency that is higher than f by a factor of the square root of 2.

7. The bouncing ball is not an example of simple harmonic motion. The ball does not follow a sinusoidal function for its position as a function of time. The daily movement of a student is also not simple harmonic motion, because the student stays at a fixed location (school) for a long period of time. If this motion were sinusoidal, the student would move more and more slowly as she approached her desk, and as soon as she sat down at the desk, she would start to move back toward home again.

9. We assume that the buoyant force acting on the sphere is negligible in comparison to its weight, even when the sphere is empty. We also assume that the bob is small compared to the pendulum length. Then, the frequency of the pendulum is

$$f = \frac{1}{T} = \frac{1}{2\pi}\sqrt{\frac{g}{L}}$$

independent of mass. Thus, the frequency does not change as the water leaks out.

11. As the temperature increases, the length of the pendulum increases, due to thermal expansion, and with a greater length, the period of the pendulum increases. Thus, it takes longer to execute each swing, so that each second according to the clock will take longer than an actual second. Thus, the clock will run slow.

13. A pulse in a long line of people is longitudinal, because the movement of people is parallel to the direction of propagation of the pulse. The speed is determined by the reaction time of the people and the speed with which they can move, once a space opens up. There is also a psychological factor, in that people may not want to fill a space that opens up in front of them too quickly, so as not to intimidate the person in front of them. The "wave" at a stadium is transverse, because the fans stand up vertically as the wave sweeps past

them horizontally. The speed of this pulse depends on the limits of the fans' abilities to rise and sit rapidly, and on psychological factors associated with the anticipation of seeing the pulse approach the observer's location.

15. A wave on a massless string would have an infinite speed of propagation because the linear mass density of the string is zero.

17. The kinetic energy is proportional to the square of the speed, whereas the potential energy is proportional to the square of the displacement. Therefore, both must be positive quantities.

19. From $v = \sqrt{F/\mu}$, we see that increasing the tension by a factor of 4 doubles the wave speed.

Problems

1. (a) 24 N toward the equilibrium position
   (b) 60 m/s² toward the equilibrium position
3. (b) 1.81 s
   (c) No, the force is not of Hooke's law form.
5. 1.23 cm
7. (a) 60 J   (b) 49 m/s
9. 2.94×10^3 N/m
11. (a) $PE = E/4$ $KE = 3E/4$
    (b) $x = A/\sqrt{2}$
13. 0.478 m
15. (a) 0.28 m/s   (b) 0.26 m/s
    (c) 0.26 m/s   (d) 3.5 cm
17. 39.2 N
19. (a) You observe uniform circular motion projected on a plane perpendicular to the motion.
    (b) 0.628 s
21. The horizontal displacement is described by $x(t) = A\cos\omega t$, where A is the distance from the center of the wheel to the crank pin.
23. 0.63 s
25. 3.14×10^6 Pa
27. (a) 11.0 N toward the left   (b) 0.881 oscillations
29. $v = \pm\omega A\sin\omega t$, $a = -\omega^2 A\cos\omega t$
31. 1.001 5
33. (a) slow   (b) 9:47 A.M.
35. (a) $L_{Earth} = 25$ cm, $L_{Mars} = 9.4$ cm,
    (b) $m_{Earth} = m_{Mars} = 0.25$ kg
37. (a) 9.00 cm   (b) 20.0 cm
    (c) 40.0 ms   (d) 5.00 m/s
39. (a) 11.4 ns   (b) 3.41 m
41. 31.9 cm
43. 80.0 N
45. (a) 30.0 N   (b) 25.8 m/s
47. 28.5 m/s
49. (a) 0.051 kg/m   (b) 20 m/s
51. 13.5 N
53. (a) Constructive interference gives $A = 0.50$ m.
    (b) Destructive interference gives $A = 0.10$ m.
55. 6.62 cm
57. (a) 588 N/m   (b) 0.700 m/s
59. 1.1 m/s
61. (a) 15.8 rad/s   (b) 5.23 cm
63. $F_{tangential} = -[(\rho_{air} - \rho_{He})Vg/L]s$, $T = 1.40$ s
67. (a) 0.50 m/s   (b) 8.6 cm

Chapter 14

Quick Quizzes

1. (c)
2. (a) 10 dB   (b) 13 dB
3. (c)
4. (b)

Conceptual Questions

1. Most speech contains frequencies below about 4 000 Hz. Thus, the spoken word is not greatly affected. However, music contains frequencies spanning the gamut of 20 Hz to 20 000 Hz, and listeners may find their enjoyment of music less than optimal.

3. The camera is designed such that it assumes the speed of sound is 345 m/s, the speed of sound at a room temperature of 23°C. If the temperature should decrease to, say, 0°C, the speed will also decrease, and the camera will note that it takes longer for the sound to make its round trip. Thus, it will think the object is farther away than it really is.

   It is of interest to note that bats use echo-sounding like this to locate insects or to avoid obstacles in front of them, something that they must do because of poor eyesight and high speed. Likewise, blue whales use this technique to avoid objects in their path. The need here is obvious because a typical whale has a mass of 10^5 kg and travels at a relatively fast speed of 20 mi/h. Thus, it takes a long time for it to stop its motion or to change direction.

5. Sophisticated electronic devices break the frequency range of about 60 to 4 000 Hz used in telephone conversations into several frequency bands and then mix them in a predetermined pattern so that they become unintelligible. The descrambler, of course, moves the bands back into their proper order.

7. A rise in temperature will increase the dimensions of the wind instrument much less than it increases the speed of sound in the enclosed air. This effect will raise the resonant frequencies that are produced by the instrument. The instrument goes sharp as the temperature increases and goes flat as the temperature decreases.

9. The echo is Doppler shifted, and the shift is like both a moving source and a moving observer. The sound that leaves your horn in the forward direction is Doppler shifted to a higher frequency, because it is coming from a moving source. As the sound reflects back and comes toward you, you are a moving observer, so there is a second Doppler shift to an even higher frequency. If the sound reflects from the spacecraft coming toward you, there is a different moving source shift to an even higher frequency. The reflecting surface of the spacecraft acts as a moving source.

11. The center of the string is a node for the second harmonic, as well as for every even-numbered harmonic. By placing the finger at the center and plucking, the guitarist is eliminating any harmonic that does not have a node at that point, which is all the odd harmonics. The even harmonics can vibrate relatively freely with the finger at the center because they exhibit no displacement at that point. The result is a sound with a mixture of frequencies that are integer multiples of the second harmonic, which is one octave higher than the fundamental.

13. The bow string is pulled away from equilibrium and released, similar to the way that a guitar string is pulled and released when it is plucked. Thus, standing waves will be excited in the bow string. If the arrow leaves from the exact center of the string, then a series of odd harmonics will be excited. Even harmonics will not be excited because they have a node at the point where the string exhibits its maximum displacement.

15. At the instant at which there is no displacement of the string, the string is still moving. Thus, the energy is present at that instant entirely as kinetic energy of the string.

17. As the child is approaching you, the sound is Doppler shifted to a frequency higher than the natural frequency of the whistle. You will momentarily hear the natural frequency just as the child comes parallel with you and is in the act of passing. As soon as the child is past, you will hear sound that is Doppler shifted to a frequency lower than the natural frequency of the whistle. As the frequency of the sound steps down from one constant value to a second constant value, the intensity of the sound varies continuously. The loudness smoothly increases to a maximum and then decreases.

19. The two engines are running at slightly different frequencies, thus producing a beat frequency between the two.

Problems

1. 5.56 km
3. 32°C
5. 1.43 km/s
7. 1.99 km
9. (a) 1.00×10^{-2} W/m²   (b) 105 dB
11. 20 km
13. 9 additional machines
15. 1.0×10^{-7} W/m²
17. (a) 1.3×10^2 W   (b) 96 dB
21. (a) 75.2-Hz drop   (b) 0.953 m
23. 595 Hz
25. 0.391 m/s
27. 19.3 m
29. 48°
31. 800 m
33. (a) 0.240 m   (b) 0.855 m
35. (a) Nodes at 0, 2.67 m, 5.33 m, and 8.00 m
        Antinodes at 1.33 m, 4.00 m, and 6.67 m
   (b) 18.6 Hz
37. At 0.089 1 m, 0.303 m, 0.518 m, 0.732 m, 0.947 m, and 1.16 m from one speaker
39. (a) 79 N   (b) 2.1×10^2 Hz
41. 19.976 kHz
43. 58 Hz
45. 3.0 kHz
47. (a) 0.552 m   (b) 317 Hz
49. 5.26 beats/s
51. 3.79 m/s toward the station,
     3.88 m/s away from the station
53. (a) 1.98 beats/s   (b) 3.40 m/s
55. 1.76 cm
57. 262 kHz
59. 64 dB
61. 439 Hz and 441 Hz
63. 32.9 m/s
65. 3.97 beats/s
67. 1.34×10^4 N
69. 1 204 Hz

CHAPTER 15

Quick Quizzes

1. d
2. b
3. c
4. b
5. a
6. c and d
7. A, B, and C
8. (b) and (d)

Conceptual Questions

1. Because of lower moisture content during winter, the air acts as a better insulator and allows larger static charges to build up on objects during winter than during summer. As a result, shocks from static electricity are more severe in winter months than in summer months.

3. The configuration shown is inherently unstable. The negative charges repel each other. If there is any slight rotation of one of the rods, the repulsion can result in further rota-

tion away from this configuration. There are three conceivable final configurations shown below. Configuration (a) is stable—if the positive upper ends are pushed toward each other, they will repel and move the system back to the original configuration. Configuration (b) is an equilibrium configuration, but it is unstable—if the lower ends are moved toward each other, the attraction of the lower ends will be larger than that of the upper ends and the configuration will shift to (c). Configuration (c) is another possible stable configuration.

(a)   (b)

(c)

5. Move an object A with a net positive charge so it is near, but not touching, a neutral metallic object B that is insulated from the ground. The presence of A will polarize B, causing an excess negative charge to exist on the side nearest A and an equal magnitude excess positive charge to exist on the side farthest from A. While A is still near B, touch B with your hand. This allows additional electrons to flow from ground through your body and onto B. With A continuing to be near but not touching B, remove your hand from B, thus trapping the excess electrons on B. When A is now removed, B is left with excess electrons causing a net negative charge. This negative charge, by means of mutual repulsion, will now spread uniformly over the entire surface of B.

7. An object's mass decreases very slightly (immeasurably) when it is given a positive charge, because it loses electrons. When the object is given a negative charge, its mass increases slightly because it gains electrons.

9. Electric field lines start on positive charges and end on negative charges. Thus, if the fair weather field is directed into the ground, the ground must have a negative charge.

11. The two charged plates create a region of uniform electric field between them, directed from the positive toward the negative plate. Once the ball is disturbed so as to touch one plate, say the negative one, some negative charge will be transferred to the ball and it will experience an electric force that will accelerate it to the positive plate. Once the charge touches the positive plate, it will release its negative charge, acquire a positive charge, and accelerate back to the negative plate. The ball will continue to move back and forth between the plates until it has transferred all their net charge, thereby making both plates neutral.

13. The electric shielding effect of conductors depends on the fact that there are two kinds of charge—positive and negative. As a result, charges can move within the conductor so that the combination of positive and negative charges establishes an electric field that exactly cancels the external field within the conductor and any cavities inside the conductor. There is only one type of gravitational charge, however—there is no "negative mass." As a result, gravitational shielding is not possible.

15. The electric field patterns of each of these three configurations do not have sufficient symmetry to make the calculations practical. Gauss's law is only useful for calculating the electric field of highly symmetric charge distributions, such as uniformly charged spheres, cylinders, and sheets.

17. No. The balloon induces charge of opposite sign in the wall, causing the balloon and the wall to be attracted to each other. The balloon eventually falls because its charge slowly diminishes after leaking to ground. Some of the balloon's charge could also be lost due to positive ions in the surrounding atmosphere, which would tend to neutralize the negative charges on the balloon.

19. When the comb is nearby, charges separate on the paper, resulting in the paper being attracted. After contact, charges from the comb are transferred to the paper so that it has the same type charge as the comb. It is thus repelled.

Problems

1. 1.1×10^{-8} N (attractive)
3. 91 N (repulsion)
5. (a) 36.8 N   (b) 5.54×10^{27} m/s^2
7. 5.12×10^5 N
9. (a) 2.2×10^{-5} N (attraction)
   (b) 9.0×10^{-7} N (repulsion)
11. 1.38×10^{-5} N at 77.5° below $-x$ axis
13. 0.872 N at 30.0° below $+x$ axis
15. 7.2 nC
17. 5.5×10^{11} N/C (away from the proton)
19. 7.20×10^5 N/C (downward)
21. 1.63×10^5 N/C
23. (a) 6.12×10^{10} m/s^2 (b) 19.6 μs
    (c) 11.8 m   (d) 1.20×10^{-15} J
25. zero
27. 1.8 m to the left of the -2.5 μC charge
33. (a) 0
    (b) $+5$ μC inside, -5 μC outside
    (c) 0 inside, -5 μC outside
    (d) 0 inside, -5 μC outside
35. 1.3×10^{-3} C
37. (a) 4.8×10^{-15} N   (b) 2.9×10^{12} m/s^2
39. (a) 858 N$\cdot$m^2/C
    (b) 0   (c) 657 N$\cdot$m^2/C
41. 4.1×10^6 N/C
43. (a) 0   (b) $k_e q / r^2$ outward
47. 57.5 N
49. 24 N/C in the $+x$ direction
51. (a) $E = 2k_e qb(a^2 + b^2)^{-3/2}$ in $+x$ direction
    (b) $E = 2k_e Qb(a^2 + b^2)^{-3/2}$ in $+x$ direction
53. (a) 0
    (b) 7.99×10^7 N/C (outward)
    (c) 0
    (d) 7.34×10^6 N/C (outward)
55. 3.55×10^5 N$\cdot$m^2/C
57. 4.4×10^5 N/C
59. (a) 10.9 nC   (b) 5.43×10^{-3} N

CHAPTER 16

Quick Quizzes

1. (b)
2. Either (a) or (b) might be true
3. (a) and (b)
4. (c)
5. (a) C decreases.   (b) Q stays the same.
   (c) E stays the same.   (d) ΔV increases.
   (e) The energy stored increases.
6. (a) C increases.   (b) Q increases.
   (c) E stays the same.   (d) ΔV remains the same.
   (e) The energy stored increases.

Conceptual Questions

1. The statement is false. A uniform field simply means that the field is constant in magnitude and direction. Any electric

field does work on a charge as that charge undergoes a displacement in the direction of the field. Thus, a change in potential occurs when one undergoes any displacement along the direction of the field, even if the field is uniform. If we define the potential to be zero at one point, it will be different from zero at neighboring points that are downhill and uphill in the field.

3. The work done in pulling the capacitor plates farther apart is transferred into additional electric energy stored in the capacitor. The charge is constant and the capacitance decreases, but the potential difference between the plates increases, which results in an increase in the stored electric energy.

5. The power line, if it makes electrical contact with the metal of the car, will raise the potential of the car to 20 kV. It will also raise the potential of your body to 20 kV, because you are in contact with the car. In itself, this is not a problem. If you step out of the car, your body at 20 kV will make contact with the ground, which is at zero volts. As a result, a current will pass through your body, and you will likely be injured. Thus, it is best to stay in the car until help arrives.

7. If two points on a conducting object were at different potentials, then free charges in the object would move, and we would not have static conditions, in contradiction to the initial assumption. (Free positive charges would migrate from higher to lower potential locations; free electrons would move rapidly from lower to higher potential locations.) The charges would continue to move until the potential became equal everywhere in the conductor.

9. The capacitor often remains charged long after the voltage source is disconnected. This residual charge can be lethal. The capacitor can be safely handled after discharging the plates by short-circuiting the device with a conductor, such as a screwdriver with an insulating handle.

11. Field lines represent the direction of the electric force on a positive test charge. If electric field lines were to cross, then at the point of crossing, there would be an ambiguity regarding the direction of the force on the test charge, because there would be two possible forces. Thus, electric field lines cannot cross. It is possible for equipotential surfaces to cross. (However, equipotential surfaces at different potentials cannot intersect.) For example, suppose two identical positive charges are at diagonally opposite corners of a square, and two negative charges of equal magnitude are at the other two corners. Then the planes perpendicular to the sides of the square at their midpoints are equipotential surfaces. These two planes cross each other, at the line perpendicular to the square at its center.

13. (a) From $Q = C(\Delta V)$, if you double the potential difference across a capacitor having a fixed capacitance, you will double the charge stored in the capacitor. (b) The energy stored in a capacitor may be expressed as $W = Q^2/2C$. Assuming a parallel-plate capacitor, the capacitance is $C = \kappa\epsilon_0 A/d$, and the stored energy is $W = Q^2 d/2\kappa\epsilon_0 A$. Thus, doubling the plate separation while holding the charge constant will double the energy stored.

15. You should use a dielectric-filled capacitor whose dielectric constant is very large. Furthermore, you should make the dielectric as thin as possible, keeping in mind that dielectric breakdown must also be considered.

17. Inserting a dielectric into the capacitor increases the capacitance of that device. (a) From $W = C(\Delta V)^2/2$, it is seen that increasing the capacitance while keeping the potential difference constant will increase the energy stored in the capacitor.
(b) From $W = Q^2/2C$, it is seen that increasing the capacitance while holding the charge on the capacitor constant decreases the energy stored in the capacitor.

Problems

1. (a) 6.40×10^{-19} J   (b) -6.40×10^{-19} J
   (c) -4.00 V
3. 1.4×10^{-20} J
5. 1.7×10^6 N/C
7. (a) 1.13×10^5 N/C   (b) 1.80×10^{-14} N
   (c) 4.38×10^{-17} J
9. (a) 0.500 m   (b) 0.250 m
11. (a) 1.44×10^{-7} V   (b) -7.19×10^{-8} V
13. (a) 2.67×10^6 V   (b) 2.13×10^6 V
15. (a) 103 V
    (b) -3.85×10^{-7} J; positive work must be done to separate the charges.
17. -11.0 kV
19. 2.74×10^{-14} m
21. 0.719 m, 1.44 m, 2.88 m. No. The equipotentials are not uniformly spaced. Instead, the radius of an equipotential is inversely proportional to the potential.
23. (a) 1.1×10^{-8} F   (b) 27 C
25. (a) 11.1 kV/m toward the negative plate   (b) 3.74 pF
    (c) 74.7 pC and -74.7 pC
27. 2.26×10^{-5} m^2
29. (a) 13.3 μC on each   (b) 20.0 μC, 40.0 μC
31. (a) 2.00 μF
    (b) $Q_3 = 24.0$ μC, $Q_4 = 16.0$ μC, $Q_2 = 8.00$ μC, $(\Delta V)_2 = (\Delta V)_4 = 4.00$ V, $(\Delta V)_3 = 8.00$ V
33. (a) 5.96 μF
    (b) $Q_{20} = 89.5$ μC, $Q_6 = 63.2$ μC, $Q_3 = Q_{15} = 26.3$ μC
35. $Q_1 = 16.0$ μC, $Q_5 = 80.0$ μC, $Q_8 = 64.0$ μC, $Q_4 = 32.0$ μC
37. (a) $Q_{25} = 1.25$ mC, $Q_{40} = 2.00$ mC
    (b) $Q'_{25} = 288$ μC, $Q'_{40} = 462$ μC, $\Delta V = 11.5$ V
39. $Q'_1 = 3.33$ μC, $Q'_2 = 6.67$ μC
41. 83.6 μC
43. 2.55×10^{-11} J
45. 3.2×10^{10} J
47. $\kappa = 4.0$
49. (a) 8.13 nF   (b) 2.40 kV
51. (a) volume 9.09×10^{-16} m^3, area 4.54×10^{-10} m^2
    (b) 2.01×10^{-13} F
    (c) 2.01×10^{-14} C; 1.26×10^5 electronic charges
55. $v = 1.67$ m/s upward, $y = 6.24$ cm above bottom plate
57. 6.25 μF
59. 4.47 kV
61. 0.75 mC on C_1, 0.25 mC on C_2
65. 50 N

CHAPTER 17

Quick Quizzes

1. d, b = c, a
2. c, d
3. b
4. b, d
5. a
6. $I_a = I_b > I_c = I_d > I_e = I_f$
7. B, B

Conceptual Questions

1. It is made possible by the very high density of charge carriers ($\sim 10^{28}$/m^3) in metallic conductors.

3. The gravitational force pulling the electron to the bottom of a piece of metal is much smaller than the electrical repulsion pushing the electrons apart. Thus, free electrons stay distributed throughout the metal. The concept of charges residing on the surface of a metal is true for a metal with an excess charge. The number of free electrons in an electrically neutral piece of metal is the same as the number of positive ions—the metal has zero net charge.

5. A voltage is not something that "surges through" a completed circuit. A voltage is a potential difference that is applied across a device or a circuit. What goes through the circuit is current. Thus, it would be more correct to say, "1 ampere of electricity surged through the victim's body." Although this current would have disastrous results on the human body, a value of 1 (ampere) doesn't sound as exciting for a newspaper article as 10 000 (volts). Another possibility is to write, "10 000 volts of electricity were applied across the victim's body," which still doesn't sound quite as exciting.

7. We would conclude that the conductor is nonohmic.

9. The shape, dimensions, and the resistivity affect the resistance of a conductor. Because temperature and impurities affect the conductor's resistivity, these factors also affect resistance.

11. The radius of wire B is the square root of three times the radius of wire A, to make its cross-sectional area three times larger.

13. Drift velocity might increase steadily as time goes on, because collisions between electrons and atoms in the wire would be essentially nonexistent and the conduction electrons would move with constant acceleration. The current would rise steadily without bound also, because I is proportional to the drift velocity.

Problems

1. 3.00×10^{20} electrons move past in the direction opposite to the current.
3. 3.0 mA
5. 1.05 mA
7. 27 yr
9. (a) n is unaffected   (b) v_d is doubled
11. 32 V, is 200 times larger than 0.16 V
13. 0.17 mm
15. (a) 30 Ω (b) 4.7×10^{-4} $\Omega \cdot$m
17. silver ($\rho = 1.59 \times 10^{-8}$ $\Omega \cdot$m)
19. 2.50 mA decrease
21. 1.98 A
23. 26 mA
25. (a) 5.89×10^{-2} Ω (b) 5.45×10^{-2} Ω
27. (a) 3.0 A   (b) 2.9 A
29. (a) 1.2 Ω
    (b) 8.0×10^{-4} (a 0.080% increase)
31. 5.00 A, 24.0 Ω
33. 18 bulbs
35. 11.2 min
37. 34.4 Ω
39. 1.6 cm
41. 295 metric tons/h
43. 26 cents
45. 23 cents
47. $1.2
49. 1.1 km
51. 1.47×10^{-6} $\Omega \cdot$m, differs by 2.0% from value in Table 17.1
53. 0.400 μA
55. (a) 667 A   (b) 50.0 km
57. 3.77×10^{28}/m^3
59. (a) 144 Ω (b) 26 m
    (c) To fit the required length into a small space.
    (d) 25 m
61. 37 MΩ
63. 0.48 kg/s
65. (a) 2.6×10^{-5} Ω (b) 76 kg

CHAPTER 18

Quick Quizzes

1. Bulb R_1 becomes brighter. Bulb R_2 goes out.
2. (b)
3. (a)

4. (a) decrease; (b) decrease; (c) increase; (d) decrease;
   (e) decrease.

Conceptual Questions

1. No. When a battery serves as a source and supplies current to a circuit, the conventional current flows through the battery from the negative terminal to the positive one. However, when a source having a larger emf than the battery is used to charge the battery, the conventional current is forced to flow through the battery from the positive terminal to the negative one.

3. The total amount of energy delivered by the battery will be less than W. Recall that a battery can be considered to be an ideal, resistanceless battery in series with the internal resistance. When charging, the energy delivered to the battery includes the energy necessary to charge the ideal battery plus the energy that goes into raising the battery's temperature due to I^2r heating in the internal resistance. This latter energy is not available during the discharge of the battery. During discharge, part of the reduced available energy again transforms into internal energy in the internal resistance, further reducing the available energy below W.

5. The starter in the automobile draws a relatively large current from the battery. This large current causes a significant voltage drop across the internal resistance of the battery. As a result, the terminal voltage of the battery is reduced, and the headlights dim accordingly.

7. An electrical appliance has a given resistance. Thus, when it is attached to a power source with a known potential difference, a definite current will be drawn. The device can be labeled with both the voltage and the current. Batteries, however, can be applied to a number of devices. Each device will have a different resistance, so the current from the battery will vary with the device. As a result, only the voltage of the battery can be specified.

9. Connecting batteries in parallel does not increase the emf. A high-current device connected to batteries in parallel can draw currents from both batteries. Thus, connecting the batteries in parallel does increase the possible current output and, therefore, the possible power output.

11. The lightbulb will glow for a very short while as the capacitor is being charged. Once the capacitor is almost totally charged, the current in the circuit will be nearly zero and the bulb does not glow.

13. The bird is at rest on a wire whose electric potential is nearly constant along its length. In order to be electrocuted, a large potential difference is required between the bird's feet. The potential difference between the bird's feet is too small to harm the bird.

15. The junction rule is a statement of conservation of charge. It says that the amount of charge that enters the junction in some time interval must equal the charge that leaves the junction in that time interval. The loop rule is a statement of conservation of energy. It says that the potential increases and decreases around a closed loop in a circuit must add to zero.

17. A few of the factors involved are as follows: the conductivity of the string (is it wet or dry?); how well you are insulated from ground (are you wearing thick rubber- or leather-soled shoes?); the magnitude of the potential difference between you and the kite; the type and condition of the soil under your feet.

19. She will not be electrocuted if she holds onto only one high-voltage wire, because she is not completing a circuit. There is no potential difference across her body, as long as she clings to only one wire. However, she should immediately release the wire once it breaks, because she will become part of a closed circuit when she reaches the ground or comes into contact with another object.

21. (a) The intensity of both lamps increases because lamp C is short-circuited and there is current (which increases) only

in lamps A and B. **(b)** The intensity of lamp C goes to zero because the current in this branch goes to zero. **(c)** The current in the circuit increases because the total resistance decreases from $3R$ (with the switch open) to $2R$ (after the switch is closed). **(d)** The voltage drop across lamps A and B increases, while the voltage drop across lamp C becomes zero. **(e)** The power dissipated increases from $\mathcal{E}^2/3R$ (with the switch open) to $\mathcal{E}^2/2R$ (after the switch is closed).

Problems

1. 4.92 Ω

3. 73.8 W; your circuit diagram will consist of two 0.800-Ω resistors in series with the 192-Ω resistance of the bulb.

5. **(a)** 17.1 Ω
   (b) 1.99 A for 4.00 Ω and 9.00 Ω, 1.17 A for 7.00 Ω, 0.818 A for 10.0 Ω

7. 9.8 Ω

9. **(a)** 0.227 A   **(b)** 5.68 V

11. 55 Ω

13. 0.43 A

15. **(a)** Connect two 50-Ω resistors in parallel, and then connect this combination in series with a 20-Ω resistor. **(b)** Connect two 50-Ω resistors in parallel, connect two 20-Ω resistors in parallel, and then connect these two combinations in series with each other.

17. 0.846 A down in 8.00-Ω resistor; 0.462 A down in the middle branch; 1.31 A up in the right-hand branch

19. **(a)** 3.00 mA   **(b)** -19.0 V   **(c)** 4.50 V

21. 10.7 V

23. **(a)** 0.385 mA, 3.08 mA, 2.69 mA
   (b) 69.2 V with c at the higher potential

25. $I_1 = 3.5$ A, $I_2 = 2.5$ A, $I_3 = 1.0$ A

27. $I_{30} = 0.353$ A, $I_5 = 0.118$ A, $I_{20} = 0.471$ A

29. $\Delta V_2 = 3.05$ V, $\Delta V_3 = 4.57$ V, $\Delta V_4 = 7.38$ V, $\Delta V_5 = 1.62$ V

31. **(a)** 12 s   **(b)** 1.2×10^{-4} C

33. 1.3×10^{-4} C

35. 0.982 s

37. **(a)** heater, 10.8 A; toaster, 8.33 A; grill, 12.5 A
   (b) $I_{total} = 31.6$ A, so a 30-A breaker is insufficient.

39. **(a)** 6.25 A   **(b)** 750 W

41. **(a)** 1.2×10^{-9} C, 7.3×10^9 K$^+$ions. Not large, only $1e/320$ Å^2
   (b) 1.7×10^{-9} C, 1.0×10^{10} Na$^+$ ions
   (c) 0.83 μA   **(d)** 7.5×10^{-12} J

43. 11 nW

45. 7.5 Ω

47. **(a)** 15 Ω
   (b) $I_1 = 1.0$ A, $I_2 = I_3 = 0.50$ A, $I_4 = 0.30$ A, and $I_5 = 0.20$ A
   (c) $(\Delta V)_{ac} = 6.0$ V, $(\Delta V)_{ce} = 1.2$ V,
   $(\Delta V)_{ed} = (\Delta V)_{fd} = 1.8$ V, $(\Delta V)_{cd} = 3.0$ V,
   $(\Delta V)_{db} = 6.0$ V
   (d) $\mathcal{P}_{ac} = 6.0$ W, $\mathcal{P}_{ce} = 0.60$ W, $\mathcal{P}_{ed} = 0.54$ W, $\mathcal{P}_{fd} = 0.36$ W,
   $\mathcal{P}_{cd} = 1.5$ W, $\mathcal{P}_{db} = 6.0$ W

49. **(a)** 12.4 V   **(b)** 9.65 V

51. $I_1 = 0$, $I_2 = I_3 = 0.50$ A

53. 112 V, 0.200 Ω

55. **(a)** $R_x = R_2 - (\frac{1}{4})R_1$
   (b) $R_x = 2.8$ Ω (inadequate grounding)

59. $\mathcal{P} = \dfrac{(144 \text{ V}^2)R}{(R + 10.0 \ \Omega)^2}$

CHAPTER 19

Quick Quizzes

1. a or c might be true

2. c

3. a

4. c

5. b

Conceptual Questions

1. No. In a uniform magnetic field, the poles of the bar magnetic experience forces that are equal in magnitude but oppositely directed. Thus, the net force exerted on the magnet by the magnetic field is zero.

3. The proton moves in a circular path upward on the page. After completing half the circle, it exits the field and moves in a straight-line path back in the direction from whence it came. An electron behaves similarly, but the direction of traversal of the circle is downward, and the radius of the circular path is smaller.

5. The magnetic force on a moving charged particle is always perpendicular to the direction of motion. There is no magnetic force on the charge when it moves parallel to the direction of the magnetic field. However, the force on a charged particle moving in an electric field is never zero and is always parallel to the direction of the electric field. Therefore, by projecting the charged particle in different directions, it is possible to determine the nature of the field.

7. The magnetic field produces a magnetic force on the electrons moving toward the screen that produce the image. This magnetic force deflects the electrons to regions on the screen other than the ones to which they are supposed to go. The result is a distorted image.

9. Such levitation would never occur. At the North Pole where Earth's magnetic field is downward, toward the equivalent of a buried south pole, a coffin would be repelled if its south magnetic pole were directed downward. However, equilibrium would be only transitory as any slight disturbance would upset the balance between the magnetic force and the gravitational force.

11. If you are moving along with the electrons, you will measure zero current for the electrons, so the electrons would not produce a magnetic field according to your observations. However, the fixed positive charges in the metal are now moving backward relative to you and creating a current equivalent to the forward motion of the electrons when you were stationary. Thus, you will measure the same magnetic field as when you were stationary, but it will be due to the positive charges presumed to be moving from your point of view.

13. A compass does not detect currents in wires near light switches for two reasons. Because the cable to the light switch contains two wires, with one carrying current to the switch and the other away from the switch, the net magnetic field is very small and falls off rapidly. The second reason is that the current is alternating at 60 Hz with increasing distance. As a result, the magnetic field is oscillating at 60 Hz, also. This frequency is too fast for the compass to follow, so the effect on the compass reading averages to zero.

15. The levitating wire is stable with respect to vertical motion — if it is displaced upward, the repulsive force weakens, and the wire drops back down. Conversely, if it drops lower, the repulsive force increases, and it moves back up. The wire is not stable, however, with respect to lateral movement. If it moves away from the vertical position directly over the lower wire, the repulsive force has a sideways component, which pushes the wire away.

In the case of the attracting wires, the hanging wire is not stable for vertical movement. If it rises, the attractive force increases, and the wire moves even closer to the upper wire.

If the hanging wire falls, the attractive force weakens, and the wire falls farther. If the wire moves to the right, it moves farther from the upper wire and the attractive force decreases. Although there is a restoring force component pulling it back to the left, the vertical force component is not strong enough to hold it up and it falls.

17. Each coil of the Slinky becomes a magnet, because a coil acts as a current loop. Because the sense of rotation of the current is the same in all coils, each coil becomes a magnet with the same orientation of poles. Thus, all of the coils attract, and the Slinky compresses.

19. There is no net force on the wires, but there is a torque. To understand this, imagine a fixed vertical wire, and a free horizontal wire (Fig. ANS Q19.19). The vertical wire carries an upward current, and creates a magnetic field that circles the wire. Each segment of the horizontal wire (of length ℓ) also carries a current that interacts with the magnetic field according to the equation $F = BI\ell \sin \theta$. Applying the right-hand rule, we see that the horizontal wire experiences an upward force on one side, and an equal downward force on the other. The forces cancel, creating a torque around the point at which the wires cross.

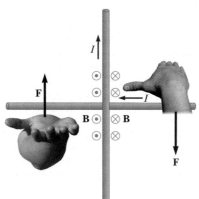

Problems

1. (a) horizontal and due east
   (b) horizontal and 30°N of E
   (c) horizontal and due east
   (d) zero force
3. (a) into the page   (b) toward the right
   (c) toward the bottom of the page
5. $F_g = 8.93 \times 10^{-30}$ N (downward), $F_e = 1.60 \times 10^{-17}$ N (upward), $F_m = 4.80 \times 10^{-17}$ N (downward)
7. 2.83×10^7 m/s west
9. 0.021 T in the $-y$ direction
11. 8.0×10^{-3} T in the $+z$ direction
13. (a) to the left   (b) into the page
    (c) out of the page   (d) toward top of page
    (e) into the page   (f) out of the page
15. 7.50 N
17. 0.131 T (downward)
19. 0.20 T directed out of the page
21. ab: 0, bc: 0.040 0 N in the $-x$ direction, cd: 0.040 0 N in the $-z$ direction, da: 0.040 0 $\sqrt{2}$ N parallel to the xz plane and at 45° with both $+x$ and $+z$ directions
23. 9.05×10^{-4} N·m tending to make the left-hand side of the loop move toward you and the right-hand side move away.
25. (a) 3.97°   (b) 3.39×10^{-3} N·m
27. 2.0×10^{-12} kg
31. 1.77 cm
33. (a) 5.00 cm   (b) 8.78×10^6 m/s
35. 20.0 μT
37. 2.0 cm
39. 20.0 μT toward the bottom of the page

41. 0.167 μT out of the page
43. (a) 4.00 m   (b) 7.50 nT   (c) 1.26 m   (d) zero
45. 4.5 mm
47. 31.8 mA
49. 2.26×10^{-4} N away from the center, zero torque
51. 1.7 N·m
53. (a) 0.500 μT out of the page
    (b) 3.89 μT parallel to the xy plane and at 59.0° clockwise from the $+x$ direction
55. 2.13 cm
57. (a) 1.33 m/s
    (b) the sign of the emf is independent of charge
59. 1.41×10^{-6} N
61. 13.0 μT toward the bottom of the page
63. (a) 53 μT toward the bottom of the page
    (b) 20 μT toward the bottom of the page
    (c) 0
65. (a) -8.00×10^{-21} kg·m/s   (b) 8.90°

CHAPTER 20

Quick Quizzes

1. b, c, a
2. The left wing tip on the west side of the airplane.
3. b
4. c
5. b

Conceptual Questions

1. According to Faraday's law, an emf is induced in a wire loop if the magnetic flux through the loop changes with time. In this situation, an emf can be induced by either rotating the loop around an arbitrary axis or by changing the shape of the loop.

3. As the spacecraft moves through space, it is apparently moving from a region of one magnetic field strength to a region of a different magnetic field strength. The changing magnetic field through the coil induces an emf and corresponding current in the coil.

5. If the bar were moving to the left, the magnetic force on the negative charges in the bar would be upward, causing an accumulation of negative charge on the top, and positive charges at the bottom. Hence, the electric field in the bar would be upward.

7. If, for any reason the magnetic field should change rapidly, a large emf could be induced in the bracelet. If the bracelet were not a continuous band, this emf would cause high-voltage arcs to occur at any gap in the band. If the bracelet were a continuous band, the induced emf would produce a large induced current and result in resistance heating of the bracelet.

11. As the aluminum plate moves into the field, eddy currents are induced in the metal by the changing magnetic field at the plate. The magnetic field of the electromagnet interacts with this current producing a retarding force on the plate, slowing it down. In a similar fashion, as the plate leaves the magnetic field, a current is induced, and once again there is an upward force to slow the plate.

13. The energy stored in an inductor carrying a current I is equal to $PE_L = \frac{1}{2}LI^2$. Therefore, doubling the current quadruples the energy stored in the inductor.

15. If an external battery is acting to increase the current in the inductor, an emf is induced in a direction to oppose the increase of current. Likewise, if we attempt to reduce the current in the inductor, the emf set up tries to support the current. Thus, the induced emf always acts to oppose the change occurring in the circuit, or it acts in the "back" direction to the change.

Problems

1. $5.9 \times 10^{-2}\,\mathrm{T \cdot m^2}$
3. $1.72 \times 10^{-6}\,\mathrm{T \cdot m^2}$, 0,
   $9.15 \times 10^{-7}\,\mathrm{T \cdot m^2}$
5. $\Phi_{B,\,net} = 0$ **(b)** 0
7. **(a)** $3.1 \times 10^{-3}\,\mathrm{T \cdot m^2}$ **(b)** $\Phi_{B,\,net} = 0$
9. 9.82 mV
11. 160 A
13. 2.7 T/s
15. **(a)** $4.0 \times 10^{-6}\,\mathrm{T \cdot m^2}$ **(b)** 18 μV
17. 11.6 μV
19. 0.763 V
21. **(a)** toward the east   **(b)** $4.58 \times 10^{-4}\,\mathrm{V}$
23. **(a)** left to right   **(b)** right to left
25. left to right
27. into the page
29. **(a)** right to left   **(b)** right to left
   (c) left to right   **(d)** left to right
31. 28.8 V
33. **(a)** 11 Ω **(b)** 76 V
35. **(a)** 60 V   **(b)** 57 V   **(c)** 0.13 s
37. 100 V
39. **(a)** 2.0 mH   **(b)** 38 A/s
43. 12 mH
45. 1.92 Ω
47. 0.140 J
49. **(a)** 18 J   **(b)** 7.2 J
51. negative ($V_a < V_b$)
53. **(a)** 20.0 ms   **(b)** 37.9 V   **(c)** 1.52 mV   **(d)** 51.8 mA
55. 1.20 μC
57. **(a)** 0.500 A   **(b)** 2.00 W   **(c)** 2.00 W
59. 115 kV
61. **(a)** 0.157 mV (end B is positive)
   (b) 5.89 mV (end A is positive)
63. **(a)** 9.00 A   **(b)** 10.8 N
   (c) b is at the higher potential   **(d)** No

CHAPTER 21

Quick Quizzes

1. c
2. b
3. b
4. b, c

Conceptual Questions

1. For best reception, the length of the antenna should be parallel to the orientation of the oscillating electric field. Because of atmospheric variations and reflections of the wave before it arrives at your location, the orientation of this field may be in different directions for different stations.
3. The primary coil of the transformer is an inductor. When an AC voltage is applied, the back emf due to the inductance will limit the current flow through the coil. If DC voltage is applied, there is no back emf and the current can rise to a higher value. It is possible that this increased current will deliver so much energy to the resistance in the coil that its temperature rises to the point at which insulation on the wire can burn.
5. An antenna that is a conducting line responds to the electric field of the electromagnetic wave—the oscillating electric field causes an electric force on electrons in the wire along its length. The movement of electrons along the wire is detected as a current by the radio and amplified. Thus, a line antenna must have the same orientation as the broadcast antenna. A loop antenna responds to the magnetic field in the radio wave. The varying magnetic field induces a varying current in the loop (Faraday's law), and this signal is amplified. The loop should be in the vertical plane containing the sight line to the broadcast antenna, so the magnetic field lines go through the area of the loop.
7. The flashing of the light according to Morse code is a drastic amplitude modulation—the amplitude is changing from a maximum to zero. In this sense, it is similar to the on-and-off binary code used in computers and compact discs. The carrier frequency is that of the light, on the order of 10^{14} Hz. The signal frequency depends on the skill of the signal operator, but it is on the order of a single hertz, as the light is flashed on and off. The broadcasting antenna for this modulated signal is the filament of the lightbulb in the signal source. The receiving antenna is the eye.
9. The sail should be as reflective as possible so that the maximum momentum is transferred to the sail from the reflection of sunlight.
11. Suppose the extraterrestrial looks around your kitchen. Lightbulbs and the toaster glow brightly in the infrared. Somewhat fainter are the back of the refrigerator and the back of the television set, while the television screen is dark. The pipes under the sink show the same weak glow as the walls until you turn on the faucets. Then the pipe on the right gets darker and that on the left develops a gleam that quickly runs up along its length. The food on the plates shines, as does human skin, the same color for all races. Clothing is dark as a rule, but your seat and the chair seat glow alike after you stand up. Your face appears lit from within, like a jack-o-lantern; your nostrils and the openings of your ear canals are bright; brighter still are the pupils of your eyes.
13. Radio waves move at the speed of light. They can travel around the curved surface of the Earth, bouncing between the ground and the ionosphere, which has an altitude that is small compared to the radius of the Earth. The distance across the lower 48 states is approximately 5 000 km, requiring a travel time of $(5 \times 10^6\,\mathrm{m})/(3 \times 10^8\,\mathrm{m/s}) \sim 10^{-2}\,\mathrm{s}$. Likewise, radio waves take only 0.07 s to travel halfway around the Earth. In other words, a speech can be heard on the other side of the world (in the form of radio waves) before it is heard at the back of the room (in the form of sound waves).
15. No. The wire will emit electromagnetic waves only if the current varies in time. The radiation is the result of accelerating charges, which can only occur when the current is not constant.
17. The resonance frequency is determined by the inductance and the capacitance in the circuit. It is not affected by the resistance in the circuit, so doubling the resistance does not alter the resonance frequency.

Problems

1. **(a)** 141 V   **(b)** 20.0 A
   (c) 28.3 A   **(d)** 2.00 kW
3. 70.7 V, 2.95 A
5. 6.76 W
9. 4.0×10^2 Hz
11. 17 μF
15. 3.14 A
17. 0.450 T$\cdot$m^2
19. **(a)** 205 mA   **(b)** 154 V   **(c)** 54.3 V   **(d)** 90.0°
   (e)

21. (a) $1.4 \text{ k}\Omega$ (b) 0.10 A (c) $51°$
    (d) voltage leads current
23. (a) 89.6 V (b) 109 V
25. 1.88 V
27. (a) 103 V (b) 150 V (c) 127 V (d) 23.6 V
29. (a) $208 \text{ }\Omega$ (b) $40.0 \text{ }\Omega$ (c) 0.541 H
31. (a) $1.8 \times 10^2 \text{ }\Omega$ (b) 0.71 H
33. $2.29 \text{ }\mu\text{H}$
35. $C_{min} = 4.9 \text{ nF}, C_{max} = 51 \text{ nF}$
37. 0.242 J
39. (a) $1\ 600$ (b) 30 A
41. (a) $1.1 \times 10^3 \text{ kW}$ (b) $3.1 \times 10^2 \text{ A}$
    (c) $8.3 \times 10^3 \text{ A}$
43. $1\ 000$ km, there will always be better use for tax money.
45. 733 nT
47. $2.94 \times 10^8 \text{ m/s}$
49. $E_{max} = 1.01 \times 10^3 \text{ V/m}, B_{max} = 3.35 \times 10^{-6} \text{ T}$
51. (a) 188 m to 556 m   (b) 2.78 m to 3.4 m
53. 60.0 km
55. (a) 18 turns   (b) 3.6 W
57. 99.6 mH
59. 1.7 cents
61. (a) resistor and inductor   (b) $R = 10 \text{ }\Omega, L = 30 \text{ mH}$
63. (a) $6.7 \times 10^{-16} \text{ T}$ (b) $5.3 \times 10^{-17} \text{ W/m}^2$
    (c) $1.7 \times 10^{-14} \text{ W}$
65. (a) 0.536 N (b) $8.93 \times 10^{-5} \text{ m/s}^2$
    (c) 33.9 days
67. $4.47 \times 10^{-9} \text{ J}$

CHAPTER 22

Quick Quizzes

1. a
2. Beams 2 and 4 are reflected; beams 3 and 5 are refracted.
3. b
4. c

Conceptual Questions

1. Sound radiated upward at an acute angle with the horizontal is bent back toward Earth by refraction. This means that the sound can reach the listener by this path as well as by a direct path. Thus, the sound is louder.
3. The color will not change, for two reasons. First, despite the popular statement that color depends on wavelength, it actually depends on the *frequency* of the light, which does not change under water. Second, when the light enters the eye, it travels through the fluid within the eye. Thus, even if color did depend on wavelength, the important wavelength is that of the light in the ocular fluid, which does not depend on the medium through which the light traveled to reach the eye.
5. A prism creates a multicolored spectrum when white light passes through it because the white light contains all visible wavelengths. Because the index of refraction varies with wavelength, the different wavelengths emerge from the prism traveling in slightly different directions. This produces the observed rainbow of colors. If only one wavelength of light is allowed to enter the prism, as in the case of the second prism, a single wavelength (and hence color) will emerge from it.
7. No, the catalog information is incorrect. The index of refraction is given by $n = c/v$, where c is the speed of light in vacuum and v is the speed of light in the material. Because light travels faster in a vacuum than any other material, it is impossible for the index of refraction of any material to have a value less than 1.
9. There is no dependence of the angle of reflection on wavelength, because the light does not enter deeply into the material during reflection—it reflects from the surface.

11. A ball covered with mirrors sparkles by reflecting light from its surface. On the other hand, a faceted diamond lets in light at the top, reflects it by total internal reflection in the bottom half, and sends the light out through the top again. Because of its high index of refraction, the critical angle for diamond in air for total internal reflection, $\theta_c = \sin^{-1}(n_{air}/n_{diamond})$, is small. Thus, light rays enter through a large area and exit through a very small area with a much higher intensity. When a diamond is immersed in carbon disulfide, the critical angle is increased to $\theta_c = \sin^{-1}(n_{\text{carbon disulfide}}/n_{diamond})$. As a result, the light is emitted from the diamond over a larger area and appears less intense.
13. The index of refraction of water is 1.333, quite different from that of air, which has an index of refraction of about 1. On the other hand, the index of refraction of liquid helium happens to be much closer to that of air. Consequently, light undergoes less refraction in helium than it does in water.
15. The diamond acts like a prism in dispersing the light into its spectral components. Different colors are observed as a consequence of the manner in which the index of refraction varies with the wavelength.
17. Light travels through vacuum at a speed of 3×10^8 m/s. Thus, an image we see from a distant star or galaxy must have been generated some time ago. For example, the star Altair is 16 lightyears away; if we look at an image of Altair today, we know only what Altair looked like 16 years ago. This may not initially seem significant; however, astronomers who look at other galaxies can get an idea of what galaxies looked like when they were much younger. Thus, it does make sense to speak of "looking backward in time."

Problems

1. 3.00×10^8 m/s
3. 114 rad/s for a maximum intensity of returning light
5. (b) 3.000×10^8 m/s
7. $19.5°$ above the horizontal
9. (a) 1.52 (b) 417 nm
    (c) 4.74×10^{14} Hz   (d) 1.98×10^8 m/s
11. (a) 584 nm   (b) 1.12
13. $111°$
15. $16.5°$
17. five times from the right-hand mirror and six times from the left
19. 0.388 cm
21. $\theta = 30.4°, \theta' = 22.3°$
23. 6.39 ns
25. $\theta = \tan^{-1}(n_g)$
27. 3.39 m
29. $\theta_{red} = 48.22°, \theta_{blue} = 47.79°$
31. (a) $\theta_{1i} = 30°, \theta_{1r} = 19°, \theta_{2i} = 41°, \theta_{2r} = 77°$
    (b) first surface: $\theta_{reflection} = 30°$, second surface: $\theta_{reflection} = 41°$
33. (a) $31.3°$ (b) $44.2°$ (c) $49.8°$
35. (a) $33.4°$ (b) $53.4°$
37. 1.38
39. $1.000\ 08$
41. (a) $10.7°$ (b) air
    (c) Sound falling on the wall from most directions is 100% reflected.
43. $27.5°$
45. $22.0°$
47. (a) $53.1°$ (b) $\geq 38.7°$
49. (a) $38.5°$ (b) ≥ 1.44
53. $24.7°$
55. 1.93
59. $\theta = \sin^{-1}(\sqrt{n^2 - 1}\sin \phi - \cos \phi)$

CHAPTER 23

Quick Quizzes

1. At C
2. c
3. (a) False   (b) False   (c) True
4. b
5. An infinite number
6. (a) False   (b) True   (c) False

Conceptual Questions

1. You will not be able to focus your eyes on both the picture and your image at the same time. To focus on the picture, you must adjust your eyes so that an object several centimeters away (the picture) is in focus. Thus, you are focusing on the mirror surface. But, your image in the mirror is as far behind the mirror as you are in front of it. Thus, you must focus your eyes beyond the mirror, twice as far away as the picture to bring the image into focus.

3. A single flat mirror forms a virtual image of an object due to two factors. First, the light rays from the object are necessarily diverging from the object, and second, the lack of curvature of the flat mirror cannot convert diverging rays to converging rays. If another optical element is first used to cause light rays to converge, then the flat mirror can be placed in the region in which the converging rays are present, and it will simply change the direction of the rays so that the real image is formed in a different location. For example, if a real image is formed by a convex lens, and the flat mirror is placed between the lens and the image position, the image formed by the mirror will be real.

5. The ultrasonic range finder sends out a sound wave and measures the time for the echo to return. Using this information, the camera calculates the distance to the subject and sets the camera lens. When the camera is facing a mirror, the ultrasonic signal reflects from the mirror surface and the camera adjusts its focus so that the mirror surface is at the correct focusing distance from the camera. But your image in the mirror is twice this distance from the camera, so it is blurry.

7. Light rays diverge from the position of a virtual image just as they do from an actual object. Thus, a virtual image can be as easily photographed as any object can. Of course, the camera would have to be placed near the axis of the lens or mirror in order to intercept the light rays.

9. We consider the two trees to be two separate objects. The far tree is an object that is farther from the lens than the near tree. Thus, the image of the far tree will be closer to the lens than the image of the near tree. The screen must be moved closer to the lens to put the far tree in focus.

11. If a converging lens is placed in a liquid having an index of refraction larger than that of the lens material, the direction of refractions at the lens surfaces will be reversed, and the lens will diverge light. A mirror depends only on reflection, which is independent of the surrounding material, so a converging mirror will be converging in any liquid.

13. This is a possible scenario. When light crosses a boundary between air and ice, it will refract in the same manner as it does when crossing a boundary of the same shape between air and glass. Thus, a converging lens may be made from ice as well as glass. However, ice is such a strong absorber of infrared radiation that it is unlikely you will be able to start a fire with a small ice lens.

15. The focal length for a mirror is determined by the law of reflection from the mirror surface. The law of reflection is independent of the material of which the mirror is made and of the surrounding medium. Thus, the focal length depends only on the radius of curvature and not on the material. The focal length of a lens depends on the indices of refraction of the lens material and surrounding medium. Thus, the focal length of a lens depends on the lens material.

17. If parallel rays strike a flat mirror, they are reflected as parallel rays. Therefore, the focal length of a flat mirror is infinite, $f \rightarrow \infty$. This is consistent with the mirror equation

$$\frac{1}{p} + \frac{1}{q} = \frac{1}{f}$$

Because the image is always as far behind a flat mirror as the object is in front of it, then $q = -p$ and the mirror equation yields $f \rightarrow \infty$.

Problems

1. on the order of 10^{-9} s younger
3. 10.0 ft, 30.0 ft, 40.0 ft
5. 0.268 m behind the mirror; virtual, upright, and diminished; $M = 0.026\,8$
7. (a) 13.3 cm in front of mirror, real, inverted, $M = -0.333$
   (b) 20.0 cm in front of mirror, real, inverted, $M = -1.00$
   (c) No image is formed. Parallel rays leave the mirror.
9. behind the worshipper, 3.33 m from the deepest point in the niche
11. 5.00 cm
13. 1.0 m
15. 8.05 cm
17. -20.0 cm
19. (a) concave with focal length $f = 0.83$ m
   (b) Object must be 1.0 m in front of the mirror.
21. 38.2 cm below the upper surface of the ice
23. 3.8 mm
25. (a) 120 cm   (b) -24.0 cm
   (c) -8.00 cm   (d) -3.43 cm
27. 20.0 cm
29. (a) 40.0 cm beyond the lens, real, inverted, $M = -1.00$
   (b) No image is formed. Parallel rays leave the lens.
   (c) 20.0 cm in front of the lens, virtual, upright, $M = +2.00$
31. (a) 13.3 cm in front of the lens, virtual, upright, $M = +\frac{1}{3}$
   (b) 10.0 cm in front of the lens, virtual, upright, $M = +\frac{1}{2}$
   (c) 6.67 cm in front of the lens, virtual, upright, $M = +\frac{2}{3}$
33. (a) either 9.63 cm or 3.27 cm
   (b) 2.10 cm
35. (a) 39.0 mm   (b) 39.5 mm
37. at distance $2|f|$ in front of lens
39. 40.0 cm
41. 30.0 cm in front of the second lens, $M = -3.00$
43. 7.47 cm in front of the second lens; 1.07 cm; virtual, upright
45. from 0.224 m to 18.2 m
47. real image, 5.71 cm in front of the mirror
49. -40.0 cm
51. 160 cm to the left of the lens, inverted, $M = -0.800$
53. $q = 10.7$ cm
55. 32.0 cm to the right of the second surface (real image)
57. (a) 20.0 cm to the right of the second lens; $M = -6.00$
   (b) inverted
   (c) 6.67 cm to the right of the second lens; $M = -2.00$; inverted
59. (a) 1.99
   (b) 10.0 cm to the left of the lens
   (c) inverted
61. (a) 5.45 m to the left of the lens
   (b) 8.24 m to the left of the lens
   (c) 17.1 m to the left of the lens
   (d) by surrounding the lens with a medium having a refractive index greater than that of the lens material

CHAPTER 24

Quick Quizzes

1. b
2. The compact disc.

Conceptual Questions

1. You will not see an interference pattern from the automobile headlights for two reasons. The first is that the headlights are not coherent sources and are therefore incapable of producing sustained interference. Also, the headlights are so far apart in comparison to the wavelengths emitted that, even if they were made into coherent sources, the interference maxima and minima would be too closely spaced to be observable.

3. The result of the double slit is to redistribute the energy arriving at the screen. Although there is no energy at the location of a dark fringe, there is four times as much energy at the location of a bright fringe as there would be with only a single narrow slit. The total amount of energy arriving at the screen is twice as much as with a single slit, as it must be according to the conservation of energy.

5. One of the materials has a higher index of refraction than water, the other lower. The material with the higher index of refraction than water appears black as it approaches zero thickness. There is a 180° phase shift for the light reflected from the upper surface, but no such phase change from the lower surface, because the index of refraction for water on the other side is lower than that of the film. Thus, the two reflections are out of phase and interfere destructively. The material with index of refraction lower than water experiences a phase change for the light reflected from both upper and lower surfaces, so that the reflections from the zero-thickness film will be back in phase, and the film will appear bright.

7. For normal incidence, the extra path length followed by the reflected ray is twice the thickness of the film. For destructive interference, this must be a distance of half a wavelength of the light in the material of the film. For a film in air no 180° phase change occurs in these reflections, so the thickness of the film must be one-quarter wavelength, which is the same as the condition for constructive interference of reflected light. This means that the transmitted light is a minimum when the reflected light is a maximum, and vice-versa.

9. Since the light reflecting at the lower surface of the air film experience a 180° phase change whereas light reflecting from the upper surface of the film does not undergo a phase change, the central spot (where the film has near zero thickness) is dark. If the observed rings are not circular, the curved surface of the lens does not have a true spherical shape.

11. For regional communication at Earth's surface, radio waves are typically broadcast from currents oscillating in tall, vertical towers. These waves have vertical planes of polarization. Light originates from the vibrations of atoms, or electronic transitions within atoms, which represent oscillations in all possible directions. Thus, light is not generally polarized.

13. Yes. In order to do this, first measure the radar-reflectivity of the metal of your airplane. Then, choose a light, durable material that has approximately half the radar-reflectivity of the metal in your plane. Measure its index of refraction, and place onto the metal a coating equal in thickness to one quarter of 3 cm, divided by that index. Sell it quick and then you can sell to the supposed enemy new radars operating at 1.5 cm, which the coated metal will reflect with extra-high efficiency.

15. If you wish to perform an interference experiment, you need monochromatic coherent light. To obtain it, you must first pass light from an ordinary source through a prism or diffraction grating to disperse different colors into different directions. Using a single narrow slit, select a single color and make that light diffract to cover both slits for Young's experiment. The procedure is much simpler with a laser because its output is already monochromatic and coherent.

17. Audible sound has wavelengths on the order of meters or centimeters, whereas visible light has wavelengths on the order of half a micrometer. Sound, therefore, diffracts around walls and doorways (roughly meter-sized apertures). Visible light diffracts only through very small angles as it passes ordinary sized objects or apertures, because $\sin\theta = m\lambda/a$ by Equation 24.11, and λ/a is extremely small.

19. Strictly speaking, the ribs do act as a diffraction grating, but the separation distance of the ribs is so much larger than the wavelength of the x-rays that there are no observable effects.

Problems

1. 1.58 cm
3. (a) 2.62 mm   (b) 2.62 mm
5. (a) 36.2°   (b) 5.08 cm   (c) 5.08×10^{14} Hz
7. (a) 55.7 m   (b) 124 m
9. 75.0 m
11. 11.3 m
13. 148 m
15. 91.9 nm
17. Any odd-integer multiple of 85.4 nm
19. 0.500 cm
21. (a) 238 nm   (b) λ will increase
    (c) 328 nm
23. 4.35 μm
25. 4.75 μm
27. No, the wavelengths intensified are 276 nm, 138 nm, 92.0 nm, . . .
29. 4.22 mm
31. (a) 1.1 m   (b) 1.7 mm
33. 1.20 mm, 1.20 mm
35. (a) 479 nm, 647 nm, 698 nm   (b) 20.5°, 28.3°, 30.7°
37. 5.49° in first order; 12.3° in second order; and 24.9° in third order
39. 44.5 cm
41. 9.13 cm
43. (a) 25.6°   (b) 19.0°
45. (a) 1.11   (b) 42.0°
47. (a) 56.7°   (b) 48.8°
49. 31.2°
53. 6.89 units
55. (a) 413.7 nm, 409.7 nm   (b) 8.6°
57. 0.156 mm
59. 2.50 mm
61. 58.1°
63. (a) 16.6 m   (b) 8.28 m
65. 1.31
67. 0.350 mm
69. 115 nm

CHAPTER 25

Answers to Quick Quizzes

1. You should choose a blue filter.

Conceptual Questions

1. The observer is not using the lens as a simple magnifier. To use a lens as a simple magnifier, the object distance must be less than the focal length of the lens. Also, a simple magnifier produces a virtual image at the normal near point of the eye, or at a point about $q = -25$ cm. With a large object distance and a relatively short image distance, the magnitude of the magnification by the lens would be considerably less than one. Most likely, the lens is part of a lens combination used as a telescope.

3. The image formed on the retina by the lens and cornea is already inverted.

5. There will be an effect on the interference pattern—it will be distorted. The high temperature of the flame changes the index of refraction of air for the arm of the interferometer in which the match is held. As the index of refraction varies turbulently, the wavelength of the light in that region will also vary turbulently. As a result, the effective difference in length between the two arms varies, resulting in a wildly varying interference pattern.

7. Large lenses are difficult to manufacture and machine with accuracy. Their large weight leads to sagging, which produces a distorted image. In reflecting telescopes, light does not pass through glass; hence, problems associated with chromatic aberrations are eliminated. Large-diameter reflecting telescopes are also technically easier to construct. Some designs use a rotating pool of mercury as the reflecting surface.

9. In order to "see" an object, the wavelength of the light in the microscope must be smaller than the size of the object. An atom is much smaller than the wavelength of light in the visible spectrum, so an atom can never be seen using visible light.

Problems

1. 30.0 cm beyond the lens, $M = -1/5$
3. Yes, the largest image dimension is only 19 mm.
5. $f/1.4$
7. $f/8.0$
9. 40.0 cm
11. 23.2 cm
13. (a) -2.00 diopters   (b) 17.6 cm
15. $+17.0$ diopters
17. $m = +3.50$
19. (a) 4.07 cm   (b) $m = +7.14$
21. (a) $|M| = 1.22$ (b) $\theta/\theta_0 = 6.08$
23. 2.1 cm
25. $m = -115$
27. $f_o = 90$ cm, $f_e = 2.0$ cm
29. (b) $-fh/p$ (c) -1.07 mm
31. (a) $m = 1.50$ (b) $m = 1.90$
33. 492 km
35. 0.40 μrad
37. 3.09 m
39. 9.8 km
41. 50.4 μm
43. 449 nm, blue
45. 98 fringe shifts
47. No. A resolving power of 2.0×10^5 is needed and that available is only 1.8×10^5.
49. (a) $+2.67$ diopters   (b) 0.16 diopters too low
51. (a) $+44.6$ diopters   (b) 3.03 diopters
53. (a) 1.0×10^3 lines   (b) 3.3×10^2 lines
55. $m = 10.7$

CHAPTER 26

Quick Quizzes

1. b
2. The answer to both questions is no.
3. a, e
4. (a) False   (b) False   (c) True   (d) False
5. a

Conceptual Questions

1. One must be careful using such phrases because events that occur at different locations and appear to occur simultaneously to one observer will not appear simultaneous to observers in motion relative to the first. This lack of simultaneity becomes more pronounced at very high speeds.

3. This scenario is not possible with light. Light waves are described by the principles of special relativity. As you detect the light wave ahead of you and moving away from you (which would be a pretty good trick—think about it!), its velocity relative to you is c. Thus, you will not be able to catch up to the light wave.

5. Present knowledge of the Universe shows that even the "fixed" stars are in motion relative to us and in motion relative to each other. Thus, the stars are not really fixed at all and do not provide us with an absolute frame of reference.

7. The light from the quasar moves at 3.00×10^8 m/s. The speed of light is independent of the motion of the source or observer.

9. For a wonderful fictional exploration of this question, get a "Mr. Tompkins" book by George Gamow. All of the relativity effects would be obvious in our lives. Time dilation and length contraction would both occur. Driving home in a hurry, you would push on the gas pedal not to increase your speed very much, but to make the blocks shorter. Big Doppler shifts in wave frequencies would make red lights look green as you approached and make car horns and radios useless. High-speed transportation would be both very expensive, requiring huge fuel purchases, as well as dangerous, because a speeding car could knock down a building. When you got home, hungry for lunch, you would find that you had missed dinner; there would be a five-day delay in transit when you watch a live TV program originating in Australia. Finally, we would not be able to see the Milky Way, because the fireball of the Big Bang would surround us at the distance of Rigel, or Deneb.

11. This is not a violation of the concepts of relativity. The theory of relativity states that the result of any measurement of the vacuum speed of light will be independent of the motion of the observer or source. It makes no claim that light must have the same speed in all media.

13. Your assignment: Measure the length of a rod as it slides past you. Mark the position of its front end on the floor and have an assistant mark the position of the back end. Then measure the distance between the two marks. This distance will represent the length of the rod only if the two marks were made simultaneously in your frame of reference.

Problems

1. (a) $t_{OB} = 1.67 \times 10^3$ s, $t_{OA} = 2.04 \times 10^3$ s
    (b) $t_{BO} = 2.50 \times 10^3$ s, $t_{AO} = 2.04 \times 10^3$ s
    (c) $\Delta t = 90$ s
3. 5.0 s
5. $c(\sqrt{3}/2)$
7. (a) 1.3×10^{-7} s   (b) 38 m
    (c) 7.6 m
9. (a) 2.2 μs   (b) 0.65 km
11. $0.950c$
13. yes, with 19 m to spare
15. (a) 39.2 μs   (b) accurate to one digit
17. 3.3×10^5 m/s
19. $0.285c$
21. $0.54c$ to the right
23. $0.357c$
25. $0.998c$ toward the right
27. (a) 54 min   (b) 52 min
29. $c(\sqrt{3}/2)$
31. $0.272c$
33. 18.4 g/cm^3
35. 1.98 MeV
37. 2.27×10^{23} Hz, 1.32 fm for each photon
39. (a) 3.10×10^5 m/s   (b) $0.758c$
41. 1.42 MeV/c
43. (a) $0.80c$ (b) 7.5×10^3 s
    (c) 1.4×10^{12} m, $0.38c$

45. $0.37c$ in $+x$ direction

47. (a) $v/c = 1 - 1.12 \times 10^{-10}$ (b) 6.00×10^{27} J
(c) $\$2.17 \times 10^{20}$

49. $0.80c$

51. (a) $0.946c$ (b) 0.160 ly
(c) 0.114 yr (d) 7.50×10^{22} J

53. (a) 7.0 μs (c) 1.1×10^4 muons

CHAPTER 27

Quick Quizzes

1. b
2. c
3. c
4. c
5. b

Conceptual Questions

1. The shape of an object is determined by observing the light reflecting from its surface. In a kiln, the objects will be very hot and will be glowing red. The emitted radiation is far stronger than the reflected radiation, and the thermal radiation emitted is only slightly dependent on the material from which the objects are made. Thus, we have a collection of objects, including the kiln walls, glowing equally with emitted radiation and only weak reflected light compared to the emitted light. This will result in indistinct outlines of the objects when viewed.

3. The "blackness" of a blackbody refers to its ideal property of absorbing all radiation incident on it. If an observed room-temperature object in everyday life absorbs all radiation, we describe it as (visibly) black. The black appearance, however, is due to the fact that our eyes are sensitive only to visible light. If we could detect infrared light with our eyes, we would see the object emitting radiation. If the temperature of the blackbody is raised, Wien's law tells us that the emitted radiation will move into the visible. Thus, the blackbody could appear possibly as black, red, white, or blue, depending on its temperature.

5. All objects do radiate energy, but at room temperature, this energy is primarily in the infrared region of the electromagnetic spectrum, which our eyes cannot detect. (Pit vipers have sensory organs that are sensitive to infrared radiation; thus they can seek out their warm-blooded prey in what we would consider absolute darkness.)

7. Most metals have cutoff frequencies corresponding to photons in or near the visible range of the electromagnetic spectrum. AM radio wave photons will have far too little energy to eject electrons from the metal.

9. We can picture higher-frequency light as a stream of photons of higher energy. In a collision, one photon can give all of its energy to a single electron. The kinetic energy of such an electron is measured by the stopping potential. The reverse voltage (stopping voltage) required to stop the current is proportional to the frequency of the incoming light. More intense light consists of more photons striking a unit area each second, but atoms are so small that one emitted electron never gets a "kick" from more than one photon. Increasing the light intensity will generally increase the size of the current but will not change the energy of the individual ejected electrons. Thus, the stopping potential remains constant.

11. Wave theory predicts that the photoelectric effect should occur at any frequency, provided that the light intensity is high enough. However, as seen in photoelectric experiments, the light must have sufficiently high frequency for the effect to occur.

13. The photons will transfer the same energy to all photoelectrons in the absorption process. However, different electrons may start from different potential energy levels and emerge from the surface with different kinetic energies. The energy of the incident photons determines only the maximum kinetic energy of the liberated electrons.

15. No. Suppose that the incident light frequency at which you first observed the photoelectric effect is above the cutoff frequency of the first metal but less than the cutoff frequency of the second metal. In that case, the photoelectric effect would not be observed at all in the second metal.

17. In a collision with an electron or other scattering particle, the photon will impart some energy to that particle. Therefore, the energy of the photon will be decreased and its wavelength increased.

Problems

1. (a) ≈ 3000 K (b) $\approx 20\,000$ K
3. (a) 999 nm
(b) Peak radiation is in the infrared region.
5. (a) 2.49×10^{-5} eV (b) 2.49 eV
(c) 249 eV
7. 2.27×10^{30} photons/s
9. (a) 2.3×10^{31} (b) $\Delta E/E = 4.3 \times 10^{-32}$
11. (a) 2.24 eV (b) 555 nm
(c) 5.41×10^{14} Hz
13. (a) 296 nm (b) 1.01×10^{15} Hz
(c) 2.71 V
15. 148 days, incompatible with observation
17. 4.8×10^{14} Hz, 2.0 eV
19. 1.2×10^2 V and 1.2×10^7 V, respectively
21. 41.4 kV
23. 0.078 nm
25. 0.281 nm
27. 1.78 eV, 9.47×10^{-28} kg·m/s
29. $70°$
31. 1.18×10^{-23} kg·m/s, 478 eV
33. (a) 1.2 eV (b) 6.5×10^5 m/s
35. (a) 1.46 km/s (b) 7.28×10^{-11} m
37. (a) $\sim 10^2$ MeV
(b) No. With kinetic energy much larger than the magnitude of the negative potential energy, the electron would immediately escape.
39. 3.58×10^{-13} m
41. (a) 15 keV (b) 1.2×10^2 keV
43. 2.1×10^{-32} m/s
45. 116 m/s
47. $\approx 5\,200$ K; clearly, a firefly is not at this temperature, so this cannot be blackbody radiation.
49. $18.2°$
51. 1.36 eV
53. 2.00 eV
55. (a) $0.0220c$ (b) $0.9992c$
57. (b) 3.72 km/s
59. (b) 2.60×10^{-16} m

CHAPTER 28

Quick Quizzes

1. d
2. (a) 9 (b) 16
3. (a) 5 values of ℓ
(b) 9 values of m_ℓ
4. b

Conceptual Questions

1. If the energy of the hydrogen atom were proportional to n (or any power of n), then the energy would become infinite

n	3	3	3	3	3	3	3	3	3	3	3	3	3	3	3
ℓ	2	2	2	2	2	2	2	2	2	2	2	2	2	2	2
m_ℓ	+2	+2	+2	+1	+1	+1	0	0	0	-1	-1	-1	-2	-2	-2
m_s	+1	0	-1	+1	0	-1	+1	0	-1	+1	0	-1	+1	0	-1

as n grew to infinity. But the energy of the atom is inversely proportional to n^2. Thus, as n grows to infinity, the energy of the atom approaches a value that is above the ground state by a finite amount, namely the ionization energy 13.6 eV. As the electron falls from one bound state to another, its energy loss is always less than the ionization energy. The energy and frequency of any emitted photon are finite.

3. The characteristic x-rays originate from transitions within the atoms of the target, such as an L shell electron making a transition to a vacancy in the K shell. This vacancy in the K shell is caused when an accelerated electron in the x-ray tube supplies energy to the K shell electron to eject it from the atom. If the energy of the bombarding electrons were to be increased, the K shell electron will be ejected from the atom with more remaining kinetic energy. But the energy difference between the K and L shell has not changed, so the emitted x-ray has exactly the same wavelength.

5. A continuous spectrum without characteristic x-rays is possible. At a low-accelerating potential difference for the electron, the electron may not have enough energy to eject an electron from a target atom. As a result, there will be no characteristic x-rays. The change in speed of the electron as it enters the target will result in the continuous spectrum.

7. The hologram is an interference pattern between light scattered from the object and the reference beam. If anything moves by a distance comparable to the wavelength of the light (or more), the pattern will wash out. The effect is just like making the slits vibrate in Young's experiment, to make the interference fringes vibrate wildly so that a photograph of the screen displays only the average intensity everywhere.

9. If the Pauli exclusion principle were not valid, the elements and their chemical behavior would be grossly different because every electron would end up in the lowest energy level of the atom. All matter would therefore be nearly alike in its chemistry and composition, since the shell structures of each element would be identical. Most materials would have a much higher density, and the spectra of atoms and molecules would be very simple, resulting in the existence of less color in the world.

11. The three elements have similar electronic configurations, with filled inner shells, plus a single electron in an s orbital. Since atoms typically interact through their unfilled outer shells, and the outer shell of each of these atoms is similar, the chemical interactions of the three atoms is also similar.

13. Each of the eight electrons must have at least one quantum number different from each of the others. They can differ (in m_s) by being spin-up or spin-down. They can differ (in ℓ) in angular momentum and in the general shape of the wave function. Those electrons with $\ell = 1$ can differ (in m_ℓ) in orientation of angular momentum.

15. The way the total energy of an atom, and in fact other energies, can be negative is through an arbitrary choice of the zero point for potential energies. The electrostatic potential energy between charged particles is arbitrarily chosen to be zero when the particles are separated by an infinite distance. When two particles having opposite signs, such as the electron and proton in a hydrogen atom, are a finite distance apart, the potential energy is less than zero, or negative. If they are sufficiently close, the magnitude of the negative potential energy may exceed the positive kinetic energy, making the total energy negative.

17. Stimulated emission is the reason laser light is coherent and tends to travel in a well-defined parallel beam. When a photon passing by an excited atom stimulates that atom to emit a photon, the emitted photon is in phase with the original photon and travels in the same direction. As this process is repeated many times, an intense, parallel beam of coherent light is produced. Without stimulated emission, the excited atoms would return to the ground state by emitting photons at random times and in random directions. The resulting light would not have the useful properties of laser light.

Problems

1. 656 nm, 486 nm, and 434 nm
3. (a) 2.3×10^{-8} N
   (b) -14 eV
5. (a) 1.6×10^6 m/s
   (b) No. $v/c = 5.3 \times 10^{-3} \ll 1$
   (c) 0.46 nm
   (d) Yes. The wavelength is roughly the same size as the atom.
7. (a) 0.212 nm   (b) 9.95×10^{-25} kg·m/s
   (c) 2.11×10^{-34} J·s
   (d) 3.40 eV   (e) -6.80 eV   (f) -3.40 eV
11. 8.22×10^{-8} N
13. (a) 0.967 eV   (b) 0.266 eV
15. (a) 122 nm, 91.1 nm   (b) 1.87×10^3 nm, 820 nm
17. 97.2 nm
19. (a) 488 nm   (b) 0.814 m/s
21. (d) $n = 2.53 \times 10^{74}$
   (e) No. At such large quantum numbers the allowed energies are essentially continuous.
23. (a) 2.47×10^{15} Hz, $f_{orb} = 8.23 \times 10^{14}$ Hz
   (b) 6.59×10^3 Hz, $f_{orb} = 6.59 \times 10^3$ Hz. For large n, classical theory and quantum theory approach each other in their results.
25. 4.42×10^4 m/s
27. (a) -122 eV   (b) 1.76×10^{-11} m
29. (a) 0.026 5 nm
   (b) 0.017 6 nm
   (c) 0.013 2 nm
31. 1.33 nm
33. $n = 3, \ell = 1, m_\ell = +1, m_s = \pm 1/2; n = 3, \ell = 1, m_\ell = 0, m_s = \pm 1/2; n = 3, \ell = 1, m_\ell = -1, m_s = \pm 1/2$
35. The table at the top of the page summarizes 15 possible states:
37. (a) 30 possible states   (b) 36
39. (a) $n = 4$ and $\ell = 2$
   (b) $m_\ell = (0, \pm 1, \pm 2), m_s = \pm \frac{1}{2}$
   (c) $1s^2 2s^2 2p^6 3s^2 3p^6 3d^{10} 4s^2 4p^6 4d^2 5s^2 = $ [Kr] $4d^2 5s^2$
41. 0.160 nm
43. L shell: 11.7 keV; M shell: 10.0 keV; N shell: 2.30 keV
45. (a) 10.2 eV   (b) 7.88×10^4 K
47. (a) -8.18 eV, -2.04 eV, -0.904 eV, -0.510 eV, -0.325 eV
   (b) 1.09×10^3 nm and 609 nm
49. (a) The four lowest energies are -10.39 eV, -5.502 eV, -3.687 eV, and -2.567 eV.
   (b) The wavelengths of the emission lines are 158.5 nm, 185.0 nm, 253.7 nm, 422.5 nm, 683.2 nm, and 1 107 nm.
   (c) 1.31×10^6 m/s
51. (a) 4.24×10^{15} W/m^2 (b) 1.20×10^{-12} J
55. (a) $E_n = (-1.49 \times 10^4$ eV$)/n^2$
   (b) $n = 4 \rightarrow n = 1$

57. **(a)** 9.03×10^{22} m/s^2 **(b)** -4.63×10^{-8} W
    (c) $\sim 10^{-11}$ s

CHAPTER 29

Answers to Quick Quizzes

1. c
2. b
3. a
4. a and b
5. b

Conceptual Questions

1. Isotopes of a given element correspond to nuclei with different numbers of neutrons. This results in a variety of different physical properties for the nuclei, including the obvious one of mass. The chemical behavior, however, is governed by the electrons. All isotopes of a given element have the same number of electrons and, therefore, the same chemical behavior.

3. An alpha particle contains two protons and two neutrons. Because a hydrogen nucleus only contains one proton, it cannot emit an alpha particle.

5. In alpha decay, there are only two final particles—the alpha particle and the daughter nucleus. There are also two conservation principles—energy and momentum. As a result, the alpha particle must be ejected with a discrete energy to satisfy both conservation principles. However, beta decay is a three-particle decay—the beta particle, the neutrino (or antineutron), and the daughter nucleus. As a result, the energy and momentum can be shared in a variety of ways among the three particles while still satisfying the two conservation principles. This allows a continuous range of energies for the beta particle.

7. The larger rest energy of the neutron means that a free proton in space will not spontaneously decay into a neutron and a positron. When the proton is in the nucleus, however, the important question is that of the total rest energy of the nucleus. If it is energetically lower for the nucleus to have one less proton and one more neutron, then the decay process will occur to achieve this lower energy.

9. Carbon dating cannot generally be used to estimate the age of a stone, because the stone was not alive to take up carbon from the environment. Only the ages of artifacts that were once alive can be estimated with carbon dating.

11. The protons, although held together by the nuclear force, are repulsed by the electrical force. If enough protons were placed together in a nucleus, the electrical force would overcome the nuclear force, which is based on the number of particles, and cause the nucleus to fission.

    The addition of neutrons prevents such fission. The neutron does not increase the electrical force, being electrically neutral, but does contribute to the nuclear force.

13. The statement is false. Both patterns show monotonic decrease over time, but with very different shapes. For radioactive decay, maximum activity occurs at time zero. Cohorts of people now living will be dying most rapidly perhaps forty years from now. Everyone now living will be dead within less than two centuries, while the mathematical model of radioactive decay tails off exponentially forever. A radioactive nucleus never gets old. It has constant probability of decay however long it has existed.

15. Since the two samples are of the same radioactive nuclide, they have the same half-life; the 2:1 difference in activity is due to a 2:1 difference in the mass of each sample. After five half-lives, each will have decreased in mass by a power of $2^5 = 32$. However, since this simply means that the mass of each is 32 times smaller, the ratio of the masses will still be $\left(\frac{2}{32}\right):\left(\frac{1}{32}\right)$, or 2:1. Therefore, the ratio of their activities will *always* be 2:1.

17. The photon and the neutrino are similar in that both particles have zero mass and zero charge. Both must travel at the speed of light and are capable of transferring both energy and momentum. They differ in that the photon has spin (intrinsic angular momentum) of $\hbar$ and is involved in electromagnetic interactions, whereas the neutrino has spin $\hbar/2$ and is closely related to beta decays.

Problems

1. $A = 2$, $r = 1.5$ fm; $A = 60$, $r = 4.7$ fm; $A = 197$, $r = 7.0$ fm; $A = 239$, $r = 7.4$ fm

3. 1.8×10^2 m

5. **(a)** 27.6 N away from the carbon nucleus
   (b) 4.16×10^{27} m/s^2 away from the nucleus
   (c) 1.73 MeV

7. **(a)** 1.9×10^7 m/s **(b)** 7.1 MeV

9. 8.66 MeV/nucleon for $^{93}_{41}$Nb, 7.92 MeV/nucleon for $^{197}_{79}$Au

11. 3.54 MeV

13. 0.210 MeV/nucleon greater for $^{23}_{11}$Na, attributable to less proton repulsion

15. 0.46 Ci

17. **(a)** 9.98×10^{-7} s^{-1} **(b)** 1.9×10^{10} nuclei

19. **(a)** 3.18×10^{-7} mol **(b)** 1.92×10^{17} nuclei
    (c) 1.08×10^{14} Bq **(d)** 8.96×10^6 Bq

21. 4.31×10^3 yr

23. **(a)** 5.58×10^{-2} h^{-1}, 12.4 h **(b)** 2.39×10^{13} nuclei
    (c) 1.9 mCi

25. $^{208}_{81}$Tl, $^{95}_{37}$Rb, $^{144}_{60}$Nd

27. $^{40}_{20}$Ca, $^{94}_{42}$Mo, $^{4}_{2}$He

29. e^+ decay, $^{56}_{27}$Co $\rightarrow$ $^{56}_{26}$Fe $+ e^+ + \nu$

31. 71.4 keV

33. 18.6 keV

35. 4.22×10^3 yr

37. **(a)** $^{30}_{15}$P **(b)** -2.64 MeV

39. **(a)** $^{21}_{10}$Ne **(b)** $^{144}_{54}$Xe **(c)** X $= e^+$, X$' = \nu$

41. **(a)** $^{13}_{6}$C **(b)** $^{10}_{5}$B

43. 1.00 MeV

45. **(a)** $^{1}_{0}$n **(b)** Fluorine mass $= 18.000\ 953$ u

47. 18.8 J

49. 24 d

51. **(a)** 8.97×10^{11} electrons **(b)** 0.100 J **(c)** 100 rad

53. 46.5 d

55. $Q = 3.27$ MeV > 0, no threshold energy required

57. **(a)** 2.52×10^{24} **(b)** 2.29×10^{12} Bq
    (c) 1.07×10^6 yr

59. **(a)** 4.0×10^9 yr
    (b) It could be no older. The rock could be younger if some ^{87}Sr were initially present.

61. 54 μCi

63. 2.3×10^2 yr

65. 4.4×10^{-8} kg/h

CHAPTER 30

Answers to Quick Quizzes

1. c
2. a
3. b
4. c, e

Conceptual Questions

1. The experiment described is a nice analogy to the Rutherford scattering experiment. In the Rutherford experiment, alpha particles were scattered from atoms, and the scattering was consistent with a small structure in the atom containing the positive charge.

3. The largest charge of a quark is $2e/3$, so a combination of only two particles, a quark and an antiquark forming a meson, could not possibly have electric charge up to $+2e$. Only

particles containing three quarks, each with a charge of $2e/3$, can combine to produce a total charge of 2e.

5. Until about 700 000 years after the Big Bang, the temperature of the Universe was high enough for any atoms that formed to be ionized by ambient radiation. Once the average radiation energy dropped below the hydrogen ionization energy of 13.6 eV, hydrogen atoms could form and remain as neutral atoms for a relatively long period of time.

7. In the quark model, all hadrons are composed of smaller units called quarks. Quarks have a fractional electric charge and a baryon number of $\frac{1}{3}$. There are six flavors of quarks; up (u) down (d), strange (s), charmed (c), top (t), and bottom (b). All baryons contain three quarks, and all mesons contain one quark and one antiquark. Section 30.12 has a more detailed discussion of the quark model.

9. Baryons and mesons are hadrons, interacting primarily through the strong force. They are not elementary particles, being composed of either three quarks (baryons), or a quark and an antiquark (mesons). Baryons have a nonzero baryon number with a spin of either $\frac{1}{2}$ or $\frac{3}{2}$. Mesons have a baryon number of zero and a spin of either 0 or 1.

11. All stable particles other than protons and neutrons have baryon number zero. Since the baryon number must be conserved, and the final states of the kaon decay contain no protons or neutrons, the baryon number of all kaons must be *zero*.

13. Yes, but the strong interaction predominates.

15. Unless the particles have enough kinetic energy to produce a baryon-antibaryon pair, the answer is *no*. Antibaryons have a baryon number of −1; baryons have a baryon number of +1; mesons have a baryon number of 0. If such an interaction were to occur, and produce a baryon, the baryon number would not be conserved.

17. Baryons and antibaryons contain three quarks, whereas mesons and antimesons contain two quarks. Quarks have a spin of $\frac{1}{2}$; thus three quarks in a baryon can only combine to form a net spin that is half-integral. Likewise, two quarks in a meson can only combine to form a net spin of 0 or 1.

19. Nuclear and coal-fired power plants are similar in that both produce high-pressure steam to drive turbines and generate electricity. They differ in the type and amount of fuel consumed, the ways they affect the environment, and in some safety issues. A moderate size nuclear-powered plant consumes a few kilograms of fissionable materials per day, whereas a comparable size coal-powered plant requires a thousand tons or more of coal each day. One byproduct of nuclear plants is thermal pollution in the waters used for cooling. In addition to thermal pollution, coal-fired plants produce emissions that cause acid rain and contribute to the greenhouse effect. Both types of plants have risks of steam explosions. In addition, nuclear-powered plants raise questions on how to dispose of radioactive waste materials.

Problems

1. $^1_0n + ^{235}_{92}U \rightarrow ^{141}_{56}Ba + ^{92}_{36}Kr + 3\,^1_0n$, 3 neutrons are released
3. 126 MeV
5. (a) 16.2 kg   (b) 117 g
7. 2.9×10^3 km ($\approx 1\,800$ miles)
9. 1.01 g
11. (a) ^8_4Be (b) $^{12}_6C$ (c) 7.27 MeV

13. 3.07×10^{22} events/yr
15. (a) 3.44×10^{30} J   (b) 1.56×10^8 yr
17. (a) 4.53×10^{23} Hz   (b) 0.622 fm
19. $\sim 10^{-23}$ s
21. $\sim 10^{-18}$ m
23. (a) conservation of electron-lepton number and conservation of muon-lepton number
   (b) conservation of charge
   (c) conservation of baryon number
   (d) conservation of baryon number
   (e) conservation of charge
25. $\bar{\nu}_\mu$
27. (a) $\bar{\nu}_\mu$ (b) ν_μ (c) $\bar{\nu}_e$ (d) ν_e
   (e) ν_μ (f) ν_μ and $\bar{\nu}_e$
29. (a) not allowed; violates conservation of baryon number
   (b) strong interaction   (c) weak interaction
   (d) weak interaction
   (e) electromagnetic interaction
31. (a) not conserved   (b) conserved
   (c) conserved   (d) not conserved
   (e) not conserved   (f) not conserved
33. (a) charge, baryon number, L_e, L_τ
   (b) charge, baryon number, L_e, L_μ, L_τ
   (c) charge, L_e, L_μ, L_τ, strangeness number
   (d) charge, baryon number, L_e, L_μ, L_τ, strangeness number
   (e) charge, baryon number, L_e, L_μ, L_τ, strangeness number
   (f) charge, baryon number, L_e, L_μ, L_τ, strangeness number
35. 3.34×10^{26} electrons, 9.36×10^{26} up quarks, 8.70×10^{26} down quarks
37. (a) Σ^+ (b) π^- (c) K^0 (d) Ξ^-
39.

Reaction	At Quark Level	Net Quarks (Before and After)
$\pi^- + p \rightarrow K^0 + \Lambda^0$	$\bar{u}d + uud \rightarrow d\bar{s} + uds$	1 up, 2 down, 0 strange
$\pi^+ + p \rightarrow K^+ + \Sigma^+$	$u\bar{d} + uud \rightarrow u\bar{s} + uus$	3 up, 0 down, 0 strange
$K^- + p \rightarrow$ $K^+ + K^0 + \Omega^-$	$\bar{u}s + uud \rightarrow$ $u\bar{s} + d\bar{s} + sss$	1 up, 1 down, 1 strange

   (d) The mystery particle is a Λ^0 or a Σ^0.
41. a neutron, udd
43. 70.45 MeV
45. 18.8 MeV
47. (a) electron-lepton and muon-lepton numbers not conserved
   (b) electron-lepton number not conserved
   (c) charge not conserved
   (d) baryon and electron-lepton numbers not conserved
   (e) strangeness violated by 2 units
49. (a) 1.5×10^{24} nuclei   (b) ≈ 0.6 kg
51. (a) 1 baryon before and zero baryons after decay. Baryon number is not conserved.
   (b) 469 MeV, 469 MeV/c
   (c) 0.999 999 4c

Index

Page numbers in italics indicate illustrations; page numbers followed by "t" indicate tables.

PHYSICAL CONSTANTS

Quantity	Symbol	Value	SI unit
Speed of light in vacuum	c	3.00×10^8	m/s
Permittivity of free space	ϵ_0	8.85×10^{-12}	$C^2/N \cdot m^2$
Coulomb constant, $1/4\pi\epsilon_0$	k_e	8.99×10^9	$N \cdot m^2/C^2$
Permeability of free space	μ_0	1.26×10^{-6} ($4\pi \times 10^{-7}$ exactly)	$T \cdot m/A$
Elementary charge	e	1.60×10^{-19}	C
Planck's constant	h $\hbar = h/2\pi$	6.63×10^{-34} 1.05×10^{-34}	$J \cdot s$ $J \cdot s$
Electron mass	m_e	9.11×10^{-31} 5.49×10^{-4}	kg u
Proton mass	m_p	$1.672\,65 \times 10^{-27}$ $1.007\,276$	kg u
Neutron mass	m_n	$1.674\,95 \times 10^{-27}$ $1.008\,665$	kg u
Avogadro's number	N_A	6.02×10^{23}	mol^{-1}
Universal gas constant	R	8.31	$J/mol \cdot K$
Boltzmann's constant	k_B	1.38×10^{-23}	J/K
Stefan-Boltzmann constant	σ	5.67×10^{-8}	$W/m^2 \cdot K^4$
Molar volume of ideal gas at STP	V	22.4 2.24×10^{-2}	L/mol m^3/mol
Rydberg constant	R_H	1.10×10^7	m^{-1}
Bohr radius	a_0	5.29×10^{-11}	m
Electron Compton wavelength	$h/m_e c$	2.43×10^{-12}	m
Gravitational constant	G	6.67×10^{-11}	$N \cdot m^2/kg^2$
Standard free-fall acceleration	g	9.80	m/s^2
Radius of Earth (at equator)	R_E	6.38×10^6	m
Mass of Earth	M_E	5.98×10^{24}	kg
Radius of Moon	R_M	1.74×10^6	m
Mass of Moon	M_M	7.36×10^{22}	kg

The values presented in this table are those used in computations in the text. Generally, the physical constants are known to much better precision.